A TEXTBOOK OF MACHINE DRAWING

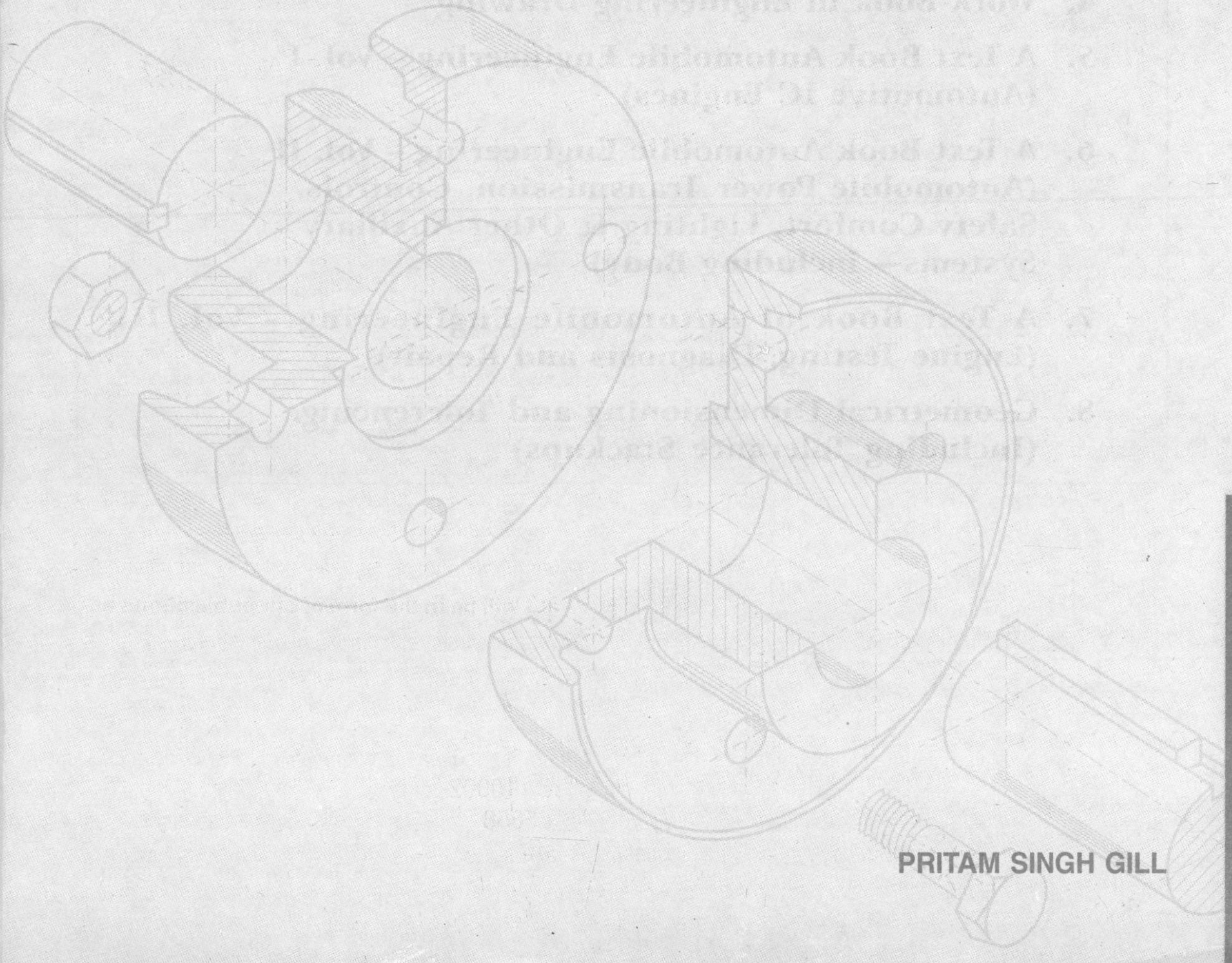

PRITAM SINGH GILL

BY THE SAME AUTHOR

1. A Text Book of Engineering Drawing
(Geometrical Drawing)
2. A Text Book of Engineering Graphics and Drafting
3. A Text Book of Graphic Statics
4. Work-Book in Engineering Drawing
5. A Text Book Automobile Engineering – Vol. I
(Automotive IC Engines)
6. A Text Book Automobile Engineering – Vol. II
(Automobile Power Transmission, Controls, Safety Comfort, Lighting & Other Auxiliary Systems— including Body)
7. A Text Book of Automobile Engineering – Vol. III
(Engine Testing, Diagnosis and Repair)
8. Geometrical Dimensioning and Tolerencing.
(Including Tolerance Stackups)

A TEXTBOOK OF

MACHINE DRAWING

For

B.Sc./B.E. Engg., Diploma in Engg., I.T.I. (Draftsman Course),
A.T.I., 10 + 2 System and other Engineering Examinations

According to IS: 696-1972
(Second Revision, Third Reprint Oct., 1984)
and SP: 46-1988)

By

PRITAM SINGH GILL

M.Sc. Engg. (Mechanical) Pb., M.E. (Prod. Engg.) Pb.,
Grad. I.E. (Mech. Engg.) India, Dip. Mech. Engg. Distinction),
M.I.E. (India), A.S.M.E., M.C.I. (India Section), M.I.S.T.E., A.M.I.A.E.

S.K. KATARIA & SONS®

Publisher of Engineering Books

4885/109, Prakash Mahal, Dr. Subhash Bhargav Lane,
Opposite Delhi Medical Association, Daryaganj, New Delhi-110002 (INDIA)
Phones: +91-11-23243489, 43551243; Mobile: +91-9871775858
e-mail: katariabook@gmail.com; katariabooks@yahoo.com
Website : www.skkatariaandsons.com

(e-books, subscriptions and bulk buying also available on our website)

Published by:

S.K. Kataria & Sons®

4885/109, Prakash Mahal, Dr. Subhash Bhargav Lane,
Opposite Delhi Medical Association, Daryaganj, New Delhi-110002
Phones: +91-11-23243489, 43551243; Mobile: +91-9871775858
e-mail: katariabook@gmail.com; katariabooks@yahoo.com
Website: www.skkatariaandsons.com

Head Office:

Opp. Clock Tower, Ludhiana (Punjab)
Ph. : 2726401

Seventeenth Revised Edition, 2001; Reprint: 2002, Reprint: 2003, Reprint: 2004, Reprint: 2005-06, Reprint: 2006-07, Reprint: 2007-08; Reprint: 2009, Eighteenth Revised Edition: June, 2013

Reprint : 2015, 2016, 2017 Reprint : 2018 Reprint: 2019 Reprint: 2020 Reprint: 2021 Reprint: 2022
Reprint: 2024

Price: ₹ 625/-

Laser Typesetting by :
PLUS COMPUTERS, Shahdara, Delhi-110093

Printed at :
PRINTED AT AL-AZIZ BOOKS, SAHIBABAD

Preface to the Eighteenth Edition

Keeping pace with the fast changing fields of Design, Drafting, Manufacturing and inspection and meet the challenges of the world wide nail biting competition of today, it becomes imperative to be fully and systematically trained in the basic fields of engineering. The culture in Design and Drafting has drastically changed and continues to be so, with the fast development of CAD systems, Manual Drafting in fast loosing its ground and is increasingly being replaced by CAD systems. It is also true that no CAD system can perform more than that its user is able to make it to do.

Sound knowledge of Engineering Drawing "Modelling and Drafting" principles is even more essential for any engineering student to be able to take full advantage of the CAD system capabilities. Clear, concise and accurate input to the CAD system being used, coupled with proper planning for the execution of design at hand is of paramount importance.

Drawing and drafting, with proper Geometric Dimensioning and Tolerancing, specified on part drawings forms a legal basis for production of component parts, of subassemblies and higher assemblies, which may be processed in different locations and by different suppliers, that come together to form a complete top level assembly of any product or machine.

Sound knowledge achieved through systematic study and training, of basic drafting principles is a must to be a good Designer, Drafter and Engineer.

It is with these aspects in mind that the author has thoroughly revised and enlarged this book so as to meet the requirements in this important field of engineering. Almost all the chapters have been revised and enlarged. The chapter on 'Geometric Dimensioning and Tolerancing' has been substantially enlarged to include enough basic information that will help the students to understand and enhance their knowledge in this important aspect of modern working drawings and component design.

The style of the book has been changed with a view to make it more clear and meaningful.

Every attempt has been made to place the text and relevant illustrations close together. An effort has been made to provide a good number of exercises that can be solved using any available CAD system.

With all these changes, the academic value of this book is expected to be sufficiently increased.

Pritam S Gill

About the Book

The main features of this text are:

- S.I. Units
- ISO and BSI standards referred.
- Predominantly used First Angle Projection.
- Third Angle Projection is intentionally used so as to train the reader in both these angles of projection to meet the expectation of world wide engineering drawing print.
- Construction and projection lines retained in many cases so as to enhance the theoretical value and facilitate its understanding and interpretation easy.
- A number of illustrative examples as each topic have been included to illustrate all relevant topic principles concerning the topic.
- Each topic is described in detail to help understand and render the relevant topic principles clear.
- Clear and well drafted illustration help easy understanding of the concepts.
- Contains enough basic material in 'Geometric Dimensioning and Tolerancing', that would help take a good start in Drawing and Drafting as used in industry.

The learning material provided in this text is designed to empower students to be good designers in the modern CAD systems. If you know what you want to accomplish, you can make the CAD system to give you the needed end result. One thing is true about all CAD systems, "Garbage in Garbage out." So learn these basic drafting principles in the book to be able to make the CAD system that you use to do what you want it to do.

This book provides a comprehensive and practical overview of Engineering Drawing and Drafting principles. It is intended to provide Drafters, Designers and Engineers with enough background information in this important field. It includes almost all topics that are necessary to provide, clear, concise and systematic information required on machine part drawings.

Contents

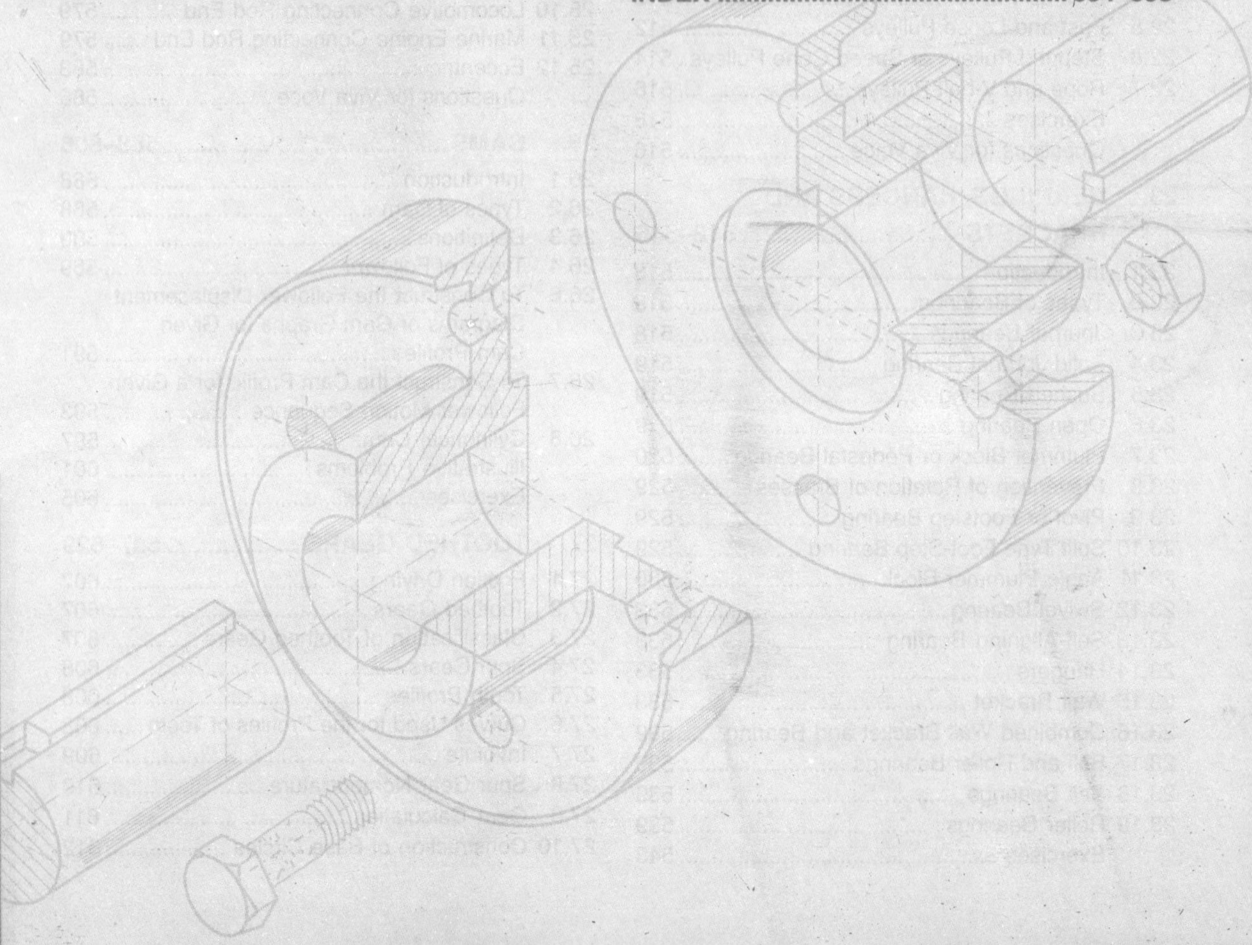

1

Manual Drafting Equipment

1.1. INTRODUCTION

Engineering drawing is entirely a graphic language. Drawing equipment and instruments are needed to record information on drawing paper or any other suitable surface. Drawing, composed of straight lines, curves, circles and arcs, is prepared with certain instruments. The quality of a drawing depends to a large extent on the quality, adjustment and care of the instruments used. Reliable and accurate equipment is essential as good results increase a students interest and enthusiasm. The correct selection and use of pencils and drawing instruments is important. The various instruments and other drafting equipment are described in this chapter. This will be helpful to students in the selection, use and care of their drafting equipments.

In selecting instruments and materials for drafting, the advice of an experienced user or a reliable manufacturer should be sought, because it is impossible for beginners to distinguish high grade instruments from those that are inferior. The best, one can afford, should be bought. With reasonable care a set of good instruments will last a life time, whereas, cheap ones will be an annoyance from the start and worthless after short use. Every student should scratch some neat mark of identification on his/her instruments.

1.2. LIST OF EQUIPMENT

The following list of equipment is a practical selection and constitutes a SET or KIT of instruments:

1. Drawing Board.
2. T-square or Mini-drafter.
3. Instrument Box containing:
    - Large-Size Compass including pen attachment and lengthening bar.
    - Large size Dividers.
    - Three Bow instruments:
        - Small Bow Pencil
        - Small Bow Divider, and
        - Small Bow Pen.
    - Two Ruling Pens
    - Box of Leads, etc.
4. Triangles or Set Squares, 45°, 30° – 60°.
5. Engineering Scale.
6. Protractor.
7. Drawing Pencils, HB, H, 2H, and 3H.
8. Drawing Paper.
9. Pencil Sharpener.
10. Sand Paper Pad.
11. Pencil Eraser.
12. Clamps, Cellotape or Drafting Tape.
13. French or Irregular Curves.
14. Clinigraph or Adjustable Set Square.
15. Dusting Cloth.
16. Drawing ink (black, water proof)
17. Tracing Paper, Cloth or Myler.
18. Erasing Shield.
19. Artgum or other cleaning eraser.

1.3. DRAWING BOARD

The drawing boards are prepared from best seasoned, good quality wood, e.g., planks of white pine, fir, oak, red cedar, mesonite, linoleum etc., soft enough to permit easy insertion and ready withdrawal of drawing pins (where these are used) and yet hard enough to stand the wear and tear of daily use. Its material must be such as to resist the warping effects of changing climatic conditions and to remain square.

To prevent warping, the board should be made of narrow strips (100 to 150 mm wide), *Fig. 1.1*, joined together accurately by groove and tongue joints and glued together edge-to-edge, rather than from one piece. The ebony edge is fitted on left side edge and has saw cuts given on it to allow for expansion and contraction. This edge is the working edge and should be perfectly straight and square with the drawing board surface. T-square head always slides against it.

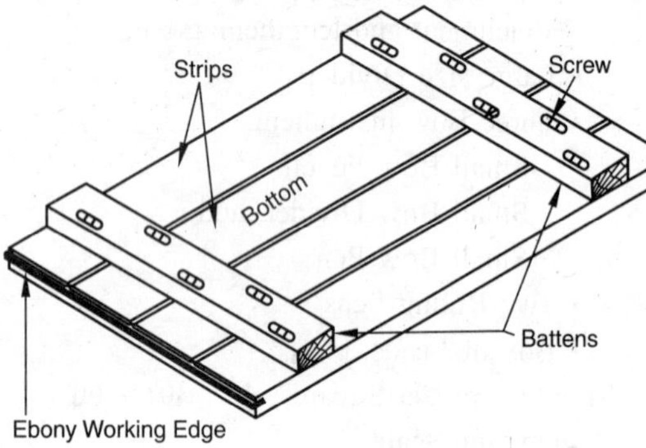

Fig. 1.1. Drawing Board

The standard sizes of drawing boards, according to BIS are given in Table 1.1.

Table 1.1. Dimensions of Drawing Boards

All dimensions are in millimeters

Designation	*To be used with sheet size*	*Size = Length × Width × Thickness (mm)*
D0	A0	1500 × 1000 × 25
D1	A1	1000 × 700 × 25
D2	A2	700 × 500 × 15
D3	A3	500 × 350 × 15

1.4. T-SQUARE

The T- square is composed of a long strip, called the blade, fastened rigidly to a shorter piece called the head, *Fig. 1.2*. The upper edge of the blade and the inner edge of the head are working edges, these should be perfectly straight and square with each other.

These are available in different sizes, i.e., T0, T1, T2, T3 of lengths 1500, 1000, 700 and 500 respectively. T-square is used primarily for drawing horizontal lines and for guiding the set-squares when drawing vertical or inclined lines.

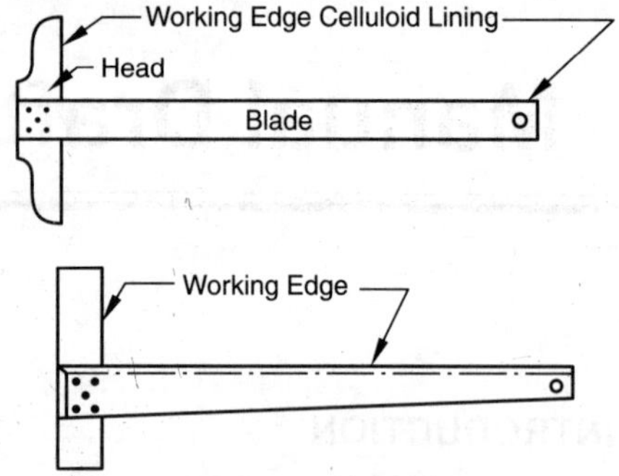

Fig. 1.2. T-Square

All of the functions of T-square can be accomplished by the correct use of a drafting machine or mini-drafter.

Figure 1.3 shows the position of the hand and pencil or open for drawing horizontal lines. The pencil point should be kept as close as possible to the T-square edge and the pencil be inclined in the direction in which the line is drawn. Horizontal lines should always be drawn from left to right. The T-square is frequently replaced with a 'mini-drafter' in class rooms, and "Drafting Machine" in drawing offices. With the increasing use of Computer Aided Drafting the use of drawing instruments is becoming absolete.

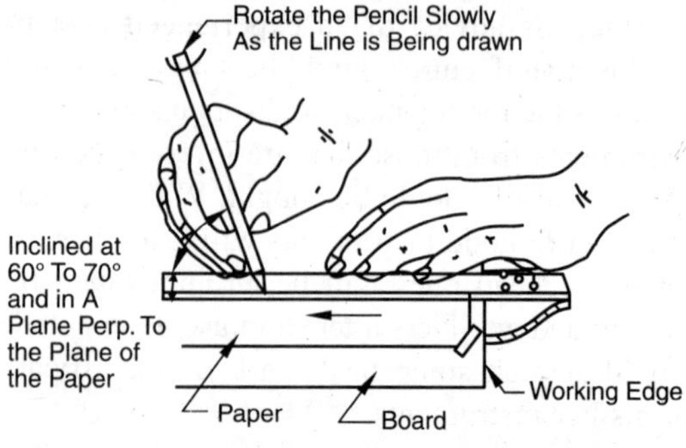

Fig. 1.3. Position of hand and pencil or pen for drawing horizontal lines

1.5. THE COMPASS OR LARGE BOW

The compass or large bow is used for drawing circles and circle arcs. It consists of two legs pivoted at the top. One leg is equipped with a steel needle attached with a screw, the other; shorter leg, is provided with a socket for detachable inserts furnished with the lead for drawing circles in pencil, *Fig. 1.4 (a)*, with a ruling pen for drawing circles in ink, *Fig. 1.4 (c)*, and with another steel needle, *Fig. 1.4 (b)*, for using it as dividers. In drawing circles of large radii a lengthening bar, *Fig. 1.4 (d)*, should be used. When a large number of concentric circles are to be drawn the needle of the compass should be put in a metal centre which has the shape of a thumbtack. For drawing pencil circles the style of pencil lead point illustrated in *Fig. 1.4* should be used because it gives more accurate result and is easier to maintain than other styles. This style of point is formed by first sharpening the lead from the outside then rubbing it on a sandpaper pad to a long flat bevel, approximately 6 mm long, and then finishing it with a sight rocking motion to reduce the width of the point. The lead in compass should preferably be of H pencil. The needle point should be adjusted approximately 10 mm beyond the end of the split sleeve, *Fig. 1.5*.

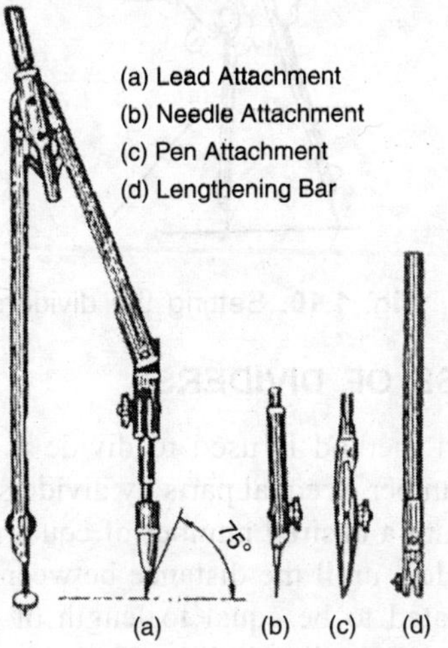

Fig. 1.4. Compass

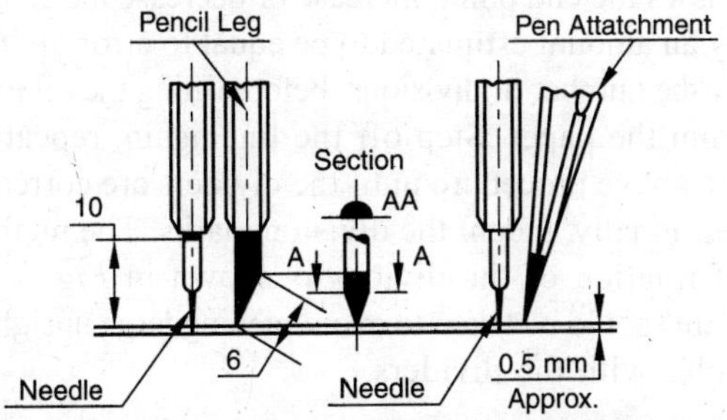

Fig. 1.5. Shaping and adjusting the compass lead to needle point

More correctly, to adjust the needle point of a large compass, the point is adjusted to pen, then replaced by the pencil leg and the pencil point adjusted to the needle point. The length of the pencil lead should be so adjusted that the instrument becomes vertically centred.

1.6. USING THE COMPASS OR LARGE BOW

Before drawing a circle, it is necessary to draw intersecting centre lines, at right angles, to locate the centre of the circle, by intersection. After the pencil point of compass has been adjusted to the length of the radius, keeping the needle point at the centre, the circle is drawn in a clockwise direction as shown in *Fig. 1.8*. The compass is usually held with the thumb and index finger of the right hand and is inclined slightly forward when drawing the circle.

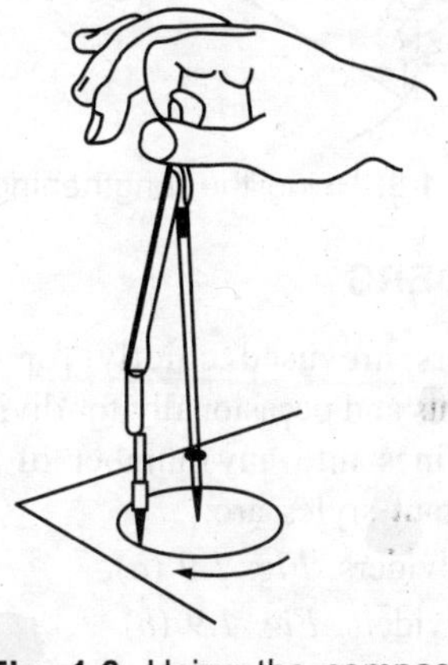

Fig. 1.6. Using the compass

When the radius of the circle, to be drawn, is more than 50 mm the legs of the compass should be bent at the knee joints to stand approximately perpendicular to the paper as shown in *Fig. 1.7*. It is particularly important to keep this in view when drawing ink circles, otherwise both nibs will not touch the paper.

For circles whose radii exceed 125 mm lengthening bar should be used to increase the capacity of the campass. While using lengthening bar, the joints must be bent to bring the legs perpendicular to the paper. Usually both hands are used to draw circles when using this method, as shown in *Fig. 1.8*, because the handle is off centre.

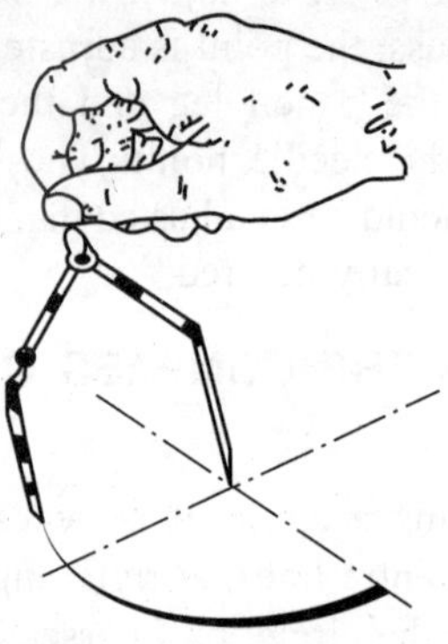

Fig. 1.7. Breaking the legs of the compass

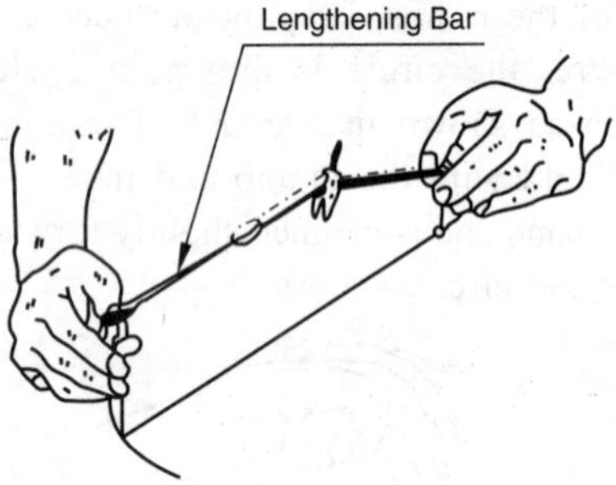

Fig. 1.8. Using the lengthening bar

1.7. DIVIDERS

The dividers are used chiefly for transferring measurements and occasionally for dividing straight or curved lines into any number of equal parts. Two important styles are:

- Plain dividers, *Fig. 1.9 (a)*.
- Bow dividers, *Fig. 1.9 (b)*.

In plain type dividers both legs have a provision of detachment of needles and a self-centring device capable of fine adjustment upto 3 mm. Bow dividers are used to hold the precise distance for transferring or dividing.

Facility in the use of dividers is essential. Further, quick and absolute control of their manipulation must be gained. To set the dividers it should be held as shown in *Fig. 1.10*. This allows an adjustment to be made quickly and easily with one hand. The second and third fingers are used to close the dividers. Practice of this motion should be continued by beginners until they can adjust the dividers to the smallest fraction.

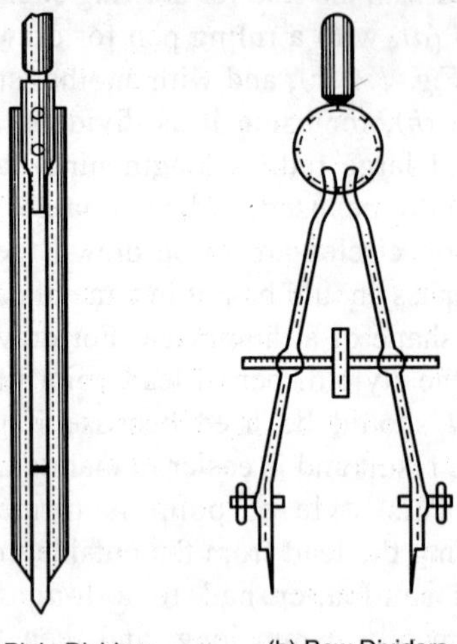

(a) Plain Dividers (b) Bow Dividers

Fig. 1.9.

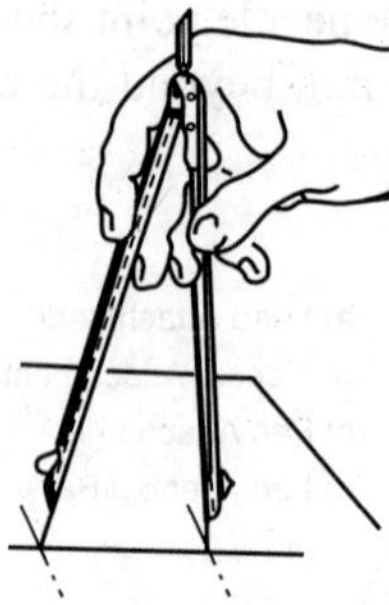

Fig. 1.10. Setting the dividers

1.8. USE OF DIVIDERS

The trial method is used to divide a line into a given number of equal parts by dividers. To divide a line, into a desired number of equal parts, open the dividers until the distance between the points is estimated to be equal to length of a division, and step off the line lightly. If the last pick mark misses the end point, increase or decrease the setting by an amount estimated to be equal to error divided by the number of divisions, before lifting the dividers from the paper. Step off the line again, repeating the above procedure until the dividers are correctly set. Finally, indent the division marks. The method of rotation of the divides is shown in *Fig. 1.11*. Care should be taken to avoid picking large unsightly holes with the dividers.

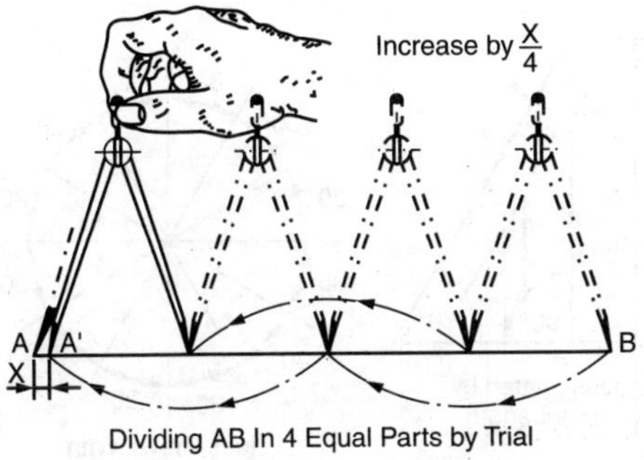

Fig. 1.11. Use of dividers

1.9. BOW COMPASSES

There are two general designs of bow instruments depending upon the adjusting screw ; the side screw and the centre screw, Fig. *1.12*. The main purpose of bow compasses is to draw small circle arcs and circles, both in ink and pencil. The spring bow comprises of two legs (or two different compasses) one with a pencil or ink point and the other with a detachable needle.

These instruments, with their even tension and positive fine adjustment mechanism, are useful in securing accurate results so essential for fine drafting.

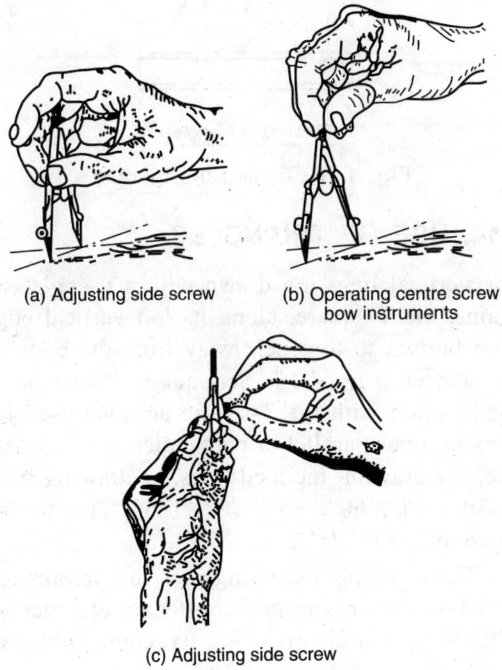

Fig. 1.12.

1.10. USE OF BOW COMPASSES

The bow pen and bow pencil are convenient for drawing circles having a radius of 10 mm or less. The needle point should be adjusted slightly longer than the marking points as in the case of the compass, *Fig. 1.13*. Chisel-edged lead, with slanted side turned outside, is used for general purposes. Whether the centre wheel or the side wheel instrument is used, the adjustment should be made with the fingers of the hand which holds the instrument, *Fig. 1.12 (a)*. When adjustment of any of the side-screw bow instruments is to be made from a large to a small radius, or vice versa, it is best to press the two legs together with the fingers of the left hand, thus relieving the pressure upon thumb screw, so that it may be spun to position quickly without any wear upon threads, *Fig. 1.12 (c)*.

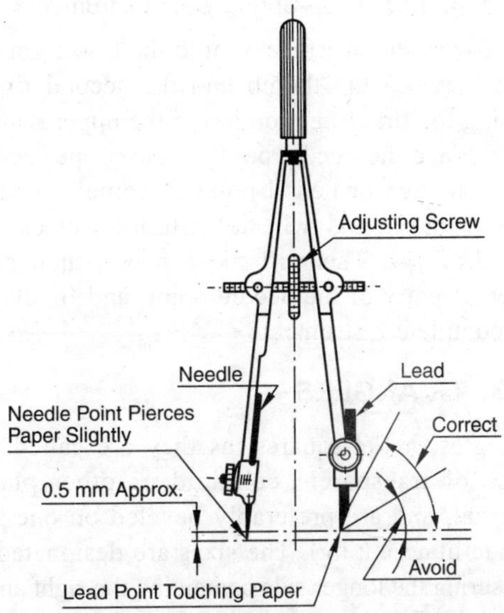

Fig. 1.13.

1.11. DROP-SPRING BOW INSTRUMENTS

These instruments are especially designed for drawing circles of small diameter. The centre rod, or needle point, remains stationary while the tube to which the pencil or pen is attached is rotated upon it and is also free to move up and down, *Fig. 1.16* (*a*). Since the needle point does not revolve it can not wear a hole in the paper.

1.12. USE OF DROP-SPRING BOW INSTRUMENTS

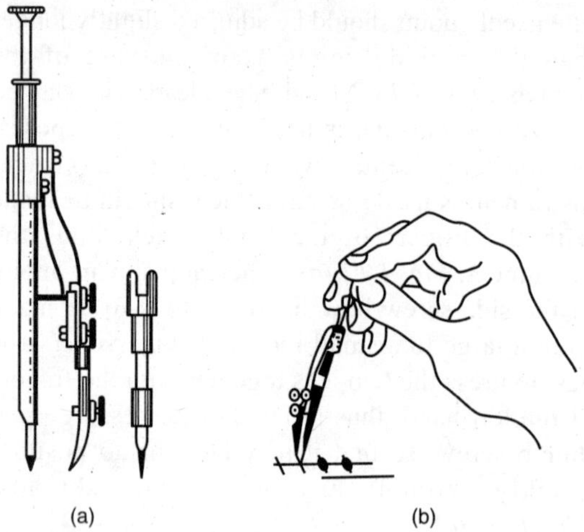

Fig. 1.14. Drop-Spring Bow Instruments

To use the instrument, hold the lower knurled head between the thumb and the second finger, placing the first finger on top of the upper knurled head. Place the needle point at the proper centre, allow the pen or pencil point to come in contact with the paper, and twirl the instrument clockwise, *Fig. 1.14 (b)*. Then lift the pen or pencil point independently of the needle point, and finally lift the complete instrument.

1.13. TRIANGLES

Triangles or Set-squares, as they are called, are made of transparent celluloid or other plastic material, and are preferably beveled on one side (to facilitate inking). The sizes are designated by measuring the longer side containing the right angle, *Fig. 1.15*. Triangles should be kept flat to prevent warping. For ordinary work, a 200 mm 45° and a 240 mm 30°-60° are the good sizes.

Through internal strain, triangles sometimes loose their accuracy. They may be tested by drawing a perpendicular and then reversing the triangles as shown in *Fig. 1.15*, and again drawing a line at the same position. Any difference in the two lines gives twice the error. When a triangle has been tested and found to be incorrect, it may be corrected by scraping, sand papering, or filing the edges where necessary.

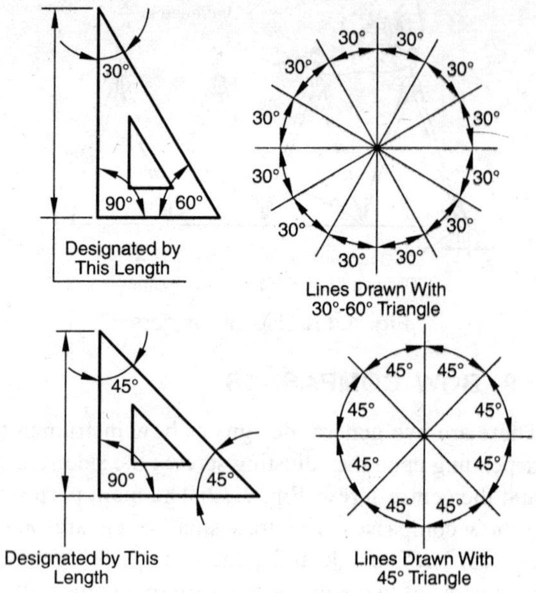

Fig. 1.15. Triangles

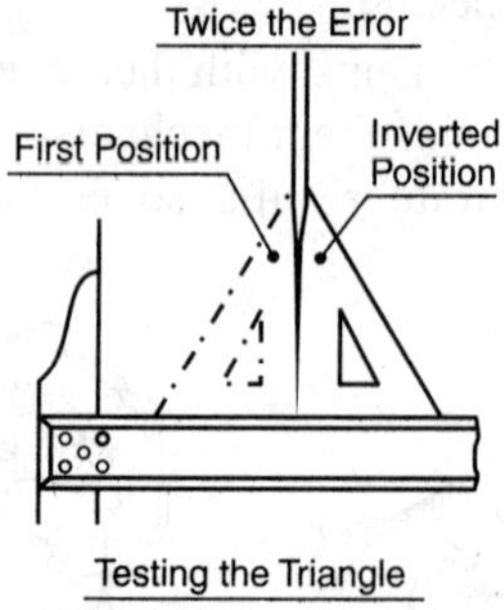

Fig. 1.16. Testing the triangle

1.14. USE OF TRIANGLES

The vertical lines are drawn with a triangle, set against the T-square, along its left vertical edge from bottom to top, i.e., away from the body of the drafter, *Fig. 1.18*. Set-squares alone or in combination with the T-square are also used for drawing lines parallel or perpendicular to another line, for drawing inclined lines, or drawing lines at 30°, 45°, 60°, 15°, 75°, 135°, etc. to the horizontal, *Fig. 1.17*.

The triangles used singly or in combination with T-square or mini-drafter offer a useful method of dividing a circle into four, six, eight, twelve or twenty-four equal parts, *Fig. 1.19*.

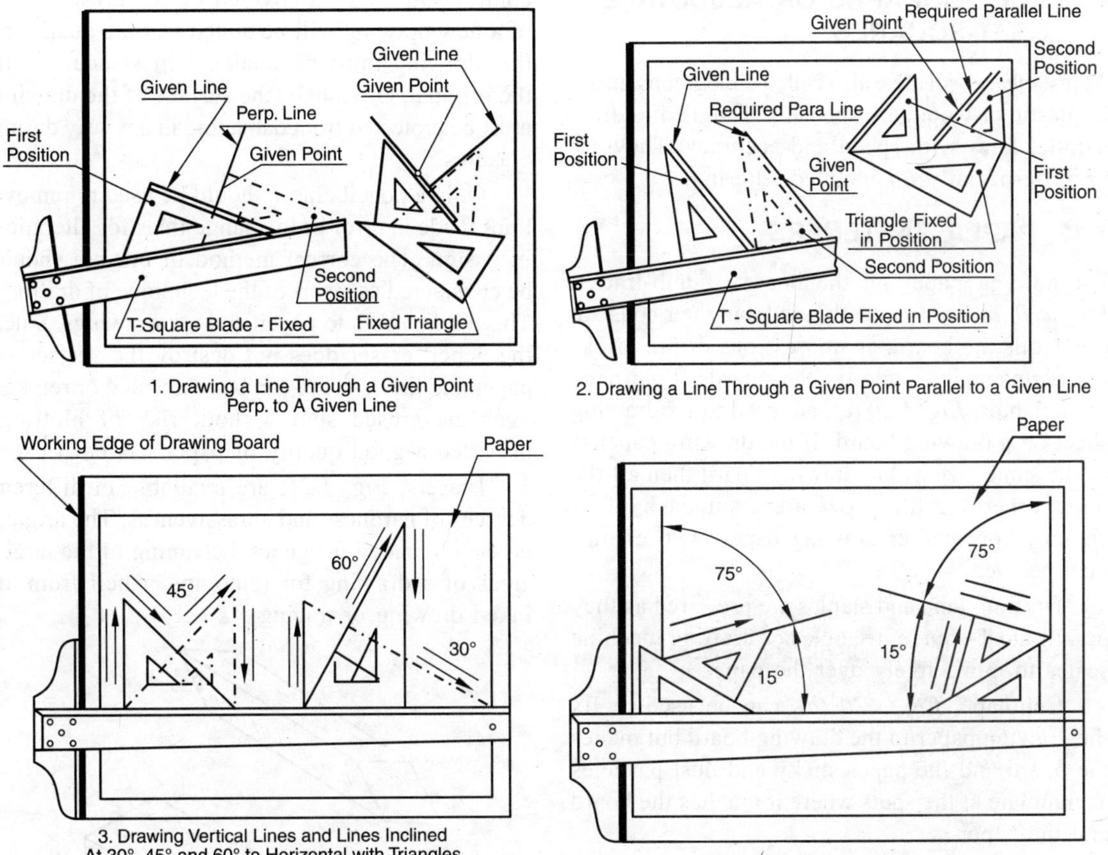

1. Drawing a Line Through a Given Point Perp. to A Given Line

2. Drawing a Line Through a Given Point Parallel to a Given Line

3. Drawing Vertical Lines and Lines Inclined At 30°, 45° and 60° to Horizontal with Triangles

4. Drawing Angles of 15° and 75° With Triangles

Fig. 1.17. Uses of Triangles

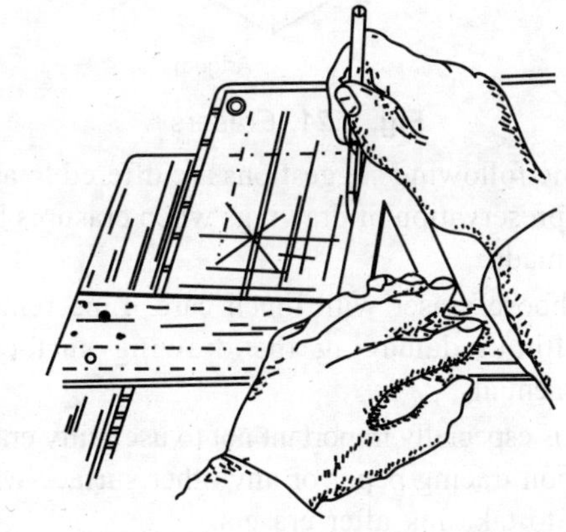

Fig. 1.18. Drawing vertical lines with triangle and T-square

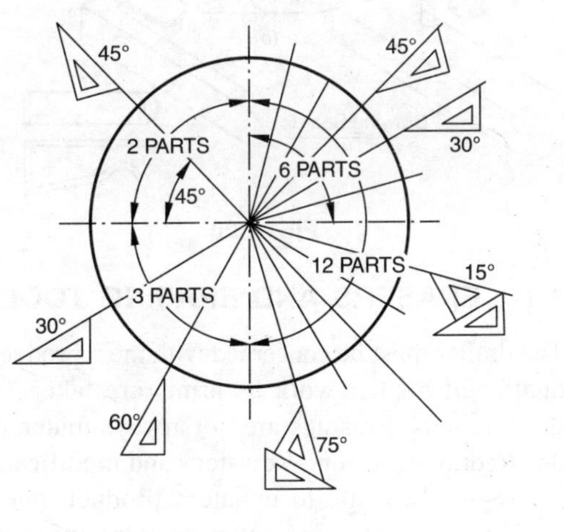

Fig. 1.19. To divide a circle into 4, 6, 8, 12 or 24 equal parts by the use of triangles

1.15. CLINOGRAPHS OR ADJUSTALE SET-SQUARES

Adjustable set-squares, also called clinographs, made of plastic or wood, are specially designed to draw parallel lines with speed and accuracy. They are useful especially for inclined lettering.

1.16. PAPER FASTENERS

For holding paper on the board, thumb-tracks, *Fig. 1.20 (a)*, are frequently used, but their constant use spoils the board, as impressions of pin pricks are permanently made on the boards. Fold-back steel clamps, *Fig. 1.20 (c)*, are used to fix drawing sheet on a drawing board. If the drawing paper is of the same size as the drawing board then all the four corners of the paper are clamped by four clamps. For smaller drawing papers two clamps may be used.

Drafting tape and staples are preferred as they permit the T-square, triangles or drafting machine scales to move freely over the paper.

Cellotape, *Fig. 1.20 (b)*, can be used to fix the drawing paper to the drawing board but makes the board and the paper sticky and dust particles accumulate at the spots where it touches the board and the paper.

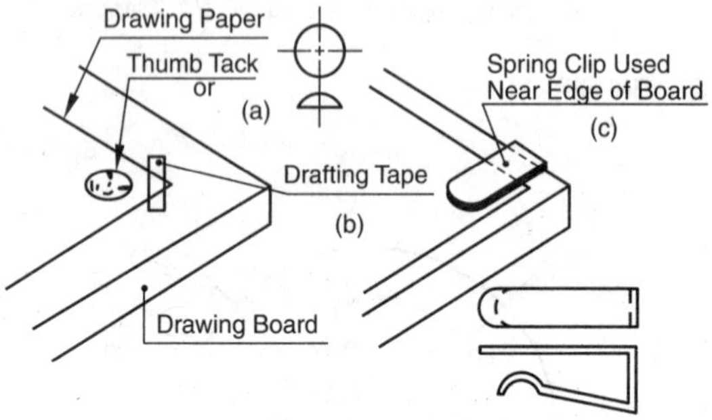

Fig. 1.20.

1.17. ERASERS AND ERASING TOOLS

The drafter must be concerned with the reproducible quality of his/her work as prints are taken from the originals. Erasings are not always undertaken due to drawing errors. Revisions and modifications are regularly made to update a product, due for example, to changes in materials, individual components, manufacturing technique, operating experience and other causes outside the drafter's control. On the area covered by the erasers, part of a new drawing will be added and the quality of this drawing must be identical in standard with the original. Obviously, the surface of the drawing must be protected from damaging in any way during erasures.

Only a pencil eraser should be used to remove lines made in error or to change lines for alteration in design. The correct method of erasing should be considered as a part of the technique of drafting. This applies both to pencil and ink drawings. Since the pencil eraser does not destroy the surface of paper and the drawing can be repenciled or reinked over the erased spot without risk of blotting, provided a good quality of paper has been used.

Erasers, *Fig. 1.21*, are available in different degrees of hardness and abrassiveness. The artgum is recommended for general cleaning of the larger areas of a drawing for removing pencil from an inked drawing or tracing.

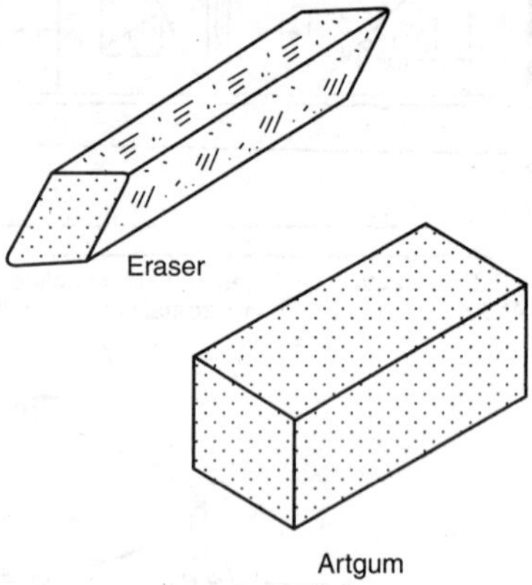

Fig. 1.21. Erasers

The following suggestions are offered to assist in the preservation of drawings when erasures have to be made:

1. Choose eraser with much care. Line removal without damaging the drawing surface is essential.
2. It is especially important not to use gritty erasers upon tracing paper or any other surface which is to take ink after erasers.
3. Better results are obtained from erasing if a hard surface, such as a triangle, is placed under the paper. If the surface has been injured by erasing, it can be improved by burnishing with the back of the finger nail.

4. When there are many lines close together and only one needs to be removed or changed, the others may be protected by using an erasing shield, *Fig. 1.22 (b)*.
5. When an ink blot is made the excess ink should be soaked by applying a blotter, or smeared with the finger if blotter is not available and not allowed to soak into the paper. When the spot is dry the remaining ink is easily erased.
6. Thoroughly erase the lines as a ghost effect can be observed with incomplete erasures when the prints are made.
7. After erasing, clean the back of the sheet as pressure on the front side of the drawing often transfers dirt containing a high proportion of graphite on to the back of the sheet.
8. Use a brush to remove the eraser crumbs from the surface of the drawing and this prevents smudging.

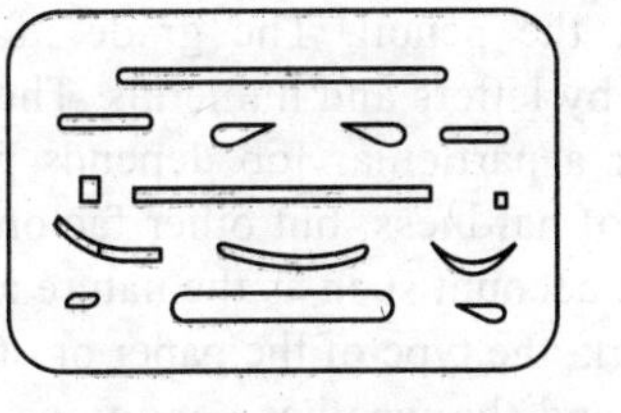

(a) Erasing Shield

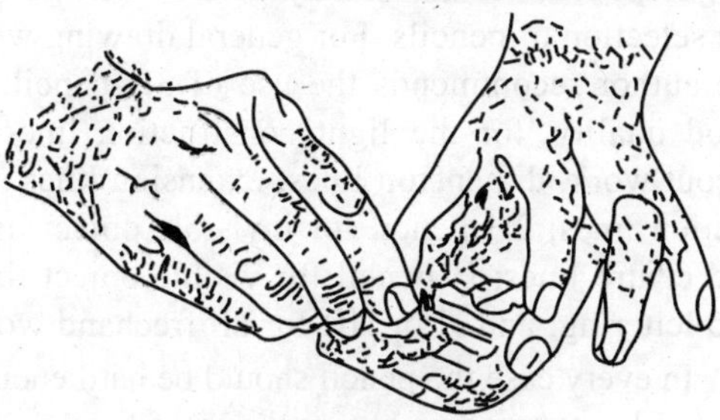

(b) Using the Erasing Shield

Fig. 1.22.

1.18. SCALES

Scale is defined as the ratio of the linear dimension of an element of an object as represented in the original drawing to the linear dimension of the same element of the object itself. Scales are made in a variety of graduations to meet the requirements of many different kinds of work. The scales adopted for drawings should have a bearing on the degree of accuracy required of scaled distances on prepared drawings.

A drawing, of an object, may be the same size as the object or it may be larger or smaller than the object. A scale with the ratio 1 : 1 is called a Full Size. A scale sswhere the ratio is larger than 1 : 1 is called an enlargement scale and it is said to be larger as its ratio increases. While a scale where the ratio is smaller than 1 : 1 is called a reduction scale and it is said to be smaller as its ratio decreases. In most cases, a drawing is made smaller than the object it represents. In general the largest scales conveniently possible, should be used. The scale of drawing should be indicated in the appropriate place, in the Title-Block. The complete designation of a scale should consist of the word 'SCALE' (or its equivalent in the language used on the drawing) followed by the indication of its ratio, as follows:

SCALE 1 : 1 for full size;

SCALE X: 1 for enlargement scales;

SCALE 1 : X for reduction scales.

The scale to be used for an engineering Drawing is the Metric Scale. The following are the scales, recommended by BIS, for use on Technical drawings:

1. Full Scale

   1 : 1

2. Reducing Scales

1 : 2	1 : 50	1 : 1000
1 : 5	1 : 100	1 : 2000
1 : 10	1 : 200	1 : 5000

3. Enlarging Scales

50 : 1	5 : 1
20 : 1	2 : 1
10 : 1	

Details that are too small for complete dimensioning in the main representation should be shown adjacent to the main representation in a separate detail view (or section) which should be drawn to a large scale.

Mostly the scales are made of box-wood in a variety of shapes. Scales are used for scaling-off measurements from scale to drawings. Opening the dividers to required measurements, by directly keeping it on the scale, and then transferring it to the required position, is quicker but this practice spoils the graduations of the scale. Some authors recommend the direct use of scale for taking a measurement. To do this, place the scale in position and mark the distance on the paper by short, light lines. The student should avoid commutative errors in the use of the scale. If a number of distances are to be set off end-to-end, all should be set-off in one setting of the scale, by adding each successive measurement to the preceding one. Since there is every possibility of making a slight error while setting a distance, if we set all distances in line, individually, such errors will accumulate to give a large error.

1.19. PROTRACTOR

This instrument is used for measuring or setting-off angles other than those obtainable with the triangles. For ordinary work the celluloid or stamped sheet metal protractor is satisfactory, *Fig. 1.23*.

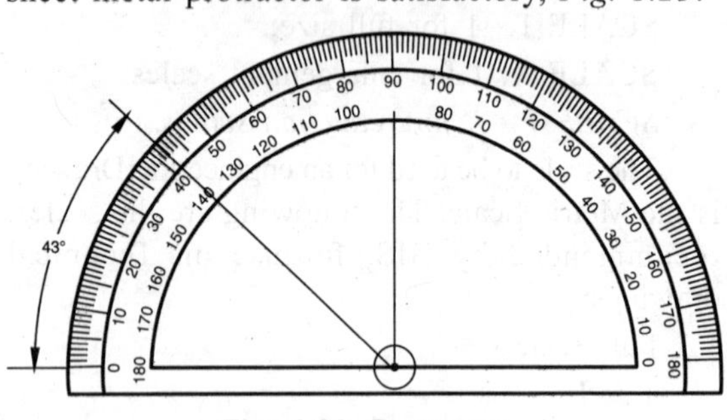

Fig. 1.23. Protractor

1.20. DRAWING PENCILS

A student of engineering and professionals should be equipped with good quality, well sharpened pencils, with leads of various degrees of hardness; e.g., 4H, 3H, 2H, H, HB, etc., descending in grades of hardness (9H, 8H, 7H, 6H, 5H, 4H, 3H, 2H, H, F, HB, B, 2B, 3B, 4B, 5B, 6B, 7B) to make eighteen grades from 9H (the hardest) to 7B (the softest), *Fig. 1.24*. Drawing pencils are hexagonal in shape, so that three planes may be in contact with the thumb, fore finger and second finger and that the pencil should not easily roll off the drawing board. Mechanical pencils, *Fig. 1.30* are now widely used. Their length is constant and they can be easily refilled with new leads. Much time is saved in sharpening, since the lead can be 'fed' from the pencil as needed, and pointing on the sand paper pad as the regular wooden pencils. It is also more economical since the lead is used to its entire length while the ordinary wooden pencil are difficult to use when shorter than about 8 cm. Drawing pencil leads are made chiefly of graphite mixed with varying quantities of clay to produce different degrees of hardness. The degree of hardness is determined by the ratio of graphite to clay in the mixture; the greater the ratio of graphite to clay the softer the pencil. The grade of a pencil is indicated by letters and numerals. The choice of a pencil for a particular job depends primarily on its grade of hardness, but other factors have to be taken into account such as the nature and accuracy of the work, the type of the paper or other drawing material, and the weather conditions etc.

Standards for the quality usually will govern the selection of pencils. For general drawing work the author recommends the use of 3H pencil, of good quality, for the light construction lines in layout work, dimension lines, extension lines and section lines; 2H pencil for invisible object lines and centre lines; H pencil for visible object lines and lettering; and H or HB for all freehand work.

In every case the pencil should be hard enough not to blur or smudge, but, at the same time, not so

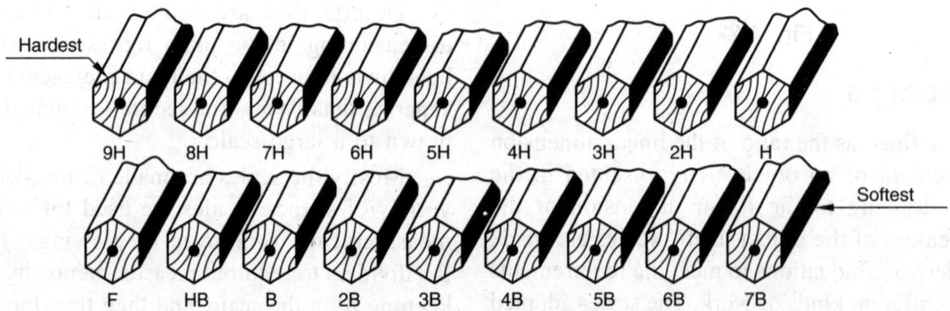

Fig. 1.24. Pencil grade chart

hard as to cut grooves in the paper under reasonable pressure. The sand paper pad should be with in easy reach and, always during use, the pencils should be kept sharp. All lines should be kept sharp. All lines should be uniform, firm and opaque. When using a conical pencil, rotate the pencil as the line is drawn, so as to keep line and pencil sharp. Positions of the lead point in relation to the straight edge for normal and very accurate work are illustrated in *Fig. 1.28*. Excessive graphite should be dusted off the drawing sheet, with a soft cloth, occasionally. Never entertain the thought that poor penciling can be corrected in inking or tracing.

1.21. POINTING THE PENCIL

In technical drawing, as in all practical work, success largely depends upon the condition of the tools employed. Without suitable and well sharpened pencils drawings are bound to be inaccurate, displeasing and characterless.

For general use a conical point is used. Whereas wedge point is more suitable for drawing straight lines, as it requires less sharpening and makes a dense line, *Fig. 1.25*. For sketching and lettering a conically mended pencil is exclusively used. When sharpening, the pencil, using a sharp pen knife, the wood at the unlettered end of the pencil, is reduced to a conical form, the strokes of the knife being straight and not 'concave' and directed towards the body, *Fig. 1.29 (a)*. About 8 to 10 mm of the lead should be exposed. The cut should be formed, including the wood and lead exposed, about 24 to 30 mm long, *Fig. 1.26*. The lead should be then shaped to a conical point on a sandpaper pad. The pad should be held stationary and the pencil moved to and fro on it and giving a rotatory motion at the same time, *Fig. 1.29 (b)*. The whole end is then lightly wiped to remove adhering specks of graphite and the end is lightly burnished on a scrap of drawing paper. The shape of the lead of the pencil, mended to sharp conical shape and also shaped to chisel shape, is illustrated in *Fig. 1.25*. While using a chisel shaped pencil point the "flat" side (actually slightly convex) of the chisel end is run along the edge of the straight edge, set-square or T-square. Ofcourse, the edge wears longer than a point but nevertheless chisel points "take getting used to" and not as widely used in the trade as conical points.

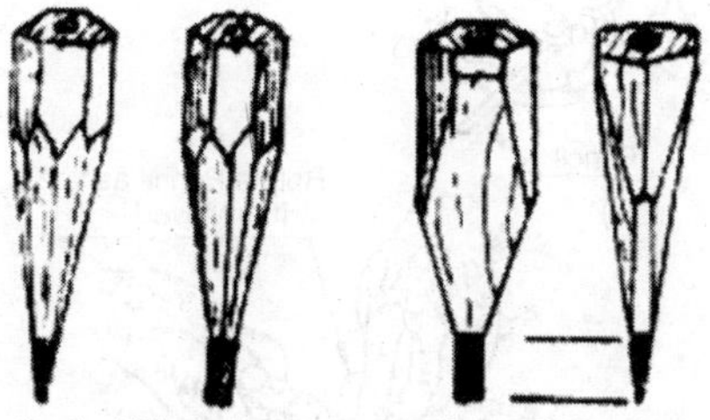

Fig. 1.25. Types of pencil point

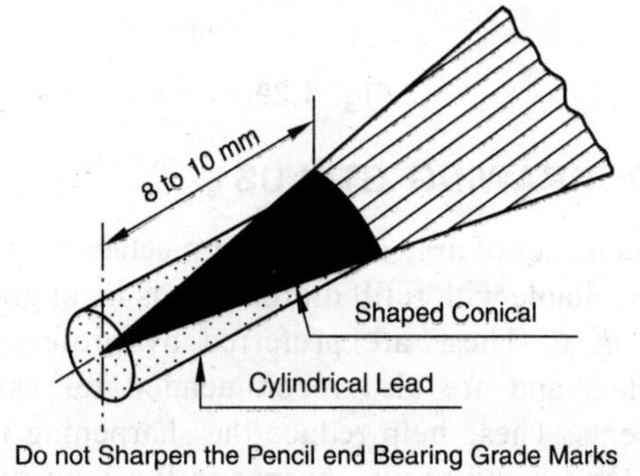

Fig. 1.26.

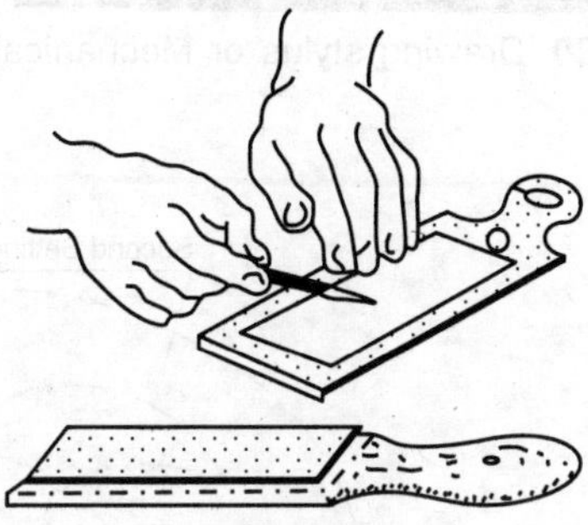

Fig. 1.27.

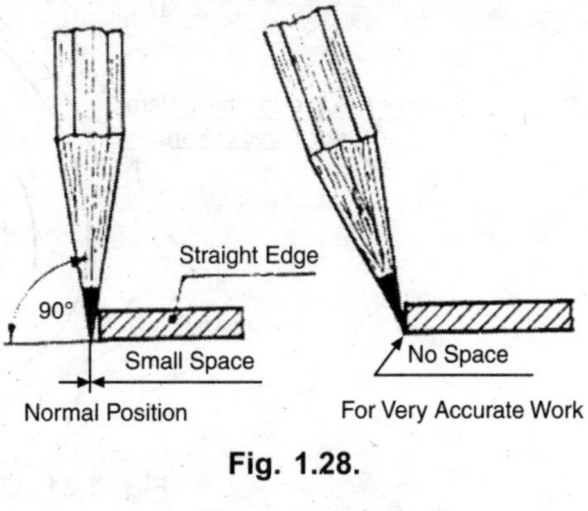

Fig. 1.28.

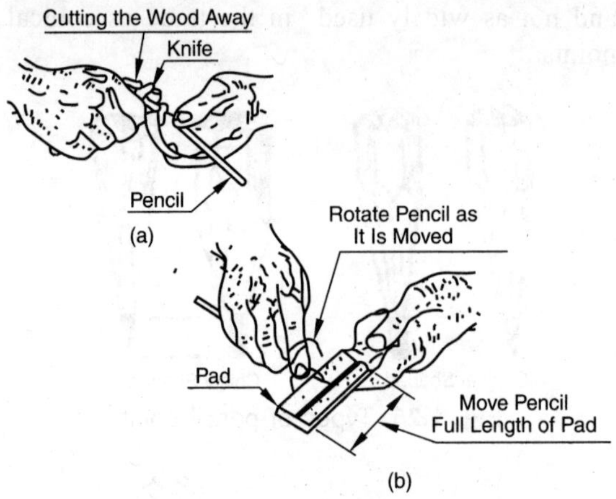

Fig. 1.29.

1.22. DRAWING STYLUS

Many makes of drawing stylus or mechanical pencil are available with refill drafting leads in all grades, *Fig. 1.30*. These are preferred by professional drafters and are also recommended for use by students. These help reduce the sharpening time, last longer and prove cheaper in the long run.

Fig. 1.30. Drawing stylus or Mecnanical pencil

1.23. FRENCH CURVES

Curved rulers called, "French Curves" or "Irregular Curves" are used for drawing curved lines other than circles or circle arcs. These are exclusively used for drawing curves where the radius of curvature is not constant. Many different forms and sizes of curves are manufacture. The best curves are made of highly transparent celluloid.

The proper size of the French Curves requires skill, especially, when the lines are to be drawn in ink. First plot the points through which the curve is to pass and then sketch a smooth, freehand line through the points.

To draw a mechanical line over a freehand line with the French curves, we need only match the various segments of the freehand curve with some part of the French curves and then draw the line with pencil, ruling pen or liner, along the edge of the curve. It is very important that the French curve should match the curve, to be drawn, for some distance at each end, beyond the segment to be drawn for any one setting of the curve, *Fig. 1.31*. If this rule is carefully observed, the successive sections of the curve will be tangent to each other, without any abrupt change in the curvature of the resulting mechanical line. The short radius part of the French curves should be kept towards the short

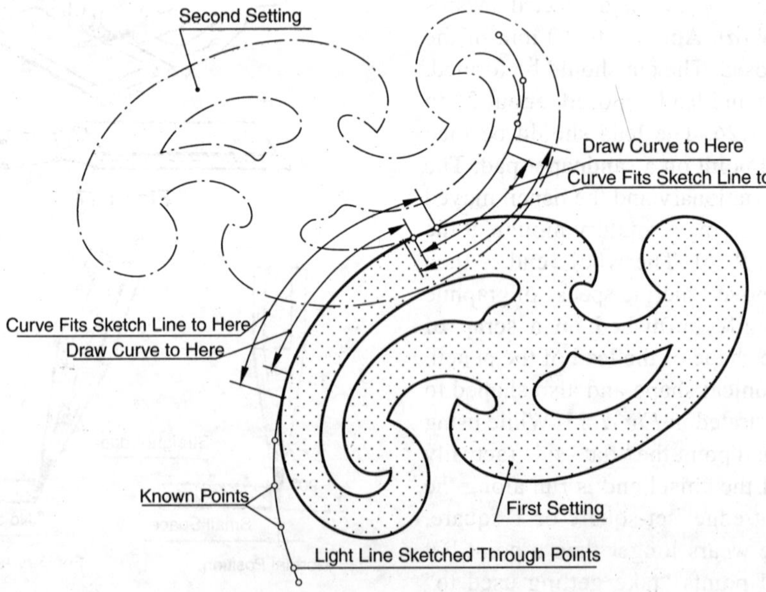

Fig. 1.31. Using the French Curves

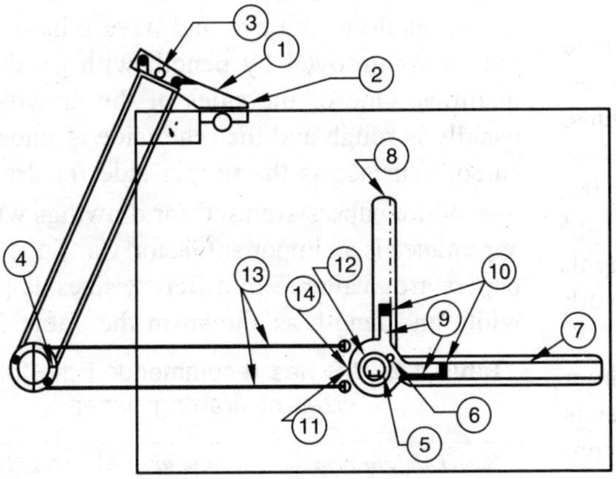

Part No.	Description
1	Clamping bracket
2	Clamping screw
3	Levelling screw
4	Centre ring
5	Protractor head
6	Protractor lockng lever
7	Big scale
8	Small sclae
9	Locating pins
10	Coin slot screws
11	Zero indicator
12	Knob
13	Parallel Arms
14	15° Dividing head

Fig. 1.32. Junior drafter

radius of the curve to be drawn. This also prevents abrupt changes in direction.

The drafter should changes his/her position with respect to the line to be drawn, when necessary, so that he/she always works on the side of the curve away from him/her.

On symmetrical curves, the same portion of the irregular curve should be used on similar parts of the curve to be drawn. These portions can be marked in pencil directly on the French curves.

1.24. DRAFTING MACHINE

Drafting machine is a special form of drawing desk. It is used to place the drawing board and is fitted with a device for changing the angle of the drawing board with the horizontal. It combines the advantages of the T-square, triangles and scales. In operation, this device keeps the two blades always parallel to their original position, no matter where they are moved on the drawing sheet fixed on the board. The two blades which may be simple straight edges with scales along them, are accurately set at right angles to each other. The blades are removable, and hence a variety of scales can be used. The adjusting head has a protractor with vernier attachment so that the blades may be set and clamped at any desired angle, *Fig. 1.32.* Considerable saving of time results from the use of this machine. Clamp the drafter bracket at about 12 cm from the top left corner of the drawing board. A simple form of the linkage to which the adjusting head and mutually perpendicular scales are attached may be separately manufactured in small, cheap and handy sizes. This when mounted on the normal drawing board, placed on a drawing desk, can serve almost the same purpose as that of a drafting machine (though it will have a much limited size range). Such a linkage with scales, and commercially being manufactured, is called a Mini-Drafter, *Fig. 1.33.* By means of this machine, horizontal, vertical and inclined parallel lines of suitable lengths anywhere on the sheet can be drawn with ease and quickly. Its use by engineering students is replacing T-squares.

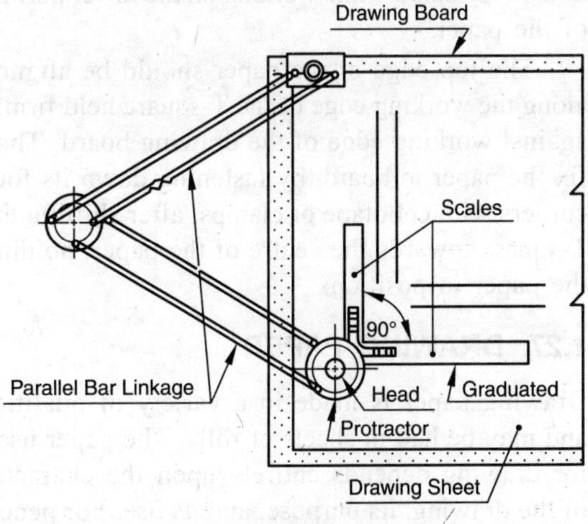

Fig. 1.33.

1.25. PREPARATION FOR DRAWING

Working on a board with its working surface horizontal is apt to be inconvenient and tiring. So boards are generally tilted by different arrangements, at an angle of 20° to the horizontal. It should be adjusted to a convenient height suitable for the concerned individual or an average height. We find that backache is avoided if the board is at a height, above the floor, sufficient to enable us do the work standing rather than sitting.

It is also important to study lighting. Without doubt the best light for a right handed drafter is natural North light. A diffused light from the front left, over the left shoulder diminates the formation of shadows cast by the hands, T-square and set squares. The instruments should be placed, within an easy reach, on the worktable or on a stand beside the table. The equipment should be well cleaned before starting to draw.

1.26. FIXING THE PAPER

As the T-square head is more rigid near the head than towards the outer end, the paper should be fastened close enough to the working edge of the board to reduce, to a minimum, any error resulting from a slight "swing" in the blade of the T-square. Further for easy manipulation of T-square and accurate preparation of drawing, the paper should be located close enough to the upper edge of the board. Keep in mind that there should be sufficient room provided below the paper for supporting the arm on the board while working on the lower portion of the paper.

The top edge of the paper should be aligned along the working edge of the T-square held firmly against working edge of the drawing board. Then fix the paper to board, by fastening down its four corners with cellotape or clamps, after shifting the T-square towards the centre of the paper, holding the paper in position.

1.27. DRAWING PAPER

Drawing paper is made in a variety of qualities and may be had in sheets or rolls. The paper used for drawing depends entirely upon the character of the drawing, its purpose, and its use. For pencil layouts and working drawings cartridge or machine made paper is largely used. In general, paper should have sufficient grain, or tooth to take the pencil, be agreeable to the eye and have a hard surface, not easily grooved by pencil, with good erasing qualities. One of the sides of the drawing paper usually is rough and the other side is smooth. The smooth surface is the proper side for drawing.

White papers are used for drawings where the appearance is an important factor. Cartridge drawing papers are available in different sizes as per their width and length as shown in the Table 1.2.

Table 1.2. BIS has recommended the following sizes of drawing paper

Sr. No.	*Designation*	*Trimmed size in mm Width × Length*	*Untrimmed size in mm Width × Length*
1.	*A0*	841 × 1189	880 × 1230
2.	*A1*	594 × 841	625 × 880
3.	*A2*	420 × 594	450 × 625
4.	*A3*	297 × 420	330 × 450
5.	*A4*	210 × 297	240 × 330
6.	*A5*	148 × 210	165 × 240

1.28. TRACING PAPER

Tracing papers are thin papers, natural or transparentized, on which drawings are traced in pencil or ink. From these traced drawings blue-prints can be prepared. Original drawings, too, can be penciled on tracing papers and blue-prints can be made directly from these drawings.

1.29. GRAPH PAPER

1. **Squared Paper:** Squares of various sizes are drawn upon it. Thick and rough graph papers may be used for freehand sketch work with advantage. The vertical and horizontal lines ruled on graph papers help to estimate dimensions and guide the sketcher for direction of drawing lines, *Fig. 1.34 (a)*.
2. **Isometric Papers:** On these papers lines inclined at 30°, towards left and right, with horizontal and also vertical lines are ruled equally spaced in all directions. This facilitates sketching of isometric drawings as the lines ruled on them conform to the directions of drawing lines in isometric drawing, *Fig. 1.34 (b)*.

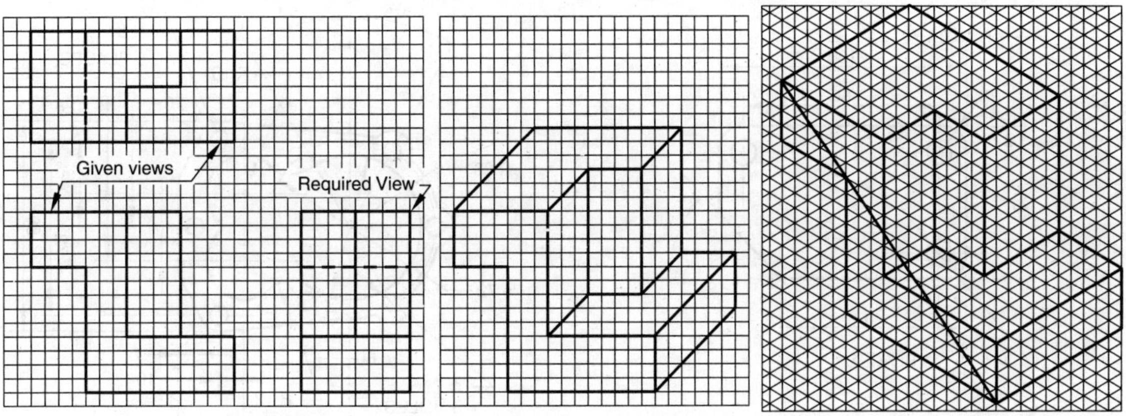

Orthographic Views (1) Oblique Views (2) Isometric Views (2)

(a) Squared paper (b) Isometric paper

Fig. 1.34.

1.30. CAUTIONS IN THE USE OF INSTRUMENTS

The following points should be carefully noted and observed when drawing:

1. Never use scale as a ruler for drawing lines.
2. Never use T-square for drawing any other lines than horizontal.
3. Never draw horizontal line with the lower edge of the T-square.
4. Never use the T-square or set-square as a guide while cutting paper with a knife edge.
5. Never use a dull pencil.
6. Never sharpen the pencil over the drawing board.
7. Never jab the driders into drawing board.
8. Never scrub a drawing all over with an eraser after finishing. It takes the life out of the lines.
9. Never start work without properly wiping off the table and instruments.
10. Never put the bow instruments away without opening to relieve the spring.
11. Never work on a table cluttered with unwanted instruments or equipment.
12. Never fold a drawing or tracing paper.

PROBLEMS 1.1 TO 1.17. *Two Dimensional (2D) Geometry of 17 different shapes is shown on the following pages as EXERCISE 1 to EXERCISE 17. Each exercise is a good candidate to practice in drawing 2D geometry and the use of drawing instruments, practicing different types of lines, dimensioning and drawing shapes with drawing instruments as well as with any CAD System available, learnt and practiced by the student. Reproduce each geometric shape; follow the drafting techniques regarding types of lines, dimensions, tangency of lines:*

1. *Using manual drafting instruments.*
2. *Using any suitable CAD system, available and learnt by the student.*

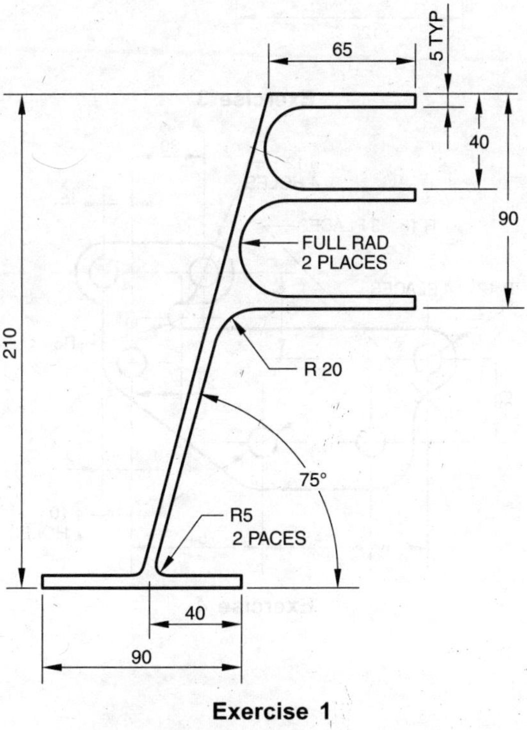

Exercise 1

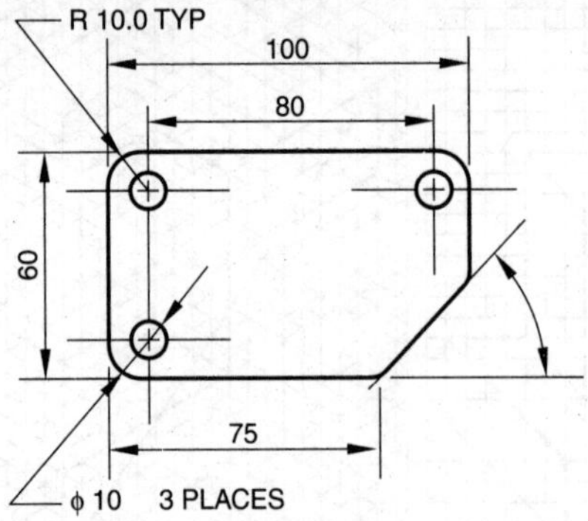

Exercise 2

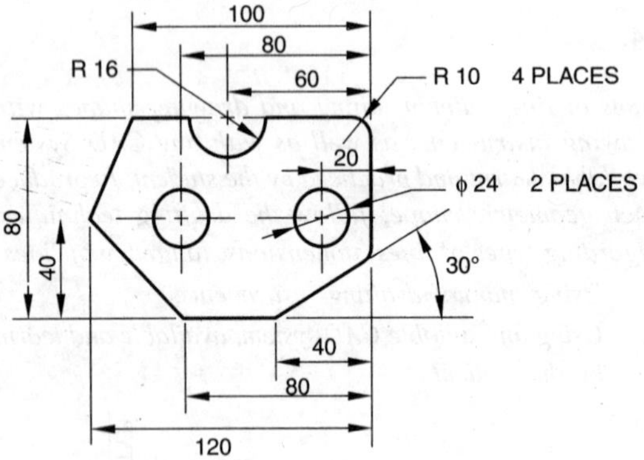

Exercise 3

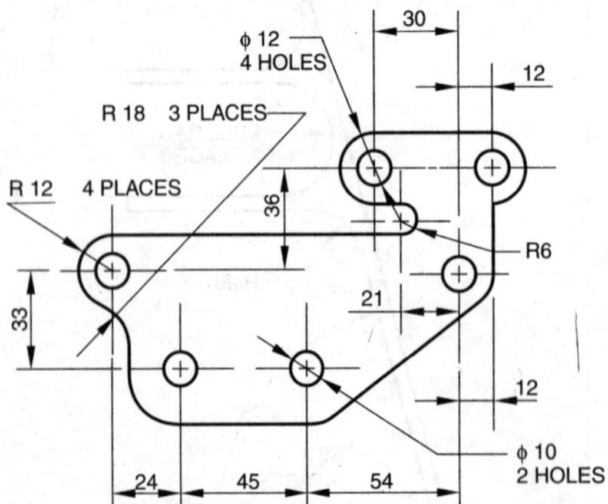

Exercise 4

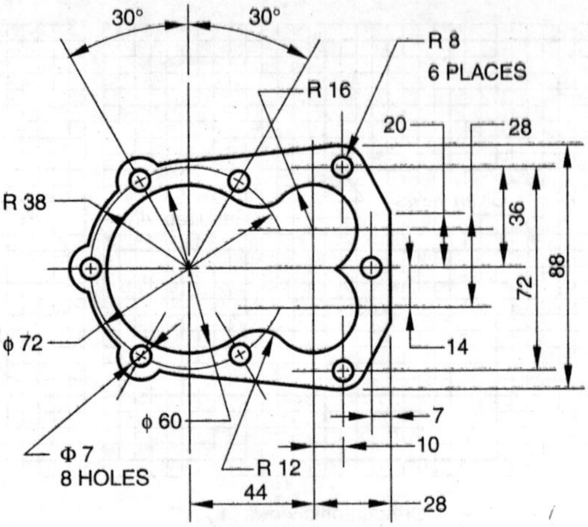

Exercise 5

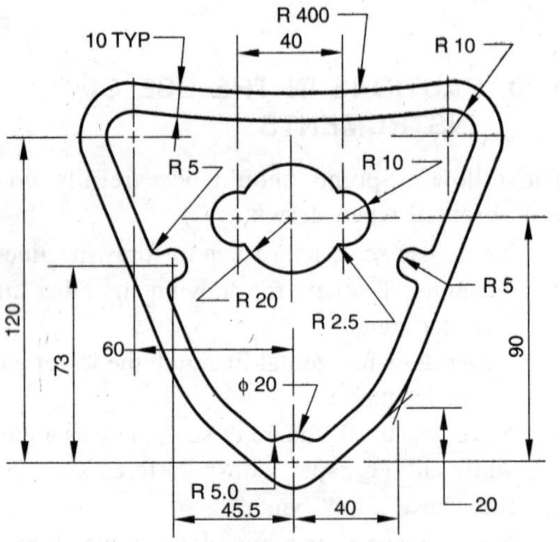

Exercise 6

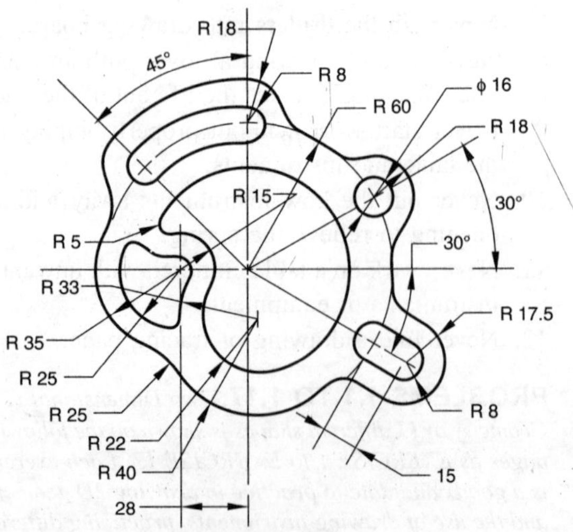

Exercise 7

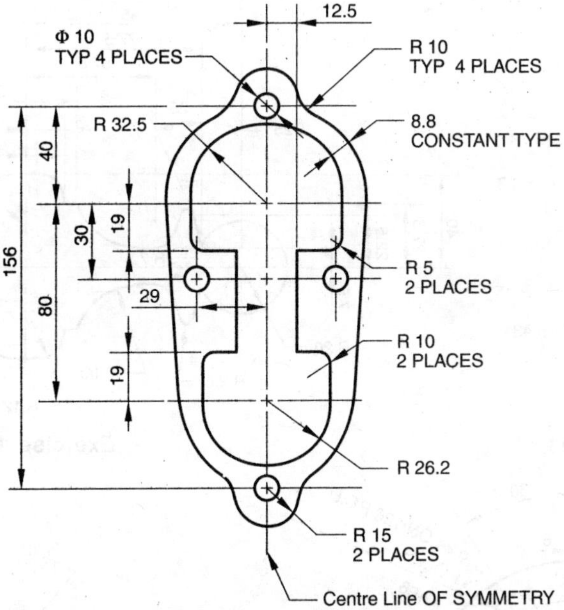

Exercise 8

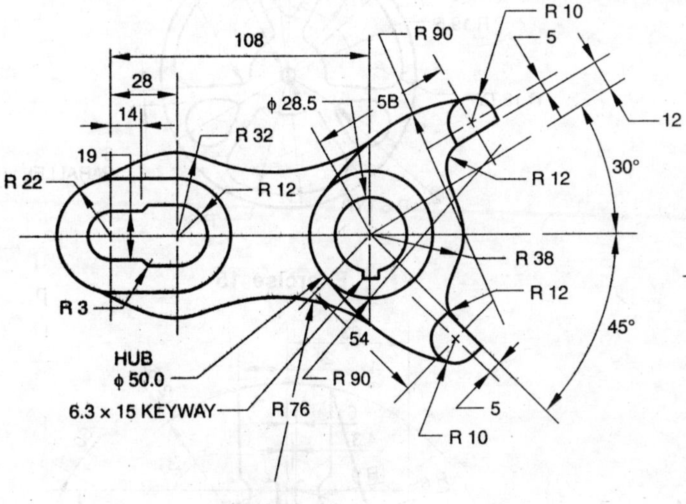

Exercise 9

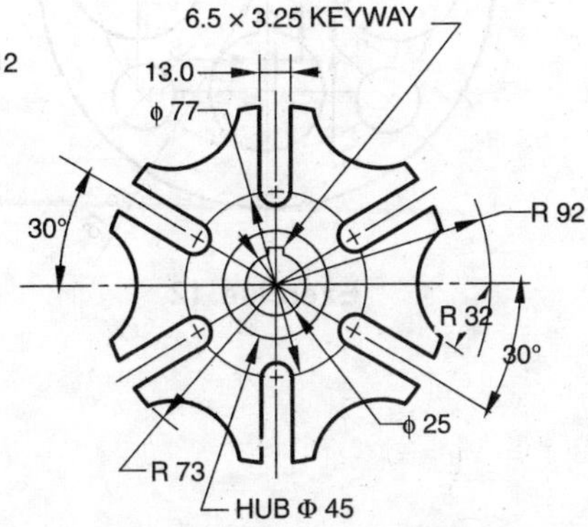

Exercise 10

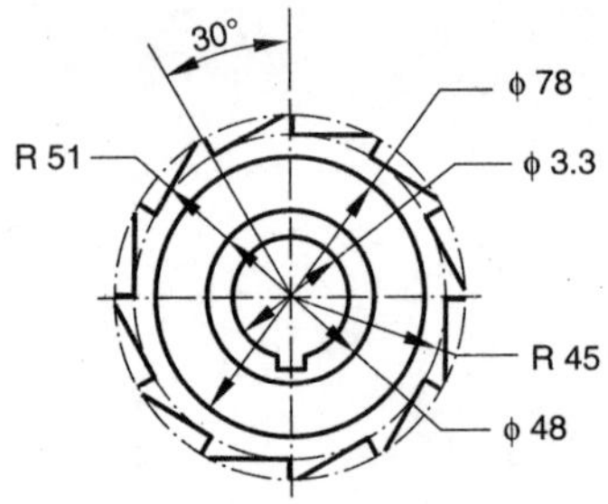

Exercise 11

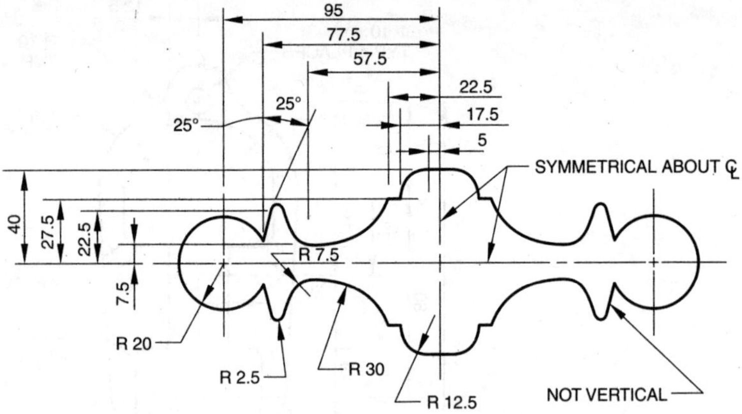

Exercise 14

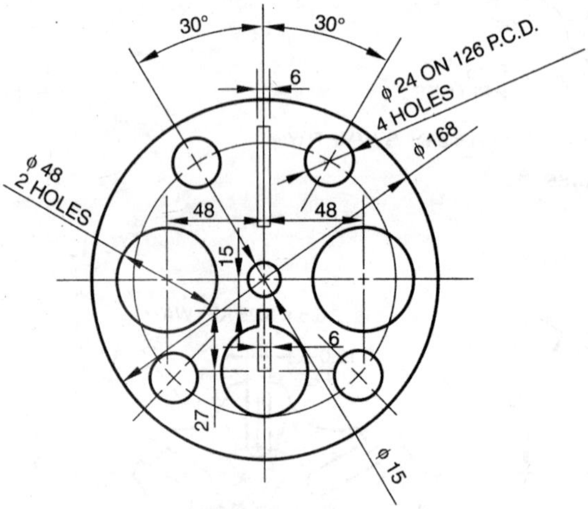

Exercise 12

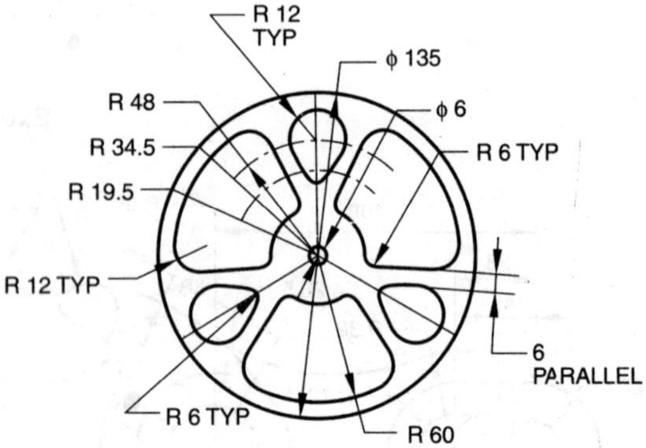

Exercise 15

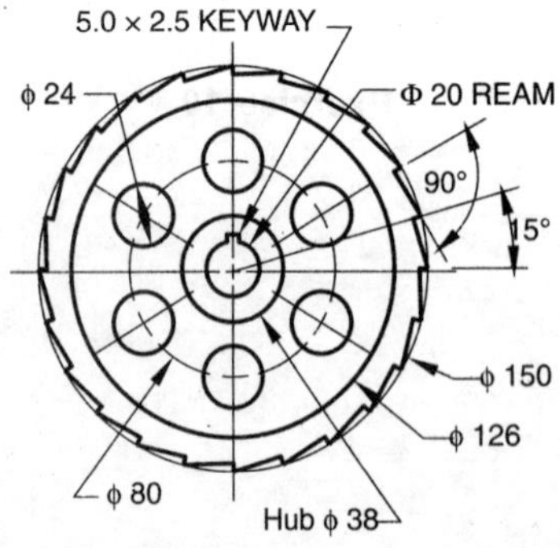

Exercise 13

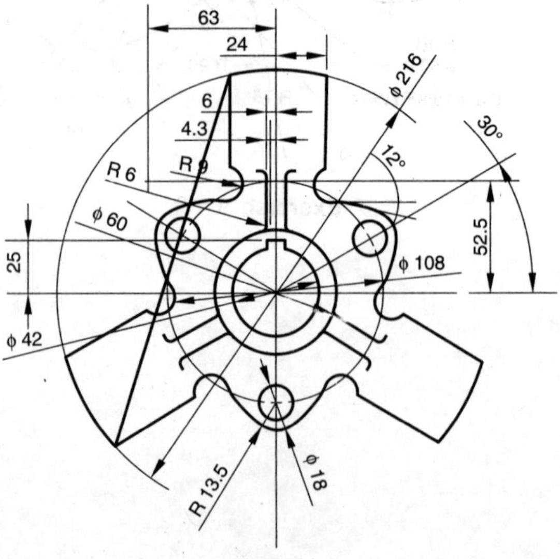

Exercise 16

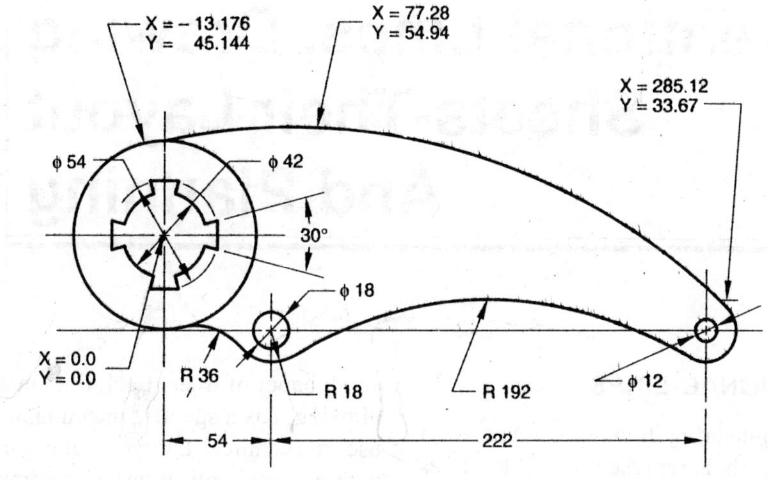

Exercise 17

EXERCISES

1. Draw a line 20 mm long and divide it into 9 equal parts by using a divider.
2. Draw a circle of 60 mm diameter and divide it into:
   (*i*) eight, (*ii*) twelve and
   (*iii*) twenty four equal parts, using T-squares and set-squares only.
3. Using set-squares and T-square only, draw triangles having the following base angles on a 90 mm base:
   (*i*) 45° and 30°, (*ii*) 60° and 75°,
   (*iii*) 15° and 75°, (*iv*) 105° and 45°, and
   (*v*) 90° and 135°.
4. Draw the following angles with a horizontal line by using the protractor:
   (*i*) 19°,
   (*ii*) 43°, 158°,
   (*iii*) 93°,
   (*iv*) 27°,
   (*v*) 142°,
   (*vi*) 158°,
   (*vii*) 167°.

QUESTIONS FOR VIVA VOICE

1. Why drawing boards are not made from one pieces?
2. Why saw cuts are provided on ebony edge of a drawing board ?
3. What are the main parts of a T-square ?
4. Why T-square is not used for drawing vertical lines ?
5. Why the Set-squares are made of transparent materials ?
6. Why cellotape is preferred to drawing pins for fixing the drawing paper to the board ?
7. Where you will use a pencil mended to conical and chisel shape and why ?
8. Why pencil should be rotated in fingers while drawing a long line ?

What is a lengthening bar and when it is used ?

9. What are the various functions of a dividers in drawing ?
10. What is a French-curve and where it is used ?
11. How you will test a T-square for accuracy ?
12. Why the needle point of a compass is slightly more in length than the lead point ?
13. How is the needle point of a compass adjusted ?
14. Why do we brake the legs of a compass ?
15. What is a bow compass and what is the difference between a bow compass and a large compass ?
16. How set-squares are checked for their accuracy
17. What are the standard sizes of drawing sheets ?
18. What is the use of, (*i*) Squared Graph, and (*ii*) Isometric Graph Papers ?

Conventional Lines, Drawing Sheets-Their Layout And Planning

2

2.1. CONVENTIONAL LINES

As the basis of engineering drawing is a line, a set of conventional symbols representing all the lines needed for different purpose may well be called an alphabet of lines. Each line, in a well executed drawing, has a specific meaning and function. For use in Technical Drawings the types of lines and their general applications, as recommended by BIS, are given in *Figs. 2.1.* (*a*), (*b*) and (*c*).

Line	Description	General Applications
A	Continuos thick	A1 Visible outlines A2 Visible edges
B	Continuous thin (straight of curved)	B1 Imaginary lines of intersection B2 Dimension lines B3 Projection lines B4 Leader lines B5 Hatching B6 Outlines of revolved sections in place
C D*	Continuous thin freehand[2] Continuous thin straight with zig-zag	C1 Limits of partial or interrupted views and sections, if the limit is not a chain
E F	Dashed thick Dashed thin	E1 Hidden outlines E2 Hidden edges F1 Hidden outlines F2 Hidden edges
G	Chain thin	G1 Centre lines G2 Lines of symmetry G3 Trajectories
H	Chain thin, thick at ends and changes of direction	H1 Cutting planes
J	Chain thick	J1 Indication of lines or surfaces to which a special requirement applies
K	Chain thin double-dashed	K1 Outlines of adjacent parts K2 Alternative and extreme positions of movable parts K3 Centroidal lines K4 Initia' outlines prior to forming K5 Parts situated in front of the cutting plane

Fig. 2.1. (a) Line Type, Application and Descritpion

*1. This type of line is suited for production of drawing by machines.

*2. Although two alternatives are available, it is recommended that any one drawing only one type of line by used.

TYPE OF LINE	APPLICATION	DESCRIPTION
Hidden Line Thin	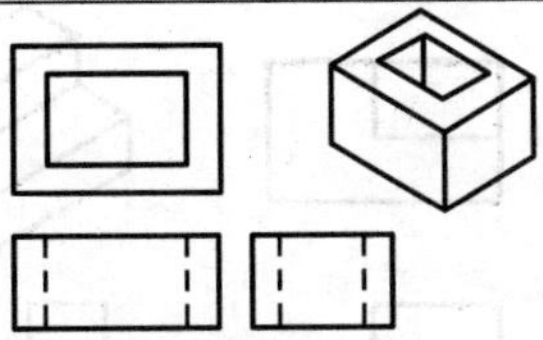	The hidden object line is used to show surfaces, edges, or corners of an object that are hidden from view.
Centre Line Thin Alternate Line and Short Dashes	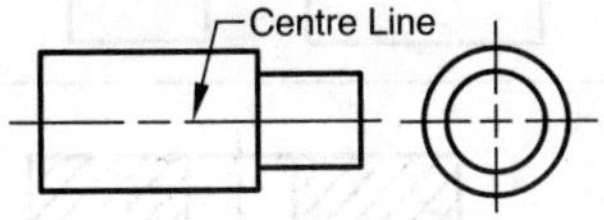	Center Lines are used to show the center of holes and symmetrical features.
Symmetry Line Center Line Thick Short Lines	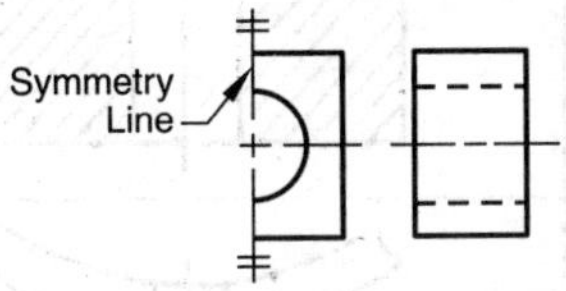	Symmetry Lines are used when partial views of symmetrical parts are drawn. It is a center line with two thick short parallel lines drawn at right angles to it at both ends.
Extension and Dimension Lines Thin Dimension Line Extension Line	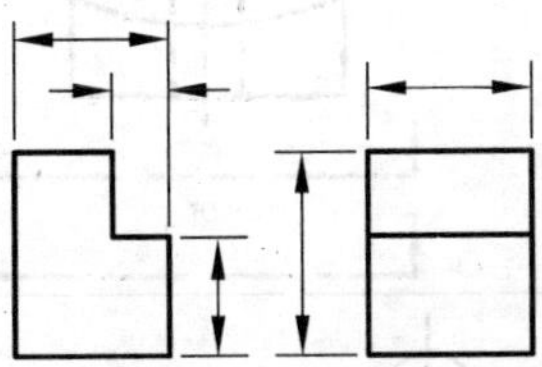	Extension and dimension lines are used when dimensioning an object.
Leaders Arrow Dot Thin	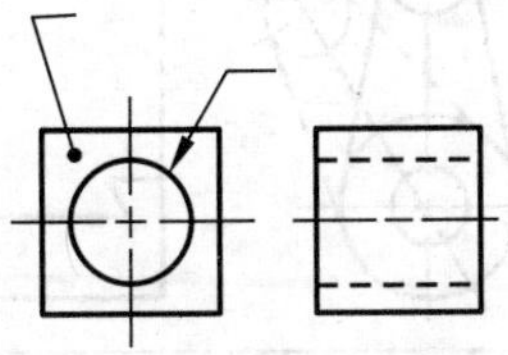	Leaders are used to indicate the part of the drawing to which a note refers. Arrowheads touch the object lines while the dot rests on a surface.
Break Lines Thin Long Break Thick Short Break	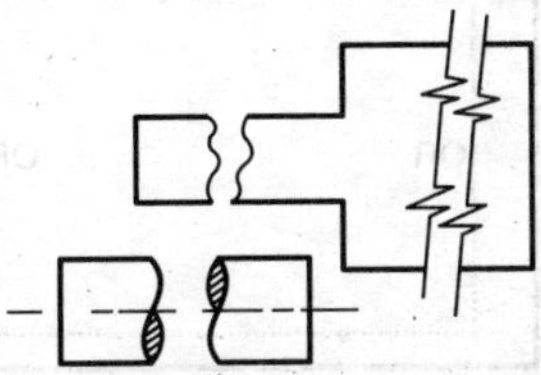	Break Lines are used when it is desirable to shorten the view of a long part.
Cutting-plane Line Thick OR	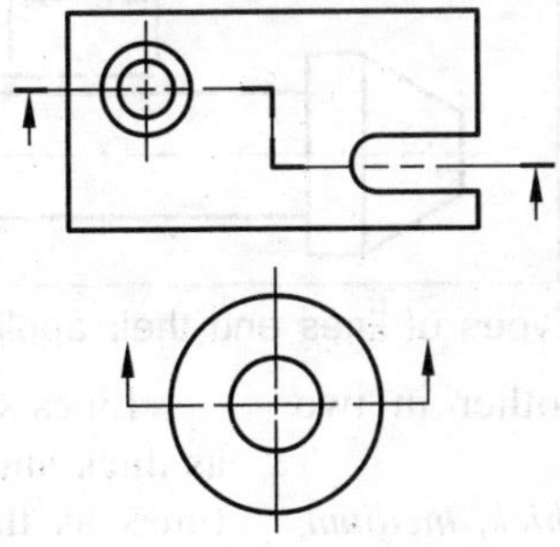	The cutting-plane line is used to designate where an imaginary cutting took place.

(Contd...)

Fig. 2.1. (b) Types of lines and their applications

TYPE OF LINE	APPLICATION	DESCRIPTION
Visible Line Thick	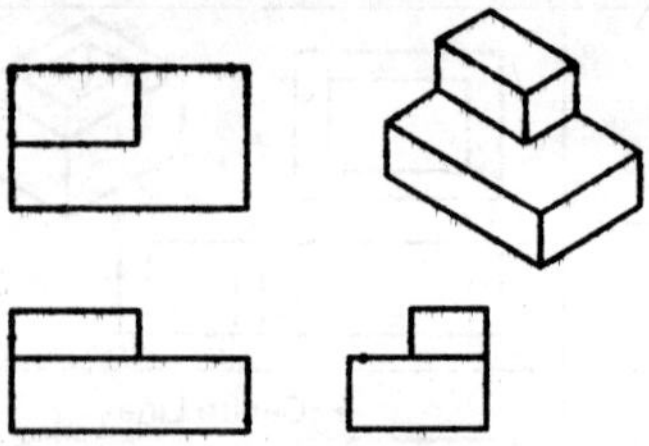	The visible line is used to indicate all visible edges of an object. They should stand out clearly in contrast to other lines so that the shape of an object is apparent to the eye
Section Lines Thin Lines	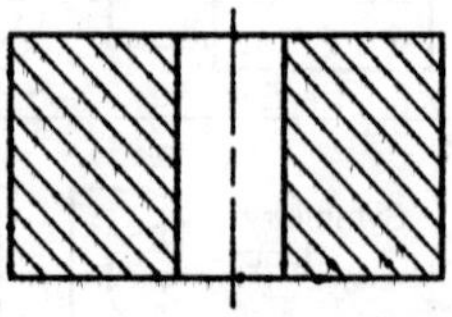	Section lining is used to indicate the surface in the section view imagined to have been cut along the cutting-plane line.
Viewing-Plane Line Thick OR	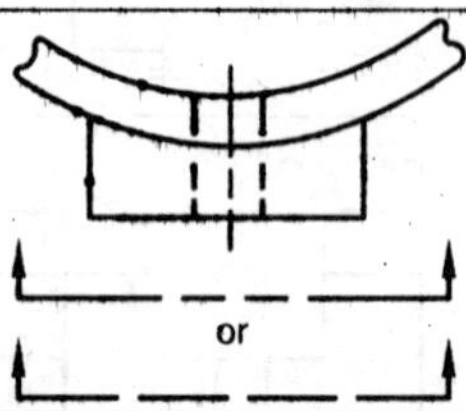	The viewing-plane line is used to indicate direction of sight when a partial view is used.
Phantom Line Thin	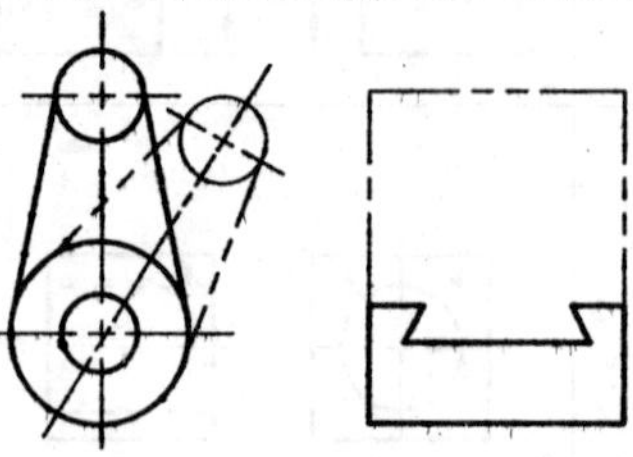	Phantom lines are used to indicate alternate position of moving parts. adjacent position of related parts, and repetitive detail.
Stitch Line Thin OR Small Dots	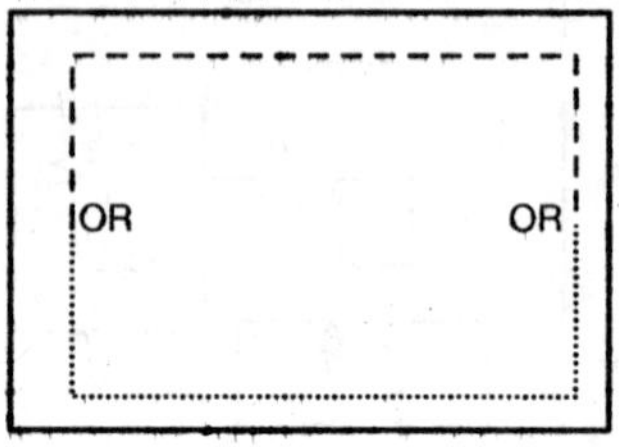	Stitch lines are used for indicating a viewing or stitching process.
Chain Line Thick	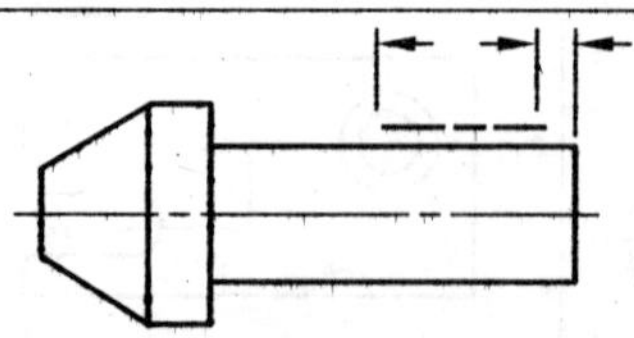	Chain lines are used to indicate that a surface or zone is to receive additional treatment or considerations.

Fig. 2.1. (c) Types of lines and their applications (ASME)

These lines differ from each other in two respects:

1. their thickness or weight, e.g., *thick, medium,* and *thin,* and
2. their *shape or construction,* depending upon their conventional use.

Lines specified *thick* should be about 3 times as thick and lines specified *medium* should be 1.5 times as thick as lines specified *thin*. The line group used on a drawing is identified from the thickest lines, i.e., visible outlines. The line group used depends upon the size of the drawing and should be chosen from one of the line groups shown in *Fig. 2.2*.

SP: 46-1988 recommends that two thicknesses of lines be used. In general, the ratio of the thick to the thin line should not be less than 2 : 1. Grading of lines is in √2 increments. The thickness of lines should be chosen, according to the size and the type of the drawing, from the following: 0.18, 0.25, 0.35, 0.5, 0.7, 1, 1.4, and 2 mm.

For all views of one piece to the same scale, the thickness of the lines should be same. Owing to difficulties in methods of reduction, the line thickness of 0.18 mm should be avoided. The minimum space between parallel lines, including hatching, should never be less than twice the thickness of heaviest line. It is further recommended that these spaces should never be less than 0.7 mm.

Typical applications of these different types of lines are shown in *Fig. 2.3* (*a*), (*b*) and *2.4*.

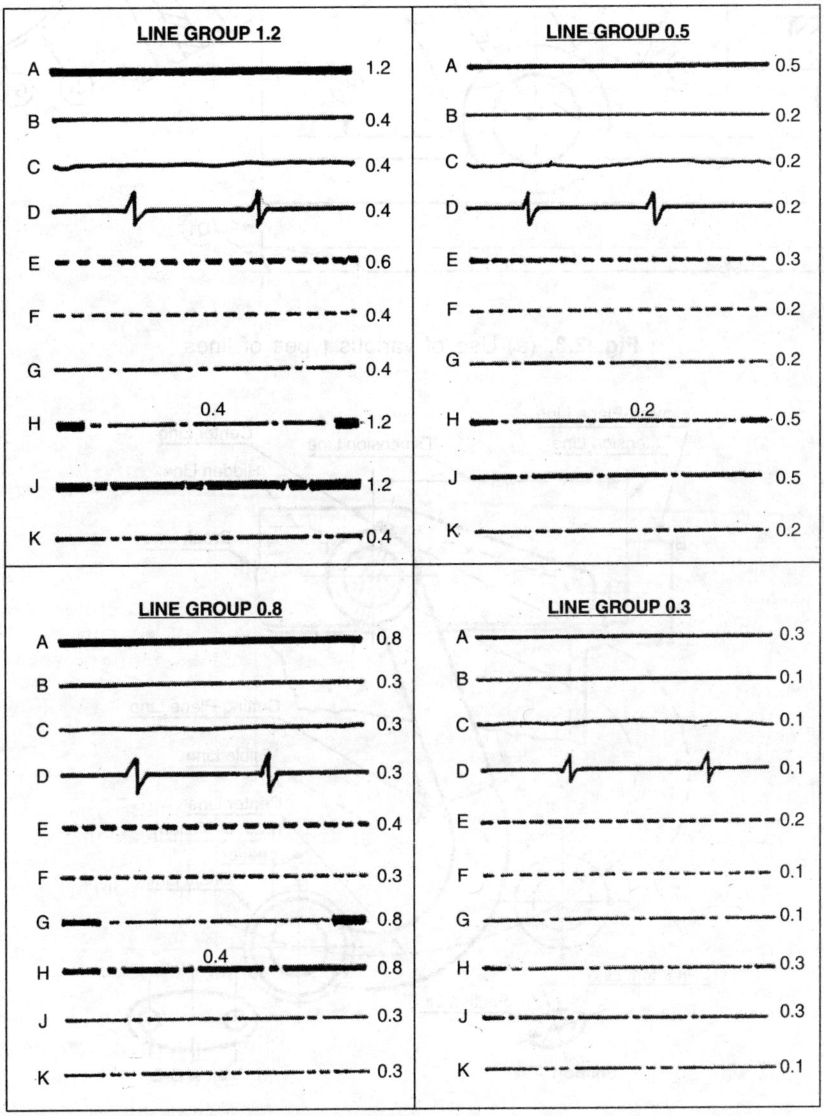

Fig. 2.2. Various line group

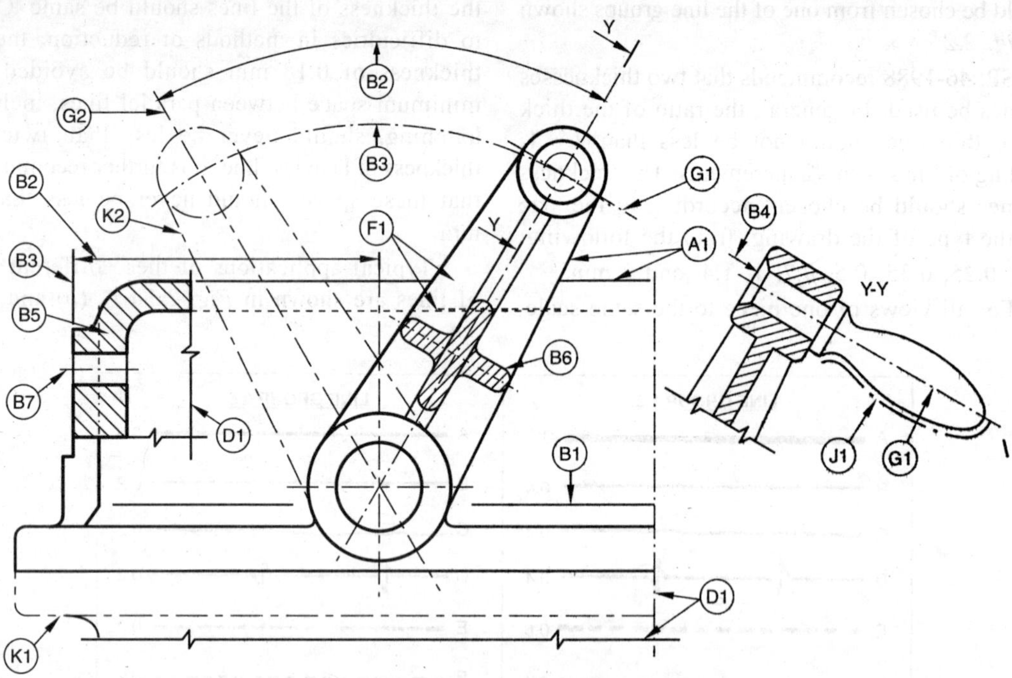

Fig. 2.3. (a) Use of various types of lines

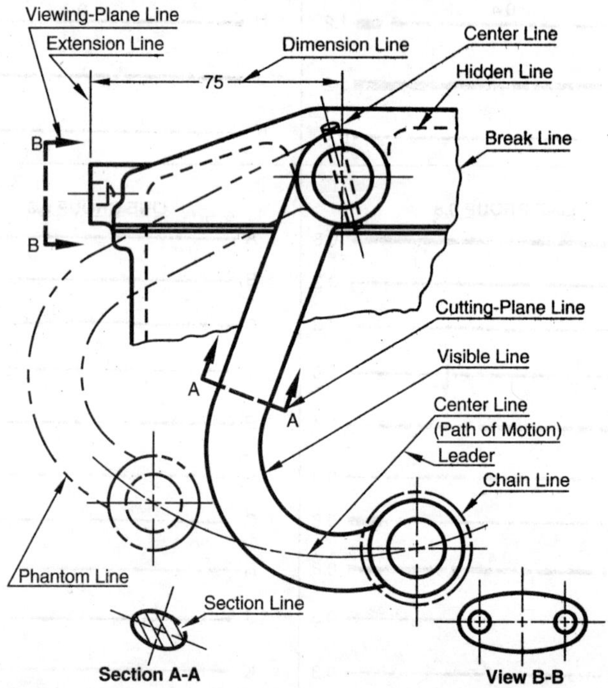

Fig. 2.3. (b) Application of lines (contd.)

1. Visible Outline or Object Line

Outline is the outstanding feature in any drawing. These are thick, continuous lines and are used to represent the outline or overall shape of an object. Every edge or surface that is visible is represented by these lines.

2. Section Lines

These are thin continuous lines, uniformly spaced and are used to show a part in section. These lines are usually drawn at an angle of 45° (may be 60° or 30°) to the horizontal. These lines extend edge-to-edge of the surface in which these are drawn.

3. Construction Lines

These are faint continuous lines used for projecting or transferring locations from one view to the other and for constructing drawings. These lines do not appear in finished drawings, except in Geometrical Drawings where these are usually not removed. These are not drawn in ink. Here in this book, in certain cases, for the purpose of illustration and to facilitate understanding of drawings for students the construction lines are retained.

4. Hidden Lines

Shape which can not be seen, when an object is looked at from an outside viewing position, can be drawn in hidden detail. It is not necessary to include all hidden detail in views in a drawing. Only sufficient of the hidden detail need be included to show quite clearly the exact shape of the object. Hidden lines may be omitted, when not required, to preserve the clarity of the drawing. When adding hidden detail lines to a drawing, the following rules should be followed:

- Hidden detail lines are medium thick lines consisting of short dashes, about 3 mm long with gaps of nearly half their length between them. The length of the dashes varies slightly in relationship to the size of the drawing.
- Hidden detail lines forming angles should be drawn with dashes touching at the angle.
- If hidden detail lines show a continuation from a visible outline, a dash should touch the visible outline.
- In a hidden detail, if a line meets a curved line, a dash should touch at the point of tangency.
- When hidden detail lines cross visible outlines either a dash should clearly cross the outline or, dashes should be drawn with a gap on each side of the visible outline. Unless this rule is observed, the hidden detail would show that the detail drawn commences at the visible outline. These are represented by evenly spaced short and uniform dashes, used to represent invisible or hidden details or parts on drawings. These dashes are about 60% thinner than the object lines.

5. Centre Lines, Locus Lines, Extreme Position of Movable Parts, Pitch Circles and Parts Situated in front of the Cutting Plane

These lines are represented by long and short dashes. The proportion for these long and short dashes is ranging from 6 : 1 to 4 : 1. In any one drawing the proportion selected should be maintained by eye judgement.

The *centre lines* are used to *centre* the symmetrical views of circular parts, to indicate centre points, axes of symmetrical parts, axes of symmetry and also for giving the location dimensions. These lines are first to be drawn in any drawing after carefully spacing the various views to be drawn. Following points need special care when drawing centre lines:

- Where two centre lines cross, the short dashes should intersect symmetrically, as shown for circle in *Fig. 2.4*.
- When a centre line coincides with either a visible or a hidden line, it should not be continued into such a line.
- No centre line should begin or end at the lines representing surfaces of the object, but should extend about 5 mm beyond such surfaces.
- Centre lines should be drawn for all parts, visible or hidden, which have these lines.
- Centre lines should begin and end with a long dash.
- Centre lines should not extend between views in an orthographic projection.
- Centre lines should not be used as dimension lines for the purpose of dimensioning, but may be extended for the purpose of dimensioning.

- Centre lines should extend for a short distance beyond the feature unless they are required for dimensioning or some other purpose.
- If an angle is formed by chain lines, then the long dashes should intersect and define the angle.

Locus lines are, particularly, used to show the paths traced out by any one point under given conditions.

Dimension Lines, Extension Lines

Leader Lines: These are discussed in chapter on Dimensioning and are thin continuous lines.

Outlines of Revolved Sections and Adjacent Parts

These lines too are thin continuous.

Cutting Plane Lines

These lines are made up of alternately a long dash and a short dash in the ratio of about 6 : 1 to 4 : 1 and are drawn as centre lines but are extra thick at ends and having arrows touching them. These are used to show the edge of the cutting plane when an object is being sectioned to show the hidden details. The arrows, attached to the ends of such lines indicate the direction in which the section is to be seen.

Break Lines

(*i*) **Short Breaks and Irregular Boundary Lines:** These are thin lines drawn freehand and are used to show a break of an object which extends over a short length of the drawing or to indicate the irregular boundary of an object. These result in a saving in space and time used for drawing without loss of detail.

(*ii*) **Long Breaks:** These are thin ruled lines having freehand zigzags at suitable intervals and are used when the brake extends over a considerable length of the drawing.

Long Chain Thick Lines

Composed of alternate long and short dashes are used to indicate surfaces which are to receive additional treatment.

2.2. DRAWING PENCIL LINES

Pencil lines should be sharp and uniform along their entire length and sufficiently distinct to fulfill their ultimate purposes. Construction lines should be drawn very lightly so that they may be easily erased. Finished lines should be drawn boldly and distinctly, so that there may be definite contrast between visible and invisible object lines, dimension lines, extension lines, section lines and centre lines. To give this contrast, which is necessary for clearness and ease in reading, object lines should be thick and very dark, invisible lines dark but not so wide, and all other lines dark and thin, *Fig. 2.5*, so that they will reproduce clearly by any reproduction method used.

When drawing a line, the pencil should be inclined, slightly, about 70° to the horizontal, in the direction in which the line is being drawn. If the pencil is rotated slowly between the fingers, as the line progresses, a symmetrical point will be maintained and a straight uniform line will be made.

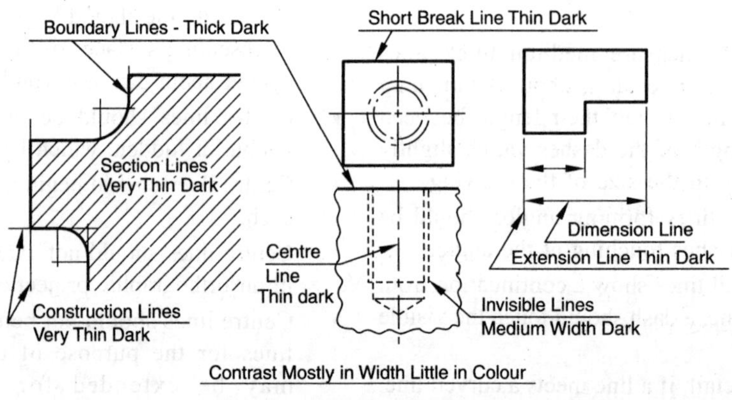

Fig. 2.5. Contrast of lines

2.3. TECHNIQUE OF DRAFTING

The technique of drafting refers not only to the good point of a finished drawing but to every movement the drafter makes from the time he/she gets the equipment out until the drawing is finished. Speed, accuracy and neatness are essential to good drafting. All these qualities will be appreciably enhanced if the tools are neatly arranged, in order, on the table, so that they are easily found and readily accessible.

To make a good pencil drawing, the original layout of the problem should be made with a 4H or 3H pencil. The lines should be firm light lines, heavy enough to be carefully seen but not so heavy as to groove the paper. When the drawing is thus completed, all construction lines over running corners, centre lines that are too long, projection lines etc. should be erased. The drawing is then made rapidly a second time with soft pencil, usually H or 2H, in the same order that one would ink a drawing. The soft pencil must be reshapened frequently so that the same weight is maintained throughout the drawing. The lines should be dense and black. T-square and set-squares should not be allowed to slide over the soft, black lines as they will smudge the lines and ruin the drawing. With soft pencil, tiny flakes of graphite will be found to break off around the line. These should be blown off as each line is drawn and thus avoid smudging, since it is these particles rather than the imbedded line which cause the trouble.

Contrast between visible outlines, hidden lines, and dimension lines etc., should be maintained by using somewhat harder pencils for the harder lines, by using the points much sharper and using a little less pressure, *Fig. 2.5*.

The contrast in line thickness, coupled with consistency in the pitch of chain lines, or lines of short dashes, is most essential in order to obtain a good proportional finish to a drawing.

2.4. TECHNIQUE OF INKING A DRAWING

In inking a drawing or making a tracing, a certain procedure should be followed to get the best results in the least time. On a good drawing there must be uniformity among lines of one kind, and contrast between lines of different kinds. To obtain this result, the drawing should be inked in the following order:

1. **Visible Lines,** all of the same weight.
   - Circles and arcs of circles and other curved lines.
   - Horizontal lines starting at the top.
   - Vertical lines starting from the left.
   - Inclined lines.
2. **Invisible Lines** in the same order as in 1 above.
3. **Centre Lines** in the same order as in 1 above.
4. **Dimension Lines** in the same order as in one above.
5. **Section lines,** light and evenly spaced.
6. **Dimensions, Arrowheads** and other lettering.
7. **Title-block** and **Border Lines.**

The importance of drawing circles and arcs first should not be overlooked, for one can always make a straight line tangent to one or two arcs, but it is extremely difficult to make an arc or a curve tangent to straight lines, particularly if one of them is just slightly out of place.

Trial lines and arcs should be drawn on a scrap paper, and adjustments made in instruments, to bring uniformity of weight or thickness between lines and arcs. Draw the arcs of desired thickness first and then later match them with the inking instrument. Correct and incorrect inking of lines and arcs over pencil lines and arcs is shown in *Fig. 2.6*.

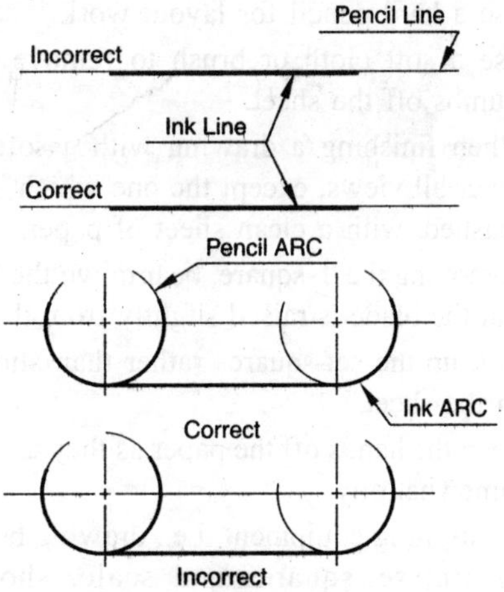

Fig. 2.6. Inking over pencil lines

Table 2.1.

Designation	*Dimensions, mm*
*A*0	841 × 1189
*A*1	594 × 841
*A*2	420 × 594
*A*3	297 × 420
*A*4	210 × 297

Wet inked lines should not be touched. Always join a wet line to a dry line. Do not fill the pen or other inking instruments over the drawing and do not leave the ink bottle uncorked.

2.5. IDEALS OF NEATNESS, ORDER AND ACCURACY

If the time spent in studying and practicing technical drawing is to be actually fruitful the student must realize from the very beginning that neatness and accuracy in detail are essential. Splashdash or perfunctory performance are useless in engineering.

Drawing must be accurate, legible and must follow standard conventions. Drawing is one link in a long chain of production. The drafter must constantly "think of about the next person", the person who is to read the drawing.

Drawing must be neat, because dirty, untidy presentation of one's ideas on paper is ill-mannered and unpleasing to a trained eye. Every care should be taken to keep the drawing clean. The following suggestions, if observed, will help to keep the drawing clean:

1. Use a hard pencil for layout work.
2. Use a soft cloth or brush to remove erasing crumbs off the sheet.
3. When finishing a drawing with a soft pencil cover all views, except the one which is being finished, with a clean sheet of paper.
4. In moving the T-square, bear down the head so that the blade is raised slightly from the paper.
5. Pick up the set-squares rather than slide them on the sheet.
6. Keep the hands off the paper as they are always somewhat oily.
7. All drafting equipment, i.e., drawing board, T-square, set-squares and scale, should be frequently cleaned.
8. Never sharpen a pencil over the drawing.
9. Never place the sandpaper pad in contact with any other drawing equipment unless it is wrapped in a wrapper.
10. Never work with the sleeves or hands resting on a penciled area while lettering, always place a piece of paper under the hand.
11. At the end of day's work, the drawing should be properly covered to protect from dust.

If the drawing is completed observing above rules, it is not necessary to scrub it with artgum and then redraw the lines, thus saving wastage of time used in scrubbing and retracing the drawing.

2.6. DRAWING PAPER

Drawing paper is made in a variety of qualities and may be had in sheets or rolls. The paper used for drawing depends entirely upon the character of the drawing, its purpose, and its use. For pencil layouts and working drawings cartridge or machine made paper is largely used. In general, paper should have sufficient grain, or tooth, to take the pencil, be agreeable to the eye, and have a hard surface, not easily grooved by pencil, with good erasing qualities. One of the sides of drawing paper usually is rough and the other smooth. The smooth surface is the proper side for drawing.

For drawings to be finished in ink the Whatman's hot pressed paper is most suited. These are available in various thicknesses, i.e., thin, medium thick and extra thick, etc. Ink is rarely applied to detail paper. Heavy, hard-surfaced detail paper is available for high grade, accurate work, while cheaper grades may be used for drawings which are not subject to rough usage. The drawing paper should not be folded.

White papers are used for drawings where the appearance is an important factor.

1. Sheet Sizes and their Designation

The basic principles involved in arriving at the size are:

1. $X : Y = 1 : \sqrt{2}$
2. $XY = 1$

where X and Y are the sides and having a surface area of 1 m^2 so that X = 0.841 m and Y = 1.189 m.

Two series of successive format size are obtained by halving along the length or doubling along the width. The areas of the two sizes are thus in the ratio of 1 : 2, *Fig. 2.7.*

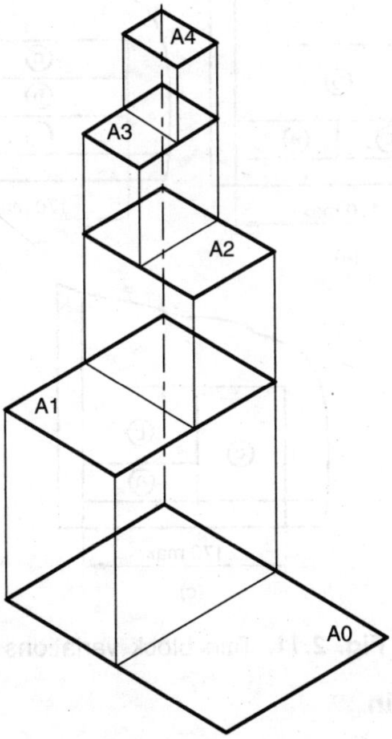

Fig. 2.7. Halving and doubling of sheets

The forms are similar to one another and hence the equation X : Y = 1 : √2 is obtained for the two sides X and Y of a format, *Fig. 2.8*, consequently the ratio between both sides is the same as that of the sides of a square to its diagonal, *Fig. 2.9.*

The preferred sizes (first choice) of the trimmed sheets as selected from the main ISO-A series are given in Table 2.2. When a sheet of greater length is needed (second choice), one of the size given in Table 2.3 should be used. These sizes are obtained by extending the shorter sides of a format of the ISO-A series of lengths that are multiples of the shorter sides of the chosen basic format.

Table 2.3.

Designation	*Dimensions, mm*
*A*3 × 3	420 × 891
*A*3 × 4	420 × 1189
*A*4 × 3	297 × 630
*A*4 × 4	297 × 841
*A*4 × 5	297 × 1051

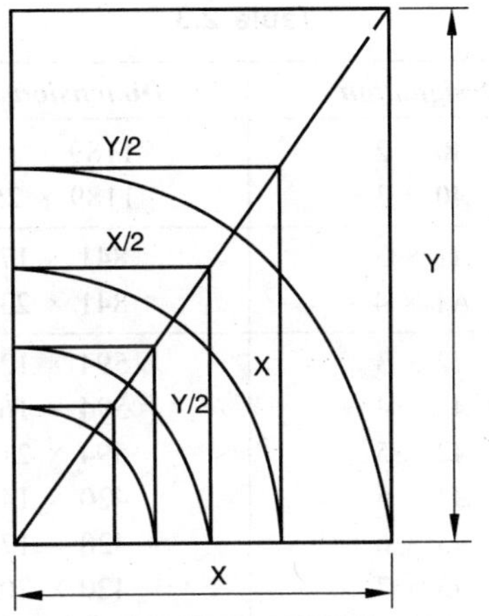

Fig. 2.8. Similarity of formats

When a very large or extra elongated sheet is essential (third choice), one of the sizes given in Table 2.4 should be used. These sizes are obtained by extending the shorter sides of the format of the ISO-A series to lengths that are multiples of the shorter sides of the chosen basic format.

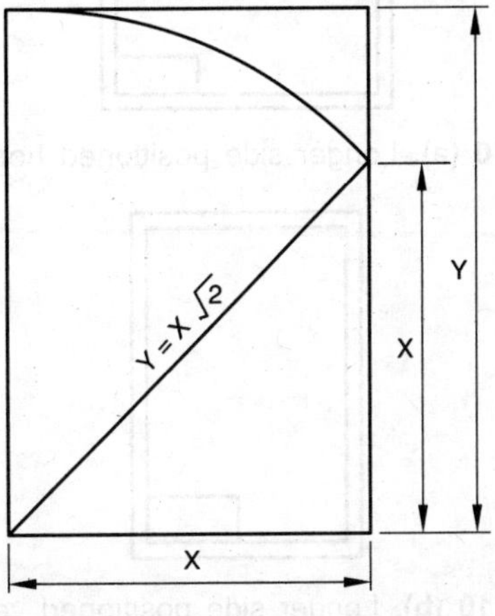

Fig. 2.9. Relationship between sides

The original drawing should be made on the smallest sheet permitting the necessary clarity and resolution. The choice of sizes should be made from the series shown in Tables 2.3, 2.4, and 2.5 in the order. Drawing sheets may be used with their longer sides positioned either horizontally, *Fig. 2.10(a)*, or vertically, *Fig. 2.10(b)*.

Since the A0 size has the area of 1 m^2, paper weights are conveniently expressed in the unit 'grams per square meter'.

Table 2.3.

Designation	*Dimensions, mm*
A0 × 2	1189 × 1682
A0 × 3	1189 × 2523
A1 × 3	841 × 1783
A1 × 4	841 × 2378
A2 × 3	594 × 1261
A2 × 4	594 × 1682
A2 × 5	594 × 2102
A3 × 5	420 × 1486
A3 × 6	420 × 1783
A3 × 7	420 × 2080
A4 × 6	297 × 1261
A4 × 7	297 × 1471
A4 × 8	297 × 1682
A4 × 9	297 × 1892

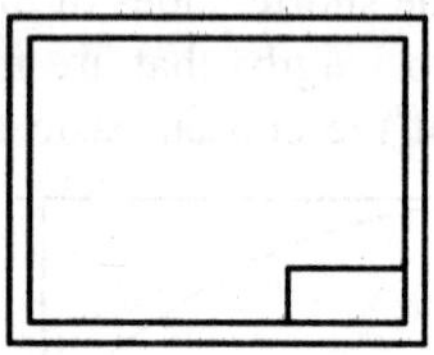

Fig. 2.10 (a). Longer side positioned horizontally

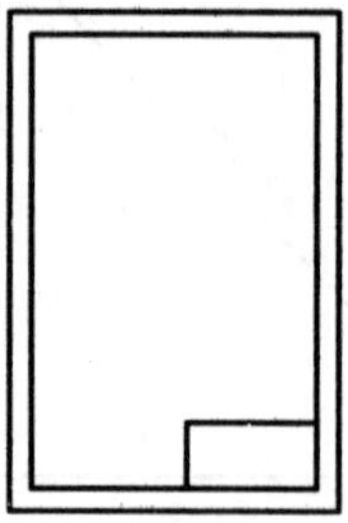

Fig. 2.10 (b). Longer side positioned vertically

2.7. LAYOUT OF DRAWING SHEETS

Every drawing office should follow a standard layout of drawing sheets which should facilitate the reading and easy identification of drawings. The arrangement followed must ensure that all necessary information is included and sufficient extra margin is also left for filing and binding of drawings whenever necessary. The general features of a drawing sheet are shown in *Fig. 2.12* as recommended by SP : 46-1988.

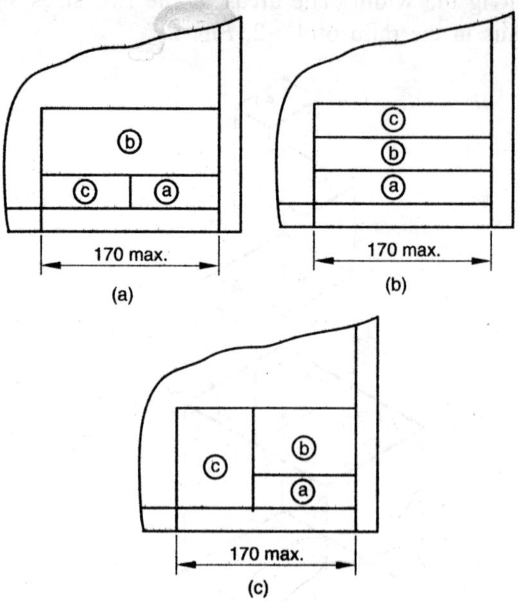

Fig. 2.11. Title block variations

1. Margin

The margin ensures that if the sheet edges become damaged by handing, the drawing within the margin enables prints to be trimmed. Prints after trimming conform to the sizes of trimmed sheets. The margins for different sizes of drawing sheets are shown in *Figs. 2.13* and *2.14*.

2. Title-Block

Title-block is an important feature in a drawing and is uniform for all drawing sheets sizes.

The position of the title-block should be within the drawing space, *Fig. 2.11 and 2.12*, such that the portion of the title-block containing the identification of the drawing is situated in the bottom right hand corner of the drawing space, both for sheets positioned horizontally (type X), *Fig. 2.9*, or vertically (type Y), *Fig. 2.10*, where it is readily seen when the prints are folded in the manner prescribed by SP : 46 - 1988. The direction of viewing the title-block should correspond, in general, with that of the drawing.

In a title-block space should be provided for the following information:

- Title of drawing;
- Drawing and sheet number;

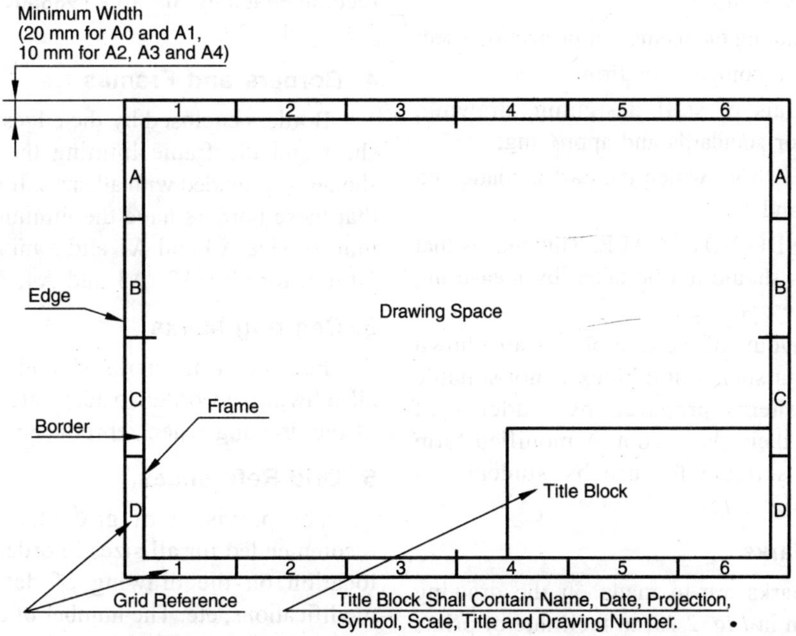

Fig. 2.12. General features of a drawing sheet

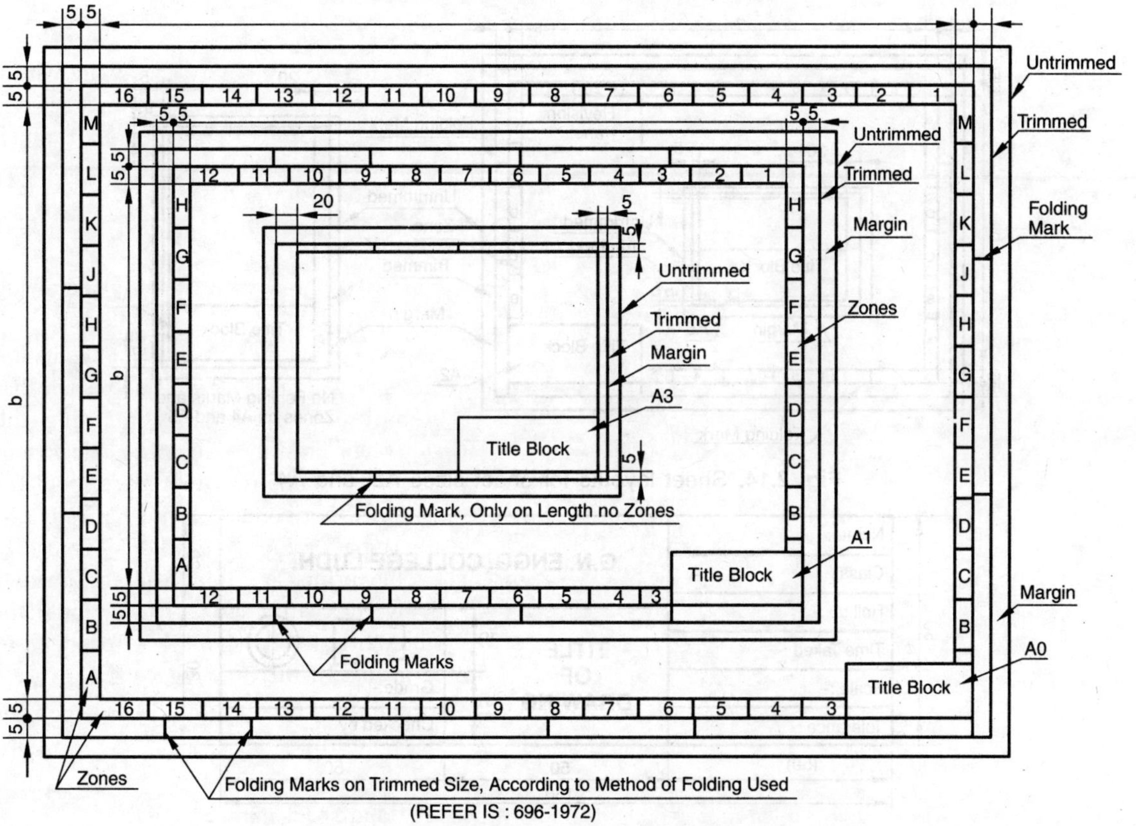

Fig. 2.13. Sheet layouts for sheet sizes A0, A1 and A3

- Scale or scales used;
- Symbol denoting the method of projection used;
- Name of the company or firm;
- Dated initials of staff designing, drawing, checking for standards and approving;
- The material from which the part is made and its finish, and
- The advice DO NOT SCALE. This means that dimensions should not be taken by measuring lines on the drawing.

Tylpical layouts of the title-blocks are shown in *Fig. 2.16*. But such a title-block is not suitable for drawing sheets prepared by students of engineering in their class room. A modified form of title-block suitable for use by students, is suggested in *Fig. 2.15*.

3. Folding Marks

Folding marks are to made on the drawing sheet, as shown in *Fig. 2.13*, to facilitate of prints in an easy manner. Two methods of folding, as recommended by SP: 46 - 1988 are shown in *Figs. 2.19* and *2.20*.

4. Borders and Frames

Borders enclosed by the edges of the trimmed sheet and the frame limiting the drawing space should be provided with all sizes. It is recommended that these borders have the minimum width of 20 mm for size A0 and A1 and a minimum width of 10 mm for size A2, A3 and A4, *Fig. 2.12.*

5. Centring Marks

Four centring marks should be provided on all drawings in order to facilitate the positioning of the drawing when reproduced or microfilmed.

6. Grid References

The provision of grid reference system is recommended for all sizes in order to permit easy location on the drawing of details, additions, modifications, etc. The number of divisions should be divisible by two and be chosen in relation to

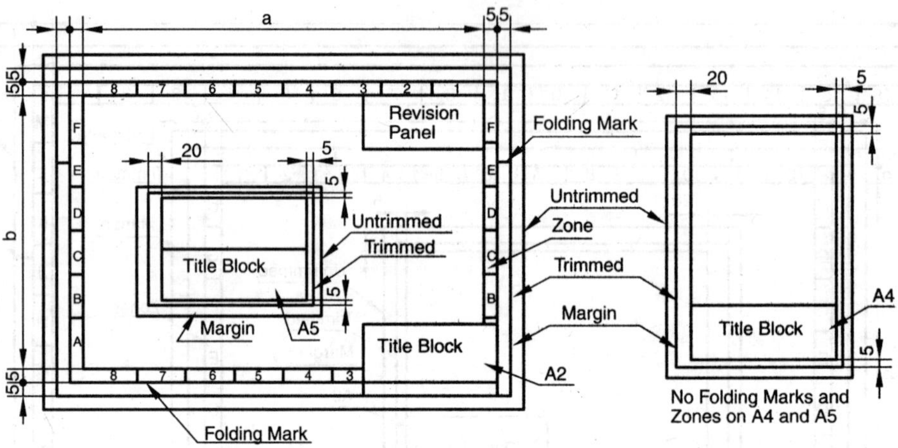

Fig. 2.14. Sheet layouts for sheet sizes A2, and A4

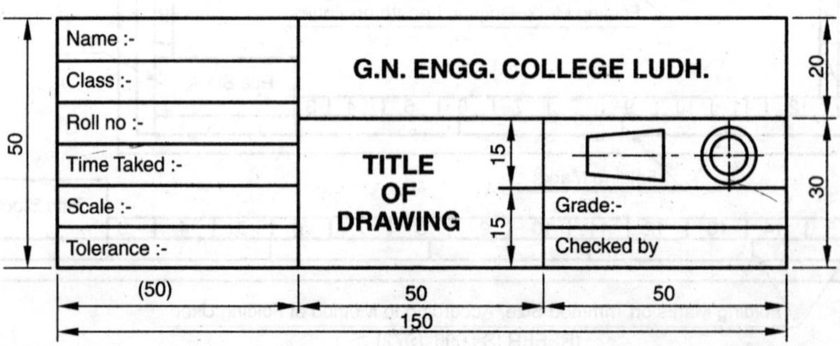

Fig. 2.15. Title-block-suggested for sheets prepared by stundents in class-room

the complexity of the drawing . It is recommended that the length of any side of the rectangles comprising the grid shall not be less than 25 mm and not more than 75 mm. The rectangles of the grid should be referred by means of capital letters along one edge and the numerals along the other edge. The numbering direction may be as shown in *Figs. 2.13* and *2.14* (IS : 696-1972) or may start at the sheet corner opposite to the title-block (SP : 46 - 1988) and be repeated on the opposite sides as shown in *Fig. 2.12*.

The sides X and Y of the drawings having a size A2 and larger should be divided in a number of equal zones as given in the following table.

Table 2.4.

		Sheet	*Sizes*	
		A0	*A1*	*A0*
No. of	*X*	16	12	8
Zones	*Y*	12	8	6

7. Multiple Sheet Drawings

Multiple Sheet Drawings marked with the same registration or identification number should be indicated by means of a sequential sheet number. In addition, the total number of sheets should be shown on sheet 1, for example:

Sheets No. = n/p

where n = sheet number, and

p = total number of sheets.

An abbreviated title-block, containing only the identification zone, may be used for all sheets after the first sheet.

2.8. ADDITIONAL INFORMATION

Where necessary, the drawing sheet may include the following additional information either in the drawing or in the title-block:

- Job or order number;
- Treatment, finish, etc.;
- Key to machining and other symbols, and legends;
- Tolerance on dimensions not individually toleranced;
- Tool and gauge, jig and fixture reference;
- Item list; and
- Any other necessary information.

It is preferred to add the above information on the top right hand corner and inside the margin and should run from top downwards.

2.9. ITEM REFERENCES

The item references should be assigned in sequential order to each component part shown in an assembly and / or each detailed item on the drawing. Further identical parts shown in the same item reference. All item references shall be shown in an item list, *Fig. 2.17* and Table 2.5.

Table 2.5. Item List

Item	*Quantity*	*Description*	*Reference*
1	1	Base	
2	1	Bottom housing	
3	1	Top housing	
4	1	Bearing	
5	1	Filling Plug	
6	2	T-bolt	
7	2	Hex.nut	
8	4	Washer	
9	2	T-bolt	
10	2	Castle nut	
11	2	Split nut	
12	1	Drain plug	

Item references should be composed of Hindu-Arabic numerals only. They may, however, be augmented by capital letters, when necessary. All item references on the same drawing shall be clearly distinguishable from all other indications. This can be achieved, for example, by:

- using characters of a larger height, for example, twice the height as used for dimensioning and similar indications;
- encircling the characters of each item reference. In that case all such circles shall have the same diameter and to be drawn with continuous thin line (Type B), *Fig. 2.18 (c)*;
- combining the above two methods.

The item references should be placed outside the general outlines of the items concerned and each item reference should be connected to its associated item by a leader line, *Fig. 2.18 (a), (b)*

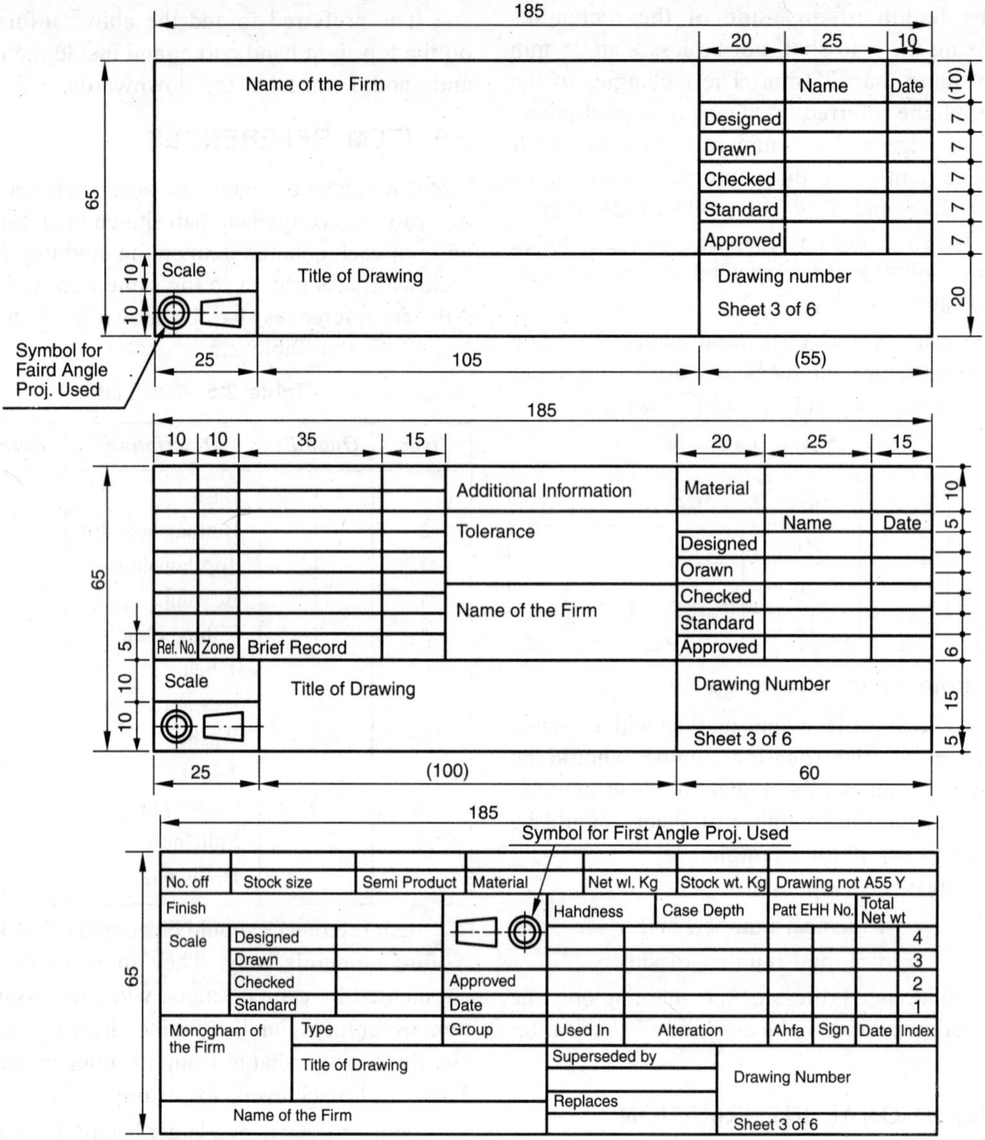

Fig. 2.16. Typical layouts of Title-block and Revision Panel

and *(c)*. Leader lines should not intersect, should be kept as small as practicable, and should be drawn at an angle to the item reference. For encircled item references, the leader line should be directed towards the centre of the circle. Item reference of related items may be shown against the same leader line, while the item references identical items need only be shown once, provided there is no risk of ambiguity.

2.10. ITEM LIST

Item lists are complete lists of the items constituting an assembly (or a sub-assembly), or of detailed parts, presented on a technical drawing. The association between the items on an item list and their representation on the relevant drawing (or on the drawings) is given by the item references. It may be included on the drawing itself or be a separate document. When included on the drawing,

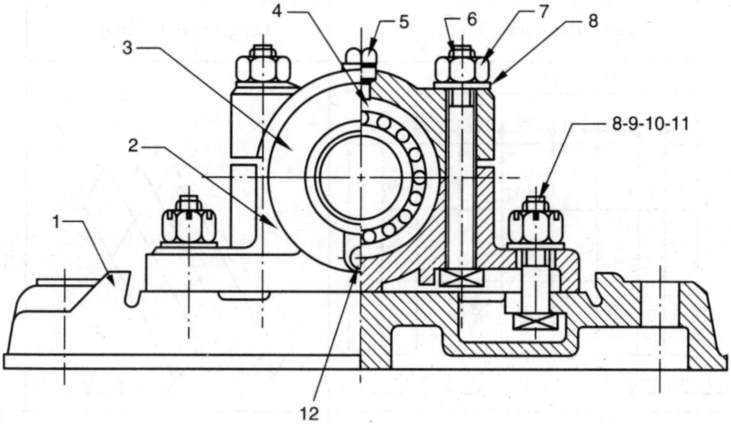

Fig. 2.17. Showing item references

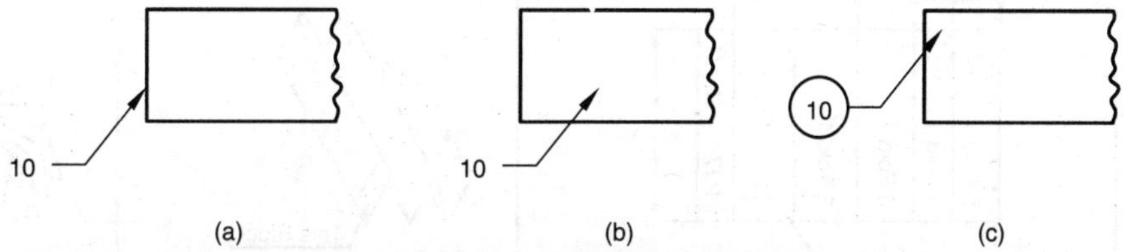

Fig. 2.18. Placing item references

the position of the item list should be such as to be read in the viewing direction of the drawing. The list may be in conjunction with the title-block and its outline be drawn with continuous thick lines. When the item list is shown on a separate document, it should be identified by the same number as that of the parent drawing but preceded by the prefix item list.

The item list should be arranged in columns by means of continuous thick or thin lines to allow information to be entered under the following headings in any suitable sequence:

- item;
- description;
- quantity;
- reference;
- material.

2.11. REVISION PANEL

It is most important that drawings should record all alterations or revisions, which are made from time-to-time. A revision table should provide space for revision number or letter, date, grid reference, brief record and dated initials of the approving authority. The revision panel may preferably be:

- a part of title-block, *Fig. 2.16*;
- at the top right hand corner, *Fig. 2.14*.

To facilitate extension, entries for revisions should begin from bottom upwards or from top downwards.

The portion of the drawing under revision should be located by referencing it in the revision panel, in relation to the zones. For example, a revision referred to in location B4 will be in the rectangle or grid made by intersections of the zones B and 4 in the drawing.

Full description can be given as: P/2B4.

which means that the detail 'P' is given in sheet No. 2, in grid obtained by the intersection of zones B and 4.

2.12. FOLDING OF DRAWING PRINTS

There are two methods of folding of drawing prints. The first method is intended for drawing prints to be filed or bound, while the second method is intended for prints to be kept individually in filing cabinet. These methods are shown in *Figs. 2.19* and *2.20*. The basic principles in each of the above methods are to ensure that:

- all large prints of sizes higher than A4 are folded to A4 sizes;
- the title-block of all the folded prints appear in topmost position; and

Sheet Size	Folding Diagram	Lengthwise Folding	Crosswise Folding
A0 841×1189	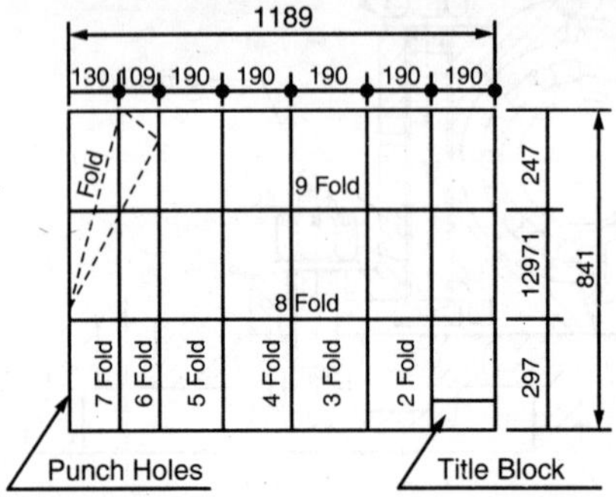	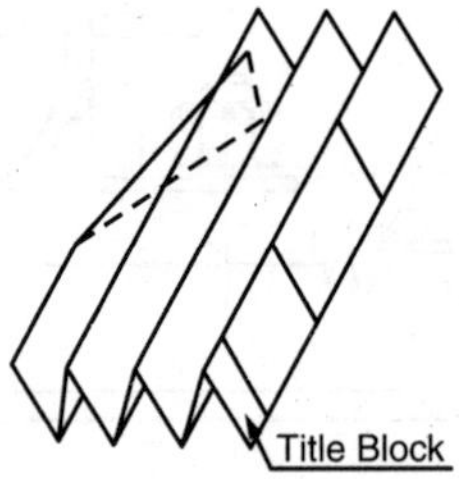	
A1 594×841	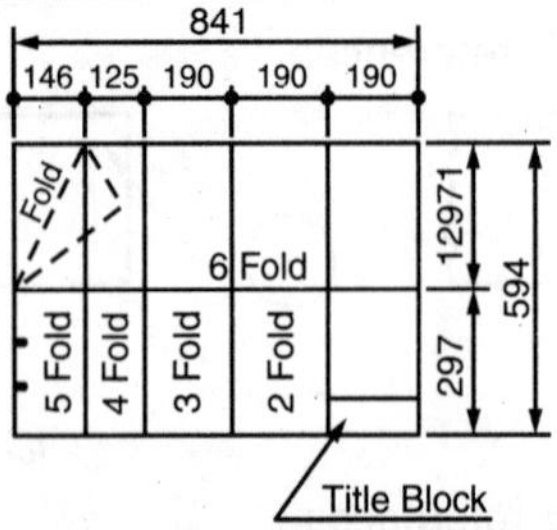	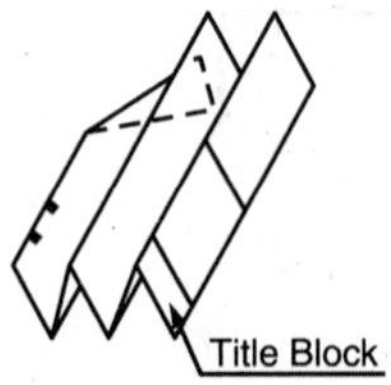	
A2 420×594	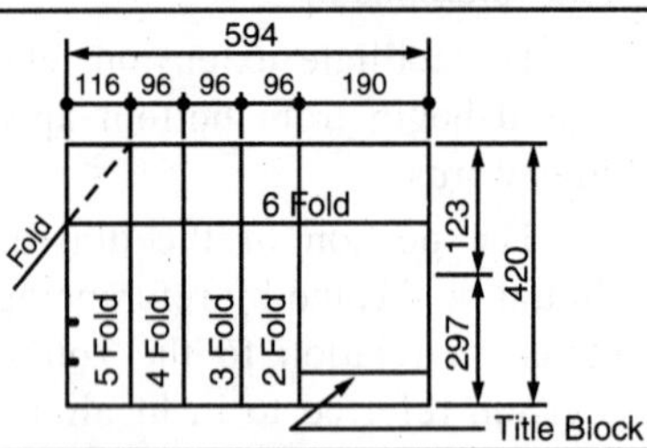	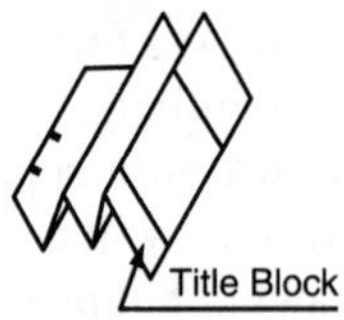	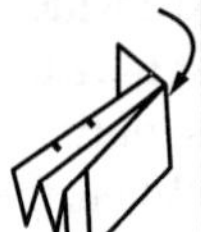
A3 297×420	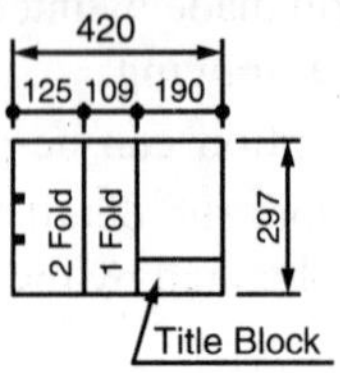		—

Fig. 2.19. Folding of prints for filing

- the bottom right corner should be outermost visible section and should have a width not less than 190 mm.

Depending on the method of folding adopted, suitable folding marks are to be introduced in the tracing sheets as a guide.

2.13. CARE AND STORAGE OF ORIGINAL DRAWING

Drawings are valuable and should be handled and stored properly to preserve them in good condition. The following simple rules, in addition to those listed in 2.5 will assist in keeping drawing in 'mint' condition.

1. Never fold drawings.
2. Do not use hands for wiping off dirt and debris; use a brush.
3. Dirt also collects on the reverse side of drawings, so keep them clean.
4. Do not roll drawings tightly since they may not lie flat during microfilming or print reproductions.

5. Do not use staples or drawing pins, instead, use tape or drawing clips.
6. When filing drawings, it is recommended that similar sizes are filed together.
7. Drawing should be stored under conditions of normal heat and humidity about 21°C and 40 to 60% relative humidity.

Sheet Designation	Folding Diagram	Lengthwise Folding	Cross Wise Folding
A0 841 × 1189	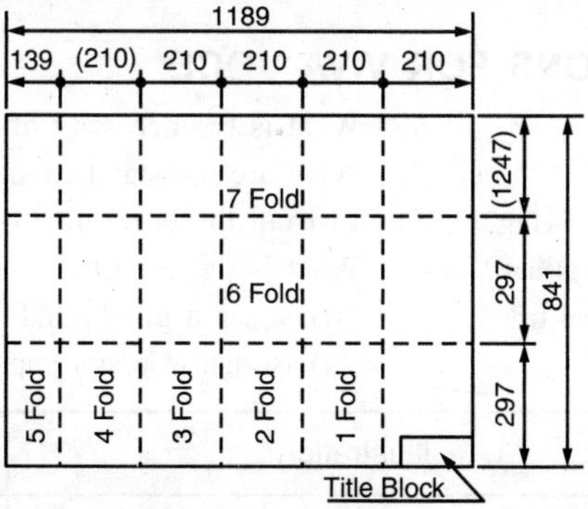	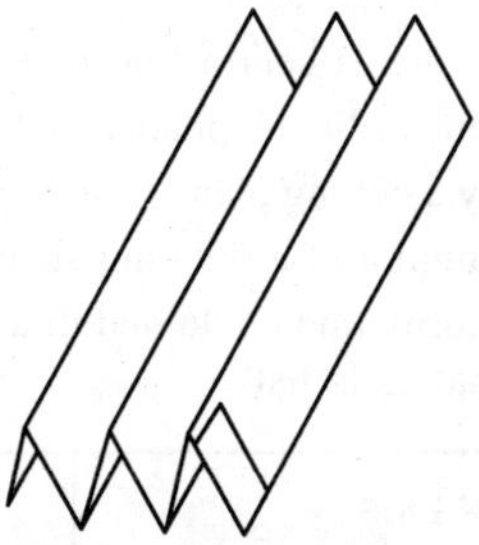	
A1 594 × 841	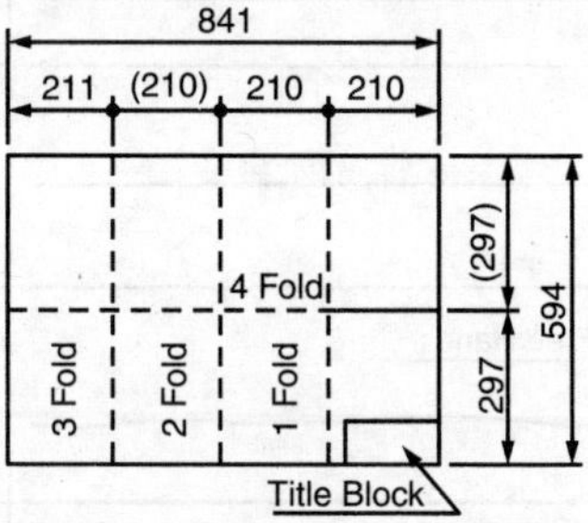	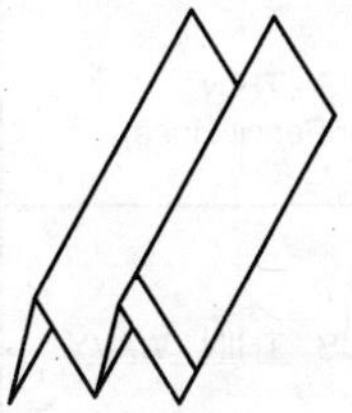	
A2 420 × 594	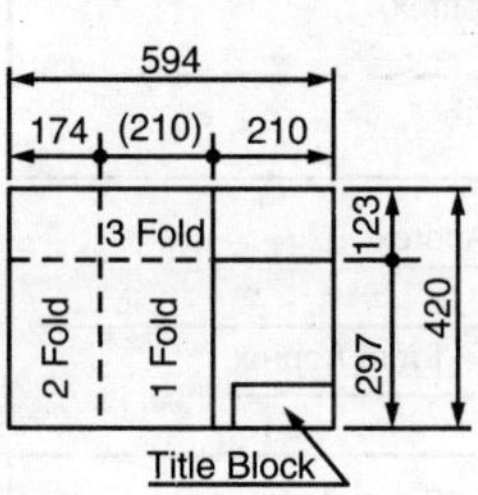	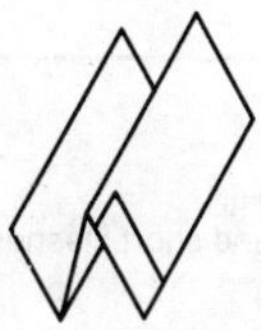	
A3 297 × 420	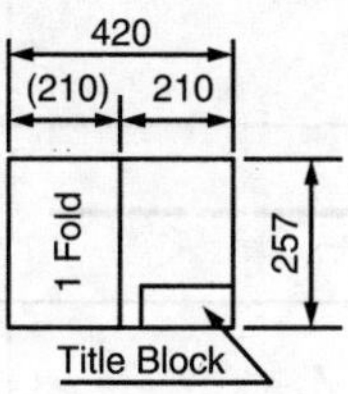		—

All Dimensions in mm

Fig. 2.20. Folding of prints to be kept individually in filing cabinet

EXERCISES

1. Divide sheet size A2 in eight equal parts and in each such division draw one of the conventional lines, allow the distance between any two cocecutive lines of a type, in its division, to be 10 mm.
2. Draw different types of conventional line, 10 time each, keeping their length as 100.
3. Redraw Fig 2.4, showing application of different types of lines, Take any convenient, proportional dimensions for it.
4. Prepare a specimen title-block:
   - as recommended by IS: 696 - 1972 and
   - for use in classroom, by students.
5. Make a complete layout of A0, A1, A2, A3, and A4 drawing sheets, showing title-block, materials list, folding marks and revision panel.

QUESTIONS FOR VIVA VOCE

1. Where, in drawing, a centre line is used?
2. What do you mean by 'Alphabet of lines'?
3. Where and why a cutting plane is drawn in drawing?
4. Why is the planning of a drawing sheet necessary?
5. List out the information contained in a typical title-block and a materials list.
6. What is the necessity of folding a drawing print?
7. What are the standard sizes of drawing sheets?
8. List out the order of inking a pencil drawing.
9. What is the use of:
   (*i*) squared graph, and
   (*ii*) isometric graph papers?

S.No.	Type of Line	Illustration	Application
A	CONTINUOUS - THICK Bold and Dark		Visible outline and edges
B	CONTINUOUS - THIN (Thin Grey for Pencil Lines)		Dimension lines, leader lines, extension lines, construction lines, outlines of adjacent parts, hatching and revolved section
C	CONTINUOUS - THIN - WAVY	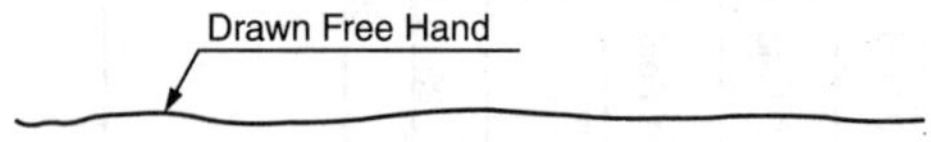	Irregular boundary lines, short break lines
D	SHORT DASHES MEDIUM	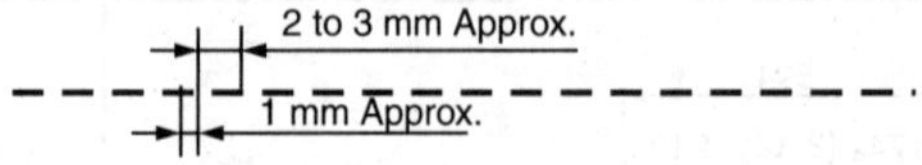	Hidden outlines and edges
E	LONG CHAIN-THIN Alternative Long and short Dashes	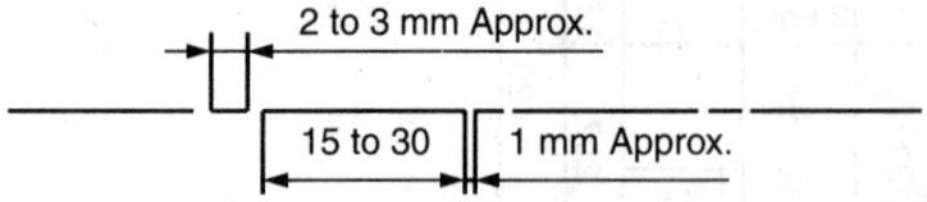	Centre lines, locus lines, extreme positions of the moveable parts, pitch circles, and parts situated infront of the cutting planes
F	LONG CHAIN THICK AT ENDS AND THIN ELSEWHERE	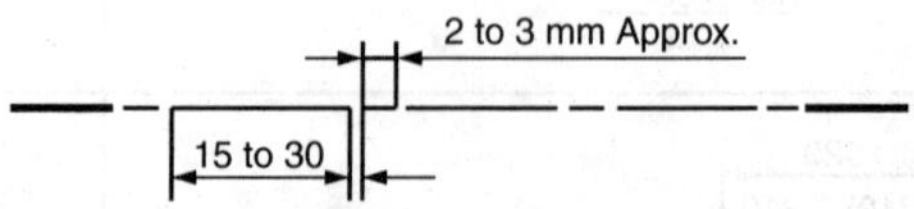	Cutting plane lines
G	LONG CHAIN-THICK		To indicate surfaces which are to receive additional treatment
H	RULED LINE AND SHORT ZIG-ZAG - THIN		Long break lines

3 Technical Lettering

3.1. INTRODUCTION

To impart to the men in the shops all the necessary information, required for complete construction of a machine or structure, the shape description, conveyed with the help of graphical language, must be accompanied by size description and instructive specifications in the form of figures, notes and dimensions. All written information on the drawing is always in the form of lettering and not in handwriting. Lettering should be done on the drawing in such a manner that it may be read when the drawing is viewed from the bottom edge, except where it is used for dimensioning purposes.

3.2. SINGLE STROKE LETTERS

Lettering is one of the "small things" which help to make or mar both the appearance and the usefulness of a drawing, regardless of the quality of the line work. The objective should be to employ uniform letters and figures which can be reproduced with reasonable rapidity and which will ensure good and legible reproductions from either pen or pencil originals. Simple freehand lettering, perfectly legible and quickly made, is an essential part of modern engineering drawing and it is conventional to print as plainly as possible. Ornament and flourish would detract from legibility and should never be used. Plain wide letters facilitate rapid reading and prevent costly misinterpretations.

The ability to letter correctly, involves:

- a perfect knowledge of the forms and proportions of the letters and a study of the direction and order of strokes which compose them;
- persistent practice in marking the set strokes;
- the development of a high degree of precision in forming and spacing of letters.

Almost anyone can learn to letter if persistent and intelligent in his/her efforts. While it is true that "practice makes perfect" there must be continuous effort to improve.

Lettering on drawing should be legible, neat in appearance and correct in style. Single stroke letters have these qualifications, therefore, they are universally used for engineering drawings. They should not be executed mechanically but should be made entirely freehand. Letters having all strokes of uniform thickness are classified as *Gothic.* The style of the letter, when the thickness of the strokes is such that it can be made with a single stroke of pencil, is called *Single Stroke Gothic*. The term '*Single stroke*' does not imply that the letter is produced with a continuous movement of the pencil or pen, but that the letter is made of one or more stems or curves, each of which is made of a single stroke.

3.3. CAPITAL AND LOWERCASE LETTERS

These letters may be of two types, *Capital* and *Small* or *Uppercase* and *Lowercase, Fig. 3.1.* Each of the two types is used in its proper place. All letters should be capitals except where the lowercase letters are accepted internationally for abbreviations.

3.4. VERTICAL AND SLANT LETTERING

Further each lettering style may be either *vertical* or *slant*, *Fig. 3.1.* In vertical lettering stems are perpendicular to the line of lettering. In slant lettering a slope of 75° to the horizontal, is recommended by *IS: 696-1972 second revision.*

Vertical letters are perhaps slightly more legible but more difficult to execute than inclined letters. Both vertical and slant letters are standard and the student should learn to execute both forms well, though they may give more attention to the style which they like and can execute better.

3.5. HEIGHT OF LETTERS

Letters are designated by their height. By this we mean the height of capital lettering only. The body of the lower case letters is 5/7 of the height of capitals. Typical proportions of uppercase and lowercase letters are shown in *Fig. 3.2*. The stems of lower case letters, such as, b, d, h, etc., which are called *ascenders,* are made the same height as capitals. The stems of lowercase letters, such as g, j, p, q, y etc., which extend downward, are known as *descenders, Fig. 3.1*. It is not desirable to grade the size of lettering with the size of the drawing. The size and weight of the lettering should be such as will produce legible prints from tracings either in pencil or in ink.

The ratio of height to width varies, but in most of the capital letters it is 7 : 5. Different sizes of letters and numerals are used for different purposes.

The height of letters is usually kept as follows:

1. Drawing number in title-block and letters denoting cutting plane or section should be – 8, 10, or 12 mm.
2. Title of drawing should be 6, 8 mm.
3. Sub-title headings should be 3, 4, 5 or 6 mm.
4. Notes, such as legends, schedules, materials list, and dimensions should be 3, 4, or 5 mm.
5. Alteration entries and the tolerances should be 2, 3 mm.

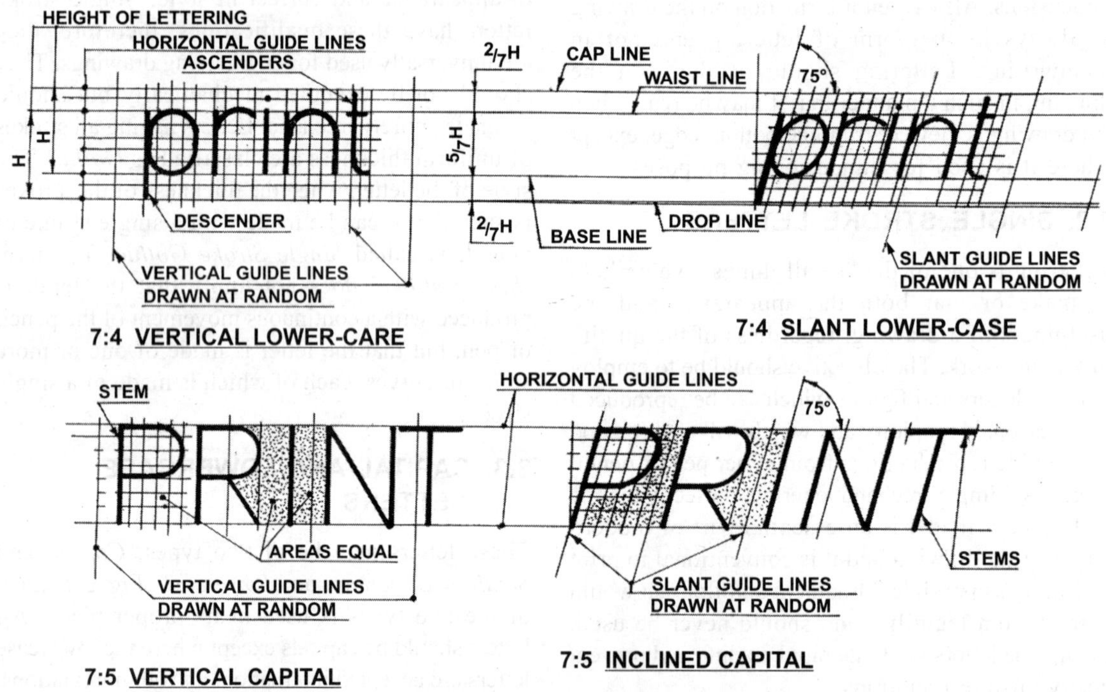

Fig. 3.1. Types of Lettering

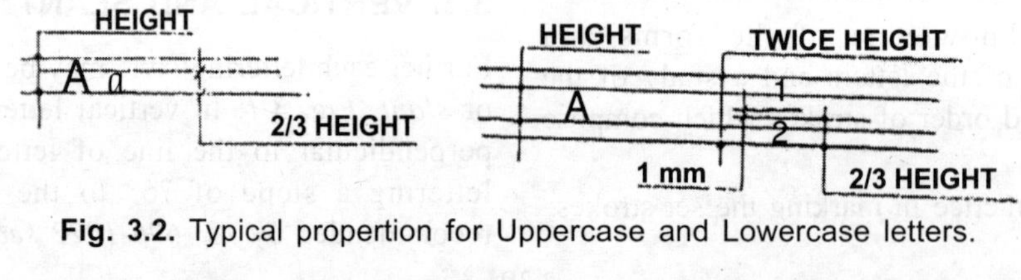

Fig. 3.2. Typical propertion for Uppercase and Lowercase letters.

3.6. GUIDE LINES

In order to obtain correct and uniform height of letters and numerals, guide lines are drawn. Vertical letters do not look pleasant if all letters are not uniformly vertical and inclined letters do not look good if all letters are not uniformly inclined. Therefore, in addition to horizontal guide lines, light vertical or inclined guide lines should be used. Vertical guide lines are not used to space the letters but instead, to assist in maintaining a uniform uprightness or uniform inclination of letters. These are to be drawn at random, *Fig. 3.3*.

Guide lines for finished pencil lettering should be drawn very lightly such that they need not be erased, as it is not possible to do so, after the lettering has been completed. A very hard pencil should be used to draw them. Whereas on inked drawings the guide line can be erased, after the ink is dry, with artgum.

In the case of capital letters two horizontal lines, and in the case of lowercase letters four horizontal line are drawn such that distance between the inner two lines is 2.5 times the distance between any two consecutive outer lines. Vertical and slant guide lines are of great help, for beginners, to maintain the proper direction of vertical or inclined stems of letters, *Fig. 3.1* and *Fig. 3.3*.

3.7. SINGLE STROKE VERTICAL LETTERS

An alphabet of vertical capital letters is shown in *Fig. 3.4* and *Fig. 3.5*. To show each unit, every rectangle has been divided into smaller squares of one unit side. The direction of movement of the pencil point to make the letter and different strokes to be executed are shown, *Fig. 3.5*. These general proportions of letters (7 : 5) and (6 : 5) are sufficiently accurate from students point of view to commence with. Horizontal strokes are made from left to right, and vertical strokes from the top downward.

Most of the strokes are natural and do not require memorizing, but the beginner must keep in mind that it is a decided mistake to practice lettering before he/she knows the proportions and order of strokes of the letter he/she is to execute. The author suggests that the beginner should consistently learn the forms of the letters by

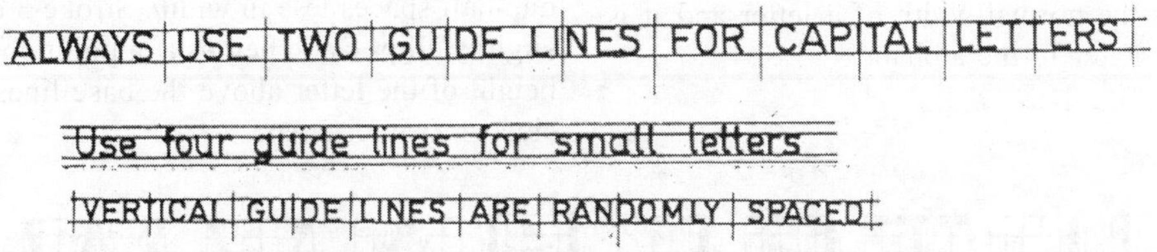

Fig. 3.3. Guide-lines for uppercase and lowercase letters

Fig. 3.4. Single stoke vertical capital letters (7.5), is : 696 (preffered). See that all letters except I, J, L. M and W are in rectangles of 7.5 and upper and lower portions of B, E, H, S, X, Y and Z are equal in area

sketching them on a squared sketch paper, taking different heights of letters, until he/she fully grasps the shapes and proportions of letters.

For purposes of close analysis and practice, letters may be classified as:

1. *straight line combinations* and
2. *curved line combinations.* They are explained below.

3.8. STRAIGHT LINE LETTERS

ILT: The ***I*** and the vertical strokes of the **L** and **T** are drawn downward with a finger movement. The horizontal strokes are drawn from left to right. The normal width of the ***L*** is 4.5 units as shown, but when it is followed in a word by a capital letter A its width should be reduced to about 4 spaces to compensate for the large area between the letters.

FEH*:** Stroke 4 of the ***E is nearly three-fifths as long as stroke 2, and is slightly above the midpoint. Stroke 3 of the ***F*** is the same as stroke 4 of the ***E***. Stroke 3 of the ***H*** is again slightly above the centre of its height.

AVW*:** Stroke 3 of the ***A is at one-third the height of the letter from the base line. The ***W*** is 1.33 times the normal width of a letter and it is the widest letter in the alphabet.

MNZ **:** There is a decided advantage in drawing the parallel strokes of the ***M*** and ***N*** before drawing the diagonal strokes, as shown. Stroke 3 and 4 of the ***M***, and stroke 2 and 3 of the ***N*** should intersect at the base line. As per SP: 46-1988 the strokes 3 and 4 of ***M*** intersect nearly two units above the base line (*M*) Fig. 3.4.

KXY **:** Strokes 1 and 2 of the ***K*** intersect at a point one-third the height of the letter from the base line. Stroke 3, extended, intersects stroke 1 at the top. The three strokes of the ***Y*** intersect a little below mid-point. The strokes 1 and 2 of the ***X*** intersect slightly above the centre.

3.9. CURVED LINE LETTERS

Refer to *Fig. 3.5.*

OCQG **:** The letters ***O, C, Q*** and ***G*** are formed with circles as bases. The ***O*** and ***Q*** are complete circles. The ***C*** and ***G*** are not complete circles and, therefore, their width is slightly less than their height. Stroke 3 of ***Q*** is a radial line making an angle of 30° with a vertical. The horizontal stroke 5 of the **G** begins at the centre of the circle and stroke 4 is vertical.

JUD*:** The *J* is a modification of the ***U but is one-half spaces less in width. Stroke 3 of the ***U*** is slightly elliptic and begins at a point one-third the height of the letter above the base line. The right side of ***D*** is circular.

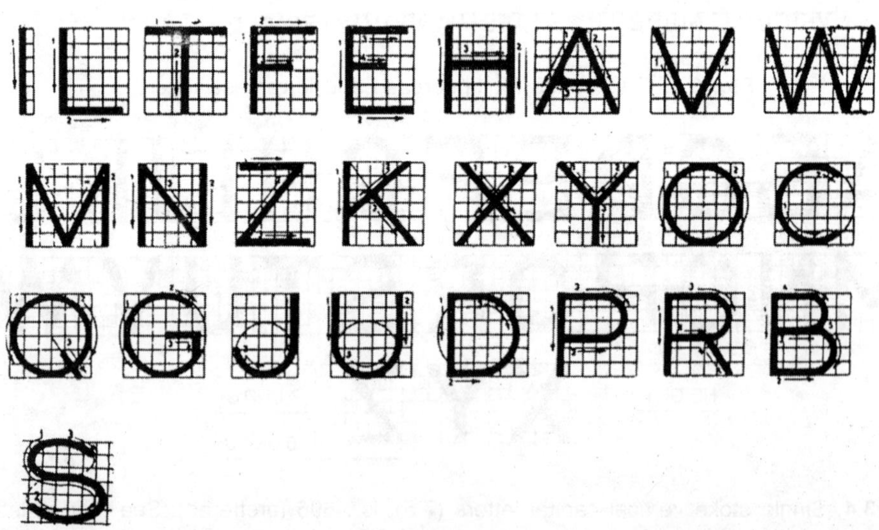

Fig. 3.5. Single stroke vertical capital letters (6.52)

PRB*:** Stroke 4 of the ***P is slightly below the centre of height of the letter, while stroke 4 of the ***R*** should be at mid-point and stroke 3 of the ***B*** is slightly above mid-height.

S **:** The upper and lower portions of the ***S*** are elliptic, the ellipses being tangent to a common vertical line on the left side. Compare the **S** with the numerals 8 and 3 in *Figs. 3.12* and *3.13*.

3.10. SINGLE STROKE INCLINED CAPITAL LETTERING

The order and direction of the strokes and also the proportions of all the slant letters are exactly the same as those for vertical letters, *Fig. 3.6.* Inclined letters can be regarded as oblique projections of vertical letters. Inclined guide lines, made independently of the widths of the letters and spaces will aid in getting a correct and uniform slant of inclined letters. The inclination is usually kept 75°. The inclined direction lines should be drawn with a special lettering triangle of 75°. Essential requirements of this lettering are:

1. Keeping to a uniform slope.
2. Having the letters full and well shaped.
3. Keeping them close together.

Figure 3.6 shows letters printed in parallelograms of proportions 6 : 5, sufficiently correct from drawing point of view. The letters

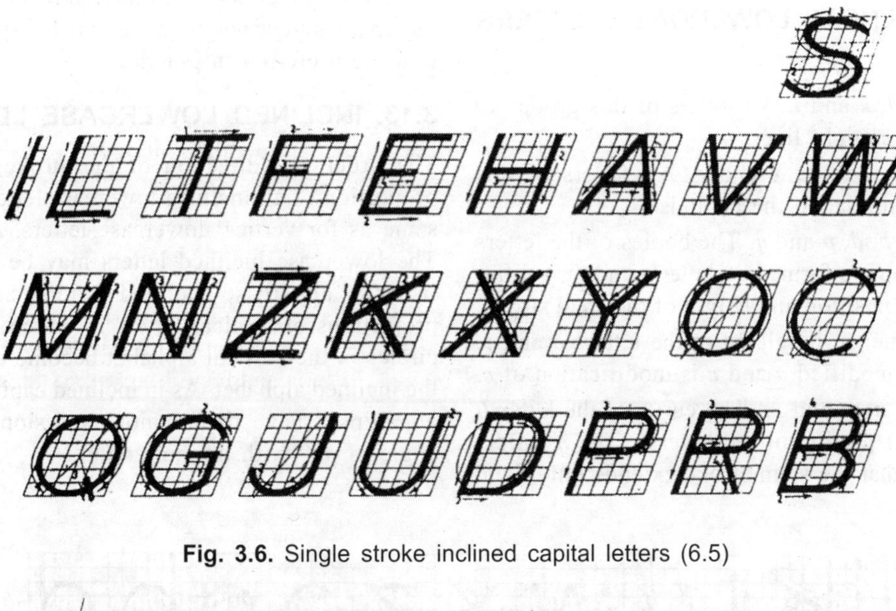

Fig. 3.6. Single stroke inclined capital letters (6.5)

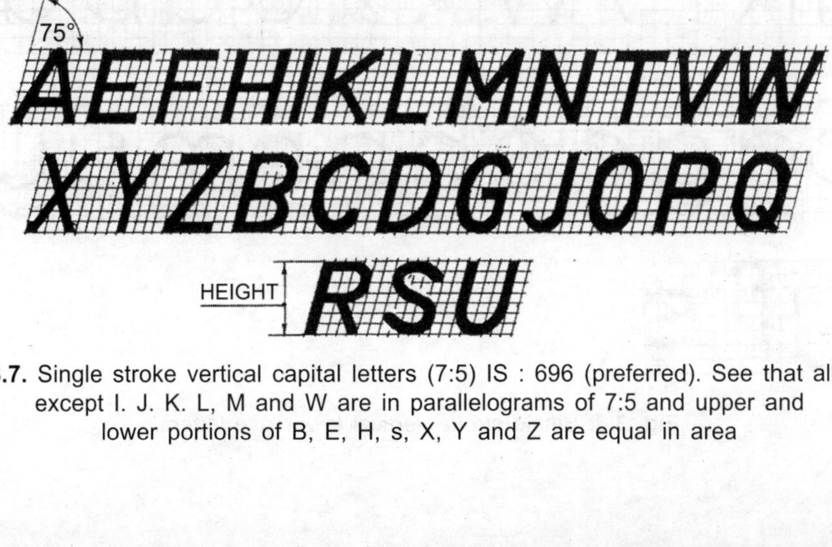

Fig. 3.7. Single stroke vertical capital letters (7:5) IS : 696 (preferred). See that all lettes except I. J. K. L, M and W are in parallelograms of 7:5 and upper and lower portions of B, E, H, s, X, Y and Z are equal in area

here too, are grouped as straight line letters and curved line letters.

These letters printed in 7 : 5 parallelograms, as recommended by SP: 46-1988, are shown in *Fig. 3.7* and preferably be printed thus.

3.11. SINGLE STROKE LOWERCASE LETTERS

These letters may be written in four guide lines. The body portion of letters is 2/3 to 5/7 of the height of related capitals. The ascenders extend to the cap line, and decenders to the drop line, *Fig. 3.1*.

Lowercase lettering may be vertical or inclined.

3.12. VERTICAL LOWERCASE LETTERS

Refer to *Fig. 3.8*.

The ***i, l, k*** and ***t***. All letters of this group are formed by straight lines.

The ***o, s, v, w, x*** and ***z***. All letters of this group are similar to the capitals.

The ***a, b, d, p*** and ***q***. The bodies of the letters in this group are formed by letter ***o*** and they differ only in the position and length of the stem stroke.

The letter ***g***, is related to the letters ***o*** and ***y***. Letter ***c*** is modified ***o*** and ***e*** is modification of ***c***.

The ***h, n, r***, and ***m***. The curve of the letter ***h*** is the upper portion of the letter ***o***. The ***n*** differs from ***h*** in that the stem stroke extends only from waist-line to base line. The letter ***r*** is a portion of the letter ***n***. The letter ***m*** consists of two modified ***n's***.

The letters ***u*** and ***y***. The letter ***u*** is inverted ***n***. The letter ***y*** is a practical combination of ***u*** and ***g***.

The letters ***j*** and ***f***. The portion of the letter ***j*** and above the base line is letter ***i***. The curve is same as of ***y***. The width of curved portion of ***f*** is smaller than other letters.

See that the body part of most of the letters in this proportion is a square of 4 units side and maximum height of letters is 6 units.

Fig. 3.9 illustrates vertical lowercase letters as recommended by: SP : 46-1988. Body part of most of these letters are formed in rectangles of 5 : 4 whereas their height is 7 units. This proportion is to be preferred in practice.

3.13. INCLINED LOWERCASE LETTERS

The order and direction of the strokes and the proportion of all inclined lowercase letters are the same as for vertical lowercase letters, *Fig. 3.10*. The lowercase inclined letters may be regarded, like the uppercase inclined letters, as oblique projections of vertical letters in which all of the circles in the vertical alphabet become ellipses in the inclined alphabet. As in inclined capital letters, all ellipses have their major axes sloping at an angle of 45° with the horizontal.

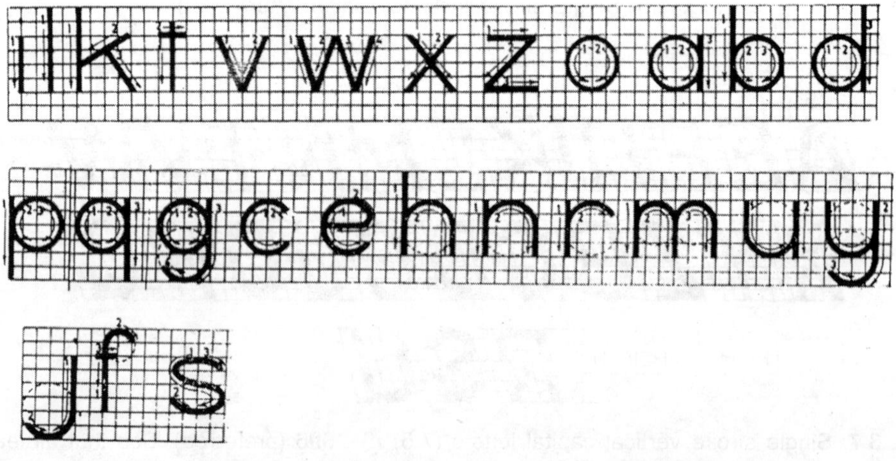

Fig. 3.8. Single stroke vertical lowercase letters

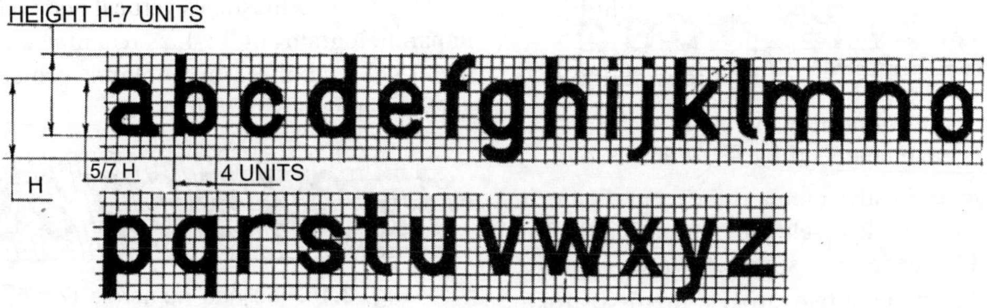

Fig. 3.9. Single stroke vertical lowercase letters (preferred)

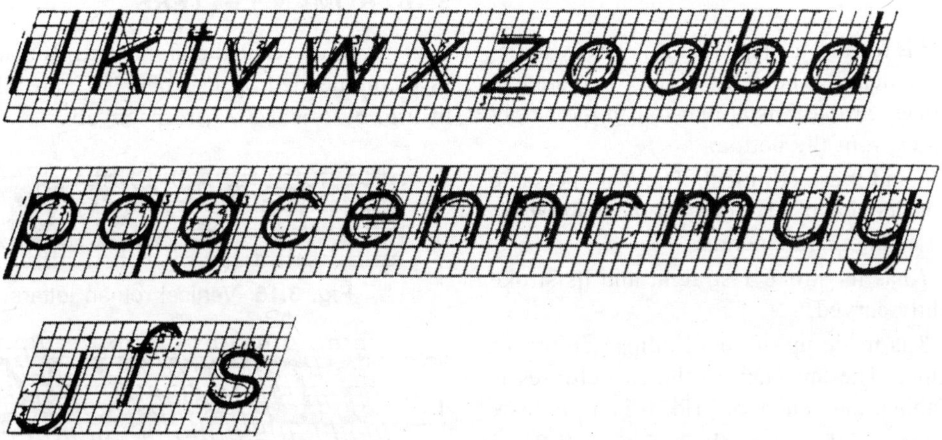

Fig. 3.10. Single stroke inclined lowercase letters

Figure 3.11 illustrates inclined lowercase letters as recommended by SP : 46 - 1988. Body parts of most of these letters are formed in parallelograms of 5 : 4 whereas their height is 7 units. See that body part of none of these letters is a circle.

3.14. VERTICAL NUMERALS

Both vertical and horizontal guide lines should be drawn for numerals too. This is particularly necessary for beginners. Since all values in engineering drawing are expressed by numerals, which should be perfectly legible, the student should make a careful study of their strokes and proportions.

All numerals except, **1**, in *Fig. 3.12*, are in rectangles of 7 : 4 and the numerals **3** and **8** are symmetrical about the centre. Numerals **6** and **9** have their loops up to exact centres of grid.

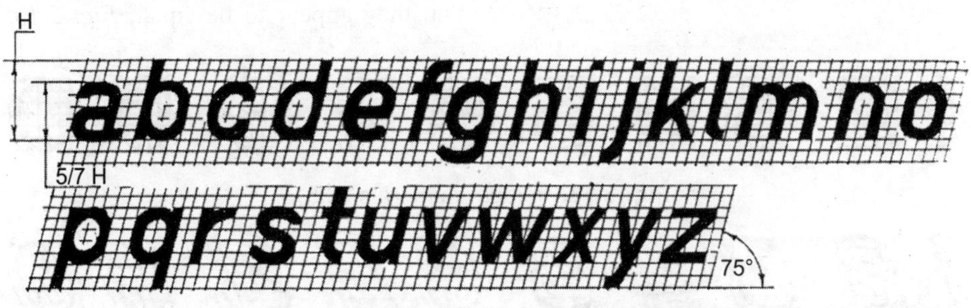

Fig. 3.11. Single stroke inclined lowercase letters

Fig. 3.12. Vertical numerals (7 : 4) as recommended by IS : 696

In *Fig. 3.13* all numerals, except **1**, are in rectangles of **6** units height and **5** units width. The numeral **1** is made by a single downward stroke.

The **2** is made in two strokes, the first stroke, which passes through the centre of the grid, the second stroke is at right angles to the end of first stroke.

The **3** is of such a shape that it is nearly three-fourth of letter **8**. The letter **4** is completed in three strokes. Stroke 3 is at one-fourth the height of the letter from the bottom.

The **6** and **9** are elliptical. The **9** is the same as the 6 upside down. The loops in the **6** and **9** are two-third as high as numerals.

The **7** has its stroke **1** straight and its stroke **2** is slightly curved.

The **8** is made up of small ellipse on top of large ellipse. The tangency of the two ellipses is slightly above the center of grid. **0** is an ellipse.

The tops of the numerals **3, 5, 6,** and **8** are narrower than the bottom to give them stability, *Fig. 3.13*, but IS : 696-1972 has adopted their symmetrical proportions above and below their centre of height, *Fig. 3.12*.

3.15. INCLINED NUMERALS

Inclined numerals also are inclined at the same angle, i.e., 75°, to the horizontal, as the inclined letters, *Fig. 3.14*. The inclined numerals may be regarded as oblique projections of vertical numerals, and the general proportions and strokes are the same.

Figure 3.15 illustrates vertical numerals printed in parallelograms of 7 : 4, as recommended by IS: 696-1972 and these are oblique projections of vertical numerals of *Fig. 3.12*.

Fig. 3.15. Inclined numerals (7 : 4) as recommended by IS : 696

3.16. ROMAN LETTERS

Specimens of vertical and inclined Roman letters, as recommended by IS : 696-1972 are illustrated in *Fig. 3.16* and *3.17*, respectively.

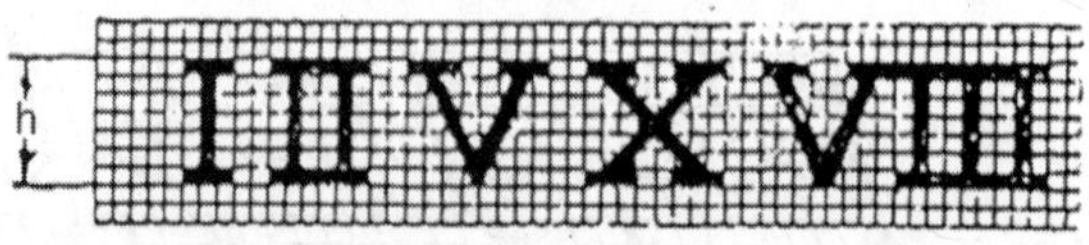

Fig. 3.16. Vertical roman letters

Fig. 3.17. Inclined Roman Letters

3.17. COMPOSITION

Composition in lettering means the arrangement and spacing of words and lines with letters of approximate size and style. Make the letters broad but closely spaced. Open spacing is hard to read. Even spacing is essential as uneven spacing spoils the entire composition.

In combining letters into words the spaces for the various combinations of letters are so arranged that they appear to be equal, *Fig. 3.18*.

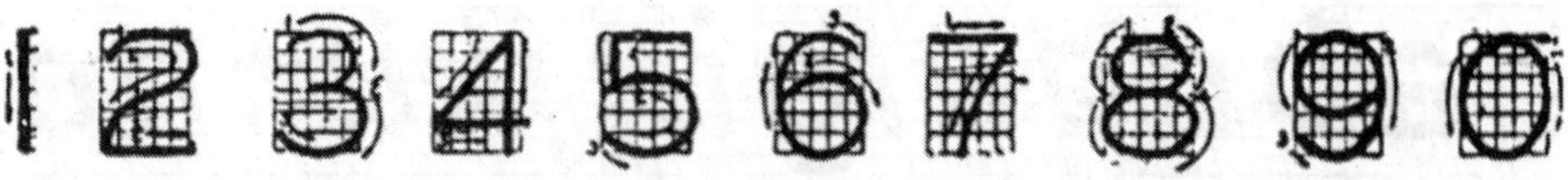

Fig. 3.13. Vertical numerals (6:5)

Fig. 3.14. Inclined numerals (6 : 5)

Fig. 3.18. Proper letter spacing

The key note of successful lettering is *"uniformity"*. Compact uniform lettering is an asset to any drawing. *'Uniform height'* is obtained by having each letter meet the top and bottom guide lines. *'Uniform weight'* by making all strokes of same thickness. *'Uniform direction'* by drawing and following direction lines, either vertical or inclined. *'Uniform shade'* by careful spacing. *'Uniformity in strength'* of lines can be acquired only by the skillful use of properly selected pencils.

In spacing words, the clear distance between them should be more than the height of the letters or they may be spaced by allowing room for the letter O between them. The space between sentences should be somewhat greater. The distance between the lines of lettering may vary from one-half the height of the capitals to 1.5 times their height, *Fig. 3.19.*

Many variations too are possible in lettering to meet design or space requirements e.g., letters may be *compressed* to be accommodated in available smaller space, or *extended* if space to be occupied by them is more than that required for normal letters, *Fig. 3.20.*

3.18. GENERAL PROCEDURE FOR LETTERING

The general procedure to make any type of lettering is suggested below:

1. Keeping in view the purpose of the lettering, decide the height of letters and draw horizontal guide lines, apart by height of letters and long enough to accommodate the required lettering.
2. Draw vertical or sloping guide lines, as the case may be, on full length of horizontal guide lines, at random.
3. Keeping in view the shape and form of letters, plot lightly, all important points which are to serve as guides for different strokes to start and end.
4. Join these points, in proper order, by light pencil strokes.
5. When the shape of letters appear to be correct and spaces in them balanced, fair out the letters and finish the work by removing reference guide lines.
6. Maintain proper distance between consecutive words in a sentence, different sentences and lines.

After a few letters are made, the pencil point will tend to become dull. To keep the lettering uniform the pencil point should be kept well sharpened, and the pencil should be continuously rotated in fingers while lettering.

While lettering in ink, excessive pressure should not be put on the pen point. Tendency to put pressure on pen point results in strokes of varying width.

Normally all vertical strokes should be drawn downward with a finger movement and all horizontal strokes to the right with a wrist movement, without turning the drawing paper. Either the position of the drawing board should be adjusted or the drafter should change his/her position such that the forearm of the hand, holding the pen, is approximately at right angle to the line of lettering. No lettering should be done if the forearm has no surface to rest upon.

Fig. 3.19. Spacing between words and sentences

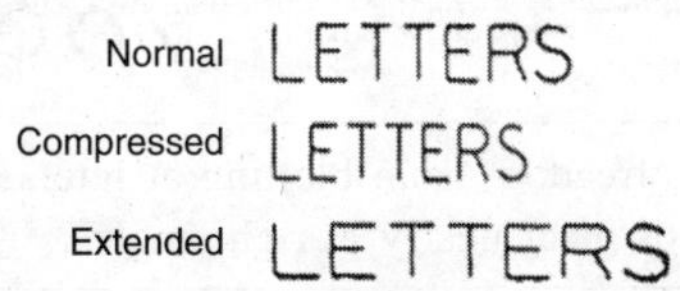

Fig. 3.20. Variations in lettering

3.19. LETTERING PENCILS

For ruling guide lines a 3H or 4H pencil, well sharpened, is best. The actual lettering should be done with a well sharpened H pencil mended to a conical point. The pencil is first sharpened to a needle point, then the point is dulled slightly by marking on paper, rotating the pencil at the same time, to keep the point symmetrical. The pencil lead should be soft enough to produce jet-black letters, yet hard enough to prevent excessive wearing down of the point, and smearing of the graphite. The soft lead containing more graphite, slides over the paper with less resistance and also produces a line dark enough to reproduce.

3.20. LETTERING WITH STENCILS

The characteristics of lettering with the aid of stencils, on technical drawings and associated document are specified in *Fig. 3.21 (a), (b), Tables 3.1* and *3.2*.

The height **h** of the capital letter is taken as the base of dimensioning. The two standard ratios for **d/h**, 1/14 and 1/10, are most economical as they result in a minimum number of line thickness as is illustrated in *Table 3.1* and *3.2*. The lettering may be inclined at 15° to the right or may be vertical. Recommended sizes of vertical *lettering A* of ratio **d/h** = 1/14 are given in *Table 3.1* and the specimen are shown in *Fig. 3.21*. Whereas recommended sizes of vertical *lettering B* of ratio

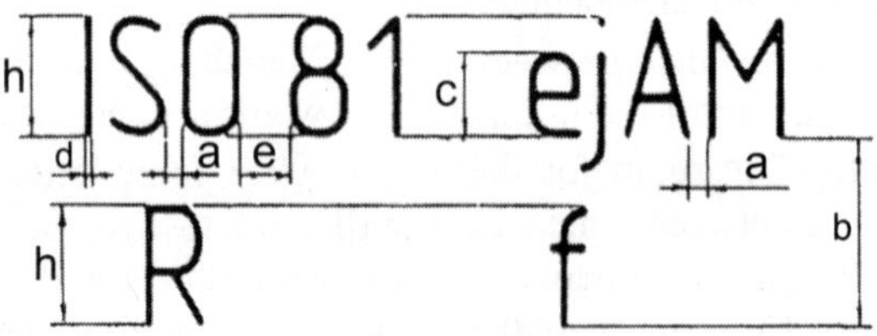

Fig. 3.21. (a)

Note: To obtain constant line-density, freedom from blotting at intersecting and case of writing the letters should be formed so that lines cross or meet nearly at right-angles.

Fig. 3.21. (b) Lettering A vertical.

1/10 are given in Table 3.2 and the specimen are shown in *Fig. 3.22.*

It should be noted that the spacing *'a'* between two characters may be reduced by half if this gives a better visual effect, e.g., LA, TV; it then equals the line thickness *'d'*.

Also to obtain constant line density, freedom from blotting at intersecting lines, and ease of writing, the letter should be formed so that lines cross or meet nearly at right angles.

LETTERING B VERTICAL

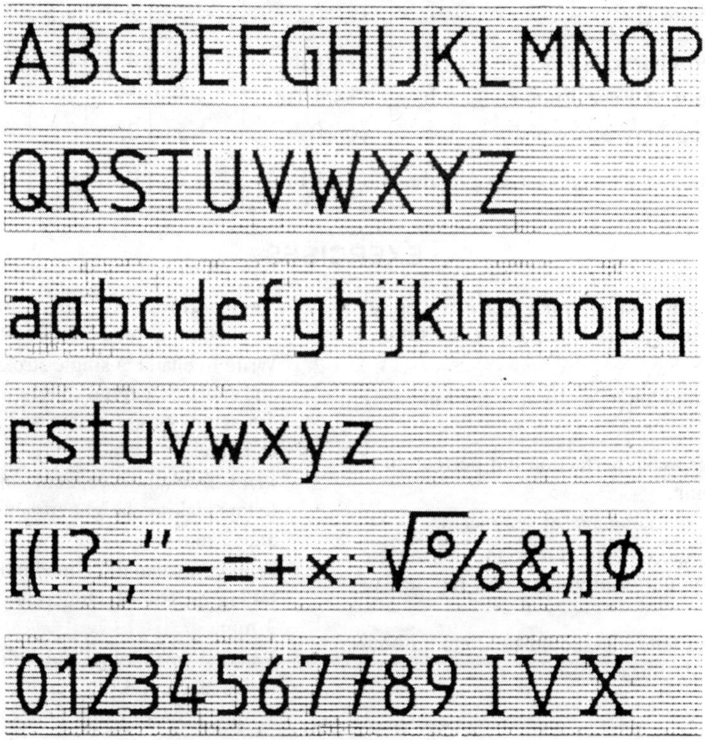

Fig. 3.22. Lettering B vertical

Table 3.1.

Lettering A (*d* = h/14)

Characteristic		*Ratio*	*Dimensions*						
Lettering height of capitals	h	(14/14)h	2.5	3.5	5	7	10	14	20
Height of lower-case letter (without stem or tail)	c	(10/14)h	—	2.5	3.5	5	7	10	14
Spacing between characters	a	(2/14)h	0.35	0.5	0.7	1	1.4	2	2.8
Minimum spacing of base lines	b	(20/14)h	3.5	5	7	10	14	20	28
Minimum spacing between words	e	(6/14)h	1.05	1.5	2.1	3	4.2	6	8.4
Thickness of lines	d	(1/14)h	0.18	0.25	0.35	0.5	0.7	1	14

Note: The spacing a between two characters may be reduced by half if this gives a better visual effect, as for example LA, VT, it then equals the line thickness d.

Table 3.2.

Lettering B (*d* = h/10)

Characteristic		*Ratio*	*Dimensions*						
Heigh t of capital letters	h	(10/10)h	2.5	3.5	5	7	10	14	20
Height of lower-case letter (without stem or tail)	c	(7/10)h	—	2.5	3.5	5	7	10	14
Spacing between characters	a	(2/10)h	0.5	0.7	1	1.4	2	2.8	4
Minimum spacing of base lines	b	(14/10)h	3.5	5	7	10	14	20	28
Minimum spacing between words	e	(6/10)h	1.5	2.1	3	4.2	6	8.4	12
Thickness of lines	d	(1/10)h	0.25	0.35	0.5	0.7	1	1.4	2

EXERCISES

1. Write freehand, in single-stroke vertical capital letters and inclined capital letter, using the ratio:
   - 6 : 5;
   - 7 : 5 and height of letters as 10 mm, the following statements:

   (*a*) Small things make perfection but perfection is no small thing.

   (*b*) Drawing is the language of engineers.

   (*c*) The main requirements of a good lettering are the legibility, uniformity, ease and rapidity of execution.

   (*d*) No freehand lettering should be done without drawing guide lines.

   (*e*) Engineering drawing is a systematic combination of different conventional lines.
2. Write freehand, a single stroke vertical capital letters and inclined capital letters using the ratios 6:5 and 7:5 and of height, 6 mm and 3 mm statements given in exercise.
3. Print freehand, vertical and inclined lowercase letters, taking height of letters as, 10 mm, 6 mm and 3 mm
4. Print vertical and inclined numerals, four times each, taking height of numerals as 10 mm, 6 mm and 3 mm.
5. Practice open type and solid type arrowheads 25 times each, taking length of the arrowhead as: 5 mm, 4 mm, 3 mm and 2 mm.

QUESTIONS FOR VIVA VOCE

1. To what shape a pencil should be mended for lettering and dimensioning and why?
2. What should be the grade of pencil used for lettering and why?
3. What are the main requirements of lettering?
4. What do you mean by single stroke letters?
5. What is Gothic style of lettering?
6. What is the importance of the concept of stability in lettering?
7. What are guide lines and why are they necessary in lettering?
8. What do you mean by height of lettering?
9. What do you understand by uniformity in lettering?
10. What do you mean by composition as used in lettering?
11. What do you mean by normal, compressed and extended lettering?
12. What is the importance of dimensioning?
13. What do you understand by the term 'Notation of dimensioning'?
14. What are end product drawings?
15. Define (*i*) A feature, (*ii*) A functional dimension, (*iii*) Datum Dimension, (*iv*) A Redundant dimension, and (*v*) An auxiliary dimension.
16. Explain with the help of simple sketches: (*i*) Progressive Dimensioning, (*ii*) Chain dimensioning, (*iii*) Parallel dimensioning, and (*iv*) Combined dimensioning.
17. What are: (*i*) Size dimensions, and (*ii*) Location dimensions?
18. Explain with the help of a simple sketch, (*i*) Aligned system and (*ii*) Unidirectional system of dimensioning.

❑❑❑

Theory of Orthographic Projections

4

4.1. INTRODUCTION

An engineer, who may be a designer, drafter or builder, is always confronted with the task of describing the shape of three dimensional objects on a two dimensional surface, *e.g.,* the drawing paper. This may be done by a written description, photograph of the object or by graphical description. The written description or the language of words, can be used to describe only very simple and standard solids, while the photographs can show only the exterior of an object. Also the object to be designed or improvement of some existing design can not be photographed. Hence the only possible media, accepted universally, by which an engineer can express his/her ideas is the language of lines or the graphical description, commonly called *"Engineering Drawing"*, which is a systematic combination of different conventional lines.

4.2. PROJECTION - TERMS

In graphic language the shape is described by *projection*, which is the image of the object, formed by *rays of light*, taken in some particular direction, from the *object* into a *picture plane*, as it appears to *an observer* stationed at the point, from or towards which the projection is made.

Depending upon the *orientation of the object, location of the point of sight, and the direction of lines of sight relative to the picture plane*, different types of projections, *e.g.,* perspective, parallel, orthographic, axonometric, oblique, etc., can be obtained.

The plane, on which the projection is taken is called the *plane of projection or picture plane.* The point, from which the observer is assumed to view the object, is called the *station point* or the *centre of projection.*

In this chapter only the orthographic system of projections will be studied in detail. Other types of projections are dealt in separate chapters.

4.3. ORTHOGRAPHIC PROJECTION

Referring to *Fig. 4.1*, if the station point (or *observer's eye) is at an infinite distance from an object* (a prism here) such that the *lines of sight (or projectors) are parallel to each other and perpendicular to the plane of projection* (or picture plane), the projection obtained, which would be of the same size and shape as the facing surface of the object, is called an orthographic projection. Note that the *object is placed with one of its faces parallel to the picture plane.*

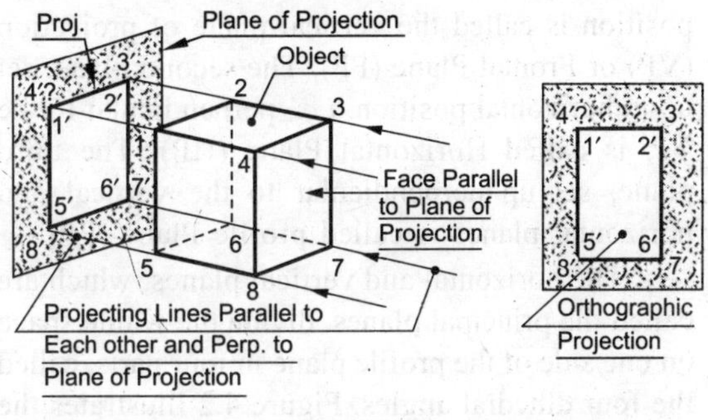

Fig. 4.1. Orthographic Projection (First Angle)

4.4. ORTHOGRAPHIC SYSTEM OF PROJECTIONS

Any object has three dimensions, *i.e.,* length, breadth (depth) and height. The problem is to represent or convey all these three dimensions, together with the other details of the object, on a sheet of drawing paper which has only two dimensions.

Orthographic system of projections is a method of representing the exact shape of a three dimensional object on a two dimensional drawing sheet in two or more views. As is clear from Fig. 4.1 that a single orthographic projection of a cube, or any other object, does not describe the form of the object, when the object is so placed that one of its faces is parallel to the plane of projection. In order to achieve a complete shape description, in such cases, it is necessary to get more than one projection, and therefore, additional planes of projection are used to project more views on them, for the object. As such, the orthographic system of projections is also called *multi-view projection* method.

These views are obtained by dropping perpendiculars from two or more sides of the object to the picture planes, generally set at right angles to each other. These picture planes are then rabated to lie them in one plane.

This method is almost universally used in engineering drawing to graphically describe the shape of objects.

4.5. CO-ORDINATE PLANES OF PROJECTION

In the orthographic projection drawing, for getting the different views of an object, three main planes are usually used. One of these set up in vertical position is called the vertical plane of projection (VP) or Frontal Plane (FP). The second plane, set up in horizontal position, *i.e.,* perpendicular to the VP, is called Horizontal Plane (HP). The third plane, set up perpendicular to the vertical and horizontal planes is called profile Plane (PP).

The horizontal and vertical planes, which are called the principal planes, divide the whole space on one side of the profile plane in four parts, called the four dihedral angles. Figure 4.2 illustrates the coordinate planes and the four dihedral angle (or quadrants). The lines of intersection of these three planes are called coordinate axes. Line of intersection of HP and VP is more commonly called "Reference Line" and is denoted by XY. The point of intersection of the three coordinate planes is called the origin.

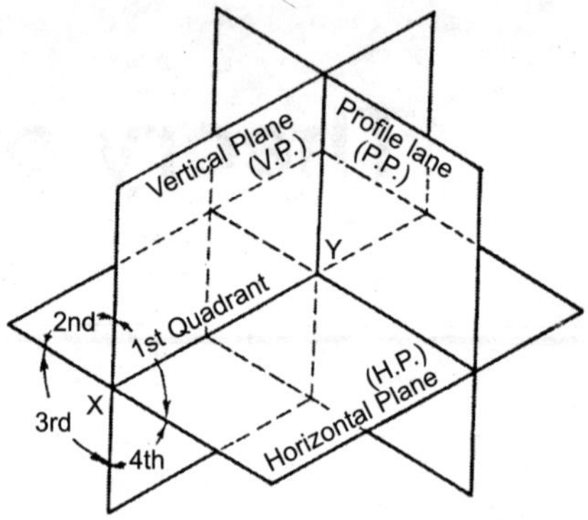

Fig. 4.2. Co-ordinate planes of projection

4.6. SYSTEMS OF ORTHOGRAPHIC PROJECTION

Orthographic views of any object can be represented by any one of the two systems of projection, *e.g.,* the *First Angle Projection* and the *Third Angle Projection.* These are named according to the quadrant in which the object is imagined to be placed, for purposes of projection. The shape and size of views are same in these two systems. The only difference lies in the relative position of the various views.

First angle projection (European) is widely used throughout Europe, the USSR, and many other countries in Asia. Third Angle Projection (American) is used in America. The current Indian and ISO Standards state that both systems of projection are equally acceptable but they should never be mixed on the same drawing.

Both these systems should be taught in Engineering Institutions, as these are used throughout the world, for the preparation of technical drawings. In India the use of First Angle Projection is in use exclusively after 1991.

4.7. FIRST ANGLE ORTHOGRAPHIC PROJECTION

Assume that a small block, 20 × 35 × 30 mm, is made and two of its corners are cut, as shown in *Fig. 4.3 (a),* in three stages.

Figure 4.3 (b) illustrates a pictorial view of the block. The block has been arranged in an arbitrary way because none of the three faces is more important than the others. Further, in order to describe the orthographic views we need to select a principal view. In the present case we have chosen the view in the direction of arrow A to be the view from the front.

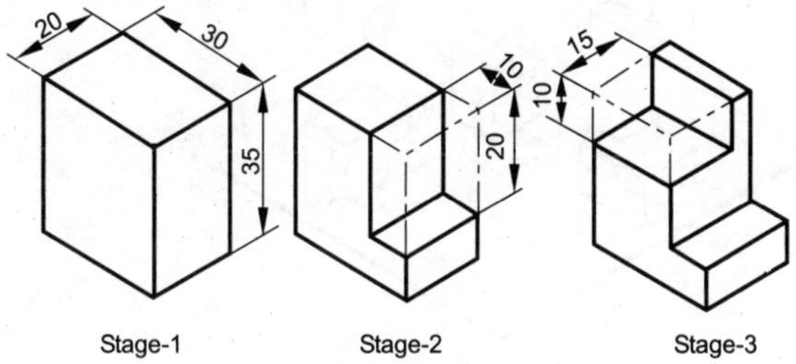

(a) Making a Small Block 35 × 30 × 20 Arbitrary Way

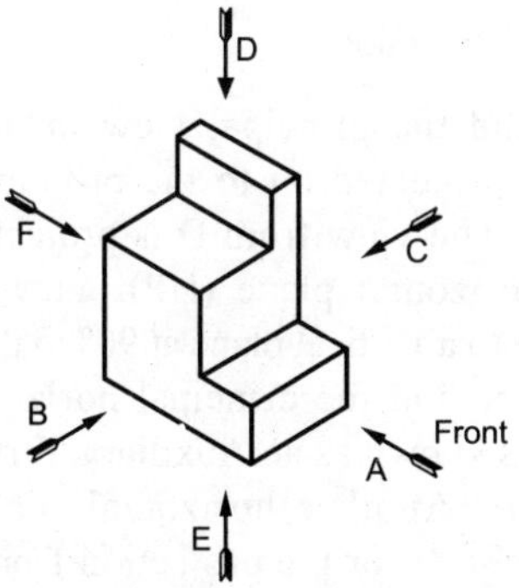

(b) Block Arranged in an Arbitrary Way

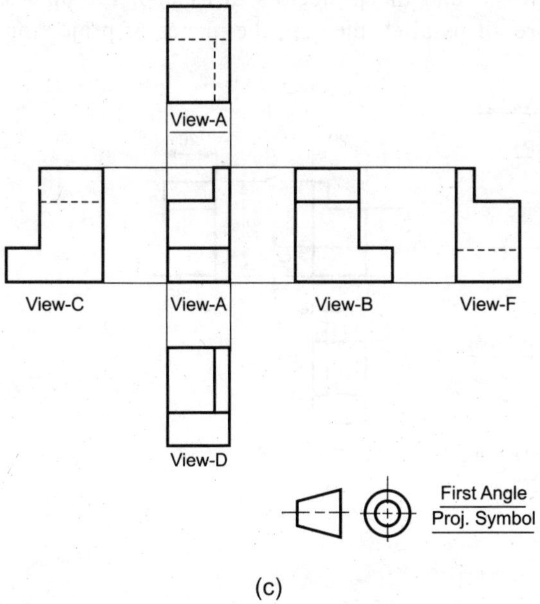

(c)

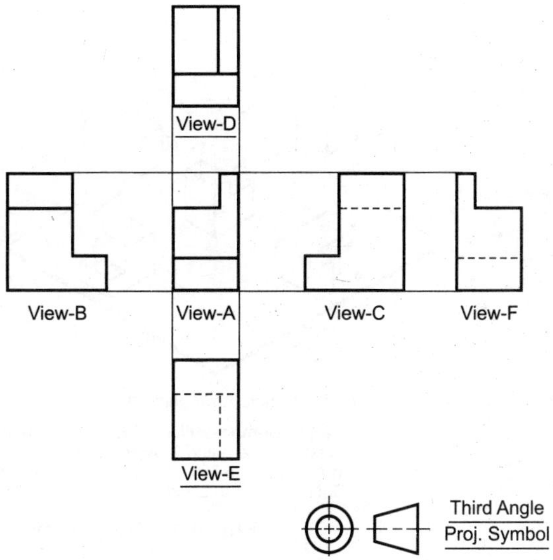

(d)

Fig. 4.3 (1)

The six arrows point to different surfaces of the block and six views will result on six different planes. The arrows themselves are positioned square to the surfaces, *i.e.,* at 90° to the surfaces. They are also at 90° or multiples of 90°, to each other. The six views are designated as follows:

- View in direction A is the view from the front,
- View in direction B is the view from the left,
- View in direction C is the view from the right,
- View in direction D is the view from above,
- View in direction E is the view from below,
- View in direction F is the view from the rear.

In first angle projection the views in the directions of arrows B, C, D, E, and F are arranged with reference to the front view as follows:

- The view from the direction B (*i.e.,* left) is placed on the right side.
- The view from the direction C (*i.e.,* right) is placed on the left.
- The view from the direction D (*i.e.,* above) is placed underneath.
- The view from the direction E (*i.e.,* below) is placed above.
- The view from F (*i.e.,* rear) may be placed on the left or on the right side, as may be found convenient, *Fig. 4.3 (d).*

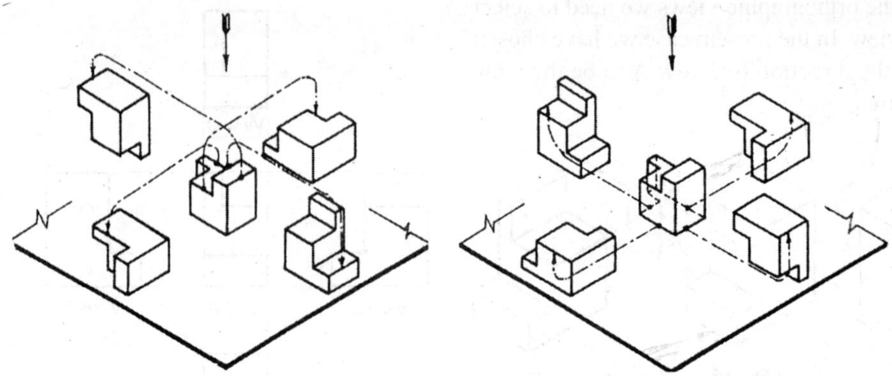

(e) First Angle Arrangement (f) Third Angle Arrangement

Simple Demonstration of First and Third Angle Projection by Placing the Block on the Drawing Board and Moving it in the Direction on the Four Chain Dotted Lines Terminating in Arrows. The View in the Direction of the Large Arrow will give the Five Views in Each Case. View-F in Each Case is Omitted.

Fig. 4.3 (2) Systems of orthographic projections

It is customary to state the projection used on orthographic drawings to remove all doubt or use the distinguishing symbol which is shown on the arrangement in *Fig. 4.3 (d)*.

If a model is made of the block in *Fig. 4.3 (b)*, which can easily be cut from polystyrene foam used in packing or soap cake, then a simple demonstration of first angle projection can be arranged by placing the block on the drawing board and observing it in the direction of the arrows.

In *Fig. 4.4 (a)*, the block is shown pictorially in the first quadrant with three of the surfaces on which views are projected. The surfaces are known as planes and the principal view in direction of arrow A is projected on to the principal vertical plane (VP). The view from D is projected on to a principal horizontal plane (HP). View B is also projected on to a vertical plane at 90° to the principal vertical plane and the principal horizontal plane. This plane is known as an Auxiliary Vertical Plane, *Fig. 4.4 (a)*. Another horizontal plane can be positioned above for the projection from arrow E, also we can position a second auxiliary vertical plane on the left for the projection of the view C. Notice that the projectors to each of the planes are all parallel, meeting the planes of projection

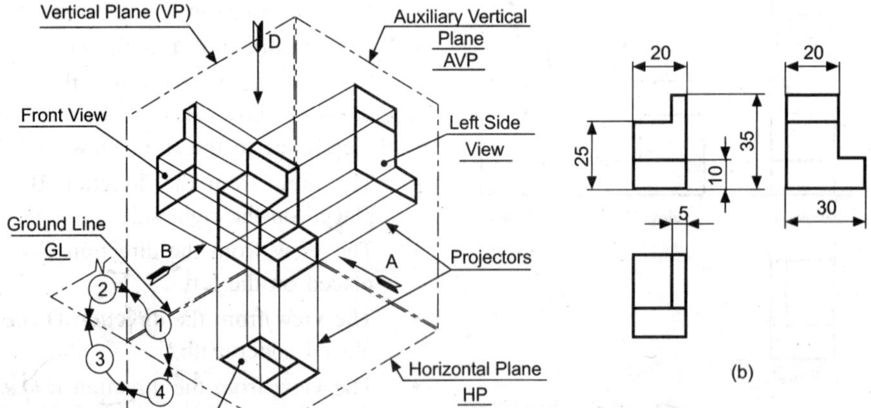

Fig. 4.4. (a) & (b)

at right angles and this is a feature of orthographic projection.

The intersection of the vertical and horizontal planes gives a line which is used as the reference line. This line is often refered to as the XY line and is useful in projection problems since it represents the position of the horizontal plane with reference to a front view and also the position of the vertical plane with reference to a plan or top view.

4.8. THIRD ANGLE ORTHOGRAPHIC PROJECTION

The difference between first and third angle projections is in the arrangement of views, *Fig. 4.3 (c)*. With reference to the front view the other views are arranged as follows:

- View B from the left side is placed on the left.
- View C from the right side is placed on the right.
- View D from the top side is placed above.
- View E from the under side is placed below.
- View F from the rear side may be placed on the left or the right, as may be found convenient.

The arrangement of views and the distinctive symbol of this projection is shown in *Fig. 4.3 (c)*.

4.9. VIEWS OF OBJECTS

To manufacture or build the objects a clear and complete description of their shape and size is necessary so that the objects may be made exactly as intended by the designers. This information is furnished by a number of systematically arranged orthographic views. Each view provides a definite information, if the view is taken in a direction perpendicular to a principal face of the object.

Every object has three principal dimensions, *e.g.*, width, height and depth, as shown in *Fig. 4.5 (a), (b)*. The front view shows only the width and height of the object, the end or side view shows height and depth, and the top view shows width and depth. No single orthographic view can show more than two of the three principal dimensions in their true sizes. A view of an object is known, technically, as a projection.

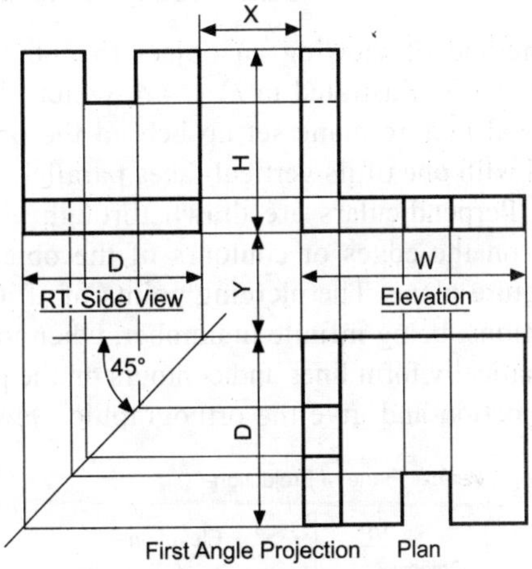

Fig. 4.5. (b) Principal dimensions of an object

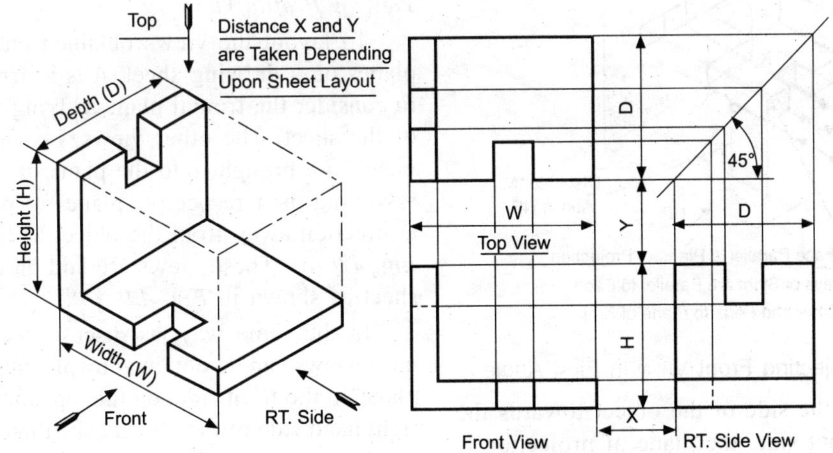

Fig. 4.5. (a) Principal dimensions of an object–Third angle projection

To describe a simple object, a drafter does not need to draw all six views, described already, and it is customary only to draw the minimum number which completely describe the object. Also, many parts do not have a definite 'Front', 'Top', or 'Side' and the orientation of the part is decided by the drafter who selects the views to give the maximum visual information to describe the part.

Traditionally, front view is also known as front elevation, side view is known as side or end elevation, and the view from above or beneath is referred to as plan.

4.10. METHOD OF OBTAINING ORTHOGRAPHIC PROJECTIONS IN FIRST ANGLE PROJECTION

The method of viewing an object to obtain the front view is illustrated in *Fig. 4.6*, which shows a vertical picture plane set up behind the object, placed with one of its vertical faces parallel to the plane. Perpendiculars are drawn through all the points on the edges or contours of the object to the picture plane. The piercing points of all these projections, being infinite in number, when joined systematically form lines and contours on the plane of projection and give the orthographic view.

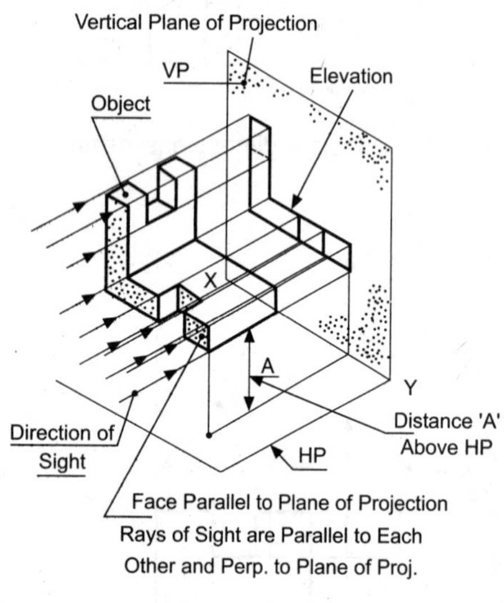

Fig. 4.6. Projecting Front View in First Angle

If we call the side of the object towards the observer as front side, the plane of projection is called the *Frontal Plane* and the orthographic view obtained on it is called the *elevation or front view*.

Figure 4.7 shows a horizontal plane placed below the object. The view obtained, by joining the piercing points of perpendiculars drawn through the various points on the object is called the *top view or plan*.

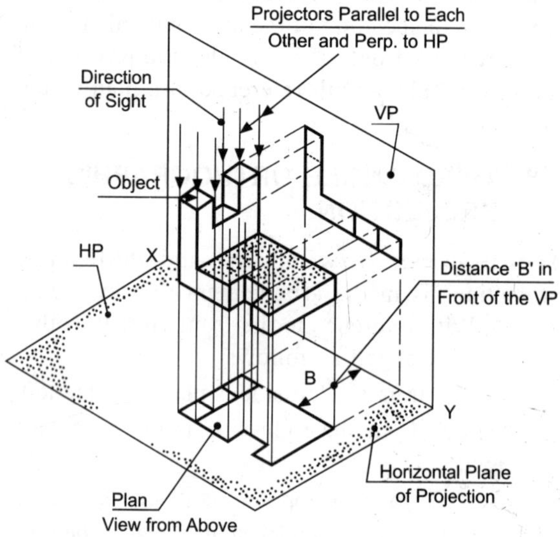

Fig. 4.7. Projecting Plan in First Angle

For simple objects two orthographic views are generally sufficient to fully describe the shape and size, but for intricate objects three or more views may be necessary.

Figure 4.8 shows a plane held perpendicular to both the frontal and horizontal planes. On this plane, which is called the *profile plane (PP)*, the view obtained is known as the *End Elevation, Side View or Profile View*.

To layout the views obtained on these three planes on a drawing sheet, it is a usual practice to consider the frontal plane as lying in the plane of the sheet. The other views, i.e., top and side views, are brought into the plane of the sheet by revolving their respective planes of projection in a direction away from the object as illustrated in *Fig. 4.9 (a)*. These views are laid on the drawing sheet as shown in *Fig. 4.9. (b)*.

In the same way, as discussed above, three more views can be obtained by placing the picture plane on the front side, on the top side, and on the right hand side of the object. The three views thus obtained are called the *Rear View, the Bottom View, and the Left End View,* respectively. These views

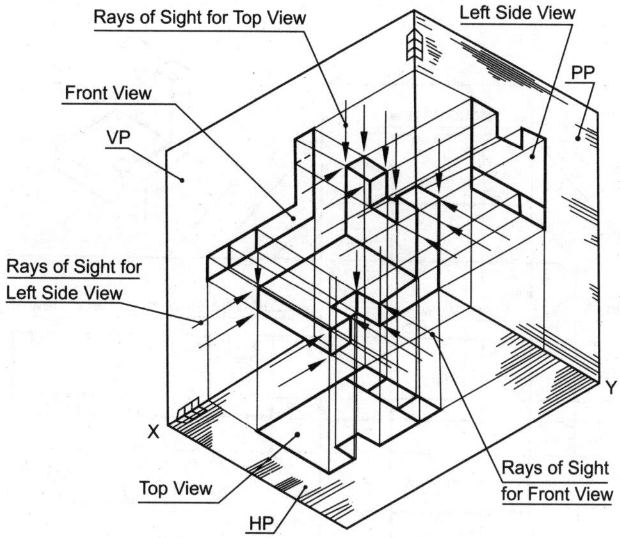

Method of Projecting Orthographic Views
as per First Angle Projection

Fig. 4.8.

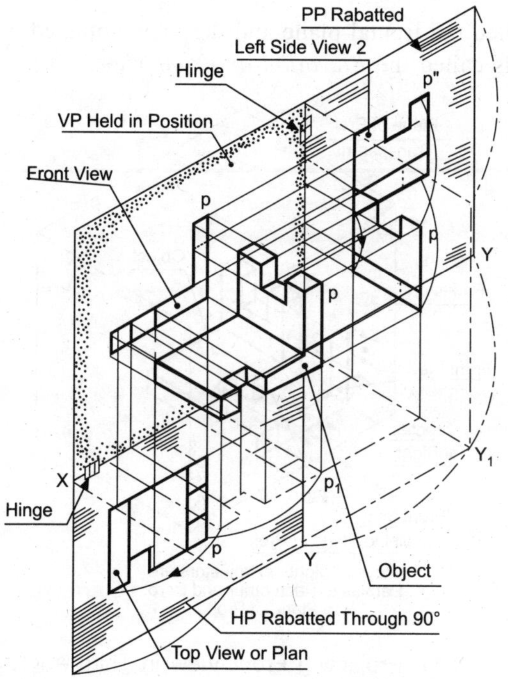

Method of Rabatting the Planes for Firts
Angle Projection

(a)

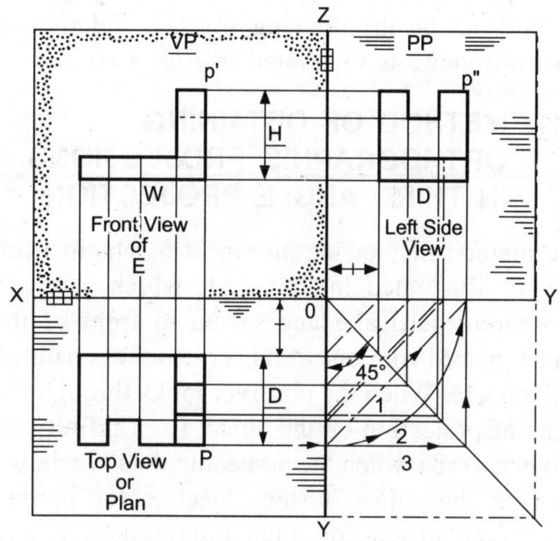

HP and PP are Rabatted to be Coplanar with the Vp
Layout of Views in First Angle Projection

(b)

Fig. 4.9.

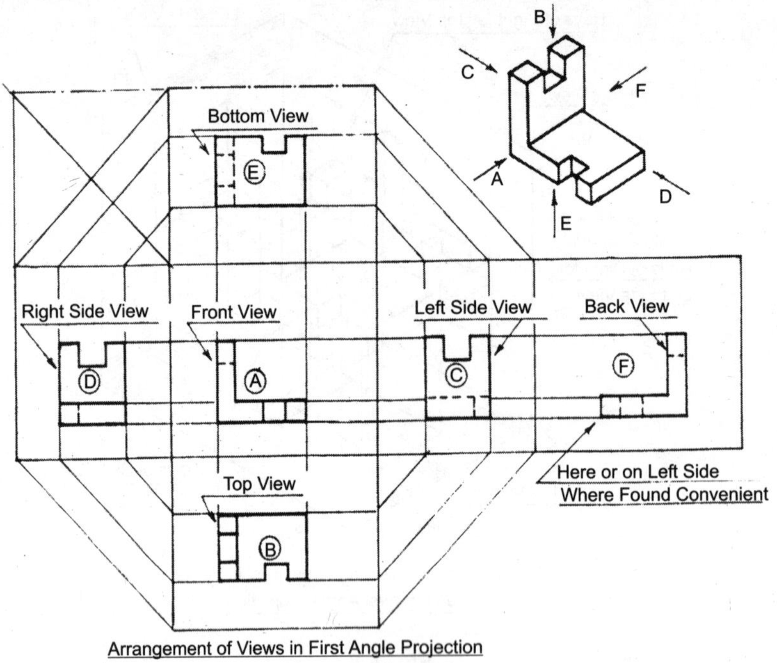

Fig. 4.10.

are laid out on the drawing sheet, in relation to the front view, as illustrated in *Fig. 4.10.*

4.11. METHOD OF OBTAINING ORTHOGRAPHIC PROJECTIONS IN THIRD ANGLE PROJECTION

The method of viewing an object to obtain front view is illustrated in Fig. 4.11, which shows a transparent vertical plane set up in front of the object, placed with one of its vertical faces parallel to the plane. When the observer looks through the plane of projection on the object from infinity, or in other words, when the perpendiculars are drawn from all the points on the object to the picture plane, the projection thus obtained is called orthographic view. This view represents all edges and contours of the object seen by the observer. In more precise terms, this view is obtained by drawing perpendiculars, called projectors, from all points on the edges or contours of the object, to the plane of projection. The piercing points of these projectors, being infinite in number, form lines on the transparent plane.

If we call the side of the object towards the observer as front side, the plane of projection is called the frontal plane and the view obtained on it is called the *Elevation or Front View.*

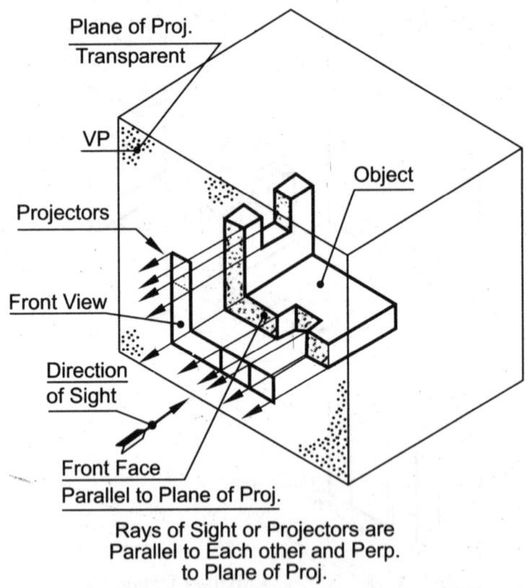

Fig. 4.11. Projecting Front View in Third Angle

Figure 4.12 shows the transparent horizontal plane placed above the object. The view obtained, by drawing perpendiculars from the object on this plane, is called the *Top View or Plan.*

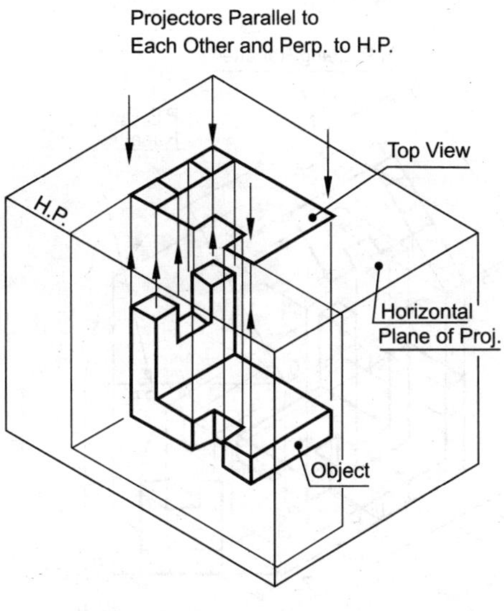

Fig. 4.12. Projecting Top View in Third Angle

For simple objects two views are generally sufficient to fully describe the shape and size, but for intricate objects three or more views are usually necessary.

Figure 4.13. (a) shows the transparent plane placed perpendicular to both the frontal and horizontal planes. On this plane, which is called the profile plane, the view obtained is known as the *End Elevation or Side View.*

It can be observed here that these three views have been obtained, imagining the object to be lying in a glass-box; or in VIIth octant.

When laying out these views on a drawing paper, it is the usual practice to consider the frontal plane as lying in the plane of the paper. The other views, *i.e.,* Top and Side views, are brought in to the plane of the paper by revolving their respective picture planes away from the object, as illustrated in *Fig. 4.13 (b), (c), (d).*

In the same way, as discussed above, three more views can be obtained by placing the picture plane on the back side, on the bottom side, and on the left hand side of the object. The three views thus obtained are called the *Rear View*, the *Bottom View*, and the *Left End View*, respectively. For laying out these views on the drawing sheet, the picture planes are revolved as illustrated in *Fig. 4.14*. The layout of all the six views on the drawing paper, is shown in *Fig. 4.15*.

4.12. VIEW ANALYSIS

We can observe the following facts in *Fig. 4.10* and *Fig. 4.15*.

1. The height of the object is shown in the front view, right side view, left side view, and rear view. All these views are in-between the same horizontal lines.
2. The width is shown in the front view, top view, bottom view, and rear view. The first three of these views do lie in-between the same vertical lines.
3. The depth is shown in the top view, right side view, left side view, and bottom view. All these views are grouped around the front view.
4. Every object line or dotted line on the drawing represents a line of intersection of two surfaces of the object or a contour.

4.13. COMPARISON OF FIRST AND THIRD ANGLE PROJECTIONS

Referring to *Fig. 4.2*, the object can be placed in any of the four quadrants and its projections on VP, HP and PP can be obtained by looking on it from front, top and side, respectively.

When the object is placed in the first dihedral angle and viewed from the front side to project the front view, the vertical plane lies behind the object. When viewed from the top, to project the top view, the horizontal plane lies below the object. When seen from left side, to project the left side view lias the profile plane again lies behind the object. Finally, when these planes are revolved such that the first quadrant opens out, as shown in Fig. 4.16 by arrows, to lie them in a single plane, the top view lias below the line of intersection of the HP and the VP, the front view lies above the line of intersection of the HP and the VP, and the left side view lies on the right side and in line with the front view. In this position, the projections obtained are called first angle projections.

For technical drawing when the system of First Angle Projection is to be used:

1. Projection planes are so set up, perpendicular to one another, such that the object is in between each plane of projection and the observer and the views are obtained as shown in *Fig. 4.4 (c), Fig. 4.9.*

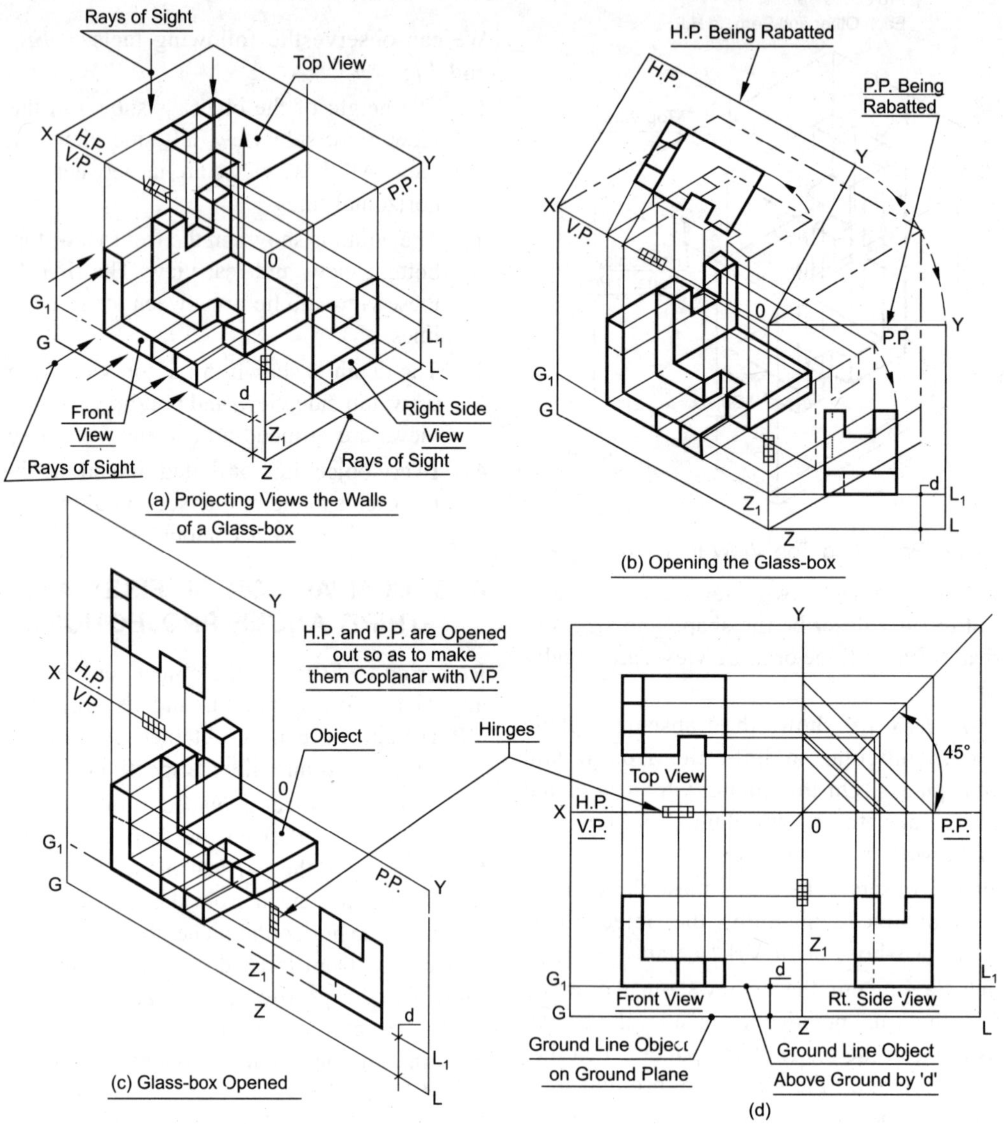

Fig. 4.13. Concept of Glass-Box (Third Angle Projection)

2. For layout of the views on the drawing sheet, planes are revolved away from the object about their edges common with the VP (*i.e.,* line of intersection of the VP with the plane of projection).
3. The points on the object, which are nearer to the observer, while looking for drawing the front view, will be projected away from the front view in the side and top views.
4. There is only one reference line XY, which is the intersection of the HP and the VP, and the intersection of the HP and the PP on rotation becomes collinear with the XY line.

When the object is placed in the third dihedral angle, *Fig. 4.17*, and viewed from front side, to project the front view, the vertical plane lies in front of the object. When viewed from top, to project the top view, the horizontal plane lies above the object. When seen from left side, to project

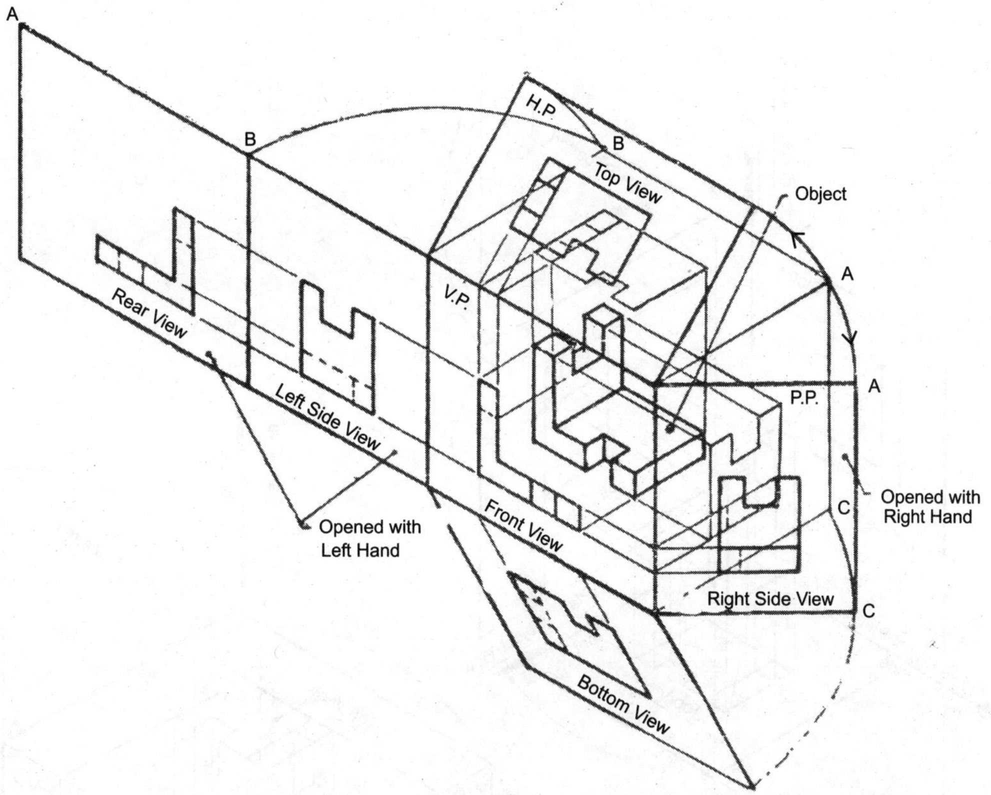

Fig. 4.14. Obtaining six views on six sides of the Glass-box (Third Angle Projection)

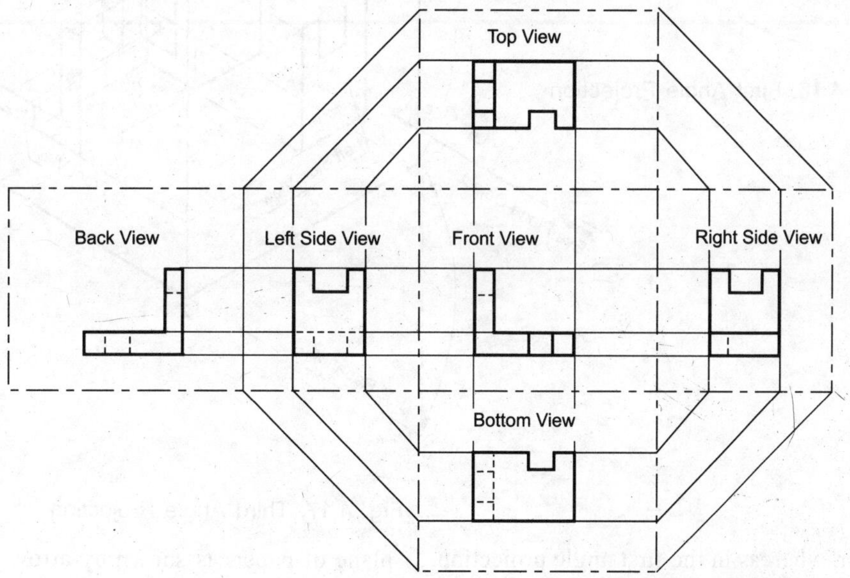

Fig. 4.15. Layout of six views after rebating the walls of the box (Third Angle Projection)

the left side view, the profile plane again lies in front of the object.

Thus we see that in third angle projection the plane of projection lies between the object and

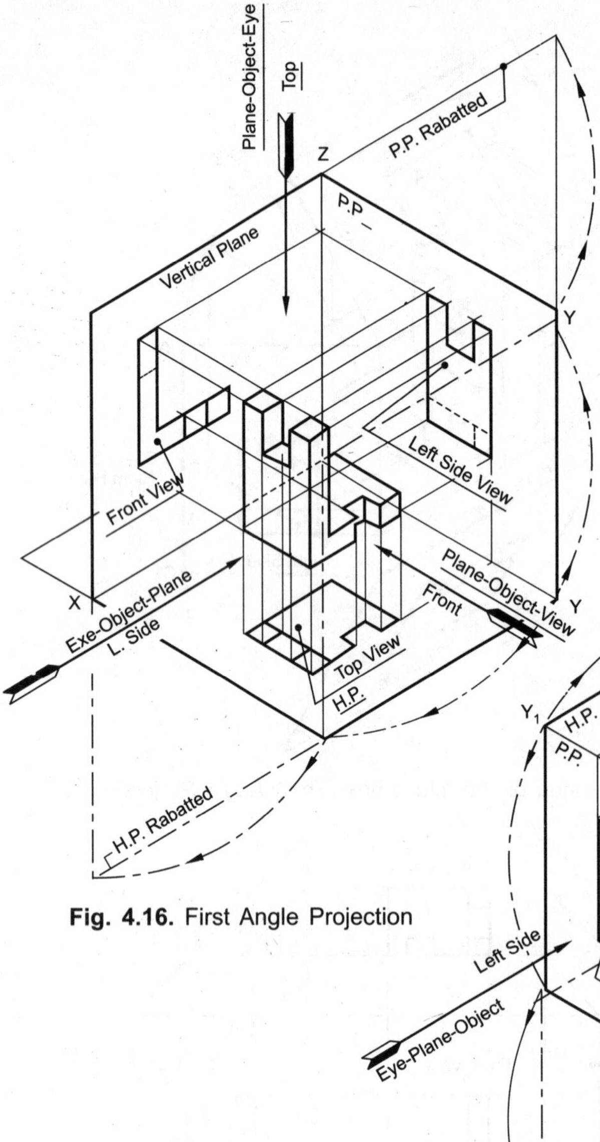

Fig. 4.16. First Angle Projection

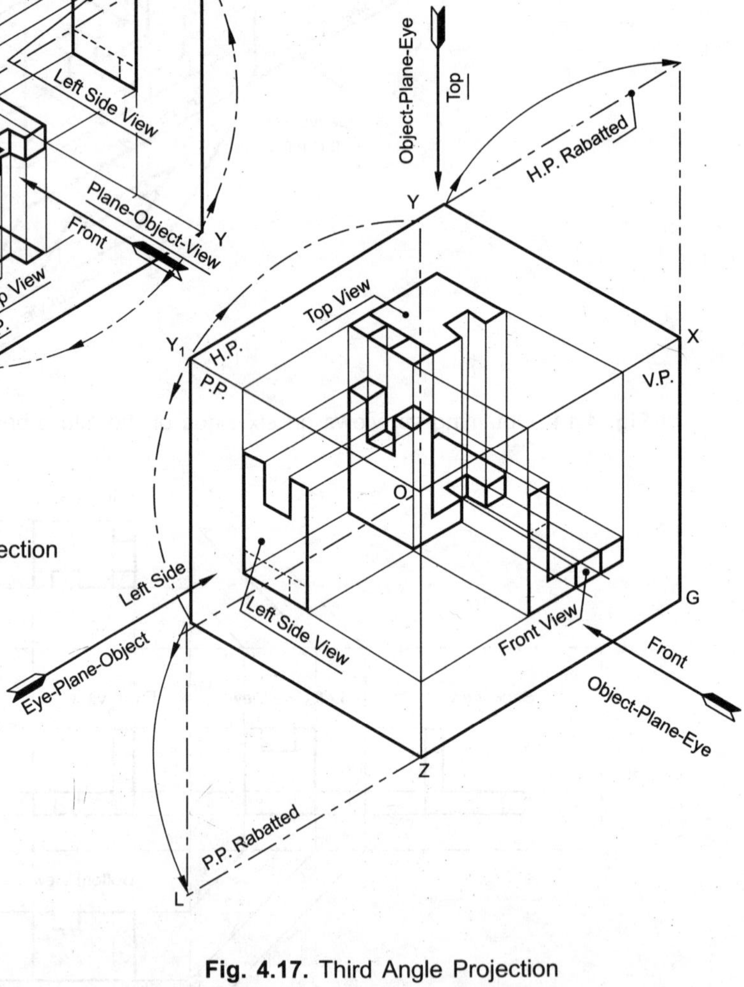

Fig. 4.17. Third Angle Projection

the observer, whereas in the first angle projection, the object lies between the projection and the observer.

For third angle projection when the planes are revolved in to position, to lie them on the plane of paper, as shown by arrows in Fig. 4.13. (*d*), the front view lies below the reference line XY, the top view lies above the reference line and above the front view, and the left side view in line with and on the left side of the front view.

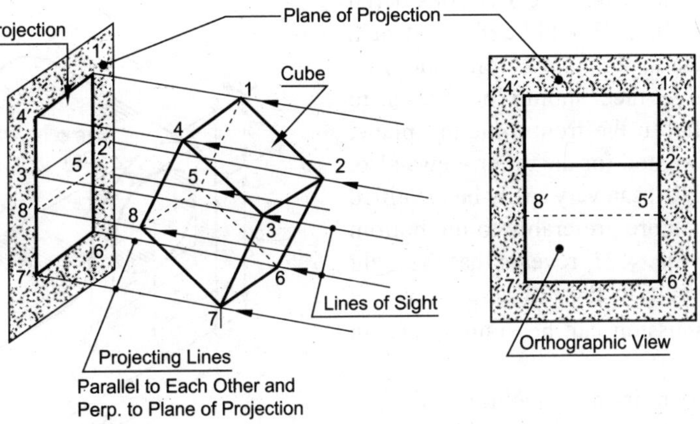

Fig. 4.18. Orientation of the object (Third Angle)

The projections thus obtained are called third angle projections. The views in third angle projection look more natural as the front view is on the front, the top view is on the top, and left side view is on the left side.

For technical drawing when the system of third angle projection is used:

1. Projection planes are set up perpendicular to one another, in a glass-box arrangement (or VIIth octant), and the object is assumed to be lying in this transparent box. The necessary views, to completely describe the shape of the object, are obtained as shown in *Fig. 4.13. (a)*.
2. For laying the views on a drawing paper, the planes are revolved away from the object, about their edges that are common with the VP (*i.e.,* about their lines of intersection with the VP).
3. The points on the object, which are nearer to the observer, while looking for drawing the front view, will be projected nearer to the front view in side and top views.
4. Actually there are two reference lines; first, the intersection of the HP and the VP, called XY line, second, the line of intersection of the VP and ground plane (AHP), called ground line GL.

4.14. ORIENTATION OF THE OBJECT

Though the theory of orthographic projections, developed in this chapter, is entirely independent of the shape of the object and its placement relative to the picture plane, yet, for natural appearance of the object drawing, the placement of the object plays an important role.

In the production shops or factories and in nature, we are accustomed to see objects in certain definite positions. When we see them placed upside down or in some other unusual placements, they look quite unfamiliar. What is true of the objects themselves, in this respect, is equally true of drawings of these objects. Hence it can be stated as a definite rule, that *objects should be placed in their normal, natural or functional positions* or as close to these positions as possible.

The most simple position of object is when its principal faces are parallel to the planes of projection. In this position the views show true sizes of faces and true lengths of lines. In any other position of the object, as shown in *Fig. 4.19* the view does not show in the true sizes of the surfaces and lines and is confusing to read.

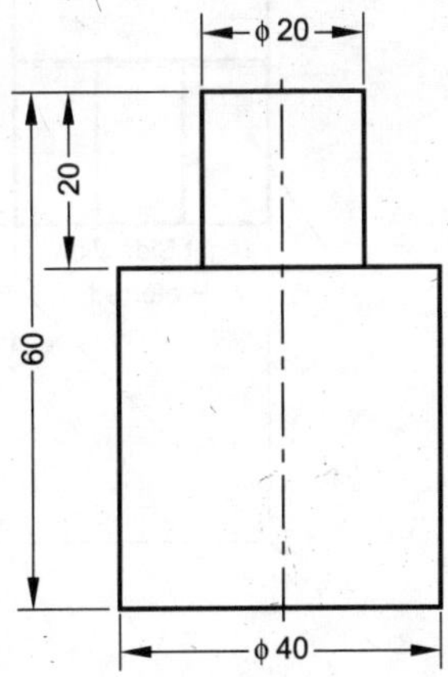

Fig. 4.19. One view drawing

As the front view of an object is considered the major view, the object should be oriented such that most of its details are on its front side, *i.e.*, the most descriptive face should be nearer to observer and parallel to the frontal picture plane. The same rule holds good for the other views also. Referring to *Fig. 4.20*, it can very easily be identified that the top view is more preferable to the bottom view. The study of *Fig. 4.21*, reveals that the right side view is preferable to the left side view.

The above discussion can be summarized in to the following rules:

1. Place the object in its most natural position.
2. Place the object with its faces parallel and near to the planes of projection.
3. Place the object with its most intricate and descriptive face on the front side.
4. Place the object (or select the views) so as to obtain the smallest number of invisible lines.

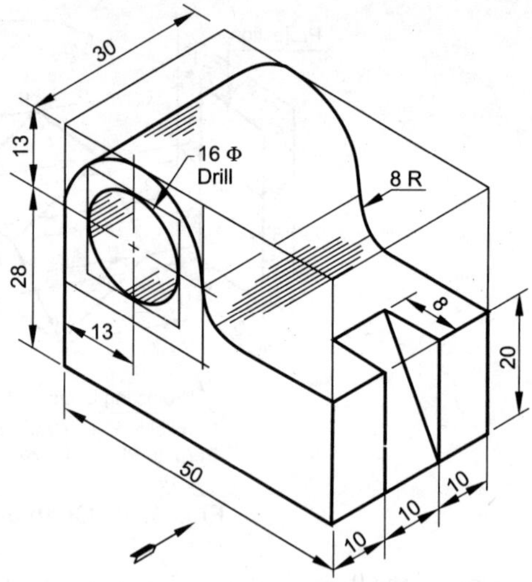

Fig. 4.20. Two view drawing

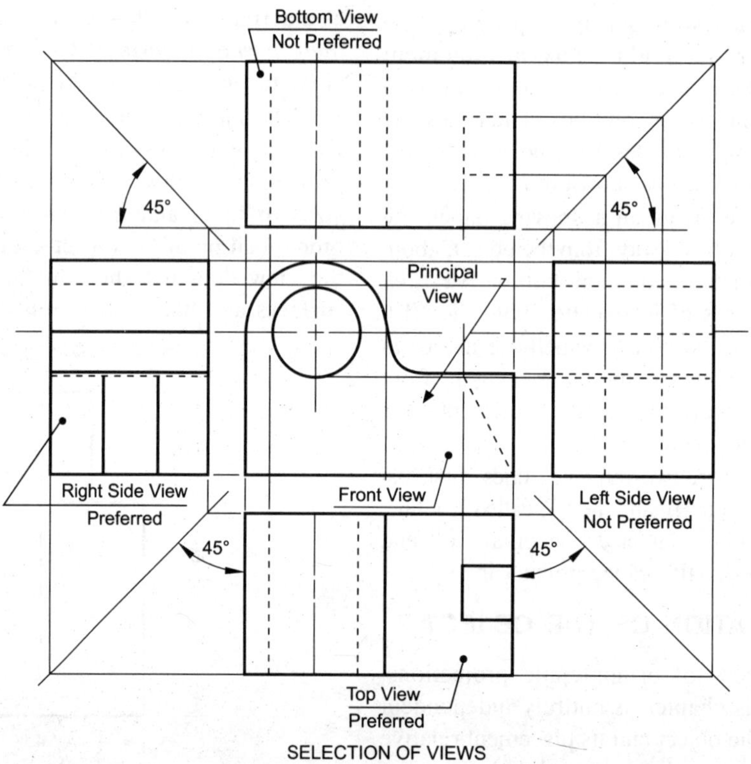

Fig. 4.21. First Angle Projection

4.15. SELECTION OF THE VIEWS

In describing the shape of an object, in the best and most economical way, it is very important to select the proper combination of views. Superfluous views and wrong combinations of views tend to confuse the reader and spoil the appearance of the drawing. If we draw all the six views shown in *Figs. 4.10* and *4.15*, we find that there is a great deal of repetition. But a multiview drawing, for use in production, or building should contain only those views, which are actually needed to describe the object completely.

Some simple objects, *e.g.,* a thin gasket or a regular cylinder, can be completely described by a single orthographic view, supplemented by a note. Such a drawing is called *'One View Drawing'*. *Fig. 4.19* shows 'One View Drawing' of a cylindrical part. Note that the notation symbol 'ϕ' is used to indicate the diameter measurement and should precede the dimension numerals.

Most objects, which are though simple and symmetrical about their axes, can not be described by a single view. A second view, which may be the top or the end view is then needed to describe the object adequately. Such drawings are called *'Two View Drawings'*. In these drawings the choice of views is determined by the descriptive value of the view. The chosen views shall have a minimum of hidden lines, because in most cases hidden features in one view will appear visible in another view, and the latter is to be chosen by the drafter, depending upon his/her personal experience, *Figs. 4.20, 21, 22,* and *24*.

When the two side views are equally descriptive, the convention is to select the right side view in third angle projection and left side view in first angle projection.

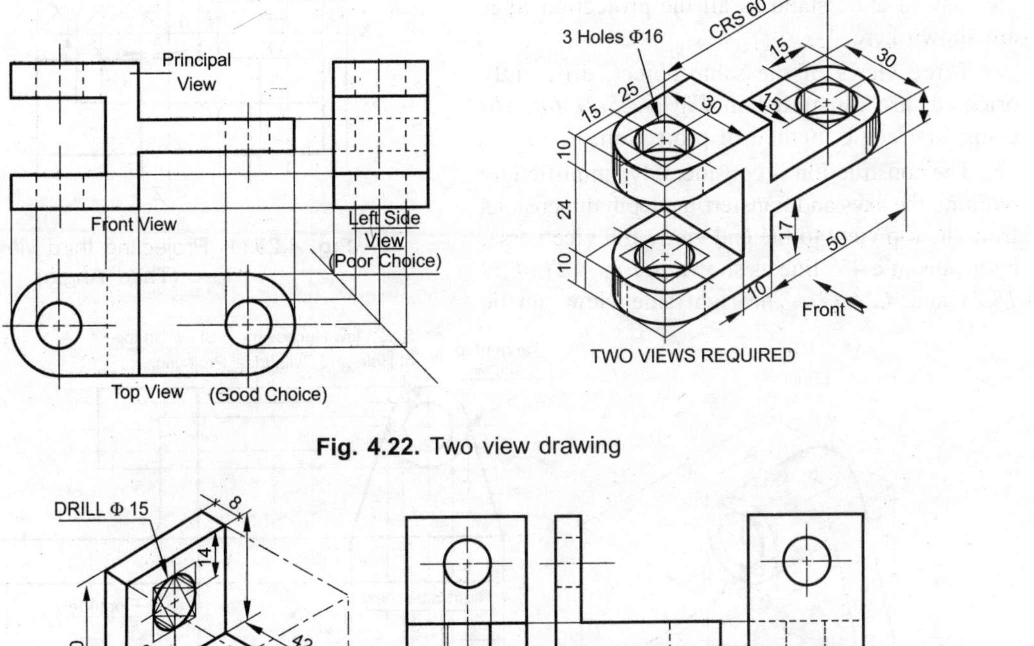

Fig. 4.22. Two view drawing

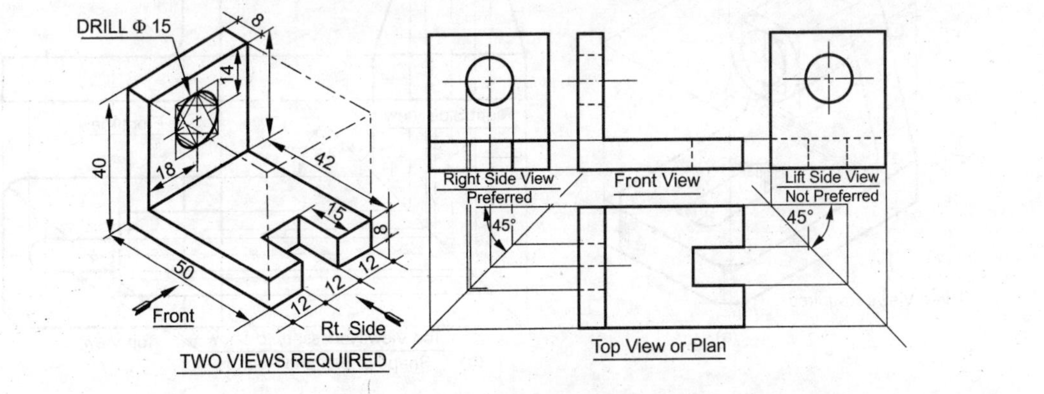

Fig. 4.23. Choice of views

As the object is usually placed with a flat face on the ground plane (or HP) and irregular contours on the top side, top view is preferred to the bottom view, *Fig. 4.20, 4.21*. When the top and side views are equally descriptive, in a two view drawing, the space available governs the selection.

When a third view is required to describe the shape of an object, the drawing is called *'Three View Drawing', Fig. 4.23.*

4.16. LAYING OUT A THREE VIEW DRAWING

A typical example of object, which require three views is shown in *Fig. 4.24 (1) (a)*. When three principal planes of the glass-box have been unfolded, the views appear as shown in *Fig. 4.24 (1)(b)* in third angle projection. The methods of projecting, from one view to the other as illustrated here, also in *Figs. 4.5 (a), (b)*, and *Fig. 4.25(1) (a)* and *(b)*, are easy to understand as all the projection lines are shown here.

Three views of the same object, differently oriented, are illustrated in *Fig. 4.25(2) (a), (b)* using first angle method of projection.

The construction is considerably simplified by omitting the axes and transferring depth dimensions from the top view to the end view, and vice versa, by means of a 45° line as shown in *Fig. 4.21, 4.22 18.23* and *4.24 (2)*. The right side view can be moved to left or right, or the top view can be moved upward or downward by shifting the 45° line accordingly. In practice, the depth dimension from the top view to the end view, or vice versa, are transferred by using the dividers as this method is faster and more accurate than the 45° line method.

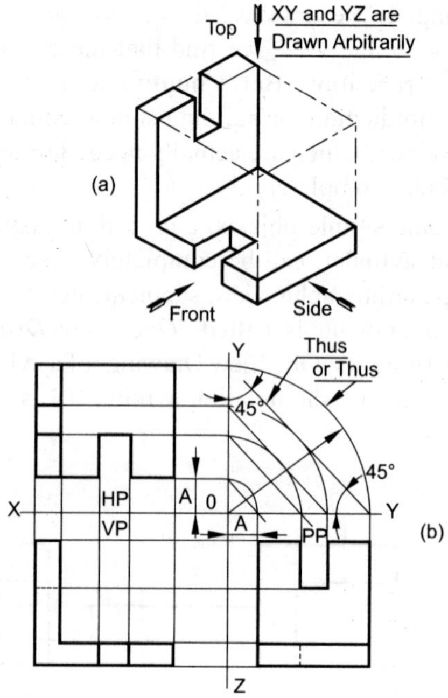

Fig. 4.25.(1) Projecting third view (Third Angle)

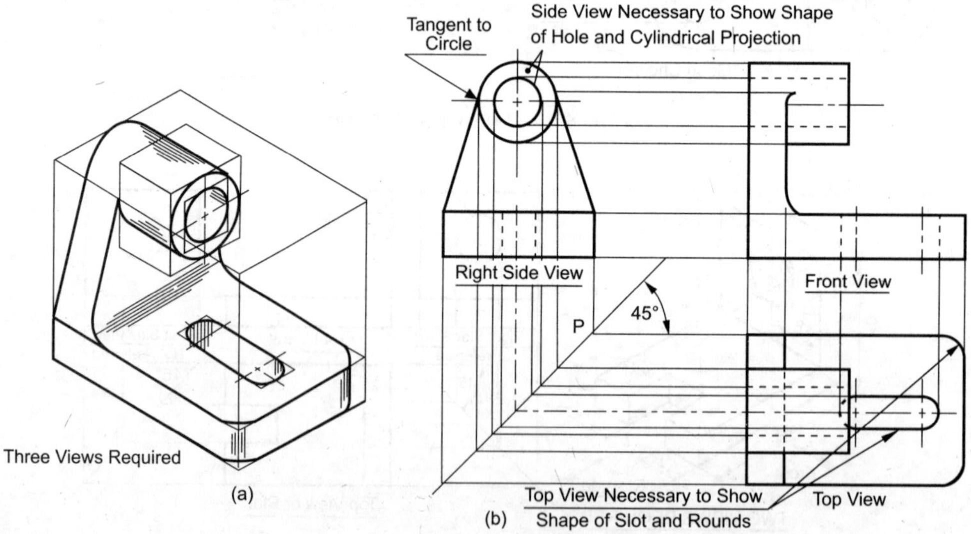

Fig. 4.24. Three view drawing and choice of view

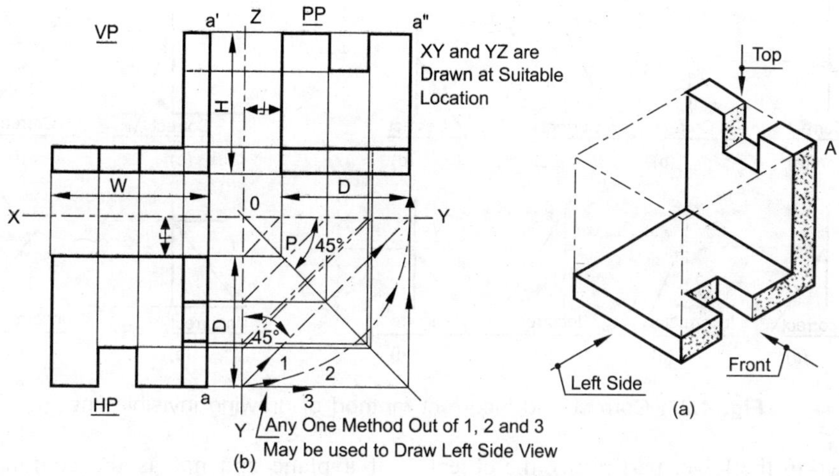

Fig. 4.25. (2) Projecting third views (First Angle)

For spacing a three view drawing, spaces A, *Fig. 4.26 (a), (b)* should be kept approximately equal and space B should be equal to or slightly greater than either of spaces A. Space C should be nearly equal and space D should be equal to or slightly greater than space C. Further adjustments, for giving dimensions and any notes on the particular drawing, be made accordingly.

4.17. INVISIBLE LINES AND ARCS

Those edges which are not visible are drawn in dotted lines. Although it is a general practice to omit those invisible lines, which tend to confuse the view, without serving any useful purpose, it is not desirable for a beginner in drafting to do so.

Figure 4.27 illustrates correct and incorrect practices in the drawing of invisible lines. The examples at (*a*), (*b*), and (*c*) show the method of drawing the dashes to secure sharp corners. Dashes should not cross an object line, as at (*d*), as this would make the invisible line appear as near the observer as the object line.

Two hidden dashes should not cross, as at (*d*). If a dash of hidden line is joined to the visible

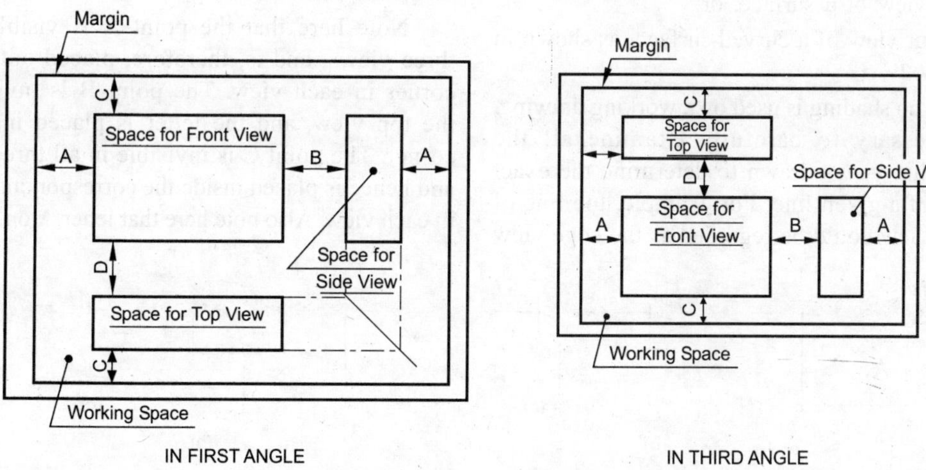

Fig. 4.26. Spacing a three view drawing

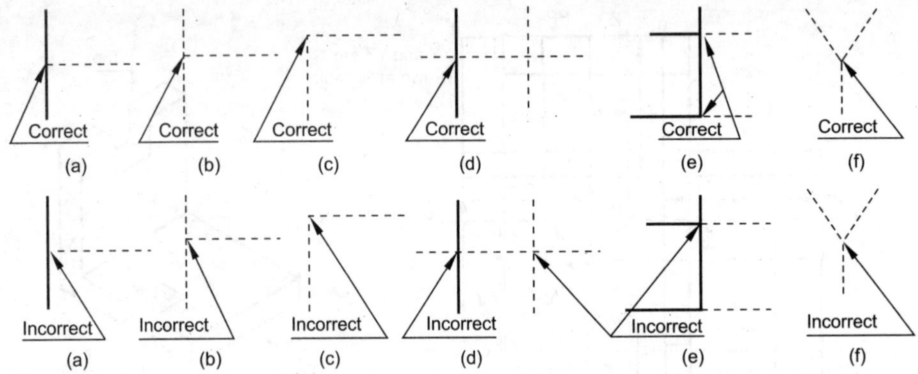

Fig. 4.27. Correct and incorrect method of drawing invisible lines

line, as shown in the lower part in (*e*), the object line would appear to continue; therefore, a gap should always be left in such cases.

Correct and incorrect practices in drawing invisible arcs are illustrated in *Fig. 4.28*. The drafter should choose views so that all features are represented by visible lines as far as this is possible. After this has been done, invisible lines should be used wherever necessary to make the drawing clear. Where they are not required, for clearness the invisible lines should be omitted.

4.18. MEANING OF LINES

An object line or a hidden line on a drawing has three possible meanings;

1. intersection of two surfaces,
2. edge view of a surface, or
3. contour view of a curved surface, as shown in *Fig. 4.29*.

Since no shading is used on a working drawing, it is necessary to carefully examine all the orthographic views drawn to determine the exact meaning of a given line. For example, top line of the front view could be regarded as the edge view of a plane and not as the contour of a curved surface if the side view is not examined. Similarly, the inside vertical line in the front view could be regarded as the edge view of a plane if we look at only the front and end views. The examination of the top view reveals that the line stands for the intersection of two surfaces.

4.19. POINTS IN ORTHOGRAPHIC VIEWS

Three orthographic views of each of the three points A, B and C are shown in *Fig. 4.30*. Here, when a point is *visible* in a given view, the numeral or letter representing the point is placed outside its corresponding corner. If the point is *invisible* the numeral or letter representing the point is placed inside its corresponding corner.

Note here that the point A is visible in all three views, and is, therefore, placed outside the corner in each view. The point B is invisible in the top view, and the letter is placed inside the corner. The point C is invisible in all three views, and hence is placed inside the corresponding corner in each view. Also note here that letter A on pictorial

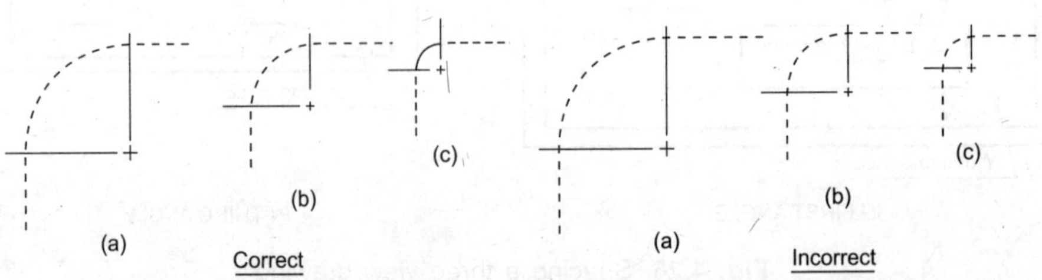

Fig. 4.28. Correct and incorrect methods of drawing invisible Arcs

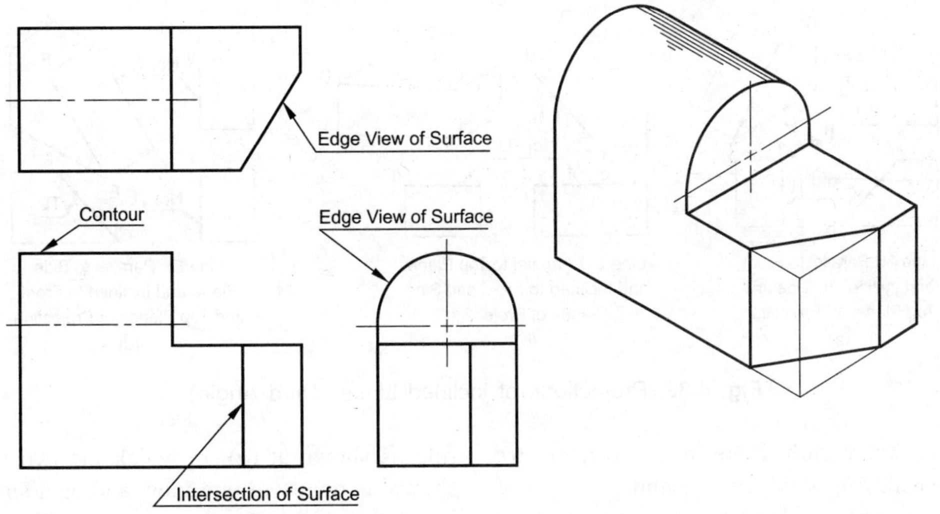

Fig. 4.29. Meaning of lines (Third Angle)

view is represented by **a′** in front view, **a** in top view and **a″** in side view.

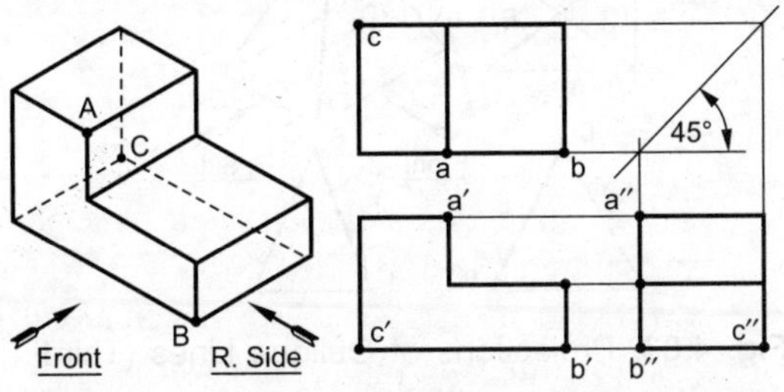

Fig. 4.30. Projections of points (Third Angle)

This numbering system, in which points are identified by the same numbers in several views, is useful in projecting known positions in two views to its unknown position in a third view, *Fig. 4.28.*

4.20. NORMAL LINES IN ORTHOGRAPHIC VIEWS

When a line is perpendicular to a principal plane of projection, it will be parallel to the adjacent planes of projection, and is called a *normal line.* Its projection on the plane to which it is perpendicular is a point. Its projection on a plane to which it is parallel, is a line projecting its true length, *Fig. 4.31 (a), (b)* and *(c).*

4.21. INCLINED LINES IN ORTHOGRAPHIC VIEWS

When a line is parallel to a principal plane of projection and inclined to adjacent planes it is an *inclined line*, and its projection on the plane to which it is parallel is true length. Its projections

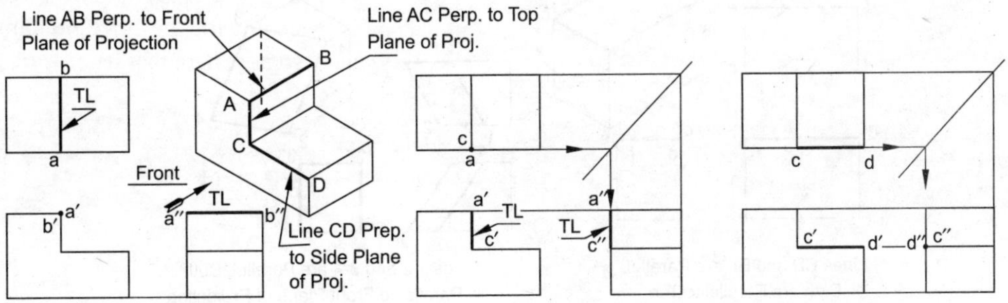

(a) (b) (c)

Fig. 4.31. Projections of Normal Lines (Third Angle)

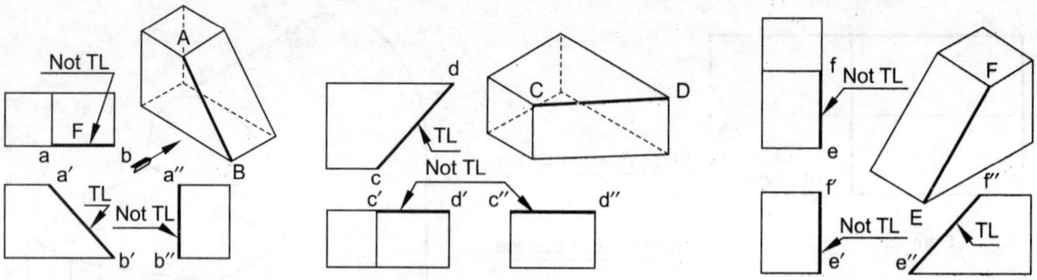

Line AB Parallel to Front and Inclined to Side and Top Planes of Projection
(a)

Line CD Parallel to Top Plane and Inclined to Front and Side Planes of Projection
(b)

Line EF Parallel to Side Plane and Inclined to Front and Top Planes of Projection
(c)

Fig. 4.32. Projections of inclined Lines (Third Angle)

on planes to which it is not parallel are foreshortened, *Fig. 4.32 (a), (b)* and *(c)*.

An inclined line will always be foreshortened in two of the three regular views and will be either vertical or horizontal in those views. In the third view it will be in an inclined position and will be shown as true length.

4.22. OBLIQUE LINES IN ORTHOGRAPHIC VIEWS

When a line is oblique to all the three planes of projection, it is an *oblique line*, and its projections in all views will be in inclined positions and also be foreshortened, *Fig. 4.33*.

4.23. PARALLEL LINES IN ORTHOGRAPHIC VIEWS

If two lines are parallel in space, their projections in any view are parallel, *Fig. 4.34*. A special case, which should not be regarded an exception to this rule, is shown at *(a)*, in which the two lines are shown as points in one view and as a single line in another view. Note that even in the pictorial drawings, the lines are shown parallel.

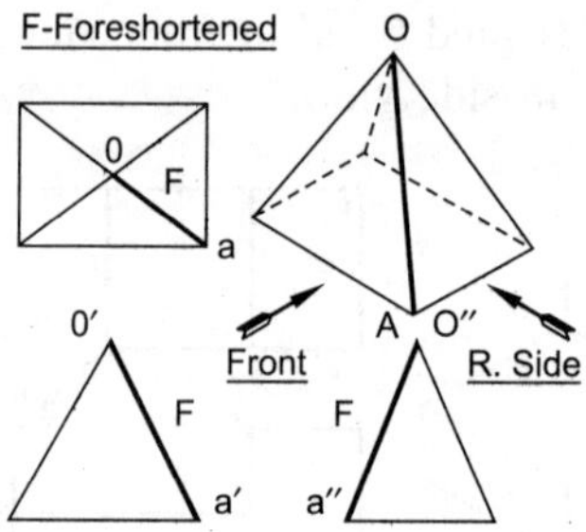

Fig. 4.33. Projections of Oblique Lines (Third Angle)

4.24. ANGLE BETWEEN LINES IN ORTHOGRAPHIC VIEWS

If an angle is in a normal plane, *i.e.,* parallel to a plane of projection, the angle will be shown true

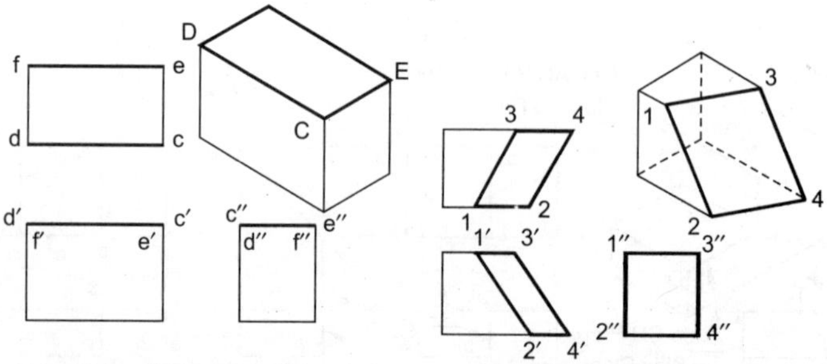

Lines CD and EF are Parallel, Both are Parallel to Top Plane of Projection

Lines 1-2 and 3-4 are Parallel, Both are Parallel to Front Plane of Projection

Fig. 4.34. Projections of Parallel Lines (Third Angle)

size on that plane of projection, while if the angle is not on a plane parallel to the plane of projection it will not project true size on that plane, *Fig. 4.35 (b).*

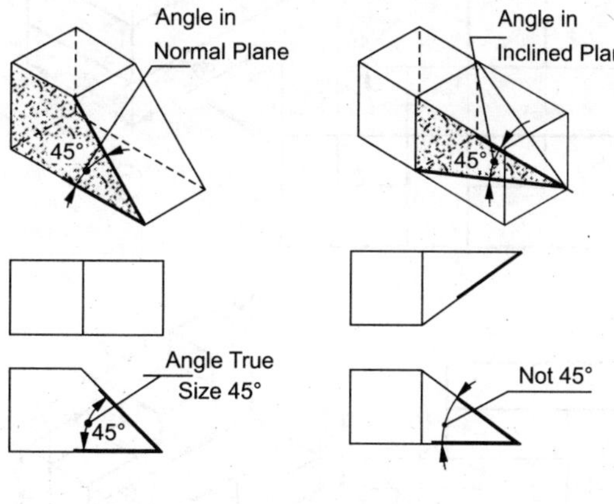

Fig. 4.35. Angles (Third Angle)

4.25. NORMAL SURFACES IN ORTHOGRAPHIC VIEWS

When a plane surface is parallel to a principal plane of projection, it is shown in its true size and shape on that plane, and as a line on each adjacent principal plane, *Fig. 4.36 (a), (b), (c).* Such a surface is called a *normal surface.* Study additional problem on simple objects having normal surfaces, shown in *Fig. 4.37.*

4.26. INCLINED SURFACES IN ORTHOGRAPHIC VIEWS

When a plane is perpendicular to a principal plane of projection and inclined to adjacent principal planes, it is an *inclined surface, Fig. 4.38 (a), (b), (c)* and is shown as a line on the plane of projection to which it is perpendicular. It will appear foreshortened on the planes to which it is inclined, the degree of foreshortening being proportional to the angle of inclination. For example, in *Fig. 4.38 (a)* the plane A is foreshortened more in the side view than in the top view as it makes a greater angle with the side plane of projection. An inclined plane will not appear true shape and size in any regular view because it is not parallel to any plane of projection. To show their true shape, a special auxiliary view is required, refer chapter 10 on *Auxiliary Views*, for detailed information. Additional problems on simple objects having inclined surfaces are shown in *Fig. 4.39*, study these problems carefully to get more insight on the treatment of inclined planes.

4.27. OBLIQUE SURFACES IN ORTHOGRAPHIC VIEWS

When a plane is oblique to all principal planes of projection it is an *oblique plane (or surface)* and its true size and shape will not be shown in any regular view, *Fig. 4.40.* An oblique surface is easily recognized by the fact that it does not appear as

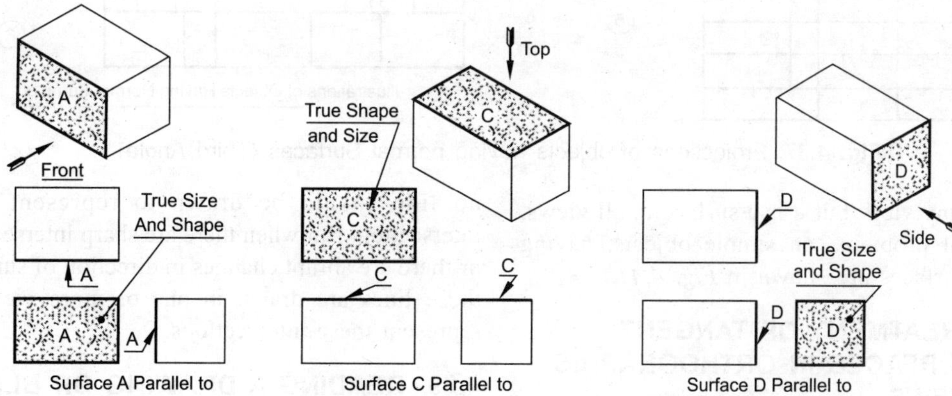

Surface A Parallel to Front Plane of Projection
(a)

Surface C Parallel to Top Plane of Projection
(b)

Surface D Parallel to Side Plane of Projeciton
(c)

Fig. 4.36. Projections of Normal Surfaces (Third Angle)

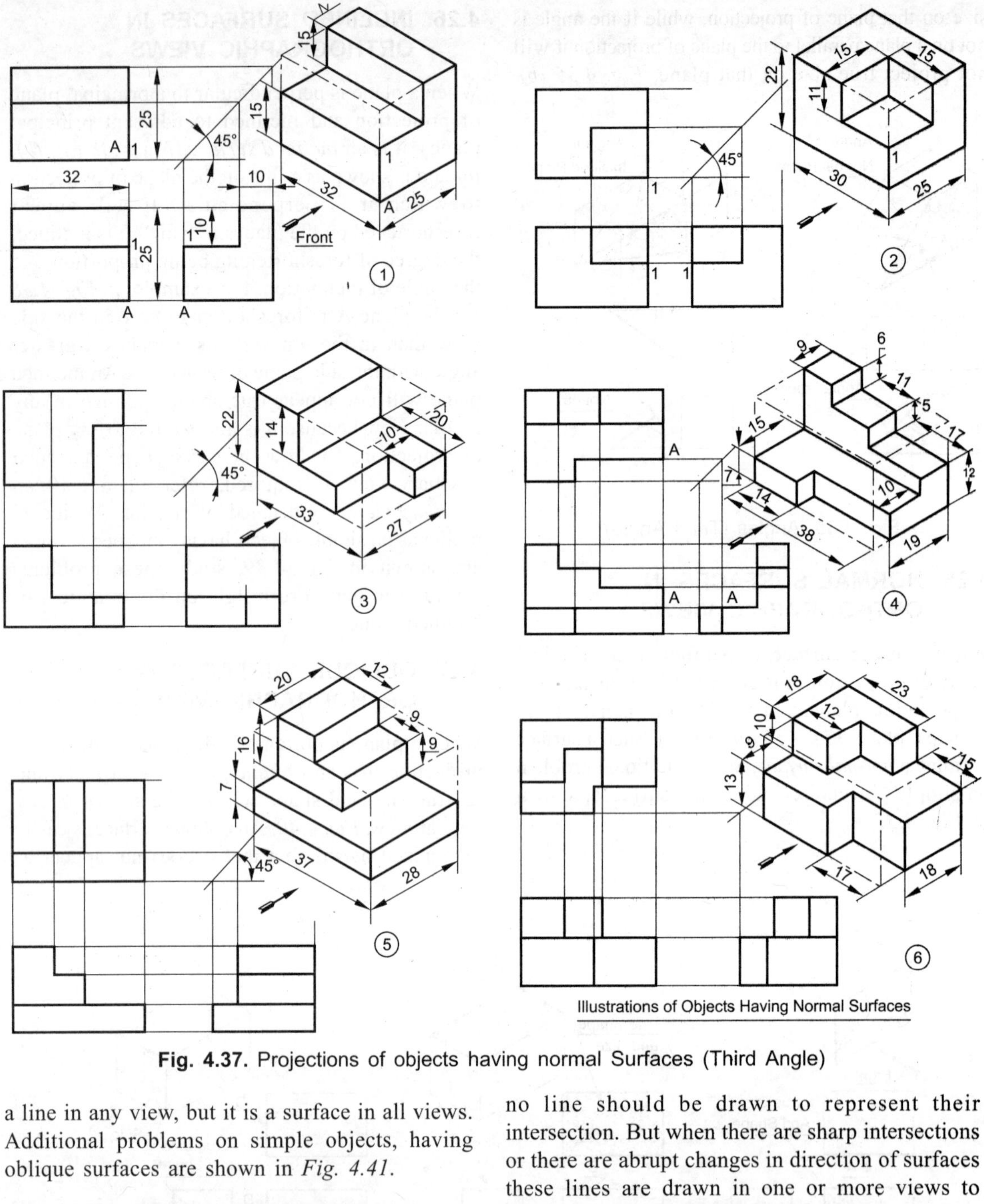

Illustrations of Objects Having Normal Surfaces

Fig. 4.37. Projections of objects having normal Surfaces (Third Angle)

a line in any view, but it is a surface in all views. Additional problems on simple objects, having oblique surfaces are shown in *Fig. 4.41*.

4.28. TREATMENT OF TANGENT SURFACES IN ORTHOGRAPHIC VIEWS

As illustrated on the pictorial drawing in *Fig. 4.42*, when a curved surface is tangent to a plane surface no line should be drawn to represent their intersection. But when there are sharp intersections or there are abrupt changes in direction of surfaces these lines are drawn in one or more views to represent these intersections.

4.29. READING A DRAWING OR BLUE-PRINT READING

Engineering drawing is the medium of expression used by engineers, designers, drafters and

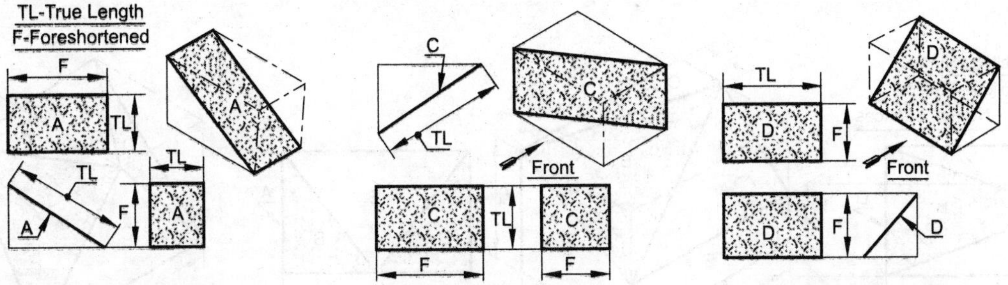

(a) Surface A Prep. to Front Plane and Inclined to Top and Side Planes of Projection

(b) Surface C Prep. to Top Plane and Inclined to Front and Side Planes of Projection

(c) Surface D Prep. to side Plane and Inclined to Front and Top Planes of Projection

Fig. 4.38. Projections of inclined Surfaces (Third Angle)

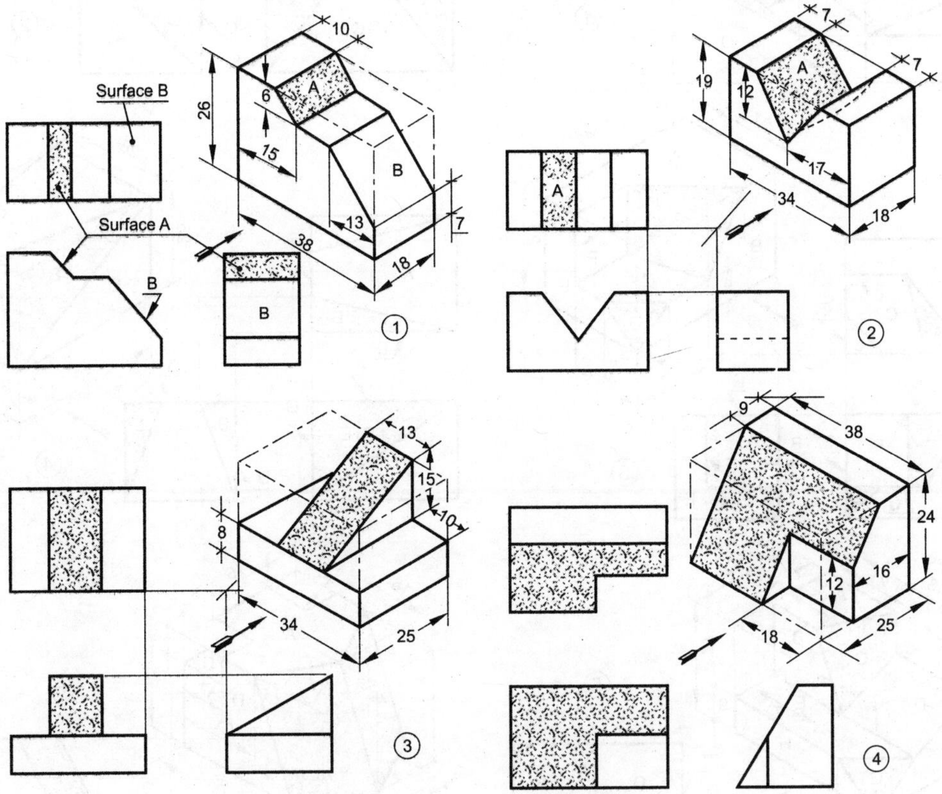

Illustrations of Simple Objects having Inclined Surfaces

Fig. 4.39. (Third Angle)

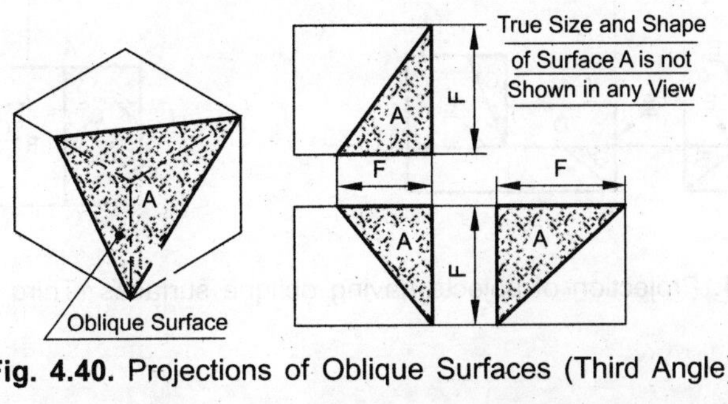

Fig. 4.40. Projections of Oblique Surfaces (Third Angle)

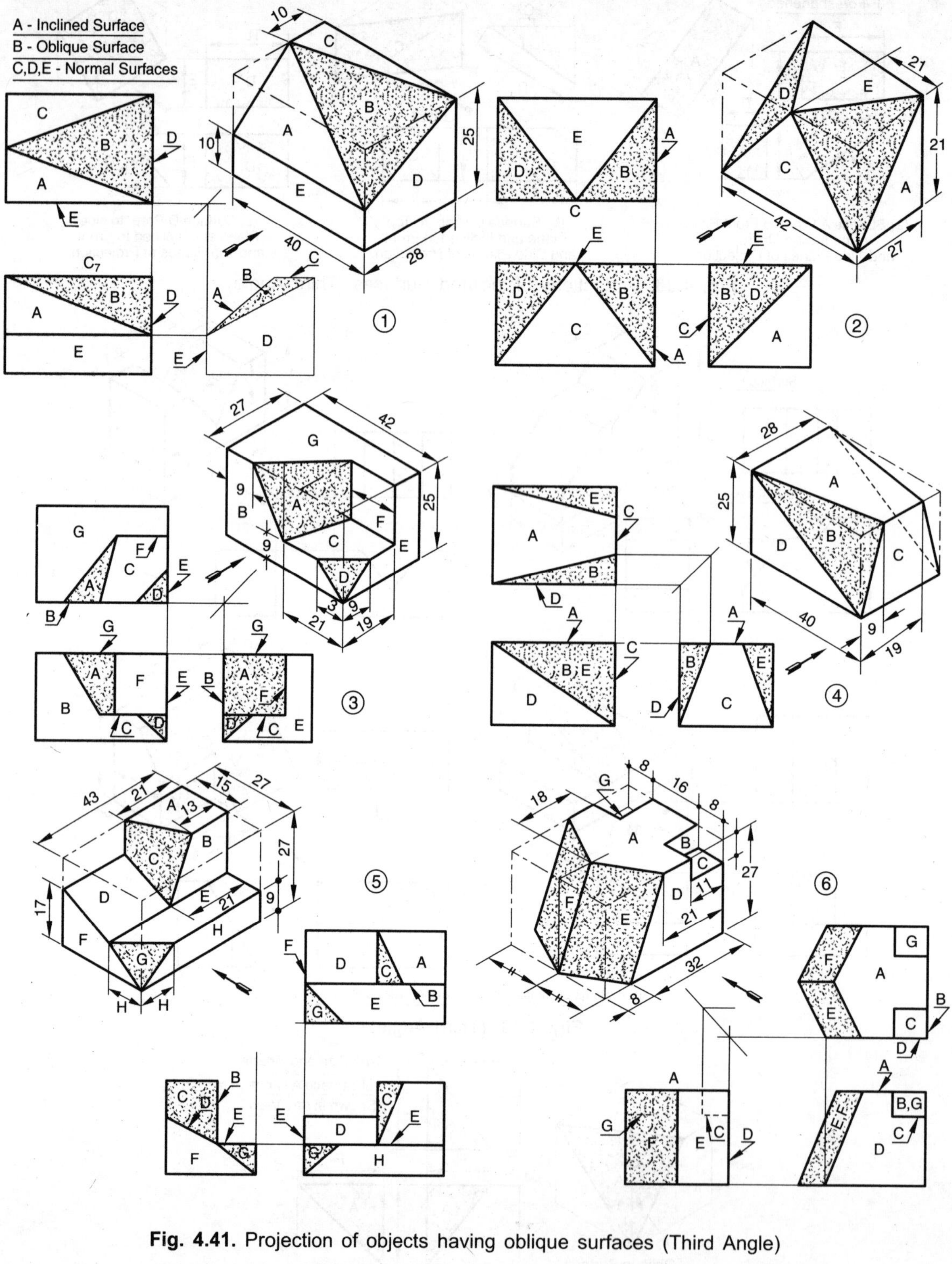

Fig. 4.41. Projection of objects having oblique surfaces (Third Angle)

technicians and each one of them would be professionally illiterate if he/she can not readily understand the graphic language used. The mental process of reading a drawing is the reverse of that used for making a drawing. The drafter has a knowledge of the shape and then makes a drawing describing that shape. The reader of a drawing is unfamiliar with the object represented, and must learn to interpret the meaning of every line in order to understand it. All prints made from the original drawing or tracing, in general are called *blue-prints or prints*. Reading a print involves the ability to visualize a three dimensional object from its related two dimensional drawing.

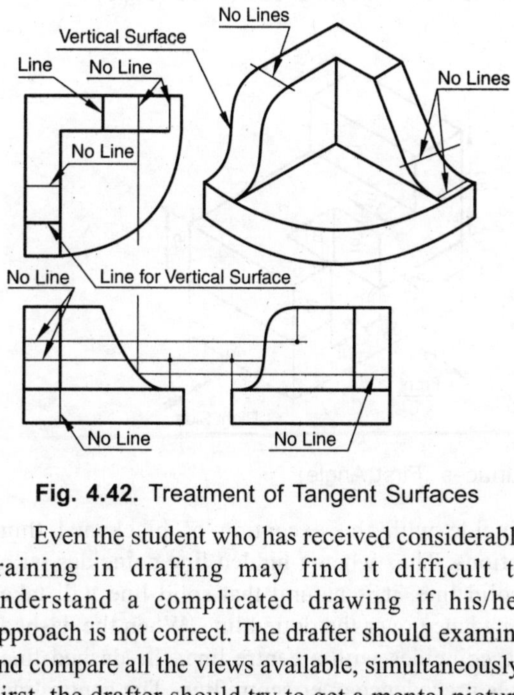

Fig. 4.42. Treatment of Tangent Surfaces

Even the student who has received considerable training in drafting may find it difficult to understand a complicated drawing if his/her approach is not correct. The drafter should examine and compare all the views available, simultaneously. First, the drafter should try to get a mental picture of the entire object without being involved in smaller details, then he/she should examine the details one by one, comparing them in different views, and finally the meaning of every line on the drawing should be determined.

To develop constructive imagination, and achieve a sound basis for reading a drawing, the principles of orthographic projection can be applied to the following types of exercises:

1. *Identification of surfaces and lines.*
2. *Missing line problems.*
3. *Possible view problems.*
4. *Missing view or third view problems.*

Only the first type of these exercises has been included in this chapter. The remaining types, which require an isometric view to be sketched to insure a correct solution to these types of problems, are not covered here.

4.30. IDENTIFICATION OF SURFACES

Having studied the fundamental principles of orthographic projections, given the pictorial view of an object, the next step towards reading a drawing and developing an orderly imagination is the identification of their corresponding orthographic projections.

Figure 4.43 shows a pictorial drawing and a three view orthographic drawing of a block. The adjacent table provides an orderly analysis of reading the three views of the block. Such a table, prepared by a student, facilitates study of a particular drawing. All lines and surfaces are numbered, at random, on the three orthographic views so that the student may take each surface designated on the pictorial drawing and identify it by a different number on each of the three views. For example, surface A on the pictorial drawing projects, as a surface in the top view and is identified by the number 3. The same surface appears as surface 7 in the front view, and a line 26 in the side view.

In *Fig. 4.44.* the surfaces of a block are numbered, in the pictorial and orthographic views, by the same latters. This will serve as a good exercise for students to identify surfaces in the orthographic views.

In *Fig. 4.45*, three orthographic views, each of 13 different blocks, shown as isometric views, are given. Study the orthographic views of each object carefully and identify, and number their corresponding pictorial views.

In *Fig. 4.46 (a)* and *(b)* 12 different blocks, with their visible surfaces numbered are given along with their orthographic views. Identify the surfaces of all these blocks in their respective orthographic views.

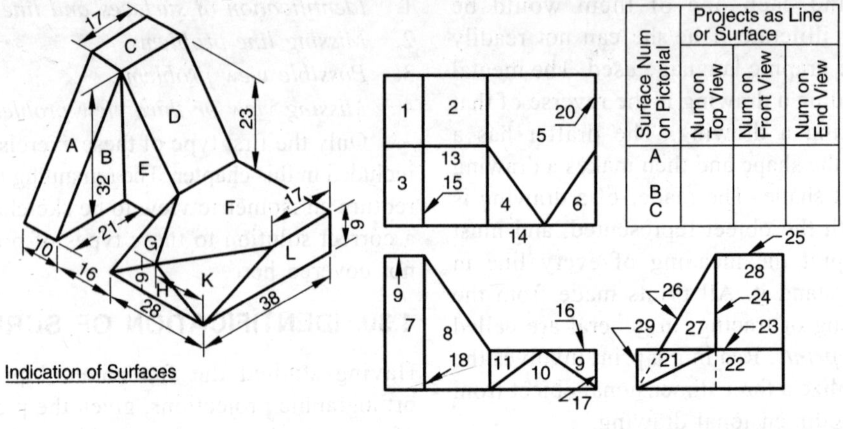

Surface Num on Pictorial	Num on Top View	Num on Front View	Num on End View
A			
B			
C			

(Columns "Num on Top View", "Num on Front View", "Num on End View" grouped under: Projects as Line or Surface)

Fig. 4.43. Identification of surfaces and interpretation of adjacent areas (Third Angle)

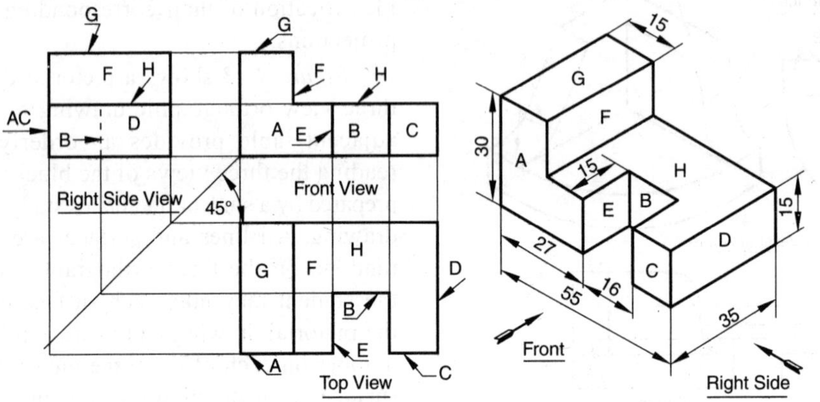

Fig. 4.44. Identification of Surfaces (First Angle)

4.31. ORDER OF PRIORITY OF COINCIDING LINES

When two or more lines of different types coincide, the following order of priority should be observed, *Fig. 4.47*:

1. Visible outlines and edges (continuous thick lines, type A);
2. Hidden outlines and edges (dashed line, type E or F);
3. Cutting planes (chain thin lines, thick at ends and changes of cutting planes, type H);
4. Centre lines and lines of symmetry (chain thin line, type G);
5. Centroidal lines (continuous thin double-dashed line, type K);
6. Projection lines (continuos thin line, type B).

Adjacent outlines of assembled parts shall coincide with the exception of blackened thin sections. Thus when a hidden line coincides with a solid line, it is natural that solid line will take precedence over the dotted line. When the dashed line coincides with a centre line, the dashed line takes precedence over centre line. The centre line is then indicated by extending the centre line at both the ends, *Fig. 4.47*.

4.32. STAGES IN PENCILING

For a balanced drawing the proper spacing of the views on a drawing sheet is very necessary. If the space left around a view or in between the views is not uniform it does not give a good appearance.

For dividing the space, measure or calculate the overall width W, height H, and depth D of the object or the assembly, *Fig. 4.48 (1)*. After drawing the margin lines on the sheet measure the

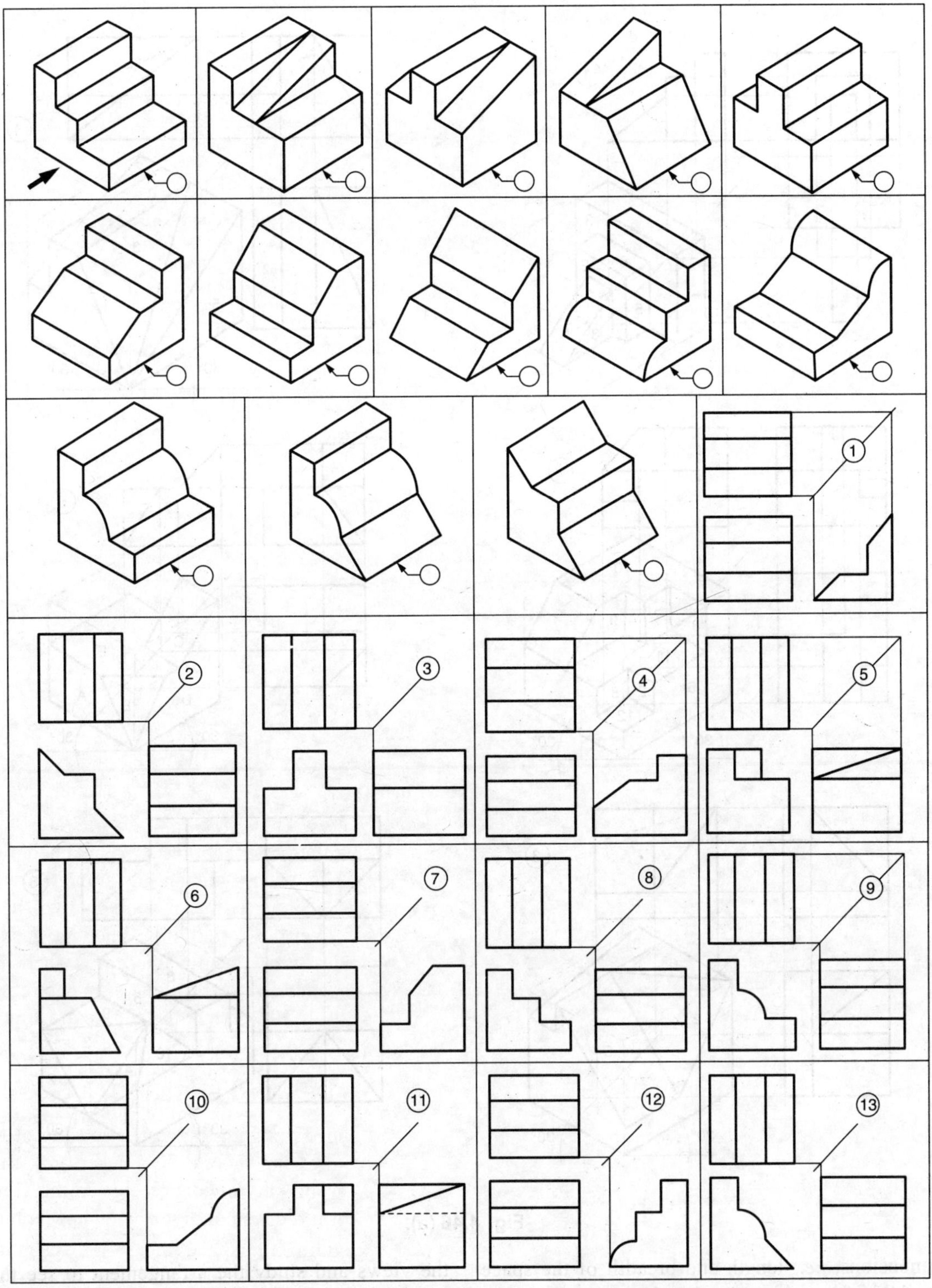

Fig. 4.45. Here on this page three orthographic views, each of 13 different objects, shown as isometric views, are given. Study the orthographic views of each object carefully and identify and number the corresponding pictorial view

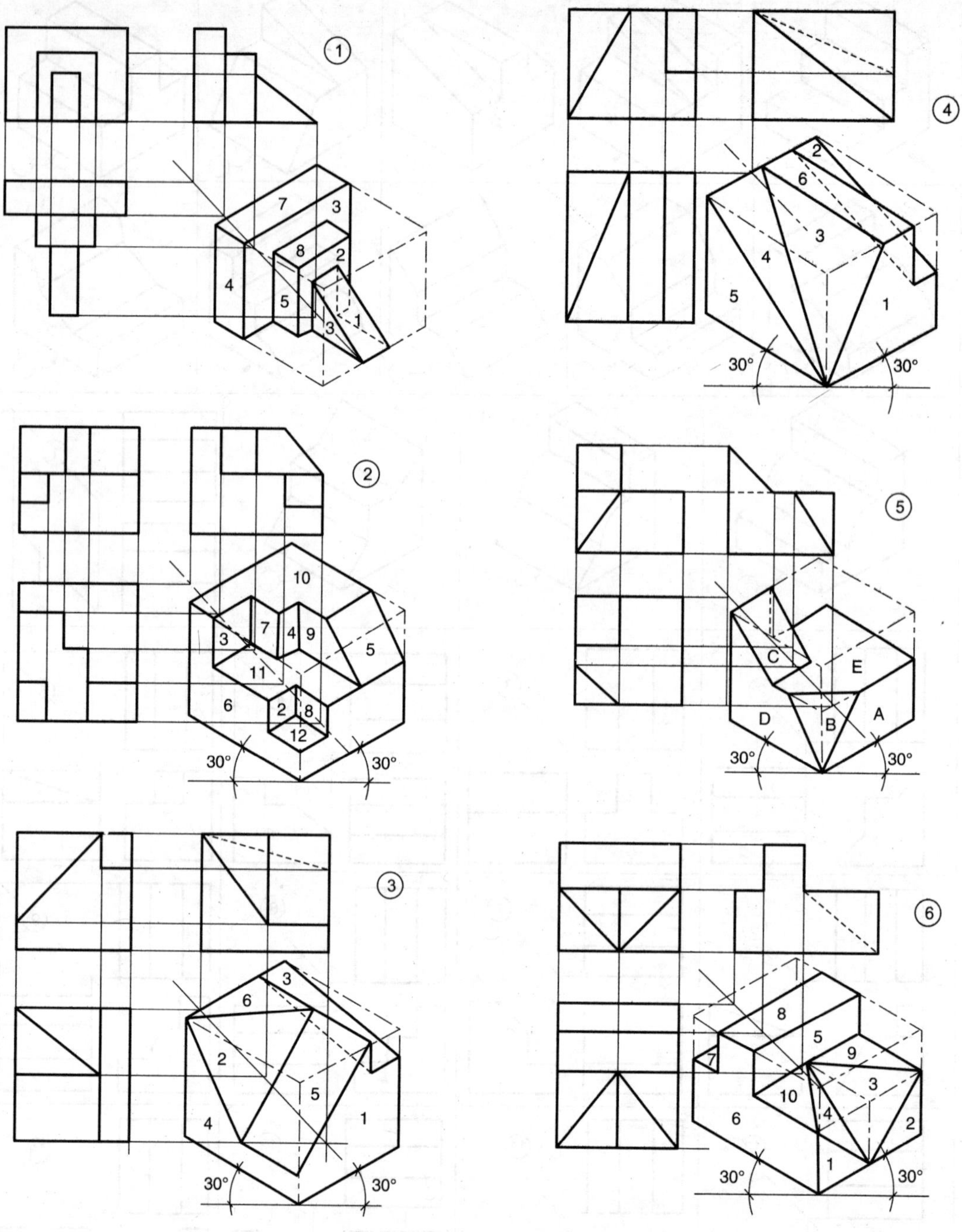

Fig. 4.46 (a).

dimensions, *i.e.*, length and breadth of the space available for making the drawing and decide the scale to be used. Block out lightly the spaces for the views and study the arrangement to see that the spaces between the views are uniform and appeal to the eye. Sufficient space should be left at the

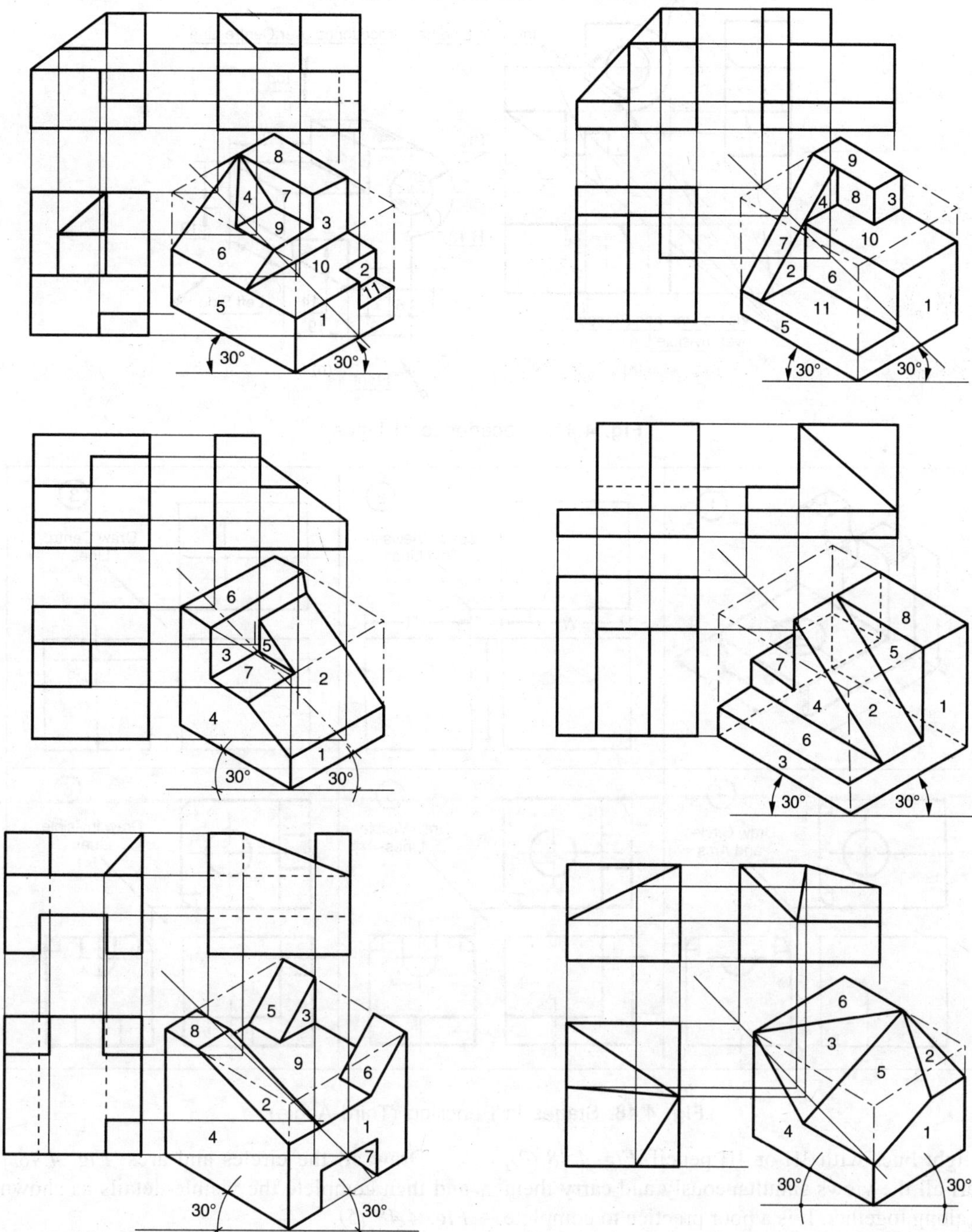

Fig. 4.46 (b).

bottom for giving the heading of the sheet and notes etc.

After selecting and determining the layout of the views, locate and draw the outlines of the views

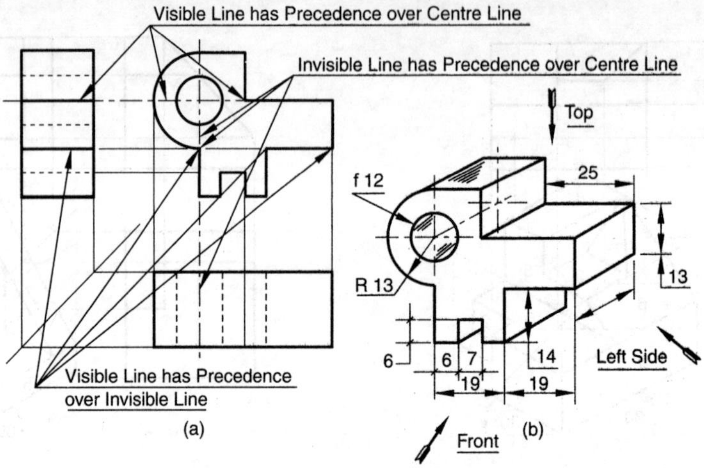

Fig. 4.47. Precedence of Lines

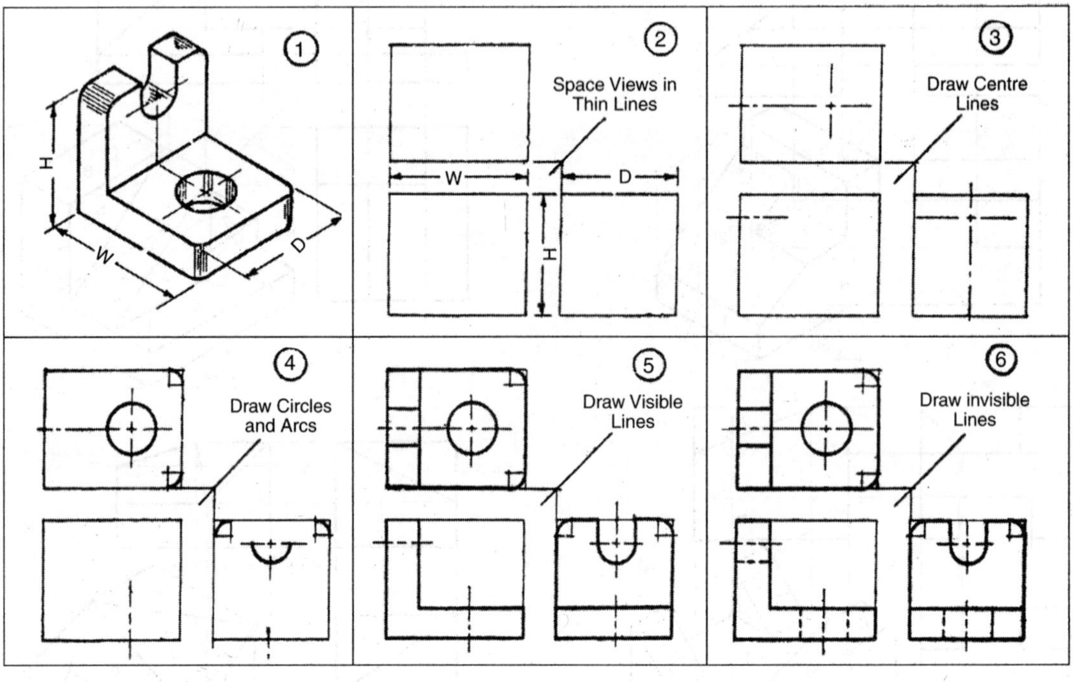

Fig. 4.48. Stages in Penciling (Third Angle)

in light lines with 3H or 4H pencil, *Fig. 4.48 (2)*. Start all the views simultaneously and carry them all along together. It is a poor practice to complete one view and then start with the other.

After the outlines, draw the centre lines for the various circular features, *Fig. 4.48 (3)*, to locate centres for various arcs. No circle or arc should be drawn before drawing the two centre lines, locating its centre.

Draw all the circles and arcs, *Fig. 4.48 (4)*, and then complete the visible details as shown in *Fig. 4.48 (5)*.

Finally add the invisible lines and finish the drawing by darkening the visible lines, *Fig. 4.48 (6)*. Then add the dimensions and notes and a suitable title-block to the drawing.

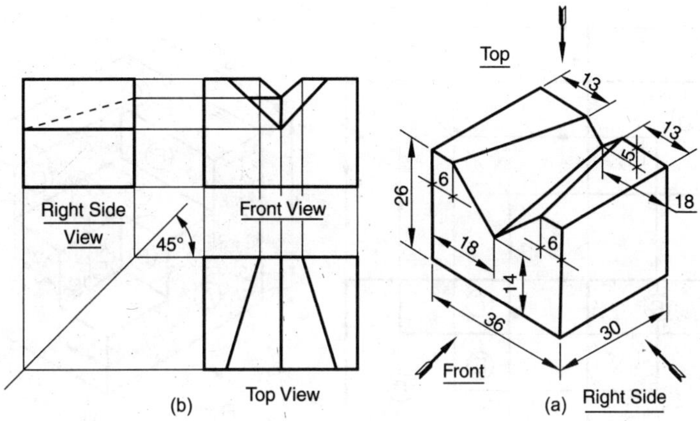

Fig. 4.46 (c). (First Angle)

EXERCISES

1. Looking in the direction of the arrow, draw the front view, right side view and top view of blocks given in *Fig. 4.20 (a), Fig. 4.21 (a), Fig. 4.22 (a)* and *Fig. 4.46 (c)*.

   Use full size scale and dimension the views completely.

- **SOLUTION:** Complete solutions of all these cases are given in the (*b*) parts of their respective figures. Study and redraw these solutions.

2. Looking in the direction as given, draw the front view, right side view, and top view of the block shown in *Fig. 4.49 (a)*. Use full size scale and dimension the views.

- **SOLUTION:** For solution refer to *Fig. 4.49 (b)*. All projection lines are shown to illustrate the method of projection.

3. Looking in the direction of the arrow, draw the front view, right side view, and top view of the *block* shown in *Fig. 4.50 (a)*. Identify the surfaces, marked by letters, by placing the corresponding letters in the orthographic views.

- **SOLUTION:** Refer *Fig. 4.50. (b)* for complete solution. Projection lines are retained to illustrate the method of projection.

4. Draw the following views of the block shown in *Fig. 4.51 (a)*:

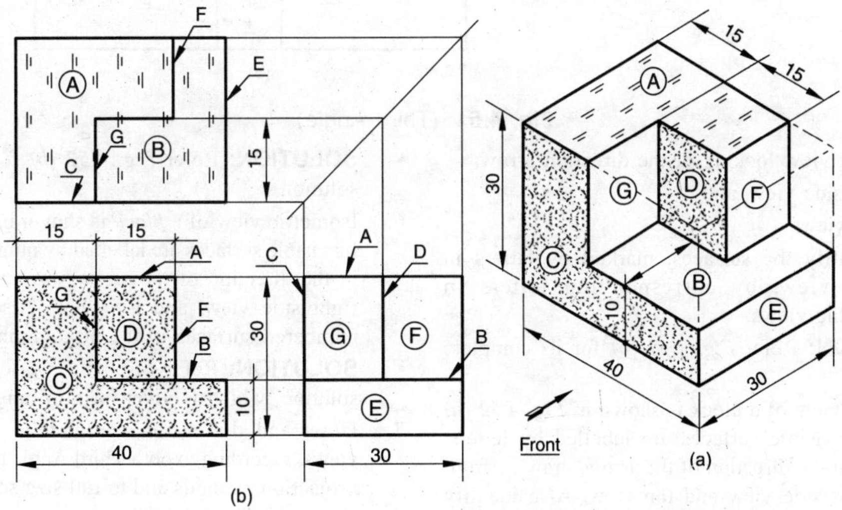

Fig. 4.49. (Third Angle)

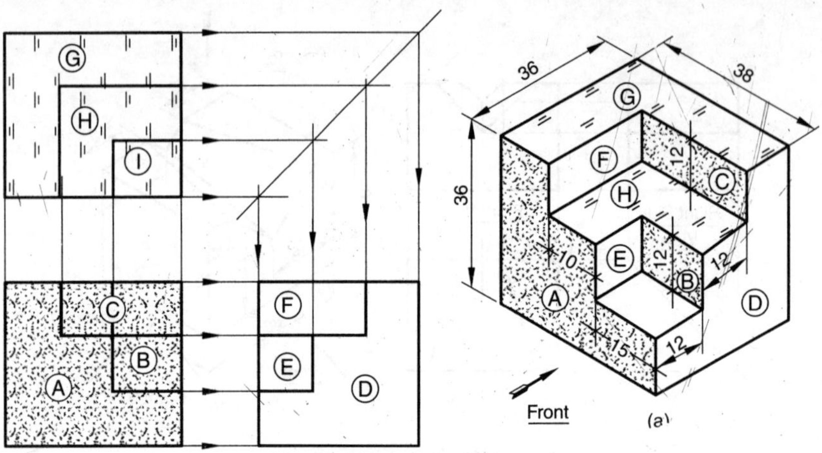

(b) (a)

Fig. 4.50. (Third Angle)

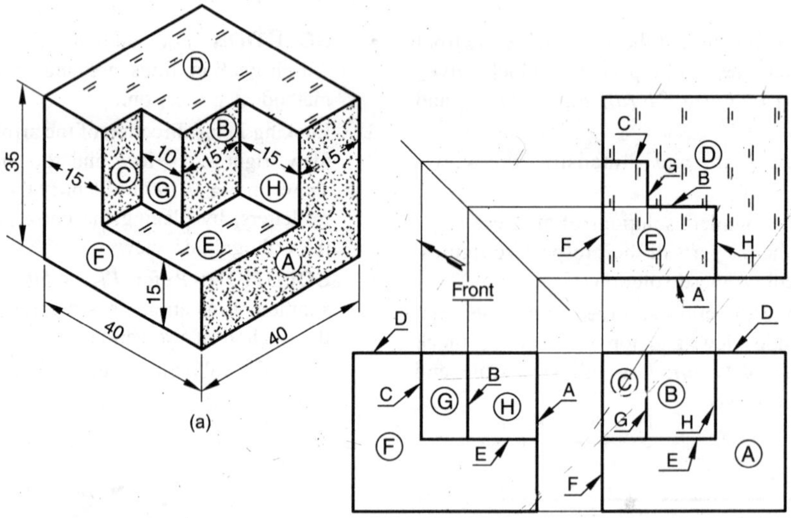

(a) (b)

Fig. 4.51. (Third Angle)

(*i*) Front view looking in the direction arrow;

(*ii*) Left side view, and

(*iii*) Top view.

Also identify the surfaces, marked by letters in pictorial view, by corresponding letters in orthographic views.

- **SOLUTION:** Refer *Fig. 4.51 (b)* for its complete solution.

5. Isometric view of a block is shown in *Fig. 4.52 (a)* where its visible surfaces are labelled by letters. Looking in the direction of the arrow, draw its front view, right side view and top view. Also identify the visible surfaces by corresponding letters in the orthographic views.

- **SOLUTION:** Refer *Fig. 4.52 (b)* for its complete solution.

6. Isometric view of a *block* is shown in *Fig. 4.53 (a)*. Its visible surfaces are labelled by numerals. Looking in the direction of the arrow, draw the front view, right side view, and top view. Also identify the numbered surfaces in the orthographic views.

- **SOLUTION:** Refer *Fig. 4.53 (b)* for its complete solution, which is given in third angle projection.

7. Draw and dimension completely, the following views, according to both Third Angle and First Angle projection methods and to full size scale, of a *block* shown in *Fig. 4.54 (a)*:

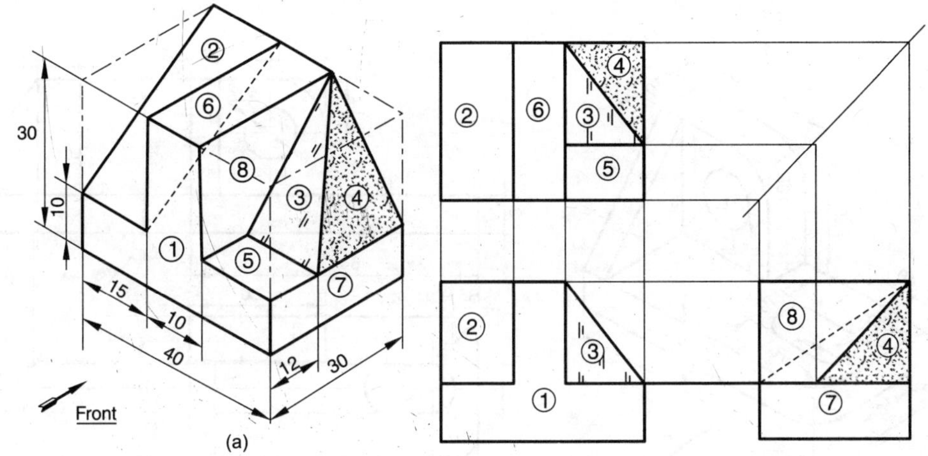

Fig. 4.52. (Third Angle)

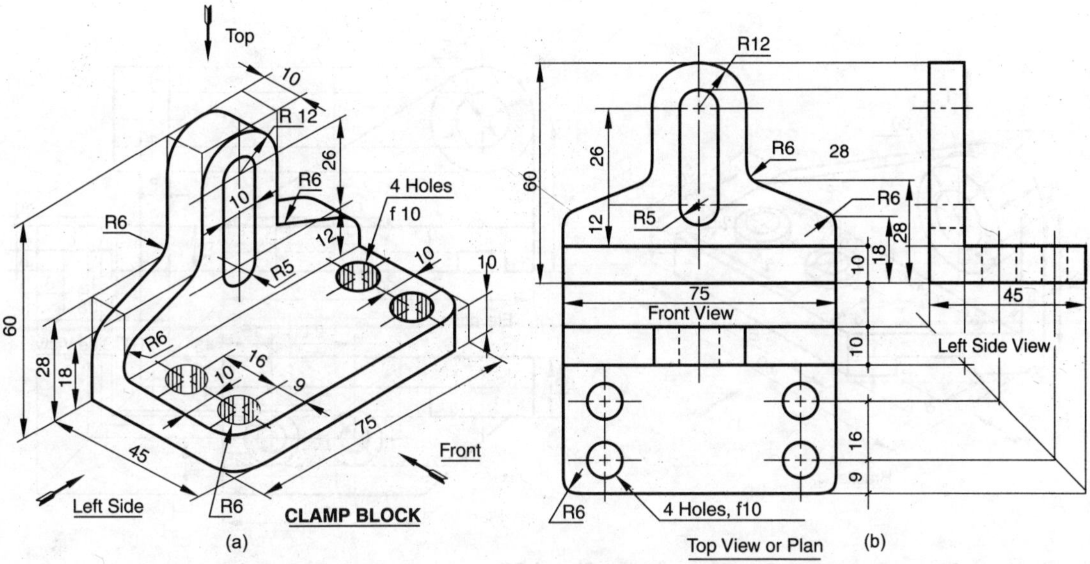

Fig. 4.53.

- Front view, looking in the direction of arrow F;
- Side view, looking in the direction of arrow S;
- Top view, looking in the direction of arrow P.

- **SOLUTION:** Refer to *Fig. 4.54 (b)* for solution in *Fig. 4.54 (b)*. All construction lines are shown here to illustrate the method of projection. The solution of the problem in third angle projection is left to the students.

8. Draw the following views of the *Clamp Block* shown in *Fig. 4.55 (a)*, both in third and first angle projections:
   - Front view looking in the direction of the arrow F,
   - Side view looking in the direction of arrow S, and
   - Top view, looking in the direction of arrow P.

- **SOLUTION:** Refer to *Fig. 4.55 (b)* for solution to this problem in first angle. All construction lines are shown to illustrate the method of projection. Its solution in third angle is left to the students.

9. Draw the following views of the *Bracket* shown in *Fig. 4.56 (a)*:
   - Front view looking in the direction of arrow F;

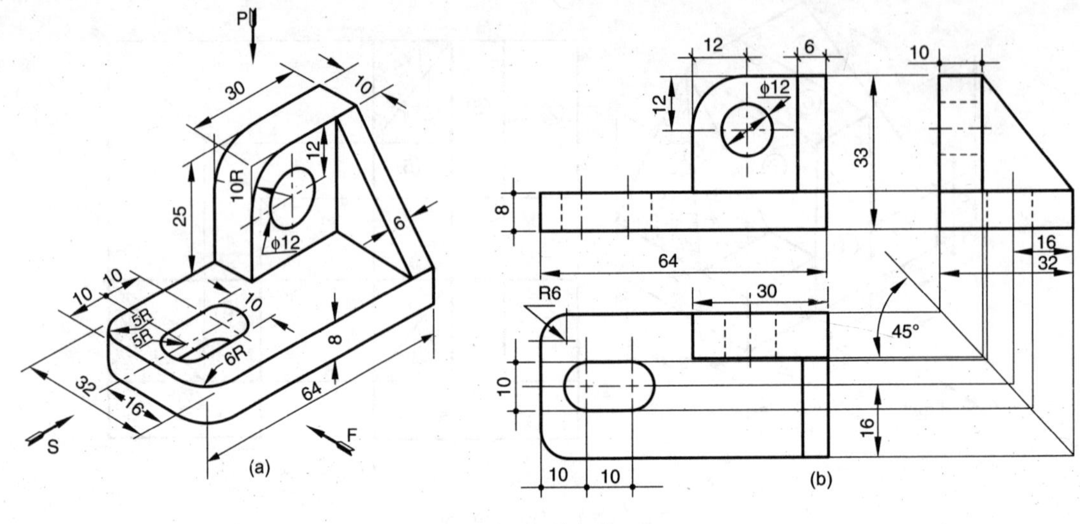

Fig. 4.54.

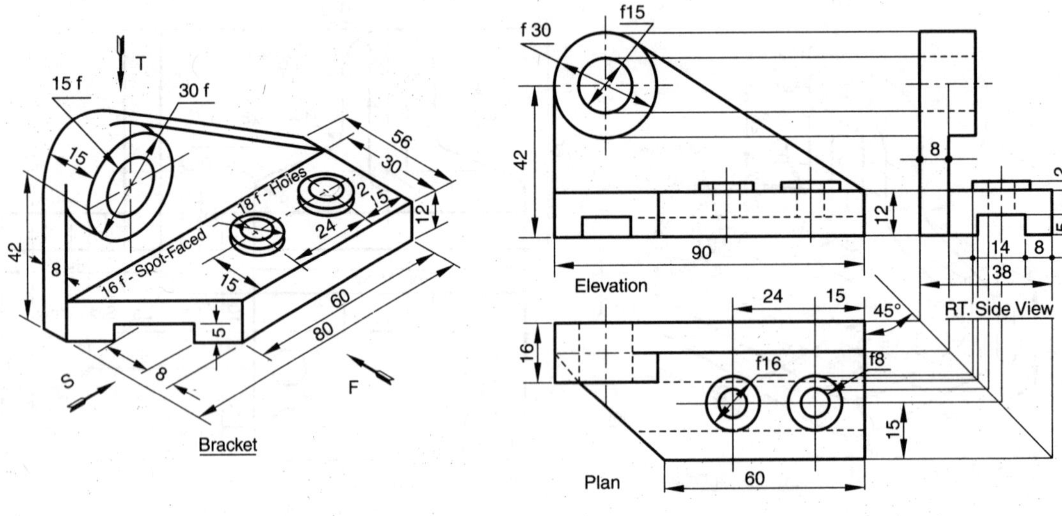

Fig. 4.55.

- Side view, looking in the direction of arrow S; and
- Top view. Use full size scale and dimension the views completely.

10. Draw the following views of the block shown in *Fig. 4.57 (a)*:
    - Front view, looking in the direction of arrow,
    - Right side view, and
    - Top view. Use full size scale and give all the necessary dimensions.

- **SOLUTION:** For solution in Third Angle refer to *Fig. 4.57 (b)*. Note the corresponding projections of points 1 and 2. Reproduce the views in First Angle and add the required dimensions.

11. Draw front view and top view of the *Support Bracket* shown in *Fig. 4.58 (a)*. Also dimension the views completely.

- **SOLUTION:** Refer to *Fig. 4.58 (b)*, for solution in Third Angle. Study these views and draw them in First Angle. Add all the required dimensions.

12. Draw the following views of the *Bracket* shown in *Fig. 4.59 (a)* in First Angle, using full size scale:
    - Front view, looking in the direction of arrow,
    - Right side view, and
    - Top view. Dimension the views completely.

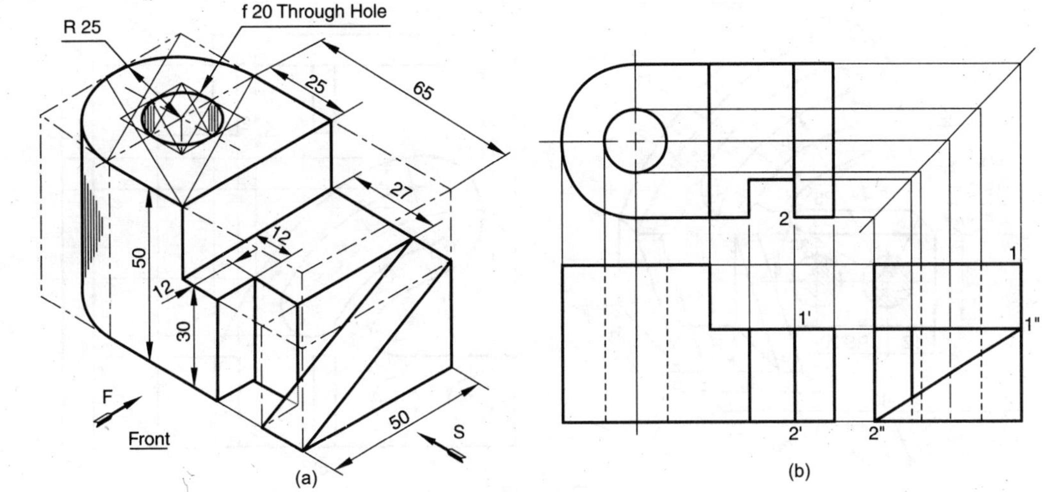

Fig. 4.56. (Third Angle)

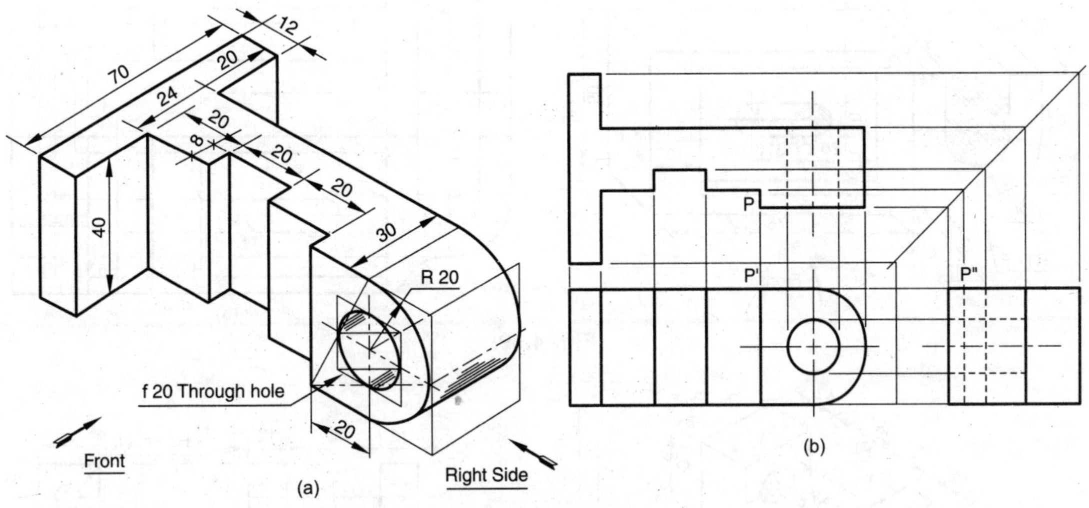

Fig. 4.57.

- **SOLUTION:** For solution, in Third Angle refer to *Fig. 4.59 (b)* Study these views carefully and then draw the required views using First Angle. Add all the required dimensions.

13. *Figure 4.60 (a)* shows oblique projection of a Support *Bracket.* Using full size scale, draw the following orthographic views;
    - Front side view, and
    - Right side view, and
    - Top view, and add all the necessary dimensions.

- **SOLUTION:** For solutions in Third Angle refer to *Fig. 4.60 (b)*. Study these views and then draw the corresponding views in First Angle. Show all the construction lines and add all the necessary dimension lines.

14. *Figure 4.61 (a)* shows isometric view of a Bracket. Draw its front view, left side view, and top view. Dimension the views. Use a full size scale.

- **SOLUTION:** For solutions in First Angle, refer to *Fig. 4.61 (b)*. Study these views carefully. Add the missing dimensions to the front view. Then draw the corresponding views in Third Angle. Show all construction lines. Give all the required dimensions.

15. *Figure 4.62 (a)* shown an isometric view of a *Block* in which the various surfaces are marked by letters A, B, C, etc. Draw its;

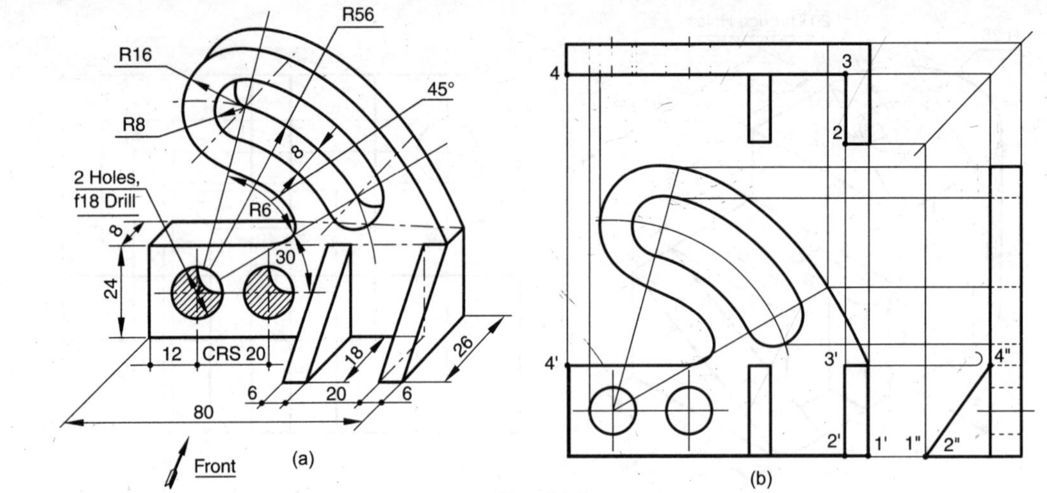

Fig. 4.58.

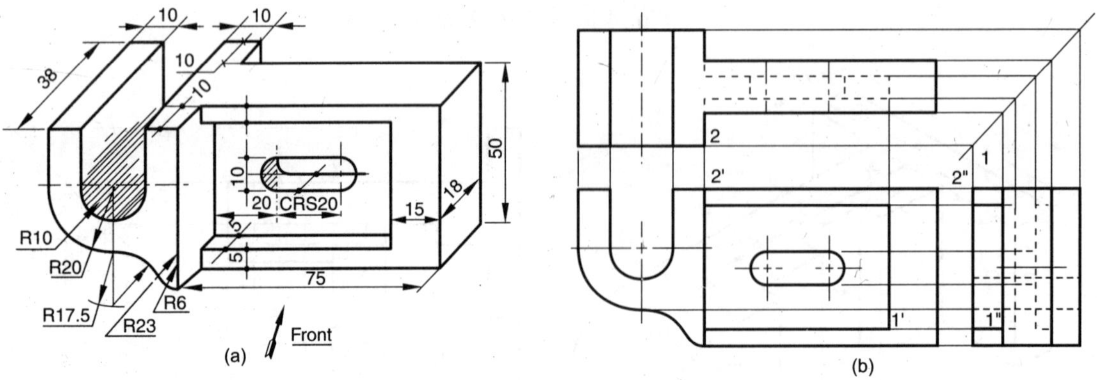

Fig. 4.59.

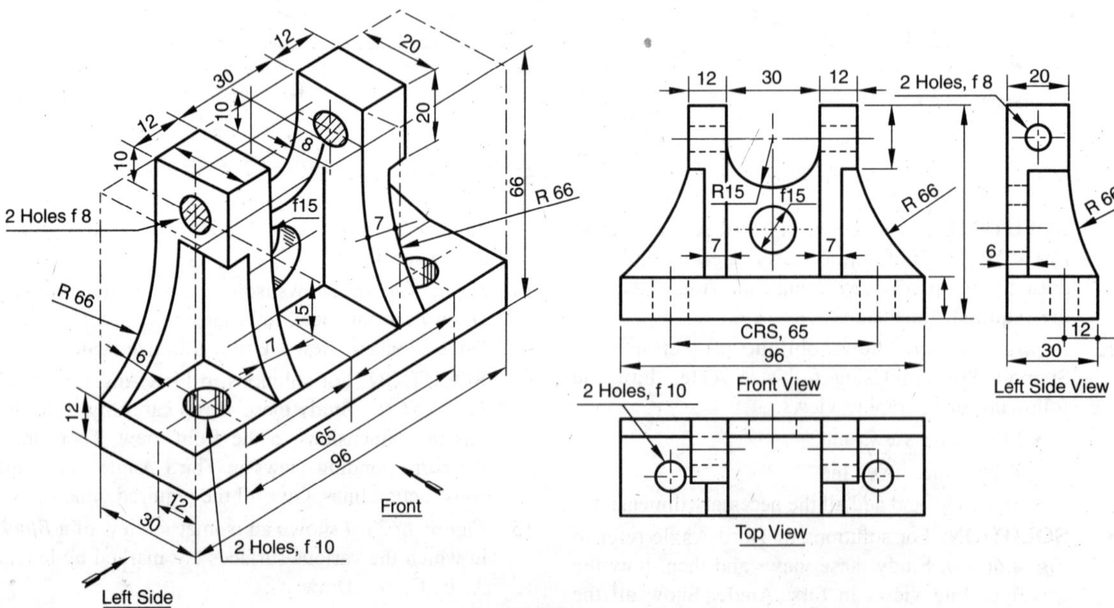

Fig. 4.60.

- Front view, looking in the direction of arrow F,
- Right side view, and
- Top view. Also identify and mark the various surfaces and their corresponding projections in the orthographic views.

- **SOLUTION:** For solution in First Angle projection refer to *Fig. 4.62 (b)*. Study these views carefully and note that a perpendicular surface projects its true shape in a view which is projected on a plane parallel to that surface. In the other two views the surface projects as a line (*i.e.,* edge view of the surface). Then draw the views in Third Angle projection. Give all the required dimensions.

16. *Figure 4.63 (a)* shows an isometric view of a Block in which the various surfaces are marked by letters A, B, C, etc. Draw its front view, right side view, and top view. Identify and mark the various surfaces and their corresponding projections in the orthographic views drawn. Use full size scale and dimension the views completely.

- **SOLUTION:** For solution in First Angle refer to *Fig. 4.63 (b)*. Identification of surfaces A and F need careful examination of the pictorial and orthographic views. Surface A is inclined to both HP and VP but is perpendicular to the Profile Plane. Therefore, it will project as a surface (rectangle) in top and front views and a line (edge) view of surface

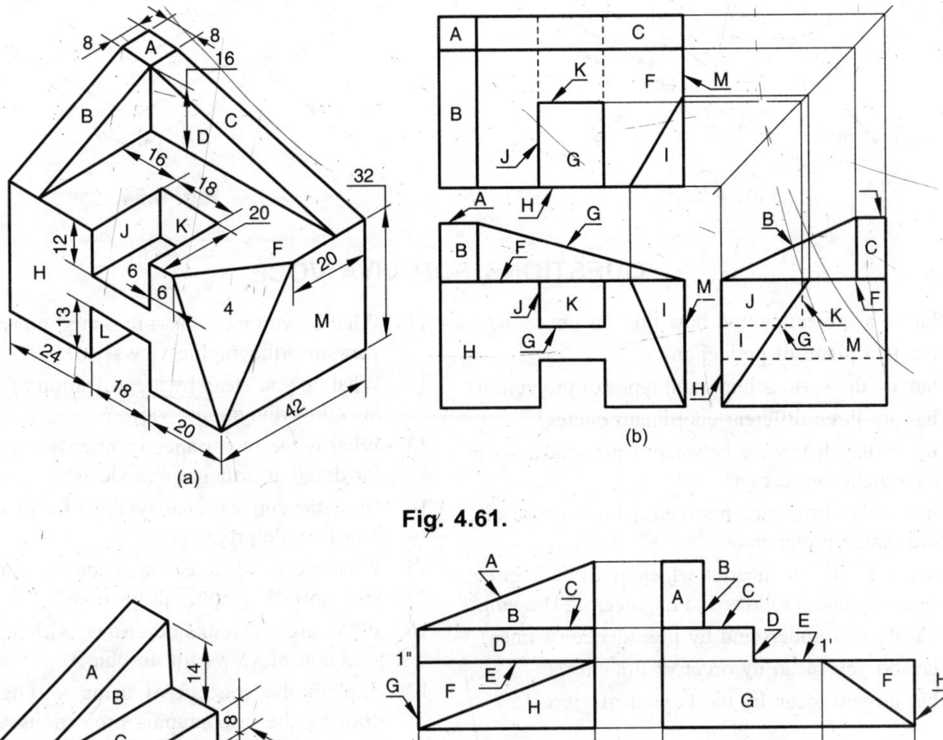

Fig. 4.61.

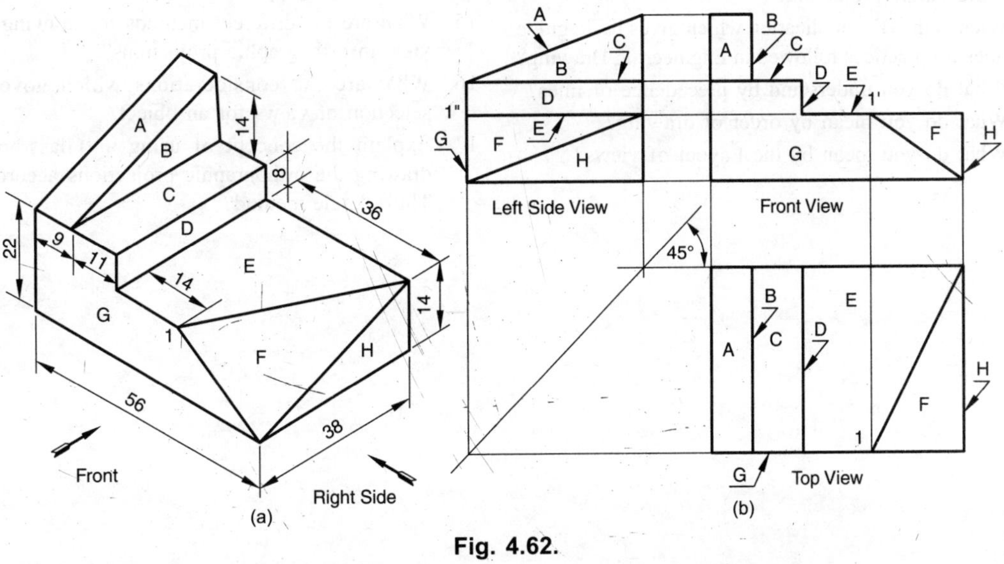

Fig. 4.62.

in side view. Surface F is an oblique surface and is inclined to all three coordinate planes of projection. Therefore, it projects as a surface (triangle) in all the orthographic views and does not project as edge view in any of them.

17. *Figure 4.64* shows a pictorial drawing of a *Block* on which the various surfaces are marked by A, B, C, etc. Draw its front view, side view and top view in First Angle. Identify and mark the projections of various surfaces by the same letters in all the orthographic views.

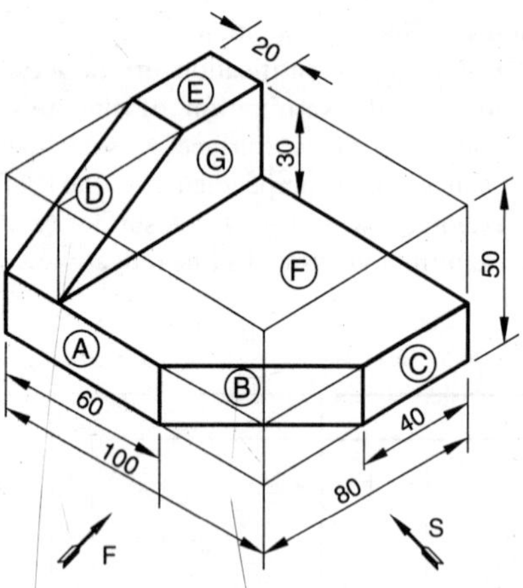

Fig. 4.63.

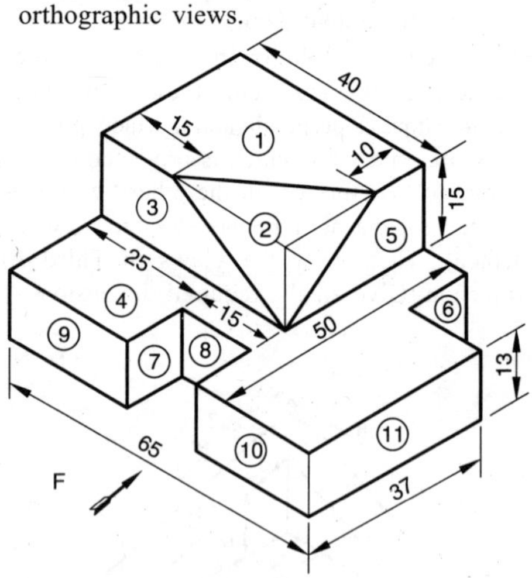

Fig. 4.64.

QUESTIONS FOR VIVA VOCE

1. What is a projection and how do you obtain it?
2. What is a plane of projection?
3. What are the various important types of projection?
4. What are three different coordinate planes?
5. What is the difference between a pictorial and an orthographic projection?
6. What is the difference between a First Angle and Third Angle projection?
7. What is the BIS publication which gives basic guide lines for practices followed in Engineering Drawing?
8. What do you understand by precedence of lines?
9. What do you mean by order of drawing?
10. What do you mean by the Layout of views?
11. What do you mean by a missing line and a missing view in orthographic views?
12. What do you mean by identification of surfaces and lines in orthographic views?
13. What is the importance of orientation of an object for drawing orthographic views?
14. Draw the conventional symbols for First Angle and Third Angle projection?
15. What are the different methods for drawing a side view in orthographic projections?
16. What are the considerations which govern the selection of views for an object?
17. Explain the concept of using a 'Glass-box' for drawing the orthographic projections according to Third Angle method?

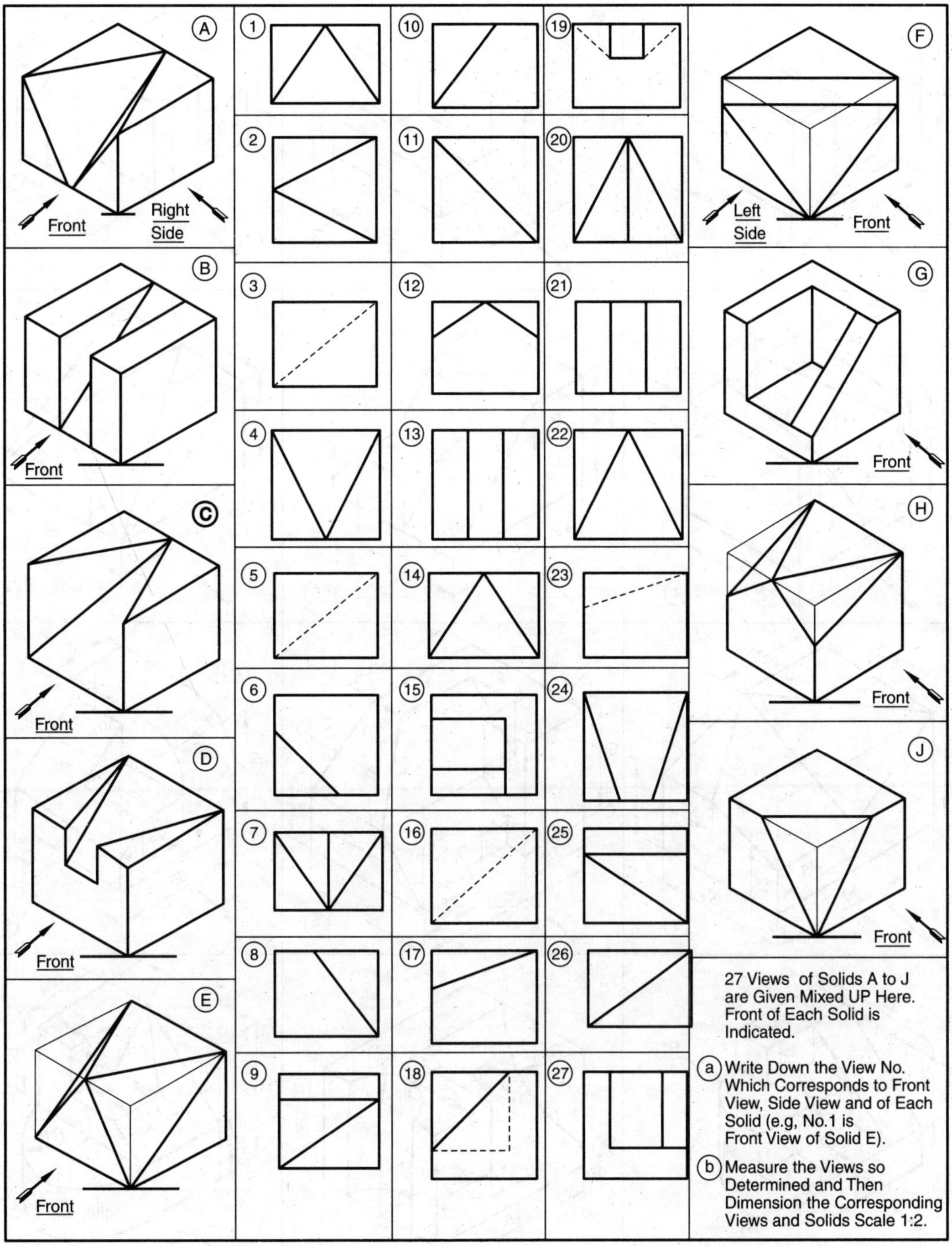
A
Front
Right Side
B
Front
C
Front
D
Front
E
Front
F
Left Side
Front
G
Front
H
Front
J
Front
1
2
3
4
5
6
7
8
9
10
11
12
13
14
15
16
17
18
19
20
21
22
23
24
25
26
27
27 Views of Solids A to J are Given Mixed UP Here. Front of Each Solid is Indicated.
(a) Write Down the View No. Which Corresponds to Front View, Side View and of Each Solid (e.g, No.1 is Front View of Solid E).
(b) Measure the Views so Determined and Then Dimension the Corresponding Views and Solids Scale 1:2.

Fig. 4.65.

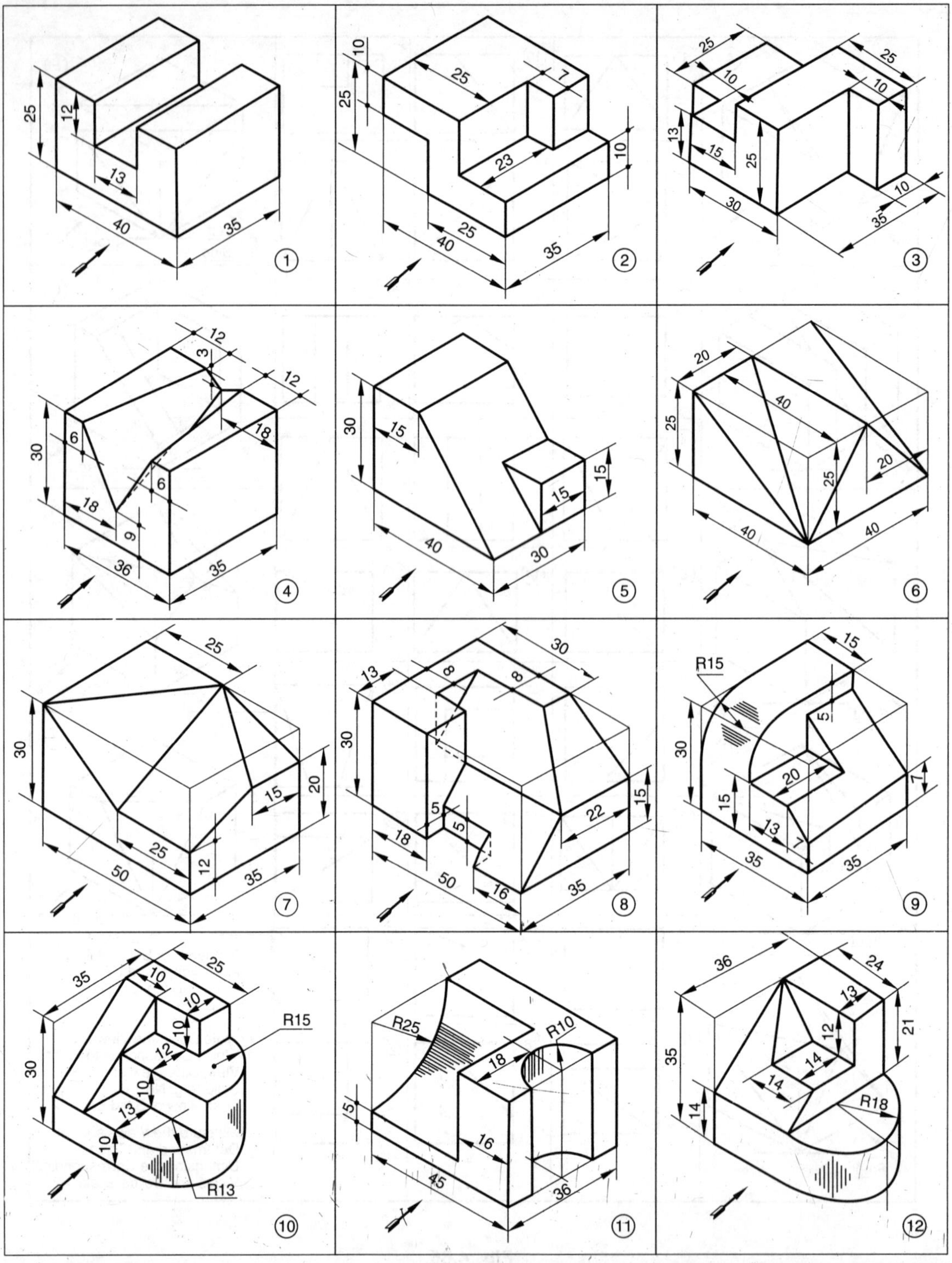

Ex. 18: Fig. 4.66. Isometric views of 12 different objects are given on this page. Front of each object is indicated; (1) Draw the front, side and top views of each object by placing corresponding letters on the orthographic of each object, (2) Name all visible surfaces of each object by placing corresponding letter on the orthographic of each object

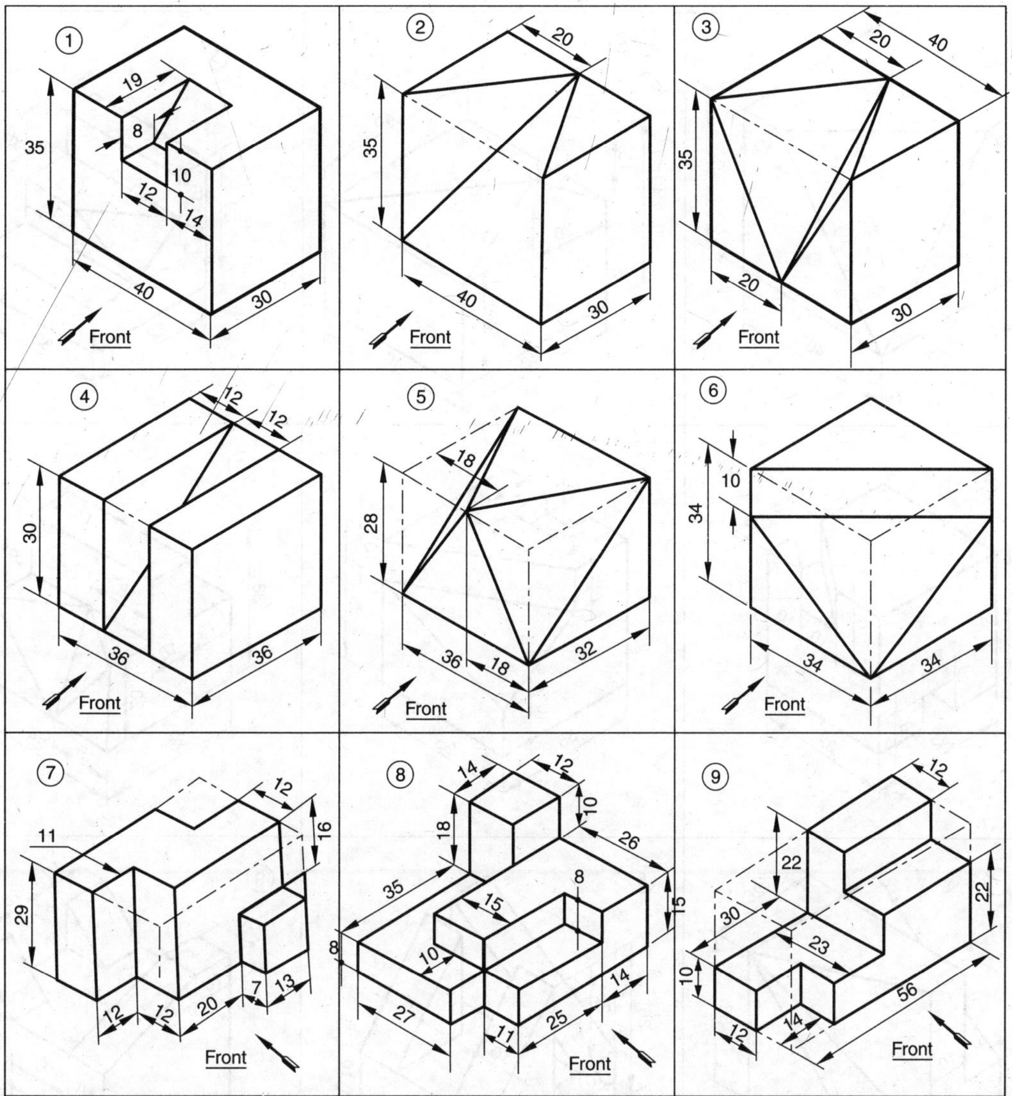

Given : Isometric Views of 11 Different Objects

(a) **Draw** : (1) Front View, in the Direction of Arrow, (2) Side View, and (3) Top View.

(b) Dimension the Views.

(c) Number all the Visible Surface in Each Case and Identify Them in the Orthographic Views Drawn.

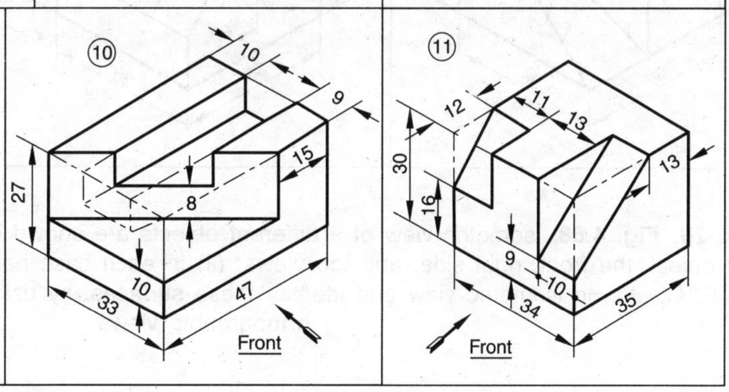

Fig. 4.67. (Ex. 19)

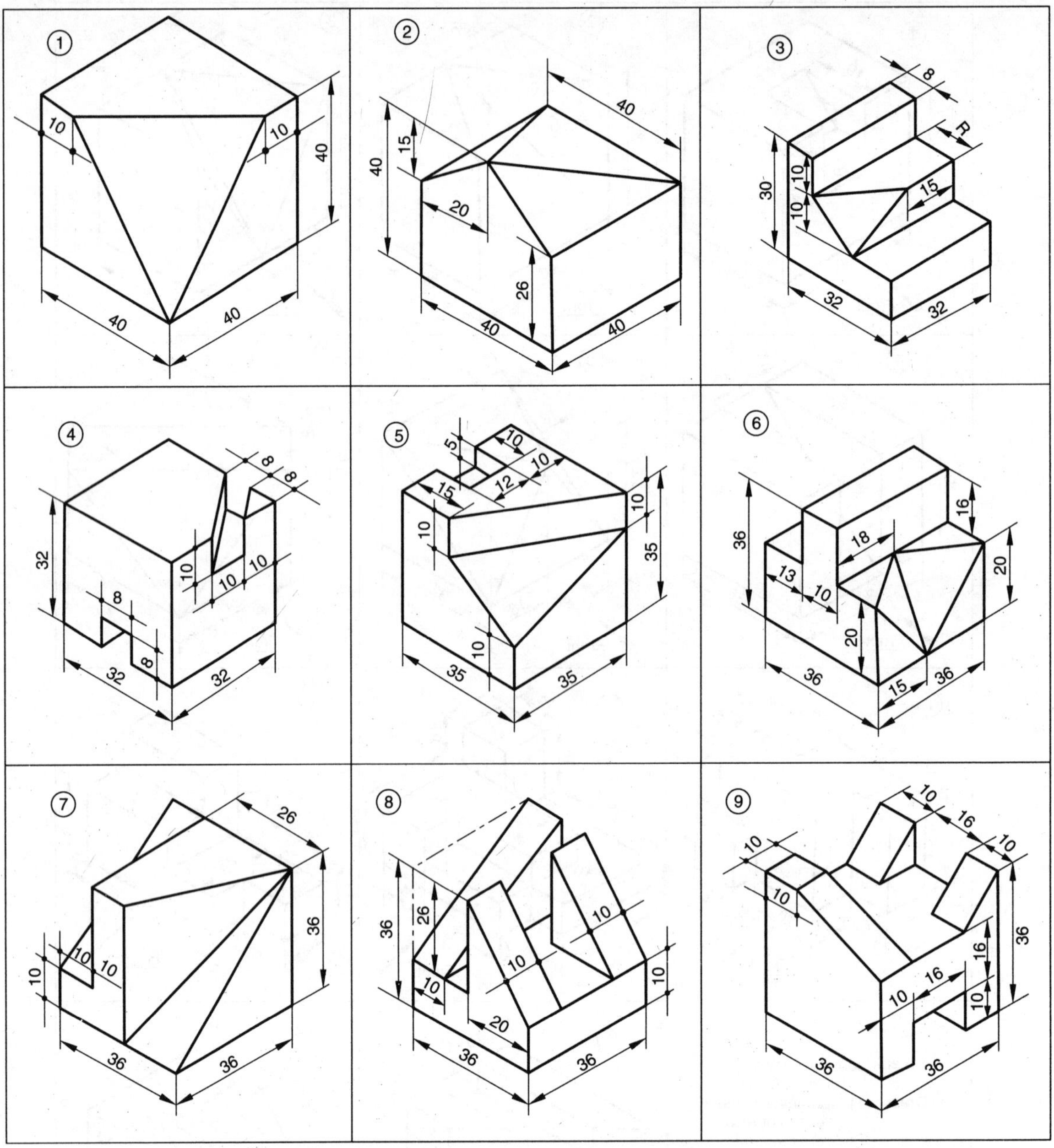

Ex. 20: Fig. 4.68. Isometric view of 9 different objects are shown on this page here; (i) Draw, for each object, the front, right side, and top views; (ii) In each case name the visible surfaces by letters A, B, C, etc., in isometric view and identify these surfaces, by using the corresponding letters, in orthographic views

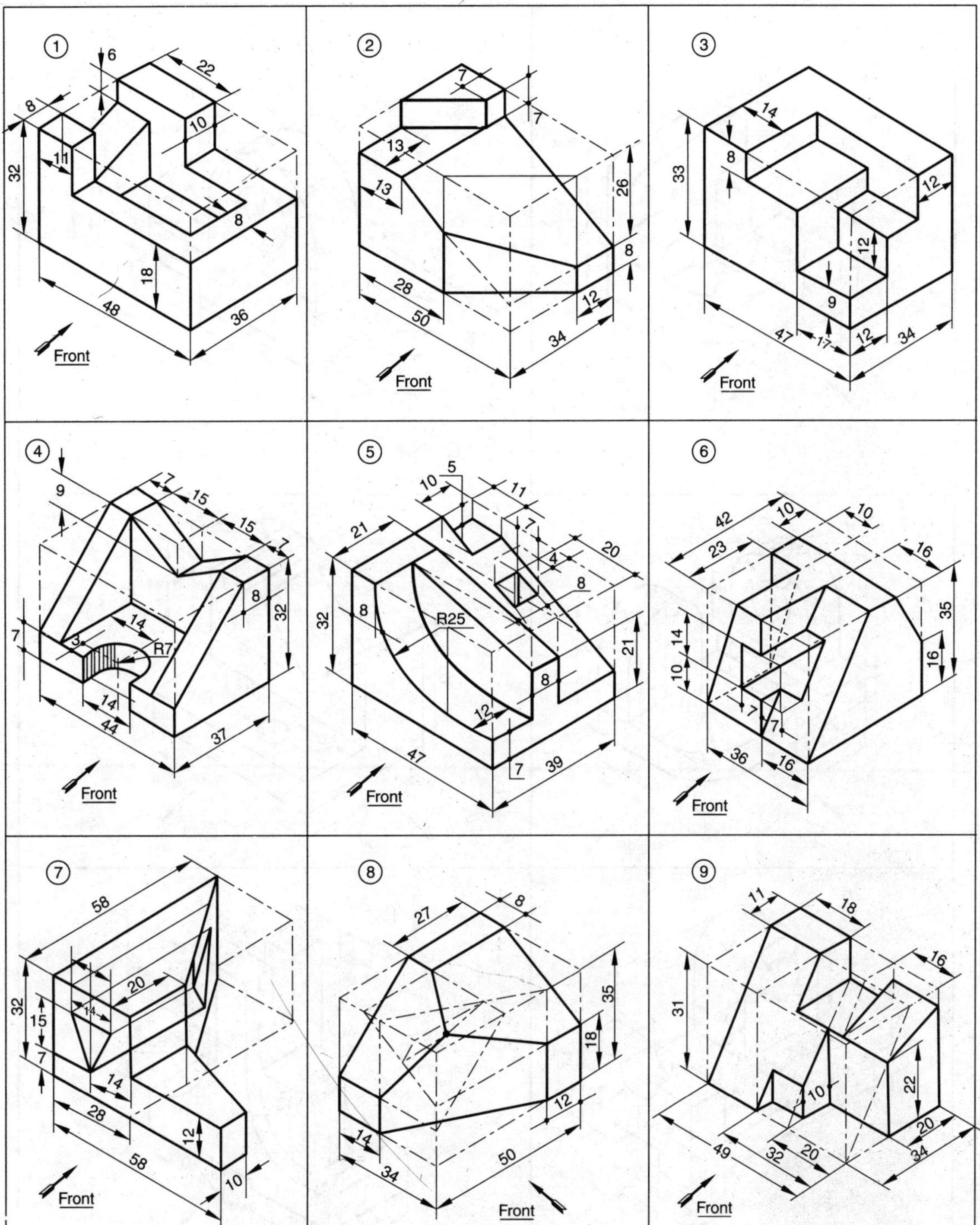

Ex. 21: Fig. 4.69. Isometric Views of a 9 different blocks are given; (1) Draw front side and top view in each case and dimensions the views, (2) Number the visible surfaces of each block and identify each such surface in orthographic projections

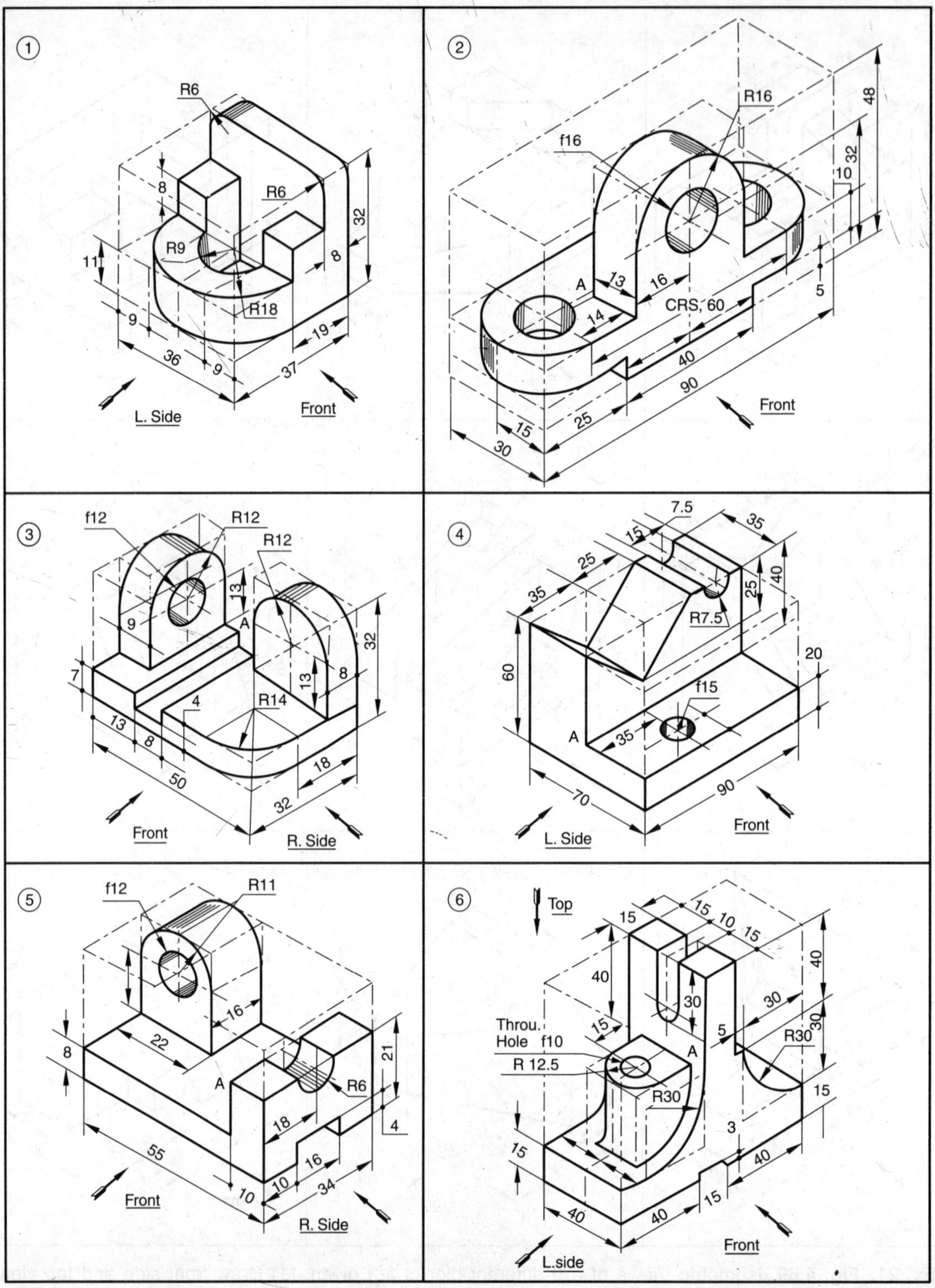

Ex. 24: Fig. 4.70. Isometric views of six different object block are given on this page; (1) Draw Front, Side, and Top views for each one of them and dimension the view; (2) Number the visible surfaces of each block and identify each surface in the corresponding orthographic views

①

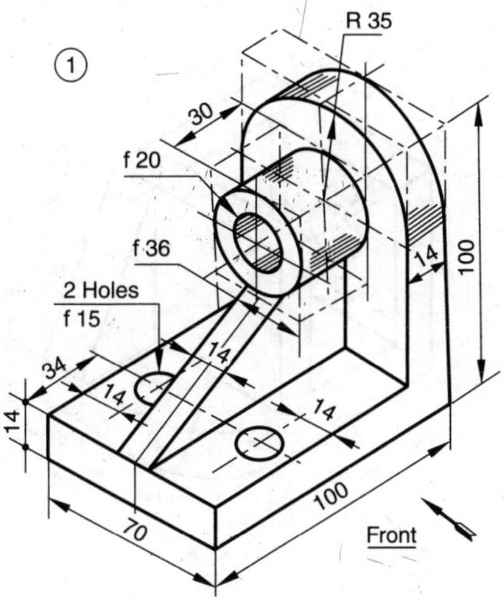

②

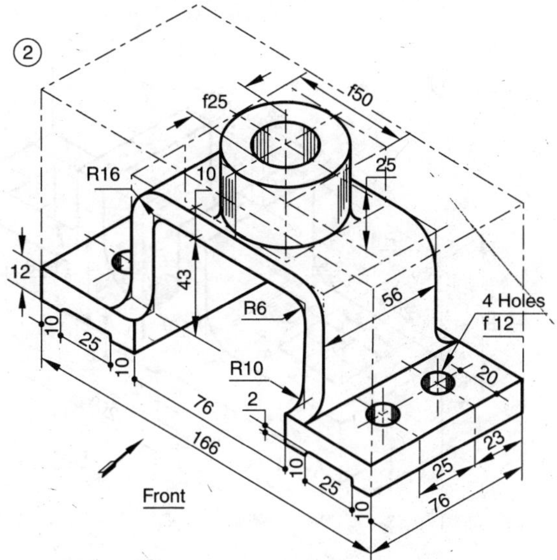

③

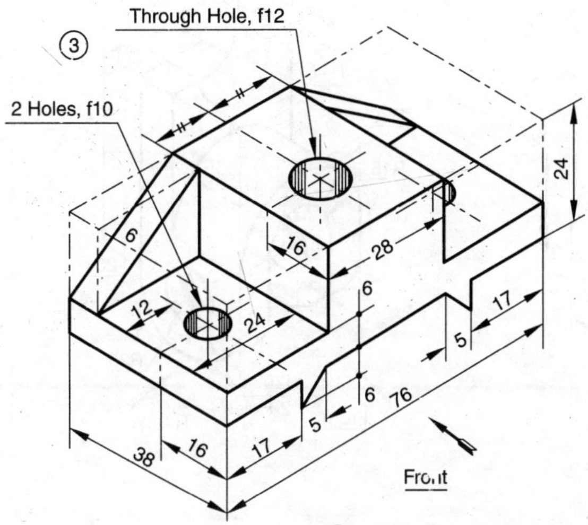

④

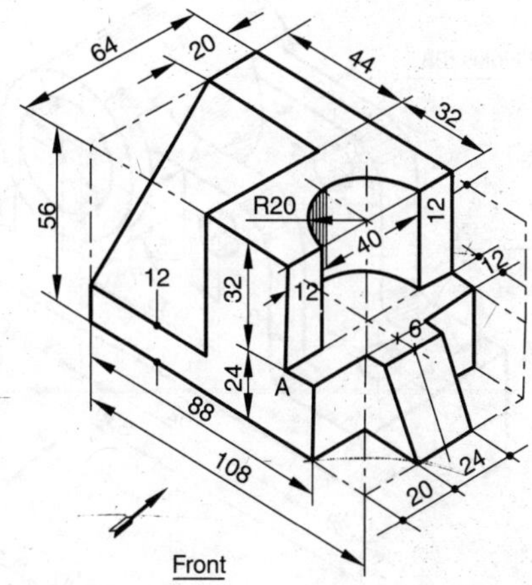

⑤

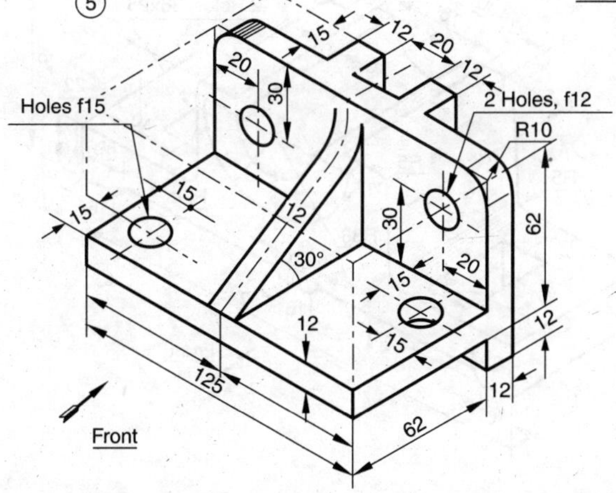

Ex. 24: Fig. 4.71. Isometric views of five different objects are given on this page; (1) Draw Front, Top and suitable Side views for each one of these and dimension the views; (2) Draw these isometric views using the CAD Systems available to you

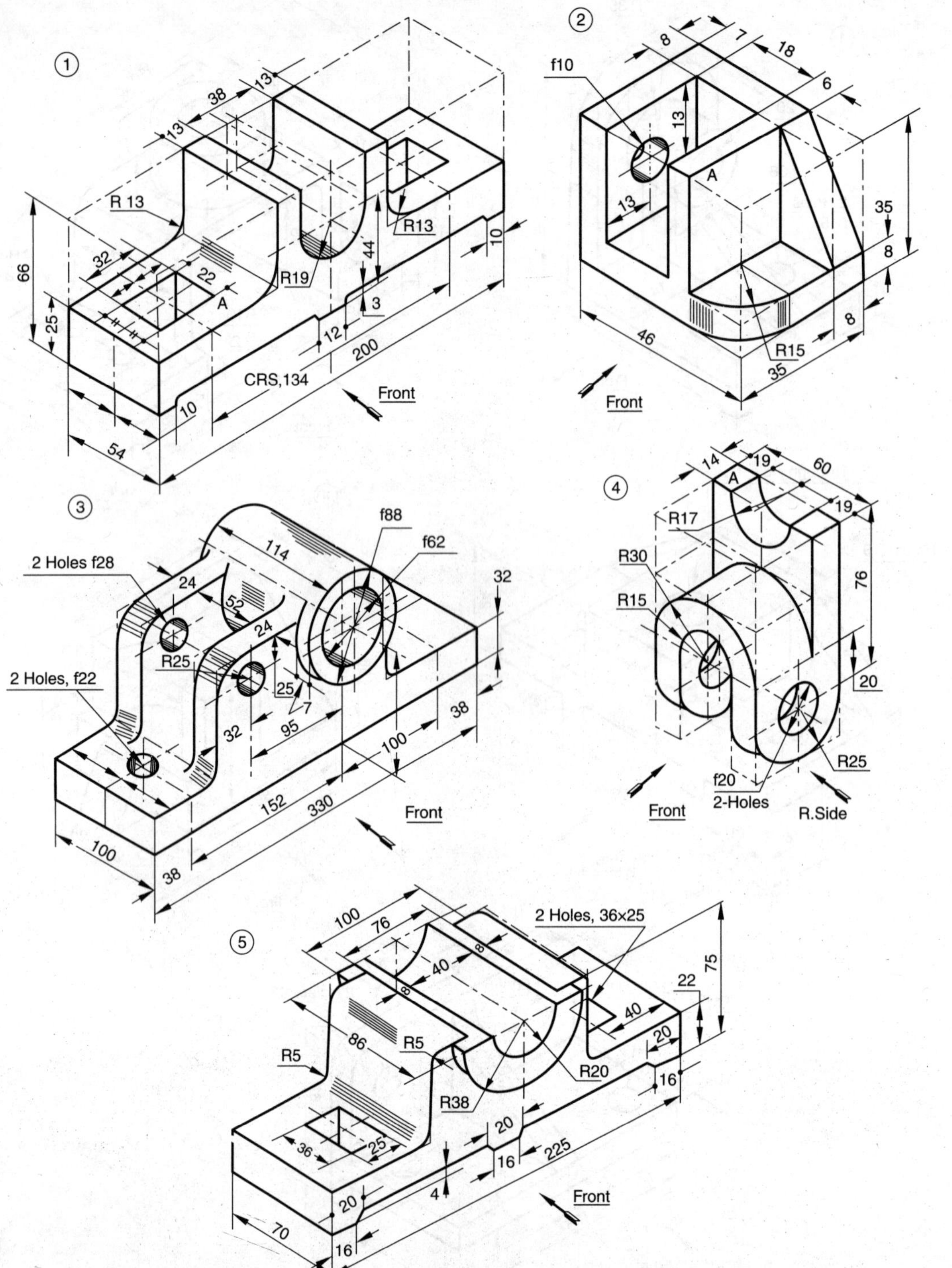

Ex. 25. Fig. 4.72. Isometric views of five different objects are given on this page; (1) Draw Front, Top and suitable Side views for each of these and dimension the views; (2) Draw these isometric views using the CAD Systems available to you

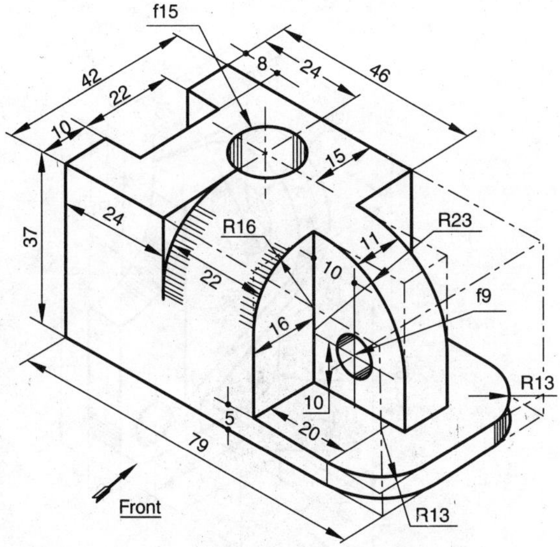

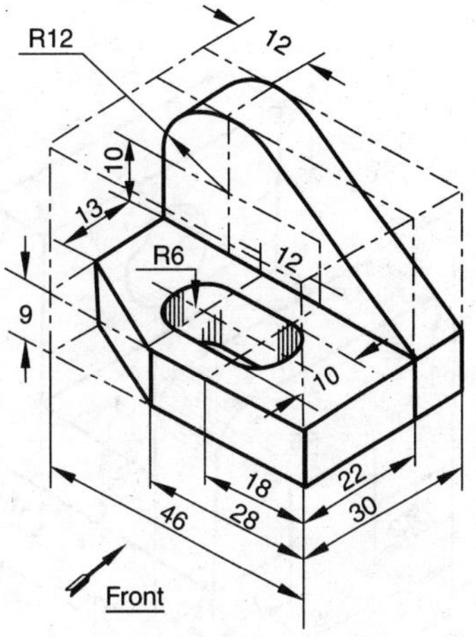

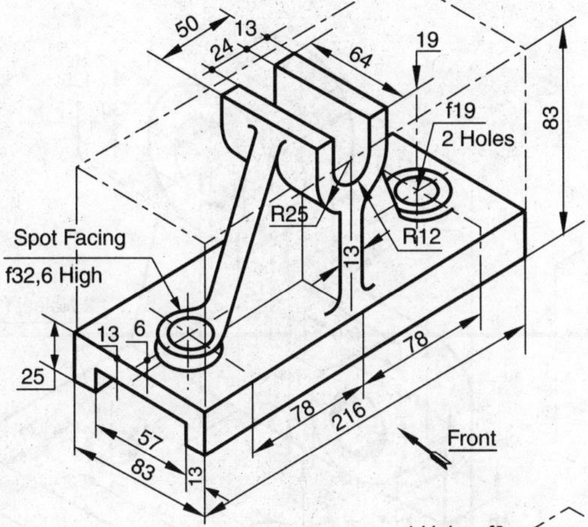

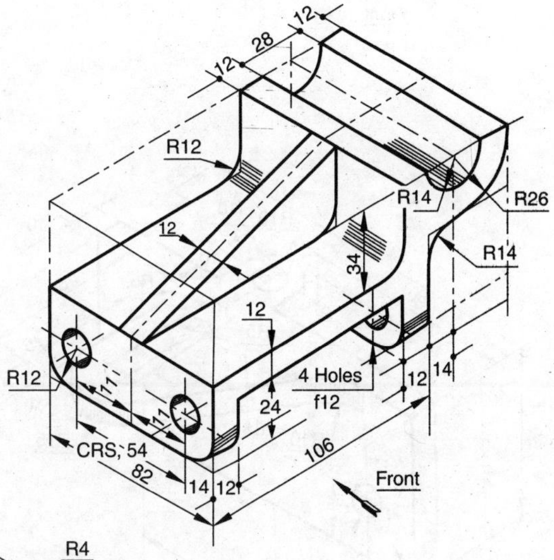

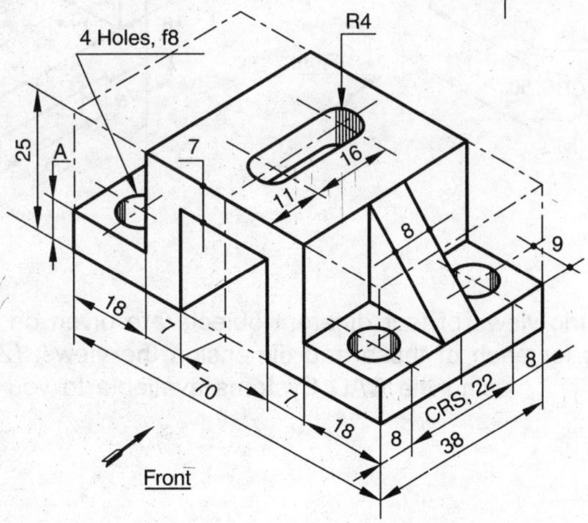

Ex. 26: Fig. 4.73. Isometric views of five different objects are given on this page; (1) Draw Front, Top and suitable Side views for each one of these and dimension the views; (2) Draw these isometric views using the CAD Systems available to you

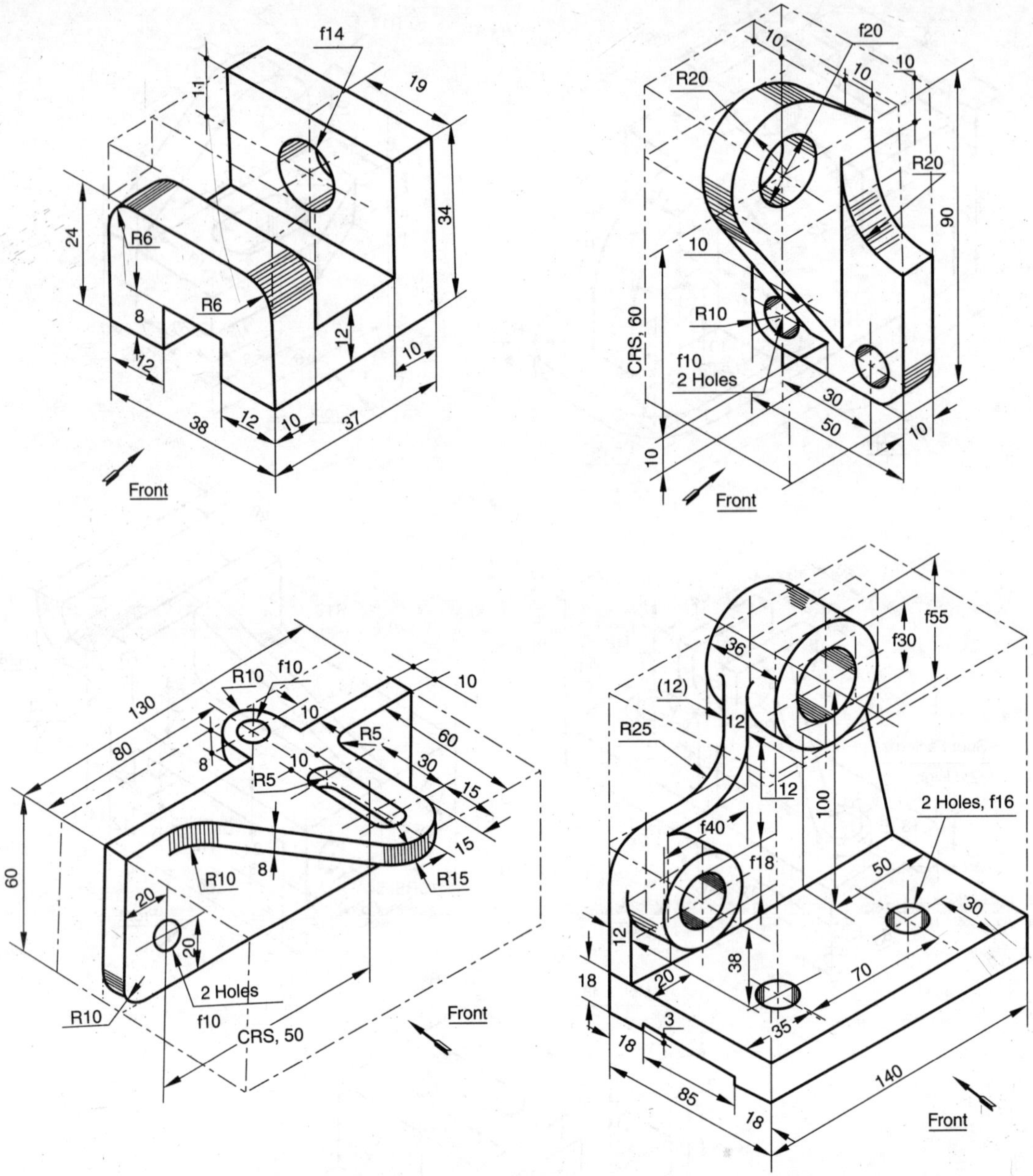

Ex. 27: Fig. 4.74. Isometric views of four different objects are given on this page; (1) Draw Front, Top and suitable Side views for each of these and dimension the views; (2) Draw these isometric views using the CAD Systems available to you

Theory of Co-Ordinate Dimensioning

5

5.1. INTRODUCTION

For convenience in using, ensuring accuracy, eliminating misinterpretations and saving time and money, in addition to shape description by orthographic views, size description on the drawings is given in the form of systematic dimensioning, standard notes, datum for measurement, geometrical dimensioning and tolerance of mating parts, machining symbols, surface finish, type of material, number required etc., arranged on a completed engineering drawing following standard practices.

In this chapter only the '*theory of conventional dimensioning*' is discussed. '*Theory of conventional tolerancing*' is discussed in chapter 13 and the '*Theory of geometric tolerancing*' is discussed in chapter 14.

ISO and National Standards are used to establish uniform rules, principles, and methods of dimensioning and tolerancing along with practices for stating and interpreting dimensioning, tolerancing, and other related requirements for use on engineering drawings and in related documents that apply to coordinate as well as geometric dimensioning methods. The reference to any standard used should be noted on the drawings or in a document referenced on the drawings.

Engineering drawing practice is very largely concerned with dimensioning and tolerancing. A student as a drafter should always bear in mind that drawings are for use, and that satisfactory dimensioning contributes to utility, facilitates mass production of interchangeable parts and simplifies inspection of finished work parts. A drafter should know how to properly represent conventional dimensioning and geometric tolerancing. Also, a technician must be able to accurately read dimensioning and geometric tolerancing. All engineering students must have a through working knowledge of the principles and methods used for specifying design requirements, on engineering drawings to be used for production of the parts.

All the dimensions, tolerances, notes, etc., necessary to provide for correct production and functioning of a part should be given directly on drawing together with any other information necessary to define the part completely in its finished form, keeping in mind the manufacturing, assembly, and inspection requirements.

5.2. DIMENSIONING TERMS AND NOTATIONS

Dimensions are applied by means of dimension lines, extension lines, or a leader line from a dimension, note, or specification directed to the appropriate feature, Fig. 5.1. A generally recognized system of terms, lines, symbols, figures and notes is used to indicate size and location of features.

1. Dimension Lines

A dimension line, with its arrowheads, shows the direction and extent of a dimension. These are thin lines, which are terminated at each end by an arrowhead, generally. Whenever possible these should be placed outside the outline of the object. A numerical value, placed above, when the dimension line is not broken, or in the dimension line when the dimension line is broken, specifies the number of units for the measurement that is indicated, *Fig. 5.1*. When numerals are in double line, one line is written above the dimension line and the other below it, *Fig. 5.27*. Dimension lines may be broken for inserting the dimension in the case of unidirectional dimensioning, *Fig. 5.23, Fig. 5.24 (b)*. A dimension line should not be broken where the feature, to which it refers, is shown broken, Fig. *5.1 (c)*, except for unidirectional method of dimensioning.

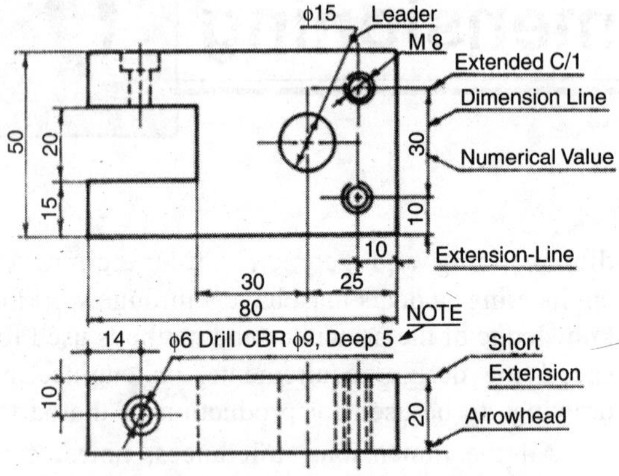

Fig. 5.1. (a) Notation of Dimensioning

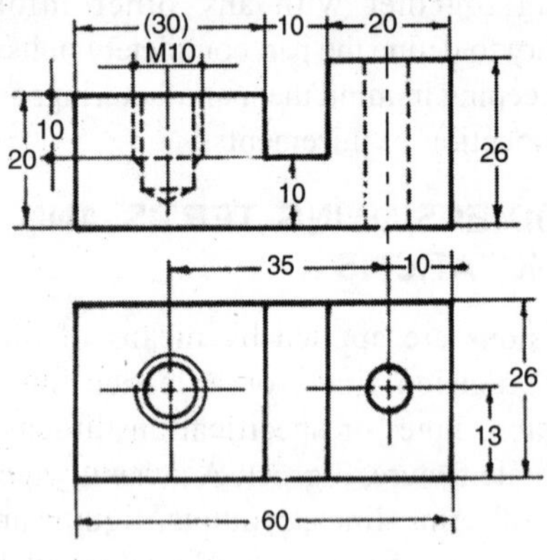

Fig. 5.1. (b)

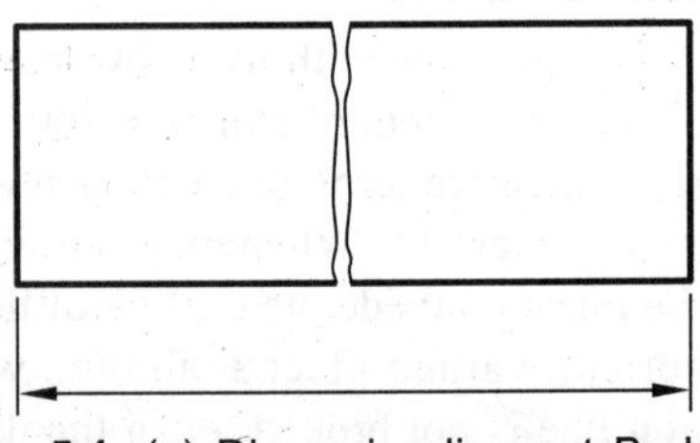

Fig. 5.1. (c) Dimension line not Broken

2. Extension Lines or Projection Lines

These are thin continuous lines extending from points, lines or surfaces to enable the dimensions to be placed outside the outline, wherever possible, to indicate the extent of a measurement given by dimension line. The line should extend about 3 mm beyond the dimension line as in *Fig. 5.1.*

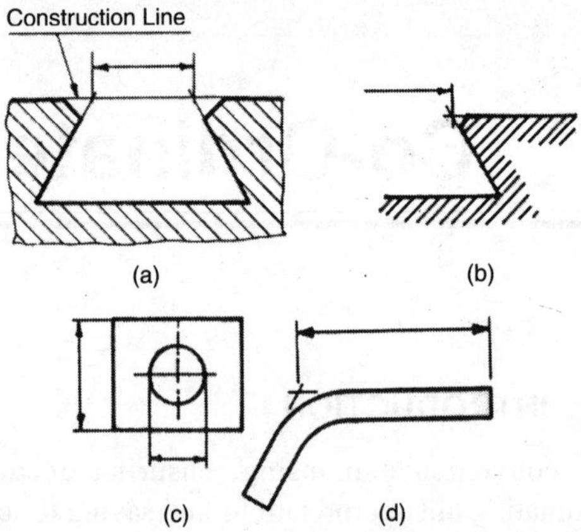

Fig. 5.2. (a),(b),(c) and (d) Drawing extension and Projection lines and locating a point

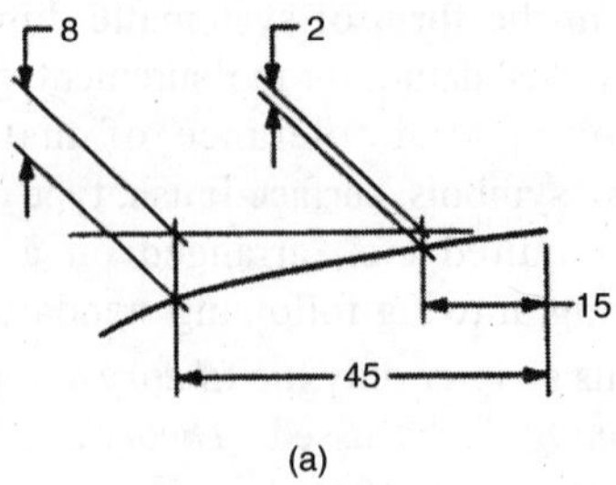

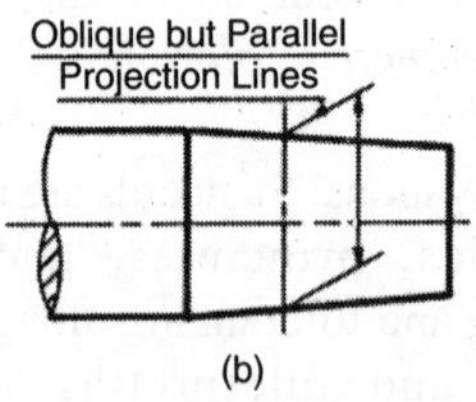

Fig. 5.3. (a) and (b) using oblique Extension lines.

In general, extension lines are drawn in a direction perpendicular to the feature to be dimensioned, *Fig. 5.2(a)*. Where space is limited, extension lines may be drawn at an oblique angle to clearly illustrate where they apply, but for preferably parallel to each other, *Fig. 5.3(a), Fig. 5.33*. Where oblique lines are used, the dimension lines are shown in the direction in which they apply, *Fig. 5.3(b)*. Intersecting construction and projection lines should extend slightly beyond their point of intersection, *Fig. 5.2(b)*. In general projection lines and dimension lines should not cross other lines unless it is unavoidable, *Fig. 5.2(c)*. Where a point is located by extension lines only, the extension lines from surfaces should pass through the point, *Fig. 5.2 (d)*.

3. Termination and Origin Indication

Dimension lines should show distinct termination, i.e., either arrowheads, or oblique strokes, or, where applicable, an origin indication. Two dimension line terminations are shown in, *Fig. 5.4 (a), (b)*, while an origin indication is shown in *Fig. 5.4(c)*.

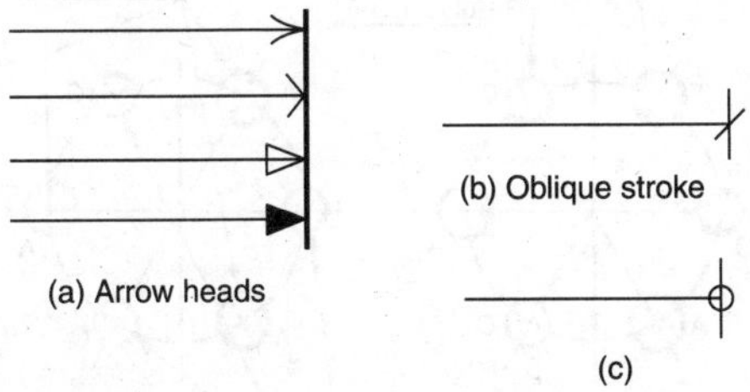

Fig. 5.4. Termination and Origin Indication

4. Arrowheads

Arroheads are used to terminate dimension lines. They should be sharp, neat and easily readable. The size of an arrowhead must be three times its spread. The construction of an arrowhead is shown in *Fig. 5.4* on enlarged scale. Preferably the space in the arrowhead should be filled. The size of arrowheads should be proportionate to the thickness of the lines of the drawing and size of drawing, as shown in *Fig. 5.5*. Good and bad practices in drawing arrowheads are shown in *Fig. 5.6 (iii)*. Adjacent arrowheads may be replaced by clearly marked dots where space is limited, *Fig. 5.6 (i)*. Instead of arrowheads, the ends of the dimension line may be marked by oblique dashes, drawn as a short lines inclined at 45°. The origin indication is drawn open circle of approximately 3 mm diameter.

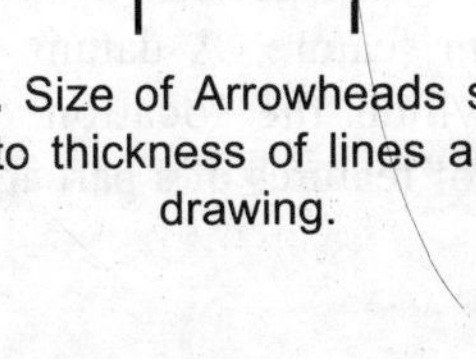

Fig. 5.5. Size of Arrowheads should be proportional to thickness of lines and size of the drawing.

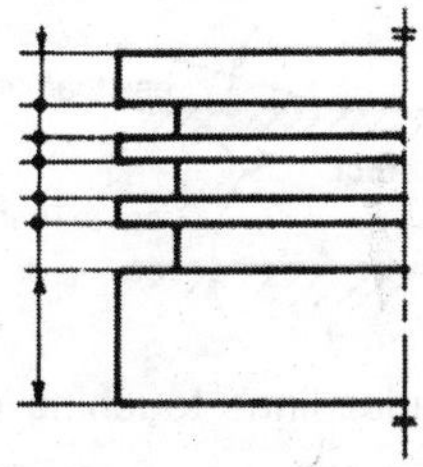

Fig. 5.6. (i) Arrowheads replaced by dots

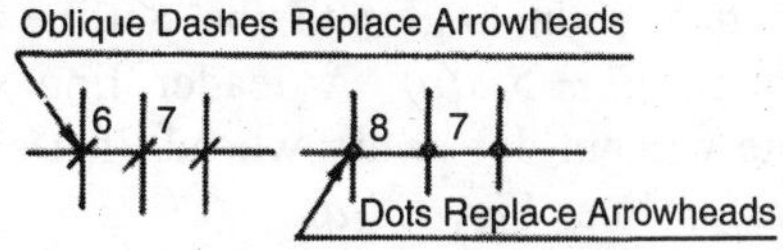

Fig. 5.6. (ii) Arrowheads replaced by Oblique dashed and dots

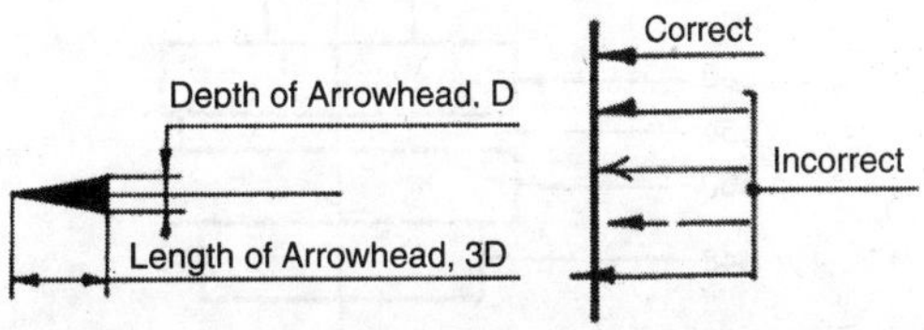

Fig. 5.6. (iii) Good and Bad practices in drawing arrowheads

5. Leaders or Pointer Lines

Leaders and pointer lines are used to link descriptive remarks to some feature, i.e., dimension, object, outline etc., of a drawing. These are thin straight lines terminated by arrowheads or dots that extend from below the note to the feature of a part to which the note applies. While arrowhead connected to a leader should always terminate at a line, *Fig. 5.8(b)*, the dot connected to a leader should be within the outline of the object, *Fig. 5.8(c)*. A leader pointing to a curve should be radial, Fig. 5.8 (*b*). Leaders should not be curved or drawn freehand. Leader may also terminate in a short horizontal bar at the bottom level of the lettering of the first or the last line of the note, *Fig. 5.1*. The main lines of most drawings are, as a rule, horizontal or vertical, so leaders are usually drawn at an angle between 30° and 60° to contrast with them, *Fig. 5.7(a)*. They should not be parallel to adjacent dimension or projection lines where confusion must arise. Two or more leaders to adjacent areas on the drawing should be drawn parallel to each other, *Fig. 5.7(b)*. Where a leader is directed to a circle or arc, its direction should be radial, *Fig. 5.7(c)*.

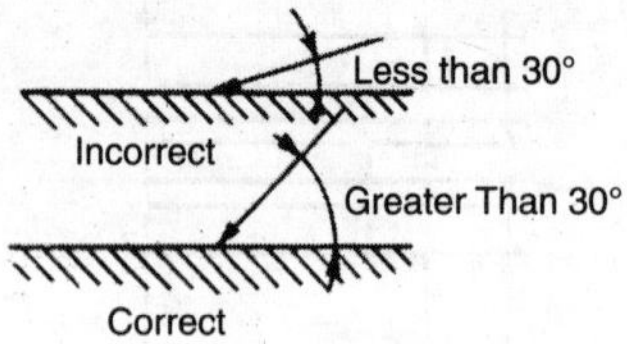

Fig. 5.7. Leader lines touching other lines

The dimension lines can sometimes be omitted and leader lines used in their place, these are positioned at right angles to the outline of the component, *Fig. 5.8(a)*. A leader line should terminate without dot or arrowhead, if it ends on a dimension line, *Fig. 5.8(d)*.

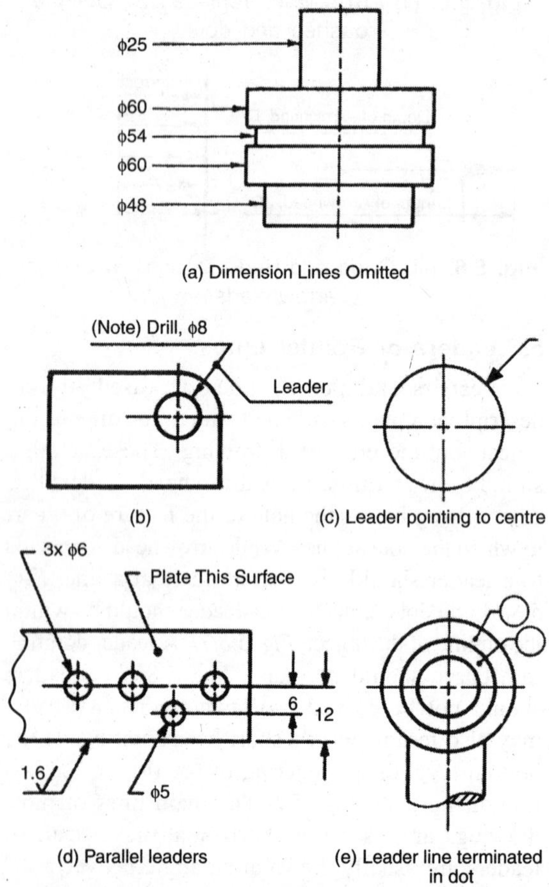

Fig. 5.8. Typical leaders

The use of common leaders should be avoided even if it means repeating the dimensions or notes, *Fig. 5.9(a)*, or using letter symbols, *Fig. 5.9(b)*.

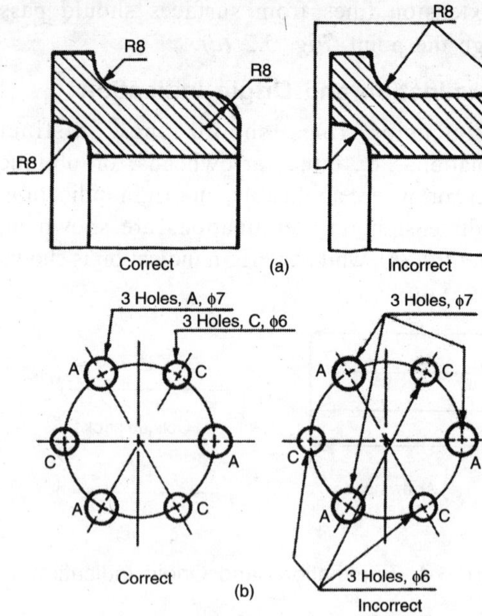

Fig. 5.9. Use of common leaders

6. Finished or End Product Drawing

It provides a complete description of the product in the condition in which it is to be used, the requirements essential to the function of the product being expressed directly on the drawing. The product may be a part ready for assembly or service, or a part for further processing.

7. Feature

It is the general term applied to an individual characteristic of a part, e.g., a cylindrical surface, shoulder, screw threads, slot, flat surface, profile, etc.

8. Functional Feature

It is feature which plays an essential role in the performance or serviceability of a part to which it belongs. It may be a location feature, e.g., a spigot that serves to locate a component in an assembly, or a working surface, e.g., the bore of a bearing.

9. Datum

It is a theoretically exact point, axis, or plane derived from the true geometric counterpart of a specified datum feature. A datum serves as the origin from which the location of geometric characteristics of features of a part are established.

10. Datum Feature

It is an actual feature of a part that is used to establish a datum.

11. Datum Feature Simulator

It is a surface of adequately precise form contacting the datum feature(s) and is used to establish the simulated datum(s). A surface plate, a gauge surface and a mandrel are some common examples. Simulated datums represent the datums during manufacture and inspection of parts.

12. Datum Target

It is a specified point, line, or area on a part used to establish a datum.

13. Dimension

It is a numerical value, expressed in appropriate units of measurement and indicated graphically on technical drawings with lines, symbols and notes to define the size, shape, location, geometric characteristic, or surface texture of a feature.

Each part has dimensions such as length, width, height, diameter, radius, etc. The drawing for the part is intended to specify a value for all dimensions necessary for the manufacture and inspection of the part relative to functional requirements. In its broadest sense, the term is not limited solely to geometric elements of the design but may also include other manufacturing and inspection criteria, e.g., mass, pressure, or capacity. However, such criteria are usually referred to under more general terms, such as drawing requirements.

Dimensions are classified according to the following types:

- **Functional Dimension:** It is a dimension, which is essential to the function of a product or space, F in *Fig. 5.39*. They must be expressed directly on the drawing. If this is done it results in the selection of datum features (surface or axis) on the basis of the functioning of the product.
- **Non-Functional Dimension:** It a dimension that is not essential to the function of the component or space, NF in *Fig. 5.39*.
- **Redundant and Auxiliary Dimensions:** Where an overall dimension is shown one of the intermediate distances is redundant and should not be dimensioned, *Fig. 5.10(a), (b)* and *(c)*. However, where redundant dimensions would provide useful information they may be added as auxiliary dimensions. Where all the intermediate dimensions are shown, the overall dimension should generally be given as an auxiliary dimension, *Fig. 5.39*. Auxiliary dimensions should not be toleranced, but should be included in brackets. Auxiliary dimensions should not govern machining operations or acceptability in any way and is derived from other values shown on the drawing.
- **Reference dimension:** A dimension, usually without a tolerance, used for information purposes only. This dimension is often a repeat of a given dimension or established from other values shown on the drawing. A reference dimension does not govern production or inspection. It is shown on a drawing with parentheses, e.g. (145).
- **Basic dimension:** A basic dimension is considered a theoretically perfect dimension. Basic dimensions are used to describe the theoretically exact size, profile, orientation, or location of a feature or datum target. These dimensions provide the basis where permissible

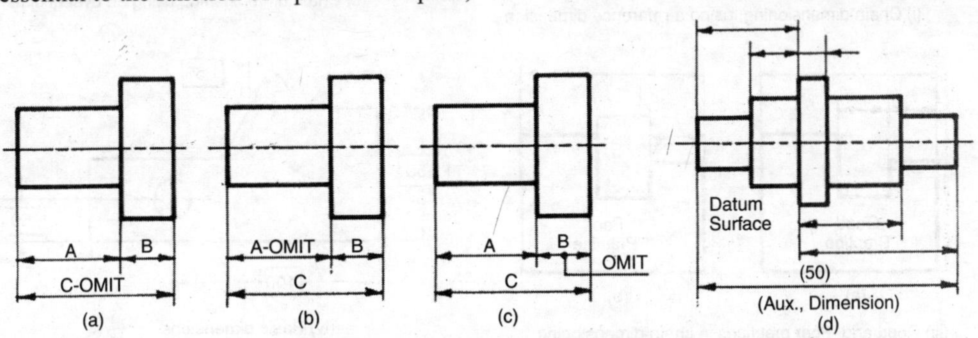

Fig. 5.10. Three possible combinations of two length dimensions (a), (b), (c) including an Auxiliary or Reference dimensions (d)

variations are established by tolerances on other dimensions, in notes, or in feature control frames. In simple terms, a basic dimension tells us where the geometric tolerance zone or datum target is located. Basic dimensions are shown on a drawing by placing a rectangle around the dimension, as shown in *Fig. 5.11 (iv)*, to indicate that the tolerances in the general tolerance note do not apply. Such dimensions may be controlled by geometric or positional tolerances or tolerances on other dimensions. They are intended to form the basis for gauges, tools, or fixtures, although it is recognized that such dimensions on tools and gauges are not exact. However, tool and gauge dimensions are made to very small tolerances, e.g., 10 times more accurate than the measured dimensions. So any variations they introduce are insignificant as compared to product tolerances. A general note may also be used to identify basic dimensions in some applications. For example, the note "*Untoleranced Dimensions Locating True Position are Basic*" indicates the dimensions that are basic.

A basic dimension symbol around a dimension is a signal to the reader to look for a geometric tolerance in a feature control frame related to the features being dimensioned.

14. Chain Dimensioning

A plan view of a twist drill stand is given in *Fig. 5.11 (i)* to illustrate chain dimensioning. Basic dimensions should not be confused with basic size.

Each of the dimensions in the chain would be subject to a manufacturing tolerance since it is not possible to mark out and drill each of the centre distances exactly. The small errors, which creep in, accumulate and the overall total of six dimensions in the chain is generally different from the auxiliary dimension shown in parenthesis (157). Chains of dimensions may be used, *Fig. 5.11 (i)*, *(ii)* provided the possible accumulation of tolerances does not endanger the functional requirements of the part. Good and bad practices in chain dimensioning are also shown in *Fig. 5.11 (iii)*.

15. Datum Dimensioning

Improved positional accuracy is obtained by dimensioning from a datum, *Fig. 5.12(a), (b)*. The datum selected is the left hand side of the drill stand. Here the overall length is not now shown as an auxiliary dimension, but as a fixed distance, which can be separately marked out, manufactured

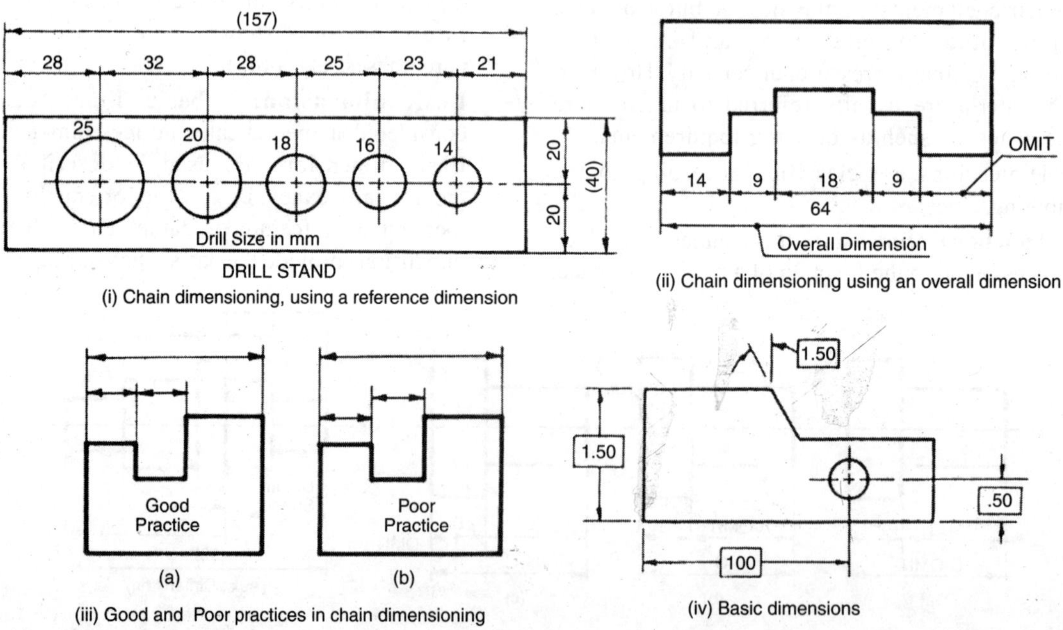

(i) Chain dimensioning, using a reference dimension

(ii) Chain dimensioning using an overall dimension

(iii) Good and Poor practices in chain dimensioning

(iv) Basic dimensions

Fig. 5.11.

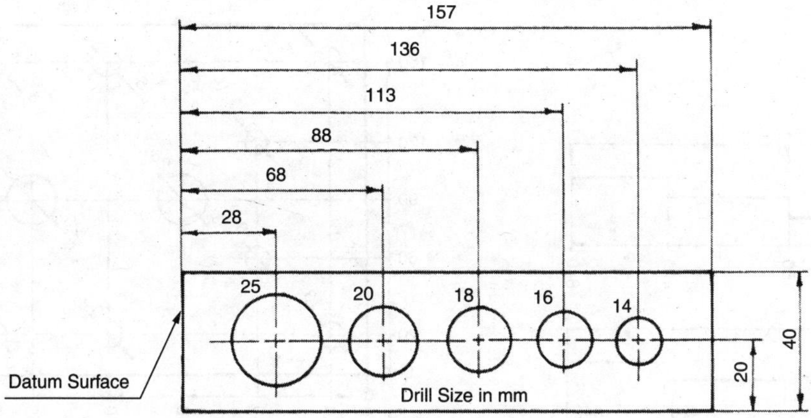

(a) Parallel dimensioning using a Datum

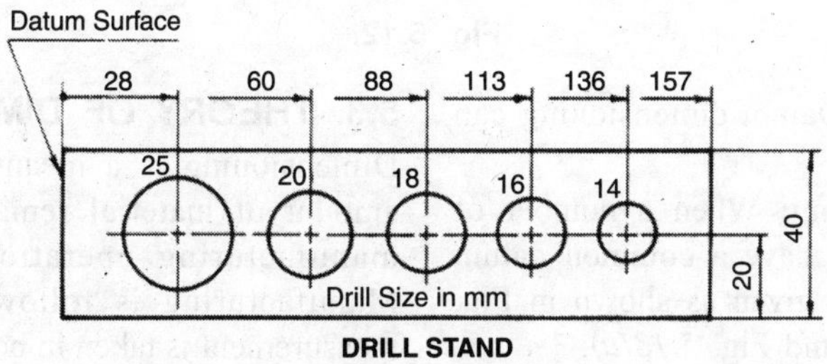

(b) Progressive dimensioning using datum in one direction

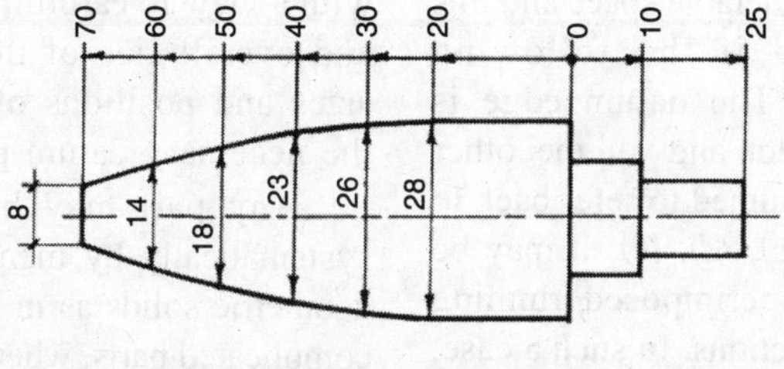

(c) Progressive dimensioning using datum in two directions

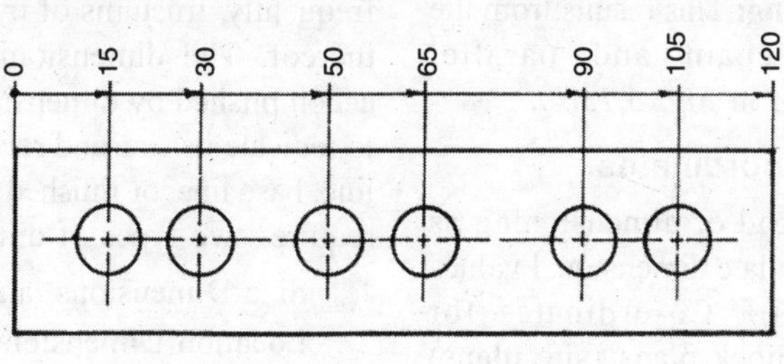

(d) Progressive dimensioning using origin

Fig. 5.12.

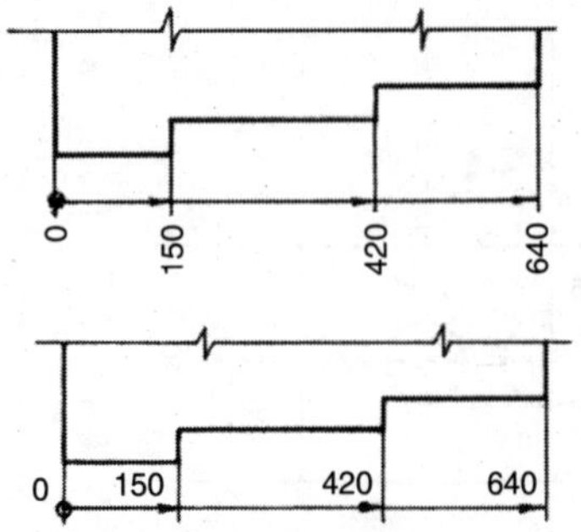

(e) Progressive or Super imposed running dimension

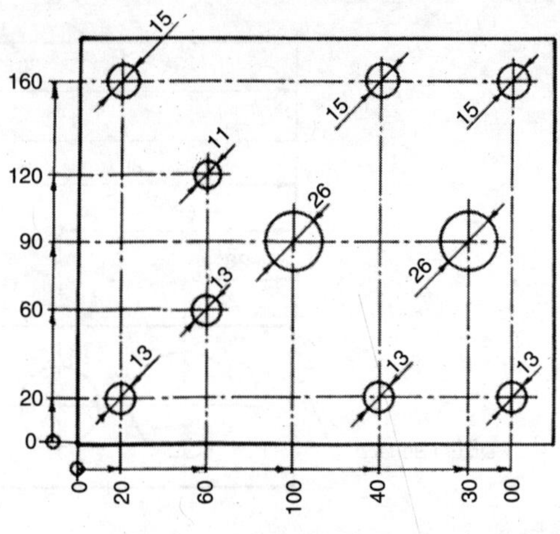

(f) Progressive dimensioning using origin for coordinates

Fig. 5.12.

and definitely checked. Datum dimensioning can be accomplished as:

- **Parallel Dimensioning:** When a number of dimensions of a part have a common datum feature these may be given as shown in Fig. *5.10(d)* Fig. *5.12(a)* and Fig. *5.13(a)*.
- **Progressive or Superimposed Running Dimensioning:** Now the spacing of the six dimensions takes up considerable space and this space can be reduced if the following convention is adopted. The datum edge is indicated by the large dot and all the other dimensions in line are assumed to refer back to this edge, Fig. *5.12(b)*, *(c)*, *(d)*, *(e)*. It may be advantageous to use superimposed running dimensioning in two directions. In such a case, the origins may be as shown in *Fig. 5.12(f)*.
- **Combined Dimensioning:** This results from the combined use of chain and parallel dimensioning as shown in *Fig. 5.13(b)*.

16. Dimensioning by Coordinates

It may be useful, instead of dimensioning as shown in *Fig. 5.12(f)*, to tabulate dimensional values as shown in *Fig. 5.14(a)*. Co-ordinates for intersections in grids on block plans (site plans) are indicated as shown in *Fig. 5.14(b)*. The coordinates for arbitrary points of reference without a grid shall appear adjacent to each point, *Fig. 5.14(c)* or in a tabular form, *Fig. 5.14(d)*.

5.3. THEORY OF DIMENSIONING

Dimensioning is a means of defining the exact amount of material remaining after a series of manufacturing operations have taken place. Manufacturing is followed by inspection and measurement is taken in both cases from functional axes or planes, known as datums.

Dimensioning, therefore, should be undertaken with a view to carefully defining the shape or form and overall size of the component, and also the sizes and positions of the various features from the necessary datum planes or axes.

Any part may be dimensioned easily and systematically by mentally dividing it into simple geometric solids as in Fig. 5.64 on page 131. Even complicated parts, when analyzed, usually, are found to be composed principally of cylinders, prisms, and, frequently, frustums of pyramids and cones or parts thereof. The dimensioning of an object may be accomplished by dimensioning each elemental form to indicate its size and relative location from a centre line, base line, or finished surface. A drawing usually requires two types of dimensions:

1. Size Dimensions, and
2. Location Dimensions.

Figure 5.15 shows a drilled plate. In addition to the size dimensions of the drilled hole, dia. or ϕ and length, we have to state the location of its axis from edges of the plate.

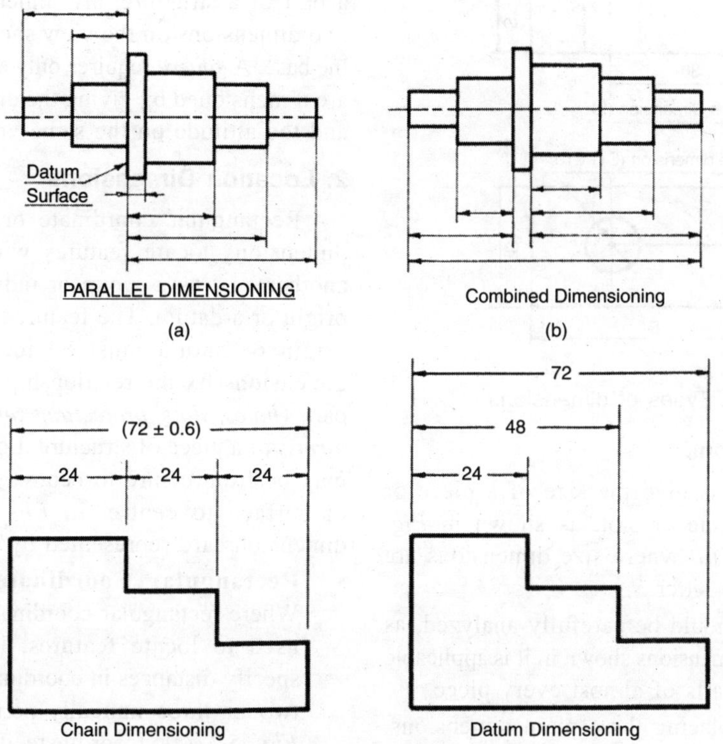

Unspecified Tolerances are ± 0.2

(c) Difference between chain and Datum dimensioning

Fig. 5.13.

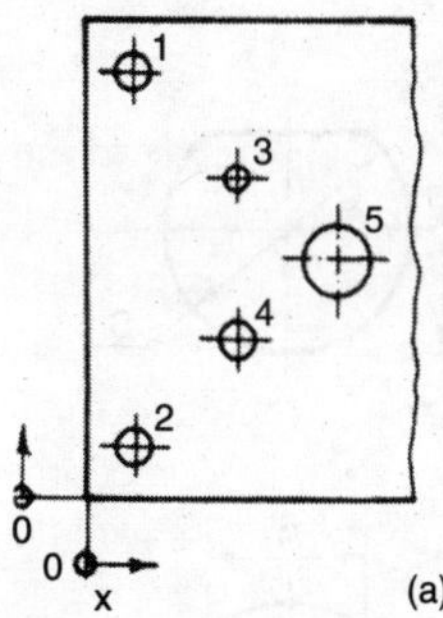

	X	Y	φ
1	20	160	15
2	20	20	15
3	60	120	9
4	60	60	15
5	100	90	15
6			
7			
8			

(a)

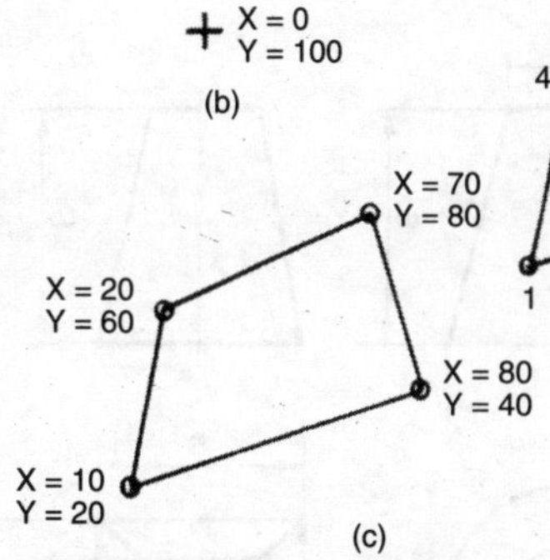

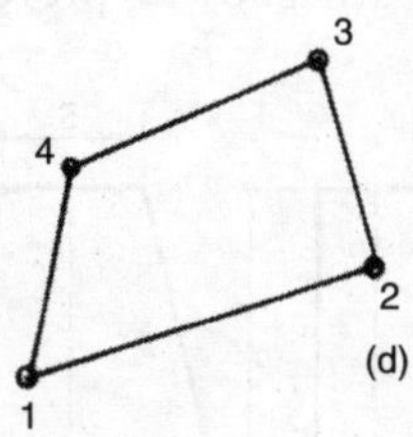

	X	Y
1	10	20
2	80	40
3	70	80
4	20	60

(d)

Fig. 5.14. Co-ordinate dimensioning

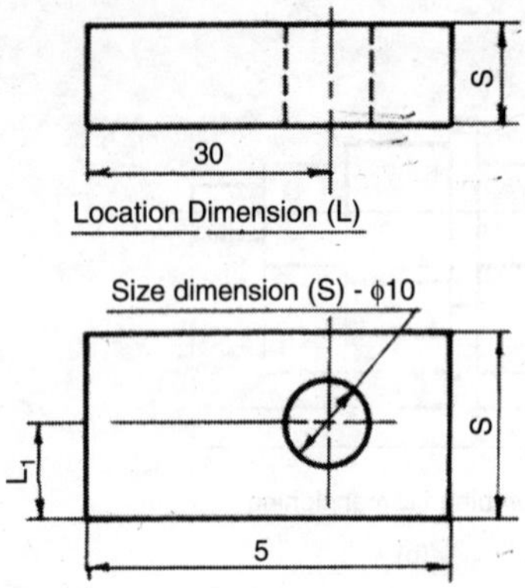

Fig. 5.15. Types of dimensions

1. Size Dimensions

Size dimensions give the size of a piece or component part, hole or slot, as shown in *Fig. 5.16* and *Fig. 5.17(a)* where size dimensions are represented by the letter *S*.

Figure 5.16 should be carefully analyzed, as the placement of dimensions shown in it is applicable to the elemental parts of almost every piece.

The rule for placing three main dimensions, (width, height, and depth) on the drawing of a prism is that two dimensions be given on a principal view and the third dimension on one of the other views. The circular *cylinder*, which appears as a bar or a shaft, requires only the diameter and length, both of which are shown, preferably, on the rectangular view. *Pyramids*, which frequently form a part of a structure, are dimensioned by giving two dimensions on the view showing the shape of the base. A *sphere* requires only the diameter. *Cones* are dimensioned by giving the diameter of the base and the altitude on the same view.

2. Location Dimensions

Rectangular coordinate or polar coordinate dimensions locate features with respect to one another, and as a group or individually, from an origin or a datum. The features that establish this origin or datum must be identified. Location dimensions fix the relationship of the component parts (*holes, slots, projections and other significant forms*) of a piece of structure. Location dimensions may be from centre to centre, surface to surface or surface to centre. In *Fig. 5.17(b)* location dimensions are represented by letter L.

- **Rectangular Coordinate Dimensioning:** Where rectangular coordinate dimensioning is used to locate features, linear dimensions specify distances in coordinate directions from two or three mutually perpendicular planes, *Fig. 5.17(c)*. Coordinate dimensioning must clearly indicate which features of the part establish these planes.
- **Rectangular Coordinate Dimensioning Without Dimension Lines:** Dimensions may be shown on extension lines without the use of

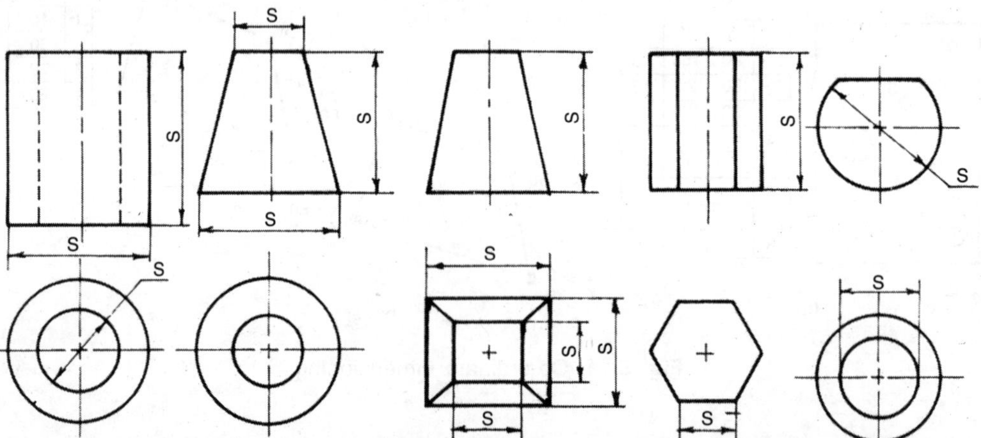

METHOD OF DIMENSIONING GEOMETRICAL SHAPES

Fig. 5.16.

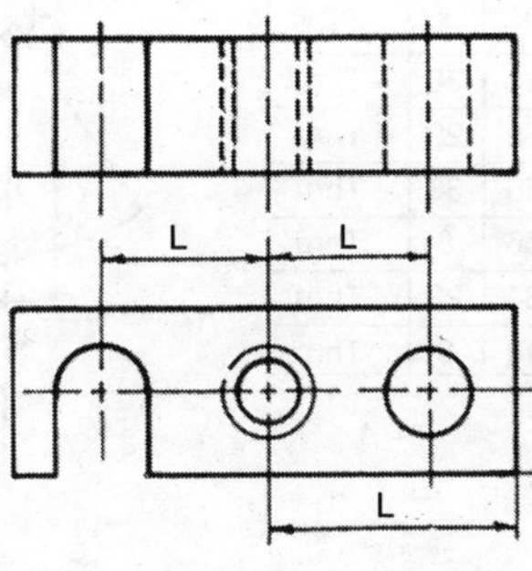

Fig. 5.17. (a) Location Dimensions (L)

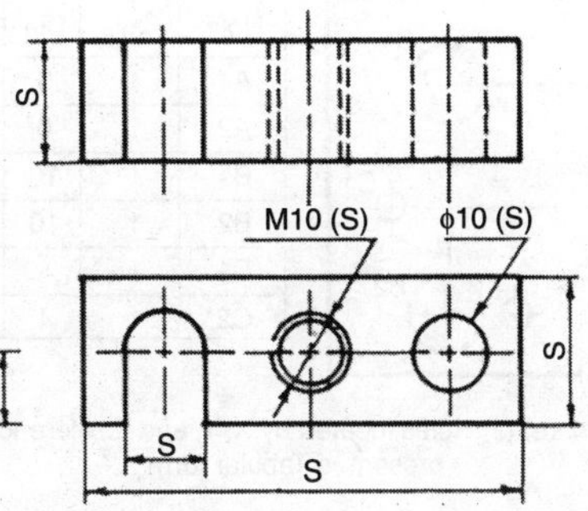

Fig. 5.17. (b) Size Dimensions (s)

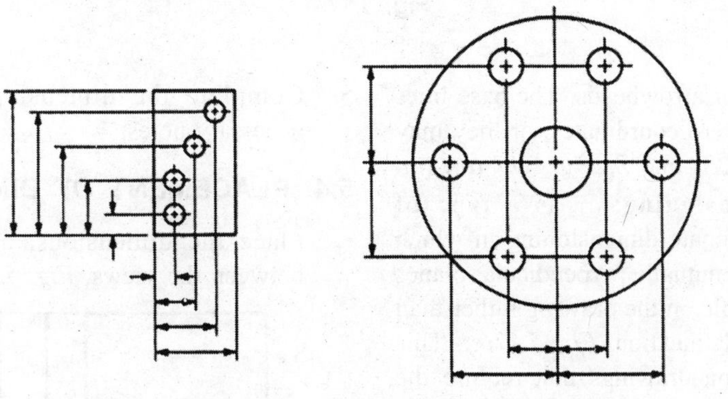

Fig. 5.17. (c) Rectangular Co-ordinate Dimensioning

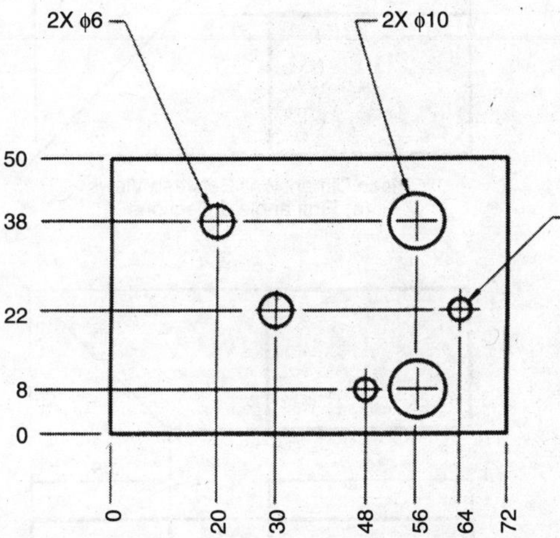

Fig. 5.17. (d) Rectangular Co-ordinate Dimensioning without dimension lines and arrowheads

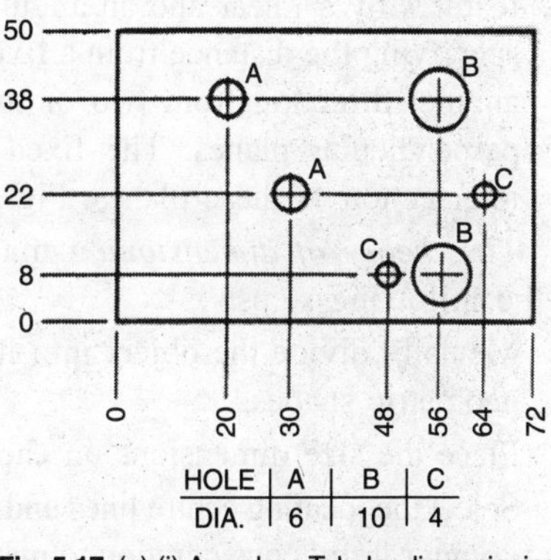

HOLE	A	B	C
DIA.	6	10	4

Fig. 5.17. (e) (i) Arrowless Tabular dimensioning, Holes are referenced by letters

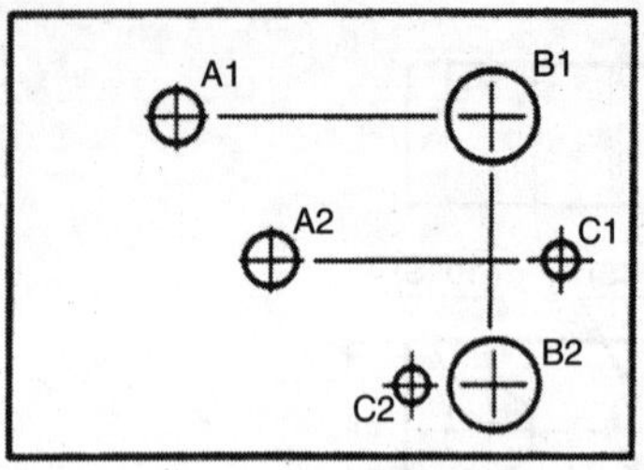

Hole	Qty	Dia	X	Y	Z
A1	1	6	20	38	Thru
A2	1	6	30	22	Thru
B1	1	10	56	38	Thru
B2	1	10	56	8	Thru
C1	1	4	64	22	Thru
C2	1	4	48	8	Thru

5.17 (e) (ii) Holes located by X, Y, and Z references presented Tabular form

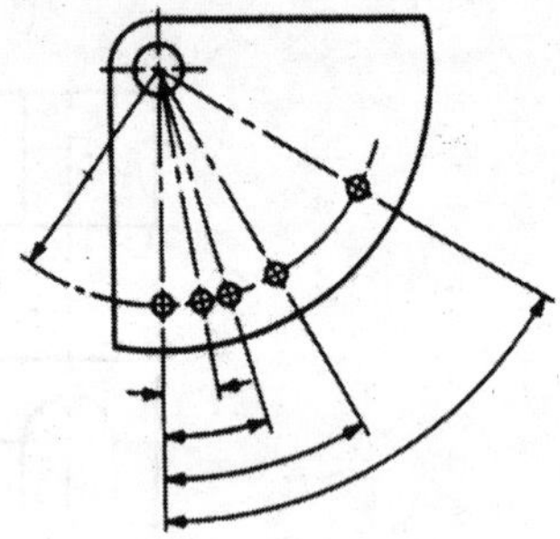

5.17 (f) Polar Co-ordinate dimensioning

Fig. 5.17.

dimension lines or arrowheads. The base lines are indicated as zero coordinates, or they may be labelled as, *X, Y,* and *Z, Fig. 5.17 (d), (e).*

- **Tabular Dimensioning:** It as a type of rectangular coordinate dimensioning in which dimensions from mutually perpendicular planes are listed in a Table on the drawing rather than on the pictorial delineation, *Fig. 5.17(e).* This method is used on drawings that require the location of a large number of similarly shaped features. Table may be prepared in any suitable manner that adequately locates the features.
- **Polar Coordinate Dimensioning:** Polar coordinate dimensioning is used to locate features by a linear and an angular dimension specifying the distance from a fixed point at an angular direction from two or three mutually perpendicular planes. The fixed point is the intersection of these planes, Fig. *5.17(f).*

The *theory of dimensioning* may be applied in the following steps:

1. Mentally divide the object into its component geometric shapes.
2. Place the size dimensions on each form.
3. Select the locating centre lines and surfaces after giving careful consideration to mating parts and to the process of manufacture.
4. Place the location dimensions so that each geometric form is located from a centre line or a finished surface.
5. Complete the dimensions by adding the necessary notes.

5.4. PLACEMENT OF DIMENSIONS

- Place the dimensions, as far as possible, between the views, *Fig. 5.18.*

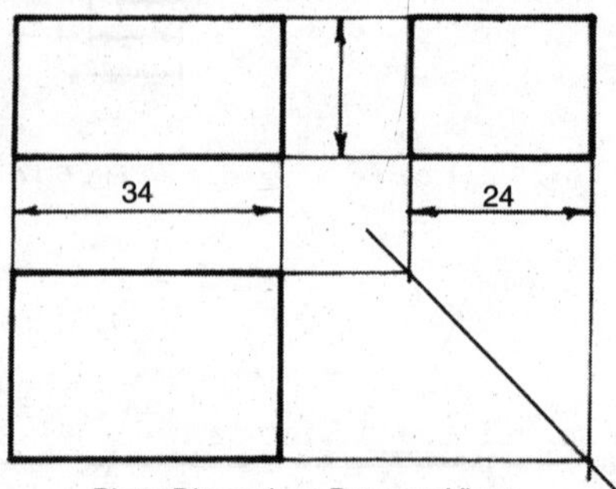

Place Dimensions Between Views
(a) First angle projections

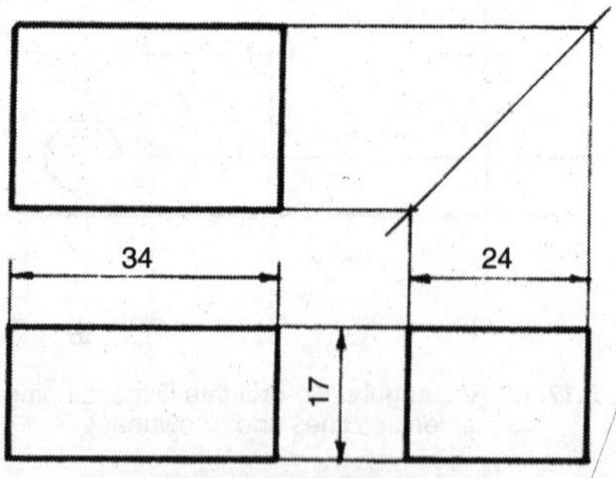

Place Dimensions Between Views
(b) Third angle projections

Fig. 5.18. Placement of Dimensions

- Dimensions must be placed where they will be most easily understood. They generally are attached to the view that shows the contour of the feature to which they apply, *Fig. 5.16*.
- The dimensions shall be placed outside a view, *Fig. 5.20*. When directness of application makes it desirable or where extension lines or leader lines would be excessively long, dimensions may be placed within the outline of a view.

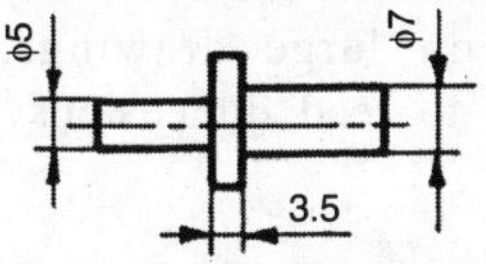

Fig. 5.19. Dimensions numerals placed beyond arrowheads

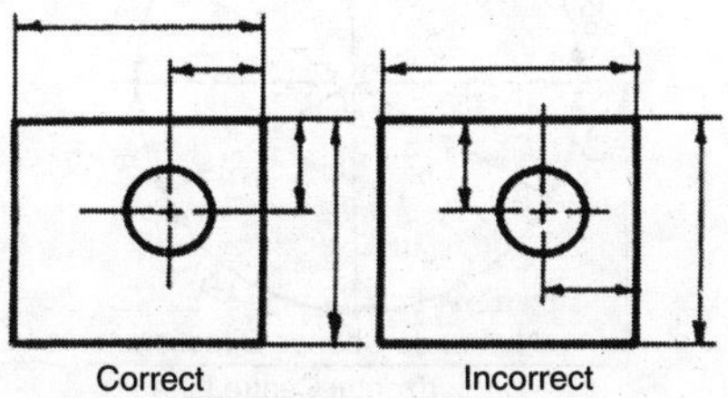

Fig. 5.20. Placement of dimensions (contd.)

- Never crowd the dimensions.
- Place the dimension lines not less than *12 mm* from the view and at least *10 mm* from each other unless availability of space restricts it.
- Dimensioning in small spaces calls for special care to ensure clearness of figures and arrowheads to avoid lack of legibility through crowding figures in to small recesses, etc. If unavoidable, due to lack of space, the dimension numerals may be placed above the extended portion of the dimension line, beyond the arrowheads, but preferably on right hand side, *Fig. 5.19*.
- Place smallest dimensions nearest the view being dimensioned, *Fig. 5.21*.
- Notes should be placed so as to be read from the bottom of the drawing with regard to the orientation of the drawing format.
- The two recommended systems of placing the dimensions are:

  (*a*) Aligned System, and

  (*b*) Unidirectional System.

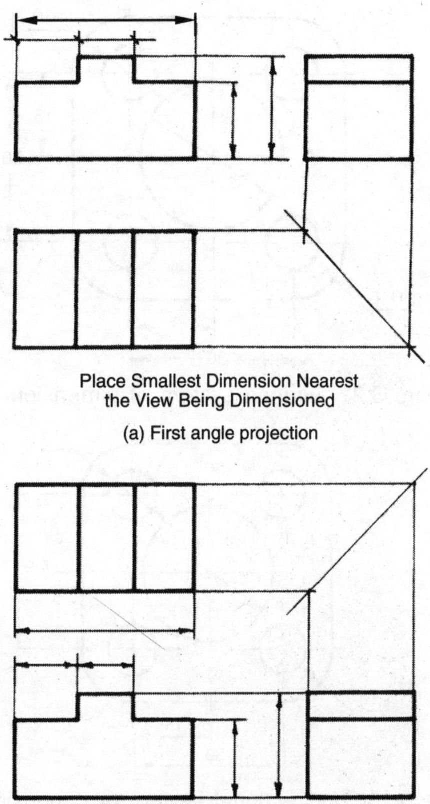

Place Smallest Dimension Nearest the View Being Dimensioned

(a) First angle projection

Place Smallest Dimension Nearest the View Being Dimensioned

(b) Third angle projection

Fig. 5.21. Placement of Dimensions

1. Aligned System

It is a system most generally used. In this, numerals for the dimension values are so placed that they may be read from the bottom or right hand edge of the drawing, *Fig. 5.22*. All dimensions are placed above the dimension lines.

An aligned expression is placed along and in the direction of the dimension line. It follows that figures should not be printed to face the left hand side of the top of the drawing. However, as shown in *Fig. 5.25(a)*, it is acceptable practice to let figures face leftwards when dimensioning angles awkwardly oriented.

Dimension lines should not run in direction included in the hatched areas, shown in *Fig. 5.26*, except in unavoidable cases.

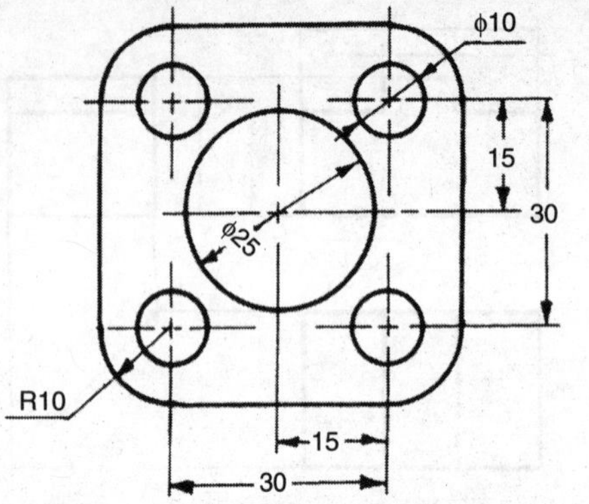

Fig. 5.22. Aligned system of dimensioning

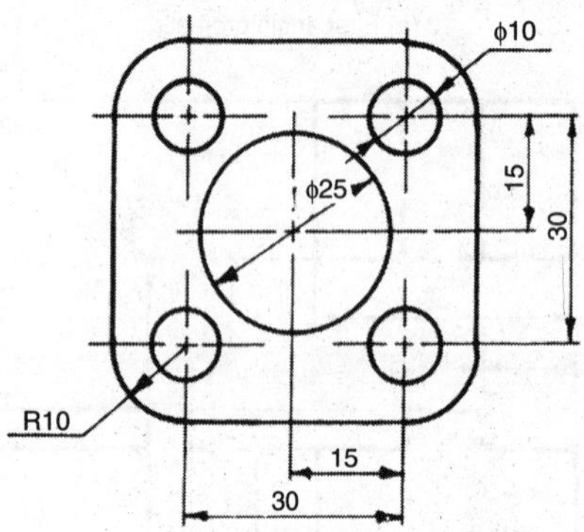

Fig. 5.23. Undirectional system of dimensioning

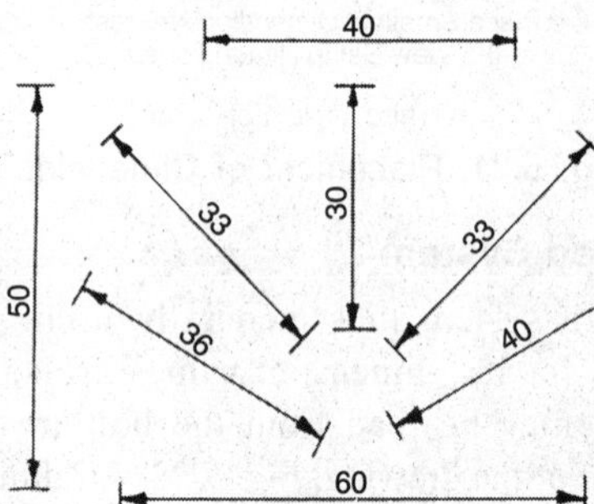

(a) Aligned system

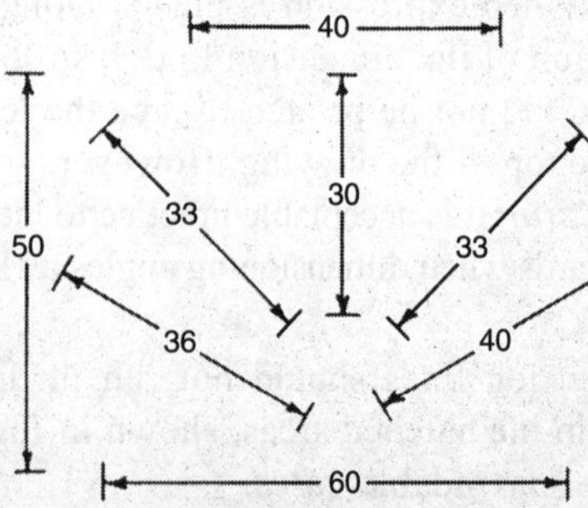

(b) Unidirectional system

Fig. 5.24. Reading horizontal, vertical, oblique dimensions

Oblique dimensions are readable from directions shown in *Fig. 5.24(a)*.

2. Unidirectional System

In this system, all the numerals for dimension values are so placed that they can be read from the bottom edge of the drawing sheet, *Fig. 5.23*, *Fig. 5.24(b)* and *Fig. 5.25(b)*. In this system there is no restriction controlling the direction of dimension lines. This system may be used with advantage, on large drawings, where it is inconvenient to read dimensions from the right hand side.

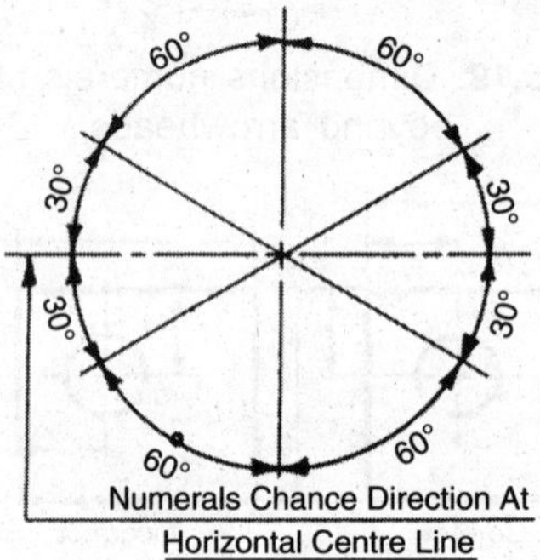

(a)

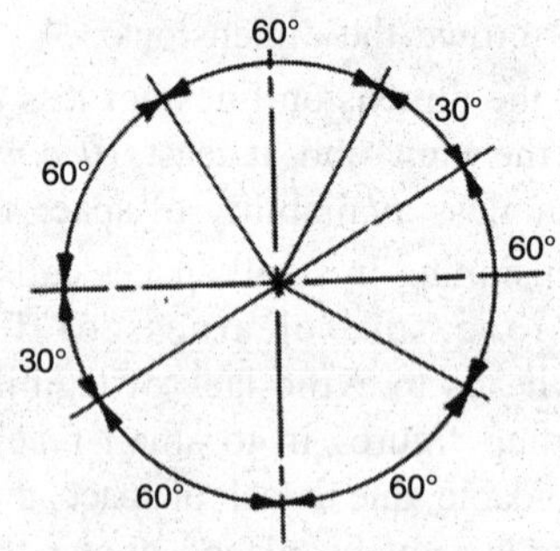

(b)

Fig. 5.25.

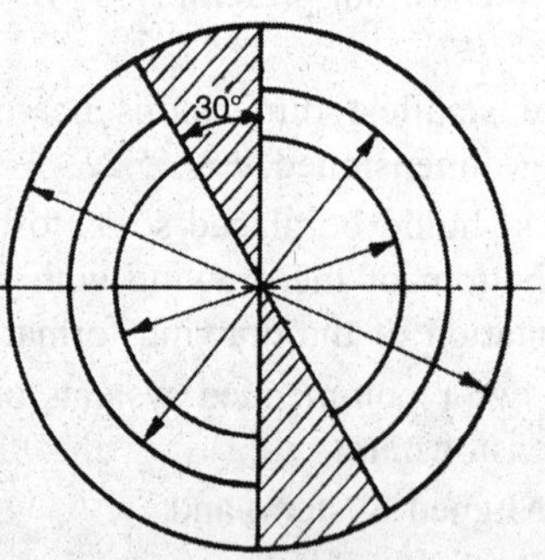

Fig. 5.26. Avoid hatched area while giving dimensions

5.5. UNITS OF DIMENSIONING

1. **SI linear units:** In the *SI system* linear dimensions, as far as possible, are expressed in one unit only; preferably in millimeters. The symbol of the unit, mm, may then be omitted provided a distinct note is added on the drawing stating the unit in which all the dimensions of the drawing are expressed. Where other units have to be shown as part of the drawing specification, e.g., *Nm* for torque or KPa for pressure, the appropriate unit symbol should then be shown with the value.

The following points should be observed when specifying millimeter dimensions on drawing:

- The decimal points should be bold, uniform and large enough to be clearly visible and should be in line with the bottom line of the figures, *Fig. 5.27.*
- When the dimension is less than unity, the zero should precede the decimal point, e.g., 0.52.
- Zero may be omitted if the decimal point is preceded by a symbol, e.g., 45.553 + .76.
- Where the dimension is a whole number, neither the decimal point nor a zero is shown.
- Where the dimension exceeds a whole number by a decimal fraction of one millimeter, the last digit to the right of the decimal point is not followed by a zero. This practice differs for tolerances expressed bilaterally or as limits.
- Neither commas nor spaces should be used to separate digits into groups in specifying millimeter dimensions on drawings.

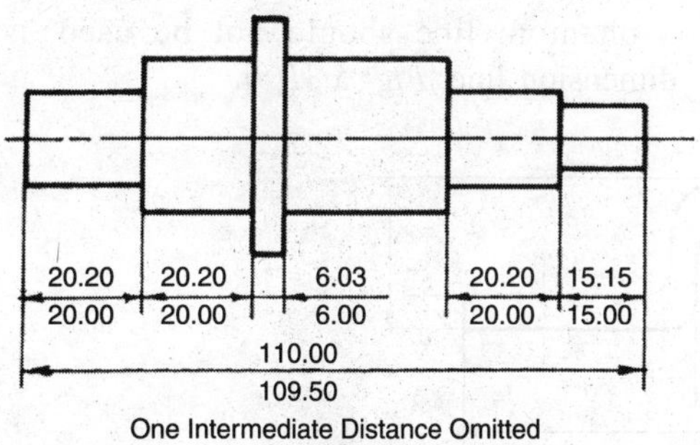

Fig. 5.27.

2. **Angular units:** Angular dimensions are expressed in either degrees or decimal parts of a degree or in degrees, minutes, and seconds. The latter dimensions are expressed by symbols: for degree °, for minutes ′, and for seconds ″. Where degrees are indicated alone, the numerical value should be followed by the symbol. When only minutes or seconds are specified, the number of minutes or seconds should be preceded by 0° or *0°0′*, as applicable, *Fig. 5.37(a).*

5.6. IMPORTANT DIMENSIONING RULES

1. Every dimension necessary for the complete definition of a finished work part, together with any other information necessary to define the part completely, should be expressed directly on the drawing.
2. Each dimension has a tolerance except for dimensions specifically identified as reference, maximum, minimum, or stock.
3. Dimensioning and tolerancing must be complete to the extent that there is full understanding of the characteristics of each feature. Neither measuring the drawing or assumption of a dimension is permitted. Exceptions include drawings such as printed wiring, templates, and master layouts prepared on a stable material. However, in these cases the necessary control dimensions must be given.
4. Dimensions must be selected and arranged to suit the function and mating relationship of a part. Dimensions should not be subject to more than one interpretation.
5. Bearing in mind the size of the actual dimensions, and the fact that there may be two numbers together, where limits of size are quoted, adequate space must be left between rows of dimensions. A spacing of about 12 mm is recommended, *Fig. 5.27.* Try and ensure that similar spacing is kept between dimension lines as this gives a neat appearance to the completed drawing.
6. It should not be necessary for any functional dimension to be deduced from other dimensions or for the drawing to be scaled.
7. There should not be more dimensions than are necessary to define the product nor should any feature be located by more than one toleranced dimension in any one direction. When an overall dimension is used, dimensioning of one of the intermediate distances may be avoided,

Fig. 5.27, except when they are required to serve the purpose as an auxiliary dimension.

8. Dimensions should be arranged to provide required information and optimum readability. Start by dimensioning the view, which gives the clearest understanding of the profile or shape of the component being dimensioned.
9. The dimensions should be quoted once only. Dimensions given in one view should not be repeated in other views, except for reference purposes.
10. Dimensions should not be placed upon a view unless the drawing becomes clearer by doing so. In general if space permits, dimensions should be placed outside the views, where possible, and between the views, as shown in *Fig. 5.20*. Also the overall dimensions which are given for surfaces that can be seen in two projected views of a component are generally best positioned between these two views.

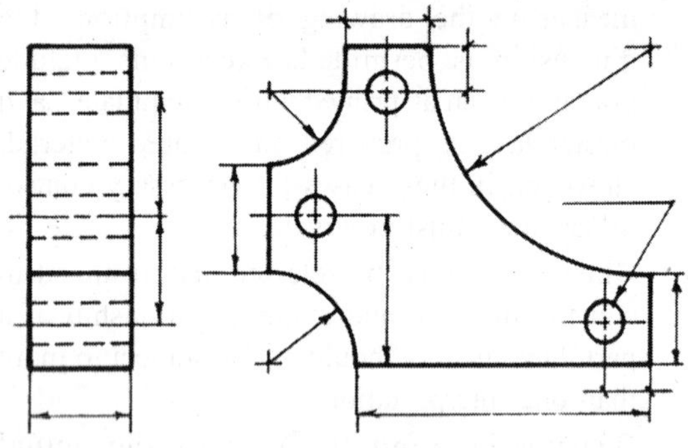

Fig. 5.28. Dimension placed on the view best showing the fact of the dimension

11. A dimension should be placed on a view best showing the effect of the dimension, *Fig. 5.28*.
12. Dimensions should be taken from visible outlines rather than from invisible lines, *Fig. 5.29*.

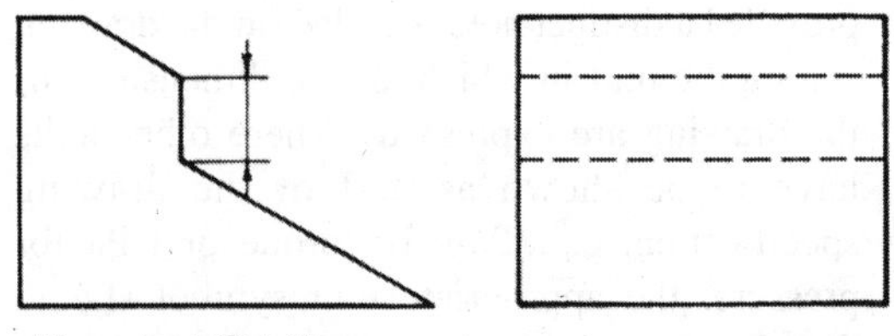

Fig. 5.29. Take dimension from visible lines

13. Dimensions for different operations, on a work part, e.g., drilling and bending, should be kept separate if permitted by its design, *Fig. 5.30*.
14. The crossing of dimension lines should be avoided as far as possible. If, however, the intersection of two dimension lines is necessary, the lines should not be broken. Dimension lines are broken only to insert the dimension figures in the unidirectional system and not in the aligned system.
15. Also intersecting projection and dimension lines should be avoided. Where unavoidable, however, neither line should be shown with a break, *Fig. 5.31(a)*.
16. The dimension line should not cut a dimension figure. The dimension values should be placed in such a way that they are not crossed or separated by any other line on the drawing.
17. An outline, a centre line, an extension line, or a phantom line should not be used as a dimension line, *Fig. 5.31(b)*.

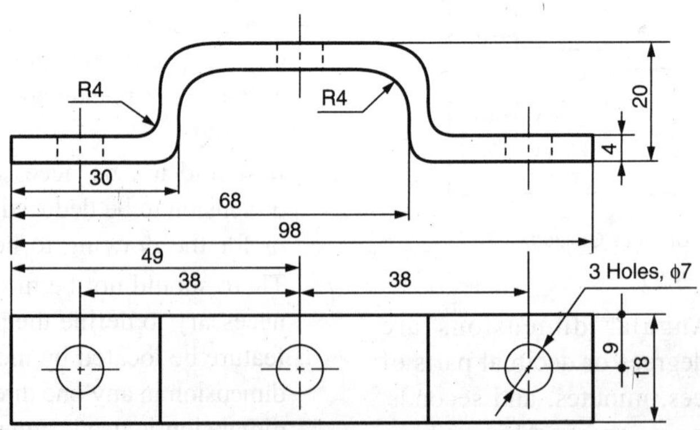

Fig. 5.30. Dimensions for different operation be kept separate

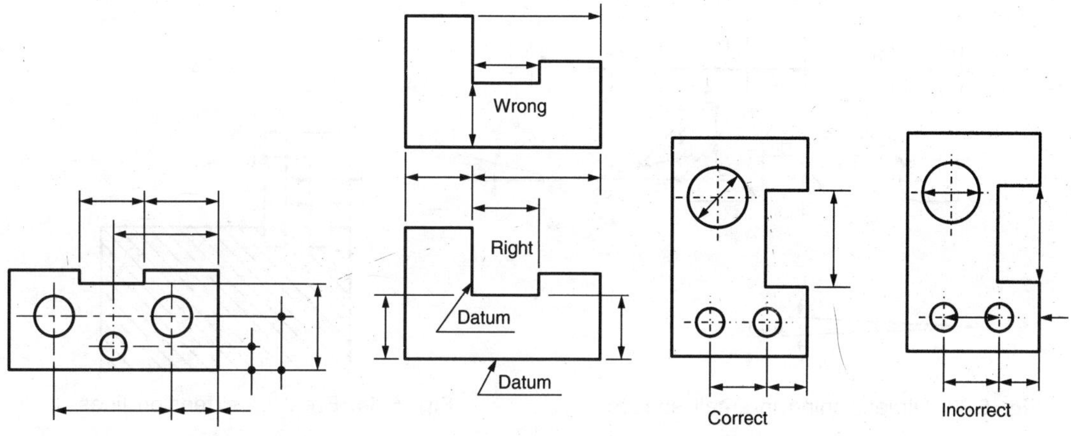

(a) Dimension lines should not be broken except for inserting numerals

(b) Correct and Incorrect practices in dimensioning

(c)

Fig. 5.31.

18. Also a dimension line should not be in line with a boundary line, *Fig. 5.31(c)*.
19. When there are several parallel dimension lines, they should be equally spaced and the figures should be staggered for clarity, as shown in *Fig. 5.32 (a), (b)*.
20. Never crowd dimensions into small spaces. Here the arrowheads should be reversed and methods shown in *Fig. 5.33* should be used. Adjacent arrowheads may be replaced by clearly marked dots, *Fig. 5.6*.
21. Locate dimension lines so that they do not cross extension lines. But if it is not possible, extension lines should be broken where they pass a dimension line, *Fig. 5.34*. Place longer dimensions outside the shorter ones so that extension lines do not cross dimension lines, *Figs. 5.1, 12, 13,* and *21*.
22. Always give locating dimensions to the centre of circles that represent holes, cylindrical projections, and bosses etc., *Figs. 5.17, 5.19, 5.35*.
23. Avoid superfluous dimensions, *Fig. 5.35*. Dimensions written as "NO" are superfluous.
24. When dimension figures appear on a sectional view, a small portion around the figures should be left unhatched, *Fig. 5.36*. This is achieved by drawing section lines after giving dimensions.
25. When an arc is used as a dimension line, for dimensioning angles, use vertex of the angle as centre and place dimension figures so that they

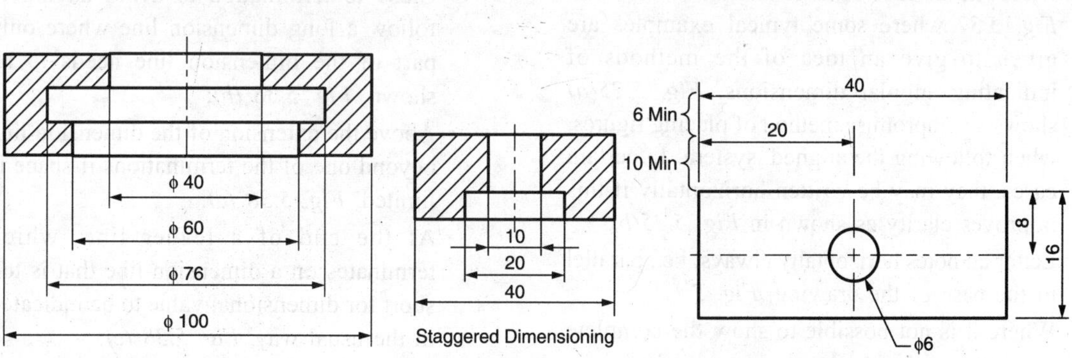

(a) Staggered dimensioning

(b) Value staggered

(c) Distance between dimension lines

Fig. 5.32.

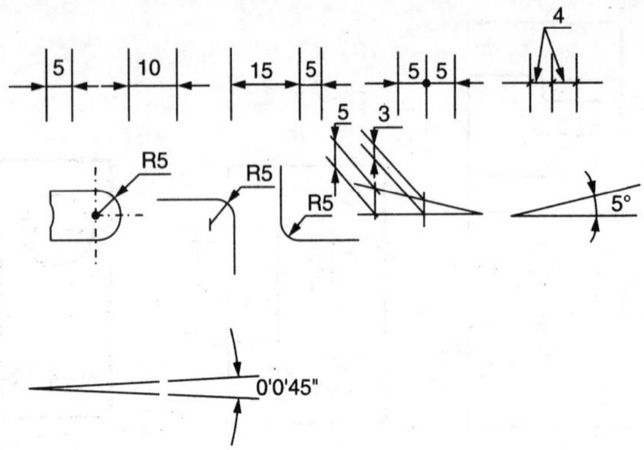

Fig. 5.33. Dimensioning in small spaces

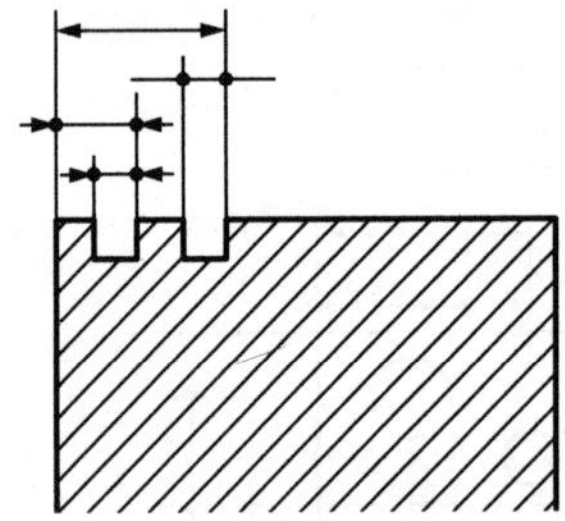

Fig. 5.34. Break in extension lines

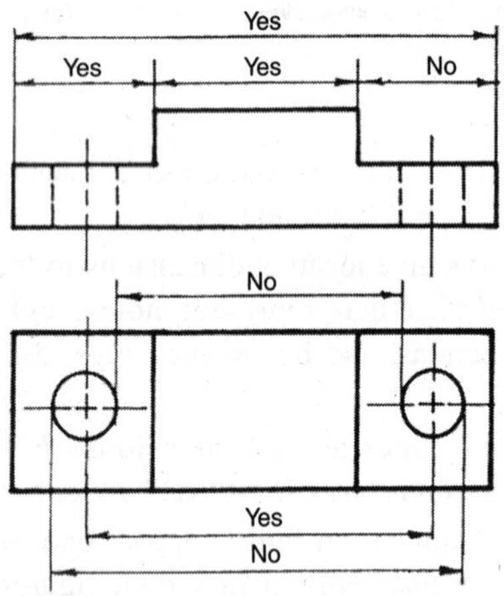

Fig. 5.35. Give locating dimensions for holes to their centres and avoid superfluous dimensions

Fig. 5.36. Figures and numerals clear of section lines

will read from the bottom of the drawing, except in the case of large angles, as shown in *Fig. 5.37* where some typical examples are given to give an idea of the methods of indicating angular dimensions. *Fig. 5.25(a)* shows a compromise method of placing figures, when following the aligned .system. In certain cases, they may be written horizontally if this improves clarity as shown in *Fig. 5.25(b)*.

26. Letter all notes horizontally always, i.e., parallel to the base of the drawing, *Fig. 5.1*.
27. Where it is not possible to show the complete outline, with in the scope of drawing, dimensioning is done as shown in *Fig. 5.38*

(a). The dimensional values can be given;

- Close to termination to avoid having to follow a long dimension line where only part of the dimension line needs to be shown, *Fig. 5.38 (b)*;
- Above the extension of the dimension line beyond one of the terminations if space is limited, *Fig. 5.38 (c)*;
- At the end of a leader line, which terminates on a dimension line that is too short for dimensional value to be indicated in the usual way, *Fig. 5.38 (c)*;
- Above a horizontal extension of a dimension line where space does not allow

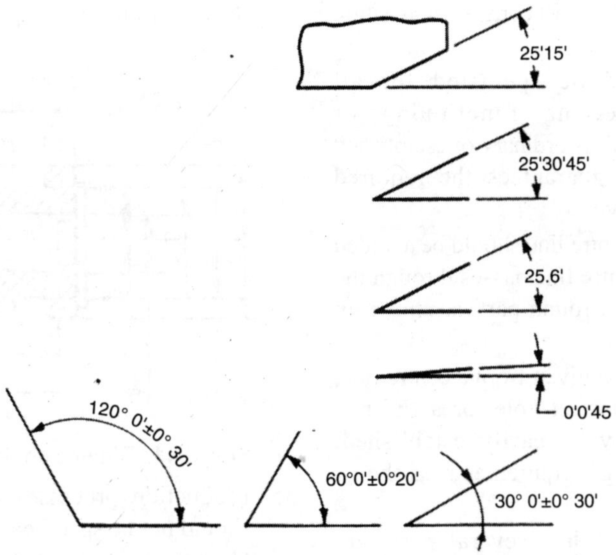

Fig. 5.37. Dimensioning Angles

placement at the interruption of a non-horizontal dimension line, *Fig. 5.38 (d).*

28. When giving limiting dimensions, the dimension line is not broken. Instead one dimension is written above and the other below the dimension line, *Fig. 5.27.*

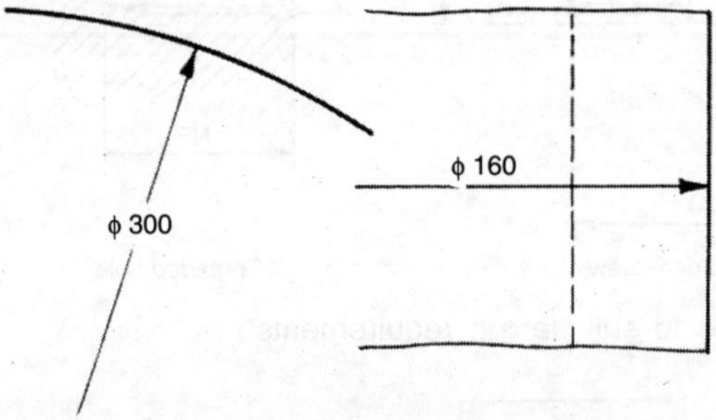

(a) Dimensioning where complete outline is not shown

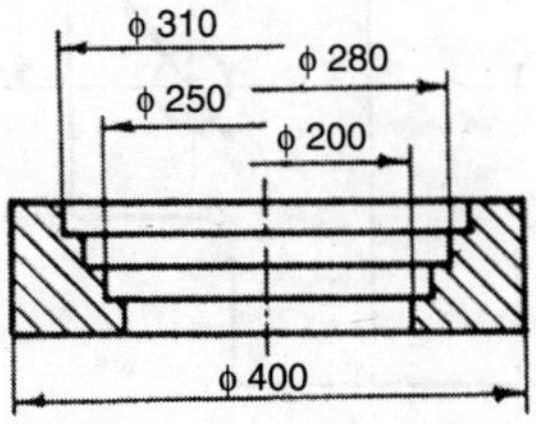

(b) Show dimension close to termination

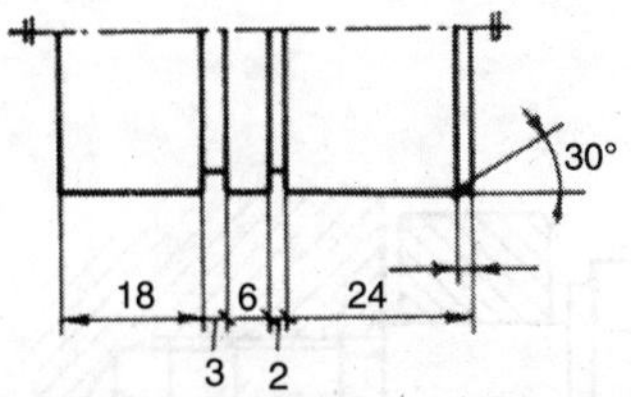

(c) Dimension value put at the end of leader line

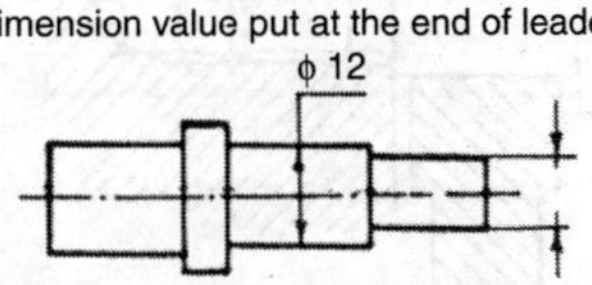

(d) Value above a horizontal extension of a dimension line

Fig. 5.38. Dimensioning in special situations

29. Dimensions which are not to scale, as when a dimension on a drawing is changed, should be clearly indicated by the addition of the letters 'NTS' or by underlining, *Fig. 5.38 (e)*, giving indications of the changes in the revision panel. If a number of dimensions are changed, the words "NOT TO SCALE" should be substituted for the usual statement of scale, or lettered on the drawing, in a prominent place.

30. The functional dimensions should be expressed directly on the drawing, *Fig. 5.39*. The application of this principle will result in the selection of datum features on the basis of the function of the product, and the method of locating it in any assembly of which it may

form a part. The non-functional dimension should be placed in the most convenient position to read.

31. Tolerances should be specified for all requirements affecting functioning or interchangeability unless ordinary or established workshop practice guarantees the required standard of accuracy.

32. Dimensioning to a centre line should be avoided except where the centre line passes through the centre of a hole, or a round part, as shown in *Fig. 5.40*.

33. Dimensions should be given from a centre line, a base line, an important hole, or a finished surface, which may be easily established, depending upon design requirements, as shown in *Fig. 5.39*.

34. **Assembled Parts:** When several parts are drawn in an assembly, the groups of dimensions related to each part should be kept separate, as much as possible, when individual part drawings are not prepared, *Fig. 5.41*.

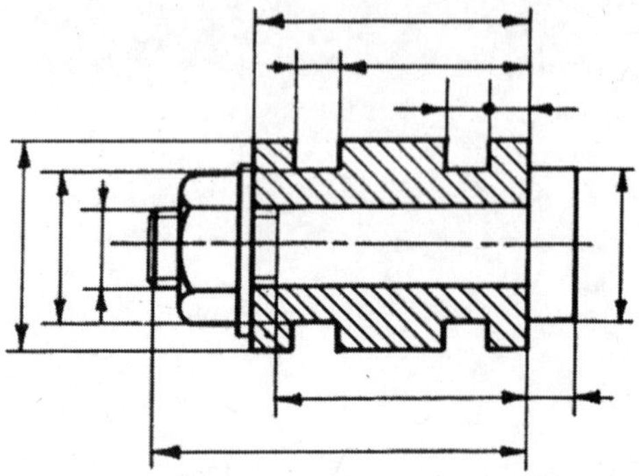

Fig. 5.41. Dimensioning Assembled Parts

35. Production processes or inspection methods should not be specified unless they are essential to ensure satisfactory functioning and interchangeability.

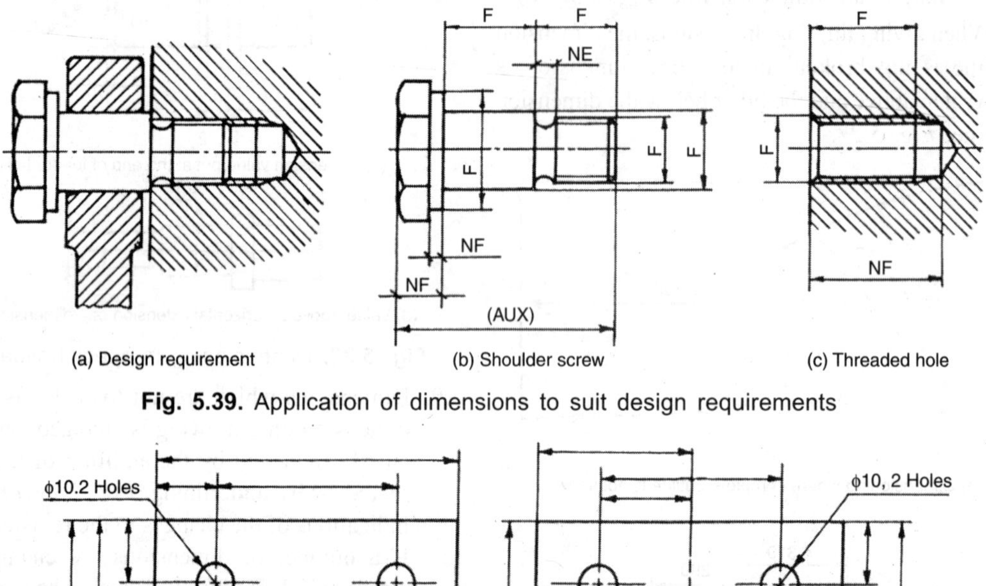

(a) Design requirement (b) Shoulder screw (c) Threaded hole

Fig. 5.39. Application of dimensions to suit design requirements

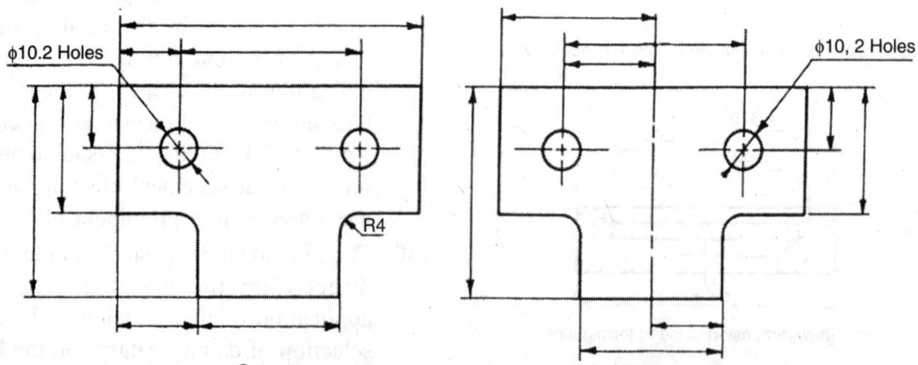

Correct Incorrect

Fig. 5.40. Dimensioning to a centre line

36. Dimensional values should be shown on drawings in characters of sufficient size to ensure complete legibility on the original drawing as well as on reproductions made from microfilms.
37. Wires, cables, sheets, rods, and other materials manufactured to gauge or code numbers should be specified by dimensions indicating the diameter or thickness. Gauge or code numbers may be shown in parentheses following the dimension.
38. A 90° angle is implied where centre lines, and lines displaying features, are shown on a drawing at right angles and no angle is specified. The tolerance for these 90° angles is the same as the general angular tolerance specified in the title-block or in a general note.
39. A 90° basic angle applies where centre lines of features are located by basic dimensions and no angle is specified. Basic dimensions are considered theoretically perfect in size, profile, orientation, or location. Basic dimensions are the basis for variations that are established by other tolerances.
40. Unless otherwise specified, all dimensions are measured at 20°C. Compensation may be made for dimensions taken at other temperatures.
41. Dimensions apply on the drawing where specificed.
42. All dimensions and tolerances apply in a free state condition except for non-rigid parts. Free state condition describes distortion of the part after removal of forces applied during manufacturing. Non-rigid parts are those that may have dimensional change due to thin wall characteristics.
43. Unless otherwise specified, all geometric tolerances apply for full depth, length, and width of the feature.

5.7. DIMENSIONING COMMON FEATURES

Various characteristics and features of parts require unique methods of dimensioning.

1. Dimensioning Diameters

- Where space permits dimensions of diameters should be placed in or on the dimension line inside the circle, *Fig. 5.43*.
- Dimensions of diameters should be placed on the most appropriate view to ensure clarity, as for instance, on a longitudinal view in preference to an end view consisting of a number of circles, *Fig. 5.42 (a)*.
- Where half views of symmetrical parts, such as cylindrical parts, are shown the portions of the dimension lines shown should be extended slightly beyond the centre line and the second arrow be omitted, *Fig. 5.42 (b)*.
- Where space is restricted, or in order to avoid superimposing of numerals, or having to follow long dimension lines, the method shown in *Fig. 5.42 (e)* may be used.

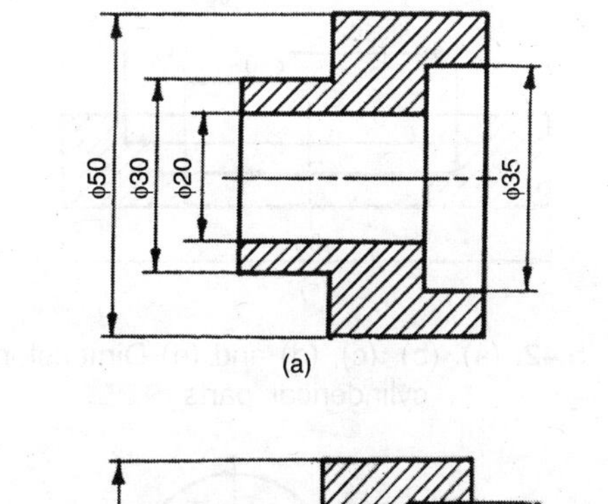

Fig. 5.42. Dimensioning cylindrical parts

(contd)...

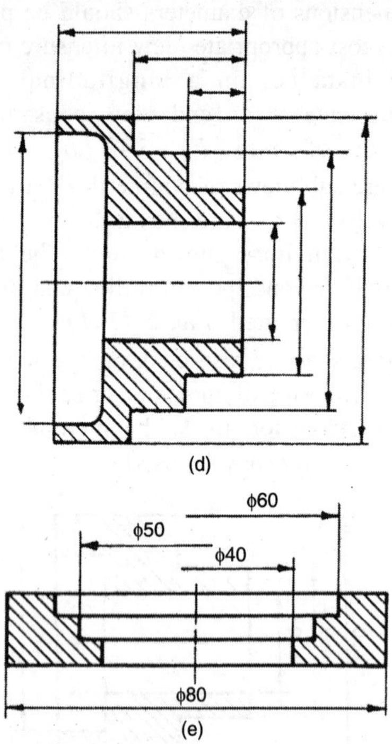

Fig. 5.42. (a), (b), (c), (d) and (e) Dimensioning cylinderical parts

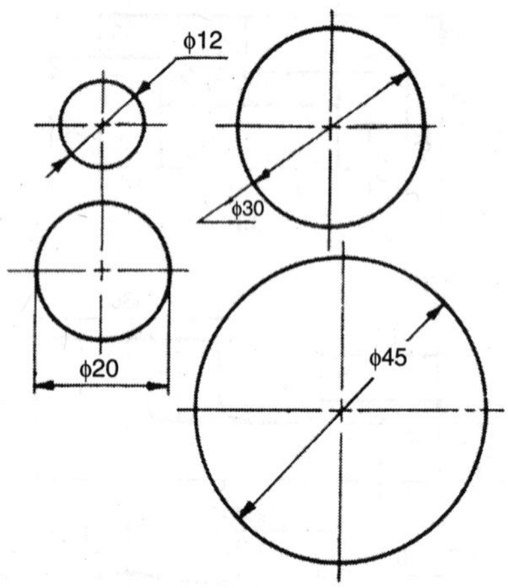

Fig. 5.43. Dimensioning circles

- A dimension indicating the diameter of a circle or cylinder should be preceded by the symbol 'ϕ'.
- Where the diameter of a spherical feature is specified, the dimension value is preceded by the spherical diameter symbol, or it should be dimensioned as in *Fig. 5.44*.

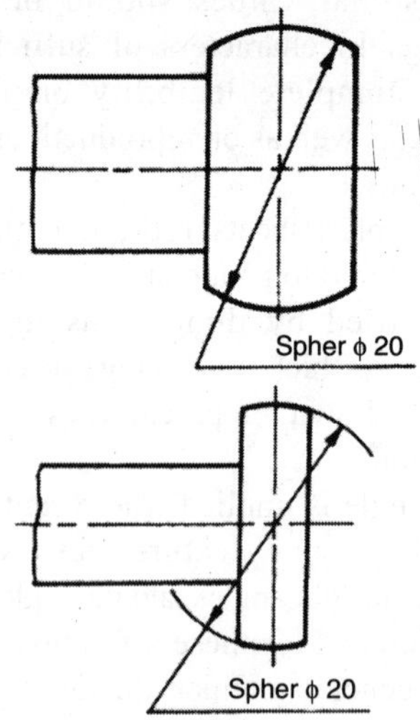

Fig. 5.44. Dimensioning spherical diameters

2. Dimensioning Radii

- Most radii are relatively small and wherever possible their dimensions should be shown outside the outline of the object by dimension lines, which pass through, or are in line with, the centre of arc. The dimension line should have one arrowhead only, and it should be touching the arc.
- The letter *R* should always precede the dimension numerals.
- A leader must be normal to arc, *Fig. 5.45*.
- Where the centre point of a radius is dimensioned, a dot or a cross should be placed at the intersection of the lines:, which locate the centre of the radius. While dimensioning small radii, the arrow may be reversed, *Fig. 5.45 (a)*. Where location of the centre is unimportant, the drawing must clearly show that the arc location is controlled by other dimensioned features such as tangent surfaces, *Fig. 5.45 (d)*.
- Where the size of the radius can be derived from other dimensions, it should be indicated by a radius arrow and the symbol *R* without an indication of the value, *Fig. 5.45 (b)*.
- Radii of arcs, which need not have their centres located, should be dimensioned as shown in *Fig. 5.46*.

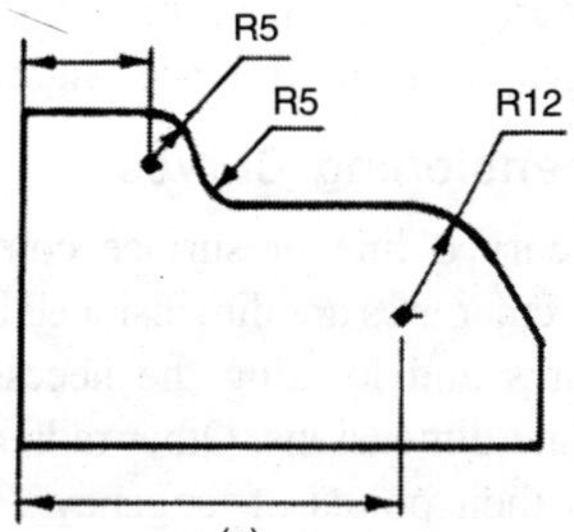

(a)

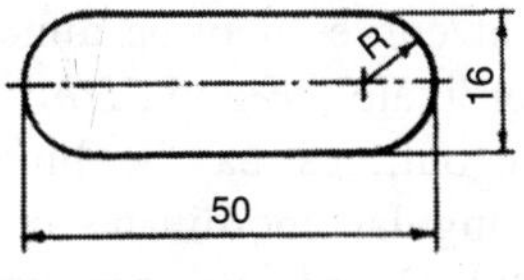

(b)

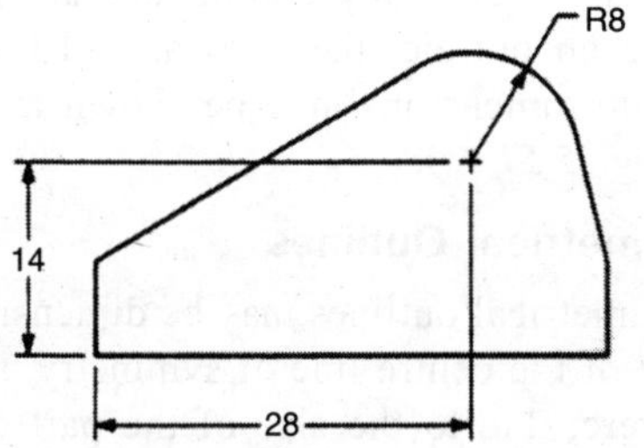

(c)

Fig. 5.45. (a), (b) and (c), Dimensioning radii of arcs which have their centre located

- When the centre of the arc falls outside the limits of the space available on drawing or interferes with another view, the radius dimension line may be foreshortened, *Fig. 5.46 (a), (b)*. That portion of the dimension line extending from the arrowhead is radial to the arc. Where the radius dimension is foreshortened and the centre is located by coordinate dimensions, the dimension line locating the centre is also foreshortened.
- Where a radius is dimensioned in a view that does not show the true shape of the radius, TRUE R is added before the radius dimension, *Fig. 5.46 (c)*.
- Where a part has a number of equal radii, a note may be used instead of dimensioning each radius separately.
- Where a spherical surface is dimensioned by a radius, the radius dimension is preceded by the symbol SR, *Fig. 5.47 (a)*.

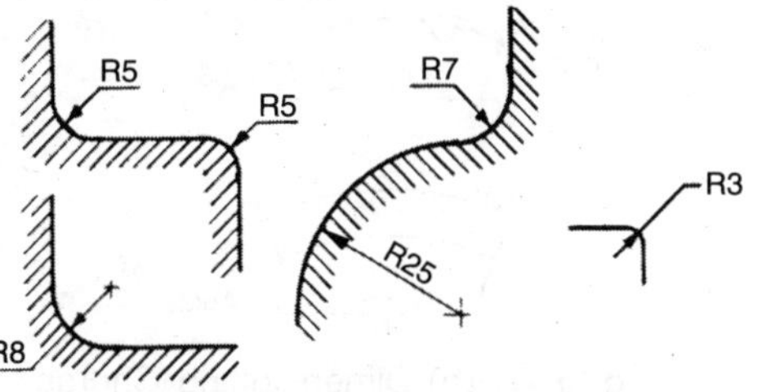

(a) Dimensioning Radii of arcs which need not have their centres located

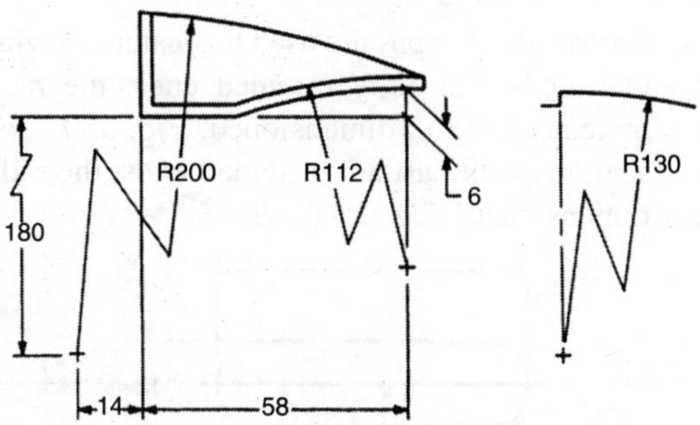

(b) Foreshortened radii

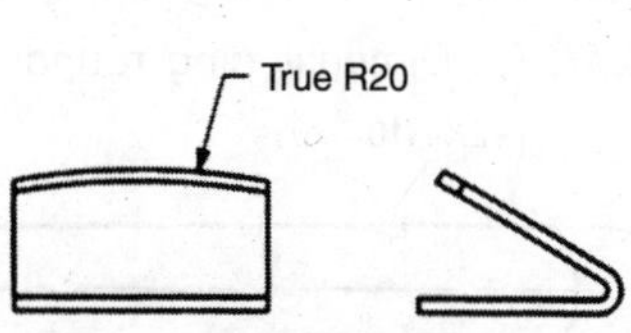

(c) True Radius

Fig. 5.46.

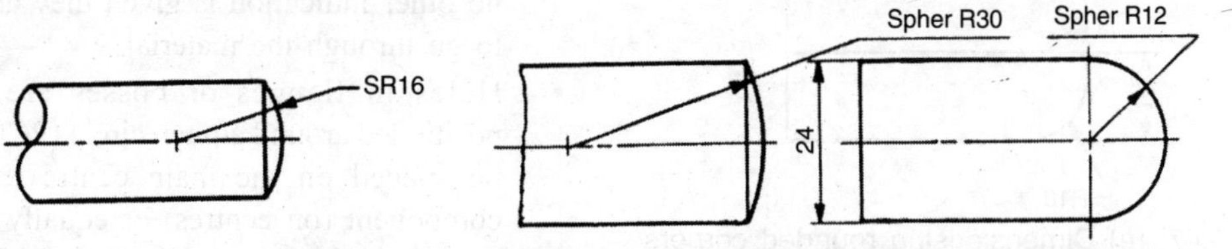

Fig. 5.47. (a) Dimensioning spherical radii

3. Dimensioning Chords, Arcs And Angles

Chords, arcs, and angles are dimensioned as shown in *Fig. 5.47 (b)*.

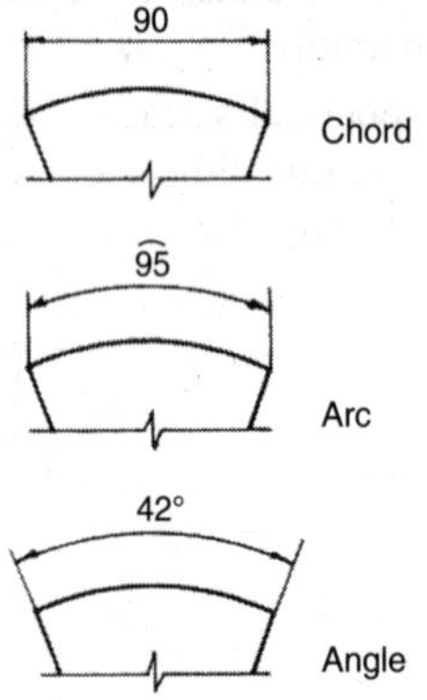

Fig. 5.47. (b) Dimensioning Chords, Arcs and Angles

4. Dimensioning Rounded Ends

Overall dimensions are used for features having rounded ends. For fully rounded ends, the radii are indicated but not dimensioned, *Fig. 5. 47 (c)*. For features with partially rounded ends, the radii are dimensioned, *Fig. 5.47 (d)*.

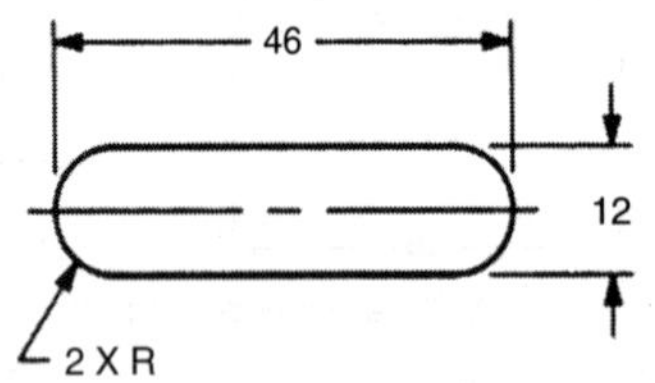

Fig. 5.47. (c) Dimensioning rounded ends

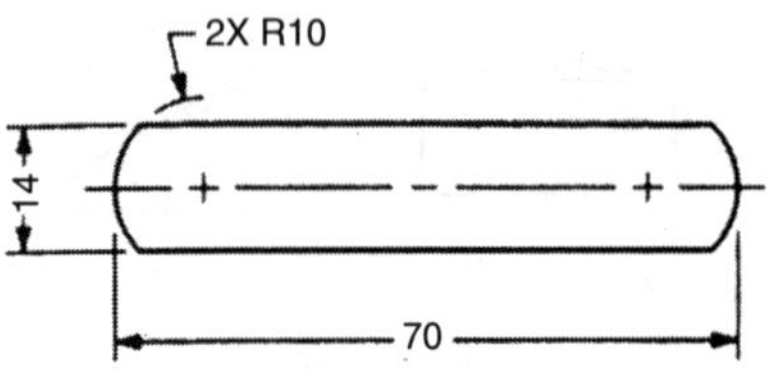

Fig. 5.47. (d) Dimensionsing partially rounded ends

5. Dimensioning Rounds

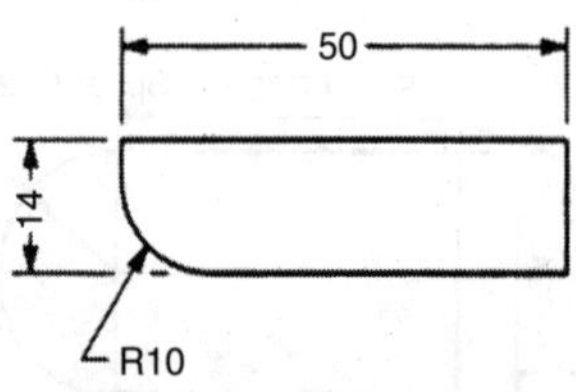

Fig. 5.47. (e) Dimensionsing rounded corners

Where corners are rounded, dimensions define the edges, and the arcs are tangent, *Fig. 5.47 (e)*.

6. Dimensioning Curves

A curved line or surface composed of two or more circular arcs are dimensioned by giving the radii of all arcs and locating the necessary centres with coordinate dimensions. Other radii are located on the basis of their points of tangency, *Fig. 5.51 (a)*.

7. Dimensioning Irregular Curves

Irregular curves may be dimensioned by the use of co-ordinates, *Fig. 5.51(b), (c)*. Circular or non-circular outlines may be dimensioned by the use of rectangular coordinates or offset method, *Fig. 5.51 (b)*. Coordinates are dimensioned from base lines. When many coordinates are required to define an outline, the vertical and horizontal coordinate dimensions may be shown in Tabular form, *Fig. 5.51 (c)*.

8. Symmetrical Outlines

Symmetrical outlines may be dimensioned on one side of the centre line of symmetry, as in the case where, due to the size of the part or space limitations, only part of the outline can be conveniently shown, *Fig. 5.51 (d)*. One-half of the outline of the symmetrical shape is shown and symmetry is indicated by applying symbol for part symmetry to the centre line.

9. Dimensioning Holes

- Different methods of dimensioning hole sizes are shown in *Fig. 5.48 (a), (b)*.
- The depth of a drilled hole, when stated in note form, refers to the depth of the cylindrical hole and not to the point of the drill.
- Where the depth dimension is not clear, as from a curved surface, the depth should be dimensioned.
- Where it is not clear that a hole goes through, the abbreviation THRU follows a dimension. If no other indication is given they are assumed to go through the material.
- Holes in flanges or bosses are generally positioned around a pitch circle (PCD) and may be spaced on the main centre lines of the component (on centres) or equally spaced off centres. Holes are usually drilled off centres to

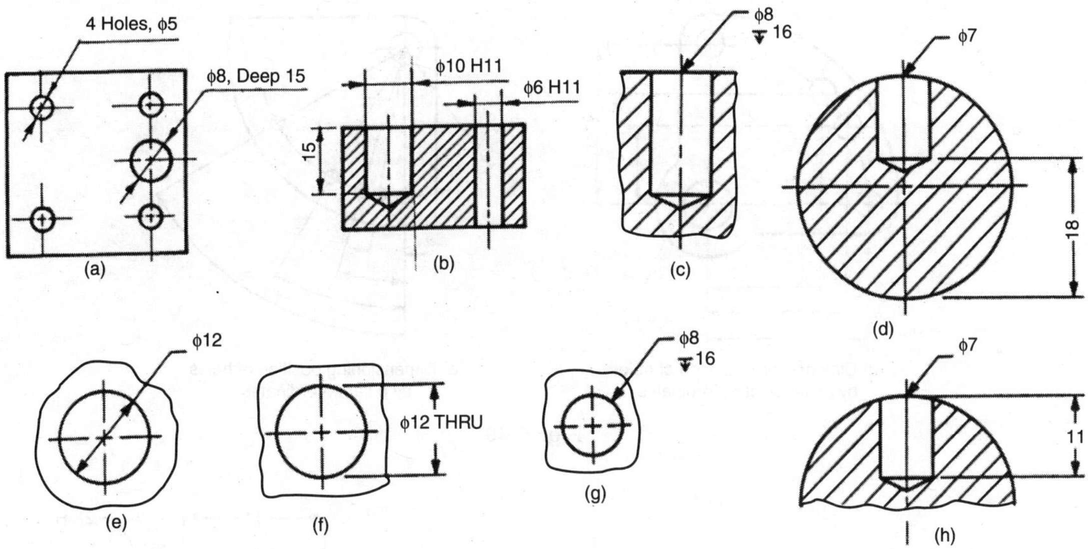

Fig. 5.48. Dimensioning holes

provide for maximum access to fixing bolts in the case of valves and pipeline findings.

- The method of positioning holes on pitch circles is shown in Fig. *5.49 (a), (b), (c), (d)*.
- Dimensions for locating the holes should be possibly given in plan view of holes, by either spacing them on pitch circles, *Fig. 5.49 (a), (b)* or by giving the rectangular coordinates, *Fig. 5.49(c)*, or by giving the centre distances, *Fig. 5.50 (a), (b)*.
- When the holes are not of equal pitch, the dimension should be given from a master hole, *Fig. 5.50 (a)* or a surface, *Fig. 5.50 (b)*, which controls the location.

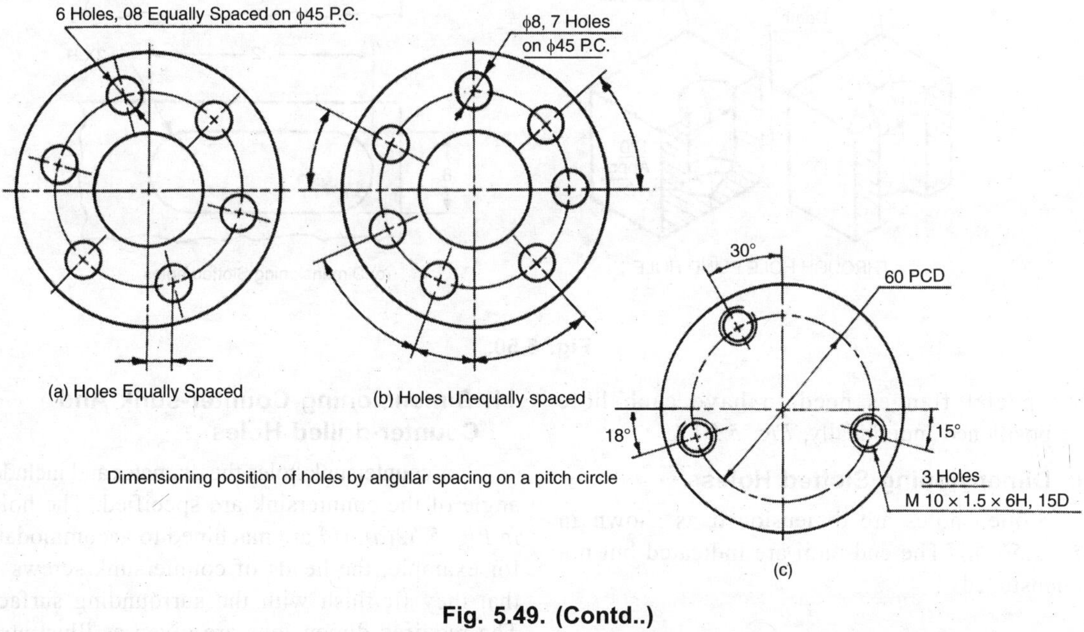

Fig. 5.49. (Contd..)

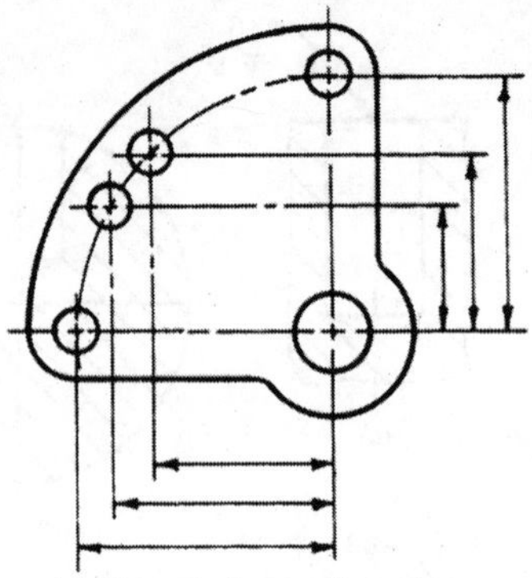

(d) Dimensioning position of holes by rectangular co-ordinates

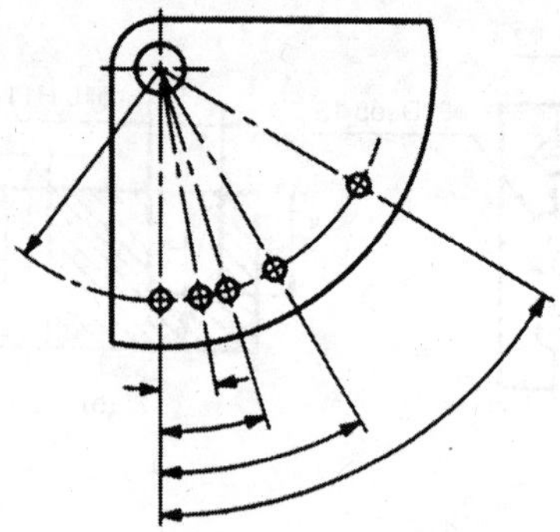

(e) Dimensioning position of holes by polar co-ordinates

Fig. 5.49.

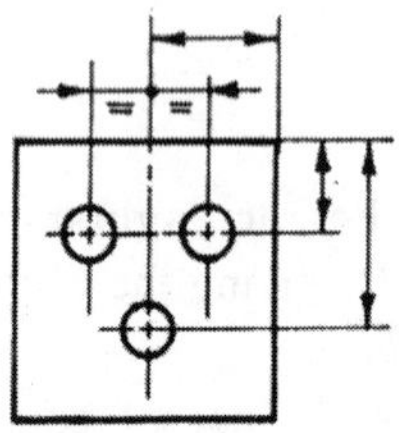

(a) Dimensioning from a Master Hole

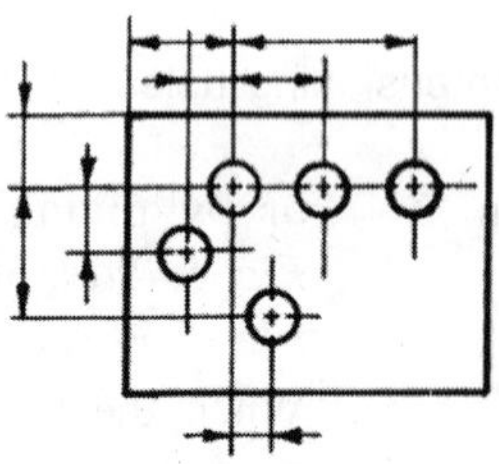

(b) Dimensioning to a surface

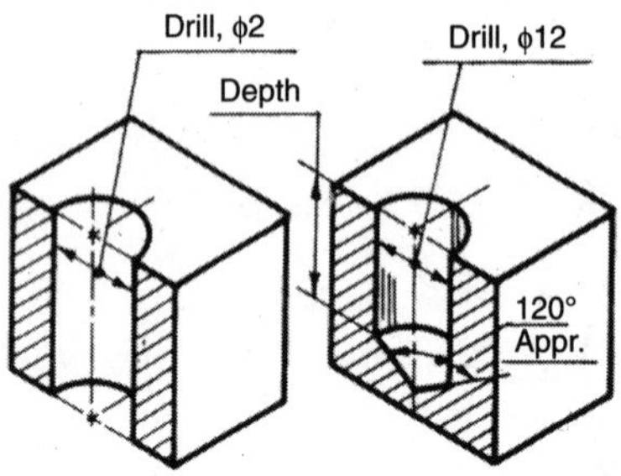

THROUGH HOLE BLIND HOLE

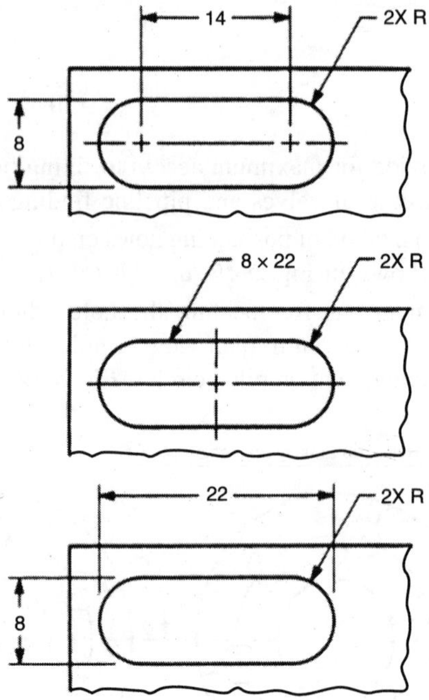

(c) Dimensioning Slotted holes

Fig. 5.50.

Special flanges need to have each hole positioned individually, *Fig. 5.50 (c)*.

10. Dimensioning Slotted Holes

Slotted holes are dimensioned as shown in *Fig. 5.50 (d)*. The end radii are indicated but not dimensioned.

11. Dimensioning Counter-sunk And Counter-drilled Holes

For countersunk holes the diameter and included angle of the countersink are specified. The holes in *Fig. 5.52(a)*, *(b)* are machined to accommodate, for example, the heads of countersunk screws so that they fit flush with the surrounding surface. The required dimensions are given as illustrated.

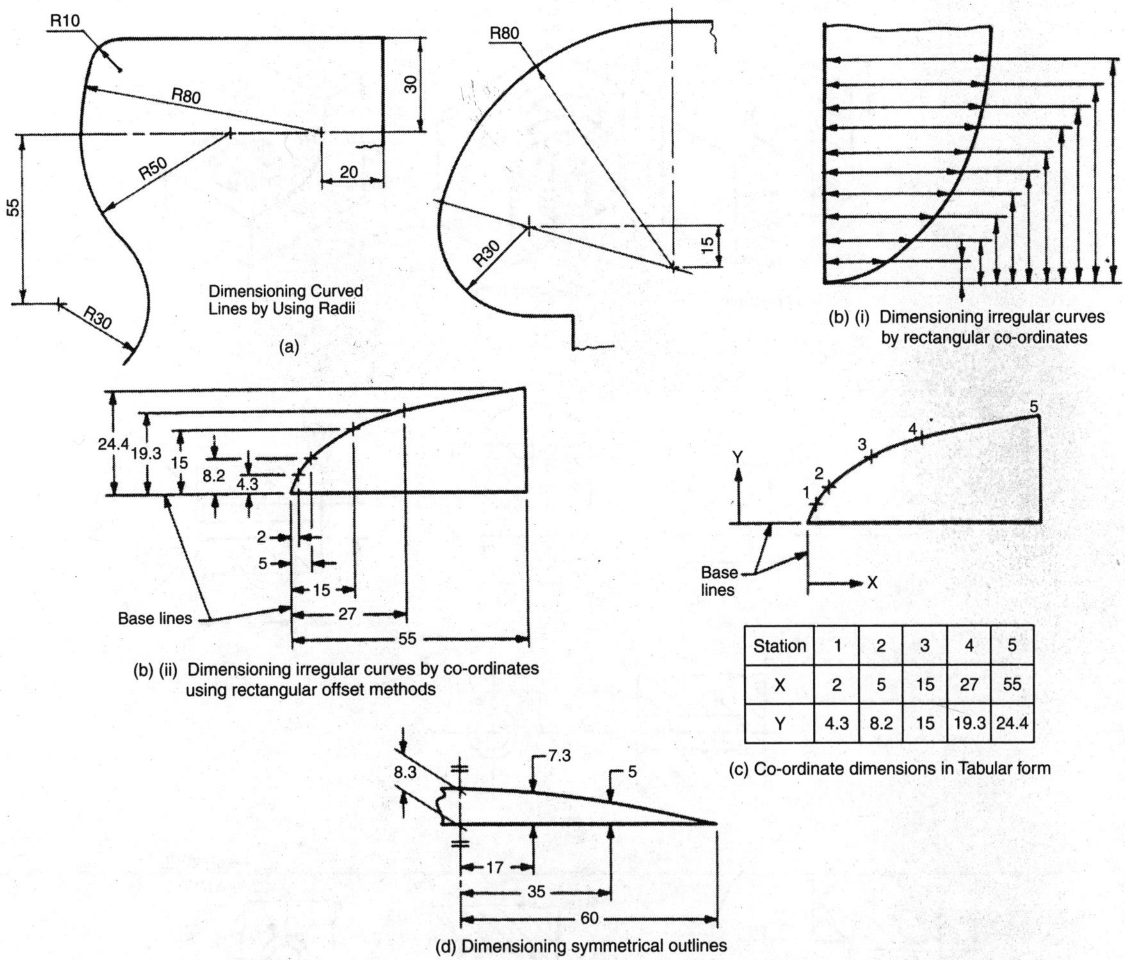

Dimensioning Curved Lines by Using Radii

(a)

(b) (i) Dimensioning irregular curves by rectangular co-ordinates

(b) (ii) Dimensioning irregular curves by co-ordinates using rectangular offset methods

Station	1	2	3	4	5
X	2	5	15	27	55
Y	4.3	8.2	15	19.3	24.4

(c) Co-ordinate dimensions in Tabular form

(d) Dimensioning symmetrical outlines

Fig. 5.51.

For counter-drilled holes, the diameter and depth of the counter-drill are to be specified. Specifying the included angle of the counter-drill is optional, *Fig. 5.52 (c)*. The depth dimension is the depth of the full diameter of the counter-drill from the outer surface of the part.

12. Dimensioning Counter-bored Holes

A counter-bore provides, for example, a recess in a casting, which is used to house a bolt head so that it does not project above the surface. The machining operation also provides an area square to the hole axis so that the bolt head is accurately seated on a smooth surface. *Fig. 5.54 (a), (b),* illustrates the method of dimensioning counter-bores. Where the thickness of the remaining material has significance, this thickness is dimensioned instead of the depth. For method of dimensioning holes having more than one counter-bore refer *Fig. 5.54 (c).*

13. Dimensioning Chamfered And Countersunk Holes On Curved Surfaces

For dimensioning a chamfered or countersunk hole on a curved surface, whether convex or concave, the diameter specified on the drawing applies at the minor diameter of the chamfer or countersink, *Fig. 5.53.*

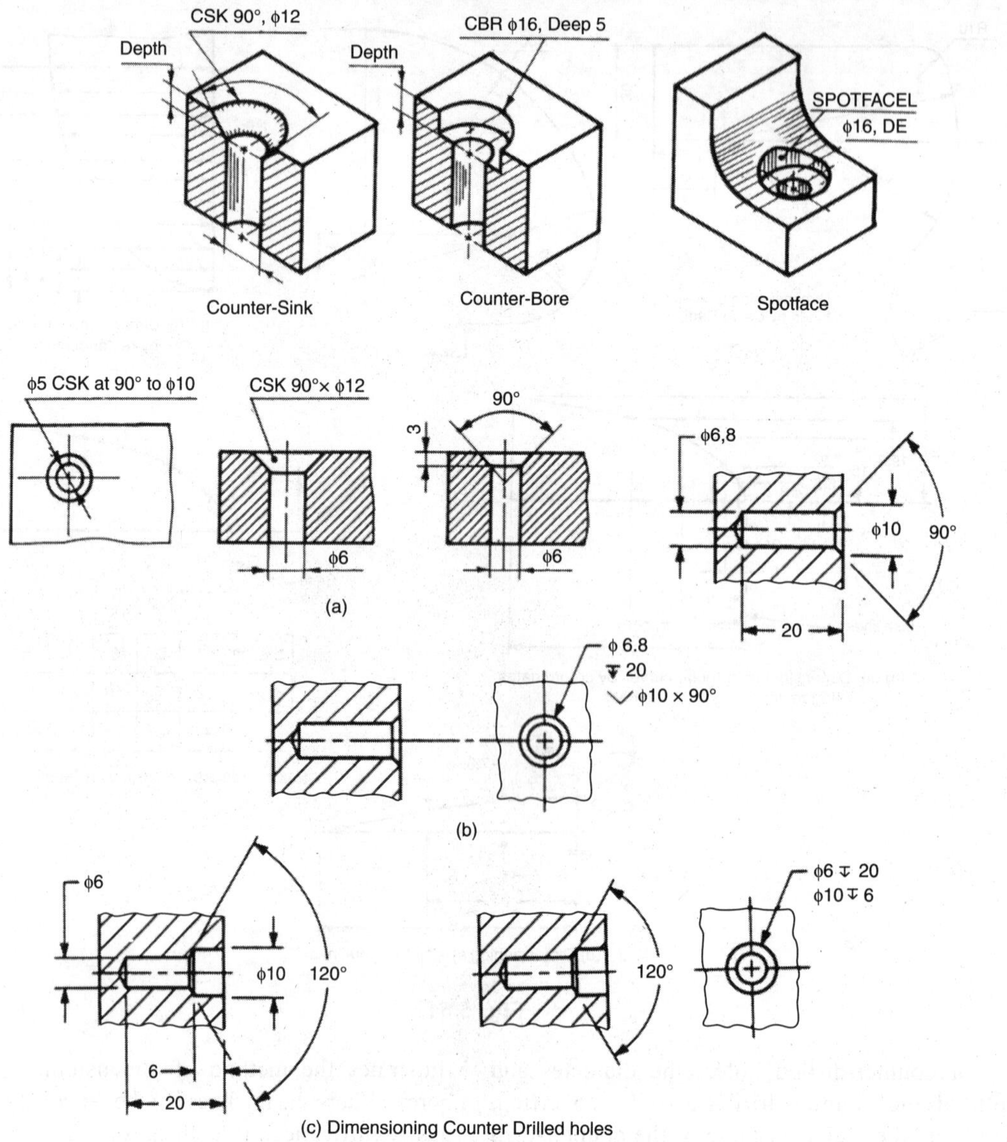

Fig. 5.52. Dimensioning Counter-Sunk Holes

14. Dimensioning Spotfaces

A spot-face is machined, for example, on a rough casting around a hole and perpendicular to the hole axis, so that a true flat surface is obtained for a nut or bolt. The depth of machining is only sufficient to ensure an accurate surface and hence the nut or bolt will project above the surrounding. The diameter of the spot faced area is specified. Either the depth or the remaining thickness of material may be specified. A spotface may be specified by a note only and need not be delineated on the drawing. If no depth or remaining thickness of material is specified, the spotface is the minimum depth necessary to clean up the surface to the specified diameter. *Fig. 5.55 (a), (b) (c)* illustrates the method of dimensioning spotfaces.

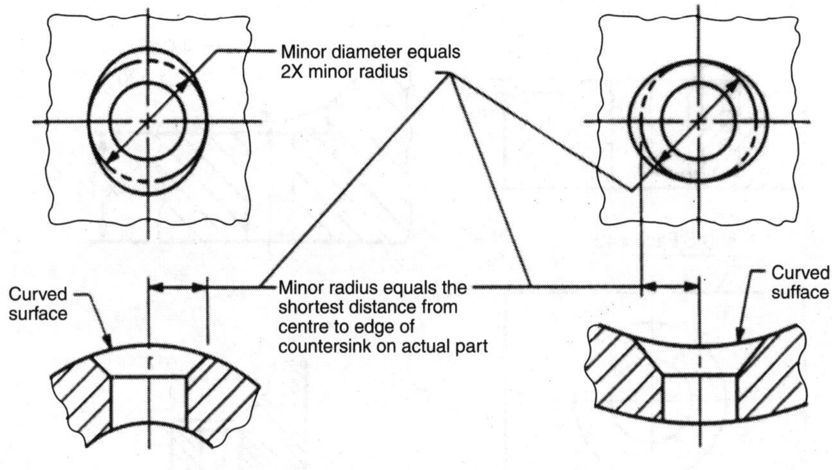

Fig. 5.53. Dimensioning a chamfered or counter-sunk joles on a curved surface

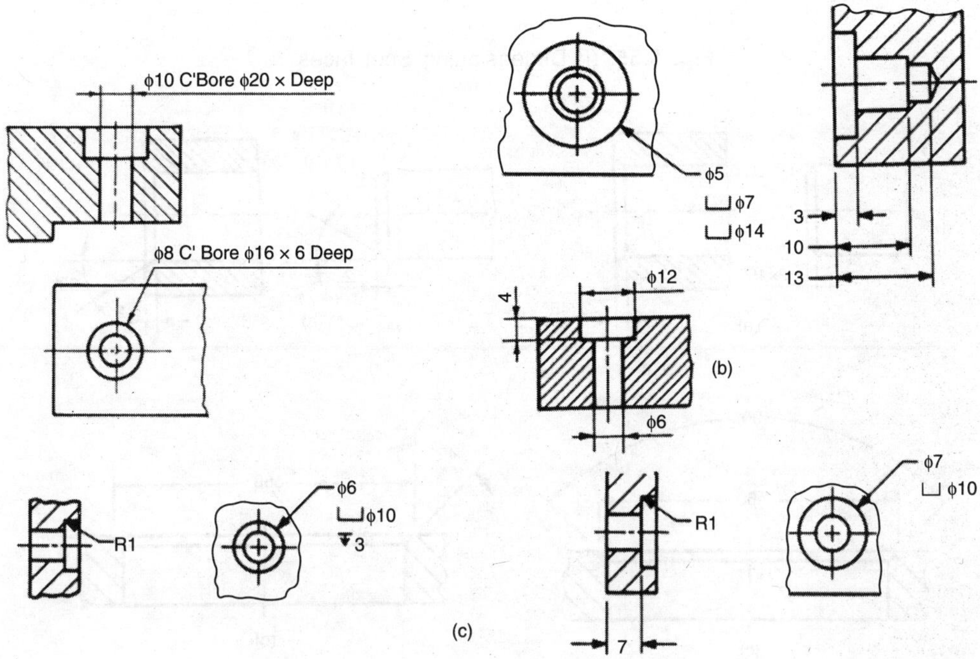

Fig. 5.54. Dimensioning Counter Board holes

15. Chamfers

Chambers are machined with a view to eliminate a sharp edge or corner. Chambers are dimensioned by a linear dimension and an angle, as shown in *Fig. 5.56 (a)* and *(b)* and not by a note and a leader. The linear dimension is the distance from the indicated surface of the part to the start of the chamfer. While dimensioning 45° chamfer angles, methods shown in *Fig. 5.55 (a)* should be used. Whereas for angles other than 45°, methods shown in *Fig. 5.55 (b)* should be used.

16. Tapered Features

Tapered features may be dimensioned by the methods shown in *Fig. 5.56*. Notes such as taper 1:10, *Fig. 5.56 (a)*, taper 1:15, *Fig. 5.56 (b)*, should

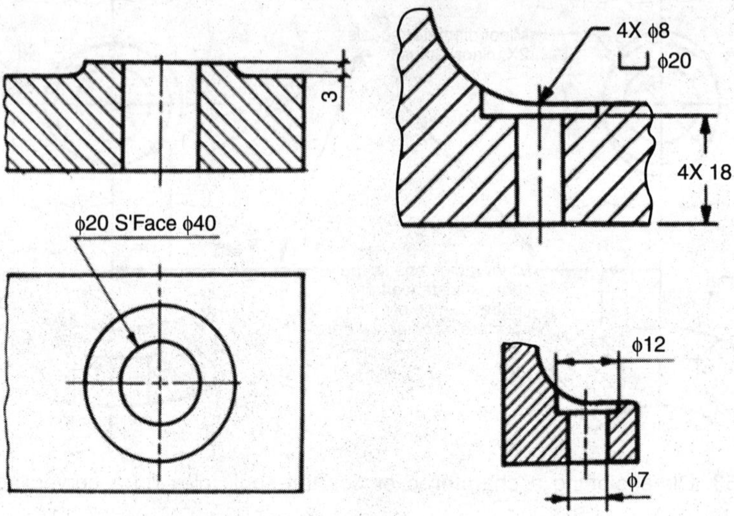

Fig. 5.55. (i) Dimensioning Spot faces

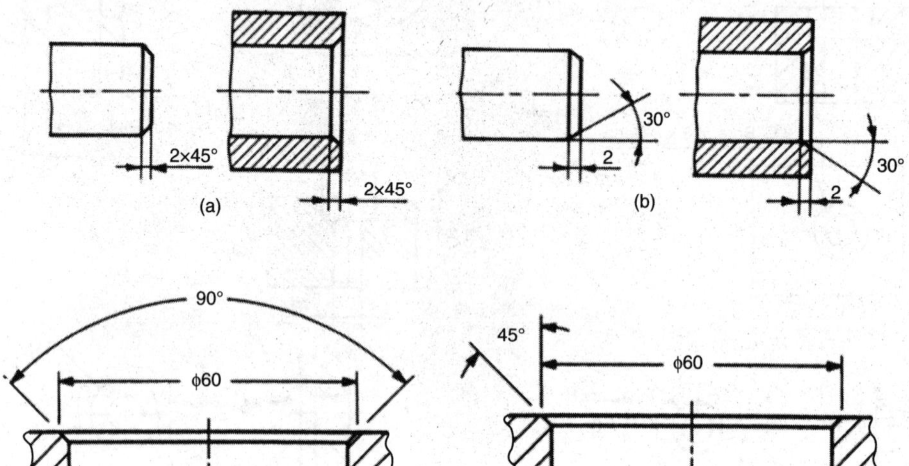

Fig. 5.55. (ii) Dimensioning internal and external chamfers

be placed parallel to the centre line, while for indication of slope, *Fig. 5.56 (c)*, be placed parallel to the envelope. If necessary, the direction of taper may be added, *Fig. 5.56 (d)*. In addition, the half angle or the taper angle is indicated, in brackets, to facilitate the machine setting, *Fig. 5.56 (a), (b)*.

17. Production Processes Or Inspection Methods

These should not be specified on drawing unless they are essential to ensure satisfactory functioning or interchangeability. This does not apply to process drawings, nor does it preclude the quoting of drill sizes.

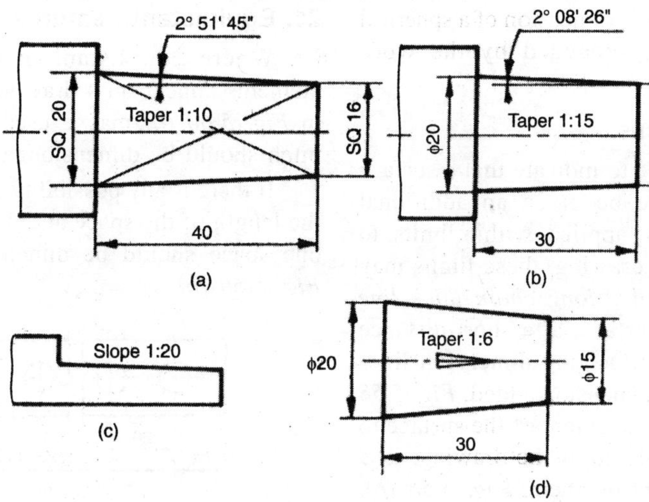

Fig. 5.56. Dimensioning Tapered features

18. Addition of Letters And Symbols

The following letters or symbols should be placed:

1. Before a dimension for diameter, the symbol 'ϕ', *Fig. 5.57(d).*
2. Before a dimension for a radius, the letter 'R', *Fig. 5.57(b).*
3. Before a dimension □ for a square, or the letters 'SQ', *Fig. 5.57 (c).*
4. Before across dimension for a hexagon, the letters 'HEX', *Fig. 5.57 (d).*

- If it clearly appears from the drawing that the dimension applies to a square section or diameter or hexagonal section the symbols may be omitted.

5. SF for spherical diameter and SR for spherical radius.

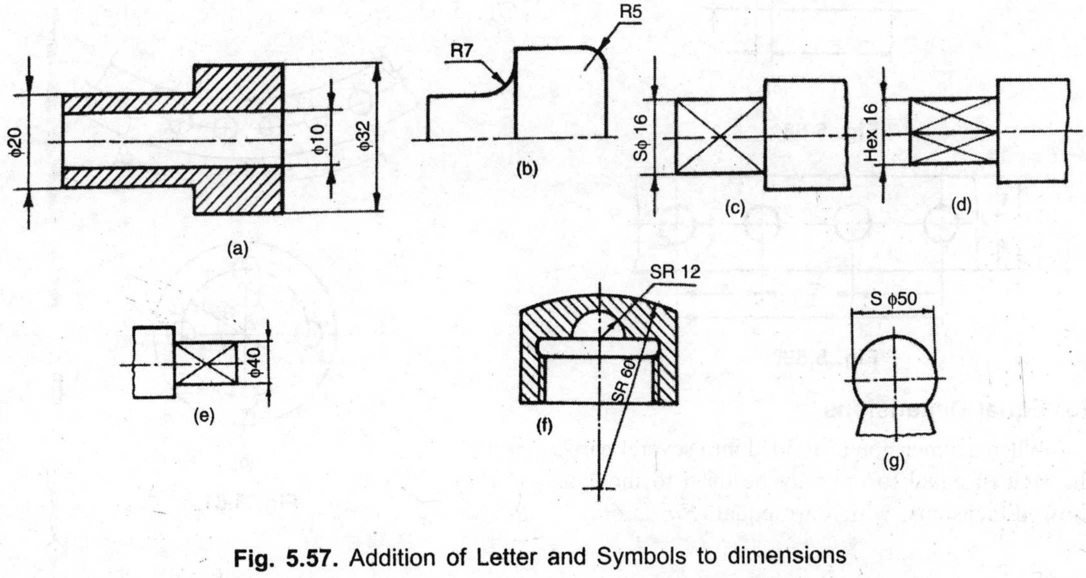

Fig. 5.57. Addition of Letter and Symbols to dimensions

6. The radius or diameter dimension of a spherical surface should be preceded by the word SPHER, *Fig. 5.47.*

19. Special Indications

When it is desirable to indicate that a surface or surface zone has to be given an additional treatment which shall be applied, within limits, to a specified area on the drawing, these limits may be defined by means of a *long chain thick line* drawn parallel to the surface, at a short distance from it, and having the locating dimension lines and the corresponding dimension added, *Fig. 5.58 (a)*. If the location and the extent of the surface to be treated appear clearly from the drawing, it is not necessary to dimension them, *Fig. 5.58 (b)*. Where the desired area is shown on a direct view of the surface, the area is section lined within the chain line boundary and appropriately dimensioned, *Fig. 5.58 (c)*.

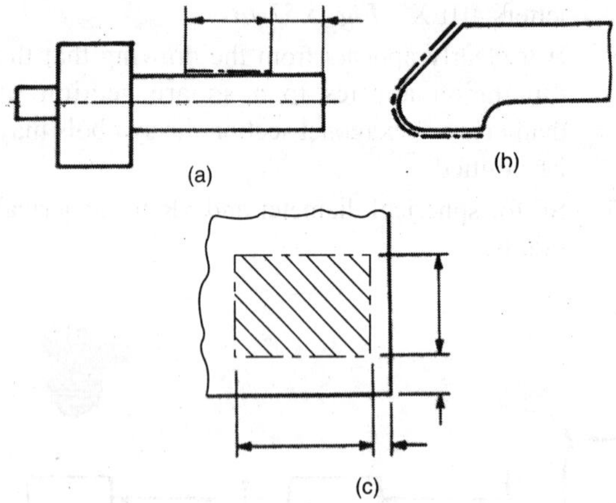

Fig. 5.58.

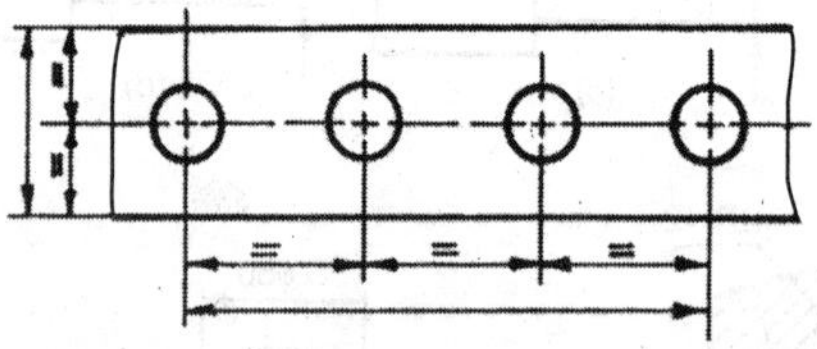

Fig. 5.59.

20. Equal Dimensions

When a dimension is divided into several parts, the sign of equal to (=) may be used to indicate those dimensions, which are equal, *Fig. 5.60.*

21. Equidistant Features

Where equidistant or regularly arranged elements appear on a drawing the method shown in *Fig. 5.60 (a)* may e used for simplicity. One pitch should be dimensioned.

If there is any possibility of confusion between the length of the space and the number of spacing one space should be dimensioned as shown in *Fig. 5.60 (b)*.

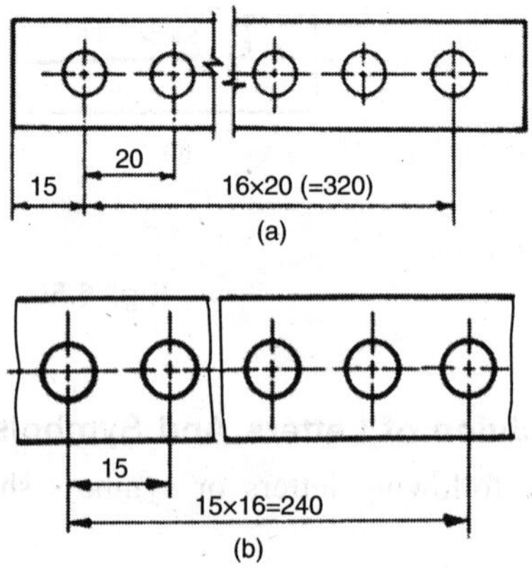

Fig. 5.60.

Angular spacings of holes and other features may be dimensioned as shown in *Fig. 5.61 (a)*. The angles of spacing may be omitted if their number is evident without confusions, *Fig. 5.61 (i)*. Circular spacing may be dimensioned indirectly by giving the number of elements as shown in *Fig. 5.62.*

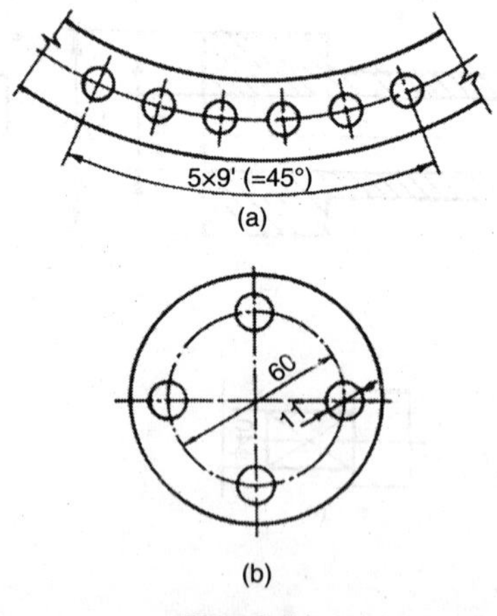

Fig. 5.61.

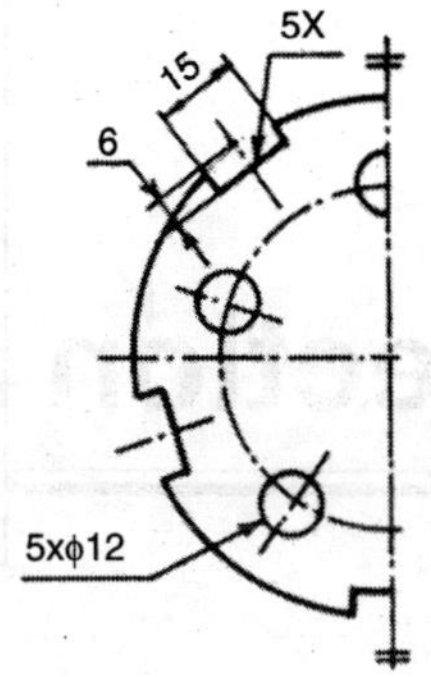

Fig. 5.62.

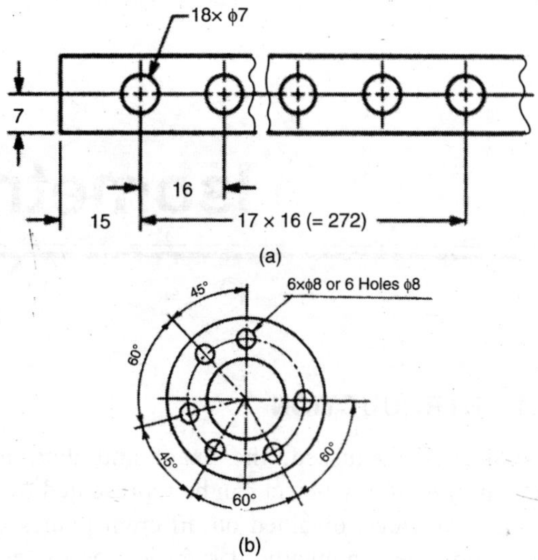

Fig. 5.63.

22. Repeated Features

If it were possible to define a quantity of elements of the same size so as to avoid repeating the same dimensional value, they may be given as shown in *Fig. 5.63 (a), (b).*

QUESTIONS FOR VIVA VOCE

1. What is the importance of dimensioning?
2. What do you understand by the term 'Notation of Dimensioning'?
3. What are end product drawings?
4. Define:
   (*i*) a Feature,
   (*ii*) a Functional Dimension,
   (*iii*) Datum Dimension,
   (*iv*) a Redundant Dimension, and
   (*v*) an Auxiliary Dimension.
5. Explain with the help of simple sketches:
   (*i*) Progressive Dimensioning,
   (*ii*) Chain Dimensioning,
   (*iii*) Parallel Dimensioning, and
   (*iv*) Combined Dimensioning.
6. What are:
   (*i*) Size Dimensions, and
   (*ii*) Location Dimensions?
7. Explain with the help of a simple sketch,
   (*i*) Aligned System, and
   (*ii*) Unidirectional System of Dimensioning.
8. What is the consideration kept in mind for dimensioning an object prepared by a number of operations ?
9. Explain with the help of simple sketches, the methods of dimensioning:
   (*i*) circles, (*ii*) arcs, (*iii*) angles, (*iv*) restricted spaces, (*v*) spherical shapes, (*vi*) holes, (*vii*) curves, (*viii*) chambers, counter-sinks, counter-bores, and spot-faces.

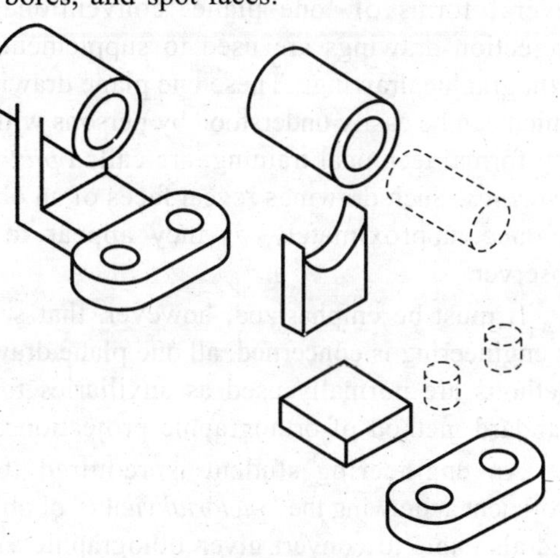

Fig. 5.64.

Mentally break the solid into smaller components the then dimension each smaller component individually so that its complete dimensions information is captured.

Isometric Projection

6

6.1. INTRODUCTION

As already discussed, the exact and detailed descripition of any object can be represented by a number of views obtained on different planes of projection, held mutually at right angles to each other, for *orthographic projections.* Such drawings can be correctly interpreted and quickly visualized only by those persons who have a good knowledge of principles used for these projections. Also their execution requires a definite exercise of constructive imagination.

A detailed study of the theory of orthographic projections will be undertaken in the next chapter.

To make the drawing more understandable, several forms of 'one plane' conventional or projection drawings are used to supplement the orthographic drawings. These one plane drawings, which can be easily understood by persons without any formal technical training, are called *pictorial drawings*. Such drawings reveal faces of an object at once, approximately, as they appear to the observer.

It must be emphasized, however, that so for as engineering is concerned, all one plane drawing methods are normally used as auxiliaries to the standard method of orthographic projections.

An engineering student is required to be proficient at drawing the *"pictorial views"* of objects and also able to convert given othographic views to pictorial views.

6.2. METHODS OF PROJECTION

Pictorial drawings are projected on a single plane of projection by the methods of:

1. Axonometric Projection,
2. Oblique Projection, and
3. Perspective or Central projection.

Out of these *three types of pictorial drawing* of an object; it can be seen that the perspective is the most realistic but it is tedious to draw, and is not used in machine drawing. The *isometric projection, a particular case of axonometric projection*, though dose not look as pleasings as the perspective, is mostly used because it is easier to draw and dimension, *Fig. 6.1.* The *oblique projection* is used, with advantage, for objects with curved features only on one face or on parallel faces. It is equally easy to draw and dimension as the isometric projection.

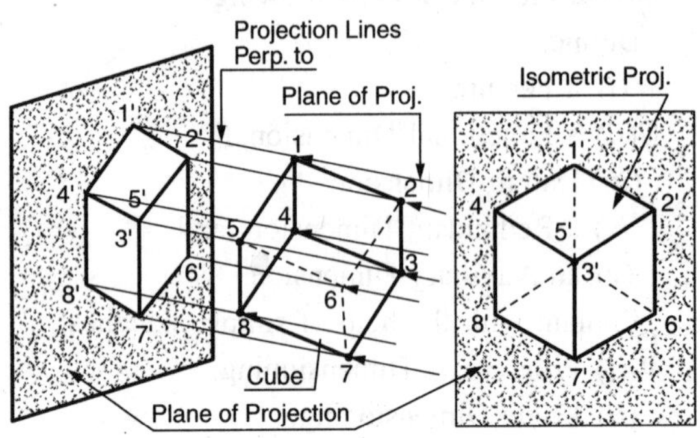

(a) Axonametrix Projection (b) Isometric Projection

Fig. 6.1.

Here, in this chapter we shall deal with the *axonometric projection with special emphasis on the isometric projection only.* The perspective and oblique projections will be dealt in separate chapters.

6.3. AXONOMETRIC PROJECTION

In axonometric projection the object, a cube for example, is placed such that its faces are inclined to the plane of projection, so that its three principal faces may be represented in one view.

When the projectors are perpendicular to the plane of projection, the resulting pictorial projection is a *normal axonometric projection*, otherwise it is an *oblique axonometric projection*. The latter is not used in actual practice, the 'axonometric projection' is generally understood to refer to normal axonometric projection. *The normal axonometric projection*, in a way, *is a branch of orthographic projection, Fig. 6.2.*

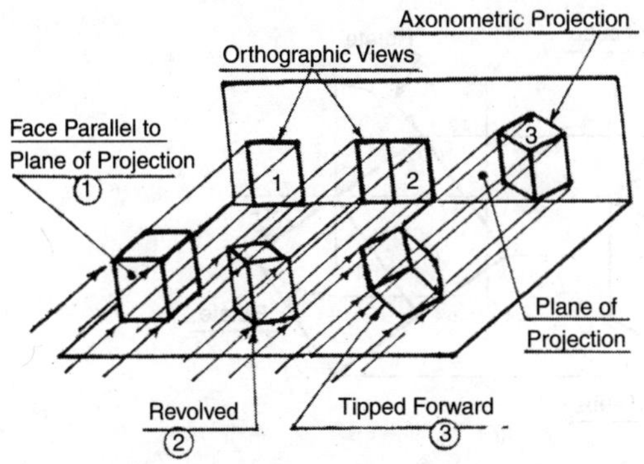

Fig. 6.2. Method of Obtaining Axonometric

6.4. TYPES OF AXONOMETRIC PROJECTION

The feature which distinguishes axonometric projection from other projections, is the inclined position of the object with respect to the plane of the projection. As the axes, or the principal edges, of the object are inclined to the plane of projection, the lengths of the lines, the sizes of the angles, and the general proportions of the projection of the object vary with the infinite number of possible positions in which the object may be placed with respect to the plane of projection.

Axonometric projection is classified as:

1. Isometric projection,
2. Dimetric projection, and
3. Trimetric projection, *Fig. 6.3*.

The theory behind each of these types is same; but the angle of projection for each type is different. *Figure 6.3* shows the three types of axonometric projection. In isometric projection, the three axes form equal angles of 120° to the plane of projection, and only one scale is needed for measurement along each of the three axes. This is the easiest type of axonometric drawing to execute.

In dimetric projection, only two of the three angles are equal, and two special foreshortened scales are required to measure distances along their respective axes, *i.e.*, one scale to do measurements along the axes that foreshorten by same amount, while the second scale for measurements along the third axis, which foreshorten by a different amount, proportionally.

Trimetric projection requires three different foreshortened scales, as all three angles between the three axes are different.

Because of the complexity of constructing dimetric and trimetric drawings, in general, only the isometric type of the axonometric projection is used in practice.

The principle involved in axonometric projection can be best explained by considering the cube, as illustrated in *Fig. 6.2*.

6.5. ISOMETRIC PROJECTION

The isometric projection is used in the preparation of such drawings in which the essential features are required to be shown, otherwise, in three orthographic views. The isometric projection of an object is a single orthometric projection, drawn with the object so oriented with respect to the

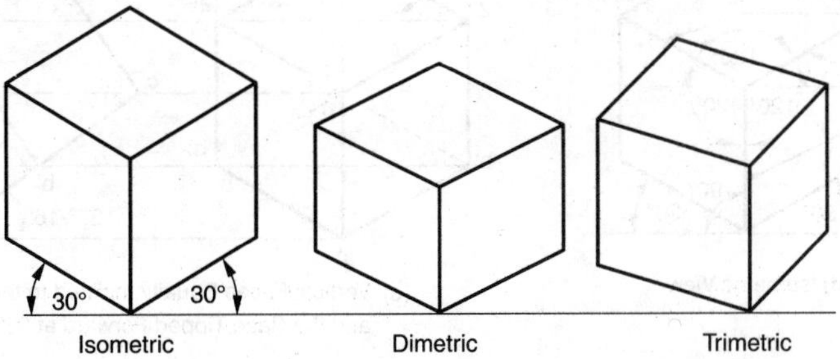

Fig. 6.3. Three types of Axonometric Projections

plane of projection, that all the three axes of the object are equally inclined to it.

The principle involved in isometric projection can be explained by considering the case of a cube. If a cube is held across the diagonal BD on any side between the thumb and the first finger and rotated on this axis, the arrangement shown in *Fig. 6.4*, (*i*) (*a*) can be obtained. Here the cube is so held that the angles in the centre, about the corner point A, are equal. The sides AB, AD, and AE appear to be equal in length, but foreshortened from their original length. In this position, although the line AE here is drawn as a vertical line, it actually slopes away from the observer. Face diagonals BD and AC are also no longer equal in length.

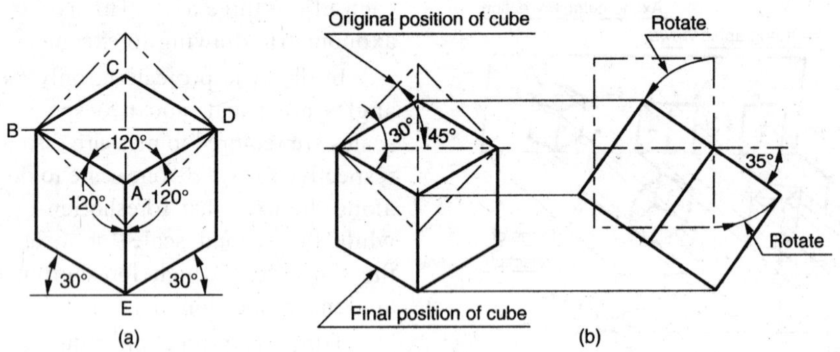

Fig. 6.4. (i) Principle of Isometric projection

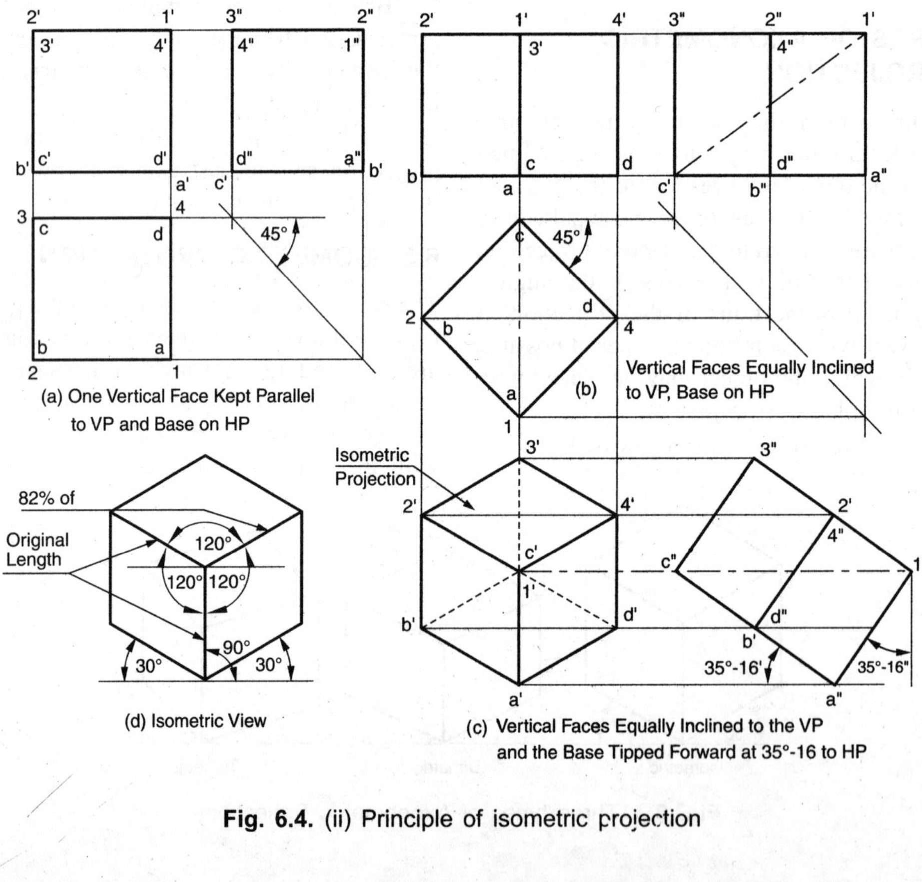

Fig. 6.4. (ii) Principle of isometric projection

By looking at the cube from the right side, in *Fig. 6.4. (i) (b)*, and following the first angle projection, it can be seen in the side view that the cube has been rotated to within **35°** 16′ of the horizontal centre line.

Geometrically, the principle of isometric projection, by rotation method, is illustrated in *Fig. 6.4. (ii) (a), (b), (c),* and *(d).*

Figure 6.4. (ii) (a) shows three orthographic views of a cube, when placed with one of its faces parallel to the principal vertical plane of projection. At (*b*) the three orthographic views of the cube are shown when it has been revolved so that the vertical faces make equal angles (*i.e.,* 45°) with the principal vertical plane of projection. At (*c*) the two views are obtained by rotating the revolved cube in such a way that all the three edges meeting at the front corner 1 are equally inclined to the vertical plane of projection. In this position of the cube the body diagonal 1C will also be perpendicular to the picture plane.

The three front edges are called the *isometric axes* and their angle of inclination with the vertical plane of projection is 35° 16′. The front view obtained in this way shows three principal faces of the cube equal in shape and is called isometric projection. Literally isometric also means *'equal measure'*. In this type of projection the angles between the projections of the front three edges, or between the *three isometric axes*, are 120° and their projected lengths are approximately 82% (actually 81.65%) of their true lengths.

It should be observed that the 90° angles of the cube, in the isometric projection, appear as 120° or 60°. Any line which is parallel to one of the isometric axes is called an *isometric line.* A line, which is not parallel to either of the isometric axes is called *a non-isometric line.* In an isometric protection of a cube the faces of a cube, or any planes parallel to them, are called *isometric planes.*

6.6. ISOMETRIC SCALE

Figure 6.5. (a) shows the isometric projection of a cube the top of which appears as a rhombus **a_1' b_1' f_1' e_1'**, the true shape of this face is represented by **a' b_1'' f_1' e_1''**,. It can be noted that all the four sides of the top square are foreshortened in isometric projection. Thus **a_1' b_1'** is the isometric length of **a_1' b_1''** and the actual measurement would show that the isometric length is 81.65% of the true length, or approximately equal to 82%.

To draw the isometric scale, as shown in *Fig. 6.5. (b),* from any point on a horizontal line draw two lines inclined at 30° and 45° respectively. Along the 45° line set out a true length scale, and from it draw vertical lines, *i.e.,* 90° to the horizontal line, to cut the 30° line, then the corresponding scale projected on 30° line is the isometric scale.

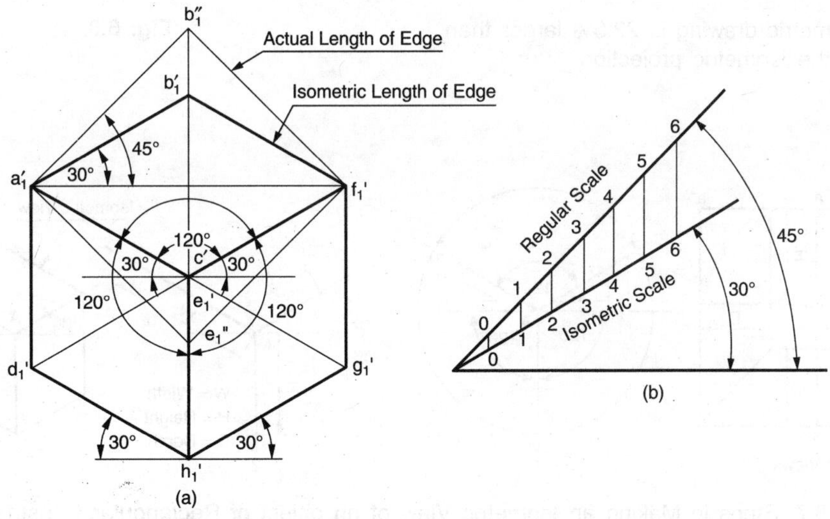

Fig. 6.5. Method of drawing isometric scale

6.7. ISOMETRIC DRAWING

As discussed above, the lines in isometric projection do not show the true lengths of the object edges, but are foreshortened by a definite amount. In making such projections, either it should be projected from the orthographic views of the object, as discussed already, or a special isometric scale should be used for transferring measurements. To avoid this tedious construction, if the foreshortening of the lengths is ignored and the true lengths are laid out along the isometric axes or along the isometric lines, the view obtained is called *isometric drawing*.

Figure 6.6 shows that the isometric drawing is slightly larger (22.5%) than the isometric projection. The variation in size does not effect the use of the regular scale, the construction of an isometric drawing is much simpler than the construction of isometric projection.

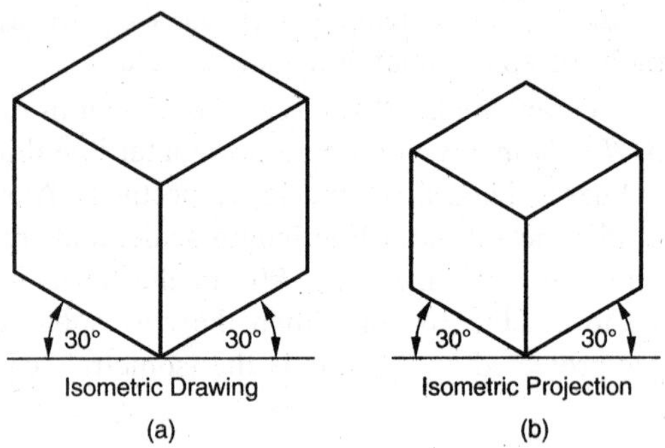

Fig. 6.6. Isometric drawing is 22.5% larger than the isometric projection

6.8. RECTANGULAR CONSTRUCTION

The steps in making an isometric drawing of a simple rectangular object are illustrated in *Fig. 6.7*. The axes are first drawn at an angle of 120° with each other, the measurements are made equal size along the axes, and the remaining lines are drawn parallel to their corresponding axes.

6.9. USE OF INVISIBLE LINES

The use of hidden lines in isometric drawing is governed by the same rule as in all other types of projection, *i.e.*, invisible lines are omitted unless they are needed to make the drawing clear. A case where invisible lines are needed is illustrated in *Fig. 6.8* in which a projecting part can not be clearly described without the use of the invisible or dotted lines.

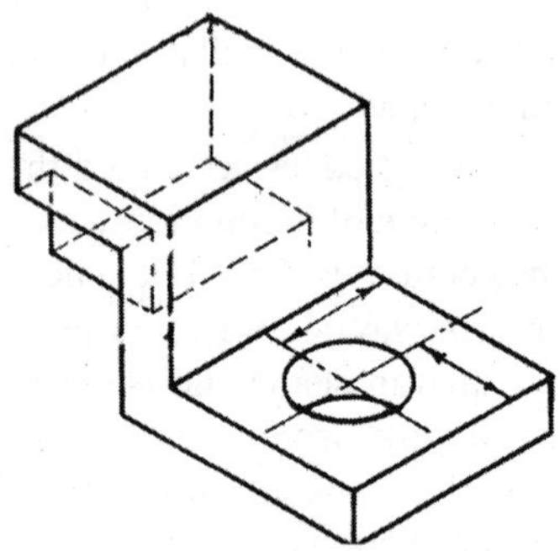

Fig. 6.8.

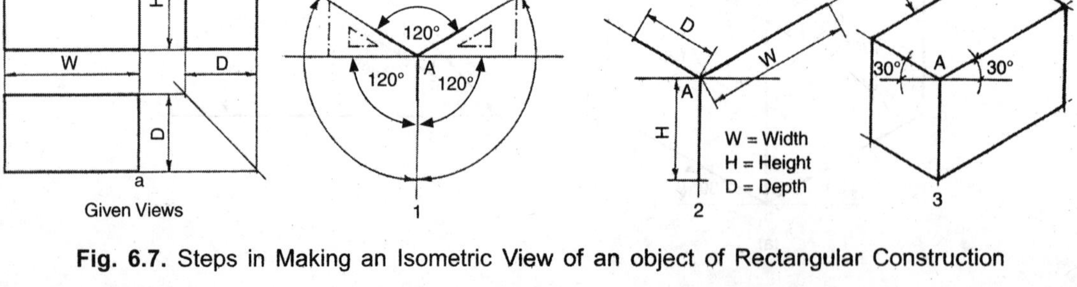

Fig. 6.7. Steps in Making an Isometric View of an object of Rectangular Construction

6.10. CENTRE LINES

The use of center lines in isometric drawing is governed by the same rule as in multi-view drawing, *i.e.*, centre lines are drawn if they are required for dimensioning, *Fig. 6.8*. In general, centre lines should be used sparingly and be omitted in case of doubt. The use of too many centre lines may produce a confusion of lines which detracts from the clearness of a drawing.

6.11. NON-ISOMETRIC LINES

Since the only lines of an object which are true length in an isometric drawing are the isometric axes or the lines parallel to them, *i.e.*, the isometric lines. The non-isometric lines can not be set off directly according to their length. For example, in the given view of *Fig. 6.9* the inclined lines AC and AB are projected true length in the top view, but since none of them is parallel to either of the isometric axes, they will not be true length in the isometric.

Such lines are drawn in isometric view by *means of box construction and offset measurement,* in which the end points of the lines are located. Point A is located after drawing the enclosing box, by means of offset measurement 'X' in step I, and construction lines are then drawn to complete rest of the construction, as in step *II*. Finally the required lines are made heavy to complete the isometric view, as in step *III, Fig. 6.9*.

It will be realized on measuring that AC is shorter and AB is longer in the isometric view than their corresponding lengths in the given orthographic top view.

6.12. OFFSET LOCATION MEASUREMENTS

As shown in *Fig. 6.10. (a), (b)* one part is located with respect to another in isometric drawing by

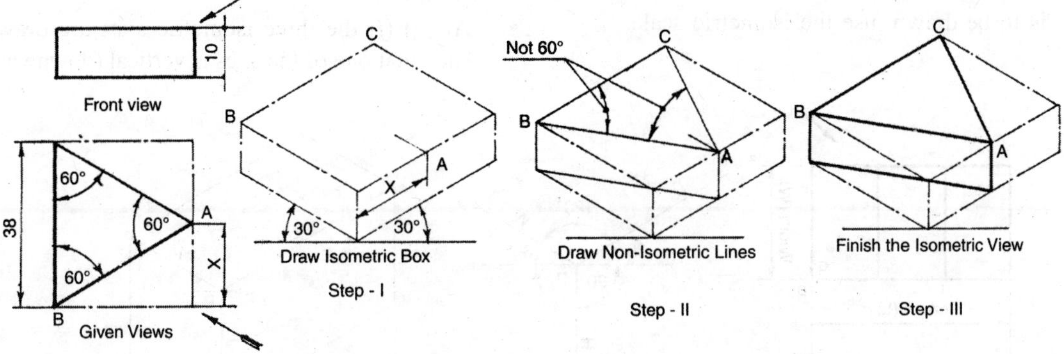

Fig. 6.9. Non-isometric lines and angles in isometric

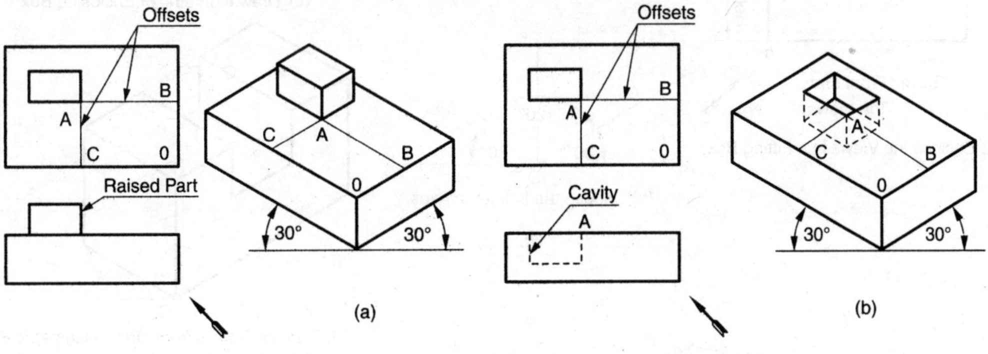

Fig. 6.10. Isometric offset location measurements

taking offset lines, like CA and BA, in the multi-view drawing and then transferring these lines to the isometric. These measurements are called offset measurements, and since they are parallel to certain edges of the main block, in the multi-view drawing, they will be parallel, respectively, to the same edge in the isometric view.

6.13. METHODS OF DRAWING AN ISOMETRIC VIEW

Some important methods of drawing isometric views are discussed below:

1. Box Method

In this method the object, to be drawn, is supposed to be enclosed in a rectangular box of the size in which it may just fit. It is drawn around the object in the given orthographic views. The different steps in making the isometric drawing or projection, with the help of this box, are as follows:

1. Draw the orthographic views of the object to the desired scale. If isometric view or projection is to be drawn, use the isometric scale.
2. Enclose the views in a rectangular box, by very light lines.
3. On the isometric axes, construct the isometric of the enclosing box.
4. Taking measurements along isometric axes and isometric lines, transfer the various details of the object, from the rectangular orthographic box, to the isometric box.
5. Finish the view by brightening the visible lines. Invisible lines are usually omitted in isometric views. Since the enclosing box and construction lines are drawn very light, they may not be erased.

Isometric of a Block, With Isometric Lines Only, By Box-Method

To make the procedure clear a very simple object, containing only isometric lines is taken and the different steps outlined above are illustrated in *Fig. 6.11*.

- At (*a*), the orthographic views of the object are shown.
- At (*b*) (*i*) the three isometric axes are drawn, such that one of the axes is vertical (downward)

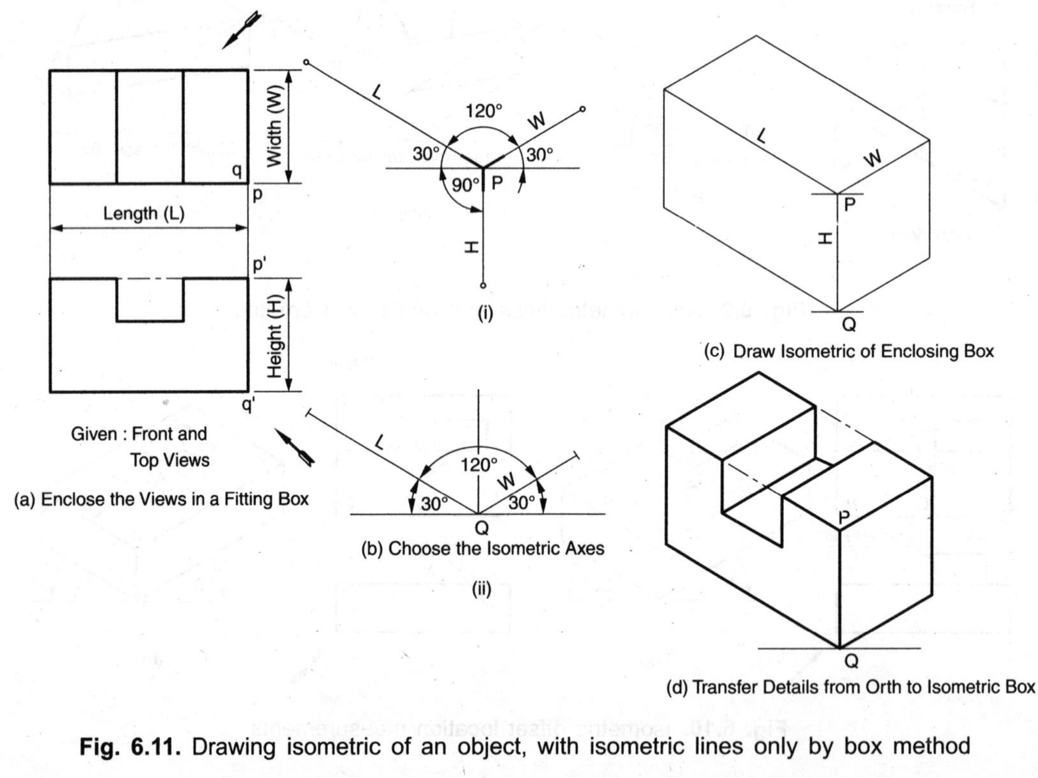

Fig. 6.11. Drawing isometric of an object, with isometric lines only by box method

and the other two making 30° angles with the horizontal.

- At (*b*) (*ii*) is shown an alternate method of locating the axes.
- Along these axes, dimensions **L, W,** and **H** are laid off, and as shown at (*c*) the isometric box is completed.
- Then, taking the measurements along the isometric lines, the details of the object are transferred from the orthographic view to the isometric box.
- Finally all the details which were drawn by light lines are brightened (made thick) to make the figure more distinct as at (*d*).

Isometric of an Object, With Non-Isometric Lines, By Box-Method

Since the non-isometric lines neither show their true lengths nor the isometric lengths in isometric view the ends of such a line should first be located and then joined to draw the isometric view of the line.

- Referring to *Fig. 6.12 (a)*, two views of an object, the edges 1-3 and 2-4 of which are non-isometric lines, are shown.
- To draw its isometric view proceed on the same lines as in the previous case.
- Construct the isometric view of the enclosing rectangular box, and then locate the points 1 and 3, in the front isometric plane, by laying off the dimensions 15 mm and 10 mm along the corresponding lines, as shown at (*b*).
- Similarly locate points 2 and 4 on the rear isometric plane. Draw lines 1-3 and 2-4 and complete the figure as shown at (*c*).
- From this figure it can be noted that the lines which are parallel on the object are parallel in the isometric view also.

In *Fig. 6.13*, the method of drawing the isometric view of a triangular pyramid, by box-method, is illustrated. Note, in this case, that all the edges, except one which is in the plane of the box, are non-isometric. Also it can be noted in this figure that the lines which are not parallel on the object are also not parallel in the isometric view.

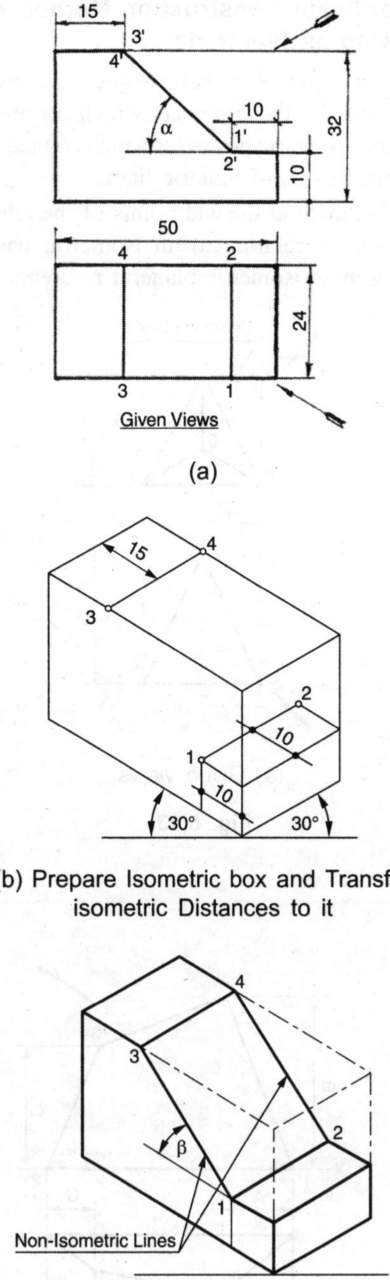

(a)

(b) Prepare Isometric box and Transfer isometric Distances to it

(c) Complete the isometric View

Fig. 6.12. Drawing isometric of an object with non-isometric lines by box-method

2. Co-ordinate Construction Method of Drawing an Isometric

- This method of making isometric drawing is adopted for those objects which are made up of non-isometric planes, or which contain a large number of non-isometric lines.
- In this method the end points of the edges are located in relation to an isometric base line lying in an isometric plane or reference plane.

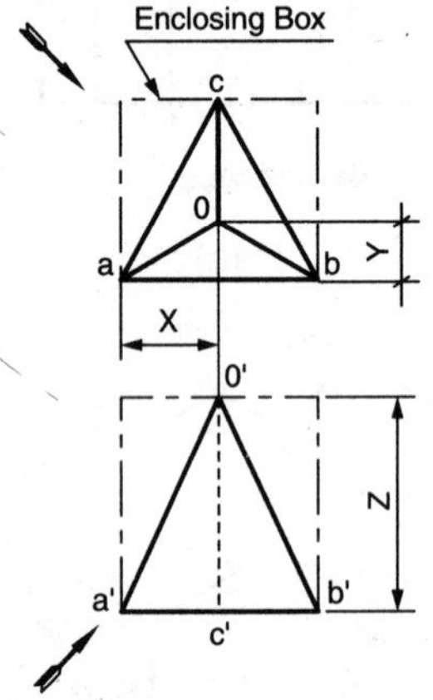

(a) Given views

Fig. 6.13.

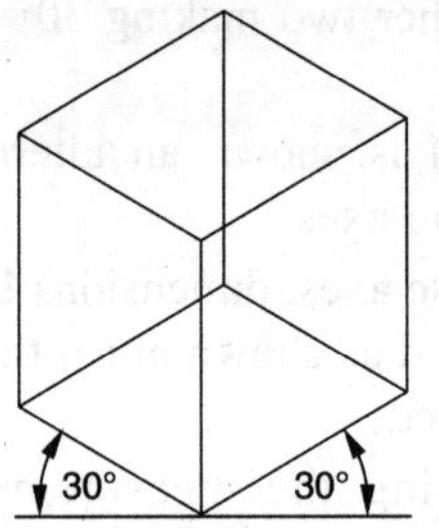

(b) Draw isometric Box

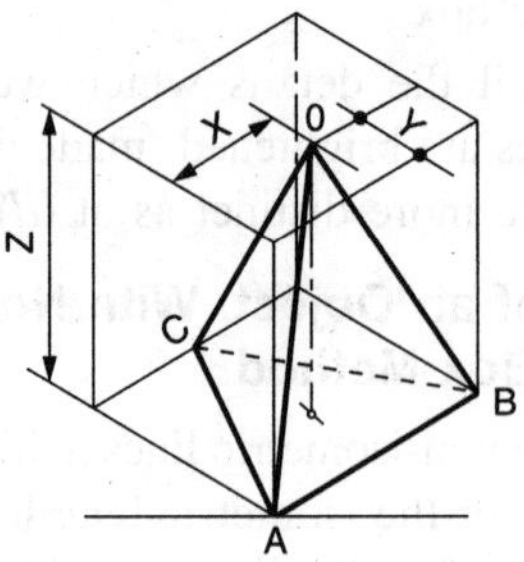

(c) Transfer detail From Orth. to Box

Fig. 6.13. Isometric of a triangular pyramid by box method

- *Figure 6.14* shows the construction of the isometric drawing of an object, in which except base all other planes are non-isometric. One of the base edges, 1-2, is made to lie on the base line and all other points are located by isometric co-ordinates.

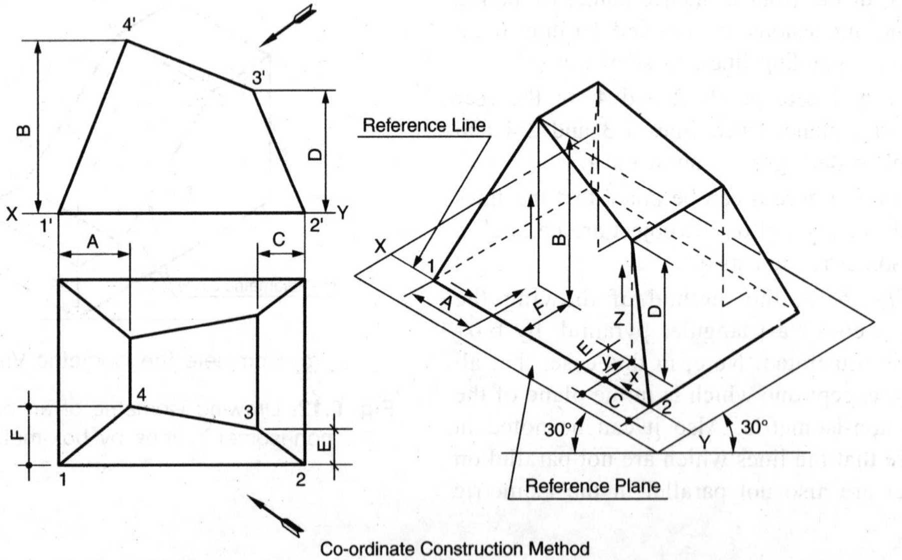

Co-ordinate Construction Method

Fig. 6.14. Use of co-ordinates in drawing an isometric

- As shown in orthographic views, the co-ordinates of point 3 are distance 'C' along X-axis, 'E' along Y-axis and 'D' along Z-axis.
- For location point 3 in the isometric view, measure distance 'C' along the X-axis and then distance 'E' parallel to Y-axis, and finally distance 'D' parallel to Z-axis.
- In the same manner the other points can also be located, and by joining these points, in proper order, the final shape of the object is obtained.

For the sake of comparison of the coordinate construction method to the box-method, the isometric view of a tetrahedron is obtained by these two methods, as illustrated in *Fig. 6.15*. The two methods are nearly alike.

3. Centre Line Layout Construction Method of Drawing an Isometric

The method is adopted for drawing the isometric view of those objects which consist of a number of circular parts lying on the same or parallel centre lines. In *Fig. 6.16* this method of obtaining isometric is illustrated.

In this method, first the various centre lines are laid out along the isometric lines, taking the main centre line as the reference line and by taking measurements along the isometric lines. The various isometric squares are drawn for the circles and

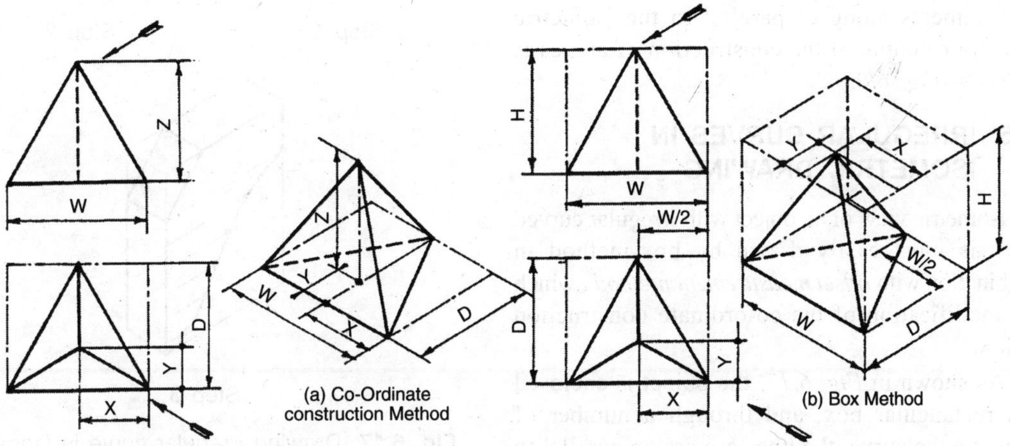

Fig. 6.15. Comparison of box and co-ordinate construction method

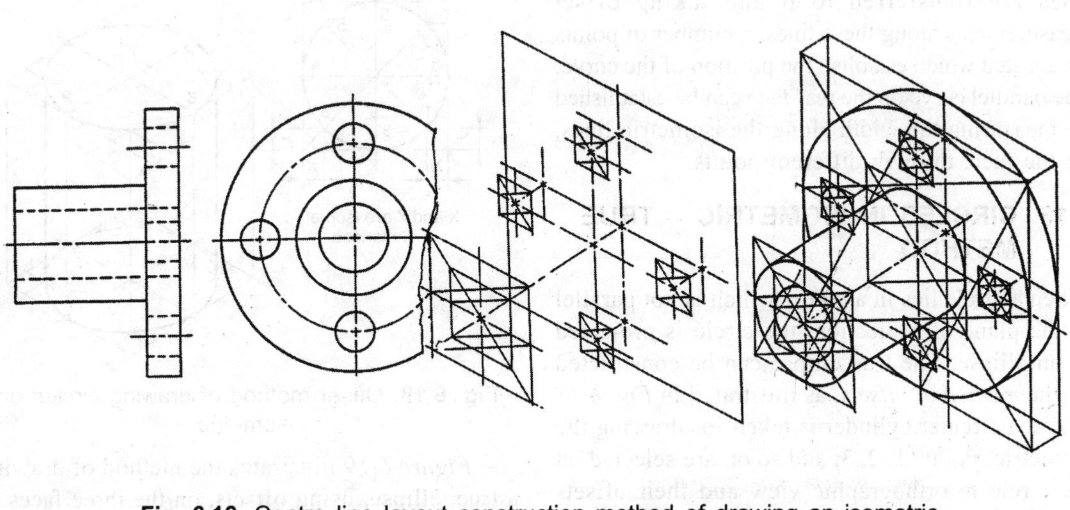

Fig. 6.16. Centre line layout construction method of drawing an isometric

the isometric view is completed by drawing dark lines for the visible edges and ellipses for the circles.

6.14. ANGLES IN ISOMETRIC VIEW

Only those angles of an object will be true size in isometric view which lie in planes parallel to the plane of projection. As almost all edges and faces of an object, usually, are inclined to the plane of projection, it follows that angles, generally, will not be projected in ture size. For example, in multi-view drawing of a cube, none of the 90° angles will be 90° in the isometric view.

Since angles can not be set off directly in isometric drawing, it is necessary to locate the end points of the lines, which subtend the angle or, in other words, they must be translated to linear measurements along or parallel to the isometric axes. An example of the construction of angles is shown in *Fig. 6.9.*

6.15. IRREGULAR CURVES IN ISOMETRIC DRAWING

The isometric view of an object with irregular curved edges is most easily drawn by box-method in combination with *offset measurement method.*, which is a modification of the co-ordinate construction method.

As shown in *Fig. 6.17,* the object is enclosed in a rectangular box, and through a number of points on the curve, the lines are drawn parallel to the line. After constructing isometric box, the parallel lines are transferred to it and taking offset measurements along these lines, a number of points are located which establish the position of the curve. The parallel curve on the rear face can be established by measuring the width along the isometric lines, on the face, through different points.

6.16. CIRCLES IN ISOMETRIC – TRUE METHOD

When a circle lies in a plane, which is not parallel to the plane of projection, the circle is projected as an ellipse. The true ellipse can be constructed by the *method of offsets*, as illustrated in *Fig. 6.18* where a circular cylinder is taken for drawing the isometric. Points 1, 2, 3, and so on are selected on the circle in orthographic view and their offsets are transferred from the othographic view to the isometric box, thus locating their corresponding positions in the isometric view. Then all the points, plotted thus, are joined by a smooth curve. If the number of points taken on the circle is sufficient, this method gives the most accurate isometric circle.

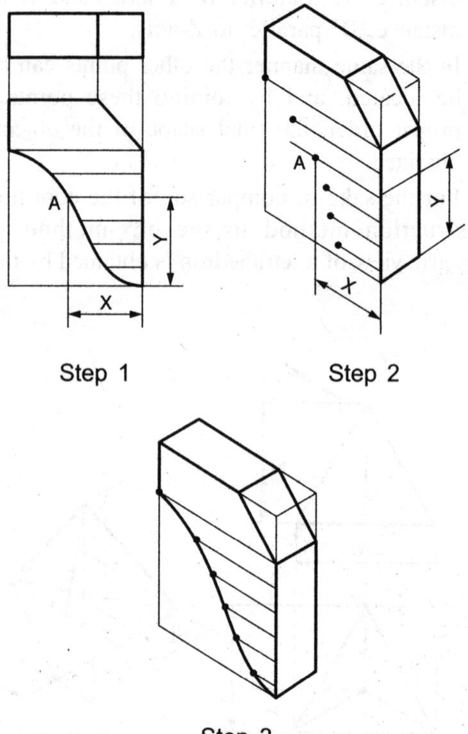

Fig. 6.17. Drawing irregular curve in isometric

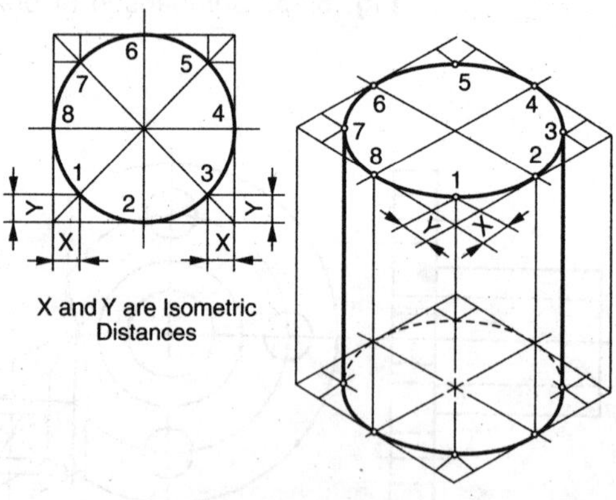

Fig. 6.18. Offset method of drawing circles on isometric

Figure 6.19 illustrates the method of drawing a true, ellipse, using offsets, on the three faces of a cube.

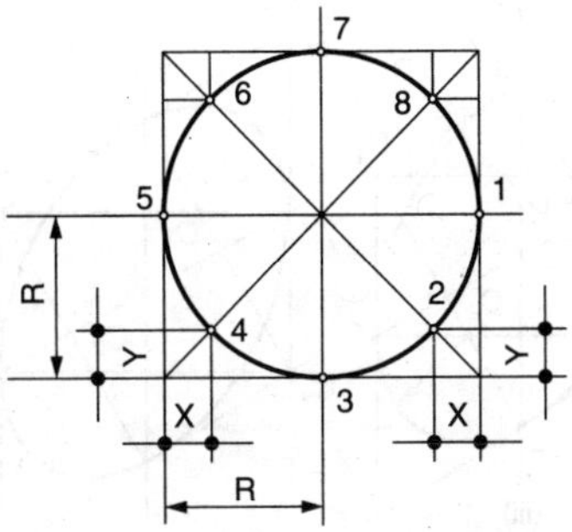

Fig. 6.19. (a) Given Circle

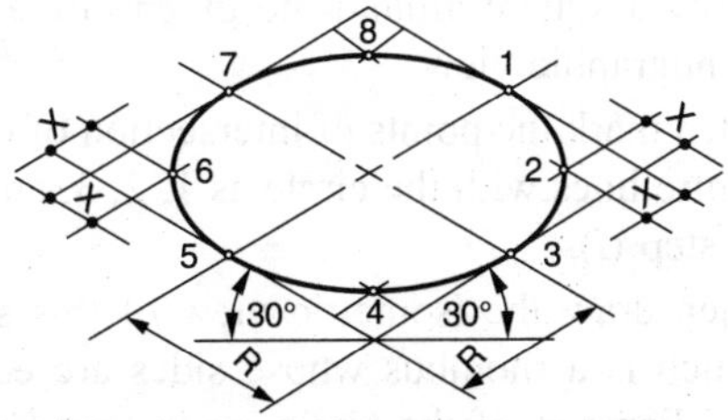

Fig. 6.19. (b) Circle in Top View

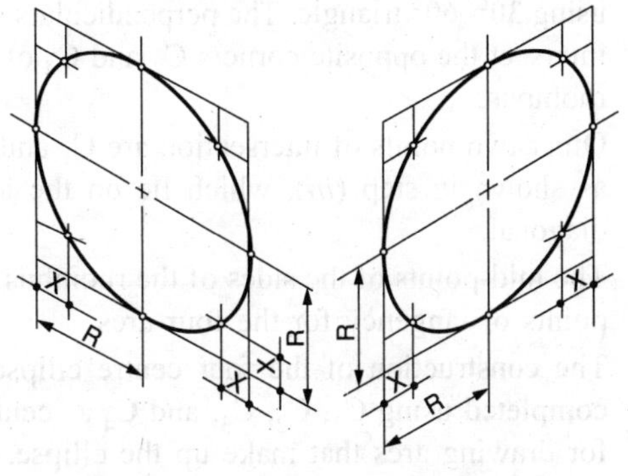

Fig. 6.19. (c) Circle in Front and Side Views

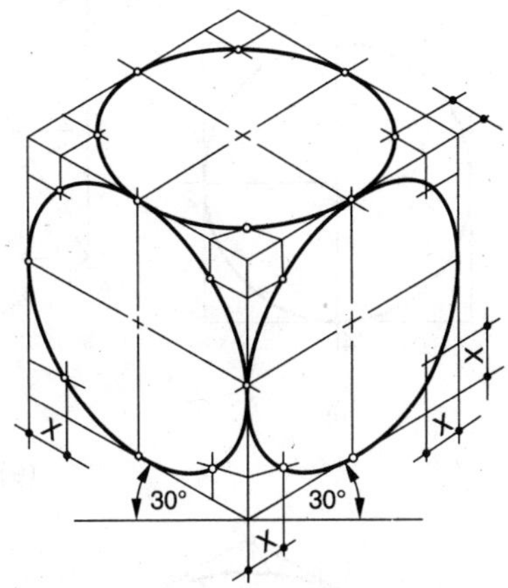

Fig. 6.19. (d) Isometric Projection of the circle on Three Faces of a Cube

Fig. 6.19. Offset Actual or ture method of drawing circles in isometric on three faces of a cube

Figure 6.20 is drawn to illustrate the coordinate construction method of drawing an isometric circle in four stages.

6.17. FOUR-CENTRE METHOD FOR DRAWING AN APPROXIMATE ELLIPSE

The true ellipse is seldom used in isometric drawing because its construction is time consuming and, in most cases approximate methods are sufficiently accurate for all isometric drawings.

The method frequently used, called the four centre method of drawing an ellipse, is illustrated in *Fig. 6.21 (a)*. It can be used only for ellipse in isometric planes.

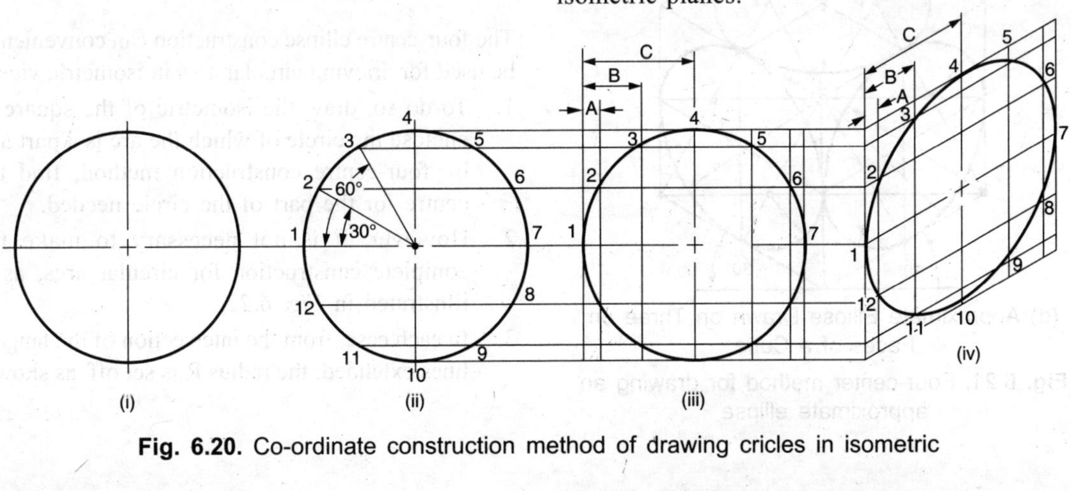

Fig. 6.20. Co-ordinate construction method of drawing cricles in isometric

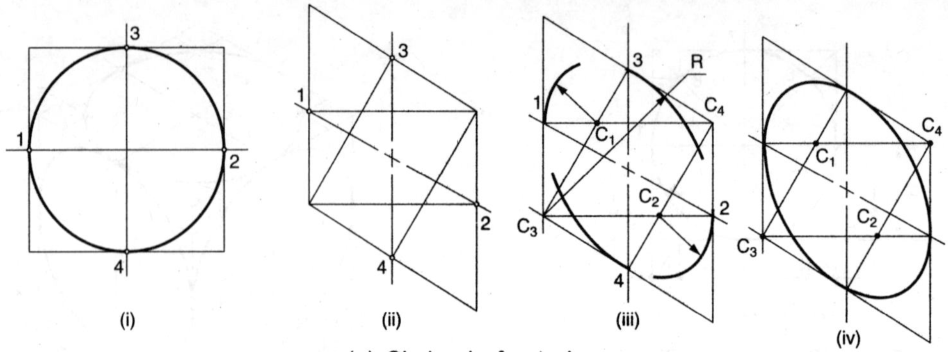

(a) Circles in front view

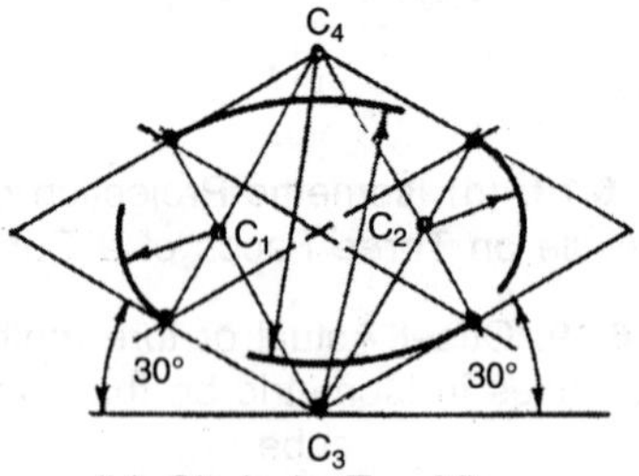

(b) Circle in Top View

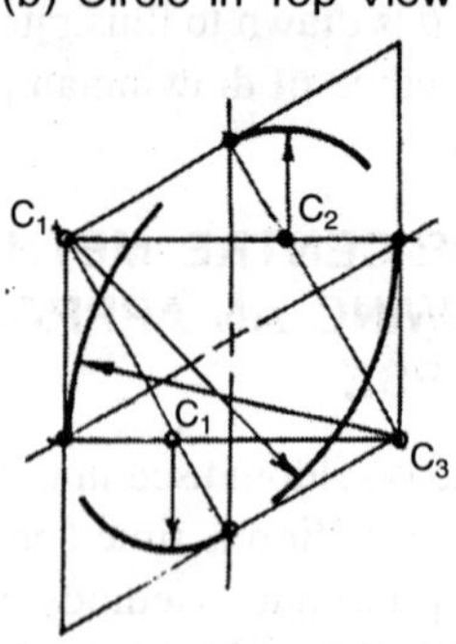

(c) Circle in Right Side View

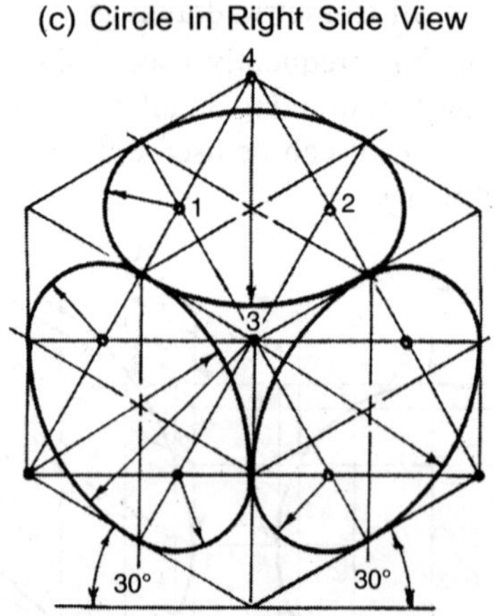

(d) Approximate Ellipse Drawn on Three on Faces of a Cube

Fig. 6.21. Four-center method for drawing an approximate ellipse

1. Draw a square around the given circle in the orthographic view.
2. Then mark the points of intersection of the two centre lines with the circle as 1, 2, 3, and 4, as in step (*i*).
3. Then draw the isometric view of this square, which is a rhombus whose sides are equal to the diameter of the circle, as in step (*ii*).
4. Then erect perpendicular bisectors to each side, using 30°–60° triangle. The perpendiculars will intersect the opposite corners C_3 and C_4 of the rhombus.
5. Other two points of intersection are C_1 and C_2 as shown in step (*iii*), which lie on the long diagonal.
6. The mid-points of the sides of the rhombus are points of tangency for the four arcs.
7. The construction of the four centre ellipse is completed using C_1, C_2, C_3, and C_4 as centres for drawing arcs that make up the ellipse.

6.18. ARCS OF CIRCLES IN ISOMETRIC

The four-centre ellipse construction can conveniently be used for drawing circular arcs in isometric views.

1. To do so, draw the isometric of the square to enclose the circle of which the arc is a part and by four-centre construction method, find the centre for the part of the circle needed.
2. However, it is not necessary to make the complete construction for circular arcs, as is illustrated in *Fig. 6.22*.
3. In each case, from the intersection of the tangent lines extended; the radius R is set off, as shown.

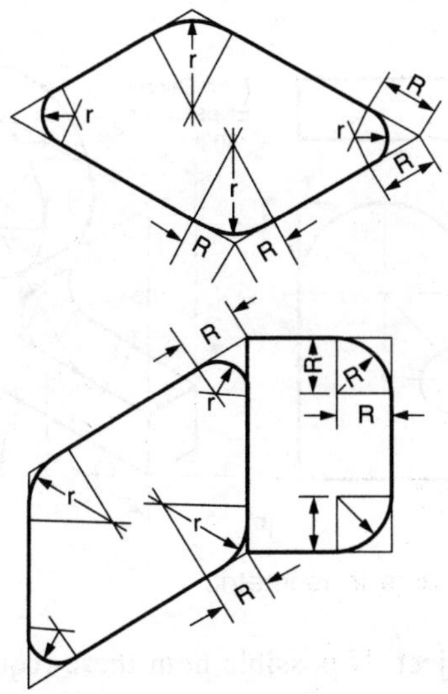

Fig. 6.22. Drawing arcs of circles in isometric

4. At each point thus plotted, perpendiculars to the lines are erected. Their point of intersection being the centre of the required arc.

6.19. SPHERE IN ISOMETRIC VIEW

The isometric drawing of any curved surface is the envelope of all lines which can be drawn on that surface. For a sphere, its great circles, *i.e.*, the circles in which the sphere is cut by any plane through the centre of the sphere, may be taken to serve as the lines on its surface.

As all great circles, except those which are either perpendicular or parallel to the plane of projection, are shown as isometric ellipses of equal major axes, it follows that their envelope is a circle whose diameter is the major axis of the ellipses.

1. Referring to *Fig. 6.23* two orthographic views of a sphere enclosed in a construction cube are shown.
2. The isometric drawing of the cube is shown in *Step 1*, together with the isometric of a great circle which lies in a plane parallel to one face of the cube, where the ends of the major axes are shown located by offset measurement *'a'*.
3. Then with radius R, half of the major axis, draw the isometric drawing of the sphere, as shown at *step 2.*
4. The diameter of this circle is 3/2 times the actual diameter of the sphere.
5. Isometric projection of the sphere is equal to the true diameter of the sphere as at *step 3.*

6.20. TANGENT CIRCLES AND ARCS IN ISOMETRIC VIEW

The four-centre method may be used for tangent isometric circles or isometric arcs only when they are tangent to each other at the midpoints of the sides of their enclosing rectangles, *Fig. 6.24 (a)*.

If the tangency points occur at any other places, the isometric circles will overlap or miss, as shown in *Fig. 6.24 (b)*, because of their departure from the true ellipse.

In such cases, either the offset measurement method be used, or some other approximation be made.

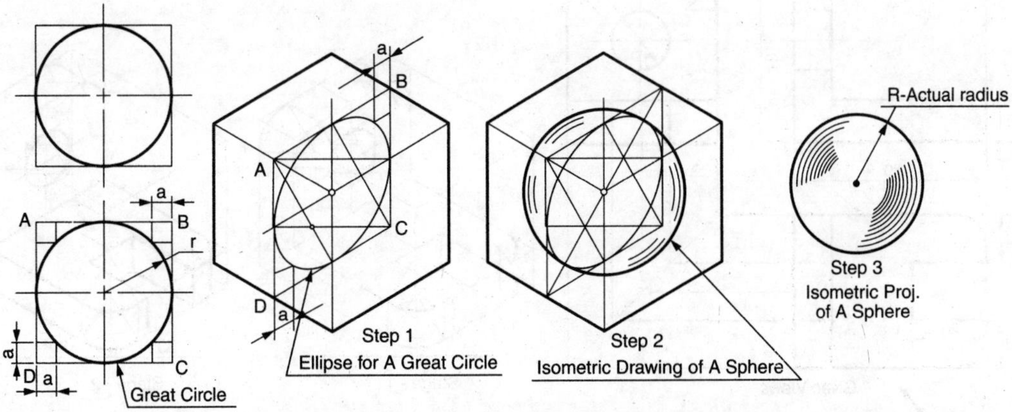

Fig. 6.23. Isometric of a sphere

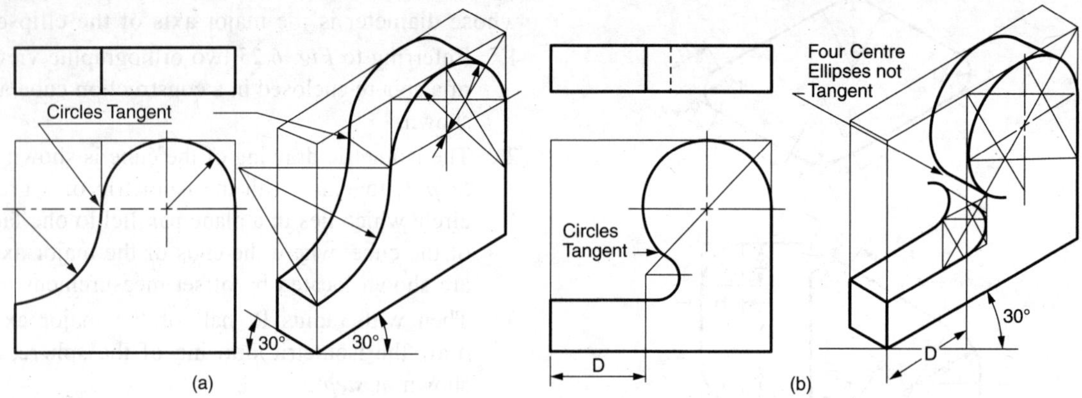

Fig. 6.24. Drawing tangent circles and arcs in isometric

Figure 6.25. illustrates the systematic approach of making an isometric view. The object is divided into a number of parts and separate isometric construction boxes are drawn for each part. The sequence and treatment of arcs in the isometric view has also been illustrated by four-centre method.

6.21. CHOICE OF POSITION OF ISOMETRIC AXES

The isometric axes may be placed in any desired position according to the requirements of the problem, but the axes are determined from the position from which the object is usually viewed, or by the position, which best describes the shape of the object. If possible both these requirements should be met.

In order to reveal certain faces containing some important details of the object, it may be placed in a position different to the ordinarily adopted vertical position, *Fig. 6.26 (a)*.

Sometimes, when it is desirable to represent a part as viewed from below, the axes may be reversed, as shown in *Fig. 6.26 (b)*.

When the object is quite long the axes are placed such that the long axis is horizontal, *Fig. 6.26 (c)*.

Whatever may be the direction of the axes, the procedure followed in constructing the isometric view remains the same.

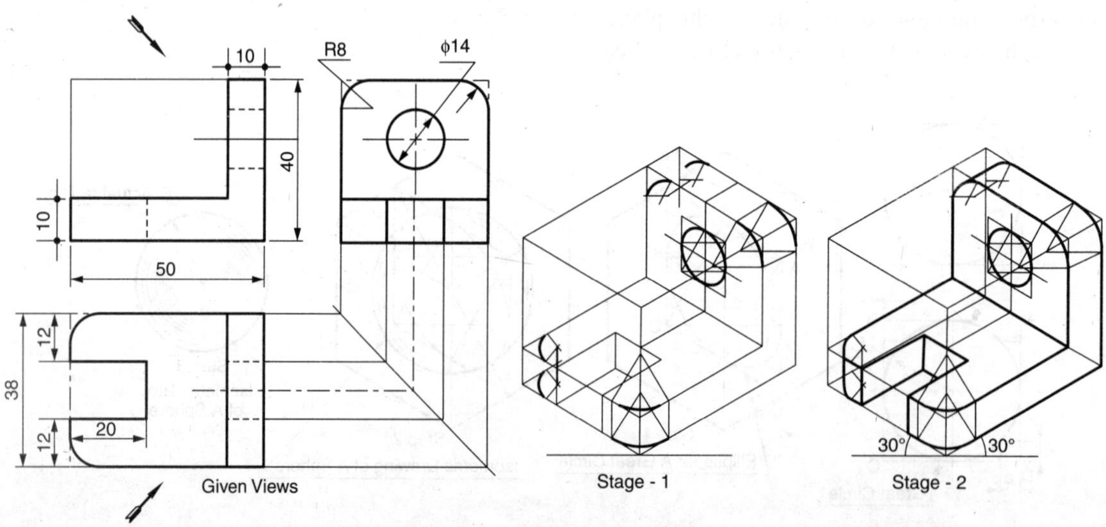

Fig. 6.25. Steps in making an isometric

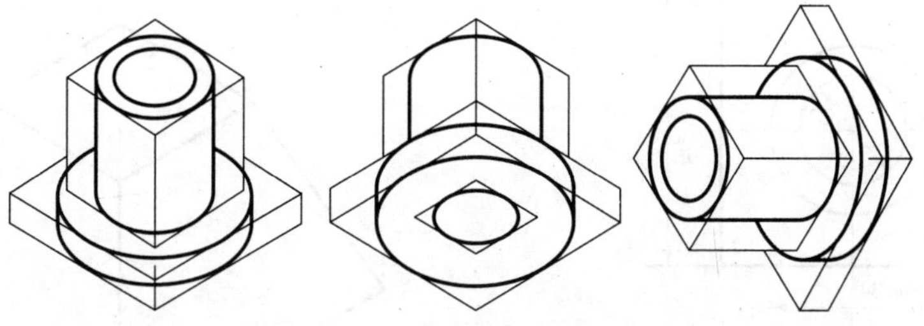

(a) (b) (c)

Fig. 6.26. Choice of position of isometric axes

6.22. ISOMETRIC SECTIONS

As the aim of isometric drawing is to represent the pictorial appearance of an object, usually outside views are used. But sometimes to show the interior details, sectional views are made. The cutting planes are taken as isometric planes and the section lining is done in the direction that gives the best effect. For a full section the face is drawn first and then the details of the object, remaining behind, are added, *Fig. 6.27*. For drawing a half section, first the object as a whole, is outlined and then the front quarter is removed, as in *Fig. 6.28*.

6.23. ISOMETRIC DIMENSIONING

The general rule for dimensioning the multi-view drawings are discussed in detail in chapter 18. All those rules hold good in isometric drawing and in addition to those the following rule must be observed.

1. All extension and dimension lines must be isometric lines, lying in isometric planes, *Fig. 6.29 (a)*. In *Fig. 6.29 (b)* is illustrated a very common error in dimensioning non-isometric line and also the correct method of dimensioning it.
2. If possible, apply the dimensions to visible surfaces.
3. Note may be lettered either on isometric guide lines or as on ordinary drawing. Only vertical type of lettering should be used, *Fig. 6.30*.
4. Dimension numerals should appear to be in plane of the surface whose dimensions these indicate. *Figure 6.29* shows the method of constructing figures on different faces of a cube. Note that the sides of the enclosing parallelogram are parallel to the pictorial axes.
5. Dimensioning examples are shown in *Fig. 6.30 (a), (b)*.

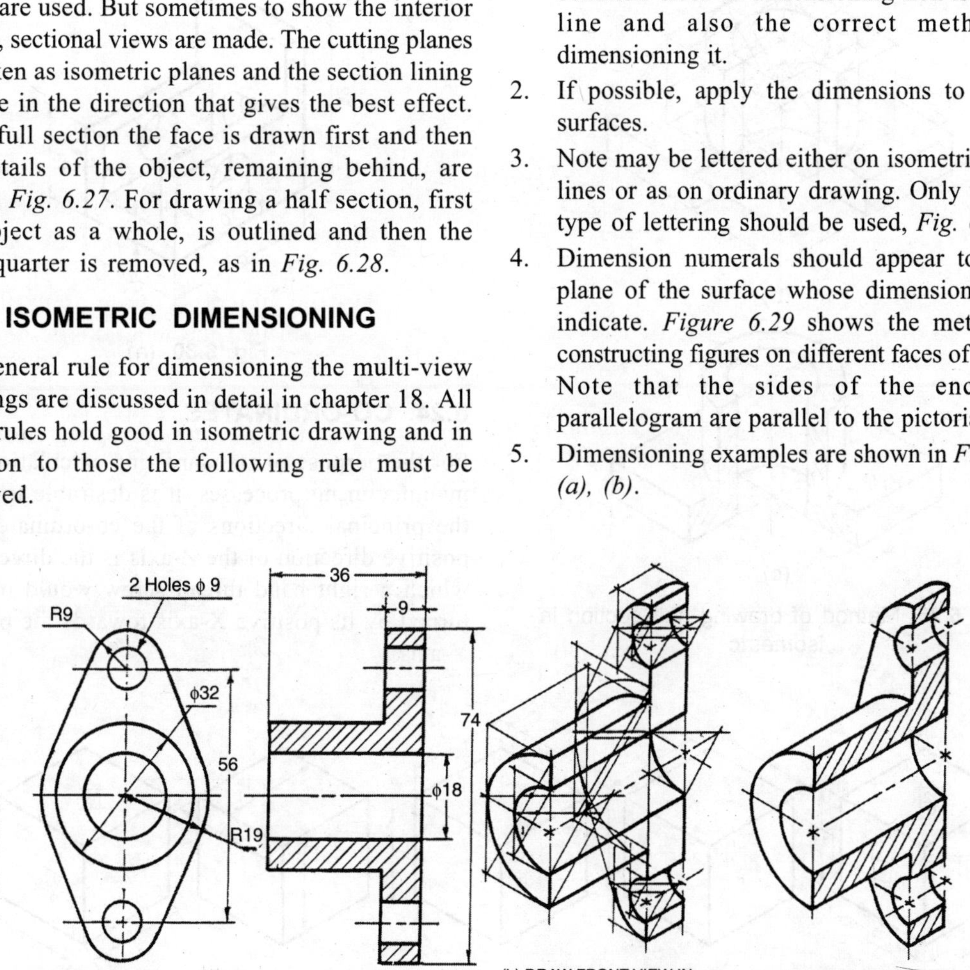

(a) GIVEN ORTHOGRAPHIC VIEWS

(b) DRAW FRONT VIEW IN ISOMETRIC FORM AND ADD. DEPTH DETAIL FROM SIDE VIEW BY BOX METHOD

(c) COMPLETE THE FULL SECTIONAL ISOMETRIC VIEW

Fig. 6.27. Method of drawing full section in isometric

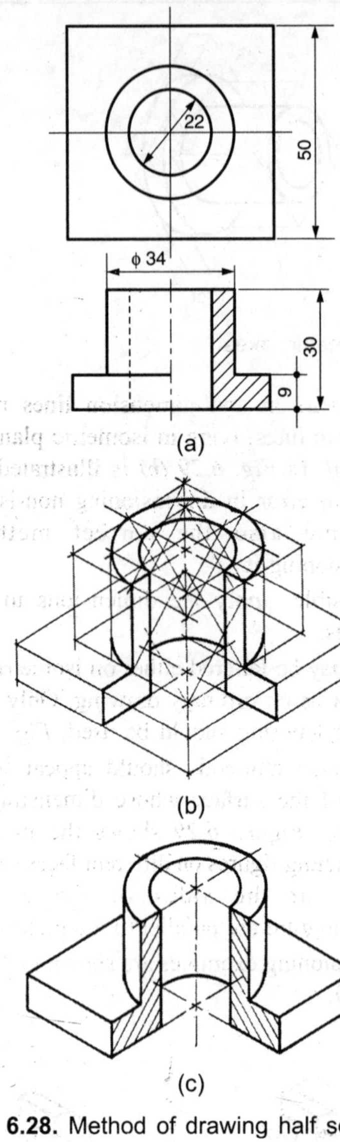

Fig. 6.28. Method of drawing half section in isometric

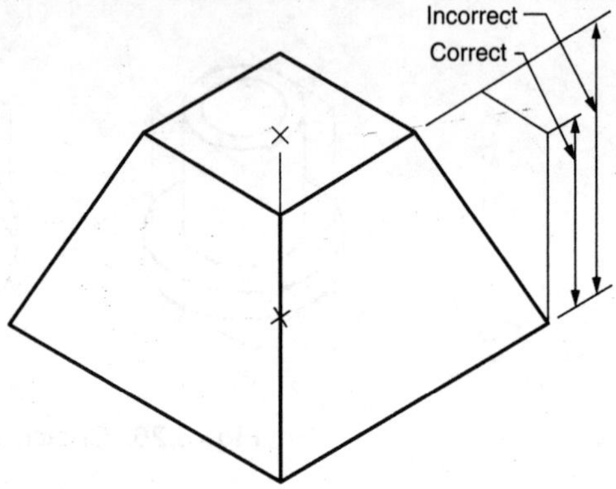

Fig. 6.29. (b)

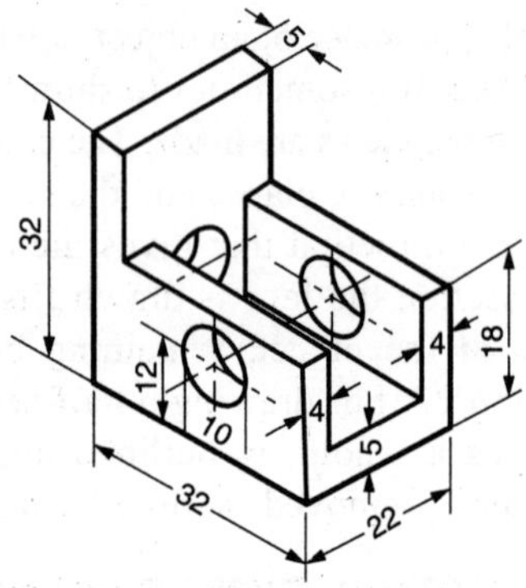

Fig. 6.30. (a)

6.24. CO-ORDINATES

For the purpose of standardizing calculations and manufacturing processes, it is desirable to define the principal directions of the co-ordinates. The positive direction of the Z-axis is the direction in which a right hand thread screw would move if turned by its positive X-axis towards the positive Y-axis.

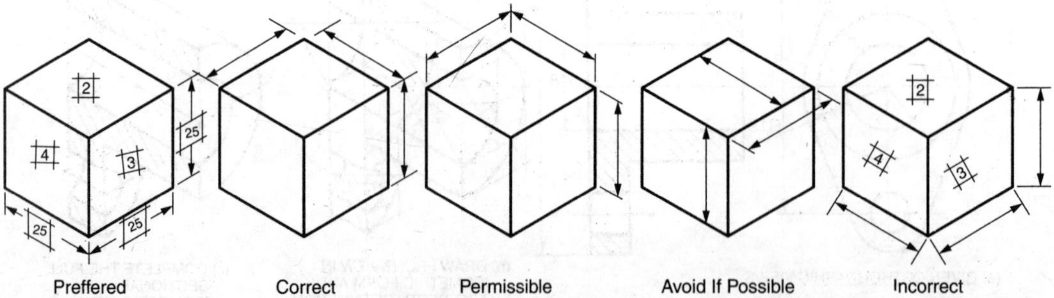

Fig. 6.29. (a) isometric dimensioning

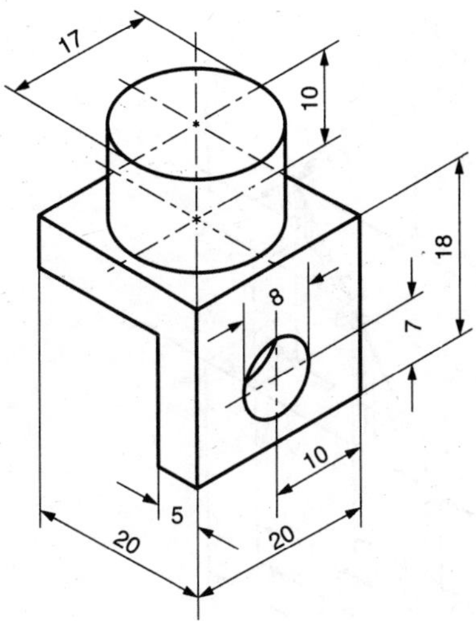

(b)

Fig. 6.30.

All co-ordinate values taken from the origin in the direction of the arrow are positive and those in the opposite direction are negative, *Fig. 6.31.*

The directions of the co-ordinates X, Y, Z, are called the principal directions and the areas enclosed by them are called the principal planes.

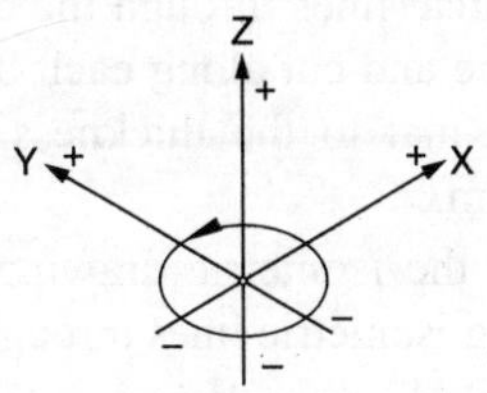

Fig. 6.31.

6.25. REPRESENTATION OF THE SYSTEM OF CO-ORDINATES

In order to provide an unambiguous representation of lines, *e.g.*, pipe bends, in isometric projection, it is necessary to show the principal planes by hatching. The planes of the side view, *i.e.*, co-ordinates Y, Z, and front view, *i.e.*, co-ordinates X, Z, should be hatched vertically and the planes of the top view, *i.e.*, co-ordinates X, Y, should be hatched at –30°, *Fig. 6.32.*

Figure 6.33 shows a bent pipe in isometric projection in the co-ordinate system. The starting point for the drawing and the dimensioning is 1(p_1) with co-ordinates X_1 = 0, Y_1 = 0, Z_1 = 0.

The section 1-2 lies on the X co-ordinate and has co-ordinates X_2 = + 50, Y_2 = 0 and Z_2 = 0.

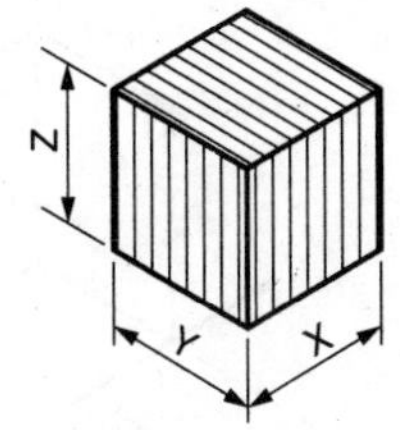

Fig. 6.32.

Section 2-3 lies in the principal plane X, Z and has dimension X_3 and Z_3 and co-ordinates X_3 = + 75, Y_3 = 0 and Z_3 = + 34. The vertical hatching, in *Fig. 6.32*, shows clearly that the plane of bending of the pipe lines in the principal plane X, Z,

Although, in the representation, section 3-4 is a continuation of 2-3, point 4 is outside the principal plane X, Z, and has dimensions X_4, Y_4, and Z_4 ; their co-ordinates are X_4 = + 104, Y_4 = + 12 and Z_4 = + 45. To show the three dimensional bending cleary in the representation, it is necessary to project the co-ordinate point 4 together with 4 onto the corresponding principal planes and to use hatching as shown in *Fig. 6.32.* Sections 4-5 and 5-6 are represnted in a similar manner whilst section 6-7 lies in the direction of the Y- co-ordinate. The co-ordinate values for the bent run shown in *Fig. 6.33* are given in Table 6.1.

Table 6.1. Co-ordinate Values For Bent Pipe run

p_1	X_1 = 0	Y_1 = 0	Z_1 = 0
p_2	X_2 = 50	Y_2 = 0	Z_2 = 0
p_3	X_3 = + 75	Y_3 = 0	Z_3 = + 34
p_4	X_4 = + 104	Y_4 = + 12	Z_4 = + 45
p_5	X_5 = + 118	Y_5 = + 62	Z_5 = + 54
p_6	X_6 = + 26	Y_6 = + 52	Z_6 = + 36
p_7	X_7 = + 26	Y_7 = + 100	Z_7 = + 36

SOLVED PROBLEMS

PROBLEM 6.1. *A cube, 25 mm edge, is placed centrally on the top of another square block, of 40 mm edge and 15 mm thick. Draw the isometric drawing of the two solids.*

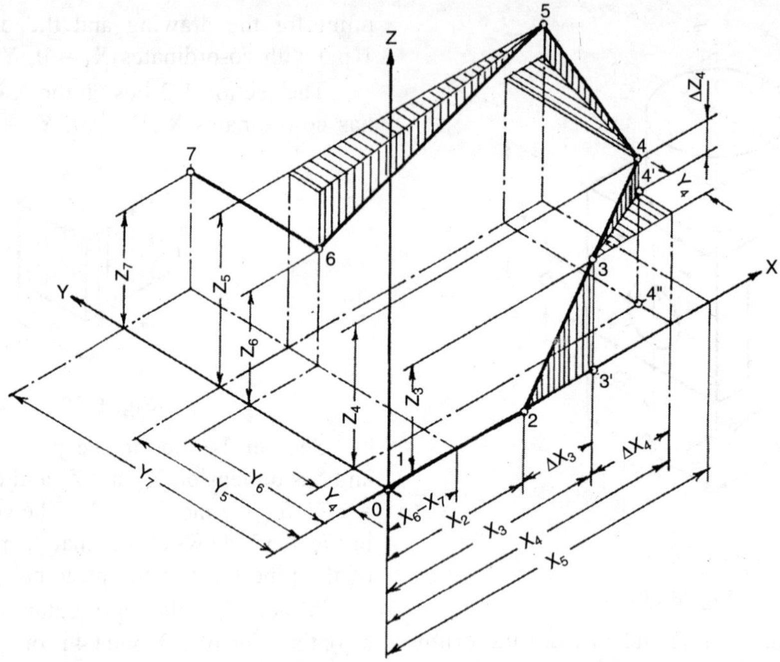

Fig. 6.33.

SOLUTION: Refer *Fig. 6.34.*

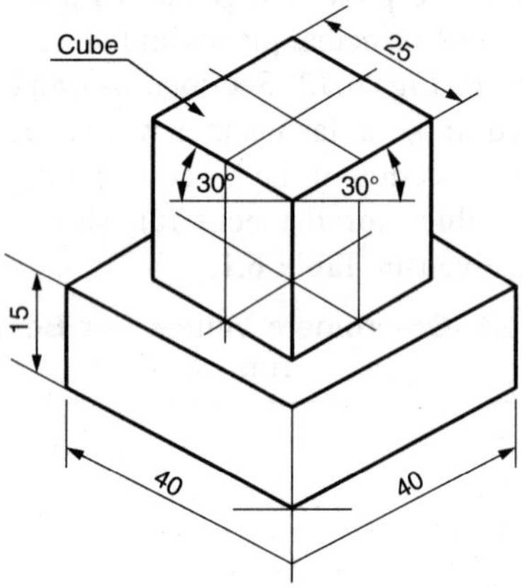

Fig. 6.34.

1. Draw the isometric drawing, using full scale, of the cube following the box-method.
2. Draw two lines at its base surface, parallel to the sides of the surface.
3. Plot points at a distance of half the length of the edge of block, *i.e.,* 20 mm, along these lines on either side of the centre.
4. Through these points draw isometric lines to represent the top of the surface of the block.
5. Draw vertical lines through the corner point of this surface and cut along each of these lines a distance equal to the thickness of the block, *i.e.,* 15 mm.
6. Complete the isometric drawing of the block by drawing isometric lines through these points.
7. Only visible lines or their visible portions need be shown in firm lines.

PROBLEM 6.2. *Draw the isometric drawing of the frustum of a right regular hexagonal pyramid, side of base hexagon is 20 mm, side of top hexagon is 10 mm and height of the frustum is 40 mm.*

SOLUTION: Refer *Fig. 6.35. (a), (b).*

1. Draw the top view of the frustum and enclose the two hexagons, that represent the top and bottom surface of the frustum, in enclosing rectangles.
2. The enclosing rectangle for the base hexagon gives width of the box is same as the height of the frustum.

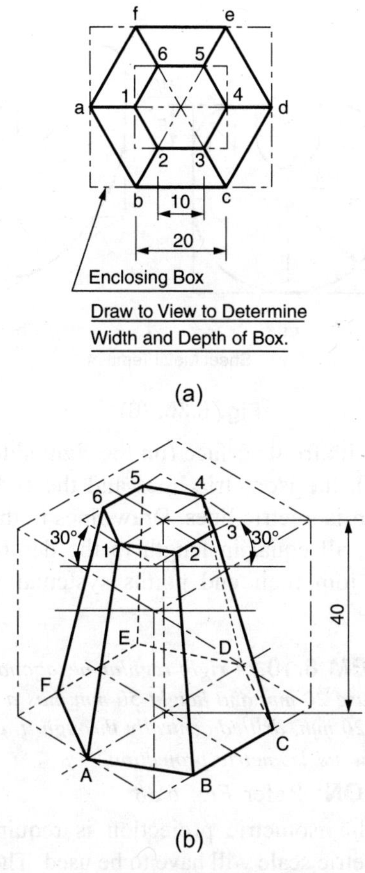

Fig. 6.35.

3. Draw the isometric of the enclosing box and locate the corners of the two hexagons on the box surfaces.
4. Join these points, as shown, to complete the isometric.

Note here that there can be more than one solution to each problem unless the viewing direction and location are defined. Also there are more than one ways of solving each problem.

PROBLEM 6.3-6.8. *The orthographic views of each of six different geometrical solids are given, along with their isometric drawings, in Fig. 6.36. Redraw all of them on A2 size drawing sheet. Time allowed is 2 hours.*

PROBLEM 6.9. *Front and top views of a V-block are shown in Fig. 6.37 (a). Draw its isometric view.*

SOLUTION: Refer *Fig. 6.37 (b)*. The depth of all the edges is same. Draw the isometric view, using isometric scale, for the enclosing box. Then

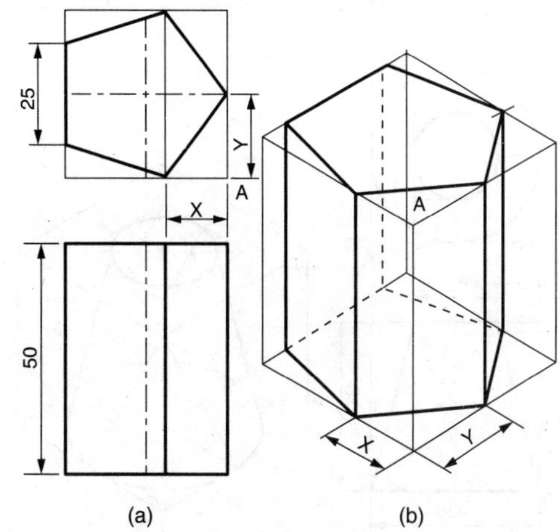

Right Regular Pentagonal Prism

Fig. 6.36. (1)

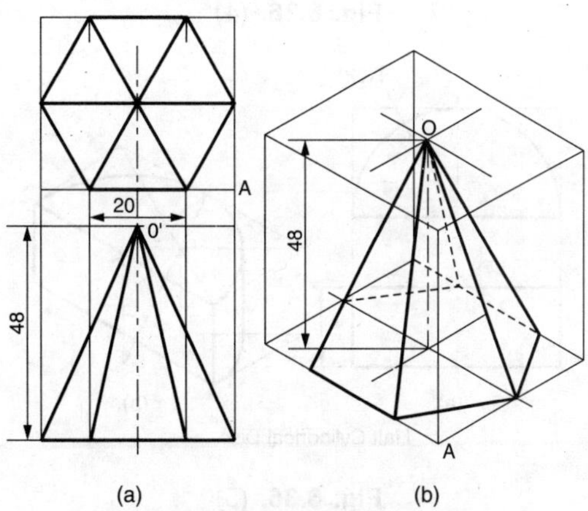

Right Regular Hexagonal Pyramid

Fig. 6.36. (2)

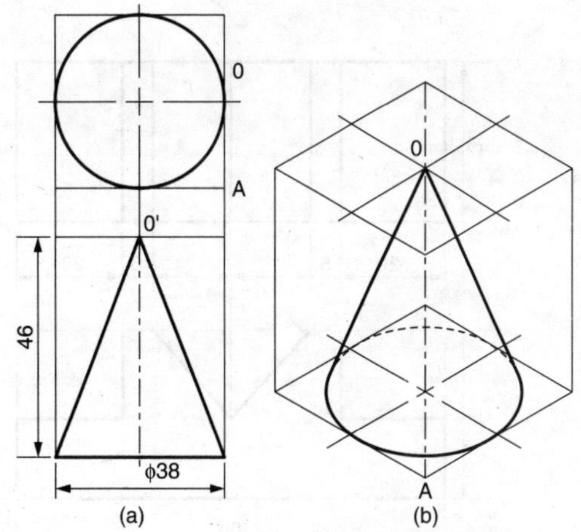

Right Circular Cone

Fig. 6.36. (3)

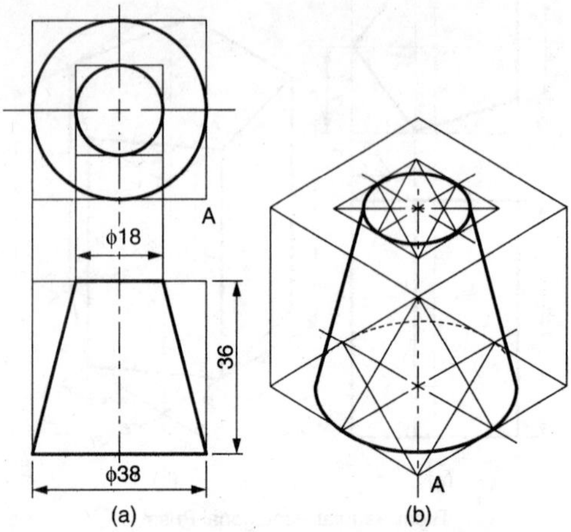

Frustum of A Cone

Fig. 6.36. (4)

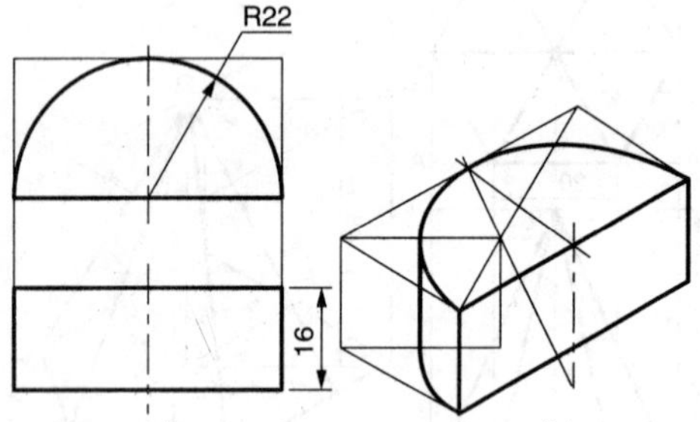

Half Cylindrical Disc

Fig. 6.36. (5)

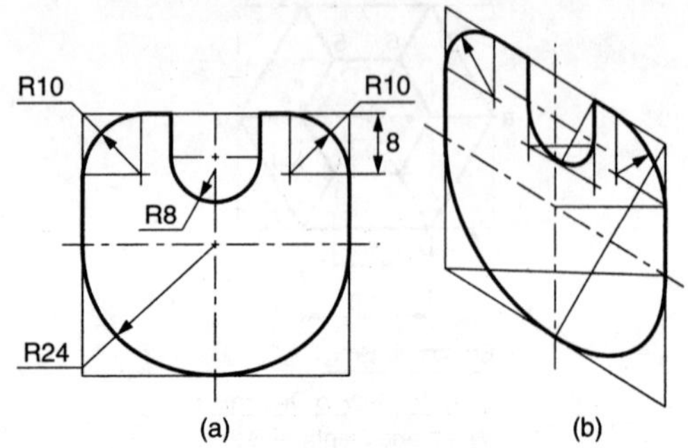

Sheet Metal Templet

Fig. 6.36. (6)

locate on its front surface (on the right side of the box here), the isometric lines and the end points of the non-isometric lines. Draw lines in the depth direction, all equal in length to the depth of the box and join their end points systematically as shown.

PROBLEM 6.10. *A right regular hexagonal prism, edge of base 20 mm and height 50 mm, has a circular hole of ϕ 20 mm, drilled centrally through it, along its axis. Draw its isometric projection.*

SOLUTION: Refer *Fig. 6.38.*

1. As the isometric projection is required, the isometric scale will have to be used. Therefore, draw the isometric scale to start with.
2. Draw the top view of the prism using the isometric scale.
3. Enclose the top view in a construction box to determine its width and depth dimensions. The

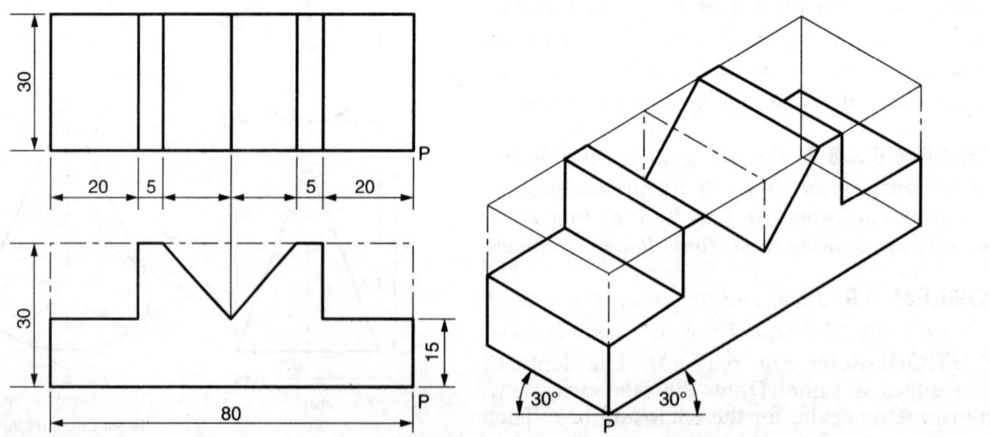

Fig. 6.37.

height of the box will be equal to the height of the prism.

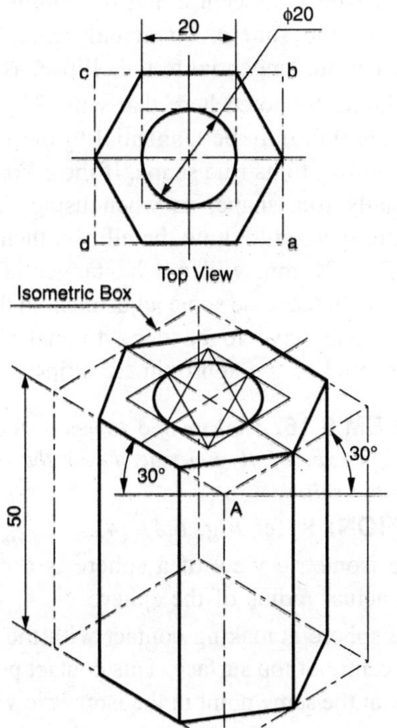

Fig. 6.38.

4. Prepare the isometric projection of the construction box and locate on it the positions of the corner points of the hexagon.
5. Add the ellipse for the hole circle in top view, as shown, and complete the isometric projection.

PROBLEM 6.11. *A right circular cone of ϕ 30 mm base and height 36 mm rests centrally on top of a square block of 48 mm side and 22 mm thick. Draw the isometric projection of the two solids.*

SOLUTION: Refer *Fig. 6.39.*

1. Draw the isometric scale to start with.
2. Draw the isometric projection of the block, using the isometric scale.
3. Draw the two centre lines on the top surface of the block.
4. Draw the rhombus symmetrically about the centre point of the top surface of the block, as shown, for the cone base.

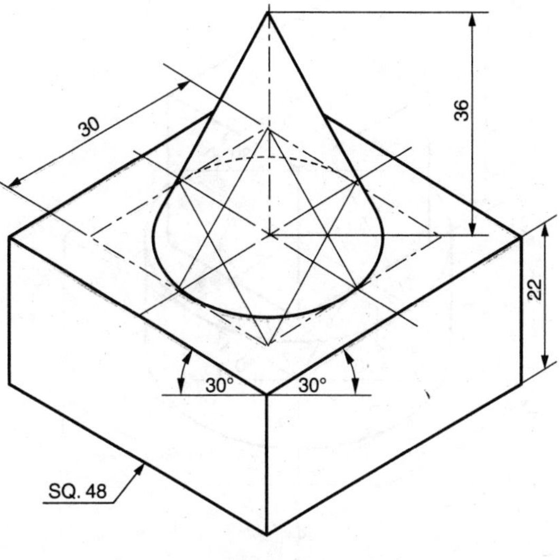

Fig. 6.39.

5. Draw the ellipse for the cone base, using the four centre method.
6. Draw a vertical line through the centre point and cut along it a distance equal to the height of the cone, using isometric scale. Join this point to the ellipse tangentially, on the two sides, as shown.
7. Finish the view by making the visible portions of lines firm.

PROBLEM 6.12. *A cube, of 30 mm edge, is placed centrally on top of a cylindrical block of ϕ 52 mm and 20 mm height. Draw the isometric drawing of the solids.*

SOLUTION: Refer *Fig. 6.40.*

1. Draw the isometric box, using full size scale, for the block, since we need to draw the isometric drawing.
2. Draw the two centre lines for its top surface and draw the ellipse for the top of the block, around this centre point, using the four-centre method.
3. Locate the corresponding points, to be used as centres of the four arcs for the base of the block, by offsetting in the vertical downward direction from their positions in the top surface ellipse.
4. Draw the visible part of the base ellipse using the principle of symmetry, as shown.
5. Draw the isometric of the cube coinciding the centre of its base surface with the centre of the top surface of the block, as shown.

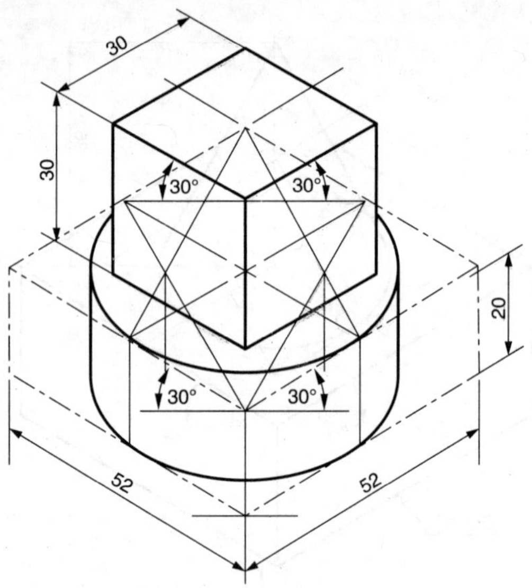

Fig. 6.40.

PROBLEM 6.13. *In Fig. 6.41 at (1), on left side, are shown two views of an object. Draw its isometric view.*

SOLUTION: Refer *Fig. 6.41. (1)*. All necessary construction lines are shown to render the solution easy to understand. The interpretation of the solution is left to the students. Note that the isometric scale will have to be used.

PROBLEM 6.14. *Two orthographic views of an object are shown in Fig. 6.41 (2) (a). Draw its isometric view.*

SOLUTION: Refer *Fig. 6.41 (2) (b)*. The interpretation of the solution is left to the reader.

PROBLEM 6.15. *Front view of a cut sphere is shown in Fig. 6.41 (3) (a). Draw its isometric view.*

SOLUTION: Refer *Fig. 6.41 (b)* and *(c)*.

1. Draw the circle for the part bottom view of the cut sphere, below the given front view.
2. Draw true ellipse, using offsets, as shown for isometric view of this circle. See that its enclosing square has a side equal to **b.** The centre of the ellipse is O_1.
3. In front view the height of the centre O of the sphere is 12 mm above the centre O_1 of the cut base.
4. Draw a vertical line through O_1 in the ellipse drawn and cut along it a distance equal to the isometric equivalent of 12 mm as shown by **a,** to locate the centre O of the sphere in isometric view.
5. Then with O as centre and the radius R = 25 mm of the sphere, on actual scale, draw a circular arc tangential to the ellipse, as shown.
6. It should be noted here that with R = 25 mm the arc will only be tangential to the ellipse if it is drawn to its true shape. If the ellipse is not actually true shape, as when using the four-centre method to draw the ellipse, then the arc of R = 25 mm will not be tangential to the ellipse. In that case some adjustment in the value of R will have to be done to make the arc tangential to the approximate ellipse.

PROBLEM 6.16. *A sphere of ϕ 40 mm rests centrally on top of a cube of 40 mm side. Draw the isometric views of the solids.*

SOLUTION: Refer *Fig. 6.41 (4)*.

1. The isometric view of a sphere is a circle of the actual radius of the sphere.
2. The sphere is making contact with the cube at the centre of top surface. This contact point will stay at the same point in the isometric view too.
3. The distance of the centre of the sphere from this contact point, which is R = 20 mm, will remain only equal to the isometric equivalent of the radius value, shown as r in isometric view.
4. Draw the isometric view of the cube, using isometric scale.
5. Locate the centre point A of its top surface by drawing two centre lines in isometric directions.
6. Draw a vertically upward line through A and cut along it a distance equal to the isometric equivalent of the radius R = 20 mm of the sphere to locate point O. Then O is the centre of the sphere.
7. Then with O as centre and R = 20 mm, the actual radius of the sphere, draw a circle to represent the sphere.
8. Finish the view by marking the visible part of the cube firm.

PROBLEM 6.17. *A hemisphere of ϕ 36 mm rests on its circular base on the top of a cube of 36 mm side. Draw the isometric projection of the solids.*

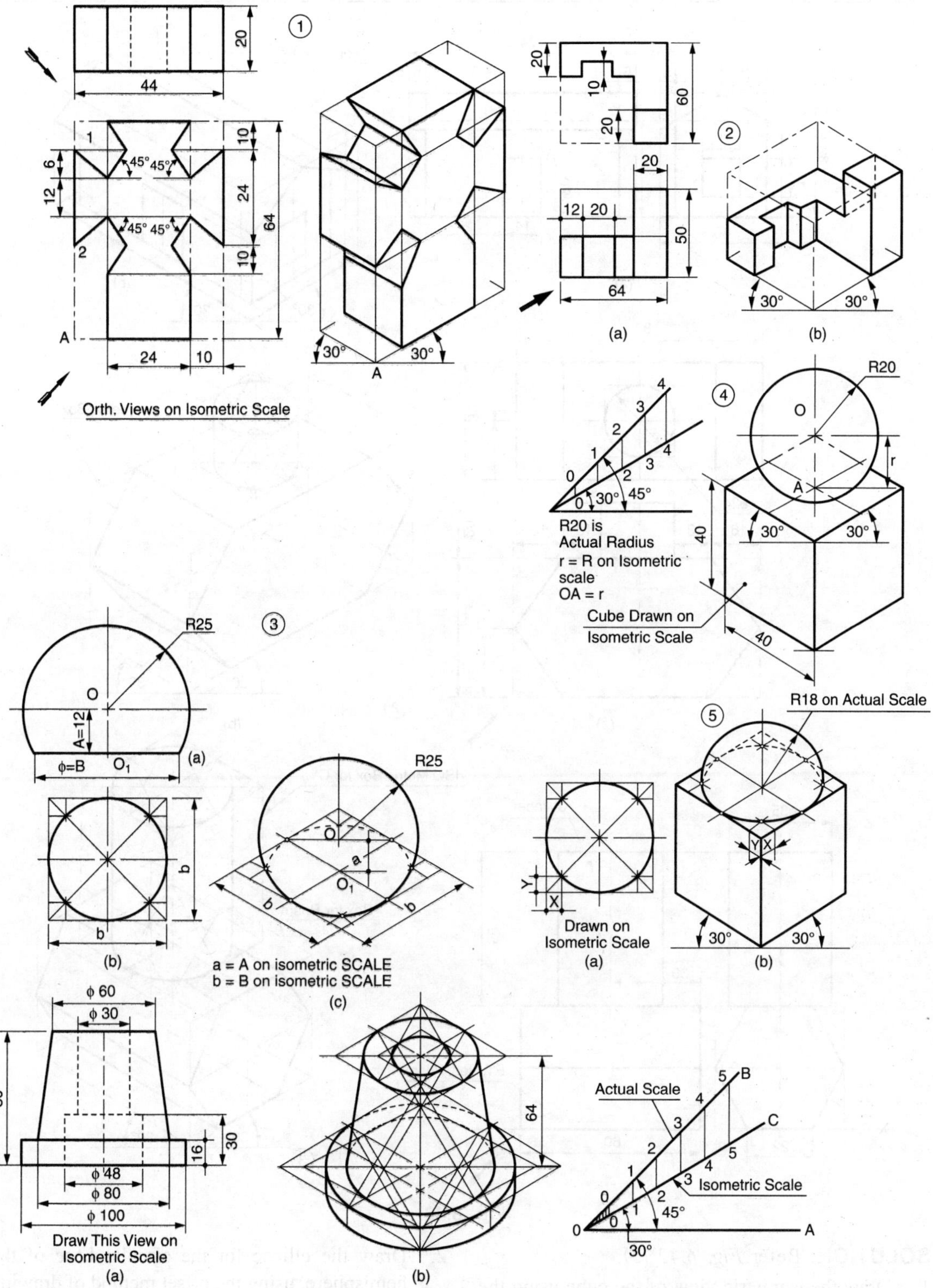
1
20
44
10
24
64
10
6
12
45° 45°
45° 45°
1
2
A
24
10
Orth. Views on Isometric Scale
30°
30°
A
2
20
10
20
60
20
12 20
50
64
(a)
30°
30°
(b)
4
R20
O
r
A
30°
30°
40
40
4
3
2
1
0
4
3
2
0
30°
45°
R20 is
Actual Radius
r = R on Isometric
scale
OA = r
Cube Drawn on
Isometric Scale
3
R25
O
A=12
ϕ=B
O1
(a)
b
b
(b)
R25
O
a
O1
b
b
a = A on isometric SCALE
b = B on isometric SCALE
(c)
5
R18 on Actual Scale
Y
X
Drawn on
Isometric Scale
(a)
Y X
30°
30°
(b)
ϕ 60
ϕ 30
80
30
16
ϕ 48
ϕ 80
ϕ 100
Draw This View on
Isometric Scale
(a)
64
(b)
Actual Scale
5 B
4
3
2
1
0
C
5
4
3
2
1
0
Isometric Scale
45°
0
A
30°
Fig. 6.41.

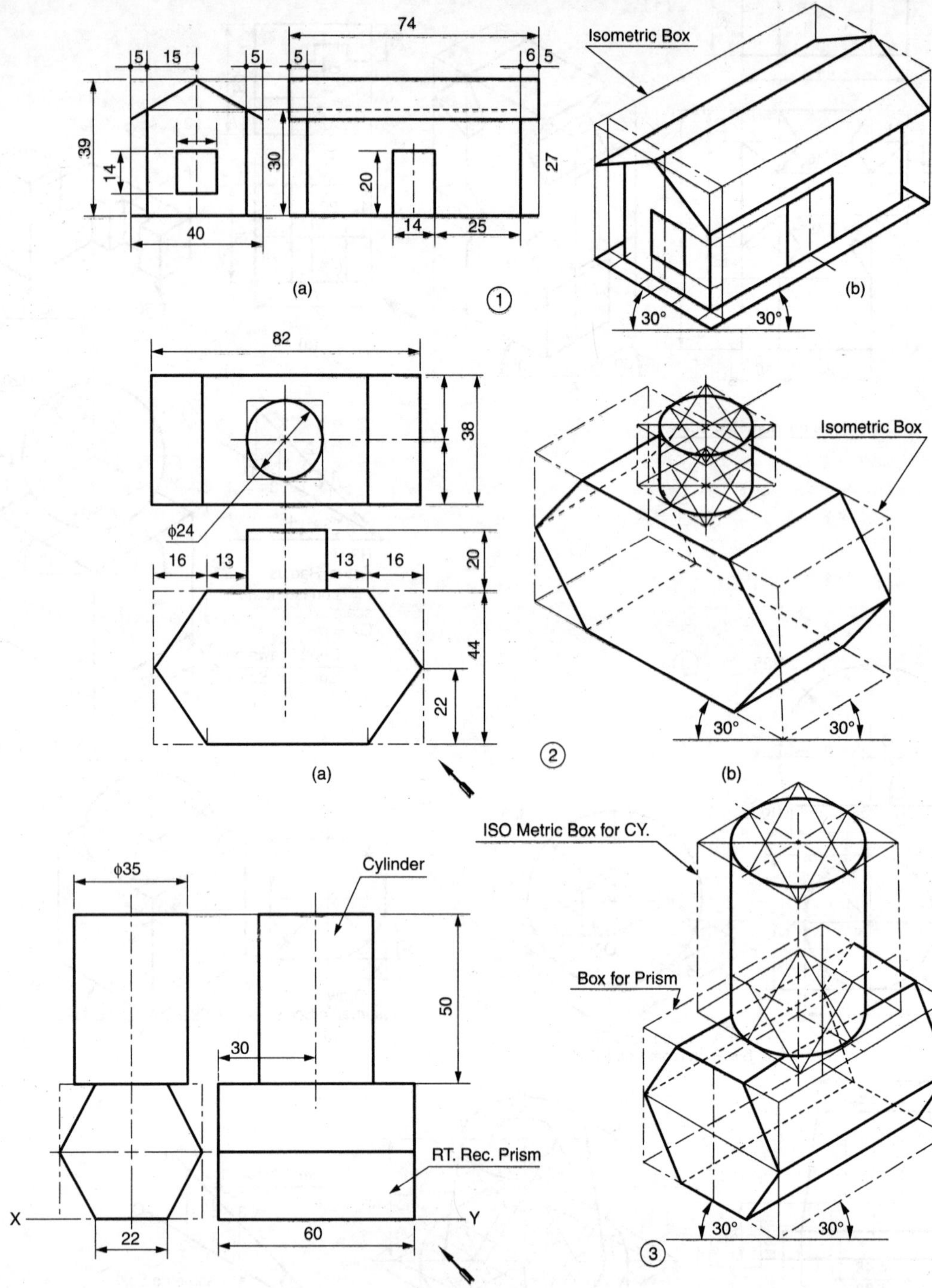

Fig. 6.42.

SOLUTION: Refer *Fig. 6.42 (5).*

1. Draw the isometric view of the cube using the isometric scale.
2. Draw the ellipse for the circular base of the hemisphere, using the offset method of drawing the true ellipse.

3. Draw a circular arc of R = 18 mm tangential to the ture ellipse, as shown.
4. In case the ellipse drawn for the base of the hemisphere is not true form, a slight adjustment in the radius value might be necessary to make the arc tangential to the eliipse.

PROBLEM 6.18. *Front view of an object is shown in Fig. 6.41. (6) (a). Draw its isometric projection.*

SOLUTION: Refer *Fig. 6.41 (6) (b).* All construction lines are retained to make the understanding of the solution easy. The arrangement consists of: a Cylindrical hole that gives its ellipse in isometric view; a frustum of a cone that given two ellipses for its top and bottom surfaces in isometric view; and a circular block. The frustum of the cone is resting centrally on top of the block. The counterbored hole is coaxial with the axis of the frustum and the block.

The interpretation of the solution is left to the reader.

PROBLEM 6.19. *Two views of a paper model of a hut are shown in Fig. 6.42 (1) (a). Draw its isomeric view.*

SOLUTION: Refer *Fig. 6.42 (1) (b).* This problem is solved by using the box and offset measurement methods in combination. Important construction lines are retained. The interpretation of the solution is left to the students

PROBLEM 6.20. *Two simplified orthographic views of an ink pot are shown in Fig. 6.42 (2) (a). Draw its isometric projection.*

SOLUTION: Refer *Fig. 6.42 (2) (b).* The problem consists of two separate parts: a hexagonal shaped body part of constant depth, and a cylindrical part placed centrally on top surface of the hexagonal shaped body part. All construction lines are retained to make the solution easy to understand.

PROBLEM 6.21. *A right regular hexagonal prism, edge of base 22 mm and 60 mm long, lies on its rectangular face on ground plane. A right circular cylinder, of ϕ 35 mm and 50 mm long, rests centrally on top rectangular surface of the prism. Draw isometric projection of the solids.*

or

Front and end views of a right circular cylinder, resting on a hexagonal prism are shown in Fig. 6.42. (3) (a). Draw the isometric projection of the solids.

SOLUTION: Refer *Fig. 6.42 (3).* All construction lines are shown to make the solution easy to understand.

PROBLEM 6.22. *Front and top views of a solid are shown in Fig. 6.43 (a). Draw its isometric view.*

SOLUTION: Refer *Fig. 6.43 (b).* All construction lines are retained to help in understanding the solution. See that the length could also be placed on the left side instead of that shown in the solution but the arrowhead indicating the direction of sight fixes the right choice as shown here.

PROBLEM 6.23. *Figure 6.44 (a) shows an orthographic view of a rivet. Draw its isometric drawing.*

SOLUTION: Refer *Fig. 6.44 (b).* The problem consists of drawing isometric view of a part sphere placed centrally on top of a cylinder. All construction

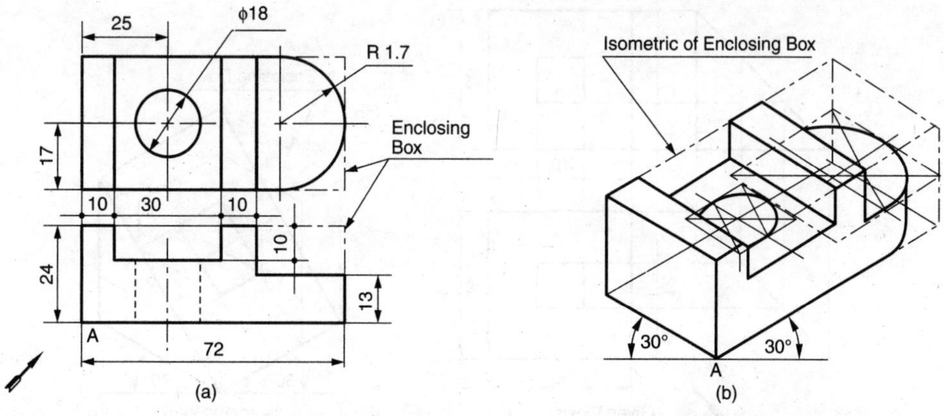

Fig. 6.43.

lines are shown and the interpretation of the solution is left to the students.

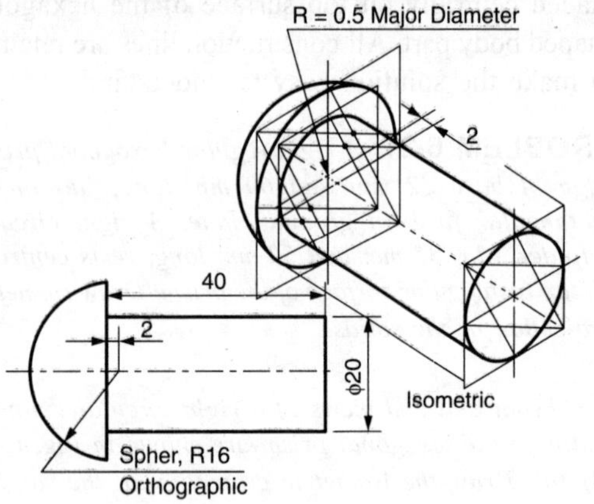

Fig. 6.44.

PROBLEM 6.24. *A right regular square pyramid is placed centrally on the top of a cube and the two solids are then placed centrally on top of a circular cylinder, as shown in Fig. 6.45 (a) in front and top views. Draw isometric projection of the solids.*

SOLUTION: Refer *Fig. 6.45 (b)*. All construction lines are shown. Also, in such problem, care should be taken to maintain the relative positions of the solids to be same as given in the orthographic views.

PROBLEM 6.25. *Figure 6.46 (a) shows three views of an object. Draw its isometric view.*

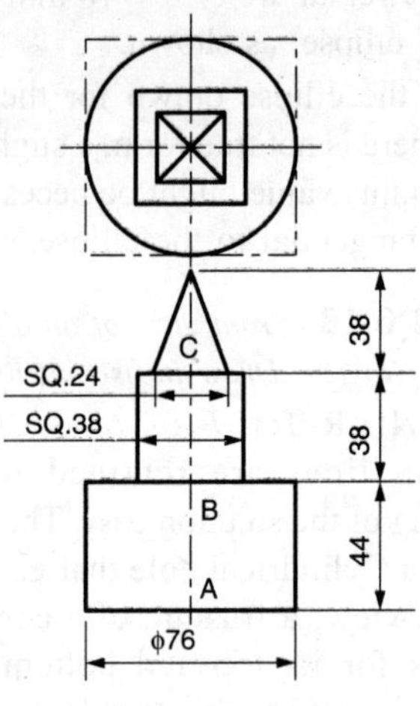

Fig. 6.45. (a)

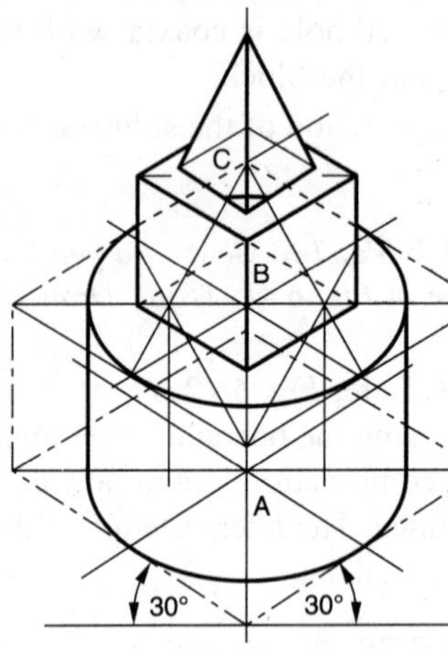

Fig. 6.45. (b)

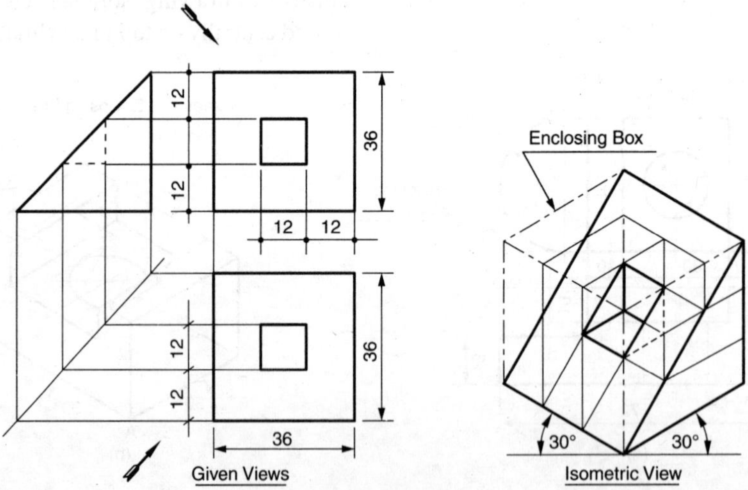

Fig. 6.46.

SOLUTION: Refer *Fig. 6.46 (b)*. All construction lines are retained. The interpretation of the solution is left to the students.

PROBLEM 6.26. *Figure 6.47 (a) shows three views of a cube cut by an oblique plane containing the adjacent face diagonals. Draw its isometric view.*

SOLUTION: Refer *Fig. 6.47 (b)*. The figure is self-explanatory.

PROBLEM 6.27. *Figure 6.48 (a) shows three orthographic views of an object. Draw its isometric view.*

SOLUTION: Refer *Fig. 6.48 (b)*. All construction lines are shown to make the solution easy to understand.

PROBLEM 6.28. *Figure 6.49 (a) shows three orthographic views of an object. Draw its isometric view.*

SOLUTION: Refer *Fig. 6.49 (b)*. Three orthographic views show that the solid consists of both the isometric and non-isometric lines that are placed systematically on the surface of a cube. All eight corners of the cube are symmetrically cut. All construction lines are shown to make the solution easy to understand.

PROBLEM 6.29. *Draw the isometric view of a hexagonal M25 nut. Show all construction lines.*

SOLUTION: Refer *Fig. 6.50 (b)*. The interpretation of the solution is left to the students.

PROBLEM 6.30. *A right regular hexagonal prism, edge of base 20 mm and height 60 mm, lies on one of its rectangular faces. A right circular cone. diameter of base 30 mm and height 35 mm rests centrally on the upper rectangular face of the prim, as shown in Fig. 6.51 (a). Draw the isometric projection of the two solids.*

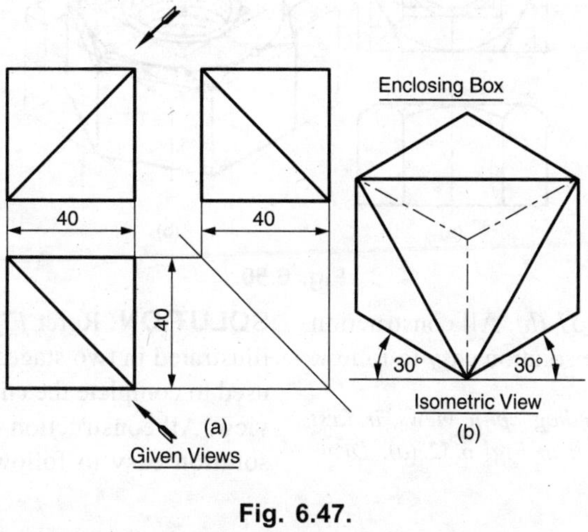

Fig. 6.47.

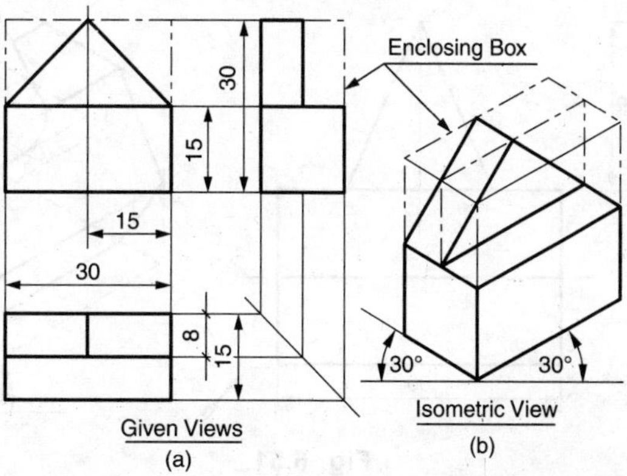

Fig. 6.48.

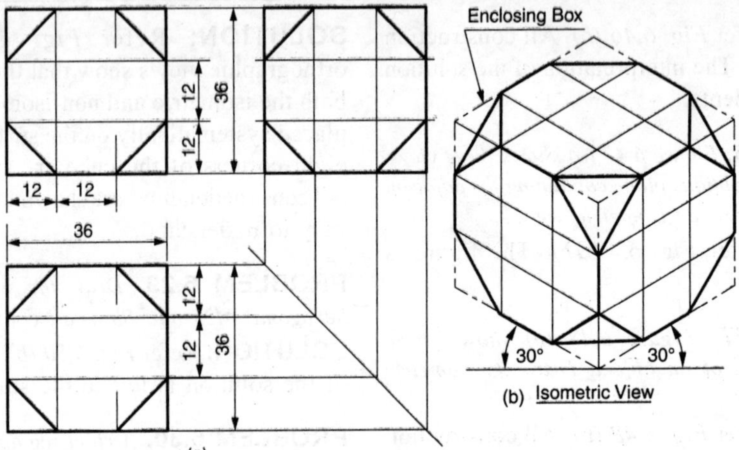

(a)

Fig. 6.49.

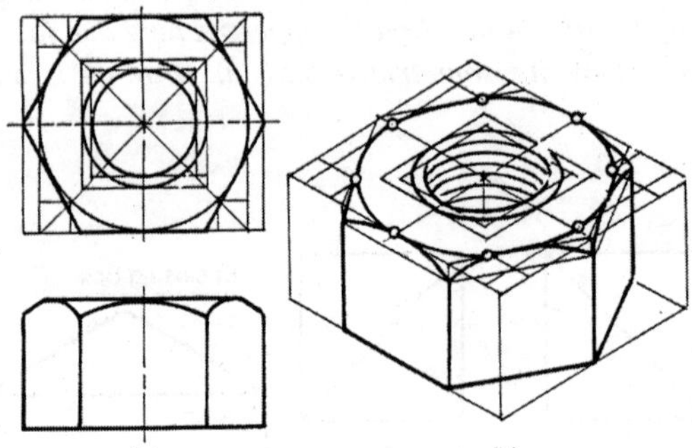

(a) (b)

Fig. 6.50.

SOLUTION: Refer *Fig. 6.51 (b)*. All construction lines are shown to make the solution easy to follow.

PROBLEM 6.31. *Three orthographic views, in first angle, of an object are shown in Fig. 6.52 (a). Draw its isometric projection.*

SOLUTION: Refer *Fig. 6.52. (b)*. The solution is illustrated in two stages. Note carefully the method used to complete the curved cut portion in isometric view. All construction lines are shown to make the solution easy to follow.

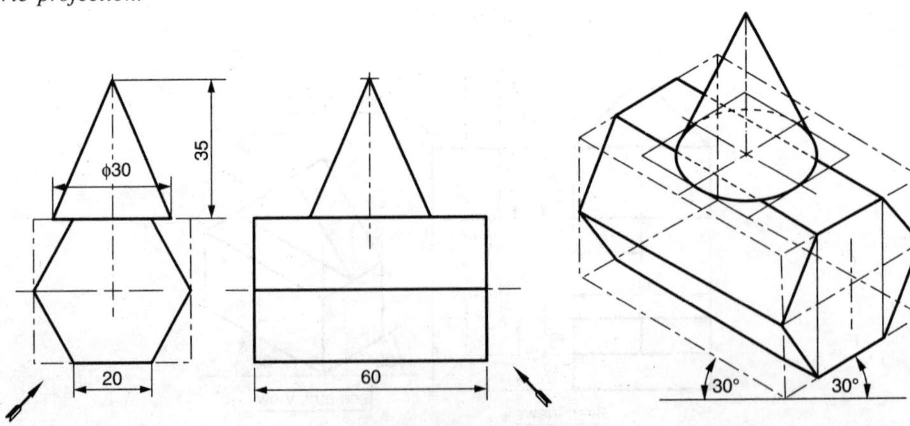

Fig. 6.51.

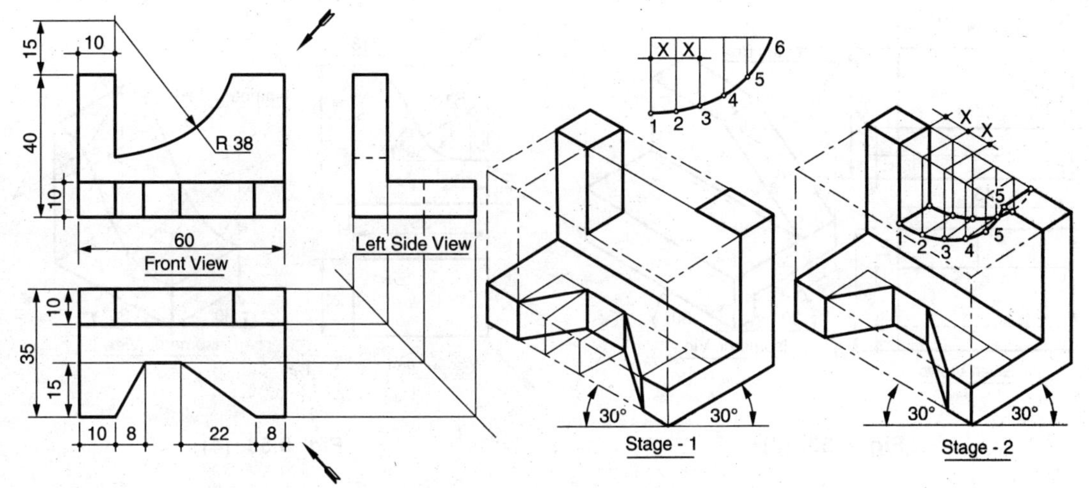

Fig. 6.52.

PROBLEM 6.32–6.37. *Two orthographic view of each of the six objects are given, in first angle, in Fig. 6.53. Draw the isometric view in each case, looking the direction of arrowhead.*

SOLUTION: Refer *Figs. 6.53* 1 to 6. The *interpretation of the solution in each case, is left to the student.*

PROBLEM 6.38 TO 6.40. *Three orthographic views of each of the four different objects are given, in first angle, in Fig 6.54. In each case some important construction lines are retained. It is required to draw the isometric drawing in each case.*

SOLUTION: Refer *Fig. 6.54. (1)* to *(4)*. The methods of drawing a hole, with its axes, (*i*) and (3), respectively, along with methods used to draw the same in the corresponding isometric drawings.

After careful study of these case the student can attempt to draw the isometric projections of these four solids on an A2 size drawing sheet.

PROBLEM 6.42. *Draw the Isometric view of an object shown in Fig. 6.55. The Orthographic views are given in third angle projection. On purpose, the required construction lines are omitted from the solution.*

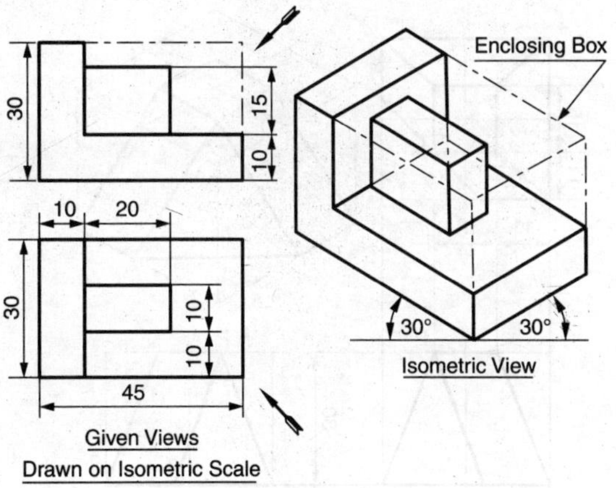

Fig. 6.53. (1)

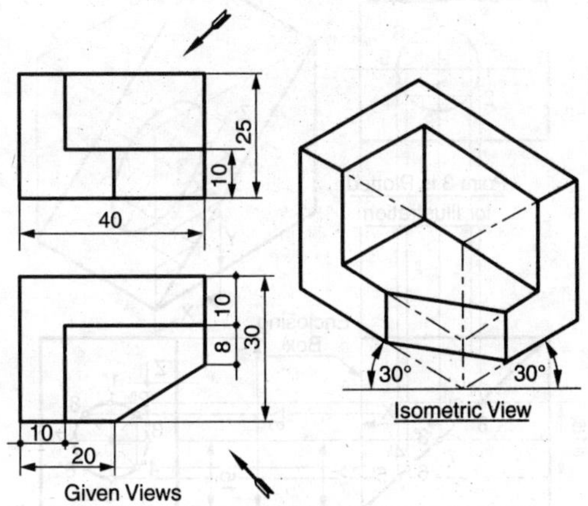

Fig. 6.53. (2)

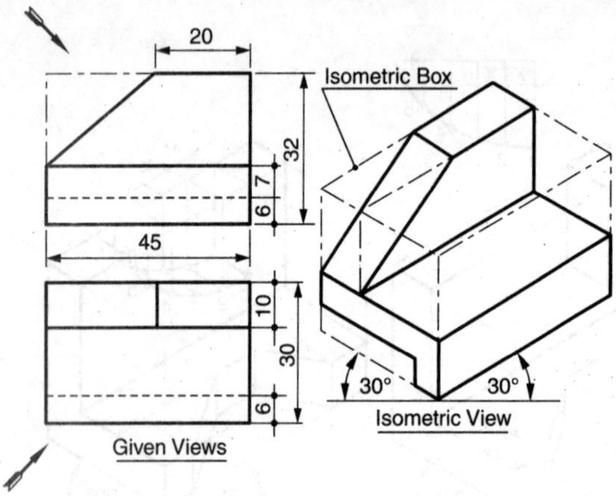

Fig. 6.53. (3)

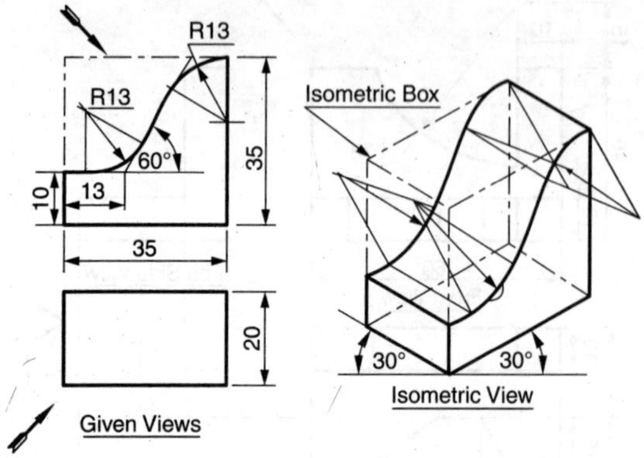

Fig. 6.53. (4)

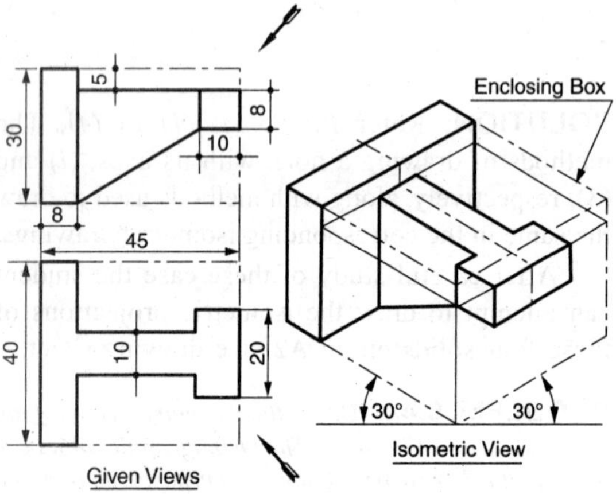

Fig. 6.53. (5)

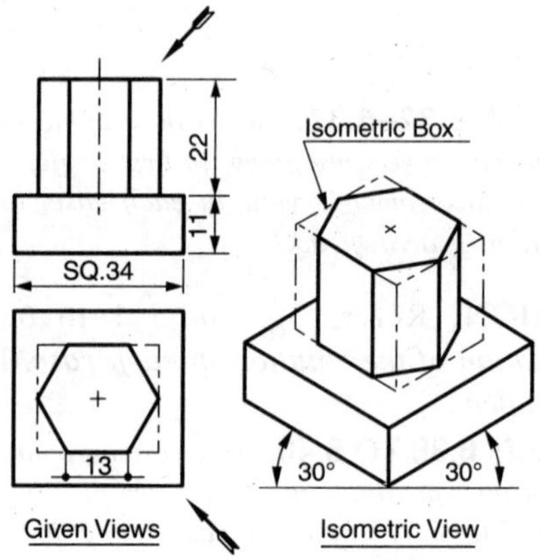

Fig. 6.53. (6)

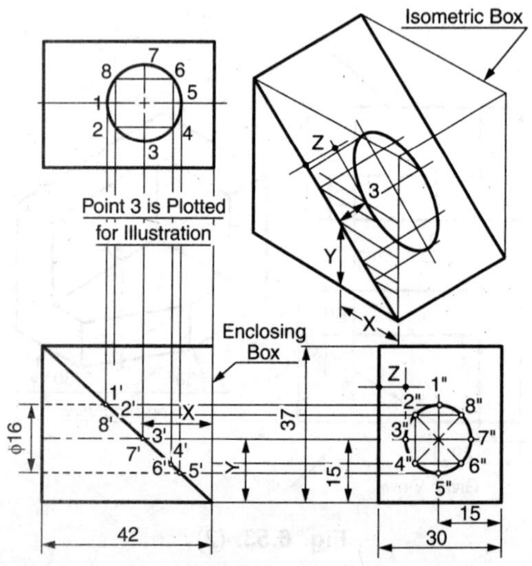

Fig. 6.54. (1)

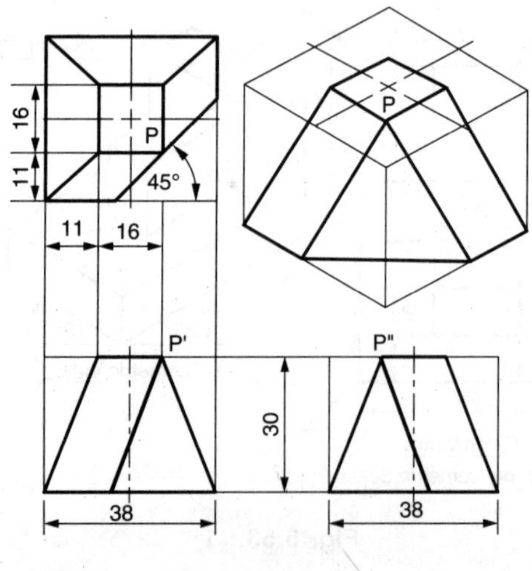

Fig. 6.54. (2)

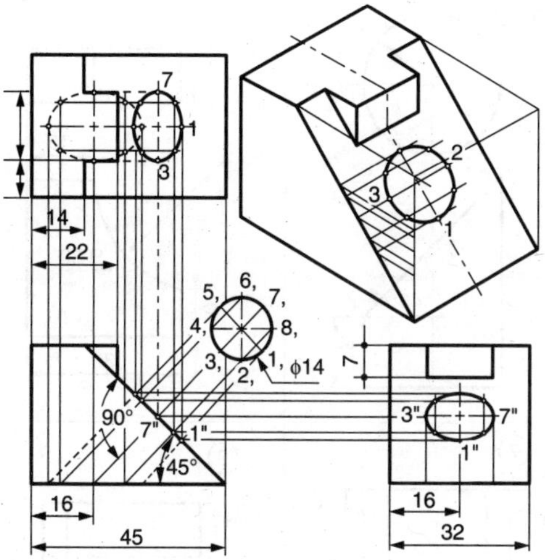

Fig. 6.54. (3)

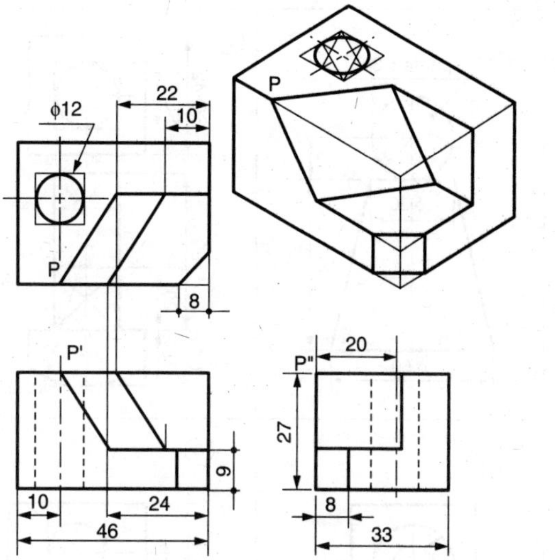

Fig. 6.54. (4)

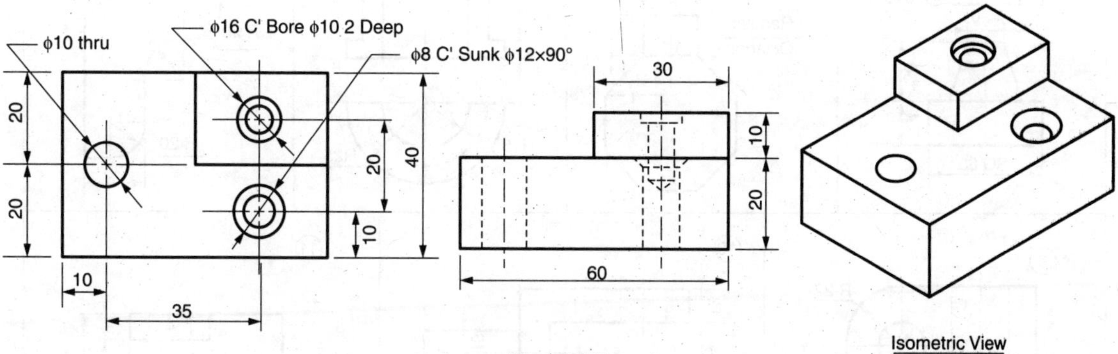

Fig. 6.55.

EXERCISES

1. *Figure 6.56* shows orthographic views of eleven different objects. Draw the isometric drawing for each one of them and give all the necessary dimensions.

   HINT. Solutions to exercise 1 and 2 of *Fig. 6.56* are given in *Fig. 6.57* and solutions to exercise 3 and 4 of *Fig. 6.56* are given in *Fig. 6.58*.

2. A cut sphere rests centrally on top of a right circular cylinder which in true is put centrally on top of a square block, as shown in *Fig 6.59*. Draw the isometric view of the solids placed in the given position.

3. Front and top views of an object are shown in *Fig. 6.60*. Draw its isometric view.

4. Front and top views of an object are shown in *Fig 6.61*. Draw its isometric view.

5. A right regular pentagonal prism, side of base 30 mm and height 60 mm, is resting on HP such that its base edges, meeting at the corner point farthest away from VP, are marking equal angles with the VP. It is cut by a section plane perpendicular the VP and inclined at 45° to the HP, such that a point P on its axis and 42 mm above the base is in the section plane. Draw the isometric drawing of the truncated portion of the prism.

   HINT: Refer *Fig. 6.62* for its solution.

6. A right regular pentagonal pyramid, edge of base 35 mm and height 65 mm, is resting on its base in HP, such that its left side base edge is perpendicular to the VP. It is cut by a section plane, perpendicular to the VP and inclined to the HP at 30°, intersecting its axis at a point 28 mm above the base. Draw the isometric drawing of the truncated pyramid.

   HINT: Refer *Fig. 6.63.* for its solution.

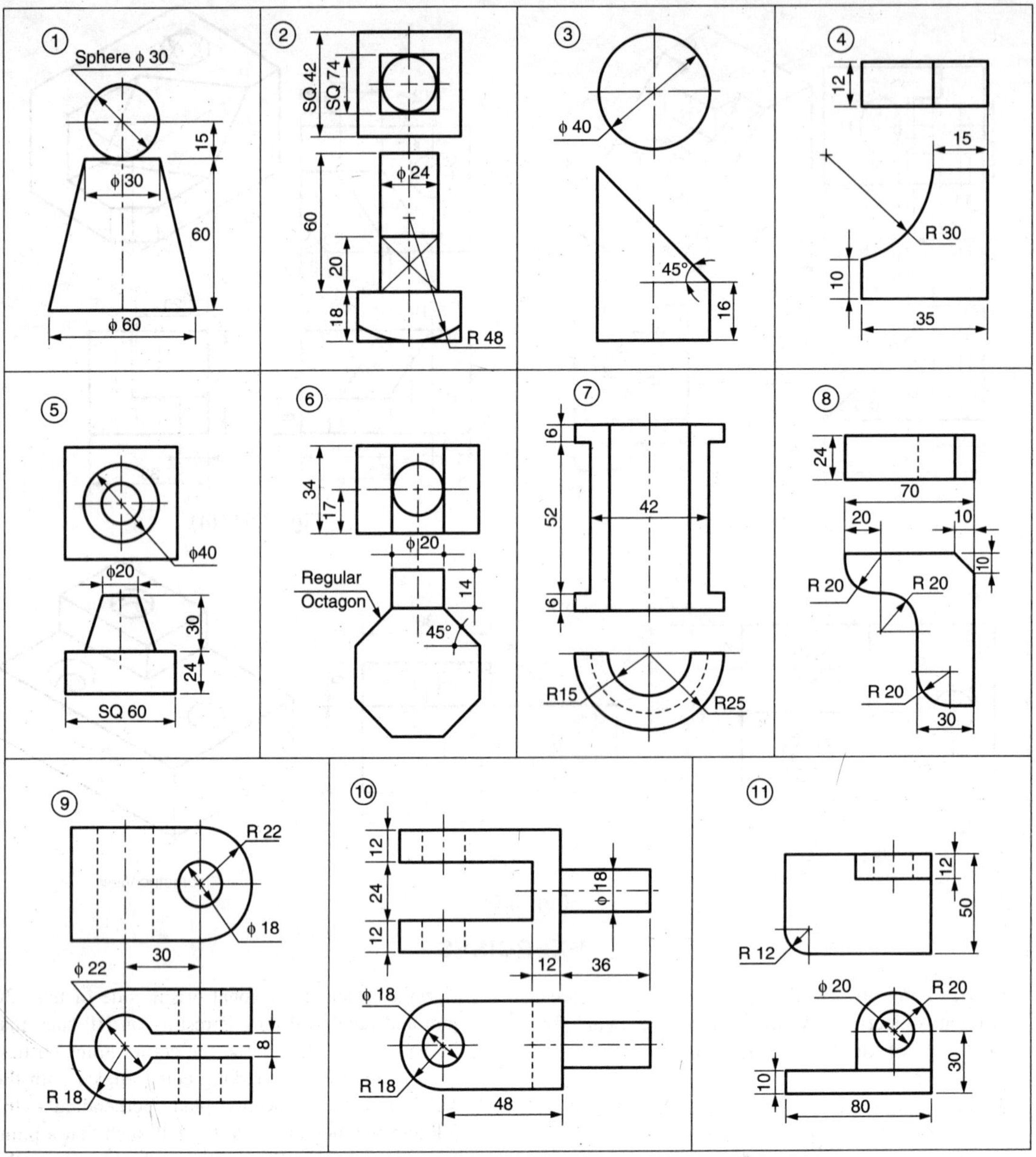

Fig. 6.56.

7. A right cylinder, diameter of base 50 mm and height 60 mm, is resting on its base in HP. It is cut by a section plane perpendicular to VP and inclined at 45° to the HP, such that a point P on its axis and 42 mm above the HP is in the section plane. Draw the isometric drawing of the truncated cylinder.

HINT: Refer *Fig. 6.64.* for its solution.

8. A right circular cone, diameter of base 52 mm and height 56 mm, is resting on its base in HP. A section plane perpendicular to VP and inclined at 30° to the HP cuts the cone, such that it passes through a point P on the axis and 26 mm above the base of the cone. Draw the isometric drawing of the truncated cone.

HINT: Refer *Fig. 6.65.* for its solution.

Any missing dimensions may be taken proportionately. Isometric view in each case is drawn using a CAD System. All these problems are good candidates for practicing on any CAD system that may be available to the students.

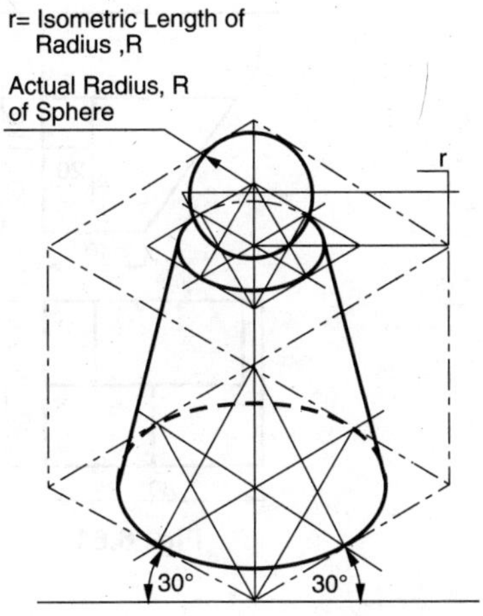

Solution to Exercise (1)

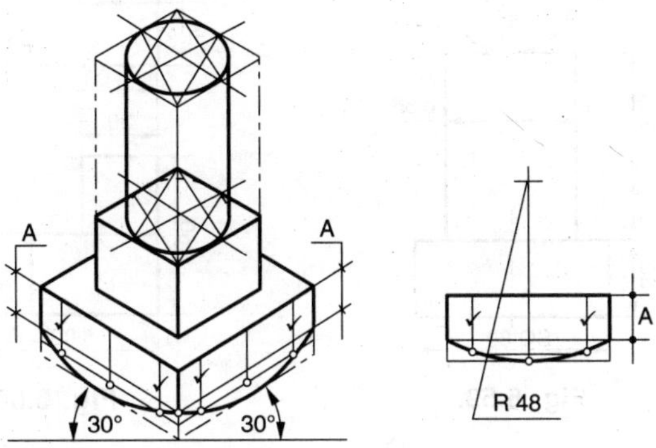

Solution To Exercise (2)

Fig. 6.57.

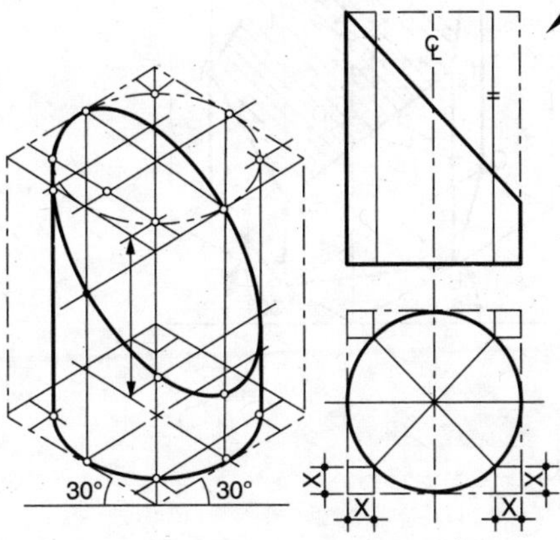

Solution to Exercise No. 3

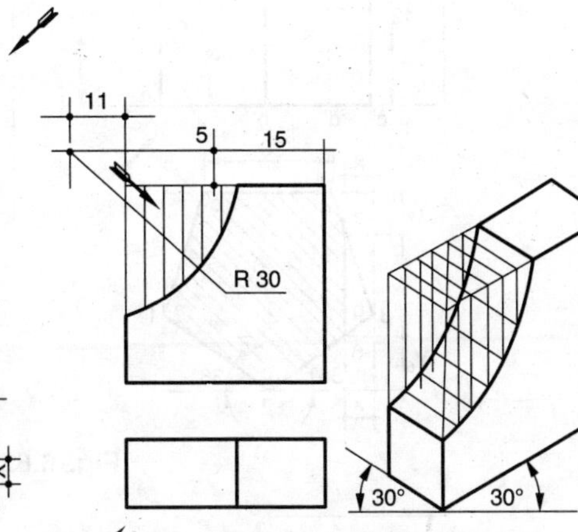

Solution to Exercise No. 4

Fig. 6.58.

How fast these isometric are drawn will certainly depend upon the capability features of the CAD System used and how well the student knows these capabilities, and the individual practice on and experience the CAD System.

Personal approach, analytical power, knowledge of drafting and aptitude surely go a long way to increase productivity.

Almost all major commands of any CAD System can be practiced by going through these exercise. Practice and persistence always help.

You can try both making a two dimensional (2D) Geometry model or a three dimensional (3D) model in each case.

Through knowledge of drafting principles goes a long way in achieving the best out of any CAD System. You will learn in the course of time that these are always more than one ways of achieving the end results with CAD. Choice of best way of doing it is learnt by experience and personal skills only.

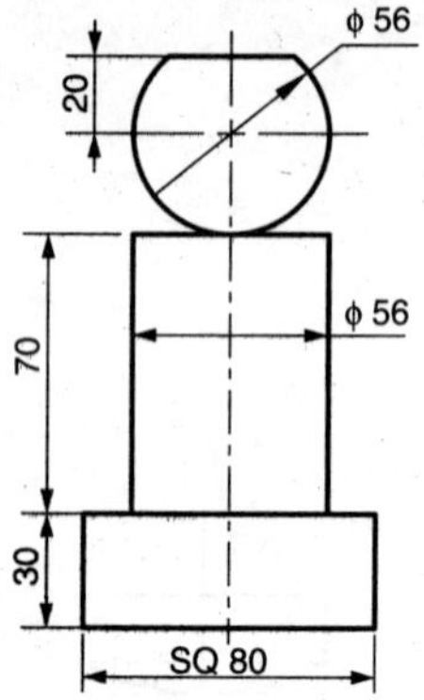

Fig. 6.59.

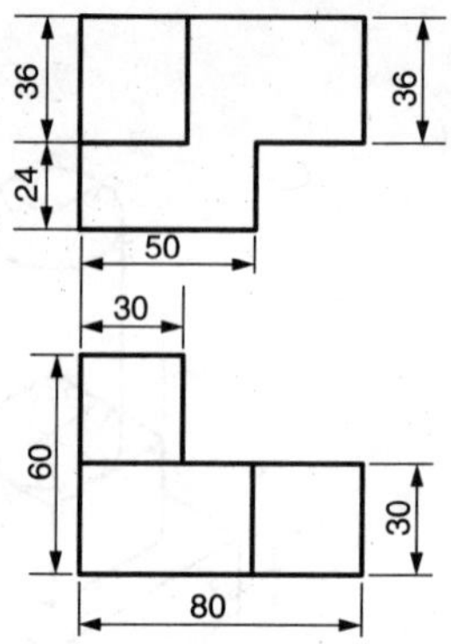

Fig. 6.60.

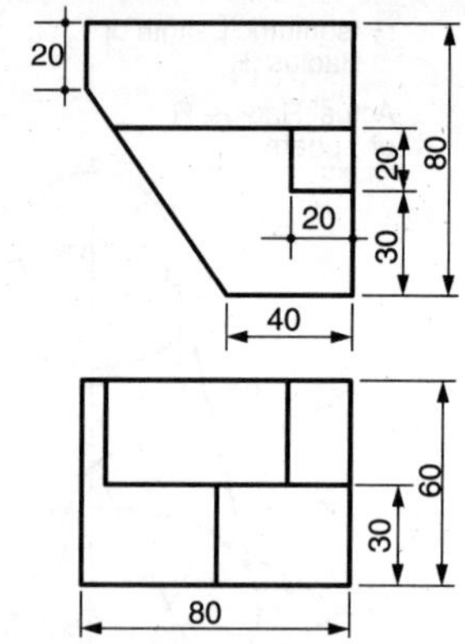

Fig. 6.61.

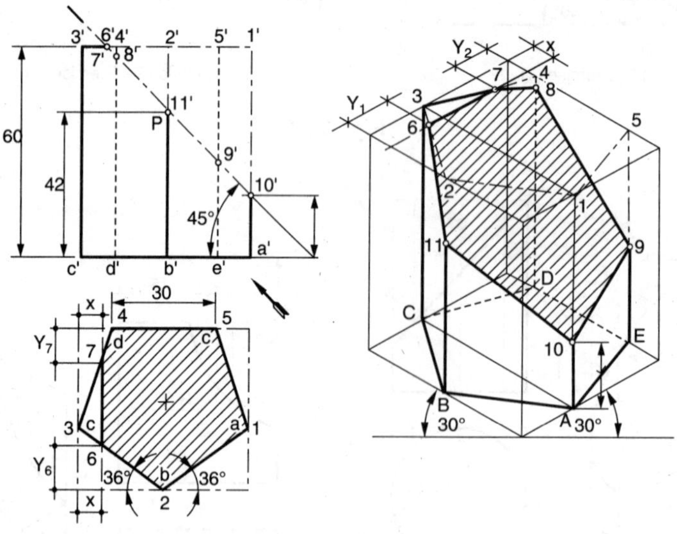

Fig. 6.62.

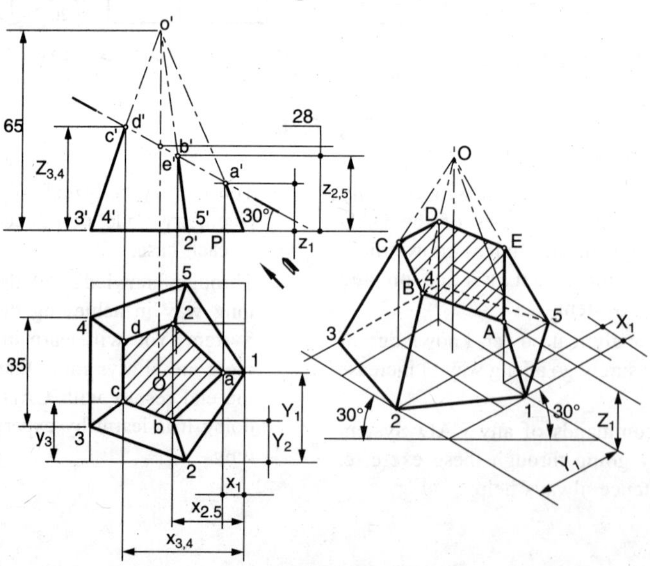

Fig. 6.63.

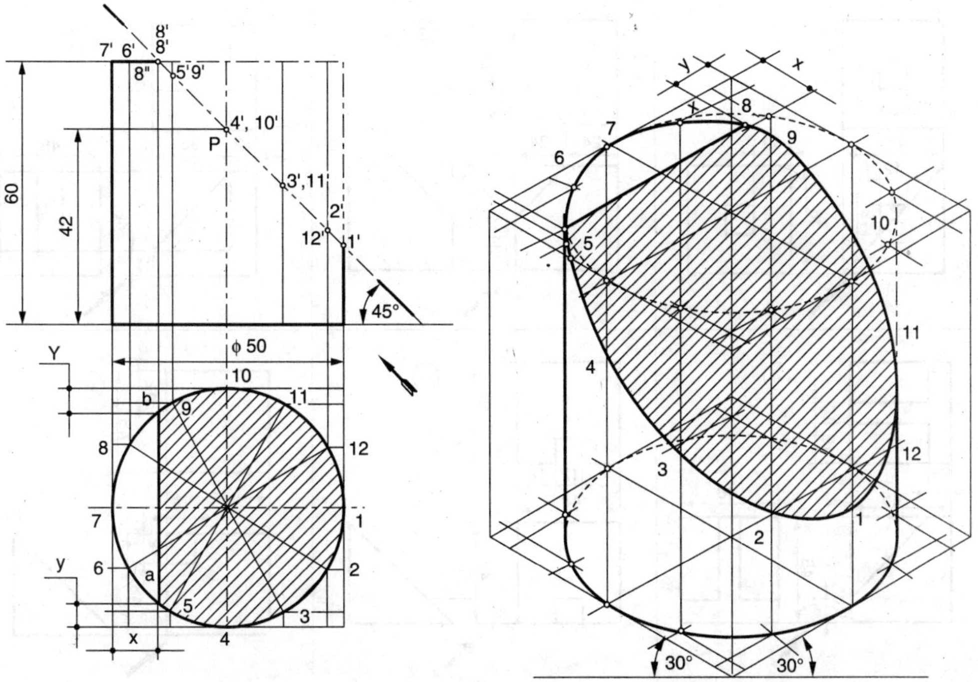

Fig. 6.64.

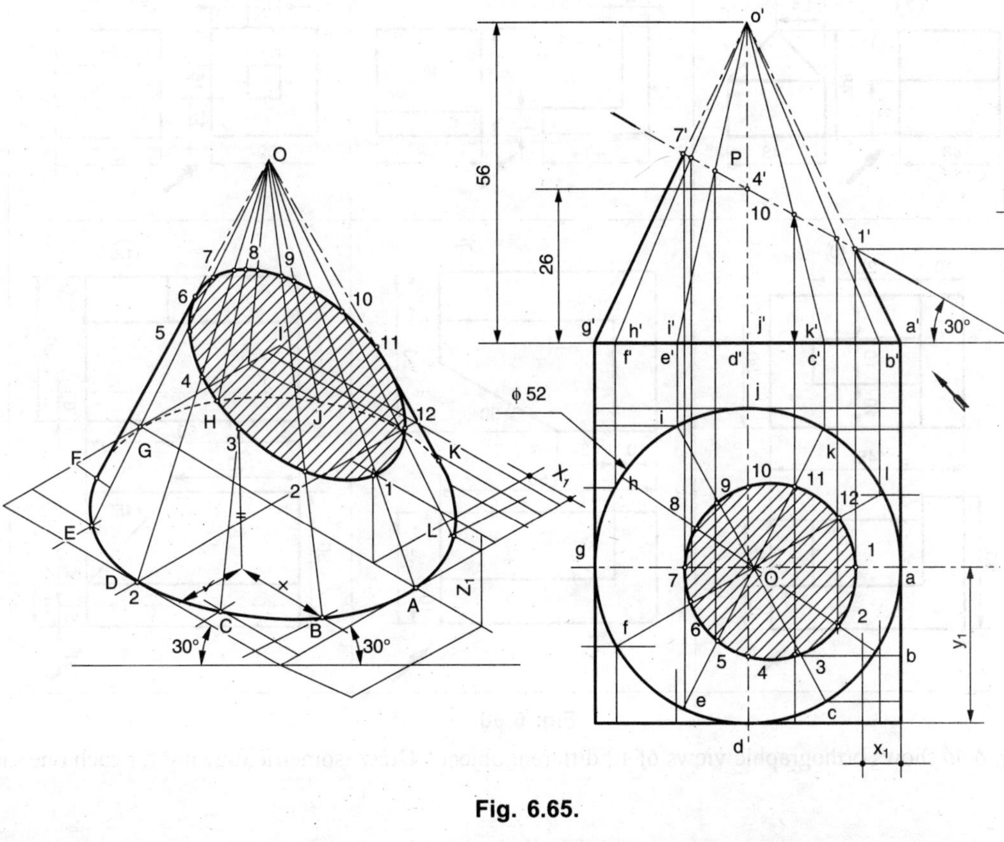

Fig. 6.65.

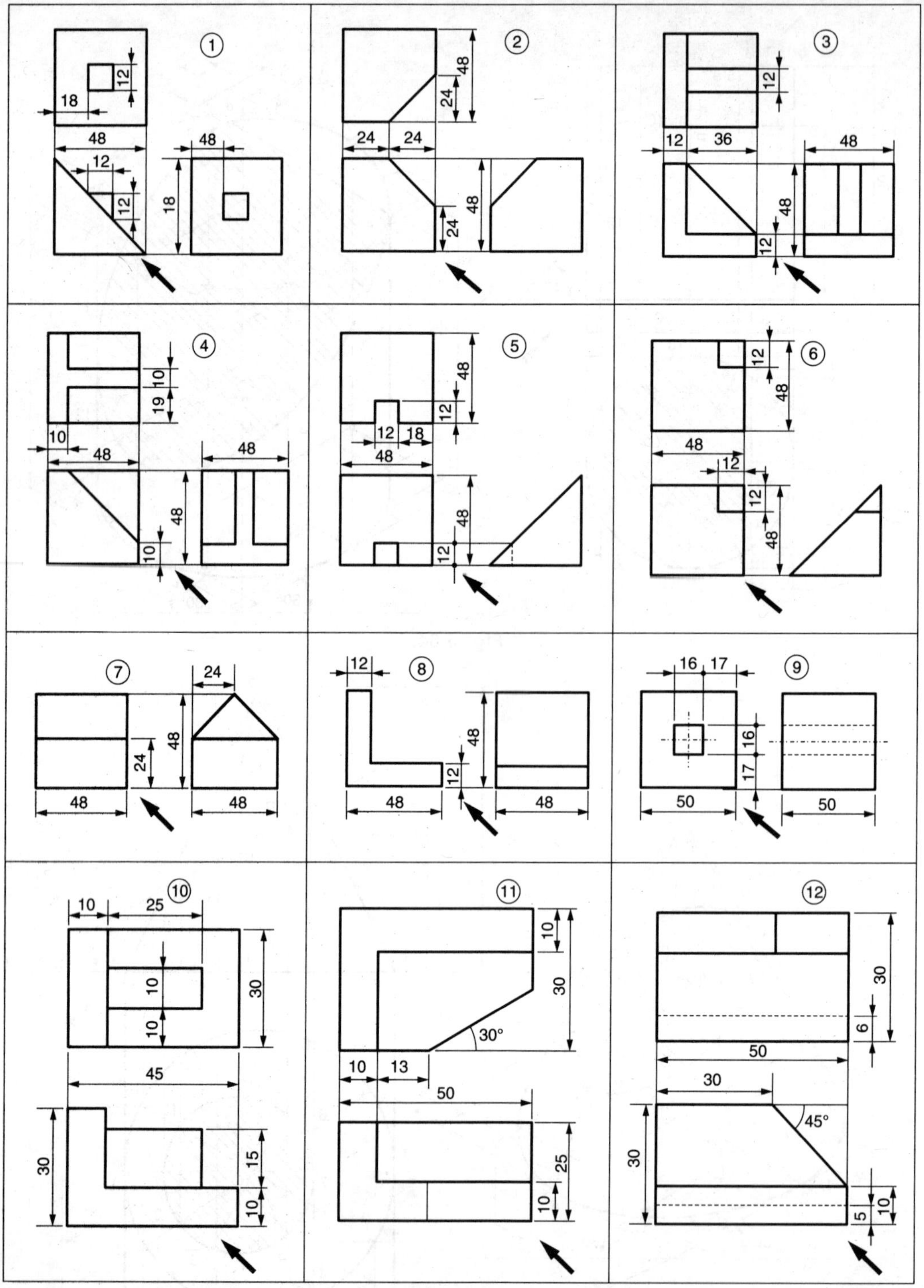

Fig. 6.66.

Fig. 6.66 shows orthographic views of 12 different objects. Draw isometric drawing for each one them.

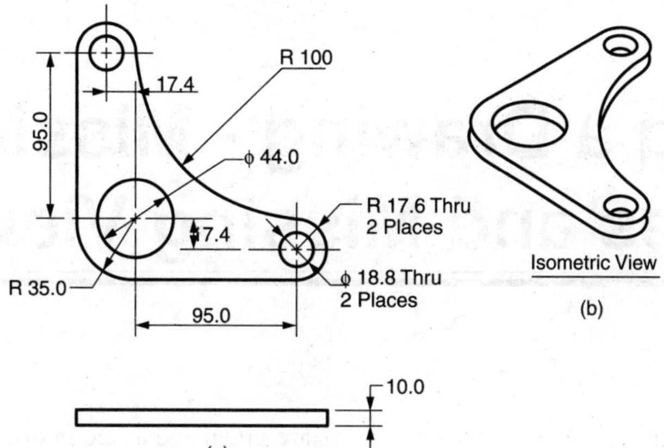

(a) (b)

Fig. 6.67.

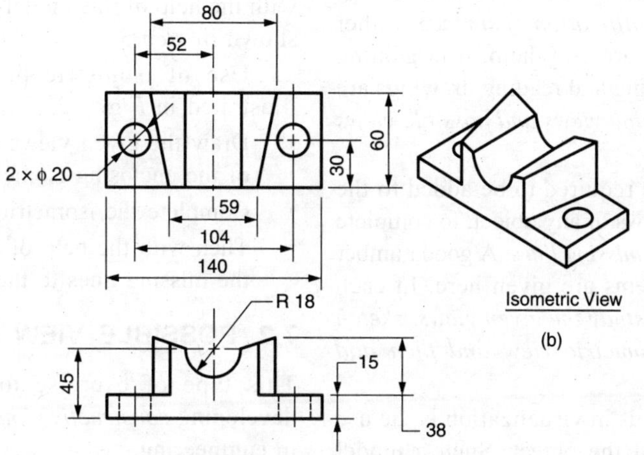

Lift Guide

(a) (b)

Fig. 6.68.

QUESTIONS FOR VIVA VOCE

1. What is a pictorial drawing?
2. What are the various methods of pictorial projection?
3. What is an axanometric projection and how can you obtain it?
4. Name the three principal types of axanometric projection. How these differ from each other?
5. What is an isometric projection? Name the various method used to obtain an isometric view of a cube.
6. Differentiate between isometric and non-isometric lines.
7. What are isometric axes? At what angles these are inclined mutually?
8. Why it is necessary to draw an isometric scale to draw an isometric view?
9. What is the difference between isometric drawing and isometric projection or view?
10. Under what circumstances hidden lines need to be drawn in an isometric view?
11. Sketch the methods to represent a circle in isometric view and name the method which is more correct?
12. By a simple sketch illustrate how an arc of a circle is drawn in an isometric view?
13. What is the isometric radius of a sphere? How do the isometric drawing and isometric view of a sphere differ in radii?

Reading a Drawing - Missing Lines and Missing Views

7

7.1. VISUALIZATION

It was pointed out in *chapter 4*, on 'Orthographic Projections', in the section on *Blue Print Reading*, that in addition to *Identification of surfaces*, other types of exercise that are very helpful in gaining experience in visualizing and reading drawings are the *missing lines, missing views and possible views problems*.

The lines that are required to be added to the given orthographic views of any object, to complete the drawing, are called *missing lines*. A good number of missing line problems are given here. In each case, the student is to *study the given views, sketch the corresponding isometric view and then add the missing line(s)*.

One of the best aids in visualization is the use of an actual model of the object. Such a model can be made to any approximate scale and of any suitable material available, *e.g.,* modelling clay, laundry soap, polystyrene, wood etc.

A typical example of the use of a model is shown in *Fig. 7.1.* Here three views of an object are given and a line in one of the views is missing. The object is to provide the missing line. Carve the model as illustrated in *steps 1 and 2.* Then, with the help of this model add the missing line as shown in *step 3.*

Use of isometric box in visualization is illustrated in *Fig. 7.2.*

- Draw the given views on the respective sides of the enclosing isometric box.
- Complete the isometric of the object.
- Then with the help of the isometric view add the missing lines to the views.

7.2. POSSIBLE VIEW PROBLEMS

This type of exercise too is very useful for developing *constructive imagination,* so important in engineering.

A view of an object, whose shape is not otherwise known, in given. The view may be front side or top view. One has to imagine and sketch or draw the object, which will have the view as given. Obviously there are a number of solutions

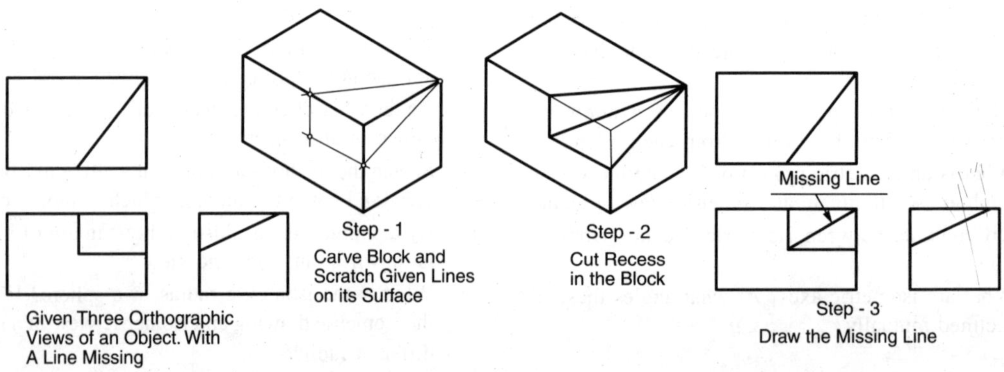

Fig. 7.1. Use of an actual model in visualization (Third Angle Projection)

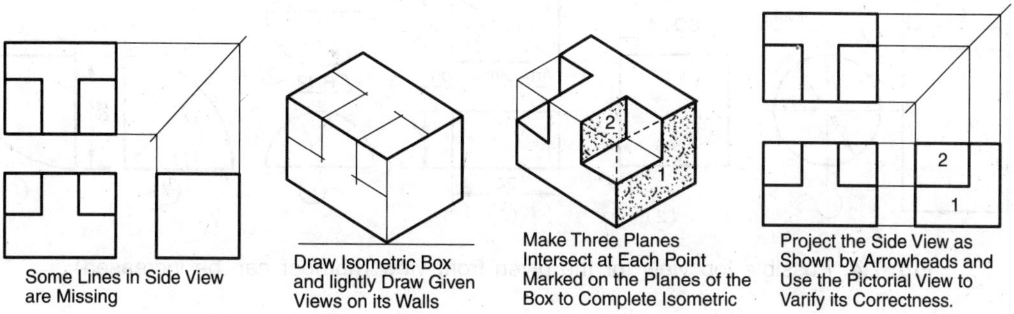

Fig. 7.2. Use of isometric box in visualization (Third angle projection)

to this type of problem. The solutions may be sketched or drawn. Brain storming will result in thinking of more and more possible solids that will go with a given view.

To form a complete concept of the true geometric shape of a part, all the areas on a given view need to be carefully examined because each closed area represents a surface on the part. In reading a drawing it has to be determined whether or not a particular area in the view represents a surface that is horizontal or vertical or inclined, or whether the surface is higher or lower than the adjacent surface. For example, five distinct objects are shown in *Fig. 7.3*, and all of them have the same top view. Here the triangular surface could be thought to be high and horizontal, sloping, or low and horizontal, but it can only be established by considering the corresponding front view. A slight consideration will show that a number of additional solids can be visualized and sketched for the same top view.

In *Figs. 7.4* and *7.5* a number of possible *top views are* drawn *for the given front view* in each case. Careful thinking will increase the number of possible views in each case.

Figure 7.6 illustrates four different solid blocks which have *different front views for the given side view*. The number of these views can further be increased.

7.3. ADDING A MISSING VIEW OR THIRD VIEW

Good practice in reading drawings can be obtained by sketching or drawing a *third view* on an object

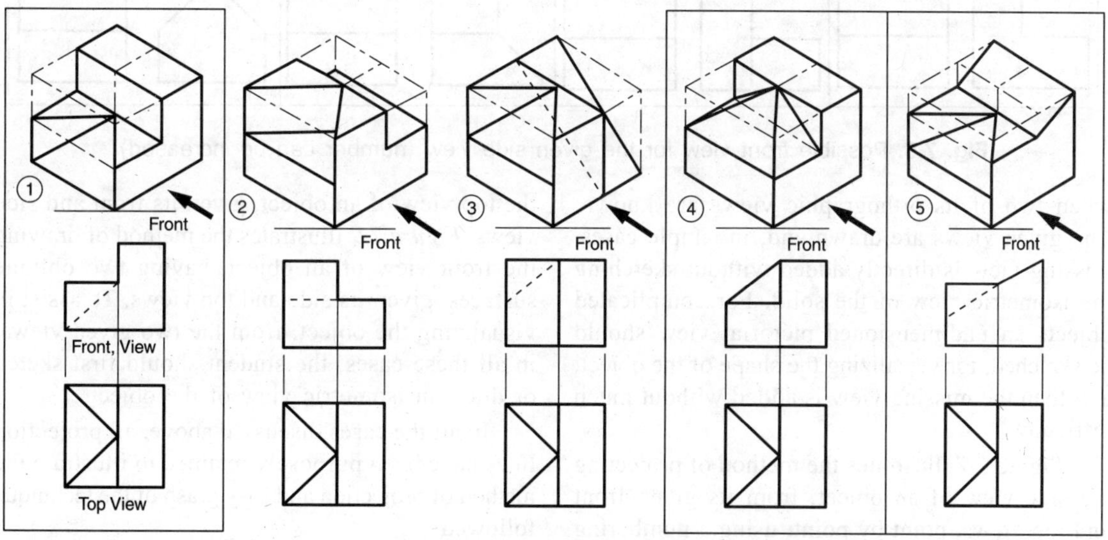

Fig. 7.3. Possible front views for the given plan (first Angle) (No. can be increased further)

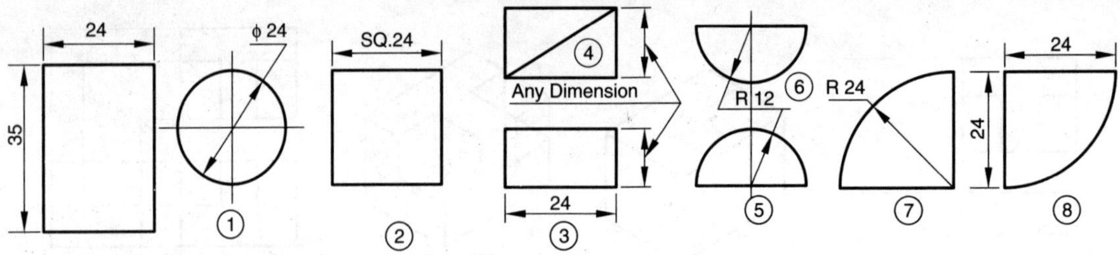

Fig. 7.4. Possible top view for the given front view (number can be increased)

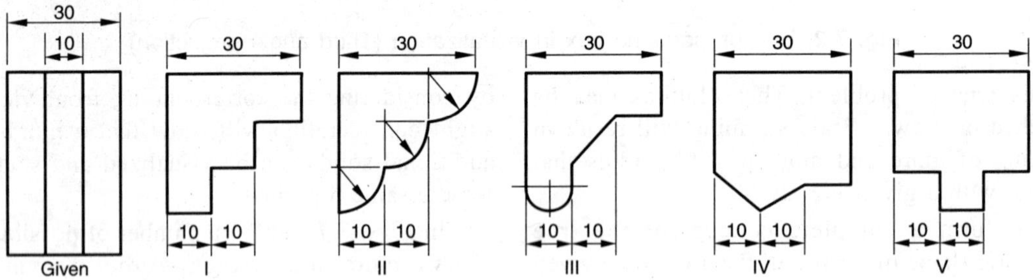

Fig. 7.5. Possible top view for the given front view (number can be increased)

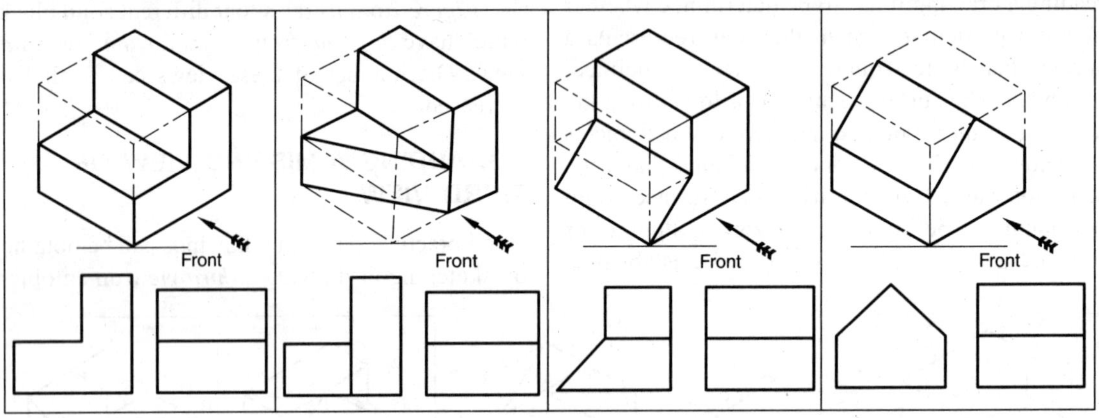

Fig. 7.6. Possible front view for the given side view (number can be increased)

when two of its orthographic views are known. The given views are drawn and, in simple cases, missing view is directly added without sketching the isometric view of the solid. For complicated objects an undimensioned pictorial view should be sketched, for visualizing the shape of the object, and then the missing view is added without much difficulty.

Figure 7.7 illustrates the method of projecting the side view of an object, from its given front and top views, point by point, using a numbering system. *Figure 7.8* illustrates the method of drawing the top view of an object, given its front and side views. *Figure 7.9* illustrates the method of drawing the front view of an object, having two oblique surfaces, given its side and top views. To assist in visualizing the object, from the two given views, in all these cases, the student should first sketch or draw an isometric view of the object.

In all the cases discussed above, all projection lines have been purposely retained to illustrate the method of projection and easy grasp of the technique followed.

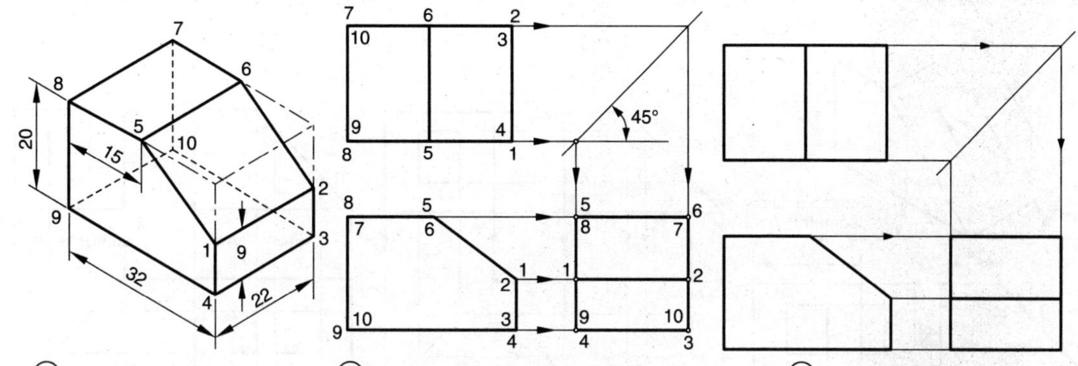

(1) Sketch the Isometric View (2) Projecting Side View Point by Point (3) Complete the Side View

Fig. 7.7. (i) Projecting side view, point by point, given the front and side views (Third Angle Projection)

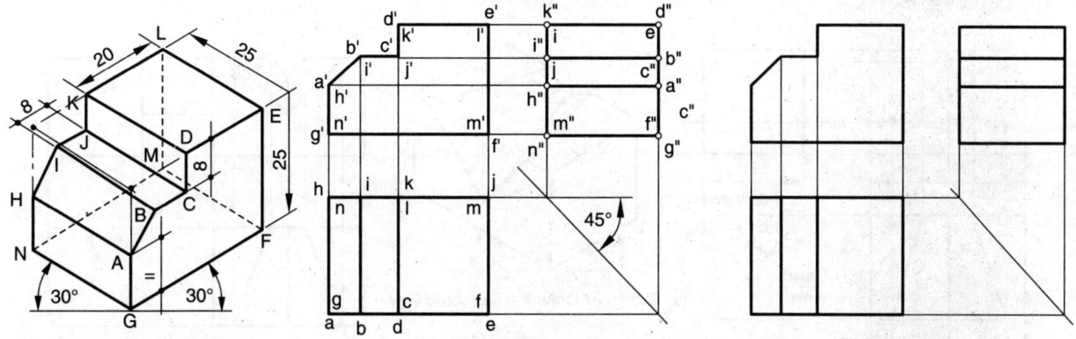

(a) Sketch or Draw Isometric View and Name its Corner Points. (b) Project Side View Point by Point. (c) Complete the Side View

Fig. 7.7. (ii) Projecting side view, point by point, given the front and side views (First Angle Projection)

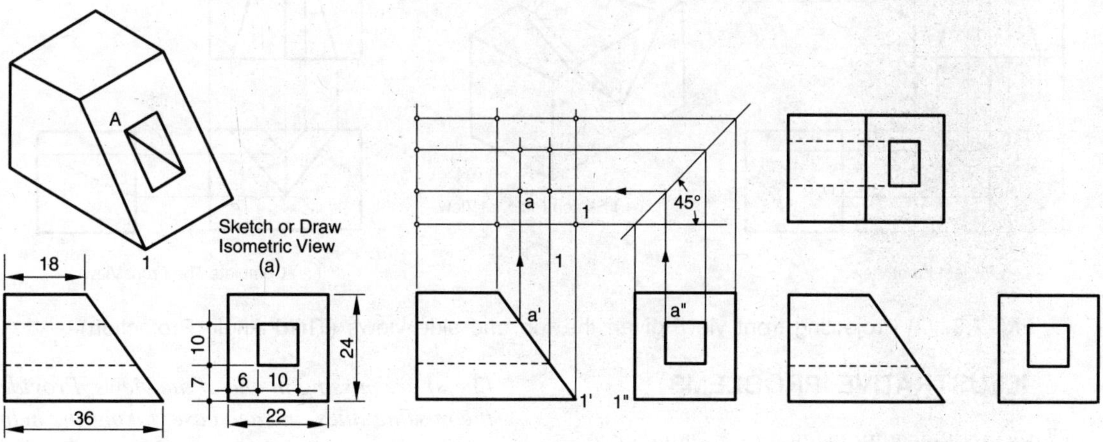

(a) Given Views (b) Project top View (c) Complete the top View

Fig. 7.8. (i) Projecting top view, given the front and side views (Third Angle projection)

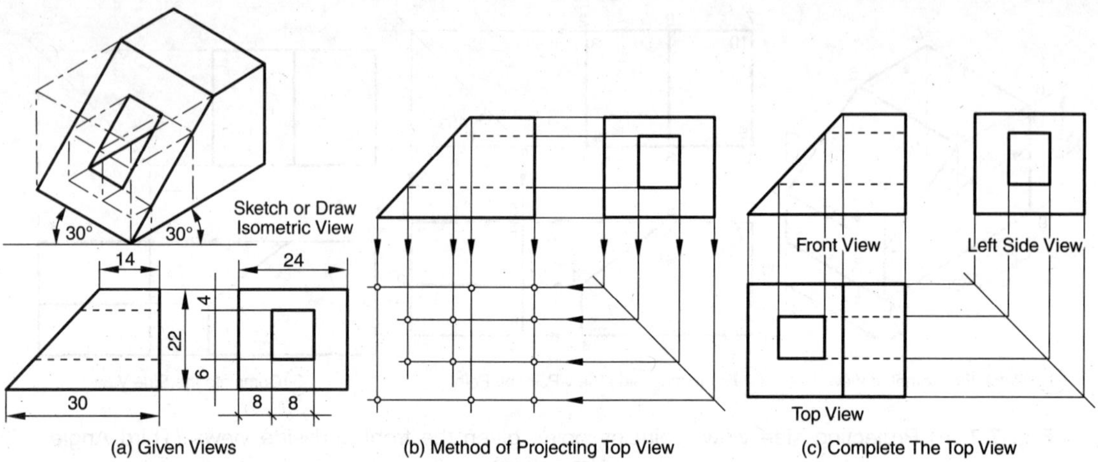

Fig. 7.8. (ii) Projecting top view, given the front and side views (First Angle projection)

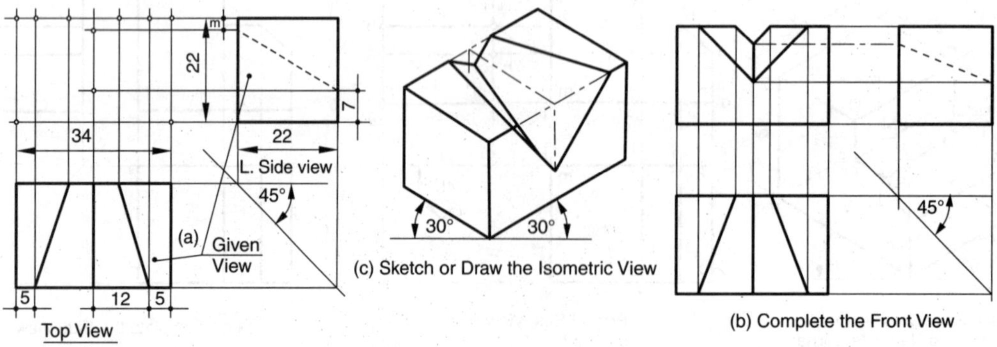

Fig. 7.9. (i) Projecting front view, given the top and side views (First Angle Projection)

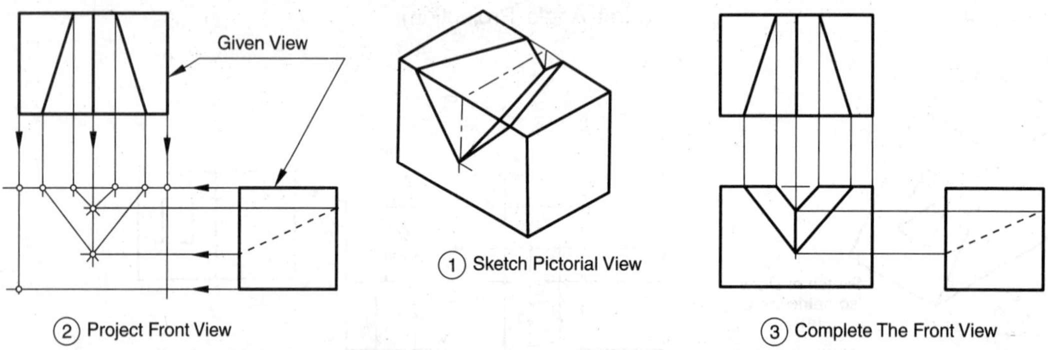

Fig. 7.9. (ii) Projecting front view, given the top and side views (Third Angle Projection)

ILLUSTRATIVE PROBLEMS

1. *Three orthographic views of a number of problems are given in Figs. 7.10 and 7.11 along with their isometric views. In each case some line(s) are missing in one of the views. Provide the missing lines, in each case, taking the help of their corresponding isometric views.*
2. *Three orthographic views of a number of problems are given in Figs. 7.12 and 7.13 In*

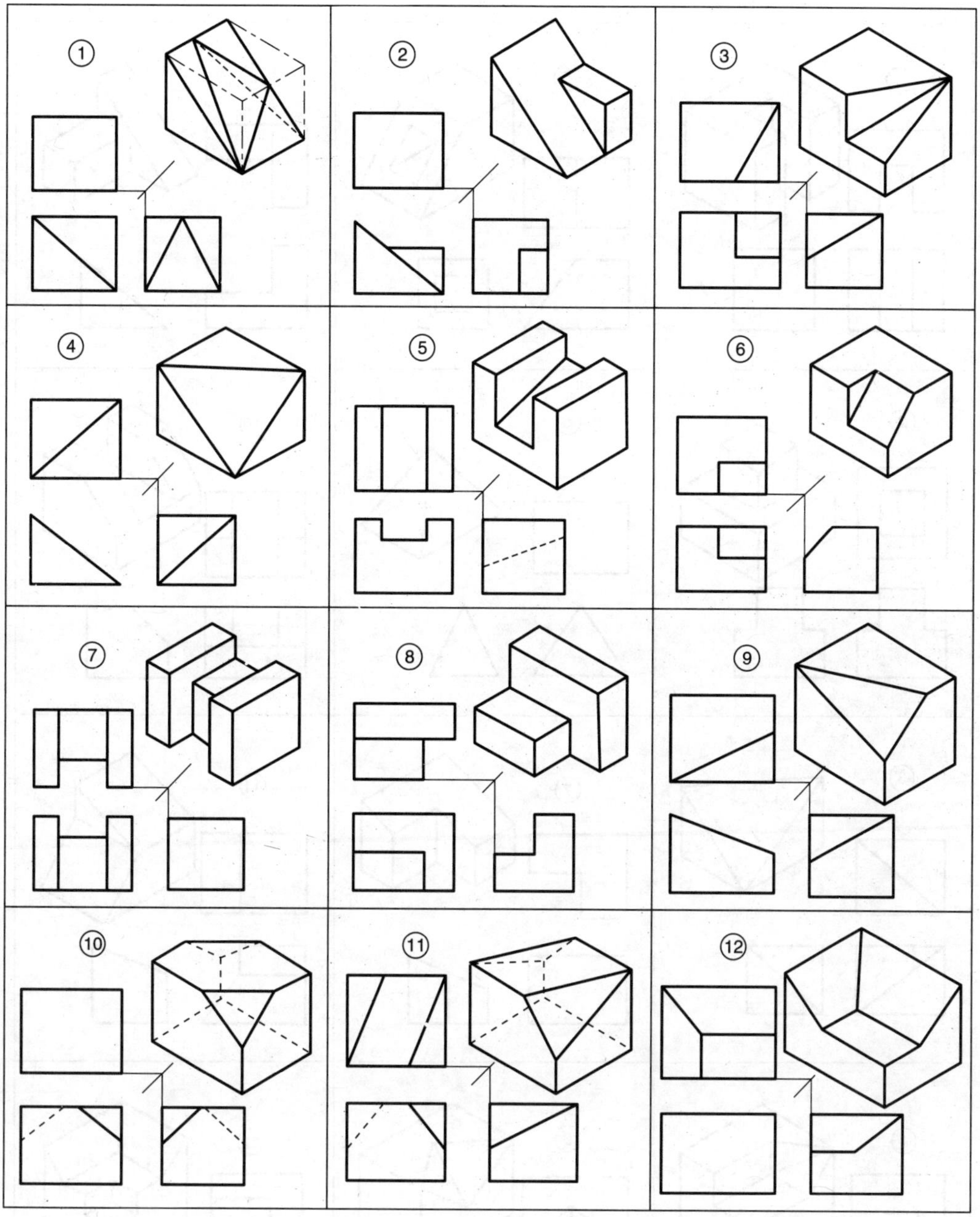

Fig. 7.10.

each case line(s) add the line (s) that are missing to the proper view.

- **HINT:** On an isometric graph paper, sketch proportionately the isometric view in each case. Then with its help add the required missing line(s). In *Fig. 7.14.* the isometric views of the problems given in *Fig. 7.12* are shown. The students can verify the correctness of the isometric views that they have sketched by comparing with these.

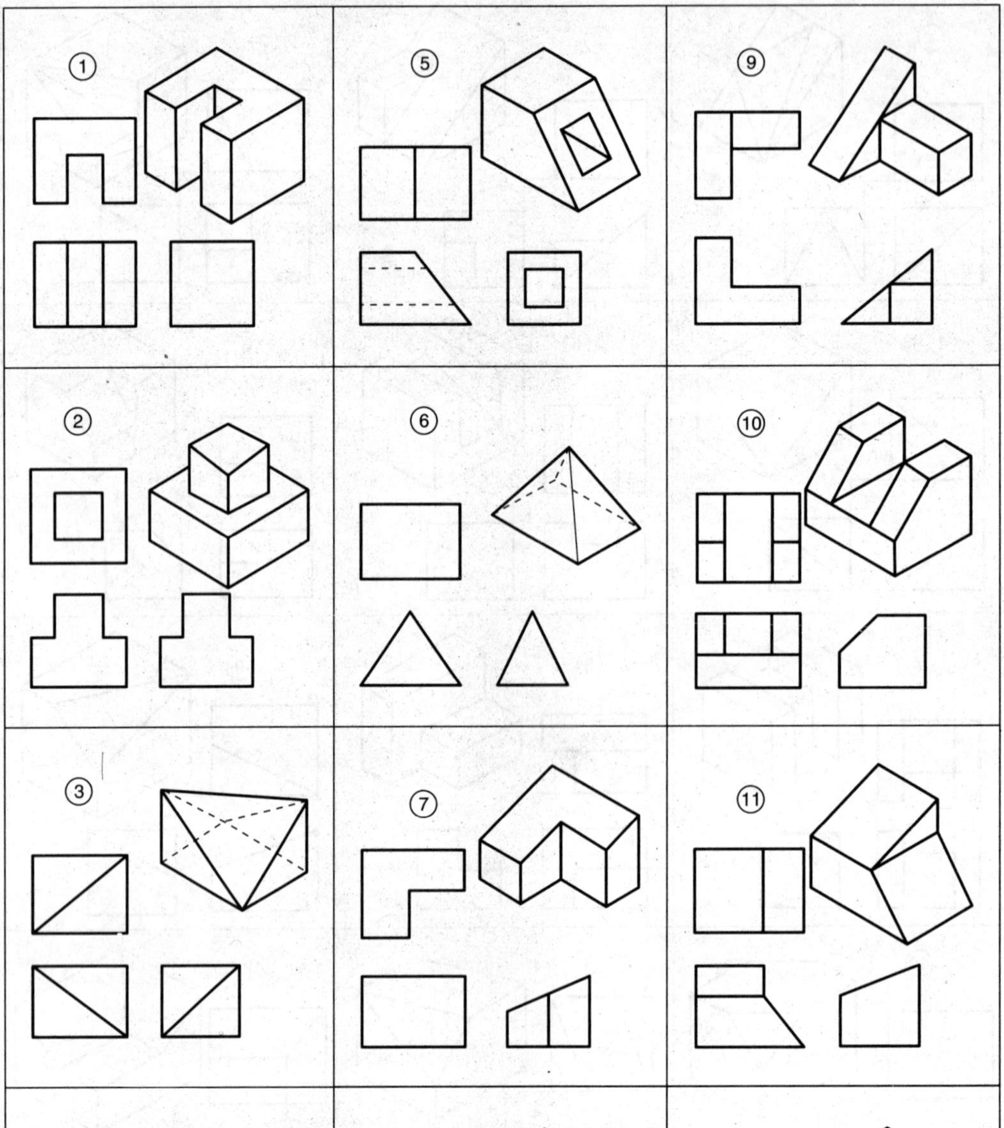

Fig. 7.11.

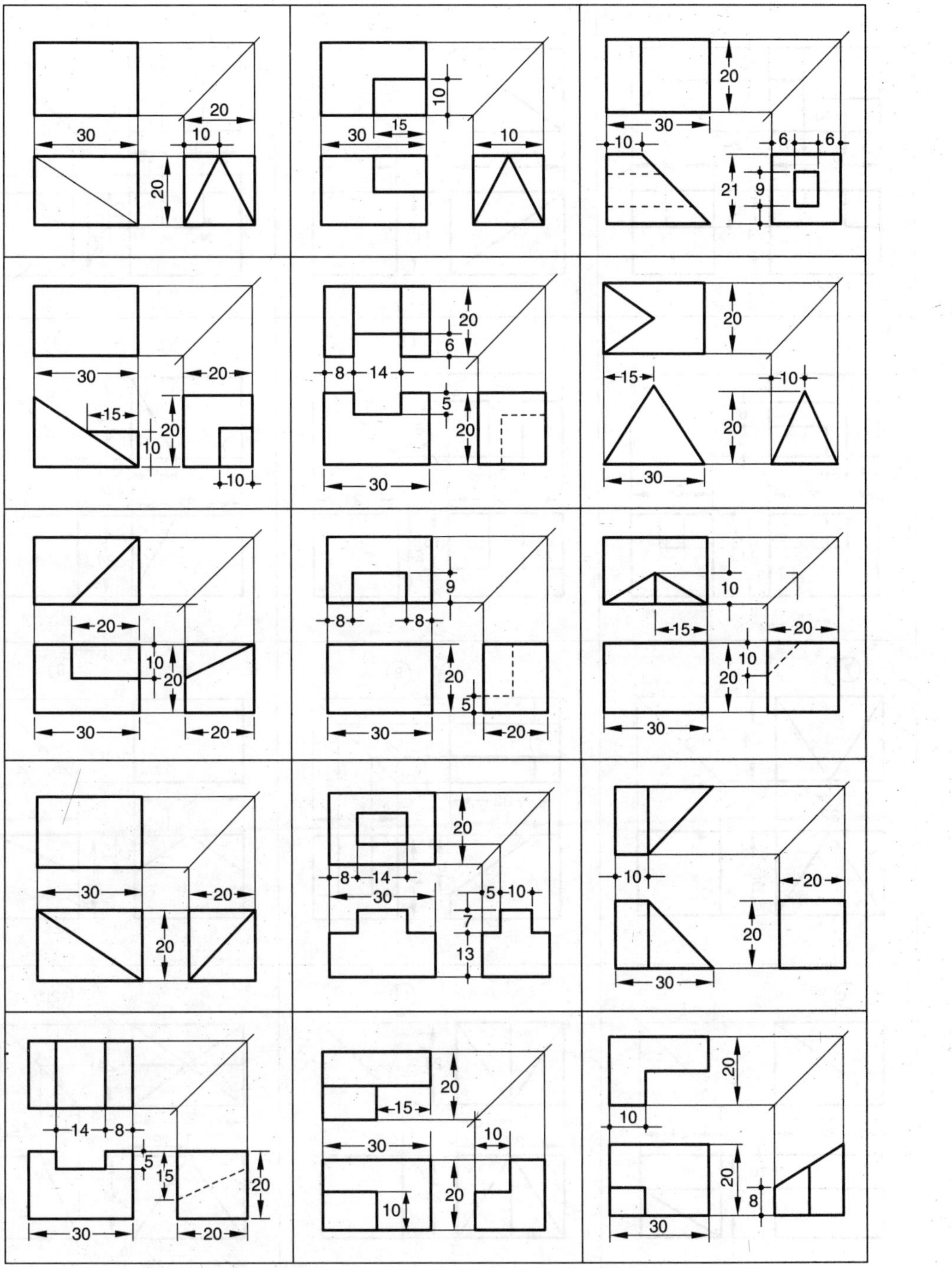

Fig. 7.12. Missing the problems (Third angle projection)

3. *A group of eight problems is given in Fig 7.15. In their orthographic views some lines are missing. Sketch proportionately, on an isometric sketch paper, the isometric view for each case. Then using the isometric views add the missing lines to the corresponding orthographic views from which these are missing.*

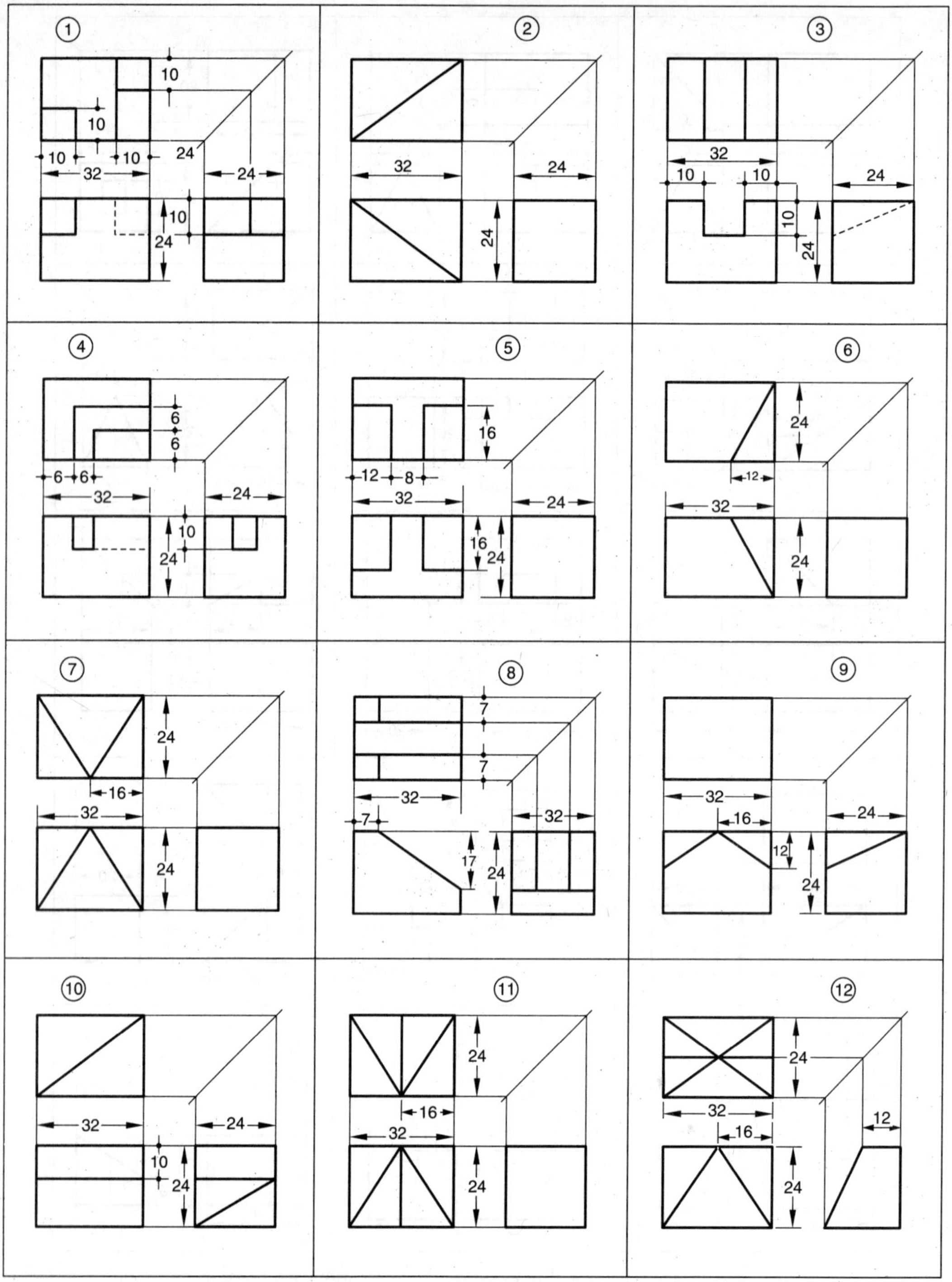

Fig. 7.13. Missing line problems (Third angle projection)

4. *Eight different blocks, with their isometric view and three orthographic views are given in Fig. 7.16. In each case some lines are missing in orthographic view(s). Reproduce the given*

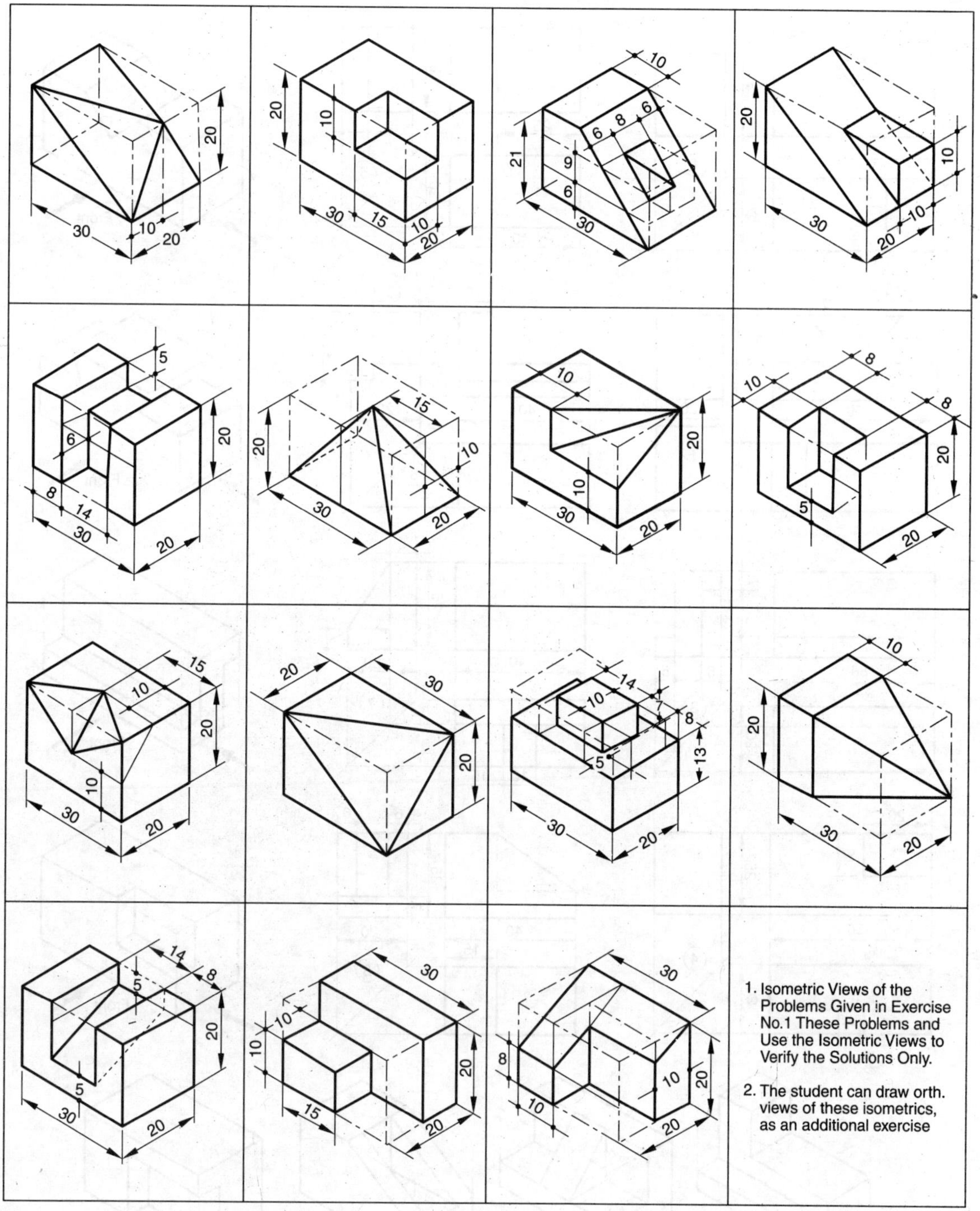

Fig. 7.14.

views, draw the isometric view, and add the missing line(s) in each case. You may also try these problems in Third Angle Projection.

5. *Front and top views of a block are given in Fig. 7.17 (a). Reproduce the given view and add the left side view.*

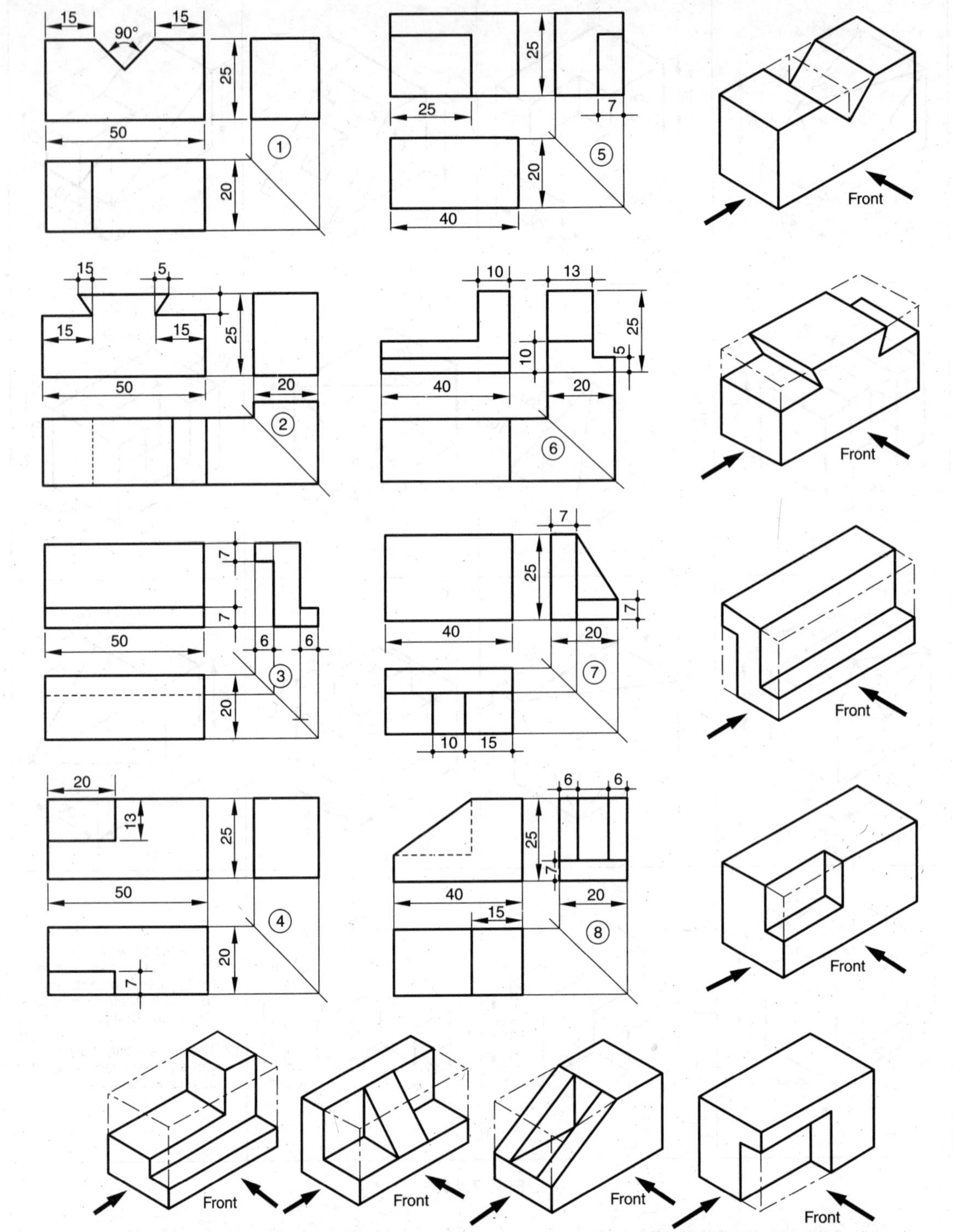

Fig. 7.15. Missing line problems (First angle projection)

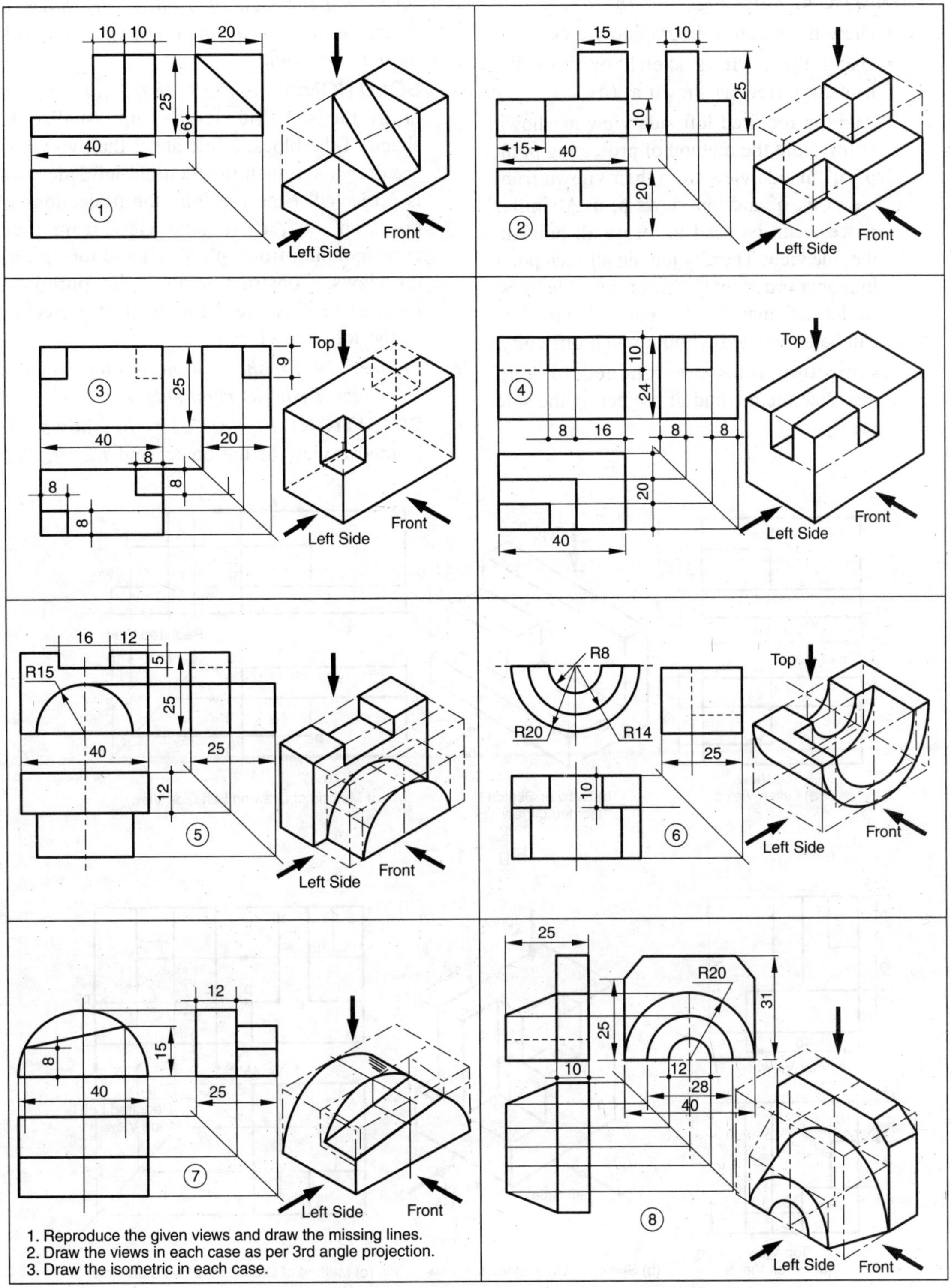

1. Reproduce the given views and draw the missing lines.
2. Draw the views in each case as per 3rd angle projection.
3. Draw the isometric in each case.

Fig. 7.16.

- **SOLUTION:** Refer *Fig. 7.17 (b), (c).*
  - Draw the given orthographic views.
  - Using these views sketch or draw the isometric view as shown at (*b*).
  - Add the required left side view as shown at (*c*). Note the method of projecting point **p″, q″** in side view from their known front views **p′, q′** and top views **p, q**. A similar process can be used to locate all point in the side view. Then by joining all such point in proper order, by visible or invisible lines, as the case may be, the required side view can be completed without much difficulty.
  - Projection lines are retained to help visualize the method of projecting the side view.

6. *Front and top view of a block are shown in Fig. 7.18 (a). Redraw the given views and add the left side view.*

- **SOLUTION:** Refer *Fig. 7.18 (b), (c)*. At (*b*) is drawn the isometric view to help visualize the shape of the block, while at (*c*) the two views are drawn and then the required left side view is projected. Note carefully the projections of points **p″, q″, r″** in side view from their corresponding front **(p′, q′, r′)** and top **(p′, q′, r′)** views. Construction lines are purposely retained to illustrate the method of projection of the left side view.

8. *Figure 7.19. (a) shows front and top views of an object. Draw its right side view.*

- **SOLUTION:** *Figure 7.19 (b)* shows the isometric view of the object and *Fig. 7.19(c)*

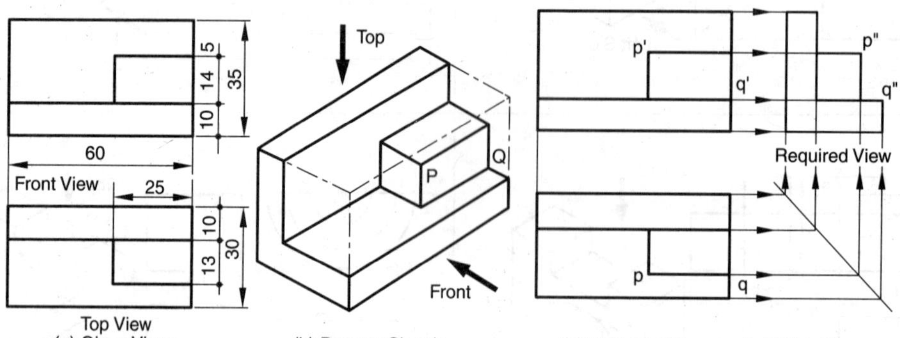

(a) Given Views (b) Draw or Sketch Isometric View (c) Method of Drawing Left Side View

Fig. 7.17.

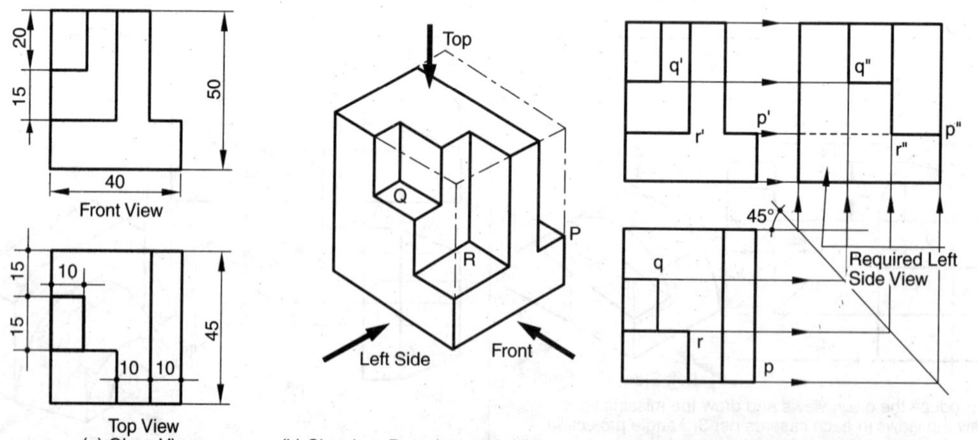

(a) Given Views (b) Sketch or Draw Isometric View (c) Method of Drawing the Left Side View

Fig. 7.18.

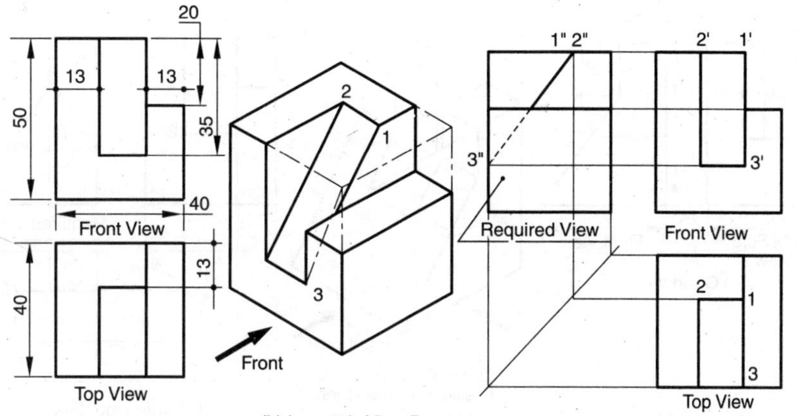

(a) Given Views (b) Isometric View Drawn to Facilitate Visualisation (c) Method of Drawing right side view

Fig. 7.19.

shows the method of drawing the required side view using the two given views. All important construction lines are retained to illustrate the method of projecting the side view.

8. *Left side view and top view of a block are shown in Fig. 7.20 (a). Draw the given views and add the front view.*

• **SOLUTION:** Draw the isometric view as shown at (*b*) and then redraw and add the required front view as shown at (*c*) in *Fig. 7.20*. All construction lines are retained to make the solution easy to follow.

9. *Front and right side view of a block are given in Fig. 7.21 (a). Redraw these views and add the top view.*

• **SOLUTION:** Draw the isometric view, as shown at (*b*) in *Fig. 7.21*. Then reproduce the given views and project from these the required top view, as illustrated at (*c*) in *Fig. 7.21*. Projection lines are retained to help visualize the solution.

10. *Right side and top views of an object are given in Fig. 7.22 (a). Reproduce the given views and add the front view.*

• **SOLUTION:** Refer *Fig. 7.22 (b)* for its isometric view. The views are reproduced and the require front view is added as shown at (*a*).

11. *In Fig. 7.23 front and top views of a block are given. Redraw these views and add the right side view.*

• **SOLUTION:** Refer *Fig. 7.23 (c)*. Here the right side view is drawn and all the projection lines are shown. The students, taking the help from isometric view drawn, should complete the side

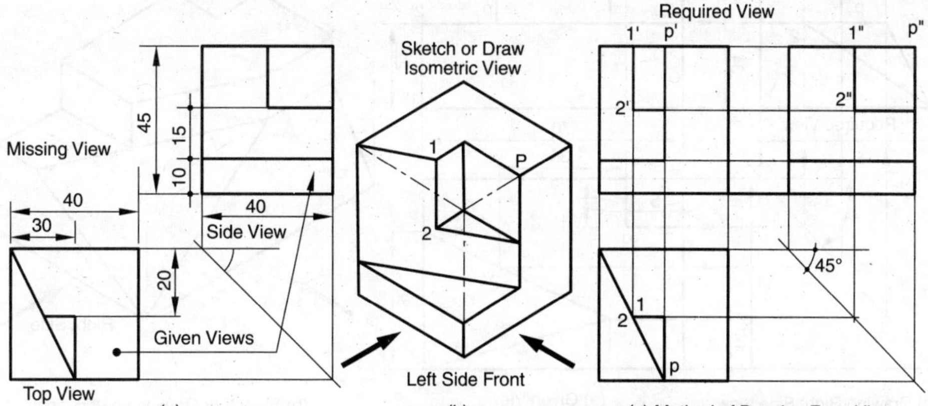

(a) (b) (c) Method of Drawing Front View

Fig. 7.20.

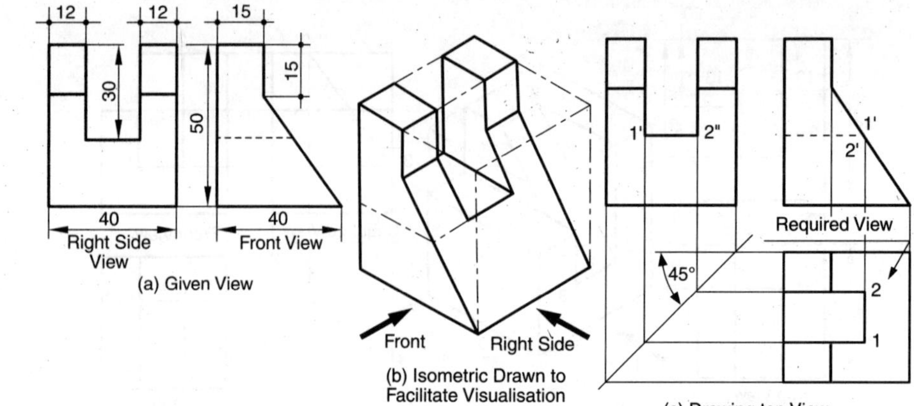

(a) Given View

(b) Isometric Drawn to Facilitate Visualisation

(c) Drawing top View

Fig. 7.21.

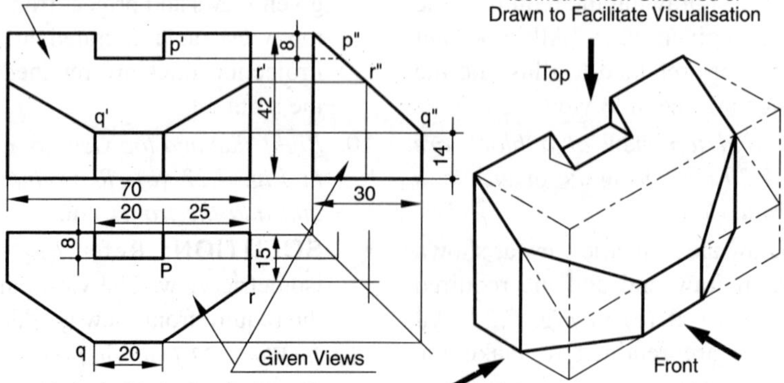

Method of Projecting the Front View

(a)

(b)

Fig. 7.22.

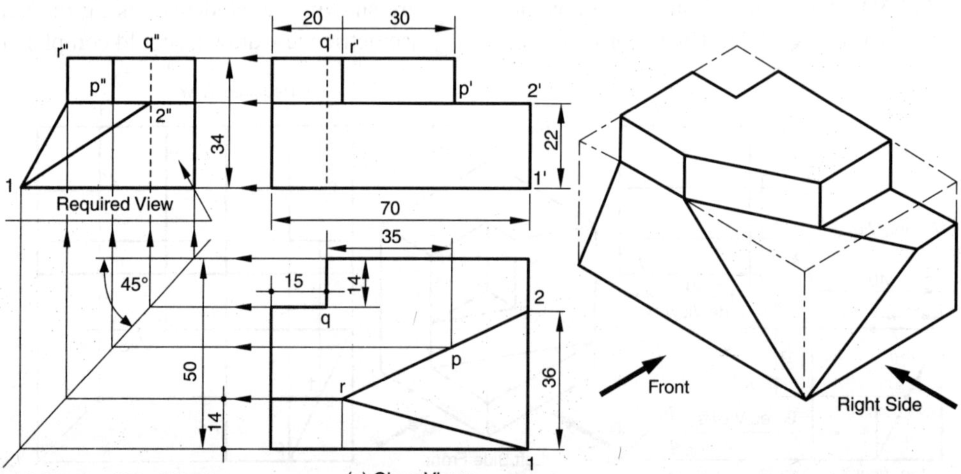

(c) Drawing Right Side View

(a) Given Views

(b) Sketch or Draw Isometric View

Fig. 7.23.

view themselves after redrawing the given views.

12. Front and right side views of a block are given in *Fig. 7.24 (a)*. Draw the given views and add the top view.

- **SOLUTION:** Refer *Figs. 7.24 (b)* for the isometric view of the block. The required top view is shown added to the given orthographic views in *Fig. 7.24 (a)*. Construction lines and arrow heads are shown to illustrate the direction of projectors and facilitate following the solution.

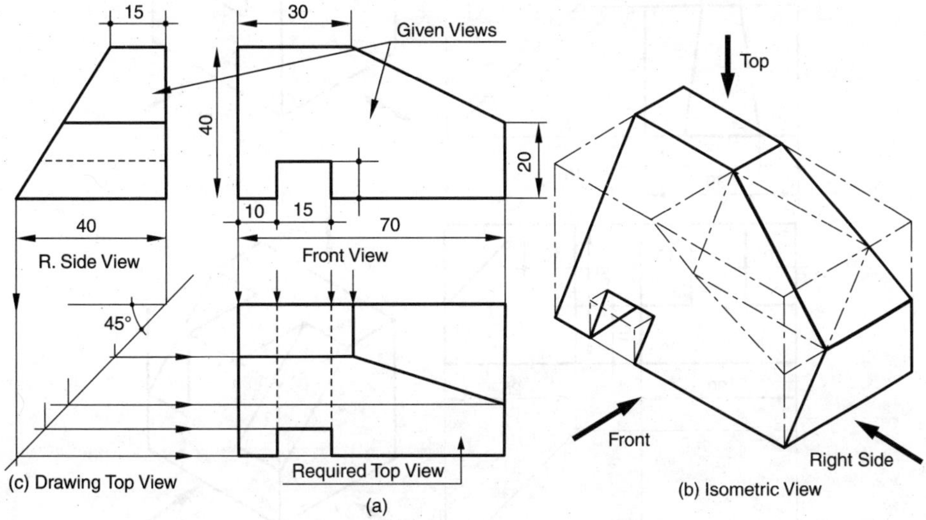

Fig. 7.24.

EXERCISES

1. Two orthographic views, each of five different blocks are given in *Fig. 7.25*. Their isometric views too are given. Reproduce the given views and add the third view in each case.
2. Two orthographic views each of twelve different blocks are given in *Fig. 7.26*. Draw the given views and add the third view in each case.

- **HINT:** Refer *Fig. 7.27* for isometric views of these blocks. With their help draw the required third view in each case.

3. Two orthographic views of a number of objects are given in *Fig. 7.28*. Sketch the isometric views of these objects.

- **HINT:** Refer *Fig. 7.29* for isometric views of these objects.

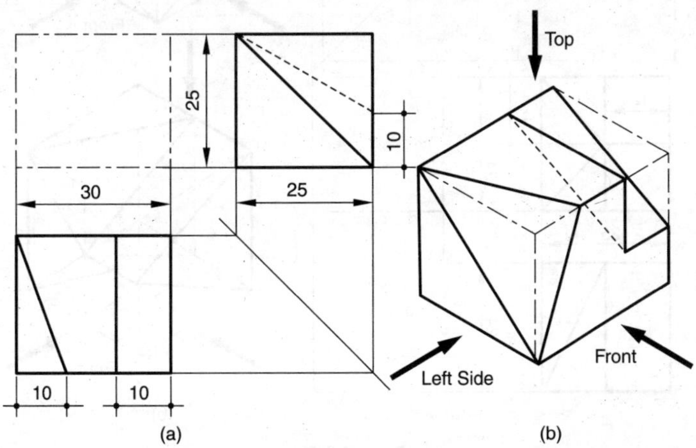

Fig. 7.25. (contd.)

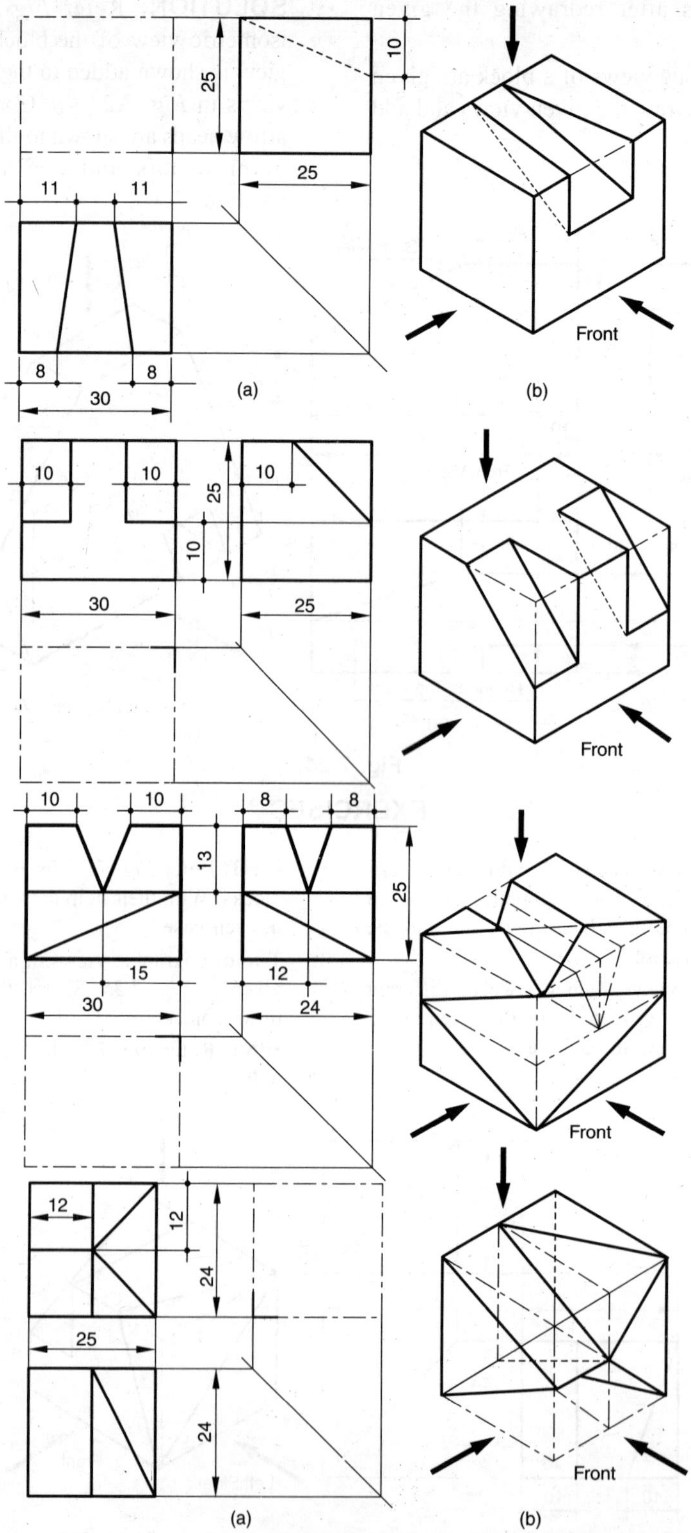
10
25
25
11
11
8
8
30
(a)
Front
(b)
10
10
10
25
10
30
25
Front
10
10
8
8
13
25
15
12
30
24
Front
12
12
24
25
24
Front
(a)
(b)

Fig. 7.25.

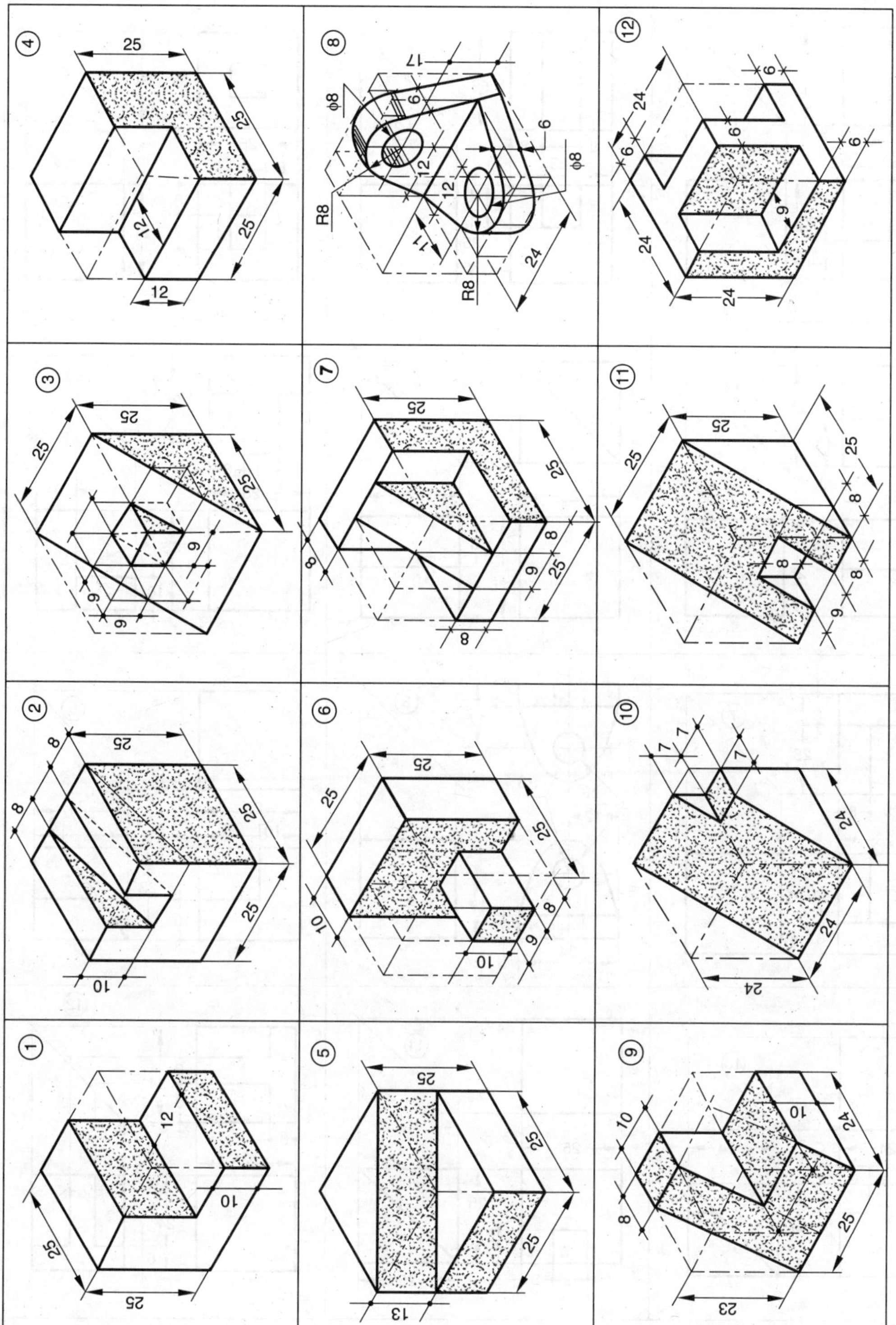

Fig. 7.26. Isometric views of the objects given in Fig. 19.28

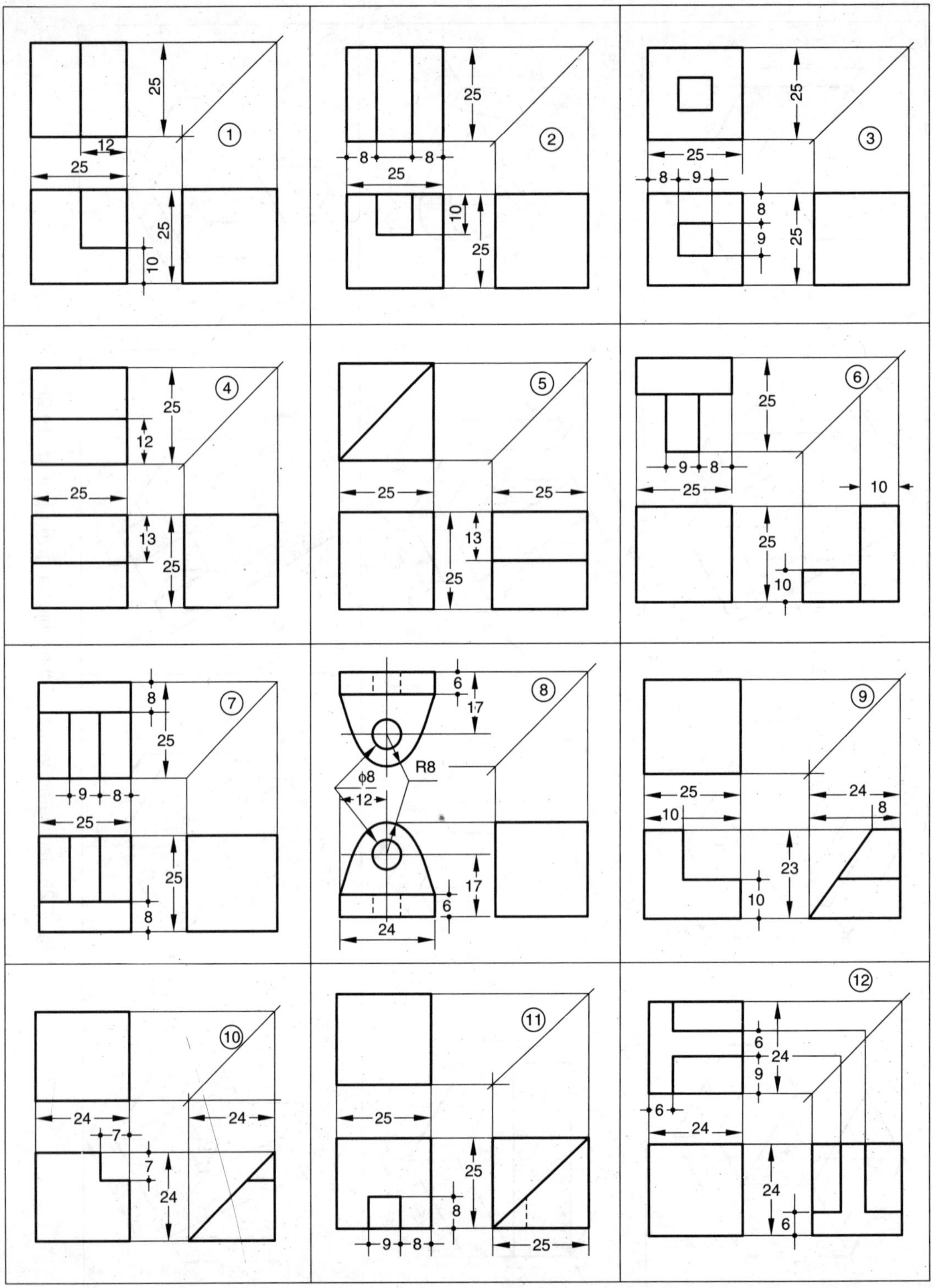

Fig. 7.27. Given two orthographic views to draw the third view

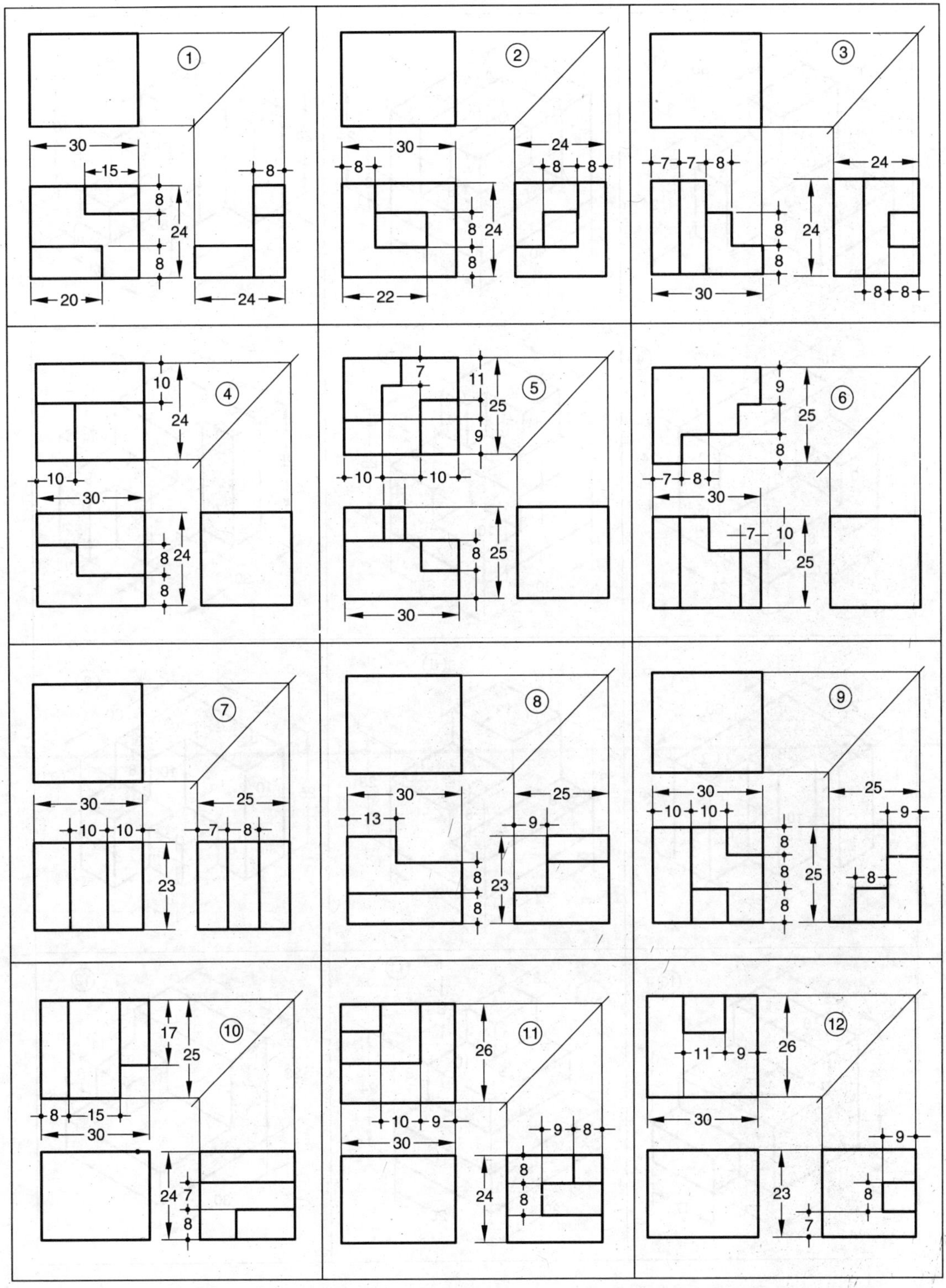

Fig. 7.28. Given two orthographic views to draw the third view

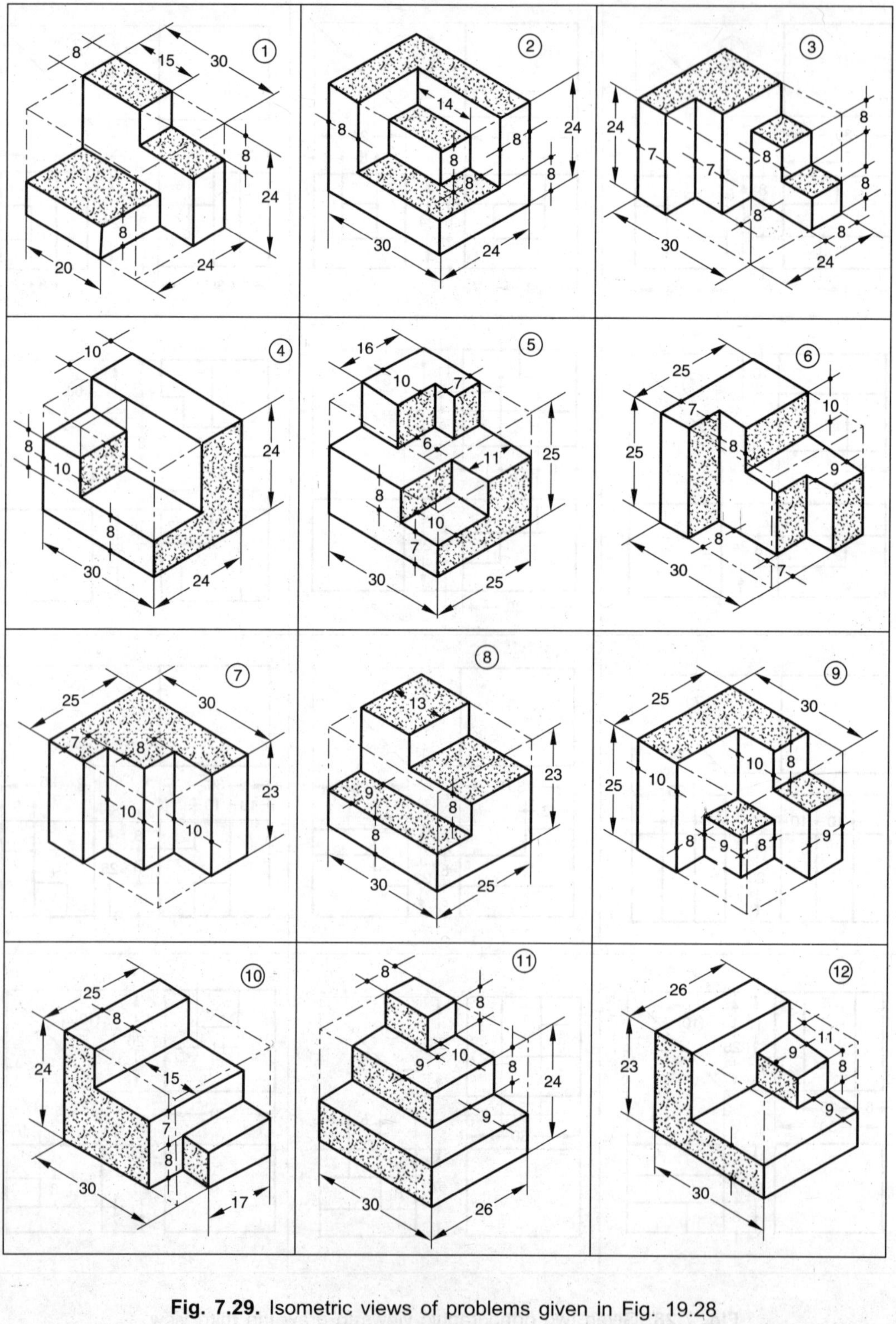

Fig. 7.29. Isometric views of problems given in Fig. 19.28

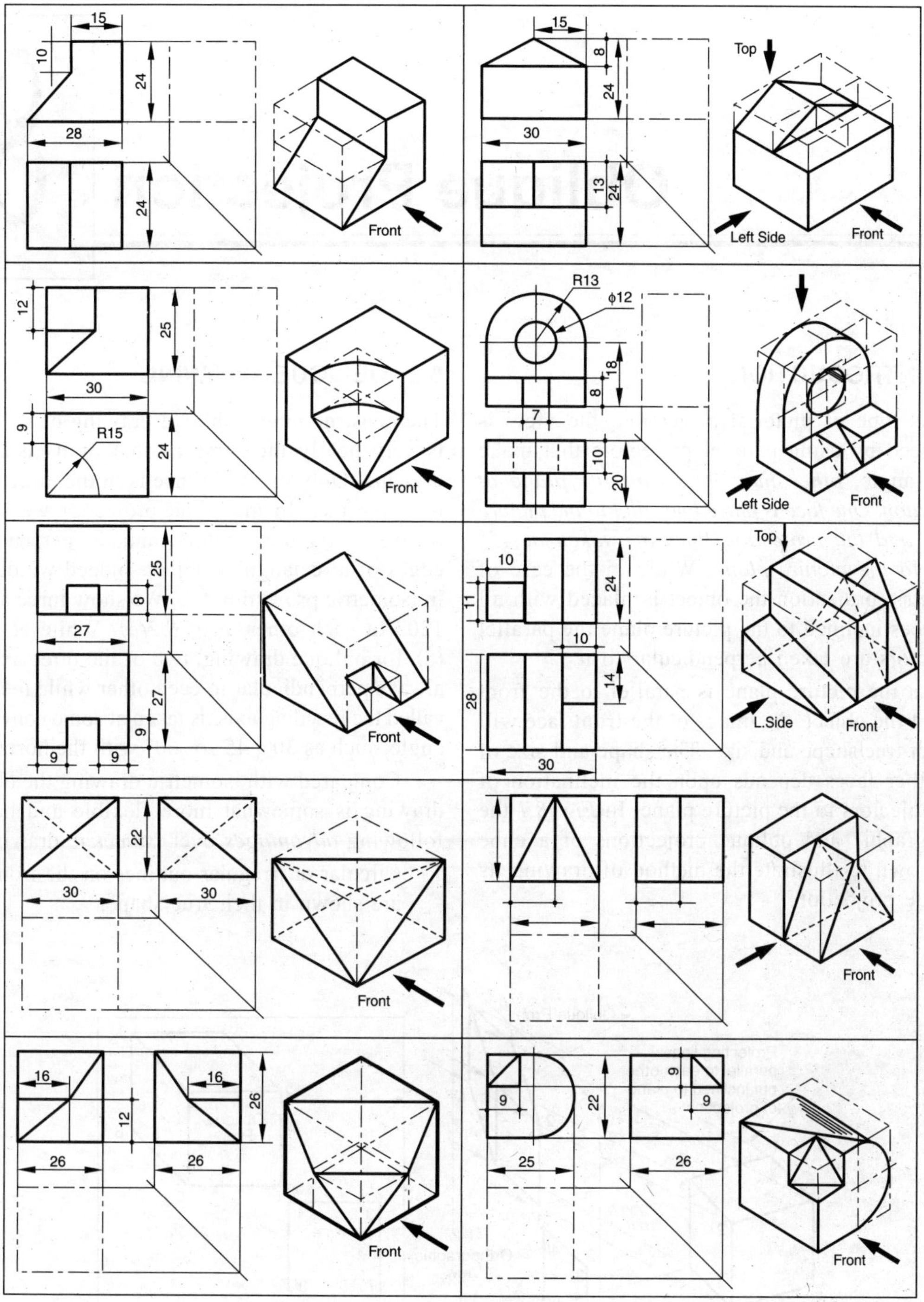

Fig. 7.30.

4. Two orthographic views, each of a number of objects are given in *Fig. 7.30*, along with their corresponding isometric views. Draw the corresponding third view in each case.

8

Oblique Projection

8.1. INTRODUCTION

In this type of pictorial projection, the view is produced by using parallel projectors that make some angle, *other than 90°, with the plane of projection. One face is placed parallel to the picture plane and the projection lines are taken at 45° (say) to the picture plane*. While in the case of isometric projection the object is placed with all the faces inclined to the picture plane the parallel projectors are taken perpendicular to it.

As the picture plane is parallel to the front face of the object, the image of the front face will project true shape and size. The shape and size of the other faces depends upon the inclination of the projectors to the picture plane. In *Fig. 8.1* the orthographic and oblique projections of a cube are shown to illustrate the method of drawing its oblique projection.

8.2. OBLIQUE DRAWING

The construction of oblique drawing of an object is very nearly the same as that of its isometric drawing. Only the difference is in the direction of the three axes. In an oblique projection we arrange three axes to represent three mutually perpendicular edges of a rectangular object. Indeed we did that in isometric projection too. We show three axes at 120° to each other, *Fig. 8.2(a)*. While at (*b*) to (*e*), for oblique drawing, two of the three axes are always perpendicular to each other while the third, called the receding axis, is taken at some convenient angle, such as 30°, 45° or 60° with the horizontal.

Compared with isometric drawing the oblique drawing is somewhat more flexible and has the following *advantages* over *isometric* drawing:

1. Circular or irregular outlines on the front face are shown in their true shape,

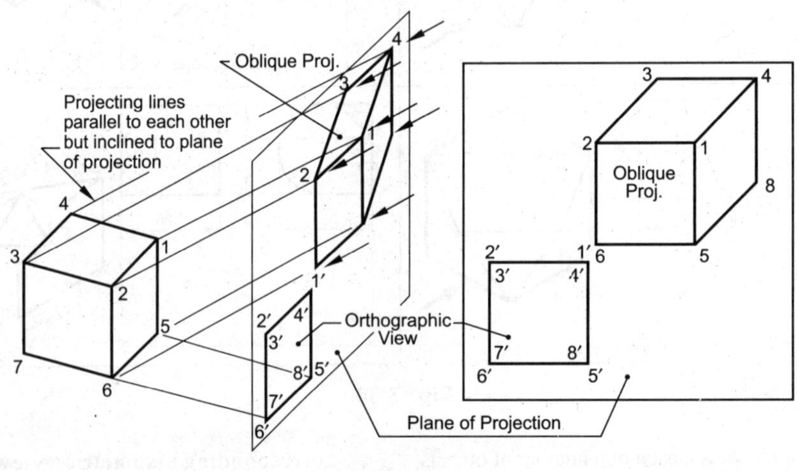

Fig. 8.1.

2. Distortion can be reduced by foreshortening along the receding axis, and
3. A greater choice is permitted in the selection of the positions of the axes. A few of the various views that can be obtained by varying the inclinations of the receding axis are illustrated in *Fig. 8.2,* 1 to 4. The selection of different angles for the oblique axis enables the drafter to emphasize more strongly the sides, back or bottom of the object at will. Usually the selection of the position of axes is governed by the character of the object.

The following rules for placing the object, when making its oblique drawing, should be observed:

- The *most irregular face, or the one containing the maximum circular outlines*, should be placed parallel to the picture plane so as to minimize distortion and simplify construction.
- If the first rule is not in clash, place the object with its longest *face facing the picture plane.*

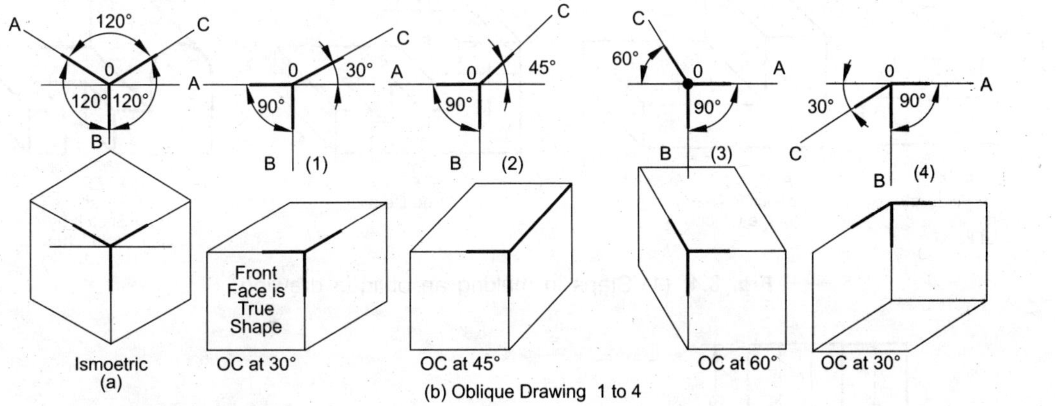

Fig. 8.2. Position of axes in isometric (a) and oblique drawings is (1) to (4)

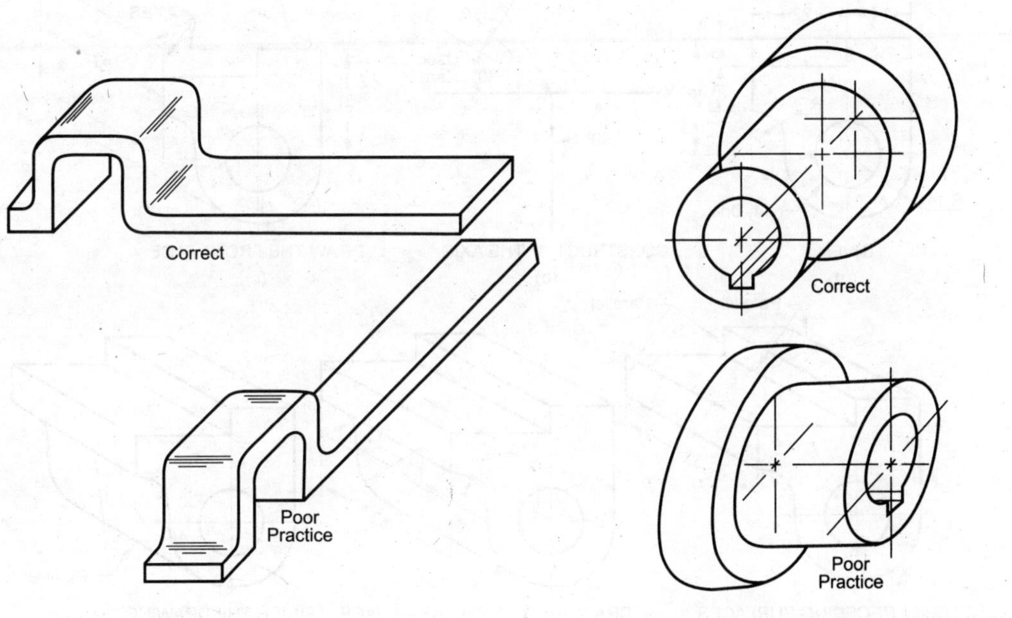

(a) Placement of a long object

(b) Placement of an object having curves or circles

Fig. 8.3. Placement of the object in oblique drawing (Orthographic views are in Third Angle)

In this position the view will be distorted to a lesser degree and its appearance will be more realistic and pleasing, *Fig. 8.3(a) (b)*.

8.3. MAKING AN OBLIQUE DRAWING

In general, the procedure for constructing an oblique drawing is the same as used for making an isometric drawing. Any of the box-method or the offset measurements method may be used for making an oblique drawing.

Figure 8.4 illustrates the different steps of construction for an oblique drawing using the box-method. The orthographic views of the object reveal that most of the irregular lines and circles are in the front view, as shown at (*a*). So this view should be kept parallel to the picture plane. The three

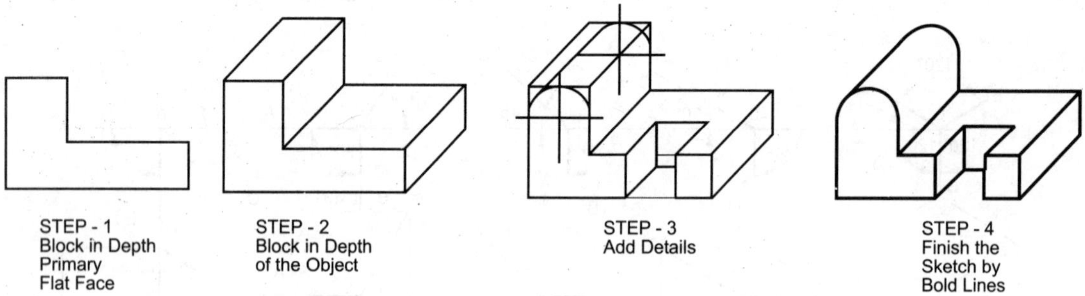

Fig. 8.4. (1) Steps in making an oblique drawing

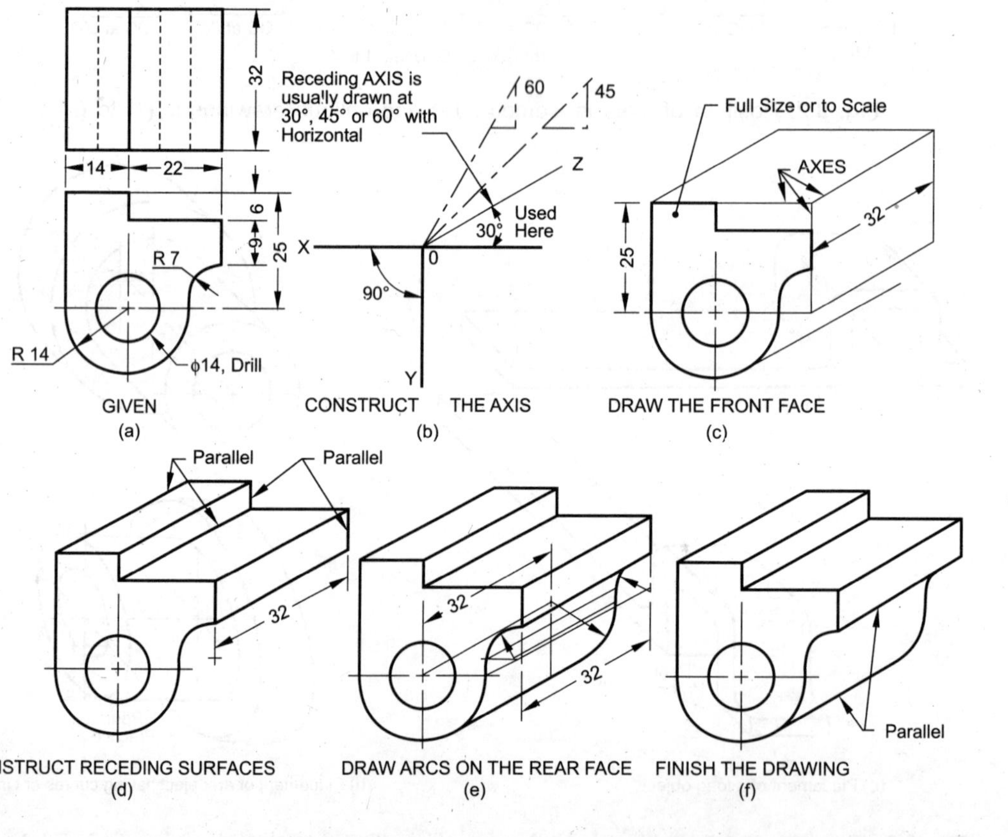

Fig. 8.4. (2) Steps in making an oblique drawing (Orthographic views in third angle)

axes that establish the perpendicular edges are drawn through O, representing the front corner, as shown at (*b*). Here, OX and OY are perpendicular to each other and OZ is set at any desired angle 30°, 45° or 60° (30° used here). On these three axes the boxes to enclose different component parts of the object are constructed. After the width, height and depth have been set off, the front face may then be laid out in its true shape and size, as shown at (*c*). The view can be completed in thin lines by drawing lines parallel to the receding axis through all the established corners, as at (*d*) and (*e*). Finally the required lines are brightened to finish the drawing, as at (*f*).

The *Centre Lines Method* of drawing an oblique projection of an object, having cylindrical construction, is illustrated in *Fig. 8.5* in three steps.

Steps in making an oblique drawing of an object, having arcs and curves, are illustrated in *Fig. 8.6 (a)* to *(d)*. The method of drawing the oblique projection of an object having a circular hole through its inclined surfaces is shown in *Fig. 8.7*.

8.4. SHORTENING THE MEASUREMENTS ALONG OBLIQUE AXIS

One of the *disadvantages* of both isometric and oblique projections is the unreal, indeed distorted, appearance of some such views that takes place due to the lack of '*vanish* or *convergence*', in the oblique receding lines. As the oblique lines in these projections are parallel and do not converge as these naturally appear on the object, the oblique drawing often presents a distorted appearance. Especially so when the dimension along the receding axis is quite long.

This drawback of the oblique drawing is partly overcome by reducing the measurements in the direction of the oblique axis, as shown in *Fig. 8.8*. The reduction may be done to any scale. But usually, these are reduced to one-half and the oblique drawing thus obtained is called Cabinet Projection, as shown in *Fig. 8.8 (c)* and *8.9 (b)*.

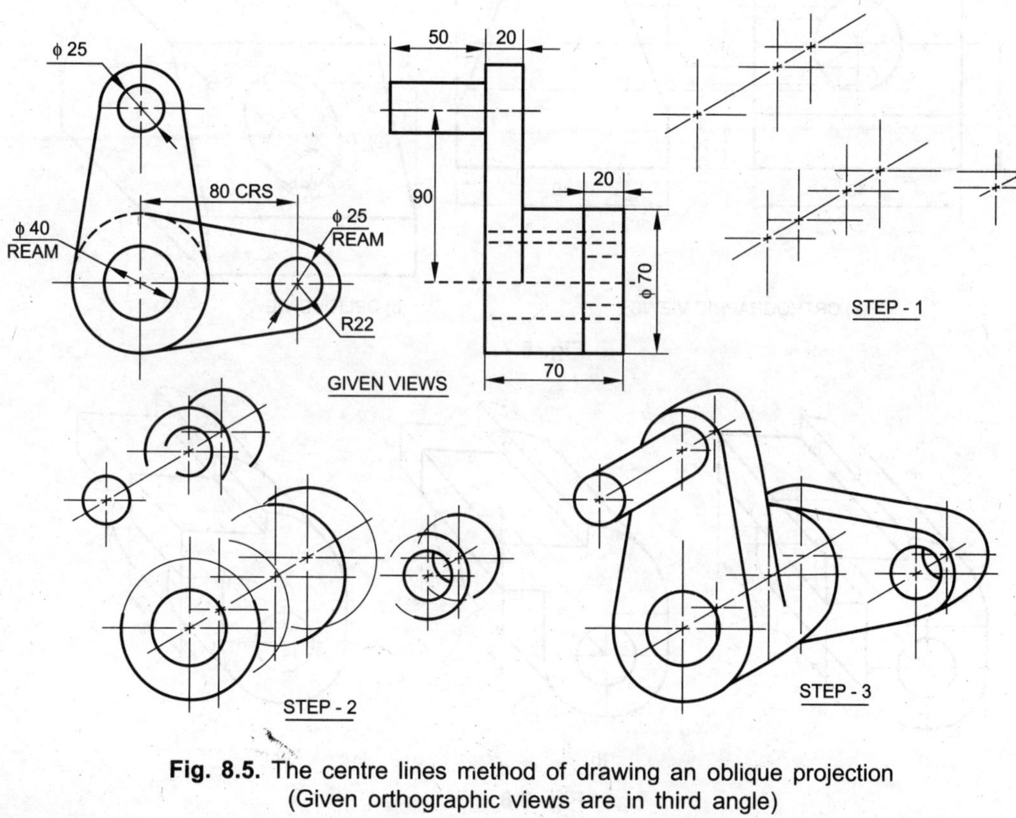

Fig. 8.5. The centre lines method of drawing an oblique projection (Given orthographic views are in third angle)

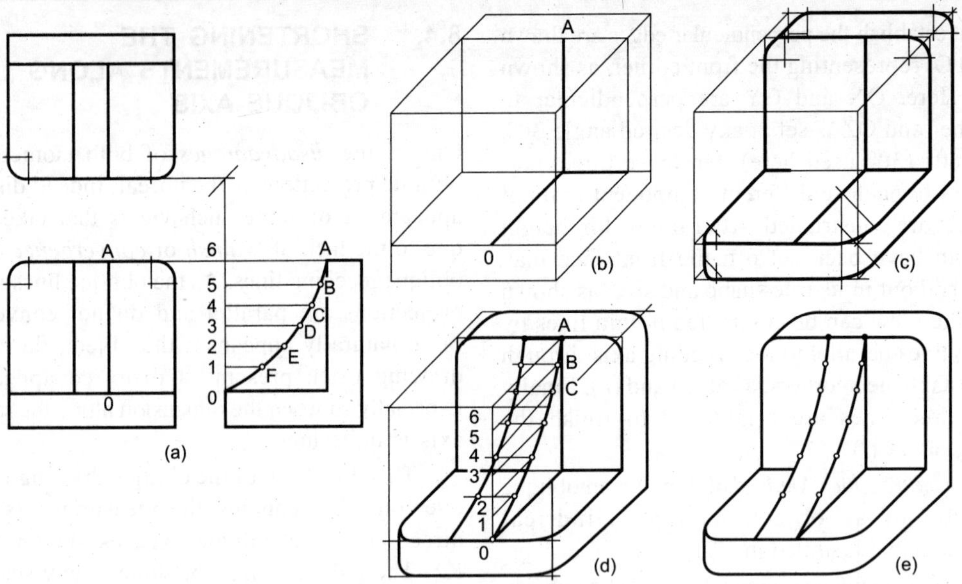

Fig. 8.6. Steps in drawing oblique projection of an object having arcs and curves by box-method (Orthographic views in third angle)

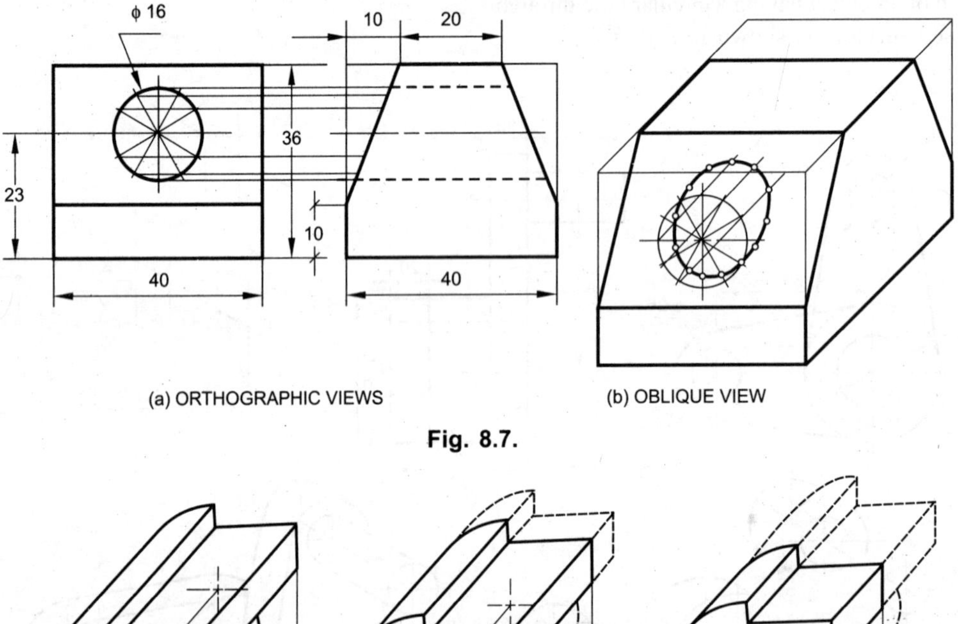

(a) ORTHOGRAPHIC VIEWS

(b) OBLIQUE VIEW

Fig. 8.7.

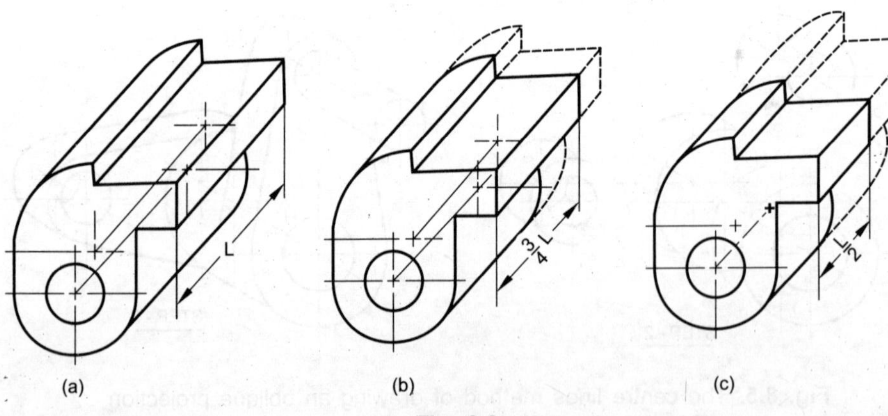

(a) (b) (c)

Fig. 8.8.

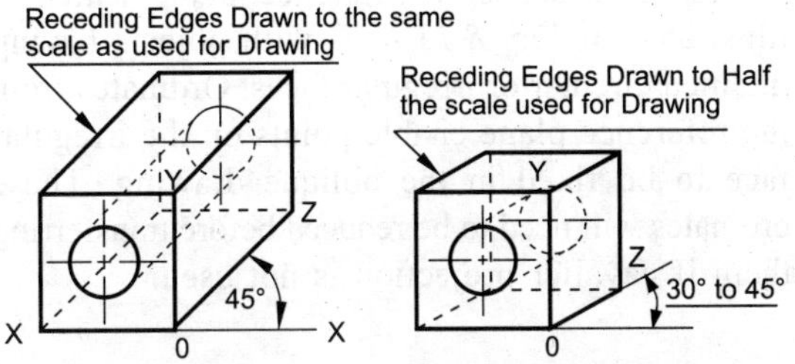

(a) Cavalier projection (b) Cabinet projection

Fig. 8.9.

8.5. CAVALIER PROJECTION

When the oblique axis is drawn at 45° to the horizontal and the receding edges are drawn to the same scale as that used for the measurements along the horizontal and vertical axes, as shown in *Fig. 8.9 (a)*, we obtain a particular type of oblique projection called '*Cavalier Projection*'.

If all the receding edges are drawn to a scale one half the scale used for measurements along the horizontal and vertical axes, the drawing is called '*Cabinet Projection*', as shown in *Fig. 8.9 (b)*. This halving of the scale for receding edges can apply irrespective of the angle selected for the oblique axis.

8.6. ANGLES, CIRCLES AND CIRCLE ARCS IN OBLIQUE PROJECTION

When the angles, circles, arcs and irregular outlines are on surfaces parallel to the plane of projection, they project their true shape and size. When located on oblique faces, the construction methods used in isometric drawing usually are applied. *Figure 8.10,* shows oblique projection of a cube with circles scribed on two faces using the Offset Measurements Method. *Figure 8.11,* shows the shape and construction of an oblique circle on the face of a cube by *Four Centre Method.*

In isometric projection two of the four centres are on the corners of the rhombus that represents the square face, on which the ellipse is drawn. In oblique projection the position of the centres depends on the angle of the receding axis. It must be noted that this approximate construction can only be used in cavalier projection, *i.e.*, the scale of all three oblique axes must be same.

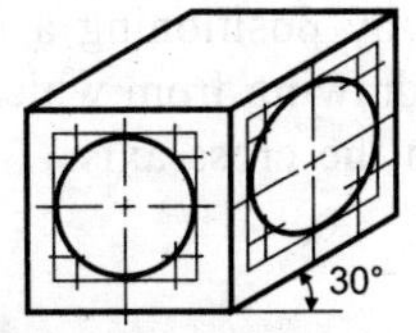

Fig. 8.10. Oblique projection of a cube with circles scribed on its two faces

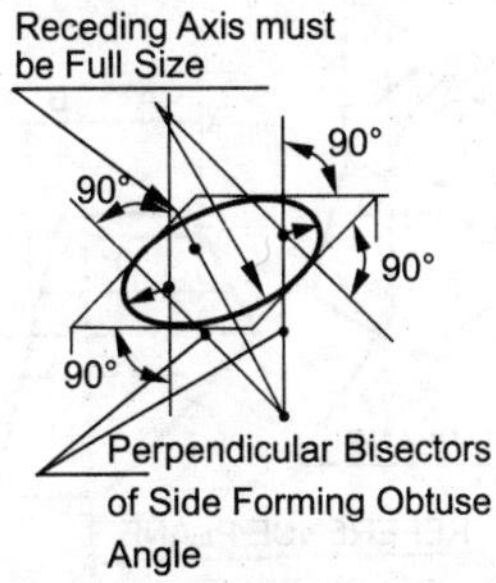

Fig. 8.11. Drawing oblique projection of a circle by four-centre method

Circles on receding planes in cabinet projection can be drawn by plotting points with ordinates as described for isometric projection. The necessary reduction in the scale on the cross axis can be made as the ordinates are transferred from the orthographic view to the oblique view.

The method of drawing the oblique projection of an object having curved portion is illustrated in *Fig. 8.6.* The method of dimensioning oblique views is shown in *Fig. 8.12.*

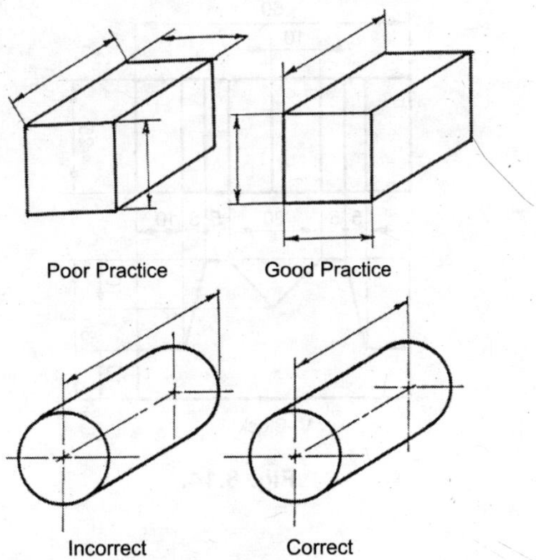

Fig. 8.12.

8.7. USE OF REFERENCE PLANES

Objects such as the Catch shown in *Fig. 8.13(a)* are best drawn by positioning a reference plane on the oblique drawing from which distances can be measured on the cross axis.

Another use of a reference plane which is illustrated in *Fig. 8.13(b)* is with objects having inclined circular or irregular faces. Ordinates from the reference plane enable points on the irregular face to be fixed in the oblique drawing. These ordinates will need to be reduced before transferring them if cavalier projection is not used.

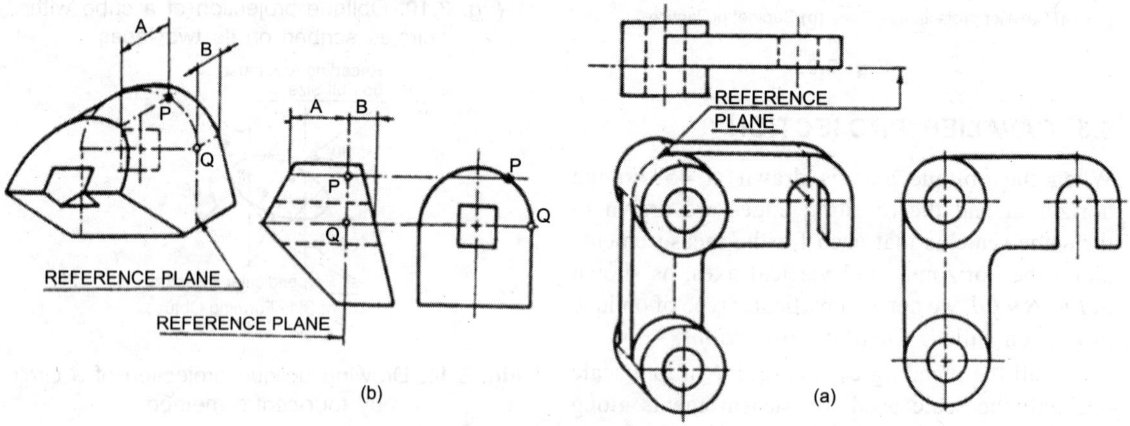

USE OF REFERENCE PLANES

Fig. 8.13.

EXERCISES

1. *Figures. 8.14, 8.15,* and *8.16* shows orthographic views of three different objects, in third Angle. Draw the oblique drawing for each one of them and dimension the views.

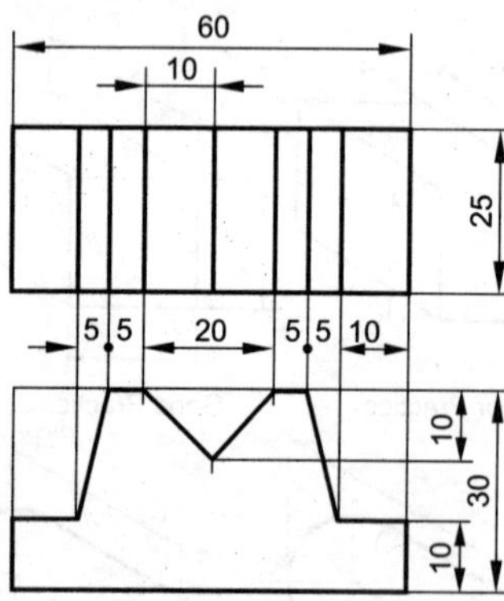

V–Block

Fig. 8.14.

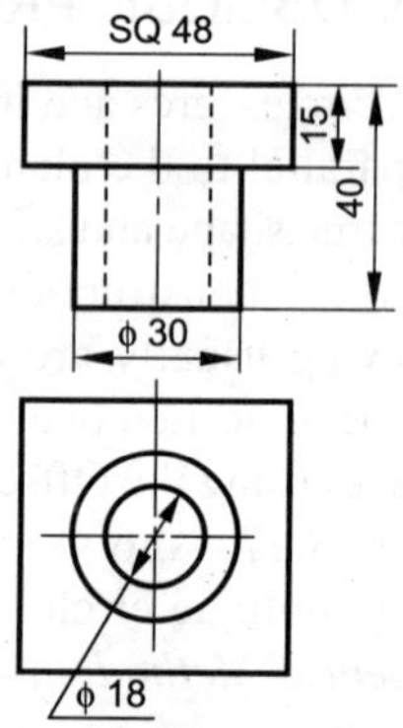

Bush

Fig. 8.15.

2. Oblique views of two different objects are shown in *Fig. 8.17.* Draw the front view, right side view and top view for each of them. Dimension the views. Use full size scale.

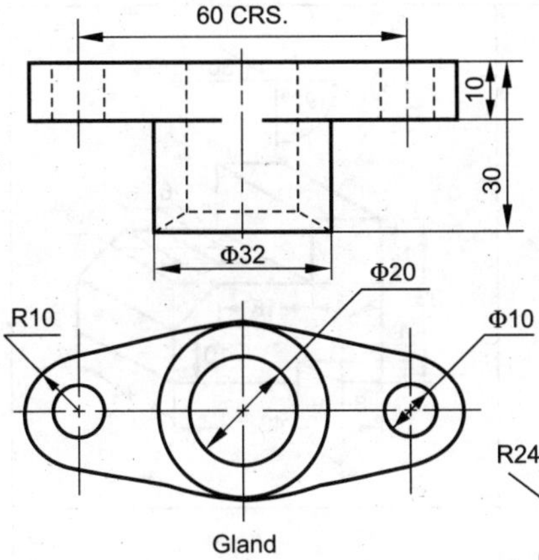

Fig. 8.16.

3. Oblique views of five different objects are shown in *Fig. 8.18*. Draw the front, right side and top views for each one of them. Dimension the views.
4. Oblique views of a group of eleven different objects is given in *Fig. 8.19*. Using an A2 size drawing sheet complete the exercise given at the bottom left hand side of these views.
5. Oblique views of a lever are shown in *Fig. 8.20*. Draw their front, right side and top views. Dimension the views completelys.

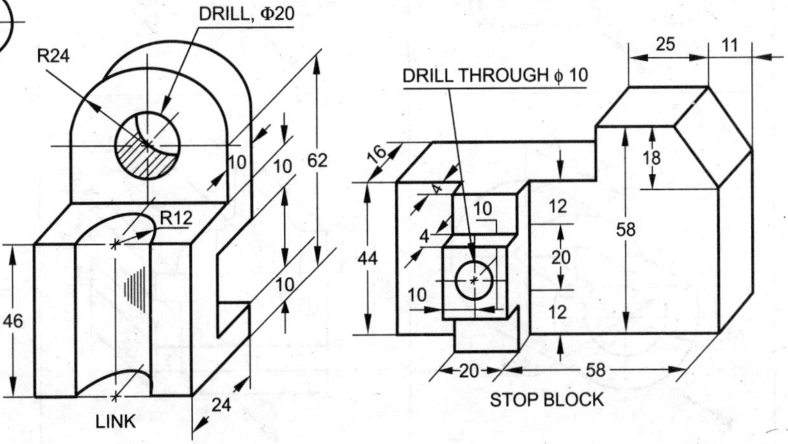

Fig. 8.17.

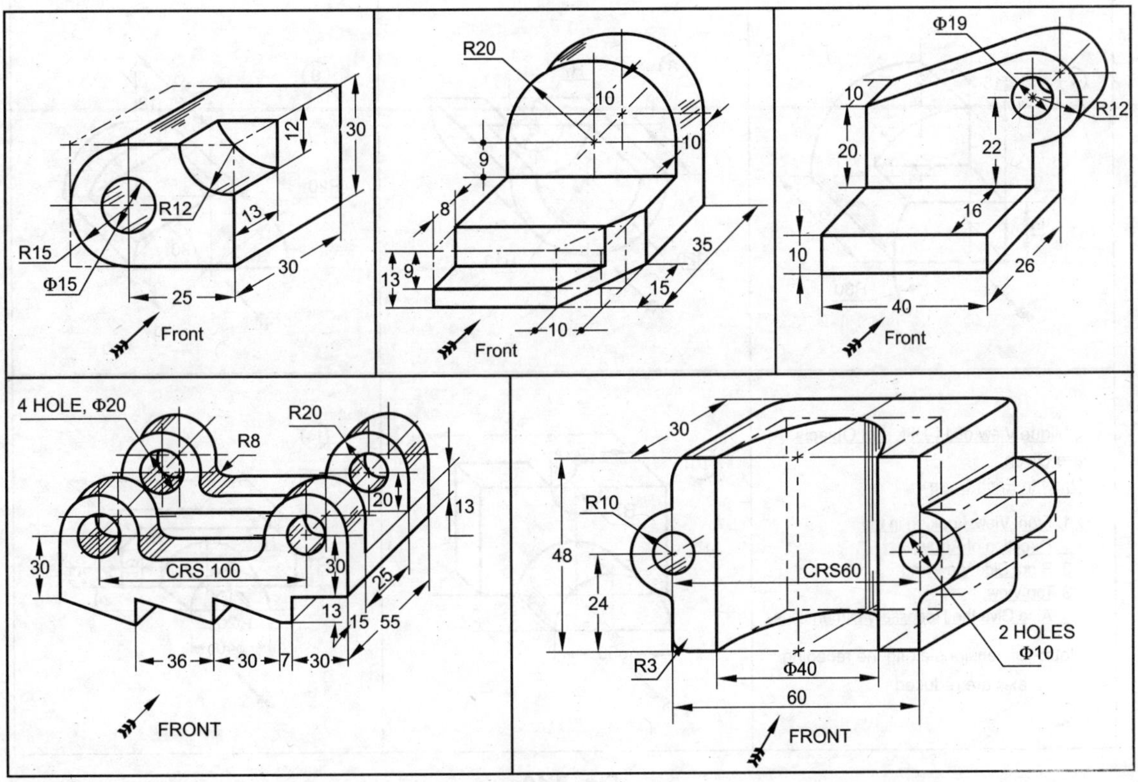

Fig. 8.18.

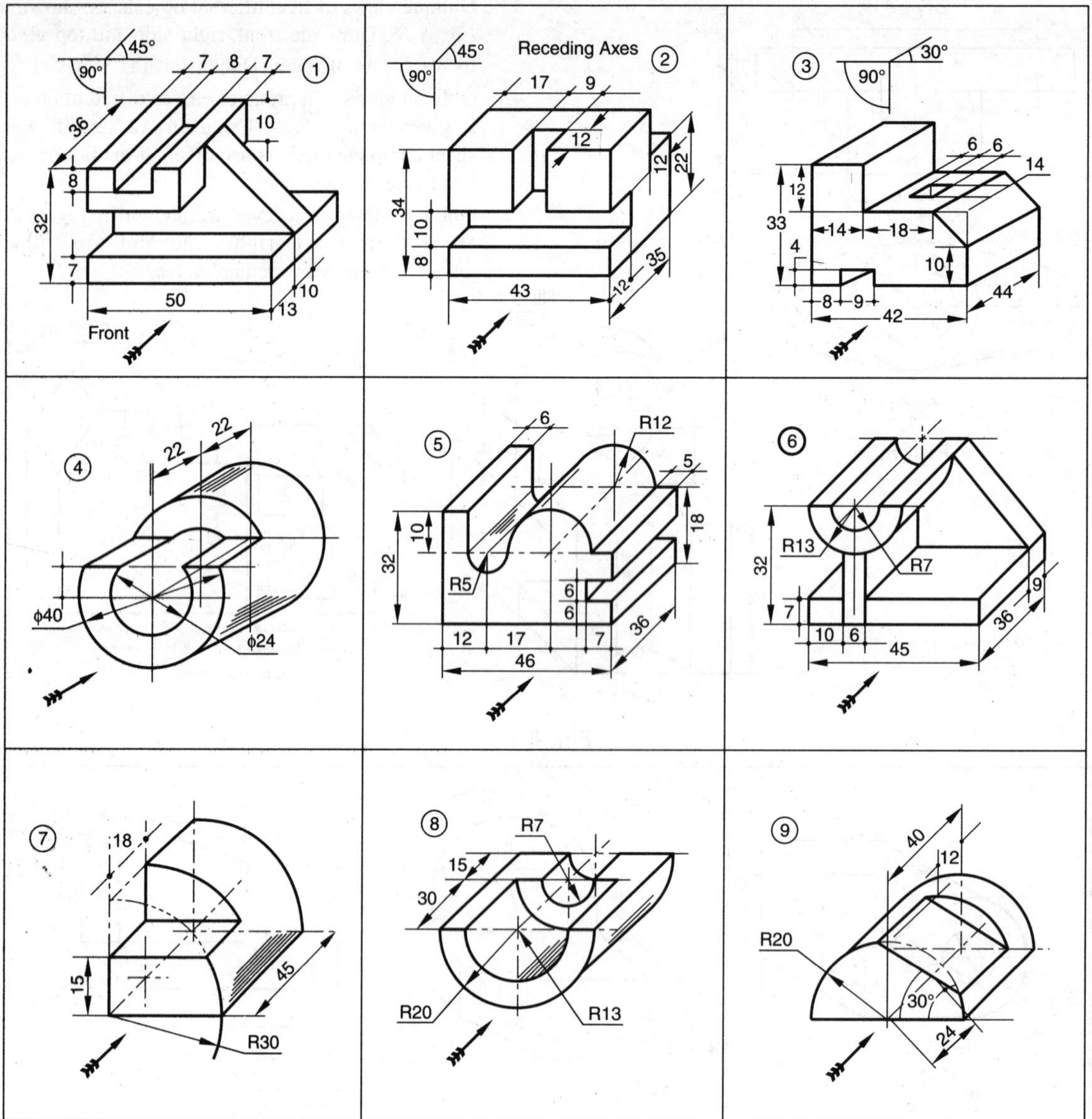

<u>Oblique View of 11 Different Objects are Given:</u>
<u>Draw in Each Case</u>

1. <u>Front View. Looking in the Direction of Arrow,</u>
2. <u>Right Side View and</u>
3. <u>Top View,</u>
   <u>Also Give the Necessary Dimensions.</u>

Note:- Dimensions along the receding axis are reduced.

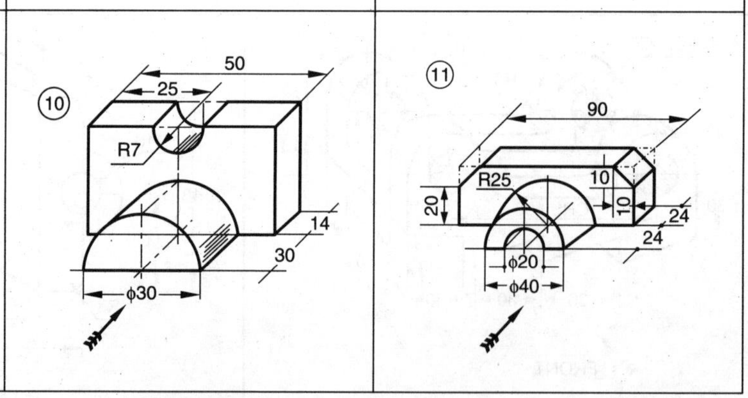

Fig. 8.19.

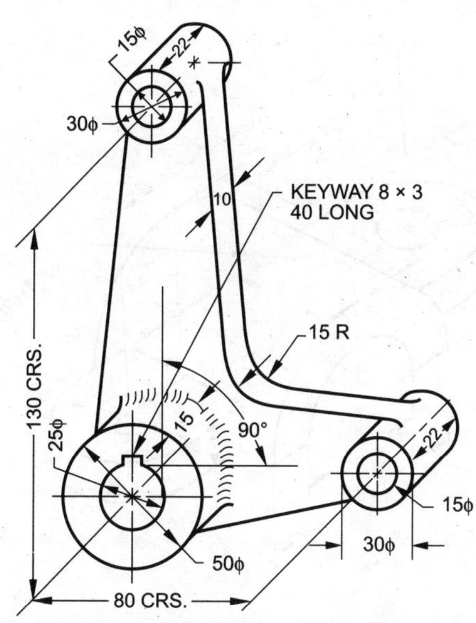

Fig. 8.20.

6. *Figure 8.22* shows two orthographic views of a simple bracket in Third Angle Projection. Draw its oblique projection.
7. Isometric views of three different objects are given in *Fig. 8.23* at 1, 2 and 3. Draw an oblique view for each one of these.
8. Orthographic views and corresponding isometric views of seven different objects are given in *Fig. 8.24* to *8.28*. Study the views and choosing a best face to keep perpendicular to the direction of sight draw a corresponding oblique views for each one of these.

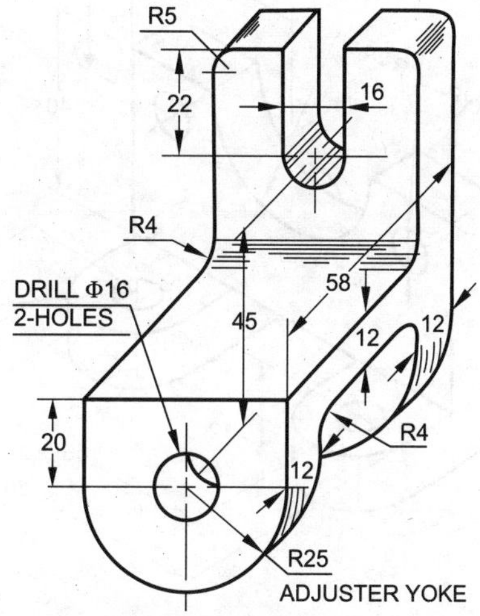

Fig. 8.21.

Note:

- All these problems also are good examples to practice different commands in any CAD System, *i.e.*, AutoCAD.
- It may be seen by drawing oblique views of the objects in these problems that for objects having circular arcs and circles in plane(s) parallel to one of the principal planes of projection on oblique view is faster and easier to draw as compared to their Isomatric views.

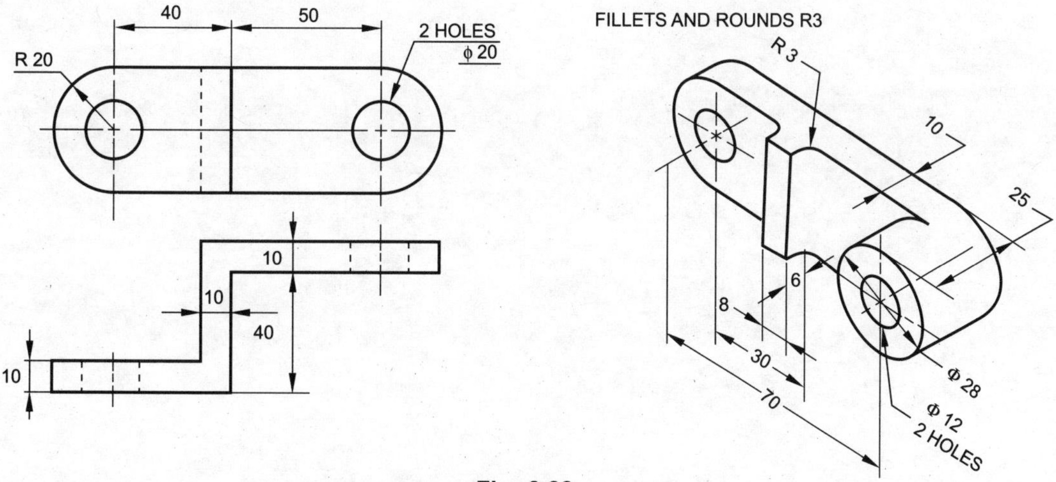

Fig. 8.22.

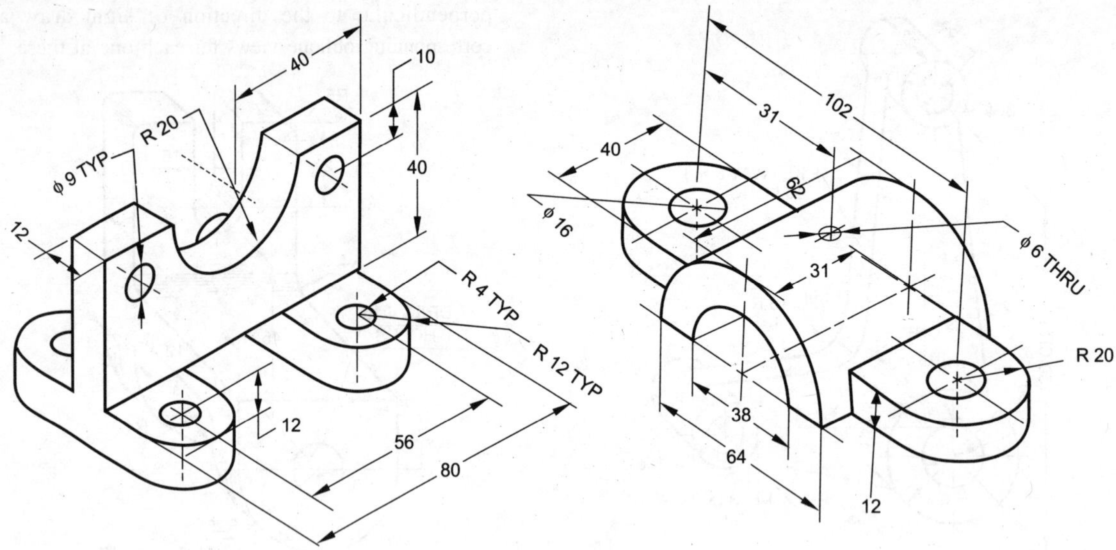

Fig. 8.23.

QUESTIONS FOR VIVA VOCE

1. What is an oblique projection?
2. For which type of an object an oblique view is preferred to an isometric view?
3. Differentiate between an oblique view and an isometric view.
4. What is a receding axis in oblique drawing? Why it is necessary to foreshorten the measurements in the direction of the receding axis?
5. Name the different methods used for oblique peojection. For what type of object each one of them is preferred?
6. What is a Cavalier Projection?
7. What is a Cabinet Projection?

❏❏❏

9

Auxiliary Projections

9.1. INTRODUCTION

Many objects are of such shape that their principal faces cannot be assumed parallel to the regular planes of projection. In other words they have certain details that are on planes that are not parallel to any of the principal planes of projection, as shown in *Fig. 9.1*. The projections of such objects on principal planes are either insufficient to describe the object or are difficult to draw and dimension. Such cases generally include objects with inclined faces of intricate shape. When it is desirable to show the true size and shape of any irregular surface, which is inclined to the principal planes of projection, a view of the surface must be projected on a plane parallel to it. The direction of sight - is assumed to be perpendicular to the plane of the surface. This imaginary projection plane is called an auxiliary plane and the orthographic view obtained on it is called an auxiliary view. These views are often found necessary in technical drawings.

The theory underlying the method of projecting principal views applies equally to auxiliary views. Ordinarily an auxiliary view is a partial view showing only an inclined surface. The reason for this is that a projection showing the entire object adds very little to the shape description. Instead, the added lines are likely to cause confusion.

A *partial auxiliary view* often is needed to complete the projection of a foreshortened feature in a principal view, *Figs. 9.1, 9.9*. Auxiliary views may also be used to find solution to a variety of problems in descriptive geometry such as for determining: (*i*) the true length and inclinations of a line, (*ii*) the point view of a line, e.g., shortest distance of a line, (*iii*) the edge view of a plane,

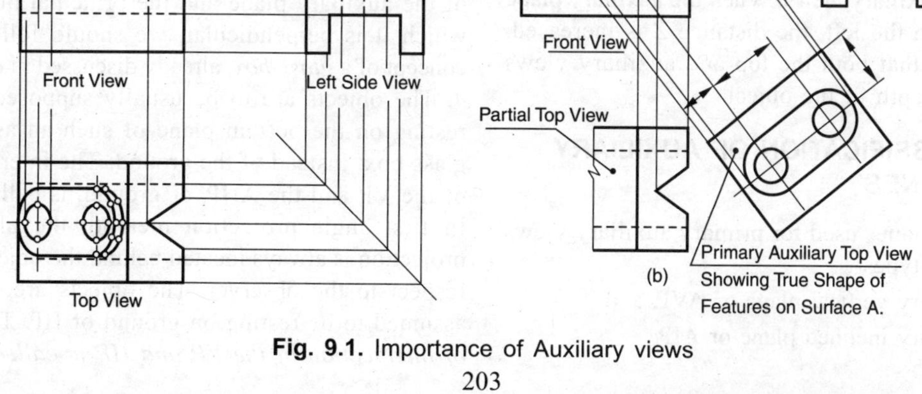

Fig. 9.1. Importance of Auxiliary views

(*iv*) the true shape and size of a plane, (*v*) distance between two skew lines, (*vi*) projections of solids, (*vii*) true shapes of sections of solids, (*viii*) curves of intersection of surfaces, etc.

9.2. AUXILIARY PLANES

As shown in *Fig, 9.2. (2)*, the object has an inclined surface which does not show in its true shape and size in the top view, nor does it show true shape in any other principal view. The auxiliary plane is assumed parallel to the inclined surface, i.e., perpendicular to the line of sight, which is at right angles to that surface. The auxiliary plane will then be perpendicular to the front plane of projection (as is the horizontal plane), to which it is assumed as hinged.

When the HP and auxiliary plane are unfolded to lie in the plane of the front view, as shown in *Fig. 9.2 (1)*, the *folding lines* represent the "*hinge*" *lines* between the views. To simplify the drawing the planes are omitted while drawing the orthographic projections, as shown in *Fig. 9.2 (3)*. As shown in *Fig. 9.5*, the folding lines may themselves be omitted in the actual drawing. As the auxiliary plane is taken parallel to inclined surface, its projection on the auxiliary plane (*i.e.*, auxiliary view) gives true shape and size. *Length dimension* of the surface is projected directly from the front view and the *depth dimension* from the top view.

It should be observed that the positions of the folding lines depend upon the relative positions of the planes in the pictorial view, *Fig. 9.2 (2)*. If the horizontal plane is moved downward, the distance *Y* is increased. If the front plane is shifted backward the distance *X* is increased but it remains equal in front and auxiliary views. When the auxiliary plane is moved to the left, the distance *Z* is increased. Note here, that both the top and auxiliary views show the depth of the object.

9.3. CLASSIFICATION OF AUXILIARY PLANES

Auxiliary planes used for primary auxiliary views are of two types:

1. Auxiliary vertical plane or AVP, and
2. Auxiliary inclined plane or AIP.

An auxiliary plane perpendicular to the HP and inclined to the VP is called *auxiliary vertical plane or AVP*. Projection on an *AVP* is called *auxiliary front view*.

It may be parallel to the vertical plane as well. In *Fig. 9.3*, if ϕ be reduced to 0° the plane will be parallel to the VP. Such a plane may be behind or in front of the VP and towards right or left of the object being drawn.

An auxiliary plane perpendicular to the VP and inclined to the HP is called *auxiliary inclined plane or AIP*. Projection on an AIP is called *auxiliary top view*.

It may be parallel to the horizontal plane. In *Fig. 9.4*, if ϕ_1 were reduced to zero, the plane will be parallel to the HP. Such a plane is called an *auxiliary horizontal plane* and it may be above or below the HP.

9.4. POSITION OF AUXILIARY PLANE

In practice, the auxiliary plane is assumed in that position which best facilitates the construction of the auxiliary view and leaves the principal views clear. $X_1 Y_1$ line should be drawn, at the desired angle to *XY*, in such a position that, on projecting the auxiliary view, there is no overlapping of views and other rules of orthographic projection are also observed. Further, on a drawing the reference line will be perpendicular to the projection lines drawn from the view from which the auxiliary view is to be projected.

In third angle projection method, the plane of projection, assumed transparent, whether auxiliary or principal plane, is to be invariably located in between the observer and the object. Therefore, for drawing the reference line or line of intersection of the auxiliary plane and the principal plane, to which it is perpendicular, we should follow the concept of a *glass-box*, already discussed in *chapter 4*. The objects are to be usually supposed to be resting on the bottom plane of such an assumed glass-box, instead of the ground. The intersection of the VP and the AHP, or ground, is called GL. In first angle projection method, the plane of projection is always located behind the object with respect to the observer. The objects are usually assumed to be resting on ground or HP. The *line of intersection of the VP and HP is called XY*.

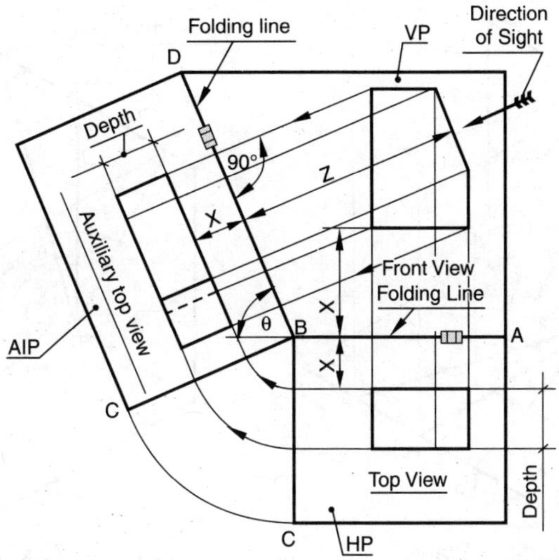

1. Position of Planes and Projections after Rabatment

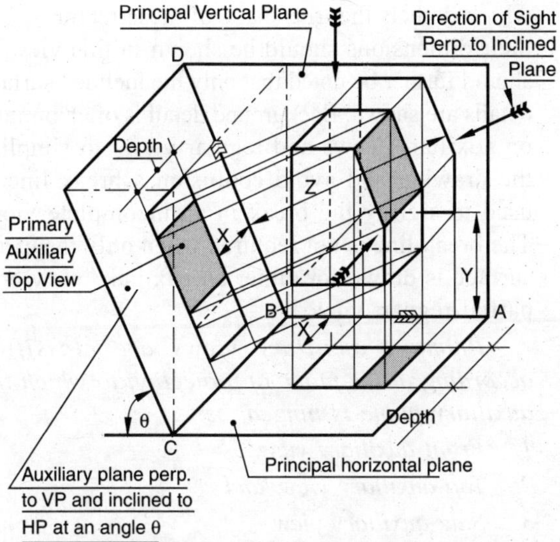

2. General Arrangement of Planes

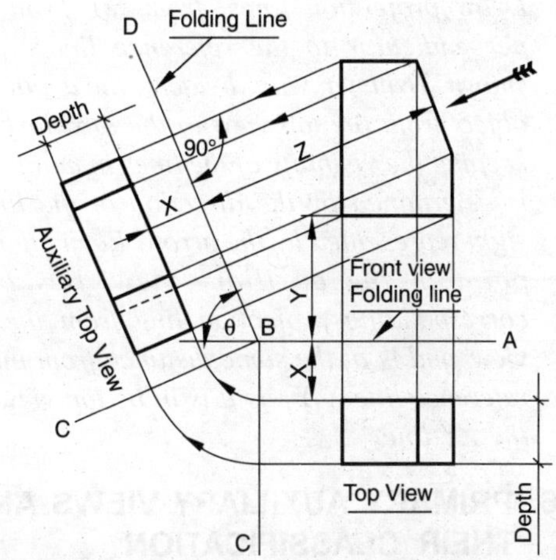

3. Method of Drawing Primary Auxiliary Top view

Fig. 9.2. (1, 2, 3) Auxiliary Plane

X_1-Y_1 denotes the intersection of the auxiliary plane and the principal plane to which it is perpendicular. As the *XY* line is normally the reference line for drawing principal view X_1Y_1 will be the reference line for drawing auxiliary view.

9.5. TO DRAW AN AUXILIARY TOP VIEW OR PLAN

Refer *Fig. 9.5*. The corner points of the object are numbered, as illustrated. The procedure draw an auxiliary view is explained in three steps:

1. *Draw the views of the object assumed to be lying on ground or HP, any convenient distance in front of the VP. Accordingly draw the XY line. Indicate the direction in which the auxiliary view is to be drawn by an arrow.*
2. *Draw auxiliary view of the reference plane (i.e. a line X_1Y_1, at any convenient distance from the front view for third angle projection and touching the front view for first angle projection method, and perpendicular to the front view of the arrow. The reference line represents the*

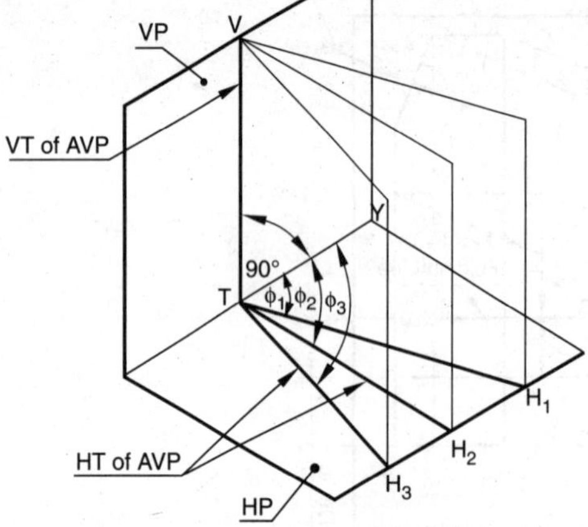

Fig. 9.3. Auxiliary Vertical Plane

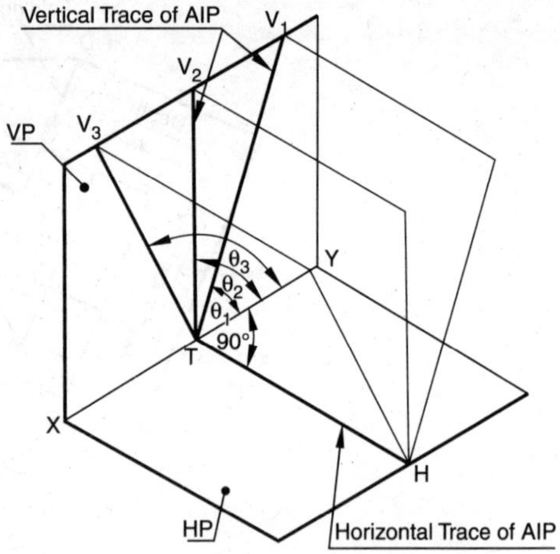

Fig. 9.4. Auxiliary Inclined Plane

auxiliary view of the reference plane, and also the line of intersection of the auxiliary plane and the reference plane.

3. *Draw projection lines, from the front view, perpendicular to the reference line X_1Y_1 as shown. Transfer, with dividers, the depth of the object from the top view to the auxiliary view, keeping the visibility of the lines in mind, which is determined by the direction of the lines of sight represented by the arrow. Each numbered point in the auxiliary view, lies on its corresponding projection line from the front view, and is at the same distance from the new reference line X_1Y_1 as it is in its top view from the XY line.*

9.6. PRIMARY AUXILIARY VIEWS AND THEIR CLASSIFICATION

Inclined surfaces of objects appear to be foreshortened and their true shape is not shown in their principal views. When such an inclined surface has important characteristics, which should be shown clearly and without distortion, a primary auxiliary view is used to completely explain the shape of the object.

When the auxiliary plane is perpendicular to a principal plane of projection, the view obtained is called a primary auxiliary view, *Figs. 9.1, 9.6, 9.7, 9.8. Fig. 9.1* clearly brings out the necessity of the primary auxiliary view. The circular features on the inclined surface A, in the front view, do not appear in their true shape on either the top or side view. The primary auxiliary view is the only view, which depicts the true shape of these features, and their dimensions should be shown in this view. It should further be noted that only the inclined surface details are shown. Background detail is often omitted on auxiliary views and regular views to simplify the drawing and avoid confusion a break line is used to signify the break in an incomplete view. The break line is not required when only the exact surface is drawn for either an auxiliary view or a partial regular view.

Primary auxiliary views are classified according to the plane of projection to which the auxiliary plane is hinged, as:

1. *Front auxiliary view,*
2. *Top auxiliary view, and*
3. *Side auxiliary view.*

When the auxiliary plane is assumed perpendicular to and, therefore, hinged to the principal HP, the projection obtained on it is called the *front auxiliary view*.

When the auxiliary plane is assumed perpendicular to and, therefore, hinged to the vertical plane of projection, the projection obtained on it is called the *top auxiliary view*.

When the auxiliary plane is assumed perpendicular to and, therefore, hinged to the profile plane, the projection obtained on it is called the *side auxiliary view*.

In practice usually the front and top auxiliary views are drawn. Orthographic auxiliary projections

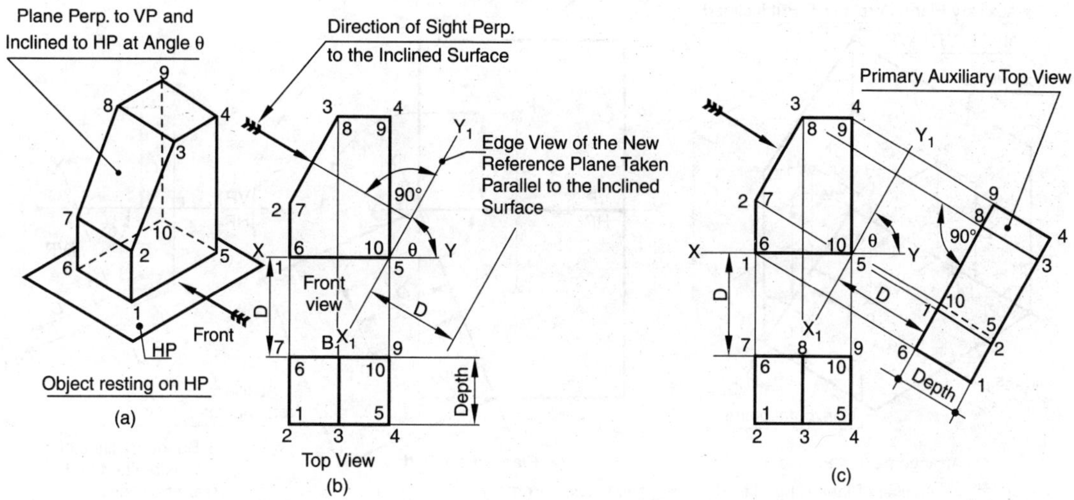

Fig. 9.5. Drawing an auxiliary view

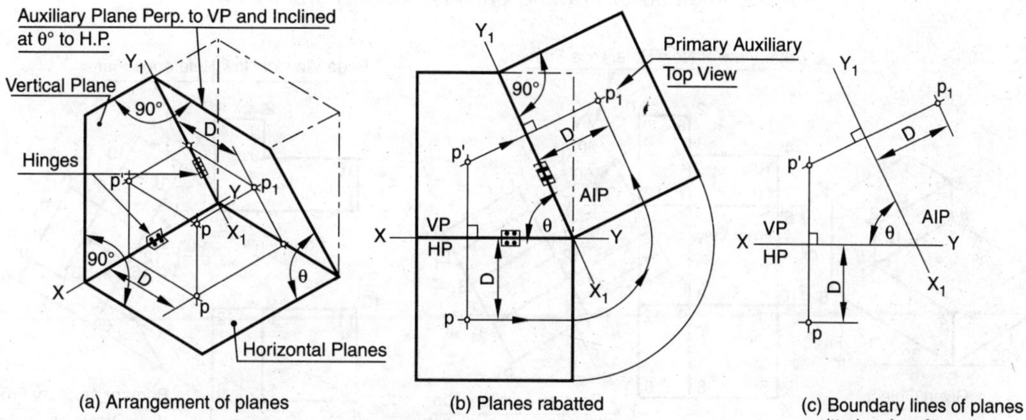

(a) Arrangement of planes

(b) Planes rabatted

Auxiliary plane hinged to front plane. It is used to draw primary auxiliary top view also called front auxiliary view.

(c) Boundary lines of planes omitted, only reference lines, XY, X,Y, retained

Fig. 9.6. Method of drawing primary auxiliary top view

are the drawing paper by rotating the auxiliary plane, through 90°, about that principal plane to which it is perpendicular and hinged, *Figs. 9.6* and *9.7*.

In *Fig. 9.8 (b)* is illustrated the method of drawing primary auxiliary front view, drawn on an auxiliary plane perpendicular to HP and inclined to VP at 60°. All the corner points on the isometric view are numbered and the same numerals are used in orthographic and auxiliary views.

In *Fig. 9.8 (a)* is illustrated the method of drawing primary auxiliary top view, drawn on auxiliary plane perpendicular to VP and inclined to HP at 60°. The figure is self-explanatory.

9.7. SECONDARY AUXILIARY VIEWS

If the auxiliary plane is oblique to all the regular planes of projection, it must be hinged to some primary auxiliary plane to which it is perpendicular. Its line of intersection with the primary auxiliary plane, to which it is perpendicular, is usually denoted by X_2Y_2. Such a plane is called a secondary auxiliary plane and a view projected on it, by taking X_2Y_2 as the reference line, is called secondary auxiliary view, *Fig. 9.9*.

In such cases two distinct operations are necessary to draw the auxiliary view. *First*, a primary auxiliary view must be drawn, and *second*; from it a secondary auxiliary view is drawn, as shown in *Fig. 9.9*.

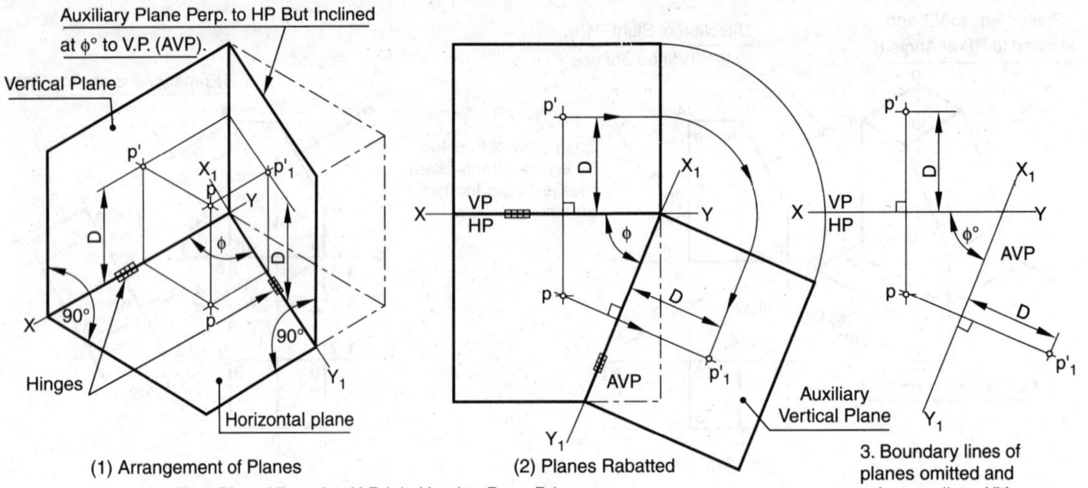

(1) Arrangement of Planes (2) Planes Rabatted

Auxiliary Plane Hinged to H.P. it is Used to Draw Primary Auxiliary Front View-also Called Top Auxiliary View.

Fig. 9.7. Method of drawing primary auxiliary front view

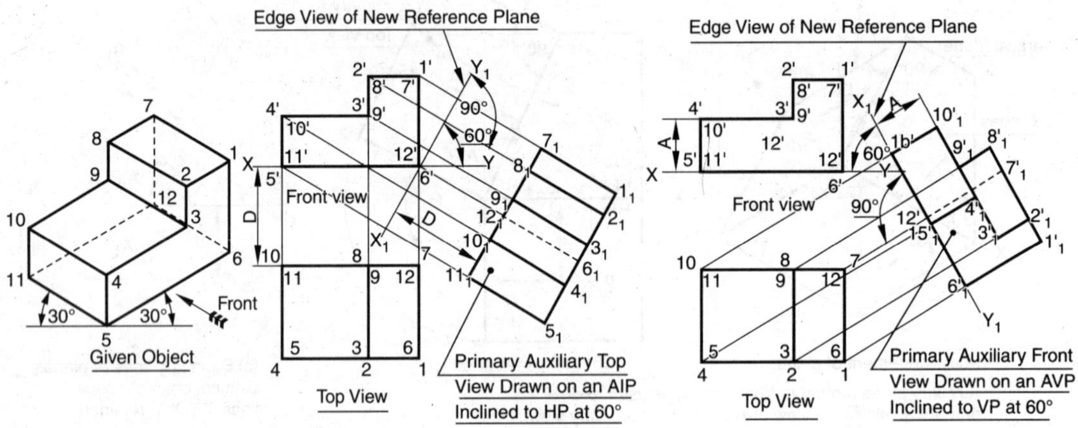

(a) (b)

Fig. 9.8.

To bring out a clear distinction between the primary auxiliary and secondary auxiliary views, let us consider the case illustrated in *Fig. 9.10*. In it at (*a*), as drawn an auxiliary view which shows the true projection and true shape of surface *A*. This view is called a primary auxiliary view because it is projected directly from the regular front view.

The surfaces *C* are inclined so that they are not perpendicular to any of the three planes of projection and appear as foreshortened in all the regular views. If a true view is required for this surface, which also shows the true shape and location of holes *M* on this, two auxiliary views, a primary auxiliary and a secondary auxiliary view, must be drawn as at (*c*).

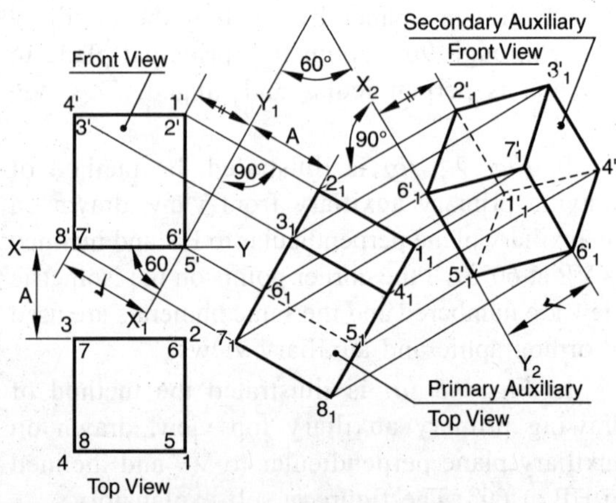

Fig. 9.9. Drawing secondary auxiliary front view

This second auxiliary view is projected from the primary auxiliary view and is known as a *secondary auxiliary view*. The view at (*b*) is also a primary auxiliary view because it is projected from one of the regular views. This view gives the true shape of inclined surface *B* but surfaces *C* on it are still distorted.

The principles of drawing the auxiliary projections are illustrated by solving some problems.

9.8. IMPORTANT OBSERVATIONS

From the preceding theory the following important observations can be made:

1. *An auxiliary front view of a point lies on a projector drawn through its top view and perpendicular to the new reference line* X_1Y_1. *The* X_1Y_1 *line can be drawn in any suitable position, fulfilling the given condition. The distance of the auxiliary front view from the* X_1Y_1 *line is equal to the distance of the previous front view from XY line. The distances of all the auxiliary front views of the same point, projected from a common top view, from their respective new reference lines are equal.*
2. *An auxiliary top view of a point lies on a projector drawn through its front view, perpendicular to the new reference line* X_1Y_1 *Its distance from the* X_1Y, *is equal to the distance of its previous top view from XY line. The distances of all the auxiliary top views of the same point projected from a common front view from their respective new reference lines are equal.*

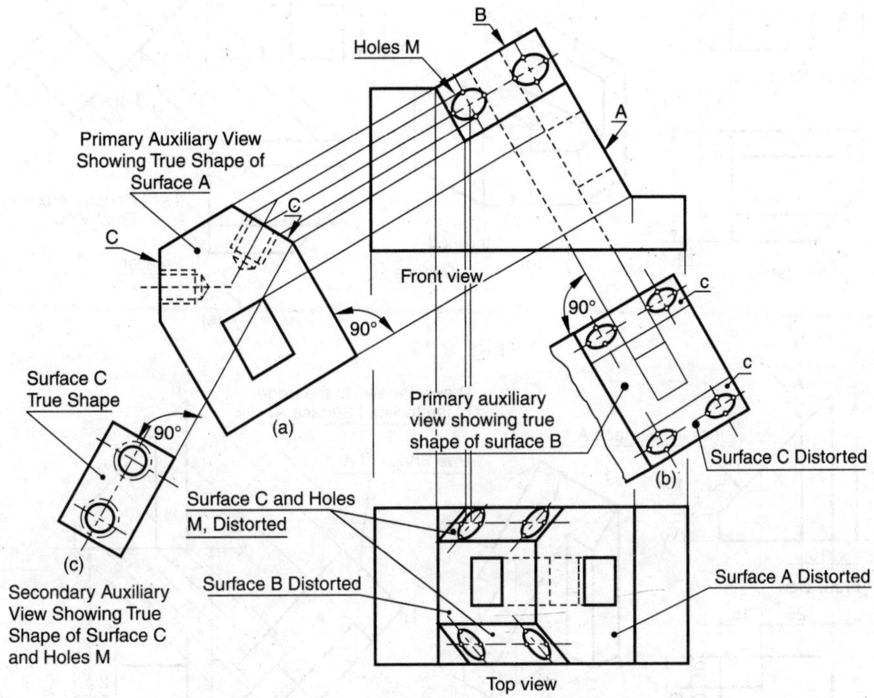

Fig. 9.10. Importance of secondary auxiliary views

ILLUSTRATIVE PROBLEMS

PROBLEM 9.1. *Two views of an object, with an inclined surface A, are given in Fig. 9.11 (a), in first angle projection. Draw its primary auxiliary front view showing true shape of the surface A.*

SOLUTION: In *Fig. 9.11 (b)* isometric view of the solid is shown to facilitate visualization, while at (*c*) is given the complete solution, with all construction lines. The solution is in first angle projection. Solve the problem using third angle projection.

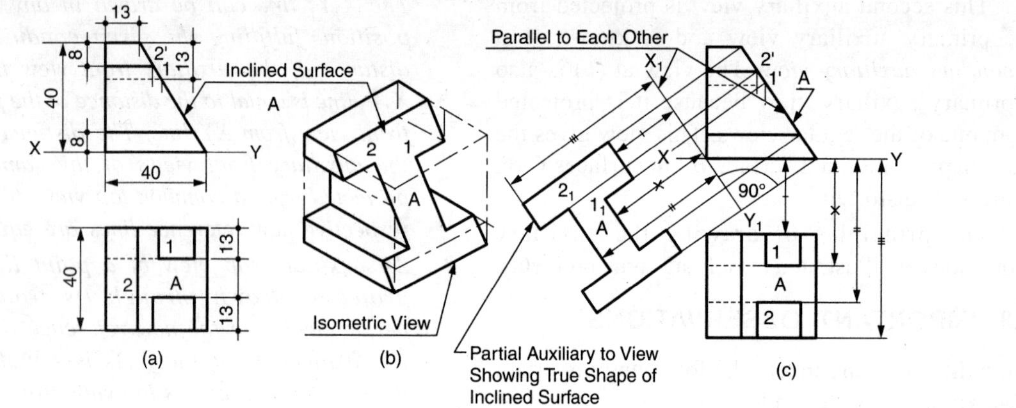

Fig. 9.11.

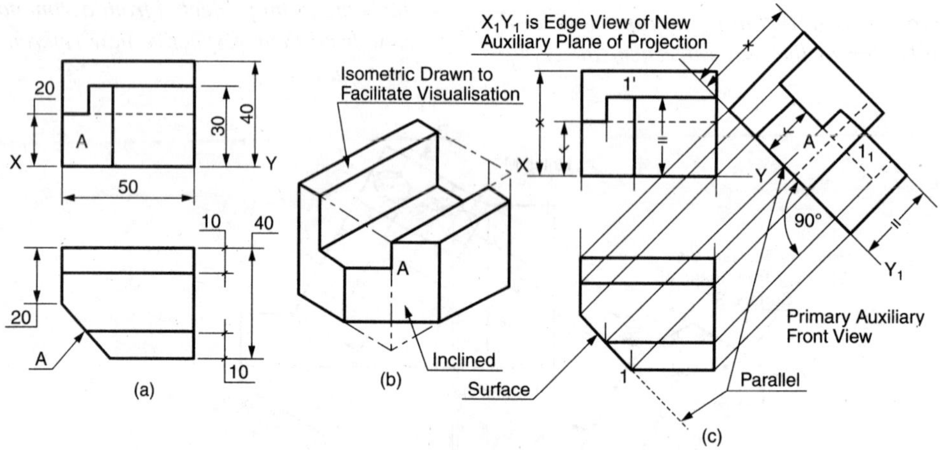

Fig. 9.12.

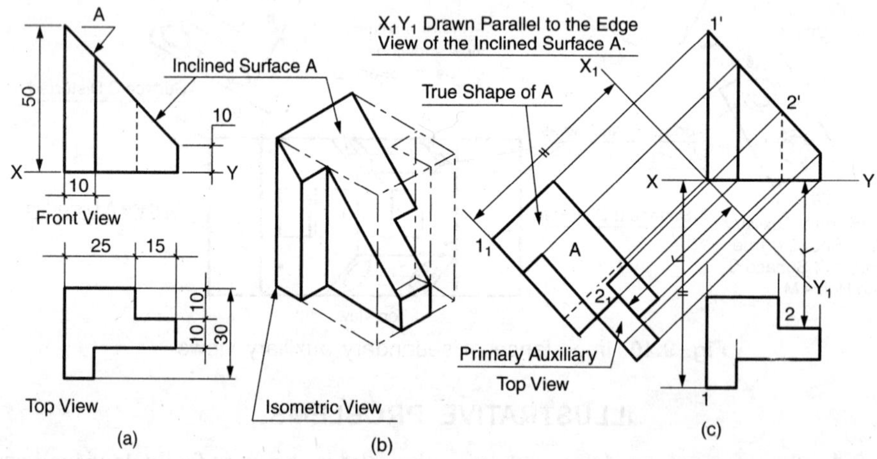

Fig. 9.13.

PROBLEM 9.2. *Two views of a block, with an inclined surface, are given in Fig. 9.12 (a), in first angle projection. Draw its partial primary auxiliary view showing true shape of the inclined surface.*

SOLUTION: In *Fig. 9.12 (b)* is shown the isometric view of the block, to facilitate visualization. At (*c*) is illustrated, with the help of points 1 and 2, the method of drawing the partial auxiliary view. Try this problem using third angle projection.

PROBLEM 9.3. *Front and top views of a block, with an inclined surface, are shown in Fig. 9.13 (a), in first angle. Draw its primary auxiliary top view showing true shape of the inclined surface.*

SOLUTION: In *Fig. 9.13 (b)* is shown isometric view of the block, to facilitate understanding, while the required primary auxiliary top view, projected on an auxiliary plane parallel to the inclined surface, is shown at (*c*). The figure is self-explanatory. You should try this problem in third angle projection method.

PROBLEM 9.4. *A right regular pentagonal pyramid, side of base 25 mm and height 55 mm, rests in HP on its base with one of its base edges perpendicular to the VP. Draw its principal views and add its primary auxiliary plan on an AIP, inclined at 45° to the HP.*

SOLUTION: Refer *Fig. 9.14*, for its solution. All construction lines are shown to facilitate understanding. It is suggested that this problem be studied after going through the projections of solids.

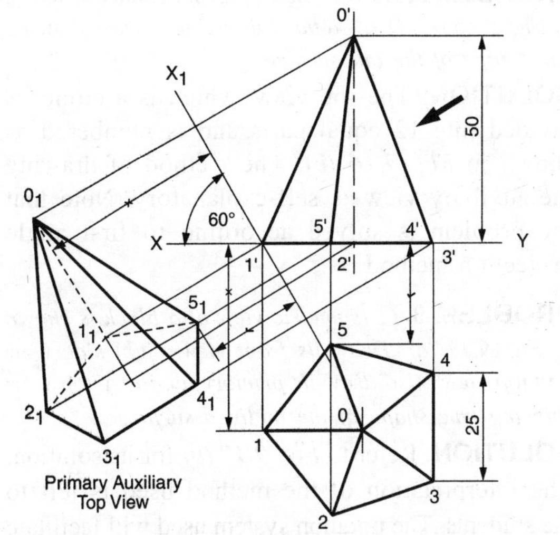

Fig. 9.14.

PROBLEM 9.5. *A right regular hexagonal pyramid, side of base 25 mm and height 60 mm, rests on its base on HP with one of its base edges parallel to the VP. An auxiliary inclined plane (AIP), inclined to the HP at 45°, cuts the pyramid, cutting its axis at a distance of 32 mm from the base. Draw its front view, top view (in section) and primary auxiliary top view, on the AIP, showing true shape of the section.*

SOLUTION: Refer *Fig. 9.15* for solution in first angle projection. All construction lines are shown to facilitate understanding. It is suggested that this problem be studied after going through the projections and sections of solids.

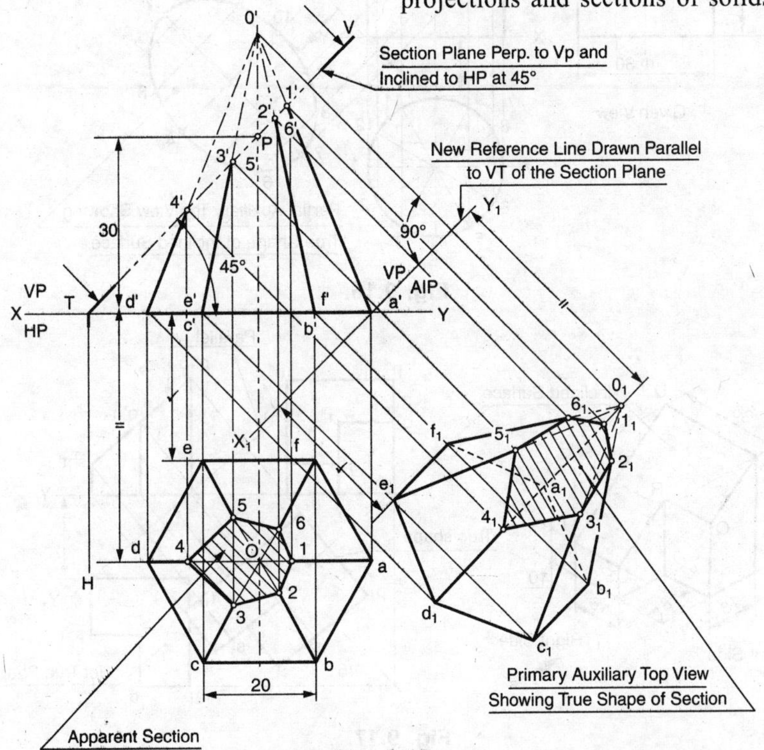

Fig. 9.15.

PROBLEM 9.6. *Front view of a cut cylinder is shown in Fig. 9.16 (a). Draw a partial auxiliary view showing true shape of the cut surface.*

SOLUTION: The top view, which is a circle, is divided into 12 equal parts and is numbered as shown in *Fig. 9.16 (b)*. The method of drawing the auxiliary view is self-explanatory. Note that the problem is solved according to first angle projection method.

PROBLEM 9.7. *Isometric view of a block is shown in Fig. 9.17 (a). Draw its front view, right side view and top view. Also draw its primary auxiliary top view showing true shape of the inclined surface.*

SOLUTION: Refer to *Fig. 9.17 (b)* for its solution. The interpretation of the method used is left to the students. The notation system used will facilitate understanding.

PROBLEM 9.8. *Isometric view of an object is shown in Fig. 9.18 (a). Draw its front view, left side view, and top view. Also draw its primary auxiliary top view, showing true shape of the inclined surface A.*

SOLUTION: Refer to *Fig. 9.18 (b)* for its solution. A suitable notation system used, and the construction lines retained, will help to understand and render the solution self-explanatory.

PROBLEM 9.9. *Isometric view of an object is shown in Fig. 9.19 (a). Draw its front view, right side view, and top view. Also draw its primary auxiliary top view showing true shape of the inclined surface.*

SOLUTION: Refer *Fig. 9.19 (b)* for solution. All construction lines are retained to make the solution self-explanatory .

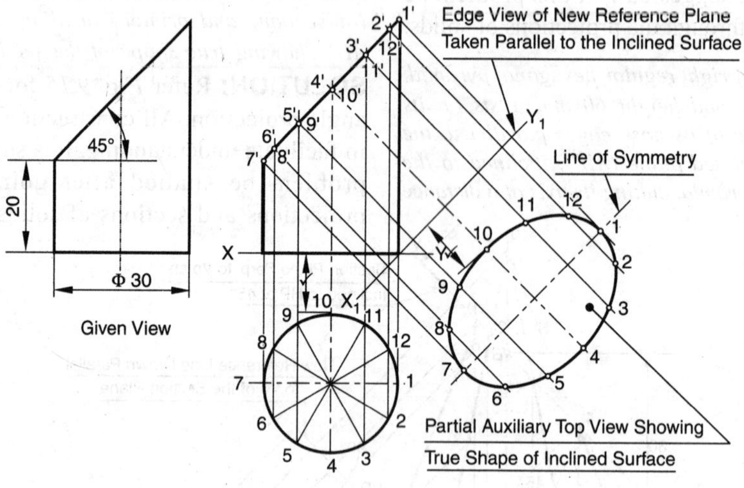

Fig. 9.16.

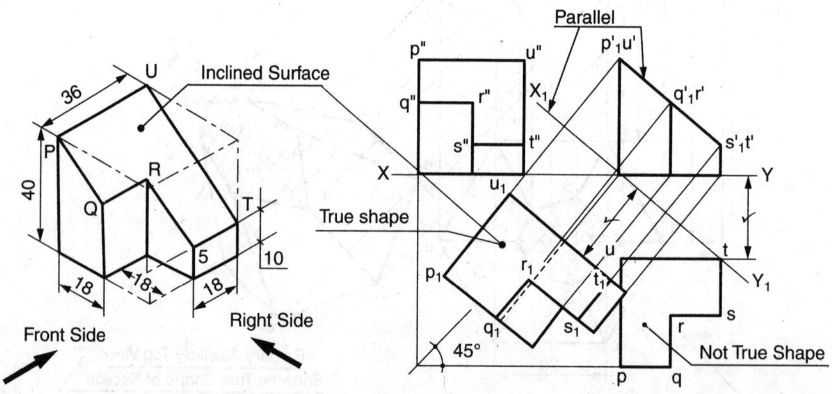

Fig. 9.17.

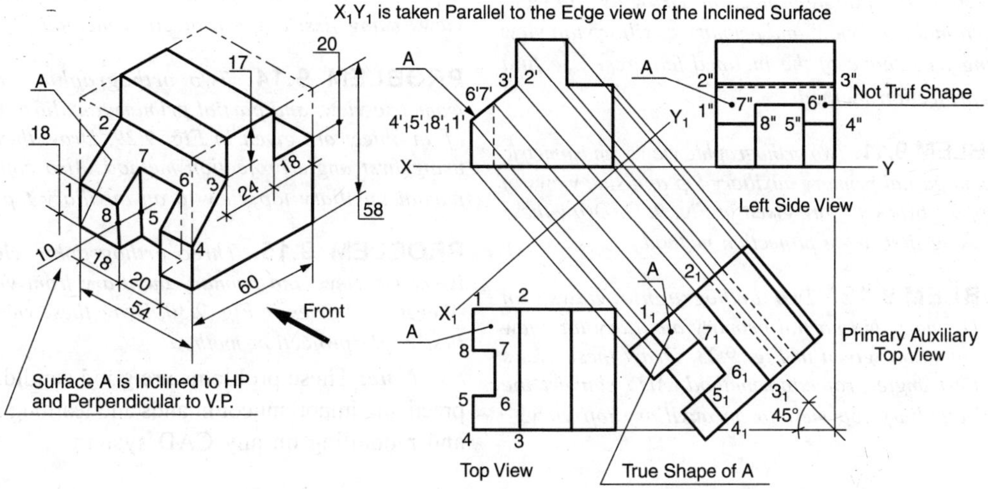

Fig. 9.18.

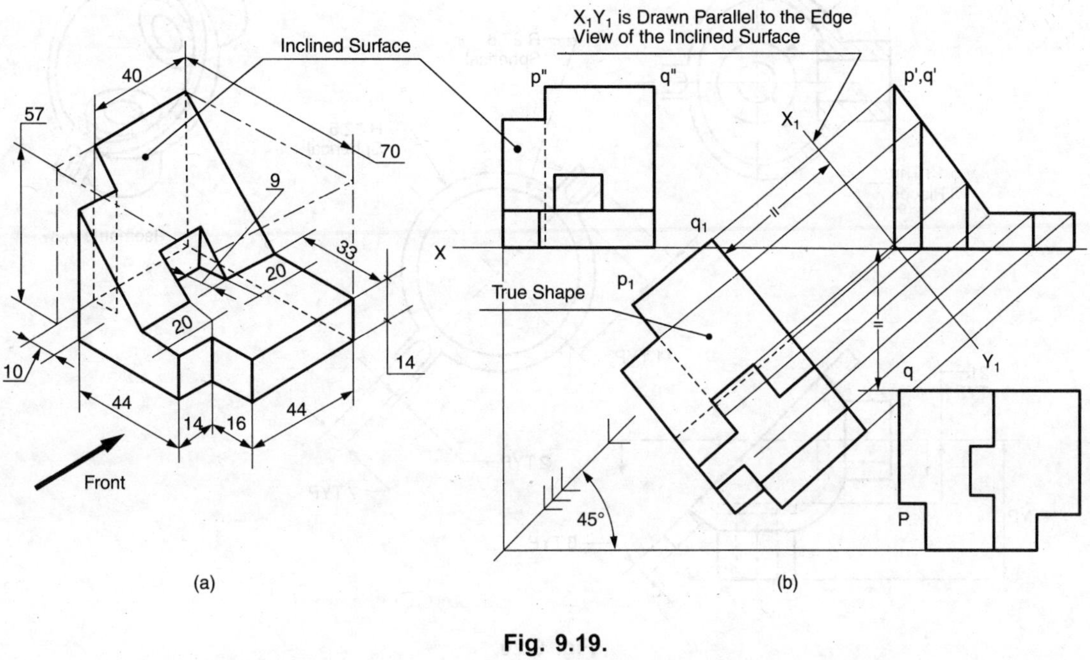

Fig. 9.19.

PROBLEM 9.10. *Two orthographic views, a partial auxiliary top view, and an isometric view are given in Fig. 9.20. in third angle projection. Draw its Top view, full sectional front view, and primary auxiliary top view showing true shape of the inclined features. Use first angle projection method.*

PROBLEM 9.11. *Two orthographic views, an isometric view, and partial primary auxiliary top and side views of a "Support-Bracket " are given in Fig. 9.22. Draw these views using first angle projection method.*

PROBLEM 9.12. *Two orthographic views, an isometric view, and partial primary auxiliary top view of an object are given in Fig. 9.23. Draw these views using first angle projection method. Also convert the partial auxiliary top view to an auxiliary top view.*

PROBLEM 9.13. *Three orthographic views, an isometric view, and primary auxiliary top view of a " Support-Bracket" are given in Fig. 9.28. Draw these views using first angle projection method.*

PROBLEM 9.14. *Two orthographic views, an isometric view, and partial primary auxiliary top view of an object are given in Fig. 9.29. Draw these views using first angle projection method. Also convert the partial auxiliary top view to an auxiliary top view.*

PROBLEM 9.15. *Three orthographic views, an isometric view, and primary auxiliary front view of a "Lever" are given in Fig. 9.30. Draw these views using first angle projection method.*

Note: These problems are good candidates for practicing important commands and learning drafting and modelling on any CAD system.

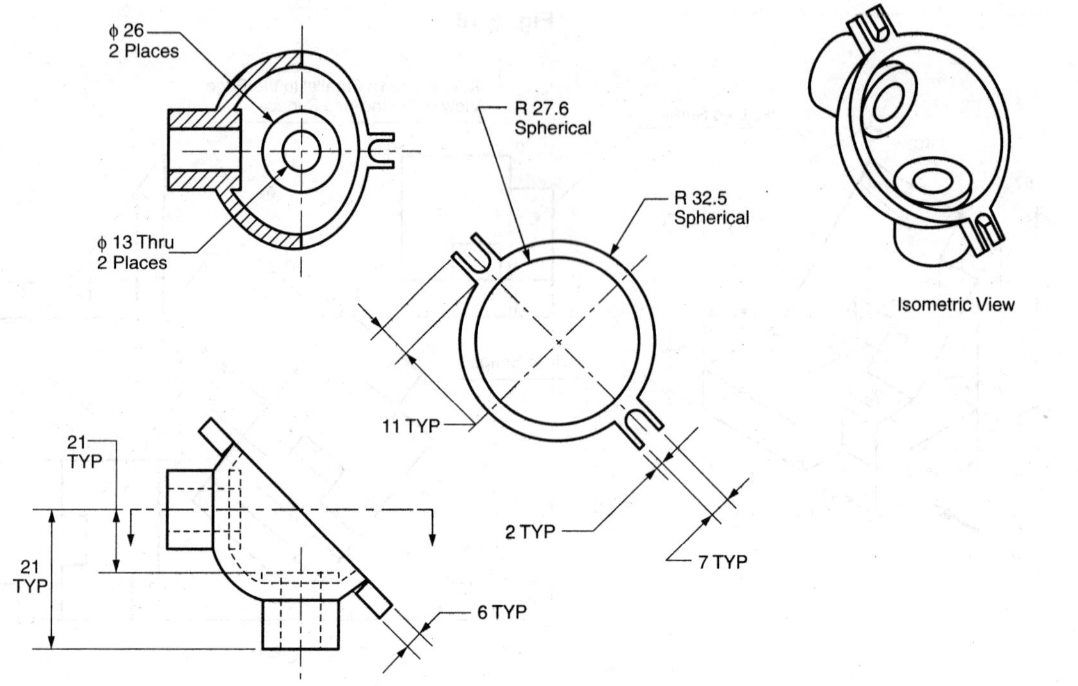

Fig. 9.20.

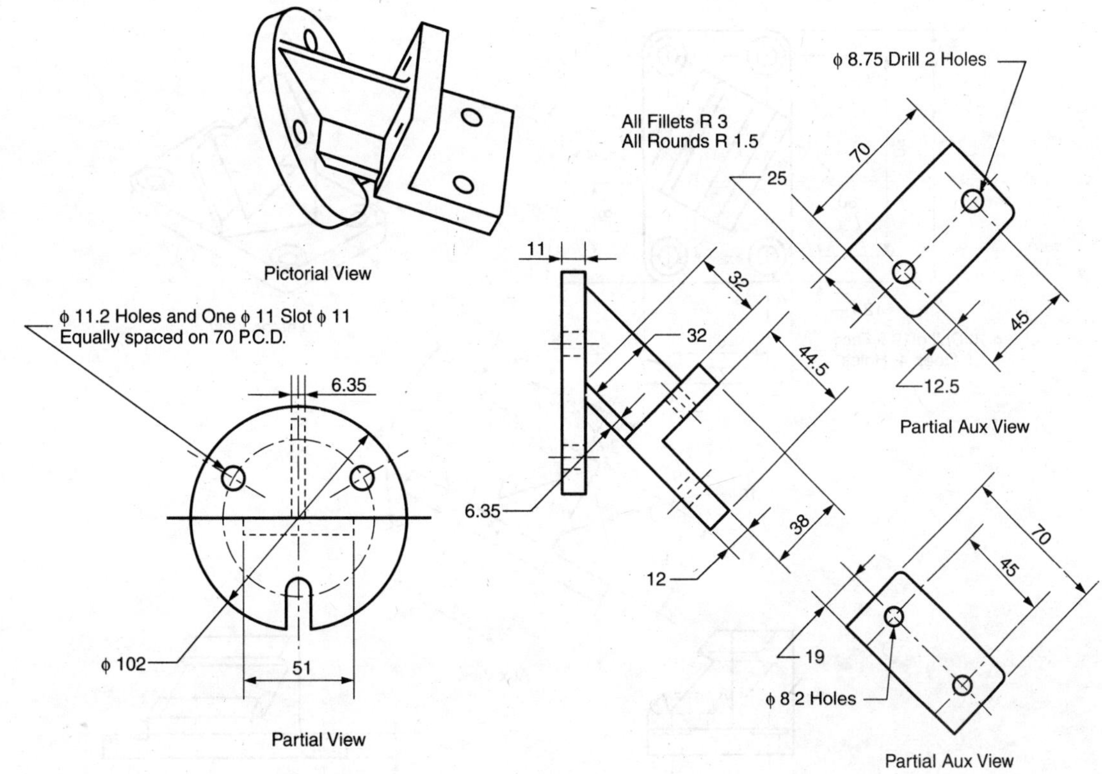

Fig. 9.21.

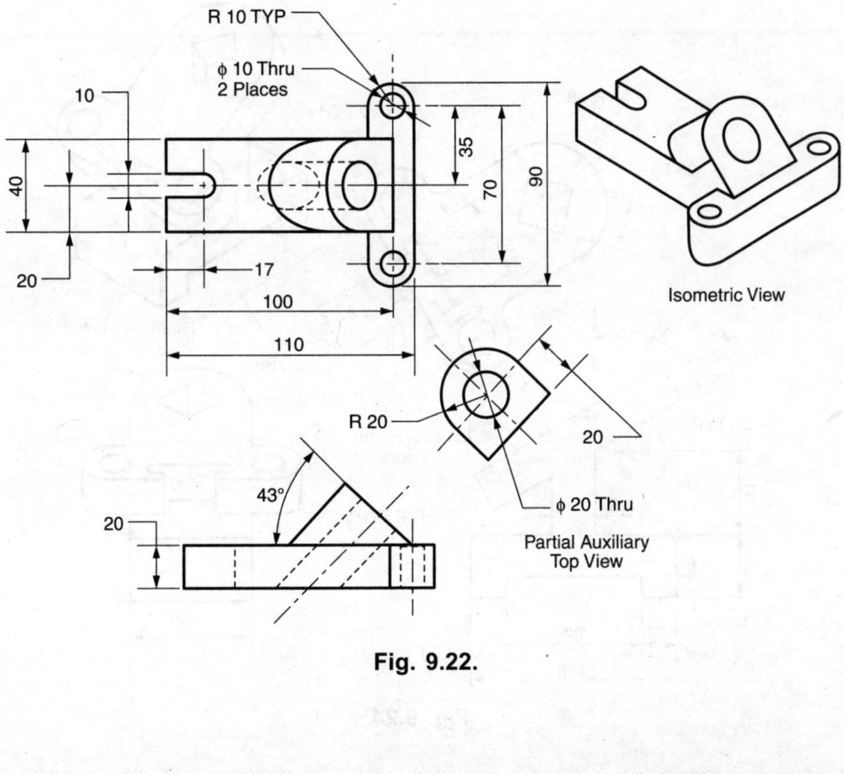

Fig. 9.22.

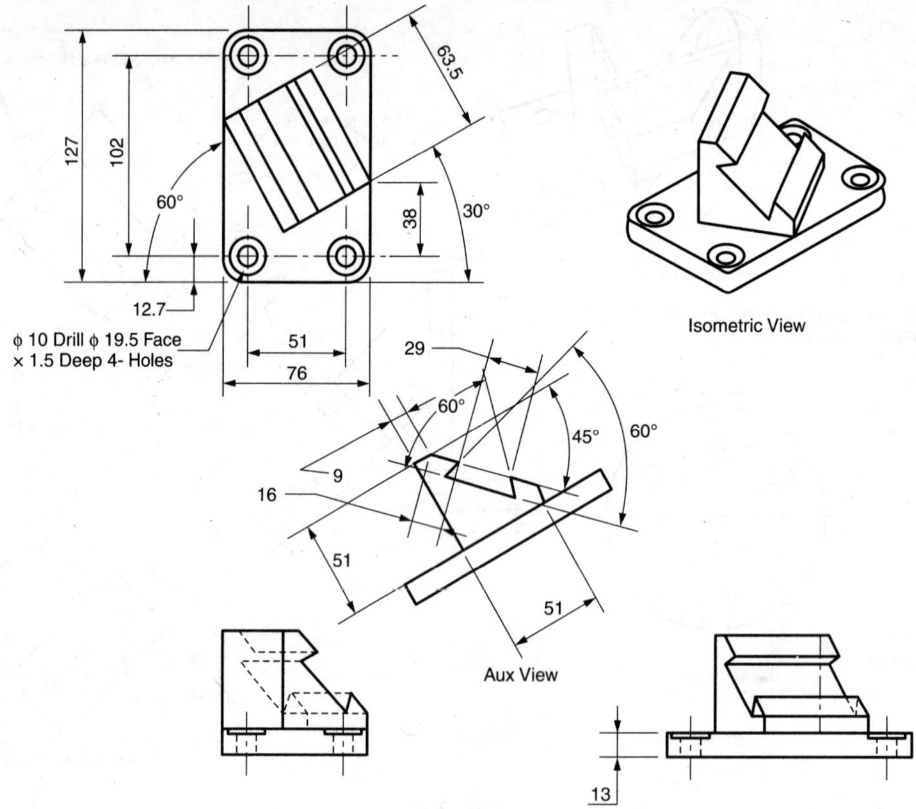

Fig. 9.23.

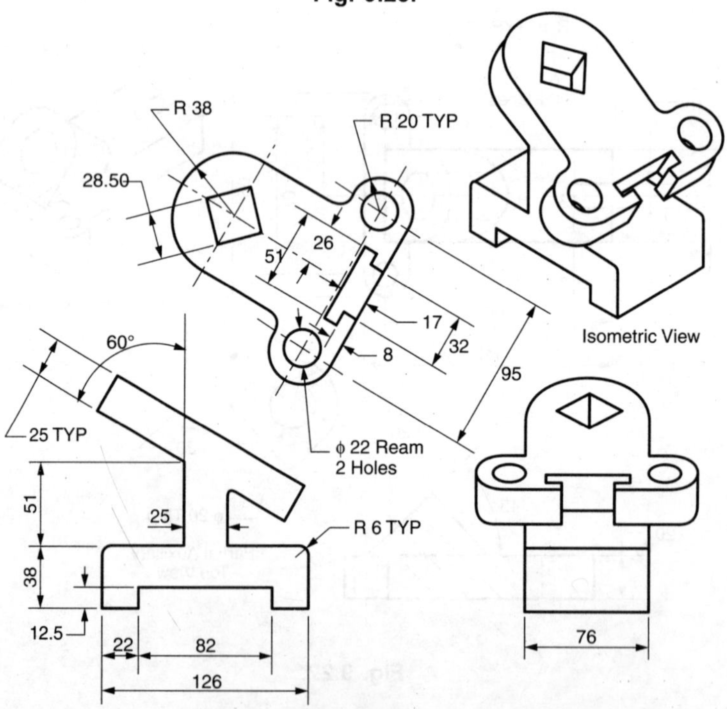

Fig. 9.24.

EXERCISES

1. Two views each of nine different objects are shown in *Fig. 9.31*. Draw partial auxiliary top view for each one of them showing true shape of the inclined surface.
2. Two views each of nine different objects are shown in *Fig. 9.31*. Draw primary auxiliary top view for each one of them showing true shape of the inclined surface.

QUESTIONS FOR VIVA VOCE

1. What are an auxiliary view and when an auxiliary view becomes necessary to be drawn?
2. What is a partial auxiliary view?
3. What is an auxiliary plane?
4. What are the two important types of auxiliary planes?
5. How would you determine the position of an auxiliary plane?
6. What are a primary auxiliary view and a secondary auxiliary view?
7. How are primary auxiliary views classified?

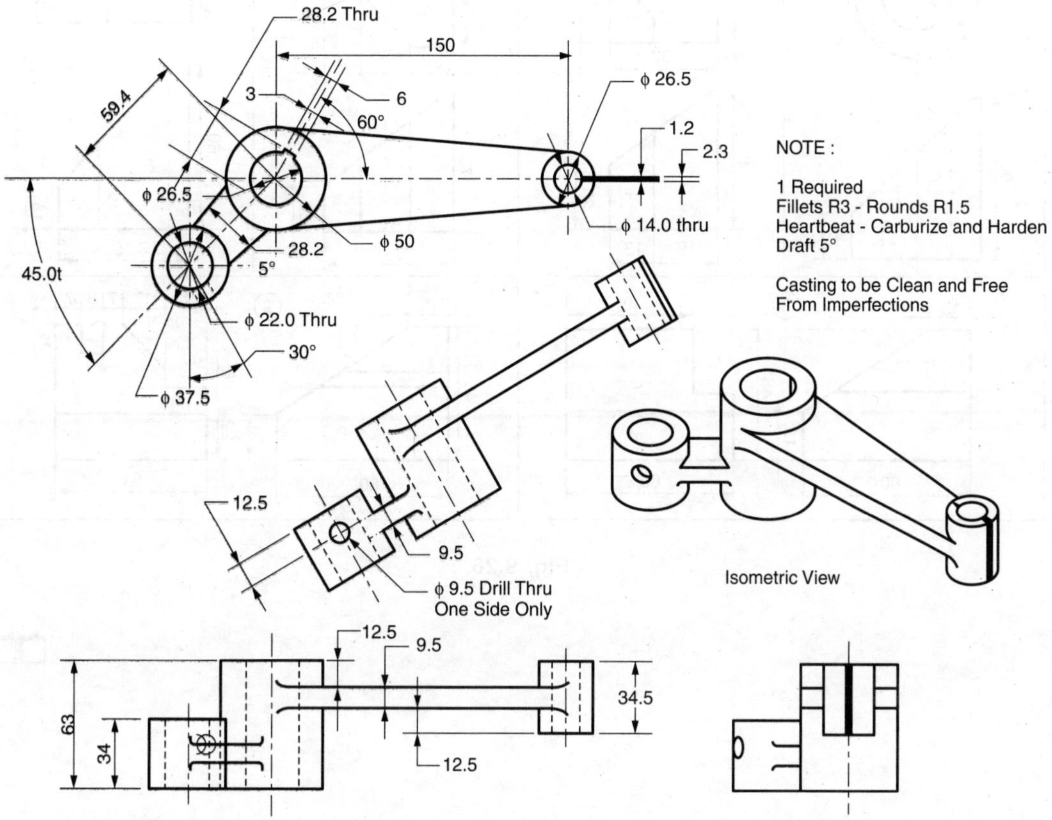

Fig. 9.25.

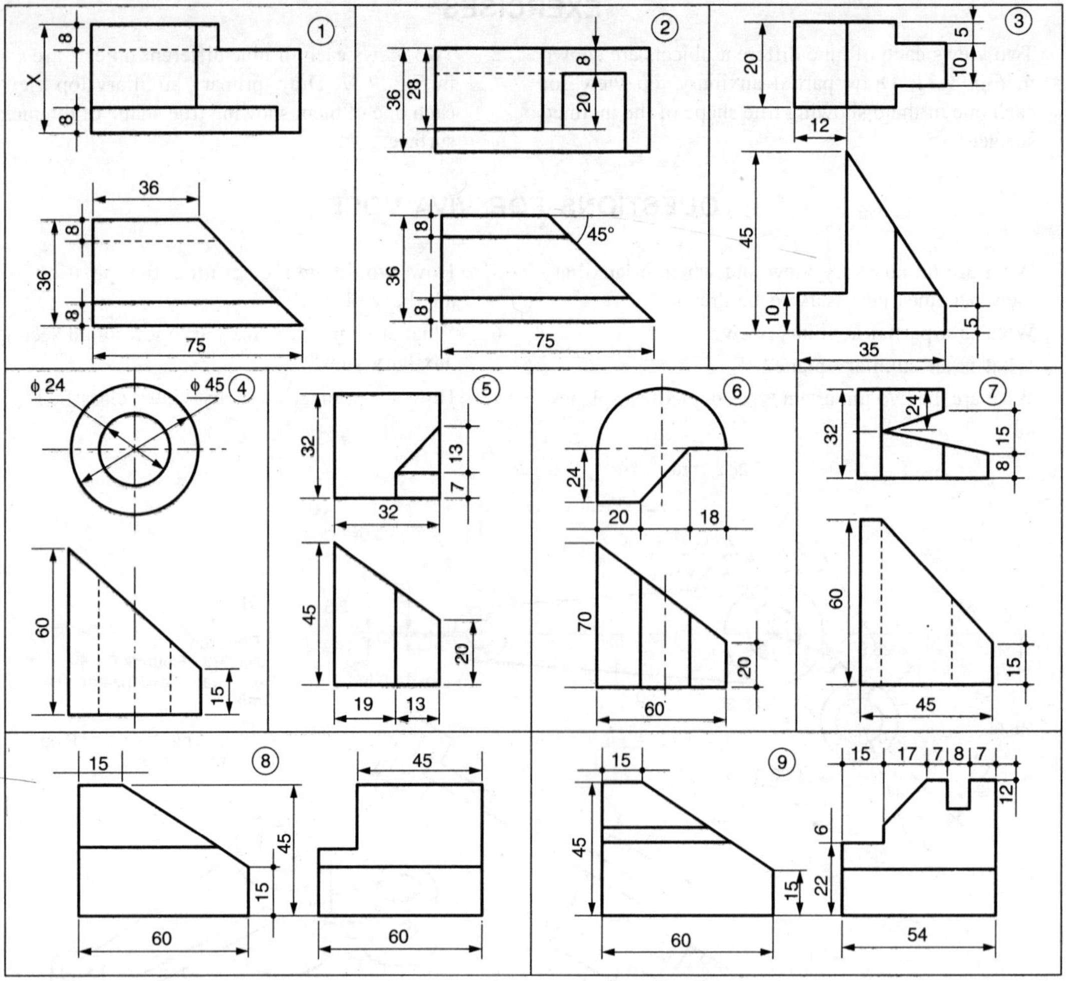
1
8
X
8
36
8
36
8
75
2
8
36
28
20
8
45°
36
8
75
3
5
20
10
12
45
10
5
35
ϕ 24
ϕ 45
4
60
15
5
32
13
7
32
45
20
19
13
6
24
20
18
70
20
60
7
24
32
15
8
60
15
45
8
15
45
15
60
45
60
9
15
45
15
60
15
17
7
8
7
12
6
22
54

Fig. 9.26.

Freehand Drafting 10

10.1. INTRODUCTION

Freehand drafting or sketching is the technique of making drawing without the use of drawing instruments. It is very convenient and quick method of shape description. Engineers find this method to be very helpful when working up new mechanisms, considering the alternate arrangements of a system being designed, and planning changes in existing designs to improve functionality or lower cost. It is necessary, therefore, for engineering students to be able to produce readable, quick and informative sketches of existing objects or new designs. Sketches can formulate record, express design and development ideas and instructions, and reduce the need for long verbal explanations. Good sketching ability enables a designer or engineer to make rapid and clean drawings of ideas or drawings that can be given to his/her dealing section for development into scale working drawings.

Sketches that are prepared developing a new design, and are the original expression of a technical thought, are sometimes called *schematic or scheme sketches*. The sketches of existing objects or mechanisms, to enable the replacement or modification of worn or broken parts, which are usually prepared away from the drawing office, are sometimes called *field sketches*.

Sketches are usually drawn freehand, but some ruling and compass work is done if it facilitates the sketch work.

Although full proficiency in freehand drawing is synonymous with complete understanding of the graphic language, and is gained only after acquiring a good knowledge and skill in drawing with instruments, some teachers advocate that a student should be conversant with freehand drawing before starting to make drawings with instruments.

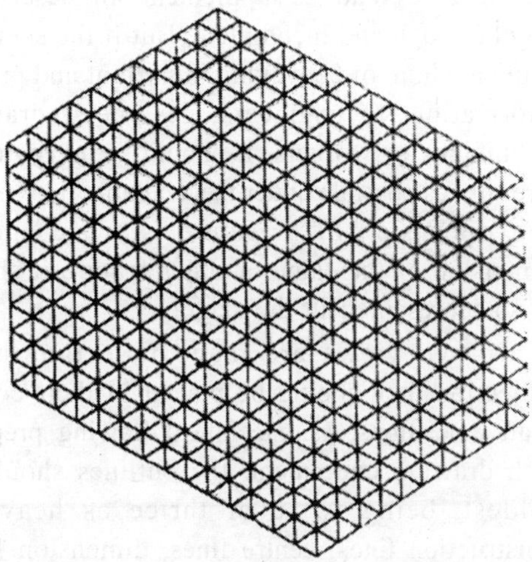

Fig. 10.1. (a) Isometric paper used for sketching.

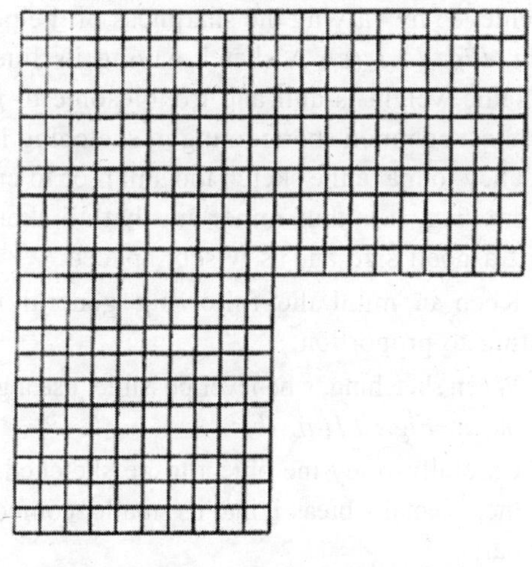

Fig. 10.1. (b) Graph paper used for sketching

10.2. PROPORTION IN SKETCHING

Sketches are not made to scale. Although produced quickly, the sketches should be in good proportion, with clear and bold outlines and the dimensioning as complete as possible. Whilst technical sketches are not scale drawings, their value is greatly enhanced if their general proportions are good. All of the various components of a sketch should be kept in proportion to those of the actual object, using eye judgment only. This technique takes a lot of practice to master. The degree of perfection for a required sketch depends upon its use. Sketches made to supplement oral description may be rough and incomplete. But if the sketch is to the medium of conveying important and precise information, to engineers, designers, drafters, technicians or workers, it should be executed as carefully as possible under the available time frame. If a dimension of a feature is twice another dimension of the same or other feature it should be so represented in the sketch.

If two views are made in orthographic projection they should be placed in their correct relative positions. As in technical drawing, prepared with drawing equipment, the outlines should be boldest, being twice or thrice as heavy as construction lines, centre lines, dimension lines, leaders, etc., while hidden lines are somewhat thinner than object lines. The variation in pencil line width is achieved by varying the sharpness of the pencil point, *Fig. 10.1 (c)*. A sketch having its lines of the same weight is dull and troublesome to read. The most common shortcoming in sketching is the tendency to make the sketch too small or to crowd the drawing. It is thus important that all sketches be of a good size and is wisely spaced.

Keep in mind the following general rules relating to proportion:

1. When sketching, whenever possible, use a graph paper, *Fig.10.1(a), (b)* .
2. Carefully study the object to be sketched, and then mentally break it into its smaller component parts.
3. Estimate the length, width and height of the largest component of the object. Comparing these dimensions to the size of the sketch paper you have, estimate the proportions that will be possible to use for the overall sketch.

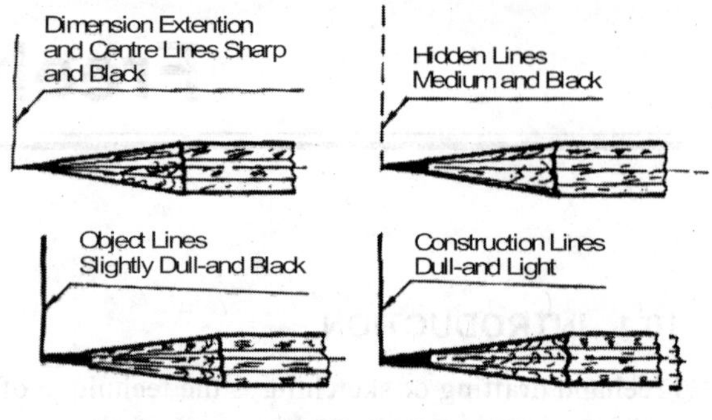

Fig. 10.1. (c) Pencil points for sketching different lilnes

4. Layout the largest component according to the proportions that you have decided in the previous step, using construction line squares and rectangles to block in irregularly shaped components.
5. Repeat steps 3 and 4 until the entire object shape is finished.

10.3. MATERIALS FOR SKETCHING

Sketching is done with minimum of equipment, which include only a lead pencil, an eraser, and a suitable paper. The pencil should be soft enough to give a good texture on a medium rough paper. Though any of the soft pencils, in the range F to 2H, can be selected but the HB grade pencil is mostly preferred. The pencil should frequently be sharpened to a conical point, which should then be rounded on some stray paper.

Depending upon the conditions of work and the purpose of drawing, a variety of papers are used for sketching. An engineer working in field may use a plain paper of his her letter pad to sketch some information. An engineer working in a design office should use a drawing sheet or sketch paper of a good quality.

To initiate a beginner in sketching, many types of squared papers are available. This helps the beginners in proportioning the object views correctly, sketching straight lines and keeping a drawing to good scale and squares. For isometric sketching, a specially ruled paper is available. Cross-section paper is of real practical value in sketching to scale as values can be assigned to squares and the squares counted to secure proportionate distances. But from an examination point of view, the author does not recommend its use to practice proportionate sketching as the examiners expect the students to do their sketches, in the examination, on plain drawing paper without any aid from instruments or squared or isometric paper. Sketching should be done freehand but with pride.

10.4. TYPES OF SKETCHES

Usually technical sketches are made of three dimensional objects and, therefore, the form of the sketch corresponds to one of the standard types of projection, used for technical drawings, i.e., *orthographic, axonometric, oblique*, and *perspective*, as shown in *Fig. 10.2*.

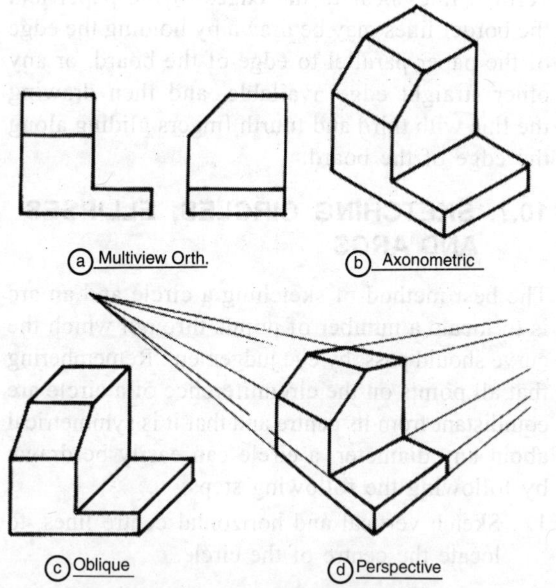

Fig. 10.2. Types of sketches

Therefore, before making a technical sketch a drafter must be familiar with the methods of projection used in technical drawing. The rules of projection, applicable to the particular type being used for making the sketch, should be kept in mind.

10.5. QUALITY OF LINES

The lines that are drawn freehand are naturally different from those drawn with instruments. The chief difference between an instrument drawing and a freehand sketch lies in the technique of making the lines, circles, curves and arcs. Freehand lines are not, and need not be, as uniform in weight and straightness as the mechanical lines, *Fig. 10.3*, but this does not imply that low standards of drafting should be adopted. But at the same time an undue emphasis on precision should not be made, as this would necessarily mean more time needed to finish a sketch.

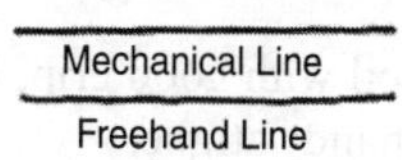

Fig. 10.3.

A character of each of the conventional lines used in freehand sketching is shown in *Fig. 10.4*. The widths of the lines are determined more by the size of the pencil point than by the pressure on the pencil point. Freehand line, just like the line drawn with instruments, should have blackness and should contrast sharply in size.

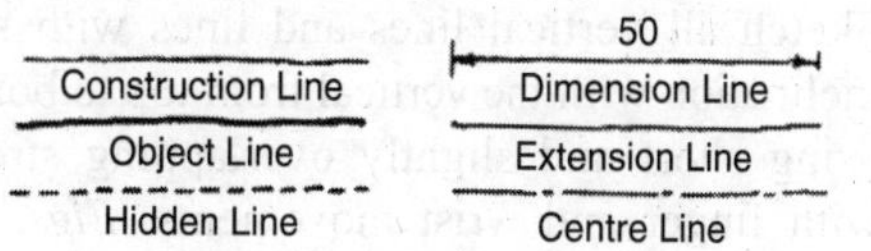

Fig. 10.4. Contrast of sketch lines

10.6. SKETCHING STRAIGHT LINES

Sketches, as with technical drawings, consist of straight and curved lines. With disciplined practice, any one can obtain the skill and become proficient in creating neat sketches involving straight and curved lines.

Before starting sketching of object views, it will be of great help to practice in sketching the lines and curves. For sketching of object views, the pencil should be held naturally, about 40 mm from the pencil point. The orientation of the pencil should be at right angles to the general direction of the line to be drawn. Here, by technique of sketching a line is meant, the freedom, snap and confidence expressed by the line. Though the technique of sketching depends more upon the natural movements of the drafter's hand holding the pencil, yet there are some suggestions, which will help in acquiring the required skill in sketching technique.

1. Place the paper in rotated position so that the horizontal edges are perpendicular to the natural

position of your sketching hand forearm, *Fig. 10.5 (a)* and *10.6 (a)*.

2. When a squared paper is being used for sketching, try to locate the main lines on the ruling of the sketch paper. Sketching straight lines on a graph paper is a simple process of guiding the pencil point using the existing graph lines.
3. Hold the pencil with good grip, but not so tight as to create hand fatigue.
4. Hold the pencil approximately 25 to 40 mm away from the pencil point.
5. Maintain a comfortable angle between the pencil and sketching strokes.
6. Sketch all horizontal lines and lines with small inclination to the horizontal from left to right, using slightly overlapping strokes, with free and easy wrist and arm motion, *Fig. 10.5 (b)*.
7. Sketch all vertical lines and lines with small inclination with the vertical from top to bottom, using short and slightly overlapping strokes, with finger and wrist movements, *Figs. 10.5 (c)* and *10.6 (b)*.
8. For drawing a long straight line, first fix the end points of the line. Add some intermediate points on the line if the line is sufficiently long. Then sketch the line by a series of overlapping strokes, keeping the eye on the point to which the line goes, rather than on the point of the pencil, *Fig. 10.5 (a) (3)*.
9. On a plain drawing paper, pencil dots can be positioned to plot the path of the line, *Fig. 10.5 (a) (3)*. In this figure the sketcher enters a series of pencil dots on the paper, which provide a basic outline as to the shape of the object. Then using a series of slightly overlapping strokes, the pencil dots are connected, *Fig. 10.5 (a)*.
10. Finger movements, with the wrist stationary, are to be used for short lines, small circles etc., whereas stiff forearm movements are more effective for long lines.
11. For any sketch it is advisable to start by sketching light centre lines for circles, hexagons and all other symmetrical figures or objects.

In *Fig. 10.5 (d)* are shown some inclined lines with the arrows indicating the direction sketching. When the inclination of the lines puts the hand in some tedious position, the paper may be turned to some convenient position. The horizontal and vertical lines near to the edges of the paper and the border lines may be drawn by holding the edge of the paper parallel to edge of the board, or any other straight edge available, and then drawing the line with third and fourth fingers gliding along the edge of the board.

10.7. SKETCHING CIRCLES, ELLIPSES AND ARCS

The best method of sketching a circle and an arc is to locate a number of points through which the curve should pass, by eye judgement. Remembering that all points on the circumference of a circle are equidistant from its centre and that it is symmetrical about any diameter, a circle can easily be drawn by following the following steps:

1. Sketch vertical and horizontal centre lines, to locate the centre of the circle.

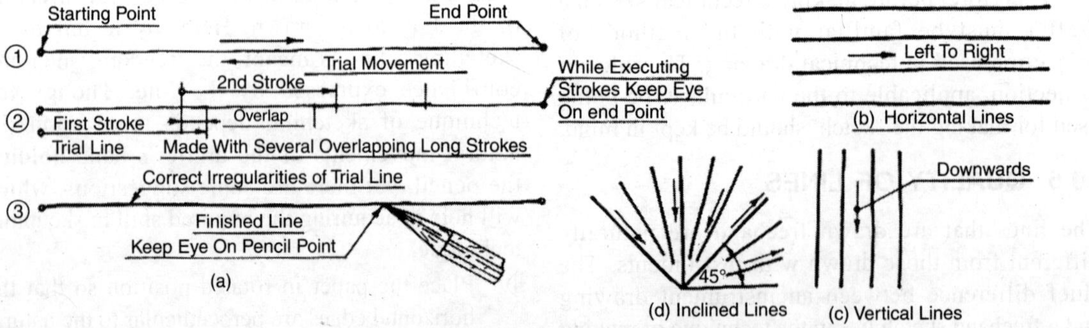

Fig. 10.5. Sketching straight lines

(a) For Drawing Horizontal Lines

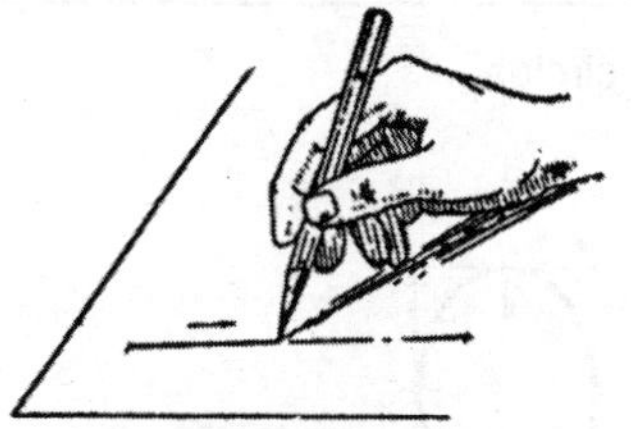

(b) For Drawing Vertical lines

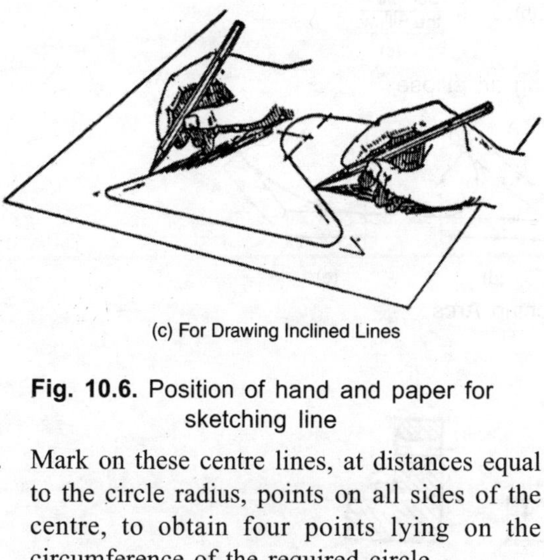

(c) For Drawing Inclined Lines

Fig. 10.6. Position of hand and paper for sketching line

2. Mark on these centre lines, at distances equal to the circle radius, points on all sides of the centre, to obtain four points lying on the circumference of the required circle.
3. Sketch a square through these four points.
4. Sketch the diagonals of the square.
5. Along these diagonals, plot four more points, one on each side of the centre, at distances of circle radius. These points too lie on the circumference of the circle.
6. Sketch the right half of the circle from top to bottom using slightly overlapping strokes. Then, similarly, sketch the left half of the circle to complete the circle passing through these eight points, as shown in *Fig. 10.7 (i) (a)*.

Another method is to draw radial construction lines lightly, then sketch short arcs across the lines at the estimated radius distance away from the centre, and finally to sketch the full circle lightly and heavy it in, as shown in *Fig. 10.7 (i)(b)*.

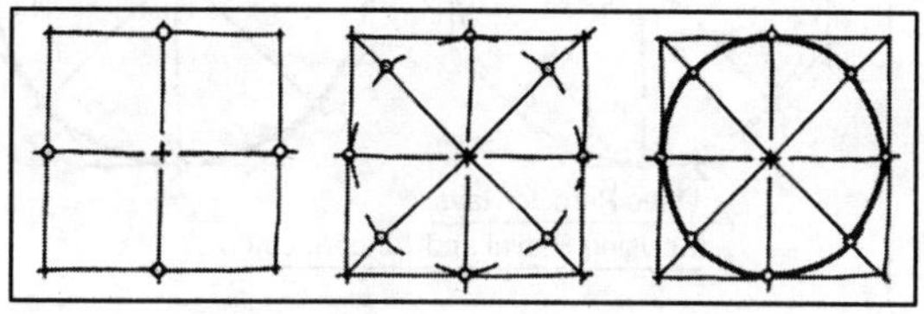

(a) Using Enclosing Square

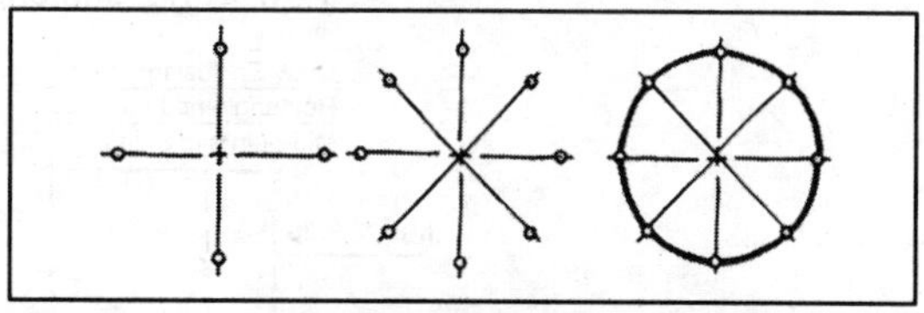

(b) Using Radial Lines

Fig. 10.7. Sketching (i) small circles

If the circle is fairly large it helps to mark off points equally spaced from the centre, as shown in *Fig. 10.7 (ii) (b)*, using a paper trammel.

Another method of sketching a large circle is shown in *Fig. 10.7 (ii) (a)*.

In sketching an ellipse, remember that an ellipse has a centre through which pass all its diameters. The longest diameter is called the *major axis* and the shortest diameter is called the *minor axis*. The ellipse is symmetrical about both of these axes, *Fig. 10.8 (a) to (c)*. The enclosing sphere for a circle becomes a rectangle for an ellipse.

Methods used to sketch arcs are simple adaptations of those used in sketching circles, as shown in *Fig. 10.9 (a)* to *(e)*. Curves should usually be drawn with the pencil on the concave side of the curve. Points of tangency should be carefully approximated and ringed. Fillets and rounds are sketched in after sketching the straight lines, *Fig. 10.10 (a)* to *(c)*. Where the sketches are to be tangential to lines at angles greater or less than 90° it is easy to use the simple preliminary constructions shown at *(d)* and *(e)*, *Fig. 10.10*. Measure off equal distances from the vertex of the angle to locate the two end points of tangency, and then sketch perpendiculars from these points

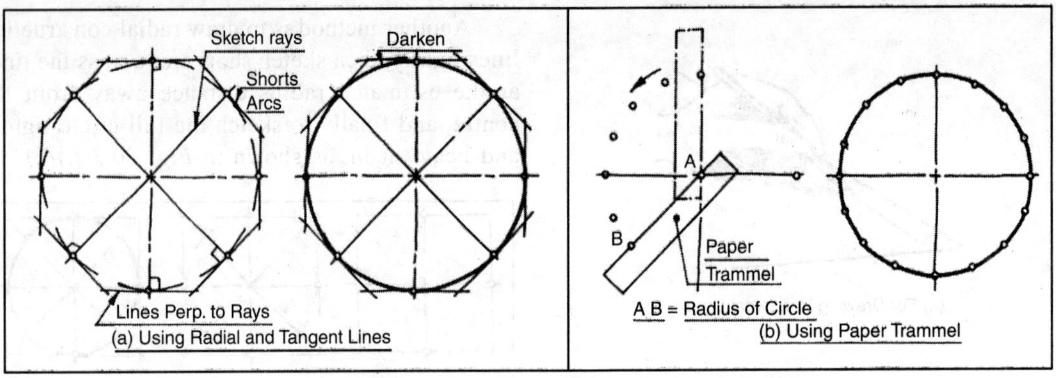

Fig. 10.7. Sketching (ii) large circles

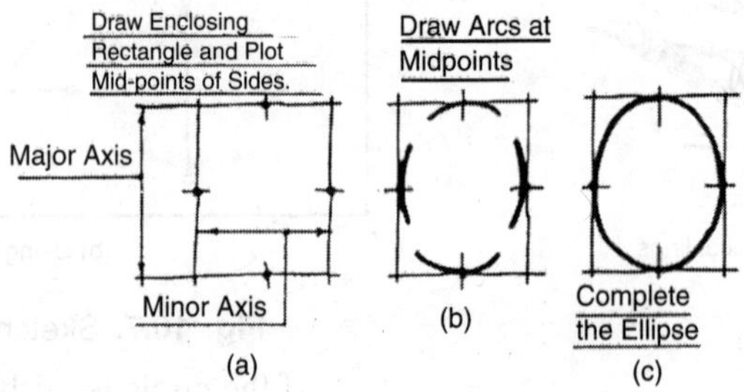

Fig. 10.8. Skeetching an ellipse

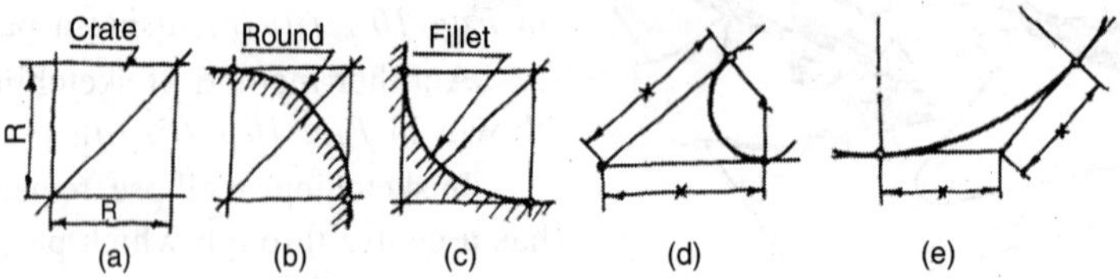

Fig. 10.9. Sketching Arcs

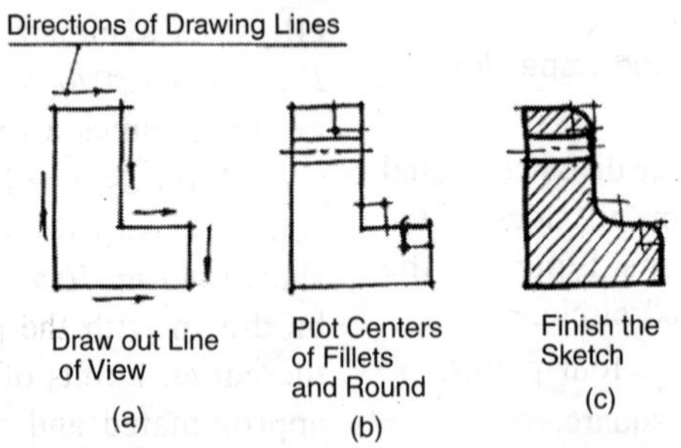

Fig. 10.10. Sketching Fillets and Rounds

to locate centres, which should be used to guide the eye when sketching the arcs.

10.8. ORTHOGRAPHIC PROJECTION

Freehand working drawings of existing objects are often required in practice for the purpose of changing some existing part, for repairing or building a lost or broken part or for replacing drawings which have been lost, damaged or found to be incorrect.

Orthographic sketching may require sketching any combination of the six principal views of the object. Usually the front, top and right side views are selected for describing an object in an orthographic sketch. In general, a combination of views that show the most detail and the least number of hidden lines should be chosen.

The procedure followed in making a sketch is almost the same as that in drawing with instruments. Though the sketch is not made to scale, yet it should be fairly proportionate, and should be according to the rules of projection, following all the conventional practices. A well-prepared sketch should give the same good appearance as a good instrument drawing. Steps in making a sketch are illustrated in *Fig. 10.11 (1)* to *(4)*, and are as follows:

1. Study the object until its shape and function are understood.
2. Determine the views which best describe the shape of the object.
3. Block in the views in light lines and space the views to give a balanced appearance to the sketch, *Fig. 10.11 (1)*.
4. Locate the details in each view in light construction lines, so that they need not be erased, *Fig. 10.11 (2)*.
5. Sketch in the required hidden lines and finish the sketch by brightening the visible lines, *Fig. 10.11(3)*.
6. Add the necessary dimensions and notes and give a suitable title to the drawing, *Fig. 10.11 (4)*.

As the proportioning is done only by eye, no scale should be indicated on the sketch.

Figure 10.12 illustrates the method of sketching orthographic views from a given isometric view, using the isometric sketch paper and the graph paper.

10.9. AXONOMETRIC SKETCHING

As for instrument drawing, may be one of three types: *isometric*, *diametric*, or *trimetric*. For the same reasons already described in isometric projection, the type most frequently used is isometric, in which length and width lines recede at 30° to the horizontal and height lines are vertical.

Theory of isometric drawing has already been discussed in detail in chapter 20. All the rules and conventions discussed there are applicable in isometric sketching. Isometric sketches are made either on plain paper or isometric sketch paper of the type. Though any of the methods used in making isometric sketching can be adopted for making

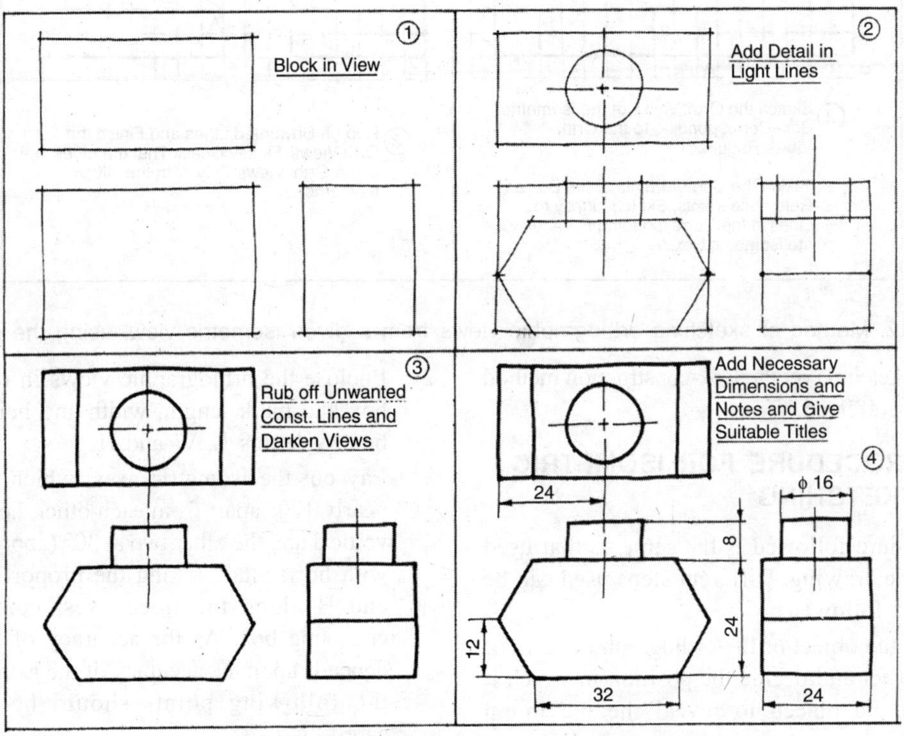

Fig. 10.11. Steps in Orthographic Sketching

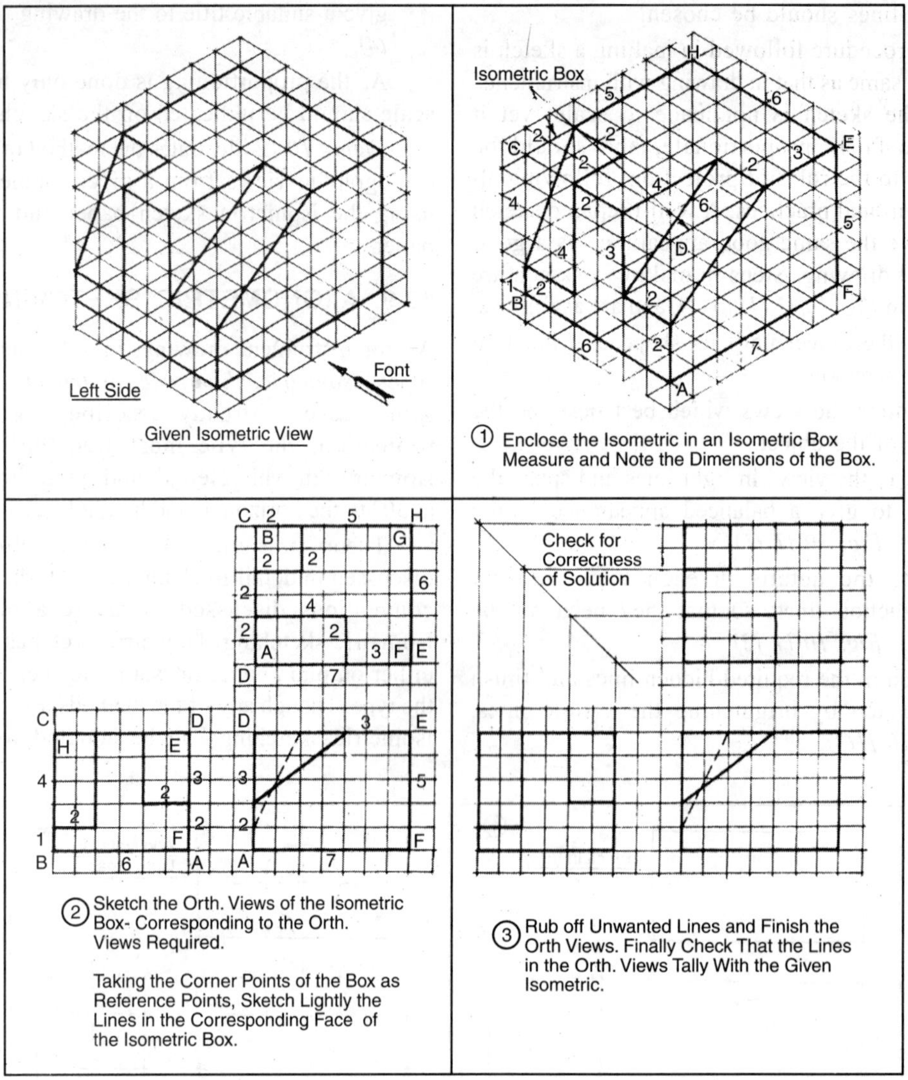

Fig. 10.12. Method of sketching orthographic views from a given isometric view, using the isometric

isometric sketches, yet the Box-construction method is mostly preferred.

10.10. PROCEDURE FOR ISOMETRIC SKETCHING

The procedure followed is the same as that used in isometric drawing. Different steps used can be outlined as followed:

1. Study the object or the orthographic views of the object and decide the position in which it should be placed to reveal the maximum details.
2. Enclose the orthographic views in rectangular box and mark length, width and height of the box by letters L, W and H.
3. Lay out the isometric axes, which should be nearly 120° apart from each other, i.e., one axis vertical and the other two at 30° (approximately) with horizontal. Taking the proportions L, W and H along the three axes, complete the enclosing box. As the accuracy of the sketch depends upon the accuracy of the isometric box, the following points should be observed carefully:

- Vertical edges must be vertical.
- Transverse lines must be parallel, may be little converging but never diverging.
- The angle of the inclined axes should never be more than 30°. A little less than 30° will give better appearance to the sketch, while greater than 30° shows the sketch as distorted.

4. Transfer the details from orthographic views to the isometric box. Take the proportions along or parallel the isometric axes only. For non-isometric lines, first locate the end points to establish the line.
5. Brighten the required lines and sketch the necessary hidden lines. In the end, add the necessary dimensions and the title.

Figure 10.13 illustrates the step-by-step construction of an isometric sketch by the "Box method". Three orthographic views of the object are shown.

10.11. SKETCHING CIRCLES AND ARCS IN ISOMETRIC

Circles and arcs appear as ellipses or parts of ellipses in isometric views. The best method to sketch the isometric circles and arcs is to first enclose them in rectangles, then to sketch the isometrics for rectangles and then locate the different points on them, to establish the curve. In *Fig. 10.15* at *(a)* sketching of circles on the three faces of a cube and an object containing semicircular arcs are shown. For irregular curves the offset construction method may be adopted.

When the object is not a simple part, and it is not convenient to enclose it in one box, it is divided into a number of parts and each part is enclosed in a separate box. *Figure 10.14* illustrates the sketching of one such object. Note carefully the proportioning of different parts and construction of holes and slots in the isometric sketch.

As the proportioning is done entirely by eye, great accuracy is unnecessary. Do not waste much time and energy on accuracy. The proportions should be approximately close to the actual distances, and the final sketch should give a realistic appearance.

10.12. OBLIQUE SKETCHING

For oblique sketching layout the orthographic front view of flat front face of the object, and show the depth lines receding at an angle, usually 45°, from the horizontal. As for oblique projection, oblique sketching is particularly useful for dealing with an object having round components on one face. It allows round components to be sketched round, rather than sketching elliptical. The following steps are usually used for oblique sketching:

1. Lay out the flat front surface of the object, using the blocking-in method.
2. Block-in the depth of the object using parallel lines.
3. Block-in the visible profile of the rear surface.
4. Add the details to true shape proportions to the front face and complete the sketch adding the depth dimensions of these details along the receding axis, showing all three dimensions for edges meeting in corner points, as shown in *Fig. 10.16*.

10.13. PERSPECTIVE SKETCHING

As for perspective projection, perspective sketching closely approximates the way a human eye actually sees an object. Two common types of perspective

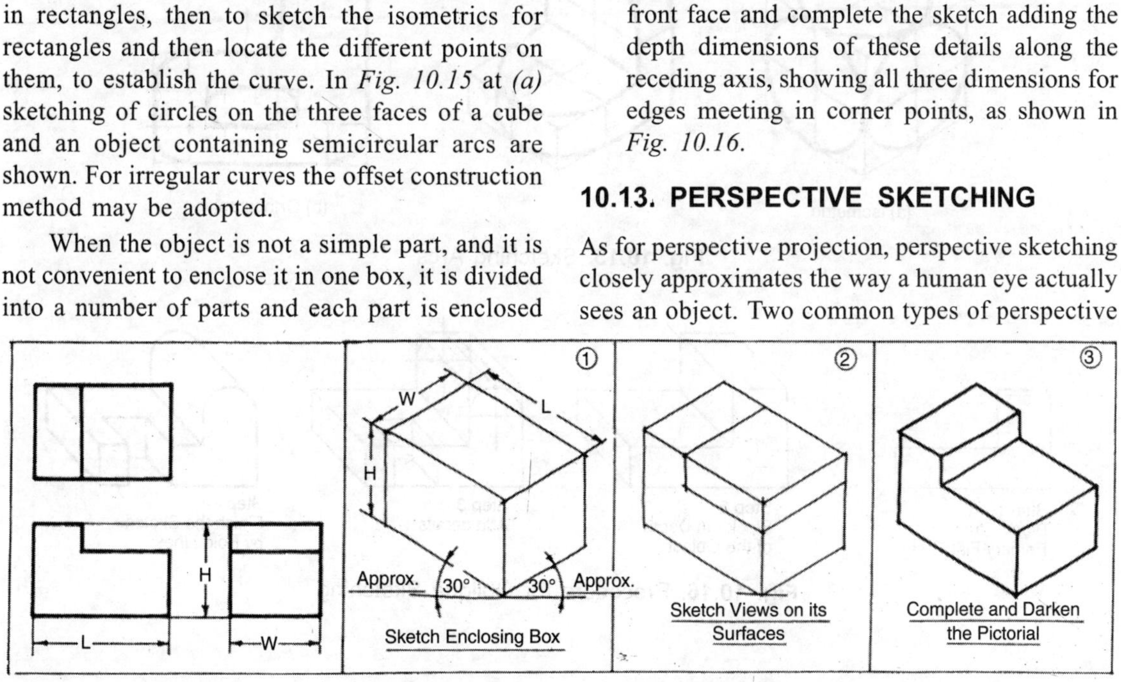

Fig. 10.13. Steps in Isometric Sketching

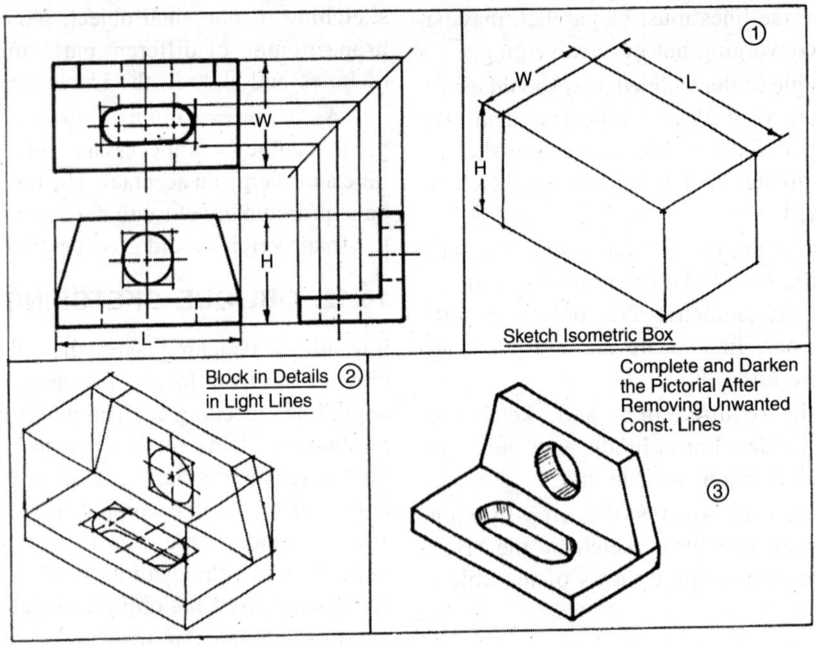

Fig. 10.14. Procedure of Isometric sketching

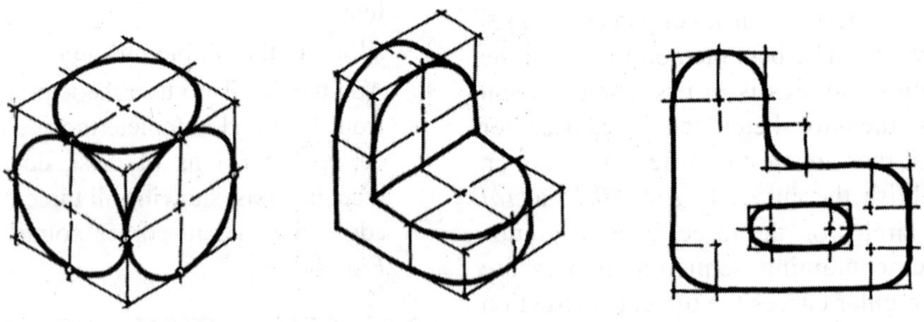

(a) Isometric

(b) Orthographic

Fig. 10.15. Sketching Arcs

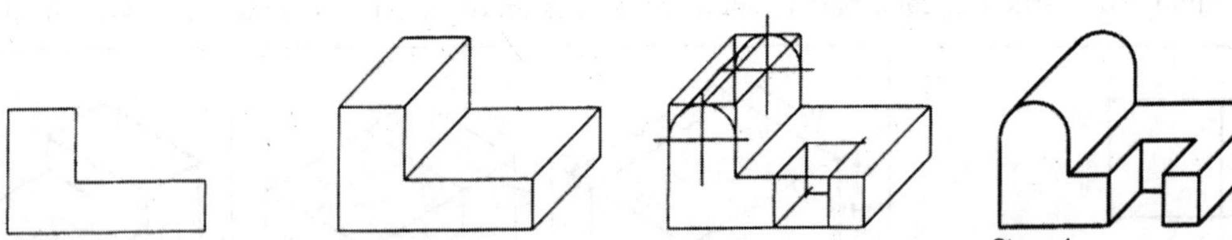

Step 1
Block - in
Primary Flat Face

Step 2
Block - in Depth
of the Object

Step 3
Add details

Step - 4
Finish the Sketch
by Bold Lines

Fig. 10.16. Procedure for Oblique Sketching

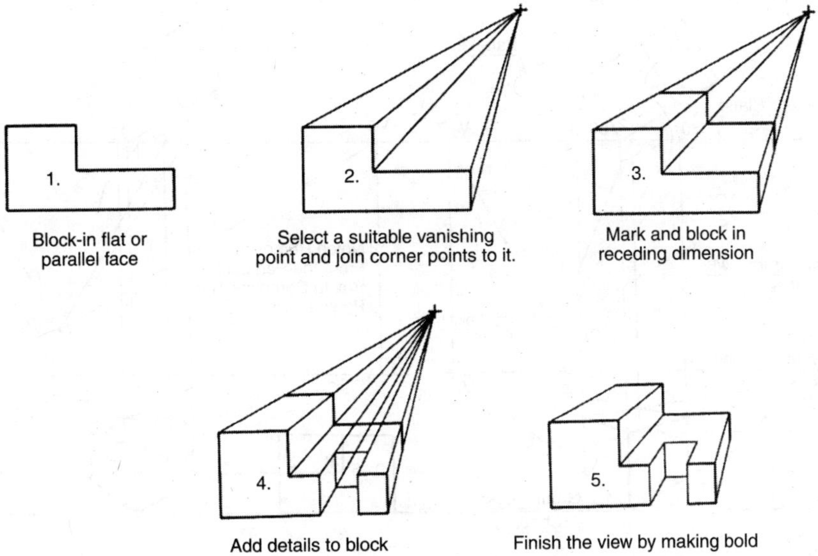

Fig. 10.17. Procedure for sketching One Point Perspective

sketches are *one-point perspective* and *two-point perspective*.

ONE-POINT PERSPECTIVE

A one-point perspective sketch is similar to an oblique sketch, except that the depth lines of the object recede to a vanishing point instead of receding parallel to one another, *Fig. 10.17*. The recommended procedure to be followed in making a one-point perspective is as follows:

1. Lay out the flat front surface of the object, to be sketched, using the lock-in method, *Fig. 10.17 (a)*.
2. Fix a suitable vanishing point for the sketch. Project all points on the front surface back to the vanishing point, *Fig. 10.17 (b)*.
3. Mark, proportionately, the depth of each object edge along all the line projectors, *Fig. 10.17 (c)*.
4. Complete the sketch by sketching the visible profile of the rear surface for each component part of the sketch, *Fig. 10.17 (d)*.

TWO-POINT PERSPECTIVE

It closely resembles an isometric sketch, except that the width and depth lines recede to the left and right vanishing points rather than receding parallel to each other, *Fig. 10.18*. The following procedure is recommended for sketching a two-point perspective sketch:

1. Layout, proportionately, the two-point perspective frame consisting of vertical height line, left and right vanishing points, and the receding lines, *Fig. 10.18 (a)*. As in perspective projection, the horizon line when positioned below the view shows the bottom of the object and when positioned above shows the top of the object in the sketch. The vanishing points must be on the horizon line.
2. Block-in the object, proportionately, estimating the length and width for proportions, *Fig. 10.18 (b)*.
3. Lay out the details, in light lines, maintaining proper proportions, *Fig. 10.18 (c)*.
4. Complete the two-point perspective sketch, *Fig. 10.18 (d)*, by adding the required visible lines at each visible corner point.

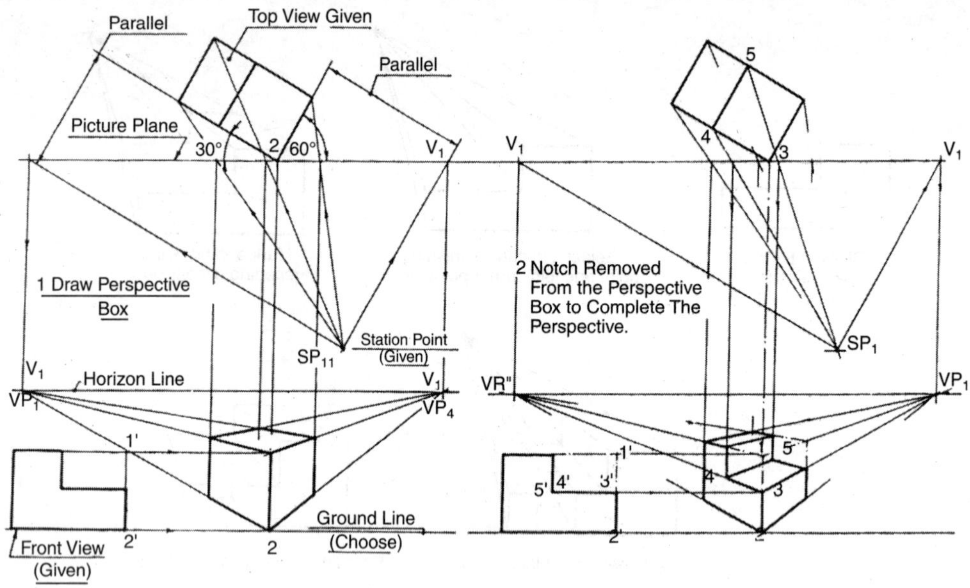

Fig. 10.18.

EXERCISES

1. On an *A2* size drawing sheet practice freehand the sketching of different lines and shapes as shown in *Fig. 10.19*. The sheet should be divided proportionately in to similar parts as is done for preparing the exercise.
2. In *Fig. 10.20* four different objects are shown sketched as isometric views on isometric sketch paper. Sketch the orthographic views for each using the squared graph paper space provided in each case.
3. On an *A4* size-drawing sheet, neatly sketch freehand the front, top and right side views of each pictorial view shown in *Fig. 10.21 to 10.28*. Use the grid provided to estimate the proportional measurements. Add all lines, including hidden lines and centerlines. You can further sketch the isometric sketches, oblique sketches, and perspective sketches of any suitable number of these problems to develop the required skill in all types of sketching.
4. In Fig. *10.29-39* eleven different objects are given as isometric views, drawn on isometric sketch paper space. For each of them sketch three orthographic views using the squared graph paper space provided for the purpose.
5. *Figures. 10.40-43* show orthographic views of six different objects, sketched on squared graph paper space for each object in each of the figures. Study the given views and sketch the isometric view, in each case, on the isometric graph paper space provided or you may use larger isometric sketch paper boxes provided.
6. Oblique views of six different objects are shown drawn on a squared graph paper space for each, in *Fig. 10.44.* Sketch the orthographic views in each case using the squared graph paper space provided. Proportions should be taken by counting the number of divisions in each edge.
7. Orthographic views of six different objects are shown sketched on graph paper space for each, in *Fig. 10.45*. Sketch, proportionately, an oblique view for each object, using the oblique graph paper space provided.

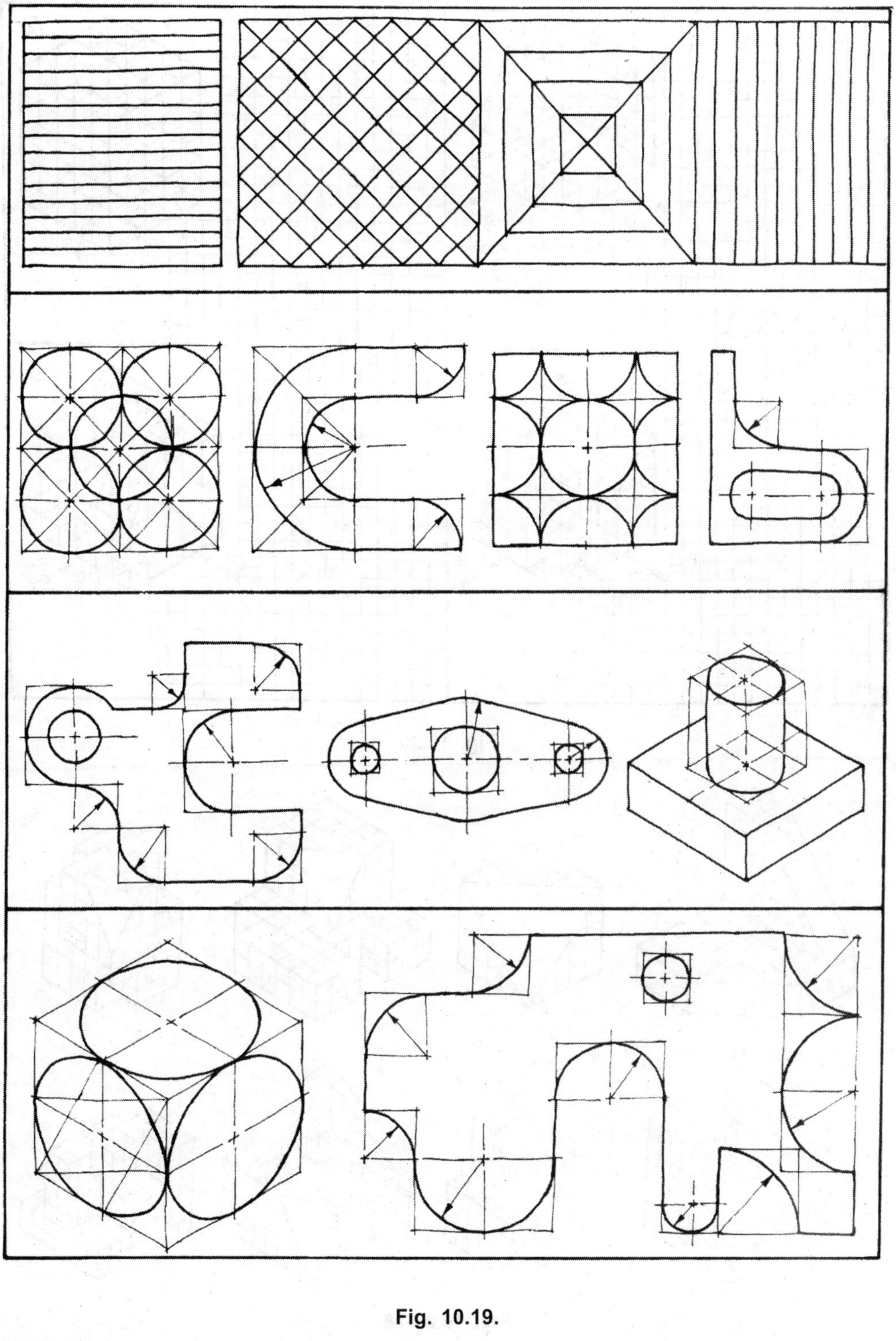

Fig. 10.19.

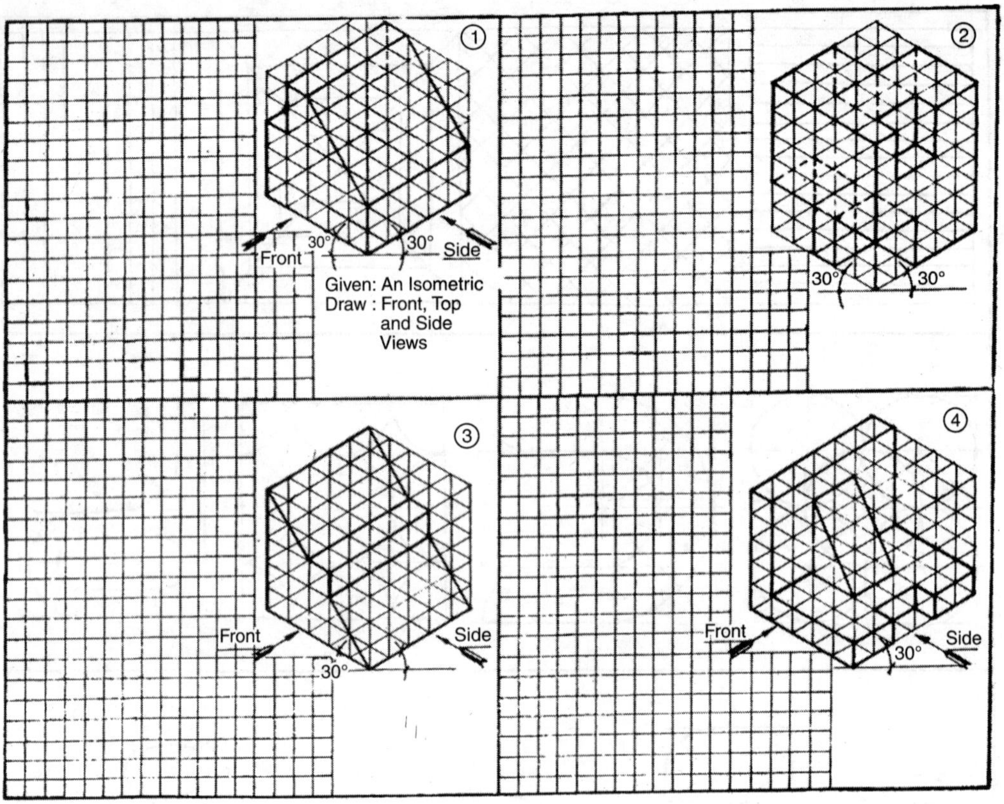

Fig. 10.20.

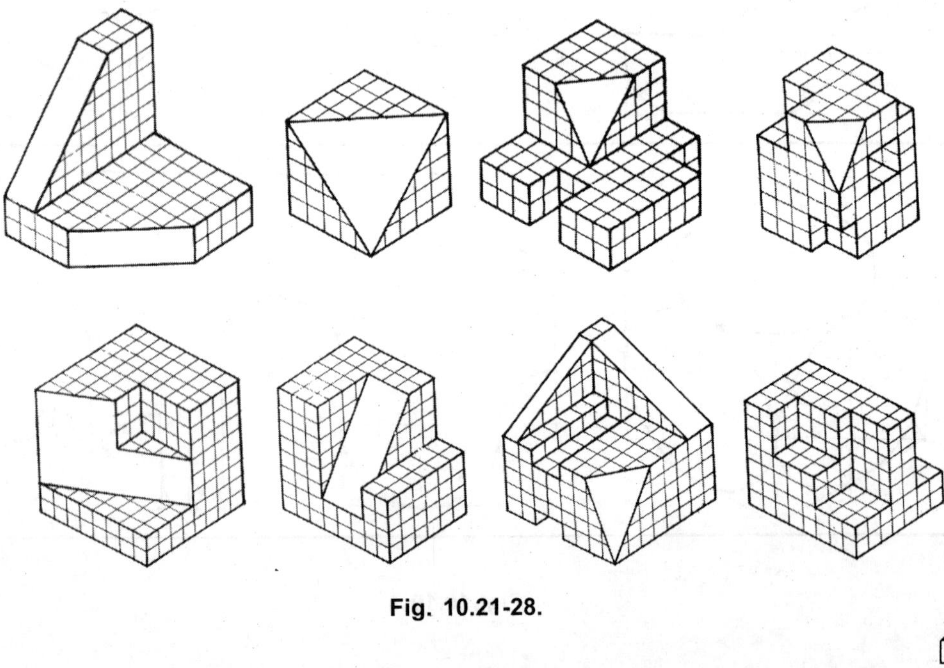

Fig. 10.21-28.

Sections and Conventions 11

11.1. INTRODUCTION

So far we have learnt that the detail of an object can be shown by drawing a limited number of carefully chosen views and showing external features of the object by firm lines and invisible detail by hidden lines. But when the interior of the object is complicated to be shown clearly and dimensioned by outside view and hidden lines, or when the component parts of a machine are drawn assembled, the method of showing the interior details by hidden lines makes the view confusing and} hard to read, *Fig. 11.1 (a)*. In such cases the object, or assembly, is assumed to be but by an imaginary plane and by assuming the front portion of the cut object (between the observer and the cutting plane) removed the view is drawn as shown in *Fig. 11.1 (b)*. Section lining or cross-hatching identifies the exposed or cut surfaces. The view thus obtained is called the '*Sectional View*' or '*View in Section*' of the remaining object. Hidden lines and details behind the cutting plane are usually omitted unless they are required for clarity. A sectional view, in most cases, replaces one of the regular orthographic views. The general rules that apply to the arrangement of orthographic views apply equally when drawing sections.

11.2. CONVENTION OF SHOWING A SECTION

A cutting plane line or a section line is used to indicate where the imaginary cutting of the object takes place. The cutting plane is passed through the most important and identifiable place on the drawing, keeping in mind the detail that we wish to expose. In most cases it may pass through the centre line of an object. In such cases there is no need of showing the position of the cutting as it can be very easily interpreted from the sectional view, *Fig. 11.2*.

If the place, from which the section is taken, is not obvious, a cutting plane line, directional arrows and identification letters are used to describe it. Referring to *Fig. 11.1*, for drawing the sectional front view, the cutting plane passes through the centre of the object. As it is the most obvious section, there is no need of showing the cutting plane line.

It should be noted here that, while drawing a sectional view, the part of the object in front of the imaginary cutting plane is supposed to be removed, *Figs. 11.2* and *11.3*. But in the other view this portion is not left out.

If the location of the section plane is not an important one, *Fig. 11.2 (d)*, or where it is necessary to distinguish between several cutting planes, *Fig. 11.2 (e)*, the position of the cutting plane(s) should be indicated by means of a thin chain line, thick at ends. The cutting plane(s) should be indicated by letter designations, e.g., capital letters. The arrows at the ends of the line(s) indicate the direction in which the oυject is viewed while drawing the section. The section should be indicated by the relevant designations, *Fig. 11.2 (e)*. The designations on the referenced sections should be placed either immediately below or above the relevant sections. But in any one drawing the references should be placed in the same way. -No other indication is necessary.

In certain cases, the parts located beyond the cutting plane line need not be drawn completely. In a sectional view, the cut surfaces are 'cross-hatched' or 'section lined' by drawing thin, equally spaced and parallel lines, generally inclined at 45° to the horizontal.

11.3. SYMBOLIC REPRESENTATION OF A CUTTING PLANE

A cutting plane employed in sectioning is indicated on a drawing, in a view adjacent to the sectional view, as shown in *Fig. 11.1 (b)*. In this view the plane appears as a line. This line is a thin chain line thick at ends, as shown in *Fig. 11.2 (a)*, *Fig. 12.3*. Arrowheads, put at its ends, indicate the direction in which the cut away object is viewed, i.e., the 'direction of sight \ as shown in *Fig. 11.1 (b)*.

Notice that the arrows always point away from the view showing the section in third angle projection and towards the view in first angle projection. Further note that part of the object, imagined to be removed, is the part opposite to the direction of the arrows. At the ends of the cutting plane line capital letters are often needed to identify the cutting plane line with the corresponding section. When the cutting plane line coincides with a centre line, the cutting plane line takes the precedence over the centre line.

It is customary to omit the cutting plane line when its position is obvious. This usually occurs in the case of symmetrical objects, in which the plane is understood to pass through the axis of symmetry. We can say that the cutting plane line is omitted unless it is needed for clarity, *Fig. 11.2 (b)*.

11.4. TYPES OF SECTION

Although the different sectional views have to e named for identification and for specifying type of view required in a drawing the names are not shown on the drawing for the same reason as these can be easily interpreted.

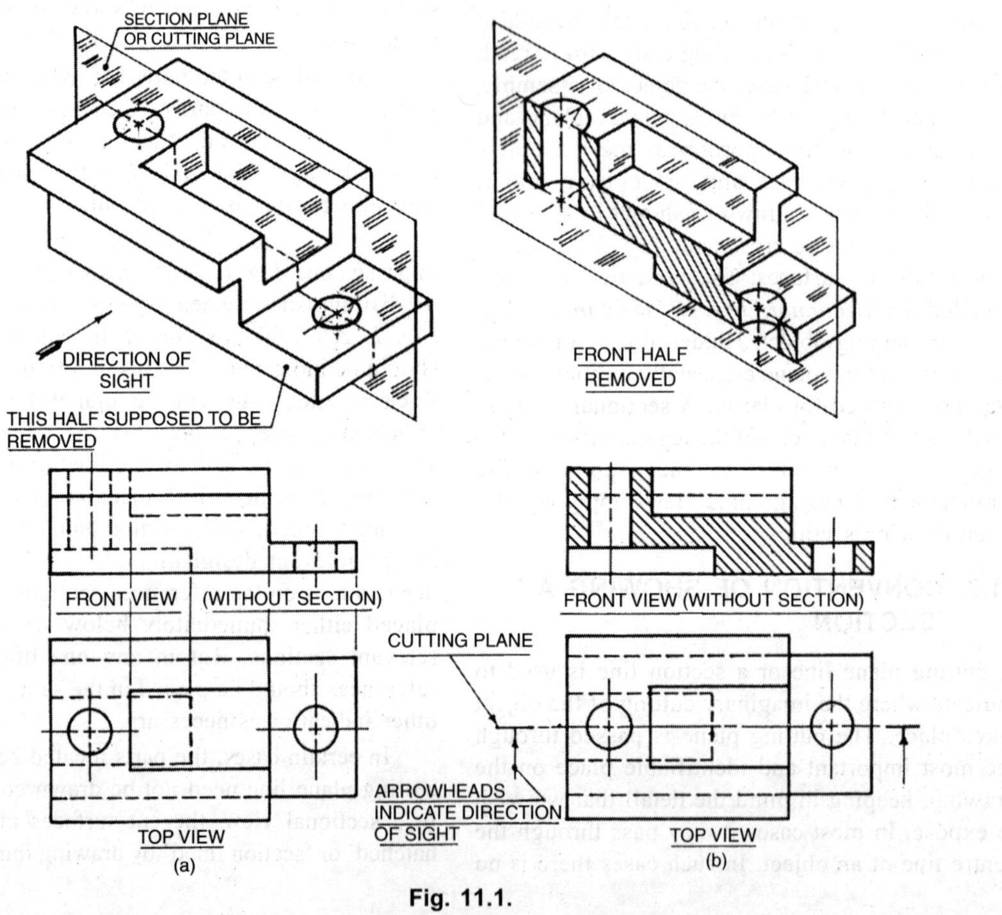

Fig. 11.1.

When the cutting plane cuts an object lengthwise, the section obtained is commonly a *longitudinal section*. When it cuts crosswise, it is called a cross section. It is designated as being either in *full section*, a half section, or a broken section. If the plane cuts entirely across the object the section represented is known as a full section, *Figs. 11.2 (c), (d) and 11.3*. If it cuts only halfway across a symmetrical object, the sectional view obtained is a half section, *Figs. 11.2 (a) (b), and 11.4*. A broken section is a partial one, which is used when less than half section is needed, Fig. 11.6. The classification of the names depends upon the character of the section, as the longitudinal and cross, and amount the view in section, as full and half sections.

1. Full Section

When the section plane passes entirely through the object, in a straight line the view obtained is called a full section of the object or the '*view full in section*'. *Figure 11.3* is an example of a full section. When the object is sectioned along its length, it is called a *longitudinal section*. When

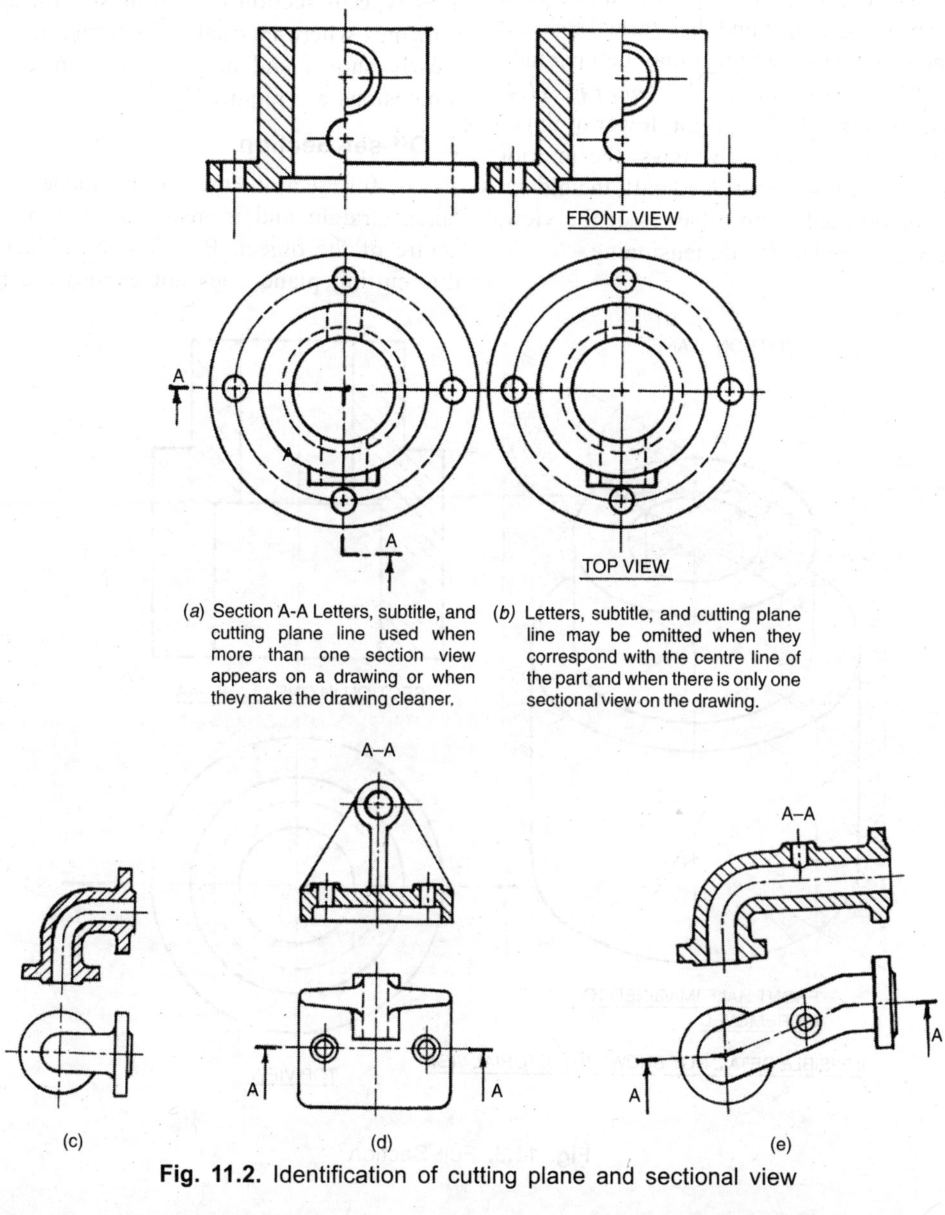

(*a*) Section A-A Letters, subtitle, and cutting plane line used when more than one section view appears on a drawing or when they make the drawing cleaner.

(*b*) Letters, subtitle, and cutting plane line may be omitted when they correspond with the centre line of the part and when there is only one sectional view on the drawing.

Fig. 11.2. Identification of cutting plane and sectional view

the object is sectioned crosswise, the section is called a *cross-section.*

This type of section is used for both detail and assembly drawings. When the cutting plane divides the object in to two identical parts, it is not necessary to indicate its location. However, the cutting plane may be identified and indicated in the usual manner to enhance clarity, *Fig. 11.3.*

2. Half Section

In symmetrical objects, to show the interior and exterior details in the same view, half sections are drawn. The cutting plane is imagined to extend half way across the object and then turned forward so that one quarter of the object may be imagined as removed, as shown in *Fig. 11.2, and Fig. 11.4.* The half section may be left, right, lower or upper half of the view. As, in such views, the interior details are visible in the sectioned half, the hidden lines may be omitted in both halves of the view, unless they are needed for dimensioning.

It should be noted that only the centre line separates the two parts of the half sectional view of a symmetrical object. As the object is only imagined as cut and no actual cutting edge exists, there is no reason for separating the two halves of the view by a solid line. Due to the same reason the change in direction of the offset cutting plane is also not shown in a sectioned view.

A half section drawing should not be used where the dimensioning of internal diameters is required. This will prevent the addition of hidden lines to the portion showing the external features. This type of section is used mostly for assembly drawings where internal and external features are clearly shown and only overall centre-to-centre dimensions are required.

3. Off-set Section

In full section, the cutting plane is usually taken straight and is made to pass through the centre of the object, But in some objects, when this cutting plane does not expose the required

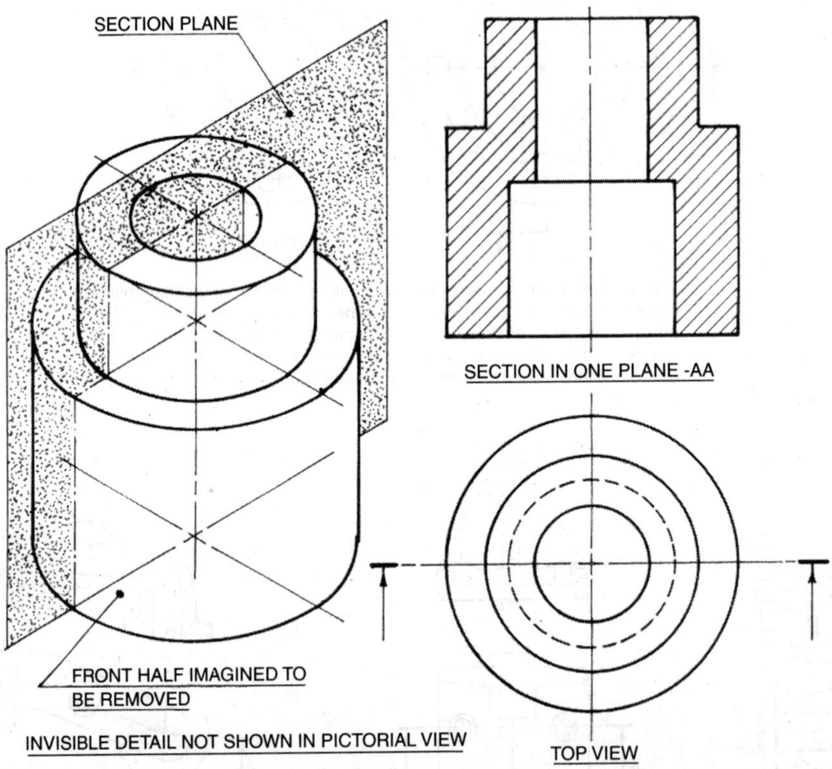

Fig. 11.3. Full Section

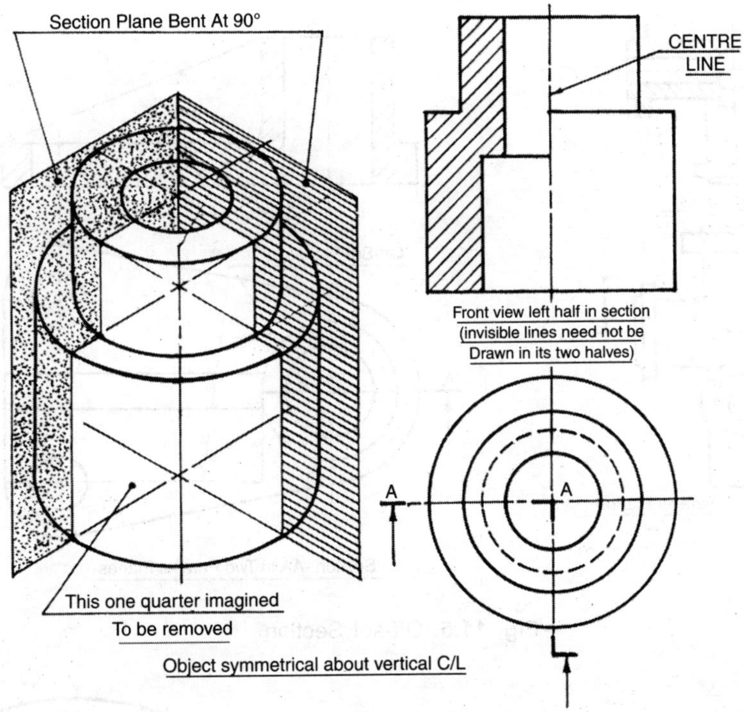

Fig. 11.4. Half Section

details, or it passes through features, which should not be sectioned, it is offset by changing the direction and is made to pass through required details that need to be shown in section. *Fig. 11.5 (a)* illustrates a section plane taken in two parallel planes. In *Fig. 11.5 (b)* the plane is offset to pass through the hub and a hole in the base of the object, while the rib is left out. If the section were continued through the centre of the object, due to sectioning of the rib, the view would have looked much heavier.

4. Broken Section

In some objects when some important interior detail must be shown, but due to the important exterior feature, that has to be shown, a full or half section is not feasible, showing a broken section is used as an alternative. The cutting plane is extended up to the required length and the part in front is imagined to be removed by breaking, a portion of the object. This gives an irregular, but thin, boundary line to the section, as shown in *Fig. 11.6*.

5. Revolved Section

As shown in *Fig. 11.7*, when the cross section of an object is revolved through 90°, to show in the view perpendicular to the cutting plane, the section is said to be a revolved section. These sections are conveniently used for showing the cross section of elongated bars, like connecting rods, arms, ribs of pulleys and so many other similar shaped features. For a revolved section a centre line is drawn through the shape on the plane to be described. The part is then imagined as rotated through 90°, and the view that would then be seen, after rotation, is superimposed on the regular view. If the revolved section does not interfere with the view on which it is revolved, then the view is not broken unless it would facilitate clearer dimensioning. When the revolved section interferes or passes through lines on the view on which it is revolved, the view is usually broken. When superimposed on the view, the outline of the revolved section is a thin continuous line.

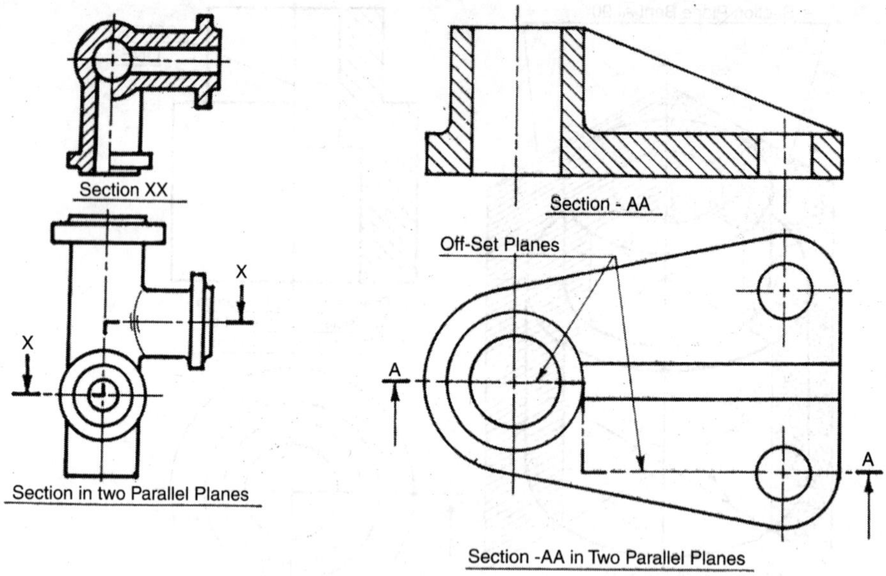

Fig. 11.5. Off-set Section

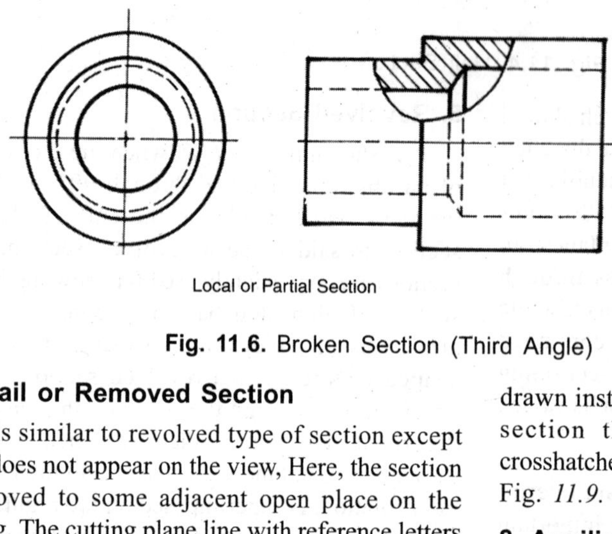

Fig. 11.6. Broken Section (Third Angle)

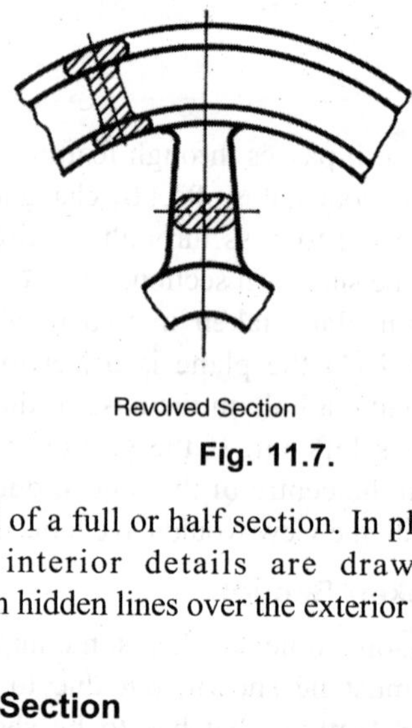

Fig. 11.7.

6. Detail or Removed Section

It is similar to revolved type of section except that it does not appear on the view, Here, the section is removed to some adjacent open place on the drawing. The cutting plane line with reference letters indicates the place of section.

Removed sections of symmetrical parts are placed on the extension of the centre line where possible. *Fig. 11.8* illustrates the detailed sections on extended centre lines. Frequently, the removed section is drawn to an enlarged scale for clarification and clearer dimensioning.

7. Phantom Section

In case of the objects, which are not completely symmetrical, a hidden or phantom section may be drawn instead of a full or half section. In phantom section the interior details are drawn and crosshatched in hidden lines over the exterior details, Fig. *11.9*.

8. Auxiliary Section

When the cutting plane is not parallel to any of the co-ordinate planes the section view obtained is called an auxiliary section. These views are used to show the interior details of the inclined features of the object, *Fig. 11.10.*

9. Aligned Section

Certain objects give an unsymmetrical and confusing section if drawn strictly according to the rules of projection. In sectioning such objects

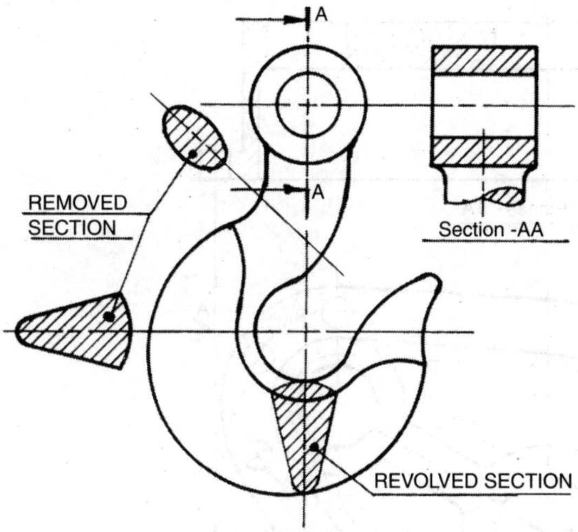

Fig. 11.8. Remove Section on extended centre lines

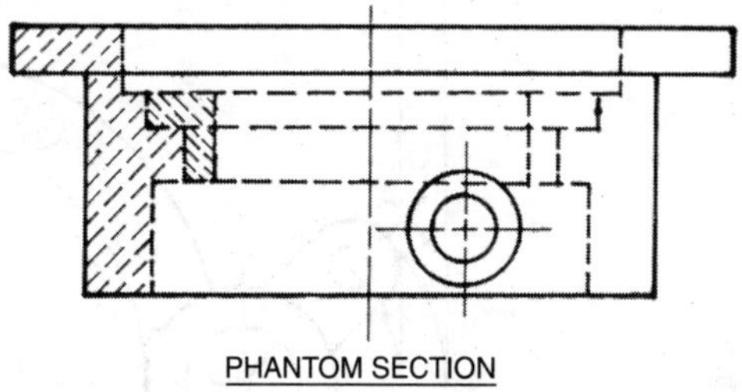

Fig. 11.9.

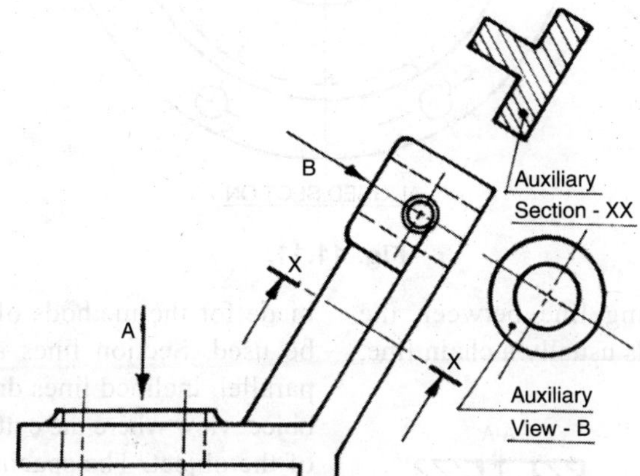

Fig. 11.10. Auxiliary Section

more details can be shown if the cutting plane is bent to pass through the required features of the object. In *Fig. 11.11* is illustrated a cutting plane A-A, bent twice to pass through the required features. While drawing the sectional view the various segments of the cutting plane are imagined as revolved until they are parallel to the plane of projection. Thus the sectional view obtained shows all the cut portions of the object in true shape. The section thus obtained is called an aligned section. For the same reasons, as already discussed, the bends in the cutting plane are not shown in the sectional view.

10. Adjacent Section

In case of adjacent sections of the same pan, parallel planes are shown side by side, section should be shown identical, *Fig. 11.12 (a)*, but may

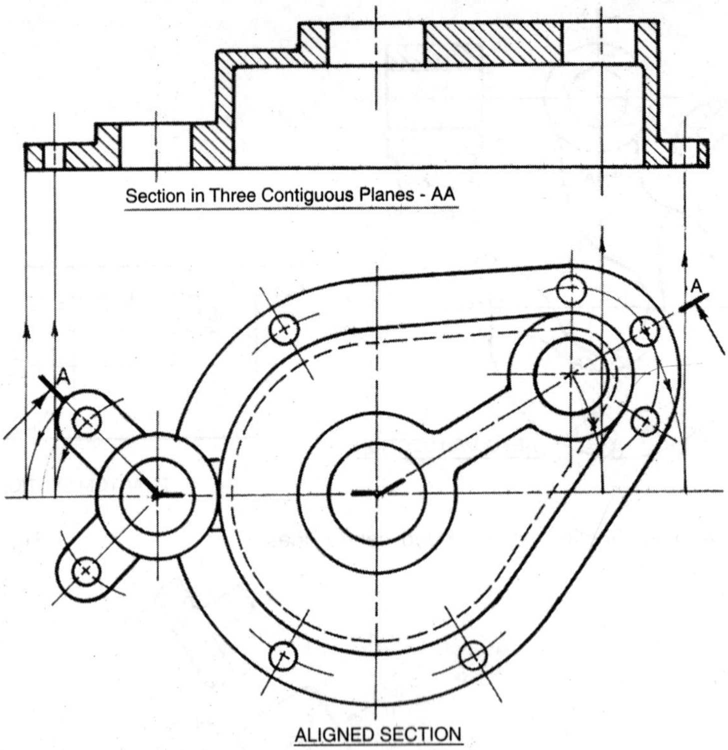

Fig. 11.11.

be offset along the dividing line between the sections. The dividing line is usually a chain line, *Fig. 11.12 (b)*.

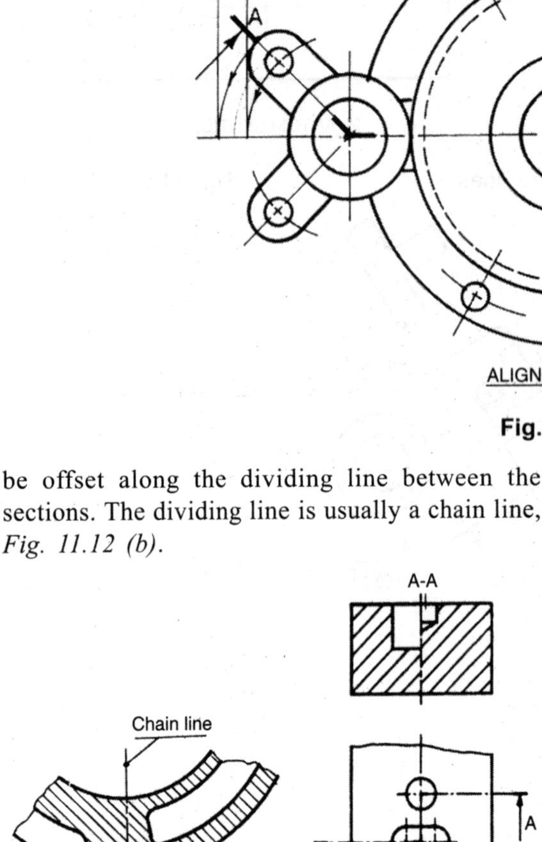

Fig. 11.12. Adjacent, section of section lining a part in different parallel planes

11.5. SECTION LINING

In the sectional view, wherever the section plane, to make the cut surface distinct, has cut the material a conventional scheme called "*section lining*" or "*cross hatching*" is employed. Allowance must be made for the methods of reproduction that are to be used. Section lines are thin, equally spaced, parallel, inclined lines drawn in those areas of the object view where the cutting plane cuts the material of the object. The spacing between lines depends upon the area to be section lined and are spaced entirely by eye judgement to give an even tint in manual drafting or setting the spacing in CAD drafting. For ordinary working drawings the spacing varies from one mm to three mm. The section lines may be inclined at any angle to the horizontal, but 45° inclination is commonly adopted. For better appearance the section lines should not be parallel to any of the main boundary lines of the area, *Fig. 11.13*. Correct and incorrect angles of section lining are illustrated in *Fig. 11.14* and some common errors in section lining are shown in *Fig. 11.15*.

If the section plane cuts the same part at different places, a section line should be continued on all the corresponding areas on the drawing in an identical manner, *Fig. 11.11*. In assembly drawings, the adjacent parts are crosshatched in opposite directions. If three parts adjoin, two of

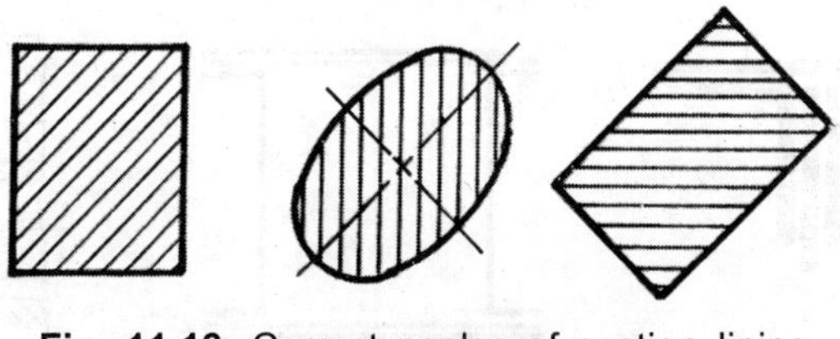

Fig. 11.13. Correct angles of section lining

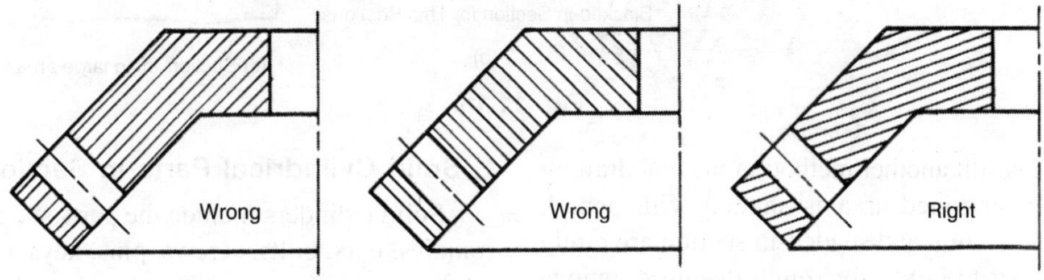

Fig. 11.14. Correct and incorrect section lining

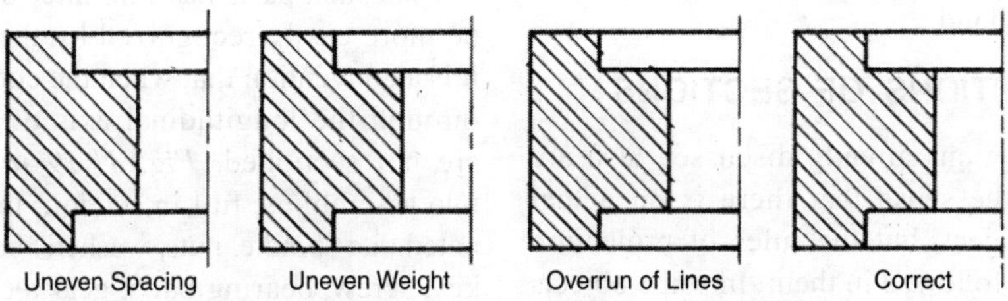

Fig. 11.15. Common errors in section lining

them can be crosshatched at 45° inclination, but in opposite directions and the third part may be crosshatched:

1. By 45° lines which should not be continuous with the lines in the adjoining parts;
2. By lines inclined at an angle other than 45° such as 30° or 60°, *Fig. 11.16 (a)*;
3. By 45° lines with the spacing in between different than that in the adjoining parts, *Fig. 11.16 (b)*.

For quick identification of the parts, each part, in every view of the assembly drawing, should be section lined in the same direction and with the same spacing in between the section lines.

Varying the type of section lines identifies sometimes different materials. The meaning of these hatchings should be clearly defined on the drawing or by reference to appropriate standards. The various types of section lines used for different materials are shown in *Fig. 11.30*.

If a dimension need be given in any sectioned area, the section lines should be discontinued from around the dimension numerals, *Fig. 11.17 (a), (b)*.

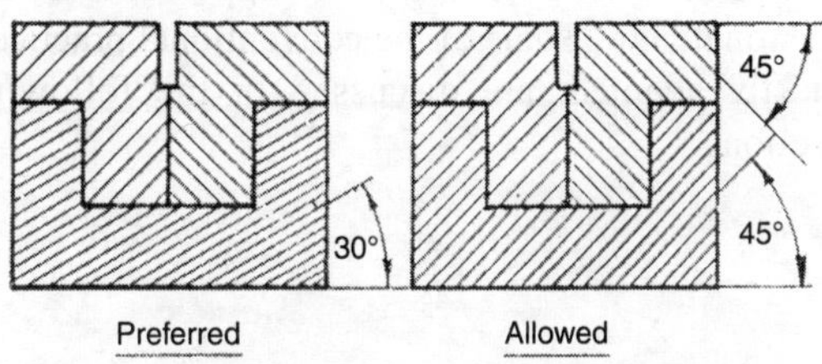

Fig. 11.16. Section lining in adjacent parts

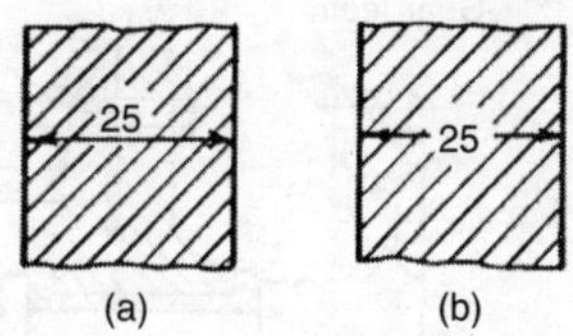

Fig. 11.17. Giving a dimension in a sectional ares

When the area to be sectioned is very small, as for thin plates and structural members, 'blacked in section' may be used. A space of not less than 0.7 mm must be left between adjacent sections of this type, *Fig. 11.18 (a), (b)*. For very large areas, an outline section limited to a zone following the contour of the hatched area is adopted to save time, *Fig. 11.18 (c)*.

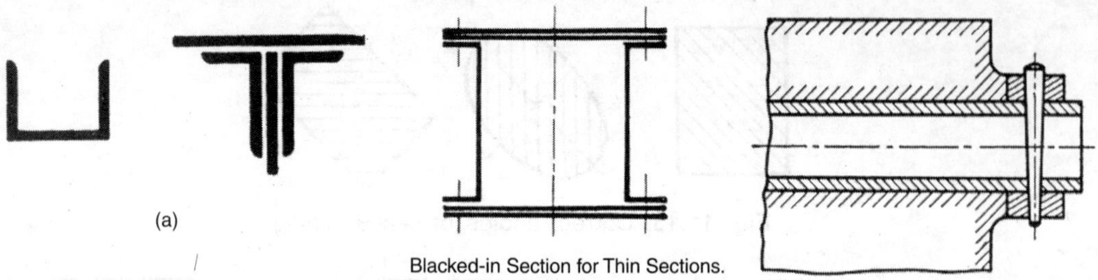

(a)

Blacked-in Section for Thin Sections.

(b)

(c) Section lining large areas

Fig. 11.18.

There is still another method in manual drafting which the sectioned area is shaded with pencil. The outline section and shaded-in section are rarely used and that too only for rough design drawings and sketches. The section lining in these methods is generally freehand.

11.6. CONVENTIONS OF SECTIONS

Although the sections already discussed also are conventions in the sense that there is no actual cutting of the object, but the rules of projection had been strictly followed in them. In some objects the deviation from the strict rules of projection results in better readability and appearance of the sectional view. Some of the conventional practices, mostly adopted, are discussed in the following sections.

1. Solid Cylindrical Parts In Section

Solid cylinders include the parts such as shafts, spindles, bars, bolts, screws, pins, keys and cotters, etc.

As such parts have no interior details and can be more easily recognized by their exterior view, it is a convention that when the cutting plane passes through the longitudinal axis of such parts, they are not sectioned. *Fig. 11.19 (a), (b)* illustrates two assemblies, full in section, in which it can be noted that spindle, nuts, washers, bolts, rivets, cotter, key, screw, bearing balls, gear teeth, taper pin and solid plug are left unsectioned. It is evident from these figures that if these parts were sectioned the drawing would have been more confusing and difficult to read.

2. Spokes and Webs of Wheels in Section

A wheel with a solid web is shown in *Fig. 11.21 (a)*, a pulley with four spokes is shown in

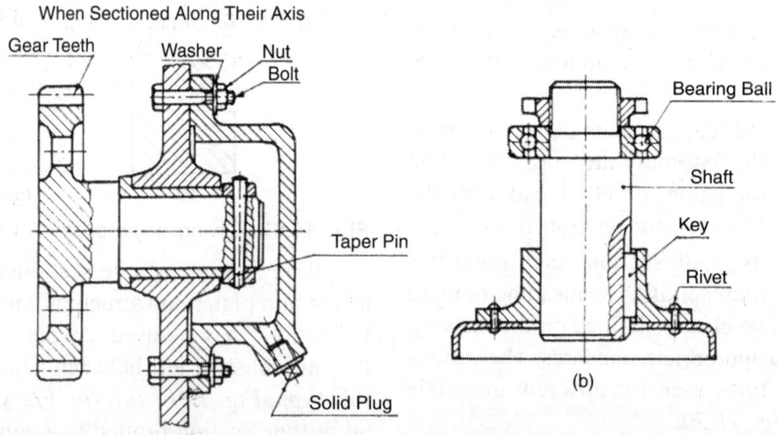

Fig. 11.19. Solid Cylindrical Parts in Section

Fig. 11.20. The most commonly accepted convention for such parts (circular wheel) is that any element, that is not solid or continuous around the axis of the part should not be sectioned. According to this practice, as the web is solid around the axis of the wheel, it is always sectioned, and as the spokes are not a solid element they are never shown in section. *Fig. 11.22* shows the preferred method of identifying spokes in section for aligned and unaligned designs. Note that the spokes are not sectioned in either case.

3. RIBS and LUGS In Section

When the cutting plane passes through a rib parallel to its larger dimension, the commonly accepted practice is to omit the section lining, to show the visible outline as if the cutting were just in front of the rib, *Fig. 11.23*. Sometimes this practice is not adequate. In *Fig. 11.21* at *(a)* the wheel has a solid web, while at (b) it has four ribs in addition to the web. When drawn in accordance with the above-discussed practice, the sectional view at (*b*) is identical with that at (*a*). In such cases the better practice is to use 'alternate cross-hatching' in which the intersection line between the web and the solid portion is made hidden line and the alternate section lines in the web are continued through the rib, *Fig. 11.23 (c)*.

11.7. TREATMENT OF HIDDEN LINES

To improve the clarity of the drawing, hidden lines, if possible, should be omitted. In *Fig. 11.24 (a)* the shape of the object is completely clear from the upper sectional front view and the hidden lines are not required. While at (*b*) if the hidden lines

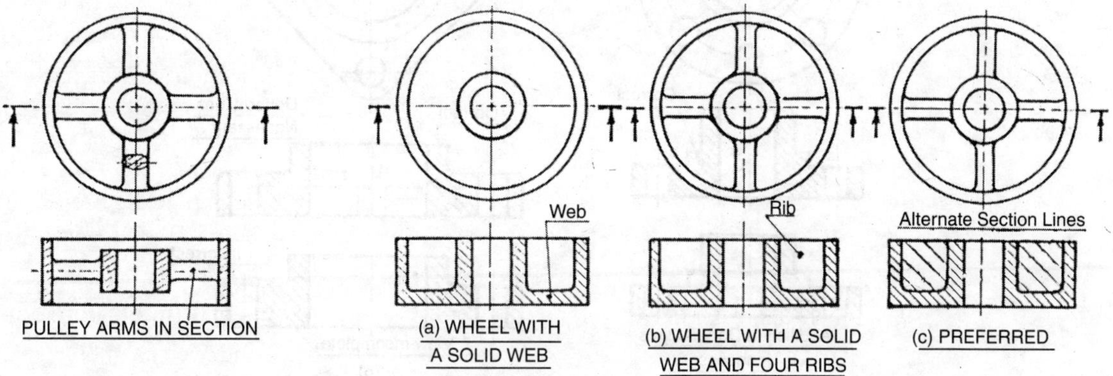

Fig. 11.20. Third Angle Projection

Fig. 11.21. Webs of wheels and Ribs in Section (Third Angle Projection) (a), (b), (c)

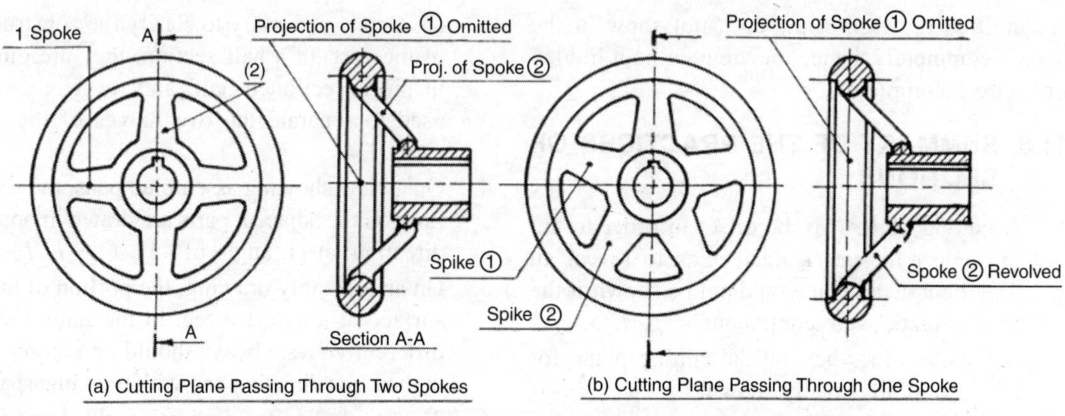

Fig. 11.22. Spoke of wheels in section (Third Angle Projection)

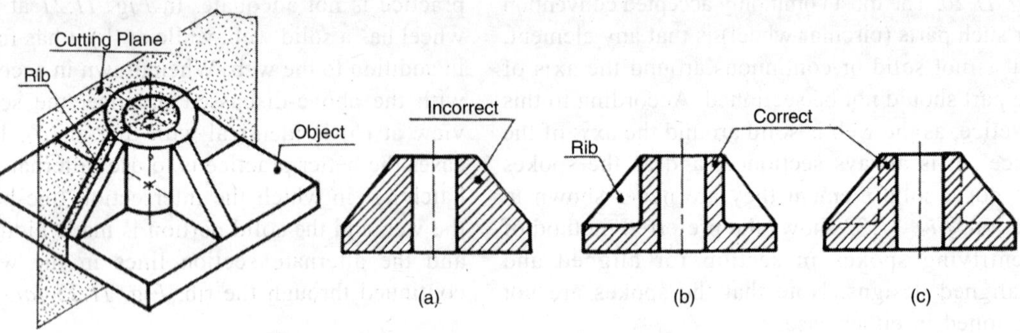

Fig. 11.23. Ribs and Lugs in Section

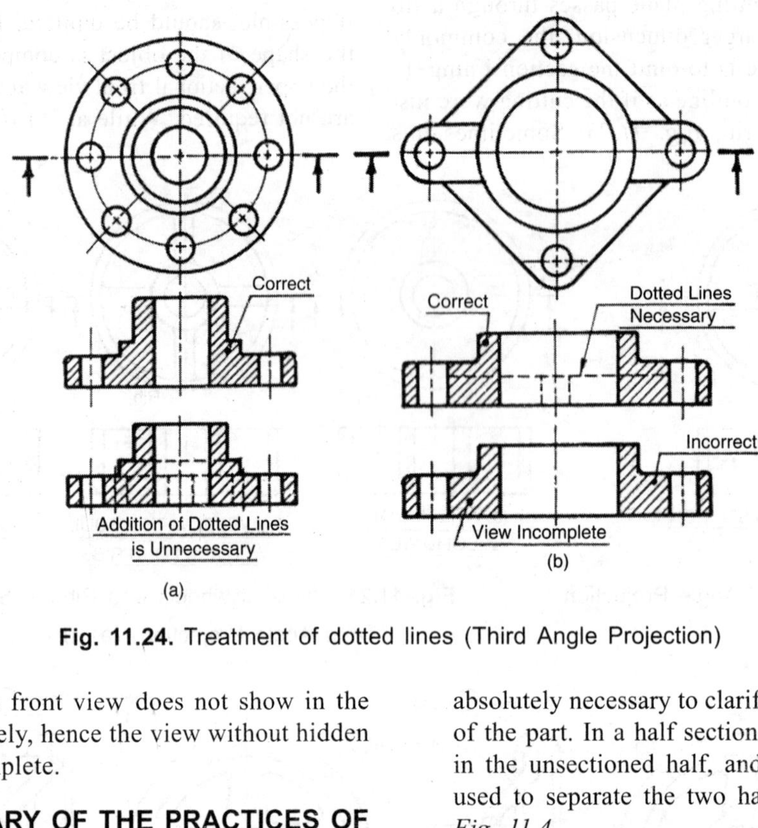

Fig. 11.24. Treatment of dotted lines (Third Angle Projection)

are omitted the front view does not show in the details completely, hence the view without hidden lines are incomplete.

11.8. SUMMARY OF THE PRACTICES OF SECTIONS

1. A cutting plane may be offset in order to cut the object in such a manner as to reveal an important detail that would not be shown if the cutting plane were continuous, *Fig. 11.5*.
2. All visible lines beyond the cutting plane for the section are usually shown.
3. Invisible lines beyond the cutting plane for the section are usually not shown, unless they are absolutely necessary to clarify the construction of the part. In a half section, they are omitted in the unsectioned half, and' a centre line is used to separate the two halves of the view, *Fig. 11.4*.
4. On a view showing assembled pans, the section lines on the adjacent parts are drawn in opposite direction, at an angle of 45°, *Fig. 11.16*.
5. On an assembly drawing, the portion of the cut surface of a single piece, in the same view or different views, always should be section-lined in the same direction and with the same spacing.
6. The symbolic line indicating the location of the cutting plane may be omitted if the location

of the plane is obvious, as for symmetrical objects.

7. On a sectioned view, showing assembled pieces, an exterior view is preferred for shafts, rods, nuts, rivets, keys, cotters, pins, bolts and so on, when the axes of these parts are in the plane of section, *Fig. 11.19 (a), (b).*

11.9. CONVENTIONAL PRACTICES

As already discussed in conventions for sections, in some objects, if the readability and clearness of a view can be improved by violating some principles of projection, these should be violated. In addition to this, to save time and effort in preparing drawings, some conventional symbols are used to convey the necessary information. These conventional practices and symbols are universally recognized and appear in the various codes of standards. Some of these, which are very common, are discussed in the following sections.

1. Half and Partial Views

When the space available is not enough to make the drawing to a suitable scale, it is allowed to make one of the top or side views as either a half or a partial view. If the front view is the exterior view the half view should be the front half of the top or the side view, *Fig. 11.25 (a)*. If the front view is a sectional view, then, rear half of the top or the side view should be drawn, *Fig. 11.25 (b), (c)*.

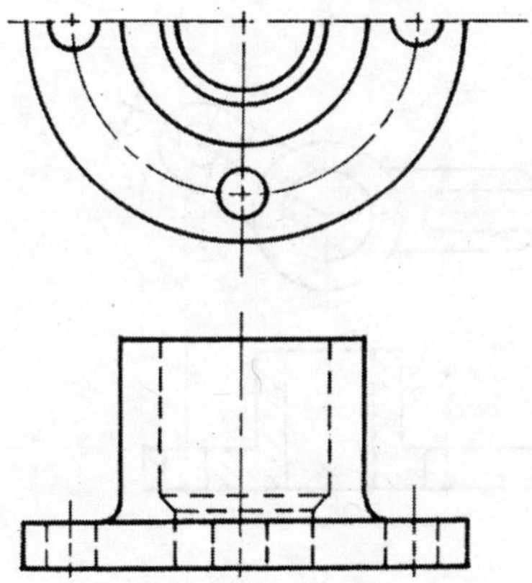

(a) Front View without Section

Fig. 11.25. ***(Contd.)***

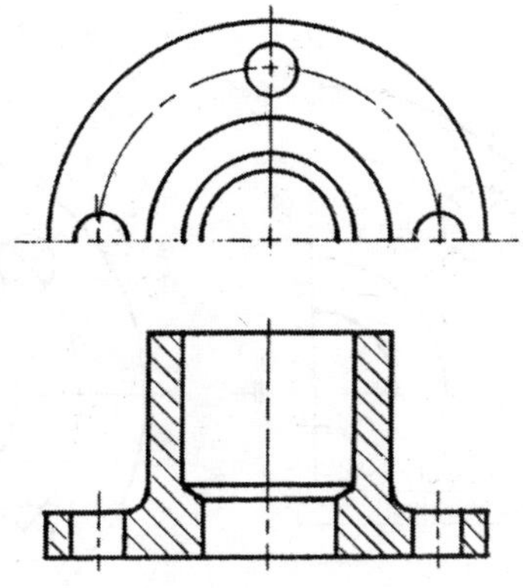

(b) Front View full Section

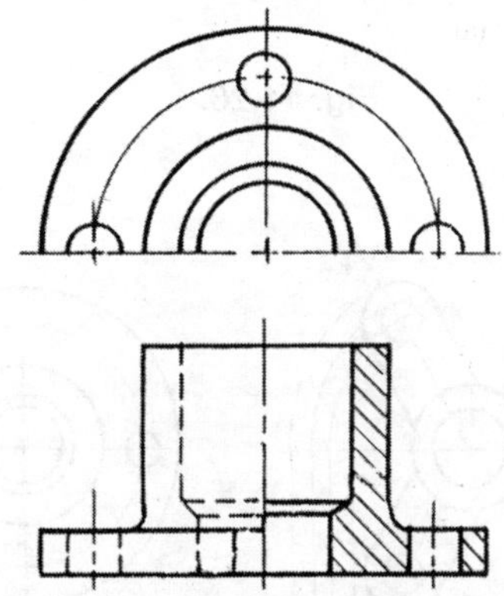

(c) Front View Half Section

Fig. 11.25. Half views (Third Angle Projection)

A partial view is one in which only part of an object is shown, *Fig. 11.26 (a)*. The line of symmetry is identified at its ends by two thin short parallel lines drawn at right angles to it. Another method is to show the lines representing the object extending a little beyond the line of symmetry, *Fig. 11.26 (b)*. In this case the short parallel lines may be omitted. However, sufficient information is given to complete the description of the object. Partial views are used because:

- They save time in drawing;
- They conserve space which might otherwise be needed for drawing the object;
- They sometimes permit the drawing to be made to a scale large enough to bring out all details clearly;
- If the part is symmetrical, a partial view (referred to as half, *Fig. 11.25*, or quarter view, *Fig. 11.26* (*a*) may be drawn on one side of the centre line.

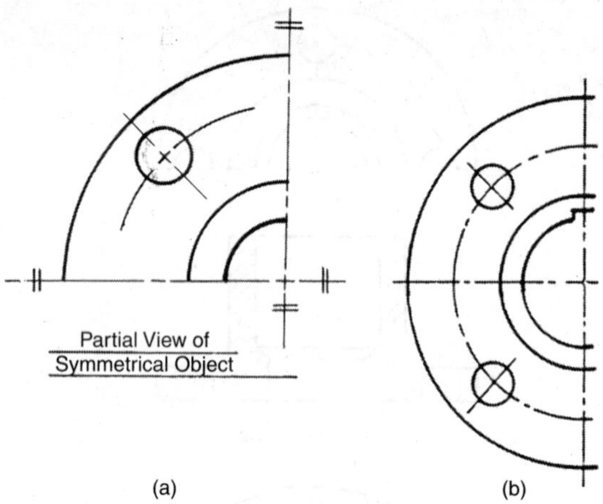

Fig. 11.26.

2. Alignment Of Parts And Holes

Two important factors which must be considered when drawing an object are the number of views to be drawn and the time required to draw them. Wherever possible attempt should be made to use time saving devices and less space for making the drawing, being sure to maintain consistency and clarity for easy reading. To simplify the representation of common features a number of conventional drawing practices are used, *Fig. 11.27 (a)* to *(g)*. Many conventions deviate from true position for the purpose of clarity; others are used to save drafting time. These conventions must be executed carefully, clarity is even more important than speed.

When the true projections of parts with inclined features, such as arms, ribs, and lugs, are drawn

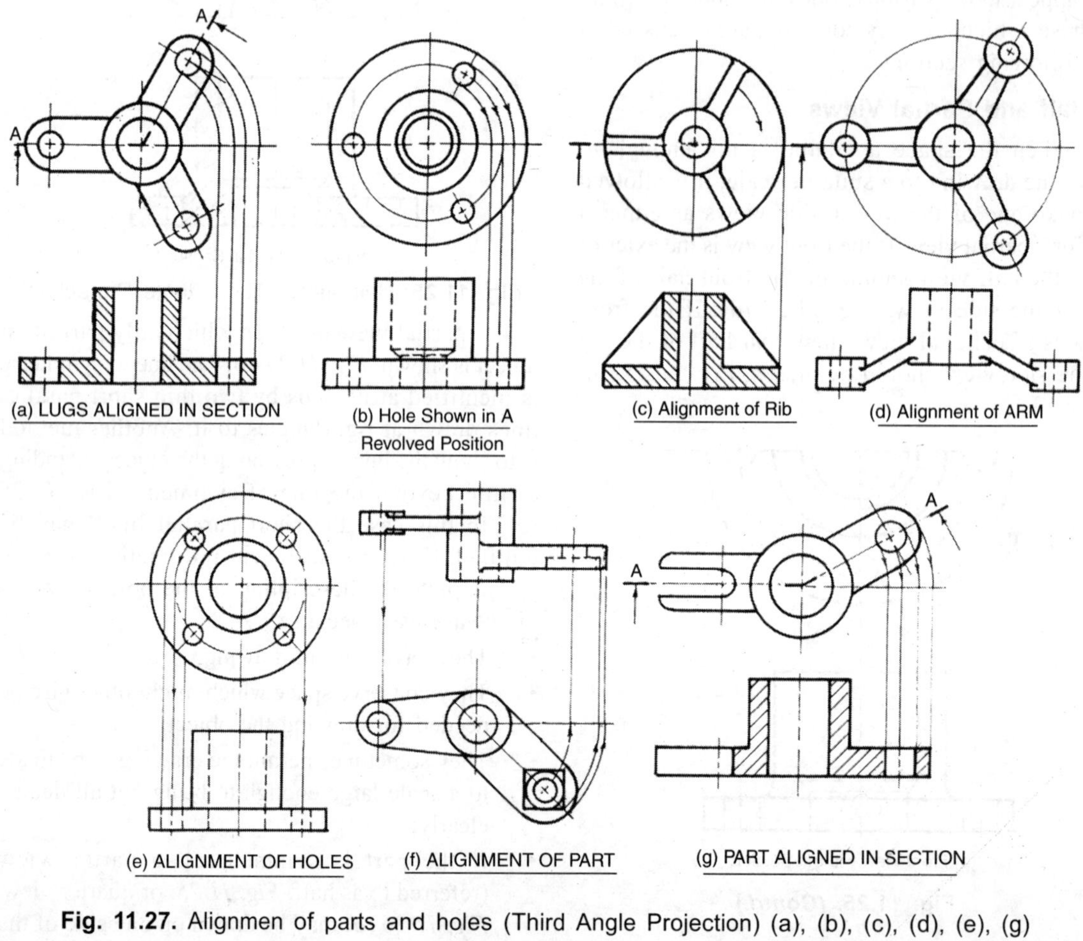

Fig. 11.27. Alignmen of parts and holes (Third Angle Projection) (a), (b), (c), (d), (e), (g)

the foreshortening of the inclined features gives a misleading appearance. In such cases the arms, ribs, and lugs are shown aligned or straightened out, *Fig. 11.27 (a), (c), (d), (f), (g)*, until they become parallel to the line of the section or projection. Drilled flanges in front view or section should show the holes at their true distance from centre, rather than the projection, *Fig. 11.27 (b), (e)*. On the same lines, in *Fig. 11.27 (c)* is shown the treatment of the radially arranged ribs.

3. Conventional Breaks

In making drawing of long parts, it is not preferred to draw the whole length of the member, if its cross-section is uniform. It can be shown to a larger scale by breaking and removing out a part, and moving the ends of the member closer. The conventional sections show the break in the member and its true length is given by giving its overall dimension. The conventional break, usually, is enough to give the complete description of the cross-section. The conventional breaks for some of the standard members are shown in *Fig. 11.28*.

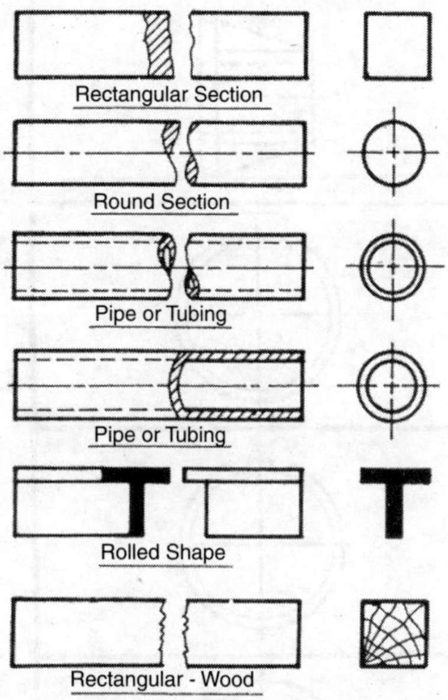

Fig. 11.28. Conventional Breaks

4. Interrupted Views

In order to save space it is permissible to show only those parts of a long object, which are sufficient for its definition. The limits of the parts retained are shown as for partial views, and the parts are drawn close to each other, *Fig. 11.29 (a), (b)*.

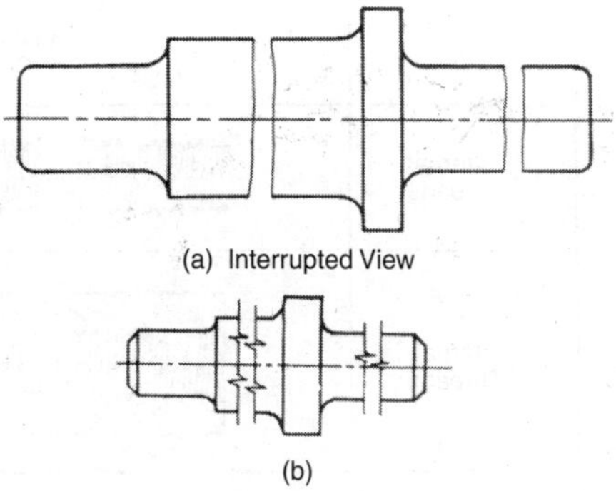

Fig. 11.29. Interrupted View

5. Space Between Views

Type	Convention	Material
Metals		Steel, cast iron copper and its alloys, aluminum and its alloys. etc.
		Lead, zinc, tin, white metal etc.
Glass		Glass
Packing and insulating internals		Porcelain stoneware, marble Slate, etc. asbestos, Fibre, felt. synthetic resin products, Paper, c or, Linoleum, rubber leather, wax, insulating and filling materials.
Liquid		Water, oil, petrol, kerosine, etc.
Wood		Wood, plywood etc.
Concrete		– –
Stacked Lamination		

Fig. 11.30. Conventional Representation of Common Material in section

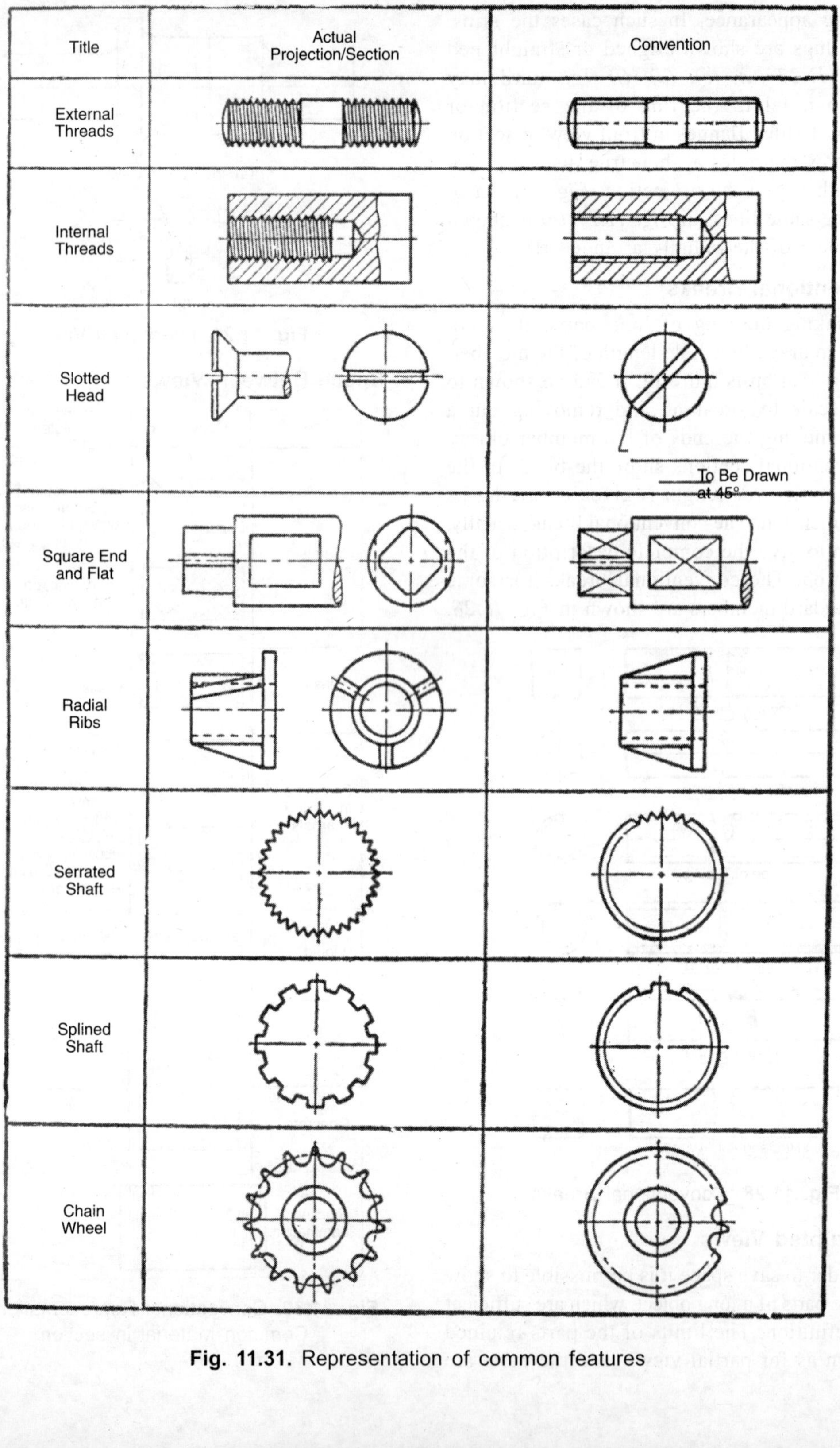

Fig. 11.31. Representation of common features

Title	Actual Projection/section	Convention
Ratchet and Pinion	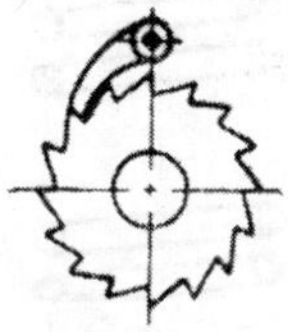	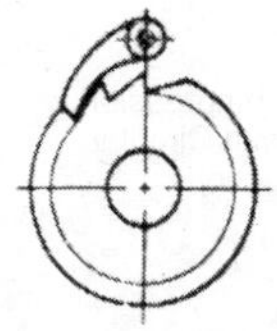
Bearings	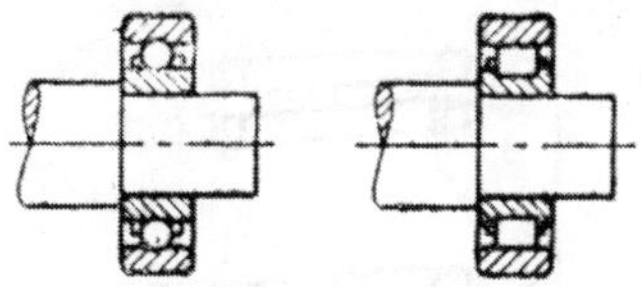	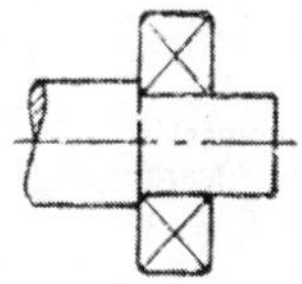
Straight Knurling	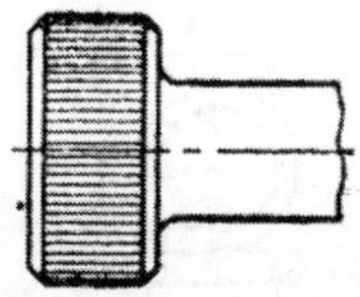	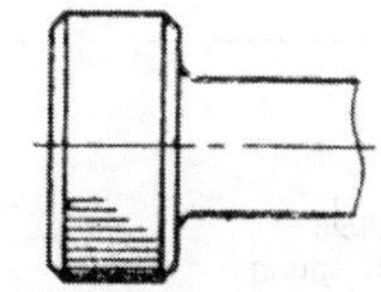
Diamond Knurling	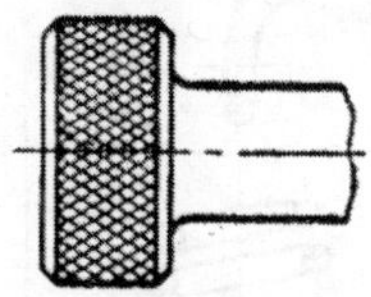	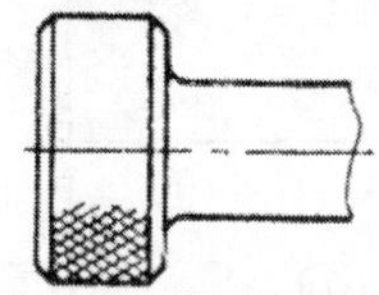
Holes on a Linear Pitch	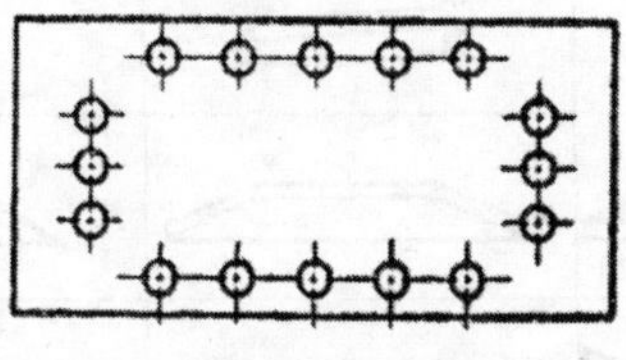	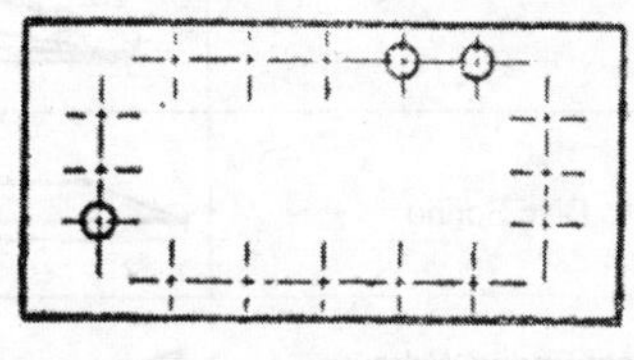
Holes on Circular Pitch	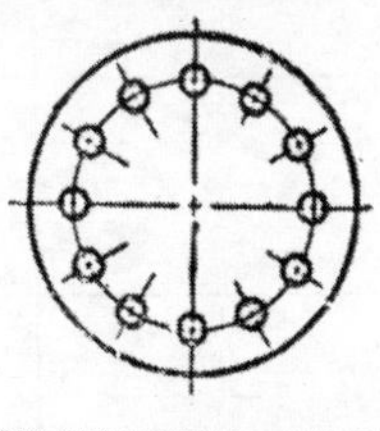	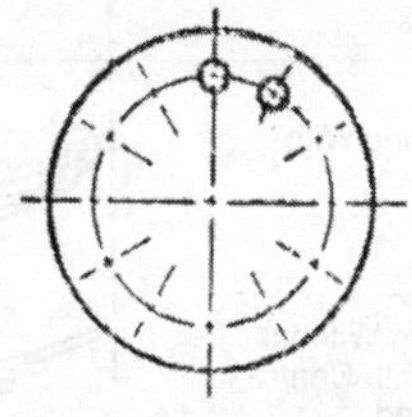
Repeated Parts	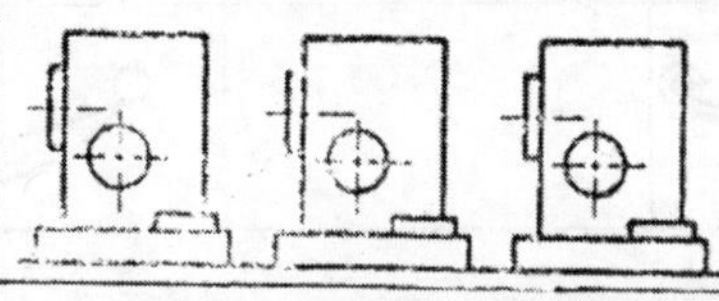	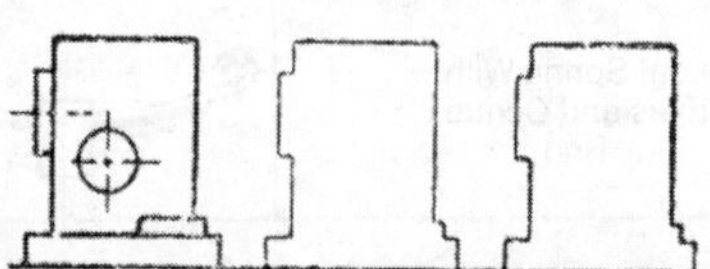

Fig. 11.32. Representation of common features

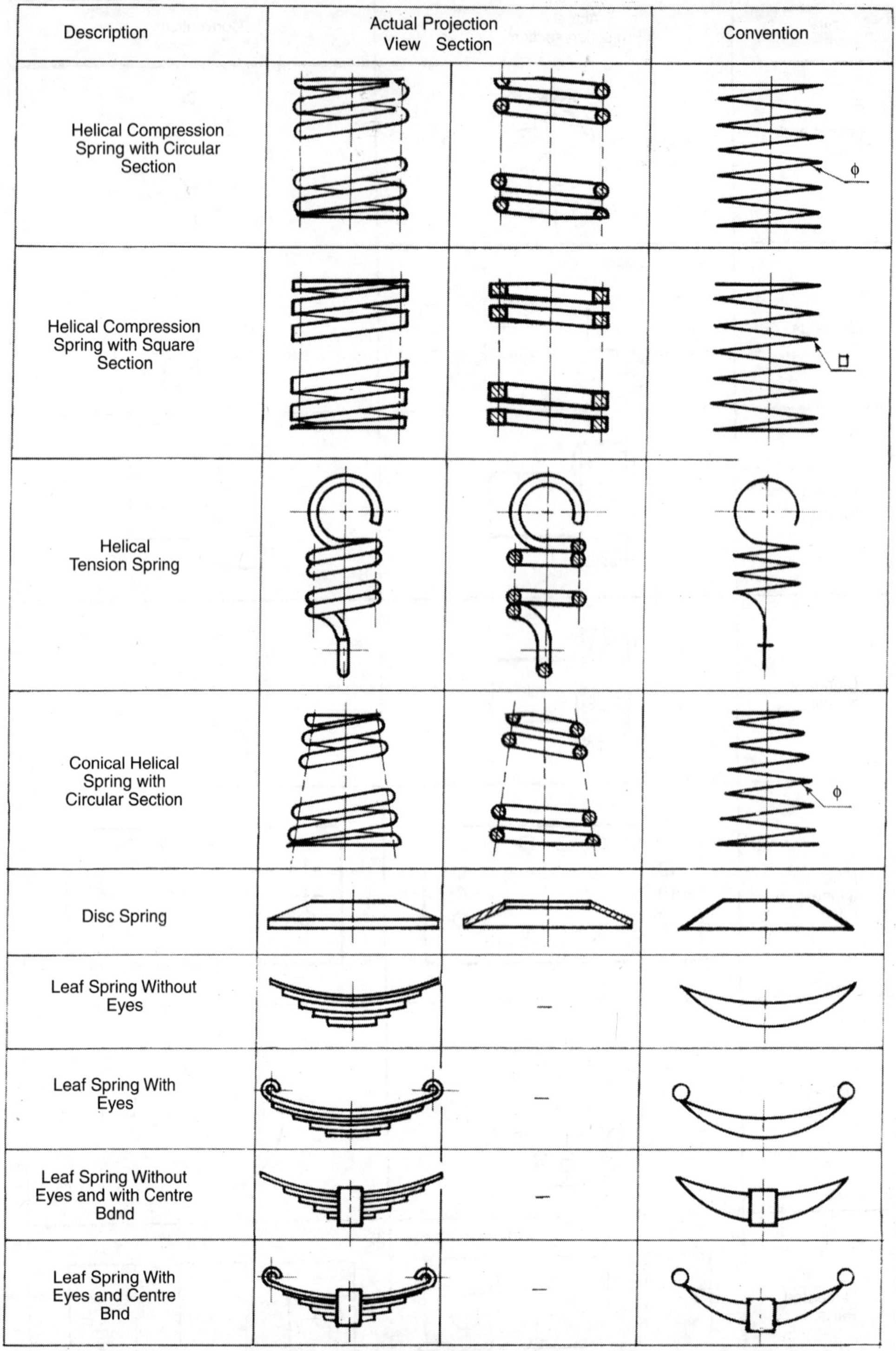

Fig. 11.33. Representation of springs

Sufficient space should be provided between views to permit placing of dimensions without crowing and to preclude the possibility of notes pertaining to one view overlapping or crowding the other views.

6. Representation Of Common Features

Conventional representation is adopted in cases where complete description of the part would involve unnecessary drawing time or space. Typical examples are shown in *Figs. 11.31*, *11.32* and *11.33*. Where the conventional representation given is not considered adequate, a more detailed view may be shown.

7. Representation of Common Materials

In view of the variety of materials used in manufacture and building of objects, it is desirable that different conventions of section Hanging, to differentiate between various materials, are made use of. Recommended conventions to indicate different materials in section are illustrated in *Fig. 11.30.*

8. Parts Located In Front of A Cutting Plane

When it is necessary to represent parts located in front of a cutting plane, these parts are represented in thin chain lines, *Fig. 11.34.*

9. Conventional Treatment of Intersections

Drawing the true section curve of two parts takes too much time and effort and adds little to the value of a drawing. When the intersections are not very important, the intersection curves may be ignored or replaced by a simple curve or straight line. *Fig. 11.35* illustrates the conventional treatment of the intersection of rectangular and circular parts with other circular parts. In *Fig. 11.35 (a)* and *(b)* the curves have been replaced by straight lines and in *Fig. 11.35 (c)*, as the curve is large, it is replaced by a circular arc. The conventional treatment of slots and holes in circular parts is illustrated in *Fig. 11.36 (a), (b)*.

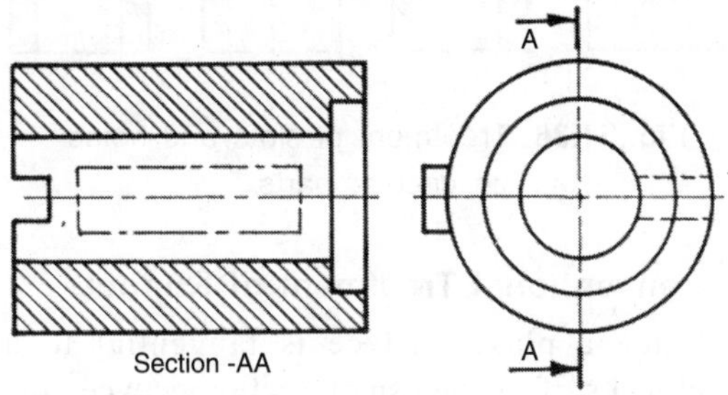

Fig. 11.34. Parts located on front of cutting plane

10. Conventional Treatment of Fillets and Rounds

Sharp corners are generally avoided in order to avoid stress concentration, to give better appearance, and for the safety of the person who handles the part. A rounded internal corner, rounded by filling in the material, is called *fillet* and a rounded external corner is called a *round.*

Though the fillets and rounds eliminate the intersection lines of intersecting surfaces, to improve the descriptive value of the view, it is a convention to represent these non-existing lines by using continuous thin lines. These lines are drawn through the intersection of surfaces, had the corner not been rounded, *Fig. 11.37.*

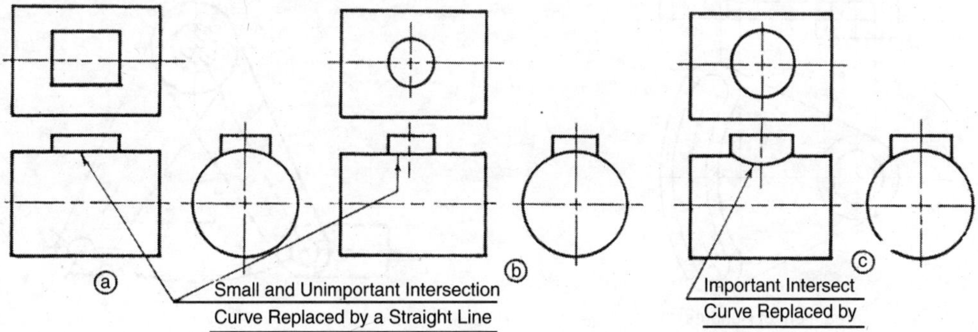

Fig. 11.35. Conventional treatment of intersections

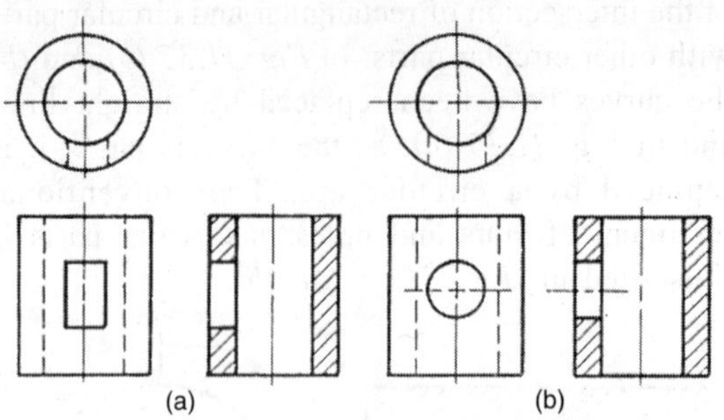

Fig. 11.36. Treatment of slots and holes in circular parts

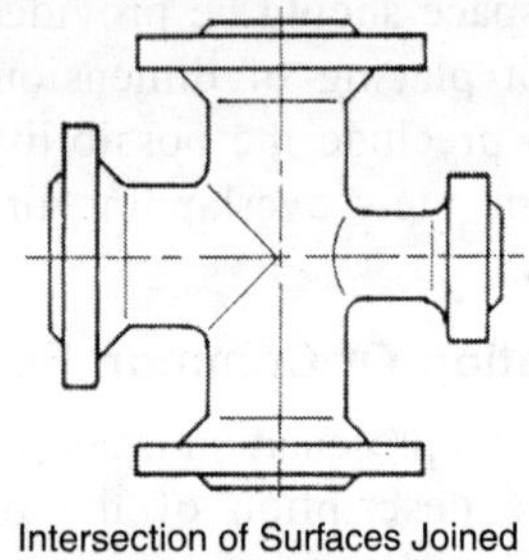

11.37. Conventional treatment of fillets and rounds

11. Conventional Treatment of Runouts

When a plane surface is tangential to a cylindrical surface, and sharp section between the two is avoided by filling in, the runout intersection of the fillet is represented as shown in *Fig. 11.38*. Runouts are usually drawn freehand, with a French curve, or with a bow compass.

12. Representation of Alternate Position of a Part

Figure 11.39 illustrates the procedure of showing the alternate position of a link. It is shown in thin dashed or phantom lines.

13. Representation of Adjacent Parts

The part, which actually does not form part of a drawing, but exists adjacent to the drawn part, on the machine, is shown in the drawing by thin lines, *Fig. 11.40*.

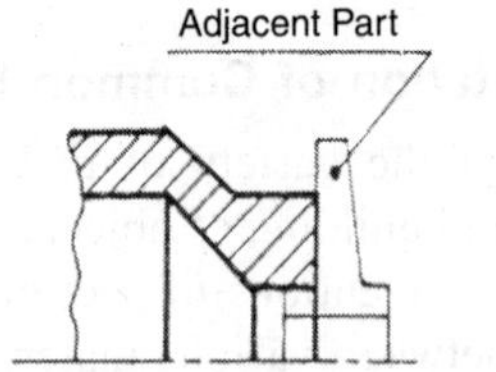

Fig. 11.40.

14. Representation pf Continuously Repeated Identical Details

To minimize labour and to save time, continuous thin lines are used to indicate the continuously repeated identical parts, shown in the Table giving conventional representation of common features, in *Fig. 11.32*.

15. Representation of Some Plane Faces

In order to avoid drawing a supplementary view, a tapered, a square, or a fiat on a cylindrical

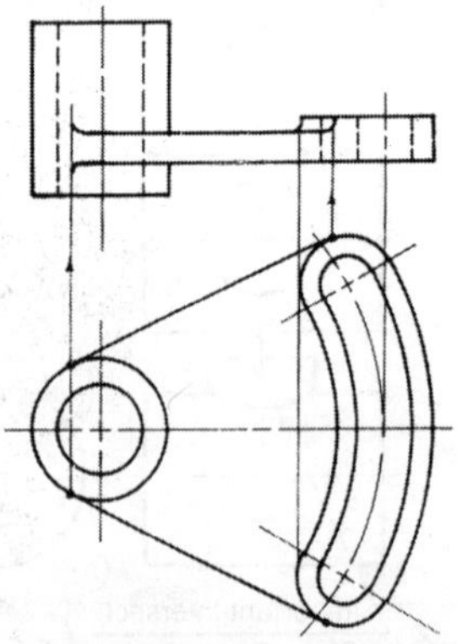

Fig. 11.38. Conventional treatment of run-out (First Angle Projection)

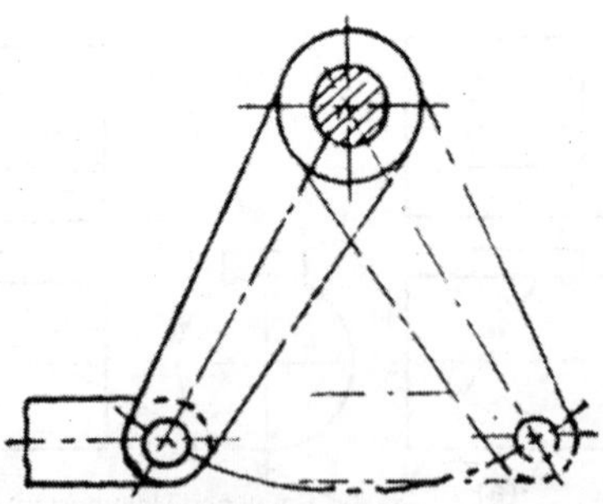

Fig. 11.39. Representation of alternate postion on a part

part may be indicated, as shown in *Fig. 11.41*, by the two diagonals drawn as continuous thin lines.

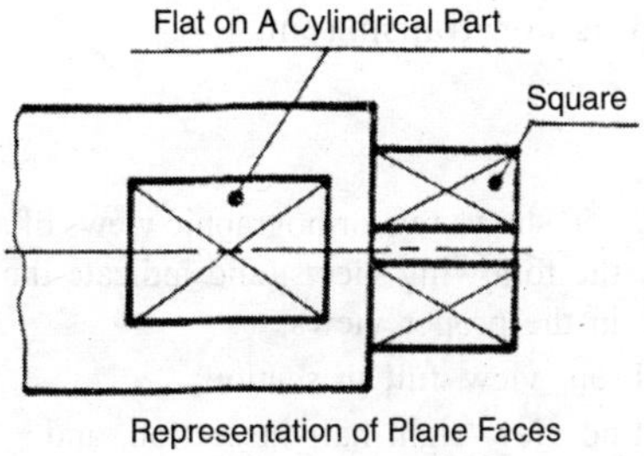

Fig. 11.41.

16. Arrangement of Successive Sections

Successive sections may be arranged in a manner similar to the examples shown in *Figs. 11.42, 11.43* as convenient for layout and understanding of the drawing.

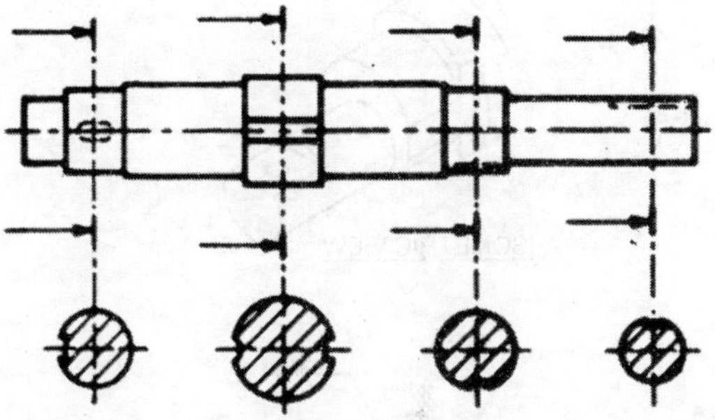

Fig. 11.42.

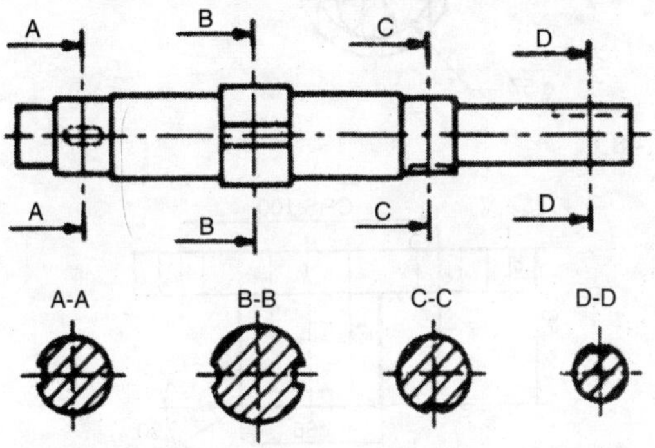

Fig. 11.43.

17. Local Views

Provided that the presentation is unambiguous it is permitted to give a local view instead of a complete view for symmetrical items. The local view should be drawn in third angle regardless of the arrangement used for the general execution of the drawing. Local views should be drawn with continuous thick lines (type *A*), and should be connected to the principal view by a centre line (type *G*), *Figs. 11.44* and *11.45*.

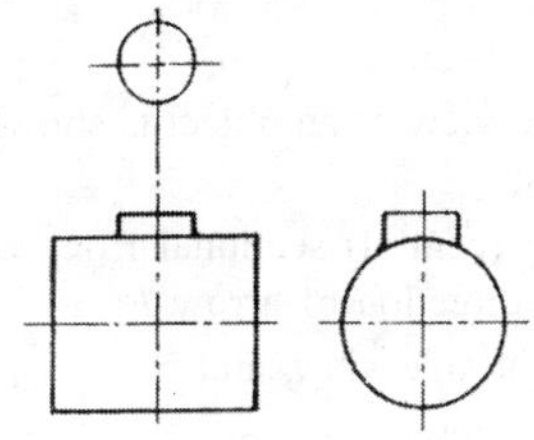

Fig. 11.44.

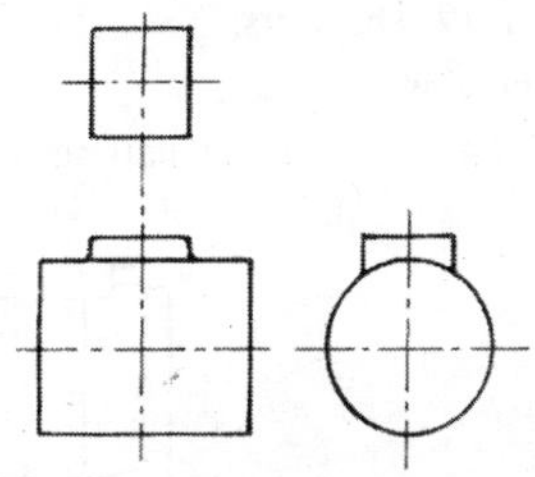

Fig. 11.45.

18. Elements On A Larger Scale

In cases where the scale is so small that details of a feature cannot be shown or dimensioned, the feature of the part may be framed by a continuous thin line (type B) and identified by a capital letter, *Fig. 11.46 (a)*. The relevant feature is then drawn to a stated larger scale accompanied by its identification letter, *Fig. 11.46 (b)*.

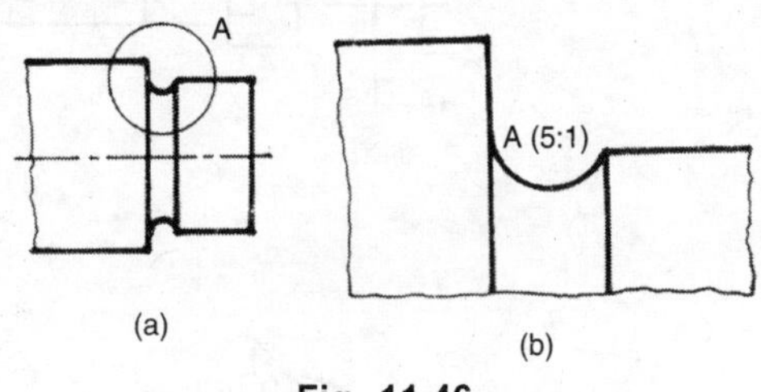

Fig. 11.46.

19. Initial Outlines

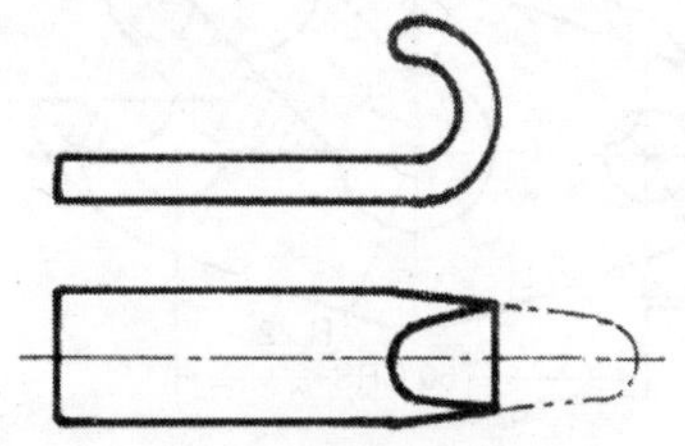

Fig. 11.47.

When it is necessary to depict the initial outline of a part, prior to forming, it should be indicated by chain, thin double dashed lines (*type K*). *Fig. 11.47.*

20. Transparent Objects

All objects made of transparent materials should be drawn as non-transparent.

EXERCISES

1. Isometric view of an object is shown in *Fig. 11.48.* Draw its;
   - Full (central) sectional front view, looking in the direction of arrow *F*;
   - Right side view, and
   - Top view.
2. A half sectional oblique view of a Gland is shown in *Fig. 11.49.* Draw its;
   - Front view;
   - Left end view–upper half in section.
3. *Fig. 11.50* shows two orthographic views of a Gland. Draw the following views, and indicate the cutting plane in the proper views;
   - Front view–full in section;
   - End view–right half in section, and
   - Top view.
4. In *Fig. 11.51* is given a piece of Brasses. Draw its sectional front view, end view, and outside view.
5. A half sectional pictorial view of a Gland is shown in *Fig. 11.52.* Draw its half sectional front view,

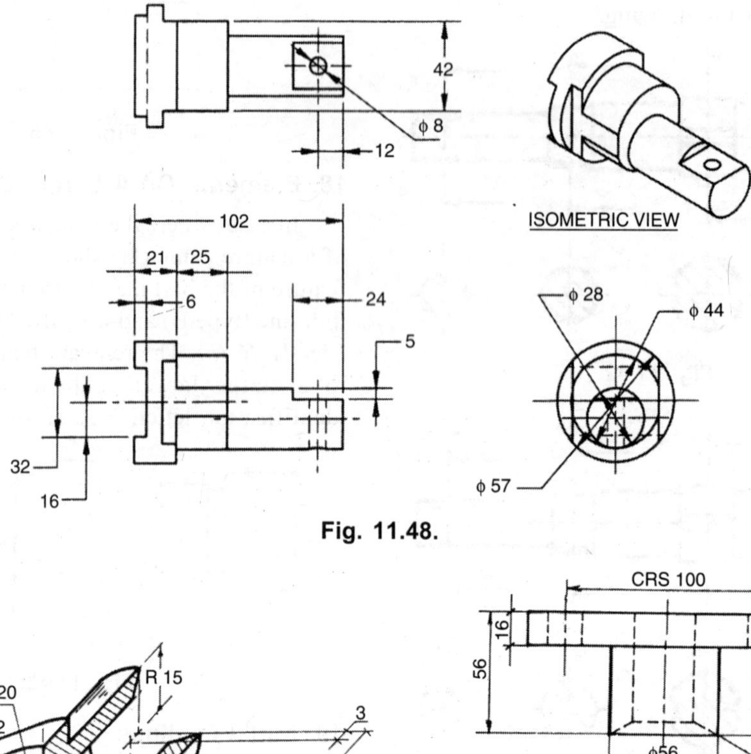

Fig. 11.48.

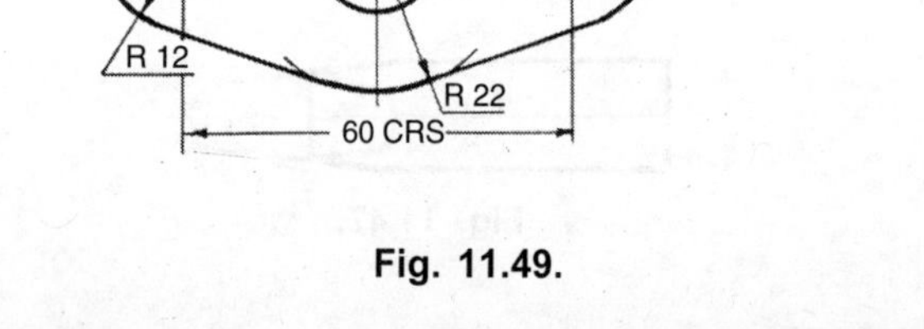

Fig. 11.49.

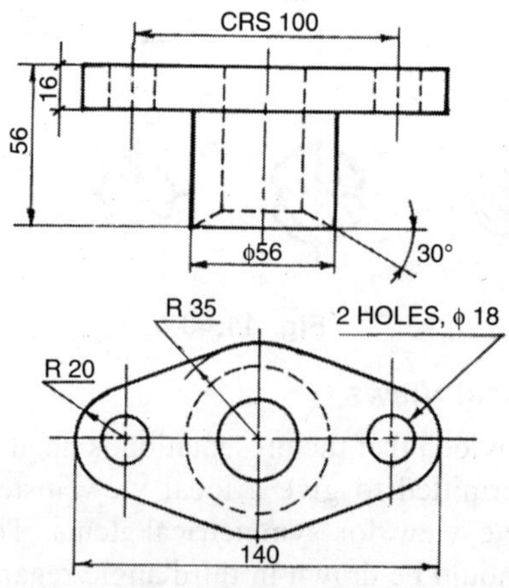

Fig. 11.50. Gland

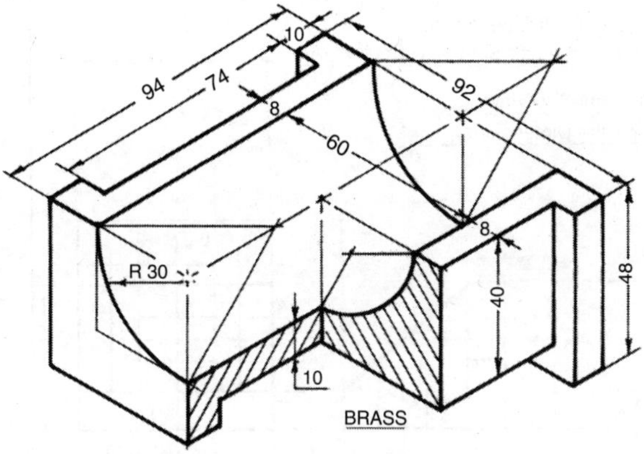

Fig. 11.51.

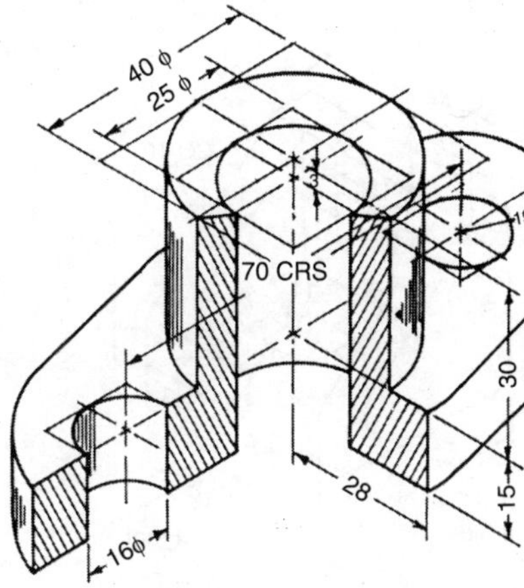

Fig. 11.52. Gland

side view looking from the left hand side, and top view.

6. Half sectional pictorial (isometric) view of a Bearing Body is shown in *Fig. 11.53*. Draw its following views;
   - Front view–right half in section,
   - Left end view–left half in section, and
   - Top view–outside.
7. From the drawing of a casting, shown in Fig. 11.54, draw the following views;
   - Front view–right half in section,
   - Full sectional end view, and
   - Top view–outside.
8. From the orthographic views of a Simple Journal Bearing, shown in *Fig. 11.55*, draw;
   - Front view–left half in section,
   - Full sectional end view, and
   - Top view–outside.
9. Two views of a Bearing Bracket are shown in *Fig. 11.56*. Reproduce its left side view and draw the front view-full in section at XX, and the top view-outside.
10. *Fig. 11.57* shows two views of a Bracket. Draw its;
    - Front view–outside,
    - Full sectional right end view–looking at A-A, and
    - Top view–outside.
11. *Fig. 11.58* shows orthographic projections of a "Support Bracket". Draw its;
    - Full sectional front view–section A-A,

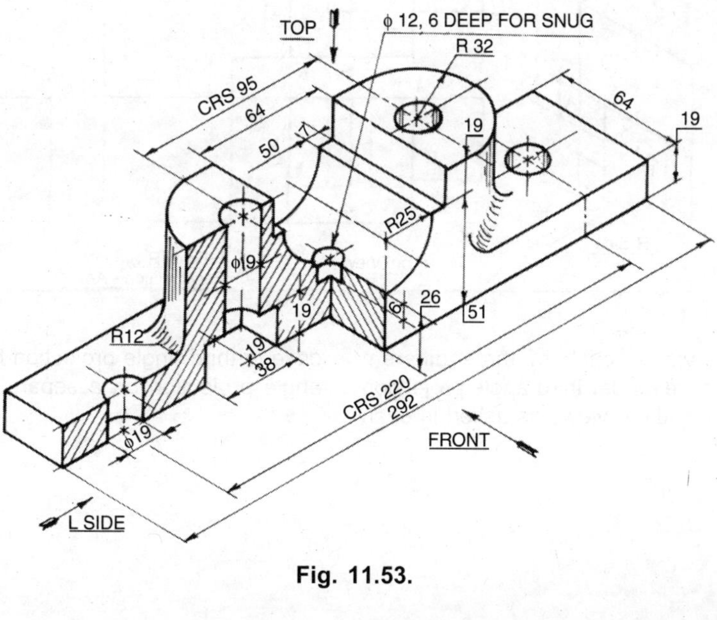

Fig. 11.53.

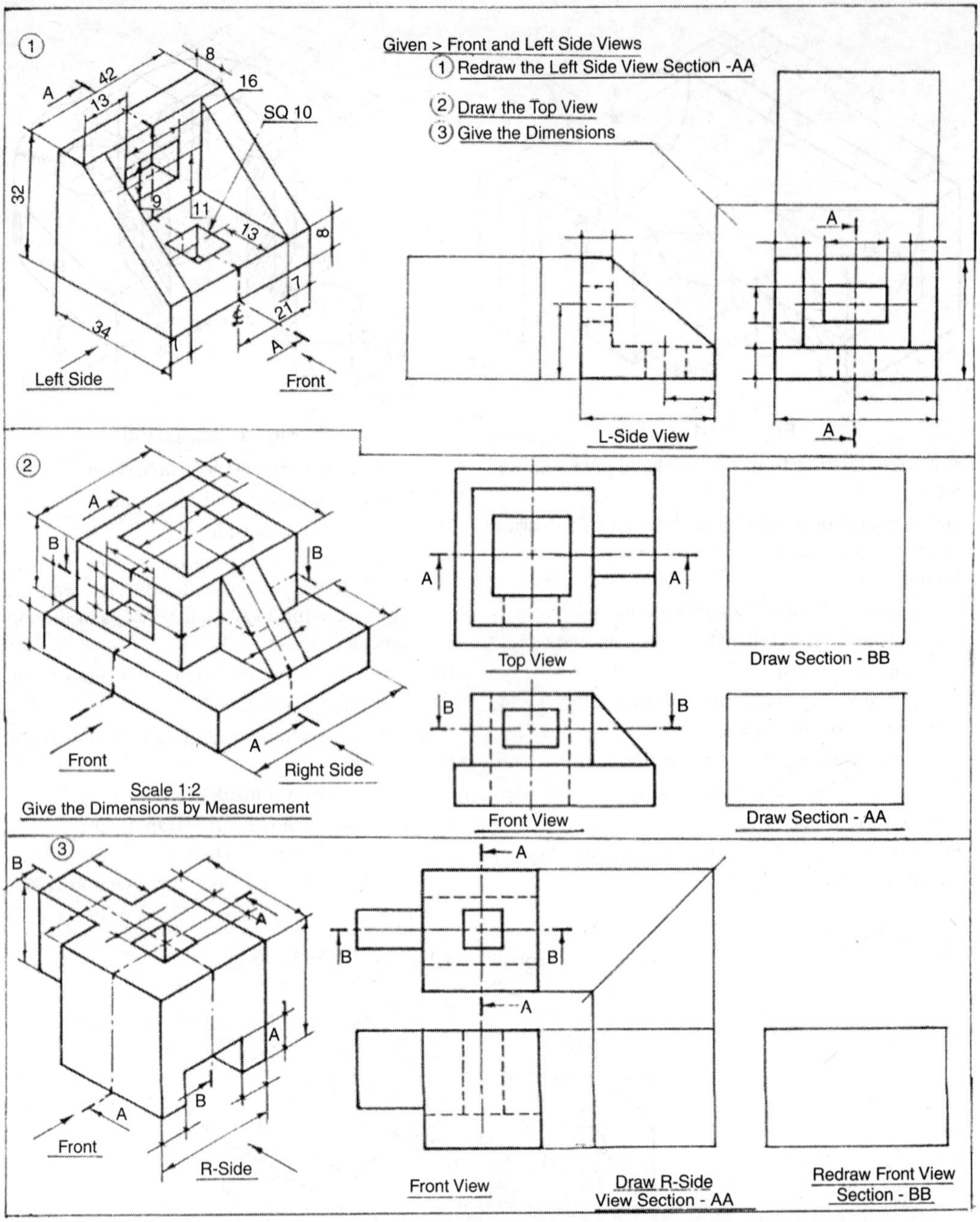

Two orthographic views. each of three different objects are given here as per third angle projection method. Draw the required views as asked in each case in third angle projection here itself and in first angle projection on a separate drawing sheet.

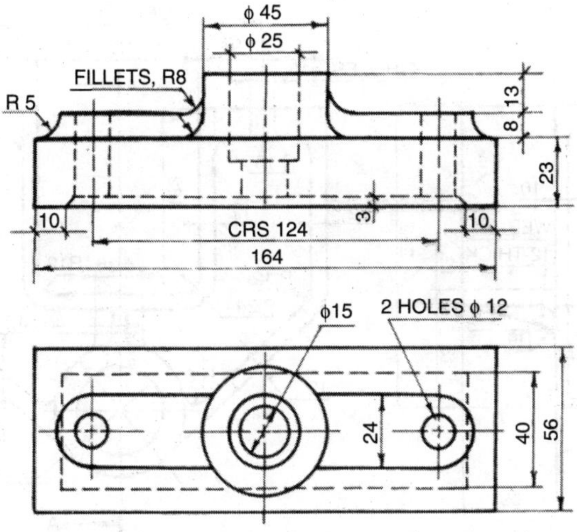

Fig. 11.54. Casting (First Angle Projection)

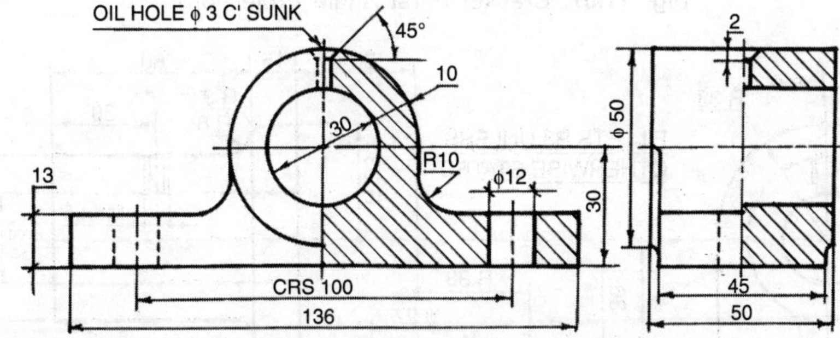

Fig. 11.55. Simple journal bearing

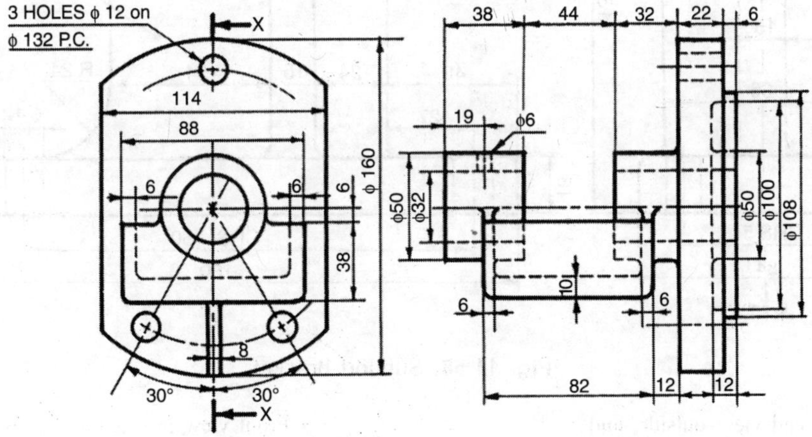

Fig. 11.56. Bearing bracket (Third Angle Projection)

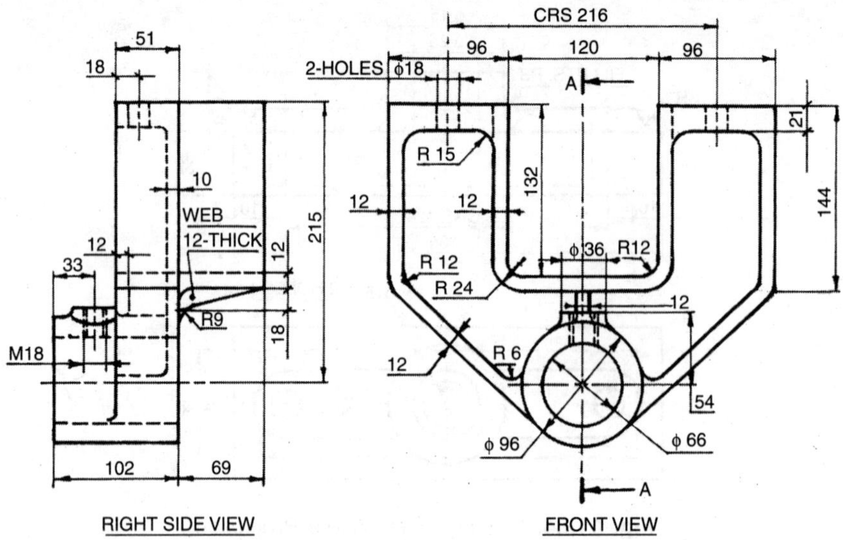

Fig. 11.57. Bracket (First Angle Projection)

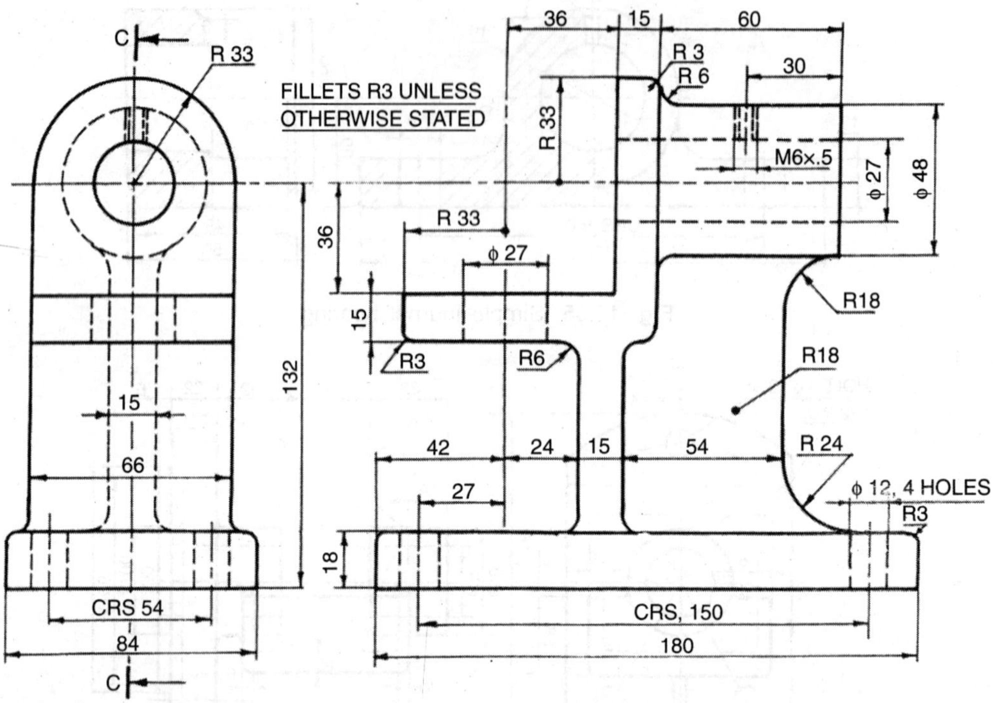

Fig. 11.58. Support bracket

- Right end view–outside, and
- Top view–outside.

12. *Fig. 11.59* shows the isometric view of an object. Using full size scale, draw its;

- Front view in section at A-A,
- Side view looking from left side, and
- Top view.

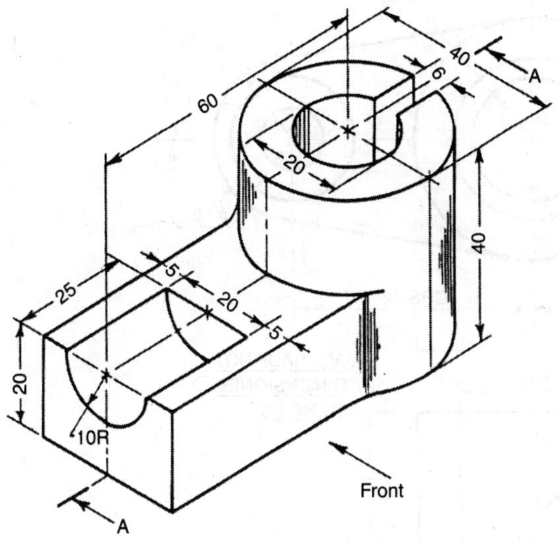

Fig. 11.59.

13. Draw the following views of an object, shown in *Fig. 11.60*;
    - Front view full in section at A-A,
    - Side view looking from left, and
    - Top view.

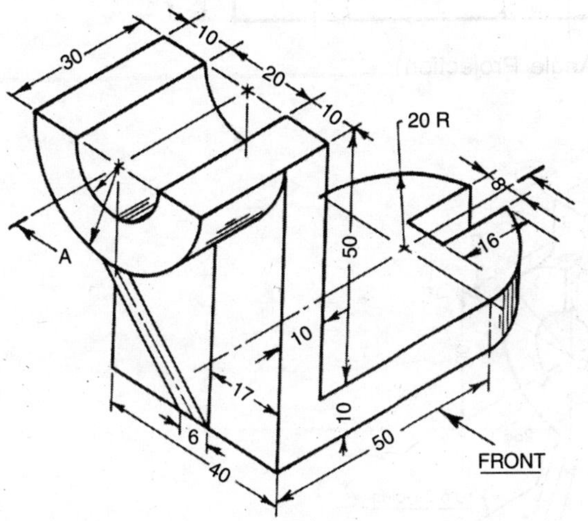

Fig. 11.60.

14. *Fig. 11.61* shows two views of an object. Draw its;
    - Top view,
    - Full sectional front view, looking at A-A, and
    - Left end view.

15. Draw the following views of a Bearing, shown in *Fig. 11.62*;
    - Front view–full in section at A-A,
    - Top view, in section at C-C.

16. Two views of a Bearing Mounting are shown in *Fig. 11.63*. Reproduce the top view. Draw its front view–full in section at X-X. Also add the leftside view–outside.

    Front view–full in section at X-X. Also add the left side view–outside.

17. Three views of a Guide Bracket are shown in *Fig. 24.64*. Draw its;
    - front view full in section at A-A,
    - left side view, and
    - top view–in section at C-C.

18. Two views of a Lever Casting are shown in *Fig. 24.65*. Draw its;
    - front view–full in section,
    - left side view–in section at A-A, and
    - top view–outside.

19. Two views of a Feed Shaft Bracket are shown in *Fig. 24.66*. Draw its;
    - front view–in section at A-A,
    - right side view–in section at C-C, and
    - top view–outside.

20. Two orthographic views, each of three different objects, are given in *Fig. 24.67*, in third angle projection. Draw the required sectional views as labeled. You may also practice drawing these sectional views in first angle projection, on a separate drawing sheet.

21. Four different objects are shown in isometric views in *Fig. 24.68*. Draw the required views for each one of them, in either the first or the third angle projection method.

22. Draw the required sectional views for each of the six objects, for each one of which two outside orthographic views are given in *Fig. 24.69*.

23. Two orthographic views of each of the six different objects are shown-in *Fig. 24.70*. Draw the required views, as labelled, for each one of them. Use third angle projection method.

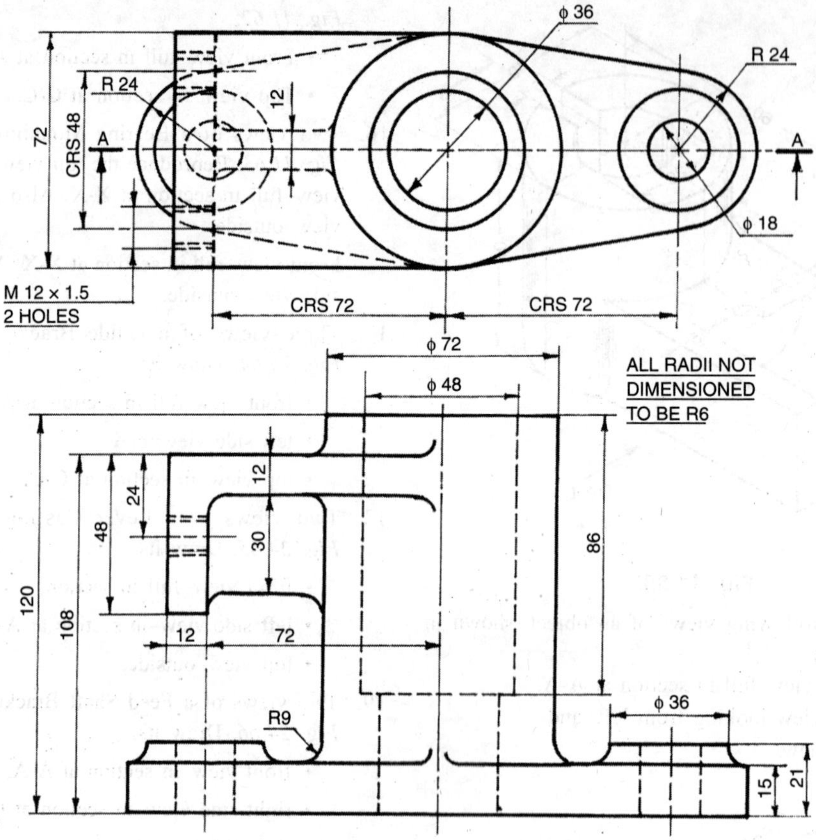

Fig. 11.61. (First Angle Projection)

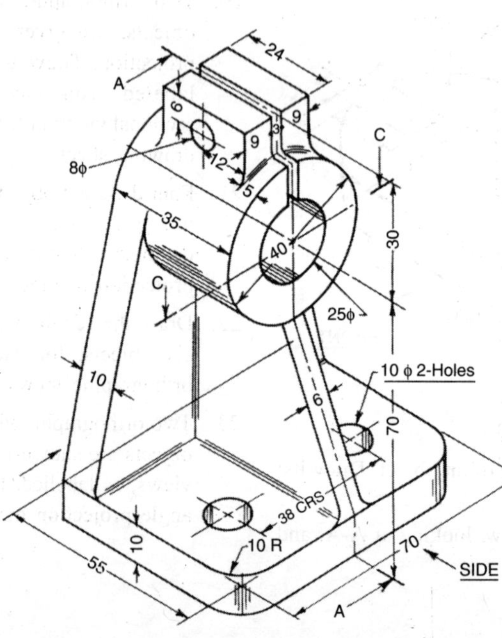

Fig. 11.62. Bearing

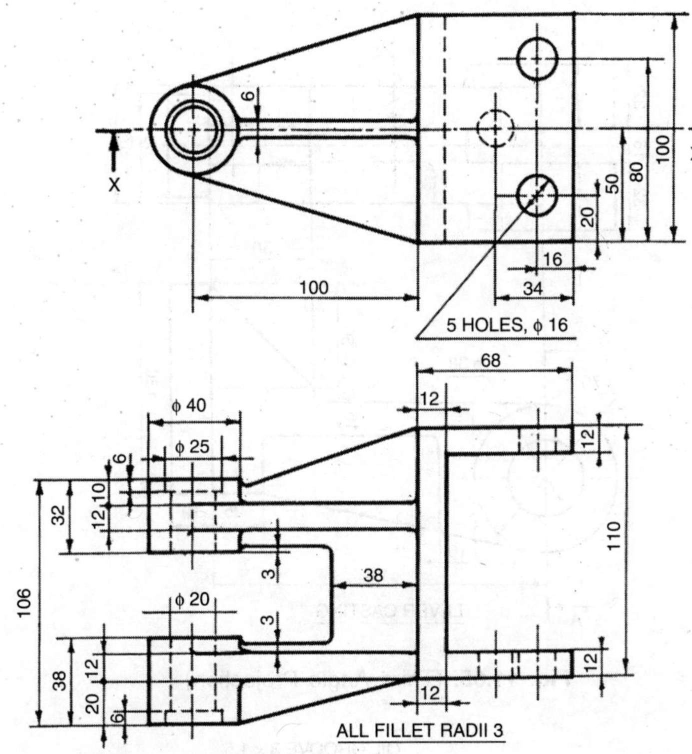

Fig. 11.63. Bearing Mounting (Third Angle Projection)

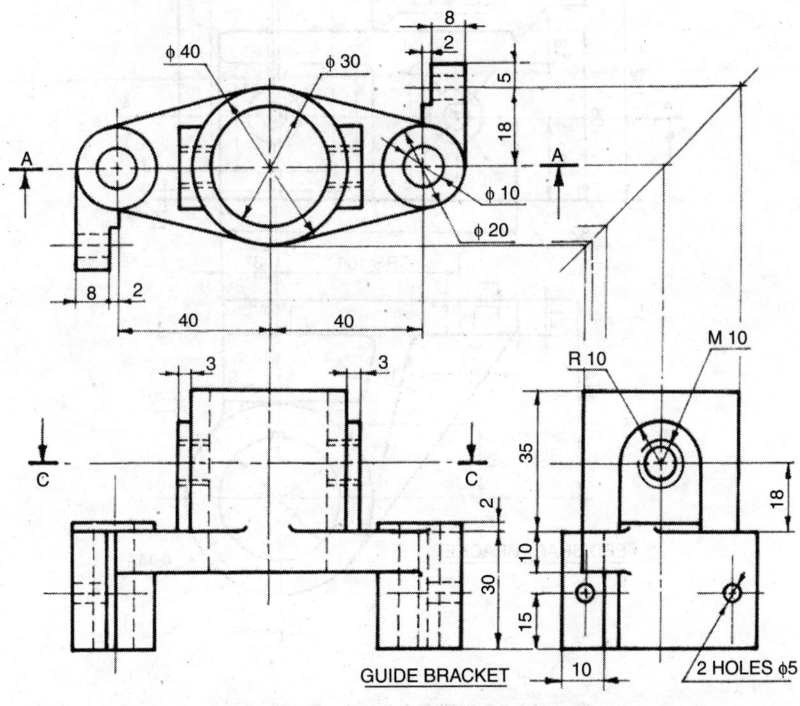

Fig. 11.64. (Third Angle Projection)

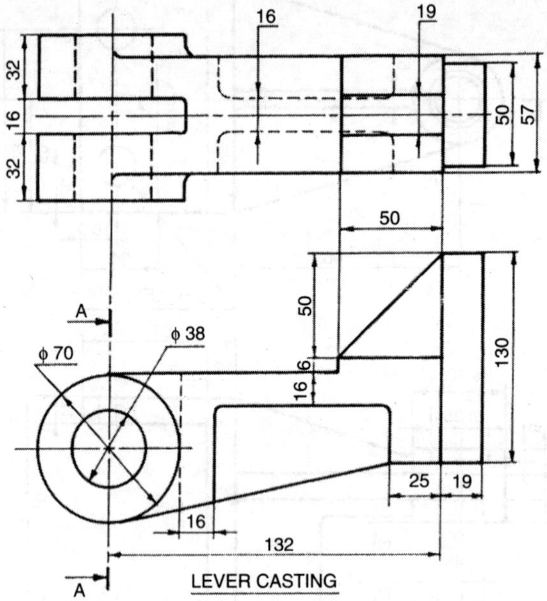

Fig. 11.65. (Third Angle Projection)

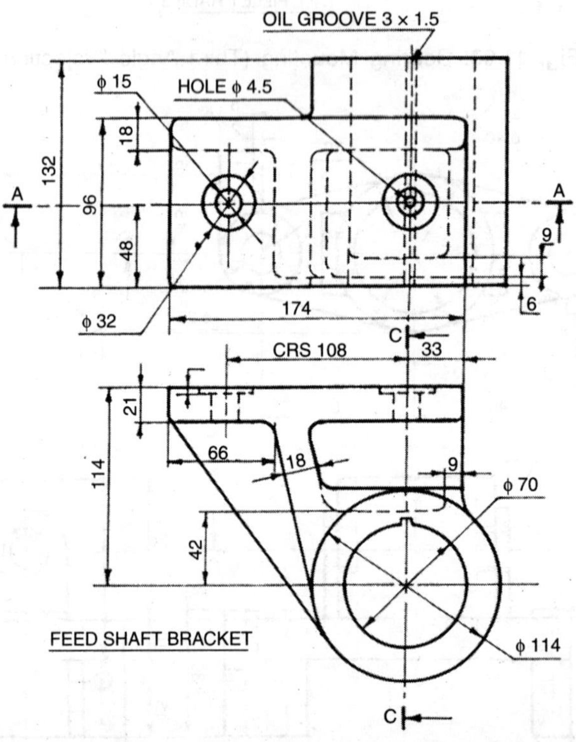

Fig. 11.66. (Third Angle Projection)

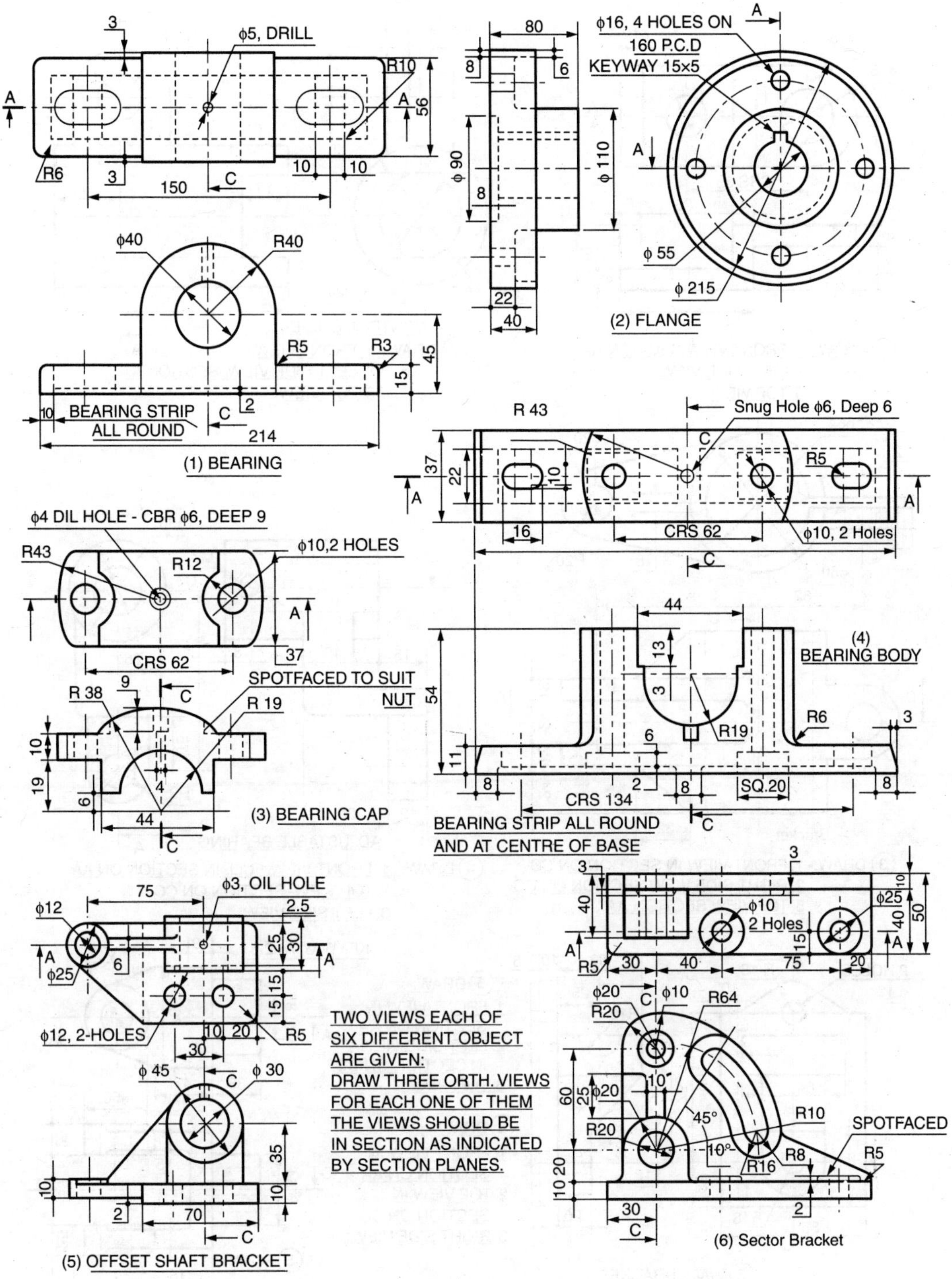
(1) BEARING
(2) FLANGE
(3) BEARING CAP
(4) BEARING BODY
(5) OFFSET SHAFT BRACKET
(6) Sector Bracket
TWO VIEWS EACH OF
SIX DIFFERENT OBJECT
ARE GIVEN:
DRAW THREE ORTH. VIEWS
FOR EACH ONE OF THEM
THE VIEWS SHOULD BE
IN SECTION AS INDICATED
BY SECTION PLANES.

Fig. 11.67.

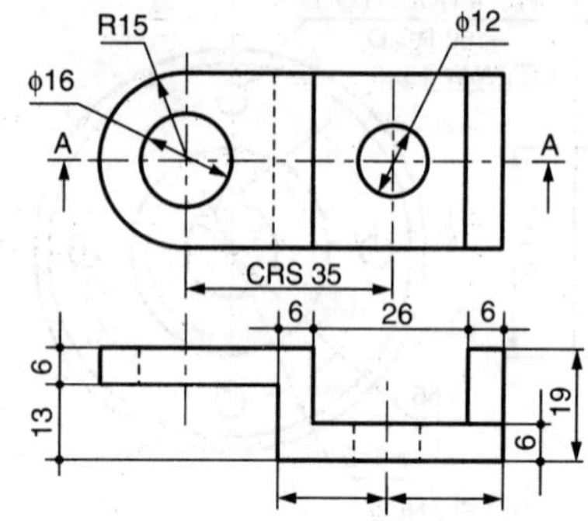

(1) DRAW:- 1. FRONT VIEW SECTION AA
2. LEFT SIDE VIEW
3. TOP VIEW

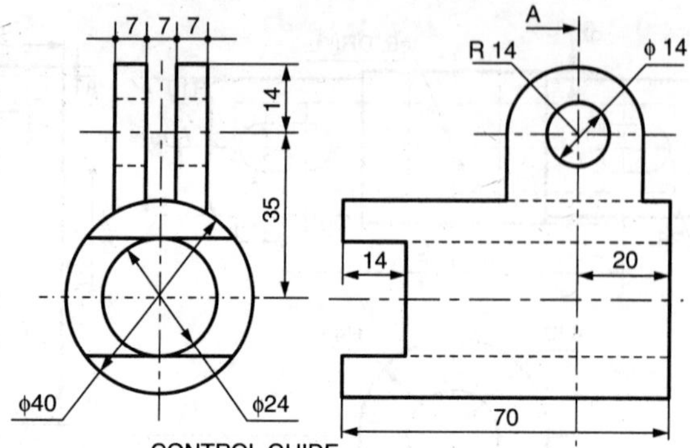

CONTROL GUIDE

(2) DRAW:- 1. FRONT VIEW
2. LEFT SIDE VIEW, SECTION AA
3. TOP VIEW

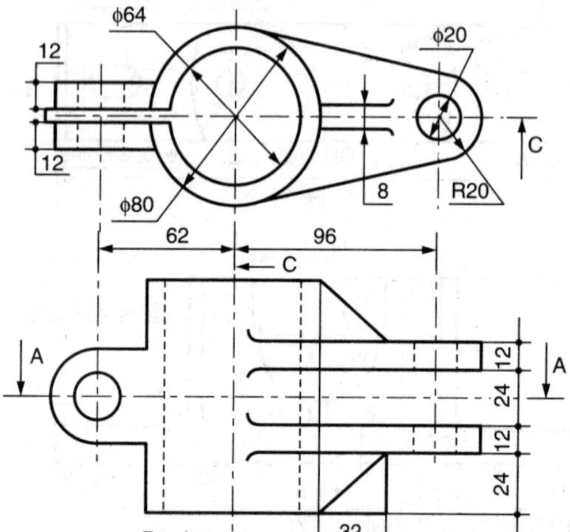

Bracket

(3) DRAW:- 1. FRONT VIEW IN SECTION ON CC
2. RIGHT SIDE VIEW SECTION ON CC.
3. TOP VIEW SECTION AA.

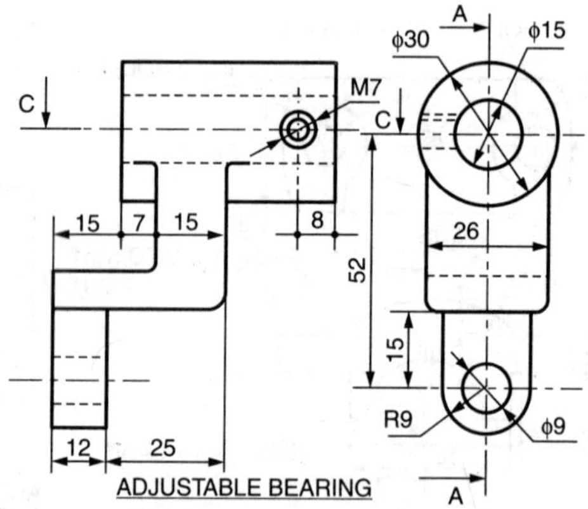

ADJUSTABLE BEARING

(4) DRAW:- 1. FRONT VIEW FULL IN SECTION ON AA
2. TOP VIEW–SECTION ON CC
3. LEFT SIDE VIEW

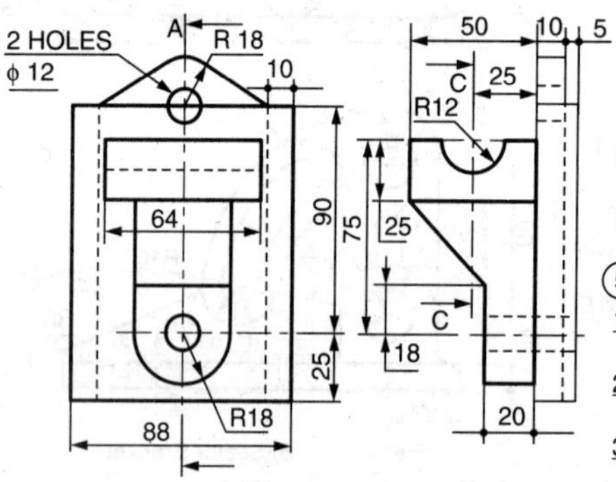

(5) WALL BRACKET

(5) DRAW:-
1. FRONT VIEW IN SECTION ON AA
2. LEFT SIDE VIEW IN SECTION ON CC.
3. TOP VIEW

(5) DRAW:-
1. FRONT VIEW IN SECTION ON AA
2. TOP VIEW IN SECTION ON CC.
3. RIGHT SIDE VIEW

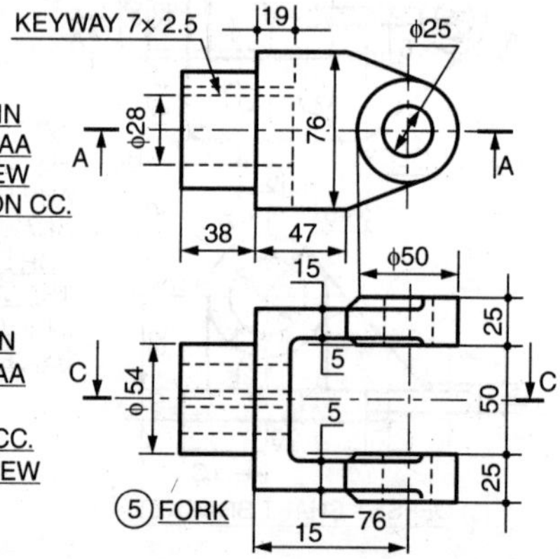

(5) FORK

Fig. 11.68.

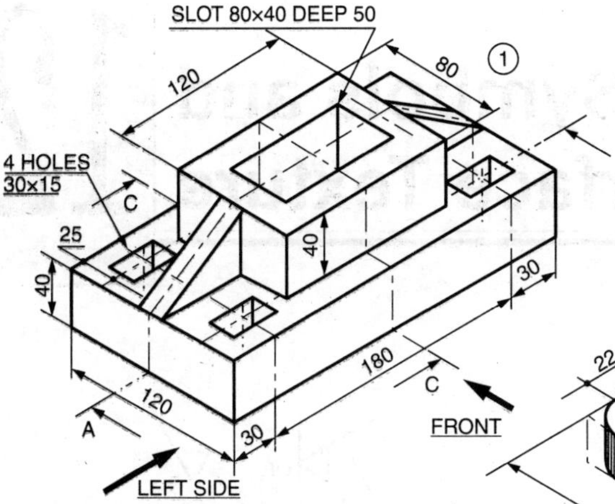

DRAW:- 1. FRONT VIEW SECTION AA.
2. LEFT SIDE VIEW SECTION CC.
3. TOP VIEW ALSO DIMENSION THE VIEWS.

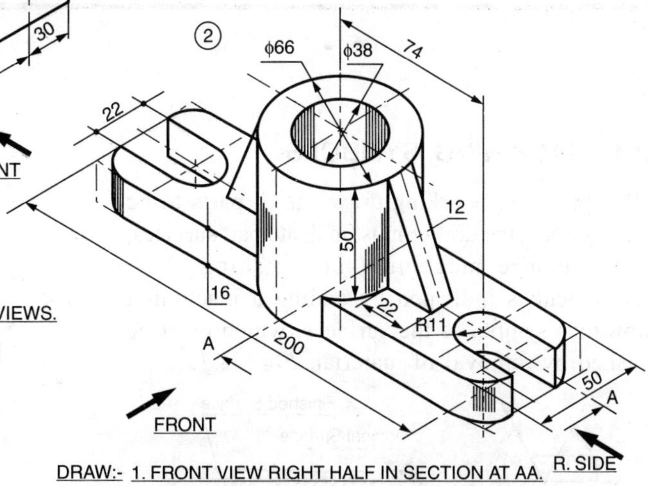

DRAW:- 1. FRONT VIEW RIGHT HALF IN SECTION AT AA.
2. RIGHT SIDE VIEW, LEFT HALF IN SECTION
3. TOP VIEW

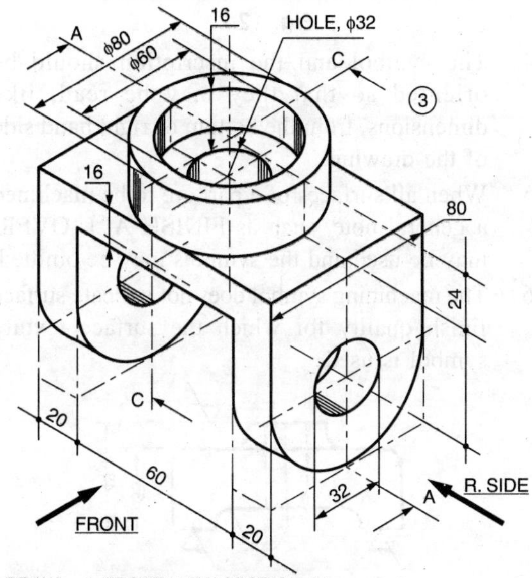

DRAW:- 1. FRONT VIEW SECTION AA.
2. R. SIDE VIEW SECTION CC.
3. TOP VIEW. ALSO DIMENSION THE VIEWS

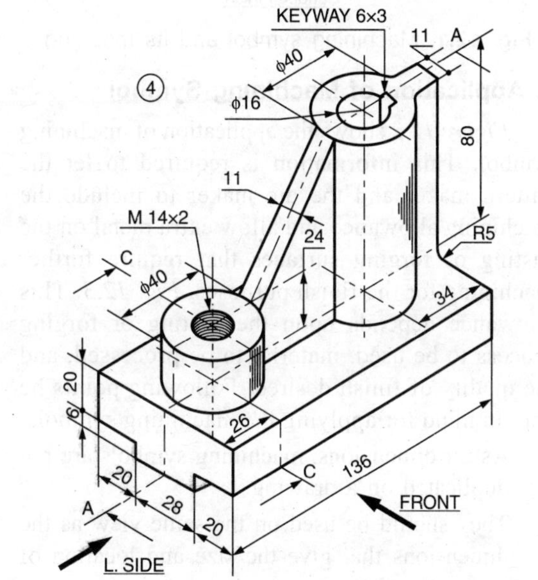

DRAW:- 1. FRONT VIEW, SECTION AA.
2. LEFT SIDE VIEW SECTION CC.
3. TOP VIEW.

Fig. 11.69.

12 Machining Symbols and Surface Texture

12.1. MACHINING SYMBOLS

When preparing working drawings of parts to be cast or *forged*, the drafter must indicate part surfaces, which require machining or finishing, This requirement is indicated by adding a machining allowance symbol to the surfaces, which must be finished by removal of material, *Fig. 12.1.*

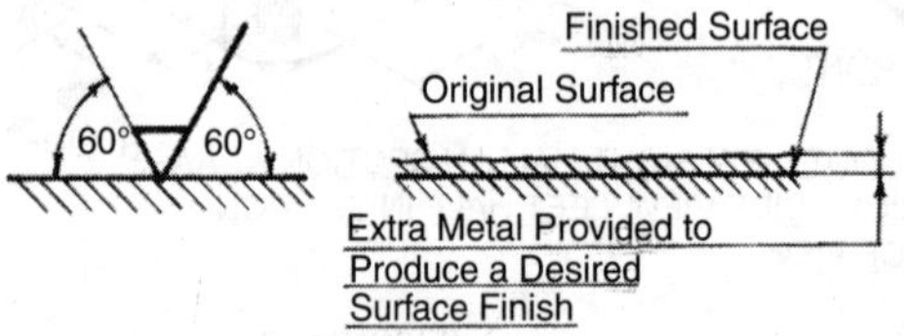

Fig. 12.1. Machining symbol and its meaning

1. Application of Machining Symbol

Figure 12.2 shows the application of machining symbol. This information is required to let the pattern maker and the die maker to include the machining allowance and allow extra metal on the casting or forging surfaces that require further machining for the finish purposes, *Fig. 12.3*. This allowance depends upon the casting or forging process to be used, material to be processed, and the quality of finish desired. Following points be kept in mind for applying the machining symbols:

1. As for dimensions, machining symbols are not duplicated on a drawing.
2. They should be used on the same view as the dimensions that give the size and location of the surface involved.
3. Place the symbol on the line representing the surface or on a leader or the extension line locating the surface.

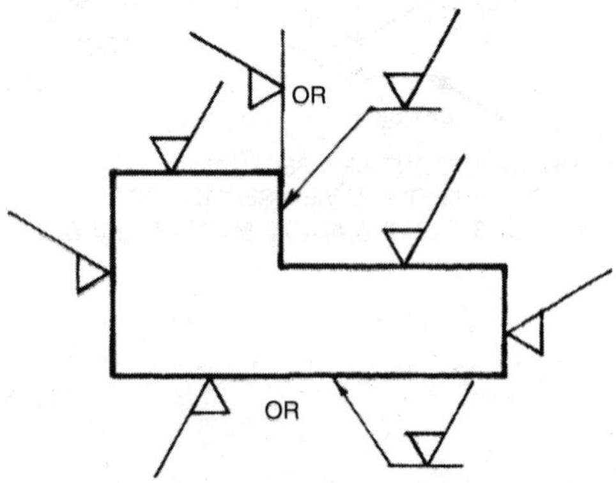

APPLICATION OF MACHINING SYMBOL

Fig. 12.2.

4. The symbol and the inscription should be oriented so that they may be read, like dimensions, from the bottom or right hand side of the drawing.
5. When all surfaces of a part are to be machined a general note, such as FINISH ALL OVER, may be used and the symbols may be omitted.
6. The machining symbol does not indicate surface finish quality for which the surface texture symbol is used.

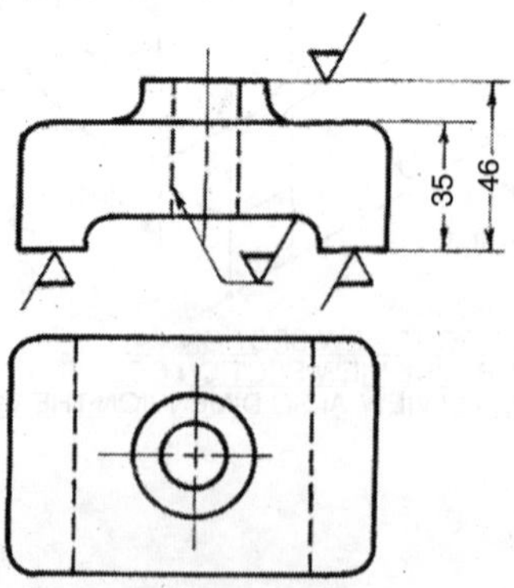

(a) Finished Cast Part

Fig. 12.3. (contd...)

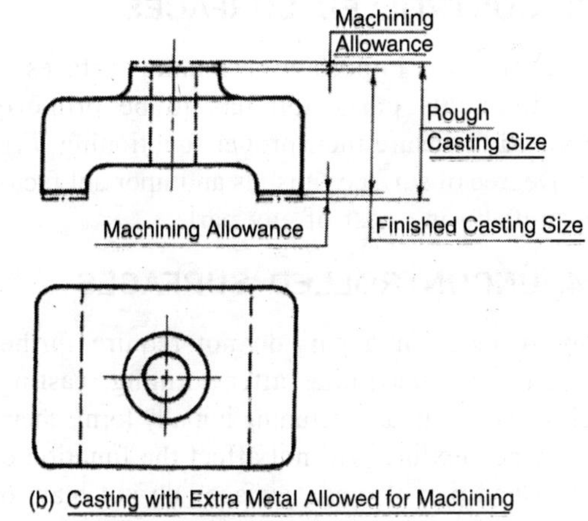

(b) Casting with Extra Metal Allowed for Machining

Fig. 12.3. Allowance for machining

2. Indication of Machining Allowance

When the value of the machining allowance needs to be specified, it is indicated to the left of the machining symbol, *Fig. 12.4*. This value is expressed in mm for metric system.

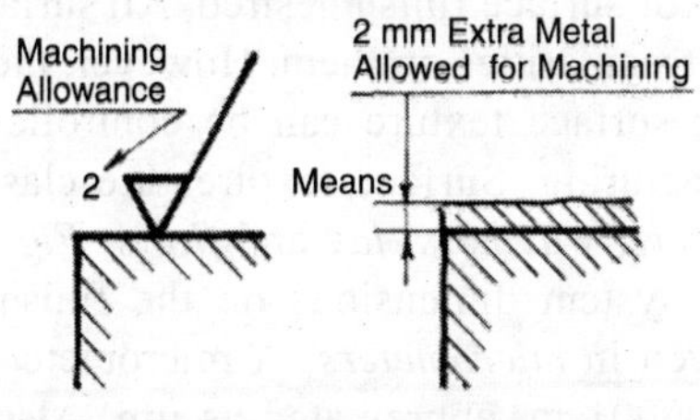

Fig. 12.4. Indication of machining allowance

12.2. SURFACE TEXTURE AND ITS IMPORTANCE

Surface texture or surface finish is the amount of geometric irregularity produced on the surface. Surface texture notes on a drawing refer to the required finish that must be machined on any particular surface. An engineer, a designer or a technician must learn to note and read the finished surfaces on a drawing.

With the advancement of technology and better materials and processes the quest of man to develop machines that can operate at much higher speeds and withstand greater loads without failure is continuously increasing. The resulting severe operating conditions at mating surfaces of the relatively moving parts of modern machinery demand that all such surfaces should be finished such that the friction, and consequently wear, should be minimum possible. The efficiency of the machines is higher and at the same time is competitive and economical. This has increased the necessity for more close control of surface quality to be exercised by designers, regardless of the size of mating part features. For a particular application machining a particular surface texture often becomes a required characteristics of its mating surfaces.

All surface finish control begins in the drawing office. The designer specifies the required surface finish to be obtained for maximum performance and service life at possible lowest cost of production with the available machinery for processing the parts. The designer bases his/her decision on best common practices followed for similar parts, testing and validation, design analysis information at hand, field service reports, finite element analysis etc. done, machine and process capabilities available in house or with the suppliers, as the case may be. In addition, the choice is influenced by other factors that include size and function of the part, loading conditions, operating speeds and conditions, characteristics of the materials of the mating parts, method of lubrication of the parts, operating temperature range, and many other conditions unique to a particular application. Surface quality specifications should be used only where required for intended operational performance of the part features, because the cost of producing a finished surface becomes greater as the quality of the surface called for is increased. As a rule the ideal surface finish is the roughest surface finish of the mating part features that will do the intended job satisfactorily.

In general, two principal reasons for surface texture control are:

- Decrease in friction, and
- Control of wear of mating surfaces.

It is a well-known fact that providing a suitable lubricating condition at the mating surfaces can very well control both these factors. The following points need to be kept in mind in taking a decision about the surface finish.

1. Whenever a continuous lubricating fluid film is required to be maintained between the mating surfaces, the surface irregularities must be controlled to a limit small enough to prevent penetration of the lubricant film under even the

most sever operating conditions. A number of parts in manufacturing and automotive industry like bearings, bushings, journals, machine ways, cylinder bores, piston, piston pins, gears, seal surfaces, and so on, are the objects where this condition must be met by controlling the surface finish.

2. Surface finish is also important to the wear service of certain mating parts that are subject to dry friction, e.g., threading and stamping dies, dry clutch plates, drum and disk brakes, cutting tool bits, rolls, etc.
3. Smooth surface finish is required on certain mating parts that require high precision in manufacturing for their intended function. Smooth finish is required on measuring instruments and gauges etc., to ensure their performance accuracy.
4. Smooth surface finish is also essential on mating surfaces of parts where leakage of fluid has to be sealed to minimum required for pressure retention, e.g., for valves in their bores, high pressure cylinders, atomizers and fuel injectors, pumps, etc.
5. Smooth surface finish is also important for visual appeal of many finished products. For this purpose surface finish control is important on forming rolls, extrusion dies, and precision casting dies so as to obtain the required surface characteristics of the finished surfaces that will not receive any further process finish.
6. Surface finish control is important for some applications of mating parts to achieve quieter operation, e.g., for power transmission gears.
7. Surface finish character is also important in situations where the mating part surfaces operate under boundary lubrication conditions. In such cases a certain amount of surface roughness will be helpful in retaining some oil and assist in maintaining a boundary lubrication layer.

To meet universal requirements for effective control of surface quality, under various types of operating conditions, a system is developed for accurately describing the surface character and finish.

12.3. CONTROLLED SURFACES

Surface finish is important on mating surfaces of all components parts and has to be properly controlled to ensure their proper functioning, *Fig. 12.5*. Degree of surface finish is an important factor in manufacturing cost of any part.

12.4. UNCONTROLLED SURFACES

Many surfaces of a part do not require further processing or machining after forming, casting, forging, stamping, or machining initially forms them. The surface texture will not affect the function of the finished part if these surfaces do not mate or contact with any other part surfaces, *Fig. 12.5.*

12.5. SURFACE FINISH CHARACTERISTICS

An absolutely smooth surface is not possible. More so in modern high volume interchangeable assembly of parts the cost of a part increases directly to the quality of surface finish desired. All surfaces have some irregularities on them. However, the amount of this surface texture can be controlled during manufacturing. Surface textures are classified as *roughness, waviness, lay* and *flaws*, *Fig. 12.6.* In metric system dimensions on the finish symbol are given in *micrometers.* A micrometer is equal to 0.000001 m, abbreviated as μm. Also 1μm is equal to 0.001 mm. Units read in micrometers are roughness height. Units read in millimeters are waviness-height, waviness-width, roughness-width, and roughness-width cut-off

1. Roughness

All smooth surfaces have some small peaks and valleys caused by machine cutting operations. These finely spaced surface irregularities are called roughness, *Fig. 12.6.* These include traverse feed marks and other irregularities within the limits of the *roughness-width cutoff.* Measurements for roughness, unless otherwise specified, apply in the direction that gives the maximum reading, usually across the lay.

2. Roughness-height

Roughness-height, also called roughness average (R_a) is an arithmetical average of the roughness deviation and is measured in micrometers

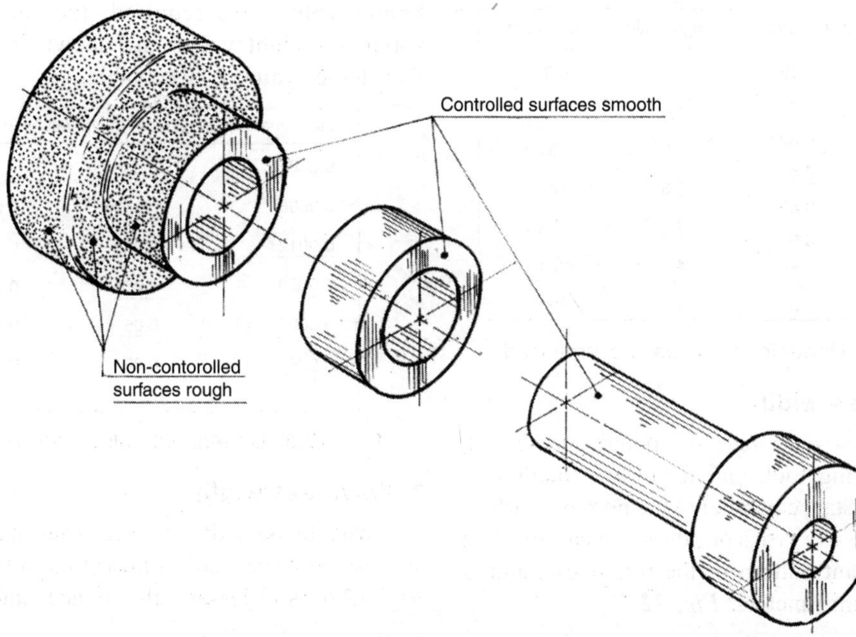

Fig. 12.5. Controlled and uncontrolled surfaces

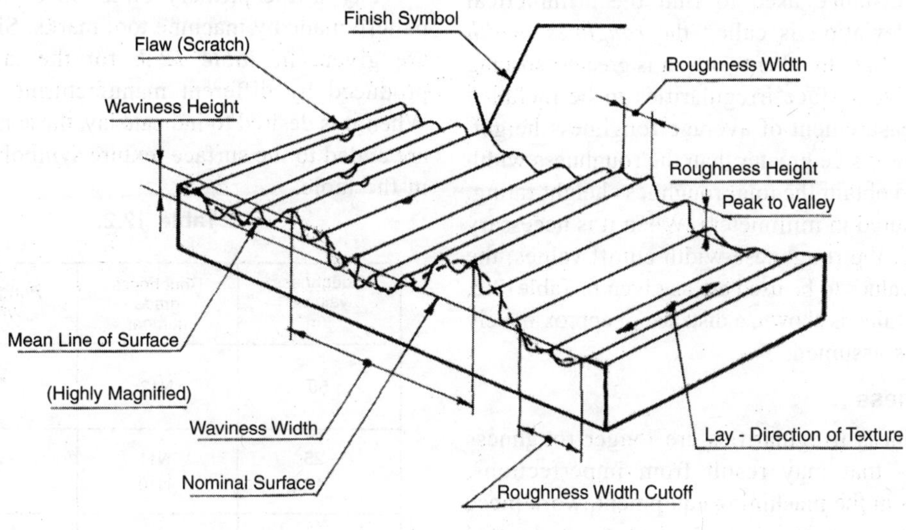

Fig. 12.6. Surface finish termonology

or roughness grade numbers, *Tables 12.1 and 12.2.* The measured distance is the variation from ridge or peak to valley in cut marks, *Fig. 12.6.* The rating value placed on the finish symbol indicates the maximum value permitted, and any lesser value is acceptable. The specifications of maximum and minimum values indicate the permissible range of roughness height permitted. Recommended roughness values are shown in *Fig. 12.7.*

Table 12.1.

Roughness value R_{α} μm	50	25	12.5	6.3	3.2	1.6	0.8	0.4	0.2	0.1	0.05	0.025
Roughness grade symbols	N12	N11	N10	N9	N8	N7	N6	N5	N4	N3	N2	N1

Recommended Roughness Average Rating Values in µm			
0.025	0.25	1.25	6.3
0.050	0.32	1.6	8.0
0.075	0.32	2.0	10.0
0.100	0.40	2.5	12.5
0.125	0.50	3.2	15.0
0.15	0.63	4.0	20.0
0.20	0.80	5.0	25.0
	1.00		

Fig. 12.7. Underlined values are preferred

3. Roughness-width

Roughness-width depends upon the processing machine, cutting tool, and feed. The roughness-width is the distance parallel to the nominal surface between successive peaks or ridges, which constitute the predominant pattern of the roughness, and is measured in millimeters, *Fig. 12.6.*

4. Roughness-width Cut-off

The distance used to find the arithmetical average deviation is called the *roughness-width cutoff Fig. 12.6.* In other words, it is greatest spacing of repetitive surface irregularities to be included in the measurement of average roughness height. It must always be greater than the roughness width in order to obtain the total roughness height rating. It is measured in millimeters. When it is necessary to indicate the roughness-width cutoff values, the standard values to be used are as given in Table12.2. When no value is shown, a distance of approximately 0.8 mm is assumed.

5. Waviness

Waviness irregularities are longer roughness variations that may result from imperfections, vibrations in the machining equipment, work piece deflections, warping, heat treatment, or strained material. Roughness may be considered superimposed on a '*wavy*' surface, and measurements for waviness, unless otherwise specified, apply in the direction that gives the maximum reading, usually across the lay, *Fig. 12.6.*

6. Waviness-height

Waviness height is the distance from the peak of the wave to its valley, and is measured in millimeters, *Fig. 12.6.* When maximum waviness height values are required, the recommended waviness-height values are shown in *Fig. 12.8.* Any lesser value is acceptable.

Recommended Waviness Height Values in µm		
0.0005	0.008	0.12
0.0008	0.012	0.20
0.0012	0.020	0.25
0.0020	0.025	0.38
0.0025	0.05	0.50
0.005	0.08	0.80

Fig. 12.8. Underlined values are preferred

7. Waviness-width

Waviness-width is the spacing between successive waves, and is measured in millimeters, *Fig. 12.6.* Any lesser value is acceptable.

8. Lay

Lay is the primary direction of the surface pattern made by machine tool marks. Six symbols are given, in *Table 12.3,* for the lay patterns produced by different manufacturing processes. When it is desired to indicate lay, these lay symbols are added to the surface texture symbols as shown in the table.

Table 12.2.

Roughness value a µm	Roughness grade number	Roughness symbol
50	N12	~
25 12.5	N11 N10	∇
6.3 3.2 1.6	N9 N8 N7	∇∇
0.5 0.4 0.2	N6 N5 N4	∇∇∇
0.1 0.05 0.025	N6 N5 N4	∇∇∇∇

9. Flaws

Flaws are infrequent irregularities that occur at random places on a machined part. The most common flaws are cracks, scratches, gauges and checks. There are no finish symbols for flaws. A supplementary note must be used on the drawing when a surface is required to be free from flaws, or the number and size of flaws is required to be .limited.

12.7. PRODUCTION PROCEDURES

To specify the proper finish of a machine part, the designer must know the equipment available and the type of surface qualities each machine produces, *Fig. 12.9*. The surface finish left by manufacturing equipment is controlled by:

1. Type of machine used,
2. Cutting speed,
3. Feed speed,
4. Condition of the cutting tool,
5. Type of the material being machined,
6. Type of cutting fluid or coolant being used,
7. Type of the abrasive surface.

Closely mating parts require a smoother finish than non-mating surfaces. The smoother the finish, costlier will be production of the part.

12.8. INDICATION OF SURFACE ROUGHNESS

Standard symbols and additional indications are used on production drawings to indicate that the surfaces are to be finished in some way so that

Table 12.3.

Symbol	Interpretation	
=	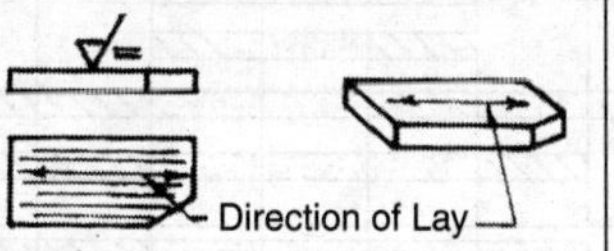	Parallel to the Plane of Projection of the View in Which the Symbol is Use
⊥	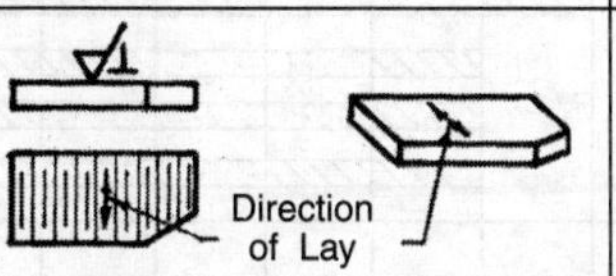	Perpendicular to the of Projection of the View in Which the Symbol is Used
X	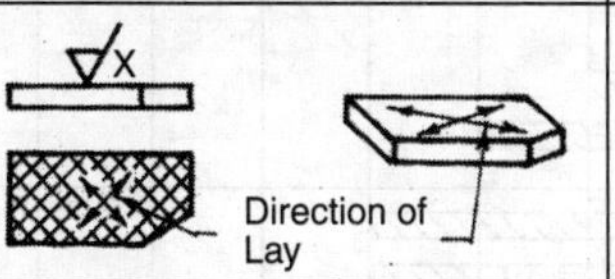	Crossed in two Slant Direction Relative to the Plane of Projection of the View in Which the Symbol is Used
M	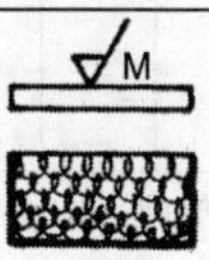	Multi-Directional
C		Approximately Circular Relative to the Center of the Surface to Which the Symbol is Applied
R	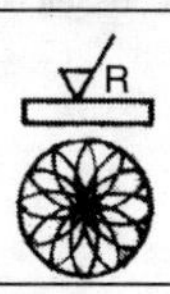	Approximately Radial Relative to the Center of the Surface to Which the Symbol is Applied

Note: Should it be Necessary to Specify a Direction of Lay not Clearly Defined by these symbols, then this must be achieved by a suitable drawing note.

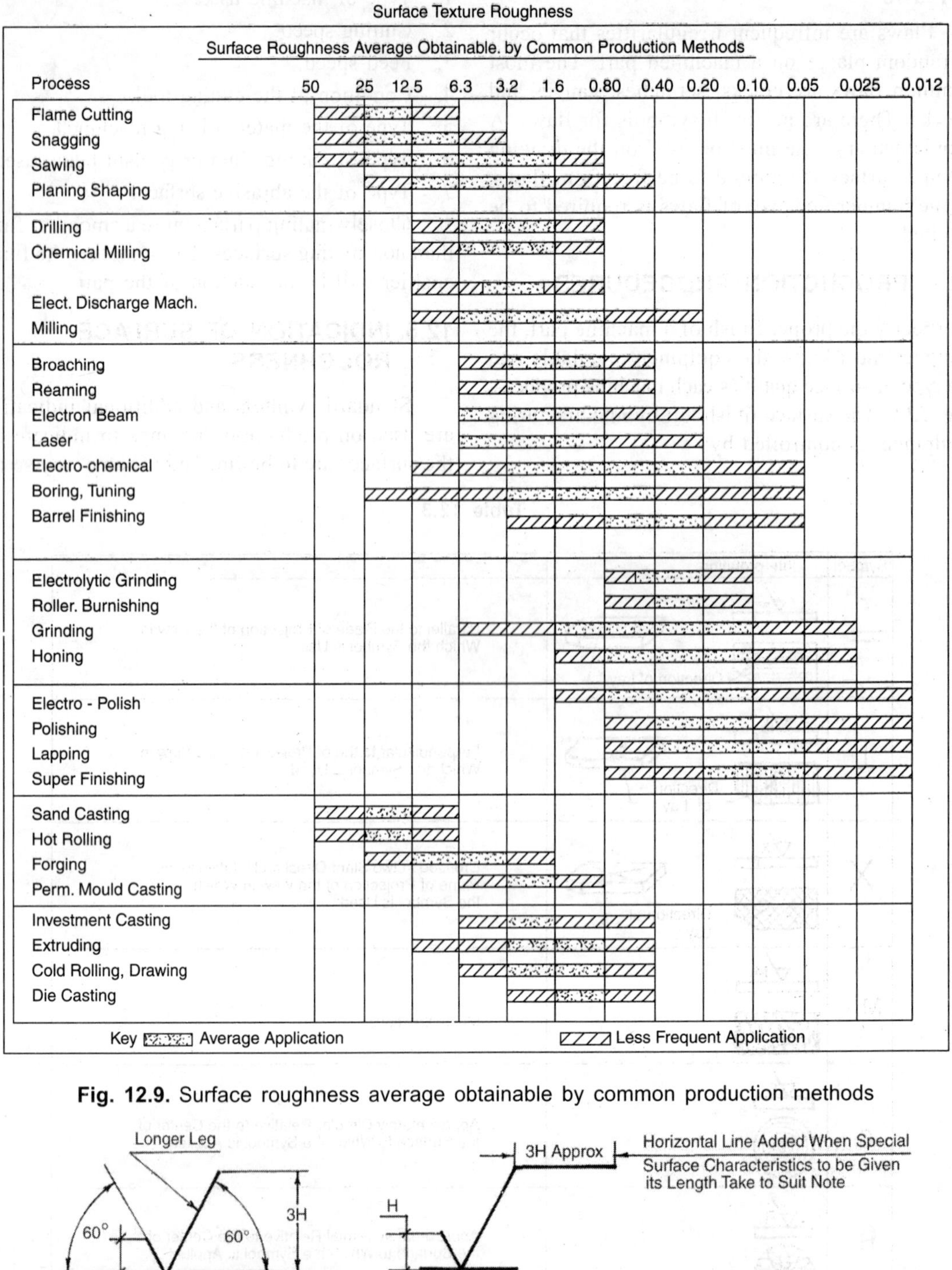

Fig. 12.9. Surface roughness average obtainable by common production methods

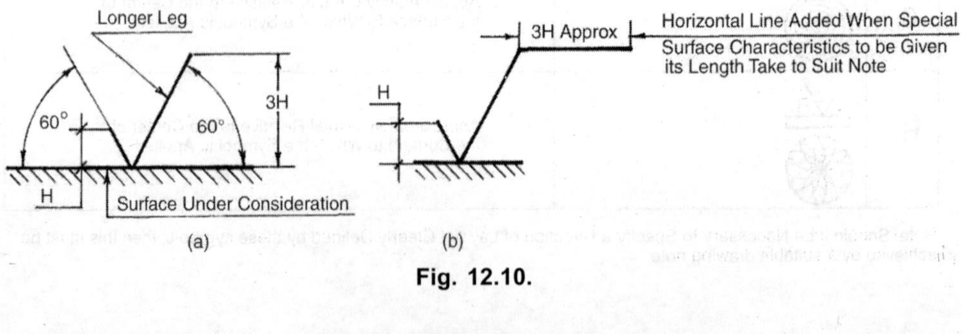

Fig. 12.10.

the roughness of the surfaces is controlled within permissible limits.

The basic symbol will be as shown in *Fig. 12.11* (*a*). It consists of two legs of unequal length inclined at 60° to the line representing the surface under consideration. It may only be used alone when it's meaning is explained by a note.

The basic symbol when used as in *Fig. 12.11* (*a*) means that the surface may be produced by any method.

If, from the surface, the removal of material by machining is required, a bar is added to the basic symbol, *Fig. 12.11* (*b*). Such a symbol when used means a machined surface with no indication of any other detail.

If the removal of material from the surface is not permitted, a circle is added to the basic symbol, *Fig. 12.11* (*c*). This symbol may also be used on a drawing concerning a production process to indicate that the surface is to be left in the state resulting from the preceding manufacturing process, whether the state be achieved by removal of material or otherwise.

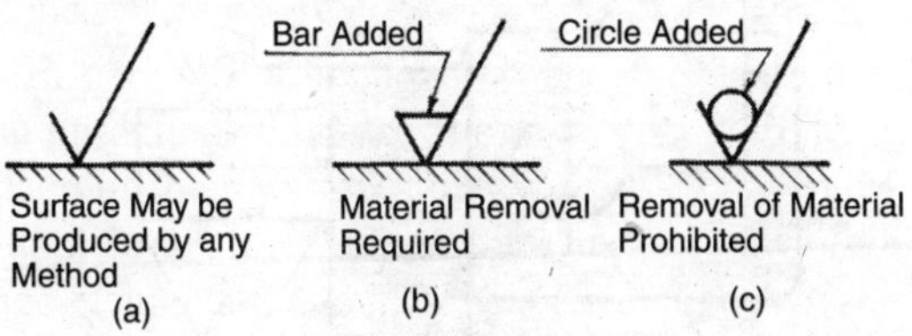

Fig. 12.11.

The value or values defining the principal criterion of roughness, Ra, are added to the symbols, *Fig. 12.12* (*a*), (*b*), (*c*). As shown in *Fig. 12.12* (*a*), it may be obtained by any production method; as in *Fig. 12.12* (*b*), it shall be obtained by removal of material by machining; and as in *Fig. 12.12* (*c*), it shall be obtained without the removal of material. The value specified represents the maximum permissible value of surface roughness. The principal criterion of roughness, Ra, may also be indicated by the corresponding roughness grade number (or symbol), *Tables 12.1 and 12.2*.

Under certain circumstances, for functional reasons, it may be necessary to specify additional special requirements concerning roughness, e.g., when the required characteristics of a surface have to be produced by a particular production method such as lapping, honing, milling, surface treatment or coating, etc.; this method should be indicated in plain language on a line added to the longer leg of any of the above symbols, *Fig. 12.13*. Unless otherwise stated the numerical value of the roughness applies to the surface roughness after surface treatment.

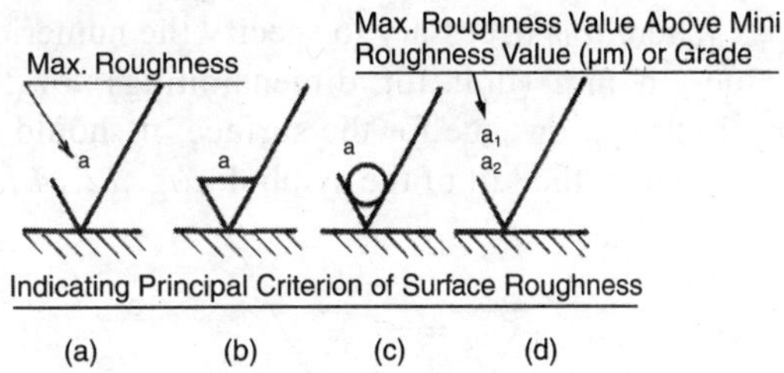

Fig. 12.12.

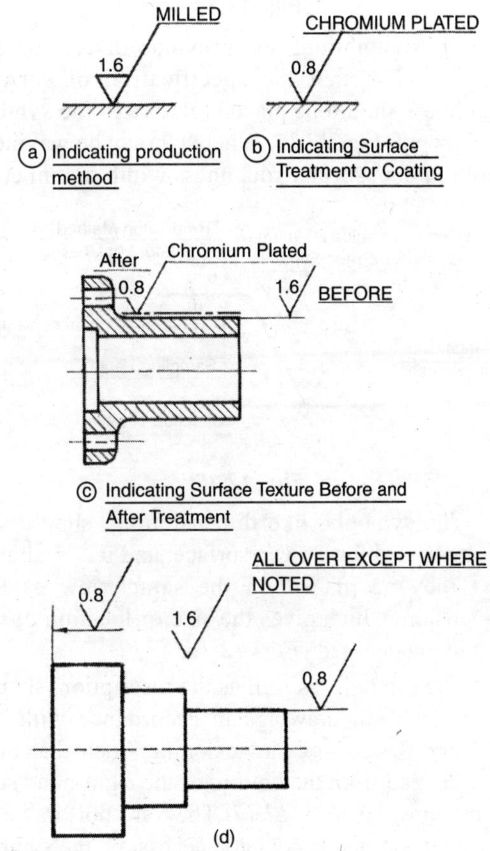

Fig. 12.13.

If it is necessary to indicate the sampling length, it should be selected from the series given in National Standard on '*Assessment of Surface*

Roughness', and stated adjacent to the symbol, *Fig 12.14 (a)*.

If it were necessary to control the direction of lay or the direction of the predominant surface pattern, it is specified by a corresponding symbol, out of the *Table 12.3*, added to the surface 'roughness symbol', *Fig. 12.14 (b)*.

Where it is necessary to specify the numerical value, in mm (unit for dimensioning), of the machining allowance for the surface, it should be indicated on the left of the symbol, *Fig. 12.14 (c)*.

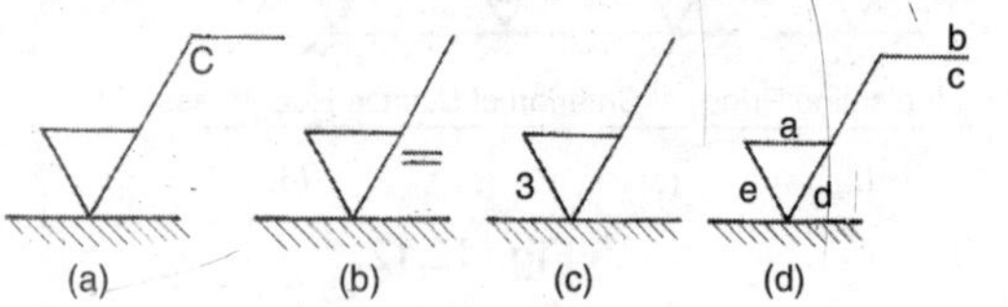

Fig. 12.14.

Thus combining the previous discussion we can establish that the specification of surface roughness should be placed relative to the symbol as shown in *Fig. 12.15*. Where (*f*) may be waviness width in mm, surface roughness width in mm etc.?

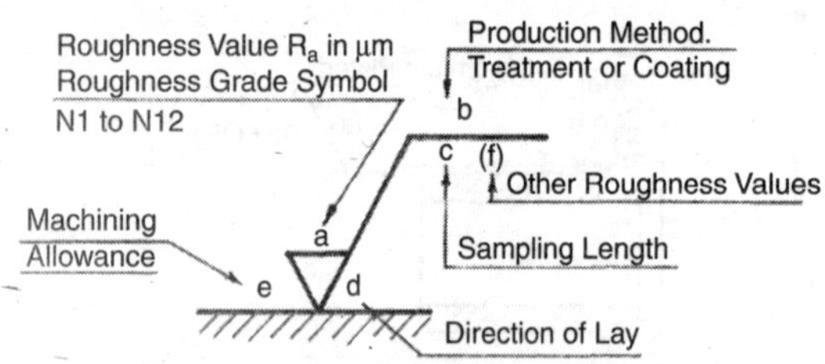

Fig. 12.15.

The symbols, like dimensioning, should not be duplicated for a given surface, and it is desirable that they be placed on the same view as the dimension, which gives the size or location of the surface concerned, *Fig. 12.16*.

The symbols, as well as the inscription, should be oriented on drawings in accordance with the '*Aligned System*' of dimensioning, such that they may be read from the bottom or the right-hand side of the drawing, *Fig. 12.17*. They are not drawn at any angle or upside down. If necessary, the symbol may be connected to the surface by a leader line terminating in an arrow. The symbol or the arrow should point from outside of the material of the component, either to the line representing the surface, or to an extension of it, *Fig. 12.17 (a), (b)*.

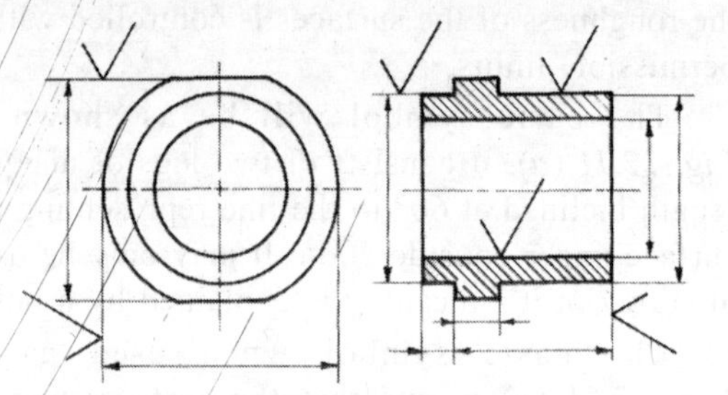

Fig. 12.16.

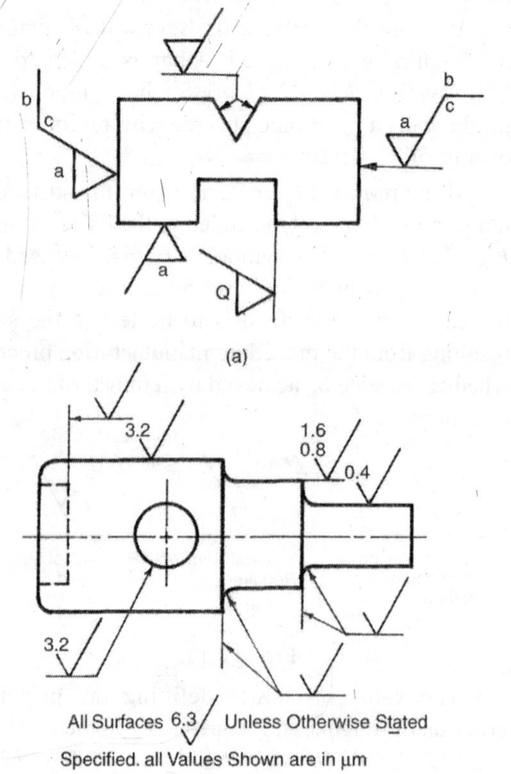

Fig. 12.17.

If the same roughness grade is required on all the surfaces of a part, it may be specified either by, (*i*) a note near the view or in the Title-Block, *Fig. 12.18 (a)*, or (*ii*) following the part number on the drawing, *Fig. 12.18 (b)*. Whereas if the surface roughness is required on the majority of the surfaces of a component, any one of may specify it the three methods indicated in *Fig. 12.19*. The symbols for the surface roughness, which are exceptions to the general symbol, are indicated on the corresponding surfaces.

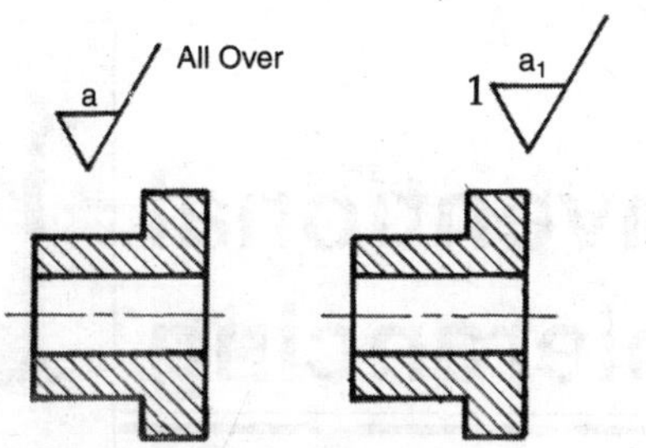

Fig. 12.18.

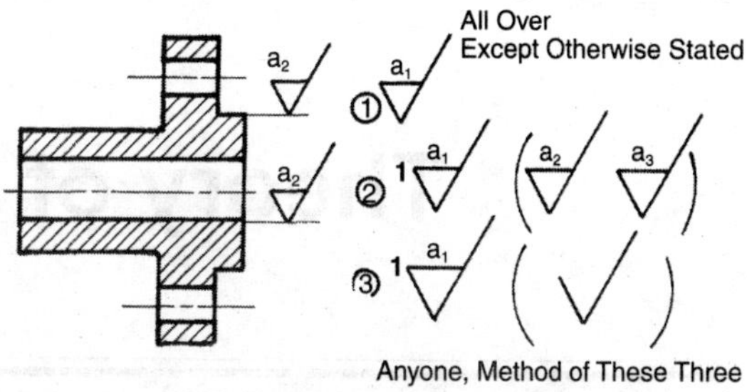

Fig. 12.19.

> **Important Note:** Should it be necessary to specify a direction of lay not clearly defined by symbols given in *Table 12.3* then a suitable note on the drawing shall achieve this.

The indications of the roughness, method of production, or machining allowance should only be given if these are necessary to ensure fitness and only for those surfaces which require these.

Indications for surface roughness are not necessary wherever the ordinary manufacturing processes by themselves ensure an acceptable surface finish.

The correct interpretation, and placement of the principal criterion of roughness, on the roughness symbol, is necessary as illustrated in *Fig. 12.20.*

Surface finishes obtainable by different machining processes are given in *Fig. 12.21* for reference.

Symbol			Meaning
Removal of Material by Machining			
Optional	Obligatory	Prohibited	
1.6 or N7	1.6 or N7	1.6 or N7	A Surface With a Max. Surface Roughness Value of 1.6 μm
3.2 0.8 / N8 N6	3.2 0.8 / N8 N6	3.2 0.8 / N8 N6	A Surface With a Max. Surface Roughness Value of 3.2 μm and a Minimum of 0.8 μm

Fig. 12.20. Symbols with indication of the principal criterion of roughness R_a

Roughness Height μm	Surface Description	Machine Processes
25.c	Very Rough	Saw or Torch Cutting, Forging or Sand Casting
12.5	Rough machining	Coarse Feeds and Heavy Cuts in Machining.
6.3	Coarse	Coarse Surface Grind, Medium Feed and Average Cuts in Machining
3.2	Medium	Sharp Tools, Light Cuts: Fine Feeds, High Speeds With Machining.
1.6	Good Finish	Sharp Tools. Light Cuts, Extra Fine Feeds. With Machining
0.8	High Grade Finish	Very Fine Feeds, Very Sharp' Tools. and' Fine Cuts.
0.4	Higher Grade Finish	Surface grinding. coarse honing, corse lapping.
0.2	Very Fine Machine Finish	Fine Honing and Fine Lapping.
0.05	Extremely smooth Machine finish	Extra Fine Honing and Lapping.

Fig. 12.21. Surface finishes obtainable by different machining operations

Theory of Conventional Tolerancing 13

13.1. INTRODUCTION

This chapter covers *general tolerancing as applied to conventional dimensioning practices*. The term *conventional dimensioning* as used here implies dimensioning without the use of geometric tolerancing, as in chapter 5. *Conventional tolerancing* applies a degree of form and location control by increasing or decreasing the tolerance magnitude.

Conventional dimensioning methods provide the necessary basic background to begin a study of geometric dimensioning and tolerancing, described in detail in *chapter 14*. Geometric tolerancing does not take the place of conventional dimensioning. However, geometric tolerancing specifies requirements more precisely than conventional tolerancing, leaving no doubts as to the intended part definition. This precision may not be the case when conventional tolerancing is used and notes on the drawing may become ambiguous.

13.2. CONVENTIONAL TOLERANCING

Not a long time ago the major of the components used to be actually mated together, and their dimensions being adjusted until the required type of fit was obtained. But with the passage of time it was realized that variations in the sizes of parts had always been present and that such variations could be restricted but not avoided. Also the slight variations in the size, which a part was originally intended to have, could be tolerated without impairment of its correct functioning. Soon it became evident that *interchangeable parts* need not be exactly identical parts, and it would rather be sufficient if the significant sizes, which controlled the part fits, lay between definite limits. Therefore, the problem of **interchangeable manufacturing** developed from the making of the mating parts to a supposedly exact size, to the holding of part dimensions between two limiting sizes, lying so closely together that any intermediate size would be acceptable and perform as intended.

The **concept of limits** means essentially that a precisely defined basic condition, expressed by a numerical value or specification, is replaced by two limiting conditions. Any resulting value of size lying between these **two limits of size** is acceptable. The difference between these two limits of size is called the **permissible tolerance**, *Fig. 13.1.* Two limiting levels enclosing a zone of acceptance or tolerance replace one standard level. In this way, a workable scheme of interchangeable manufacture, that is indispensable to mass production, has been established. Without effective size control by the designers and engineers, interchangeable manufacturing cannot be obtained.

All present day competitive manufacturing requires **quantity production** and interchangeability for many closely fitting parts. The production of these mating parts to an exact dimension, although theoretically possible, is economically unfeasible since the cost of a product rapidly increases as an absolute correct size is approached.

Also, due to numerous advantages of **mass production and simplification** it is not possible for a single industrial facility to manufacture all the components of an assembly unit. Interchangeable manufacturing allows parts manufactured in widely situated facilities to be brought together for assembly of the unit.

Interchangeable assembly manufacturing is

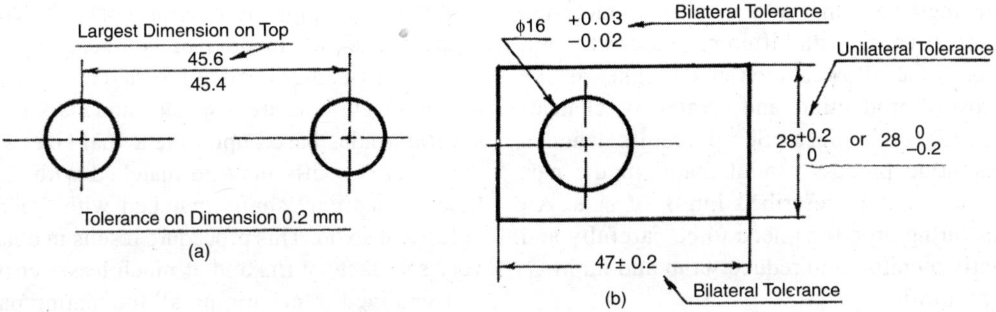

Fig. 13.1. Dimensions with tolerance

a term that is used to describe a method of part manufacture in which units and the machine as a whole may be assembled without individual fitting, while during repairs any part or unit of the machine may be replaced without any fitting whatsoever. A strictly interchangeable part is one, which can be assembled without selection, adjustment or supplementary operations.

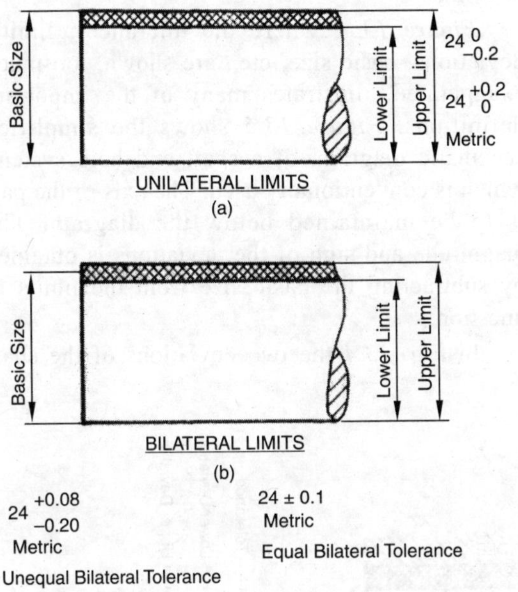

Fig. 13.2. Unilateral and Bilateral limits

The dimensions of any part of an assembly unit are established in designing the part, usually as a result of computation or practical judgement, experience, and best common practices followed on similar parts. These dimensions are then suitably rounded off to fix their **basic** or nominal sizes. The dimensions actually obtained in manufacturing a part are called **actual sizes**. Thus, for the sake of convenience, a basic size is ascribed to a dimension and each of its two limits is defined by its deviation from this size. The magnitude and sign of the deviation are obtained by subtracting the basic size from the limit in question. *Fig, 13.4*, which illustrates these definitions, is in practice replaced by a schematic diagram similar to *Fig. 13.5* for the sake of simplicity.

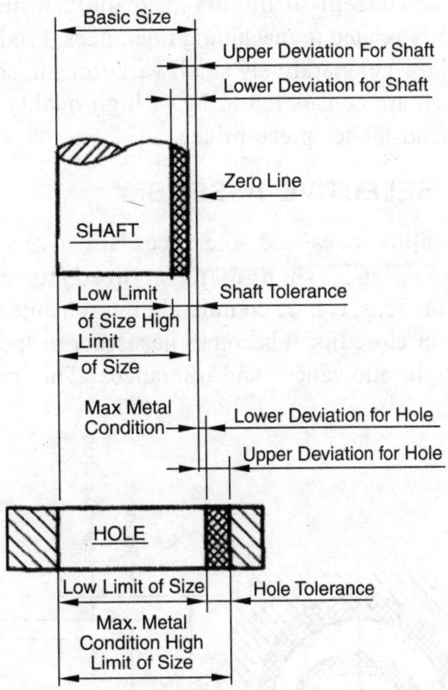

Fig. 13.3. (a) Deviation and maximum material condition

Tolerancing is thus an extension of dimensioning, **Tolerance** on a dimension is the algebraic difference between the high and low limits of size, *Fig, 13.1*. It has to be allowed in order to cover the reasonable imperfections in workmanship and the inevitable inaccuracy of manufacturing processes. It varies with different grades of work. The closer the allowed tolerance, higher is the piece cost of production, and greater the chances of rejection of a piece. Scrap results when a manufacturing process cannot maintain the part dimensions within prescribed limits of size. All manufacturing processes need to be carefully and constantly monitored to reduce scrap and improve company profits.

In general, tolerances should be chosen with due regard to performance, possible difficulties in manufacture, inspection and assembly. The designer specifies the allowance error that may be tolerated for a given dimension and still permit the satisfactory part function. *As a greater accuracy costs more money, the designer needs to specify the largest possible tolerance that will allow the intended part function and life of the mating parts.*

The **concept of quality** in manufacturing is primarily related to machining tolerances. Products that allow comparatively small variations in shape and size are considered to be of high quality and command higher piece prices.

13.3. SELECTIVE ASSEMBLY

When allowances and tolerances are given, the mating parts, when manufactured to print specifications, can be completely interchangeable. Also, for close fits, it becomes necessary to specify very tight allowances and tolerances. This might result in very high production costs. To cut this high cost, either manual or computer-controlled selective assembly is often resorted to. In selective assembly, all parts are inspected and classified into several grades based upon the actual sizes so that the "small" shafts may be matched with "small" holes, "medium" shafts matched with "medium" holes, and so on. This procedure results in obtaining very satisfactory fits and at much lesser cost than that involved in machining all the mating parts to very accurate and tight dimensions.

13.4. DEFINITIONS CONCERNING TOLERANCING AND LIMITS SYSTEM

The ISO (metric) system of limits and fits for mating parts is approved and adopted for general use worldwide. It establishes the designation symbols used to define specific dimensional limits on drawings.

Figure 13.4, where the tolerances, limits, deviations, basic size, etc., are shown, too much exaggerated, illustrates many of the important definitions. *Figure 13.5* shows the simplified schematic diagram, of a shaft and hole system, which is conventionally used. The axis of the part is to be maintained below the diagram. The magnitude and sign of the deviations is obtained by subtracting the basic size from the limits in question.

In *Fig. 13.4* the two deviations of the shaft

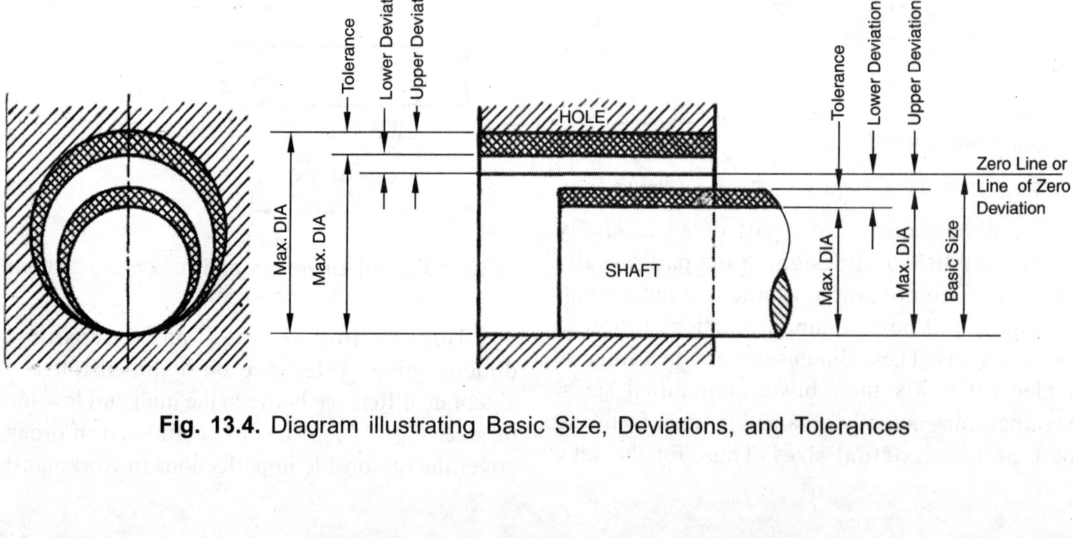

Fig. 13.4. Diagram illustrating Basic Size, Deviations, and Tolerances

are negative and those of the hole are positive. It is assumed that the shaft touches at the bottom of the hole and then other dimensions are considered. *A line is assumed to be there in between the upper tolerance on shaft and lower tolerance on hole.* All tolerances and deviations are referred to from this *datum line*. Further in the system of limits and fits we are simply interested in the tolerance on shafts and holes and not in their sizes. Therefore, for the conventional diagram the axis of shaft and hole is maintained very much below the diagram.

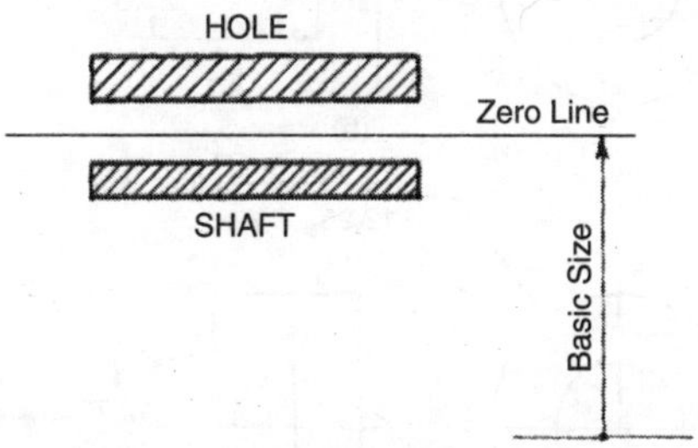

Fig. 13.5. Simplified Schematic diagram of a shaft and hole Clearance Fit

The following basic terms are defined here as these are used frequently in dimensioning of drawings. The understanding of their exact meanings is paramount in order that there should be no ambiguity in the precise interpretation of tolerancing methods described in text.

1. Size of Dimensions

It is the number expressing, in any particular unit, the numerical value of a dimension. In theory it is impossible to produce a part to an exact size, because, if measured with sufficient accuracy every part would be found to have a slightly different size, in general. However, for purposes of discussion and interpretation, a number of distinct sizes for each 'dimension have to be recognized.

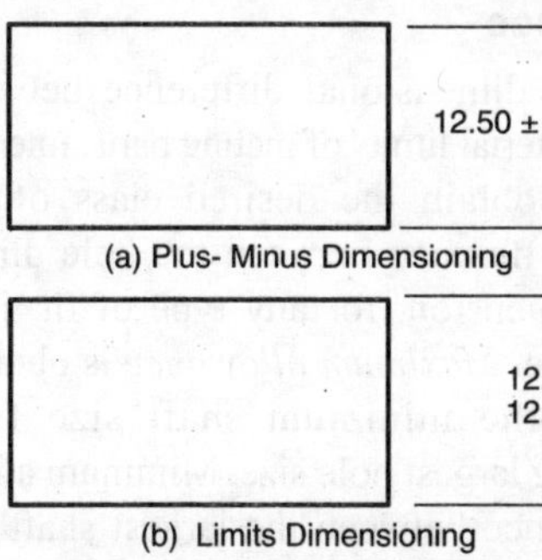

Fig. 13.6. Limit Dimensions

2. Nominal Size

The nominal size of a dimension or part is the size by which it is referred as a matter of convenience. It is a close approximation to a standard size, without any specified tolerance, used for the purpose of general identification such as stock size or thread diameter. In *Fig. 13.7*, the nominal size of both hole and shaft is 40 mm.

3. Basic Size

It, has got to be ascribed for the sake of convenience and is the size of a part in relation to which all limits of variation are determined by the application of the allowance and the tolerance. It will be same for the male and female parts of the fit on dimensions which do not control mating features, or when no allowance is applicable, the basic size, which is also the design size, is specified on the drawing, *Fig. 13.6 (a)*.

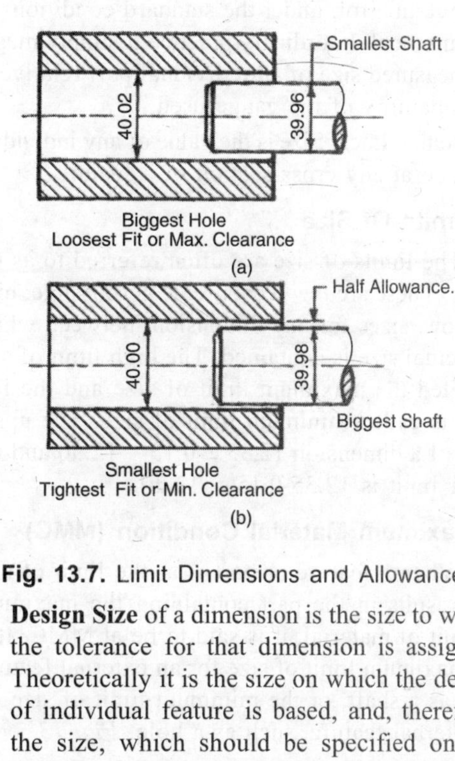

Fig. 13.7. Limit Dimensions and Allowance

- **Design Size** of a dimension is the size to which the tolerance for that dimension is assigned. Theoretically it is the size on which the design of individual feature is based, and, therefore, the size, which should be specified on the drawing. For dimensions of mating features it is derived from the basic size by the application of the allowance. When there is no allowance, it is identical to the basic size.
- **Specified Size** is the size identified on the drawing when associated with a tolerance. The

specified size is usually identical to the design size or, if no allowance is involved, to the basic size.

The basic size is sometimes specified on a drawing without a tolerance, and is enclosed in a rectangular frame to indicate that tolerances expressed in the general tolerance note do not apply. In this case variations are governed by geometric tolerances.

It will be same for the mating parts of the fit Often, basic size and nominal size of a part or dimension are used with the same sense.

4. Actual Size

It is the size of a part after manufacturing as may be found by actual measurement. The term includes the actual mating size and the actual sizes. The actual size of a dimension is the value that would be obtained on an individual measurement, without an error, under the standard conditions of measurement. In ordinary practice it simply means the measured size of an individual part relative to the capability of the gauge used.

Actual Local Size is the value of any individual distance at any cross-section of a feature.

5. Limits Of Size

The limits of size are often referred to as the limits. These are the two extreme permissible, high and low, sizes for any dimension, between which the actual size is contained. The high limit of size is called the maximum limit of size and the low limit is called minimum limit of size. The upper limit of a dimension 12.35 ± 0.15 = 12.50 and the lower limit is 12.35-0.15 = 12.20.

6. Maximum Material Condition (MMC)

When a feature or part is at the limit of size that results in the part containing the maximum amount of material, it is said to be at MMC. It is the maximum limit of size for an external feature, such as a shaft, or the minimum limit of size for an internal feature, such as a hole, *Fig. 13.3 (b), 13.3 (c).*

7. Least Material Condition (LMC)

When a feature or part is at the limit of size that results in the part containing the minimum amount of material, it is said to be at LMC. It is the minimum limit of size for an external feature, e.g., a shaft, *Fig. 13.3 (b)*, or the maximum limit of size for an internal feature, e.g., a hole, *Fig. 13.3 (c).*

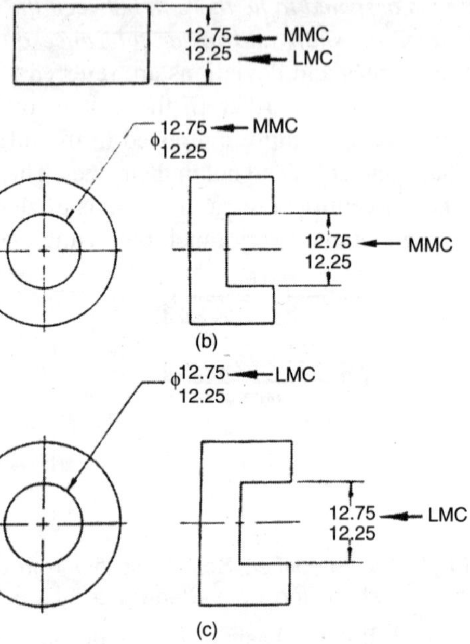

Fig. 13.3. Material Conditions

8. 'Go Limit'

The designation is applied to that of the limits of size, which corresponds to the maximum metal condition, i.e., the upper limit of a shaft and lower limit of a hole. This is the limit of size checked by the 'Go' gauge.

9. 'Not Go Limit'

This designation is applied to that of the two limits of size, which corresponds to the minimum material condition, i.e., the lower limit of a shaft and the upper limit of a hole. This is the limit of size checked by the 'Not Go' gauge.

10. Allowance

It is the dimensional difference between the maximum material limits of mating parts, intentionally provided to obtain the desired class of fit. An intentional difference between the hole dimension and shaft dimension, for any type of fit, is called the **allowance**. *Maximum allowance* is obtained by subtracting the minimum shaft size from its corresponding largest hole size. Minimum allowance is the difference between the largest shaft and the smallest hole size, *Fig. 13.13*, and *13.13 (a)*.

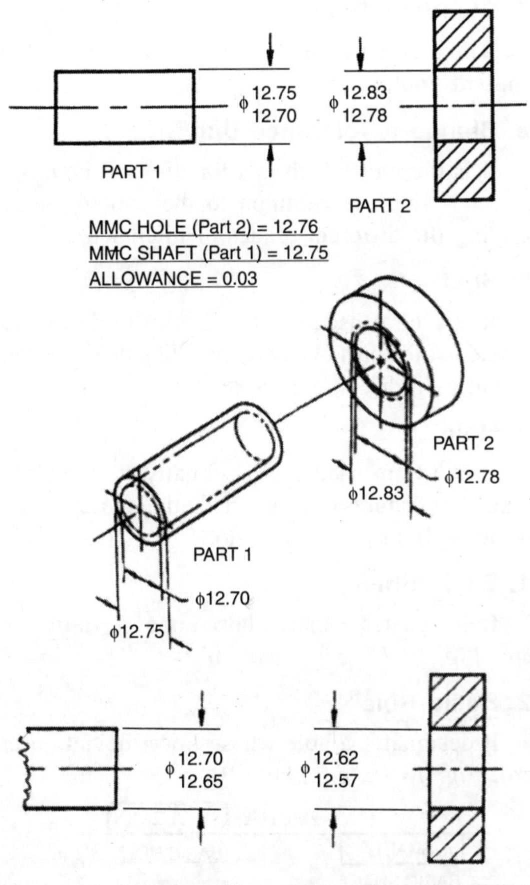

Fig. 13.13. (a) Allowance

Allowance is the minimum *clearance* (*positive allowance*), or maximum interference (*negative allowance*) between mating parts. It can be considered to be the tightest possible fit between parts.

Allowance may be positive or negative depending on the type of fit between the shaft and the mating hole. Positive allowance denotes a clearance and negative allowance denotes interference. Allowance can be calculated as: (*MMC Hole -MMC Shaft*) = *Allowance*.

11. Deviation

It is defined as the algebraic difference between a size (actual, maximum, minimum etc.) and .the corresponding basic size. For the sake of convenience, a basic size is ascribed to the part and each of the two limits of size is defined by its *deviation from the basic size*. Magnitude and sign of the deviation is obtained by subtracting the basic size from the limit of size in question. *Fig. 13.4* and *Fig, 13.3 (a)* illustrate these definitions and *Fig. 13.5* shows the simplified schematic diagram that is conventionally used.

In this simplified schematic diagram, the axis of the shaft is to be maintained below the diagram. In the example illustrated, the two deviations of the shaft are negative and those of the hole are positive.

- **Actual Deviation.** It is the algebraic difference between the actual size and the corresponding basic size.
- **Upper Deviation**. It is the algebraic difference between the maximum limit of size and the corresponding basic size, *Fig. 13.8, Fig. 13.3.*
- **Lower Deviation.** It is the algebraic difference between the minimum limit of size and the corresponding basic size, *Figs. 13.8, 13.3 (a)* and *13.4.*

12. Zero Line

In a graphical representation of limits and fits, it is a straight line to which the deviations are referred. The zero line is the line of zero deviation and represents the basic size, *Fig. 13.8*. By

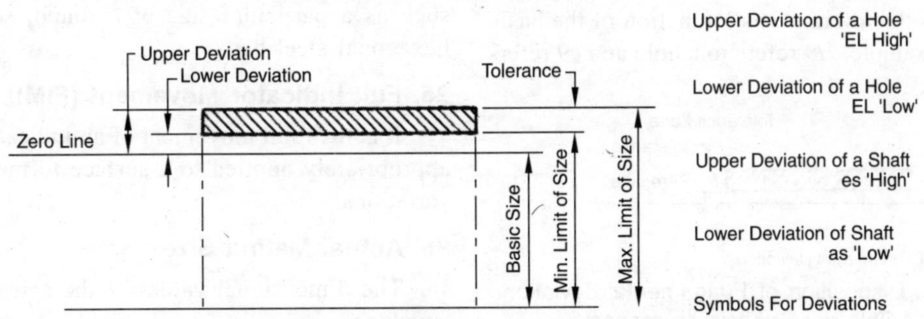

Fig. 13.8. Basic size and its deviations

convention, when the zero line is drawn horizontally, positive deviations are shown above and negative deviations below it.

13. Tolerance

The permissible variation of a size is called tolerance. The dimension 12.35 ± 0.15 is referred to as *plus-minus dimensioning. The tolerance of this dimension is {(12.35 + 0.15) – (12.35 – 0.15)}= 12.50 – 12.20 = 0.30.* It is equal to the algebraic difference between upper and lower deviations and has an absolute value without sign, *Fig. 13.4.* A tolerance is not given to dimension values that are identified as reference, maximum, minimum, or stock sizes.

14. Tolerance Zone

In a graphical representation of tolerance, it is the zone bounded by the two limits of size of the part. It is defined by its magnitude and position in relation to zero line, *Fig. 13.9.*

15. Fundamental Deviation

It is that one of the two deviations which is conventionally chosen to define the position of the tolerance zone in relation to the zero line, *Fig. 13.9.*

16. Grade of Tolerance

In a standard system of limits and fits, groups of tolerances are considered as corresponding to the same level of accuracy for all basic sizes. It is the name given to one standard series of tolerances calculated according to a certain law in terms of the basic size. *Tolerance grade* is an indication of the magnitude of the tolerance. It will be seen, from the table which follows, that the lower the grade, the finer will be the tolerance. In the *ISO System of Limits and Fits* a number symbol designates the tolerance as a function of the basic size, for example, *H9* refers to a hole and *e9* refers to a shaft.

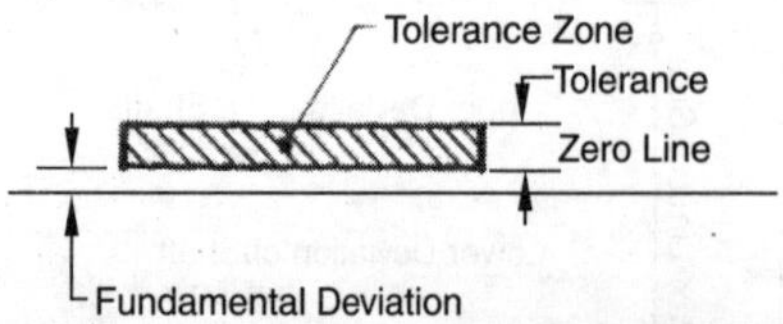

Fig. 13.9. Disposition of Fundamental deviation and Tolerance zone with respect to the Zero Line

17. Standard Tolerance

It is a tolerance belonging to any one of the standard grades.

18. Standard Tolerance Unit

It is the unit which is a function of the basic size and which is common to the two formulae defining the different grades of tolerance.

19. Shaft

It is a term used conventionally to designate all external features of a part, including those, which are not cylindrical, e.g., a key.

20. Hole

It is a term used conventionally to designate all internal features of a part, including those, which are not cylindrical, e.g., a slot.

21. Basic Shaft

It designates a shaft whose upper deviation is zero, *Fig. 13.10,* e.g., shaft 'h'.

22. Basic Hole

It designates a hole whose lower deviation is zero, *Fig. 13.10,* e.g., hole 'H'.

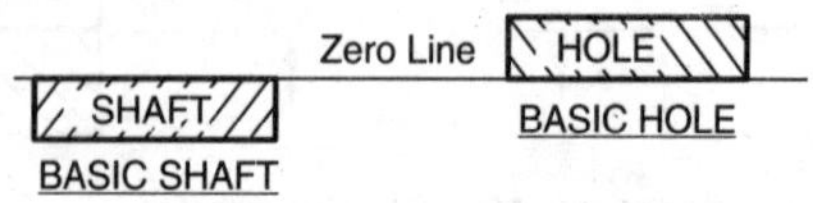

Fig. 13.10. Graphical representation of Basic Shaft and Basic Hole

23. Reference Temperature

The standard reference temperature is 20°C for industrial measurements and, consequently, for dimensions defined by the system.

24. Stock Size

It is the commercial or premanufactured size, such as a particular size of a round, square, or hexagonal steel bar.

25. Full Indicator Movement (FIM)

It is the total movement of an indicator where appropriately applied to a surface to measure its variations.

26. Actual Mating Size

The dimensional values of the actual mating envelope.

27. Dimension Origin Symbol

This symbol is used to indicate that a toleranced dimension between two features originates from one of these features, *Fig. 13.11 (b)*.

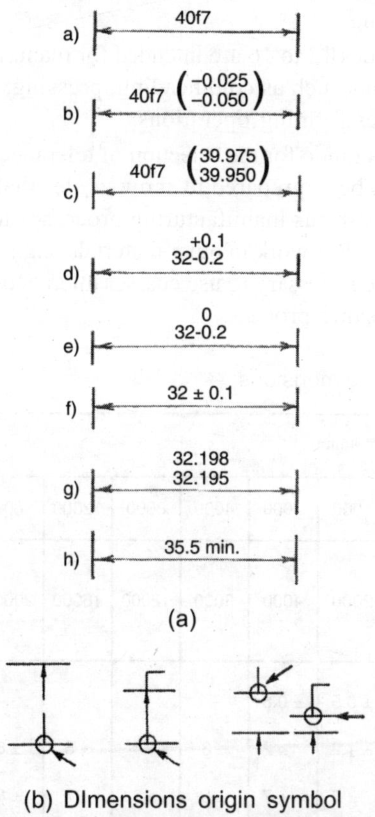

Fig. 13.11. Indication of the components of a linear dimension

13.5. UNILATERAL AND BILATERAL LIMITS

When tolerance distribution is on only one side of basic size, it is referred to as **unilateral tolerance,** *Figs. 13.1 (b), 13.2 (a)*. If permissible variation is both plus and minus, it is referred to as a **bilateral tolerance,** *Figs. 13.1 (b), 13.2 (b)*.

Figure 13.2 (a) shows an example of unilateral limits, where the maximum and minimum limits of size are disposed on the same side of the basic size. This system is preferred since the basic size is used for the GO limit gauges, and changes in the tolerance magnitude affect only the size of the other gauge dimension, i.e., NOT GO gauge. The inspection departments use these gauges to inspect and check finished products. *Figure 13.2 (b)* shows an example of bilateral limits. Note that in this case the limits are disposed above and below the basic size.

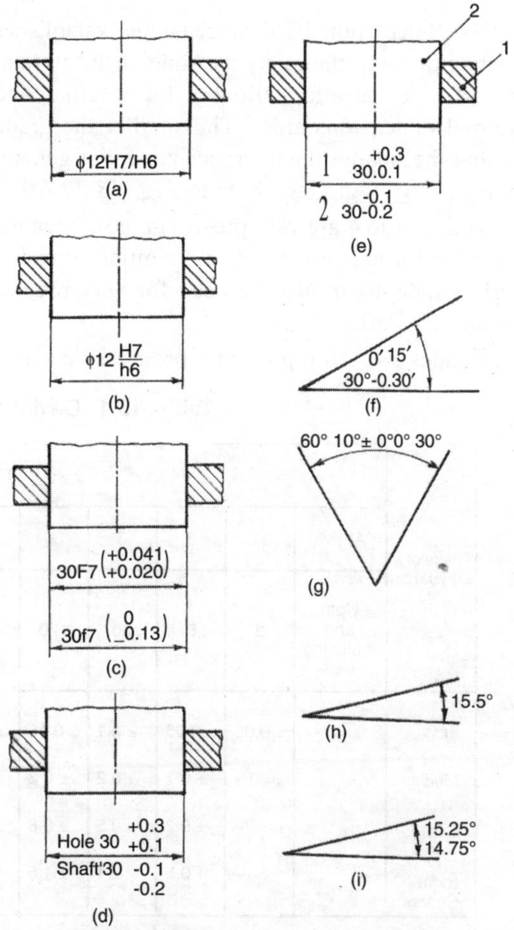

Fig. 13.12. Dimensioning assembled parts, including the tolerance on individual parts, (*a*), (*b*), (*c*), (*d*), (*e*)

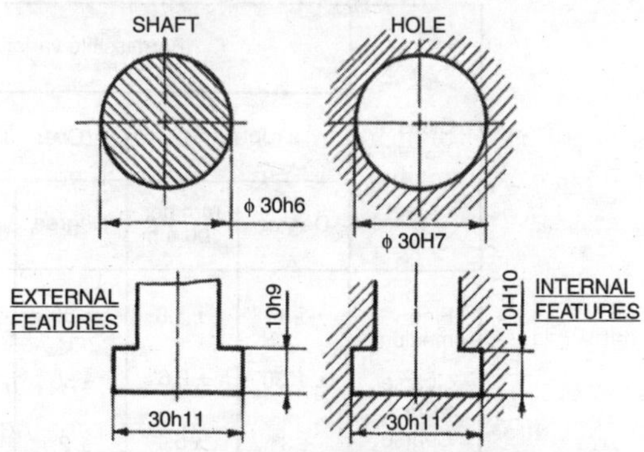

Fig. 13.12. (f) Tolerance dimensions for external and internal features

13.6. SYMBOLS FOR TOLERANCES, DEVIATIONS, AND FITS

13.6.1. Fundamental Tolerances

An "International Tolerance Grade" establishes the magnitude of the tolerance zone or the amount of part size variation allowed for internal and external dimensions alike. The smaller the grade number the smaller the tolerance zone. For general applications IT grades, refer to *Fig. 13.20 (a)*.

Grades 1 to 4 are very precise grades, intended primarily for gauge making and similar precision work. Grade 4 can also be used for very precise production work.

Grades 5 to 16 represent a progressive series suitable for cutting operations, e.g., turning, boring, grinding, milling, drilling, planing, and shaping, etc. Of these Grade 5 is the most precise grade, obtainable by fine grinding and lapping, while Grade 16 is the coarsest grade for rough sawing and machining.

Grades 12 to 16 are intended for manufacturing operations such as cold heading, pressing, rolling, and other forming operations.

As a guide for the selection of tolerances, Table 13.3 has been prepared to show grades that can be held by various manufacturing processes for work in metals. For work in other materials, e.g., plastics, it may be necessary to use coarser tolerances grades for the same process.

Table 13.1. Deviations for linear dimensions

Class of Devuation		Rangel of nominal dimensions											
	Above	0.5	3	6	30	120	315	1000	2000	4000	8000	12000	16000
	Upto and Including	3	6	30	120	315	1000	2000	4000	8000	12000	16000	20000
Fine		± 0.05	± 0.05	± 0.1	± 0.15	± 0.2	± 0.3	± 0.5	± 0.8	–	–	–	–
Medium		= 0.1	± 0.1	± 0.2	± 0.3	± 0.5	± 0.8	± 1.2	± 2	± 3	± 4	± 5	± 6
Coarse		–	± 0.2	± 0.5	± 0.8	± 1.2	± 2	± 3	± 7	± 5	± 6	± 7	± 8
Extra Coarse		–	± 0.5	± 1	± 1.5	± 2	± 3	± 4	± 6	± 8	± 10	± 2	± 12

Table 13.2 (a). Deviations for angular dimensions

(All dimersions mm)

Class of Deviation	Permissible variations on length of shorter side or angle							
	Upto 10		Over 10 to 50		Over 50 to 120		Over 120	
	Degree	mm per 100 mm	Degree	mm per 100 mm	Degree	mm per 100 mm	Degree	mm per 100 mm
Fine medium	± 1	± 1.8	± 30′	± 0.9	± 20′	± 0.6	± 10′	± 0.3
Coarse extra	± 1°30′	± 2.6′	± 50′	± 1.5	± 25′	± 0.7	± 15′	± 0.4
Coarse	± 3°	± 5.1	± 2°	± 3.5	± 1°	± 1.8	± 30′	± 09

Table 13.2 (b). Deviations for Radii and Chamfers

(All dimersions mm)

Class of Deviation	Range of Dimensions					
	Above	0.5	3	6	30	120
	Upto and including	3	6	30	120	315
Fine medium		± 0.2	± 0.5	± 1	± 2	± 4
Coarse extra coarse		± 0.2	± 1	± 3	± 4	± 8

A fundamental deviation establishes the position of the tolerance zone with respect to the basic size. "Tolerance position letters" express fundamental deviations. Capital letters are used for internal dimensions and Lowercase letters are used for external dimensions.

Tolerance is denoted by two symbols, a letter symbol and a number symbol called the grade, *Table 13.4. Fig. 13.20* shows the graphical illustration of tolerance sizes or fundamental deviations for letter symbols. It is also evident that these letter symbols represent the degree of

Table 13.3. (a) Possible accuracy from industrial forming equipment

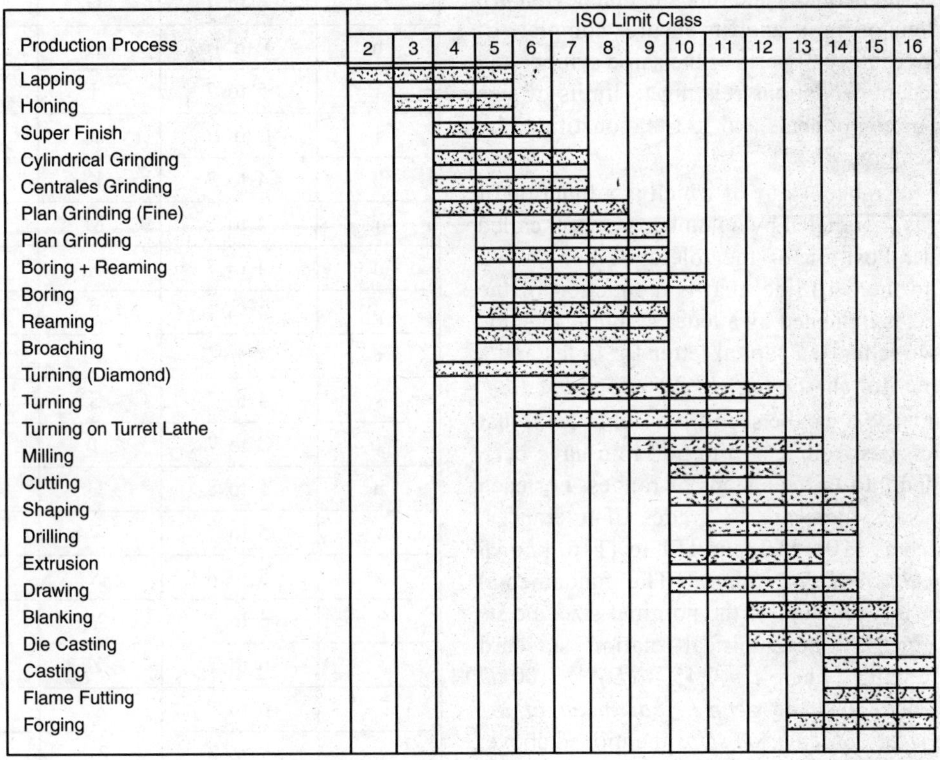

Table 13.3. (b)

Grade	IT5	IT6	IT7	IT8	IT9	IT10	IT11	IT12	IT13	IT14	IT15	IT16
Values	7i	10i	16i	25i	40i	64i	100i	160i	250i	400i	640i	1000i

closeness of the tolerance zone (*positive or negative*) to the basic size. In order to satisfy the usual requirements, both of individual parts and of fits, the system provides for any given basic size, a whole range of tolerances with respect to the line of zero deviation, called the zero line.

In India *BIS 919-1963* describes **"Limits, Fits and Tolerances"**. The system comprises of suitable combinations of **18 Grades of Tolerances,** also called *Grades of Accuracy of* ***manufacture***, and 28 **Fundamental Deviations** indicated by letter symbols of both holes and shafts in diameter steps of up to 500 mm. The various 28 deviations are represented by capital letters *A. B, C, CD, D, E, EF, F, FG, G, H, J, JS, K, M, N, P, R, S, T, U, V, X, Y, Z, Z4, ZB, ZC* for *holes* and corresponding lowercase letters, *a* to *zc*, for shafts.

Innumerable **fits,** ranging from extreme interference to those of extreme clearance, may be obtained by a suitable combination of fundamental tolerances and fundamental deviations. This system of limits and fits ensures complete fit interchangeability of parts of the same symbol and makes recommendations relating to limits of size of parts or components and to fits, constituted by their assembly.

The tolerance, value of which is a function of the size, is designated by a number symbol, called the grade. Position of the tolerance zone, with respect to the zero line, that is a function of the basic size, is indicated by a letter symbol (in some cases, two letters), a capital letter for holes and a small letter for shafts, *Fig. 13.20*, and *Table 13.4*.

Similarly, it can be seen from Table 13.5, that the basic sizes from 1 mm to 500 mm have been subdivided into thirteen steps or ranges. For each nominal step, there are 18 grades of tolerances, designated as IT01, IT0, and IT1 to IT16, known as '*Fundamental Tolerances*'. The fundamental tolerance is a function of the nominal size and its unit is given by the empirical relation, standard tolerance unit *i*, where $i = 0.45 \times (D)^{1/3} + .001D$ microns and *D* is the *geometrical mean of the limiting values of the basic steps* mentioned above, in millimeters. This relationship is valid for *grades 5 to 16* and *nominal sizes from 3 to 500 mm*. For grades below 5, and for sizes above 500 mm, there are other empirical relations for which refer to *BIS.—919-1963*. Relative magnitudes of different grades tolerances for grades 5 to 16 in terms of tolerance unit *i*, for sizes up to 500 mm, are given in *Table 13.3 (b)*.

Table 13.4.

Shafts		*Holes*	
Latter symbol	*Grades*	*Letter symbol*	*Grades*
a	9, 11	A	9, 11
b	8, 9, 11	B	8, 9, 11
c	8, 9, 11	C	8, 9, 11
d	5 to 11	D	6 to 11
e	5 to 9	E	5 to 10
f	4 to 9	F	5 to 9
g	4 to 7	G	5 to 7
h	1 to 16	H	1 to 16
i	5 to 7	I	6 to 8
j	1 to 16	J	1 to 16
k	4 to 7	K	5 to 8
m	4 to 7	M	5 to 8
n	4 to 7	N	5 to 11
p	4 to 7	P	5 to 9
r	4 to 7	R	5 to 8
s	4 to 7	S	5 to 7
t	5 to 7	T	5 to 7
u	5 to 8	U	6 to 7
v	5 to 7	V	6 to 7
x	5 to 8	X	6 to 7
y	6 to 7	Y	7
z	6 to 7	Z	7 to 8
z_a	6 to 7	Z_A	7 to 8
z_b	7 to 8	Z_B	8 to 9
z_c	7 to 8	Z_C	8 to 9

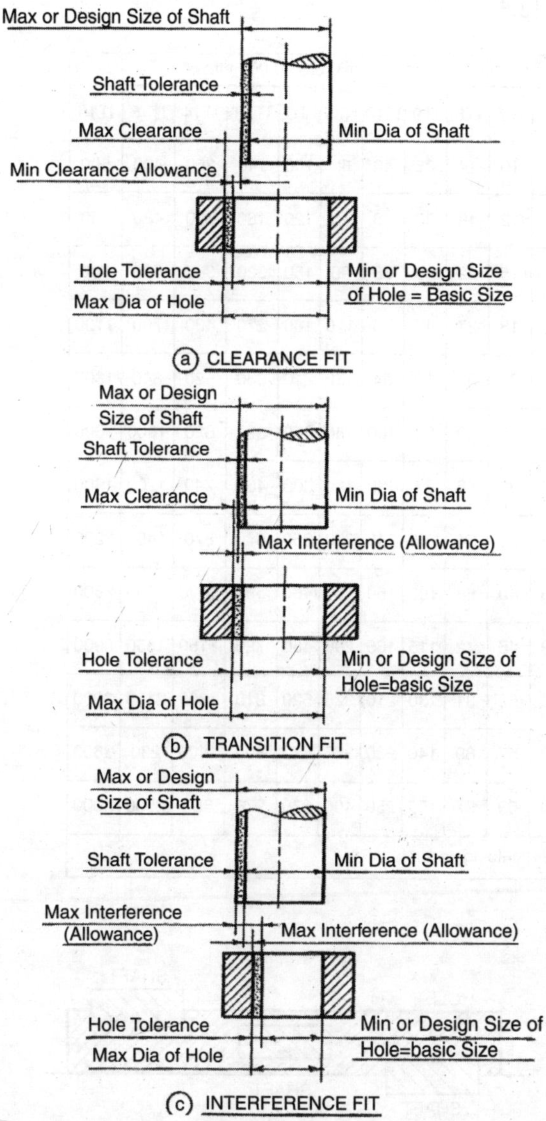

Fig. 13.13.

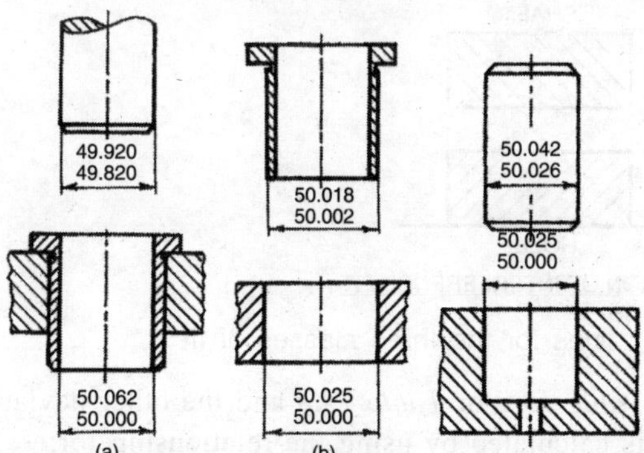

Fig. 13.14.

Thus the fundamental tolerance value for different grades (IT) may be obtained either from *Table 13.5* or calculated from the relationship given in the *Table 13.3 (b)*.

- **Metric Tolerance Symbol:** It is established by combining IT grade number and the tolerance position letter. This symbol identifies the actual maximum and minimum limits of the part. The toleranced sizes are thus defined by the basic size, of the part followed by the symbol composed of a letter and number, *Fig. 13. 21(a)*.

The tolerance size is thus defined by its basic value followed by a symbol composed of letter (in some cases two letters) and a number, e.g. 50 g7.

PROBLEM 13.1. *Calculate the fundamental tolerance for a shaft of ϕ100 mm and grade 7.*

SOLUTION: The shaft size 100 mm lies in the basic step 80 to 120 mm and the geometric mean for the step is:

$$D = \sqrt{(80 \times 120)} = 98 \text{ mm}$$

The tolerance unit,

$$i = 0.45 \times (98)^{1/3} + 0.01 \times 98$$
$$= 2.172 \text{ microns}$$

For grade 7, as per *Table 13.3 (a)*, the tolerance value is:

$$16i = 16 \times 2.172 = 35$$

microns, as is given in *Table 13.5*.

- **Metric Fit Symbol:** A fit is indicated by the basic size common to both mating components, followed by a symbol corresponding to each component, with the internal part symbol preceding the external part symbol, *Fig. 13.21 (b)*. A fit is indicated by the basic size common to both the mating components, followed by symbols corresponding to each component, the hole being quoted first, e.g. *50 H8/g7*.

13.6.2. Fundamental Deviations

The symbols used in *Fig. 13.20* for the fundamental deviations for the shaft and the hole are as follows:

	Shaft	Hole
Upper deviation	es	ES
Lower deviation	ei	EI

Table 13.5.

Dia. steps in mm	Tolerance grades, values of tolerance in microns, 1 micron = 0.001 mm																	
	IT01	ITO	IT1	IT2	IT3	IT4	IT5	IT6	IT7	IT8	IT9	IT10	IT11	IT12	IT13	IT14	IT15	IT16
To and Incl. 3	0.3	0.5	0.8	1.2	2	3	4	6	10	14	25	40	60	100	140	250	400	600
Over 3 To and Incl. 6	0.4	0.6	1	1.5	2.5	4	5	8	12	18	30	48	75	120	180	300	480	750
Over 6 To and Incl. 10	0.4	0.6	1	1.5	2.5	4	6	9	15	22	36	58	90	150	220	360	580	900
Over 10 To and Incl. 18	0.5	0.8	1.2	2	3	5	8	11	18	27	43	70	110	180	270	430	700	1100
Over 18 To and Incl. 30	0.6	1	1.5	2.5	4	6	9	13	21	33	52	84	130	210	330	520	840	1300
Over 30 To and Incl. 50	0.6	1	1.5	2.5	4	7	11	16	25	39	62	100	160	250	390	620	1000	1600
Over 50 To and Incl. 80	0.8	1.2	2	3	5	8	13	19	30	46	74	120	190	300	460	740	1200	1900
Over 80 To and Incl. 120	1	1.5	2.5	4	6	10	15	22	35	54	87	140	220	350	540	870	1400	2200
Over 120 To and Incl. 180	1.2	2	3.5	5	8	12	18	25	40	63	100	160	250	400	630	1000	1600	2500
Over 180 To and Incl. 250	2	3	4.5	7	10	14	20	29	46	72	115	185	290	460	720	1150	1850	2900
Over 250 To and Incl. 315	2.5	4	6	8	12	16	23	32	52	81	130	210	320	520	810	1300	2100	3200
Over 315 To and Incl. 400	3	5	7	9	13	18	25	36	57	89	140	230	360	570	890	1400	230	3600
Over 400 To and Incl. 500	4	6	8	10	15	20	27	40	63	97	155	250	400	630	970	650	2500	4000
Incl. Means Inculding																		

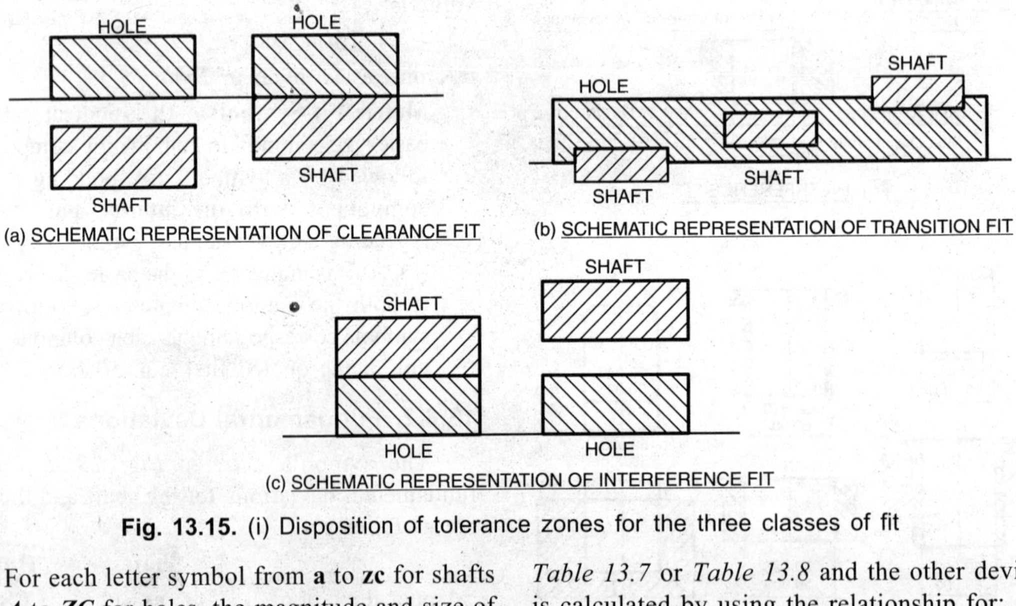

Fig. 13.15. (i) Disposition of tolerance zones for the three classes of fit

For each letter symbol from **a** to **zc** for shafts and ***A*** to ***ZC*** for holes, the magnitude and size of one of the two deviations may be obtained from *Table 13.7* or *Table 13.8* and the other deviation is calculated by using the relationship for;

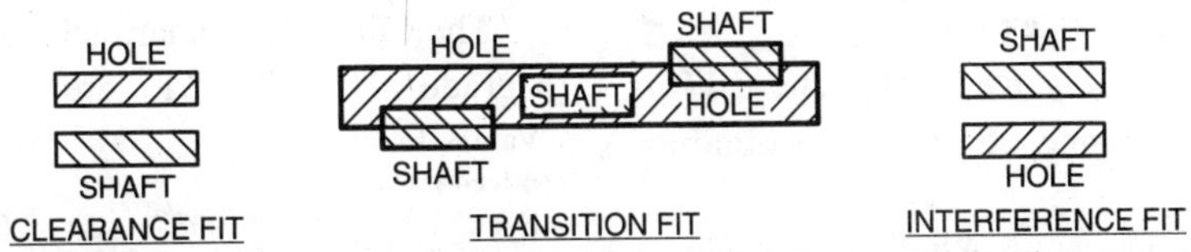

Fig. 13.15. (ii) Disposition of tolerance zones for the three classes of fit

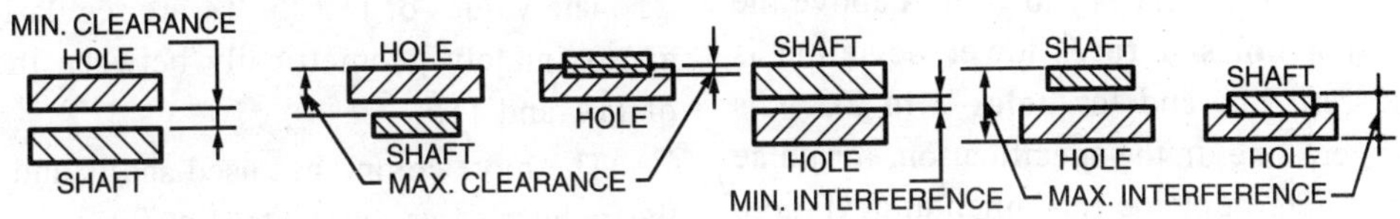

Fig. 13.16.

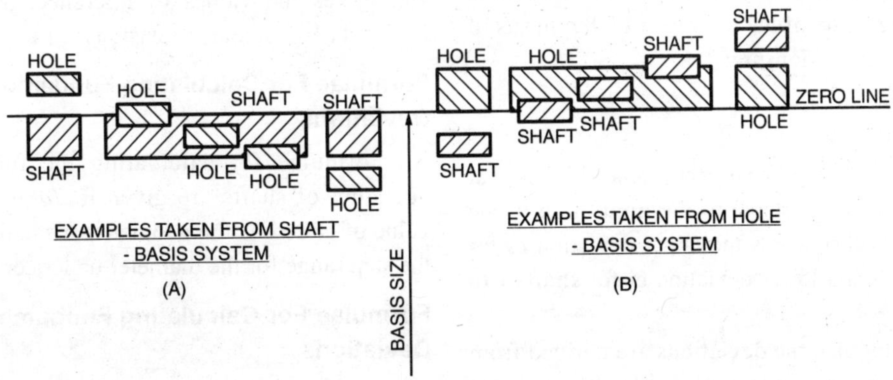

Fig. 13.17.

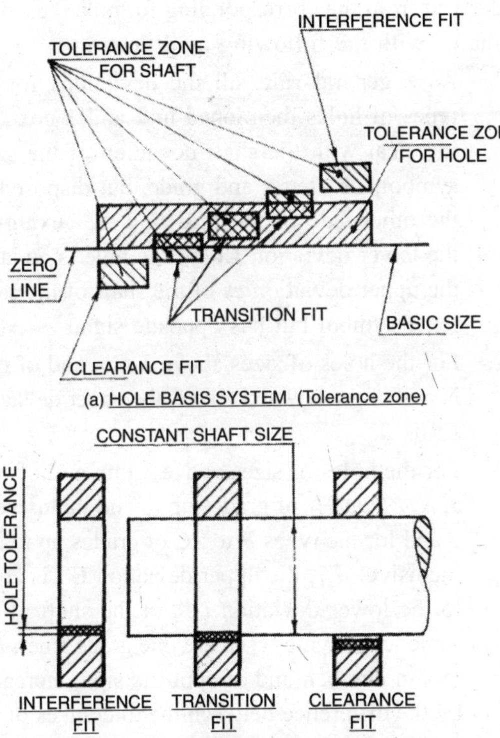

Fig. 13.18.

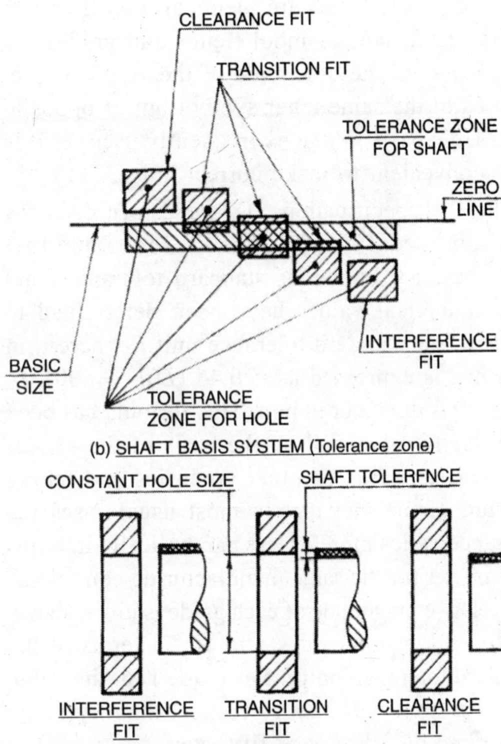

Fig. 13.19.

Shafts **ei = es - IT**

Holes **EI = ES – IT**, where **IT** is the fundamental tolerance of the grade obtained from *Table 13.5*.

For **shafts *a*** to ***h*** the upper deviation is below the zero line and for **shafts *j*** to ***zc*** it is above the zero line. For **holes *A* to *H*** lower deviation is above the zero line and for **holes *J* to *ZC*** it is below the zero line. In the specification, formulae are given to determine the fundamental deviation. The other deviations may be derived directly using the absolute value of the tolerance **IT** by means of the algebraic relationship:

ei = es – IT

es = ei + IT.

The deviation given in the *Table 13.7* is that corresponding in principal to the limit close to the zero line, in other words the upper deviation **es** for **shafts a** to **h** and lower deviation **ei** for **shafts j to zc.**

For holes also the deviations are derived from those of the corresponding shafts. The general rule is that the hole limits are identical with the shaft limits of the same symbol (letter and grade) but distributed on the other side of the zero line, i.e. EI = es of the same letter symbol but of opposite sign. Hole basis system is frequently used, as it is very convenient to make correct holes.

In BIS specifications, 18 grades of tolerances are designated: IT01, IT0, IT1 to IT16, Table 13,3 (*a*). These are known as standard tolerances and their numerical values have been determined in terms of the standard tolerance unit ***i***, where *i*, in microns, is expressed as = $0.45\ (D)^{1/3} + .001\ D$, where *D* is diameter in mm, This formula has been empirically calculated on the basis of the former national standards of other countries and taking account of the fact that in most usual cases the tolerance varies more or less parabolically in terms of diameter for the same manufacturing conditions. The relative magnitude of each grade is given above, in table form, for grades 5 to 16, in terms of the standard tolerance unit ***i***. For grade IT6, the value 10*i* and above it the tolerance magnitude is multiplied by 10 at each fifth step.

The values of standard tolerances corresponding to grade IT01, IT0, IT1 are:

Values in microns	IT01	IT0	IT1
For *D* in mm	0.1 + .008	0.5 + .012D	0.8 + .020D

The values of IT2 to IT4 are regularly scaled approximately geometrically between the values of IT1 and IT5.

The most commonly used shafts and holes in the industries are indicated in *Table 13.4*. *Table 13.5* gives the values of tolerance grades from IT1 to IT16 for diameter steps up to 500 mm.

Formulae For Calculating Fundamental Shaft Deviations

Formulae for calculating the fundamental deviation of shafts are given in *Table 13.9*. The value of D here is the geometric mean diameter of the step range for the diameter under consideration.

Formulae For Calculating Fundamental Hole Deviations

The fundamental deviations for holes are derived from the corresponding formulae, used for shafts, with the following modifications:

1. As a general rule, all the deviations for the types of holes mentioned in 2 and 3 next, are identical with the shaft deviation of the same symbol, i.e., letter and grade, but disposed on the other side of the zero line. For **example,** the lower deviation **EI** for the hole is equal to the upper deviation **es** of the shaft of the same letter symbol but has opposite sign.
2. For the holes of sizes above 3 mm and of type N and of grade 9 and above, the upper deviation ES is 0.
3. For the holes of sizes above 3 mm and of type *J*, *K*, *M*, and *N* of grades up to and inclusive of 8 and for the types *P* to *ZC* of grades up to and inclusive of 7, the upper deviation **ES** is equal to the lower deviation *i.e.*, of the shaft of the same letter symbol but are one grade finer, *i.e.*, less in number, and of opposite sign, increased by the difference between the tolerances of the two grades in question.

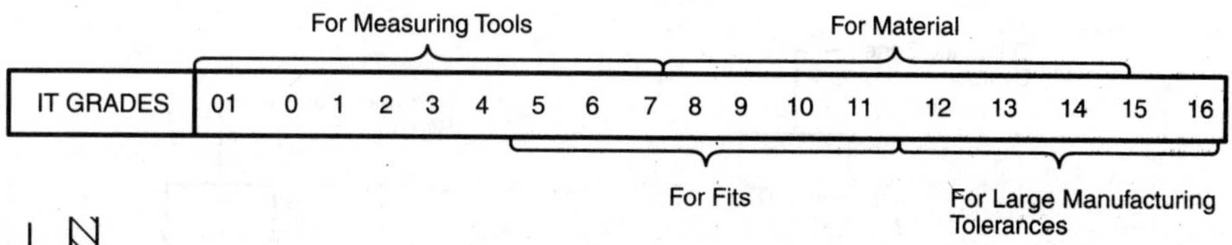
For Measuring Tools
For Material
IT GRADES
01 0 1 2 3 4 5 6 7 8 9 10 11 12 13 14 15 16
For Fits
For Large Manufacturing Tolerances

(A)

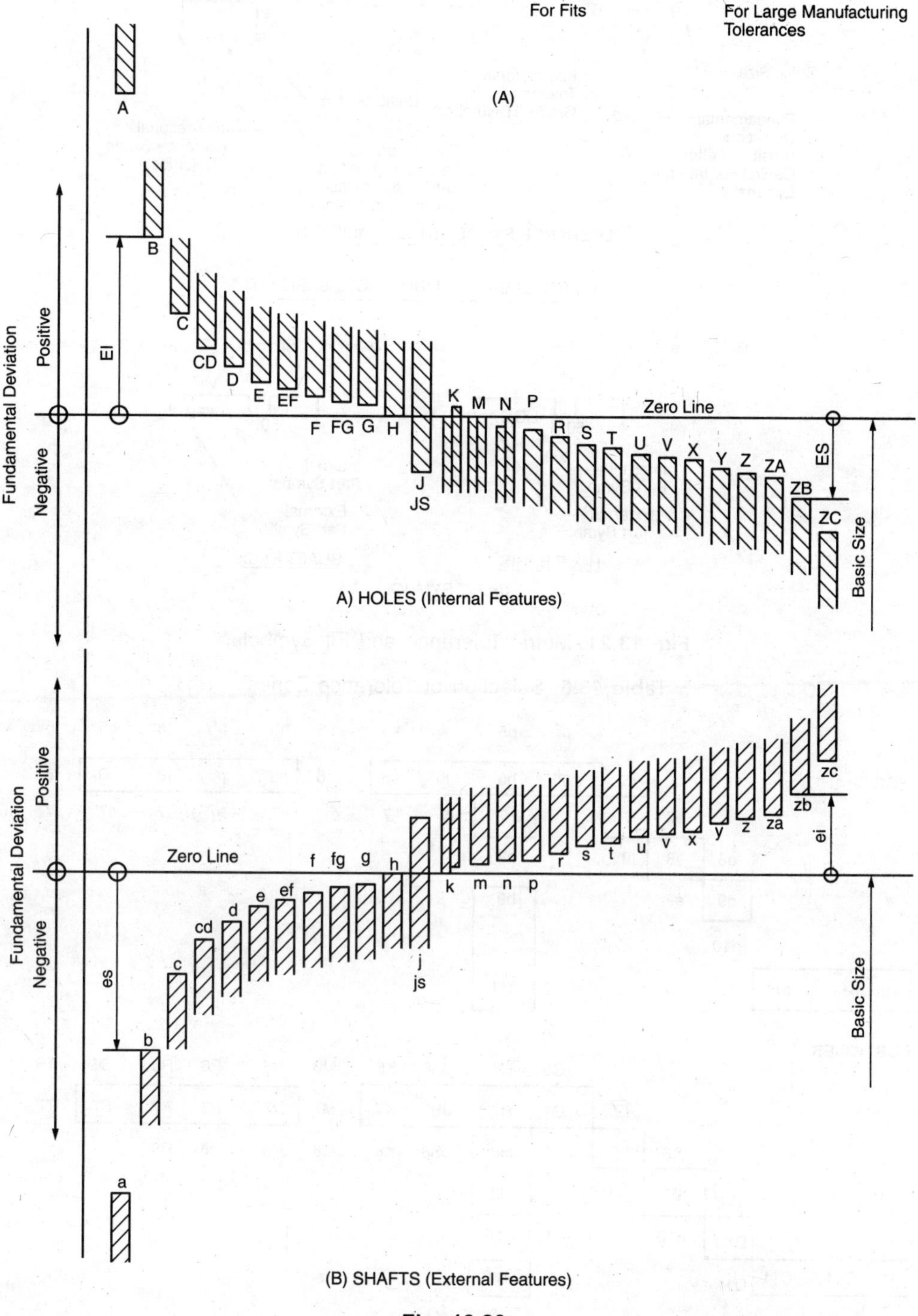
Fundamental Deviation
Positive
Negative
EI
A B C CD D E EF F FG G H J JS K M N P R S T U V X Y Z ZA ZB ZC
Zero Line
ES
Basic Size
A) HOLES (Internal Features)
Fundamental Deviation
Positive
Negative
es
a b c cd d e ef f fg g h j js k m n p r s t u v x y z za zb zc
Zero Line
ei
Basic Size
(B) SHAFTS (External Features)

Fig. 13.20.

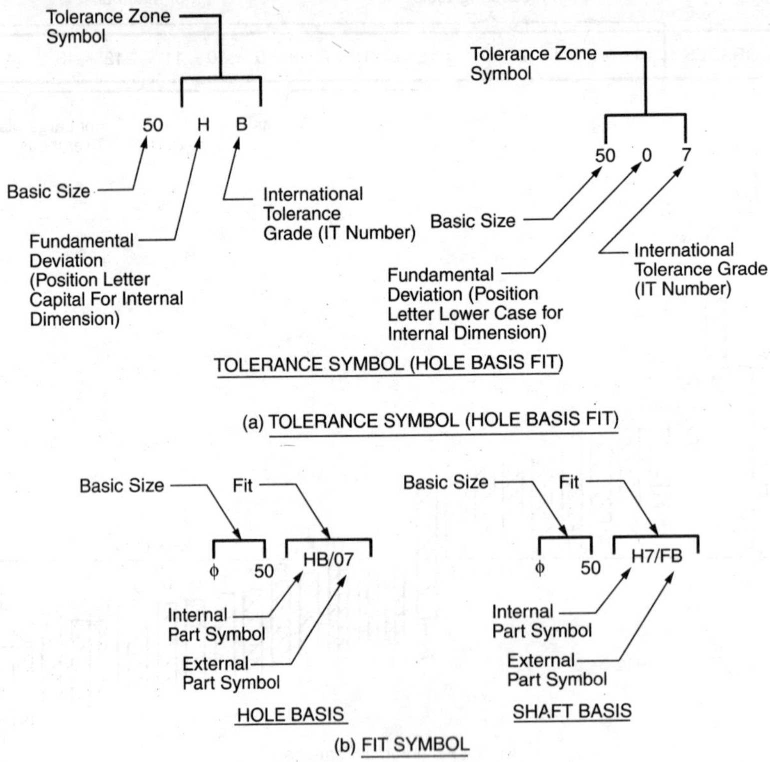

Fig. 13.21. Metric Tolerance and Fit Symbols.

Table 13.6. Selection of Tolerance Zones

FOR SHAFTS

a	b	c	d	e	f	g	h	j/js	k	m	n	p	r	s	t	u
						g5	h5	j5	k5	m5	n5	p5	r5	s5	t5	
					f6	g6	h6	j6	k6	m6	n6	p6	r6	s6	t6	
				e7	f7		h7	js7	k7	m7	n7	p7	r7	s7	t7	u7
			d8	e8	f8		h8									
			d9	e9			h9									
			d10													
a11	b11	c11					h11									

FOR HOLES

A	B	C	D	E	F	G	H	JS	K	M	N	P	R	S	T
						G6	H6	JS6	K6	M6	N6	P6	R6	S6	T6
					F7	G7	H7	JS7	K7	M7	N7	P7	R7	S7	T7
				E8	F8		H8	JS8	K8	M8	N8	P8	R8		
			D9	E9	F9		H9								
			D10	E10			H10								
A11	B11	C11	D11				H11								

Note: Deviations js and JS may be replaced by corresponding deviations j and J

PROBLEM 13.2. *Calculate the fundamental deviations for the following shaft sizes:*

(a) 30 $e8_t$

(b) 50 g6, and

(c) 40 m6.

SOLUTION: From Table 13.9 the deviations for shafts can be obtained.

(*a*) The upper deviation **es** for the shaft ***e*** is;

$= -\ 11\ D^{0.41}$

The value for *D* is

$= \sqrt{(80 \times 30)} = 23.24$ mm.

$\therefore$ ***es*** $= -\ 11 \times 23.24^{0.41} = -\ 40$ microns, as given in Table 13.7.

Table 13.7. Fundamental deviations for shafts of types a to zc of sizes upto 500 mm. (cont.)

Fundamenta deviation in microns															(1 micron - 0.001 mm)
Diameter step in mm		Upper deviation (es)								js	Lower deviation (el)				
		a	b	c	d	e	f	g	h		j			k	
Over	upto	All grades									5.6	7	8	4 to 7	≤ 3 > 7
–	*3	–270	–140	–60	–20	–14	–6	–2	0		–2	–4	–6	–0	–0
3	6	–270	–140	–70	–30	–20	–10	–4	0		–2	–4	–	+1	–0
6	10	–280	–150	–80	–40	–25	–13	–5	0		–2	–4	–	+1	0
10	14	–290	–150	–95	–50	–32	–16	–6	0		–3		–	+1	0
14	18											–6			
18	24	–300	–160	–100	–65	–40	–20	–7	0		–4	–8	–	+2	0
24	30														
30	40	–310	–170	–120	–80	–50	–25	–9	0		–5	–10	–	+2	0
40	50	–320	–180	–130											
50	65	–340	–190	–140	–100	–60	–30	–10	0	±IT/2	–7	–12	–	+2	0
65	80	–360	–200	–150											
80	199	–380	–	–170	–120	–72	–36	–12	0		–9	–15	–	+3	0
100	120	–410	220	–180											
			–240												
120	149	–460	260	–200											
140	160	–520	–280	–220	–145	–85	–43	–14	0		–11	–18	–	+3	0
160	180	–580	–310	–230											
180	200	–660	–340	–240											
200	225	–740	–380	–260	–170	–100	–50	–15	0		–13	–21	–	+4	0
225	250	–820	–420	–280											
250	280	–920	–480	–300	–190	–110	–56	–17	0		–16	–26	–	+4	0
280	315	–1050	–540	–330											
315	355	–1200	–600	–360	–210	–125	–62	–18	0		–18	–28	–	+4	0
355	400	–1350	–680	–400											
400	450	–1500	–769	–440	–230	–135	–68	–20	0		–20	–32	–	+5	0
450	500	–1650	–840	–480											

* The deviation of shafts of types a and b are not provided for diameters upto 1 mm.

** For types js in the particular grades 7 to 11, the two symmetrical deviations ± IT/2 may possibly be rounded if the IT value in microns is odd value, by replacing it by the even value immediately below.

Table 13.7. Fundamental deviations for shafts of types a to zc of sizes upto 500 mm.

Fundamenta deviation in microns															(1 micron - 0.001 mm)
Diameter steps in mm		Lower deviation (es)													
		m	n	p	r	s	t	u	v	x	y	z	za	ab	zc
Over	upto	All grades													
–	3	+2	+4	+6	+10	+14	–	+18	–	+20	–	+26	+32	+40	+60
3	6	+4	+8	+12	+15	+19	–	+23	–	+28	–	+35	+42	+50	+80
6	10	+6	+10	+15	+19	+23	–	+28	–	+34	–	+42	+52	+67	+97
10	14	+7	+12	+18	+23	+28	–	+33	–	+40	–	+50	+64	+90	+130
14	18								+39	+45	–	+60	+77	+180	+150
18	24	+8	+15	+22	+28	+35	–	+41	+47	+54	+63	+73	+98	+136	+188
24	30						+41	+48	+55	+64	+75	+88	+118	+160	+218
30	40	+9	+17	+26	+34	+43	+48	+60	+68	+80	+94	+112	+148	+200	+274
40	50						+54	+70	+81	+97	+114	+136	+180	+242	+325
50	65	+11	+20	+32	+41	+53	+66	+87	+102	+112	+144	+172	+226	+300	+405
65	80				+48	+59	+75	+102	+120	+146	+174	+210	+274	+360	+480
80	100	+13	+23	+37	+51	+71	+91	+124	+146	+178	+214	+258	+335	+445	+585
100	120				+54	+79	+104	+144	+172	+210	+254	+310	+400	+525	+690
120	140				+63	+92	+122	+170	+202	+248	+300	+365	+470	+620	+800
140	160	+15	+27	+43	+65	+100	+134	+190	+228	+280	+340	+415	+535	+700	+900
160	180				+68	+108	+146	+210	+252	+310	+380	+465	+600	+780	+1000
180	200				+77	+122	+166	+236	+274	+350	+425	+520	+670	+880	+1150
200	225	+17	+31	+50	+80	+130	+180	+258	+310	+385	+470	+575	+740	+960	+1250
225	250				+84	+140	+196	+284	+340	+425	+520	+640	+820	+1050	+1320
250	280	+20	+34	+56	+94	+158	+218	+315	+385	+475	+580	+710	+920	+1200	+1500
280	315				+98	+170	+240	+350	+425	+525	+650	+790	+1000	+1300	+1700
315	355	+21	+37	+62	+108	+190	+268	+390	+475	+590	+730	+900	+1150	+1500	+1900
355	400				+114	+208	+294	+435	+530	+660	820	1000	1300	+1650	+2100
400	450	+23	+40	+68	+126	+232	+330	+490	+595	+740	+920	+1100	+1455	+1850	+2400
450	500				+132	+252	+360	+540	+660	+820	+1000	+1250	+1600	+2100	+2600

(*b*) The upper deviation ***es*** for the shaft ***g*** is ;

$= -2.5\ D^{0.34}$

The value of D = 38.73 mm.

Hence ***es*** $= -2.5 \times 38.73^{0.34} = -9$ microns, which is again same as given in *Table 13.7.*

(*c*) The lower deviation ***ei*** for the shaft ***m*** is ;

= + (**IT7** – **IT6**).

From the *Table 13.5* the size 40 is in the step 30 to 50, hence the geometric mean diameter D = 38.73 mm, as before.

Tolerance unit $i = 0.45 \times (D)^{1/3} + 0.001$

$D = 0.45 \times (38.73)^{173} + 0.001 \times 38.73$

= 1.58 microns

The fundamental tolerance for grade 7, from the *Table 13.3 (a)* is 16*i*, *i.e.*, 1.58 × 16 = 25 microns. Hence **ei** = 25 (IT7) –16 (IT6) = + 9 microns, same as given in Table 13.7.

The fundamental tolerance for grade 6 is 10i, from *Table 13.3 (a)*, or 16 microns, from *Table 13.5*.

Table 13.8. Fundamental deviations for holes of types **A** to **ZC** for sizes upto 500 mm (contd.)

Fundamental deviation in microns																				(1 micron = 0.001 mm)
Diameter steps mm		Lower deviation (EI)																		P to ZC
		A*	B*	C	D	E	F	G	H	JS+	J			K		M		N		
Over	Upto	All grades									6	7	8	≤8	>8	≤ 8+*	>8	≤8	>8	≤7
-	3*	+270	+140	+60	+20	+14	+6	+2	0	+IT/2	+2	+4	+6	0	0	–2	–2	–4	–4	Same deviation as for grades > 7 + Δ
3	6	+270	+140	+70	+30	+20	+10	+4	0	±IT/2	+5	+6	+10	-1+Δ	–	-4+Δ	-4Δ	-8+Δ	0	
6	10	+280	+150	+80	+40	+25	+13	+5	0	±IT/2	+5	+8	+12	-1+Δ	–	-6+Δ	-6Δ	-10+Δ	0	
10 14	14 18	+290	+150	+95	+59	+32	+16	+6	0	±IT/2	+6	+10	+15	-2+Δ	–	-7+Δ	-7	-12+Δ	0	
18 24	24 30	+300	+160	+110	+65	+40	+20	+7	0	±IT/2	+8	+12	+20	-2+Δ	–	-8+Δ	-8	-15+Δ	0	
30 40	40 50	+310 +320	+170 +180	+120 +130	+80	+50	+25	+9	0	±IT/2	+10	+14	+24	-2+Δ	–	-9+Δ	-9	-17+Δ	0	
50 65	65 80	+340 +360	+190 +200	+140 +150	+100	+60	+30	+10	0	±IT/2	+13	+18	+28	-2+Δ	–	-11+Δ	-11	-20+Δ	0	
80 100	100 120	+380 +410	+220 +240	+170 +180	+120	+72	+36	+12	0	±IT/2	+16	+22	+34	-3+Δ	–	-13+Δ	-13	-23+Δ	0	
120 140 160	140 160 180	+460 +520 +580	+260 +280 +310	+200 +210 +230	+145	+85	+43	+14	0	±IT/2	+18	+26	+41	-3+Δ	–	-15+Δ	-15	-27+Δ	0	
180 200 225	200 225 250	+660 +740 +820	+340 +380 +420	+240 +269 +280	+170	+100	+50	+15	0	±IT/2	+22	+30	+47	-4+Δ	–	-17+Δ	-17	-31+Δ	0	
250 280	280 315	+920 +1050	+480 +540	+300 +330	+190	+110	+56	+17	0	±IT/2	+25	+36	+55	-4+Δ	–	-20+Δ	-20	-34+Δ	0	
315 335	335 400	+1200 +1350	+600 +680	+360 +400	+210	+125	+62	+18	0	±IT/2	+29	+39	+60	-4+Δ	–	-21+Δ	-21	-37+Δ	0	
400 450	450 500	+1500 +1650	+760 +840	+440 +480	+230	+135	+68	+20	0	±IT/2	+33	+43	+66	-5+Δ	–	-23+Δ	-23	-40+Δ	0	

- The deviation of holes, of types A and B , in grades >8 are not findamentryupto 1 mm.

\+ For the hole of type in Grades 7 and 11, the two symmetrical ± deviations IT/2 may possible be rounded. If the value in microns is an odd v alue replace it by the even v alue immmediately below.

\+ Special case: For the hole M6 ES = 9 from 250 to 315 (instead of –11).

Table 13.8. Fundamental deviations for holes of types **A** to **ZC** for sizes upto 500 mm (contd.)

Fundamental in microns														**(1 micron=0.001 mm)**					
Diameter		**Upper deviations (ES)**																	
Steps in mm		**P**	**R**	**S**	**T**	**U**	**V**	**X**	**Y**	**Z**	**ZA**	**ZB**	**ZC**	**Δ in microns**					
Over	**upto**	**>7**												**3**	**4**	**5**	**6**	**7**	**8**
-	3	-6	-10	-14	-	-18	-	-20	-	-26	-32	-40	-60	Δ = 0					
- 3	6	-12	-15	-19	-	23	-	-28	-		-42	-50	-80	1	1.5	1	3	4	6
6	10	-15	-19	-23	-	-28	-	-34	-	42	-52	-67	-97	1	1.5	2	3	6	7
10	14	-18	-23	-28	-	33	-	-40	-	50	-64	-90	-130	1	2	3	3	7	9
14	18						-39	-45	-	-60	-77	-109	-150						
18	24	-22	-28	-35	-	-41	-47	-54	-63	-73	-93	-136	-188	1.5	2	3	4	8	12
24	30				-41	-48	-55	-64	-75	-8	-18	-160	-218						
30	40	-26	-34	-43	-48	-60	-68	-80	-94	-112	-148	-200	-274	1.5	3	4	5	9	14
40	50				-54	-70	-81	-97	-114	-136	-180	-242	-325						
50	65	-32	-41	-53	-65	-87	-102	-122	-144	-172	-226	-300	-405	2	3	5	6	11	16
65	80		-43	-59	-75	-102	-120	-146	-174	-210	-274	-360	-480						
80	100	-37	-51	-71	-91	-124	-146	-178	-214	-258	-335	-445	-585	2	4	5	7	13	19
100	120		-54	-79	-104	-144	-172	-210	-254	-310	-400	-525	-690						
120	140		-63	-92	-122	-170	-202	-248	-300	-365	-470	-620	-800						
140	160	-43	-65	-100	-134	-190	-228	-280	-340	-415	-535	-700	-900						
160	180		-68	-108	-146	-210	-252	-310	-380	-465	-600	-780	-1000						
180	200		-77	-122	-166	-236	-284	-350	-425	-520	-670	-880	-1150	3	4	6	9	17	26
200	225	-50	-80	-130	-180	-256	-310	-385	-470	-575	-740	-960	-1250						
225	250		-84	-140	-196	-284	-340	-425	-520	-640	-820	-1050	-1350						
250	280		-94	-158	-218	-315	-385	-474	-580	-710	-920	-1200	-1550	3	4	7	9	20	29
280	315	-56	-98	-170	-240	-350	-425	-525	-650	-790	-1000	-1300	-1700						
315	355		-108	-190	-268	-390	-475	-590	-730	-900	-1150	-1500	-1900	4	5	7	11	21	32
355	400	-62	-114	-208	-294	-435	-530	-650	-820	-1000	-1300	-1650	-2100						
400	450		-126	-232	-330	-490	-595	-740	-920	-1100	-1450	-1850	-2400	4	5	7	13	23	34
455	500	-68	132	-252	-360	-540	-660	-820	-1000	-1250	-1600	-2100	-2600						

* In determining K.M.N upto grade 8 and P to ZC upto grade 7 take the Δ values from the column on the right.
Example: For P7 from diameters 18 to 30 mm Δ = 8, hence ES = – 14.

Table 13.9.

Upper Deviation (es)

Shaft designation	In microns (for D in mm)
a	$= -(265 + 1.3D)$ for $D <= 120$ $= -3.5D$ For $D = 120$
b	$= -(140 + 0.85D)$ for $D < = 160$ $= -1.8D$ for $D > 160$
c	$= -52D^{0.2}$ for $D < = 40$ $= -(95 + 0.8D)$ for $D > 40$
d	$= -16D^{0.44}$
e	$= -11D^{0.41}$
f	$= -5.5D^{0.41}$
g	$= -2.5D^{0.34}$
h	$= 0$
$j5$ to $j8$	no formula

Lower Deviation (ei)

Shaft Designation	In microns (for D in mm)
$k4$ to $k7$	$= 0.6(D)^{1/3}$
k for grades <= 3 and >= 8	$= 0$
m	$= +(\text{IT} - \text{IT6})$
n	$= +5D^{0.34}$
p	$= +\text{IT7} + 0 + 5$
r	= geometric mean of values for p and s
s	$= +\text{IT8} + 1$ to 4 for $D <= 50$ $= +\text{IT7} + 0.4\text{D}$ for $D > 50$
t	$= \text{IT7} + 0.63D$
u	$= +\text{IT7} + D$
v	$= +\text{IT7} + 1.25D$
x	$= +\text{IT7} + 1.6D$
y	$= +\text{IT7} + 2D$
z	$= +\text{IT7} + 2.5D$
za	$= +\text{IT8} + 3.15D$
zb	$= +\text{IT9} + 4D$
zc	$= +\text{IT }10 + 5D$

For Js = the two deviations are equal to ± IT/2

Problem 13.3. *Calculate the fundamental deviations for the sizes :*

1. 40 D9,

2. 65 F8.

SOLUTION: From *Table 13.9*, the deviations for holes can also be obtained.

1. The lower deviation **EI** for the hole D is given by $16 \times D^{0.44} = 16 \times (38.73)^{0.44} = 80$ micron, where $D = \sqrt{(30 \times 50)}$ 38.73 mm.
2. Lower deviation for the hole F is $= 5.5 \times D^{0.41}$, where $D = \sqrt{(50 \times 80)} = 63.246$ mm. Hence **EI** = 30 microns, as given in *Table 13.8*.

13.7. METHODS OF TOLERANCING

Tolerances should be specified for all requirements critical to functioning and interchangeability whenever it is doubtful that ordinary or established workshop techniques and manufacturing equipment may be relied upon to achieve a satisfactory standard of accuracy. Indicating such tolerances on drawings does not necessarily imply the use of any particular method of production, measurement or gauging.

Units of the deviations should be expressed in the same units as the basic size, If two deviations relating to the same dimension have to be shown, both should be expressed in the same number of decimal places, *Fig, 13.11 (a)*, except if one of the dimensions is zero, *Fig, 13.11 (b)*, The decimal points should be vertically in line, The size of letters and numerals are same as that of the basic size.

Indication of the Components of A Linear Dimension

1. **ISO Symbols:** The components of the toleranced dimension should be indicated, using ISO Symbols, in the following order :
   - **the basic size**, and
   - **the tolerance symbol**.

   If in addition to the symbols, *Fig. 13.11 (a)*, it is necessary to express the values of the deviations, *Fig. 13.11 (b)*, or the limits of size, *Fig. 13.11 (c)*, the additional information should be shown in brackets.

2. **Permissible Deviations :** The components of the toleranced dimension should be indicated in the following order, Fig. 13.11 (*d*), (*e*), (*f*);
   - the basic size, and
   - the values of the deviation.

   If one of the two deviations is zero, this should be expressed by the digit zero, *Fig. 13.11 (e).* If the tolerance is symmetrical in relation to the basic size, the value of the deviations should be indicated once only, preceded by the sign ±, *Fig. 13.11 (f).*
3. **Limits of Size:** The limits of size may be indicated by an upper and lower dimension, Fig. *13.11 (g)*. If a dimension needs to be limited in one direction only, this should be indicated by adding min. or max. to the dimension, Fig. *13.11(h)*.
4. **General Tolerancing Note:** General tolerance notes greatly simplify the drawing and save much labour in its preparation, *Fig. 13.22*. The use of general notes on a drawing is given below:

- Tolerance on all dimensions ± 0.5.
- Tolerance except where otherwise specified ± 0.02.
- Tolerance on dimensions 140 mm and below ± 0.25.
- Tolerance on cast thickness ± 10%.
- Dimensional deviations for untoleranced dimensions: Coarse class according to BIS: 2102-

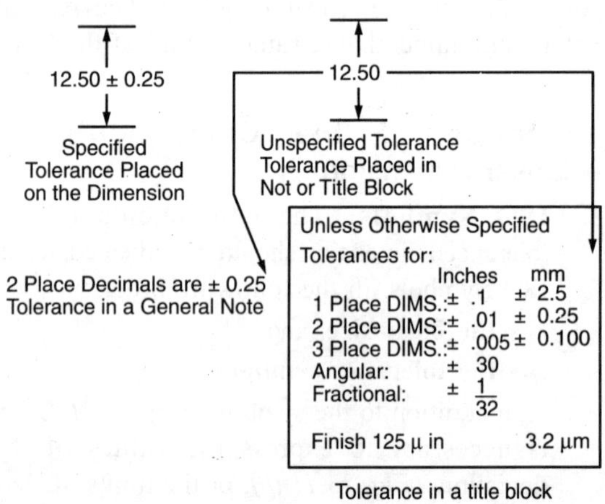

Fig. 13.22.

'*Allowable deviations for dimensions -without specified tolerances*'.

Order Of Indication of Deviations And Limits of Size

The upper deviation or the upper limit of size should be written in the upper position and the lower deviation or the lower limit of size in the lower position, irrespective of whether a hole or a shaft is toleranced.

Indication, of Tolerances on Drawings of Assembled Parts

1. **Symbols :** The tolerance symbol for the hole should be placed before that for the corresponding shaft, *Fig. 13.12 (a),* or above it, *Fig. 13.12 (b)*, the symbols being preceded by the basic size indicated once only. If it is also **ISO** necessary to specify the numerical values of the deviations, they should be written in brackets, *Fig. 13.12 (c)*. But for the sake of simplicity, dimensioning with only one dimension line may be used, *Fig. 13.12 (d)*.
2. **Values By Digits:** Dimensions for each of the components of the assembled parts, should be preceded by the feature name, *Fig. 13.12 (d)*, or the item reference, *Fig. 13.12 (e)*, of the components. The dimension for the hole being placed, in both these cases, above that for the shaft.

Indication of Tolerances on Angular Dimensions

The rules given for the indication of tolerances on linear dimensions are equally applicable to angular dimensions, except that the units of the basic angle and the fractions thereof, as well as deviations should always be indicated, *Fig. 13.12 (f), (g), (h)*. If the angular deviation is expressed in either minutes of a degree or seconds of a minute of a degree, the value of the minutes or seconds shall be preceded by 0° or 0° 0' as applicable.

13.8. GENERAL TOLERANCES FOR LINEAR AND ANGULAR DIMENSIONS

The permissible machining variations in linear dimensions, without tolerance indications, which are applicable for all machining processes with chips like turning, milling, etc., and with chips like punching, drawing, embossing, pipe bending, etc., but not applicable for production methods like forging, casting, welding, flame cutting, etc., as recommended by BIS, are given in *Table 13.1*. Those recommended for angular dimensions are given in *Table 13.2*.

13.9. FITS BETWEEN MATING PARTS

When two mating parts are to be assembled, the relation resulting from the difference between their sizes before assembly is called a Fit.

Any machine, as well as its mechanisms and units, is assembled of parts arranged in a definite relationship to each other. These parts, corresponding surfaces of these parts, and the dimensions of these surfaces are called **mating parts, mating surfaces**, and **mating dimensions,** respectively.

Mating surfaces are classified as **enveloped or external mating surfaces** and **enveloping or internal mating surfaces**. Their dimensions are correspondingly called **enveloped** and **enveloping dimensions**. The difference between the dimensions of the enveloping and enveloped surfaces determines the nature of the association or fit between them. The enveloping surface of a cylindrical part is a hole, and the enveloped surface is a shaft. For the sake of simplicity, and in view of the particular importance of the cylindrical parts with circular section, only these are referred to explicitly. It should be understood, however, that recommendations for this type of components apply equally well to other sections. Thus all internal surfaces of other shapes will also be referred to as holes, and external surfaces of their mating parts as shafts, e.g., a keyway and a key.

Fit is the general term used to signify the range of tightness or looseness resulting from the condition of a specific, combination of allowances and tolerances in the design of mating parts. There are a large number of recommended fits. The selection of a suitable fit, for any specific application, is made by the designer mating *Tables* such as; *13.3, 13.4, 13.6, 13.8, 13.9, 13.10, 13.11* and *13.12*. The designer must also consider the accuracy that can be achieved with the available industrial manufacturing or forming equipment, as shown in *Table 13.3*.

13.10. DEFINITIONS CONCERNING FITS

Important definitions concerning fits are given below:

1. **Clearance:** In a fit, this is the difference between the sizes of the hole and the shaft, before assembly, when this difference is positive, *Fig. 13.13, 13.14 (a), 13.16.*
2. **Interference:** This is the difference between the sizes of the hole and the shaft, before assembly, when this difference is negative, *Fig. 13.13, 13.14 (c).*
3. **Clearance Fit:** It is a fit, which always provides a clearance, *Fig. 13.14 (a)*. In this case the tolerance zone of the hole is entirely above that of the shaft, *Fig. 13.15 (a)*.
4. **Transition Fit:** It is a fit, which may provide either a clearance or an interference, *Fig. 13.15 (b)*. Here the tolerance zones of the hole and the shaft overlap.
5. **Interference Fit:** It is a fit, which always provides an interference, *Fig. 13.15 (c)*. Here the tolerance zone of the hole is entirely below that of the shaft.
6. **Minimum Clearance:** In a clearance fit, it is the difference between the minimum size of the hole and the maximum size of the shaft, *Fig. 13.16 (a)*.
7. **Maximum Clearance:** In a clearance or transition fit, it is the difference between the maximum size of the hole and the minimum size of the shaft, *Fig. 13.16 (b)*.
8. **Minimum Interference:** In an interference fit, the arithmetical difference between the maximum size of the hole and the minimum size of the shaft before assembly, *Fig. 13.16 (c)*.
9. **Maximum Interference:** In an interference or transition fit, the arithmetical difference between the minimum size of the hole and the maximum size of the shaft before assembly, *Fig. 13.16 (4)*.

13.11. SYSTEMS OF FITS

Two of the most commonly used methods of applying fits are the **hole basis system** and the **shaft basis system**. In hole basis system of fits each of the fits has a basic hole, *Fig. 13.17 (a)*. In

Table 13.10. Commonly used Fits

I-For Holes							
Types of Fit	Class of shaft	With Holes H6*	H7	H8	H9	H11	Remarks
CLEARANCE	a b					a11 b11	Large clearance fit and widely used
	c					c11	Stock running fit
	d		d8*	d8*, d9,d10	d8*, d9,d10	d9	Loose running fit
	e	e7*	e8	e8 e9*			Easy running fit
	f	16*	17	17,18*			Normal running fit
	g	g5*	g6				Close running fit or sliding fit, also spigol and location fit
	h	h5*	h6	h7,h8*,h9		h11	Precision sliding fit. Also fine spigot and location fit.
TRANSITION	js	js5*	js6	js7*			Push fit for very accurate location with easy assembly and disassembly
	k	k5*	k6	k7			Light keying fit (true transition) for keyed shafts, non-running locked pins etc.
	m	m5*	m6*	m7*			Medium keying fit
	n	n5*	n6	n7*			Heavy keying fit (for tight assembly mating surfaces)
INTERFERENCE	p	p5*	p6	p7*			Light press fit with easy dismentting for non-ferrous parts, standard press with easy dismentling for ferrous and non-ferrous parts assembly
	r	r5*	r6	r7*			Medium drive fit with easy dismentling for ferrous parts assembly. Light drive fit with easy dismentling for non-ferrous parts assembly
	s	s5*	s6*	s7*			Heavy drive fit ferrous parts permanent or semi-permanent assembly standard press fit for non-ferrosus parts.
	t	t5*	t6*	t7*			Force fit on ferrous parts for permanent assembly
	u			u7*			Heavy force fit or shrink fit.

* Second preference fits.

Table 13.10. (Concluded)

II-For Shafts							
Types of Fit	Class of shaft	With Shafts					Remarks
		h5*	h6	h8*	h9	h11	
CLEARANCE	A B					A11 B11	Large clearance fit widely used
	C					C11	Stock running fit
	D			D9*		D9	Loose running fit
	E			E8*			Easy running fit
	F			F7*			Normal running fit
	G		G6*	G7			Close running fit or sliding fit, also spigot and location fit
	H	H6*	H7	H8	H9	H11	Precision sliding fit also fine spigot and location fit.
TRANSITION	JS	JS6*	JS7*	JE8*			Push fit for very accurate location with easy assembly and disassembly
	K	K6*	K7	K8*			Light keying fit (true transition) for keyed shafts, non-running locked pins etc.
	M	M6*	M7*	46*			Medium keying fit
	N	N6*	N7*	N8*			Heavy keying fit (for tight assembly mating surfaces)
INTERFERENCE	P	P6*	P7				Light press fit with easy dismentting for non-ferrous parts, standard pressfit with easy dismentling for ferrous and non-ferrous parts assembly
	R	R6*	R7				Medium drive fit with easy dismentling for ferrous parts assembly. Light drive fit with easy dismentling for non-ferrous parts assembly
	S	S6*	S7				Heavy drive fit ferrous parts permanent or semi-permanent assembly standard press fit for non-ferrosus parts.
	T	T6*	T7				Force fit on ferrous parts for permanent assembly

* Second preference fits.

Table 13.11

UNIT HOLE SYSTEM I.S.O																																				
			DIFFERENC IN MEASUREMENT OF NOMINAL DIAMETER IN MICRONS																																	
			VERY ACCURATE WORK							ACCURATE WORK											ORDINARY WORK										ROUGH WORK					
			HOLE	SHAFT						HOLE	SHAFT										HOLE	SHAFT									HOLE	SHAFT				
OVER	TO			CLEAR FIT		TRANS. FIT			INT. FIT.		CLEAR FIT					TRANS. FIT				INT. FIT.		CLEAR FIT					TRANS. FIT					CLEAR FIT				
1	3	Go	0	-3	0	+4	-	+7	+11	0	-20	-14	-7	-13	0	+6	-	+9	+9	+16	0	-20	-14	-7	0	0	+7	-	-	+15	0	-270	-140	-60	-20	0
		Not go	7	-8	-5	-1	-	+2	+6	+9	-45	-28	-16	-10	-7	-1	-	+2	+2	+9	+14	-60	-39	-21	-14	-9	-2	-	-	+5	+60	-330	-200	-120	-80	-60
3	6	Go	0	-4	0	+4	-	+9	+13	0	-30	-20	-10	-4	0	+7	-	+12	+12	+20	0	-30	-20	-10	0	0	+9	-	-	+20	0	-270	-140	-70	-30	0
		Not go	-8	-9	-5	-1	-	+4	+8	-12	-60	-38	-22	-12	-6	-1	-	+4	+4	+12	+18	-78	-50	-28	-18	-12	-3	-	-	+8	+75	-345	-215	-145	-105	-75
6	10	Go	0	-5	0	+4	+7	+12	+16	0	-40	-25	-13	-5	0	+7	+10	+15	+15	+24	0	-40	-25	-13	0	0	+10	+16	+21	+25	0	-280	-150	-80	-40	0
		Not go	-9	-11	-6	-2	+1	+6	+10	-15	-76	-47	-28	-14	-9	-2	+1	+6	+6	+15	+22	-98	-61	-35	-22	15	-5	+1	+6	+10	+90	-370	-240	-170	-130	-90
10	18	Go	0	-6	0	+5	+9	+15	+20	0	-50	-32	-16	-6	0	+8	+12	+18	+18	+29	0	-50	-32	-16	0	0	+12	+19	+25	+30	0	-290	-150	-95	-50	0
		Not go	-11	-14	-8	-3	+1	+7	+12	-18	-93	-59	-34	-17	-11	-3	+1	+7	+7	+18	+27	-120	-75	-43	-27	-18	-6	+1	+7	+12	+110	-400	-260	-205	-150	-110
18	30	Go	0	-7	0	+5	+11	+17	+24	0	-65	-40	-20	-7	0	+9	+15	+21	+21	+35	0	-65	-40	-20	0	0	+13	+23	+29	+36	0	-300	-160	-110	-65	0
		Not go	-13	-16	-9	-4	+2	+8	+15	-21	-117	-73	-41	-20	-13	-4	+2	+8	+8	+22	+33	-149	-92	-53	-33	-21	-8	+2	+8	+15	+130	-430	-290	-240	-195	-130
30	50	Go	0	-9	0	+6	+13	+20	+28	0	-80	-50	-25	-9	0	+11	+18	+25	+25	+42	0	-80	-50	-25	0	0	+15	+27	+34	+42	0	-310	-170	-120	-80	0
		Not go	–16	-20	-11	-5	+2	+9	+17	-25	-142	-89	-50	-25	-16	-5	+2	+9	+9	+26	+39	-180	-112	-64	-39	-25	-10	+2	+9	+17	+160	-480	-340	-290	-240	-160
50	80	Go	0	-10	0	+6	+15	+24	+33	0	-100	-60	-30	-10	0	+12	+21	+30	+30	+51	0	-100	-60	-30	0	0	+18	+32	+41	+50	0	-340	-190	-140	-100	0
		Not go	+19	-23	-13	-7	+2	+11	+20	-30	-174	-106	-60	-29	-19	-7	+2	+11	+11	+32	+46	-220	-134	-76	-46	30	-12	+2	+11	+20	+190	-550	-390	-340	-290	-190
80	120	Go	0	-12	0	+6	18	+28	+38	0	-120	-72	-36	-12	0	+13	+25	+35	+35	+59	0	-120	-72	-36	0	0	+20	+38	+48	+58	0	-380	-220	-170	-120	0
		Not go	+22	-27	-15	-9	+3	+13	+23	-35	-20	-126	-71	-34	-22	-9	+3	+13	+13	+37	+54	-260	-159	-90	-54	-35	-15	+3	+13	+23	+220	-630	-480	-400	-340	-220
120	180	Go	0	-14	0	+7	+21	+33	+45	0	-145	-85	-43	-14	0	+14	+28	+40	+40	+68	0	-145	-85	-43	0	0	+22	+43	+56	+67	0	-480	-260	-200	-145	0
		Not go	+25	-32	-19	-11	+3	+15	+27	-40	-245	-148	-83	-39	-25	-11	+3	+15	+15	+43	+63	-305	-185	-106	-66	-40	-18	+3	+15	*27	+250	-830	-560	-480	-395	-250
180	250	Go	0	-15	0	+7	+24	+37	+51	0	-170	-100	-50	-15	0	+16	+33	+46	+46	+79	0	-170	-100	-50	0	0	+25	+50	+63	+77	0	-660	-340	-240	-170	0
		Not go	+29	-35	-20	-13	+4	+17	+31	-48	-285	-172	-96	-44	-29	-13	+4	+17	+17	+50	+72	-355	-215	-122	-72	-46	-21	+4	+17	+31	+290	-110	-710	-570	-450	-290
250	315	Go	0	-17	0	+7	+27	+43	+57	0	-190	-110	-56	-17	0	+16	+36	+52	+52	+88	0	-190	-110	-56	0	0	+26	+56	+72	+86	0	-920	-480	-300	-190	0
		Not go	-32	-40	-23	-16	+4	+20	+34	-52	-320	-191	-108	-49	-32	-16	+4	+20	+20	+56	-81	-400	-240	-137	-81	-52	-26	+4	+20	+34	+320	-1370	-860	-650	-510	-320
315	400	Go	0	-18	0	+7	+29	+46	+62	0	-210	-125	-62	-18	0	+18	+40	+57	+57	+98	-0	-210	-125	-62	0	0	+29	+61	+78	+94	0	-1200	-600	-360	-210	0
		Not go	-36	-43	-25	-18	+4	+21	+37	-57	-350	-214	-119	-54	-36	-18	+4	+21	+21	+62	+89	-440	-265	-151	-89	-57	-28	+4	+21	+37	+360	-1710	-1040	-760	-570	-360
400	500	Go	0	-20	0	+7	+32	+50	+67	0	-230	-135	-68	-20	0	+20	+45	+63	+63	+108	0	-230	-135	-68	0	0	+31	+68	+86	+103	0	-1500	-760	-440	-230	0
		Not go	-40	-47	-27	-20	+5	+23	+40	-63	-385	-232	-131	-60	-40	-20	+5	+23	+40	+68	97	-480	-290	-165	-97	-63	-32	+5	+23	+40	+400	-2050	-1240	-880	-630	-400

shaft basis system of fits each fit has a basic shaft, *Fig. 13.17 (b)*.

The hole basis system is the system of fits in which design size of the hole is the basic size and the allowance is applied to the shaft, *Fig. 13.17 (a), 13.18*. The design size of the shaft is obtained by subtracting the allowance from the basic size of the hole. Tolerances are then applied to each part separately. The lower deviation of hole is zero. The letter symbol for this situation is 'H'. The shaft sizes are varied according to different types of fit required. From the production point of view, this system is economical as a single drill or reamer size can be used to produce a variety of fits by merely altering the shaft limits. Shafts can be accurately produced to size by turning or grinding. **For example**, for a ϕ 25 H8/f7 fit, which is a preferred hole basis clearance fit, the limits for the hole and shaft will be as follows:

Hole limits = ϕ 25.000 – ϕ 25.033

Shaft limits = ϕ 24.959 – ϕ24.980

Minimum clearance = smallest hole – largest shaft

= 25.000 – 24.980

= 0.020

Maximum clearance = Biggest hole – Smallest shaft

= 25.033 – 24.959

= 0.074

If a ϕ 25H7/s6 preferred hole basis interference fit is required, the limits for the hole and shaft will be as follows:

Minimum interference = Largest hole – smallest shaft

= 25.021-25.035

= – 0.014

Maximum interference = smallest hole – largest shaft

= 25.000 – 25.048

= – 0.048

In the shaft basis system, *Fig. 13.18*, the design size of the shaft is the basic size and the allowance is applied to the hole. The design size of the hole is obtained by adding the allowance to the basic size of the shaft. Tolerances are then applied to each part separately. The upper deviation of the shaft is zero. The letter symbol used for this situation is '*h*'. The shaft size remains constant and the mating holes are varied to produce the required grade of fit.

The application of either system depends upon many conditions, such as, the nature of the product, the manufacturing methods, the condition of the raw material, etc.

A hole basis system is the extensively used system. This is because a hole is more difficult to produce than a shaft due to the fixed character of the hole producing tools. The shaft basis system should only be used where it will convey unquestionable economic advantages, i.e., where it is necessary to be able to mount several parts with holes having different deviations on a single shaft of drawn steel bar without machining the latter. The designers should decide on the adoption of either system to secure general interchangeability.

Tolerances for holes and shafts are identical with those for a basic hole system. However, the basic size becomes the maximum shaft size. **For example,** for a ϕ16 *C* 11/ *h* 11 fit, which is a preferred shaft basis clearance fit, the limits for the hole and shaft will be as follows:

Hole limits = ϕ 16.095 – ϕ 16.205

Shaft limits = ϕ 15.890 – ϕ 16.000

Minimum clearance = smallest hole - largest shaft

= 16.095- 16.000

= – 0.095

Maximum clearance = largest hole – smallest shaft

= 16.205 – 15.890

= 0.315

A shaft dimension to a certain class of fit with a hole in the hole basis system differs from the shaft giving the same fit in the shaft basis system.

BIS recommend a unilateral hole basis system, but a full range of holes suitable for use on shaft basis, unilateral or bilateral, is also included. Certain variations in the sizes of the shafts and holes will be allowed during manufacture. If these variations are set in such a manner that any shaft will make an assembly with any hole and vice versa, the method is called an "interchangeable system".

13.12. CLASSIFICATION OF FITS

Depending upon the respective positions of the tolerance zone of the hole or the shaft the fits may be of three general types : clearance fit, transition fit, and interference fit, depending upon the actual limits of the hole or shaft. *Fig. 13.13* illustrates the three types of fit. *Fig. 13.14* illustrates a clearance fit, and *Fig. 13.15* illustrates the disposition of tolerance zones in various cases. It is very obvious from *Fig. 13.14* and *Fig. 13.15* that the setting of tolerance value is not sufficient to define particular limits. The position of the tolerance zone relative to the basic size of the feature must also be specified.

1. Clearance Fits

Clearance fit results in a positive clearance over the whole range of the tolerance. Shafts **a** to **g** produce a **clearance fit** with the basic bole *H*. Clearance fits are intended for accurate assembly of parts and bearings. The parts can be assembled by hand because the hole is always larger than the mating shaft.

Figure 13.14 (a) shows a spindle, which slides freely in a bush. The basic size of the fit is 50 mm. Here selected ISO fits , hole basis, are used. If an H9 fit is chosen for the hole and a d10 fit for the shaft then, from the ISO chart, the tolerances quoted are;

+ 62 for the H9 hole and
0
– 80 for the d10 shaft
– 180

These tolerances are given in units of one thousand of millimeter. So the upper limit of hole will be 50.062 and lower limit for the hole is the same as the basic size of 50.000.

The shaft upper limit = 50.000 – 0.080
= 49.920.

The shaft lower limit = 50.000 – 0.180
= 49.820.

The allowance for the hole = 50.062 – 50.000
= 0.062.

The tolerance for the shaft is = 49.920 – 49.820
= 0.100.

The minimum clearance or allowance is
= 50.000 – 49.920
= 0.080.

2. Transition Fits

Transition fits have limits of size indicating that either a clearance or interference may result when mating parts are assembled. Transition fits are a compromise between clearance and interference fits. **Shafts js** to **n** produce transition fit with the basic **hole *H***. They are used for applications where accurate location is important, but either a small amount of clearance or interference is permissible.

Figure 13.14 (b) shows a bushing, which is required to fit into a casting, and some interference can be tolerated. For this particular application, a H7 fit has been selected for the hole and a k6 fit for the shaft. The tolerances quoted are ;

+ 25 for the hole and
0
+ 18 for the shaft.
+ 2

The upper limit for the hole will be
= 50.000 + 0.025 = 50.025

The lower limit for the hole will be
= 50.000 + 0 = 50.000

The hole tolerance is
= 50.025 – 50.000 = 0.025.

The upper limit for the bushing will be
= 50.000 + 0.018 = 50.018.

The lower limit for the bushing will be
= 50.000 + 0.002 = 50.002.

The bush tolerance is
= 50.025 – 50.002 = 0.023.

The maximum interference between the bushing and the hole is
= 50.018-50.002 = 0.016

The maximum clearance between the bush and the hole is
= 50.018 – 50.000 = 0.18.

3. Interference Fits

Interference fits have limits of size so prescribed that an interference always results when the mating parts are assembled. The hole is always smaller than the shaft. **Shafts p** to **u** produce interference fit with the **basic hole H**. Interference fits are for permanent assemblies of parts, which require rigidity and alignment, e.g., dowel pins and bearings in castings. Parts are usually press fitted together using an arbor press.

Figure 13.14 (c) shows a dowel pin that is required to be inserted in a base. This is an example of a light press fit with a small minimum interference. For this application a **H7 fit for the hole** and a **p6 fit for the shaft** has been chosen. In the ISO Table for fits, the tolerances quoted are

+25 for the hole and

0

+ 42 for the mating shaft.

+ 26

The combination gives the following dimensions :

The upper limit for the hole will be

= 50.000 + 0.025 = 50.025.

The lower limit for the hole will be

= 50.000 + 0 = 50.000.

The hole tolerance is

= 50.025 – 50.000 = 0.025.

The upper limit for the dowel pin will be

= 50.000 + 0.042 = 50.042.

The lower limit for the dowel pin will be

= 50.000 + 0.026 = 50.026.

The tolerance on the dowel pin is

= 50.042 – 50.026 = 0.016.

The maximum interference between the dowel pin and the hole is

= 50.042 – 50.000 = 0.042.

The minimum interference between the dowel pin and the hole is

= 50.026 – 50.025 = 0.001.

Tolerances and deviations for both holes and shafts offer a wide range of fits of which many of the possible combinations may not be of practical use. Majority of common engineering requirements may be satisfied on the basis of a restricted selection of tolerance grades resulting in economy and ease of standardization, yet leading to universally recommended fits. Commonly used fits are given in *Table 13.10*.

13.14. GUIDE FITS FOR SELECTION OF FTS

This guide gives recommendations in engineering problems concerned with the mating of a shaft and a hole. The recommendations are also applicable to non-cylindrical fits.

This guide also gives the representative usage of various classes and grades of fit. The examples are only of an illustrative character and they do not specify any design details.

Much experience is necessary in order to select the appropriate tolerance for a particular application. Fits may be selected on the hole basis system or the shaft basis system.

An engineer must be familiar with the manufacturing processes involved and bear in mind such factors as the length of contact of mating parts, running speed, lubrication, temperature, humidity, surface finish, type and quality of the material used for each part, also the forces and loading which are present. Practical experience is indispensable in the selection of fits to guarantee design performance.

The selection of machine tools to give the required degree of accuracy must also be given serious consideration, since it is of little use quoting tolerances on drawings, which cannot be produced and maintained for economical production in the shop. It should be appreciated that the cost of production to maintain the increased accuracy required for each dimension increase at a steeper rate. For all applications, the manufacturing tolerance chosen should be the largest possible that will permit the intended functional performance in operation.

Innumerable number of combinations of shafts and holes may be used for a fit, as are given in the National Standard on Limits and Fits. As a common practice the hole basis system of fits is followed. However, in certain cases when products are made from bright drawn bars, shaft basis system may also be employed, and actually which system should be chosen depends upon many conditions.

Some characteristics of various grades of hole basis system, unilateral, from H5 to H11 are discussed below:

- *H5* - precision boring, fine internal grinding or honing can obtain this grade.
- *H6* - This grade can be obtained by precision boring and honing or by hand reaming.
- *H7* - grinding, broaching or careful reaming can produce this.
- *H8* - boring or machine reaming can produce this grade.

- *H*9 - This grade is mostly used for noncircular fits and can be obtained by boring and reaming.
- *H*10 - This grade is not used for diameter fits. It is used for milled widths and unimportant parts or drilled holes.
- *H*11 - This grade too is never used in fits as it is very coarse. It is useful only for coarse drilled or punched holes.

We will now study the cases of three types of fits one by one.

1. Clearance Fit

This type of fit results in a positive clearance over the whole range of the tolerance. *Shafts **'a'** to **'h'*** produce a clearance fit with the basic holes 'H7'.

Shafts *'a'*, *'b'* and *'c'*: These fits give a large clearance and are not widely used.

Shaft *'d'* : This fit is available in grades 8 to 11 and is generally suitable for loose running fits as in case of Plummer block bearings and loose pulleys, etc.

Shaft *'e'*: This fit is available in grades 7, 8, and 9. It is best suited for general loose clearance fits such as properly lubricated bearings requiring appreciable clearance. In the finer grades it may be used on large, high speed, heavily loaded bearings such as turbogenerator and large electric motor bearings etc.

Shaft *'f'*: This fit is available in grades 6,7 and 8. It is widely used for normal grease lubricated or oil lubricated bearings where no substantial temperature differences are encountered, e.g., in gearbox shaft bearings, bearings for small electric motors and pumps, etc.

Shaft *'g'*: This fit is expensive to manufacture as the clearance is small and is, therefore, not recommended as a running fit except in precision equipment where shaft loadings are very light, e.g., in bearings for accurate link work, for piston and slide valves etc., and for the spigot and socket location fits.

Shaft *'h'*: This fit exists in grades 5 to 11 and is widely used for non-running parts. It is useful for normal location and spigot fits. In finer grades, it is used as precision sliding fit.

2. Transition Fits

These may result in either clearance fit or interference fit depending upon the actual value of the individual tolerances of the mating features. **Shafts *'j'* to *'n'*** produce transition fit with the *basic hole 'H7'*.

Shaft *'j'* : This is available in grades 5 to 11 and is a transition fit averaging a slight clearance. It is recommended for location fits requiring less clearance than is given with the 'h' shaft and where a slight interference is permissible. For example, in coupling spigots and recesses, gear rings clamped to steel hubs, etc.

Shaft *'k'* : This is the true transition fit averaging virtually no clearance. Available in grades 5 to 11, it is best suited for location fits where a slight interference can be tolerated as with the object of diminating vibration.

Shaft *'m'*: This is available in grades 5, 6 and 7. It averages a slight interference, although appreciable force is required if extremes are encountered.

Shaft *'n'* : This transition fit gives clearance only in extreme sizes. Available in grades 5, 6 and 7. It is recommended for tight assembly fits. The true interference would result with the finest grade shaft (n5) with the hole H6.

3. Interference Fits

These fits result in positive interference over the whole range of tolerance and can be obtained by shafts *'p'* 'to 'z' with the basic hole 'H7'.

Shaft *'p'*: This gives a true interference fit with holes of grades 6 or 7 but with holes of grade 8 transition fit will be obtained. The amount of interference is not excessive but is sufficient to give nonferrous parts a light press fit, which can be dismantled when required. It is the standard press fit for steel, cast iron or brass to steel assemblies, e.g. fixing a bush on a gear. For light M alloys (very elastic materials) the amount of interference is too small for getting satisfactory fits.

Shaft *'r'* : This combination gives a medium fit on ferrous parts and a light drive fit on non-ferrous parts, which can be easily dismantled. An interference fit is obtained with the holes H8 with diameters over 100 mm, and with smaller diameters a transition fit results.

Shaft *'s'* : This is used for permanent or semi permanent assemblies of steel and cast iron members, the amount of interference being sufficient to provide considerable gripping force. When using elastic materials, e.g., light alloys, it is equivalent to the 'p' shafts in ferrous metals. Typical applications in this type of fit are collars pressed on to shafts, valve shafting, etc.

Shafts '*t*', '*u*' : No general recommendations can be made about these fits as these involve an increasing degree of interference.

Shafts '*x*', '*y*', '*z*', '*v*': These fits give a very large interference with the basic hole and are, therefore, not recommended for fits.

In order to obtain the most economical manufacture consistent with satisfactory functional performance and quality, it is recommended that as wide a tolerance should be provided as is possible for the required fit and function. In the allocation of tolerances to the hole and shaft members of a fit, it is generally advisable to give a larger tolerance to the hole than the shaft due to the fact that a hole is more difficult to produce than a shaft. The exception is in the case of very large sizes where the effects of temperature play a large role.

Recommendations for selection of tolerance zones for general purposes are given in *Table 13.6.*

Whenever possible the tolerance zones shall be chosen from the corresponding symbols for shafts and holes as indicated in *Table 13.6*. The first choice shall preferably be made from the symbols enclosed in the frames.

Commonly used fits for holes and shafts are given in *Table 13.7*. *Table 13.8* gives the tolerances for different types of work as per the unit hole system ISO. *Table 13.12* illustrates some sample situations and can serve as a guide for selecting the fits in parallel cases. For complete coverage of the various types of fits and tolerances the national standards should be used for production environment applications.

The use of the tables for deciding various fits is illustrated by the following solved problems.

PROBLEM 13.4. *Find , (a) the basic size, (b) type of fit, and (c) tolerance of a 45 mm diameter shaft rotating at a normal speed in its bearing. Neglect temperature effects.*

SOLUTION: (*a*) Basic size = 45 mm.

(*b*) Type of fit. From the given condition, i.e., shaft rotating at normal speed in its bearing, referring to *Table 13.7* or *13.8* on selection of fits, we see that the fit may be *H6/f6* or *H7/J7* or *H8/f8*. Let us say that here the fit adopted is *45 H8/f8* where H8 means an *'H' hole to tolerance grade 8, and f8 means an f shaft to tolerance grade 8.*

(*c*) Tolerance : To establish the tolerances we can follow any one of the following two methods:

METHOD 1: Referring to Table 13.8 unit hole basis ISO, as 45 mm lies in diameter step 30 mm to 50 mm, we can prepare the following table.

Over	To		Hole	Shaft
			H8	f8
30	50	Go	+.000	–.025
		Not Go	+.039	–.064

∴ In Go Limit

(*a*) Upper limit of shaft = – .025

(*b*) Lower limit of hole = + .000

In 'Not Go' Limit

(*a*) Lower limit of shaft = – 0.64

(*b*) Upper limit of hole = + 0.39

$$\therefore\ 45\text{ H8} = 45^{+.039}_{+.000}$$

$$\text{and } 45\text{ f8} = 45^{-.025}_{-.064}$$

METHOD 2: To find out the tolerance for the shaft and hole pair designated 45 H8/f8, from the *Table 13.8* we see that 45 mm lies in diameter step 30 to 50 mm.

$$\therefore \quad D = \sqrt{(30\times 50)} = \sqrt{1500}$$

$$= 10\sqrt{15} = 38.67 \text{ or } 38.7 \text{ mm.}$$

∴ Value of tolerance unit i

$$= 0.45\sqrt[3]{D} + .001\,D + .001\,D$$

$$= 0.45\sqrt[3]{38.7} + .001\times 38.7$$

$$= 0.45 \times 3.38 + 0.387 = 0.00156 \text{ mm}$$

Now for hole of type '*H*' and grade '8' the fundamental tolerance IT8 = 25 i (from Table of fundamental deviations in *Table 13.3*)

$$= 25 \times 0.0156 = 0.039 \text{ mm.}$$

(This can be directly read from *Table 13.8*, against the given grade and diameter step). For the holes 'H' the fundamental deviation = 0, (from BIS : 919 formulas for shaft and hole deviation).

$$\therefore \text{ the hole limits are } 45^{+.039}_{+.000}$$

For the quality '8' shaft, the fundamental

tolerance, as for the hole, but with -ive sign = $25i$

$= -.039$ mm

Also for 'f' shaft the fundamental deviation (from BIS : 919, table giving formulae for shaft and hole deviation);

$$= -5.5\ D^{0.41}$$
$$= -5.5 \times 38.7^{0.41}$$
$$= -5.5 \times 4.545$$
$$= -25 \text{ microns}$$
$$= -0.025 \text{ mm.}$$

$\therefore$ Upper deviation = 0.039 mm, and

Work tolerance = 0.039 mm

$\therefore$ Lower deviation = – 0.025 – 0.039

= – 0.064 mm

$$\therefore \quad 45\ f8 = 45^{-.025}_{-.064}$$

PROBLEM 13.5. *Fix the limits of tolerance and allowance for a 25 mm diameter shaft and hole pair designated H8/d9. Comment on this type of fit represented.*

SOLUTION:

METHOD 1: To find tolerance for 25 H8/d9, we can prepare the following table by referring to *Table 13.8* of unit hole system ISO.

Over	To		Hole H8	Shaft d9
18	30	Go	+0.000	–0.065
		Not Go	+0.033	–0.117

$\therefore$ In 'Go' Limit

(*a*) Upper Limit of shaft = – 0.065

(*b*) Lower Limit of hole = + 0.000

In 'Not Go' Limit

(*a*) Lower Limit of shaft = – 0.117

(*b*) Upper Limit of hole = + 0.033

+ 0.033

$\therefore$ 25 H8 = 25

+ 0.000

– .065

$$\text{and } 25\ d9 = 25^{-.065}_{-.117}$$

METHOD 2: From Table 13.8 we note that 25 mm lies in diameter step 18 to 30 mm and that for hole of quality '8', the fundamental tolerance IT8 = 33 microns.

Also for the 'H' hole the fundamental deviation = 0

$$\text{The hole limits are : } 25\ H8 = 25^{+.033}_{+.000}$$

Also from *Table 13.8*, for the quality '9' shaft the fundamental tolerance IT9 = 52 microns with negative sign.

For 'd' shafts the fundamental deviation from the table of formulae for shaft and hole deviations given in BIS: 919

$$= -16\ D^{0.44} = -16 \times (23.2)^{0.44} = -65 \text{ microns}$$

Here $D = \sqrt{18 \times 30} = 23.2$ mm.

$\therefore$ The limits for shaft 25 d9 are 25.000 – 0.065

= 24.935 mm, and

25.000 – 0.052 – 0.065 = 25.000 – 0.117

= 24.83 mm. – .065

or 25 d9 = 25 – 0.117

$\therefore$ Tolerance = 24.935 – 24.833 – 0.052 mm.

By referring to Table 13.7, on selection of fits, we see that the designation H8/d9 represents a fit which is suitable for loose running fits, e.g., loose pulleys. Plummer blocks, etc.

PROBLEM 13.6. *A ϕ 50 mm shaft is to be supported in oil lubricated journal bearing consisting of a bushing ϕ100 mm fitted into a housing. Determine the working dimensions for;*

- *bushing bore,*
- *bushing, and*
- *the shaft.*

Also calculate the maximum and minimum clearance interference.

SOLUTION:

1. Select the type of fit to be used based on the intended function of the parts. Referring to *Table 13.9*, the fits to be employed are selected as :
   - for the bushing and the housing H7/p6 . an interference fit,
   - for the shaft and the bushing H7/f7, a clearance fit for normal running
2. Obtain the tolerances on the linear dimensions of the parts. From *Table 13.5*, the fundamental tolerances (IT) for different grades based on size are:

Table 13.12.

H11-c11 Slack Running Fit	Used to give flexibility under load, easy assembly for a close fit at elevated working temperatures	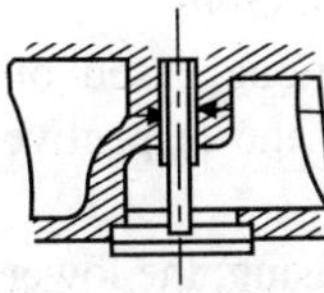
H9-d10 Loose Running Fit	Used for gland seals, Loose pulleys and very large bearing	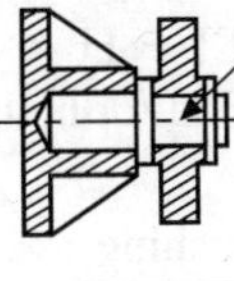
H9-e9 Easy Running Fit	Used for widely separated bearing or several bearing in line.	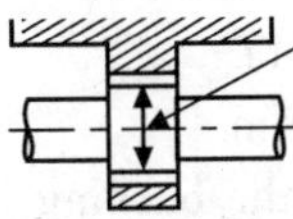
H8-f7 Normal Running Fit	Suitable for applications requiring a good quality fit heat that is easy to produce.	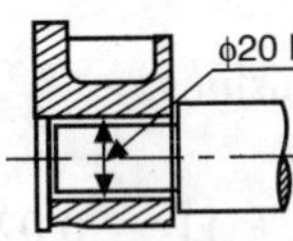
H7-g6 Sliding and Location Fit	Not normally used for continuously running bearing unless load is slight suitable for precision sliding and location	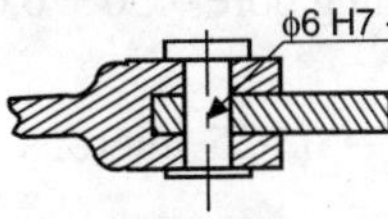
H7-h6 Location Fit	Suitable for many non-running assembles.	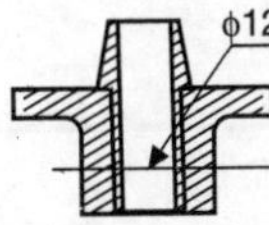
H7-k6 Push Fit	Used for location fit when a slight interference of one part relative to the other is an advantage.	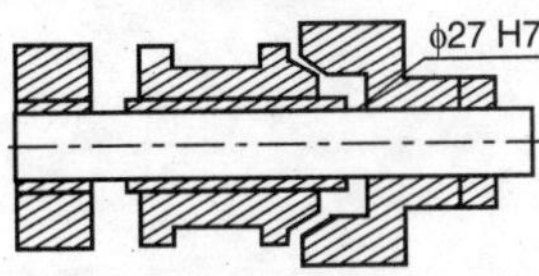
H7-n6 Tight Assembly FIt	Used when the degree of clerance that can result from a H7-k6 fit is not acceptable	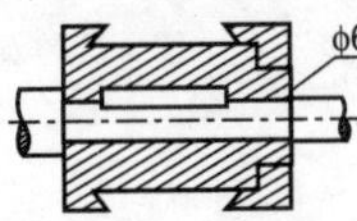
H7-p6 Press Fit	Ferrous parts are not over strained during assembly or dismentling.	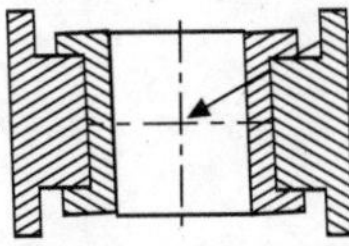
H7-s6 Heavy Press Fit	Mainly used for permanent assemblies.	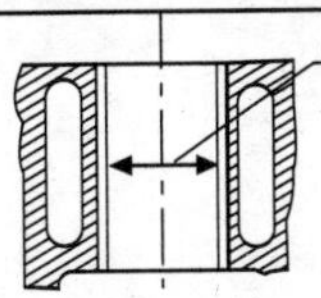

- for ϕ 100 and grade 6 = 22 microns,
- for ϕ 100 and grade 7 = 35 microns, and
- for ϕ 50 and grade 7 = 25 microns.

3. Obtain the fundamental deviations based on the type of hole/shaft and then the respective sizes. From *Table 13.8*:

- For a hole of type *H, i.e.*, housing; the lower deviation, EI = 0.000
  the upper deviation, ES = EI + IT = 0 + 0.035 mm. Hence the dimension for housing bore = 100 + 0.035
- For a shaft of type p, i.e., bushing :
  lower deviation, ei = 0.037 microns
  upper deviation, es = ei + IT
  = 0.037 + 0.022 = 0.059 mm.
  Hence the outside bore of the bushing
  = 100 + 0.059
- for a hole of type H, i.e., bushing:
- lower deviation, EI = 0.000
- upper deviation, ES = EI + IT = 0.025 microns Hence the bushing bore = 50 + 0.025
- For a shaft of type f:
- Upper deviation, es = – 0.025 microns
- lower deviation, ei = es – IT
  = – 0.025 – 0.025 = –0.050 microns.

Hence, the shaft diameter = 50 – 0.050

4. Calculate the interference/clearance amounts.
Between bushing and housing:

- Maximum interference = 100.00 – 100.059
  = – 0.059 mm
- Minimum interference = 100.035 – 100.037
  = – 0.002 mm

Between bushing and shaft:

- Maximum clearance = 50.025 – 49.050
  = 0.075 mm
- Minimum clearance = 50.000 – 49.075
  = 0.025 mm

Notes:

1. The two examples discussed above make it clear that the two methods are equivalent, to each other and for selecting limits and fits, the reference to *Table 13.6, Table 13.7* and Table 13.8 serve the purpose.
2. Limits, Fits, and Tolerance can be decided using the National Standards on Limits, Fits and Tolerance.

❒❒❒

14 Geometric Dimensioning And Tolerancing

14.1. INTRODUCTION

An engineering drawing of a production part conveys information from the designer to the manufacturing and inspection. It must contain all information necessary for the part to be correctly manufactured. All the dimensions, tolerances, notes, etc., necessary to provide for correct production and functioning of a part should be given directly on drawing together with any other information necessary to define the part completely in its finished form, keeping in mind the manufacturing and inspection requirements. The information must also enable the inspector to determine precisely whether the finished parts are acceptable. The drawing must also indicate any permissible variations for size and shape in the form of limits or tolerances. Therefore, each drawing must convey three essential types of information, e.g.,

- The material to be used, and these are usually covered by separate specifications or supplementary documents, and the drawing need only make reference to these.
- The size or dimensions of the part. Linear, circular, and angular dimensions specify size. Tolerances on these size dimensions may be applied directly, or may be specified by means of a general note, or occasionally by reference to other documents.
- The shape or geometrical characteristics are indicated on orthographic views of the part drawing, supplemented by dimensions.

In this chapter important principles of geometrical tolerancing together with the use and application of symbols, which are internationally agreed by ISO, are explained. In addition, the necessary definitions of important terms are given. Typical methods of applying geometrical tolerances to technical drawings according to ISO: 1101-1983, and BIS: 8000 Parts 1 to 4, are explained. In addition examples are included to illustrate the method of indication and interpretation of geometrical tolerancing symbols and characteristics.

14.2. GEOMETRIC DIMENSIONING AND TOLERANCING (G.D.&T.)

The concept of tolerancing is a relatively new development. It has only been since the early 1900's that tolerances were placed on drawings. Before that every thing used to be built exact or 'perfect'. When the parts were not 'perfect' and did not fit, those were filed and hand fitted until they did fit. Workers were considered as craftsmen.

As time went on, it became known that parts did not have to be exact. They could vary and could be off perfect size within some limits. So some variation or tolerance is allowed on every dimension on the drawing. This is how plus/minus or coordinate system of tolerancing, as we know today, was arrived at. This plus/minus system worked quite effectively until about the late 1930's. At this time a fellow by the name Stanley Parker from the Royal Torpedo Factory in Scotland came up with the idea of Positional Tolerancing System. This system located features with round tolerance zones rather than the traditional square zones of plus/minus tolerancing system. The idea caught on worldwide and resulted in various National and International Standards on the concept. Geometric Dimensioning and Tolerancing has developed immensely over the past half century, and is, without a doubt, the most widely accepted system of dimensioning and tolerancing worldwide.

The newer methods of engineering tolerancing consist primarily of geometric and positional tolerancing, use of theoretically exact surfaces or selected portions of a surface, e.g., use of datums, datum targets, and more precise interpretations of linear and angular tolerances. This provides designers and drafters with a means of expressing permissible variations in a very precise manner. Further more, the methods and symbols used are internationally agreed by ISO and, therefore, break language barriers.

The national and ISO Standards concerning this concept are written as standards not necessarily as instructional textbooks. It is difficult, for an untrained person, to understand the technical tolerancing terms and Jargon. The text presented here, on "**Geometrical Dimensioning** and **Tolerancing**" is only an interpretation of National and International Standards.

The figures and examples are used to help the reader understand the principles and practices of geometric tolerancing. Some of the figures show added detail to emphasize a point, and some figures are left incomplete by intent to avoid complexity. All numerical values used are illustrative only.

These concepts are extremely useful for the specification of precise information on engineering drawings, but cannot be used to the exclusion of traditional methods of coordinate dimensions and tolerances. They can only complement them by clarifying requirements that might otherwise be misunderstood. To build an understanding of Geometric Dimensioning and Tolerancing that will be strong and lasting, one must begin by establishing a solid understanding of the fundamentals of its language.

The establishment of a system of *'Geometric Dimensioning and Tolerancing'* is the result of contributed work of experts from many countries and is accepted as the worldwide authority on this subject. It deals with the techniques for the control of geometric characteristics such as **Straightness, Flatness, Parallelism, Profile, Cylindricity, Angularity, Perpendicularity, Runout, Concentricity,** and **Position.** It also includes precise interpretation of coordinate and angular tolerances, datum systems, projected tolerance zone and other related concepts. The techniques for geometric dimensioning and tolerancing have been perfected to the point where they can be used advantageously for interchangeable mass manufacture. Geometric dimensioning and tolerancing is a *dual purpose designation system:*

1. It is a *set of standard symbols,* which are used to define part features and their tolerance zones. ISO and National Standards document the symbols and their interpretation.
2. It is *a philosophy of defining parts based on how they function.* Geometric dimensioning helps readily capture the design intent by providing the designer with sophisticated tools to say what he/she means to express. As a result, manufacturing or inspection can understand more clearly the design requirements of the part.

When dealing with technical design and development problems, a drafter has to be able to properly represent conventional dimensioning and geometric tolerancing. A technician must be able to accurately read dimensioning and geometric tolerancing. Generally, a design drafter converts engineering sketches or instructions into formal drawings using proper standards and techniques. Only after acquiring adequate practice and experience, a drafter, designer, or engineer begins to implement geometric dimensioning and tolerancing on the research and development of new products or the revision of existing products.

An important concept to remember about geometric dimensioning is that the dimensions on a drawing define as to what size and shape the part must be, to function as intended in design. This dimensioning philosophy is a powerful design tool that can make a substantial difference in product cost by providing most liberal tolerances that allow the required function to be realized. On the whole, if properly executed this design philosophy results in better communication and can add up to substantial savings in any company's operating costs.

Geometric Tolerancing is intended primarily for the more advanced drafter and for those studying engineering design in technical schools and colleges. However, it will be equally useful in industry, where it should find place in the personal library of engineers, designers, drafters, inspectors and those working in the shops and tool rooms.

14.3. WHEN TO USE GEOMETRIC TOLERANCING

It is not necessary to use geometric tolerances for every feature on a part drawing. The function and relationship of a particular feature on a part dictates the use of geometric tolerancing. Geometric tolerancing should he used in circumstances where even the best applied conventional dimensioning and tolerancing is inadequate and does not ensure that parts would assemble and function as intended after manufacture. In most cases it is to be expected that, if each feature on the part meets all dimensional tolerances, form variations will be adequately controlled by the accuracy and precision of the manufacturing process and equipment used. This is further supplemented by the partial degree of control exercised by the measuring and gauging procedures used in inspection of the parts. If there is any doubt about the accuracy of such control, a geometric tolerance *of form, orientation, or position* must be specified. This is often necessary when parts are of such size or shape that bending or other distortion is likely to occur.

Geometric control also becomes necessary when errors of shape or form must be held within limits other than might ordinarily be expected from the manufacturing process, and as a means of meeting functional or interchangeability requirements. Geometric tolerancing is used with advantage for drawings of component parts of an assembly that are manufactured by subcontracting the parts to shops of widely varying equipment and experience, in widespread locations, where possible manufacturing variations are not known. On the other hand, if the same parts are to be manufactured and assembled in a workshop where the method of production has proved to produce parts and assemblies of satisfactory quality, the same degree of tolerancing may not be necessary. In such cases complete and explicit statements of requirements, provided by geometric tolerancing ensures that the designer's instructions are understood without ambiguity.

Many problems, in industry, stem from faulty *communications*. In today's worldwide competitive mass production market, it is not enough to make drawings that can be understood. Instead, we must make drawings that cannot possibly be misunderstood. To meet this end the use of *Geometric Dimensioning and Tolerancing is an effective way out.*

14.4. SHORTCOMINGS OF COORDINATE TOLERANCING

Coordinate tolerancing is a dimensioning system where a part feature is located or defined by means of rectangular dimensions with given tolerances. Coordinate tolerancing has three shortcomings:

1. **Square tolerance zones**
2. **Fixed size tolerance zones**
3. **Ambiguous instructions for inspection**

There are many problems with our old tolerancing system, e.g., lack of specified datums, lack of maximum and minimum material concepts, non-uniform interpretation, etc. But probably one of the most apparent and easiest to see is the manner in which holes, in parts, are located.

In the coordinate system, we locate holes with dimensions that have plus and minus tolerance. It is done in a manner shown in *Fig. 14.1 (a)*. To keep the example simple only the distance between the hole centres are considered, and the distances to the edges are not considered at this stage. This method of tolerancing allows the axis of the hole to move plus something and minus something in the X and Y direction. This creates a square tolerance zone in which the holes must lie, *Fig. 14.1 (b)*.

Notice that if the holes are located with a plus-minus square zone this would allow manufacturing to place the holes anywhere inside this square zone. This in effect allows the holes to be off location more in a diagonal direction than across the flats. *Fig. 14.1 (c)*. This is not logical. We do not want to allow the holes to be off location more in one direction than another, i.e., the hole location tolerance should be the same in any direction we measure.

Locating holes (or any features for that matter) should be similar to shooting at a bull's eye at an archery or rifle range. We aim for the exact centre and afterwards check to see how much we are off the centre or basic position. The tolerance zones should thus be round and not square. When we use round tolerance zones the holes will be off location the same in any direction we measure as we wish.

Another advantage of round tolerance zones over the square tolerance zones is that in some cases we may get additional tolerance for manufacturing without sacrificing function. The reason is that in plus-minus system the designer allows manufacturing to locate the holes anywhere

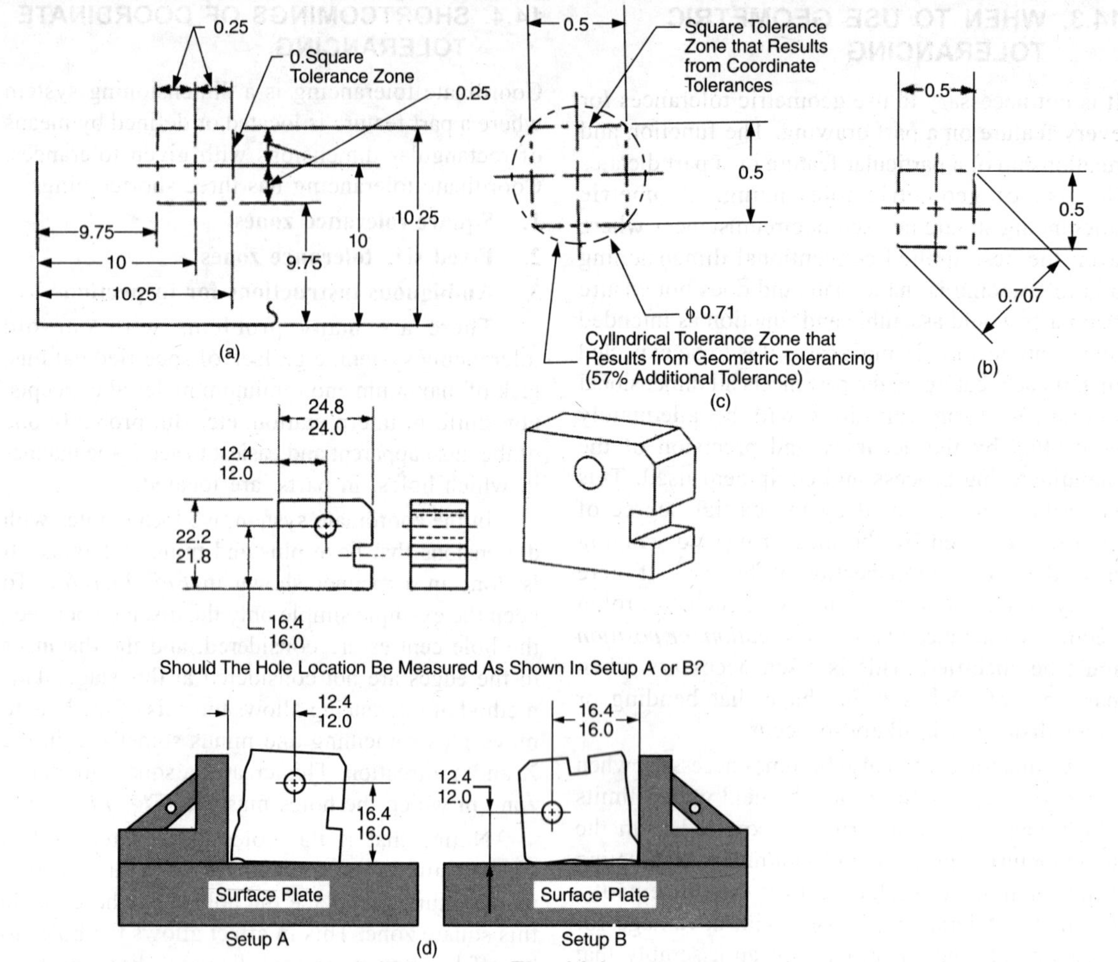

Fig. 14.1.

in the square zone. If the axis of the hole be located in the corner of a *0.5 mm* square zone it is possible that the hole could be approximately *0.707 mm* off the basic centre if measured in a diagonal direction as shown in *Fig. 14.1 (c)*.

If the designer made calculations on the two mating parts and determined that the parts would work if the holes were off their basic location by *0.707 mm* in the diagonal direction, why not let the holes be off the basic location in any direction by *0.707 mm*. A *0.5 mm* square zone converts to an approximate *0.707 mm* diameter zone. This change results in an increase of 57% additional tolerance while still assuring the assembly of the parts.

A third major shortcoming of coordinate tolerancing is that it has ambiguous instructions for inspection. *Fig. 14.1 (d)* shows two logical methods an inspector could use to set up the part for inspecting the holes. Different ways to hold the part for inspection will result in:

- Two inspectors can get different measurements from the same part.
- Good parts may be rejected or, worse yet, bad parts can be accepted as good ones.

The problem is that the drawing does not communicate to the inspector the sequence in which the surfaces should touch the gauging equipment while setting up for inspection.

Due to these shortcomings the use of coordinate dimensioning is rapidly diminishing in industry. However, coordinate tolerancing is not totally obsolete; it does have some useful applications on engineering drawings, e.g., for size dimensions, chamfer dimensions, and radius dimensions.

14.5. ADVANTAGES OF GEOMETRIC DIMENSIONING AND TOLERANCING

Modern geometric dimensioning and tolerancing is rapidly being adopted as the universal engineering drawing language, since it allows the designer to more clearly express the design intent and improve the understanding of the manufacturer. Some of the advantages offered by G.D.&T. are:

- It *results in improved communications*, at all levels, by providing a common language to design, manufacturing, and quality control. It enforces uniformity in drawing specifications and interpretation, and results in reduced controversy, guesswork and assumptions. Helps in the development of sound customer and vendor relationships.
- Use of *G.D.&T. results in improved product designs*. It provides a designer with *sophisticated tools* to express precisely the *design intent*. Also it takes into consideration the *part function* at the design stage and makes use of functional dimensioning philosophy to *establish part tolerances based upon functional requirements*.
- The use of G.D.&T. results in increased manufacturing tolerances on products because:
  1. Under certain conditions, it provides an additional tolerance, called 'bonus', for manufacturing. This additional tolerance usually will result in a significant saving in production cost.
  2. The use of functional dimensioning, the tolerances are assigned to a part based upon its functional requirements. Study of the part functional requirements will help to determine reasonable functional tolerances on the part and this usually will result in larger tolerances since the designer will not be tempted to assign tighter tolerances for lack of confidence in his/her design.

14.6. GAUGING AND MEASURING PRINCIPLES

The specification of a tolerance on a dimension, or a geometric dimensioning tolerance of form, orientation, or position, does not necessarily imply the use of any particular method of production, measurement, or gauging. Geometric tolerancing does not require the use of functional gauging. Functional gauging techniques are shown only to help the reader understand the principles and concepts. The gauges here are shown with no tolerances to avoid complexity. Gauge tolerances are usually 10% of the product tolerance with another 5% allowable for wear.

One of the best ways of analyzing the effects of tolerances and limits is to establish a gauging or measuring principle that will determine whether or not the part feature falls within its assigned tolerance zone. This method of interpretation is used here. But the gauging or measuring equipment shown in this text is intended to illustrate a principle, and not to illustrate a complete gauge design, nor it means to indicate the only suitable method available for the purpose.

It is also a common practice for manufacturers to use methods in accordance with the drawing specifications. For example, if the diameter of a part is specified without a form tolerance, a caliper-type measurement, such as one made with a micrometer, will be sufficient to meet the drawing requirement. Yet a manufacturer may elect to provide a ring gauge, which, in addition to checking the maximum size, may also be used to check for roundness. In addition, if the ring gauge were of sufficient length, it would also control straightness or cylindricity within the same limits.

Any reference to gauging is included for interpretative and explanatory purposes only and is not intended to dictate actual gauging technique for the manufactured part.

To properly understand and interpret the tolerancing procedures, it is necessary to understand the precise meaning of the terms used. Some of these terms are already explained in Chapters 5 and 18 while the rest of them are explained in this chapter.

14.7. FUNDAMENTAL RULES OF DIMENSIONING AND TOLERANCING

- Dimensioning and tolerancing should:
  1. Clearly define a size or location of each surface line, point, or feature.

2. Be selected and arranged to suit the function and mating relationship of a part.
3. Not be subject to more than one interpretation.

- A drawing should define a part without specifying manufacturing methods, e.g., only the diameter of a hole should be shown without indicating whether it is to be drilled, reamed, punched, or made by any other operation.
- It should not be necessary to specify the geometric shape of a part feature, unless some particular precision is required.
- Lines or features that appear to be straight imply straightness; those that appear to be round imply; circularity; those that appear to be parallel imply parallelism; those that appear to be square imply perpendicularity; centralines imply symmetry; and features that appear to be concentric about common centreline imply concentricity.
- It is not necessary to add angular dimensions of 90° to corners of rectangular parts or to specify that the opposite sides are parallel. However, if a particular; departure from the illustrated form is precision permissible, or if a certain degree of form is required, these must be specified.
- If a slight departure from the true geometric form or position is permissible, or if a certain degree precision of form is required, these must be specified.
- If a slight departure from the true geometric form or position is permissible, it should be exaggerate pictorially in order to show clearly where dimensions apply. *Figure 14.2,* shows some example.

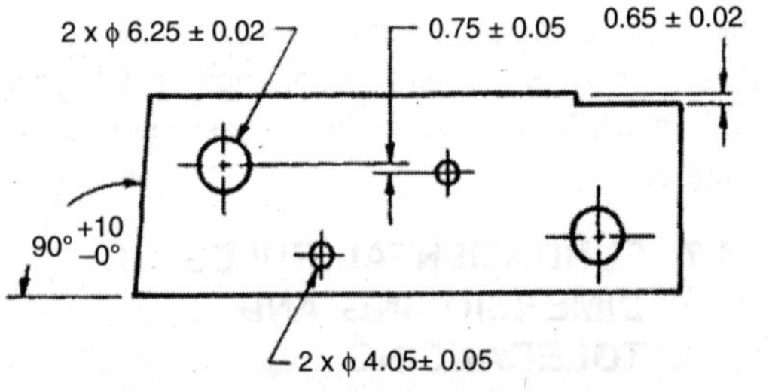

Fig. 14.2. Exaggeration of small dimension

- Dimensions that are not to scale should be underlined

14.8. LINEAR DIMENSIONS WITHOUT DATUMS

Datums are theoretically exact geometrical references, such as a line, edge or plane, to which toleranced features are related by dimensions. When datums are not specified linear dimensions are intended to apply on a point-to-point basis, either between opposing points on the indicated surfaces or directly between the points marked on the drawing.

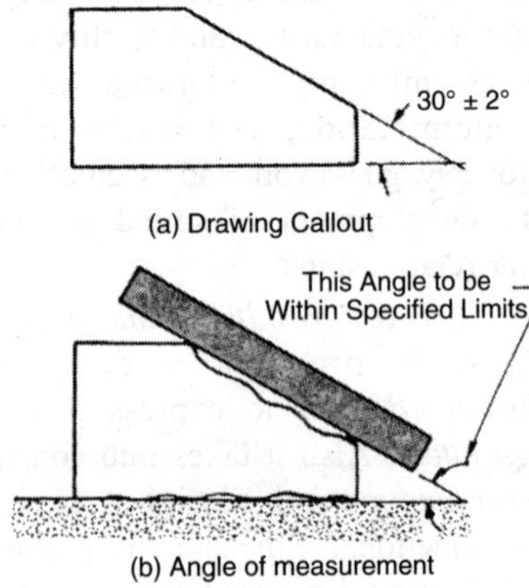

Fig. 14.3. Angular dimensions

- *Figure 14.3,* shows a simple part with an angular dimension between two surfaces. If two plane-gauging surfaces are brought into contact with the surfaces, the angle between them must be within the angular limits specified on the drawing.
- *Figure 14.4* shows a circular feature with a diameter shown as ϕ 20.25 ± 0.05. This means that the diameter, when measured between any two opposing points around the circumference, such as a-a, b-b, c-c, shall be within the specified limits of size.

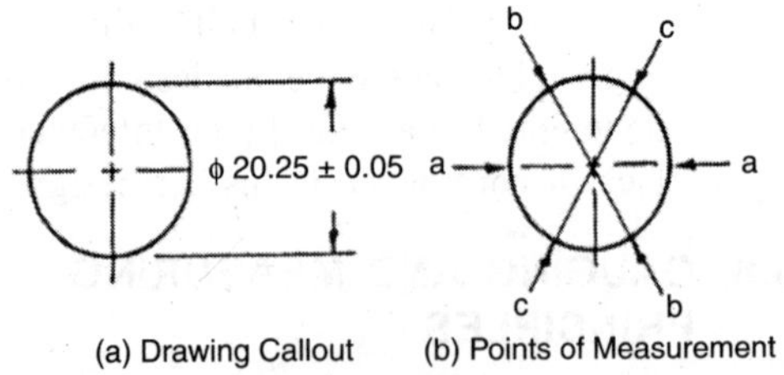

Fig. 14.4.

- The diameter of a cylindrical part, *Fig. 14.5*, applies at any opposing points along its length, such as *a-a*, *b-b*, *c-c*, or *d-d*.

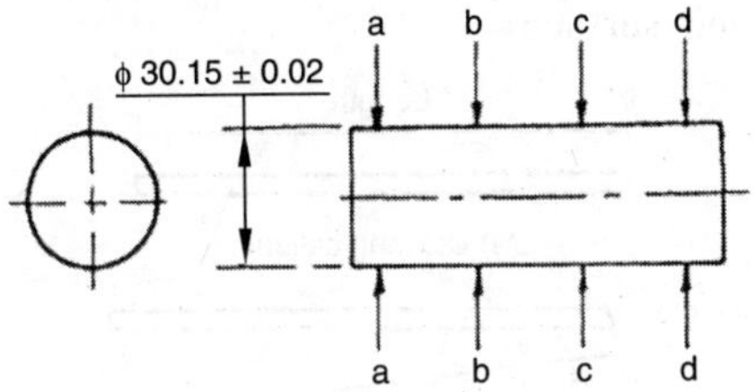

Fig. 14.5.

- This rule applies irrespective of the fact whether or not they're any obstructions in between, as shown in *Fig 14.6 (a)* where ϕ 18.24 ± 0.10 mm applies at *a-a*, *b-b*, or *c-c*. However, if there is any doubt in such cases, as when the surfaces are widely separated, it is preferable to repeat the dimension.

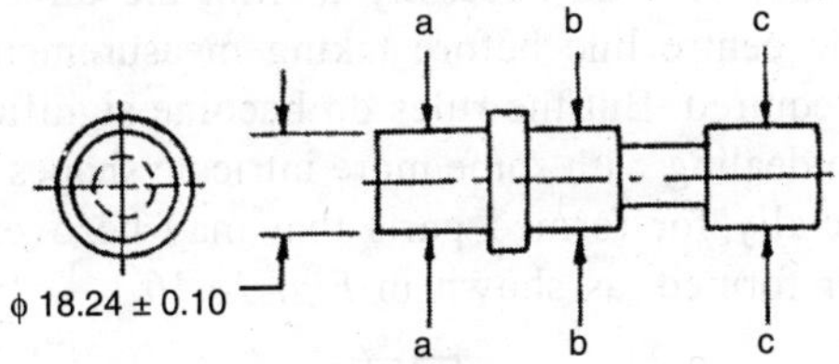

Fig. 14.6. (a)

- When applying the rule to rectangular parts, measurements of thickness are made normal to the centreline between the surfaces. For parts of uniform thickness, it is equivalent to making measurements normal to the surface. A sufficient number of such measurements are made at various points on the surface to ensure that thickness limits are met for the entire surface. This rule applies to bent or bowed parts too. *Fig 14.6 (b)*.

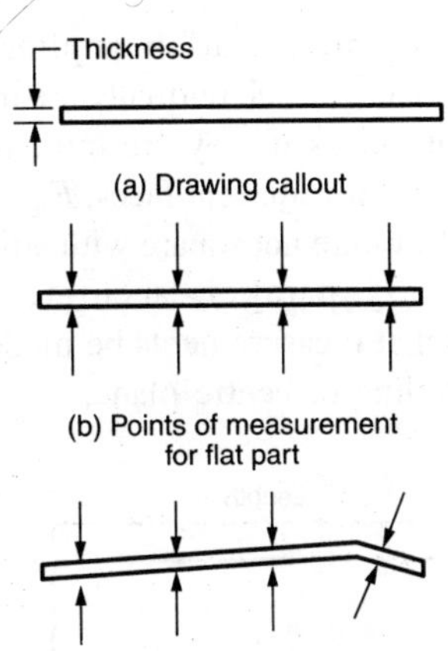

(a) Drawing callout

(b) Points of measurement for flat part

(c) Points of measurement for bent or bowed parts

Fig. 14.6. (b)

- If the thickness is not uniform, i.e., when surfaces are not parallel, measurements are made normal to theoretical centreline, *Fig. 14.7*, and these may not be truly normal to the surface.

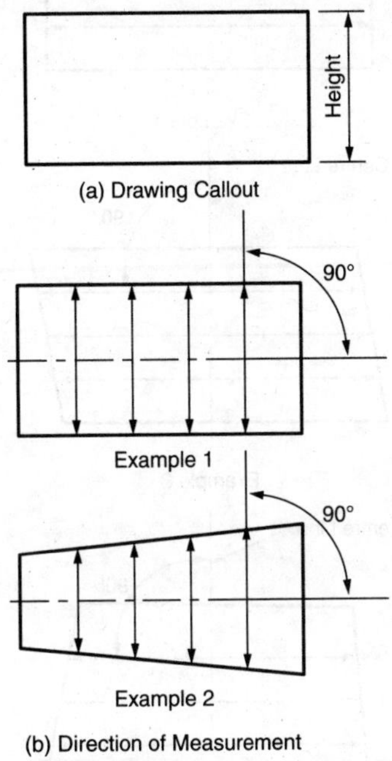

(a) Drawing Callout

Example 1

Example 2

(b) Direction of Measurement

Fig. 14.7. Thickness measurements

- The same interpretation applies to length measurements, which normally are made normal to the end faces if they are mutually parallel, and square with adjacent faces, *Fig. 14.8*. When the end faces are not square with adjacent faces, or mutually parallel, accurate measuring requires that measurements be made normal to the centreline or centre plane.

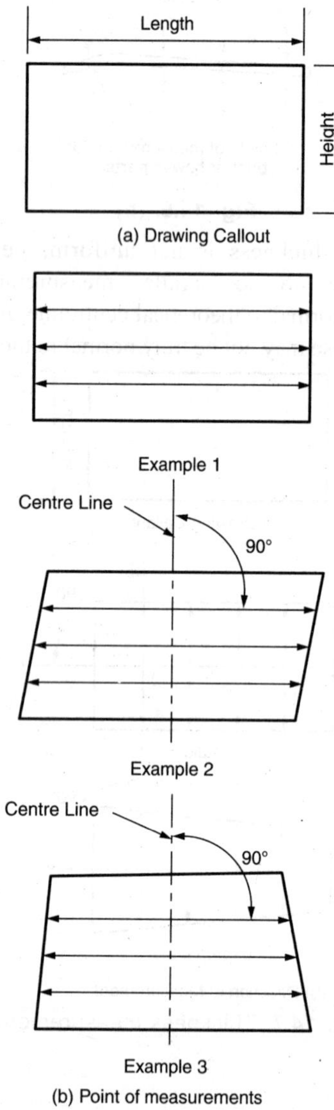

Fig. 14.8. Length measurements

In spite of the above rules, measurements must be kept within the boundary of the part and should not be made to a point in space, *Fig. 14.9*, in an attempt to measure in a direction normal to the faces. Very thin parts are thus measured parallel with their surface.

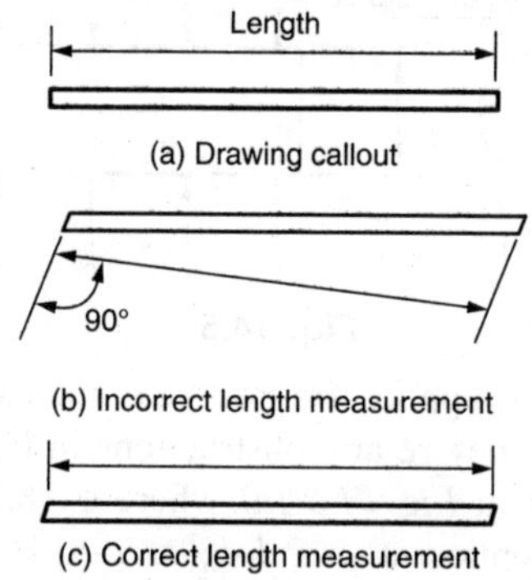

Fig. 14.9. Measuring length of thin parts

In most cases, this kind of precision measurement and necessity to find the direction of the centre line before taking measurement is not required. But the rules do become significant when dealing with some more intricate shapes and especially for formed parts that may be over or under formed, as shown in *Fig. 14.10*.

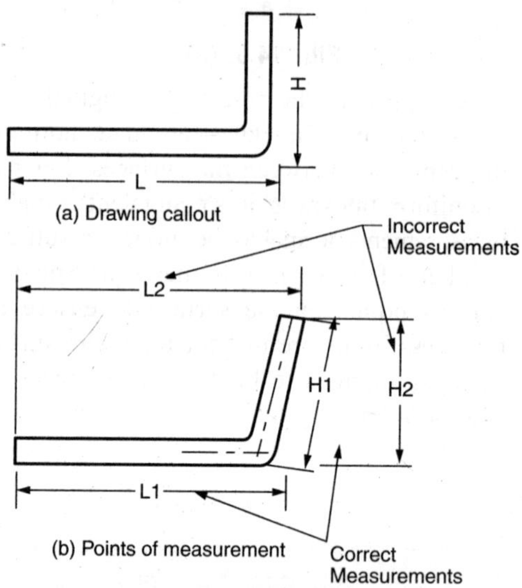

Fig. 14.10. Measurement of formed parts

The points of measurements for measurement of location dimensions, e.g., a hole in a part as shown in *Fig. 14.11*, the dimension L applies to a measurement from the axis of the hole to a corresponding point on the edge of the part, perpendicular to the edge; as at "*a*". When the part is off square, dimension D would be measured as shown at "*b*", which is the shortest distance between the axis and a point on the edge of the part.

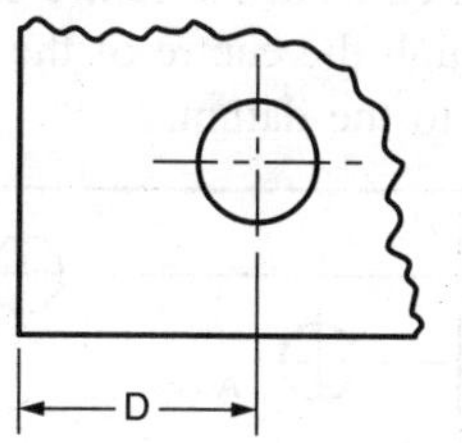

(a) Drawing callout

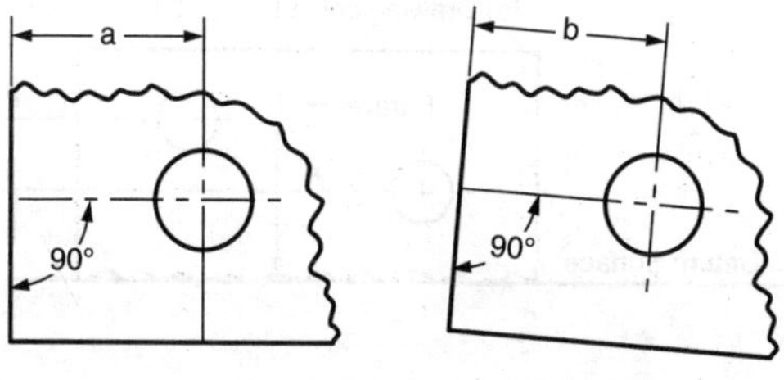

(b) Points of measurements

Fig. 14.11. Measurement of location

When a single dimension controls several features, e.g., a series of holes as shown in *Fig. 14.12*, the dimension applies individually from the axis of each hole to the corresponding point on the edge. In case the part is bowed, the series of holes will be located on a correspondingly curved centre line, so that the location of each hole will meet the drawing limits when measured at 1,2, and 3.

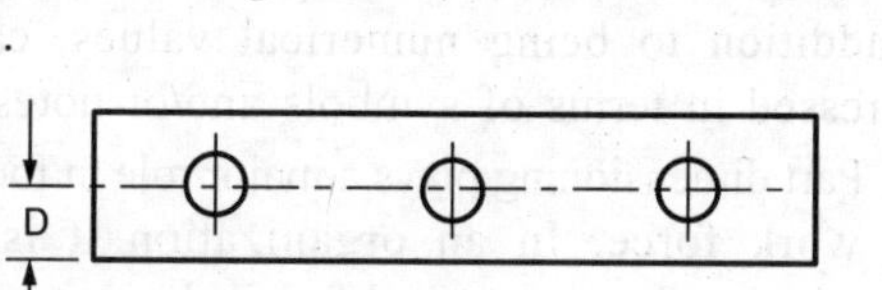

(a) Drawing callout

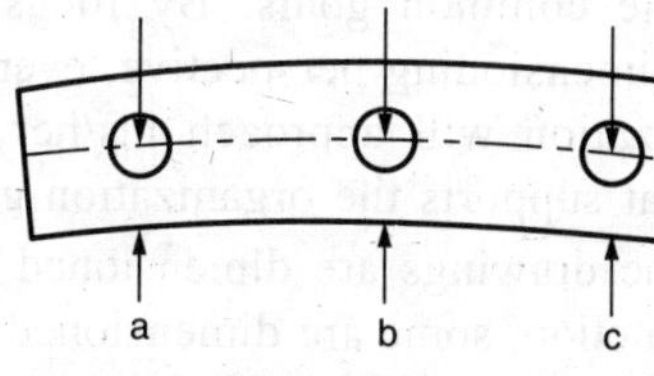

(b) Points of measurement if part is bowed

Fig. 14.12. Measurement of bowed parts

14.9. LOCATION DIMENSIONS WITH DATUM

When location dimensions originate from a feature or surface indicated as a datum, measurement is made from the theoretical datum, and not from the actual feature or surface of the part. For cases where a curved centre line, as shown in *Fig. 14.12*, would not meet functional requirements or where the position of the hole, *Fig. 14.11*, would be required to be measured parallel to the base. Referring the dimension to a datum feature, *Fig. 14.13*, can easily specify such requirements.

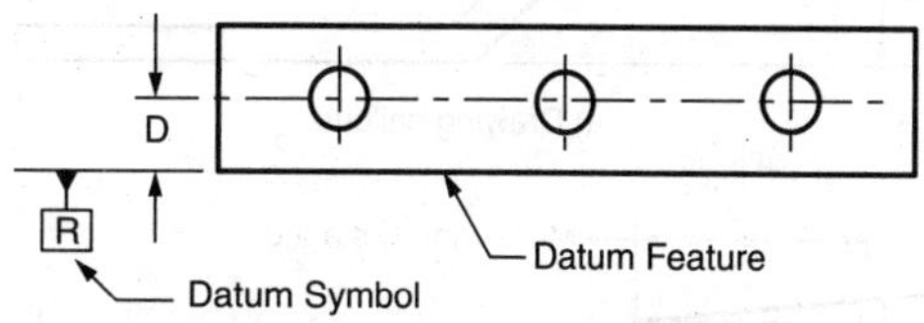

Note: Datum Feature R Applies to all Dimensions Originating Form This Surface

(a) Drawing callout

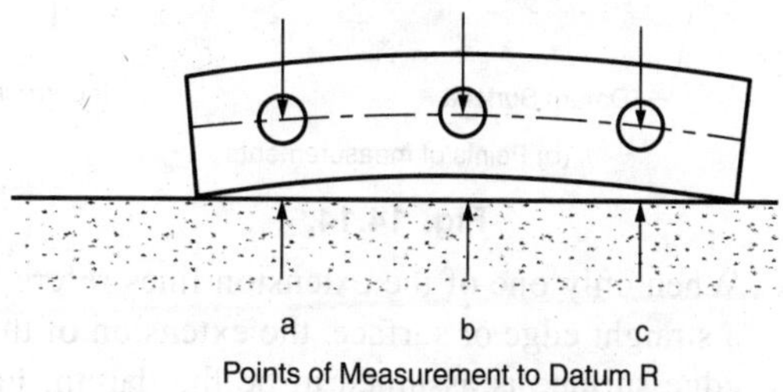

(b) Interpretation if part is bowed

Fig. 14.13. Dimension referred to a datum

14.10. ASSUMED DATUMS

There are often cases where the basic rules for measurements on a point-to-point basis cannot be applied, because the originating points, lines, or surfaces are offset in relation to the features by the dimensions. In such cases, it becomes necessary to assume a suitable datum. The datum assumed is usually the theoretical extension of one of the lines or surfaces involved.

The following general rules cover three types of dimensioning practices that are commonly encountered. But it should be kept in mind that, if any doubt is likely to exist, the required datum feature should be properly identified on the drawing. This is especially important in cases where the dimensional tolerances are small compared to possible form variations.

1. When a dimension refers to two parallel edges or planes, the larger edge or surface is assumed to be the datum feature. For example, if the surfaces of the part shown in *Fig. 14.14* are not truly parallel, as shown in the *Points of Measurement view*, dimension D would be acceptable so long as the top surface is within the size limits when measured at points 1 and 2. It does not need to be within the size limits if measured along the surface extension at the opposite end.

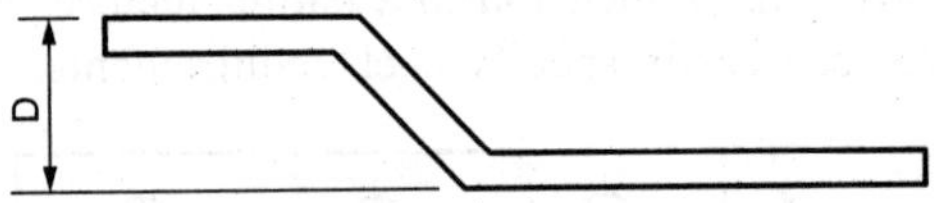

(a) Drawing callout

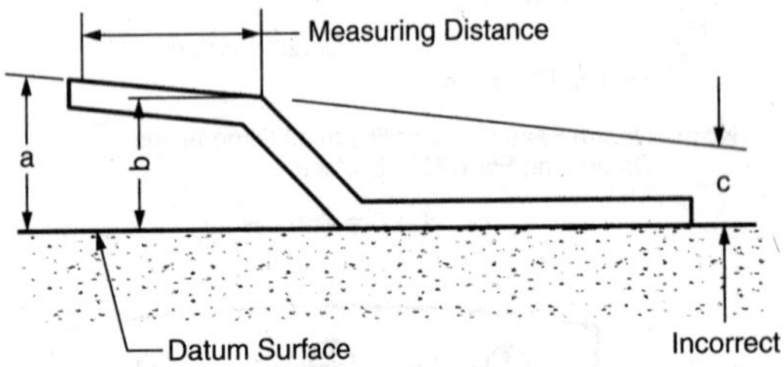

(b) Points of measurements

Fig. 14.14.

2. When only one of the extension lines refers to a straight edge or surface, the extension of that edge surface is assumed to be the datum. For example, in *Fig. 14.15*, measurement of dimension *X* is made to the datum surface at "*x*" in the Points of Measurement view.

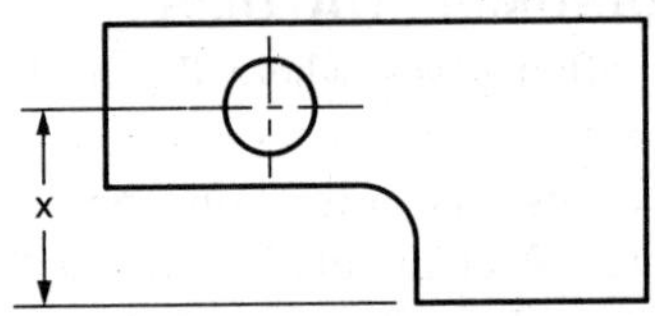

(a) Drawing callout

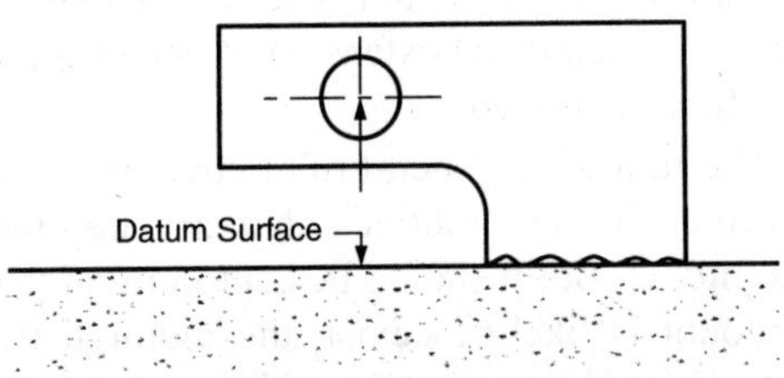

(b) Point of measurements

Fig. 14.15.

3. When both extension lines refer to offset points, rather than edges or surfaces, it should generally be assumed that the datum is a line running through one of these points, and parallel to the line or surface to which it is dimensionally related. For example, in *Fig. 14.16*, dimension *X* is measured from the centre of the hole *B* to a line through the centre of the hole *A*, which is parallel to the datum.

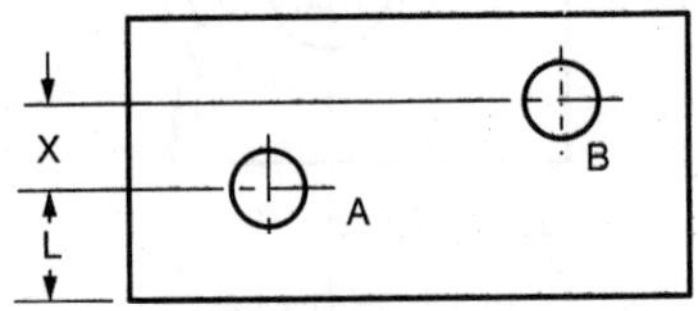

(a) Drawing callout

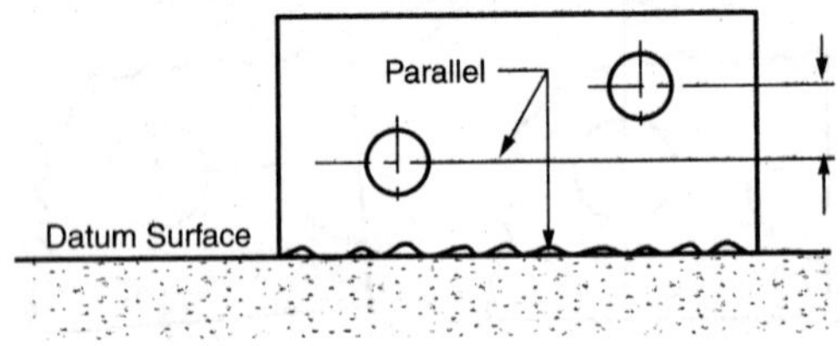

(b) Points of measurements

Fig. 14.16.

14.11. PURPOSE OF DIMENSIONS

Basically dimensions define the finished part and describe what the part must be in order to function as intended. Dimensions specify part features in terms of **size, location, orientation**, and **form.** Often, dimensions are numbers that define the geometric characteristics of the part. But dimensions, in addition to being numerical values, can be expressed in terms of symbols and/or notes.

Part dimensioning plays a major role in focusing the work force. In an organization it is very important to have the workforce always working towards the common goals. By focusing on a common dimensioning perspective, everybody in the organization will approach his/her job in a manner that supports the organization goals.

If some drawings are dimensioned based on product function, some are dimensioned based on manufacturing capabilities, and some based on preferred inspection techniques, then the workforce will not have a common goal to work to and will not be focused on a common goal.

The method used as a basis for establishing dimensions and tolerances is controversial in industry. Usually engineering wants to use product function as the main criteria in establishing dimensions and tolerances. Often, manufacturing would like to use the process capability as the main criteria for establishing dimensions and tolerances. The main intent behind this is to keep the cost as low as possible. However, it is important to keep in mind that the simplest and cheapest component design does not always result in the lowest overall costs. The lowest overall costs are achievable only in the context of the system design. Product function should be the primary focus in establishing dimensional relationships and tolerances on drawings. Using the product function as the basis for dimensioning has the following advantages and disadvantages:

Advantages

- Primary focus is on the product functional performance.
- Dimensioning established functional relationships.
- Tolerances are dependent variables.
- Provides flexibility in process selection for processing in manufacture.
- The functional effects of a product change request can be easily evaluated.
- Results in larger tolerances on dimensions without affecting the part function.

Disadvantages

- The process, intended to be used for manufacture, may not be capable.
- It requires process stacks.

Product function is the primary focus for determining component tolerances. However, there are some instances where process capability becomes a primary consideration for determining component tolerances. The choice depends upon the type of dimension that is being toleranced.

14.12. TYPES OF DIMENSIONS

All part dimensions can be classified into the following four categories:

1. Customer or end user sensitive dimensions.
2. Customer robust dimensions.
3. Safety related dimensions.
4. Non-system related dimensions.

Loss function concept is useful to understand these types of dimensions. Simplified loss function concept is briefly explained here. A loss function curve is a graph that shows the effect that variation of a dimension has on the product function. A loss function curve can be parabolic, as shown in *Fig. 14.18*, or it can be straight or curved. The shape of the loss function curve signifies a trend about the effect that variation of a dimension has on a product function. The shape of the loss function curve can be determined through extensive testing. However, in many instances, the shape of the loss function curve can be estimated using prior product experience, engineering principles, and logic. When classifying dimensions, the shape of the loss function curve often can be estimated.

Customer Sensitive Dimensions

A customer sensitive dimension is a dimension with a tolerance that has a direct relationship to a product system function such that the system function continually degrades as the dimension varies from its nominal or target value. The "target value" is the value to which the dimension is intended to be produced. The loss function curve shown in *Fig. 14.18* illustrates an example of the effect of that variation has on the product function.

The limits for a customer sensitive dimension are established by extensive testing of the product. The amount of tolerance for the dimension is a compromise between the level of allowable loss of performance and the cost to reduce the variation of the part dimension. On a customer sensitive dimension, even though a tolerance is shown on the drawing, manufacturing is required to continuously monitor the process and try to reduce the variation for this dimension. Whenever a dimension has the loss function curve of the type shown in *Fig. 14.18*, it will be referred to as a customer sensitive dimension. Customer sensitive dimensions are usually identified on engineering drawings.

Customer Robust Dimensions

A customer robust dimension is a dimension with a tolerance that has a small impact on the product system function. The part can be produced

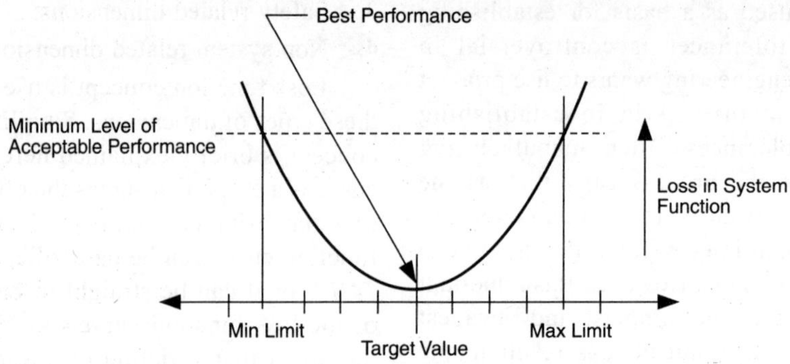

Fig. 14.18. Loss function curve for a safety related customer sensitive dimension

anywhere between the limits of the dimension and the system will function about the same. The customer will not be able to tell the difference even when the part is at the maximum or minimum end of the tolerance zone.

Whenever a dimension has a typical loss function curve of the type shown in *Fig. 14.19*, it is referred to as customer robust dimension. Such dimensions are not usually identified on engineering drawings. When this type of dimension is shown on a drawing, manufacturing can produce parts anywhere within the maximum and minimum limits without affecting the customer. There is no target value from the product function standpoint. Therefore, manufacturing should ensure that the process is capable, but not use much resources trying to reduce variation around a target value.

Safety Related Dimensions

A safety related dimension is one for which the consequences of its variation may produce an unsafe condition for the customer. The print specification ensures a safe product, but each specification should be analyzed for its effect on customer safety in case its variation exceeds the print specification. Analysis is required to consider the consequence if a part that is made outside its specification is used. If the consequences of the variation result in an unsafe condition for the customer, the dimension is a safety related dimension. The following precautions should then be taken:

- Engineering should identify the dimension as a safety related dimension.
- Manufacturing should take precautions to ensure that a part with an out of print safety related dimension is never produced or used, i.e., 100% inspection, error proofing, and documentation control be exercised.

Dimensions of this type can have various forms of loss function curves. The shape of the loss function curve is not the key issue; instead the consequence of the variation is the focus on a safety related dimension.

Non-system Related Dimensions

A non-system related dimension is a dimension that has little or no impact on the product function.

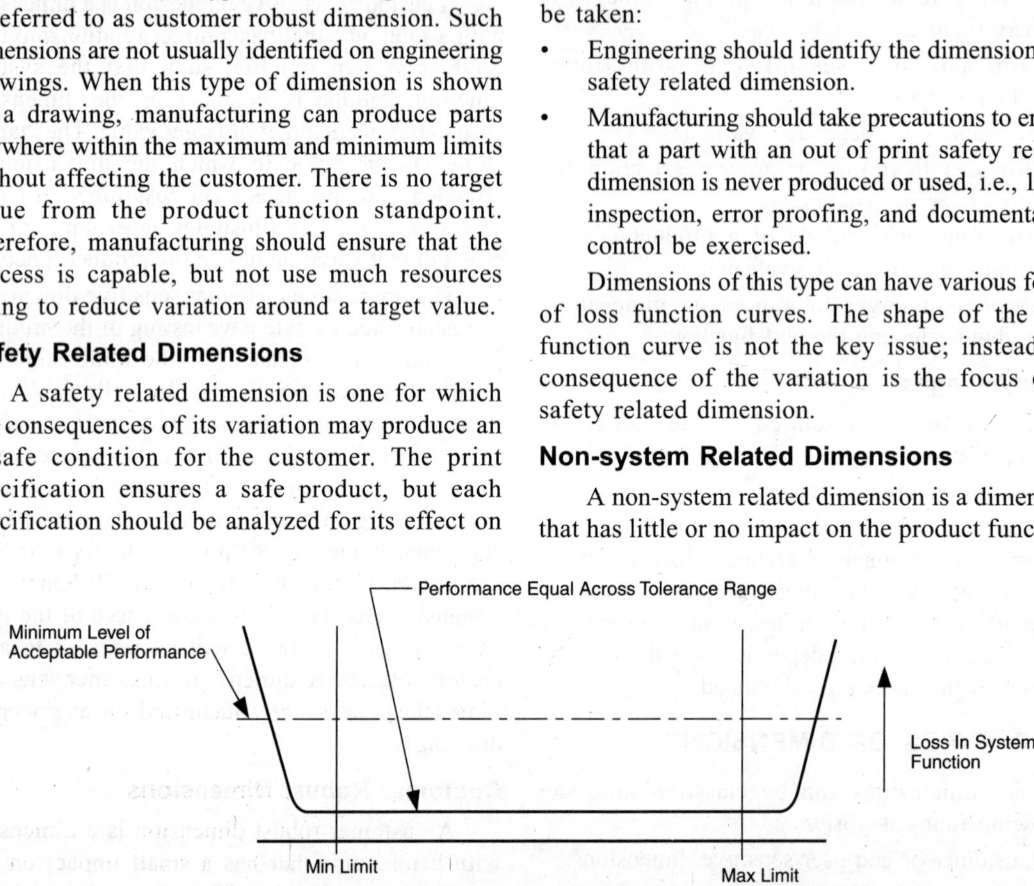

Fig. 14.19. Loss function curve for a customer robust dimension

The loss function curve usually looks like the curve shown in *Fig, 14.20*. The flat portion of the curve extends well beyond the limits of the print specification. The system is not sensitive to variation in this dimension and could be produced well outside the print limits and the part still function properly. For this type of dimension, the process capability is often the primary consideration in determining the tolerance. The manufacturing engineer should determine how much tolerance is needed for the base method used to make the part. Non-system related dimensions are not identified, in any special way, on the part drawing. However, the design team should be aware of which dimensions are non-system related on each part.

For the four types of dimensions, the basis for establishing tolerances, and the impact of manufacturing on the part, we can conclude that:

- For customer sensitive dimensions the tolerance is based on functional requirements and we have to plan to reduce the variation about the target value.
- For customer robust dimensions the tolerance is based on functional requirements and we have to ensure 100% conformance.
- For safety related dimensions tolerance is based on functional requirements and we have to plan to ensure 100% conformance with provisions to further ensure that a part out of specifications never gets into the product.
- For non-system related dimension the tolerance is based on both the functional requirements and the capability of the base process to be used. We have to plan to ensure 100% conformance.

14.13. PRINCIPLES OF FUNCTIONAL DIMENSIONING

Functional dimensioning philosophy used three basic principles:

1. All dimensions on a drawing should be clear, concise, consistent, and based on the system function. This is the first principle of the functional dimensioning philosophy. The use of geometric tolerances enables the dimensions to be clear and concise. The dimensions are based on the system function. Every part is a component in one or more product systems. When dimensioning a part, it must be recognized that the dimensions are truly ensuring that the part will fulfill the system requirements. The organization will be able to focus on the dimensional relationships that are important to the product function.
2. The second principle of functional dimensioning is that the dimensions should specify the largest possible tolerance the system function will allow. This principle states that the component tolerances are based on the system requirements. Each dimension should be analyzed to determine which classification it belongs to. Then the large possible tolerance for the dimension should determined. If a dimension has no effect on a product system, the amount

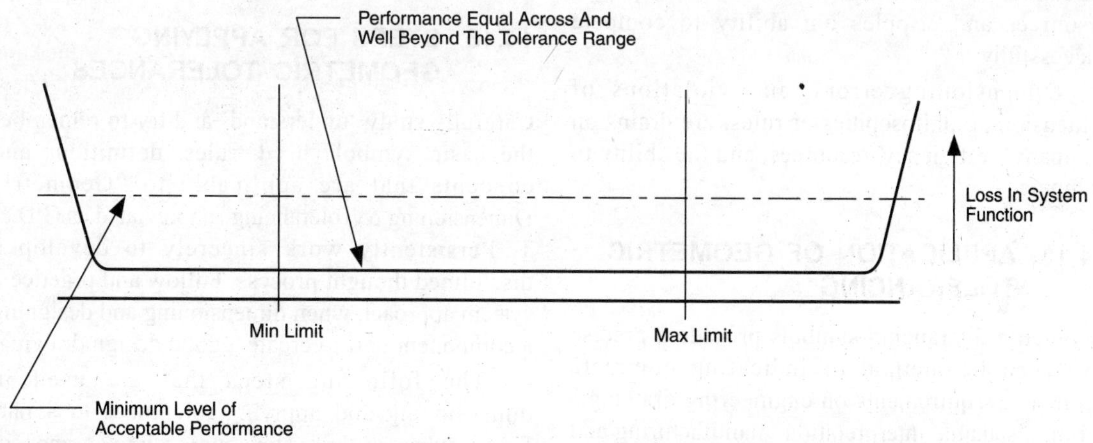

Fig. 14.20. Loss function curve for a non-system related dimension

of tolerance for the dimension should be based on what the effect will be on the product. The relationship between a tolerance and the system function should be analyzed for each dimension.

3. The third basic principle is that the dimensions should define the part without specifying the manufacturing processes. The method of producing the part feature is left to the process engineer. Specifying dimensions without specifying processes has two major advantages:
   - The process engineer has flexibility to process the part in the most economical method that will produce the end results required. This reduces production costs.
   - The designer will be focusing on part function and will use the system function requirements in specifying the part features This results in better product designs.

14.14. COST OF POOR DIMENSIONING

Poor dimensioning causes enormous waste in the organization. The waste costs the organization! causing confusion among many people who rely on the drawing's dimensions.

Confusion results in unnecessary meetings, arguments, and waste. It also results in scrap by way of parts that do not function as intended,! well as parts that would function fine, but are rejected for not meeting unnecessary tolerances. Scrap costs in terms of material expense, labour and lost productivity. Time is lost in meetings to resolve the conflicting issues, waste, and missed schedules. Poor dimension drains on the company's resources and cripples out ability to compete successfully.

Dimensioning errors, and violations of dimensioning philosophies or rules, are drains on company's efficiency, resources, and the ability to compete.

14.15. APPLICATION OF GEOMETRIC TOLERANCING

Geometric tolerancing symbols provide a precise and accurate method of indicating complete geometric requirements on engineering drawings, for unmistakable interpretation, manufacturing and quality requirements. Form tolerances limit the deviations of an individual feature from its ideal geometrical form.

Orientation, location and runout tolerances limit the deviations of the mutual orientation and/or location of two or more features. For functional reasons, one or more features may be indicated as a datum. If necessary, a geometric tolerance should be specified to the datum feature in order to ensure that the datum feature is sufficiently exact for its purpose.

As the symbols are internationally agreed upon, language barriers do not limit interpretation. Geometric tolerancing should be used in circumstances where even the best applied conventional dimensioning and tolerancing is inadequate and does not ensure that parts would assemble and function satisfactorily after manufacture.

Geometric tolerancing is used with advantage for drawings of component parts that are to be manufactured in widespread locations. In such cases, complete and explicit statement of requirements, provided by geometric tolerancing ensures that the designer's instructions are understood without ambiguity.

Geometric tolerance applies, always, to the whole extent of the toleranced feature unless otherwise specified, e.g., 0.02/50 indicates that a tolerance of 0.02 is permitted for an extent of 50 at any place of the toleranced feature.

Indication of geometric tolerances does not necessarily imply the use of any particular method of production, measurement or gauging.

14.16. STEPS FOR APPLYING GEOMETRIC TOLERANCES

Carefully study 'understand' and try to remember the basic symbols used, rules, definitions and concepts that are applicable to "Geometric Dimensioning & Tolerancing" abbreviated as G.D.& T. Persistently work sincerely to develop a disciplined thought process. Follow and practice a system approach when dimensioning and designing a component part to create a good design drawing.

The following steps that are used in dimensioning and applying tolerances to a part form a subset of the system approach to component design. Basically these steps are the essence of dimensioning a part.

- Specify the datums for dimensioning and measurements.
- Establish the dimensional relationships between part features.
- Determine the "dimension type" for each dimensional relationship between features,
- Carefully select the proper modifier for each geometric tolerance.
- Establish an appropriate numerical value for each dimensional relationship.

14.17. BALANCED DESIGN

There are two basic components that are important to a design:

1. Functional performance, and
2. Cost of production.

Functional design elements comprise of performance, appearance, durability, and reliability. A product must function and perform as intended so as to be successful in the market. Also, making a product involves several kinds of costs that include manufacturing, quality control, assembly, testing, validation, etc.

When making designs, engineering departments can strive for the best possible functional performance. But this may drive the cost elements too high and result in poor profits. Designs that strive for function at the expense of cost are poor designs. On the other hand, designing for lower cost by sacrificing performance, appearance, durability, or reliability is not a right option either. Substandard function eventually results in low sales and result in poor profits, too. Designs that strive for low cost at the expense of functional performance are also poor designs.

A good design is one that is manufacturable, durable, easy to assemble and service, attractive in appearance, and protects product function. A good design is one that is balanced as regards to functional performance and cost of production. A "balanced design" is one that defines a product and optimizes product function and product cost elements.

14.18. GENERAL DEFINITIONS

Feature

A feature is a general term used for a physical portion of a part, e.g., a surface, hole, pin, tab or slot. Think of a feature as a part surface. The part shown in *Fig. 14.21* has seven features.

Axis of Feature

Axis of feature is a straight line that coincides with the centre plane of the true geometric counterpart of the specified feature.

Centre Plane of Feature

Centre plane of feature is a plane that coincides with the centre plane of the true geometric counterpart of the specified feature.

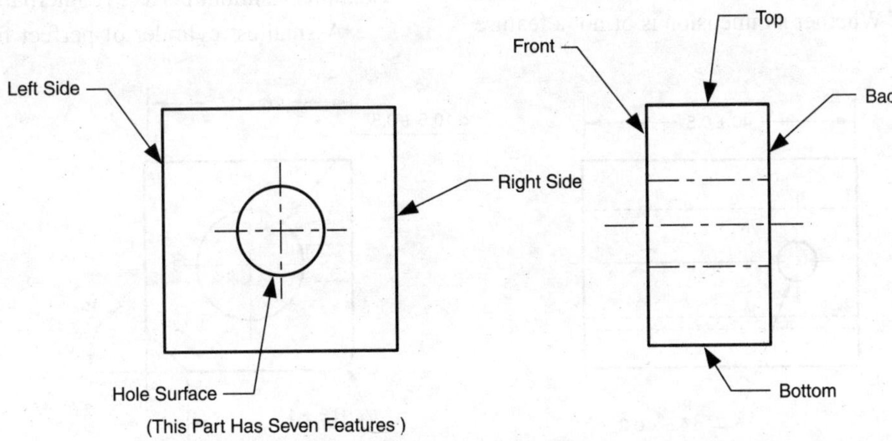

Fig. 14.21. Features

Feature of Size

A *feature of size*, abbreviated as FOS, is one cylindrical or spherical surface, or a set of two *opposed* elements or opposed parallel surfaces, associated with a size dimension. An axis, median plane or centre-point can be derived from a feature of size.

Each feature of size consists of features (surfaces). A cylindrical FOS contains one feature, the cylindrical surface. A planar FOS contains two features, the two parallel planar surfaces.

Internal and External Features of Size

Features of size are of two types – external FOS and internal FOS. External features of size are comprised of part surfaces (or elements) that are external surfaces, e.g., a shaft diameter or the overall height or width of a planar part. In *Fig. 14.22*, the dimensions 40 ± 0.5, and 26 ± 0.5 are size dimensions for external features of size.

An internal feature of size is comprised of part surfaces (or elements) that are internal part surfaces, e.g., a hole diameter or the width of a slot. In *Fig. 14.22*, the 4.5 ± 0.2 diameter hole and the 10.5 ± 0.3 spherical diameter depression are size dimensions for internal features of size.

Feature of Size Dimension

When a dimension is associated with a feature of size, it is referred to as a feature of size dimension. When a dimension is not associated with a feature of size it is referred to as a non-feature of size dimension. In *Fig. 14.22* there are four features of size dimensions and three non-features of size dimensions. Whether a dimension is or not a feature of size dimension is an important concept in geometric tolerancing.

Actual Local Size

Actual local size is the value of any individual distance at any cross-section of a feature of size. It is a two-point measurement, taken with a measuring instrument like a caliper or micrometer, taken at a point along the cross-section of the part. A FOS may have several different values of actual local size.

Resultant Condition

Resultant condition is the variable boundary generated by the collective effects of a FOS's specified MMC or LMC material condition, the geometric tolerance for that material condition, the size tolerance, and the additional geometric tolerance derived from the feature's departure from the specified material condition.

Resultant Condition Size

Resultant condition size is the actual value of the resultant condition boundary.

Actual Mating Envelope (AME)

The term "*actual mating envelope*" is defined according to the type of feature of size being considered, as follows:

1. The actual mating envelope **for an external feature of size** is a similar perfect feature counterpart of the smallest size that can be circumscribed about the feature so that it just touches the surface at the highest points. For example, a similar perfect counterpart could be:
   - A smallest cylinder of perfect form,

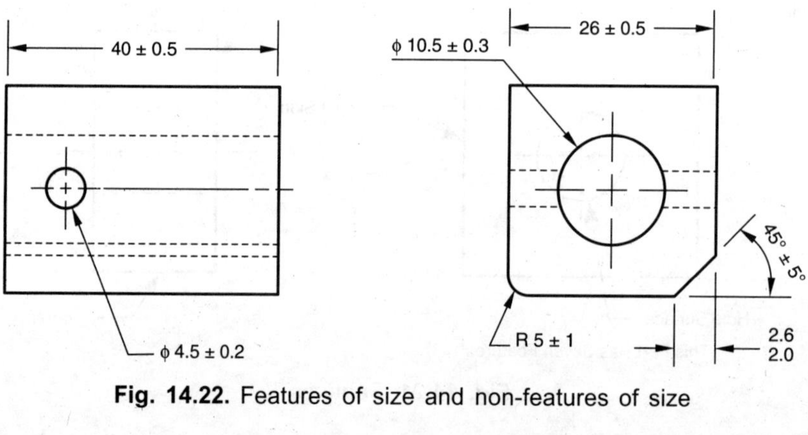

Fig. 14.22. Features of size and non-features of size

- Two parallel planes of perfect form at a minimum separation, that just contacts the highest points of the surfaces. Actual mating envelope (AME) is a variable value. It is derived from an actual part, *Fig. 14.23.*

2. The actual mating envelope (AME) **for an internal feature of size** is a similar perfect feature counterpart of the largest size that can be inscribed within the feature so that it just touches the surface at its highest points.

For example, a similar perfect feature counterpart could be a largest cylinder of perfect form or two parallel planes of perfect form at maximum separation that just contact(s) the highest points of the surfaces. AME is a variable value. It is derived from an actual part, *Fig. 14.24.*

If a feature of size is controlled by an orientation or positional tolerance, with the MMC or LMC modifier used, the AME is relative to the appropriate datum(s).

Full Indicator Movement (FIM)

The total movement of an indicator when appropriately applied to a surface to measure its variations.

True Geometric Counterpart

A true geometric counterpart is the theoretically perfect boundary (virtual condition or actual mating envelope) or best-fit (tangent) plane of a specified datum feature.

True Position

True position is the theoretically exact location of a feature established by basic dimensions.

14.18.1. Applicability of Material Conditions

Applicability of RFS, MMC, and LMC is limited to features subject to variations of size. They may be datum features or other features whose axes or centre planes are controlled by geometric tolerances. The ability to specify tolerances for a part feature at material conditions is an important concept in geometrical tolerancing. A geometric tolerance can be specified to apply at the largest

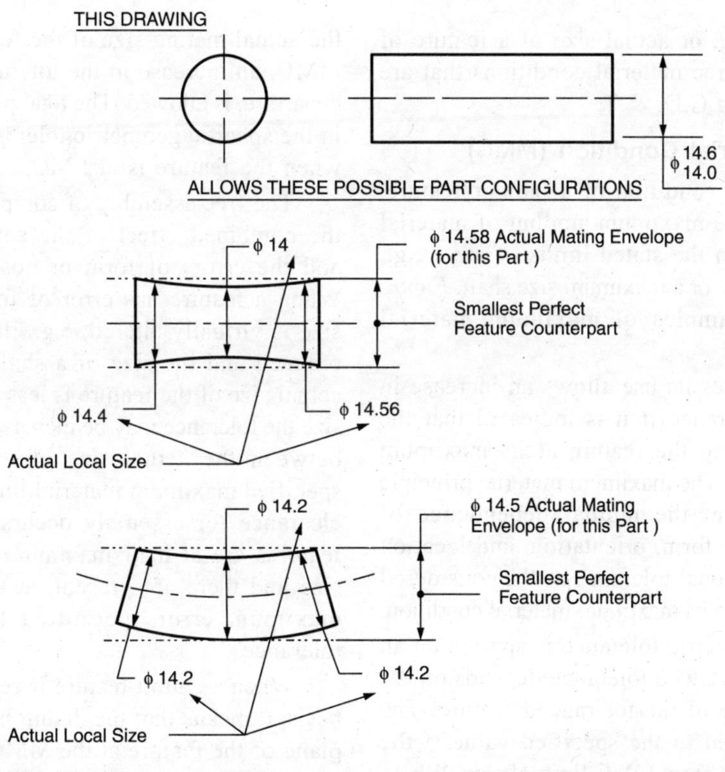

Fig. 14.23. Actual mating envelope of an External Feature of size

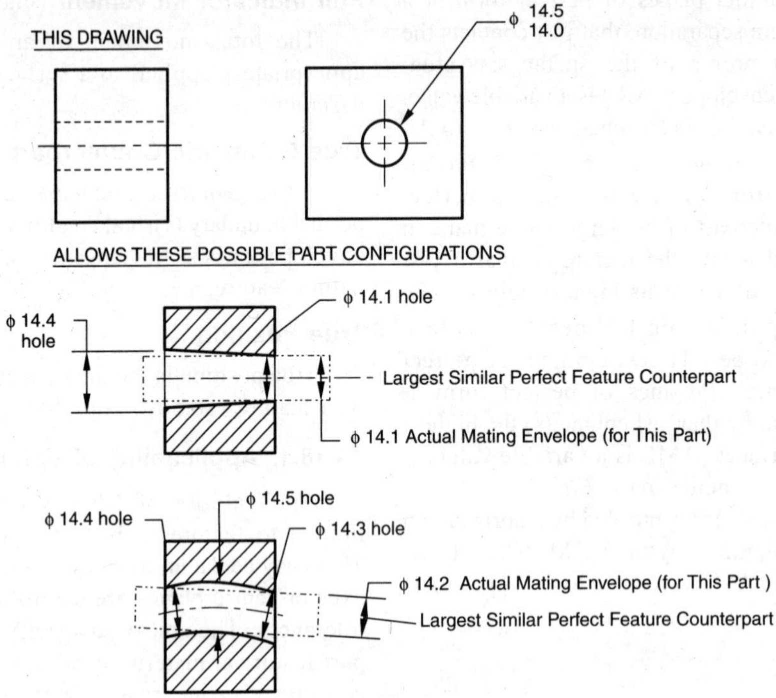

Fig. 14.24. Actual mating envelope of an internal feature of size

size, smallest size, or actual size of a feature of size. There are three material conditions that are commonly used in G.D. & T.

Maximum Material Condition (MMC)

MMC is that condition of a part or feature, which contains the maximum amount of material everywhere within the stated limits of size, e.g., minimum size hole, or a maximum size shaft. *Figure 14.25* shows examples of maximum material condition.

In certain cases its use allows an increase in the specified tolerance if it is indicated that the tolerance applies to the feature at its maximum material condition. The maximum material principle takes into account the mutual dependence of tolerances of size, form, orientation, and location and permits additional tolerance as the considered feature departs from its maximum material condition.

Where a geometric tolerance is applied on an MMC basis, the allowed tolerance depends on the actual mating size of the toleranced feature. The tolerance is limited to the specified value if the feature is produced at its MMC limit of size. Where the actual mating size of the feature departs from MMC, an increase in the tolerance, equal to such departure, is allowed. The total permissible variation in the specific geometric tolerance is at maximum when the feature is at LMC.

The free assembly of components depends on the combined effect of the actual finished sizes and the errors of form or position of the parts. When a feature has error of form or position its size is virtually altered, e.g., the size of a hole is reduced and the size of a shaft increased. If the actual size of the feature is less than the maximum size the tolerance may be increased by the difference between the actual size of the feature and its specified maximum material limits. The minimum clearance for assembly occurs when the mating features are at their maximum material limits of size and there are present, at the same time, the maximum errors permitted by the geometric tolerances.

When a datum feature is referenced on MMC basis, it means that the datum is the axis or centre plane of the feature at the MMC limit. Where the actual mating size of the datum feature has departed

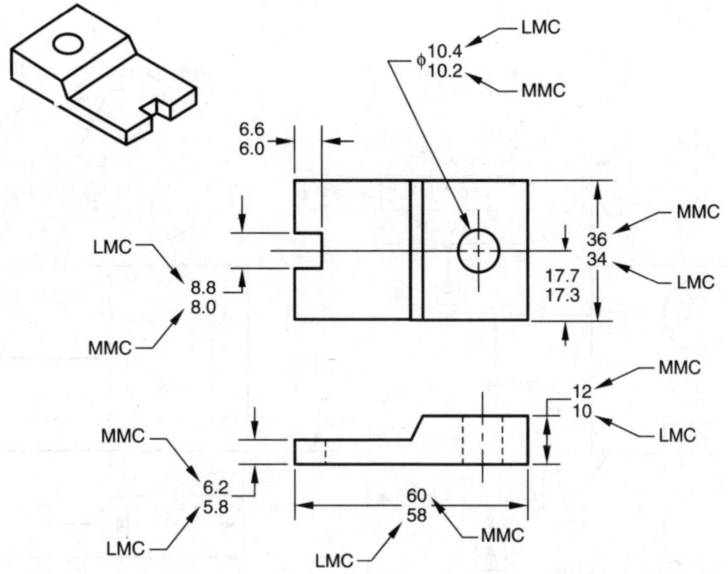

Fig. 14.25. Maximum and minimum material conditions of size

from MMC, a deviation is allowed between the axis or centre plane and the axis or centre plane of the datum.

Where a tolerance of position or orientation is applied on a zero tolerance at MMC basis, the tolerance is totally dependent on the actual mating size of the toleranced feature. No tolerance of position or orientation is allowed if the feature is produced at its MMC limit of size; and in such a case, the feature must be located at true position or be perfect in orientation, as the case may be. Where the actual mating size of the feature departs from MMC, a tolerance, equal to such departure, is allowed. The total permissible variation in position or orientation is at maximum when the feature is at LMC.

When an increase in geometric tolerances assists manufacture they should, in all cases be established by functional acceptance.

The indication that the tolerance value applies at the maximum material condition is shown by the symbol M placed after the tolerance value, the datum letter or both, *Fig. 14.26.*

The position of the symbol M decides whether the maximum material principle is to be applied to the toleranced feature, the datum feature or both of these items. If a letter does not identify the datum, the maximum material principle applied to the datum is indicated in the third compartment of the tolerance frame, *Fig. 14.27 (a), (b).*

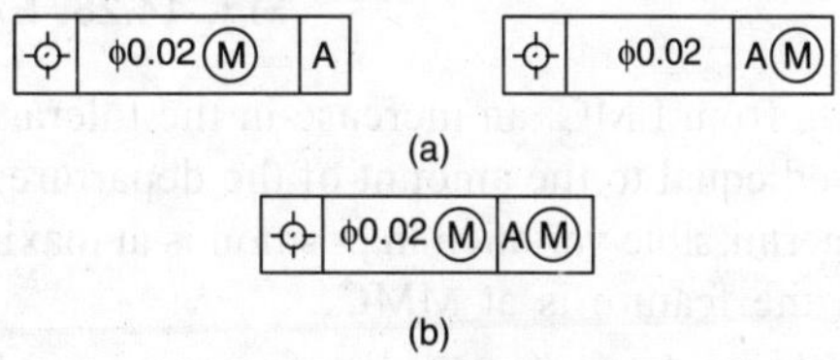

Fig. 14.26.

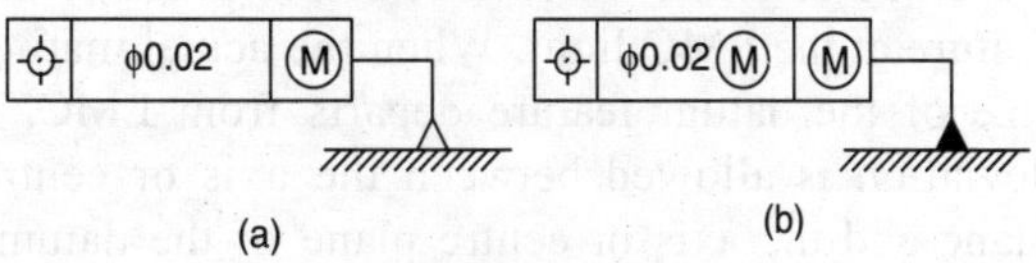

Fig. 14.27.

Least Material Condition (LMC)

Least material condition is the condition in which a feature of size contains the least amount of material everywhere within the stated limits of size, e.g., the smallest shaft diameter or the largest hole diameter. *Fig. 14.28* shows examples of the least material condition.

Where a positional tolerance is applied on a LMC basis, the allowed tolerance is dependent on the actual mating size of the feature under consideration. The tolerance is limited to the specified value if the feature is produced at its LMC limit of size. When the actual mating size

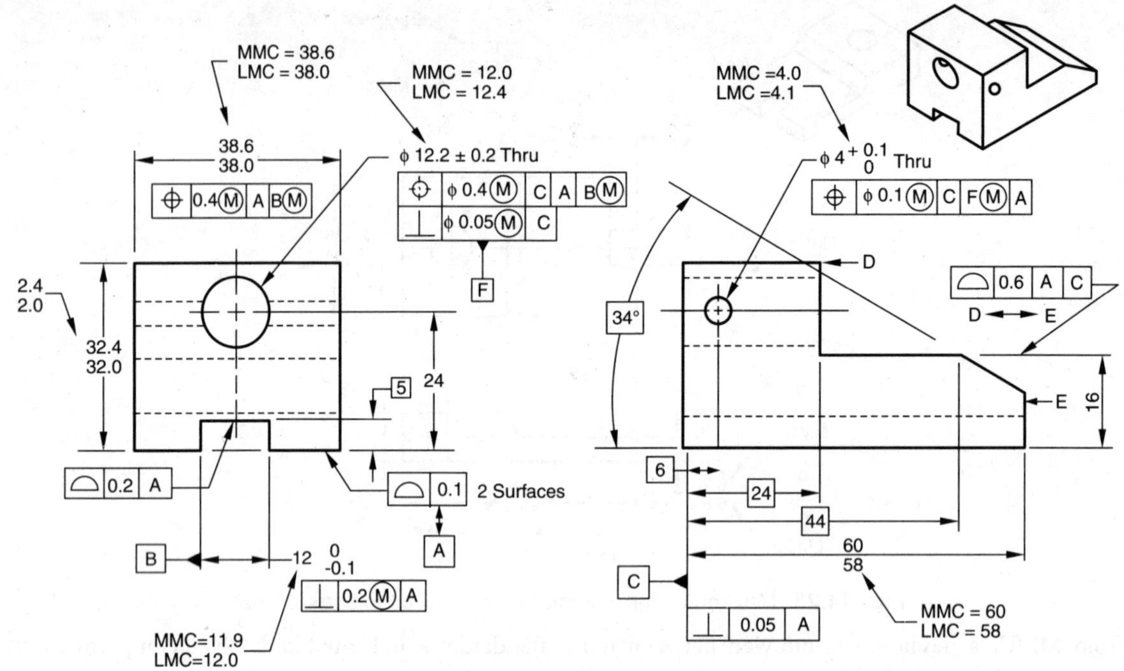

Fig. 14.28. Examples of MMC and LMC

departs from LMC, an increase in the tolerance is allowed equal to the amount of the departure. The total permissible variation in position is at maximum when the feature is at MMC.

When a datum feature is referenced on a LMC basis, the datum is the axis or centre plane of the feature at the LMC limit. When the actual mating size of the datum feature departs from LMC, a deviation is allowed between the axis or centre plane and the axis or centre plane of the datum.

When a tolerance of position or orientation is applied on a zero tolerance at LMC basis, the tolerance is totally dependent on the size of the considered feature. No tolerance of position or orientation is allowed if the feature is produced at its LMC limit of size, and the feature must be located at its true position or be perfect in orientation, as applicable. When the actual mating size of the feature under consideration departs from the LMC, a tolerance is allowed equal to the amount of the departure. The total permissible variation in position or orientation is at maximum when the feature is produced at MMC.

Every feature of size has a maximum and a least material condition. Limit dimensions directly specify the maximum and least material condition of a feature of size. When a drawing has plus-minus dimensions, the material condition may have to be derived from the dimensions, *Fig. 14.26.*

Regardless of Feature Size (RFS)

Regardless of feature size is the term that indicates a geometric tolerance applies at any increment of size of the feature, within its size tolerance. It means that the geometric tolerance applies at whatever size the part is produced. There is no symbol for RFS because it is the default condition for all geometric tolerances.

Each material condition is used for different functional reasons. Geometric tolerances are often specified to apply at MMC when the function of a FOS is assembly. Geometric tolerances are often specified to apply at LMC to ensure a minimum distance on a part. Geometric tolerances are often specified to apply at RFS to ensure symmetrical relationships.

Geometrical tolerances specified at maximum or least material condition have a significant cost advantage over geometric tolerances specified regardless of feature size.

Projected Tolerance Zone (P)

A projected tolerance zone is where the tolerance zone is projected above the part surface. A projected tolerance zone is commonly used to limit the perpendicularity of a hole to ensure assembly with the mating part.

It is sometimes necessary to tolerance the external projection of a component feature; *Fig. 14.29* shows a typical example of a **'Projected Tolerance Zone.**

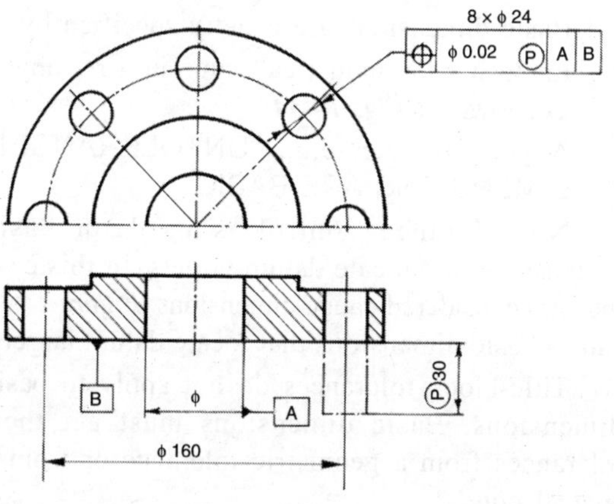

Fig. 14.29. Projected tolerance zone application

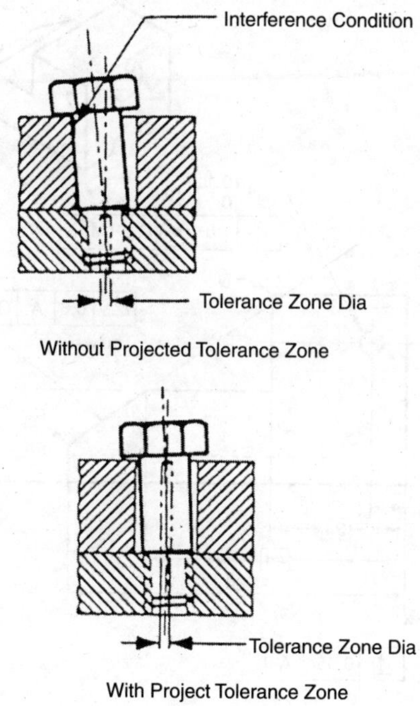

Fig. 14.30. Fixed Fastener Interference Condition

When dimensioning threaded holes (or press-fit holes), consideration must be given to variation in perpendicularity of the hole relative to the mating face of the assembly. The squareness error of the fastener (or press fit pin) may result in an interference conditions with the mating part, *Fig. 14.30*. An interference condition can occur where a position tolerance is specified for the hole, and the hole is tipped within the position tolerance zone. When the fastener is placed in the hole, the orientation of the fastener may result in an interference condition near the head of the fastener. This condition is common with fixed fastener applications.

Radius

A radius is a straight line extending from the centre to the surface of an arc or a circle. The symbol for a radius is "R". When the "R" symbol is used, it creates a zone defined by two arcs (the minimum and maximum radii). The part surface must lie within this zone, as shown in *Fig. 14.31*. The part surface may have flats or reversals within the zone.

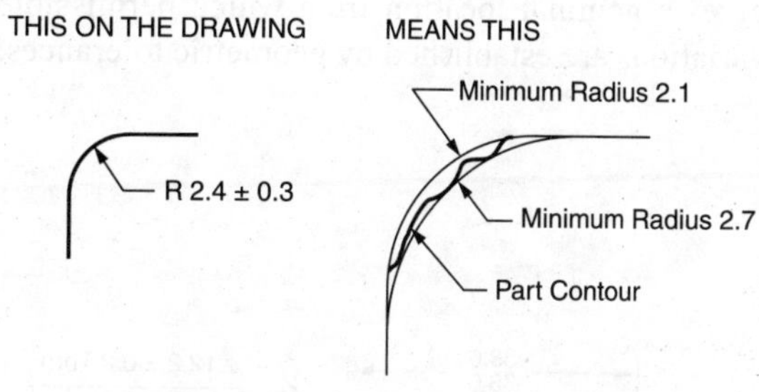

Fig. 14.31. Specifying a radius

Controlled Radius

A controlled radius is a radius with no flats or reversals allowed within the tolerance zone. The symbol for a controlled radius is "CR". When the "CR" symbol is used, it creates a zone defined by two arcs (the minimum and maximum radii). The part surface must be within the crescent-shaped tolerance zone' and be an arc without flats or reversals, as shown in *Fig. 14.32*.

A controlled radius is a stringent requirement. It should only be used in special cases. For example, when the part stresses are very high and reversals in the surface with a radius would produce higher additional stresses.

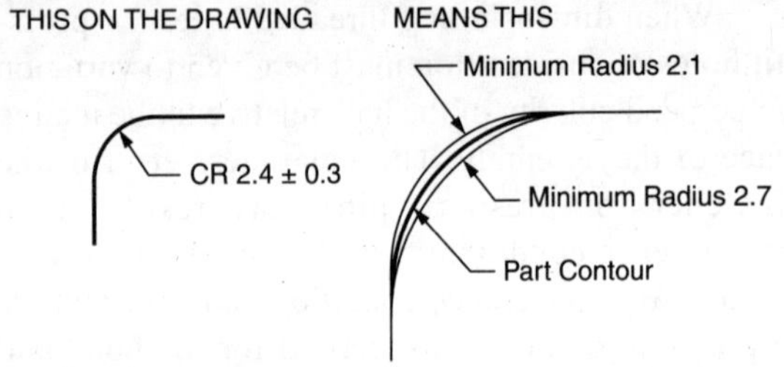

Fig. 14.32. Specifying a controlled radius

Basic Dimension

A basic dimension is a numerical value used to describe the theoretically exact size, true profile, orientation, or location of a feature or datum target. It is the basis from which permissible variations are established by tolerances on other dimensions, in notes, or in feature control frames. There are two uses for basic dimensions:

1. To define a theoretically exact location, orientation, or true profile of a part feature, and locate the tolerance zone.
2. To define datum targets.

A basic dimension used to define part features, gives a nominal location from which permissible variations are established by geometric tolerances. Basic dimensions used to describe part features must be accompanied by geometric tolerances to specify amount of tolerance the part feature can have, as shown in *Fig. 14.33*. A basic dimension designates a goal, and the geometric tolerance and the added tolerance specifies the amount of acceptable variation from the goal.

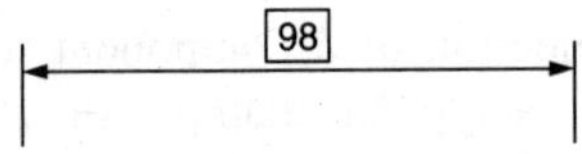

Fig. 14.33. Basic dimension symbol

Basic dimensions are usually specified by:

- Enclosing the numerical value in a rectangle, as shown in *Fig. 14.34*.
- A general note, e.g., "UNTOLERANCED DIMENSIONS ARE BASIC".

No tolerance control is used on basic dimensions that locate datum targets. In this case they are considered gauge dimensions. *Figure 14.35* shows basic dimensions that locate datum targets.

Title-block tolerances do not apply to basic dimensions. Basic dimensions must get their tolerances from a geometric tolerance or from a special note.

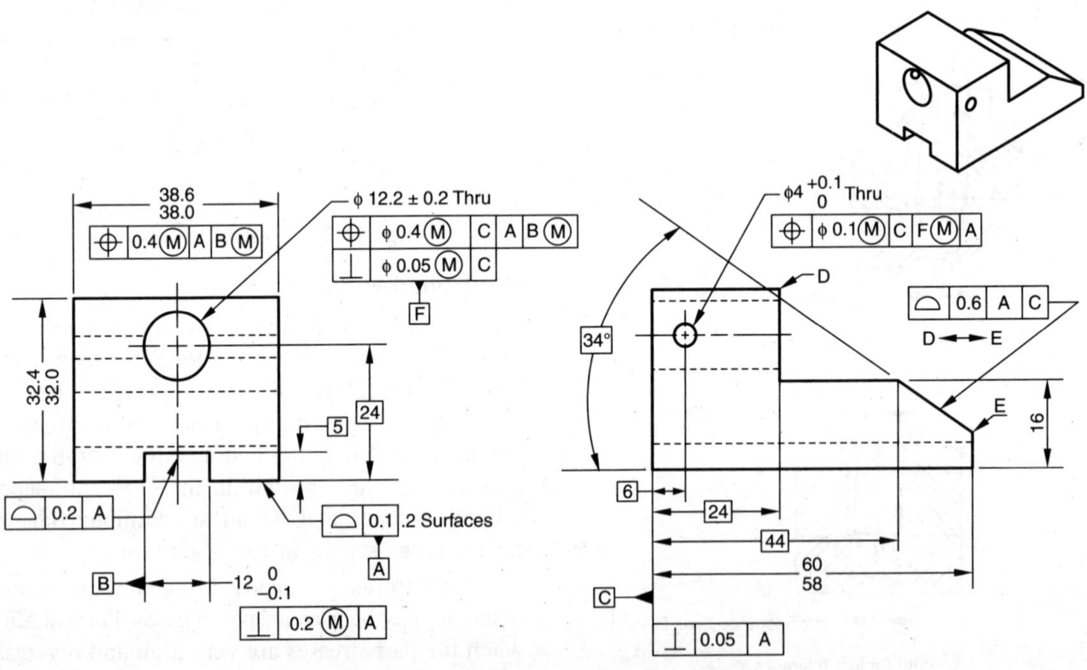

Fig. 14.34. Basic dimensions examples

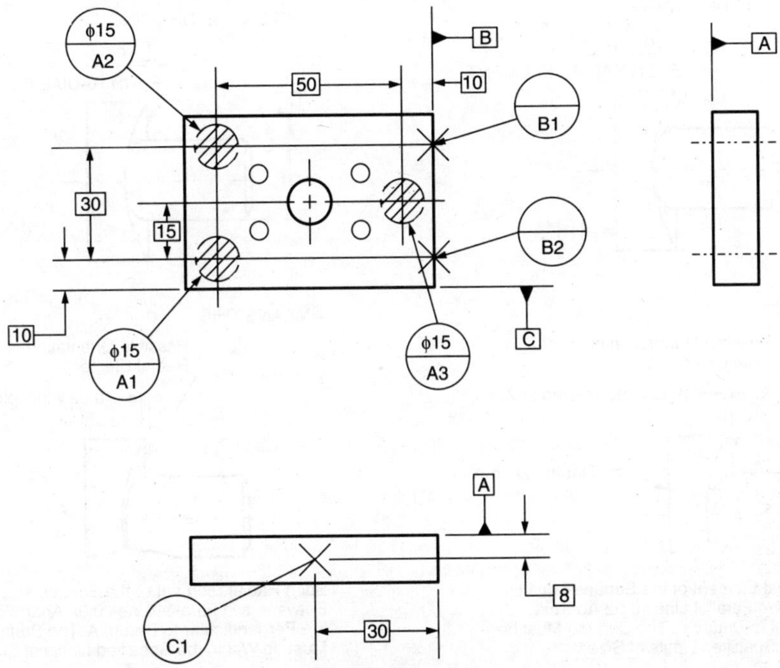

Fig. 14.35. (a)

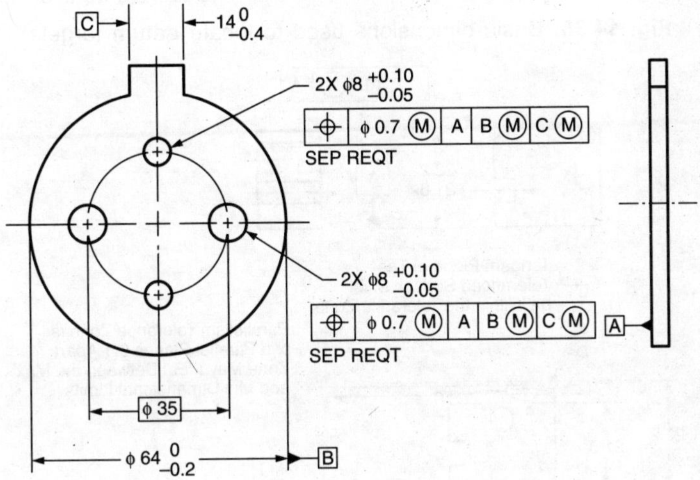

Fig. 14.35. (b) Multiple patterns of features, separate requirements

Reference Dimension

It is a dimension, usually without tolerance, used for information purposes only. A reference dimension is a repeat of a dimension or is derived from other values shown on the drawing or on related drawings. It is considered auxiliary information and does not govern production or inspection operations.

Tangent Plane

A theoretically exact plane derived from the true geometric counter part of the specified feature surface. Method of specifying a Tangent Plane is illustrated in *Fig 14.36.*

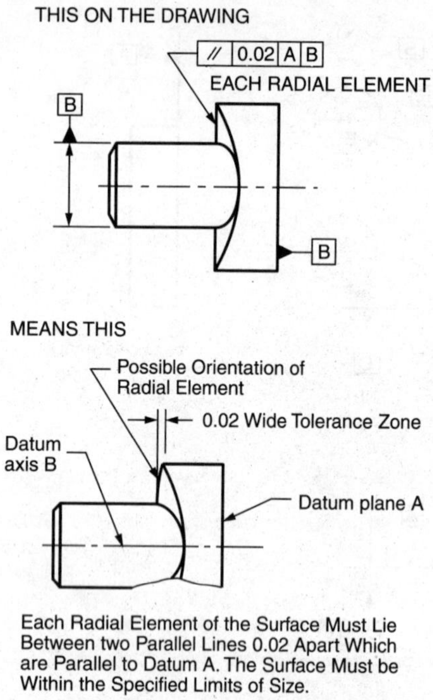

(c) Specifying parallel for a radial element of a size

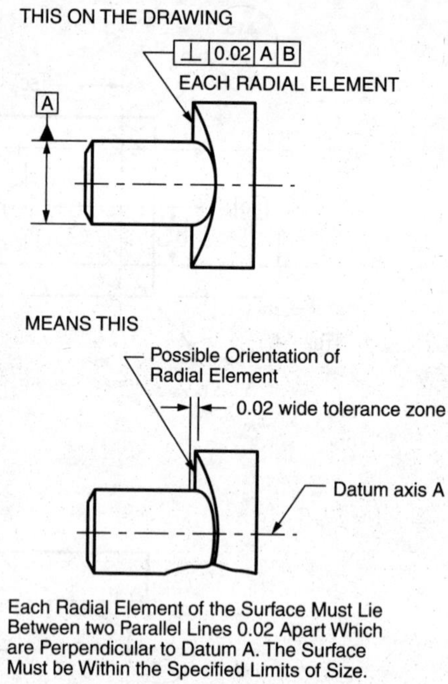

(d) Specifying perpendicularity for a radial element of a surface

Fig. 14.35. Basic dimensions used to locate datum targets

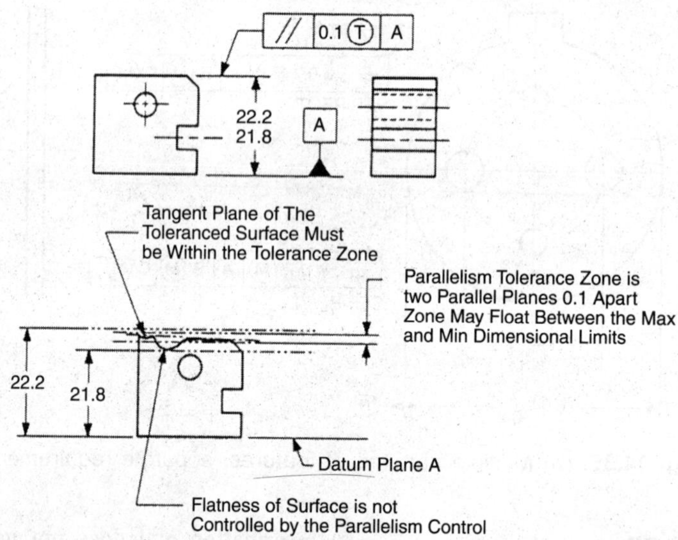

Fig. 14.36. (a) Parallelism with the tangent plane modifier

14.19. SYMBOLS

This section establishes the symbols for specifying geometrical characteristics and other dimensional requirements on engineering drawings. All symbols should be clear, legible, and meet the reproducibility requirements of national and international standards.

In case of situations where the desired geometric requirements cannot be completely conveyed by the standard symbology used, a note may be used to describe the requirement, either separately or to supplement a geometric symbol, *Fig. 14.35 (b), 14.35 (c),* and *14.35 (d).*

Fig. 14.36 (*b*) shows common dimensioning symbols and their proportions as used on drawings.

Modifiers

In the language of geometric tolerancing there is a set of eight common symbols called "modifiers". These modifiers communicate additional information about the drawing or tolerancing of a part. These modifiers are shown in *Table 14.1.*

The first two modifiers, **MMC** and LMC are explained already. These are placed inside the feature control frame.

The next two modifiers, the "*Projected Tolerance Zone*" *"P"* modifier and the "tangent plane" modifiers, too, are placed inside the feature control frame. The projected tolerance zone modifier changes the location of the tolerance zone, so that it exists above the part. It projects the tolerance zone, so that it exists above the part.

The *"Tangent Plane" "T"* modifier denotes that only the tangent plane established by the high points of the controlled surfaces must be within the tolerance zone. The tangent plane modifier can be used with the parallelism, perpendicularity and angularity controls. When the tangent plane modifier is used in parallelism callouts, the flatness of the toleranced surfaces is not controlled.

Figure 14.36 (*a*), shows a parallelism application that uses the tangent plane modifier. For this case the following conditions apply:

- The tolerance zone is between two parallel planes.
- The tangent plane established by the high points of the surfaces must be within the 0.1 parallelism tolerance zone.
- The flatness of the toleranced surface is not controlled by the parallelism control.

The diameter symbol, is used two ways:

- Inside a feature control frame to denote the shape of the tolerance zone.

  Fig. 14.37 illustrates the method of specifying a tangent plane.
- Outside the feature control frame to simply replace the word "diameter".
- *Fig. 14.37* illustrates the method of specifying a tangent plane.

Table 14.1. Modifiers

TERM	SYMBOL
AT MAXIMUM MATERIAL CONDITION	Ⓜ
AT LEAST MATERIAL CONDITION	Ⓛ
PROJECTED TOLERANCE ZONE	Ⓟ
FREE STATE	Ⓕ
TANGENT PLANE	Ⓣ
DIAMETER	Ø
SPHERICAL DIAMETER	SØ
RADIUS	R
SPHERICAL RADIUS	SR
CONTROLLED RADIUS	CR
REFERENCE	( )
ARC LENGTH	⌒

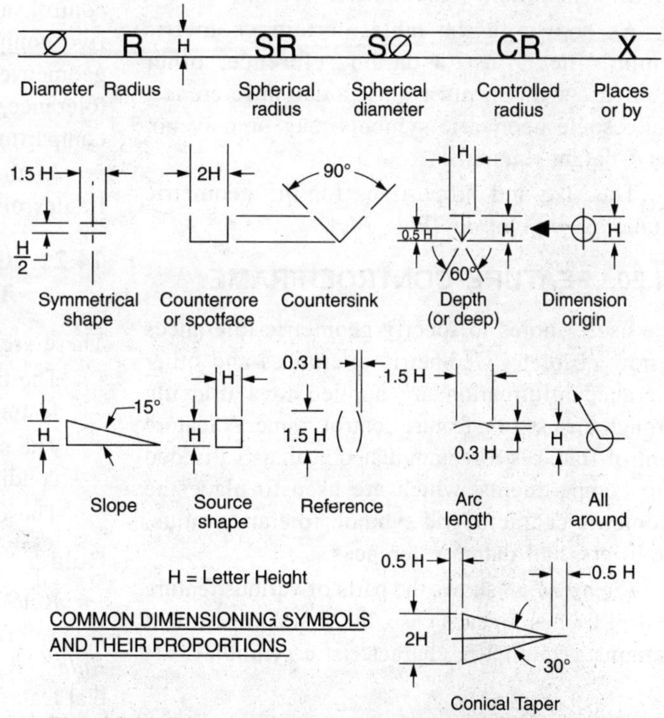

Fig. 14.36 (b).

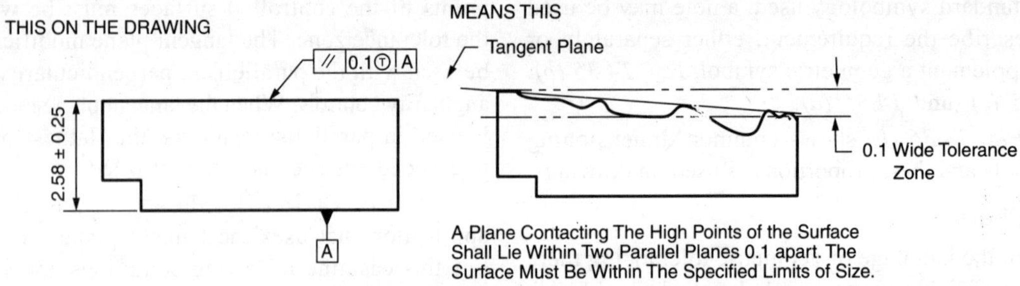

Fig. 14.37. Specifying a tangent plane

The *"Radius" "R"* and *"Controlled Radius"* "CR" modifiers are always used outside the feature control frame.

The modifier *"For Reference" "( )"* is simply the method of denoting that information is for reference only and is not to be used for manufacturing or inspection. To designate a dimension or other information as reference, the reference information is enclosed in parentheses.

Geometric Characteristic Symbols

There are fourteen geometric characteristic symbols used in the language of geometric tolerancing. They are shown in *Table 14.2*. The symbols are divided into five categories: *form, profile orientation, location, and runout.*

As shown in the table, certain geometric symbols never use a datum reference, other geometric symbols always use a datum reference, while some geometric symbols may or may not use a datum reference.

The size and proportions for the geometric symbols are given in *Table 14.3*.

14.20. FEATURE CONTROL FRAME

The use of notes to specify geometric tolerances is now obsolete. Geometric tolerances and other tolerance information are applied to a drawing through the use of a feature control frame. A feature control frame is a rectangular box that is divided into compartments, which are used to place the geometric characteristic symbol, tolerance value, modifiers, and datum references.

Figure 14.38 shows the parts of various feature control frames. In each case, the first compartment contains a geometric characteristic symbol.

The second compartment always contains the tolerance information. The tolerance value specified is always a total value. The tolerance portion may contain several bits of information:

- If the tolerance value is preceded by a diameter symbol ϕ, the shape of the tolerance zone is a cylinder.
- If a diameter is not shown in front of the tolerance value, the shape of the tolerance zone is either parallel planes or parallel lines, or a uniform boundary in the case of a profile tolerance.

When specifying a non-datum related control, a feature control frame will have only two compartments. When specifying a datum related control, a feature control frame may have up to five compartments: the first compartment for a geometric characteristic symbol, second for the tolerance information, and up to three additional compartments for datum references.

Methods of placement, of the feature control frames on a drawing, are shown in *Fig. 14.39*.

14.21. GEOMETRIC DIMENSIONING AND TOLERANCING RULES

There are two rules:

- The first rule establishes default conditions for features of size.
- The second rule establishes a default material condition for feature control frames.

These rules are described here:

Rule 1

Rule 1 is a key concept in geometric tolerancing. It is referred to as the *"Individual Feature of Size Rule"*. Rule 1 is a dimensioning rule to ensure that features of size will assemble with one another.

Table 14.2. Geometric Characteristic Symbol

CATEGORY	CHARACTERISTIC	SYMBOL	USES A DATUM REFERENCE
FORM	STRIGHTNESS	—	NEVER
	FLATNESS	⏥	
	CIRCULARITY (ROUNDNESS)	○	
	CYLINDRICITY	⌭	
PROFILE	PROFILE OF A LINE	⌒	SOMETIMES
	PROFILE OF A SURFACE	⌓	
ORIENTATION	ANGULARITY	∠	ALWAYS
	PERPENDICULARITY	⊥	
	PARALLELISM	//	
LOCATION	POSITION	⌖	
	CONCENTRICITY	◎	
	SYMMETRY	⌯	
RUNOUT	CIRCULAR RUNOUT	↗	
	TOTAL RUNOUT	⌰	

It states that, *"where only a tolerance of size is specified, the limits of size of an individual feature establish the limits to which variations in its form, as well as its size, are allowed"*. For features of size, where only a tolerance of size is specified, the surfaces shall not extend beyond a boundary or envelope of perfect form at MMC.

Often Rule 1 is paraphrased as *"perfect form at MMC"* or the *"envelope rule"*.

There are thus two components of Rule 1:

1. The envelope principle, and
2. The effect on the form of a FOS as it departs from MMC.

When Rule 1 applies, the maximum boundary or envelope of an external feature of size is its MMC. The minimum boundary or envelope for an internal FOS is its MMC. To determine whether the two features of size will assemble, the designer can compare these maximum material conditions of the features of size.

When a pin diameter is at MMC, the pin must have perfect form. This means that the pin should be perfectly straight and round. This will allow the pin to fit through a boundary equal to its MMC. If the size of the pin is less than its MMC, the pin could contain form error, i.e., errors in straightness and roundness, equal to the amount the pin departs from MMC.

Figure 14.40 illustrates the effects of Rule 1 on an external and an internal FOS of a cylindrical shape.

The form of a FOS is controlled by its limits of size, as described below:

- The surfaces of a FOS shall not extend beyond a boundary or envelope of perfect form at MMC.
- When the actual local size of a FOS has departed from MMC toward LMC. The form is allowed to vary by the same amount.
- The actual local size of an individual feature of size must be within the specified tolerance of size.

Table 14.3.

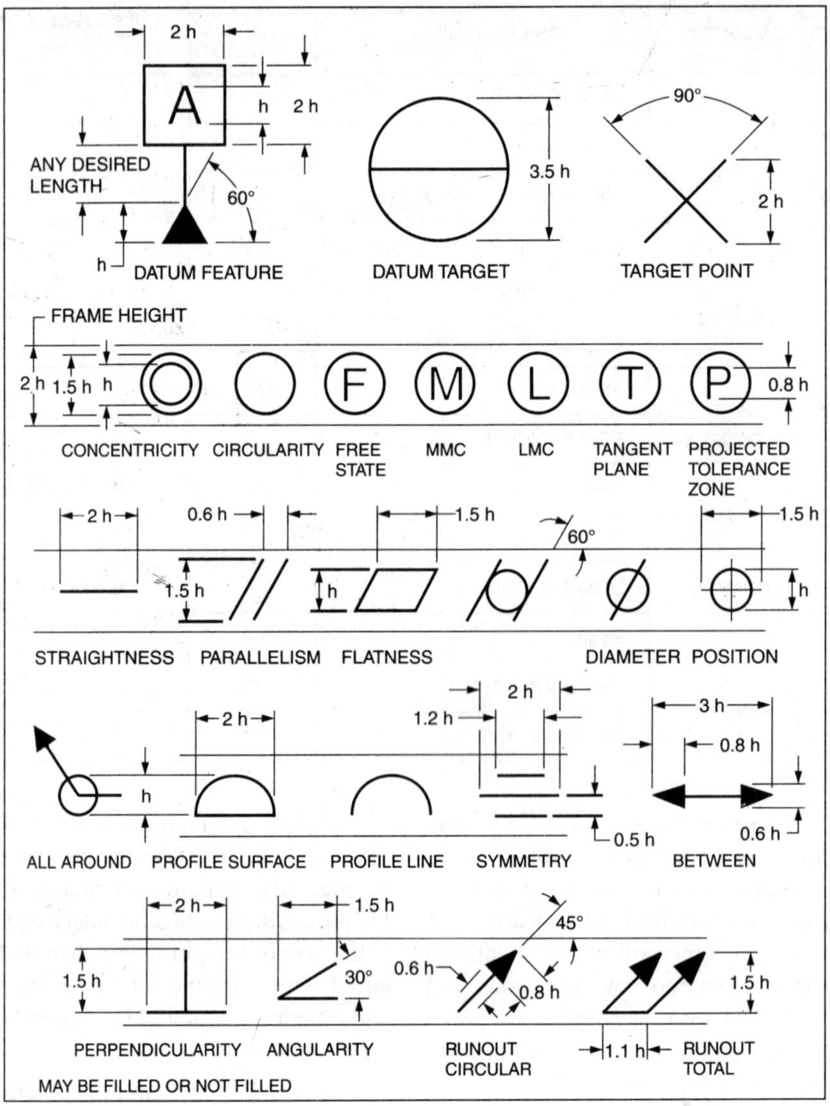

FORM AND PROPORTIONAL OF GEOMETRIC TOLERANCING SYMBOLS

- There is no requirement for a boundary of perfect form at LMC. When a FOS is produced at LMC, it can vary from true form by the amount allowed by the MMC boundary or envelope.

Figure 14.41, shows the effects of Rule 1 on a planar FOS. Here, the words "perfect form" mean perfect flatness, straightness, circularity, and cylindricity.

When the feature of size is produced at MMC, it is required to have perfect form. When the FOS produced is not at MMC, then a form error is allowed. For example, when the height of the part produced is at 10.7, then a form error equal to the amount of departure from MMC, i.e., (10.8 – 10.7) = 0.1 would be allowed. When the part is produced at LMC, a form equal to the amount of departure from MMC, i.e., (10.8 – 10.2) = 0.6 will be allowed. Thus, if the part height were 10.2 everywhere, the flatness of the bottom of the block would be limited to 0.6.

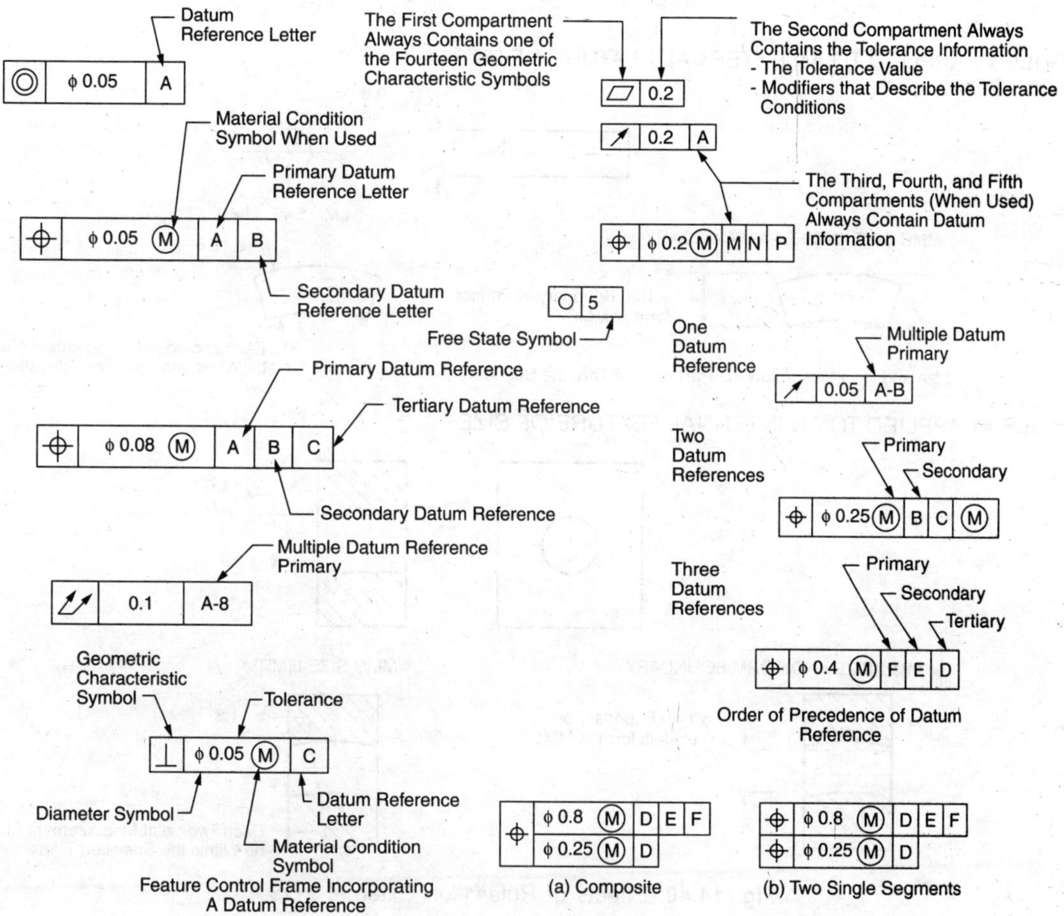

Fig. 14.38. Feature control frames

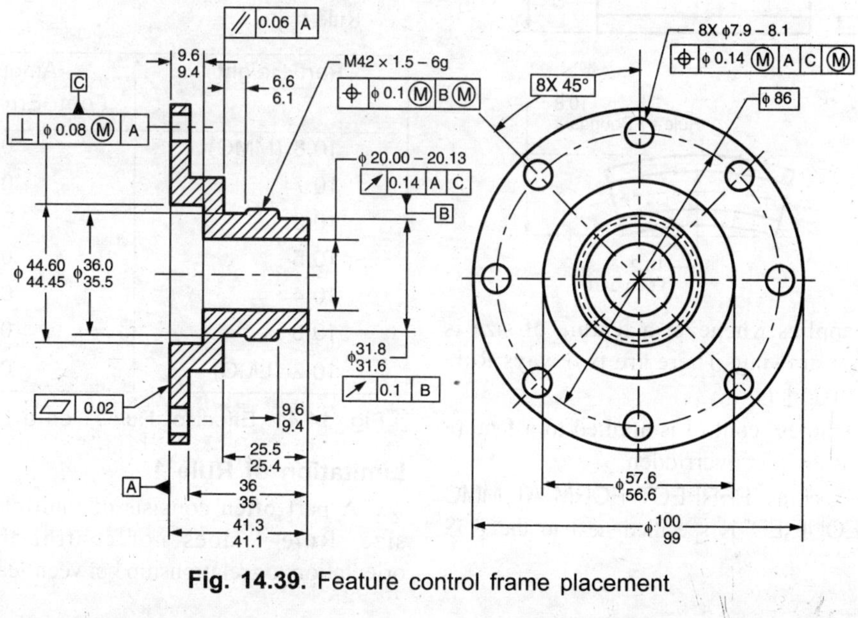

Fig. 14.39. Feature control frame placement

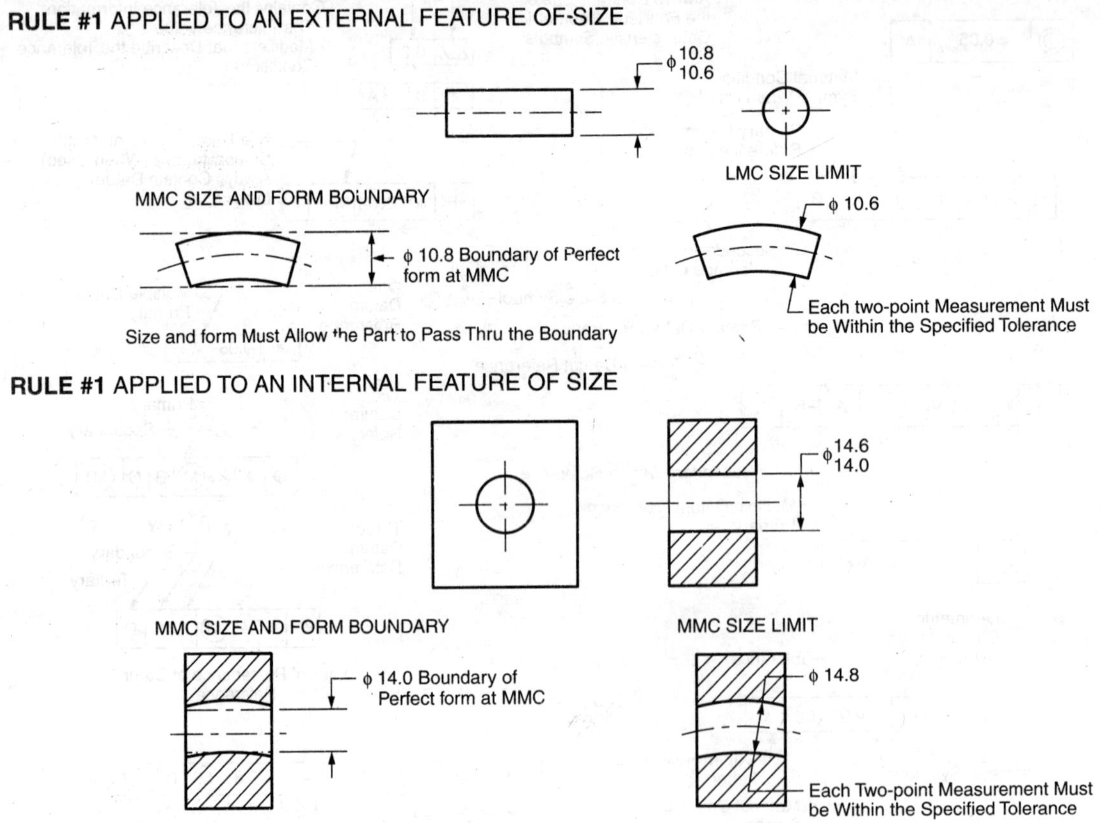

Fig. 14.40. Effects of Rule#1 on External FOS

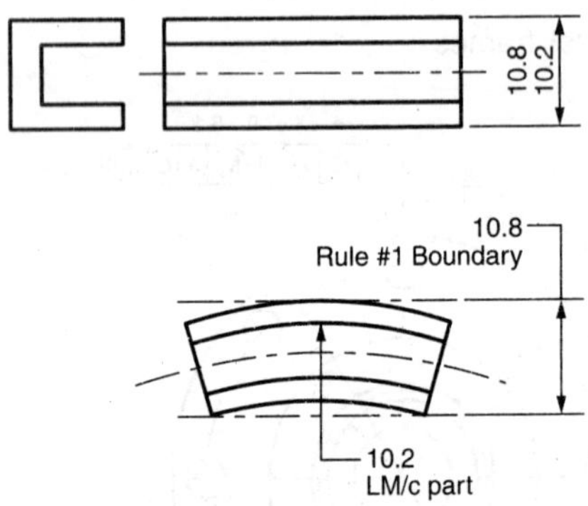

Rule 1 applies whenever a feature of size is specified on a drawing. There are two ways Rule 1 can be overridden:

1. If a straightness control is applied to a feature of size, Rule 1 is overridden.
2. If a note such as "PERFECT FORM AT MMC NOT REQUIRED" is specified next to the FOS dimension, it exempts the FOS dimension from Rule 1.

Part Height	Amount of Form Error Allowed
10.8 (MMC)	0
10.7	0.1
10.6	0.2
10.5	0.3
10.4	0.4
10.3	0.5
10.2 (LMC)	0.6

Fig. 14.41. Effect of Rule#1 on a Planar FOS

Limitation of Rule 1

A part often consists of multiple features of size. Rule 1 does not control the location, orientation, or relationship between features of size.

Features of size shown perpendicular, symmetrical, or coaxial must be controlled for location or orientation to avoid incomplete drawing specifications. Often, implied 90° angles are covered by a general angular tolerance note or by an angular tolerance in the drawing title-block.

In *Fig 14.42*, there are four features of size, marked *S*1, *S*2, *S*3, and *S*4. Rule 1 applies independently to each FOS. Rule 1 does not control angles between these features of size, i.e., *A*1, *A*2, and *A*3.

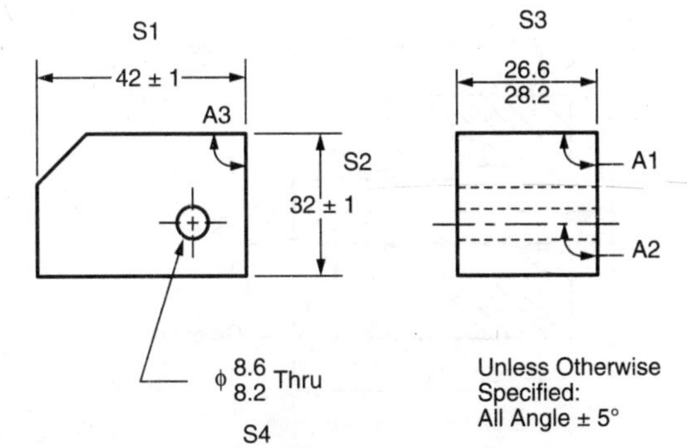

Fig. 14.42. Interrelationship between features of size

There are two exceptions to Rule 1:

1. Rule 1 does not apply to a FOS on a part subject to free-state variation in the unconstrained condition. Rule 1 does not apply to stock sizes, e.g., bar stock, tubing, sheet metal, or structural shapes.

The standards for these items govern the surfaces that remain in the as-finished condition on the finished part.

Inspecting A FOS

When inspecting a FOS that is controlled by rule 1, both its **size** and **form** need to be verified. The MMC size and Rule 1 boundary or envelope can be verified by with a GO gauge. A GO gauge is intended to fit into or fit over for an internal FOS and external FOS, respectively. A GO gauge is made to the MMC limit of the FOS and has a perfect form. It verifies the MMC limit and Rule 1 form envelope of a FOS. To fully verify the Rule 1 effects, a Go gauge must be at least as long as the FOS it is used to verify. *Figure 14.43,* shows examples of Go gauges for a pin and a hole.

The minimum size i.e., LMC, of a FOS can be measured with a NO-GO gauge. A No-Go gauge is not intended to fit into or over a FOS. It is made to the LMC limit of the FOS. A No-Go gauge makes a two-point check, e.g., a caliper or snap gauge can be used as a No-Go gauge. The two-point check is made at various points along the cross-section to ensure that the FOS does not violate the LMC limit. *Figure 14.43,* shows examples of NO-GO gauges of a pin and a hole.

Rule 2

Rule 2 is called the **"all applicable geometric tolerances rule"**. It states: RFS applies, with respect to the individual tolerance, datum reference, or both, where no modifying symbols is specified. MMC or LMC must be specified on the drawing where required. Certain geometric tolerances always apply RFS and cannot be modified to MMC or LMC.

Where a geometric tolerance is applied on an RFS basis, the tolerance is limited to the specified value regardless of the actual size of the feature.

14.22. VIRTUAL / RESULTANT CONDITION, INNER AND OUTER BOUNDARY

Depending upon its function, a FOS is controlled by a size tolerance and one or more geometric controls. Various material conditions, e.g., MMC, LMC, or RFS, may also be applied. In each case, consideration must be given to the collective effects of the size, specified material condition, and geometric tolerance of the FOS.

1. Consideration must be given to the collective effects of MMC and applicable tolerances in determining the clearance between parts and in establishing gauge feature sizes.
2. Consideration must be given to the collective effects of LMC and applicable tolerances in determining guaranteed area of contact, thin wall conservation, and alignment hole location in establishing gauge feature sizes.

The terms that apply to these conditions are virtual condition, inner boundary, and outer boundary:

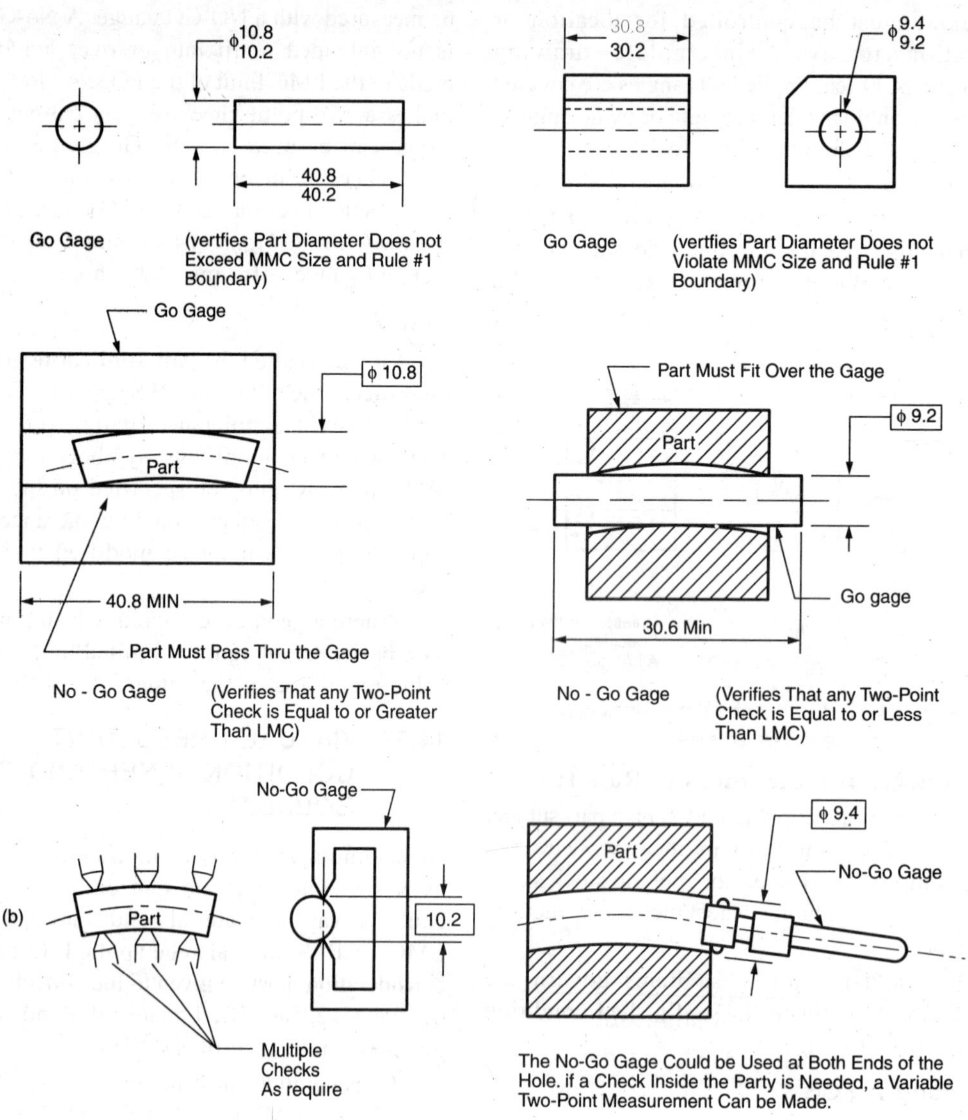

Fig. 14.43. Go Gagure and No-gauge Examples

Virtual Condition (VC)

Virtual condition (VC) is a constant worst-case boundary generated by the collective effects of a feature of size specified at MMC, or LMC and the geometric tolerance for that material condition. The VC of a FOS includes effects of the size, orientation, and location for the FOS. The virtual condition boundary is related to the datums that are referenced in the geometric tolerance that is used to determine the VC.

Datum Features At Virtual Condition

A virtual condition exists for a datum feature of size where its axis or centre plane is controlled by a geometric tolerance. In such cases, the datum feature applies at its virtual condition even though it is referenced in a feature control frame at MMC or LMC. Where a virtual condition equal to the maximum material condition or the least material condition is the design requirement, a zero tolerance at MMC or LMC is specified.

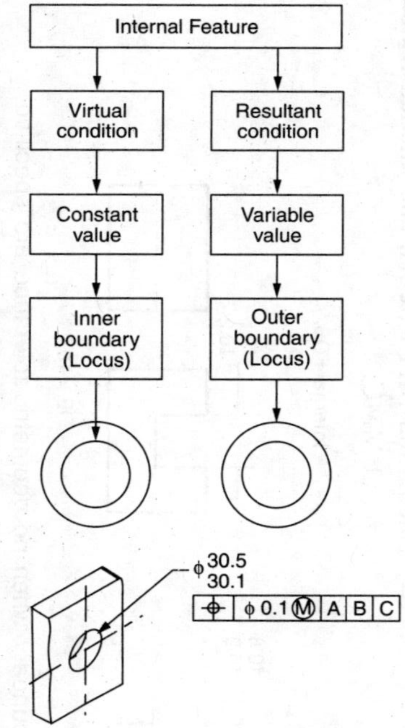

	ϕ HOLE	ϕ TOL	V COND	R COND
LMC	30.5	0.5		31.0
	30.4	0.4		30.8
	30.3	0.3	30.0	30.6
	30.2	0.2		30.4
MMC	30.1	0.1		30.2

The virtual condition of an internal feature is a constant value equal to its maximum material condition size MINUS its applicable tolerance of location

The resultant condition of an internal feature is a variable value equal to its actual mating envelope size PLUS its applicable tolerance of location

(1)

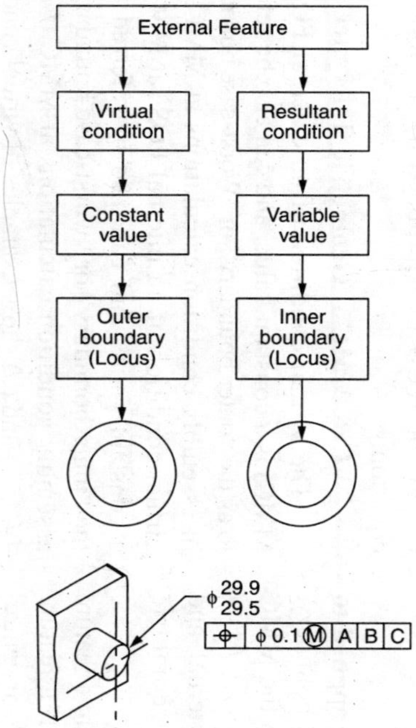

	ϕ PIN	ϕ TOL	V COND	R COND
MMC	29.9	0.1		29.8
	29.8	0.2		29.6
	29.7	0.3	30.0	29.4
	29.6	0.4		29.2
LMC	29.5	0.5		29.0

The virtual condition of an external feature is a constant value equal to its least material condition size PLUS its applicable tolerance of location

The resultant condition of an internal feature is a variable value equal to its actual mating envelope size MINUS its applicable tolerance of location

(2)

MMC CONCEPT — VIRTUAL AND RESULTANT CONDITION

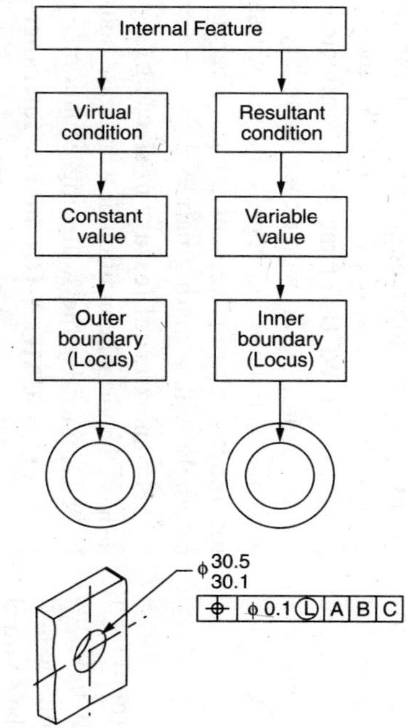

	ϕ HOLE	ϕ TOL	V COND	R COND
LMC	30.5	0.1		30.4
	30.4	0.2		30.2
	30.3	0.3	30.6	30.0
	30.2	0.4		29.8
MMC	30.1	0.5		29.6

The virtual condition of an internal feature is a constant value equal to its least material condition size PLUS its applicable tolerance of location.

The resultant condition of an internal feature is a variable value equal to its actual mating envelope size MINUS its applicable tolerance of location.

(3)

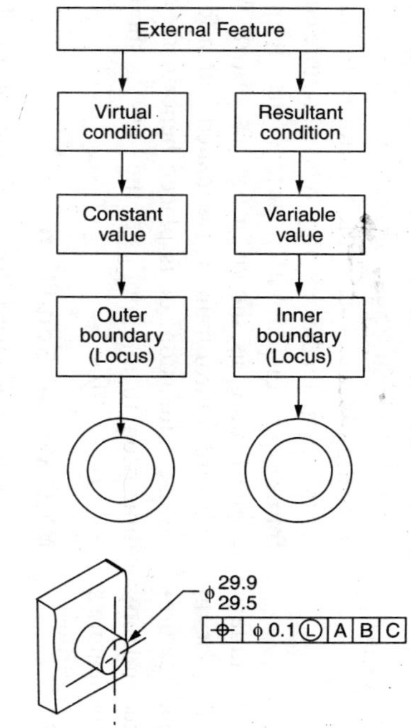

	ϕ PIN	ϕ TOL	V COND	R COND
MMC	29.9	0.5		30.4
	29.8	0.4		30.2
	29.7	0.3	29.4	30.0
	29.6	0.2		29.8
LMC	29.5	0.1		29.6

The virtual condition of an external feature is a constant value equal to its least material condition size MINUS its applicable tolerance of location.

The resultant condition of an external feature is a variable value equal to its actual mating envelope size PLUS its applicable tolerance of location.

(4)

LMC CONCEPT — VIRTUAL AND RESULTANT CONDITION

EXAMPLES

Virtual Condition Size

Virtual condition size is the actual value of the virtual condition boundary.

Inner Boundary (IB)

Inner boundary is a worst-case boundary generated by the smallest feature (MMC for an internal feature and LMC for an external feature) minus the stated geometric tolerance and any additional tolerance, if applicable, from the feature's departure from its specified material condition.

Outer Boundary (OB)

It is a worst-case boundary generated by the largest feature (LMC for an internal feature and MMC for an external feature) plus the stated tolerance, and any additional tolerance, if applicable, from the feature's departure from its specified material condition.

Worst-case Boundary (WCB)

Worst-case boundary is a general term to refer to the extreme boundary for a FOS that is the worst-case for assembly. Depending upon the part dimensioning, a worst-case boundary can be a virtual condition, inner boundary, or outer boundary.

FOS Boundary Conditions

If there are no geometric controls applied to a FOS, the WCB is the outer or inner boundary. The outer or inner boundary is equal to the MMC boundary as described by Rule 1. *Figure 14.44* shows the worst-case boundary of external and internal features of size when no geometric tolerances are specified.

If a feature control frame is applied to a feature (a surface), it does not affect its WCB. If a feature control frame is applied to a FOS (an axis or centre-plane), it does affect its WCB.

To determine whether a geometric control is applied to a feature (a surface) or a FOS (an axis or centre-plane), observe the location of the feature control frame on the drawing. When a feature control frame is associated with a surface, then it applies to the feature, as shown in *Fig. 14.45 (a)*. When a feature control frame is associated with a feature of size dimension, or is placed beneath the FOS dimension, then it applies to the FOS, as shown in *Fig. 14.45 (b)*.

MMC Virtual Condition

When a geometric tolerance that contains MMC modifier in the tolerance portion of the feature control frame is applied to a FOS, the virtual condition (or WCB) of the FOS is affected. The virtual condition or worst-case boundary is the extreme boundary that represents the worst-case for functional requirements, e.g., clearance or assembly with a mating part.

For an **external feature of size,** e.g., a pin or a shaft, the virtual condition or worst-case boundary is determined by the following equation:

VC = MMC + Geometric Tolerance

For an **internal feature of size,** e.g., a hole, the VC or WCB is determined by the following equation:

VC = MMC – Geometric Tolerance

The virtual condition of an external FOS (at MMC) is a constant value, and can also be referred to as the "outer boundary" or 'Worst-case boundary" in assembly condition calculations at MMC. The virtual condition of an internal FOS (at MMC) is a constant value, and can also be referred to as the "inner boundary" or "worst-case boundary" in assembly condition calculations at MMC. *Figure 14.45(c)* and *(d)* show examples of virtual condition calculations, for external and internal features of size, at MMC.

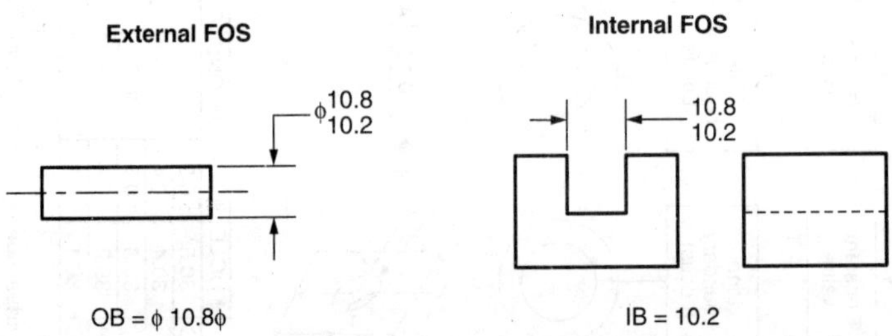

Fig. 14.44. Worst-case boundary when no geometric tolerance are specified

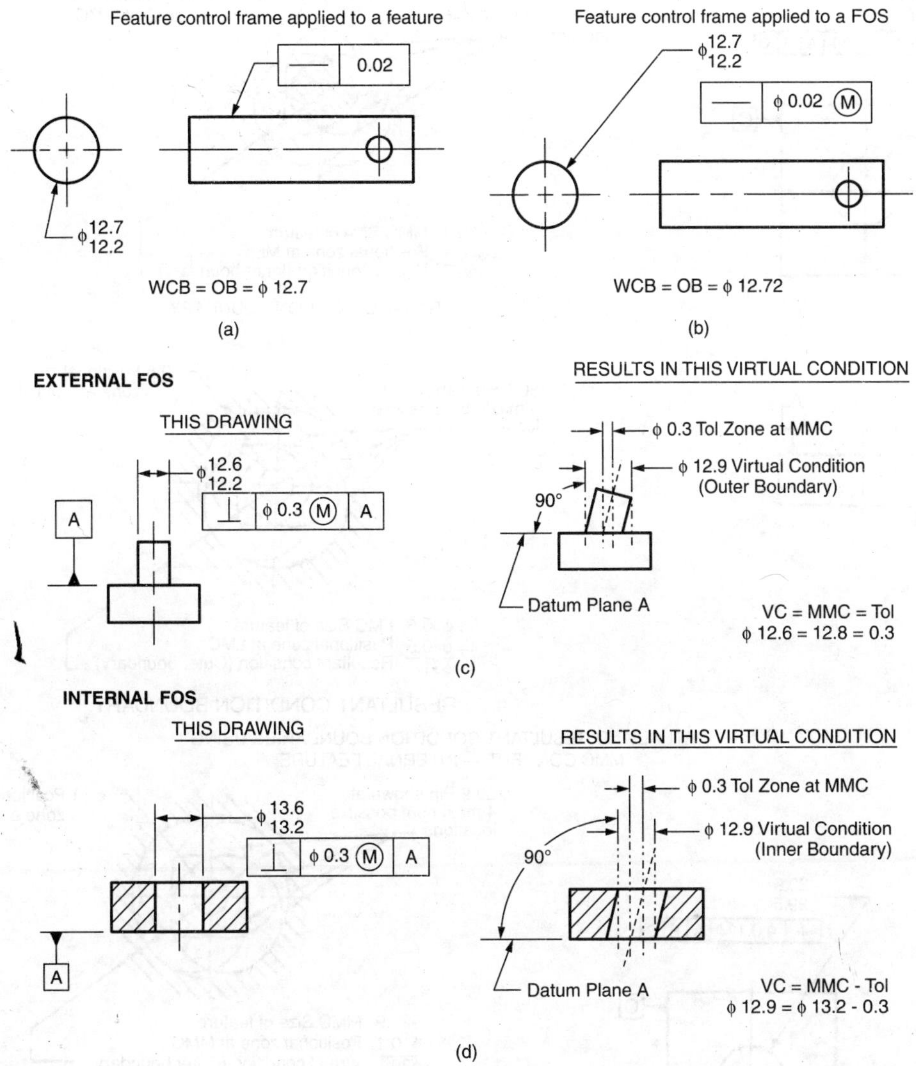

Fig. 14.45. MMC Virtual condition examples

LMC Virtual Condition

When a geometric tolerance that contains LMC modifier in the tolerance portion of the feature control frame is applied to a FOS, the virtual condition or worst-case boundary of the FOS is affected. The virtual condition (WCB) is the extreme boundary that represents the worst-case for functional requirements, e.g., wall thickness, alignment of features, minimum machine stock on a part.

For an **external feature of size,** e.g., a pin or a shaft, the virtual condition or worst-case boundary is determined by the following equation:

VC = LMC – Geometric Tolerance

For an **internal feature of size,** e.g., a hole, the VC or WCB is determined by the following equation:

VC = LMC + Geometric Tolerance

The virtual condition of an external FOS (at LMC) is a constant value, and can also be referred to as the "inner boundary" or "worst-case boundary" in calculations. The virtual condition of an internal FOS (at LMC) is a constant value, and can also be referred to as the "outer boundary" or "worst-case boundary" in calculations. *Fig 14.46* (*a*) and

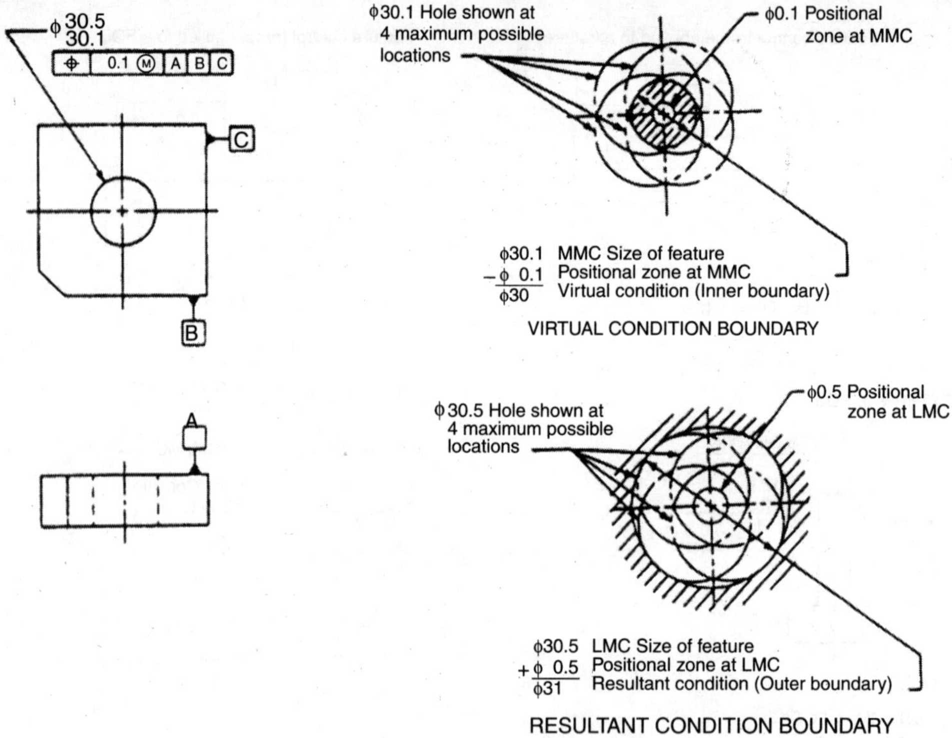

EXAMPLE 1 : VIRTUAL AND RESULTANT CONDITION BOUNDARIES USING MMC CONCEPT — INTERNAL FEATURE

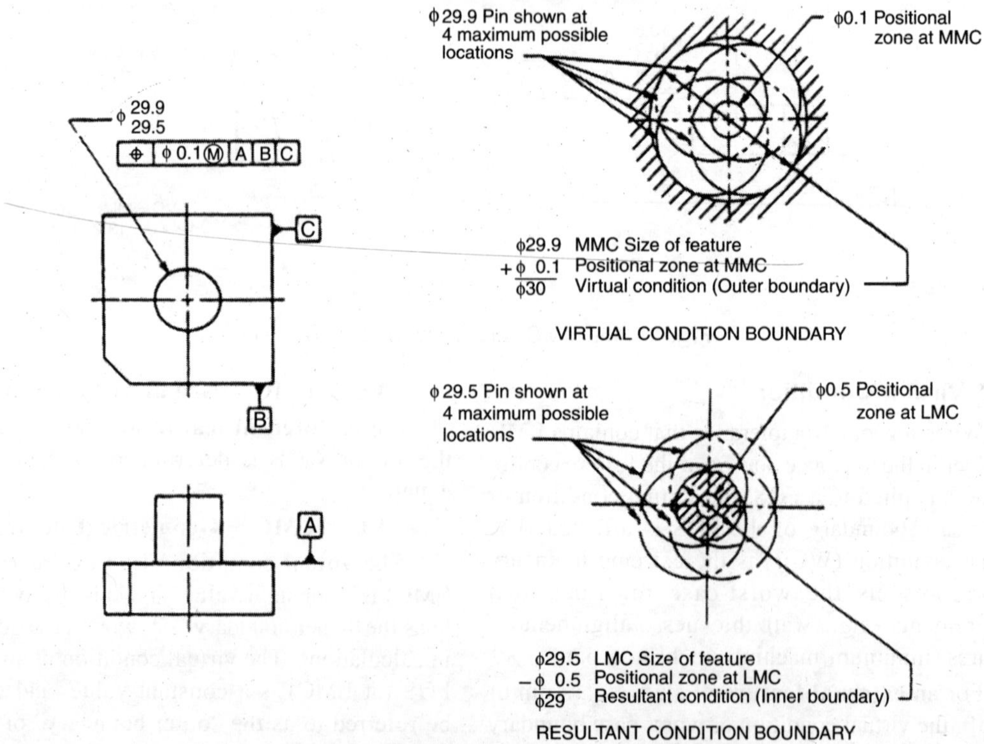

EXAMPLE 2 : VIRTUAL AND RESULTANT CONDITION BOUNDARIES USING MMC CONCEPT — EXTERNAL FEATURE

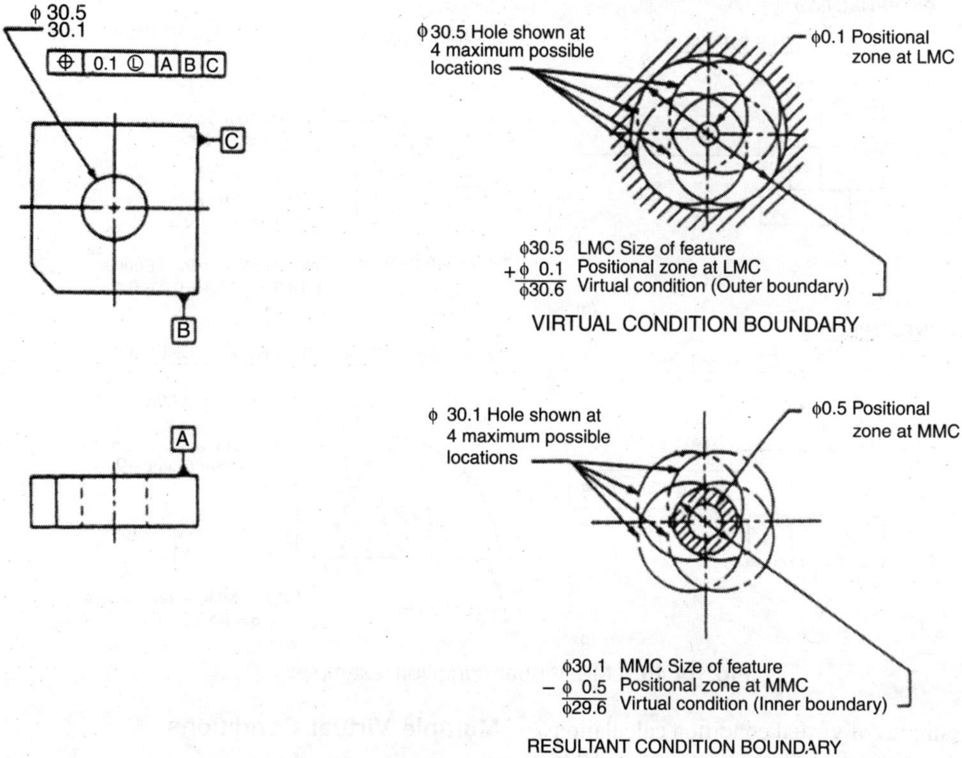

EXAMPLE 3 : VIRTUAL AND RESULTANT CONDITION BOUNDARIES USING LMC CONCEPT — INTERNAL FEATURE

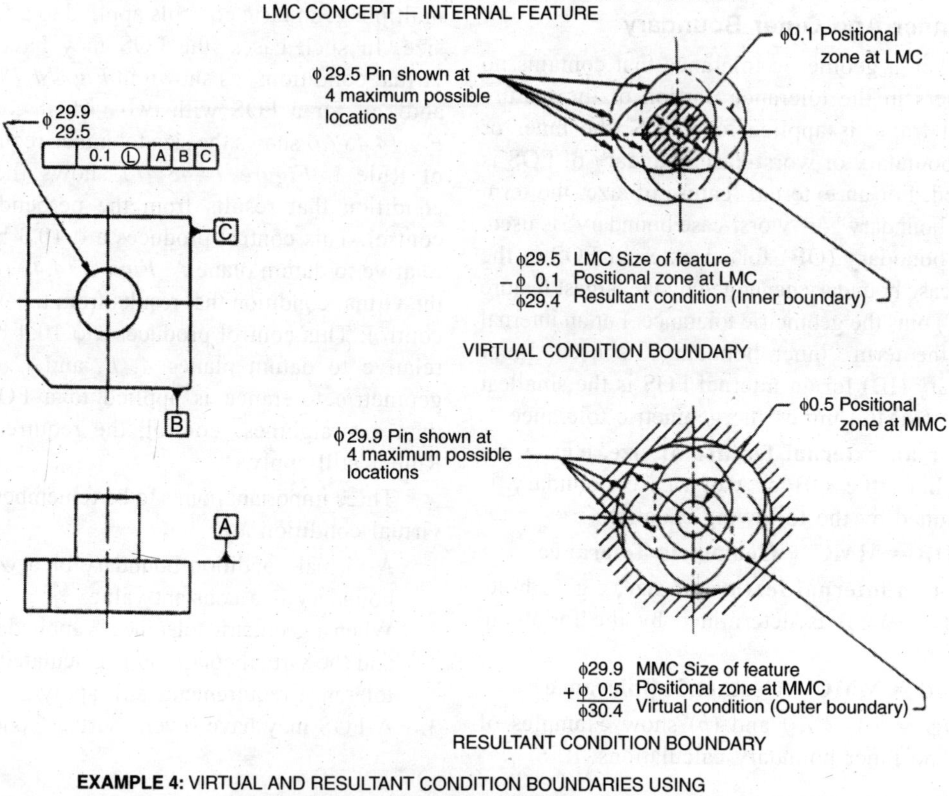

EXAMPLE 4: VIRTUAL AND RESULTANT CONDITION BOUNDARIES USING LMC CONCEPT — ENTERNAL FEATURE

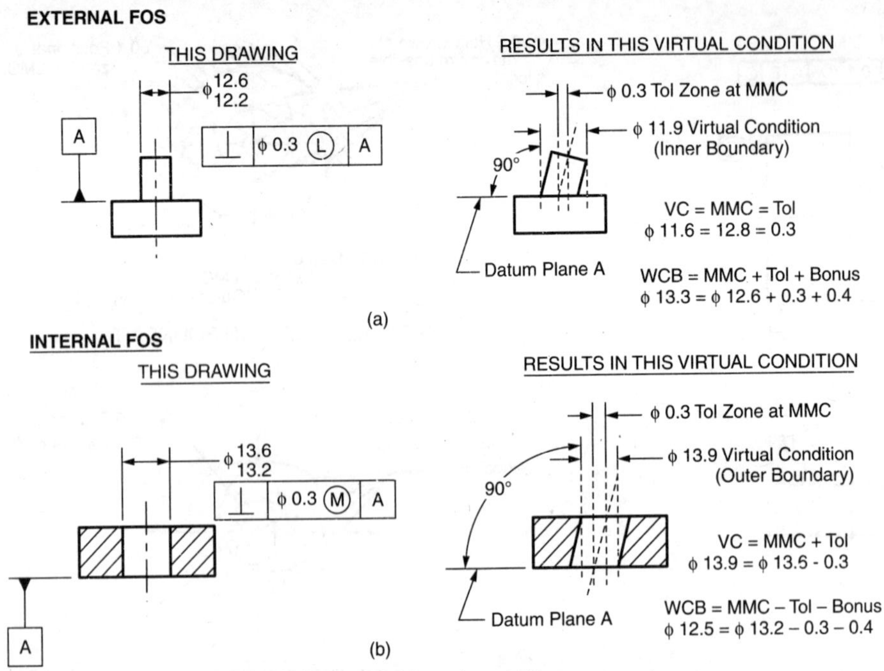

Fig. 14.46. LMC Virtual condition examples

(*b*) show examples of virtual condition calculations, for external and internal features of size, at LMC.

RFS Inner And Outer Boundary

When a geometric tolerance that contains no modifiers in the tolerance portion of the feature control frame is applied to a FOS, the inner or outer boundary or worst-case boundary of FOS is affected. For an external feature of size, the term "outer boundary" or "worst-case boundary" is used. Outer boundary (OB) for an external FOS is the worst-case boundary generated by the largest feature (MMC) plus the geometric tolerance. For an internal FOS, the term "inner boundary" is used. Inner boundary (IB) for an internal FOS is the smallest feature (MMC) minus the geometric tolerance.

For an external feature of size, e.g.; a pin or a shaft, the OB or worst-case boundary is determined by the following equation:

OB = MMC + Geometric Tolerance

For an **internal feature of size,** e.g., a hole, the IB or WCB is determined by the following equation:

IB = MMC - Geometric Tolerance

Figure 14.47 (a) and (*b*) show examples of outer and inner boundary calculations, RFS.

Multiple Virtual Conditions

On complex drawings, it is common to have multiple geometric controls applied to a feature of size. In such cases, the FOS may have several virtual conditions, as shown in *Fig. 14.48 (a)*, (*b*), and (*c*), for a FOS with two virtual conditions. *Fig 14.48 (a)* shows the size tolerance requirements of Rule I. *Figure 14.48 (b)* shows the virtual condition that results from the perpendicularity control. This control produces a φ 10.3 boundary relative to datum plane A. *Figure 14.48 (c)* shows the virtual condition that results from the positional control. This control produces *a* φ 10.4 boundary relative to datum planes *A*, *B*, and *C*. When a geometric tolerance is applied to a FOS, other than a straightness control, the requirements of Rule 1 still apply.

Three important points to be remembered about virtual condition are:

1. A virtual condition boundary or a worst-case boundary is a constant value.
2. When a geometric tolerance is applied to a FOS, and the virtual condition is calculated, the size tolerance requirements still apply.
3. A FOS may have several virtual conditions.

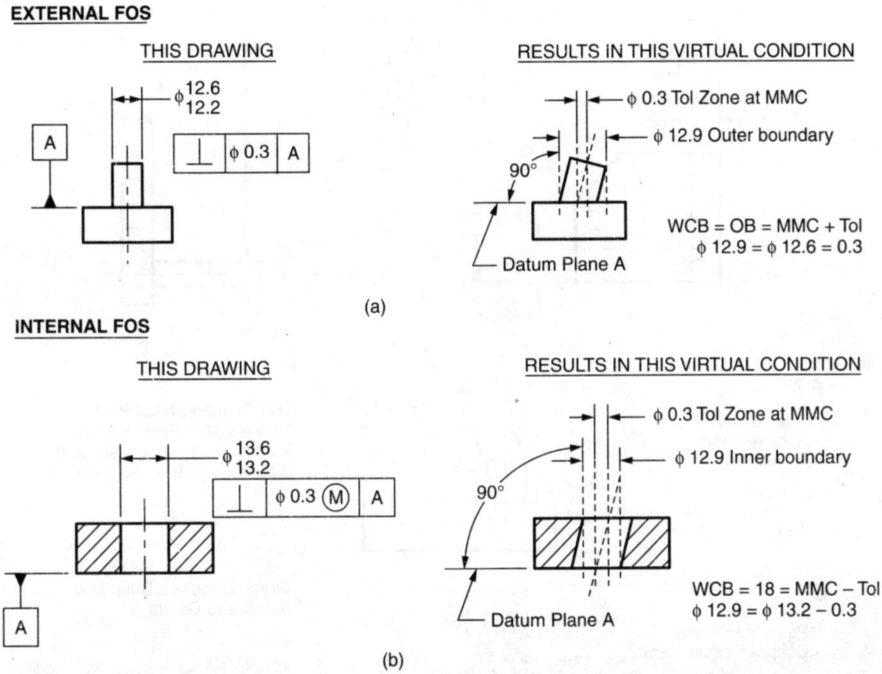

Fig. 14.47. RFS Inner and Outer Boundary examples

14.23. BONUS TOLERANCE

Bonus tolerance is an additional tolerance for a geometric control. Whenever a geometric tolerance is applied to a FOS, and it contains a MMC or LMC modifier in the tolerance portion of the feature control frame, a bonus tolerance is permissible.

Bonus tolerance is an important concept in machining parts. Bonus tolerances can reduce manufacturing costs significantly. A brief introduction to bonus tolerance concept is presented here.

When the MMC or LMC modifier is used in the tolerance portion of the feature control frame, it means that the stated tolerance applies when the feature of size is at its maximum material condition or the least material condition, as the case may be. When the actual mating size of the FOS departs from MMC or LMC, an increase in the stated tolerance, equal to the amount of the departure, is permitted. This increase, or extra tolerance, is called the bonus tolerance. *Figure 14.49*, shows how bonus tolerance is calculated in a straightness control application. When the MMC modifier is used, it means that the geometric characteristic can be verified with a fixed gauge.

A functional gauge is a gauge that is built to a fixed size (the virtual condition) of a part feature; a part must fit into, or onto, the gauge. A functional gauge does not provide a dimensional measurement; it only indicates if the part is or is not to the print specifications.

Figure 14.49, the functional gauge is designed to the virtual condition of the feature of size. Since the gauge opening is not variable, the thinner the washer becomes, the more straightness tolerance it can have and still pass through the gauge.

Although a straightness control is used for the example in *Fig. 14.49*, the bonus tolerance concept applies to any geometric control that uses the MMC or LMC modifier in the tolerance portion of the feature control frame.

Figure 14.50 shows how to determine the amount of bonus tolerance permissible in an application. The MMC modifier in the tolerance portion of the feature control frame denotes that a bonus tolerance is permissible. The maximum amount of bonus tolerance permissible is equal to the difference between the MMC and the LMC of the toleranced FOS. Important points to be kept in mind regarding Bonus Tolerance are:

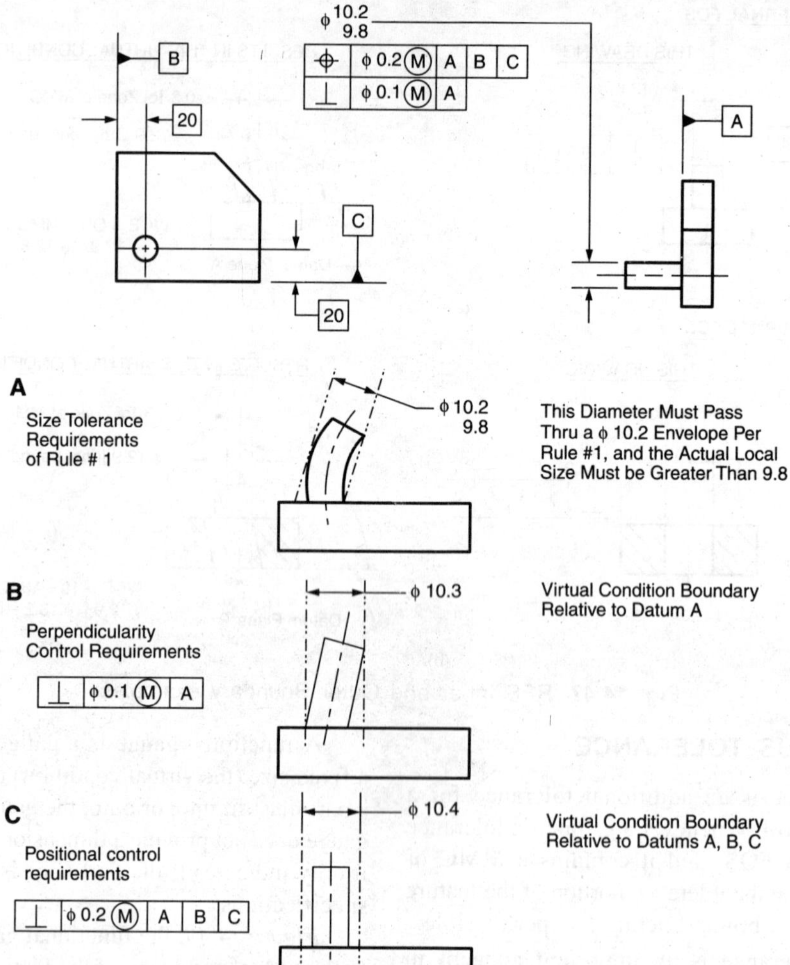

Fig. 14.48. Multiple virtual conditions

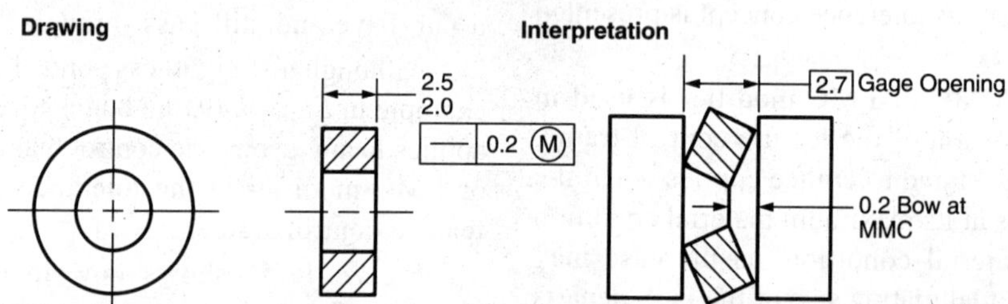

Feature AME	Straightness Tolerance	Bonus Tolerance	Total Tolerance
2.5 MMC	0.2	0.0	0.2
2.4	0.2	0.1	0.3
2.3	0.2	0.2	0.4
2.2	0.2	0.3	0.5
2.1	0.2	0.4	0.6
2.0 LMC	0.2	0.5	0.7

Note:
Maximum Bonus Occurs When the AME is at LMC

Fig. 14.49. Bonus tolerance examples

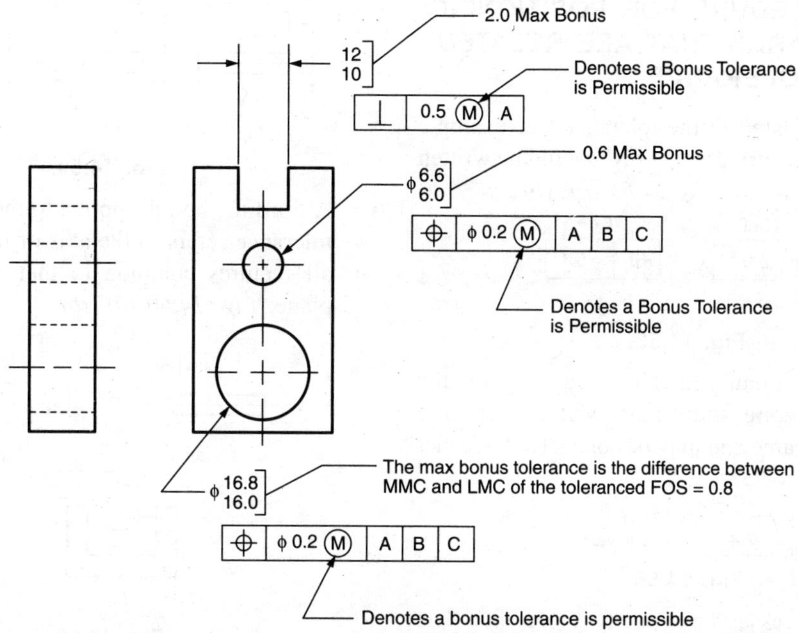

Fig. 14.50. Bonus tolerance

- Bonus tolerance is an additional tolerance for a geometric control tolerance.
- Bonus tolerance is only permissible when a MMC or LMC modifier is shown in the tolerance portion of a feature control frame.
- Bonus tolerance comes from the FOS tolerance.
- Bonus tolerance is the amount the actual mating size departs from MMC or LMC.

14.24. RELATED CHARACTERISTICS

A datum system described later, in this chapter, may include reference to one (*Fig. 14.51*) or more features for which the following procedures are used:

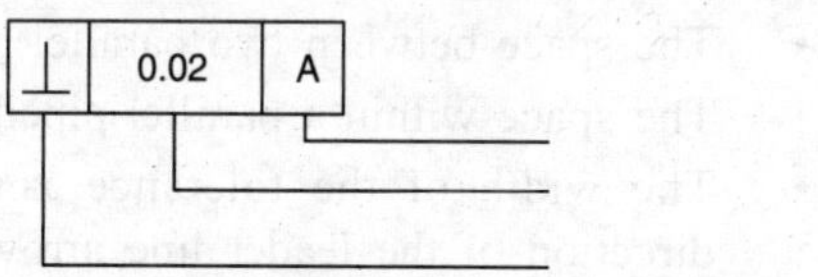

Fig. 14.51.

- Reference may be made to more than one feature, by placing a hyphen between the two letters, that designate a single datum established by the two datum features, *Fig. 14.52.*
- Where the datum system consists of a number of datum features the datum letters are placed together in the third compartment of the tolerance frame. No significance to order is implied, *Fig. 14.53.*

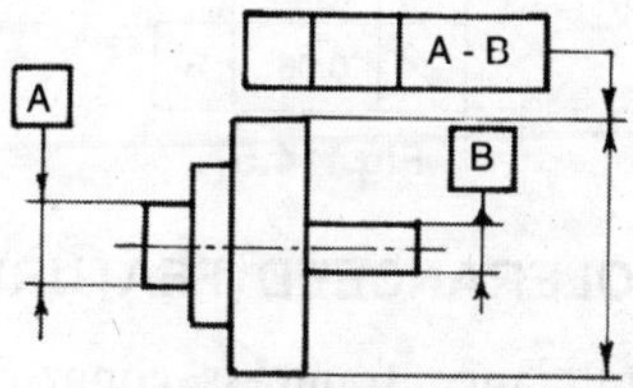

Fig. 14.52.

Fig. 14.53.

- Where the application of the datum features is required in a particular order, the designated letters are quoted in separate boxes, *Fig. 14.54.*

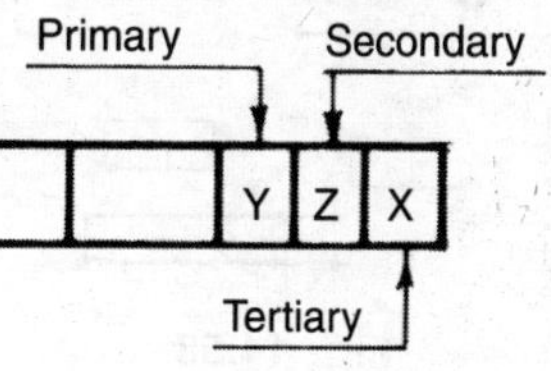

Fig. 14.54.

14.25. PROCEDURE FOR POSITIONING REMARKS THAT ARE RELATED TO TOLERANCE

- Remarks related to the tolerance, for example '6 holes', '2 surfaces' or '6X' should be written above the frame, *Fig. 14.55 (a), (b)*.

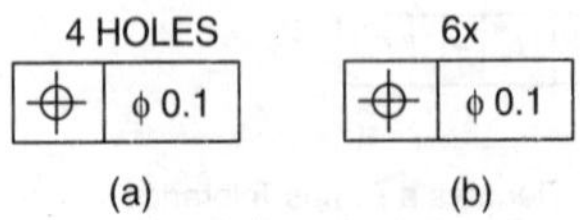

Fig. 14.55.

- Indications qualifying the feature within the tolerance zone should be written near the tolerance frame and may be connected by leader line, *Fig. 14.56*.

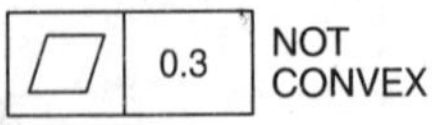

Fig. 14.56.

- If it is necessary to specify more than one tolerance characteristic for a feature, the tolerance specification should be given in tolerance frames positioned one under the other, *Fig. 14.57*.

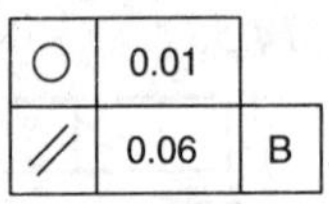

Fig. 14.57.

14.26. TOLERANCEED FEATURES

- The tolerance frame is connected to the toleranced feature by a leader line terminating with an arrow on the outline of the feature or an extension of the outline, *Fig. 14.58*.

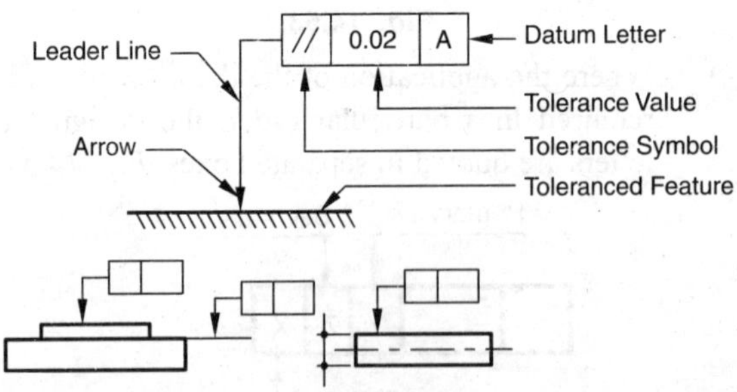

Fig. 14.58.

- A second method is to extend a dimension line when the tolerance refers to the axis or median plane of the dimensioned features, *Fig. 14.59*.

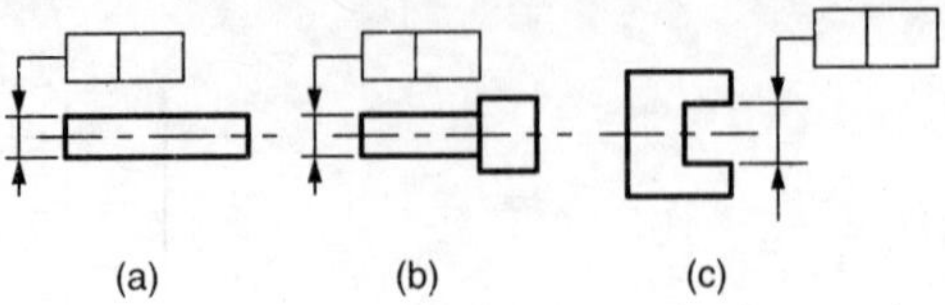

Fig. 14.59.

- A third method is applied to the axis when the tolerance refers to the axis or median plane of all features common to that axis or median plane, *Fig. 14.60 (a), (b)*.

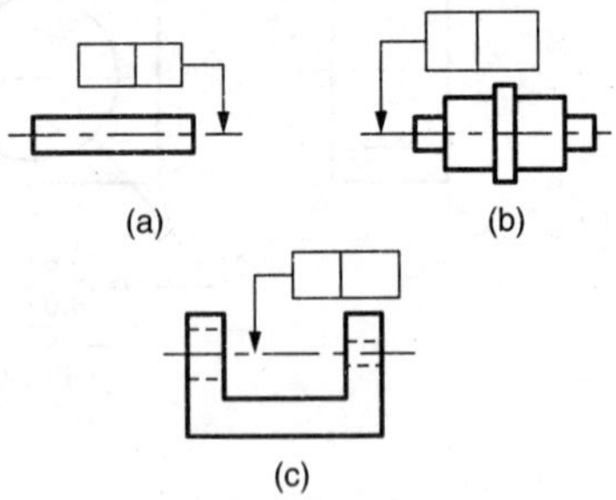

Fig. 14.60.

14.27. TOLERANCE ZONES

A geometric tolerance of a feature, that may be a point, line, axis, surface, or median plane, specifies the tolerance zone within which a feature is required to be contained. According to the characteristic, which is to be toleranced, the tolerance zone is one of the following:

- The area within a circle.
- The area between two concentric circles.
- The area between two parallel lines.
- The space within a sphere.
- The space within a cylinder.
- The space within two coaxial cylinders.
- The space between two equidistant surfaces.
- The space between two parallel planes.
- The space within a parallel-piped.
- The width of the tolerance zone is in the direction of the leader line arrow joining the symbol frame to the toleranced feature unless the tolerance zone is preceded by the sign ϕ, *Fig. 14.61 (a), (b), (c)*.
- Individual tolerance zones of the same value can be applied to more than one separate feature, *Fig. 14.62*.

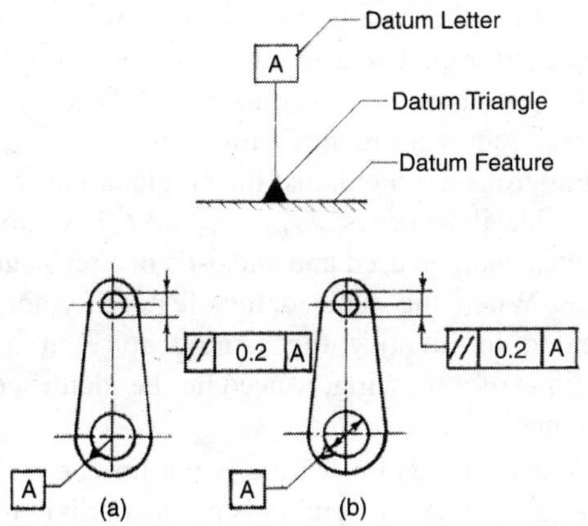

Fig. 14.61.

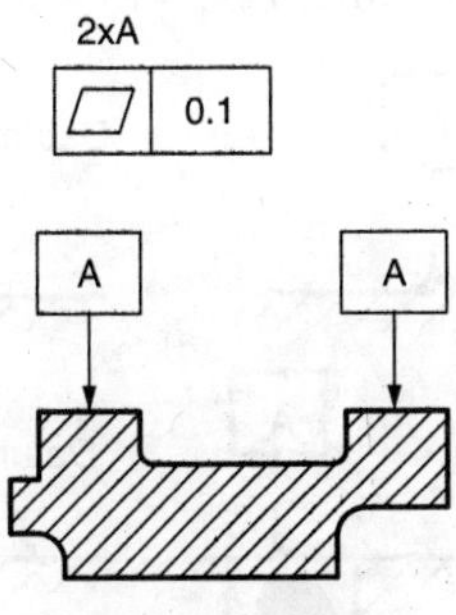

Fig. 14.62.

14.28. DATUMS AND DATUM SYSTEM

A designer/drafter should be able to read and understand the information related to datums on drawings. The **datum system** constitutes an important part of the language of geometric dimensioning and tolerancing. A datum system is a set of symbols and rules that are used to communicate to the drawing user as to how the dimensional measurements are to be made for inspection and quality control functions. It serves two basic purposes:

1. To allow the designer to specify the sequence in which the part is to contact the inspection equipment for the measurement of a dimension.
2. It allows the designer to specify the part surface that are to contact the inspection equipment for the measurement of a dimension.

The datum system provides important Benefits:

- It helps in making repeatable dimensional measurements, without any ambiguity.
- Helps in communicating functional relationships of part(s).
- Helps in taking the measurements as intended in design; by the designer.

Implied Datums

An implied datum is an assumed plane, axis, or point from which a dimensional measurement is made. Implied datums represent an old concept that resulted from coordinate system of dimensioning and toleracing. *Figure 14.68* show an example of implied datums. Here, the bottom and left sides of the block are considered implied datums.

Implied datums have two major **shortcomings**:

1. They do not clearly communicate to the drawing user as to which part surfaces are to contact the inspection equipment for taking measurements. This decision is left to the inspector, who makes his/her own assumptions.
2. They do not communicate the sequence in which the part should be brought in contact with the measuring equipment. Each inspector may assume a different sequence that can produce different results for the part measurements, as shown in *Fig. 14.68.*

Two consequences of using implied datums are:

- Good parts are rejected.
- Bad parts are accepted.

A part that is to print specifications when measured following the proper set-up may be rejected when measured from the wrong implied datums, or when using the wrong sequence in which the part contacts the measuring equipment. Also, a part that is actually out of print specification when measured properly may pass inspection when measured from the wrong surfaces or the wrong surface set up sequence.

Datum

A datum is a theoretically exact point, axis, or plane derived from the true geometric counterpart of a selected datum feature. A datum is the origin from which, the location of or geometric characteristics of features of a part are established. Datums are theoretical reference planes, points, or axes and are simulated by the inspection equipment.

Datum Feature

A datum feature is an actual feature of a part that is used to establish a datum. Datum features are part features, and they exist on the part being geometrically toleranced.

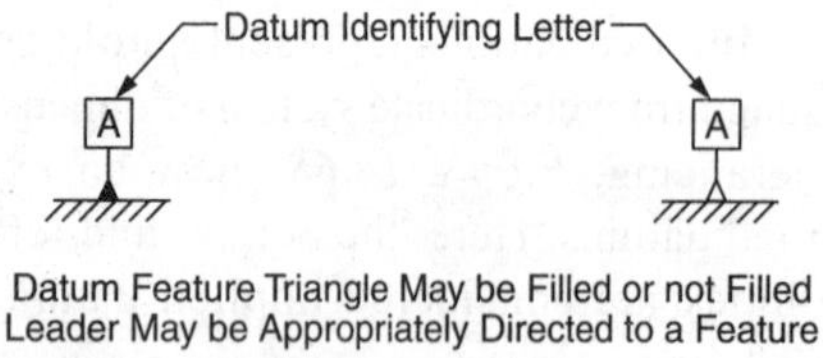

Fig. 14.63. (a)

Datum Feature Symbol

Symbolic representation of a datum feature consists of a capital letter enclosed in a square frame and a leader line extending from the frame to the considered feature, terminating in a triangle. The triangle may be filled or not filled, as shown in *Fig. 14.63 (b)*.

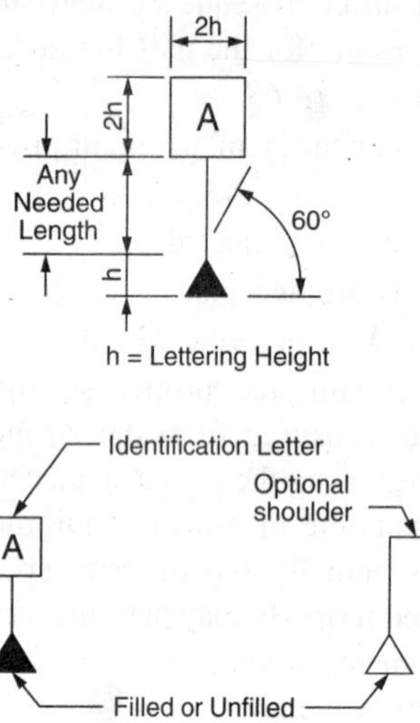

Fig. 14.63. (b)

The method of attaching this symbol to the part feature, being geometrically toleranced, determines whether it designates a planar datum or a FOS datum. *Figure 14.70* shows different ways of displaying the datum feature symbol to denote a planar datum.

When a toleranced feature is related to a datum, a letter, shown in the tolerance frame, indicates it. A capital letter enclosed in a square datum box and connected to a solid or blank equilateral datum triangle indicates a datum, *Fig. 14.63 (b)*.

Each datum feature is identified by a different capital letter enclosed in a box. The use of letters O, M, Q, and I are not recommended. When datum features requiring identification on a drawing are so numerous as to exhaust the single alpha series, the double alpha series, AA through AZ, BA through BZ, etc., shall be used and enclosed in a rectangular frame. When the same datum feature symbol is repeated to identify the same feature in other locations of a drawing, it need not be identified as reference.

- When the datum feature is the line or surface itself the datum symbol is on the outline of the feature or an extension of the outline, but not a dimension line, *Fig. 14.64 (a), (b)*.

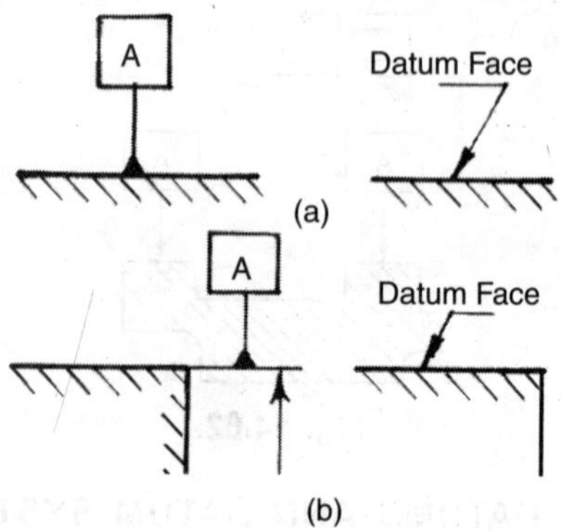

Fig. 14.64.

- A datum symbol connected to a dimension line refers to the axis or median plane of the dimensioned feature, *Fig. 14.65 (a), (b)*. When the datum symbol is on the axis or median plane, the datum is understood to be the axis or median plane of the whole component.
- When the datum is on a common axis or median plane of two or more features the datum symbol is indicated on the axis or median plane, *Fig. 14.66*.
- When the tolerance frame can be connected in a clear and unambiguous manner with the datum feature by a leader line and triangle the datum identification letter may be omitted, *Fig. 14.67*.

Datum Selection

Datum features are selected on the basis of part function and assembly requirements. The datum features are often the features that orient and locate

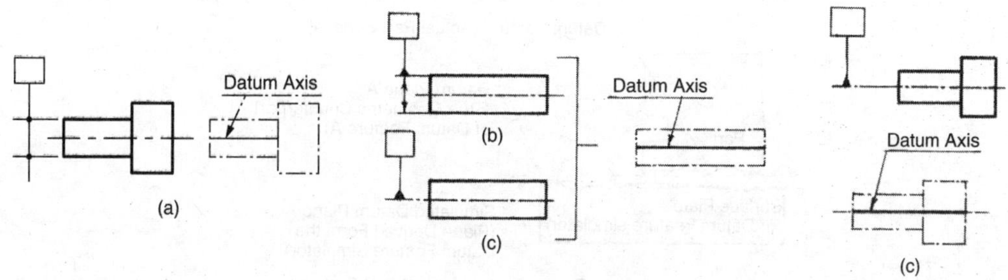

Fig. 14.65.

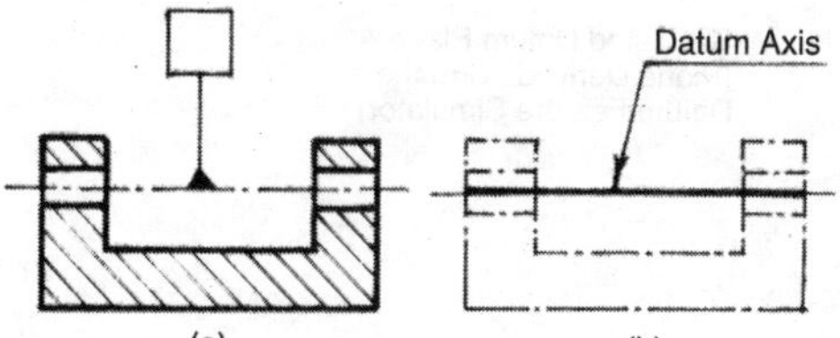

Fig. 14.66.

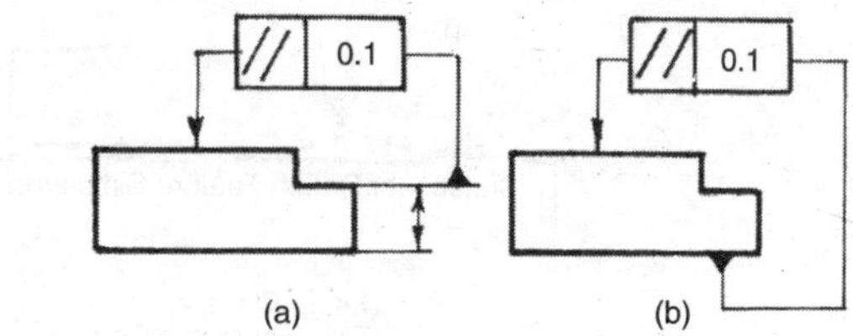

Fig. 14.67.

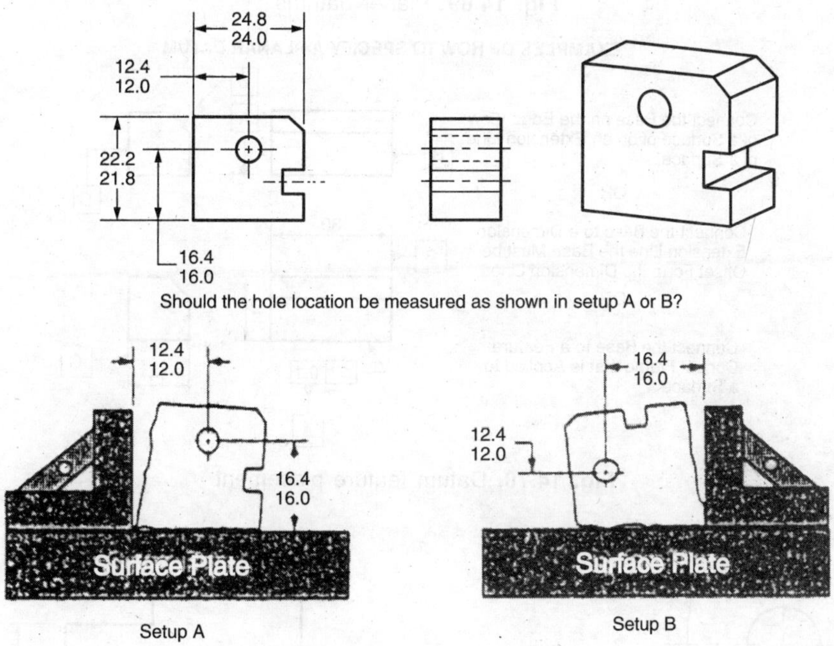

Fig. 14.68. Implied datums

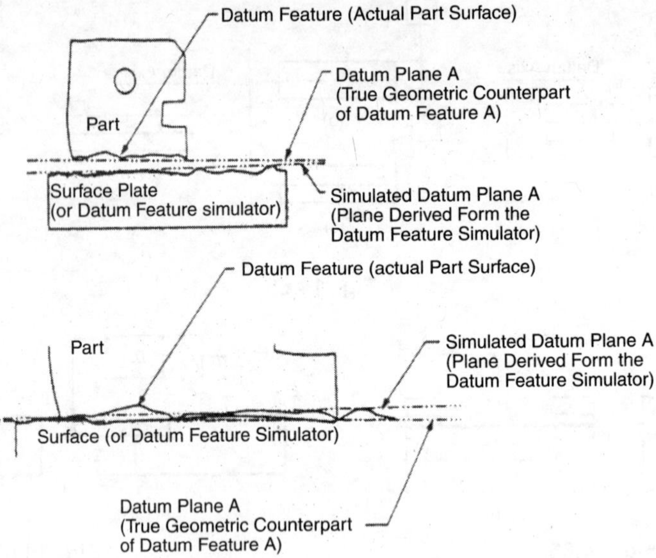

Fig. 14.69. Planer datums

EXAMPLES OF HOW TO SPECIFY A PLANAR DATUM

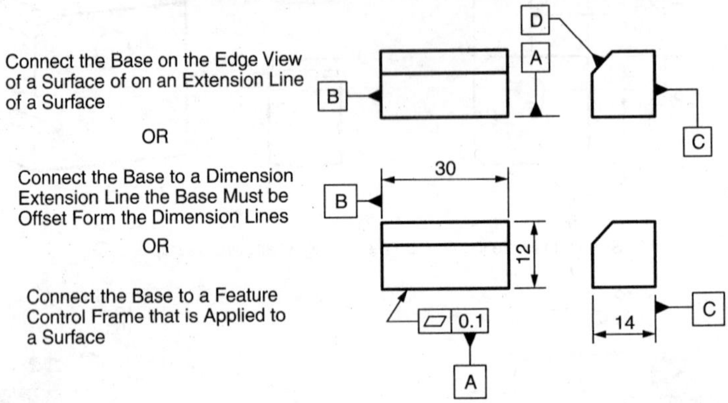

Fig. 14.70. Datum feature placement

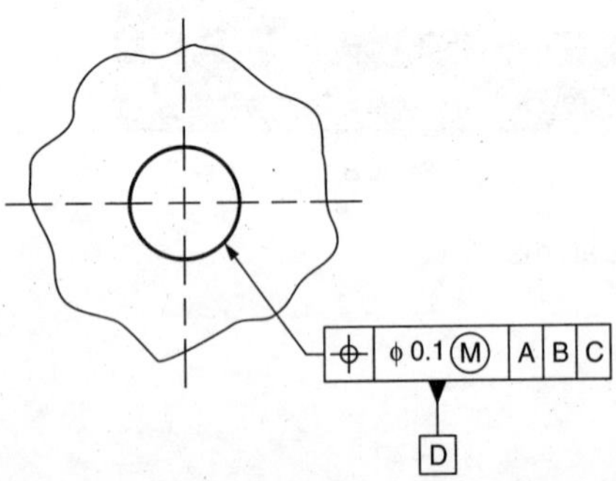

Fig. 14.71. Placement of datum feature symbol in conjunction with a feature control frame

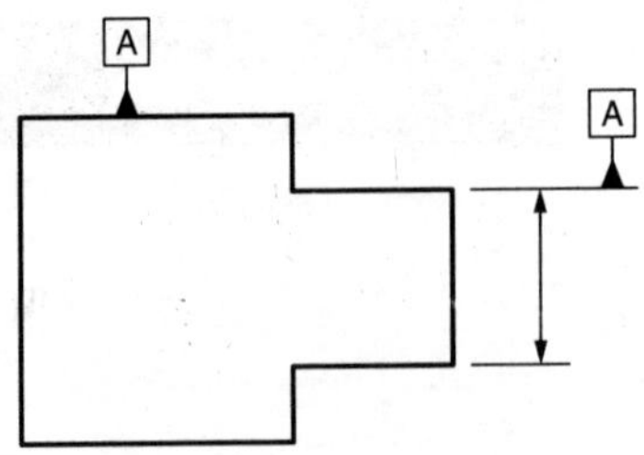

Datum Feature Symbols on a Feature Surface and an Extension Line

Fig. 14.72.

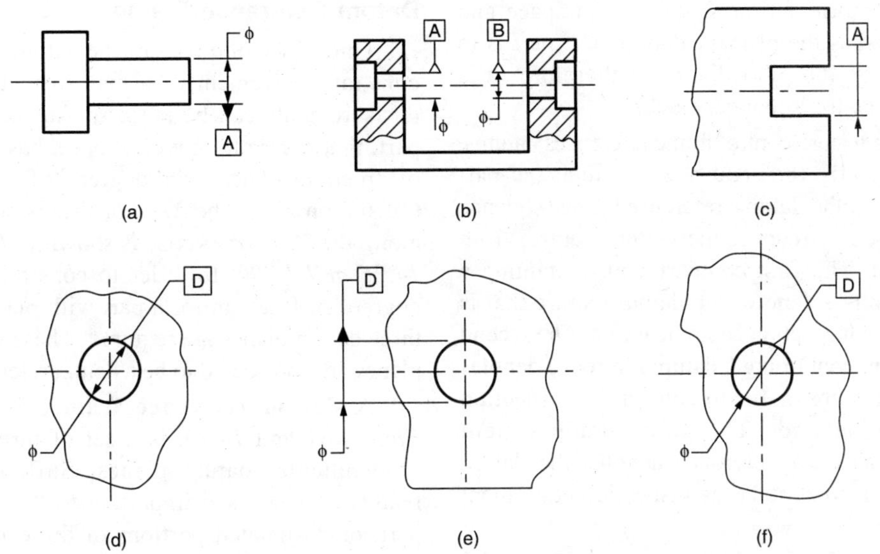

Fig. 14.73. Placement of datum feature symbols on features of size

the part in its assembly. For example, the part in *Fig. 14.74* mounts on surface *A* and is located by diameter *B*. For assembly, the four holes need to be located relative to the features that mount and locate the part with respect to the mating part. So, the surface *A* and the diameter *B* are designated as datum features.

Referencing Datums In Feature Control Frames

After datums, the drawing must also communicate when and how the datums should be used. This is done through the use of feature control frames. When feature control frames are specified, they reference the datums to be used for part measurement. For example, in *Fig. 14.74*, the bolt holes are toleranced, relative to the datums, through the use of a geometric tolerance. As the part is to be clamped against surface A, surface A will

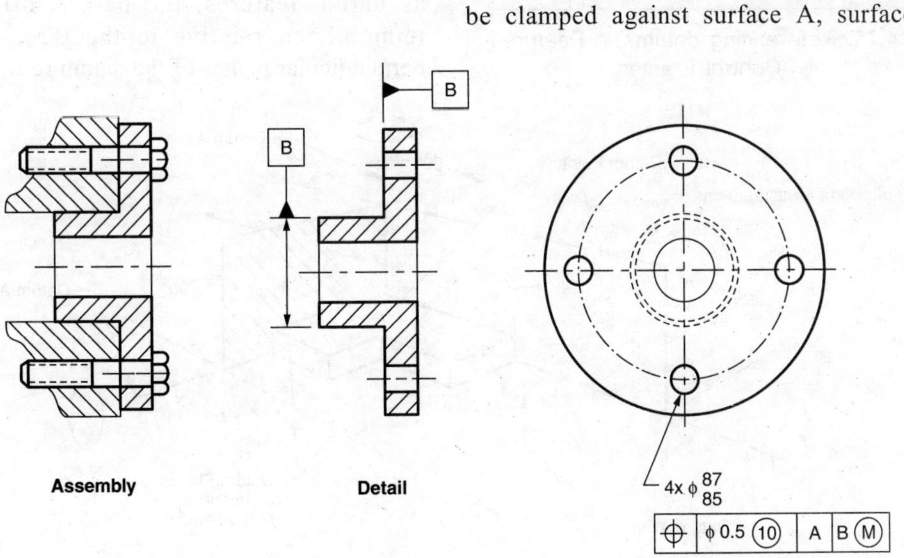

Fig. 14.74. Datum selection

establish the orientation of the part in space and is referenced as the primary datum. The part is to be located by the pilot diameter; therefore, it is referenced as the secondary datum.

When a feature control frame references datums, it has to specify the sequence in which the part should contact the datums referenced. The sequence is determined by reading the feature control from left to right. The first compartment containing a datum reference denotes the datum feature that is to contact the inspection equipment first. The second compartment containing a datum reference denotes the datum feature that is to contact the inspection equipment second. The third compartment containing a datum reference denotes the datum feature that is to contact the inspection equipment third.

Figure 14.75 shows an example of the order of interpreting the datum sequence in a feature control frame.

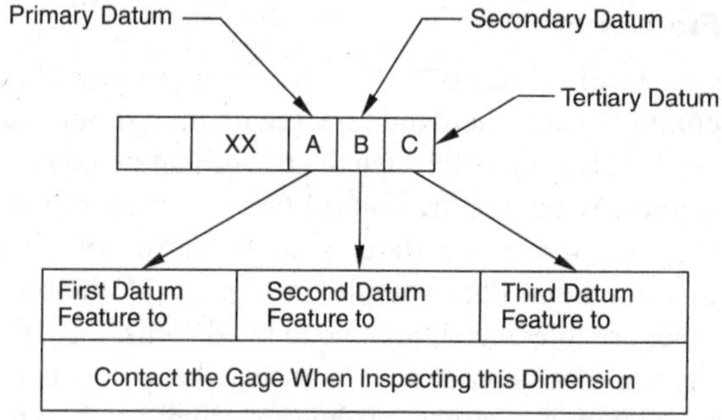

Fig. 14.75. Referencing datums in Feature Control Frames

Datum Reference Frame

One of the purposes of the datum system is to constrain movement of a part so that repeatable measurements can be made during inspection. A part, that is free to move in space, has six degrees of freedom. These six degrees of freedom are translation along the *X*, *Y*, or *Z*-axis, and rotation along the *X*, *Y*, or *Z*-axis, as shown in *Figs. 14.84, 14.76 and 14.80*. In order to constrain these six degrees of freedom for a part with planar datums, three datum planes are required. These three datum planes are considered to be a datum reference frame.

A datum reference frame, as shown in *Fig. 14.81* and *14.76*, is a set of three mutually perpendicular datum planes. Sufficient datum features, those most important to the design of a part, or designated portions of these features are chosen to position the part in relation to the datum reference frame. This reference frame exists in theory only and not on the part. Therefore, it becomes necessary to establish a method of simulating the theoretical reference frame from the actual features of the part. This simulation is accomplished by positioning specifically identified features in contact with appropriate datum simulators, in a stated order of precedence, to restrict motion of the part and to relate the part adequately to the datum reference frame.

Where features of a part have been identified as datum features, the part is oriented and immobilized relative to the three mutually-perpendicular planes of the datum reference frame

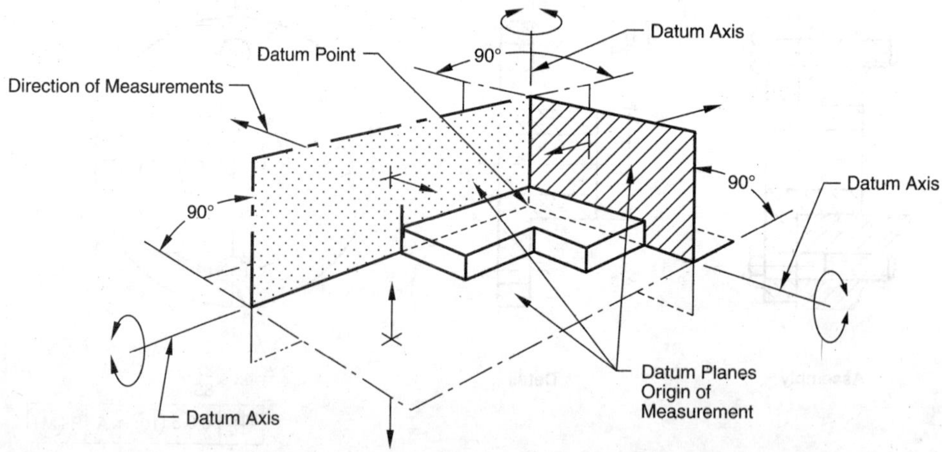

Fig. 14.76. Datum reference frame

in a selected order of precedence. This results in making the geometric relationships that exist between the features easy to measure, *Fig. 14.83*.

As measurements cannot be made from a true geometric counterpart that is theoretical, a datum is assumed to exist in and be simulated by the associated processing equipment. For example, machine tables and surface plates, though are not true planes, are of such quality that the planes derived from them are used to simulate the datums from which measurements are taken and dimensions verified, *Fig. 14.69*.

The planes of the datum reference frame are simulated in a mutually perpendicular relationship to provide direction as well as the origin for related dimensions and measurements. Thus, when the part is positioned on the datum reference frame, by physical contact between each datum feature and its counterpart in the associated processing equipment, dimensions related to the datum reference frame by a feature control frame or note are thereby mutually perpendicular.

This theoretical reference frame constitutes the three-plane dimensioning system used for datum referencing. The planes of a datum reference frame have zero perpendicularity tolerance to each other by definition, *Fig. 14.78, 14.81, and 14.83*.

The 90° angles between datum planes are basic, whereas the 90° angles between actual part surfaces have tolerance. The angular tolerance of the part surfaces may be specified on the drawing, in title-block tolerances, or in a general note. *Fig. 14.87* shows a part where the part surfaces, that are datum features, are not exactly 90° to each other. A

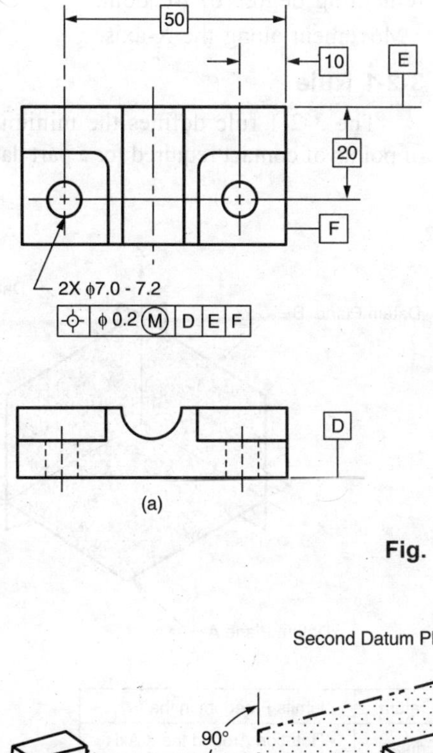

(a)

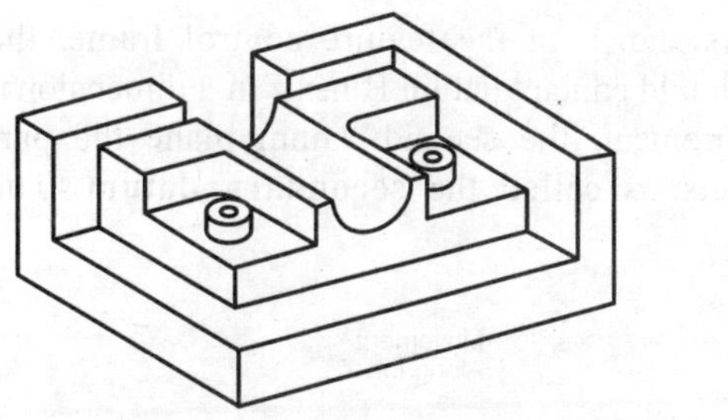
(b)
Part Where Datum Features are Plane Surfaces

Fig. 14.77.

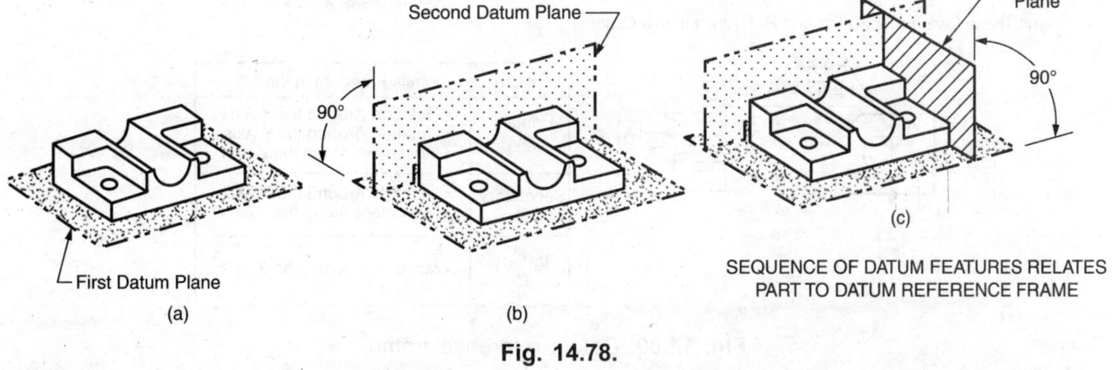

(a) (b) (c)

SEQUENCE OF DATUM FEATURES RELATES PART TO DATUM REFERENCE FRAME

Fig. 14.78.

measurement of this part will produce different results based on which part surface touches the datum reference frame first, second, and third.

A feature control frame **is read from left to right**. Here, datum A is referenced first, datum B is referenced second, and datum C is referenced third. This means that the part should contact datum A first. As such, this plane is called the **primary datum.** The primary datum plane establishes the orientation of the part to the datum reference frame. To stabilize the part at least three points contact must be established by the part with the primary datum. The primary datum restricts three degrees of freedom:

- Movement along the Z-axis,
- Rotation around the X-axis, and
- Rotation around the Y-axis.

The other three degrees of rotation remain unrestricted *Fig. 14.80.*

As shown in the feature control frame, the part should contact datum B next. In a dimensional measurement, the second datum plane the part contacts, is called the **secondary datum.** The secondary datum locates the part, restricting part movement, within the datum reference frame. The part may have a line contact with the secondary datum. Therefore, it needs a minimum of two-point contact with the secondary datum plane. It then restricts two additional degrees of freedom:

- Rotation around the Z-axis, and
- Movement along the Y-axis.

It thus leaves only one degree of freedom for the part to move.

The feature control shows that the part should contact datum C last. The third datum plane, that the part contacts, is called the **tertiary datum.** The tertiary datum locates the part, restricting its movement, within the datum reference frame. The part may have a single point contact with the tertiary datum plane. The tertiary datum restricts the last remaining degree of freedom:

- Movement along the X-axis.

3-2-1 Rule

The 3-2-1 rule defines the minimum number of points of contact required for a part datum feature

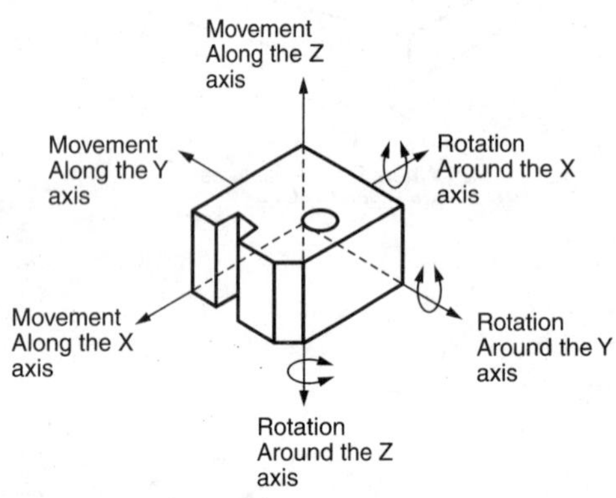

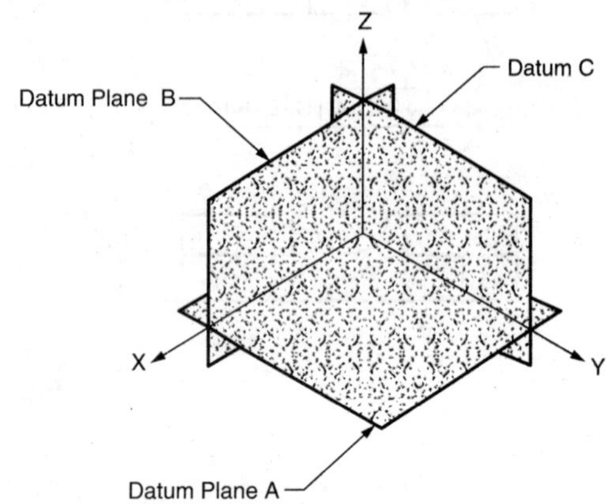

With These Datums Specified and Reference in this Order

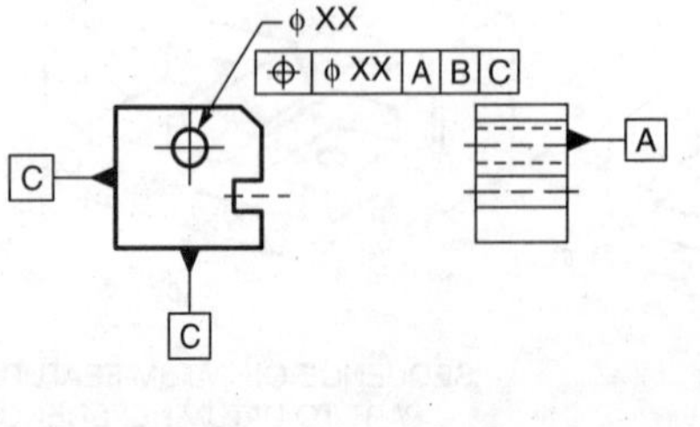

Datum	Limits Freedom in the
Primary A	• Rotation Around the X Axis • Rotation Around the Y Axis • Movement Along the Z axis
Secondary B	• Rotation Around the Z Axis • Movement Along the Y axis
Tartiary C	• Movement Along the X axis

Fig. 14.80. Datum reference frame

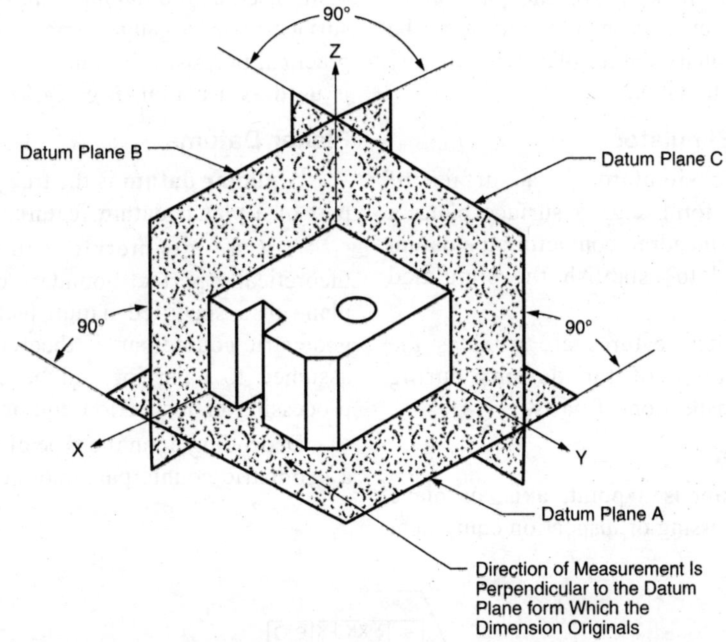

Fig. 14.81. Datum reference frame

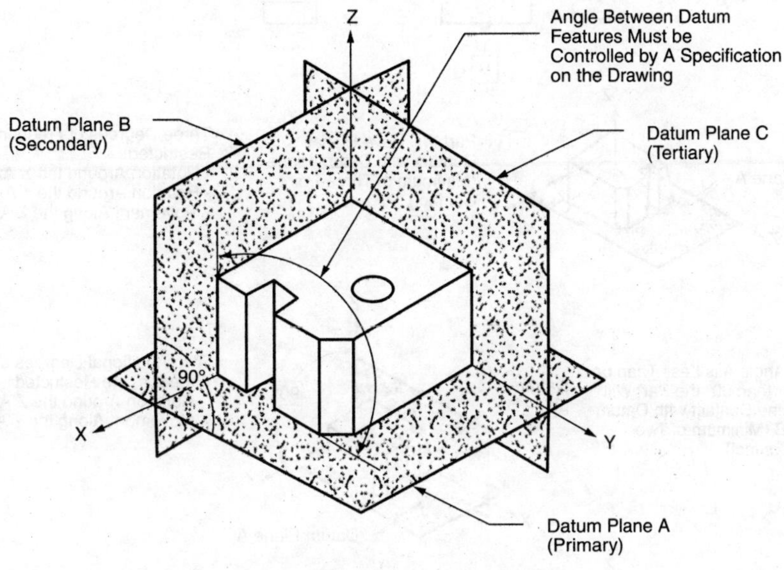

Fig. 14.82. Datum features

with its primary, secondary, and tertiary datum planes. This rule applies on a part with all planar datums. The primary datum feature should have at least three points of contact with its datum plane. The secondary datum feature should have at least two points of contact with its datum plane. The tertiary datum feature should have at least one point of contact with its datum plane.

For ideal stability of a part on a surface at least three points of contact are required. Form error on a planar datum feature can cause one-point contact of a part on a primary datum plane.

This will result in "rocking" of the part on its datum plane. In such a case, the part must be stabilized on its primary datum plane, before any measurements can be taken.

Datum Feature Simulator

Datum feature simulator is a surface of adequately precise form, e.g., a surface plate, a gauge surface, or a mandrel, contacting the datum features) and used to establish the simulated datum(s).

Simulated datum features are used as the practical embodiment of the datums during manufacture and inspection of parts.

Simulated Datum

Simulated datum is a point, axis, or plane established by processing or inspection equipment being used as the datum feature simulator, e.g., a surface plate, a gauge surface, or a mandrel. For practical purposes, in industry, a simulated datum is used as a datum, *Fig. 14.77, 14.78.*

Planar Datums

A planar datum is the true geometric counter part of a planar datum feature.

A **true geometric counterpart** is the theoretically perfect boundary or best-fit tangent plane of a specified datum feature. Since a true geometric counterpart is theoretical, the datum is assumed to exist in, and be simulated by, the processing or inspection equipment being used.

Depending upon the type of the datum feature, a geometric counterpart may be:

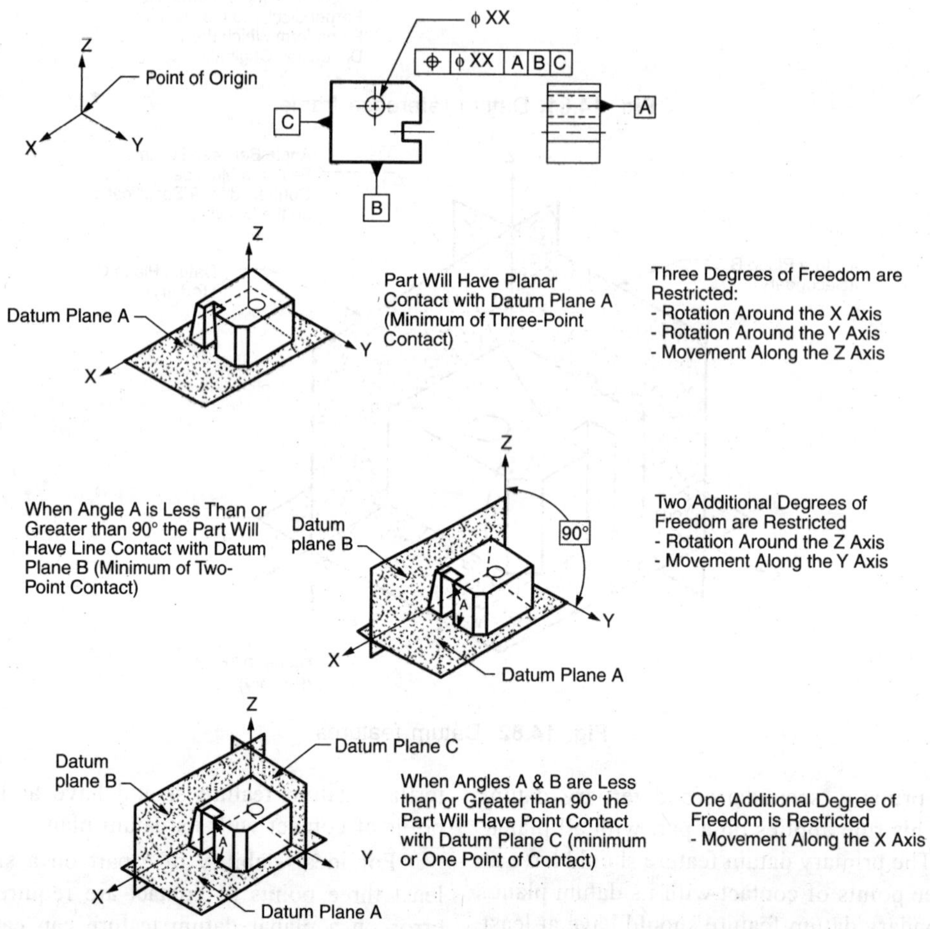

Fig. 14.83. Primary, Secondray, and Tertiary Datums

- A tangent plane contacting the high points of a surface.
- A maximum material condition boundary.
- A least material condition boundary.
- A virtual condition boundary.
- An actual mating envelope
- A worst-case boundary.

A mathematically defined contour.

Datum Related Versus FOS Dimensions

Only dimensions that are related to a datum reference frame, through geometric tolerances, should be measured in a datum reference frame. When a dimension is not associated to a datum reference frame, with a geometric tolerance, there is no specification on how to locate the part in the datum reference frame. As shown in *Fig 14.84*, the hole locations are related to the datum reference frame with datum planes D as primary, E as secondary, and F as tertiary datums. The overall dimensions are FOS dimensions. Therefore, during inspection of the hole dimensions, the part should be mounted in the datum reference frame D-E-F. Whereas, during inspection of the overall dimensions, the part need not be mounted in the datum reference frame.

Inclined Datum Features

An inclined datum feature is a datum feature that is at an angle other than 90°, relative to the other datum features. On parts with inclined datum features, i.e., surfaces, at angles other than 90°, the reference frame will contain planes at the basic angle of the part surface. The part shown in *Fig. 14.85* has a surface at an angle of 60° to the datum feature A. This surface is designated as datum surface C. The datum reference frame would have three perpendicular planes. However, the inclined datum feature would have its datum feature simulator oriented at the basic angles shown on the drawing. For this type of datum reference frame, the angle of the surface must be specified as basic, and the surface is controlled by a geometric tolerance.

Multiple Datum Reference Frames

In certain cases, the functional requirements of a component part may require more than one datum reference frame, to completely define its functional relationships, *Fig. 14.86*. The datum reference frames may be at right angles or at angles other than 90°. A datum plane can be used in more than one datum reference frame.

Coplanar Datum Features

Coplanar surfaces are two are more surfaces that are on the same plane. Similarly, coplanar

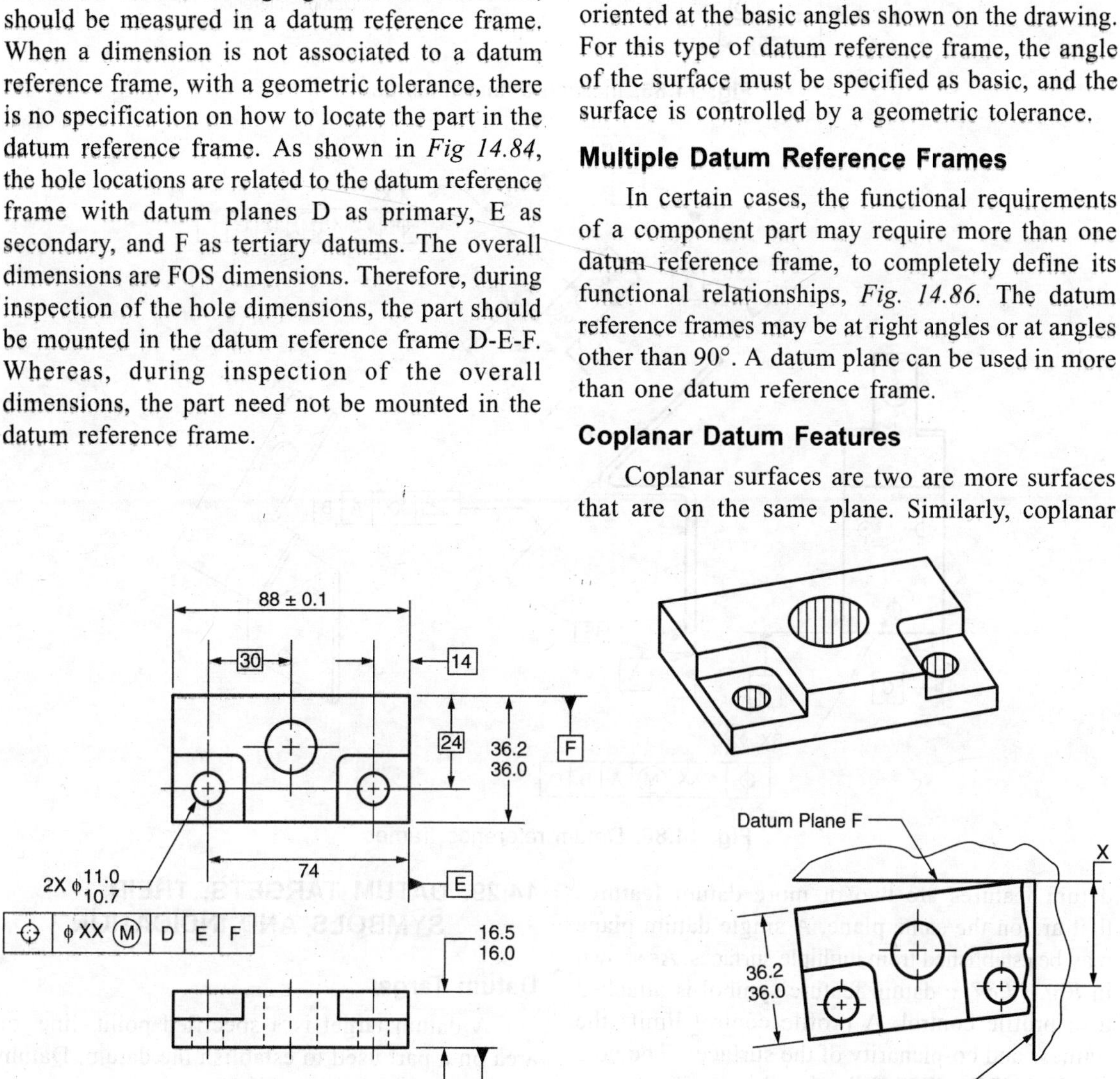

Fig. 14.84. Datum related versus feature of size dimensions

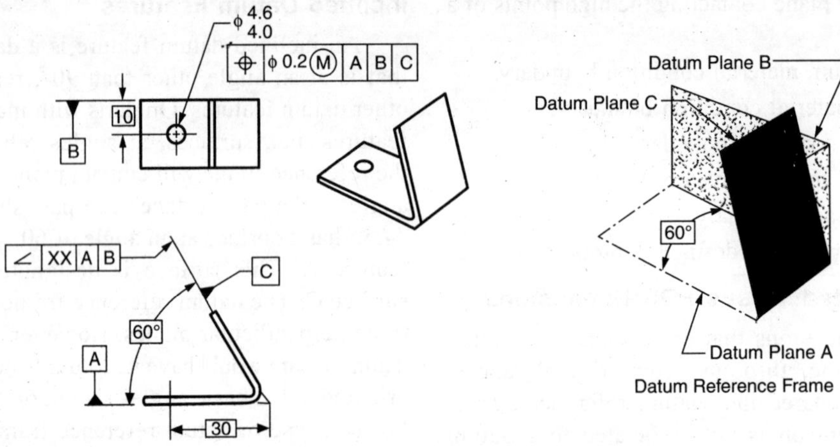

Fig. 14.85. Inclined datum features

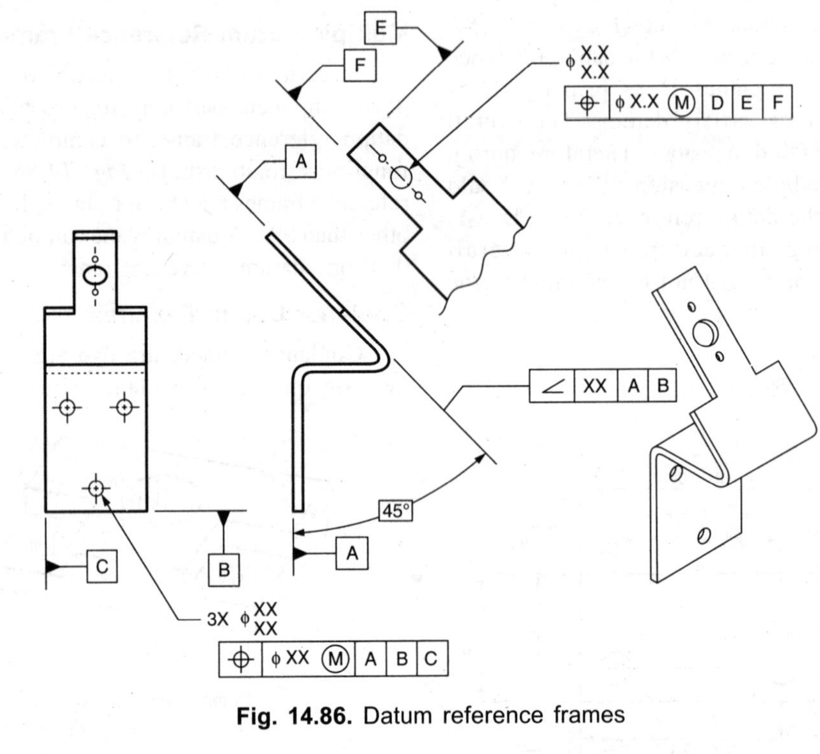

Fig. 14.86. Datum reference frames

datum features are two or more datum features that are on the same plane. A single datum plane may be established from multiple surfaces. As shown in *Fig. 14.87*, a datum feature symbol is attached to a profile control. A profile control limits the flatness and co-planarity of the surfaces. The note "TWO SURFACES", following this profile control, also denotes that datum feature A is comprised of two surfaces.

14.29. DATUM TARGETS, THEIR SYMBOLS AND INDICATION

Datum Target

A datum target is a specified point, line, or area on a part used to establish the datum. Datum targets are symbols that describe the shape, size and location of gauge elements that are used to establish datum planes, axes, and points. Datum targets are shown on the part surfaces on a drawing, but actually they do not exist on a part. Datum targets describe gauge elements. The gauge elements only contact a portion of the part surface. Datum targets can be specified to simulate a point, line or area contact on a part. The use of datum targets allows a stable and repeatable relationship for a part with its gauge.

Datum targets should be specified on parts where it is not practical to use an entire surface as a datum feature. Surfaces produced by forging, casting or sheet metal fabrication, irregularly shaped parts, plastic parts, and weldments may be subjected to warping, twisting or bowing; and will not necessarily be flat. Often the-part will rock, wobble, or not rest in the same position on a full datum plane. It is therefore, impractical to designate, on entire surface, as a functional datum because accurate and repeatable measurements cannot be made from the entire surface. Therefore, in order to define a practical datum plane, selected points or areas are indicated on the drawing. Manufacturing processes and inspection utilizes these combined points or areas as datums.

Datum Target Symbols

A datum target application uses two symbols:

- A datum target symbol, and
- A symbol that denotes which type of gauge element is to be used.

The datum target symbol is shown in *Fig. 14.88 (a)*. The symbol is divided into two parts with a horizontal line. The lower half denotes the datum reference letter and the target number associated with that datum; the upper half contains gauge element size information when applicable. The leader line from the symbol specifies whether the datum target exists on the surface shown or on the hidden surface of the part. If the leader line is solid, the datum target exists on the surface shown, e.g., datum targets *B*l and *B*2 in *Fig. 14.88.* If the leader line is dashed, it denotes that the datum target exists on the hidden surface of the part, e.g., datum targets *A*1, *A*2, and *A*3 in *Fig. 10.87.* When a datum target symbol is used on a drawing, it is often accompanied by the datum symbol.

Three symbols used to denote the type of gauge element in a datum target application are the symbols for a target point, a target line, and a target area. All symbols appear on the drawing view that most clearly shows the relevant surface, *Fig. 14.87.* Basic dimensions are used to describe the location of datum targets. This assures that there will be minimum variation between gauges. If the datum target is:

1. **A point:** It is indicated by an X-shaped symbol, consisting of two mutually perpendicular lines. The symbol is shown and dimensioned on the plan view of the surface to which it is being applied, *Fig. 14.89 (a)*. Where this type of view is not available, the symbol can be shown and dimensioned in two adjacent views, Fig. 14.89 (*b*). A datum target point is often simulated in a gauge with a spherical-tipped gauge pin, as shown in *Fig. 14.90*. Here, the datum plane *C*

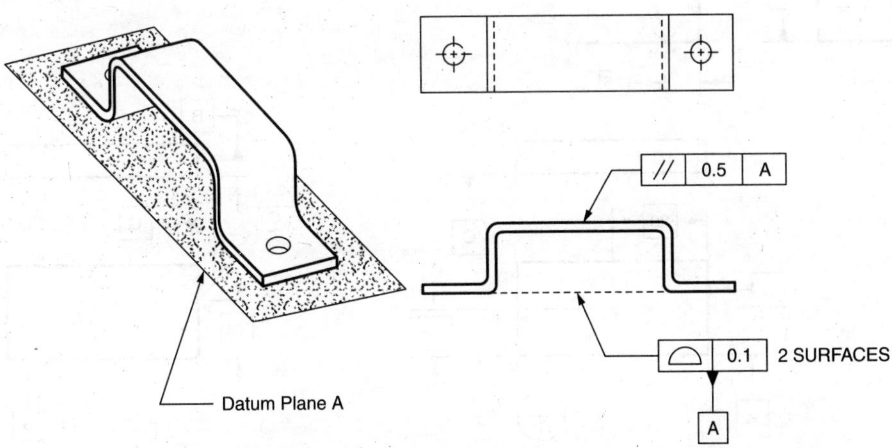

Fig. 14.87. Coplanar datum features

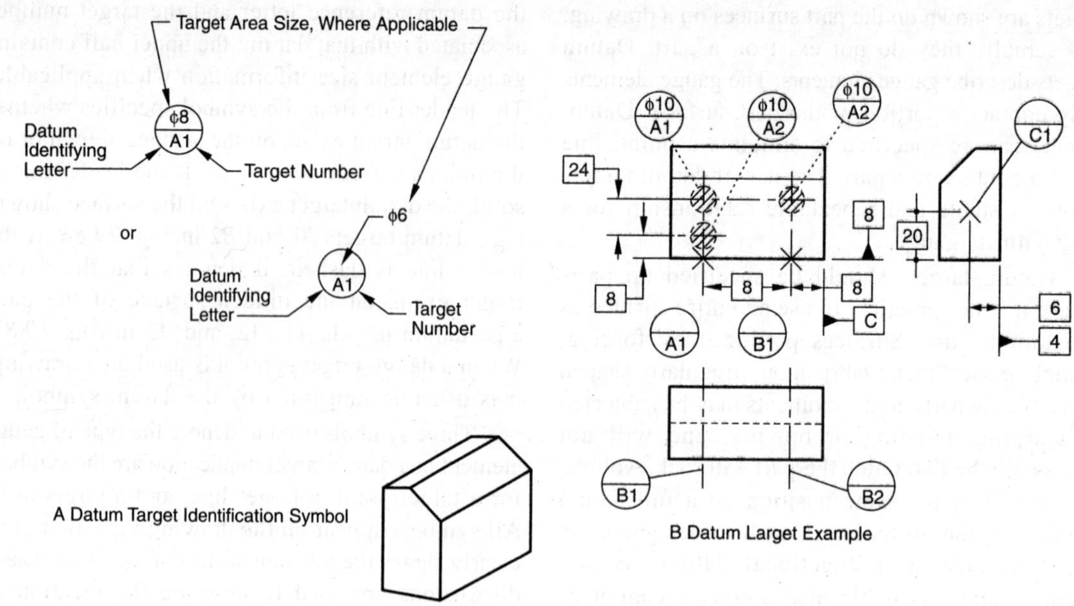

Fig. 14.88. Datum target symbols

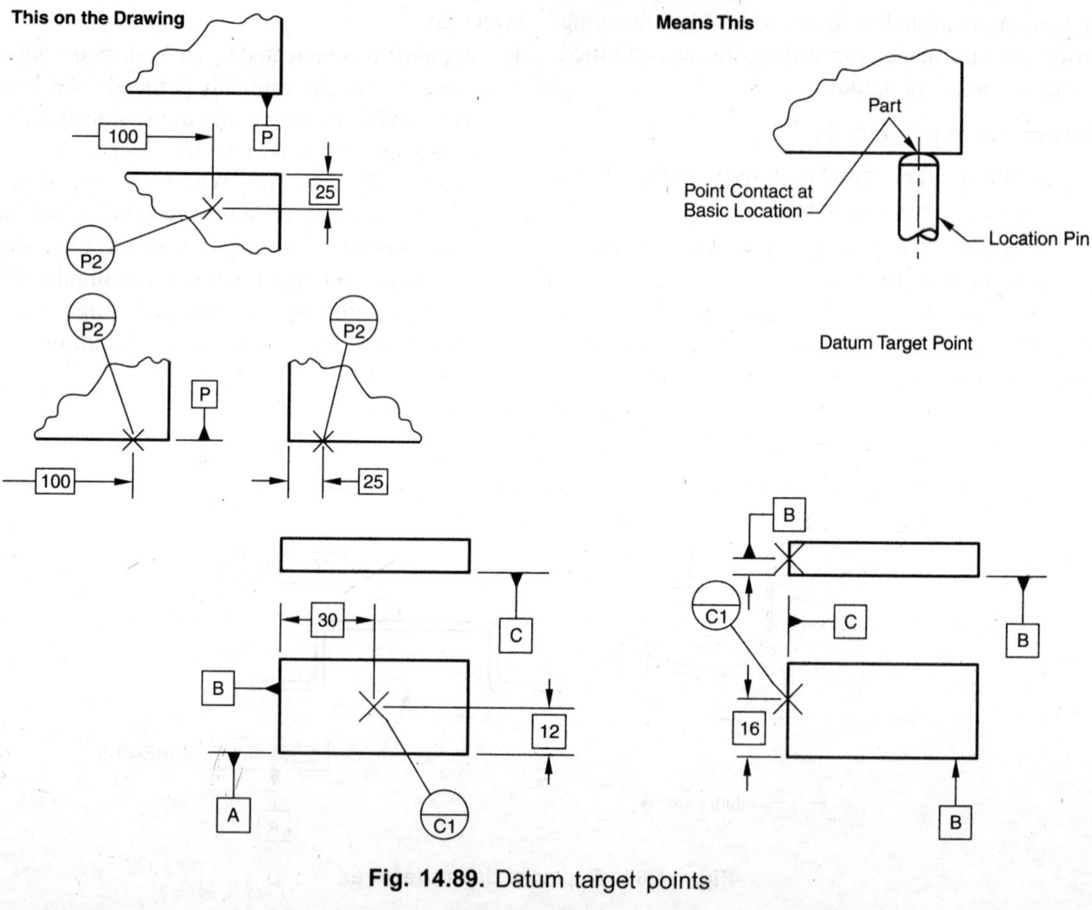

Fig. 14.89. Datum target points

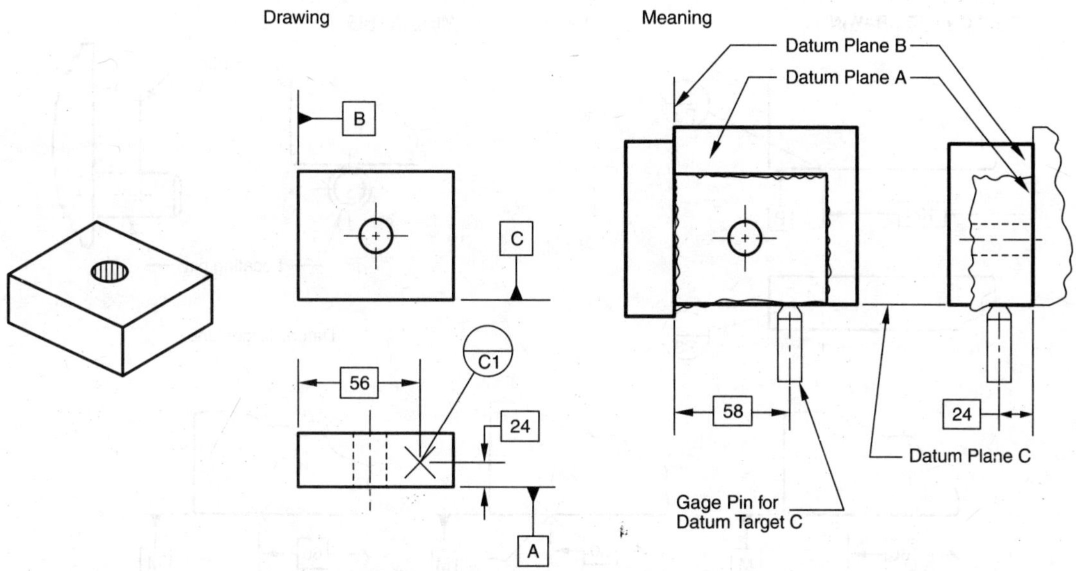

Fig. 14.90. Datum target points

is across the top of the spherical-tipped pin and is perpendicular to both datum planes A and B. The actual part surface may be above or below the datum plane.

2. **A line:** A datum target line can be specified by three different ways, as shown in *Fig. 14.91. (a)*, *(b)*, and *(c)*:
   - A phantom line on the plan view of a surface,
   - An X on the edge view of a surface, and
   - A combination of a phantom line, and an X.

   Basic dimensions should be used to locate the datum targets relative to other targets and datums. Datum target lines are simulated cylindrical gauge pins, *Fig. 14.92.*

3. **An area:** Where it is preferred to use a gauge element that represents an area of contact on a part surface, a datum target area is specified. A datum target area is designated by drawing the outline of the gauge for simulating the datum target area on the part surface. It is indicated by a hatched area, surrounded by a thin, double dashed chain line, as shown in *Fig. 14.88 (a), 14.93.*

The shape, size, and location of the area are described with basic dimensions. If the target area is circular, the diameter may be specified in the upper half of the datum target symbol as shown in *Fig. 14.88 (b).* Where it is not possible to show the circular target area, the method shown in *Fig. 14.88 (c)* may be used. The use of datum target areas is shown in *Fig. 14.89.* The flat tipped gauge pins are used to simulate the area datum targets specified on the drawing.

The leader line between the datum target symbol and the datum target is shown dashed when the datum target area at which the gauge pin contacts the surface is invisible, *i.e.*, on the rear surface of the part with respect to the viewing direction.

Datum Target Applications

Figures 14.94 to 14.99 give some practical applications of datum targets. It can be interpreted from the illustrations that:

- Datum targets *A*1, *A*2, and *A*3 establish datum *A*.
- Datum targets *B*1 and *B*2 establish datum *B*.
- Datum target *C*1 establishes has datum *C*.

Although there are only three datum target symbols, they can be used in a number of ways and on a number of different types of parts. Three different datum target applications that are representative of most common cases are:

- Creating a partial reference frame from partial surfaces.

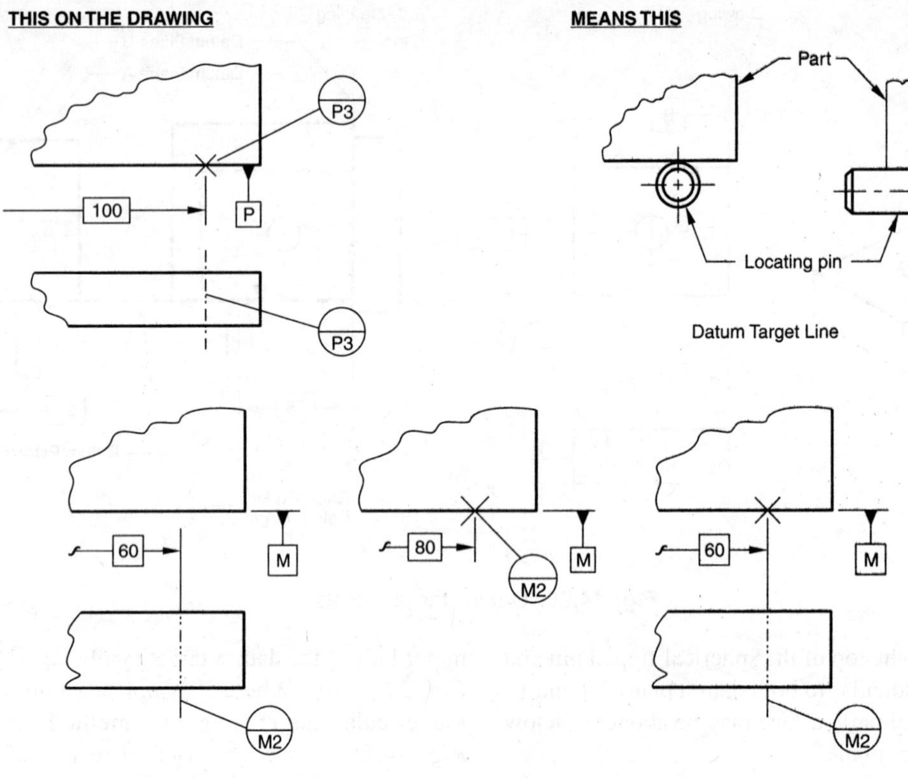

Fig. 14.91. Symbol for datum targer lines

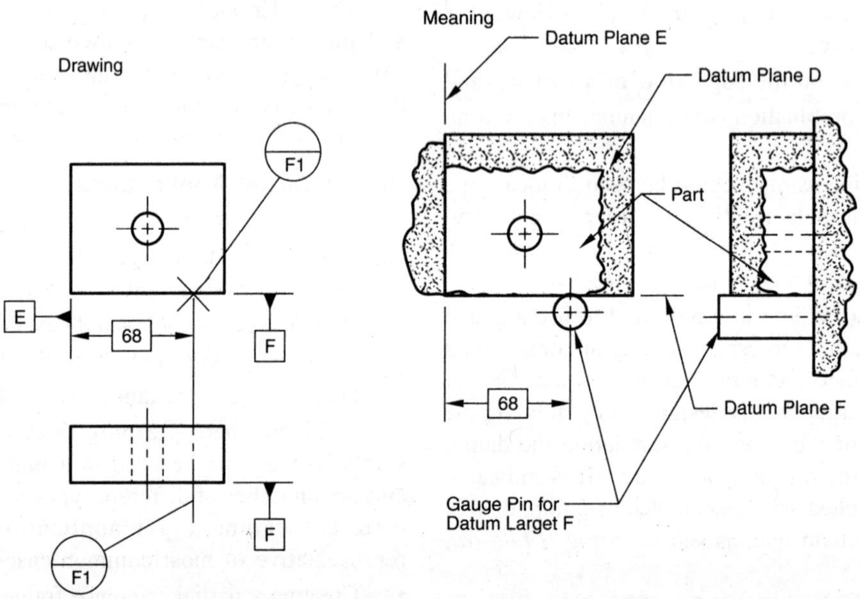

Fig. 14.92. Datum target lines

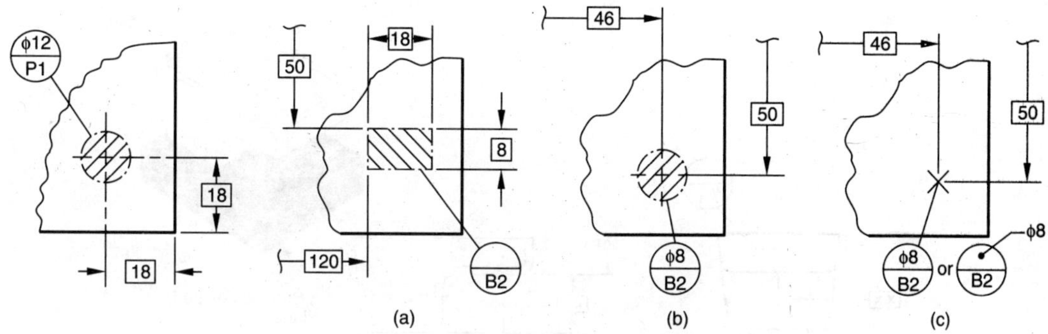

Fig. 14.93. Datum target areas

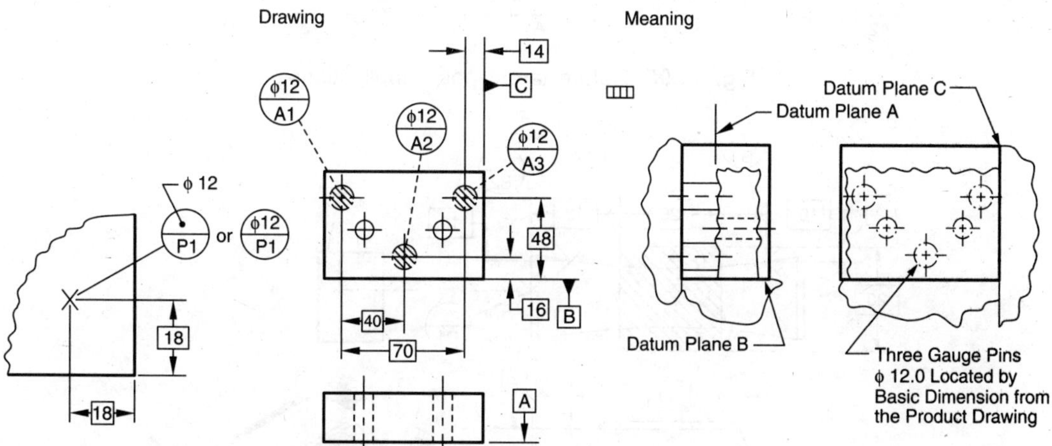

Fig. 14.94. Datum target applications

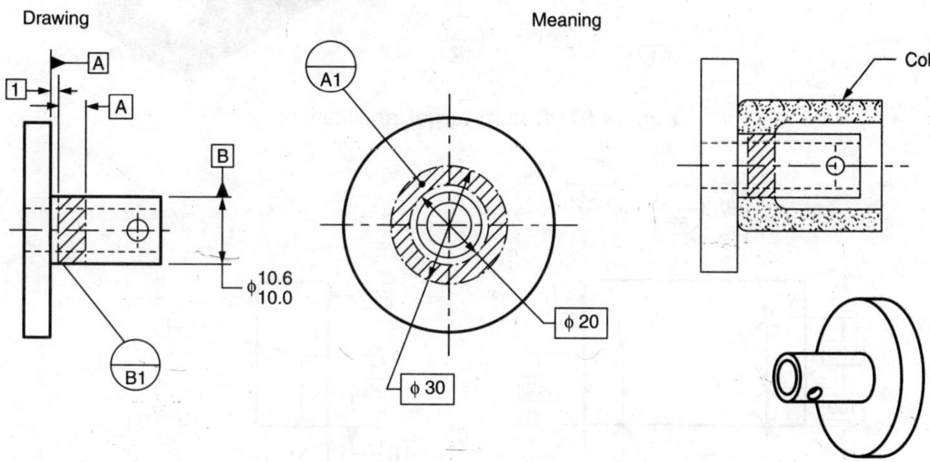

Fig. 14.95. Datum target area applications

- Creating a partial datum reference frame from offset surfaces.
- Creating a complete datum reference frame from irregular surfaces.

When using datum targets to establish a complete datum reference frame, three requirements should be met:

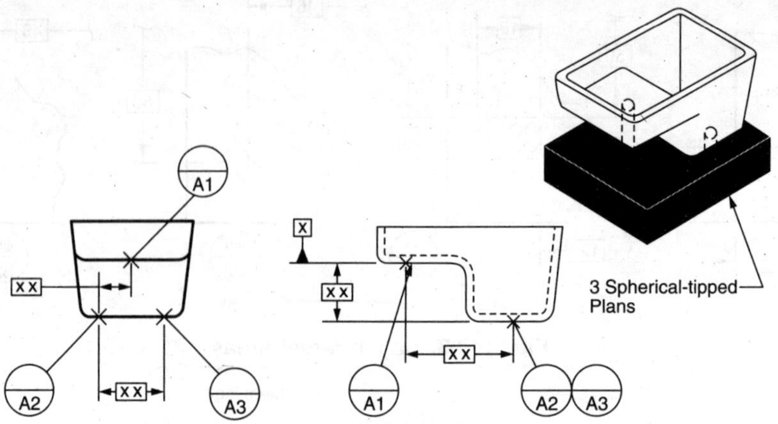

Fig. 14.96. Datum target point applications

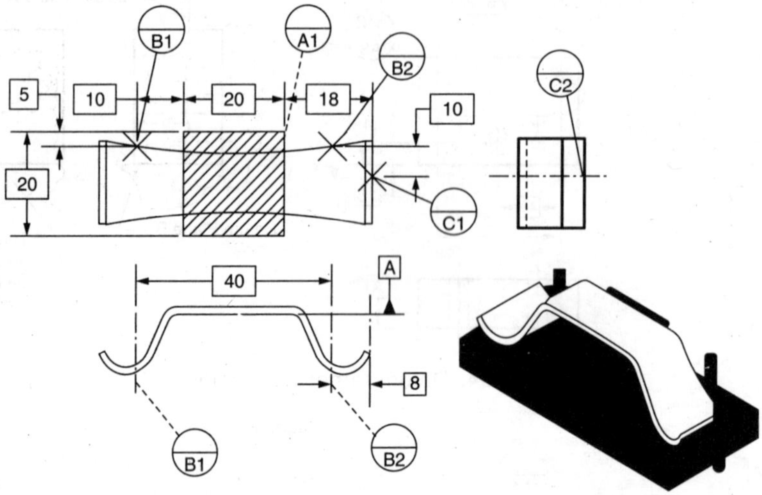

Fig. 14.97. Datum target applications

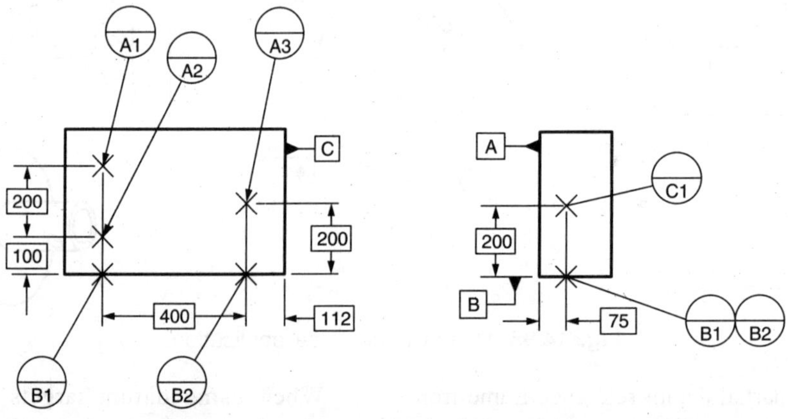

Fig. 14.98. Dimensioning datum targets

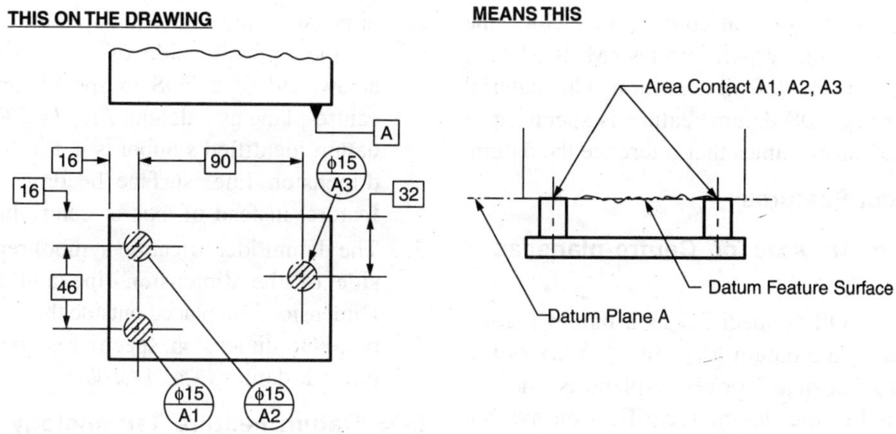

Fig. 14.99. Primary datum plane established by three datum target areas

- Basic dimensions should be used to (define) locate the datum targets.
- The datum reference frame must restrain the part in all six degrees of freedom.
- The part dimensioning must ensure that the part will rest in the gauge in only one orientation and location so as to ensure repeatable measurements.

In certain applications, it may be desired to create a datum reference frame by using a partial surface of the datum feature. *Figure 14. 95* show an application that uses datum targets to define gauge elements that contact partial surfaces of the datum features. Datum targets Al and Bl designate gauge elements that contain a portion of the surfaces they contact. When a datum target area (or line) is shown on a cylindrical surface like datum target Bl, it wraps around the diameter of the part.

In some other cases, it may be necessary to create a datum reference frame by using points (or lines, or areas) from offset surfaces of a part, as shown in *Fig. 14.96* that uses datum targets. Gauge elements that contact the offset surfaces are used to establish the datum reference frame. Here, datum target *A*l is on one surface, and datum target *A*2 and *A*3 are on another surface, A basic dimension is used to define the offset of the datum targets (gauge pins). A profile tolerance may be used to define the offset tolerance for the part surfaces. A datum symbol may be attached to either of the surfaces where the datum targets are shown.

When datum target points, lines, or areas are used to establish a primary datum plane, it is a good practice to specify them as far apart, as is possible, so as to effectively stabilize the part on the gauge. A secondary and a tertiary datum are required when the primary datum is established with datum targets. *Figure 14.98* shows the method of dimensioning datum targets.

In some cases, it may be necessary to create a datum reference frame on a part with irregular surfaces, as shown in *Fig. 14.92* that uses datum targets. Gauge elements that contact the points on the irregular surfaces are used to establish the datum reference frame. Datum target A1 is an area target, and it establishes datum plane A. Datum targets Bl and B2 are line targets, and they establish datum plane B. Datum target Cl is a line target, and it establishes datum plane C. Basic dimensions are used to define the size and location of the datum targets. A profile tolerance can be used to define the tolerance for the irregular part surfaces. A datum symbol may be attached to either of the planes where the datum targets are shc:vn.

14.30. FEATURE OF SIZE DATUM SPECIFICATIONS

When a feature of size (FOS) is used as a datum feature, it usually gives an axis or a centre-plane as the datum. Diametrical and planar features of size may be used as datum features. When a diameter is used as a datum feature, it results in a datum axis, and when a planar FOS is used as a datum feature, it results in a datum centre-plane. Also, when a FOS is used as a datum feature, the drawing must indicate the material condition at which the true geometric counterpart applies (MMC, LMC, or RFS). RFS is the default condition. The material condition for a FOS datum feature is specified in the feature control frames that reference the datum.

FOS Datum Features

Specifying An Axis or Centre-plane as a Datum

When a FOS is specified as a datum feature by associating the datum identification symbol to the FOS, it gives an axis or centre-plane as a datum. Remember that the datum identification symbol must be associated to a FOS to specify an axis or centre-plane as a datum. There are five common ways to specify an axis or centre-plane as a datum:

1. The datum identification symbol touching feature surface of a diameter to specify the axis as a datum, *Fig. 14.100 (a)*.
2. The datum identification symbol touching the beginning of a leader line of a FOS to specify its axis datum, *Fig. 14.100 (b)*.
3. The datum identification symbol touching a feature control frame to specify an axis or centre-plane as datum, *Fig. 14.100(c)*.
4. The datum identification symbol in line with dimension line and touching the extension line on the opposite side of the dimension line arrowhead of a FOS to specify an axis or a centre-plane as a datum, *Fig. 14.100 (d)*. If the datum identified symbol is not in line with the dimension line, surface becomes the datum feature, instead of axis or centre-plane.
5. The datum identification symbol replacing one side of the dimension line and arrowhead. Dimension line placed outside the extension line of a FOS dimension, specifies an axis or centre-plane a datum, *Fig. 14.100(e)*.

FOS Datum Feature Terminology

We have already defined the common terms used here. When a diameter is designated as a datum feature, placing the part in a datum feature simulator, that may be gauge elements, derives datum axis. *Figure 14.101.* illustrates the terminology used to describe the various elements in a datum axis application. The datum feature is the surface of the part. The gauge element that holds the part datum feature is the datum feature simulator. The datum feature simulator is considered to be the true geometric counterpart of the datum feature. The axis of the gauge element that holds

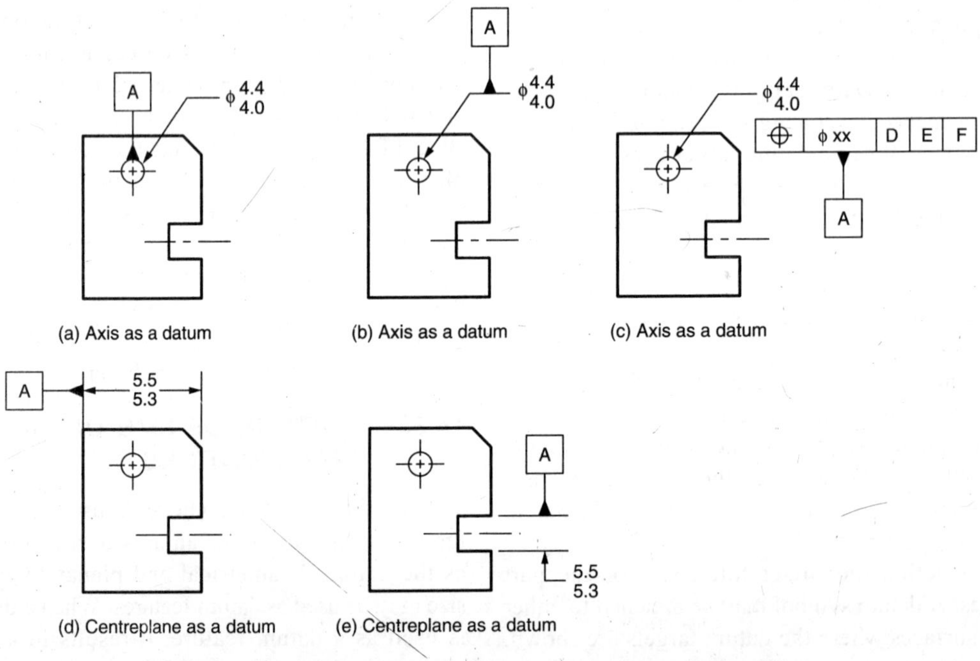

Fig. 14.100. Specifying an Axis or Centre-plane as a datum

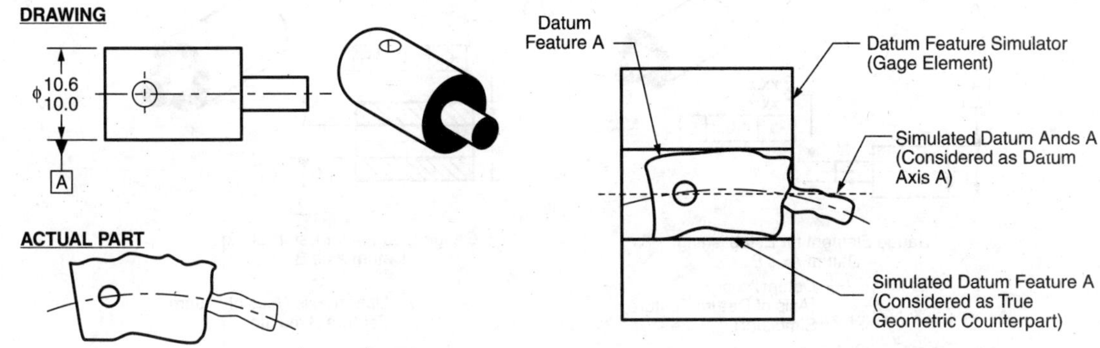

Gage Element for Establishing Datum Axis A (RFS)

Fig. 14.101. Datum terminology

the part datum feature is the simulated datum axis and is considered to be the datum axis. The datum axis becomes the origin of measurements for dimensions that are related to the datum, in dimensioning the part.

Referencing Feature of Size (FOS) Datum Features

When using FOS datum features on drawings, the material condition used to establish the datum axis or centre-plane must be specified. This is done through feature control frames that reference the FOS datum features. As has already been discussed, the feature control frame indicates the material condition (MMC, or LMC) and the datum sequence (primary, secondary, or tertiary). When a FOS datum feature is referenced in a feature control frame, and no material condition modifier is shown, it is automatically referenced at the default RFS condition, as per Rule #2.

FOS Datum Feature Applications Referenced RFS

Datum selection and referencing are dependent on the part function. Most commonly, the features that orient and locate the part in its assembly are chosen as the datum features for the part being dimensioned.

1. Datum Axis RFS Primary

When a diameter is specified as a datum feature and referenced in a feature control frame at RFS, it establishes a datum axis through physical contact between the inspection equipment and the datum feature. A datum feature simulator envelopes the specified diameter. Holding devices that are adjustable in size, e.g., a precision jaw-chuck, collet, or centring device, are used as the datum feature simulator, that is the true geometric counter-part. The datum feature simulator holds the part securely on the datum feature. The axis of the datum feature simulator becomes the datum axis and establishes the orientation of the part being inspected. *Figure 14.94 (a)* and *(b)* show examples of an external and internal diameter that is used to establish a primary datum axis referenced RFS.

When a FOS is referenced RFS and is used to establish a primary datum axis or centre-plane, its length should be sufficient to establish a repeatable relationship (orientation) with the gauge.

2. Datum Centreplane RFS Primary

When a parallel plane FOS is specified as a datum feature and is referenced in a feature control frame at RFS, it establishes a datum centre-plane, through physical contact between the inspection equipment, being used, and the datum feature. A datum feature simulator holds the part securely on the datum feature. The centre-plane of the datum feature simulator becomes the datum centre-plane, and establishes the orientation of the part. An adjustable gauge element is required to simulate the true geometric counterpart of the datum feature. The part is held by the gauge and oriented by the gauge. *Fig. 14 103 (a)* and *(b)* illustrates the use of an external and internal FOS for establishing a primary datum centre-plane, referenced RFS. It has to be kept in mind that specifying a FOS as a primary datum is comparatively expensive and

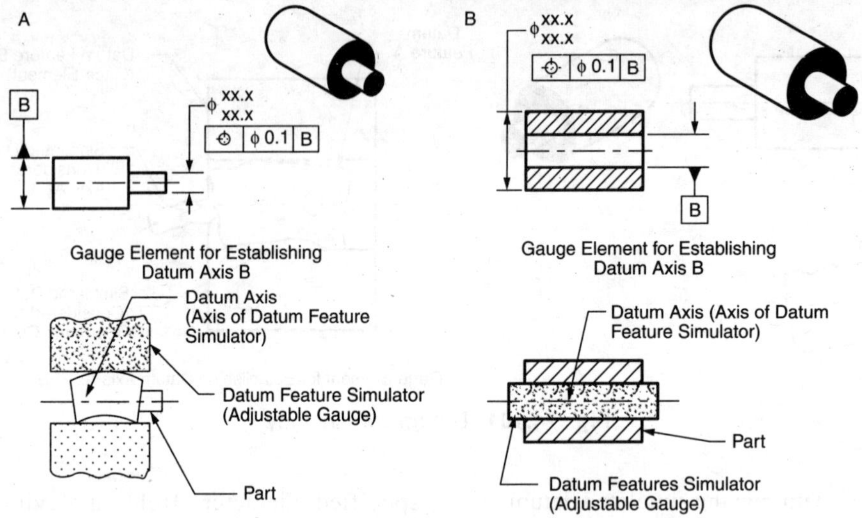

Fig. 14.102. Primary datum RFS

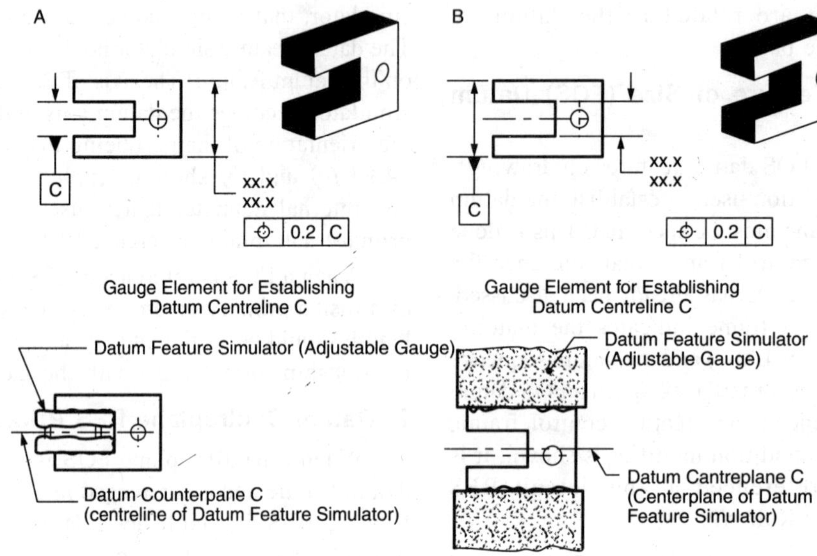

Fig. 14.103. Primary datums RFS

should only be used when the part function demands it.

3. Datum Axis Referenced RFS Secondary

When a part is oriented by a surface and located by a diameter, the surface and diameter are specified as datum features, as shown in *Fig. 14.104*. When the face is specified as primary and the diameter as secondary (RFS), the following conditions apply:

1. The part should have at least three-point contact with the primary datum plane.
2. The secondary datum feature simulator should be adjustable in size.
3. The datum axis is the axis of the datum feature simulator.
4. The datum axis should be perpendicular to the primary datum plane.
5. The second and third datum planes have to be associated with the datum axis.
6. In case an angular and/or rotational relationship has to be precisely controlled, a tertiary datum has to be specified.

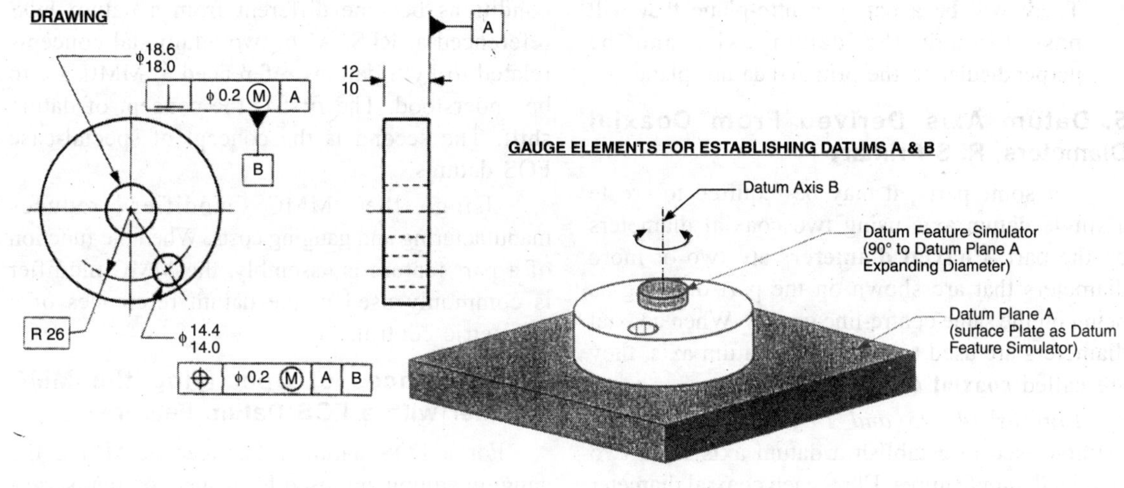

Fig. 14.104. Datum axis RFS secondary

4. Datum Axis Referenced RFS Secondary and Datum Centre-plane RFS Tertiary

For an application where a part surface is to be oriented by a surface, located by a diameter, and has an angular relationship relative to a FOS, most commonly, the surface, diameter, and FOS are specified as datum features, as shown in *Fig. 14.105.* When the face is specified as primary, the diameter as secondary (RFS), and the slot as tertiary (RFS), the following conditions apply:

- The part should have at least three-point contact with the primary datum plane.
- There will be a datum axis perpendicular to the primary datum plane.

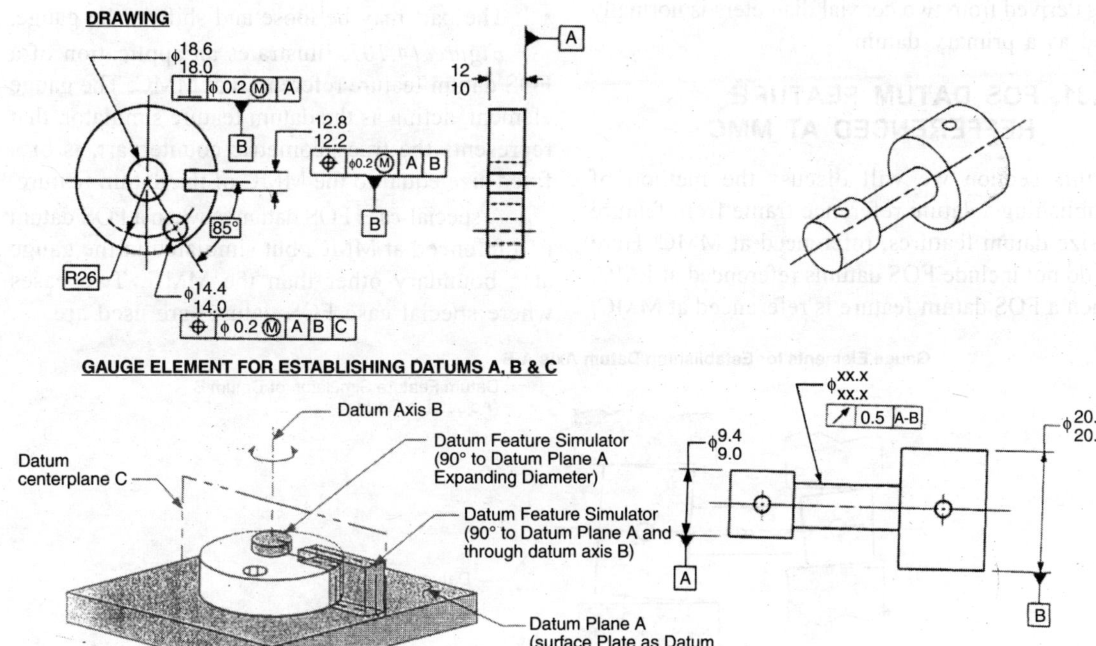

Fig. 14.105. Datum axis RFS secondary, Datum centre-plane RFS Teritary

- There will be a datum centre-plane that will pass through the datum axis, and be perpendicular to the primary datum plane.

5. Datum Axis Derived From Coaxial Diameters, RFS Primary

For some parts, it may be required to create a single datum axis using two coaxial diameters on the part. **Coaxial diameters** are two or more diameters that are shown on the part drawing as being on the same centre-line or axis. When coaxial diameters are used to establish a datum axis, they are called **coaxial datum features.**

Figures 14.105 and 14.106, illustrate the method used to establish a datum axis from two coaxial datum features. Here, each coaxial diameter is specified as a datum feature. The datum reference letters, separated by a dash, are then entered in one compartment of the feature control frame. The datum axis is established by the two coaxial diameters on the part and is simulated by contacting the high points, on both datum features, simultaneously. Coaxial diameters are frequently used when both datum features have an equal role in locating the part in its assembly. A single datum axis derived from two coaxial diameters is normally used as a primary datum.

14.31. FOS DATUM FEATURE REFFERENCED AT MMC

In this section we will discuss the method of establishing a datum reference frame from feature of size datum features, referenced at MMC. Here we do not include FOS datums referenced at LMC. When a FOS datum feature is referenced at MMC, conditions become different from a datum FOS referenced at RFS. Also, two additional concepts related to FOS datums referenced at MMC are to be understood. The first is the concept of datum shift. The second is the concept of special-case FOS datums.

Using the MMC modifier reduces manufacturing and gauging costs. When the function of a part feature is assembly, the MMC modifier is commonly used in the datum references of a geometric control.

Consequences of Specifying the MMC Modifier with a FOS Datum Feature

For a FOS datum referenced at MMC, the gauging equipment, used for inspection that serves as the datum feature simulator is of fixed size. The datum axis or centre-plane is the axis or centre-plane of the gauge element. The size of the true geometric counterpart of the datum feature is determined by the specified MMC limit of size or, in certain cases, its MMC virtual condition. Referencing a FOS datum at MMC has two effects on the part gauging:

- The gauge becomes fixed in size.
- The part may be loose and shift in the gauge.

Figure 14.107, illustrates an application of a FOS datum feature referenced at MMC. The gauge element, acting as the datum feature simulator, that represents the true geometric counterpart, is of a fixed size equal to the MMC of the datum feature.

A special-case FOS datum is when a FOS datum is referenced at MMC, but simulated in the gauge at a boundary other than the MMC. Two cases where special case FOS datums are used are:

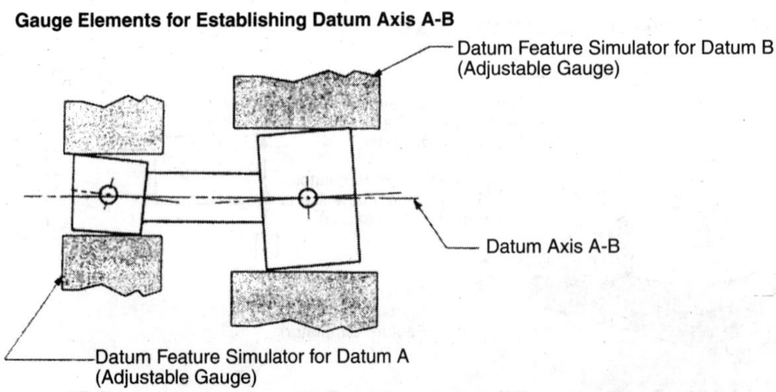

Fig. 14.106. Datum axis derived from coaxial diameters RFS primary

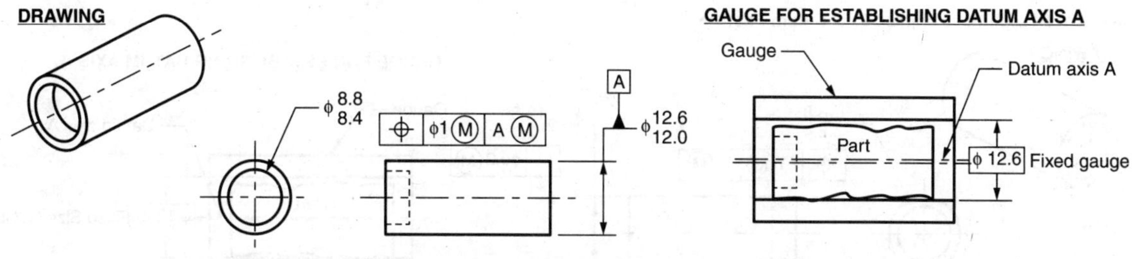

Fig. 14.107. Effects of using the MMC modifier

- When a straightness control is applied to a FOS datum feature.
- When secondary or tertiary datum features of size in the same datum reference frame are controlled by a location or orientation control with respect to this higher-ranking datum.

Figure 14.108, shows application of the two cases for special-case FOS datums.

Datum Shift

Datum shift is the allowable movement, or looseness, between the part datum feature and the gauge element. As has already been mentioned, when a FOS datum feature is referenced at MMC, the gauge element simulating the perfect feature counterpart is fixed in size. But the part datum feature can vary within its allowed size limits. Thus there may be some looseness between the part and the gauge element. This allows the datum shift and results in additional tolerance for the part, as shown in *Fig. 14.109*

When a part datum feature is at MMC, there is no datum shift allowed. As the actual mating envelope of the part datum feature departs from MMC towards LMC, a datum shift becomes available. The amount of datum shift is equal to the amount the datum feature departs from MMC, as shown in the chart. The maximum amount of datum shift possible is equal to the difference between the gauge size and the LMC of the datum feature.

Datum shift is similar to **bonus.** Like bonus, datum shift allows an additional tolerance under certain conditions. The actual amount of datum shift can be added to the stated tolerance in the feature control frame, as shown in the chart in *Fig. 14.110*. The datum shift concept also applies when LMC modifier is used. To summarize:

- Datum shift can result in an additional tolerance for a geometric tolerance.
- Datum shift is only allowed when a MMC modifier is shown in the datum portion of the feature control frame.
- Datum shift results when the actual mating envelope (AME) of the datum feature departs from the MMC.
- Maximum allowable datum shift is the difference between the gauge size, for the datum, and the LMC size of the datum feature.

Datum shift is not available on features of size that are gauged simultaneously, e.g., there is no datum shift between the holes of a pattern of holes.

The datum shift concept also applies when using special-case FOS datums. Datum shift can increase the amount of additional tolerance permissible for a special-case FOS datum, as shown in the chart in *Fig. 14.111* for various actual mating sizes of the datum feature. When the datum feature is at MMC, there can be datum shift available. The amount of datum shift available is equal to the difference between special-case FOS datum and MMC when the datum feature is at MMC. In the case of a special-case FOS datum, datum shift only exists if the datum feature is straighter than its permissible straightness error. Thus we conclude:

- For a special-case FOS datum referenced at MMC, datum shift may be available when the datum feature is at MMC.

The worst-case boundary datum shift concept also applies when the LMC modifier is used in datum feature specification.

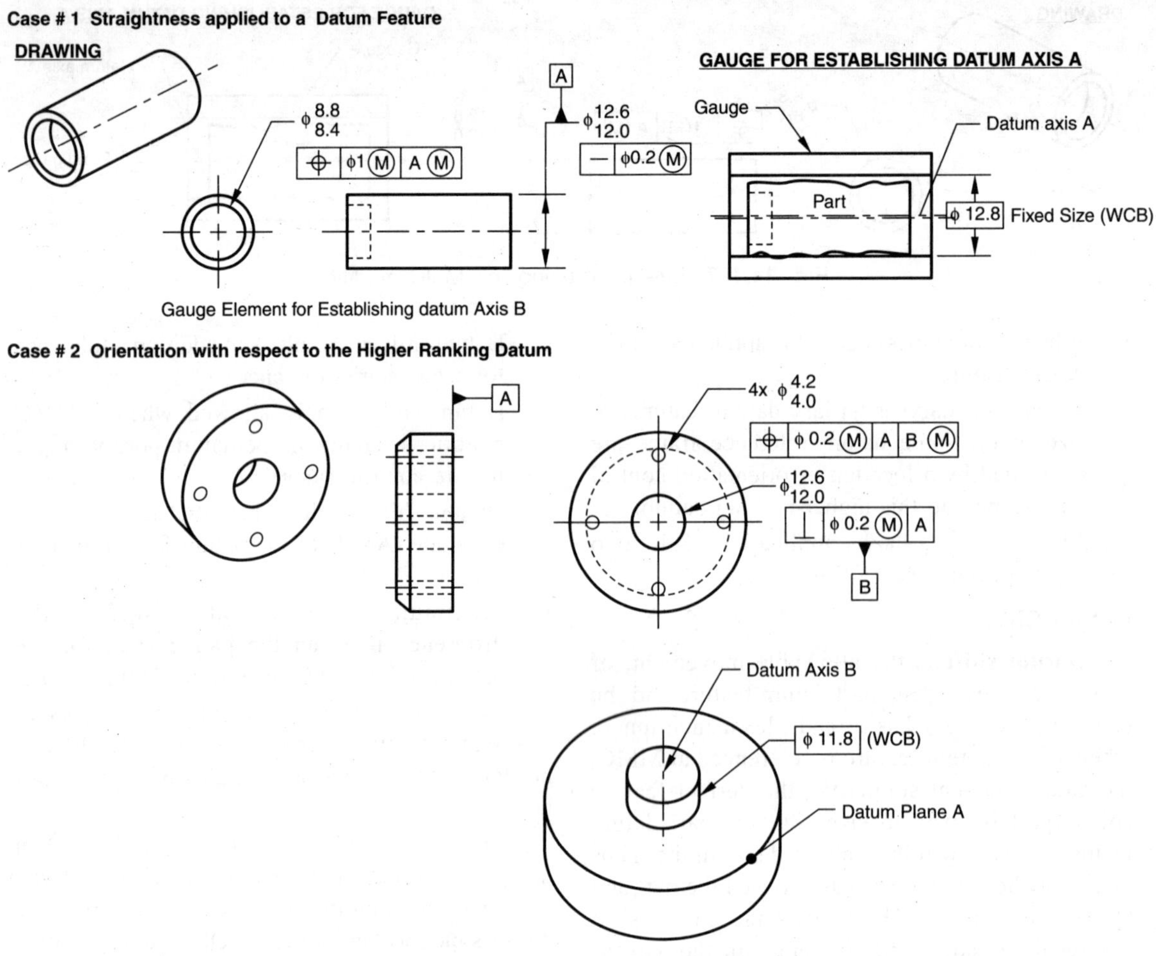

Fig. 14.108. Special-case FOS datum Examples

FOS Datum Feature Application When Referenced at MMC

Datum Axis MMC Primary

When a diameter is specified as a datum feature and referenced in a feature control frame as primary at MMC, a fixed gauge element can be used as the datum feature simulator. The size of the fixed gauge element is equal to the MMC or, in certain cases, worst-case boundary, of the datum feature. The datum axis is the axis of the datum feature simulator. Depending upon the actual mating size of the datum feature, a datum shift may be available. Examples of an external and internal diameters specified as a datum feature, referenced as primary at MMC, are shown in *Figs. 14.112 (a)* and *(b)* respectively.

Datum Centre-Plane MMC Primary

When a FOS consisting of parallel planes is specified as a datum feature and referenced in a feature control frame as primary at MMC, a fixed gauge element can be used as the datum feature simulator. The size of the fixed gauge element is equal to the MMC or, in certain cases, worst-case boundary of the datum feature. The datum centre-plane is the centre-plane of the datum feature simulator. Depending upon the actual mating size of the datum feature, a datum shift may be available. Examples of an external and internal FOS specified as a datum feature, referenced as primary at MMC, are shown in *Fig. 14.113 (a)* and *(b)*, respectively.

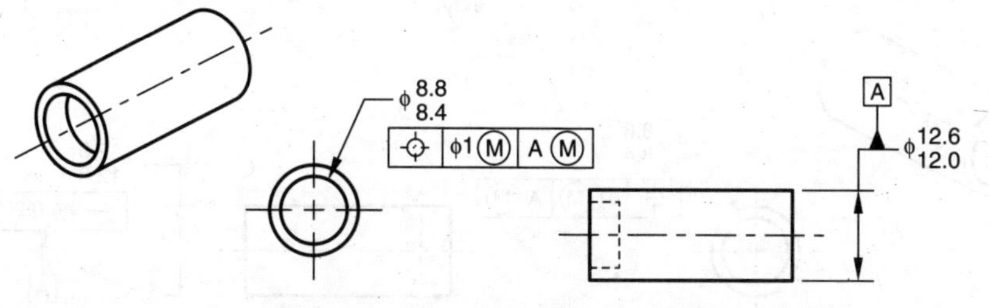

Gauge Element for Establishing Datum Axis A

	AME of Datum Feature A	Diametral Datum Shift Possible
MMC	φ 12.6	0.0
	φ 12.4	0.2
	φ 12.2	0.4
LMC	φ 12.0	0.6

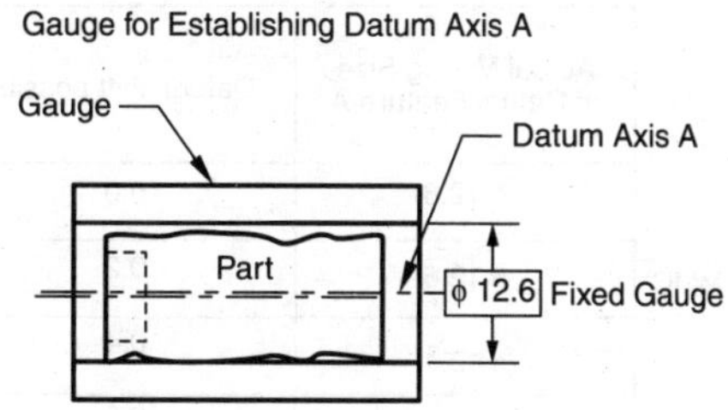

Fig. 14.109. Datum shift

Positional Tolerance	+ 1.0
Possible Bonus Tolerance	+ 0.4
Possible Shift Tolerance	+ 0.6
Total Possible Tolerance	2.0

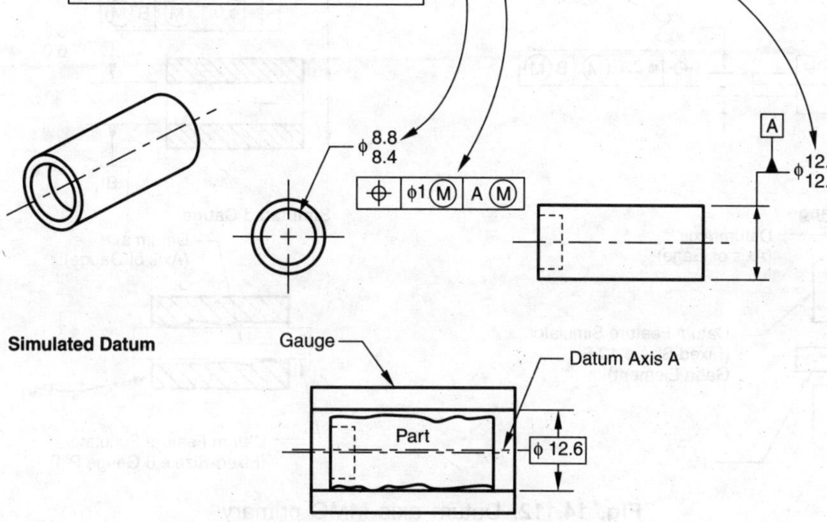

Fig. 14.110. Datum shift examples

Datum Axis MMC Secondary

When a part is oriented by a surface and located by a diameter, the surface and diameter are normally specified as datum features, as shown in *Fig. 14.114.*

When the datums are referenced with face as primary and the diameter as secondary at MMC, the following conditions are applicable:

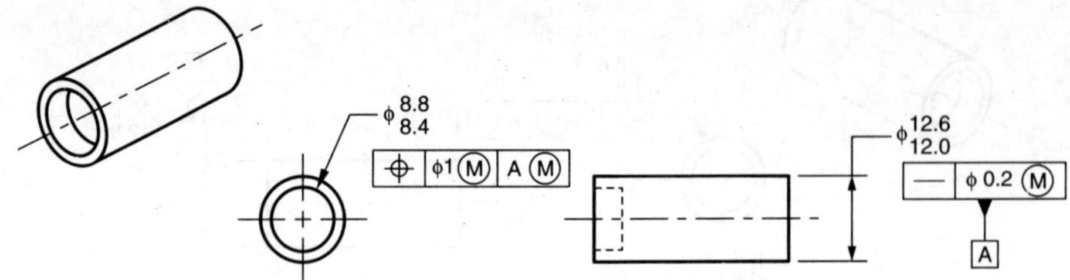

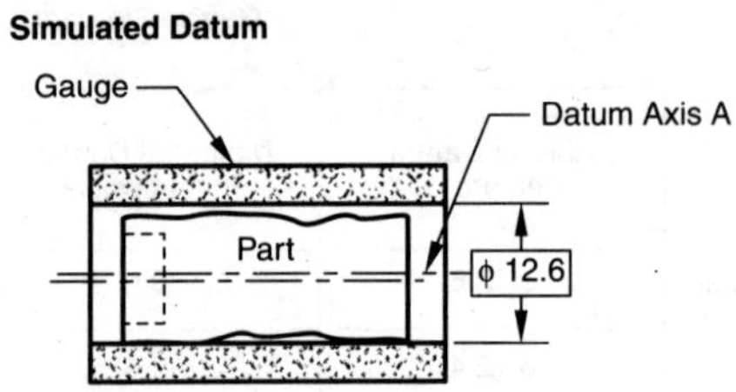

	Actual Mating Size of Datum Feature A	Datum shift possible
	ϕ 12.8	0.0
MMC	ϕ 12.6	0.2
	ϕ 12.4	0.4
	ϕ 12.2	0.6
LMC	ϕ 12.0	0.8

Datum Shift = Gauge Size – Actual Matching Size

Fig. 14.111. Datum Shift for special-case FOS Datums

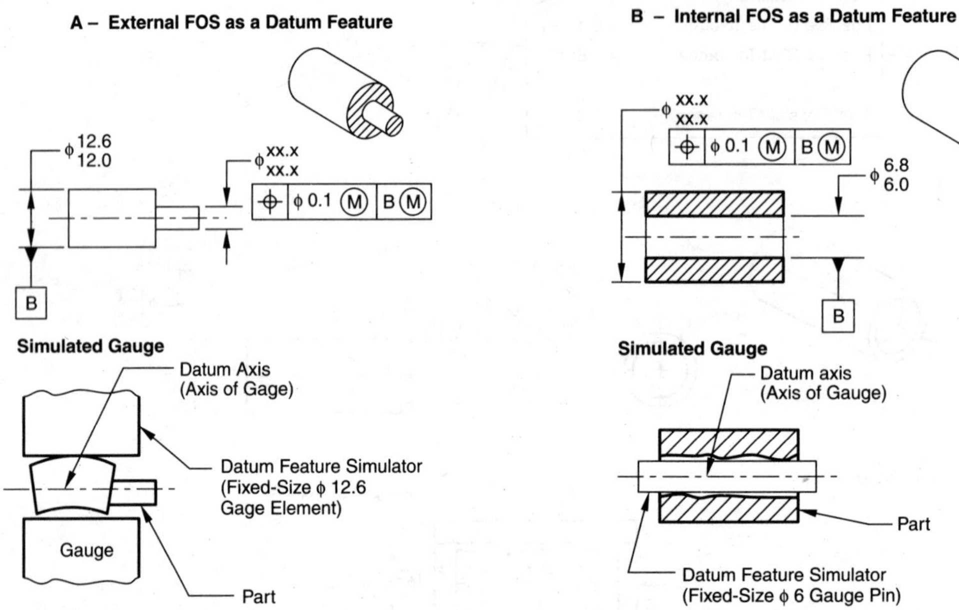

Fig. 14.112. Datum axis MMC primary

- The part should have at least three-point contact with the primary datum plane.
- The datum feature simulator will be a fixed size. Here, the worst-case boundary of datum feature *B*.
- The datum axis is the axis of the datum feature simulator.
- The datum axis will be perpendicular to the primary datum plane.

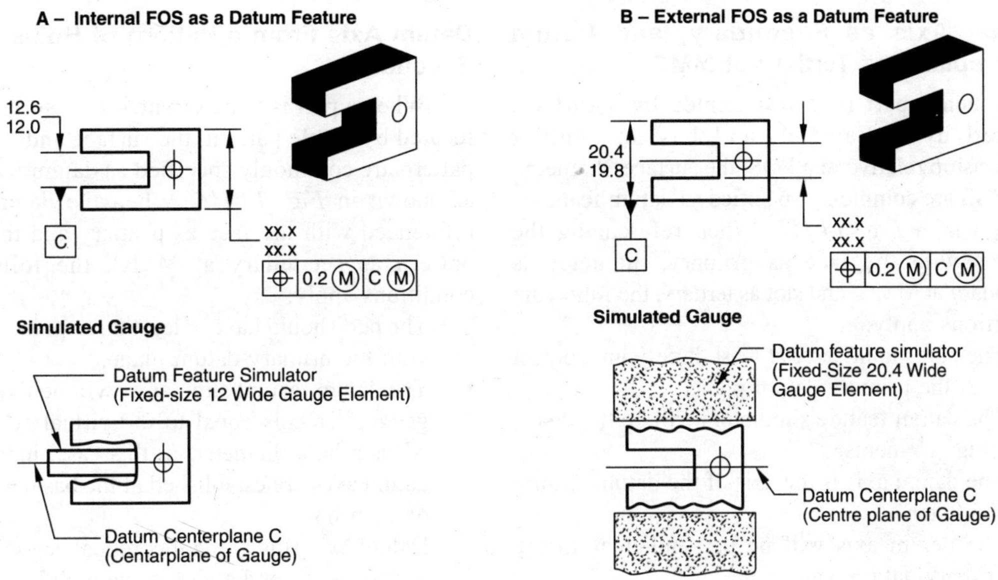

Fig. 14.113. Datum centre-plane MMC pirmary

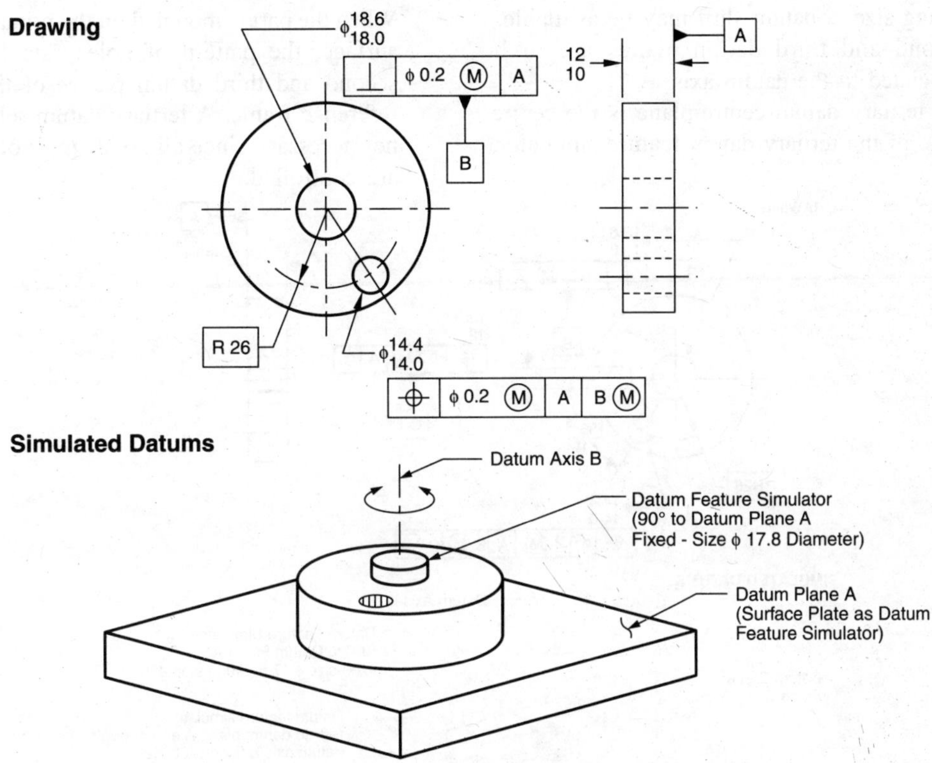

Fig. 14.114. Datum centre-plane MMC secondary

- Depending upon the datum feature's actual mating size, a datum shift may be available.

If angular relationship is to be specified, to assure proper relationship of part features, a tertiary datum becomes necessary.

Datum Axis as Secondary, and Datum Centreplane as Tertiary at MMC

When a part is to be oriented by a surface, located by a diameter, and has an angular relationship relative to a FOS, the surface, diameter, and FOS are commonly specified as datum features, as shown in *Fig. 14.115*. When referencing the datums with the face as primary, diameter as secondary at MMC, and slot as tertiary, the following conditions apply:

- The part should have at least three-point contact with the primary datum plane.
- The datum feature simulators will be fixed size gauge elements.
- The datum axis is the axis of the datum feature simulator.
- The datum axis will be perpendicular to the primary datum plane.
- Depending upon the datum feature's actual mating size, a datum shift may be available.
- Second and third datum planes are to be associated to the datum axis.
- The tertiary datum centre-plane is the centre-plane of the tertiary datum feature simulator.

Datum Axis From a Pattern of Holes, MMC Secondary

When a part is to be oriented by a surface and located by a hole pattern, the surface and the hole pattern are commonly specified as datum features, as shown in *Fig. 14.116*. When the datums are referenced with the face as primary and the hole pattern as secondary at MMC, the following conditions apply:

- The part should have at least three-point contact with the primary datum plane.
- The datum feature simulators will be fixed size gauge elements equal to the virtual condition of each hole diameter. In this case, individual datum axes are established at the basic location of each hole.
- Datum axis B is at the theoretical centre of the pattern. A second and third datum planes exist from the datum axis.
- When the part is mounted on the primary datum surface, the pattern of holes establishes the second and third datum planes of the datum reference frame. A tertiary datum reference is not necessary since all six degrees of freedom are controlled.

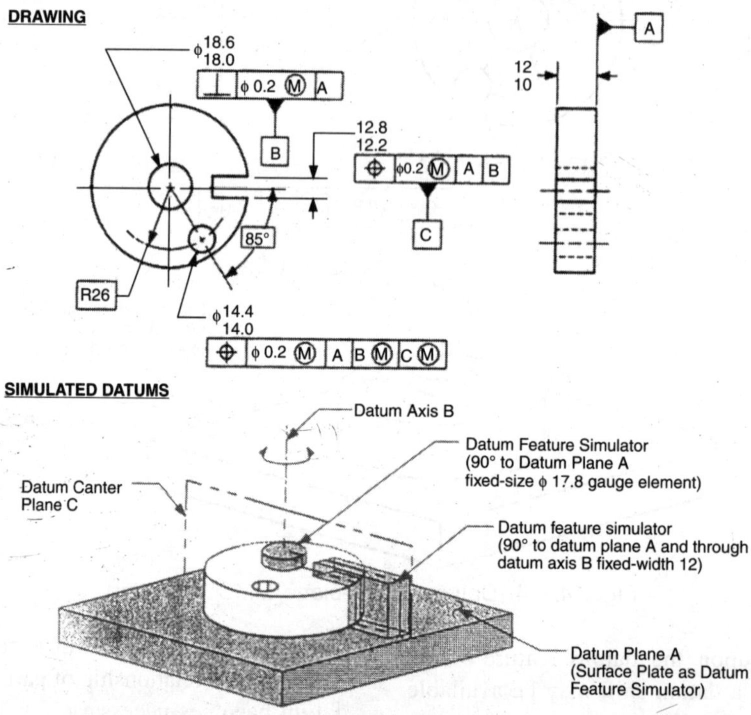

Fig. 14.115. Datum axis MMC secondary, datum centre-plane tertiary

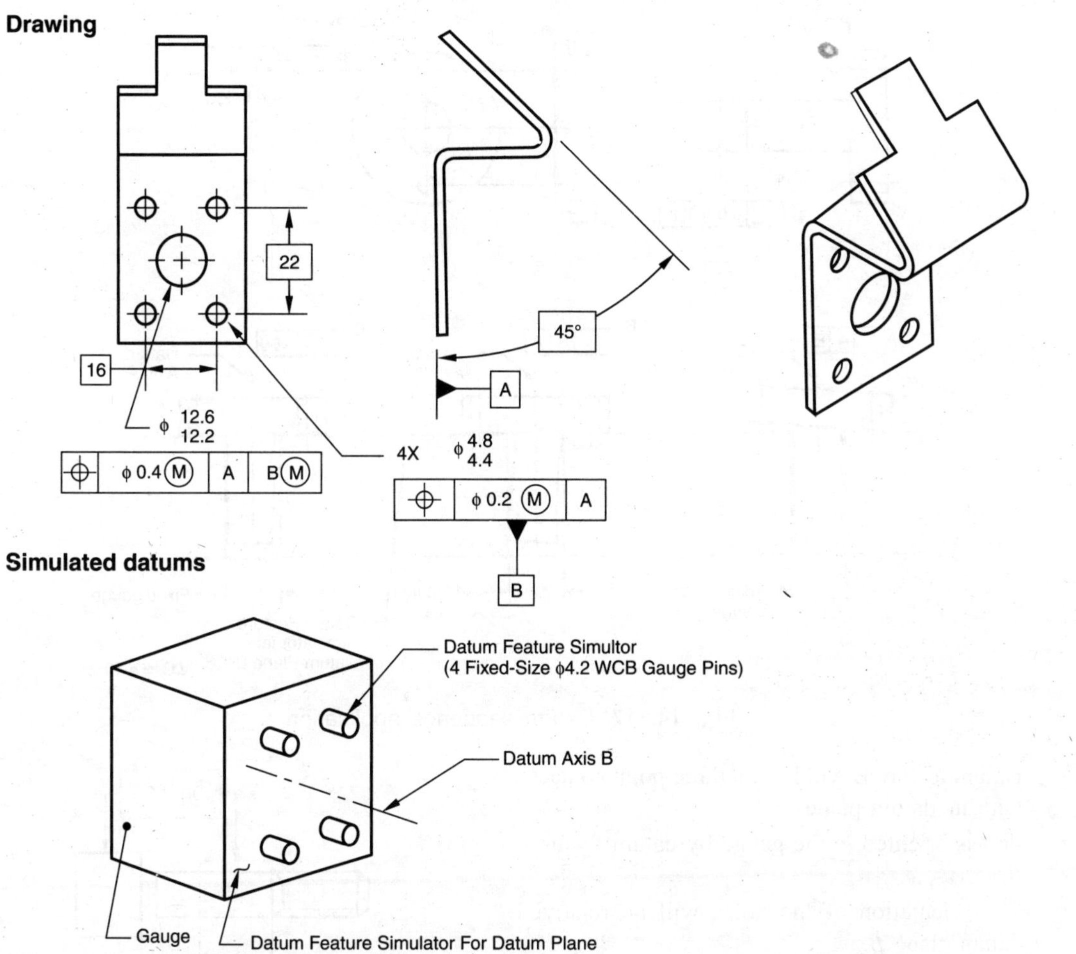

Fig. 14.116. Datum axis from a pattern of Holes MMC secondary

- Depending upon the actual mating size and actual location and orientation of the datum features, datum shift may be available.

When using a pattern of holes as a datum feature, the spacing between the holes and the square ness of the holes needs to be specified. This is often done, by using a tolerance of position control with a single datum reference.

The holes must be referenced at MMC in feature control frames that reference them.

14.32. DATUM SEQUENCE

When interpreting datum reference frames that involve both features of size datum features and planar datum features, the datum sequence plays an important role in the final tolerances for the part being dimensioned, as shown in *Fig. 14.117*. Here, the datum reference frame is shown completed in three different possible sequences.

At (*A*), the datum sequence used is A primary and *B* secondary at RFS. For this case:

- An adjustable gauge is required.
- No datum shift is allowed on datum feature A.
- Part is oriented in the gauge by datum feature *A*.
- The datum feature *B* will have a minimum of one point of contact with its datum feature simulator.
- The orientation of the holes will be relative to datum axis *A*.

At (*B*), the datum sequence is *B* primary and A secondary at RFS. For this case:

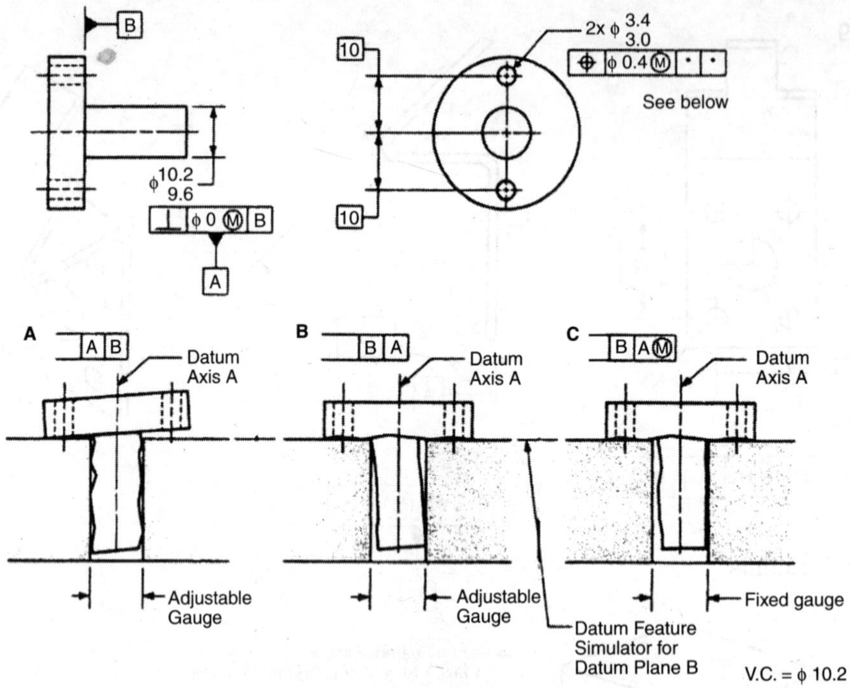

Fig. 14.117. Datum sequence application

- Datum feature B will have a three-point contact with its datum plane.
- Part is oriented in the gauge by datum feature *B*.
- The orientation of the holes will be relative datum plane *B*.
- An adjustable gauge is required, and no datum; shift is permissible on datum feature *A*.

At (*C*), the datum sequence is *B* primary and A secondary at MMC. For this case:

- When simulating datum *A*, a fixed gauge is allowed, and the gauge will allow datum shift.
- Datum feature *B* will have a three-point contact with its datum plane.
- Part is oriented in the gauge by datum feature *B*.
- Orientation of the holes will be relative to datum plane *B*.

14.33. RESTRICTIVE SPECIFICATIONS

Figure 14.118 shows a tolerance applied to a restricted part of a component.

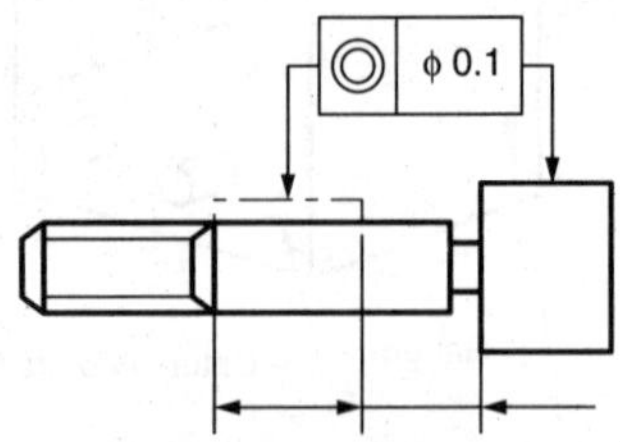

Fig. 14.118.

14.34. THEORETICALLY EXACT DIMENSIONS

These dimensions are identified by enclosure in a rectangular box, e.g.,

60 EQUISPACED 45° ϕ40

and are commonly known as "Boxed Dimensions". They define the true position of a hole, slot, boss, profile, etc.

Boxed dimensions are never individually toleranced but are always accompanied by a positional or zone tolerance specified within the tolerance frame referring to the feature, *Fig. 14.119 (a), (b)*.

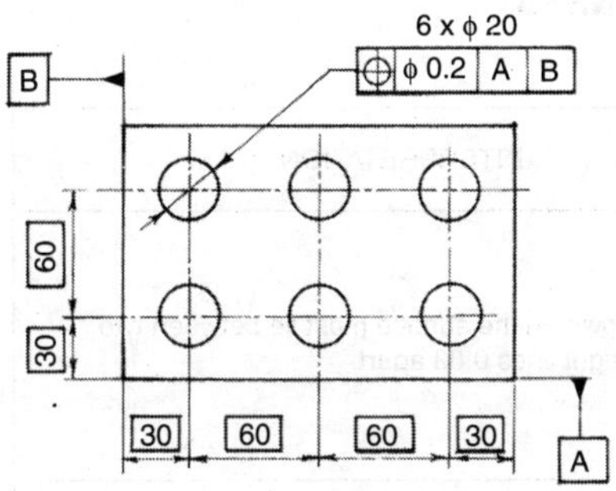

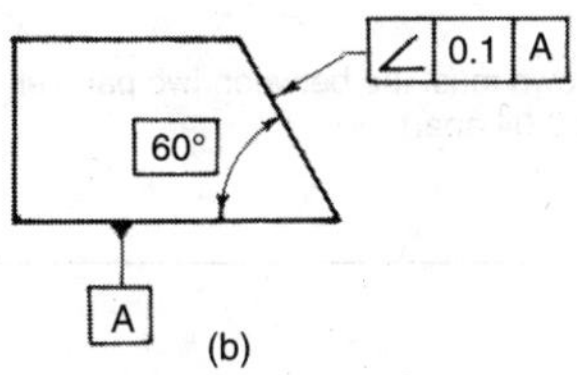

Fig. 14.119.

14.35. INTERPRETATIONS OF GEOMETRIC CONTROLS

1. Straightness

Straightness tolerance controls:

(*a*) The Straightness of a line on a surface.

(*b*) The Straightness of a line in a single plane.

(*c*) The Straightness of a line that is the axis of a feature(*s*).

The tolerance zone is the area between two parallel straight lines in the indicated plane containing the line or axis in the case of (a) and (*b*), and it is a cylinder, unless controlled in two planes, for (*c*).

The tolerance value is the specified distance between two parallel straight lines or the diameter of the cylinder, as shown in *Table 14.3.*

2. Flatness

Flatness tolerance controls the deviation of a surface from a true plane. It is the space between two parallel planes. Flatness is an unrelated characteristic and does not control the parallelism or squareness of a surface in relation to other features, *Table 14.4.*

3. Circularity (Roundness)

It is the condition where a feature is a continuous curved surface, any point of its surface being a constant distance from its centre or axis. Roundness can only be truly measured by a roundness-measuring instrument. The symbol for roundness controls the **circularity and sphericity.** Where sphericity is intended the word 'sphere* is shown above the tolerance, *Table 14.5.*

A roundness tolerance zone is the annular space between two co-planar circles concentric with each other. The tolerance value is equal to the distance apart the two co-planar circles and controls the errors of form of the profile between them. For a cylinder or cone the plane is perpendicular to the axis and for a sphere the plane intersects the maximum diameter, unless otherwise stated.

A roundness tolerance is an unrelated characteristic, each cross-section must be considered independently.

4. Cylindricity

Cylindricity is a combination of roundness, Straightness, and parallelism applied to the surface of a cylinder. Cylindricity tolerance controls the roundness, Straightness and parallelism of a surface. The tolerance zone is the angular space between two cylinders coaxial with each other. Its value is the radial distance between the cylinders, *Table 14.6.* The checking of cylindricity presents considerable difficulties. It is recommended that the individual characteristics, which comprise cylindricity, are toleranced separately.

5. Concentricity

Concentricity is the relation between two cylinders or circles that have a common centre or axis. Concentricity tolerance controls the deviation of the centre or axis from its true position, i.e., the axis or axis of the datum feature. The tolerance zone is the area contained by a circle or a cylinder within which the deviation lies. Concentricity tolerance is a particular case of positional tolerance in which the toleranced feature and the datum feature are circles or cylinders. The tolerance value is the diameter of the tolerance zone, *Table 14.7.*

6. Squareness

It is that condition where a line, plane or surface lies at 90° to another line, plane or surface. The tolerance zone is the space between two parallel lines or surfaces; or it can be the space contained

Table 14.3. STRAIGHTNESS

INDICATION ON DRAWING	TOLERANCE ZONE	INTERPRETATION
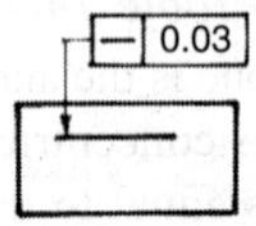	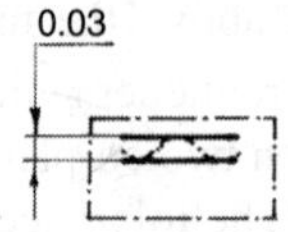	The line shown on the surface must lie between two parallel straight lines 0.03 apart.
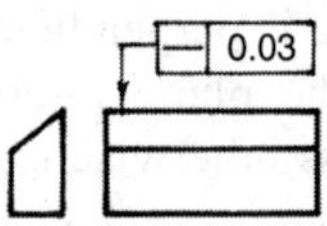	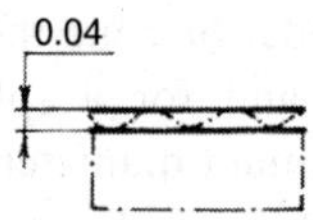	The edge shown must be between two parallel straight lines 0.04 apart.
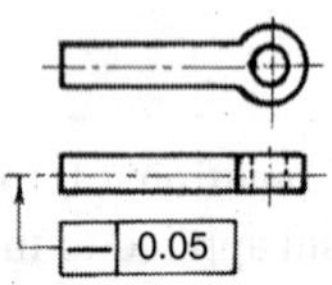	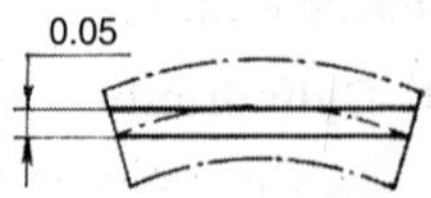	The axis of the whole part must lie between two parallel straight lines 0.05 apart.
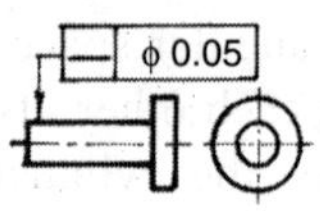	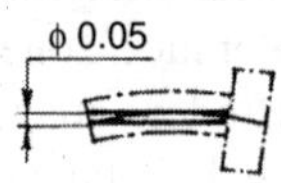	The axis of the dimensioned cylinder must lie in a tolerance zone 0.05 diameter
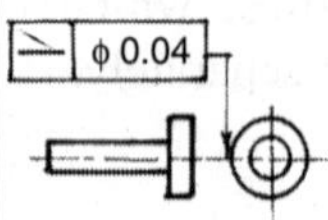	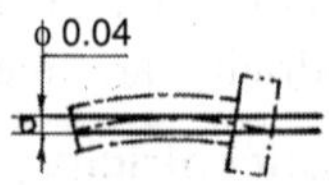	The axis of the whole part must lie in a cylinderical tolerance zone of 0.04 diameter.
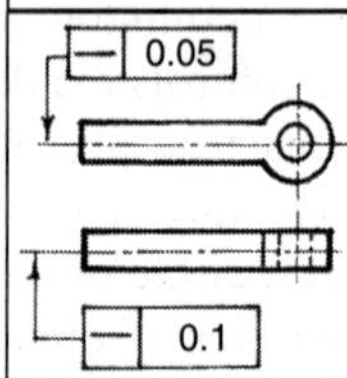	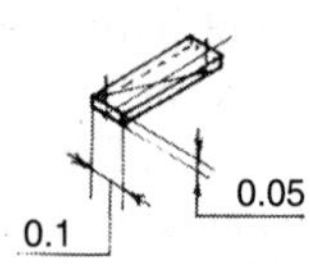	The axis of the whole part must lie in a boxed zone 0.4 wide in the horizontal plane and 0.05 wide in the vertical plane.

GEOMETRIC TOLERANCE	EXAMPLE AS SHOWN ON DRAWING	MEANING TOL. ZONE SHAPE
— STRAIGHTNESS (OF AN AXIS)	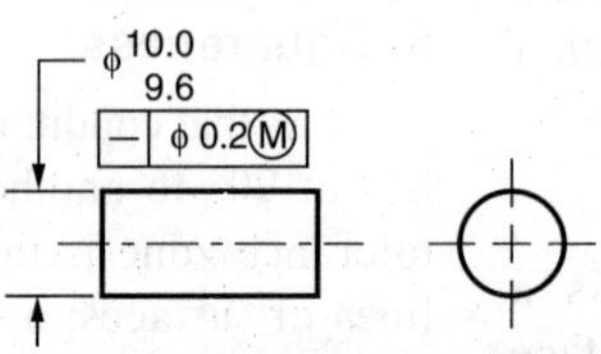	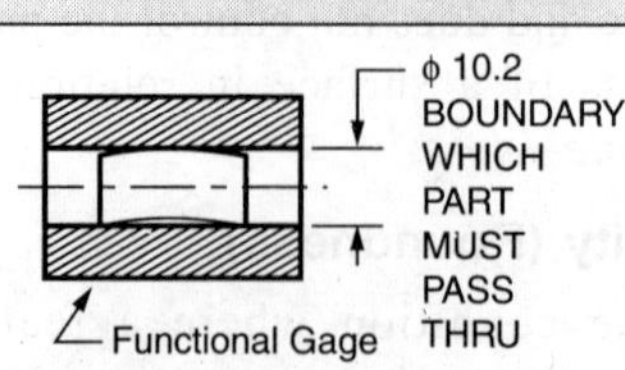

DATUMS USED	MMC OR RFS MODIFIERS	BONUS TOL.	OVERALL BOUNDARY	COMMENTS
NO	Ⓜ SHOWN IN EXAMPLE OTHERWISE RFS IMPLIED	YES WHEN Ⓜ IS USED	AFFECTED 10.0 + 0.2 = 10.2	• PART MUST ALSO BE WITHIN SIZE LIMITS • TOL. VALUE MAY BE GREATER THAN SIZE LIMITS

GEOMETRIC TOLERANCE	EXAMPLE AS SHOWN ON DRAWING	MEANING TOL. ZONE SHAPE
—— STRAIGHTNESS (OF A SURFACE ELEMENT)	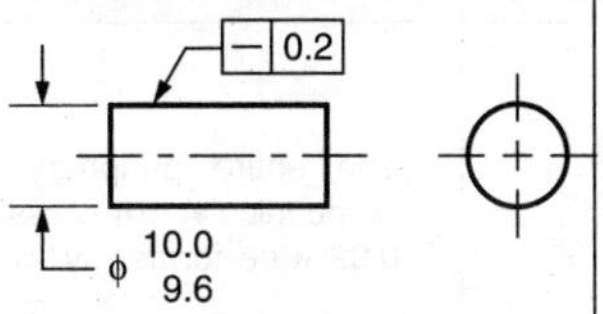	

DATUMS USED	MMC OR RFS MODIFIERS	BONUS TOL.	OVERALL BOUNDARY	COMMENTS
NO	NO ALWAYS RFS (IMPLIED)	NO	NO AFFECTED	• TOL. VALUE MUST BE LESS THAN THE SIZE TOL. • TOL. APPLIES TO EACH LINE ELEMENT

Table 14.4. FLATNESS

INDICATION ON DRAWING	TOLERANCE ZONE	INTERPRETATION
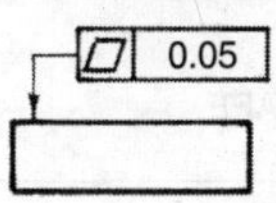	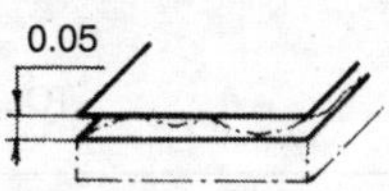	*Example* *General case* The surface must lie between two parallel planes 0.5 apart.
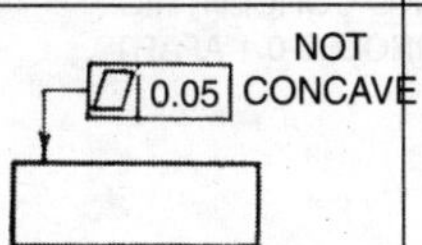	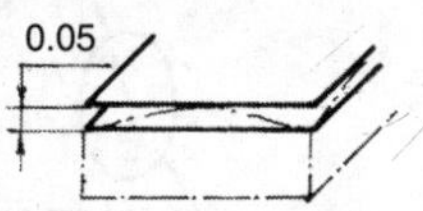	*Indications qualifying the feature* The surface must lie between two parallel surfaces 0.05 apart and must not be concave. **Note:** If convexity is not required an appropriate note should be added to the tolerance frame.

GEOMETRIC TOLERANCE	EXAMPLE AS SHOWN ON DRAWING	MEANING TOL. ZONE SHAPE
▱ FLATNESS	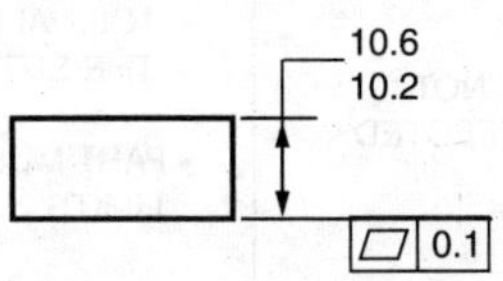	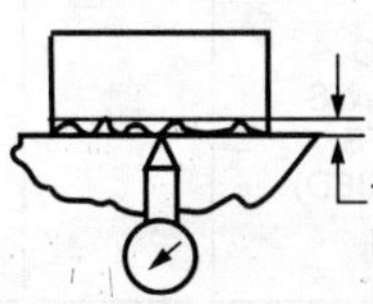

DATUMS USED	MMC OR RFS MODIFIERS	BONUS TOL.	OVERALL BOUNDARY	COMMENTS
NO	NO ALWAYS RFS (IMPLIED)	NO	NOT AFFECTED	• TOL. VALUE MUST BE LESS THAN THE SIZE TOL. • PART MUST ALSO BE WITHIN SIZE LIMITS

Table 14.5. CIRCULAIRY

INDICATION ON DRAWING	TOLERANCE ZONE	INTERPRETATION
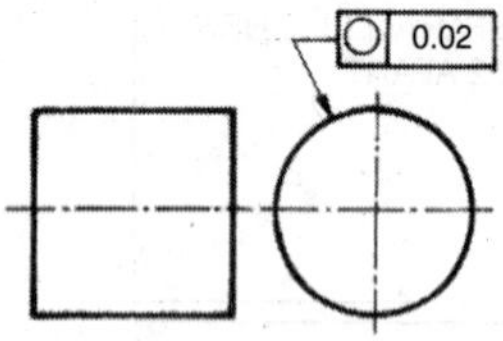	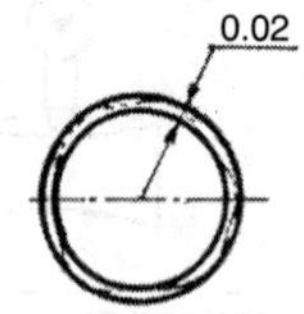	The entire periphery at the considered cross-section perpendicular to the axis must lie in an annular tolerance zone 0.02 wide, formed by two concentric circles.
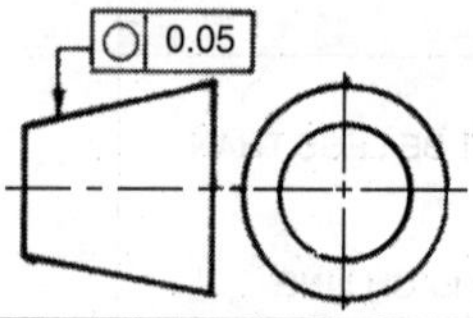	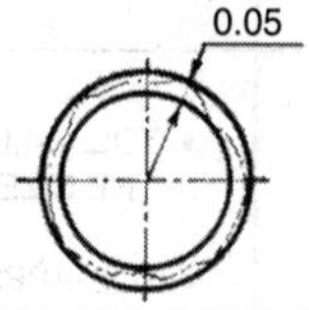	The entire periphery at the considered cross-section perpendicular to the axis must lie in an annular tolerance zone 0.05 wide, formed by two concentric circles.
SPHERE 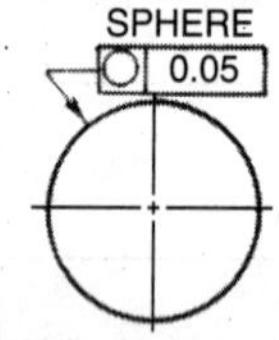	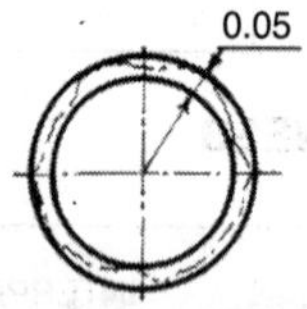	The entire periphery at any section of maximum diameter must lie in two concentric circles. a radial distance 0.01 apart in the plane of the section.

GEOMETRIC TOLERANCE	EXAMPLE AS SHOWN ON DRAWING	MEANING TOL. ZONE SHAPE
○ CIRCULARITY	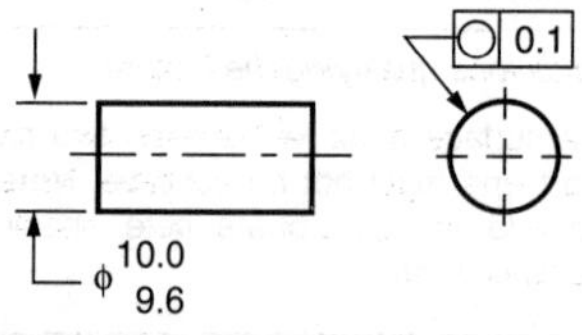	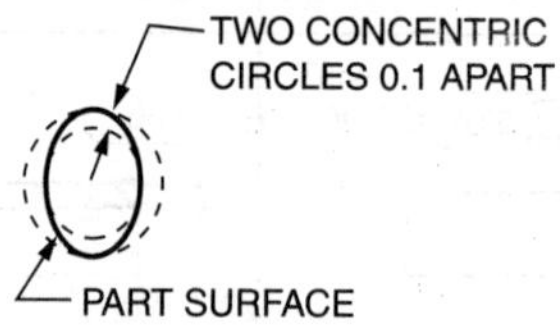

DATUMS USED	MMC OR RFS MODIFIERS	BONUS TOL.	OVERALL BOUNDARY	COMMENTS
NO	NO ALWAYS RFS (IMPLIED)	NO	NOT AFFECTED	• TOL. VALUE MUST BE LESS THAN THE SIZE TOL. • PART MUST ALSO BE WITHIN SIZE LIMITS

Table 14.6. CYLINDERICITY

INDICATION ON DRAWING	TOLERANCE ZONE	INTERPRETATION
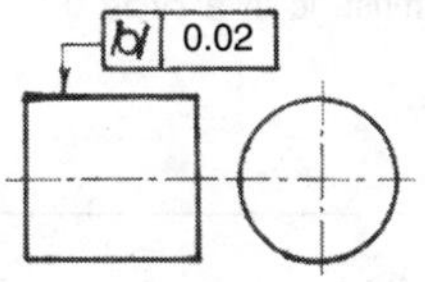	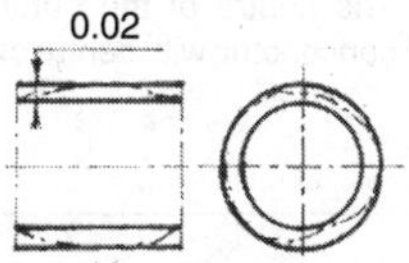	The entrie periphery of the part must lie in an annular tolerance zone 0.02 wide, formed by two concentric cylinders.

GEOMETRIC TOLERANCE	EXAMPLE AS SHOWN ON DRAWING	MEANING TOL. ZONE SHAPE
 CYLINDRICITY	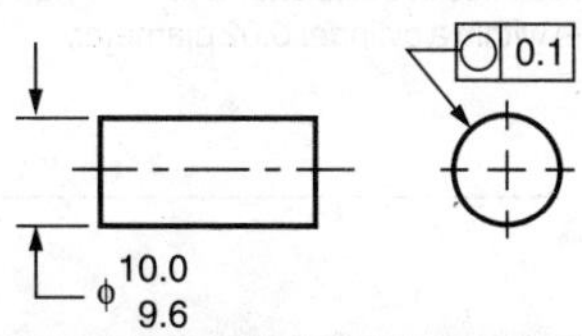	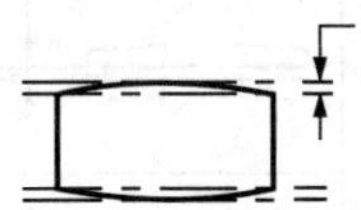

DATUMS USED	MMC OR RFS MODIFIERS	BONUS TOL.	OVERALL BOUNDARY	COMMENTS
NO	NO ALWAYS RFS (IMPLIED)	NO	NOT AFFECTED	• TOL. VALUE MUST BE LESS THAN THE SIZE TOL. • PART MUST ALSO BE WITHIN SIZE LIMITS

within a cylinder. Tolerance zones are perpendicular to datum features, *Table 14.8.* Carefully note the significance of the leader line in relation to the feature it controls.

7. Parallelism

It is the condition whereby two parallel lines or surfaces are, always, separated by a uniform distance. Parallelism tolerance controls the parallelism of two lines or surfaces and the tolerance zone is the distance between them, *Table 14.9.* It is important to note the position of the leader line in relation to the feature it controls.

8. Angularity

It defines a position between two lines, planes or surfaces that are not parallel or perpendicular to each other, *Table 14.10.* Angularity tolerance controls this relationship. The tolerance zone is the area between two parallel lines, or planes or the space within a cylinder of diameter equal to the tolerance value, which is at a specified angle to the datum feature. In the case of a circular tolerance zone the tolerance value is preceded by the symbol ϕ.

The tolerance value is the specified distance between the lines, or planes, or the diameter of the cylinder.

9. Position

A positional tolerance controls the location of one feature from another or from a datum, *Table 14.11.* True position also imply parallelism and squareness of the tolerance zones. The tolerance value is the specified deviation from the true position.

Fig. 14.7. CONCENTRICITY

INDICATION ON DRAWING	TOLERANCE ZONE	INTERPRETATION
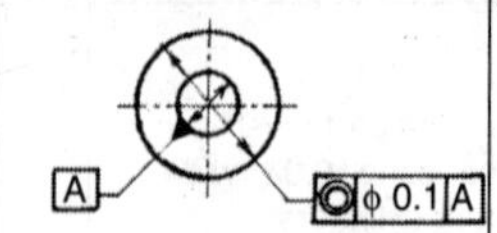	CIRCLE φ 0.1 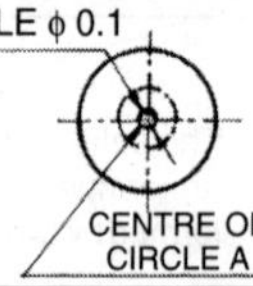 CENTRE OF CIRCLE A	The centre of the outer circle must lie in a circle 0.1 diameter concenctric with centre of circle.
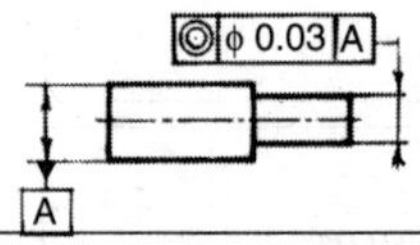	CYLINDER 0.03 DIA 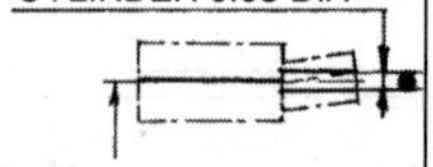	The axis of the right hand cylinderical portion must lie in a cylinder 0.03 diameter coaxial with the left hand portion.
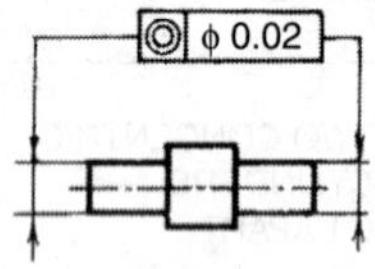	CYLINDER 0.02 DIA. 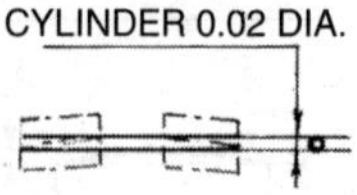	The axes of the left hand and the right hand cylindrical portions must lie within a cylinder 0.02 diameter.
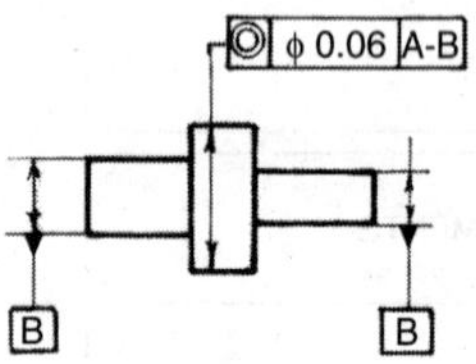	CYLINDER 0.06 DIA. 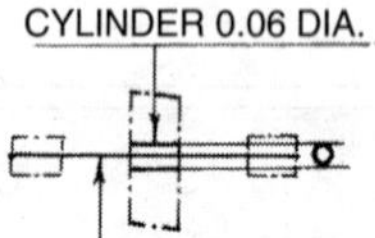 DATUM AXIS (MEAN AXIS OF A-B)	The axis of the central portion must lie within a cylinder 0.06 diameter coaxial with the mean axis of the end projections.

GEOMETRIC TOLERANCE	EXAMPLE AS SHOWN ON DRAWING	MEANING TOL. ZONE SHAPE
 CONCENTRICITY	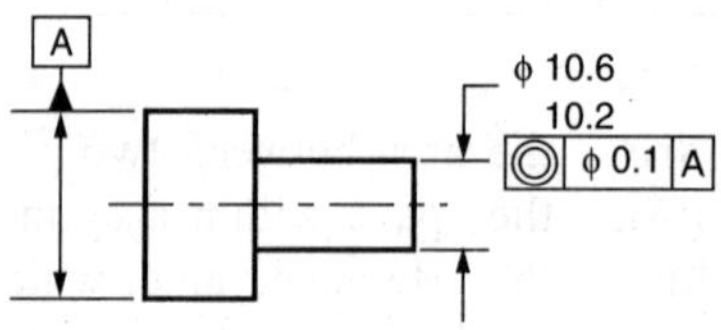	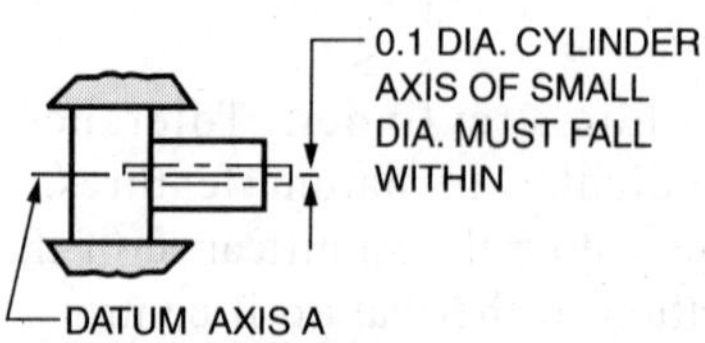

DATUMS USED	MMC OR RFS MODIFIERS	BONUS TOL.	OVERALL BOUNDARY	COMMENTS
YES	NO ALWAYS RFS (IMPLIED)	NO	AFFECTED 10.6 + 0.1 = 10.7	• Involves a complex analysis of the surface to determine the axis location (It is not a direct gate reading) • Should only be used in high speed rotating

10. Symmetry

It is the condition where a feature is divided into two identical parts by a line or plane. Symmetry tolerance controls the area between two parallel lines or planes that are parallel to the datum feature, and the value is the distance between them, *Table 14.12*. Symmetry also implies perpendicularity with the plane of the drawing.

Table. 14.8. SQUARENESS

INDICATION ON DRG.	TOLERANCE ZONE	INTERPRETATION
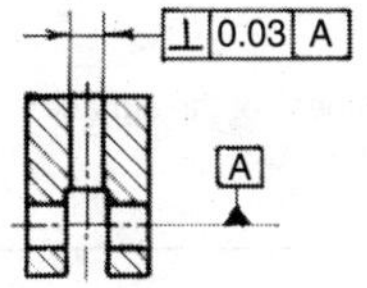	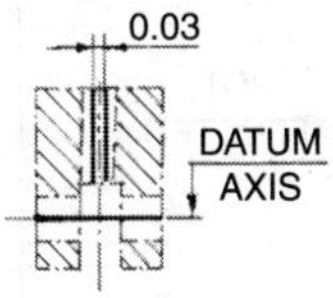	The axis of the vertical hole must lie between two planes 0.03 apart which are perpendicular to common axis of the two horizontal holes.
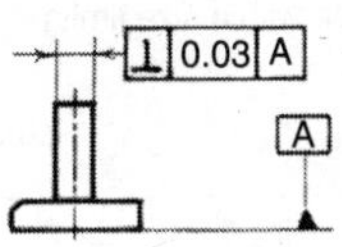	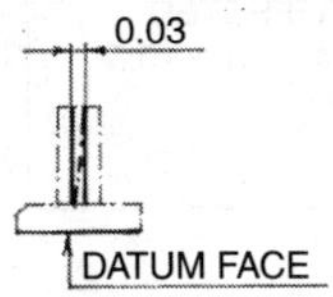	The axis of the pillar must be contained between two straight lines 0.03 apart perpendicular to the datum surface and in the plane shown.
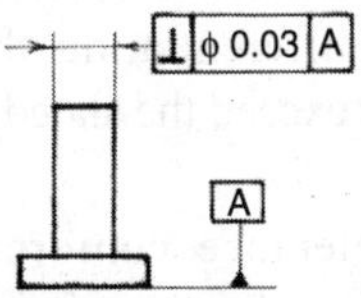	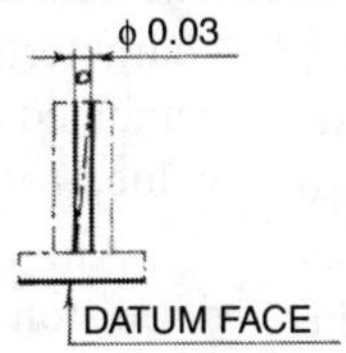	The axis of the pillar must be contained in a cylinder 0.03 diameter, the axis of which is perpendicular to the datum.
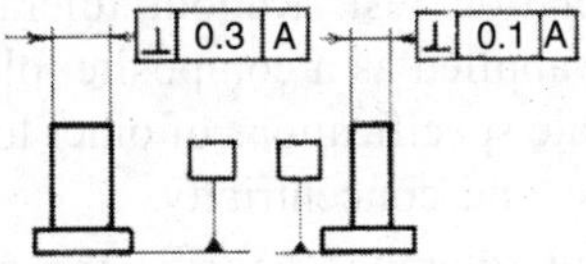		The axis of the pillar must lie in a boxed tolerance zone 0.1 by 0.3 which is perpendicular to the datum.
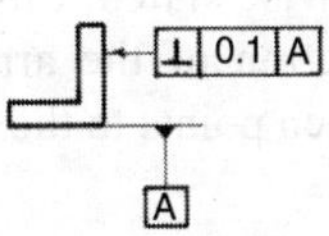	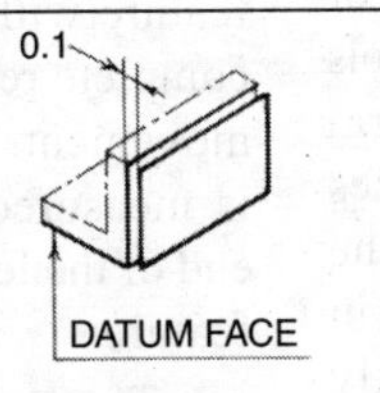	The right hand face must lie between two parallel planes 0.1 apart which are perpendicular to the datum.

GEOMETRIC TOLERANCE	EXAMPLE AS SHOWN ON DRAWING	MEANING TOL. ZONE SHAPE
⊥ PERPENDICULARITY (SURFACE TO SURFACE)	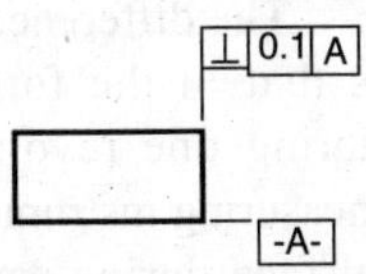	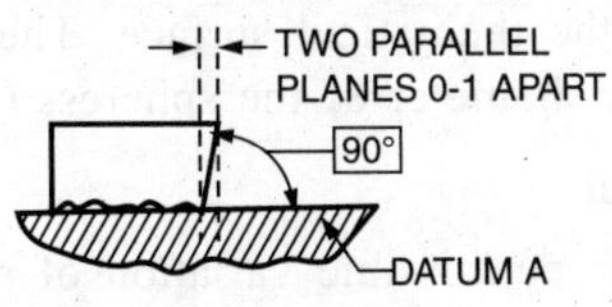
⊥ PERPENDICULARITY (Diameter TO SURFACE)	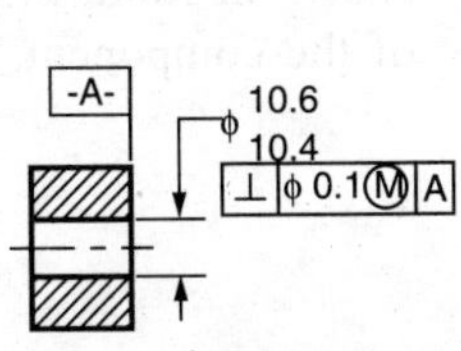	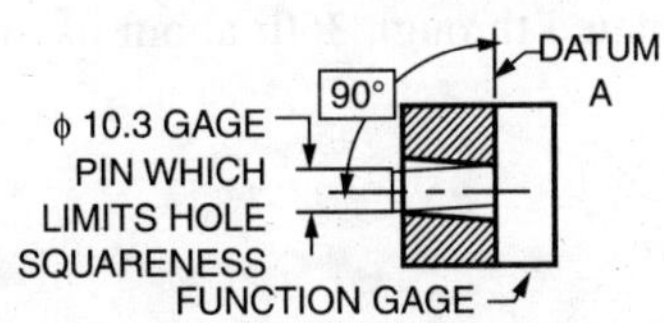

DATUMS USED	MMC OR RFS MODIFIERS	BONUS TOL.	OVERALL BOUNDARY	COMMENTS
YES	NO ALWAYS RFS (IMPLIED)	NO	NOT AFFECTED	• Also controls flatness of the surface.
YES	Ⓜ SHOWN IN EXAMPLE OTHERWISE	YES WHEN Ⓜ IS USED	AFFECTED 10.4 – 0.1 = 10.3	• Hole must also be within size limits.

11. Profile of a Line

A profile tolerance controls the ideal contour of the feature that is defined by boxed dimensions. The tolerance zone is assumed to be equally disposed about the true contour, and the value is equal to the diameter of circles that determine the zone, *Table 14.13 (a)*. If the controlled profile includes a sharp corner, then the inner boundary of the tolerance zone finishes at the intersection, and the outer boundary of the tolerance zone extends to their intersection, *Table 14.13 (b)*.

12. Profile of a Surface

Boxed dimensions define the theoretical or perfect form of the surface. Profile tolerance controls the deviation of a surface from its perfect form and the tolerance zone is the space two surfaces that envelope spheres of diameter equal to the tolerance value. The centres of the spheres lie in the theoretical surface thus giving an equally disposed bilateral tolerance. A unilateral tolerance zone may be specified in which case the surfaces of the spheres touch the theoretical surface. Both cases give a tolerance zone with a constant width normal to the theoretical surface. The tolerance value is the diameter of the spheres, *Fig. 14.23*.

13. Runout

It is the permissible variation of position of any point fixed on a surface, which occurs when the part is rotated through 360° about its own axis. The resultant indications include errors of other characteristics without differentiating them. The combined errors must not exceed the stated tolerance value, *Table 14.15*.

Although runout tolerance conforms to the definition of a geometric tolerance, it is in class apart from other geometric tolerances, demanding a practical relation test. Runout tolerance may sometimes be applied as a composite tolerance in place of separate specifications of other tolerances, e.g., roundness and concentricity.

The runout tolerance zone is the maximum permissible variation of position of the considered feature with respect to any point fixed during one complete revolution about the datum without axial movement. Unless otherwise stated, this variation is measured in the direction of the arrow at the end of the leader line, which points to the toleranced feature.

The tolerance value is the full indicator movement during one complete revolution at the fixed point without axial movement.

14. Total Runout

The difference between simple and total runout is that in the former, one measurement is taken during one revolution, while in total runout the measuring instrument is moved along the component surface during several revolutions. In both cases, shown in *Table 14.16*, there is no axial movement of the component.

Table 14.9. PARALLELISM

INDICATION ON DRG	TOLERANCE ZONE	INTERPRETATION
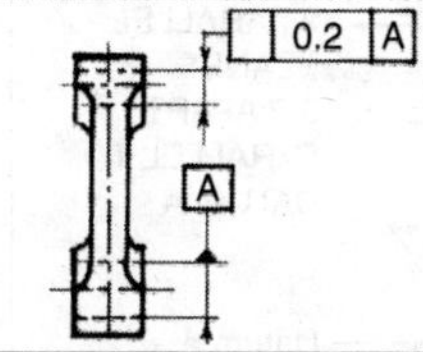	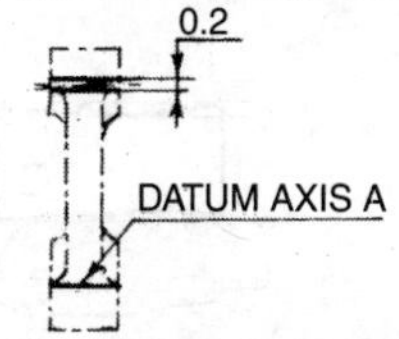	The upper axis must lie between two lines 0.2 apart which are parallel to the datum axis and lie in the vertical plane.
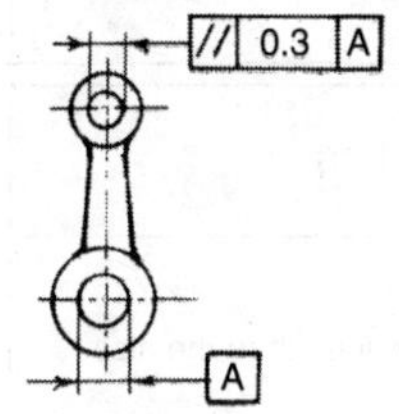	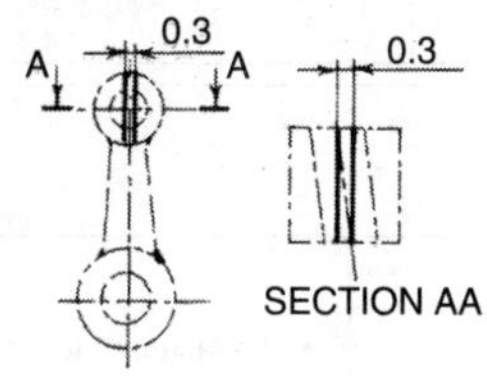	The upper axis must lie between two lines 0.3 apart which are parallel to the datum axis and lie in the horizontal plane.
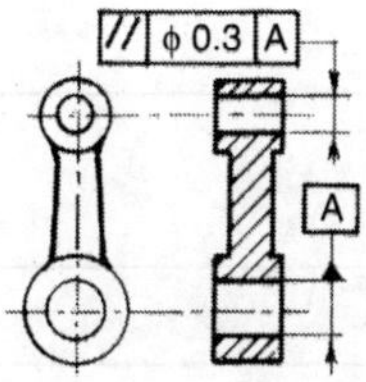	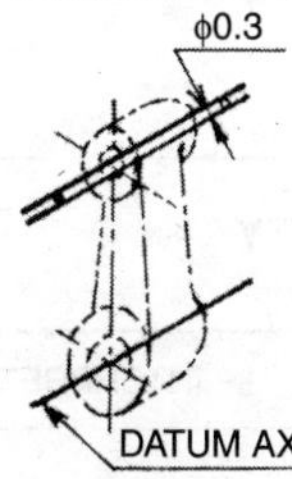	The upper axis must lie in a cylindrical tolerance zone 0.3 diameter with its axis parallel to the datum.
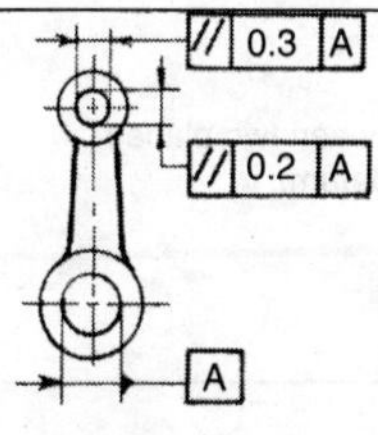	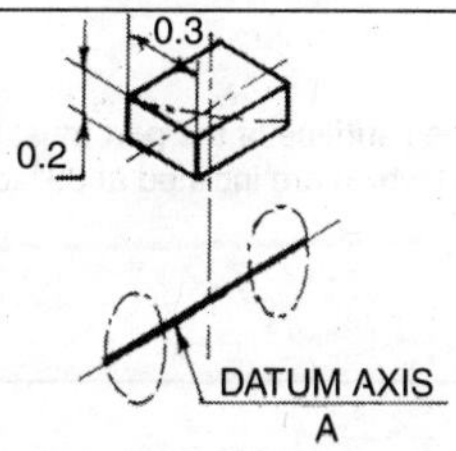	The upper axis is contained in a parallel-piped having a width of 0.3 in the horizontal plane and a width of 0.3 in the vertical plane, its side being parallel to the datum .
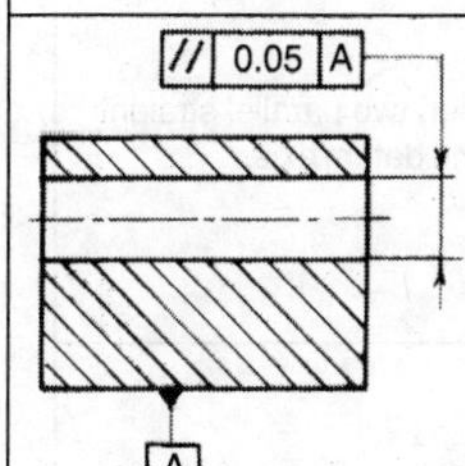	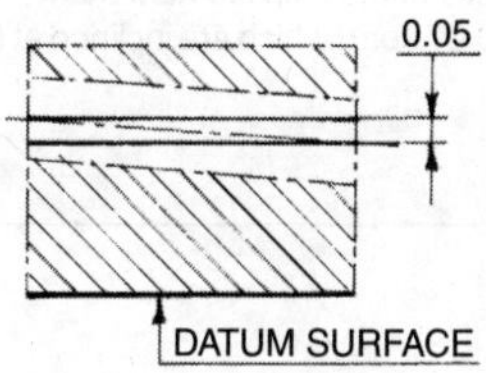	The axis of the hole must lie between two planes 0.05 apart parallel to the datum.
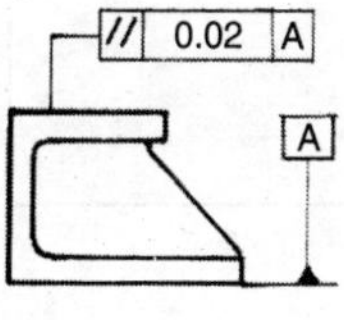	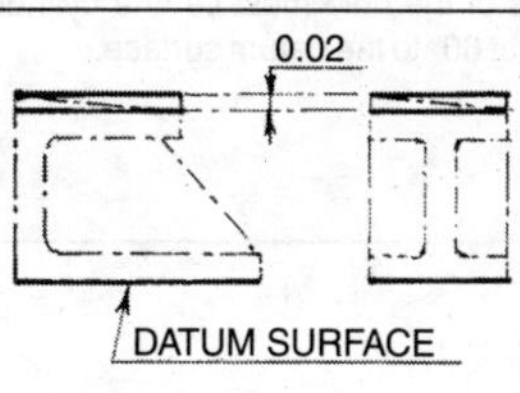	The upper surface of the part must lie between two planes 0.02 apart parallel to the datum

GEOMETRIC TOLERANCE	EXAMPLE AS SHOWN ON DRAWING	MEANING TOL. ZONE SHAPE
// PARALLELISM (SURFACE TO SURFACE)	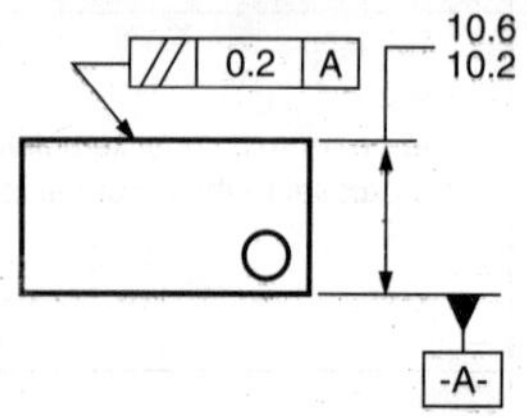	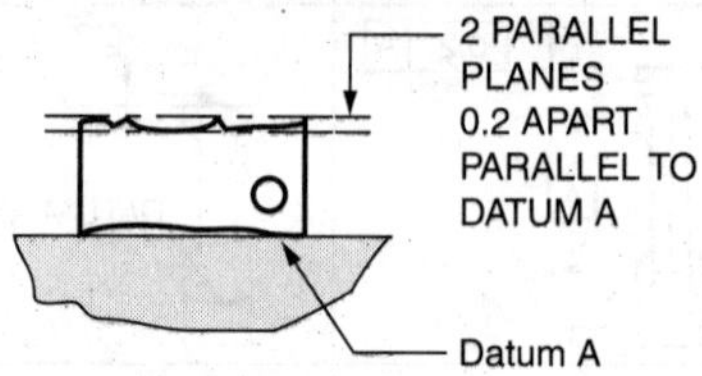

DATUMS USED	MMC OR RFS MODIFIERS	BONUS TOL.	OVERALL BOUNDARY	COMMENTS
YES	NO ALWAYS RFS (IMPLIED)	NO	NOT AFFECTED	• Parallelism tol. must be less than the size tol.

Table 14.10. ANGULARITY

INDICATION ON DRG.	TOLERANCE ZONE	INTERPRETATION
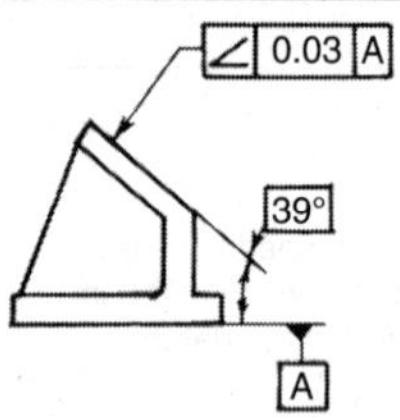	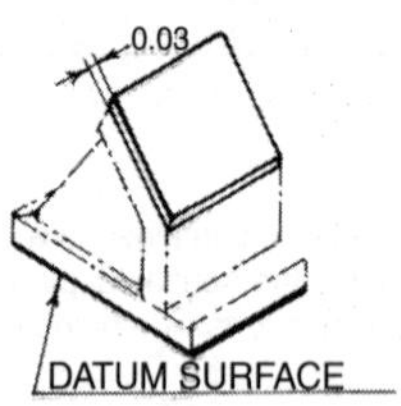	The inclined surface of the part must lie between two planes 0.03 apart which are inclined at 39° to the datum.
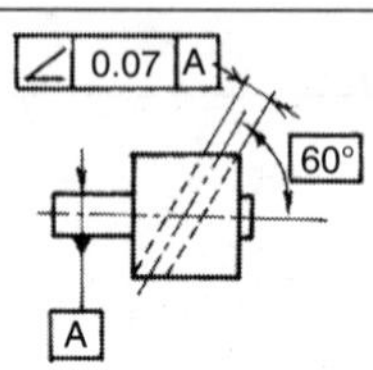	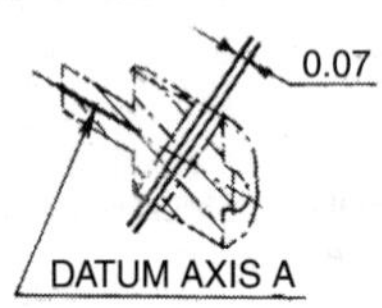	The axis of the inclined hole must lie between two parallel straight lines 0.07 apart which are inclined at 60° to the datum axis.
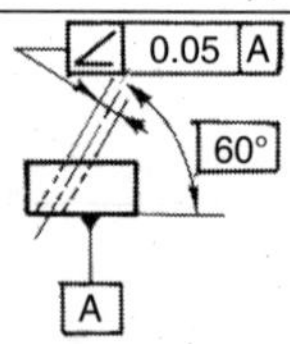	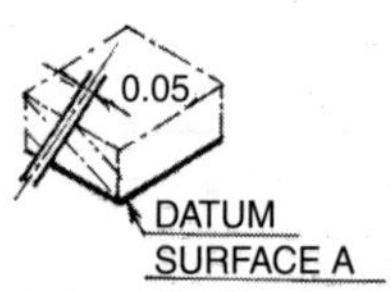	The axis of the hole must lie in a cylinder 0.05 diameter which is inclined at 60° to the datum surface.

GEOMETRIC TOLERANCE	EXAMPLE AS SHOWN ON DRAWING	MEANING TOL. ZONE SHAPE
∠ ANGULARITY	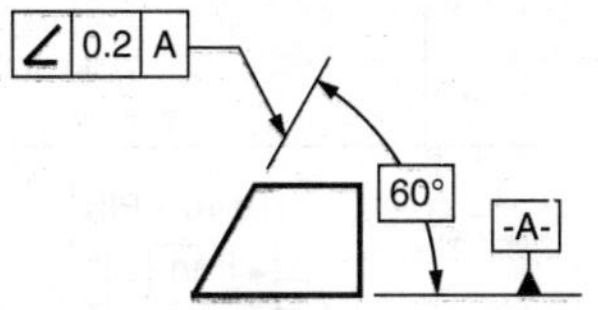	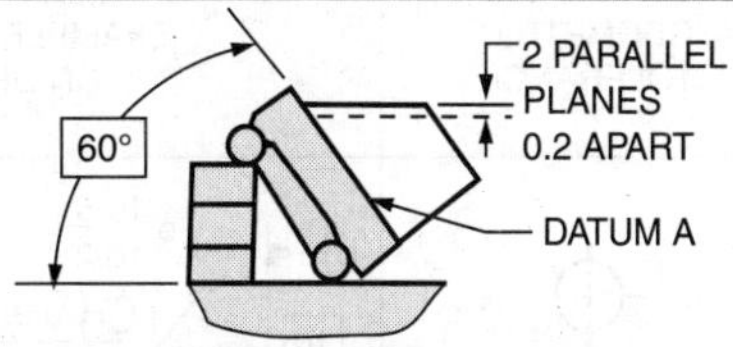

DATUMS USED	MMC OR RFS MODIFIERS	BONUS TOL.	OVERALL BOUNDARY	COMMENTS
YES	Ⓢ IMPLIED UNLESS Ⓜ IS SHOWN	YES WHEN Ⓜ IS USED	NOT AFFECTED	• A Basic dimension must be used from the tolerance feature to the datum referenced.

Table 14.11. POSITION

INDICATION ON DRG.	TOLERANCE ZONE	INTERPRETATION
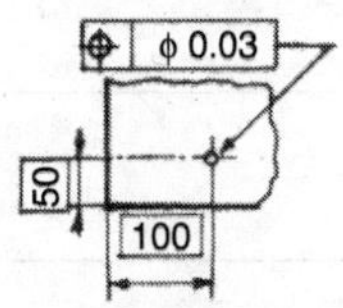		The point must lie in a circle 0.03 diameter in the plane of the surface which has its center at the specific point of intersection. If the point has to be located in three dimensions (instead of two dimensions as in the example) the tolerance will be sphere.
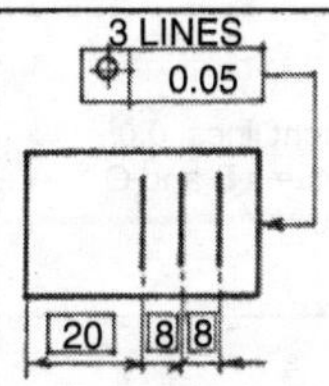	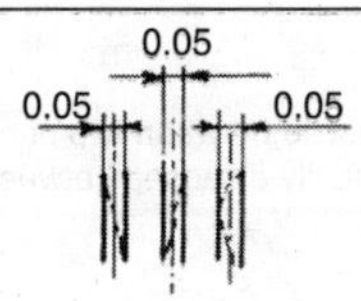	The position of each line lies between two parallel straight lines on the surface, 0.05 apart, which are symmetrically disposed about the specified true position of the axis of the hole.
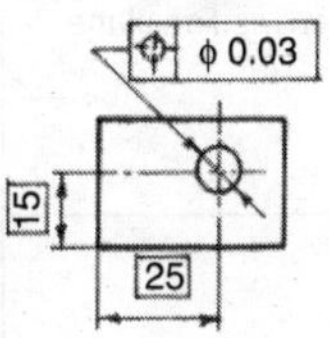	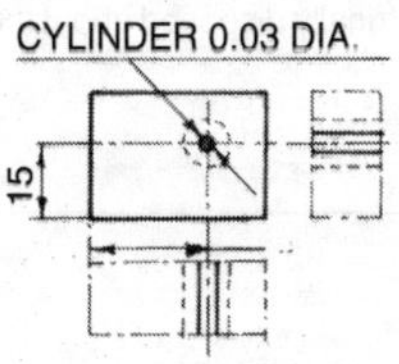	The axis of the hole must be contained in a cylinder 0.03 diameter with its axis in the specified true position of the axis of the hole.
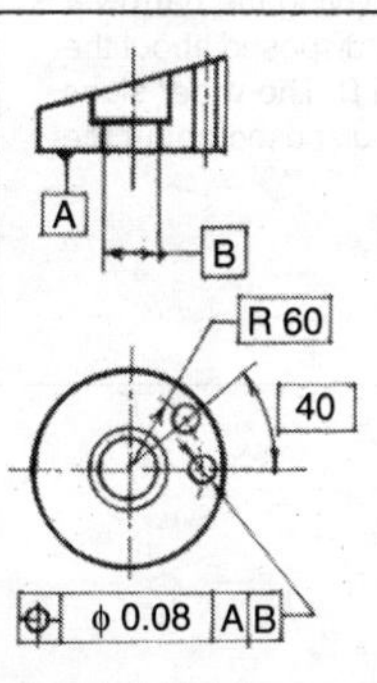	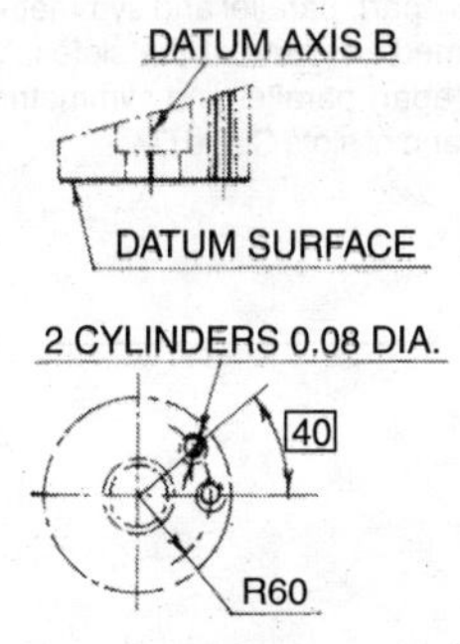	The axis of the two holes must lie in cylinders 0.08 diameter with their axes in the specified true positions of the holes in relation to datum surface A and the axis of the central hole B. T he positional tolerance also control the squareness of the two holes in relation of the datum surface A.

Table 14.11. POSITION

<table>
<tr><th>GEOMETRIC TOLERANCE</th><th>EXAMPLE AS SHOWN ON DRAWING</th><th>MEANING TOL. ZONE SHAPE</th></tr>
<tr>
<td>
POSITION</td>
<td>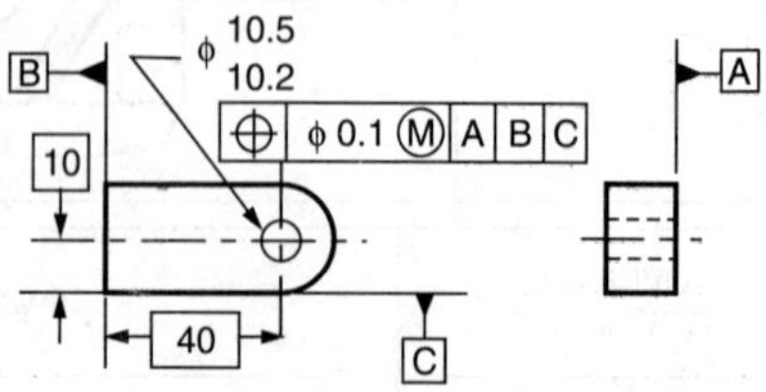</td>
<td>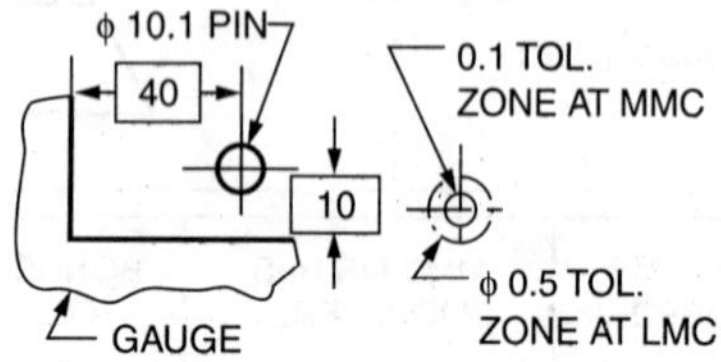</td>
</tr>
</table>

DATUMS USED	MMC OR RFS MODIFIERS	BONUS TOL.	OVERALL BOUNDARY	COMMENTS
YES	Ⓜ SHOWN IN EXAMPLE Ⓢ MAY BE SPECIFIED	YES WHEN Ⓜ IS USED	AFFECTED 10.2 – 0.1 = 10.1	• Allows round tolerance zones • Ⓜ is used when function of part feature is assembly • Basic dimensions must be used to establish the true position

Table 14.12. SYMMETRY

<table>
<tr><th>INDICATION ON DRAWING</th><th>TOLERANCE ZONE</th><th>INTERPRETATION</th></tr>
<tr>
<td>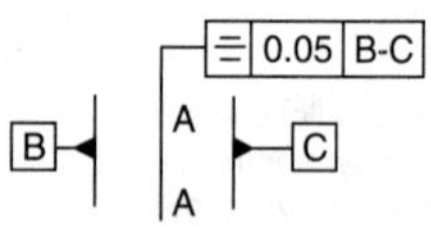</td>
<td>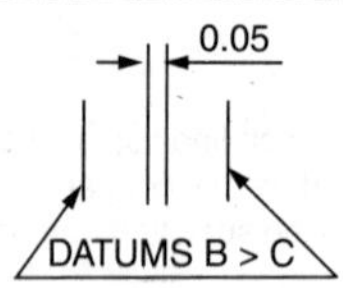</td>
<td>The line AA must lie between two parallel straight lines, 0.05 apart, symmetrically disposed between the datums B and C.</td>
</tr>
<tr>
<td>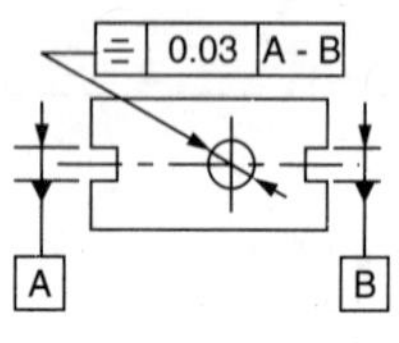</td>
<td>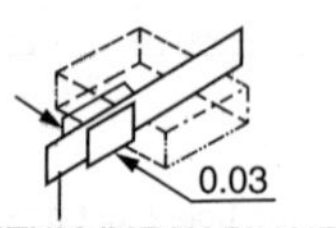DATUM (MEAN PLANE OF WIDTH AT B)</td>
<td>The axis of the hole must lie between two parallel planes. 0.03 apart, symmetrically disposed about the common median plane of the slots.</td>
</tr>
<tr>
<td>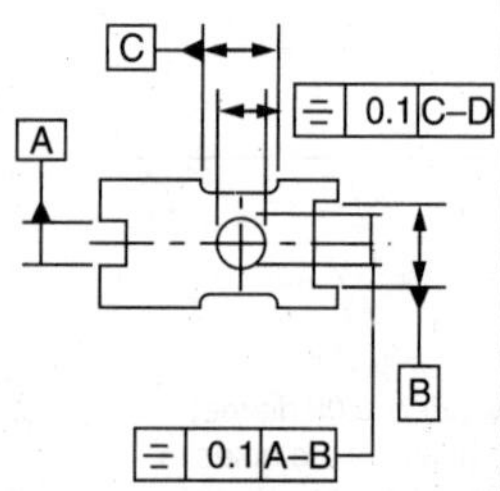</td>
<td>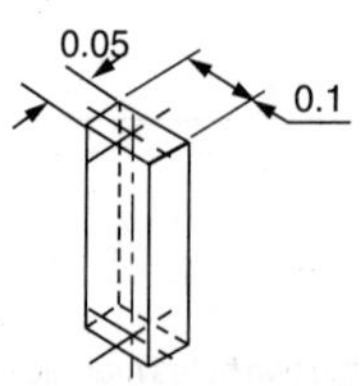</td>
<td>The axis of the hole must lie within a box having the narrower sides 0.05 apart, parallel and symmetrically disposed about the common median plane at the slotes A and B. The wider sides being 0.1 apart parallel and symmetrically disposed about the median plane of slots C and D.</td>
</tr>
</table>

Table 14.13. PROFILE OF A LINE

INDICATION ON DRAWING	TOLERANCE ZONE & INTERPRETATION
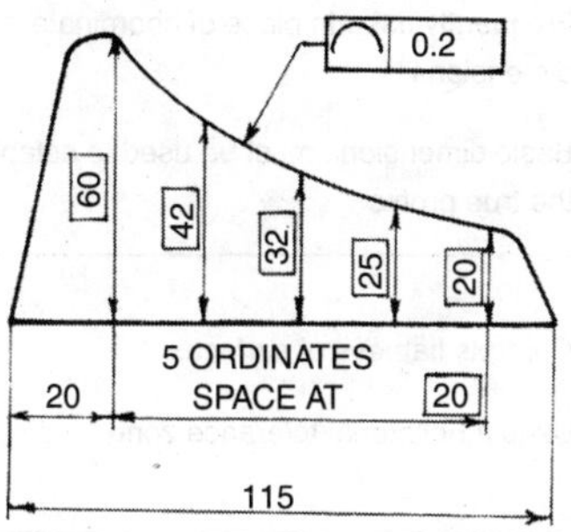	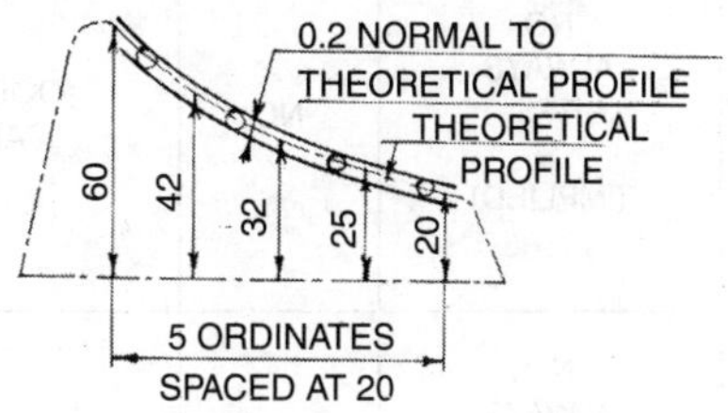 In any section parallel to the plane of projection of the drawing, the actual profile must lie between two lines which envelop a series of circles of 0.2 diameter with their centres on the theoretical profile.

Table 14.14. PROFILE OF A SURFACE

INDICATION ON DRAWING	TOLERANCE ZONE	INTERPRETATION
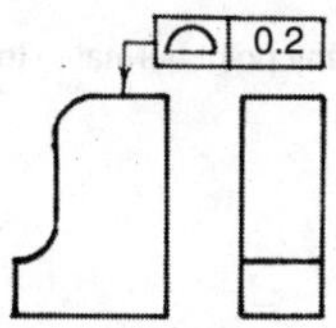	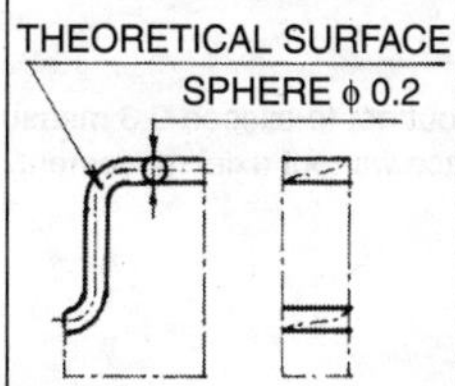	The curved surface of the part must lie between two surfaces which envelop a series of spheres of 0.2 diameter with their centres on the surface having the correct geometric shape.

GEOMETRIC TOLERANCE	EXAMPLE AS SHOWN ON DRAWING	MEANING TOL. ZONE SHAPE
⌓ PROFILE OF A SURFACE	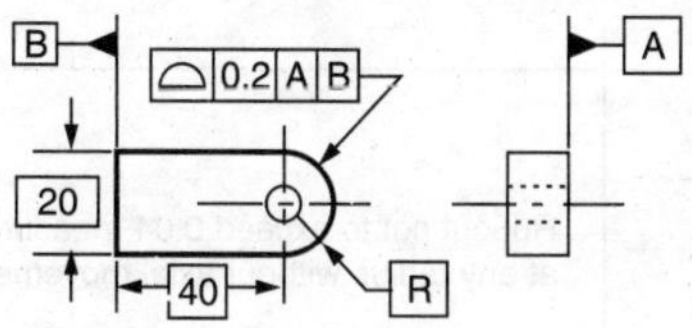	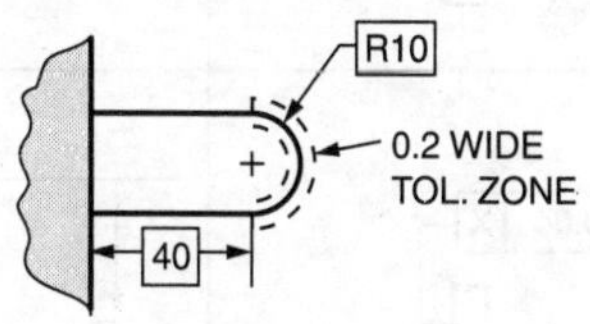
⌓ PROFILE OF A SURFACE (COPLANAR)	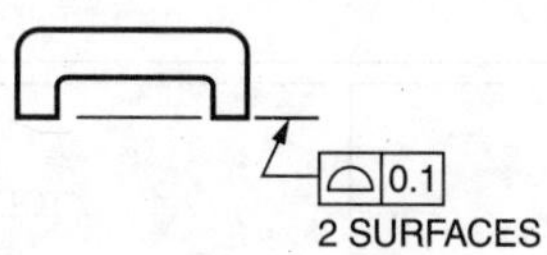	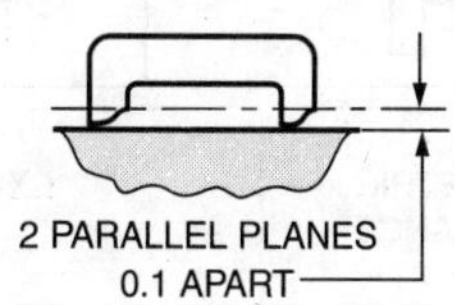

DATUMS USED	MMC OR RFS MODIFIERS	BONUS TOL.	OVERALL BOUNDARY	COMMENTS
YES	NO ALWAYS RFS (IMPLIED)	NO	DOES NOT APPLY	• Tol. zone can be bilateral or unilateral • Frequently used in place of coordinate dimensions • Basic dimensions must be used to establish the true profile
NO	NO ALWAYS RFS (IMPLIED)	NO	DOES NOT APPLY	• Controls flatness of surfaces • Uses a unilaterial tolerance zone.

Table 14.15. RUNOUT

INDICATION ON DRAWING	ROTATION	INTERPRETATION
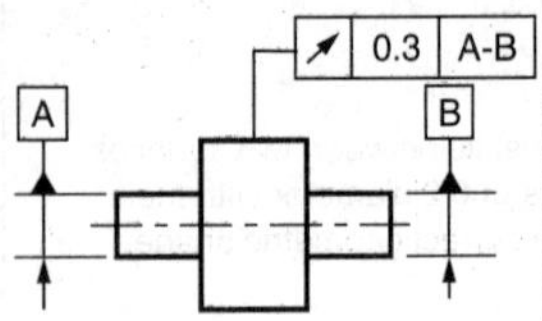	ROTATE PART ABOUT DATUM AXIS 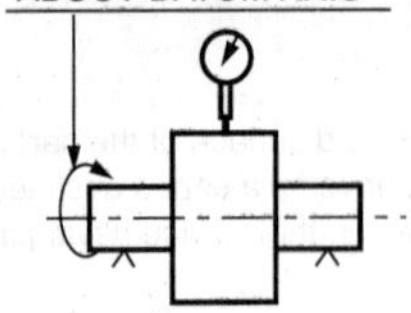	Runout not to exceed 0.3 measured at any point normal to the surface without axial movement.
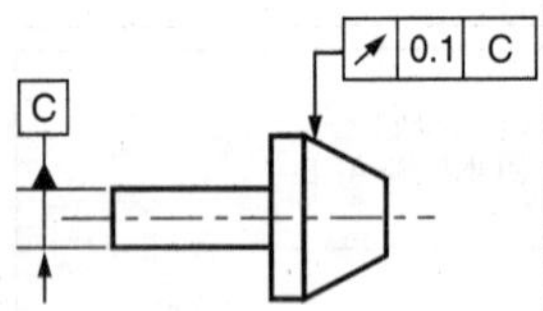	ROTATE PART ABOUT DATUM AXIS 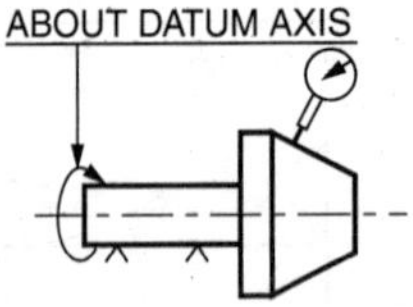	Runout not be exceed 0.1 measured at any point normal to the surface, without axial movement.
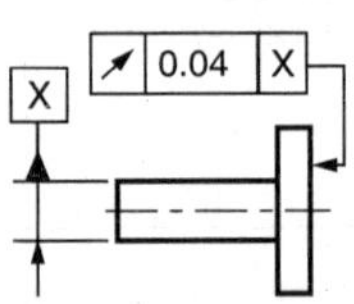	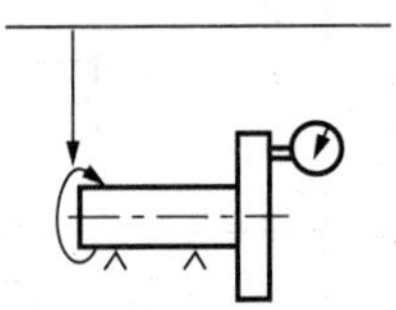	Runout not to exceed 0.04 measured parallel to the datum shaft at any radius without axial movement.

GEOMETRIC TOLERANCE	EXAMPLE AS SHOWN ON DRAWING	MEANING TOL. ZONE SHAPE
 CIRCULAR RUNOUT	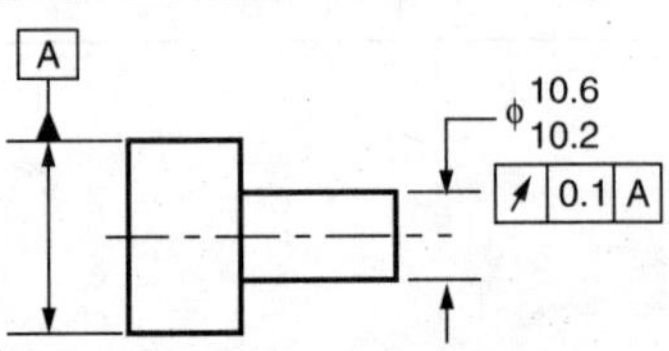	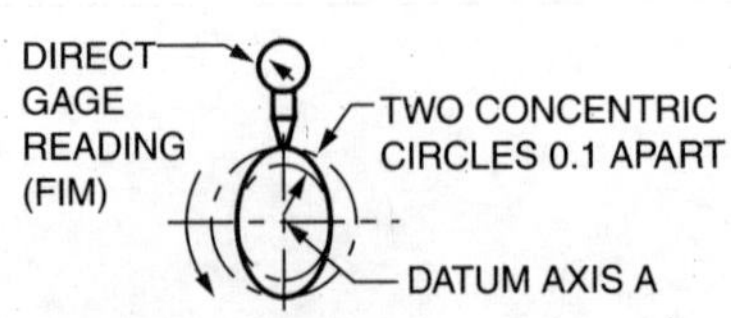

DATUMS USED	MMC OR RFS MODIFIERS	BONUS TOL.	OVERALL BOUNDARY	COMMENTS
YES	NO ALWAYS RFS (IMPLIED)	NO	AFFECTED 10.6 + 0.1 = 10.7	• APPLIES TO EACH CIRCULAR ELEMENT INDEPENDENTLY • A COMPOSITE CONTROL WHICH CONTROLS ROUNDNESS AND AXIS OFFSET

Table 14.16. TOTAL RUNOUT

INDICATION ON DRAWING	ROTATION	INTERPRETATION
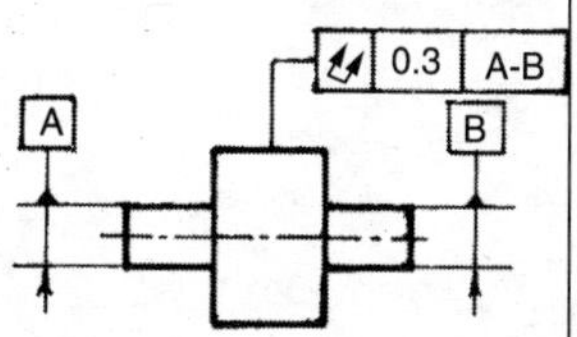	ROTATE PART ABOUT DATUM AXIS 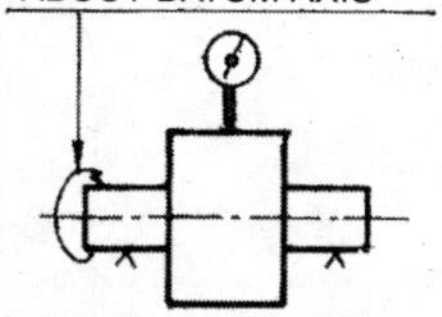	Total runout not to exceed 0.3 measured at any point normal to the surface whilst moving the measureing instrument along the feature when the part is rotated through a series of revolutions about its own axis without axial movement.
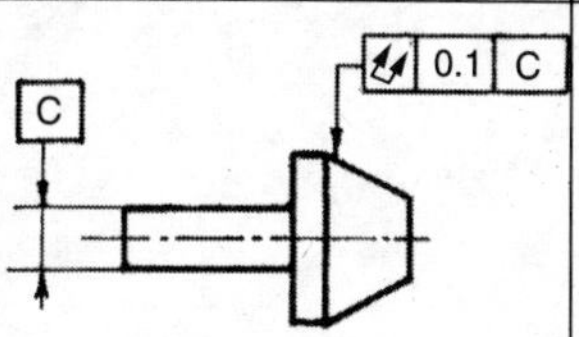	ROTATE PART ABOUT DATUM AXIS 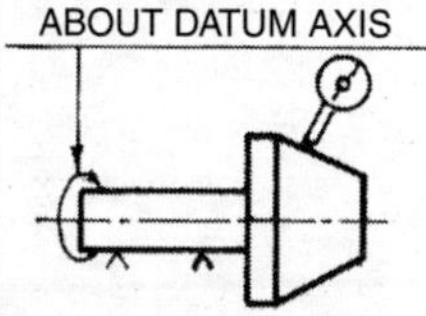	Total runout not to exceed 0.05 measured at any point parallel to the surface whilst moving the measureing instrument along the feature when the part is rotated through a series of revolutions about its own axis without axial movement.
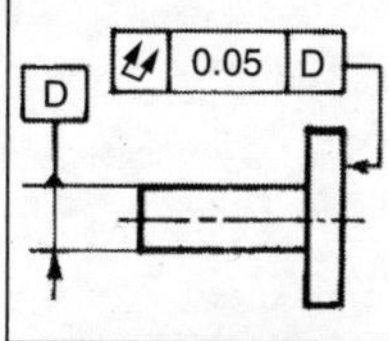	ROTATE PART ABOUT DATUM AXIS 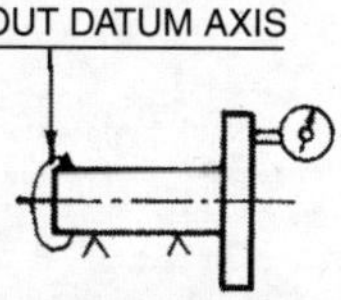	Total runout not to exceed 0.1 measured at any point normal to the surface whilst moving the measuring instrument along the feature when the part is rotated through a series of revolutionas about its own axis without axial movement.

GEOMETRIC TOLERANCE	EXAMPLE AS SHOWN ON DRAWING	MEANING TOL. ZONE SHAPE
 TOTAL RUNOUT	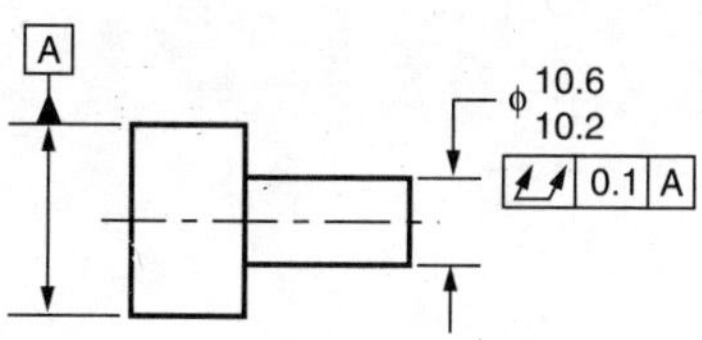	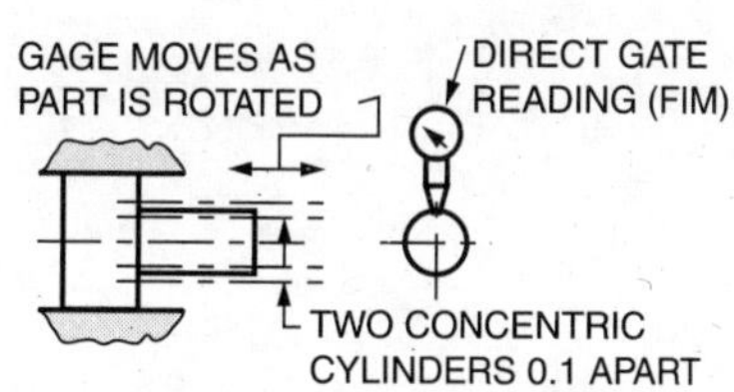

DATUMS USED	MMC OR RFS MODIFIERS	BONUS TOL.	OVERALL BOUNDARY	COMMENTS
YES	NO ALWAYS RFS (IMPLIED)	NO	AFFECTED 10.6 + 0.1 = 10.7	• APPLIES TO ENTIRE SURFACE SIMULTANEOUSLY • A COMPOSITE CONTROL WHICH CONTROLS STRAIGHTNESS, TAPER & OFFSET

This text only provides the basic necessary information about the datums, datum systems and various geometric controls. For a detailed knowledge and insight to G.D.E.T refer to the **"Geometric Dimensioning and Tolerancing"** including tolerance stade up by the Author.

❑❑❑

Detail And Assembly Drawing –Wooden Joints

15

15.1. INTRODUCTION

Engineering drawing is a universally accepted language, used by engineers, to describe accurately and clearly the form, size, finish, method of manufacture (without specifications), and construction of an object.

A technical professional can use the graphic language as a powerful means of communication with others, trained in this graphic language, for conveying technical information. Through its use any person trained in this field can visualize the shape and size of the object represented in this form. Irrespective of language barriers, the drawings can be effectively used any where in the world, irrespective of their place of origin. Without internationally standardized engineering drawing principles,. it would have been impossible to produce objects such as aircrafts, automobiles, locomotives, ships to watches, etc., each requiring a large number of component parts. Through its use the engineers develop and record their ideas and transmit the same to those who are entrusted the responsibility of executing designs prepared by them.

A technical drawing must be clear, unmistakable in meaning and interpretation. In a number of dealings, with suppliers, these drawing constitute an official document. Hence, an engineer should possess good knowledge, not only in preparing a correct drawing but also to read the drawings correctly.

Engineering drawings are made up of lines, circles, arcs and curves, with separate views arranged in a logical standard system of projection. To these views, dimensions and notes and other required information and directions for manufacturing are added. The drawings prepared by an engineer may be original idea sketches and layout drawings or they may be working drawings.

Since an engineer may frequently have to prepare preliminary design drawings and to direct the making of the working drawings which follow, he/she should have a complete understanding of the use of drawing instruments, CAD system(s) being used, planning and layout of drawing sheets, principles of projection, lettering, Geometric Dimensioning & Tolerancing, Sketching, Sections and Conventions, Surface Finishes, Drafting Practices, Drawing Reproduction and storage methods, Drawing Conformance Standards, etc.

Since engineering drawing is the foundation upon which all product designing and subsequent manufacture are based, it is one of the most important single branches of study in an engineering institution. Every engineering student must know how to make drawings, expressing himself/herself clearly to one familiar with it, and how to read a drawing made by anybody else. He/she must be able to interpret every detail for correctness and visualize clearly in space. To do so he/she must know its basic theory, composition and the accepted conventions. As its principles are essentially the same throughout the world, an engineer trained in the practices of one nation can readily adopt himself/herself to the practices of another nation.

Study of machine drawing mainly involves learning to sketch machine parts and to make working drawing and assembly drawings. Machine Drawing deals with the application of principles of Geometrical Drawing to the representation of machines and their components. Two of the many types of drawings are:

- Those which show what the designer requires in the finished product, and

- Those, which are to be used in shops for the actual manufacture of the component parts.

It is the former that are generally discussed to impart training of executing engineering drawings in the class rooms and, generally, the second type forms the part of a specialized training programs.

15.2. WORKING DRAWINGS

A working drawing of an object is one, which completely describes the object represented. Working drawings are used by engineers and drafters to convey their ideas to those who are to execute their designs. Such drawings should describe the objects so completely that those who are to erect, construct, or manufacture the object represented will need no additional information. Further the necessary information given should be in the simplest manner possible and according to conventions adopted by the respective professions and trades. A working drawing should include a representation of the object, dimensions including the required GD &T, showing the size, explanatory notes or indications of the materials to be used, the methods by which they are to be finished, and a descriptive title. For example, working drawing of an ARBOR is shown in *Fig. 15.1*.

In engineering, most objects are made of several component parts. More complicated the product the more parts there are likely to be. If there are two or three simple parts that make up the product then it is possible to put them all together on one drawing in their correct relative position, fully detailed and dimensioned. One complete drawing may be used in the shop to make these components and fit them together. In most cases, this is not possible because of the number and complexity of the components. When however, this is the case each component is drawn in detail on a separate sheet, or if the components are small, several are drawn on the same sheet. These are called "Detail Drawings" and with the drawing of the assembled parts, form a "set of drawings". As such the working drawings are classified as "Detail Drawings" and "Assembly Drawings".

15.3. SET OF WORKING DRAWINGS

A complete set of working drawings for a machine consists of detail drawing sheets, giving all the necessary information required for the production of individual parts in the shop, and an assembly drawing showing the location of each component part in the finished (assembled) machine, *Fig. 15.2 (a)* and *(b)*.

15.4. DETAIL DRAWINGS

In general, a detail drawing is one, which describes a single part, or several parts individually. The drawing should furnish complete information, according to standard conventions, regarding shape and size of each individual part, along with other specifications necessary for the purpose of manufacturing the part, *Fig. 15.2 (a)*. Such a drawing usually carries the following information:

- Shape description of the part minimum possible number of orthographic views, including special sections and auxiliary or pictorial views if necessary, to be taken to describe the part completely,
- Size of description of the part,
- Material specification,
- Necessary manufacturing processes,
- Grades of surface finish,
- Limit of accuracy, and
- Number of parts required.

Since a machinist will ordinarily make one part at a time, it is advisable to detail each piece, regardless of size, on a separate individual standard size-drawing sheet.

15.5. ONE VIEW DRAWINGS

Many parts, such as shafts, bolts, studs, and washers may require only one properly dimensioned view. In the case of each of these parts, a note can describe the complete shape of piece without sacrificing clarity.

15.6. ASSEMBLY DRAWINGS

An assembly drawing is one that shows the complete structure of a machine and the relative disposition and working relationship of its component parts. The purpose of an assembly drawing is to represent the proper working relationships of the several component parts of a structure or mechanism, *Fig. 15.2 (b)*. It usually gives only a few main dimensions and much of the hidden detail is usually

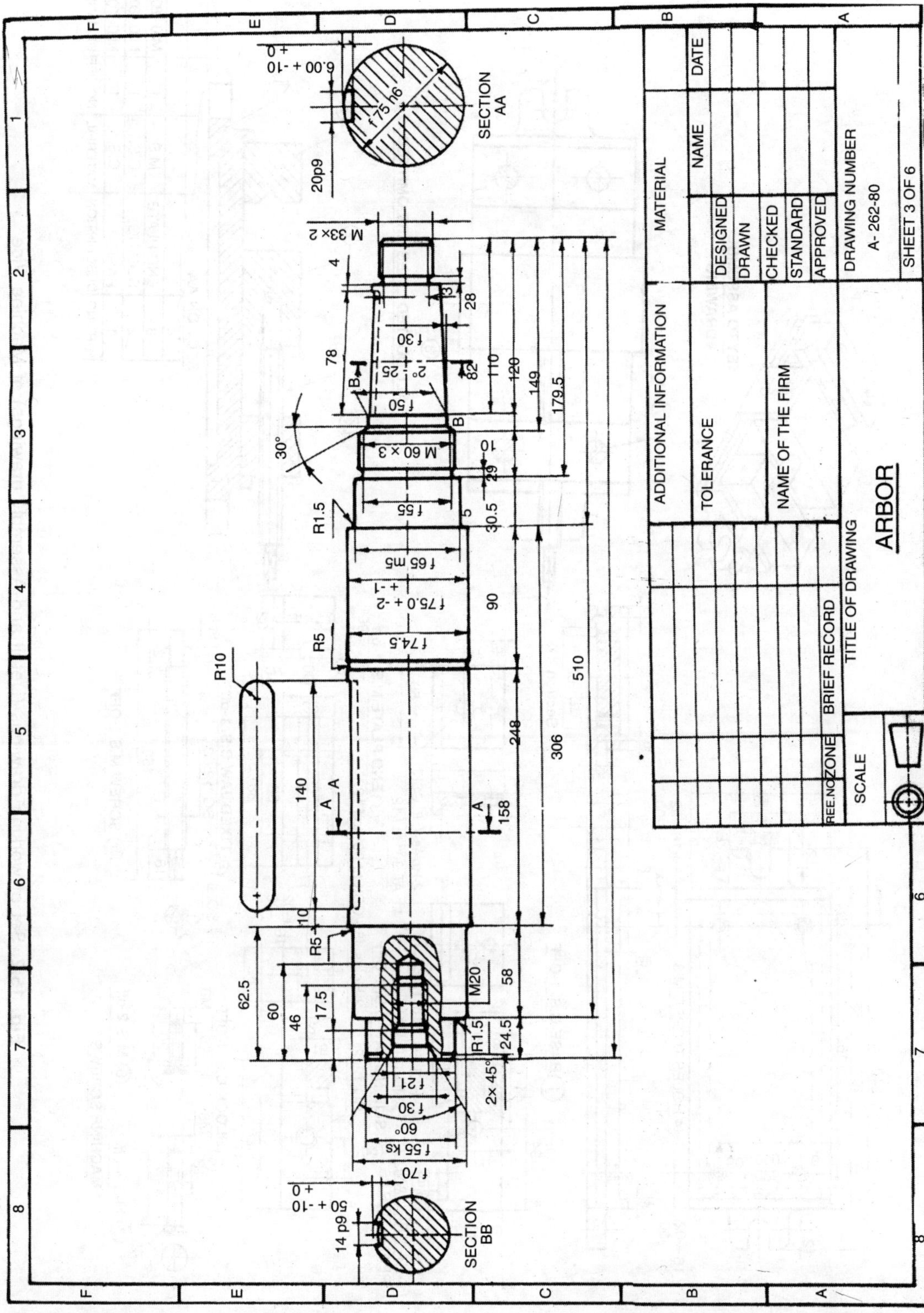

Fig. 15.1. Working drawing of ARBOR

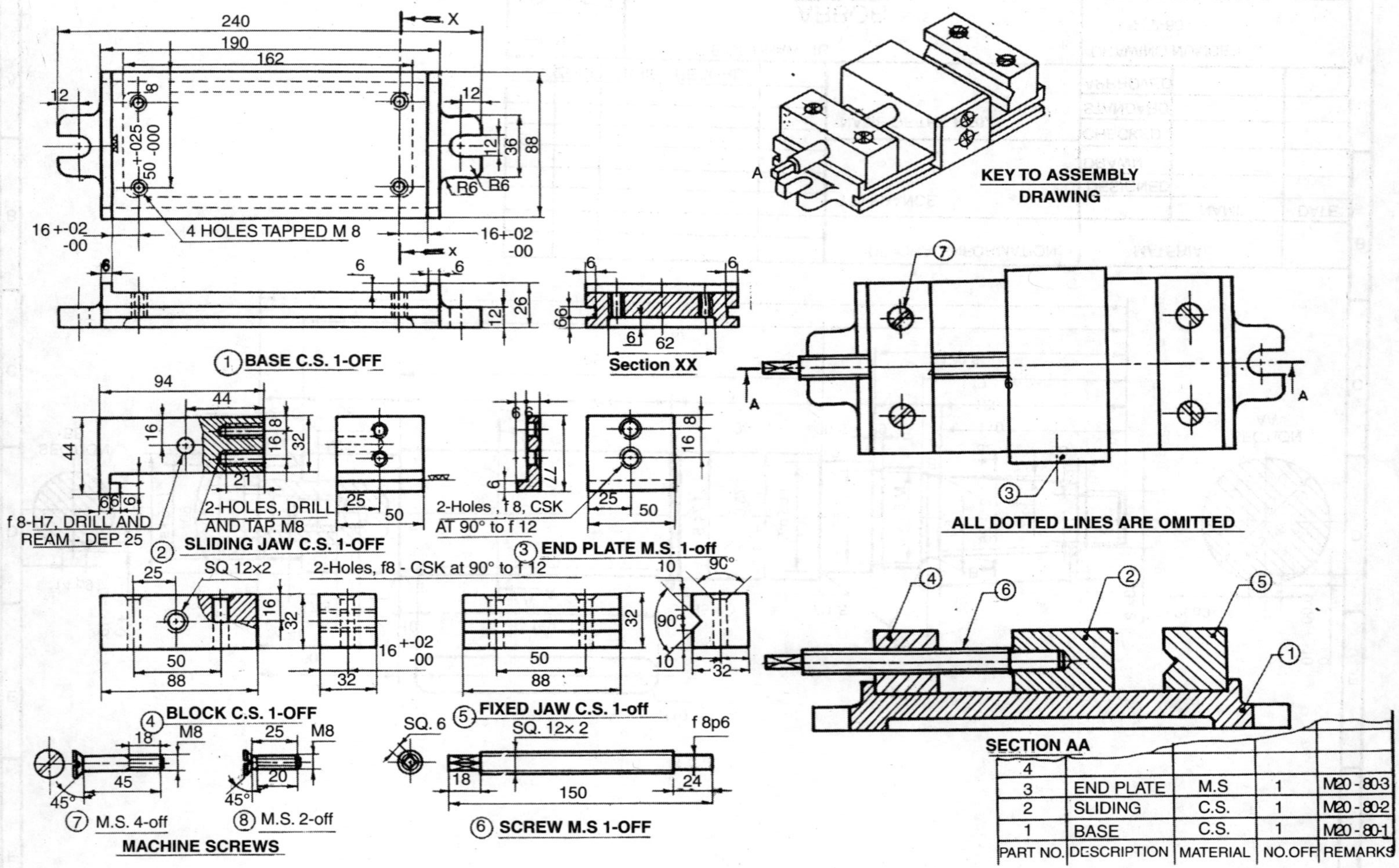

PART NO.	DESCRIPTION	MATERIAL	NO.OFF	REMARKS
4				
3	END PLATE	M.S	1	M20 - 80-3
2	SLIDING	C.S.	1	M20 - 80-2
1	BASE	C.S.	1	M20 - 80-1

Fig. 15.2. Set of working drawings (Detail and Assembly drawings) of Machine Vice

omitted. An assembly drawing usually contains the following information:

- Main view to show the best advantage of the assembly,
- Overall dimensions and centre to centre distances indicating relationship among the parts and the machine as a whole,
- Identification of individual parts, and
- Necessary sections or pictorial views, if required, for better understanding.

For an engineering drawing student, examples dealing with assembly drawings are necessarily quite simple and usually present all the details on one drawing sheet. The student is then required to assemble the parts and draw certain orthographic views.

Very often details are presented in a haphazard manner, their position on the sheet bearing little relation to their final position in the assembly drawing. If the details assemble into a familiar object then there is no difficulty. Faced with unfamiliar details, however, the best method is to draw a rough sketch, fitting the parts together by studying the dimensions carefully to find where the pieces fit. Only when every part is fitted together should a start be made on the drawing paper.

After the drawing is completed and a few dimensions added, the individual components are then denoted by leader lines having numerals or letters at their ends, which should be well clear of the out line and the dimensions, *Fig. 15.2 (b)*.

These leaders should be then tabulated in a materials and parts list, which is usually placed at the bottom right corner close to the title block, *Fig. 15.2 (b)*, and should include those of the following headings, which are necessary:

1. Item or part number,
2. Description of part,
3. Quantity required,
4. Material, and
5. Cross references to other drawings.

Assembly drawings have many specific users; hence there are many types.

1. Design Assemblies

The first stage in the development of an idea for a machine or structure is to prepare freehand sketches and to make the calculation required to determine the feasibility of the design. From these sketches the designer prepares a layout or design drawing, which gives the relationships of the mating parts and affords an accurate analysis of the design to be worked out. Assembly drawing is the natural vehicle of design. The design drawing is executed with instruments. It may include any auxiliary views or sections, which are helpful, and generally is not dimensioned. The completed design assembly is used as a basis for the preparation of complete detailed and assembly drawings.

2. General Assemblies

Frequently a design drawing is traced, with some modifications, to produce the final assembly. The details and assembly may be drawn on the same sheet in the representation of small structures, but usually the assembly is drawn on a separate sheet.

- **Sectioning:** In order to show the working relationships of interior (invisible) parts, it is generally advantageous to use sectioning more often than in detail drawings.
- **Views:** Since most structures are symmetrical about a centreline or a central plane, it is possible to expose nearly all parts by full section, and one main view is frequently sufficient. However, any additional views or sections may be used. No effort should be made to describe completely the forms of parts, since this Information is shown on the detail drawings. The views chosen should be those which show most clearly how the parts fit together and the mechanism functions.
- **Dimensions:** When dimensions are necessary, only those applying to the structure as a whole are given. These may include overall dimensions, dimensions between important centres, and any others necessary to show the relationships of the parts. Detail dimensions should be omitted.
- **Hidden Lines:** The use of hidden lines in necessary in assembly drawings is the same as that in all other types of drawing. Hidden lines should be used only when their use results in added clearness.
- **Standard Parts:** Standard parts, such as rolled sections, ropes, chains, cables, bearings, pipes, bolts, nuts, screws, pins, nails, keys, tapers,

wires, etc., are represented symbolically by approximate methods. The dimensions for these commonly used elements have been standardized, and the drafter must obtain their sizes from company standards, manuals, from handbooks, BSI pamphlets or from manufacturer's tables.

Such parts are not shown on detail working drawings unless the parts are to be altered for use, but are drawn conventionally on assembly drawing and are listed, according to standard tables, in the parts-lists.

3. Working Drawing Assemblies

The details of simple mechanisms are sometimes dimensioned completely in the assembly drawings, *Fig. 15.3 (c)*. This is a common practice in drawings of valves, jigs and fixtures.

4. Group Assemblies

It is impossible to show the relationships of all the parts in some complicated structures in a single assembly drawing. Generally, such mechanisms are divided into natural units, or groups of parts.

5. Installation Assemblies

An assembly, which is made for the specific purpose of showing how to erect or install a structure, is an installation assembly. This type of drawing is often called an outline assembly, because it shows only the out-line and the relationships of exterior surfaces.

6. Check Assemblies

After all the detail drawings of a structure have been finished, it may be necessary to make a **check assembly.** Such a drawing is an assembly of all details drawn accurately, to scale, in order to check each one, graphically. After the check assembly has served its purpose it may be converted into a general assembly drawing of any type desired. Therefore, the assembly drawing usually may be employed to check against the detail drawings and the general computations.

7. Exploded Assemblies

Exploded pictorial assembly drawings are used frequently in the parts list section of company catalogs and instruction manuals. Those with very little experience in reading multi-view drawings easily understand drawings of this type.

8. Pictorial Assembly

These are used with prospective customers who cannot read a projection drawing, or when it is desired to show the appearance of an object.

15.7. WOODEN JOINTS

The object of all wooden joints is to fix one member in an economical way, by way of minimum amount of labour and material wastage necessary, involving least number of operations, maintaining the proper strength of the joint. There are number of joints which are used in joinery. Out of these all, only a few important joints are considered here for the sake of exercise in **Detail and Assembly Drawings.**

EXERCISES

1. Prepare detail and assembly drawings of the wooden joints shown in *Fig. 15.4 (a) to (h)*, shown in the form of exploded assemblies.
2. Pictorial assembly drawings of three wooden joints are shown in *Fig. 15.5*. Draw detail and assembly drawings for one of these.
3. Detail drawings of a three-piece joint are shown in *Fig. 15.6*, in third angle. Assemble the parts and draw its three orthographic views.

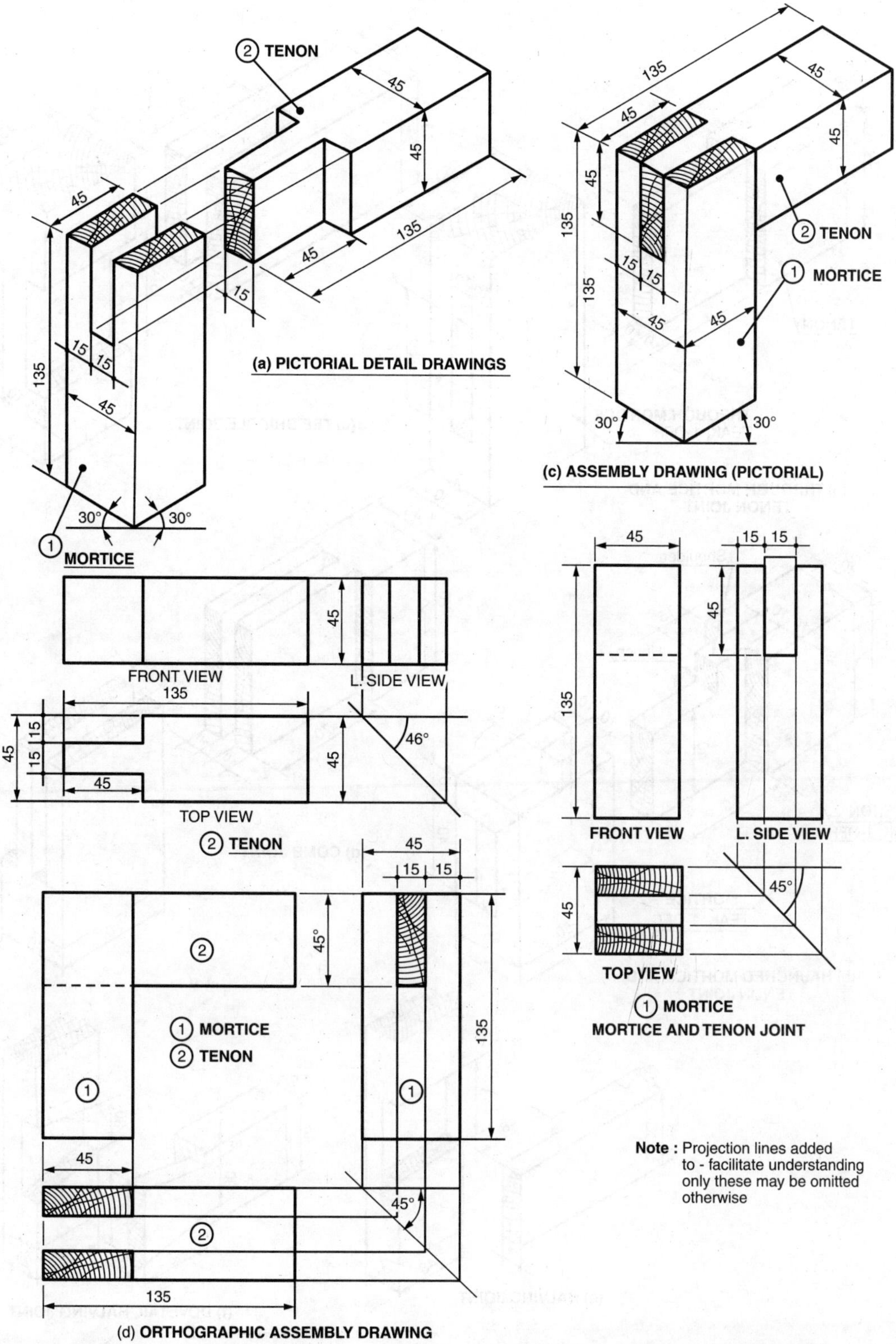

Fig. 15.3. Detail and assembly drawings of 'Mortice and Tenon' wooded joint

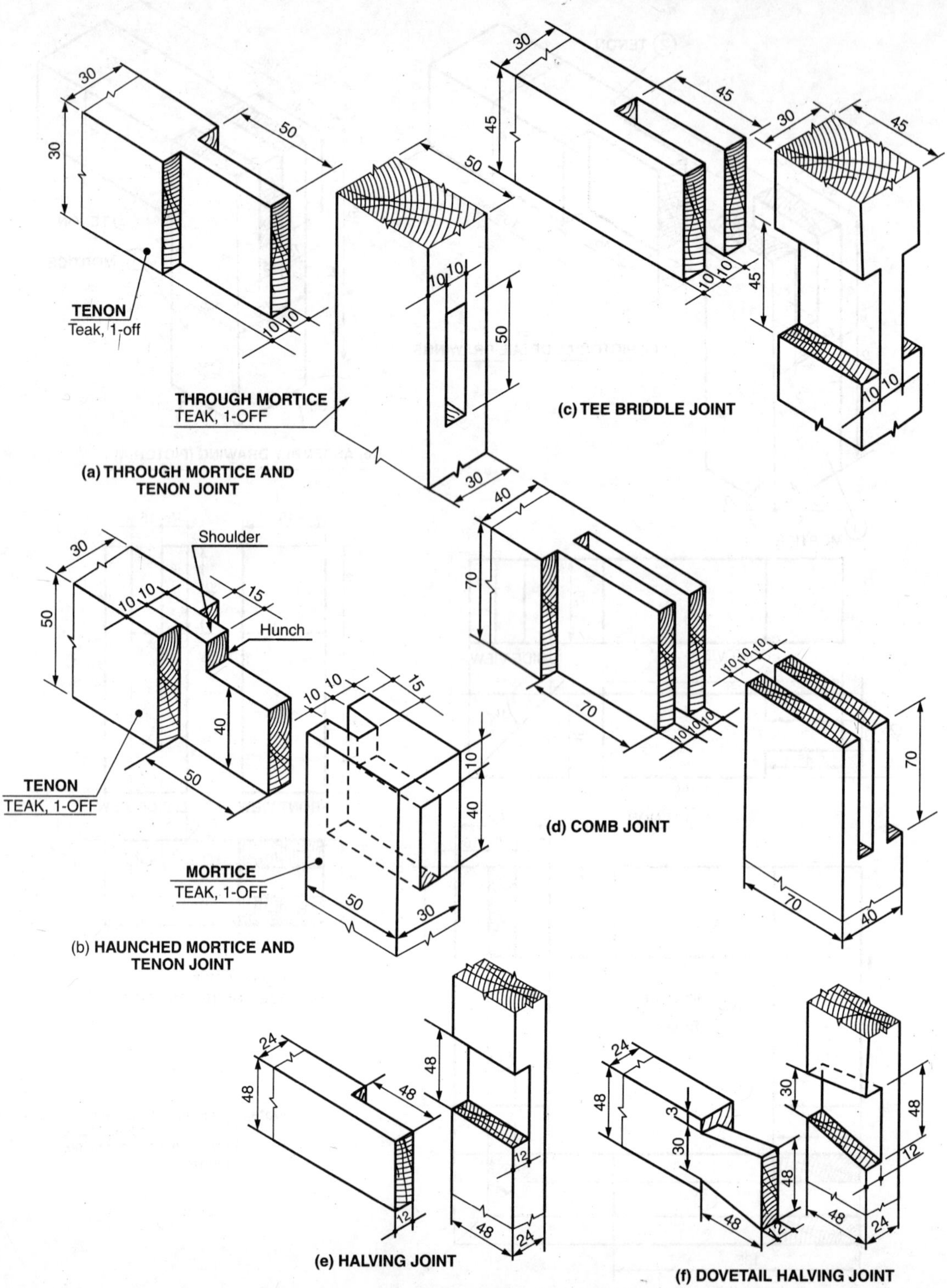
TENON
Teak, 1-off
THROUGH MORTICE
TEAK, 1-OFF
(a) THROUGH MORTICE AND TENON JOINT
(c) TEE BRIDDLE JOINT
Shoulder
Hunch
TENON
TEAK, 1-OFF
MORTICE
TEAK, 1-OFF
(b) HAUNCHED MORTICE AND TENON JOINT
(d) COMB JOINT
(e) HALVING JOINT
(f) DOVETAIL HALVING JOINT

Fig. 15.4. (cont...)

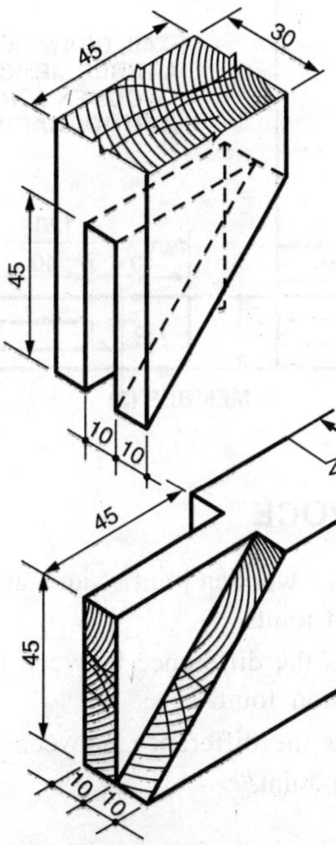

(g) MITRE FACED BRIDDLE JOINT

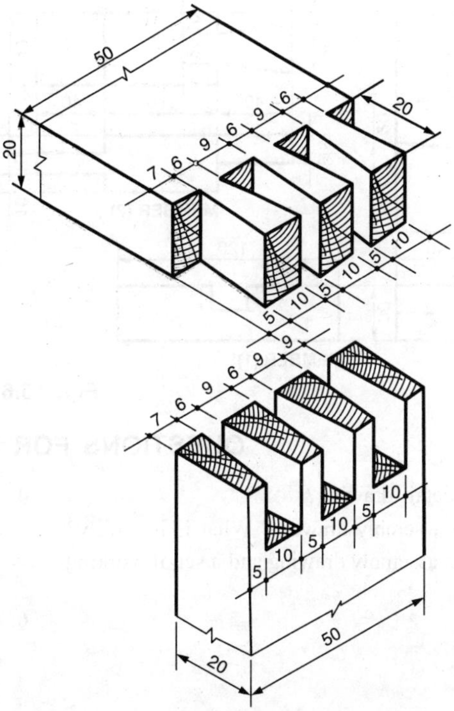

(h) LAPPED DOVETAIL JOIN

Fig. 15.4.

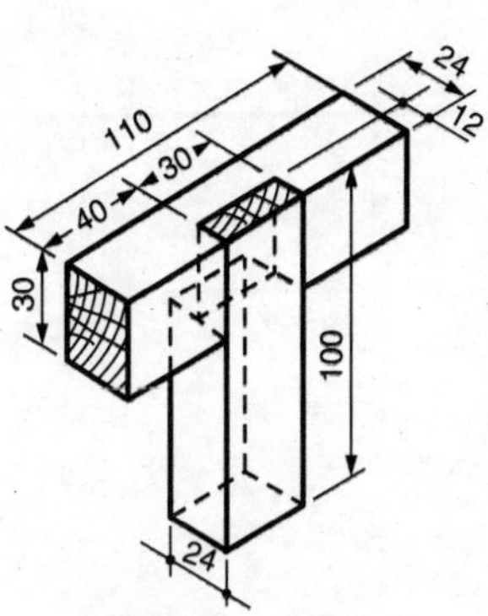

(a) TEE HALVING JOINT

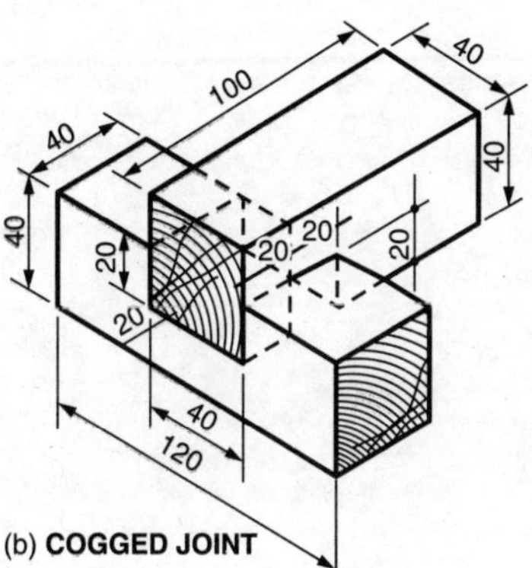

(b) COGGED JOINT

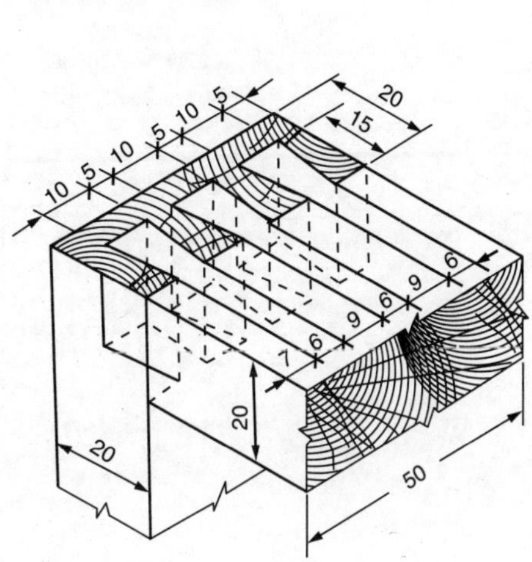

(c)

Fig. 15.5.

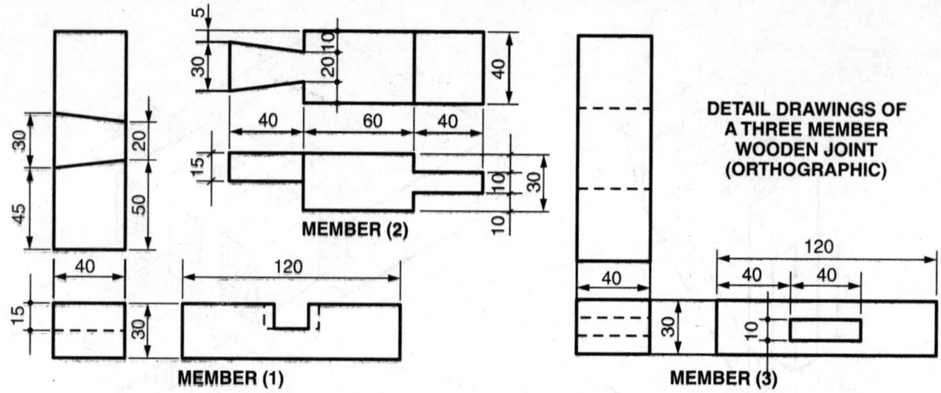

Fig. 15.6.

QUESTIONS FOR VIVA VOCE

1. What is a detail drawing?
2. What is an assembly drawing? What is its utility?
3. What are an assembly drawing and a set of working drawings?
4. What is a wooden joint? Name at least five types of wooden joints.
5. What is the difference between an "Open Mortice and Tenon Joint"?
6. What is the difference between a "Halving and a Briddle Joint"?

16 Rivets And Riveted Joints

16.1. FASTENERS

The devices used for holding together two parts of a machine or structure are called 'Fasteners or Fastenings'. These fasteners vary in kind and use, from ordinary nails and glue to screwed pieces, rivets, keys and cotters etc., varying in size for use in locomotives to watches.

16.2. CLASSIFICATION OF FASTENERS

Fasteners are divided into two main categories:

1. Temporary fasteners, and
2. Permanent or semi-permanent fasteners.

Temporary fasteners allow easy unfastening of the parts connected, without injuring the fastening element, whenever required. Whereas with semi-permanent fasteners the parts when joined once, cannot be separated without breaking the joining element. Temporary fasteners are widely used in joining the machine parts and other engineering products where frequent dismantling is required. Nuts and bolts, screws, studs, keys, cotters and pins are examples of temporary fasteners. Whereas riveting, welding, soldering and brazing etc., are permanent (semi-permanent) fastenings.

16.3. RIVETS

Rivets are short cylindrical pieces of ductile metal with a head, formed during manufacture, at one end and a tail on the other end, *Fig. 16.1*. A head is formed on the other (tail) side also after the rivet has been put in place through the aligned rivet holes in the two plates to be joined. They are usually employed to fasten together pieces of sheet metal or fasten plates to rolled sections or rolled sections to each other. In structures, boiler work, ship building, etc., where dimensions are reasonably large, the rivets are solid cylinders, usually of mild steel, but in lighter sheet metal fabrication, as in the aircraft industry, etc.; they are often semi-tubular or hollow (*Fig. 16.3*) and made from various metals and alloys, including copper, brass, aluminium, duralumin, etc.

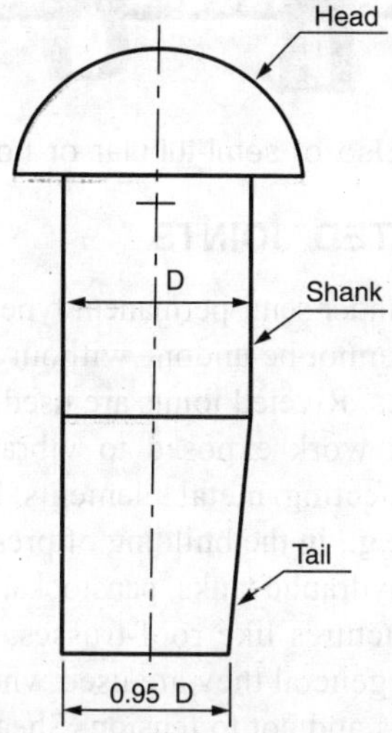

Fig. 16.1. Rivet

Rivet is designated by giving the type of head, shank diameter and length, e.g., a Snap Head Rivet of ϕ 10 mm and 50 mm long is designated as Snap Head Rivet 10 × 50.

The riveting process has been displaced in many branches of engineering by welding, but there is still a great deal of it being done for many industrial applications.

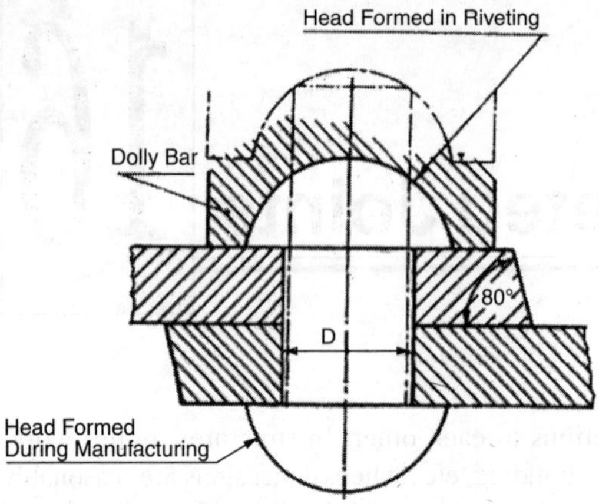

Fig. 16.2. Riveting

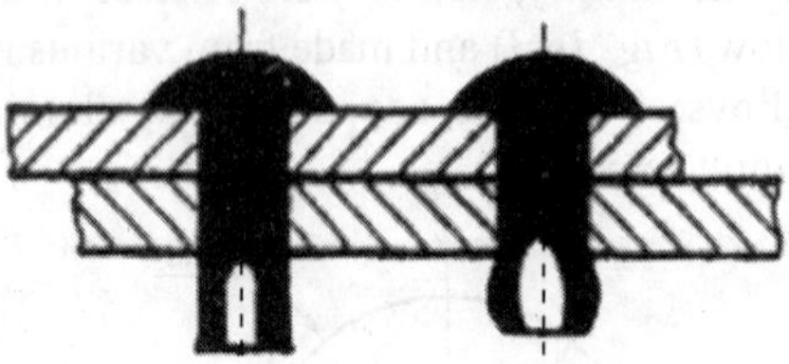

Fig. 16.3. Use of semi-tubular or hollow rivets

16.4. RIVETED JOINTS

These come under semi-permanent type of fasteners, as the joint cannot be undone without chipping off the rivet heads. Riveted joints are used in structural and machine work exposed to vibrational loads and for connecting metal elements having poor weldability, e.g., in the building of pressure vessels like boilers, hydraulic tanks, penstocks, air receivers etc., and structures like roof-trusses, bridges and tower etc. In general they are used when subjected to shear stress and not to tension. Shearing divides the material into parts, which separate from each other by sliding at the surface of separation.

16.5. SHAPES OF RIVET HEADS

The various important shapes of rivet heads used in engineering are given in *Fig. 16.5*. Each shape has some specific use. In structural works only a few kinds of rivet heads are usually used, e.g., the snap, button, flat-top counter sunk heads. For boilers and tanks most commonly snap head, conoidal head or ellipsoidal head are used. These have to be very substantial. The exact method of locating the centre of the arc for drawing the head of a snap head type rivet is illustrated in *Fig. 16.5. (a)*, whereas easy drawing proportions for snap head rivet are given in *Fig. 16.4*.

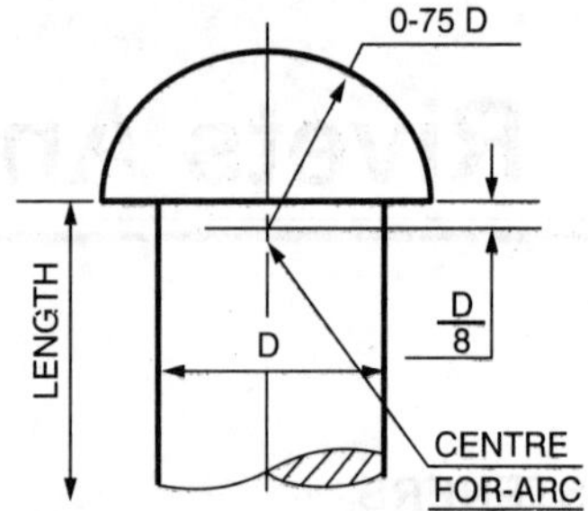

Fig. 16.4. Easy drawing proportions for snap head rivets

Snap head are usually employed in machine riveting, conical heads or conoidal heads are mainly used where restricted space permits only hand hammering. Counter-sunk heads are mainly used in ship building, where flush surfaces are required. It is generally considered that snap heads are most easily shaped, that pan heads have maximum strength and that counter-sunk heads are least reliable.

16.6. RIVET DIAMETERS

The diameter of a rivet when cold (d) is slightly smaller (about 1 to 2 mm) than the corresponding hole diameter (D), for when the rivet is heated it expands and the closed rivet (after riveting) is to completely fill the hole. When, in this chapter, the diameter (D) is referred to, it means the diameter of the closed rivet or that of the hole it fills. The length of rivet is the length of the rivet excluding its head. The length should accommodate the thickness of the plates to be fastened and should provide enough material for forming the second head at the tail end. The length of the rivet required to form a head varies according to the head shape. For snap and conical heads about 1.5 D is sufficient.

The diameter (D) is determined from the thickness of the plates to be connected. If '*t*' is the thickness of the plates then by Prof. Unwin's formula $D = 6\sqrt{t}$ where 't' is in mm or $D = 1.9\sqrt{t}$ where '*t*' is in cm. This empirical rule is applicable only when plates and rivets are of substantial size. Calculated by the use of this rule the approximate values of D for various values of t may be tabulated as given in *Table 16.1*.

16.7. RIVETING

In the plates, to be connected, holes for the rivets are punched and reamed, or drilled little larger than the diameter of a rivet. Punching is liable to damage the plates especially when of hard material. On the best the plates are clamped together by a few bolts along the seam so that the remaining holes may be drilled and reamed through both plates without any danger of trouble arising through holes being out of alignment. A burr is usually left after every drilling operation, and on a riveted joint may aid shear. Burrs are usually removed by light counter sinking. The rivets are then heated to red hot and inserted into the holes after aligning the holes in the plates. The tail side of the rivet is then forged to the shape of head. The riveting may be done by hand hammering or by machine, or hydraulic power depending upon the condition of work. When the work is being done in the shop, machine riveting is done, whereas in the field, where it is not convenient to carry the machine, hand riveting is done. For light work 'cold riveting' is resorted to. *Fig. 16.2*, illustrates the riveting process.

16.8. RIVET TESTING

A rivet is considered fit if it withstands the following two tests:

1. Bend Test

The rivet shank is bent cold by hammering until the tail side touches the shank and does not show any fracture on the outside of the bend.

2. Flatening Test

The rivet head when flattened by hammering should spread until the diameter is 2.5 times the diameter of the shank, without showing any cracks at the edges.

Table 16.1.

Thick of Plate, t mm	8	9	10	11	12	14	16	18	20	22
Dia. of Finished rivit, D mm	17	18	19	20	21	22	24	25	27	28

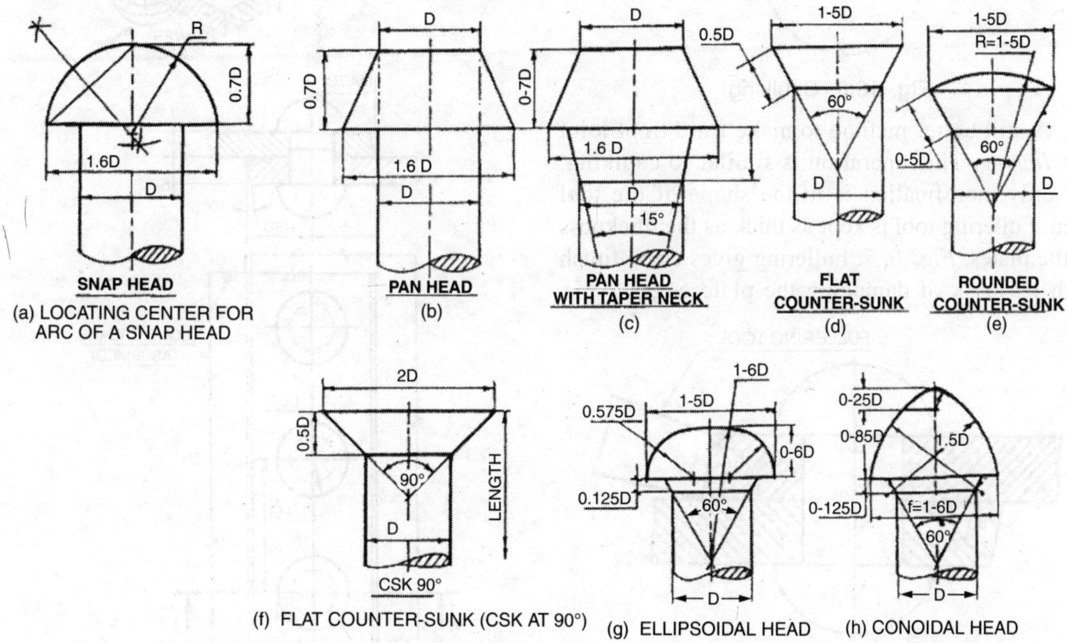

Fig. 16.5. Types of rivet heads

16.9. CAULKING AND FULLERING

However closely rivets are pitched, and however carefully riveting is performed, it is not always possible to achieve a perfectly leak-proof joint. Whereas in the construction of pressure vessels, like tanks and boilers, the joints are required to be very tight to prevent the leakage of fluids. To achieve this the edges of the plates and the rivet heads are burred down by a blunt tool of the shape of a flat cold chisel about 5 mm thick at the end and 40 mm in width, and the edge having an angle of 10° (the angle of the beveled edge of the plates). It is generally hammered either by hand or by compressed air. This process is called caulking, *Fig. 16.6*, and the tools used is called caulking tool. Both internal and external caulking is done in many cases.

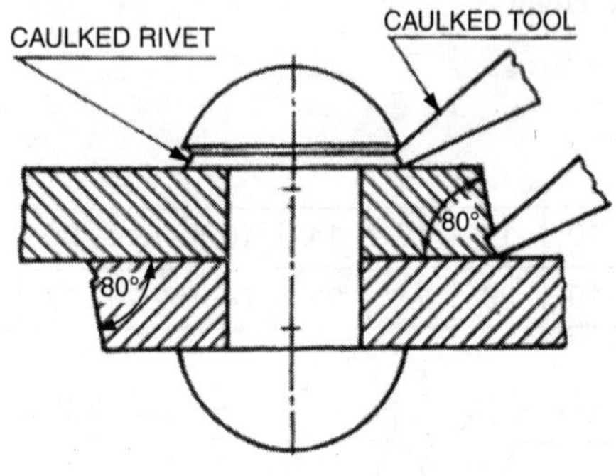

Fig. 16.6. Caulking

A still better method to make fluid tight joint is *fullering*. This operation is similar to caulking, the only modification is in the shape of the tool used. Fullering tool is kept as thick as the thickness of the plates, *Fig. 16.7*. Fullering gives clean finish with less risk of damaging the plate edges.

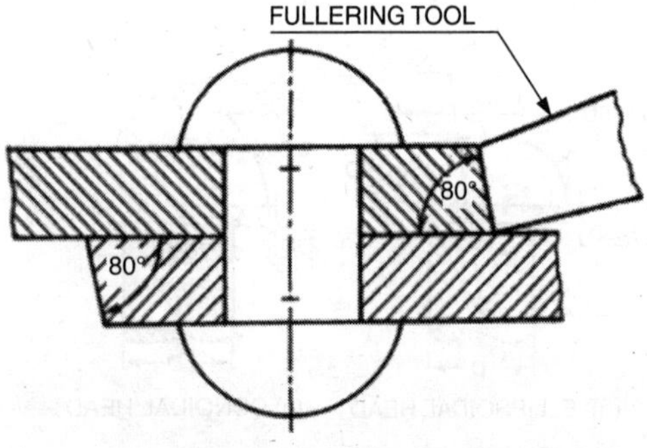

Fig. 16.7. Fullering

16.10. DEFINITIONS

The following definitions, in connections with riveted joints, are important:

1. **Pitch (P):** It is the centre to centre distance between two adjacent rivets in the same row, measured parallel to the seam or edge of the plate or rolled section. Usually $P = 3D$, where 'D' is the diameter of the rivet hole or the rivet after the riveting process is done.
2. **Diagonal Pitch:** It is a transverse measurement between centres of rivets in adjacent parallel rows, on the same plate, as indicated in Fig. *16.9 (b)*.
3. **Margin:** It is the least distance from a rivet hole centre to the nearest edge of the plate. It is usually kept equal to 1.5*D*.
4. **Back Pitch:** It is the distance between rows of rivets measured perpendicular to the seam, *Fig. 16.9 (b)*.

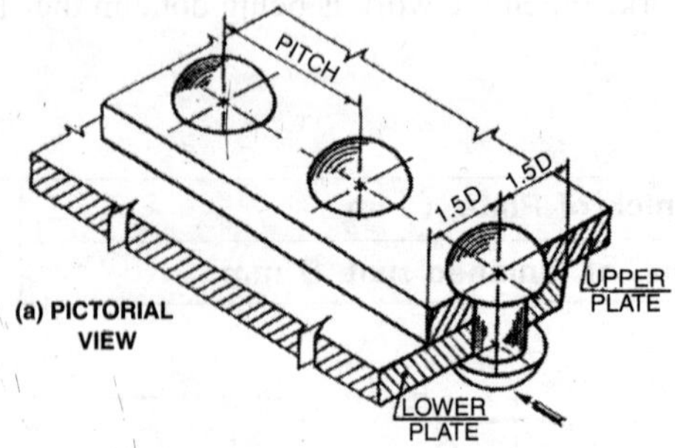

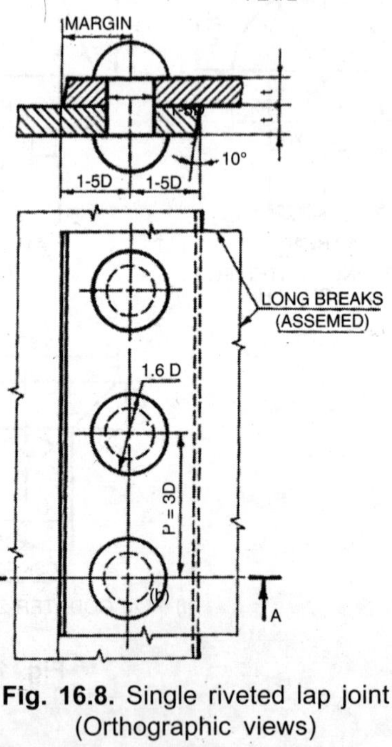

Fig. 16.8. Single riveted lap joint (Orthographic views)

16.11. TYPES OF RIVETED JOINTS

Riveted joints may be divided into two types depending upon the manner in which the plates are held in relation to each other:

1. Lap Joints; and 2. Butt Joints

1. Lap Joints

When the edges of the two plates to be connected, overlap each other and all rivets pass through both the plates, a lap joint is formed, *Figs. 16.8, 9, 10.* The distance by which the plates overlap each other is called the *overlap*. Depending upon the rows of rivets, the joints may be further classified as:

Single riveted lap joint – one row of rivets, *Fig. 16.8.*

Double riveted lap joint – two rows of rivets, *Fig. 16.9, 16.10.*

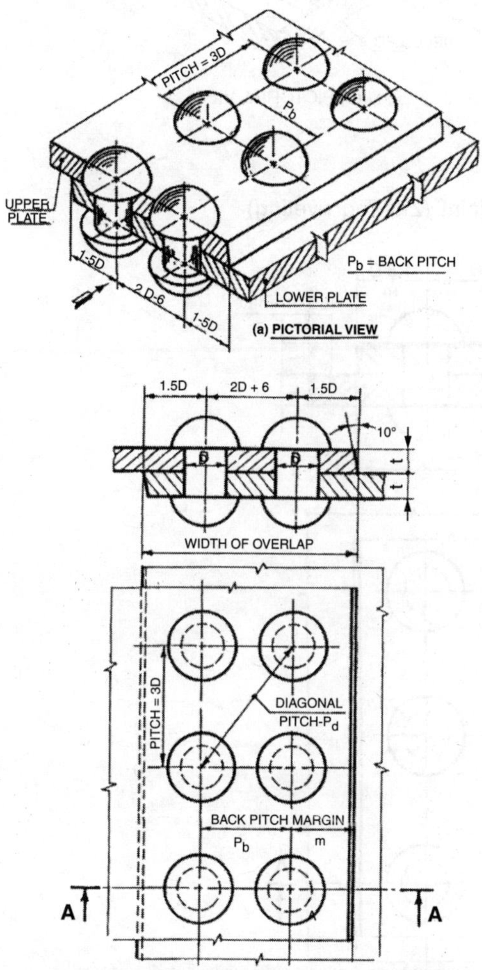

Fig. 16.9. (b) Double riveted lap joint (Chain riveting)

Tripple riveted lap joint – three rows of rivets, *Fig. 16.11, 16.12.*

Figure 16.8 (a) shows the pictorial view of a part of single riveted lap joint, and *Fig. 16.8 (b)* shows the two orthographic views of the joint in First Angle. *Figure 16.9* shows a double riveted lap joint. It can be observed that the rivets in two rows are opposite to each other. This type of riveting is called *chain riveting*. Whereas *Fig. 16.10* shows another double riveted lap joint in which the rivets in two rows are staggered. This type of riveting is called *zig-zag riveting* and is most commonly used. It can be noted from these figures that the distance between rows in chain riveting is usually taken 2D + 6 mm while in zig-zag it is 2D only.

Figure 16.11, shows two orthographic views, in First Angle, of a tripple riveted lap joint with chain type of riveting. *Figure 16.12*, shows two views of a tripple riveted lap joint with zig-zag type of riveting.

In each of the sectional views in riveted joints, though the cutting plane passes through the centres of some rivets, it is a convention not to show these rivets in section.

In cylindrical boiler work single or double riveted lap joints are often used for circular girth seams, as the pressure exerted in this direction is only one-half of that carried by the longitudinal joints for which double strapped are used – double and tripple ordinarily, but quadruple for higher pressures.

2. Butt Joints

When in a riveted joint the edges of the main plates, to be connected, butt against each other and lie in the same plane, the line of junction being covered by *cover plates* or *cover straps*, it is called a but joint. In this case each rivet passes though the cover plates and one of the main plates. Butt joints are usually used on longitudinal seams of cylindrical boilers etc.

Depending upon the number of rows of rivets, on each plate, the joint may be further classified as;

Single riveted butt joint, *Fig. 16.13*, one row of rivets in each plate.

Double riveted butt joint, *Fig. 16.14 (a), (b)* two rows of rivets in each main plate.

Tripple riveted butt joint – three rows of rivets in each plate.

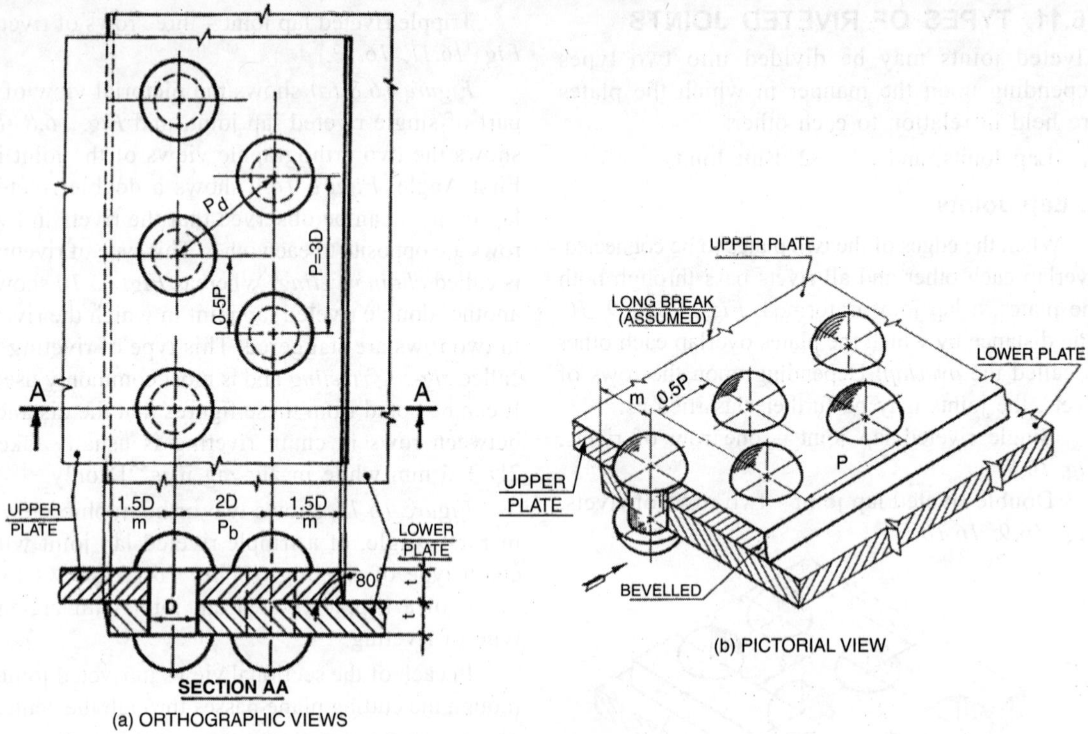

Fig. 16.10. Double riveted lap joint (Zig-Zag riveting)

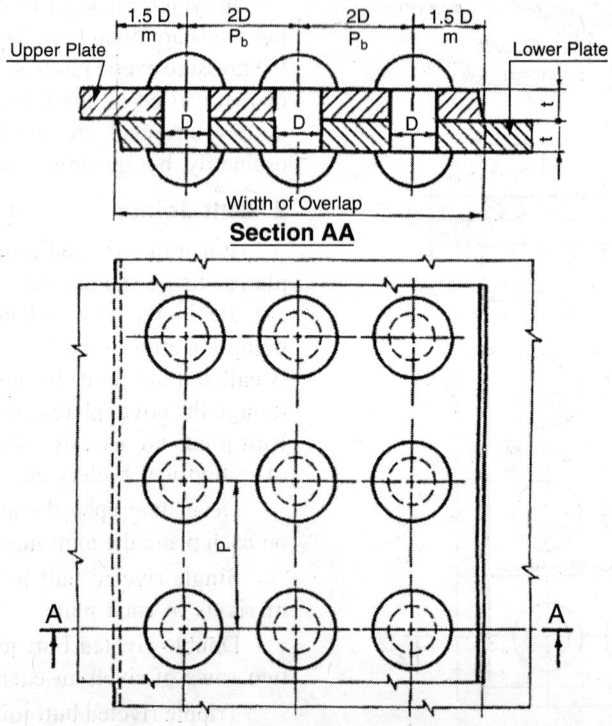

Fig. 16.11. Tripple riveted lap joint chain riveting

It can be noticed in *Fig. 16.14* that rivets in the corresponding rivet rows on the two main plates should be in line and not staggered.

Depending upon the number of cover plates used, a butt joint can further be designed as; (1) Single strap butt joint when only one strap is provided, which may be on either side of the main plates, *Fig. 16.15*, and (2) Double strap butt joint with two cover straps, one on each side of the main plates, *Fig. 16.16*. The straps may be equal or unequal to each other in width and length.

16.12. THICKNESS OF COVER PLATES (t_1)

When one cover plate is used, as shown in *Fig. 16.15*, the thickness is made more that the thickness of main plates usually t_1 = 1.125 t. When two cover plates are used t_1 = 0.625 t, *Fig. 16.16*.

Figure 16.13 shows two views of a single riveted double strap butt joint. It can be noted that there is one row of rivets in each plate and two rows in whole joint, while in the single riveted lap joint there is only one row of rivets shared by both the plates.

16.13. CHAIN AND ZIG-ZAG RIVETING

In *Fig, 16.14 (a)* are shown the two views of a double riveted butt joint. There are two rows of rivets on each plate to be joined. The rivets in the two rows are opposite to each other. This type is called the 'Chain-Riveting'.

In *Fig. 16.14 (b)* is shown another double riveted butt joint having the rivets in the adjacent rows, on the same plane, triggered. This type is called 'Zig-zag Riveting'. Proportions are shown in figures.

16.14. BOILER JOINTS

In the construction of boilers, the circumferential joint or ring joint is usually made double riveted lap point while the longitudinal joint is made double or tripple riveted butt joint. Zig-zag riveting is considered more effective in making the joint leak proof.

In the construction of boilers and other pressure vessels the main plates in lap joint and cover plates in butt joint are first bevelled at 80° angle and then rolled to the proper curvature of the boiler shell. But in the drawing, as we show only a part of the plates, the curvature is ignored.

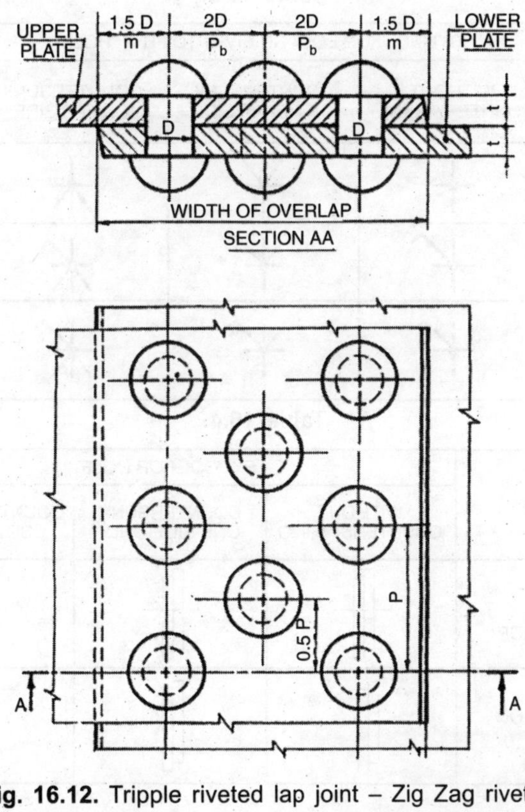

Fig. 16.12. Tripple riveted lap joint – Zig Zag riveting

16.15. CONVENTIONAL REPRESENTATION OF HOLES, BOLTS AND RIVETS

1. Representation on Projection Planes Normal to Their Axes

In order to represent holes, bolts and rivets on projection planes parallel to their axes, the symbols, represented in thick lines, given in *Tables 16.2* and *16.3*, shall be used. The symbol for holes shall be without a dot in the centre.

2. Representation on Projection Planes Parallel to Their Axes

In order to represent holes, bolts and rivets on projection planes parallel to their axes, the symbols shown in *Tables 16.4* and *16.5*, shall be adopted. Only the horizontal dash of these symbols shall be represented in thin line, while all other parts shall be represented in thick line.

In both the above cases, to distinguish bolts from rivets, the designation of bolts shall always being with a prefix showing the type of screw thread e.g. the designation of bolts with metric screw thread is M 12 × 50, whilst that for a rivet is ϕ 12 × 50.

Table 16.2.

HOLE	SYMBOL FOR HOLE			
	WITHOUT COUNTERSINKING	COUNTERSUNK ON NEAR SIDE	COUNTERSUNK ON FAR SIDE	COUNTERSUNK ON BOTH SIDES
DRILLED IN THE WORKSHOP				
DRILLED ON SITE	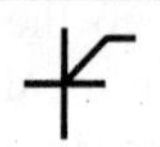	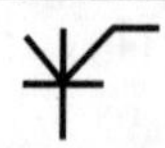		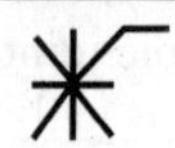

Table 16.3.

BOLT OR RIVERT	SYMBOL FOR BOLT OR RIVET TO FIT IN HOLE			SYMBOL FOR RIVET TO FIT IN HOLE COUNTERSUNK ON BOTH SIDE
	WITHOUT COUNTERSINKING	COUNTERSUNK ON NEAAR SIDE	COUNTERSUNK ON FAR SIDE	
FITTED IN THE WORKSHOP	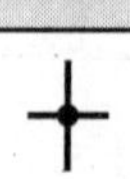			
FITTED ON SITE	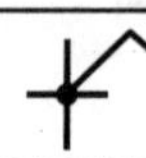			
FITTED ON SITE AND HOLE DRILLED ON SITE				

Table 16.4.

HOLE	SYMBOL FOR HOLE		
	WITHOUT COUNTERSINKING	COUNTERSUNK ONE SIDE ONLY	COUNTERSUNK ON BOTH SIDES
DRILLED IN THE WORKSHOP	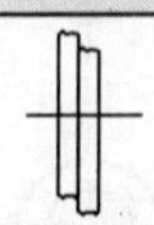	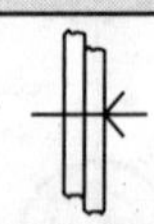	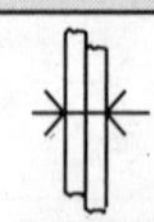
DRILLED ON SITE	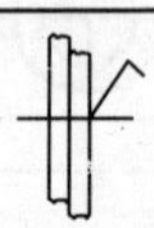	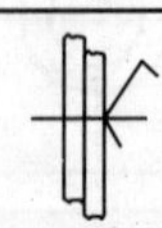	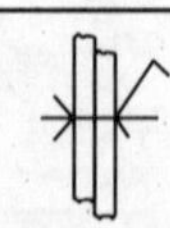

Table 16.5.

Bolt or rivet	Symbol for bolt or rivet to fit in hole		Symbol for rivet to fit in hole countersunk on both sides	Symbol for bolt with designated nut position
	Without countersinking	Countersunk on one side only		
Fitted in the workshop	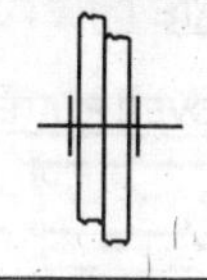	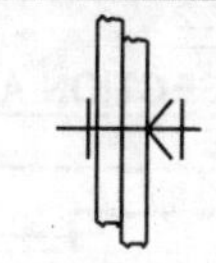	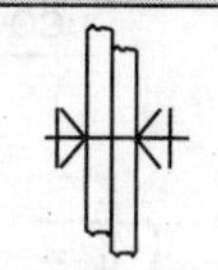	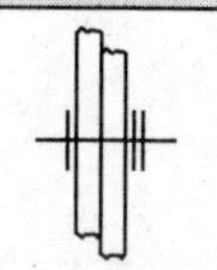
Fitted on site	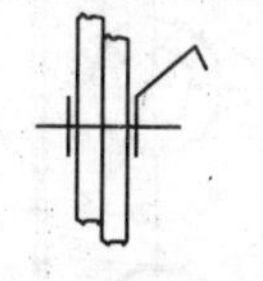	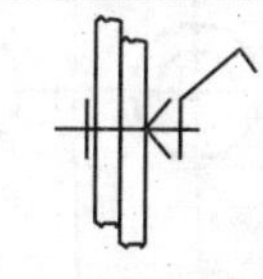	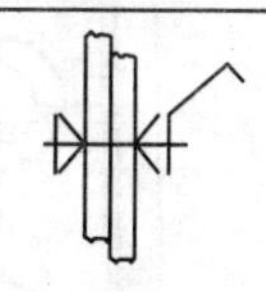	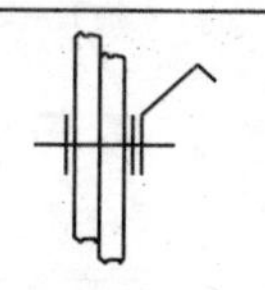
Fitted on site and hole driled on site	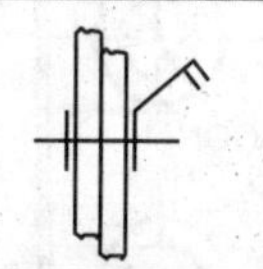	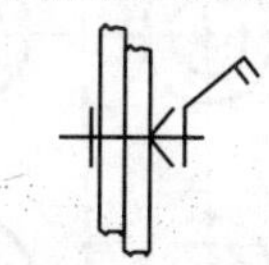	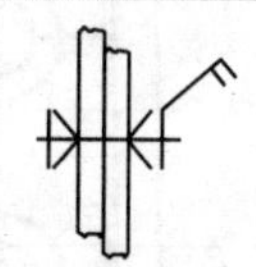	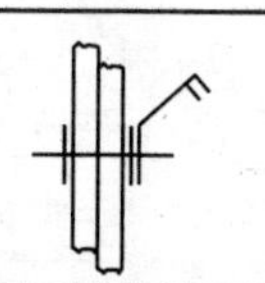

$t_1 = 0.625\ t$

t = THICKNESS OF MAIN PLATE
t_1 = THICKNESS OF COVER PLATE

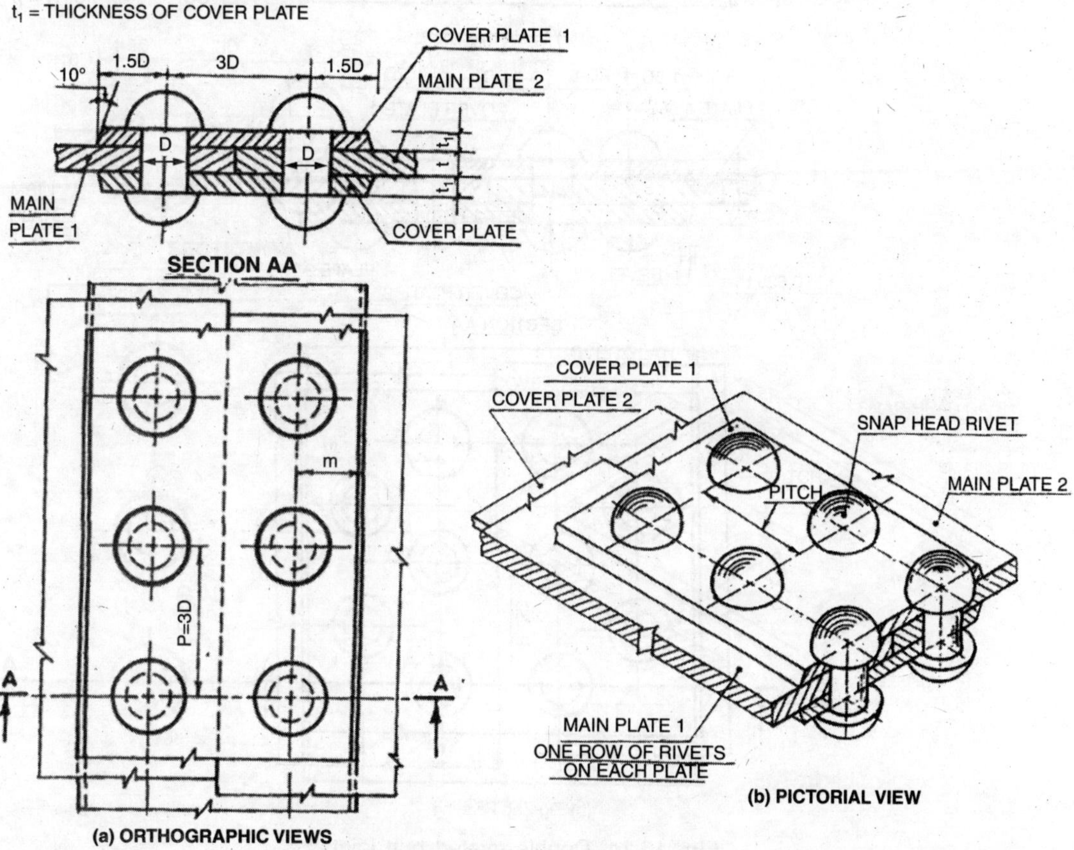

(a) ORTHOGRAPHIC VIEWS

(b) PICTORIAL VIEW

Fig. 16.13. Single riveted butt joint

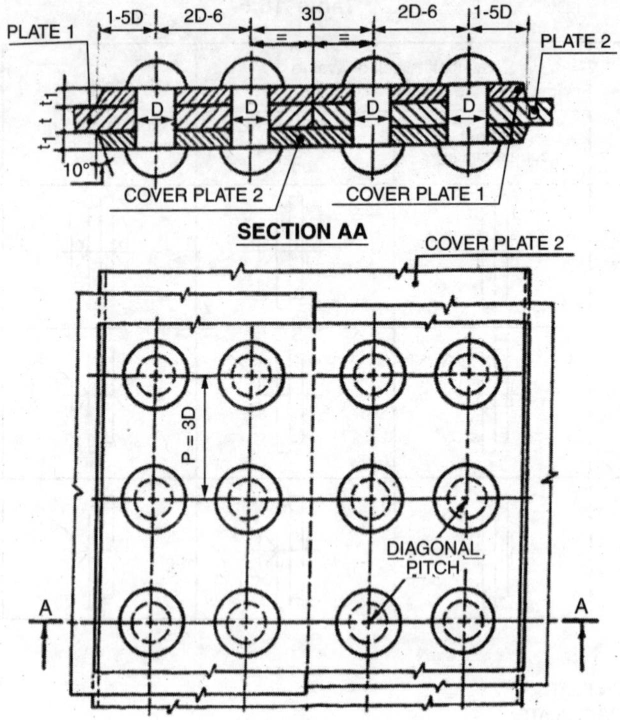

(a) Chain riverting

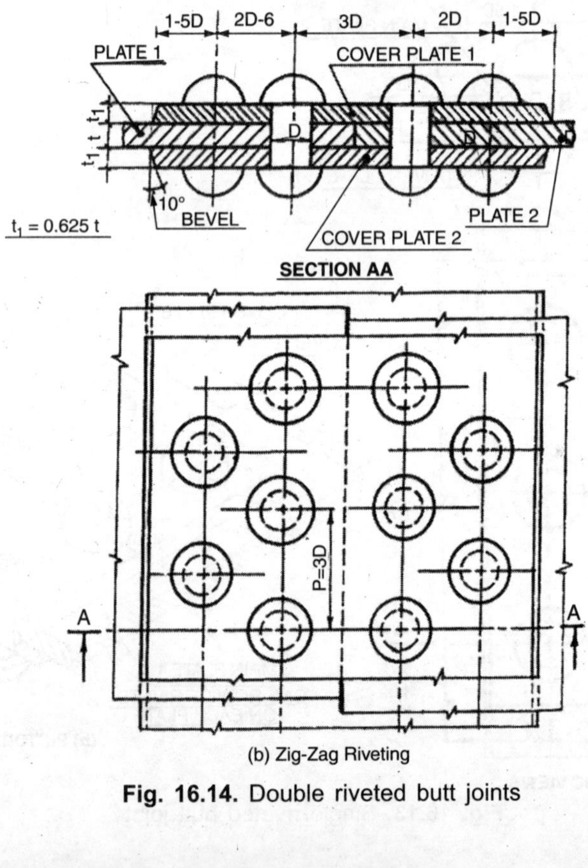

(b) Zig-Zag Riveting

Fig. 16.14. Double riveted butt joints

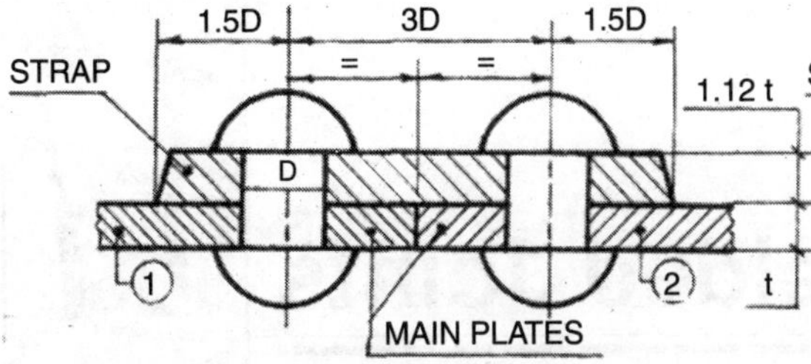

Fig. 16.15. Single Strap Butt Joint

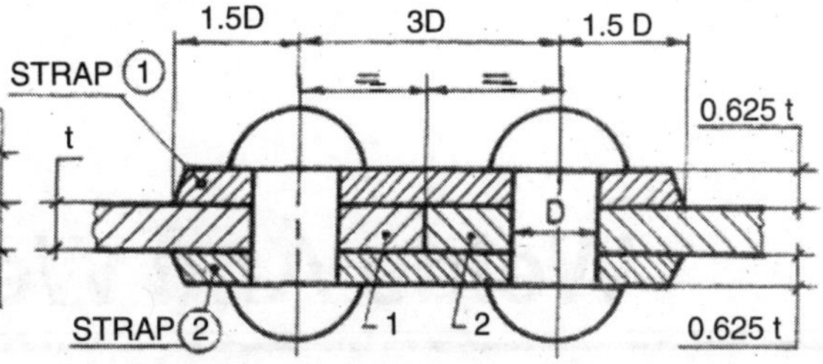

Fig. 16.16. Double Strap Butt Joint

EXERCISES

1. Sketch freehand at least four types of rivet hands and give their proportions. Take diameter for rivets as 25 mm.
2. Sketch two views of a single riveted lap joint for 16 mm thick plates. Show at least three rivets in top view.
3. Sketch two views of a double riveted lap joint for 9 mm thick plates when it has:
   (i) Chain riveting and (*ii*) Zig-zag riveting.
4. Sketch two views of a single riveted butt joint, with two cover plates, for 9 mm thick plates.
5. Sketch two views of a double riveted butt joint, with two cover straps, when the riveting is, (*i*) Chain type and (*ii*) Zig-zag type. Take thickness of plates as 9 mm. Show at least two pitch lengths.
6. What do you mean by caulking and fullering? Describe each by a suitable sketch.
7. Draw, to a full size scale, a double riveted butt joint with double cover plates, showing the zig-zag arrangement of rivets in the rows. Take the thickness of the butting plates as 12 mm and the nominal diameter D of the rivet as 20 mm. Fully dimension the drawing very clearly indicating also the dimensions of the cover plates.
8. Give dimensioned sketch of a double riveted zigzag butt joint with two cover straps for 12 mm thick main plates. Assume thickness of cover plates as 8 mm and pitch of rivets as three times the rivet diameter.
9. Give neat freehand orthographic sketches of the following riveted joints showing on them the pitch of the rivets in terms of the diameter of the rivets for the plate thickness in MM:
   (*a*) Double riveted lap joint-rivets zig-zag.
   (*b*) Double riveted butt joint-rivets zig-zag.
10. Show by means of neat sketches the difference between the chain riveting and zig-zag riveting. What is the advantage of one type over the other?
11. Draw to a full scale, two orthographic views of a triple riveted lap joint using rivets in zig-zag arrangement. Take $t = 12$ mm, $D = 20$ mm. Give all other necessary dimensions.
12. Show with the aid of proportionate sketches, three different ways on which two plates may be connected at right angles to each other.
13. Define the terms, pitch, diagonal pitch, overlap, margin, and distance between rows as used in riveted joints. Illustrate these by suitable sketches.

QUESTIONS FOR VIVA VOCE

1. What is a fastening?
2. What are the permanent and temporary fastenings? Give examples.
3. What is a rivet and of which metal it is usually made?
4. How many types of riveted joints are there?
5. What is the difference between a lap and butt riveted joint?
6. Why the ends of the overlapping plates are bevelled? State the usual angle of bevelling.
7. Why is caulking and fullering done on riveted joints?
8. What is the function of a cover plate as used in riveted joints?
9. What type of rivet you will suggest for use in: (*i*) Shipbuilding (*ii*) Structural work, and (*iii*) The field conditions, and why?

Welds And Welded Joints 17

17.1. WELDING

Welding is a process of joining together two or more pieces of metal, whose contacting surfaces are raised to a plastic or fusion state by the application of heat, with or without added metal and with or without the application of pressure. A weld is one of the permanent joints.

The primary importance of welding is the uniting of pieces of metal so that they will operate properly as a unit to support the loads to be carried. In order to design and build such a structure, to be economical and efficient, a basic knowledge of welding is essential. *Figure 17.1 (a)* illustrates many basic welding terms.

With the availability of better equipment and advanced techniques in welding, welding has made manufacturing process possible at previously unheard of temperatures and pressures. Welding enables the production of leak proof joints that will withstand high pressures and temperatures. Welding has greatly displaced riveting in shipbuilding and structural work.

Welding, as compared to riveting, is cheaper, stronger and simpler to execute. It can be performed at site and allows a considerable freedom in design.

Here, in this chapter, we shall describe the method of representing, on drawings, information concerning the type, size and position of welds in fusion-welded joints using the gas, arc or resistance welding processes.

17.2. TYPES OF WELDED JOINTS

There are five basic types of welded joints, classified according to the position of the members being joined, as shown in *Fig. 17.1 (b)*. These joints are named as Butt joint, Corner joint, Tee joint, Lap joint, and Edge joint.

17.3. ELEMENTARY WELDING SYMBOLS

The fillet, plug or slot, butt and groove are the basic types of arc and gas welds. The groove welds are further classified as V, bevel, U and J-butt. Whereas spot, seam, stitch, projection or upset and flash welds are the basic resistance welds.

The introduction of welding symbols on a drawing enables the designer to indicate clearly the type and size of weld required to meet the design requirements. It is becoming important for the designer to indicate the required type of weld correctly. Points that must be made clear are the type of weld, the joint preparation, the weld size, the root gap (if any) and the degree of penetration required. These points can be clearly indicated on the drawing by the welding symbol. Welding symbols are a shorthand language.

The various categories of welds are characterized by a symbol. The symbol, in general, is similar to the shape of the welds to be made and it, however, does not specify the process to be employed. Elementary symbols are shown in *Table 17.1*.

The vertical portions of the symbols for single-bevel butt, double-bevel butt, single-J butt, double-J butt and fillet welds should be shown in the left hand side of the symbol, irrespective of the orientation of the weld metal. The symbols, as given in *Table 17.1*, shall apply to the respective welds regardless of the number of runs deposited. The profiles of all welds are of user's standard proportions. Single-U or V joints are used up to 12 mm thick plates and double-U or V joints are used for plates having thickness above 12 mm.

Table 17.1.

DESIGNATION	ILLUSTRATION	SYMBOL
Butt weld between plates with raised edges (melted down completely).		╯╰
Square butt weld		‖
Single-V bult weld		V
Single-bevel butt weld		|/
Single-V butt weld with broad root race		Y
Single-bevel butt weld with broad root face		|/ with root
Singel-U butt weld (parallel or stoping sides)		U with stem
Single-J butt weld		J with stem
Backing run; back or backing weld		◡
Filter weld	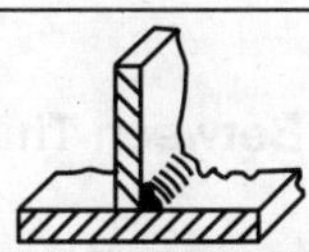	◺
Filter weld; or Slot weld		⊓
Spot weld		○
Seam weld	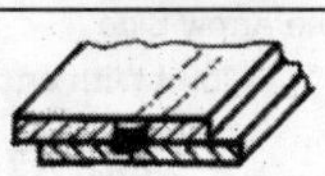	⦵

Elementary symbols can be completed by, adding to them a supplementary symbol characterizing the shape of the external surface of the weld. The supplementary symbols recommended by BIS are gives in *Table 17.2*. The absence of a supplementary symbol means that the shape of the weld surface does not need to be indicated precisely. Examples of combinations of elementary and supplementary symbols are given in *Table 17.3*.

Table 17.2.

SHAPE OF WELD SURFACE	SYMBOL
(a) Flat (usually flush)	—
(a) Convex	⌒
(a) Concave	‿

Table 17.3.

DESIGNATION	ILLUSTRATION	SYMBOL
Flat(flush) single-V butt weld		V̄
Convex double-V butt weld		X with convex arcs
Concave fillet weld		◺ with concave arc
Flat (flush) single-V butt weld with flat (flush) backing run		V̄ with flat backing run

When necessary, combinations of elementary symbols can be used as illustrated in *Table 17.4*. Though it is not forbidden to associate several symbols, it is better to represent the weld on a separate sketch, when symbolization becomes too difficult.

17.4. POSITION OF THE SYMBOLS ON DRAWINGS

The symbols form only part of a complete method of representation of welds on drawings, *Fig. 17.2*. Which comprises in addition to the symbol itself:

1. An arrow line per joint.
2. A dual reference line consisting of two parallel lines; one continuous and one dashed. The dashed line can be drawn, either above or beneath the continuous line. For symmetrical welds, the dashed line is unnecessary and should be omitted.

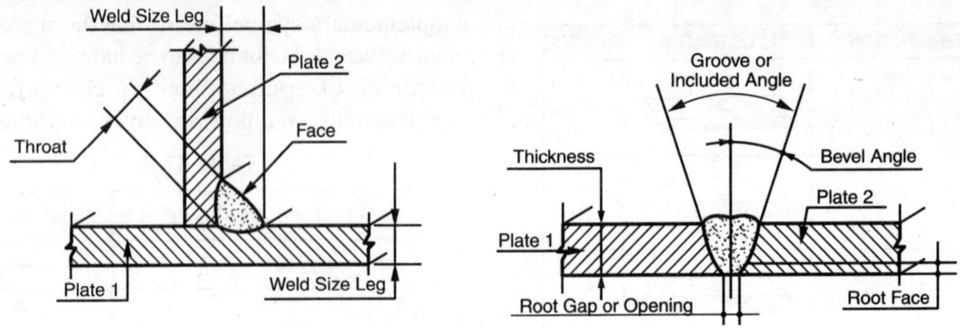

(a) Fillet Weld (b) Groove Weld

(a) Basic welding terms

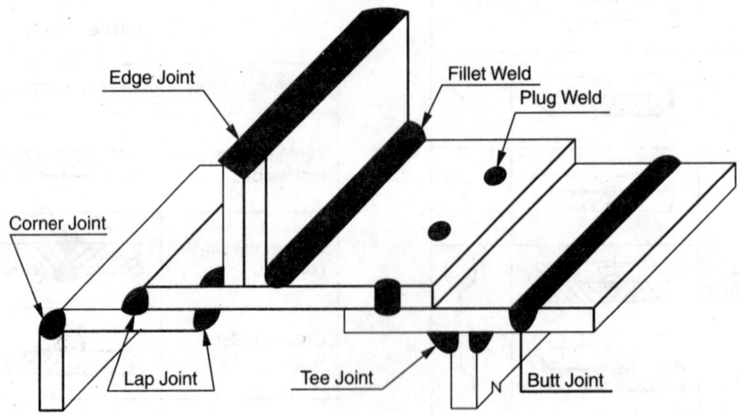

(b) Types of welded joints

Fig. 17.1.

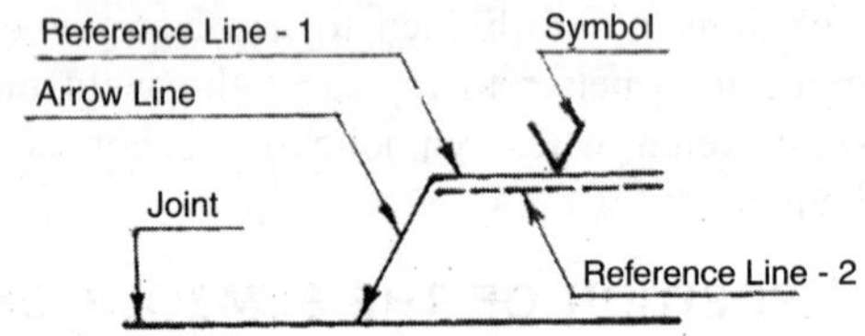

Method of Representation of Symbols on Drawing

Fig. 17.2.

3. A certain number of dimensions and conventional and signs.

Following rules define the location of welds by specifying:

(*i*) The position of the arrow line;
(*ii*) The position of the reference line; and
(*iii*) The position of the symbol.

1. Relation Between The Arrow Line and The Joint

In *Figs. 17.3* and *17.4*, meaning of the terms; (*a*) arrow side of the joint; and (*b*) other side of the joint is explained.

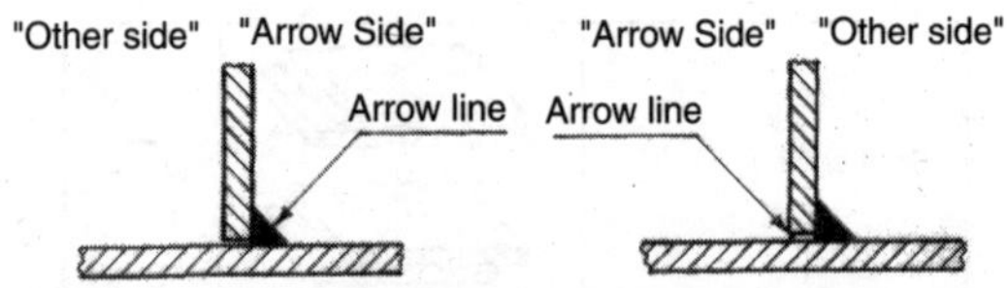

(a) Weld on the Arrow Side (b) Weld on the Other Side

T-joint With one Fillet Weld

Fig. 17.3.

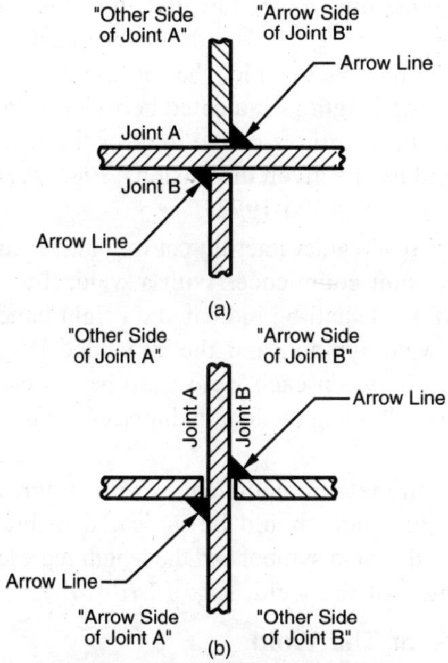

Cruciform Joint with two Fillet Welds

Fig. 17.4.

2. Position Of The Arrow Line

The position of the arrow line with respect to the weld is generally of no special significance, *Fig. 17.5 (a), (b)*. However, in the case of welds which require only one plate to be prepared, the arrow line shall point towards the plate, which is prepared, *Fig. 17.5 (c) and (d)*. The arrow line joins one end of the continuous reference line, such that it forms an angle with it, and an arrow head touching the weld seam.

3. Position of the Reference Line

The reference line shall preferably be drawn parallel to the bottom edge of the drawing or if it is not possible then perpendicular.

4. Position of the Symbol With Regard to the Reference Line

The symbol is to be placed either above or beneath the reference line in accordance with the following regulations:

(*i*) The symbol is placed on the continuous line side of the reference line if the weld (weld face) is on the arrow side of the joint, *Fig. 17.6 (a)*.

(*ii*) The symbol is placed on the dashed line side if the weld (weld face) is on the other side of the joint, *Fig. 17.6 (b)*.

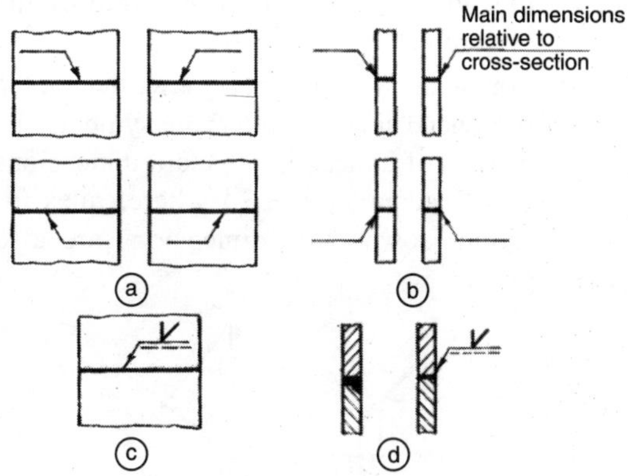

POSITION OF ARROW LINE

Fig. 17.5.

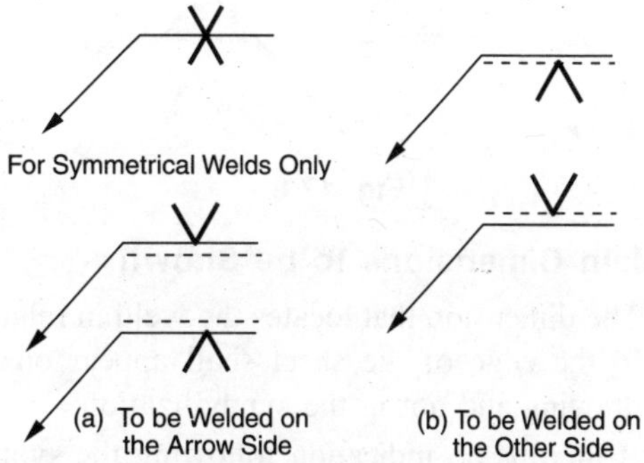

POSITION OF THE SYMBOL ACCORDING TO THE REFERENCE LINE

Fig. 17.6.

17.5. DIMENSIONING OF WELDS

1. General Rules

(*i*) Each weld symbol may be accompanied by a number of dimensions.

(*ii*) These dimensions are written, as shown in *Fig. 17.7* such that;

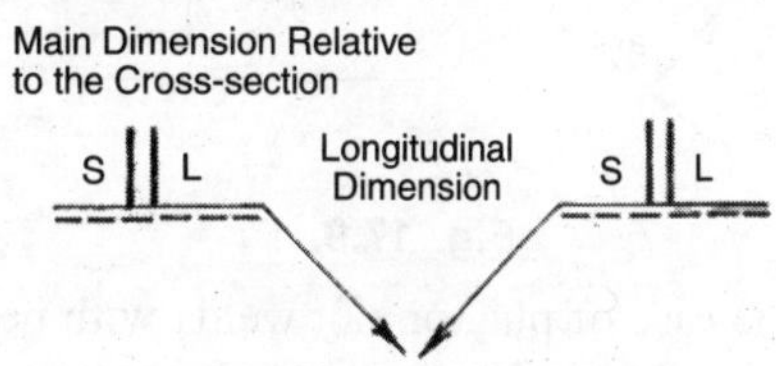

DIMENSIONING THE WELDS

Fig. 17.7.

(*a*) The main dimensions pertaining to the cross-section of the weld are written on the left hand side of (i.e., before) the symbol.

(*b*) Longitudinal dimensions are written on the right-hand side (i.e., after) the symbol. The method of indicating the main dimensions is defined in *Table 17.5* where, rules for setting down these dimensions are also given.

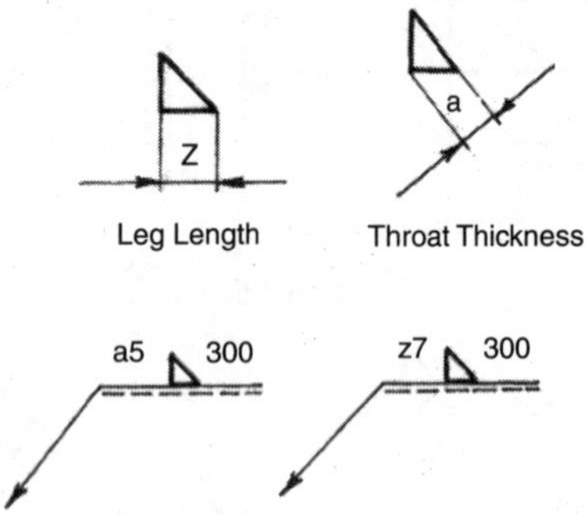

Fig. 17.8.

2. Main Dimensions to be Shown

(*i*) The dimension that locates the weld in relation to the edge of the sheet shall appear on the drawing and not in the symbolization.

(*ii*) If there is no indication following the symbol it means that the weld is to be continuous over the whole length of the work piece.

(*iii*) Unless otherwise indicated, butt welds are to have complete penetration.

(*iv*) For the fillet welds, there are two methods to indicate dimensions, *Fig. 17.9*. Therefore, the letters a or z shall always be placed in front of the value of the corresponding dimension.

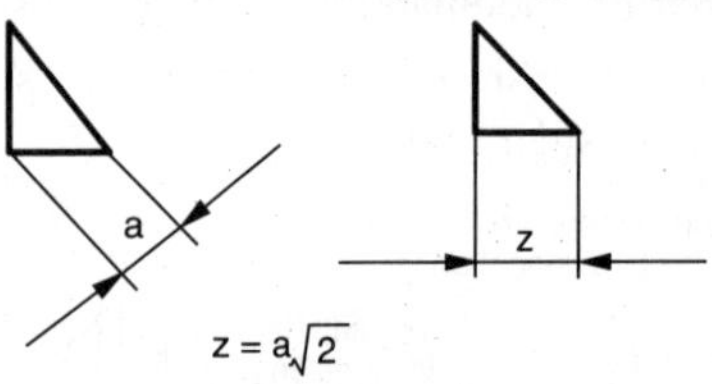

Fig. 17.9.

(*v*) In the case of plug or slot welds with bevelled edges, it is the dimension at the bottom of the hole, which shall be taken into consideration.

3. Length of Welds

Dimensions of weld lengths, weld sizes and pacing of intermittent welds should be given in mm. Unless otherwise indicated all welds shall be continuous along the length of the weld, or the length of the weld to which the basic symbol applies shall be the length as indicated between the abrupt changes in directions of the joint or the length as indicated by the given dimensions, *Fig. 17.10 (a), (b)* (as per IS : 696-1972).

Where a regular intermittent welding is required and the joint commences with a weld, the length of each weld shall be indicated on right hand side of the weld symbol, and the unwelded length of the joint between each weld shall be indicated in brackets following the length dimension, *Fig. 17.10 (c)*.

When the joint commences with an unwelded length the same should be indicated in brackets between the weld symbol and the length representing the length of the weld, *Fig. 17.10 (d)*.

4. Size of The Weld

The size of fillet, spot, seam and stich welds shall be indicated by inserting a dimension on the left hand side of the weld symbol, *Fig. 17.10 (a)* for equal welds, and Fig. *17.10 (b)* for unequal welds. The size should be indicated against each weld symbol unless there is a general note on the drawing stating that all welds are of any particular size.

In the case of single-V, single-U, single-J or bevel butt welds, where it is required that the penetration should be only partial, the vertical depth of penetration should be indicated by a dimension added to the left-hand side of symbol, *Fig. 17.11 (a)* and *(c)*.

In the case of double-V, double-J or double bevel welds, where only partial penetration is required; the same procedure is to be followed. However, in this case, the dimension should be added above and below the reference line in order to conform to the disposition of the symbol, *Fig. 17.11 (d)*. If the joint is asymmetrical in transverse cross-section, indication should be provided by the fact that the dimensions above and below the reference line should be dissimilar, *Fig. 17.11 (e)*.

17.6. WELD CONTOURS

The nominal contours for fusion, fillet and butt welds, should be as shown in *Fig. 17.12* and

Table 17.4. (Contd)...

DESIGNATION SYMBOL	ILLUSTRATION	REPRESENTATION	SYMBOLIZATION EITHER	SYMBOLIZATION OR
Butt weld between plates with raised edges and backing run	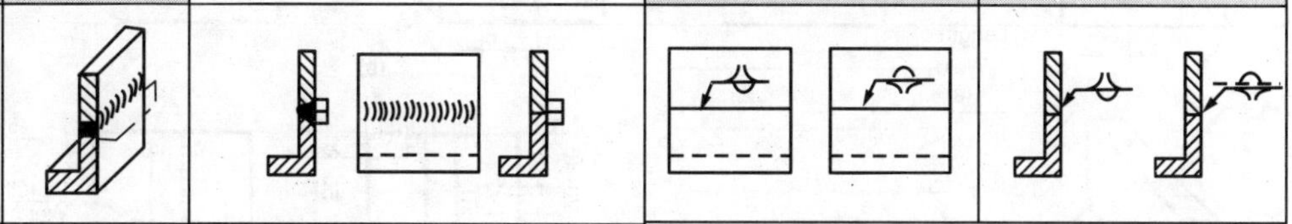			
Square butt weld ‖ Welded from both sides	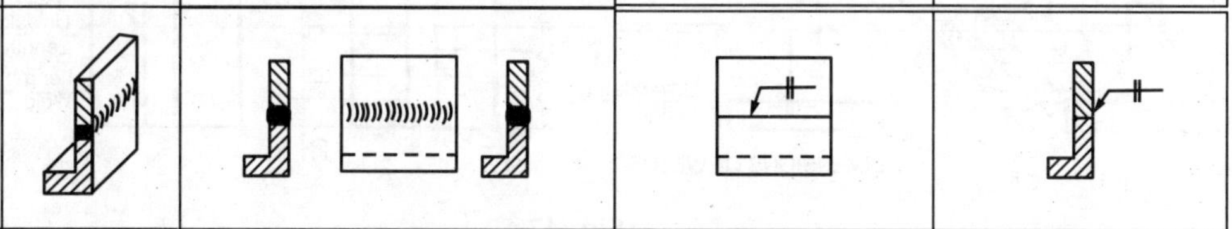			
Single-V butt weld V and	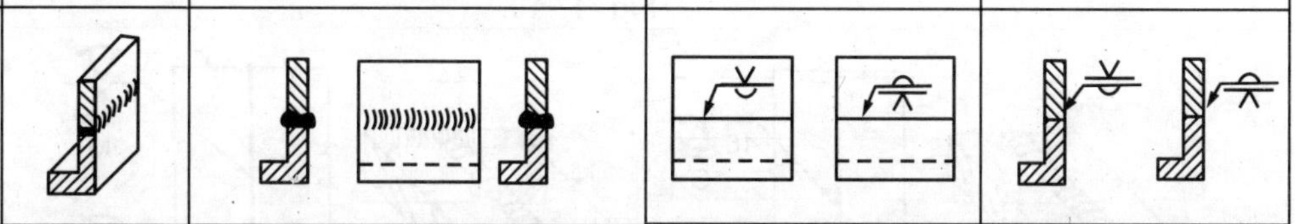			
backing run	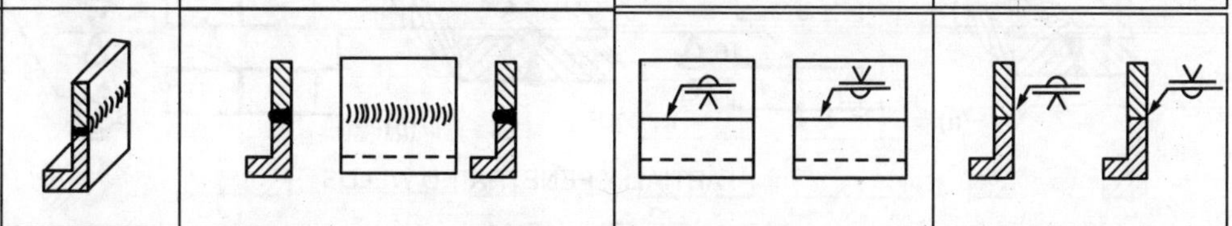			
Double-V but weld (X-weld)	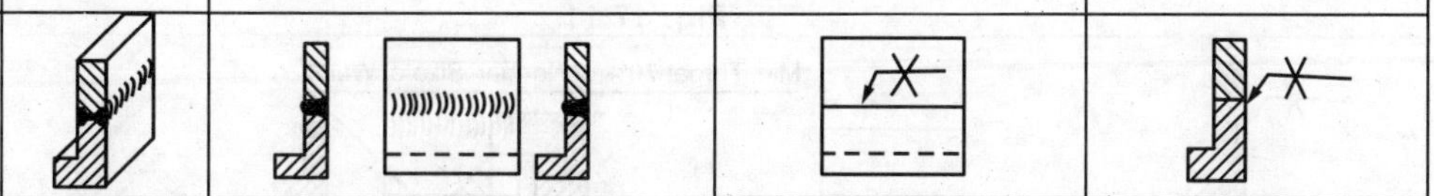			
Double bevel butt weld (K-weld)	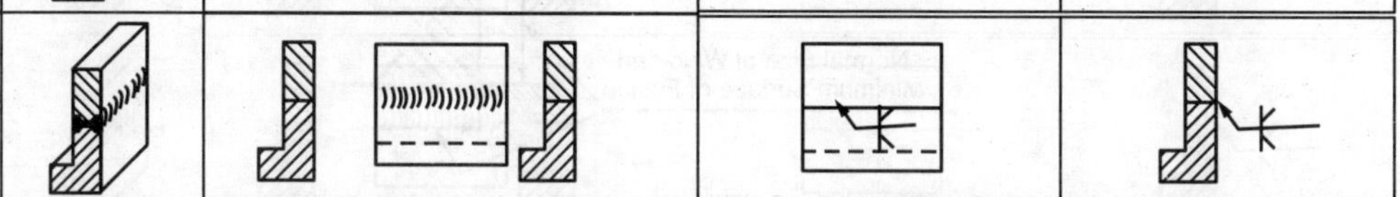			
	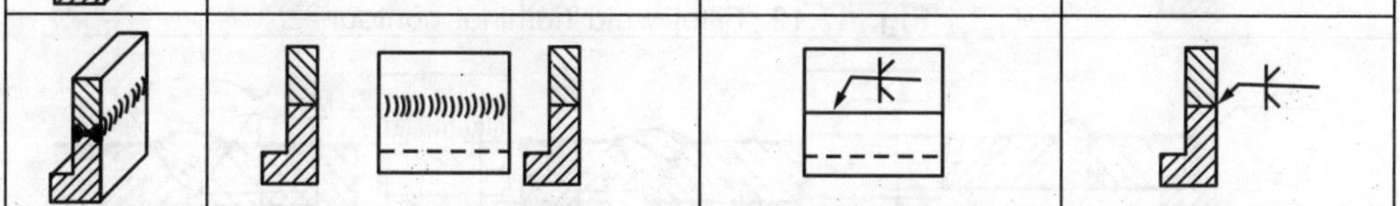			
Double-V butt weld Y with broad root face —	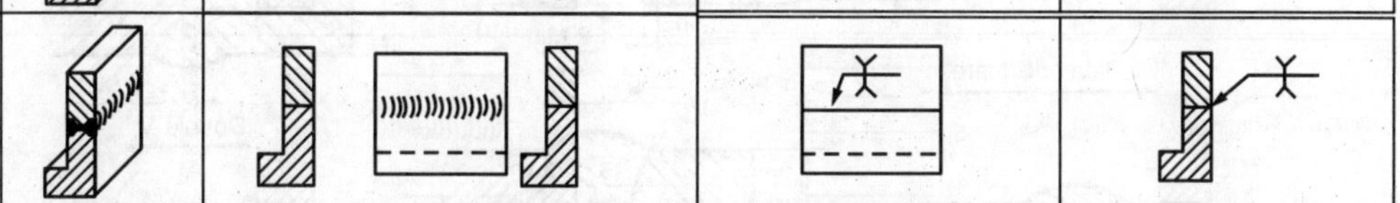			
Double bevel but weld with broad root face	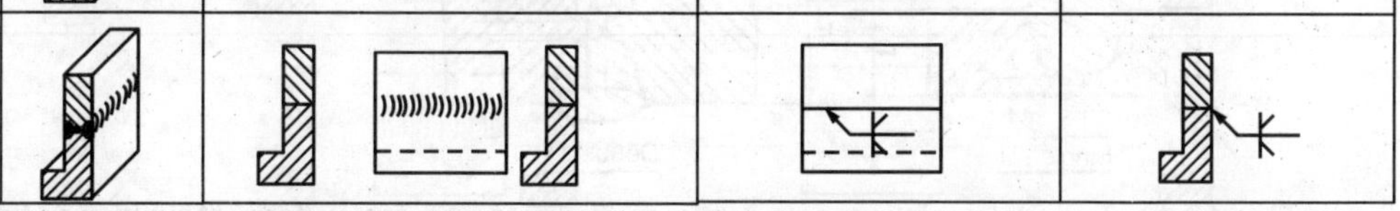			

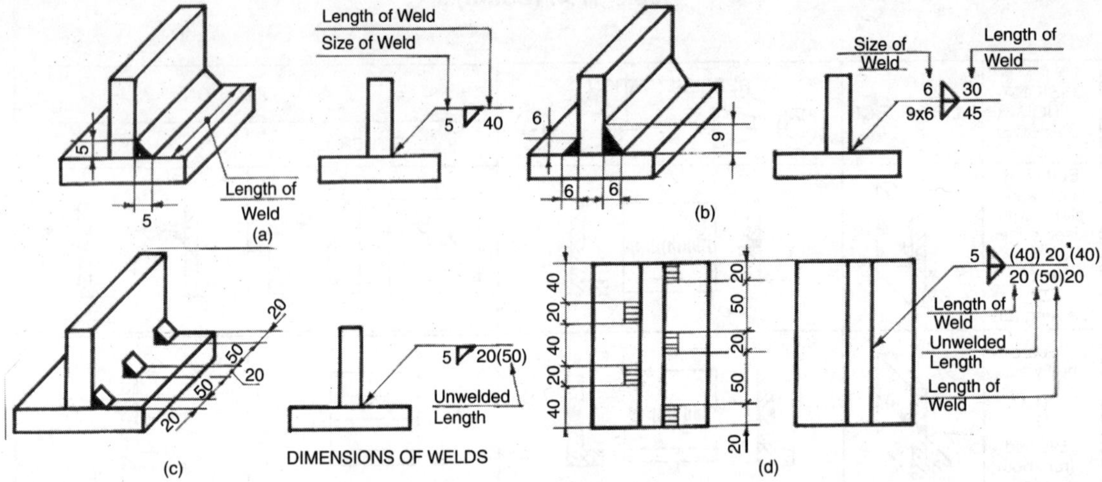

DIMENSIONS OF WELDS

Fig. 17.10.

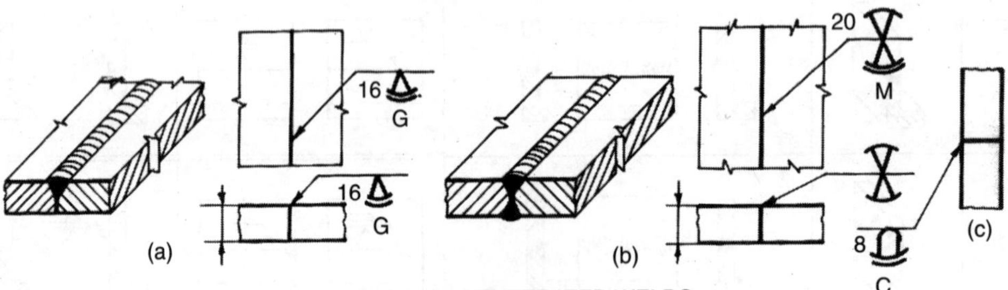

PARTIALLY PENETRATED WELDS

Fig. 17.11.

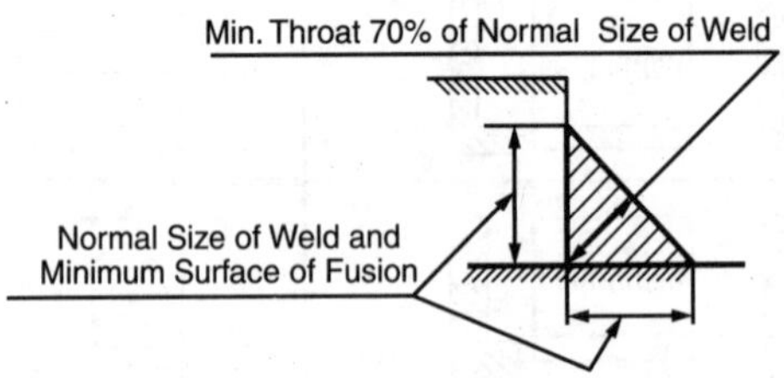

Fig. 17.12. Fillet weld nominal contour

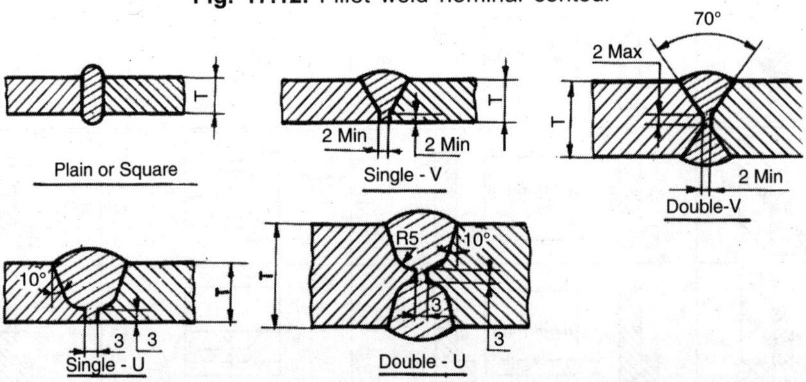

Fig. 17.13. T = Plate thickness and minimum dimension for surface of weld, i.e. finish weld nowhere below line of normal plate surface

17.13. Plain or square butt joint is used for plates of thickness about 2 mm. Single-V butt joint is used for plates of thickness from 8 to 15 mm. The edges of plate to be jointed are bevelled to form an included angle of 70° to 90°. Double-V butt joint is used for plates more than 16 mm in thickness. Single-U butt joint is used for plates of 16 mm to 20 mm thickness. Double-U butt joint is used for plates of thickness above 20 mm.

17.7. SURFACE CONTOUR AND FINISH OF WELDS

Finishing of welds other than cleaning shall be indicated by suitable contour (flush, convex, concave) and finish symbols (chipping C, machining M and grinding G), *Fig. 17.14 (a), (b)*. Where a weld is required to have approximately flush surface without recourse to any method of finishing, a straight line should be added below the symbol to indicate it, *Fig. 17.14 (c)*.

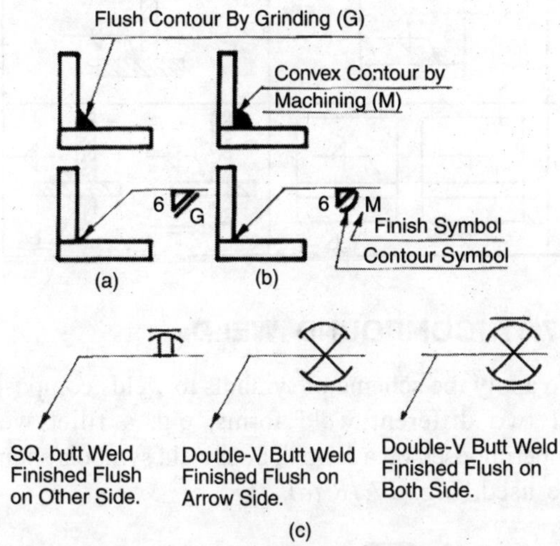

Fig. 17.14.

17.8. WELD ALL ROUND

In order to simplify the application of the scheme to case where it is required that a weld should be made all round the joint, a circle should be placed at the 'elbow' connecting the arrow to the reference line, *Fig. 17.15 (a), (b)*.

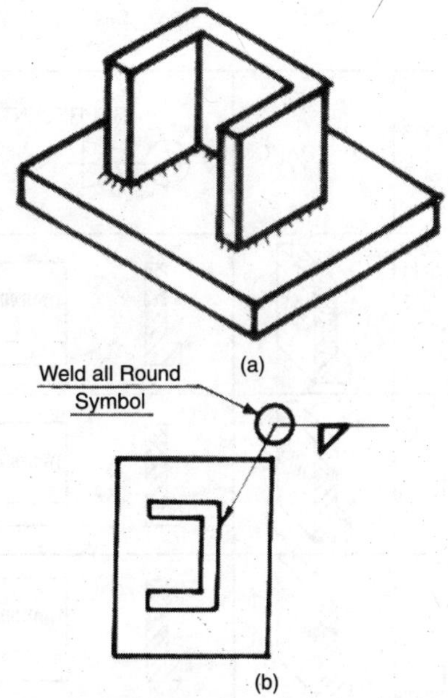

Fig. 17.15.

17.9. SITE WELD

When some of the welds in any particular welded structures are required to be made on site, during erection or assembly, they should be designated by a filled-in-circle at the 'elbow' connecting the arrow to the reference line, *Fig. 17.16*, to differentiate them from the shop welds.

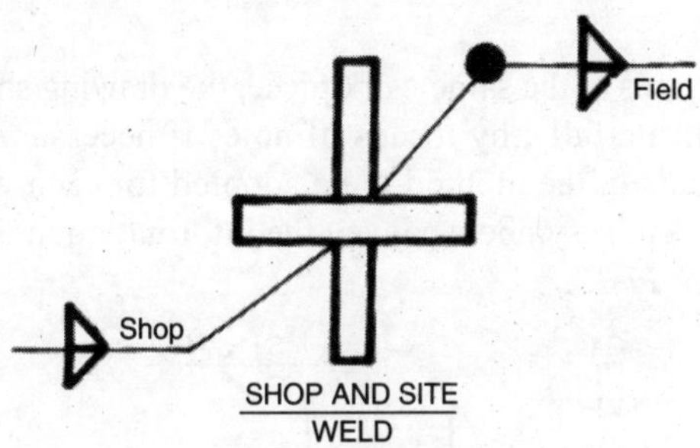

Fig. 17.16.

17.10. METHOD OF WELDING

Where all welds shown on drawing are to be deposited by the same process, a note specifying the processes may be made on the drawing. When different processes are to be used for different

Table 17.4. (Contd.)

DESIGNATION SYMBOL	ILLUSTRATION	REPRESENTATION	SYMBOLIZATION: EITHER	SYMBOLIZATION: OR
Double-U butt weld	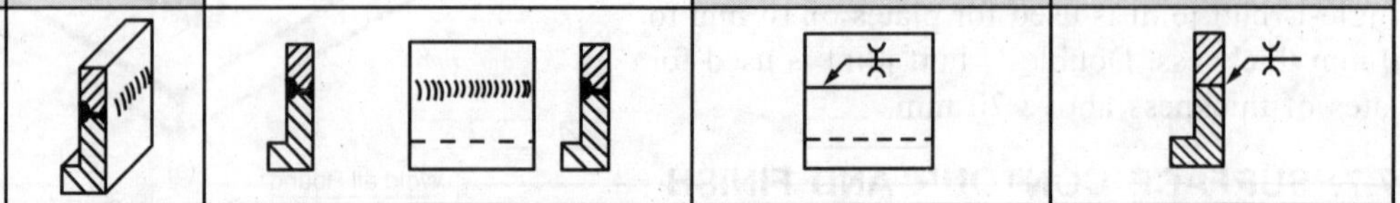			
Double-J butt weld	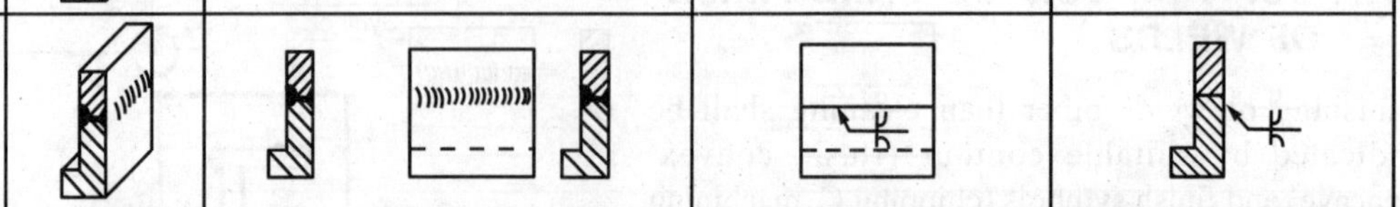			
Single-V / Single-U / Butt welds	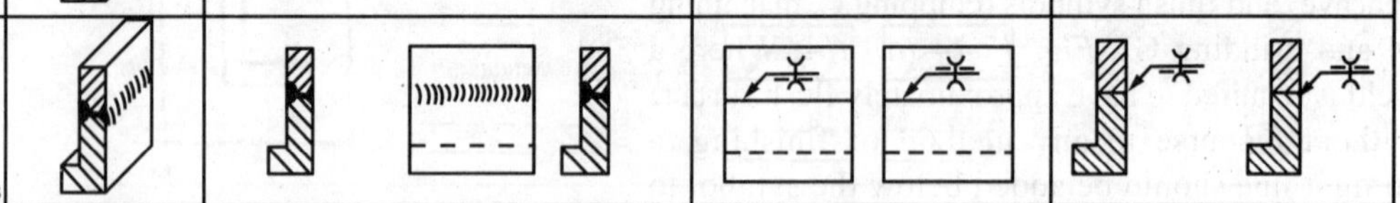			
Fillet weld and Fillet weld	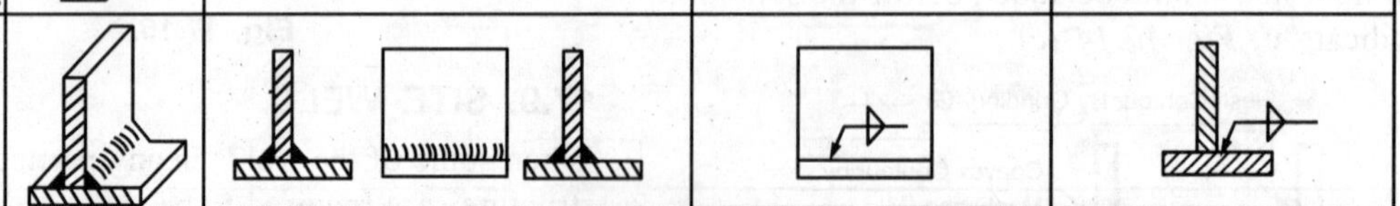			
	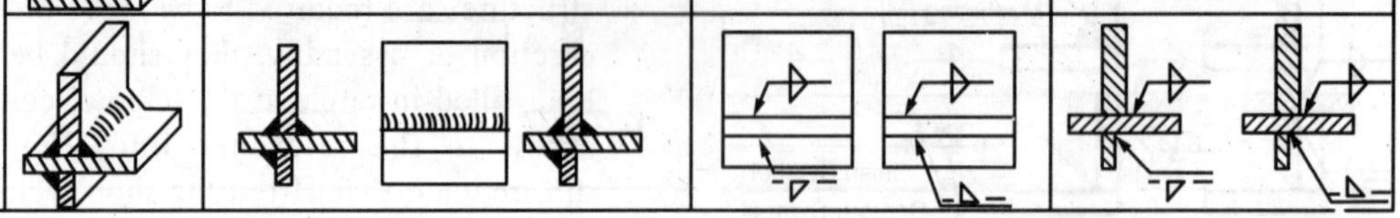			

portions of the same work piece, the drawing should indicate fully, by means of notes if necessary, the details of the method to be adopted for each weld, unless the symbols convey the information in them, *Fig. 17.17.*

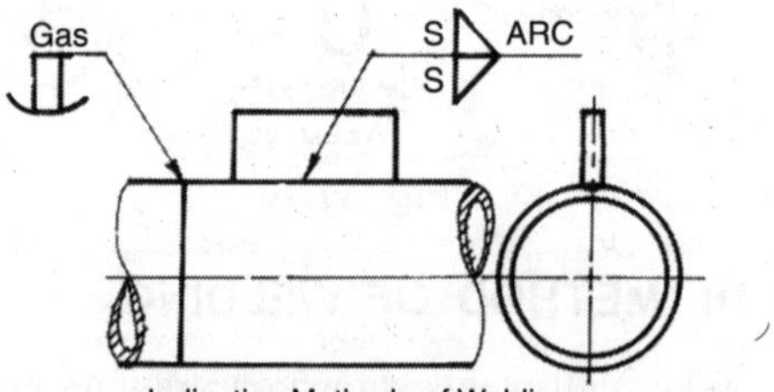

Indicating Methods of Welding

Fig. 17.17.

17.11. COMPOUND WELD

To apply the scheme of symbols to welds composed of two different weld forms, e.g. a fillet weld superimposed on a J-weld, both weld symbols should be used, *Fig. 17.18 (a), (b).*

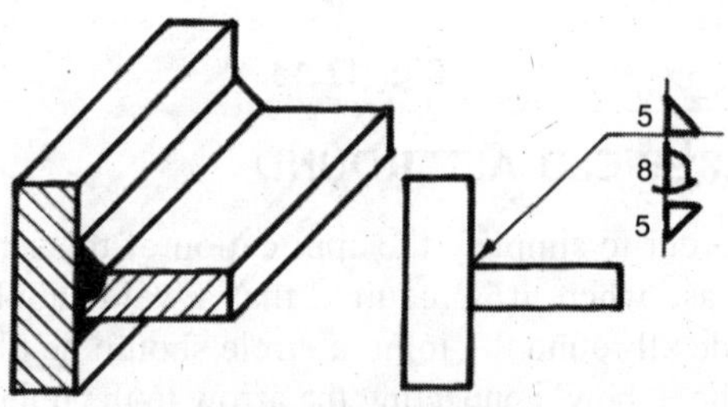

(a) Compound Weld (b)

Fig. 17.18.

Table 17.5.

DESIGNATION OF WELDS	DEFINITION		INSCRIPTION
Butt weld	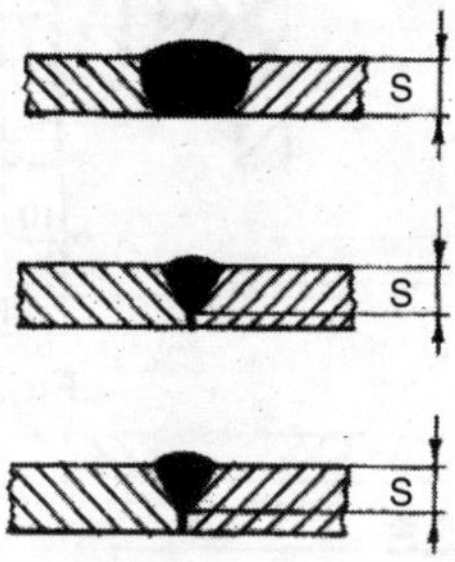	S : minimum distance from the surface of the part to the bottom of the penetration, which can not be greater than the thickness of the thinner part.	V S‖ S Y
Butt weld between plates with raised edges		S : minimum distance from the external surface of the weld to the bottom of the penetration.	S‖
Continuous fillet weld	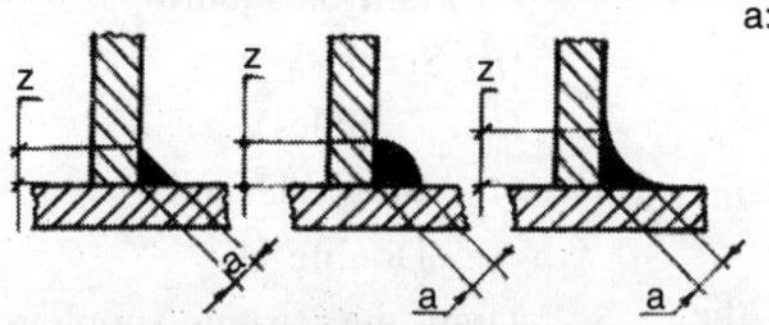	a: minimum distance from the external surface of the weld to the bottom of the penetration. z: side of the largest isoscetes triangle that can be inscribed in the section.	a ◺ z ◺
Intermittent fillet weld	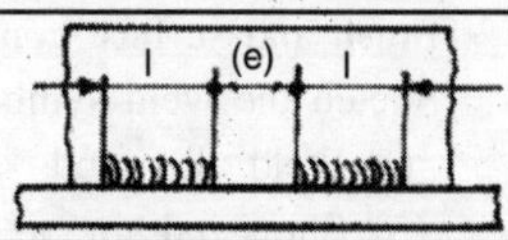	l : length of weld (without end craters) (e) : distance between adjacent weld element n : number of weld elements.	a◺ n + l(e) z◺ n + l(e)
Staggered intermittent fillet weld	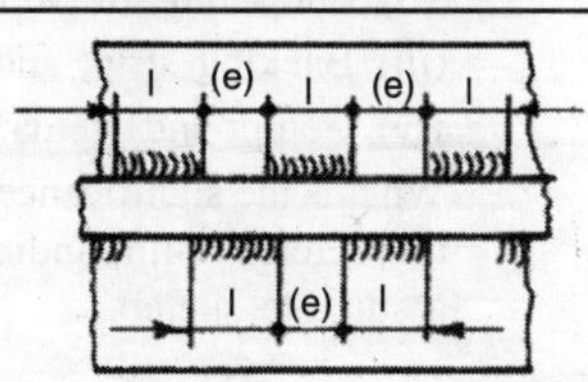	l, (e), n, a and z : all have the same meaning as given above	a n × l (e) a n × l (e) z n × l (e) z n × l (e)
Plug or slot weld	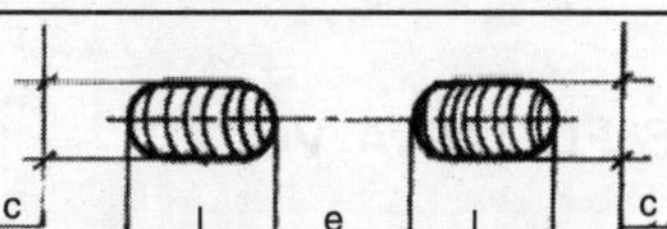	l, (e) and n : as above c : Width of slot.	c ⊓ n × l(e)
Seam weld	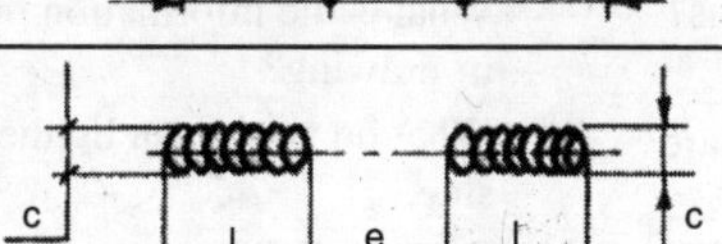	L,(e), and n : as above C : width of weld	C ⊖ n × l(e)
Plug weld	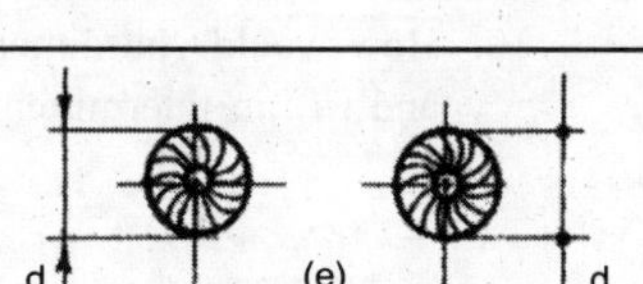	n : number of weld elements (e) : spacing d : diameter of hole	d ⊓ n × (e)
Spot weld	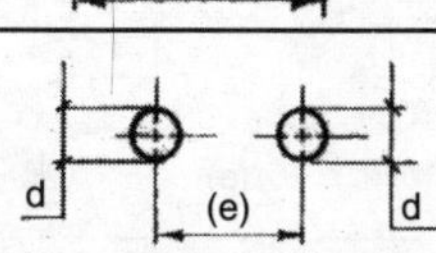	n : as above (e) : spacing d : diameter of spot	d ○ n × (e)

17.12. PROJECTION WELD

Full details of the projections should be given either on the drawing or in the welding procedure sheet (WPS). Spacing between welds should be measured centre-to-centre and should be shown on the right hand side of the symbol. Overall length of a series of projection welds should be dimensioned on the drawing. The total number of welds should be indicated in brackets above and below the symbol.

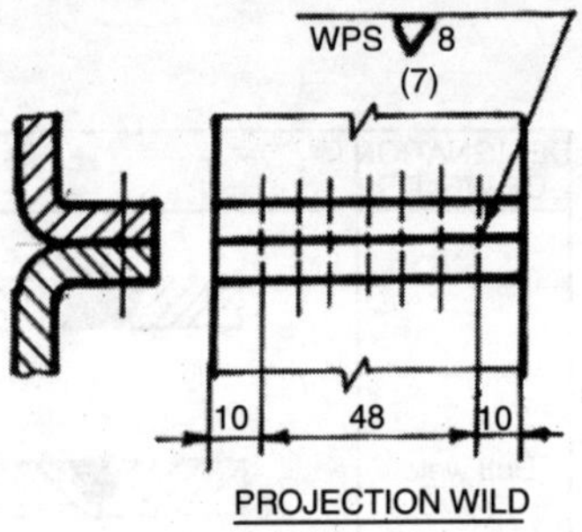

Fig. 17.19.

EXERCISES

1. Sketch neatly, taking any suitable proportions, following welded joints:
   (*a*) Butt joint (*b*) Lap joint
   (*c*) Tee joint (*d*) Corner joint, and
   (*e*) Edge joint
2. Draw a neat, sketch of 'standard location of elements of welding symbol" according to BIS.
3. Draw the 'cross-sectional representation' and the symbol of the following forms of welds:

1.Fillet	10. Plug or slot
2.Single-V butt	11. Stitch
3.Double-V	12. Flash
4.Square butt	13. Single-J butt
5.Single bevel butt	14. Double-J butt
6.Single-U butt	15. Butt resistance
7.Double-U butt	16. Backing strip
8.Stud	17. Projection
9.Bead	18. Tube

4. Sketch neatly, the nominal contours for fusion butt and fillet welds, indicating their proportions:
   (*i*) Plain or square
   (*ii*) Single-V
   (*iii*) Double-V
   (*iv*) Single-U and
   (*v*) Double-U
5. Draw the various supplementary symbols for weld finish and surface contour.
6. Sketch the weld symbols to indicate:
   (*i*) Weld all round
   (*ii*) Shop and site weld
   (*iii*) Backing strip, and
   (*iv*) Compound welds.
7. What is the significance of the position of arrowhead and reference line indication of welds? Illustrate by a suitable sketch.

QUESTIONS FOR VIVA VOCE

1. What is welding and what type of fastening it is?
2. What is fusion welding?
3. Why the edges of various parts to be welded are prepared?
4. What are the advantages of welded joints over riveted joints?
5. What are the fundamental types of welded joints?
6. What to YOU understand by, (*i*) length of a weld, (*ii*) Size of a weld?
7. What is the information required to represent a weld in drawing?
8. What do you mean by the arrow-side and the other-side.
9. What is WPS?
10. How would you dimension: (*i*) a continuous weld, and (*ii*) an intermittent weld in a drawing?

Screw Threads And Threaded Fasteners

18

18.1. INTRODUCTION

A screw thread is the functional element used on a bolt, stud, set-screw or any other threaded piece or component. A screwed piece consists essentially of a cylinder of round core along which a projection, or thread, winds its way in the form of a helix or, it can be said that a screw thread is formed by cutting helical groove on a cylindrical surface (outside or inside). All screwed pieces fall under the category of temporary fasteners, already discussed in chapter 17.

Screw threads and fastening by screwed pairs have frequent occurrence in the machine parts. The usual forms, methods of representation, and proportions of screw threads are of prime importance in machine drawing and hence should be clearly understood. Threaded pieces are always used in pairs to hold different machine parts together temporarily.

The principal uses of threads are, (*i*) for fastening, (*ii*) for adjusting, and (*iii*) for transmitting power.

As a helix is the curve of the thread elements it is logical to begin with a study of its generation and true projection.

18.2. HELIX

It is the curve generated by a point moving on the surface of a cylinder or cone in circumferential direction, at a constant angular speed, and with a simultaneous uniform rate of advance in an axial direction. The amount of this advance for one revolution is called lead of the helix. Helices occur in practice as screw threads, springs, worm gears, propeller blades and cylindrical cams. For practical purposes a screw may be considered to be wedge in the form of a helix, so that the input motion is a rotation while the output remains translation.

1. Drawing a Helix

The construction and development of a cylindrical helix, for one turn, is illustrated in *Fig. 18.1 (a)* and *(b)* respectively. Since the motion of the generating circle point is uniform in both angular and axial directions, when the projection of the point has moved from 1 to 2 on the circle, in the top view, it will move parallel to the axis from 1 to 2 in the front view. The circle in top view and the pitch of the helix in front view is to be divided in equal parts, 12 here.

If the cylinder were to be developed, the helix would become the hypotenuse of a right triangle whose base is the circumference of the cylinder and whose altitude is the lead of the helix. The angle between the hypotenuse and the base of the triangle is known as the helix angle.

18.3. THREAD TERMS AND NOMENCLATURE

In a study of threads and their representation on a drawing, it is necessary to define a few important terms used in specifying any thread, *Figs. 18.2, and 18.3.*

1. Screw Thread

It is a ridge of uniform cross-section produced by forming, on the outside or inside surface of a cylinder or a cone, a continuous helical groove, such that the distance measured parallel to the axis, between two corresponding points on its contour is proportional to their relative angular displacement about the axis.

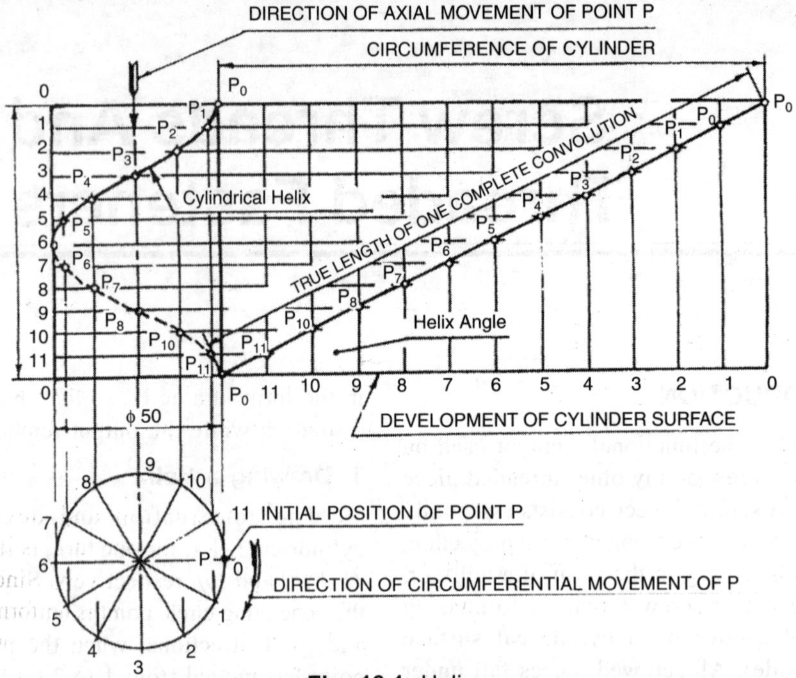

Fig. 18.1. Helix

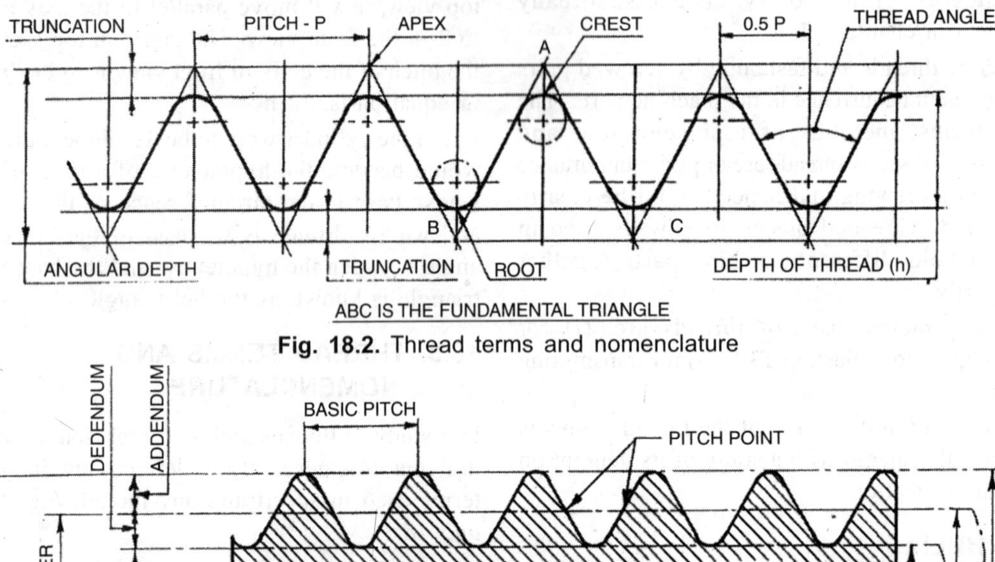

Fig. 18.2. Thread terms and nomenclature

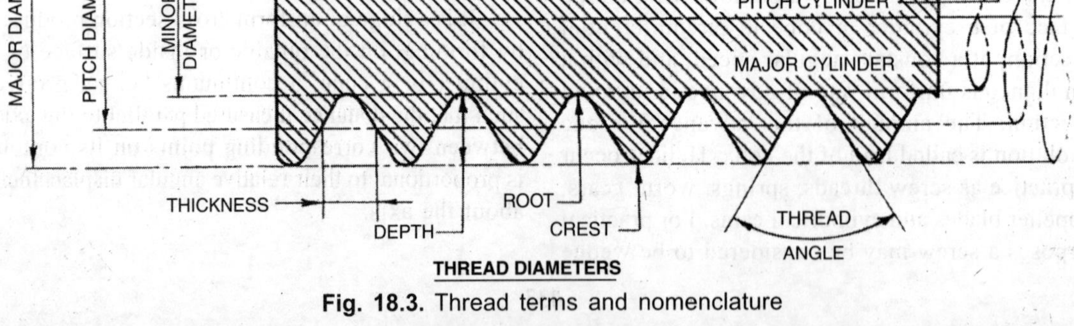

Fig. 18.3. Thread terms and nomenclature

A parallel screw thread is formed on the surface of a cylinder, whereas a tapered screw thread is formed on the surface of a cone, *Fig. 18.4*.

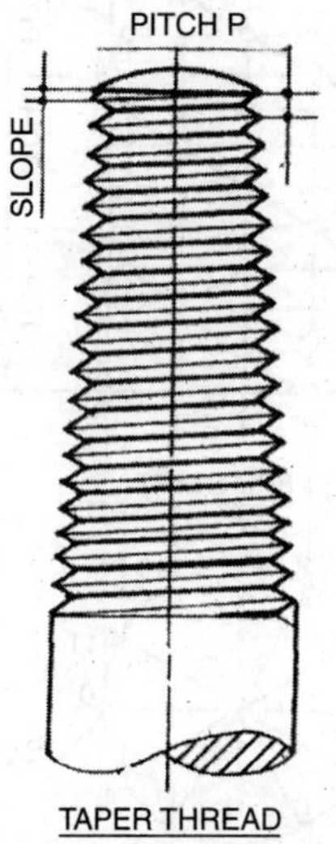

Fig. 18.4

2. External Thread

A thread on the outside surface of a member such as bolt, stud or screw.

3. Internal Thread

A thread on the inside surface of a member such as a threaded hole or a nut.

4. Crest

The top surface joining the two sides of a thread, or, that part connecting adjacent flanks at the top of thc groove.

5. Root

The bottom surface joining the sides of the two adjacent threads, or, the part which connects adjacent flanks at the bottom of the groove.

6. Angle of Thread

It is the angle between the flanks, measured in an axial plane, *Fig. 18.2*.

7. Flank

It is the surface between the crest and the root of a thread .

8. Helix Angle

It is the angle, which the helix makes at any point with a plane perpendicular to the axis. The tangent of this angle is equal to the lead divided by the circumference of the helix cylinder, *Fig. 18.1*.

9. Fundamental Triangle

It is the triangle of which two sides represent the form of the theoretical thread with sharp crests and roots having the same basic pitch and flank angle as the actual thread, and whose third side or base is parallel to the generator of the cylinder or cone on which the thread is formed. The Apex is the sharp corner of the fundamental triangle opposite to its base.

10. The Height (or Depth) of the Fundamental Triangle

It is the distance measured perpendicular to the thread axis, from its apex to its base.

11. The Truncation

It is the distance measured perpendicular to the axis, between the major and minor (or core) cylinders and adjacent apex of the fundamental triangle.

12. Depth of Thread

It is the distance measured perpendicular to the axis, between crest and root.

13. Theoretical Depth

It is the depth of the thread, which could be possible for perfect action.

14. Pitch

It is the distance measured axially between corresponding points on the consecutive thread forms in the same axial plane and on the same side of the axis. For any particular type of uniform thread, pitch remains constant for full length of thread. For this purpose we can describe pitch as a 'crest to crest' measurement taken parallel to the axis of the screw and taking adjacent thread forms.

15. Lead

It is the distance measured parallel to the axis between a point on a thread form and a corresponding point on the same thread helix after one complete revolution, *Fig. 18.5 (a), (b), (c)*. In other words it is axial distance a screw thread advances in one revolution.

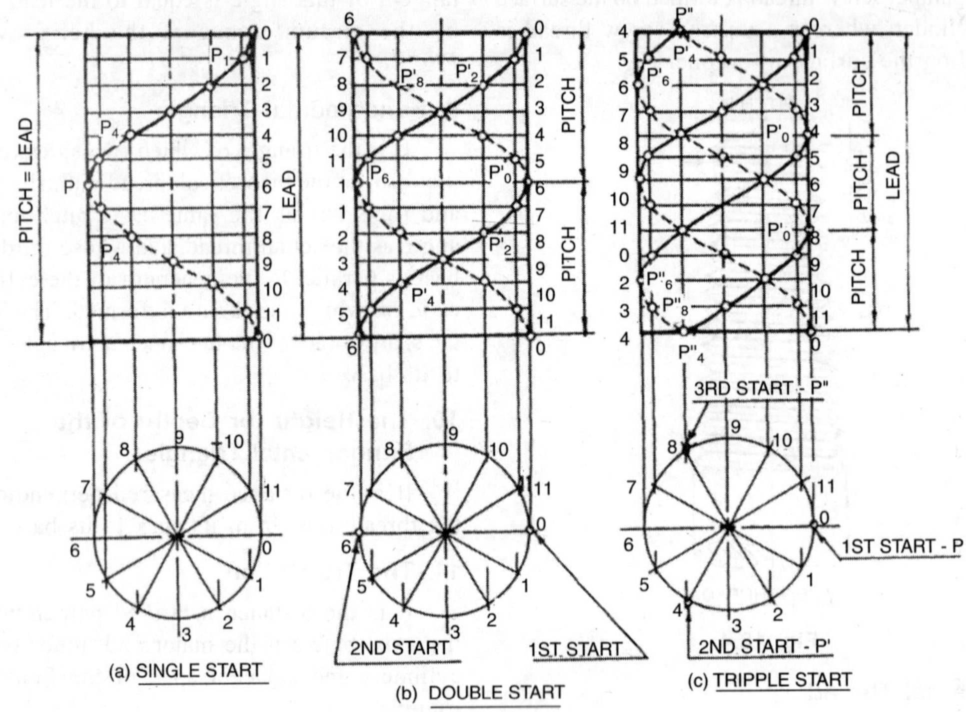

Fig. 18.5. Single and Multiple start helices

In an ordinary standard nut and bolt having single start threads, the axial advance of the nut in one complete turn, when the bolt is held secure, is termed the lead of its thread. On single start threads pitch and lead are equal. On double start threads lead is twice the pitch, so on, *Figs. 18.5, 18.6*. In general the multiple thread is determined by counting the thread ends on the end face of a screw or a nut. At (*a*), in *Fig. 18.5* helix starts from points 0 only and, therefore, pitch and lead are equal. At (*b*), in Fig. 18.5, helix starts from points 0 and 6, therefore, it is a double start thread helix and lead is twice the pitch. At (*c*), in *Fig. 18.5*, helix starts from points 0, 4 and 8, therefore, it is a triple start thread. In this case lead is thrice the pitch.

16. Slope

Slope of a thread is equal to half its lead. In case of single start threads slope is equal to half the pitch, and for double start threads it is equal to pitch, *Figs. 18.18, 18.19*.

DIAMETERS

17. Major Diameter

Also called outside or nominal diameter, is the largest diameter of a screw thread, external or internal, *Fig, 18.3*. The screw is specified on this diameter.

On a parallel thread this is the diameter of a cylinder imagined as co-axial with the axis of the screw. The cylinder just touches the crests of an external thread or the roots of an internal thread.

18. Minor Diameter

Also called core or root diameter, is the smaller diameter of external or internal threads, This on a parallel thread, is the diameter of an imaginary cylinder, which just touches the roots of an external thread, or crests of an internal thread.

19. Pitch Diameter

Also called effective diameter is the diameter of an imaginary cylinder, on a cylindrical screw thread, the surface of which would pass through the thread at such points as to make equal the width of the thread and the width of the space between the threads.

In a nut and bolt assembly, it is the diameter at which the ridges on the bolt are in complete touch with the ridges of the corresponding nut.

20. Thread Profile

It is the shape of a thread on a section plane containing the axis of thread.

21. Single Thread

When all the threads on a member are built on a single helix, *Fig. 18.5 (a)*, it is termed 'single thread'. It has got only one starting point. In other words a single start screw thread has one continuous ridge (with a groove on each side of it).

22. Double Thread

Two equally spaced threads are wrapped around the cylinder on two parallel helices, *Fig. 18.5 (b)*. It has two starting points. Its lead is twice the pitch, *Fig. 18.16*.

23. Multiple Thread

If a quick-acting thread is required the lead must be large. This means that the thread depth will be large and the bar stock on which the thread is cut will be weakened, To avoid this, multiple threads are used when a quick advance is required in a screwed pair, Two or more than two threads are cut side by side around the cylinder on an equal number of parallel helices, *Fig. 18.5. (b)*. Pitch remains fixed for a particular type of thread form on a cylinder of fixed diameter but the lead changes in multistart threads. Here again the distance between corresponding points on adjacent threads is called the pitch, and the lead is the axial movement during one revolution. *Figure 18.16* (*a*) and (*b*) shown double start Vee and square threads, respectively, A common example is the thread on a fountain pen cap. If nothing else is mentioned with a thread it is taken as single thread.

24. Right-hand and Left-hand Threads

Take any nut and bolt. Hold the bolt firmly, in left hand, and rotate the nut clockwise, by right

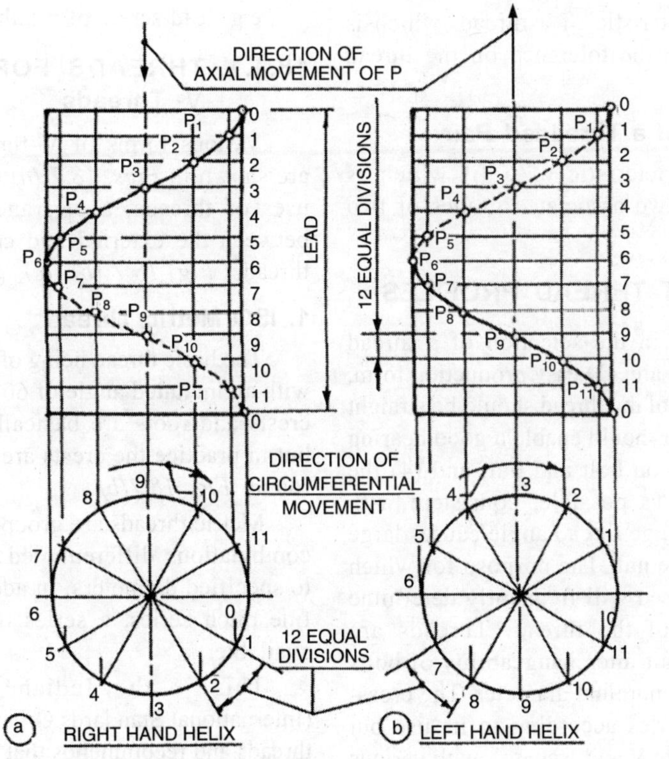

Fig. 18.6. Direction of Helices

hand, the nut will screw on the bolt if the threads are right handed. Rotate the nut anti-clockwise it will screw off the bolt. It is abbreviated as RH.

A left-hand screw thread, when assembled with a stationary mating thread, moves towards the observer when rotated clockwise, i.e. it screws off. It is abbreviated as LH.

A right hand thread advances into a threaded hole when turned clockwise; a left hand thread advances when turned counterclockwise. The thread slope, *Fig. 18.6*, can easily distinguish them. A thread is always considered to be right-hand if it is not otherwise specified. A left-hand thread is always marked out on a drawing, *Figs. 18.12, 18.13 (b)*.

25. Depth of Engagement

It is the radial distance by which the thread forms of two mating threads overlap each other. The radial distance between the basic major diameter of the external threads and the minimum minor diameter of the internal threads represents 100% depth of engagement.

26. Grade or Class of a Thread

It is the characteristic of a thread, which is determined between the tolerance on the thread and its basic size.

27. Grade of fit of a Threaded Pair

It is that characteristic of a fit which is determined by the two associated grades of two mating threads.

18.4. DIFFERENT THREAD PROFILES

Many factors come in the selection of a thread profile. Among them are: (1) easy production form, (2) the measurement of the thread should be straight forward, (3) the form should enable a good bearing between the threads on bolt and nut, and (4) the thread be as sharp as possible, consistent with required strength. Large thread angle causes large bursting force on the nut. The purpose for which the screw is to be used will frequently determine the shape profile of the thread. Threads are standardized to permit interchangeability of bolts and nuts of the same nominal diameter. The cross-section of thread varies according to its use but the two main kinds are V and 'square', with various modifications.

When the threads have a 'V' cross-section it is called a V thread, and when it has square cross-section it is called a square thread. All types of V thread have inclined flanks, making an angle between them. Whereas the flanks of square threads are perpendicular to thread axis and parallel to each other. These two basic types have their advantages and disadvantages over one another, as follows:

1. V–threads are more suitable for fastening purposes as they offer greater frictional resistance to motion than the square threads.
2. V– threads are stronger than the square threads. They have twice the amount of the internal resisting power, resisting the shearing action, at the root of the thread than a square thread.
3. V– threads can be cut easily by a tap, a die or on a machine.
4. V– threads due to high resistance frictional force, which they offer, are not suitable for use where power is to be transmitted. Whereas, square threads are ideal for power transmission as they offer less friction to relative rotation, e.g. lead-screw of a lathe, screw jacks etc.

18.4.1. THREADS FOR FASTENING — V- Threads

Various forms of V–threads in common use are shown in *Figs. 18.7 (a)* to *(f)*. In the practical uses of threads, a clearance must be provided between the external and corresponding internal threads, *Fig. 18.7 (b), (c), (f)*.

1. ISO Metric Thread

The basic thread being of symmetrical 'V' form with an included angle of 60° between flanks, The crests and roots are basically flat, *Fig. 18.7 (a)*, but in practice the crests are flat and the roots are round, *Fig. 18.7(b)*.

Metric threads are grouped into diameter pitch combinations differentiated by the pitch applied to specified diameters. In addition to a coarse and fine pitch series, a series of constant pitches is available.

This is the Indian Standard for ISO (International Standards Organization) metric screw threads and recommends that the coarse pitch series be used for the vast majority of general-purpose applications.

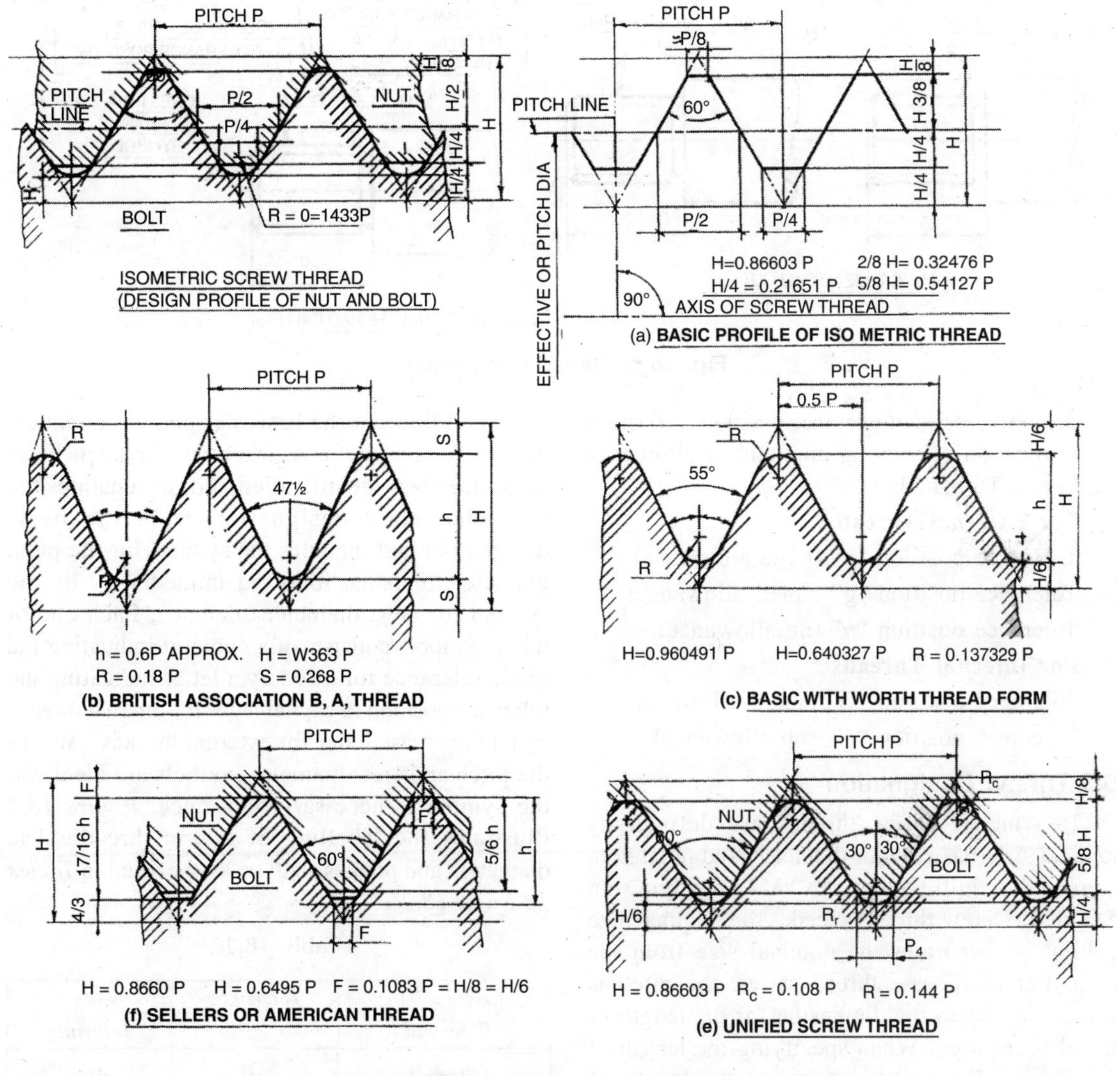

Fig. 18.7. V-Thread forms for fastening

For each of the two main thread elements, pitch diameter and crest diameter, there are numerous tolerance grades. The number of the tolerance grade reflects the tolerance size. For *example:* Grade 4 tolerances are smaller than Grade 6 tolerances, Grade 8 tolerances are larger than Grade 6 tolerances.

In each case, grade 6 tolerances should be used for medium quality and/or length of engagement applications. The tolerance grades below grade 6 are intended for applications involving fine quality and/or short length of engagement. Tolerance grades above grade 6 are intended for coarse quality and/or long periods of engagement.

In addition to tolerance grade, positional tolerance is required. The positional tolerance defines the maximum material limits of the pitch and crest diameters of the external and internal threads and indicates the relationship to the basic profile.

In conformance with current coating (or plating) thickness requirements and the demand for ease

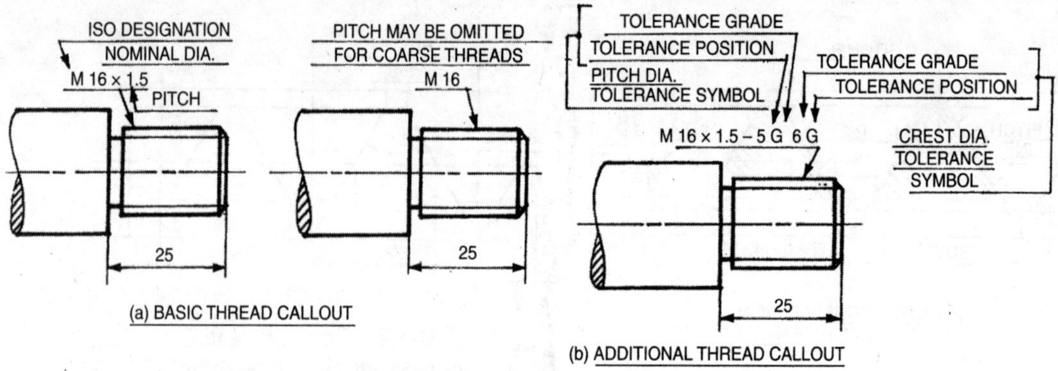

Fig. 18.8. Thread designation

of assembly, a series of tolerance positions reflecting the application of varying amounts of allowance has been established:

For External Threads:

Tolerance position "*e*" (large allowance)

Tolerance position "*g*" (small allowance)

Tolerance position "*h*" (no allowance)

For Internal Threads:

Tolerance position "*G*" (small allowance)

Tolerance position "*H*" (no allowance)

ISO Thread Designation

ISO metric screw threads are defined by nominal size (basic major diameter) and pitch both expressed in millimeters. An "M" specifying an ISO metric screw thread precedes the nominal size and an "×" separates the nominal size from the pitch. For the coarse thread series, the pitch is shown only when the dimension for the length of thread is required. When specifying the length of thread, an "×" is used to separate the length of thread from the rest of the designation. For external threads, the length of thread may be given as a dimension on the drawing.

For example, a 10 mm diameter, 1.25 mm pitch, fine series thread is expressed as M 10 × 1.25. A 10 mm diameter, 1.5 mm pitch, coarse series thread is expressed as M10; the pitch is not shown unless the length of thread is required. If the latter thread was 25 mm long and this information was required on the drawing, the thread call out would be M 10 × 1.5 × 25.

In addition to the basic designation complete designation for an ISO metric screw thread includes tolerance class identification, a dash separates the tolerance class designation from the basic designation and includes the symbol for the pitch diameter tolerance followed immediately by the symbol for crest diameter tolerance. Each one of these symbols consists of a numeral indicating the grade tolerance followed by a letter indicating the tolerance position (a capital letter for internal threads and a lowercase letter for external threads). Where the pitch and crest diameter symbols are identical, the symbol is necessary only once. *Figure 18.8* illustrates the labelling of metric threads. The diameters and pitches shall be as specified in *Tables 18.1* and *18.2*.

Table 18.3.

Classes of Fit Extermal	*Tolerance*	*Class Internal*
Close	5H	4g
Medium	6H	6g
Free	7H	8g

2. British Standard Whitworth (B.S.W.) Thread

The BSW basic thread form is a symmetrical V-thread in which the angle between the flanks, measured in an axial plane, is 55°, *Fig. 18.7 (c)*. Thread is rounded equally at crests and roots by circular arcs blending tangentially with the flanks. The basic sizes have been standardized by British Standards Association. These threads are used on bolts and screwed fastenings for several purposes.

Table 18.1.

DIAMETER			PICHES													
PREFEREED	CHOICE 2	CHOICE 3	COARSE	FINE										
			6	4	3	2	1.5	1.25	1	0.75	0.5	0.35	0.25	0.2
0.25			0.075											
0.3			0.08											
	0.35		0.09											
0.4			0.1											
	0.45		0.1											
0.5			0.125											
	0.55		0.125											
0.6			0.15											
	0.7		0.175											
0.8			0.2											
	0.9		0.225											
1			0.25											0.2
	1.1		0.25											0.2
1.2			0.25											0.2
	1.4		0.3											0.2
1.6			0.35											0.2
	1.8		0.35											0.2
2			0.4										0.25	
	2.2		0.45										0.25	
2.5			0.45									0.35		
3			0.5									0.35		
	3.5		0.6									0.35		
4			0.7								0.5			
	4.5		0.75								0.5			
5			0.8								0.5			
		5.5									0.5			
6			1							0.75				
		7	1							0.75				
8			1.25						1	0.75				
		9	1.25							0.75				
10			1.5					1.25	1	0.75				
		11	1.5						1	0.75				
12			1.75				1.5	1.25	1					
	14		2				1.5	1.25	1					
		15					1.5		1					
16			2				1.5		1					
		17					1.5							
	18		2.5			2	1.5		1					
20			2.5			2	1.5		1					
	22		2.5			2	1.5		1					
24			3			2	1.5		1					
		25				2	1.5		1					
		26					1.5							
	27		3			2	1.5		1					
		28				2	1.5		1					
30			3.5		(3)	2	1.5							
		32				2	1.5							
	33		3.5		(3)	2	1.5							
		35					1.5							
36			4		3	2	1.5							
		38					1.5							
	39		4		3	2	1.5							
		40			3	2	1.5							
42			4.5	4	3	2	1.5							
	45		4.5	4	3	2	1.5							
48			5	4	3	2	1.5							
		50			3	2	1.5							
	52		5	4	3	2	1.5							

Table 18.2.

Diameter				Piches											
Prefereed	Choice 2	Choice 3	Coarse	Fine											
				6	4	3	2	1.5	1.25	1	0.75	0.5	0.35	0.25	0.2
		55			4	3	2	1.5							
56			5.5		4	3	2	1.5							
		58			4	3	2	1.5							
	60		5.5		4	3	2	1.5							
		62			4	3	2	1.5							
64			6		4	3	2	1.5							
		65			4	3	2	1.5							
	68		6		4	3	2	1.5							
		70			4	3	2	1.5							
72				6	4	3	2	1.5							
		75			4	3	5	1.5							
	76			6	4	3	2	1.5							
		78					2								
80				6	4	3	2	1.5							
		82					2								
	85			6	4	3	2								
90				6	4	3	2	1.5							
	95			6	4	3	2								
100				6	4	3	2								
	105			6	4	3	2								
110				6	4	3	2								
	115			6	4	3	2								
	120			6	4	3	2								
125				6	4	3	2								
	130			6	4	3	2								
		135		6	4	3	2								
140				6	4	3	2								
		145		6	4	3	2								
	150			6	4	3	2								
		155		6	4	3									
160				6	4	3									
		165		6	4	3									
	170			6	4	3									
		175		6	4	3									
180				6	4	3									
		185		6	4	3									
	190			6	4	3									
200		195			6	4	3								
					6	4	3								
		205			6	4	3								
	210				6	4	3								
		215			6	4	3								
220					6	4	3								
		225			6	4	3								
		230			6	4	3								
		235			6	4	3								
	240				6	4	3								
		245			6	4	3								
250					6	4	3								
		255			6	4	3								
	260				6	4									
		260			6	4									
		270			6	4									
		275			6	4									
280					6	4									
		285			6	4									
		290			7										

3. British Association (B.A.) Standard Thread

The basic thread form is a symmetrical V-thread in which angle between flanks are 47.5°, *Fig. 18.7 (d)*. These threads are to supplement BSW series and have fine pitches. They are used on screws for precision work.

4. Unified Screw Thread

From the above explanations it can be noticed that the principal screw thread systems in the United Kingdom are the BSW and BA. The American National Thread System developed from the Seller's thread has been in use in U.S.A. and Canada. These three countries came to an arrangement for a common screw thread system with the included angle of 60° and designated as United Screw Thread in the year 1949, and were accordingly published in the National Standards.

This thread, shown in *Fig. 18.7 (e)* has an included angle of 60°. The thread on the bolt is rounded off at the crest and root and the thread in the nut is rounded off at the root but the crest is left flat.

5. American Standard Thread

Also called Seller's thread, is American National Thread used for general purposes, e.g., on bolts, nuts, screws and tapped holes. It has flat tops and bottoms. The angle of 'V' is 60°, *Fig. 18.7 (f)*. The flat crests can stand more rough usage than the sharp "V" threads.

18.4.2. THREADS FOR POWER TRANSMISSION

1. Square Threads

The sides or flanks of these threads are normal to the axis and are parallel to each other, *Fig. 18.9 (a)*. The pitch of a square thread is often taken as twice that of a BSW thread of the same diameter. In this type of thread there is no bursting action on the nut, as in V-threads, and the threads transmit the force almost parallel to the axis. The thread is, therefore, more suited for driving purposes or transmission on motion. It is, however, weaker than V- thread in resisting shearing action at the bottom of thread. It gives larger movement of nut or screw per revolution. Mechanisms of machine tools, valves, spindles, vice screws, jacks etc. are generally provided with square threads. Because of it's cutting and disengaging of split nut difficulties it has been largely replaced by the Acme thread. A square thread is designated by the letter SQ followed by nominal diameter and pitch, the two being separated by the '*x*'. For example a square thread of nominal diameter 30 mm and pitch 6 mm is designated as:

SQ 30 × 6

Square threads are available in three series as:

Fine series 14 to 640 mm

Normal series 22 to 300 mm

Coarse series 22 to 400 mm

2. Acme Thread

It is a modified form of square thread, *Fig. 18.9 (b)*. It is much stronger than square thread and can be easily produced. Acme thread is frequently used on screw cutting lathes, brass valves, cocks and bench vices. When used in conjunction with a spit nut, as on the lead screw of a lathe, the tapered sides of the thread facilitate the ready engagement of the two halves of the nut when required. It has a 29° thread angle and it is a close adaptation of the 29° worm thread but is slightly shallower.

3. Buttress Thread

It is used where power is to be transmitted in one direction only, *Fig, 18.9 (c)*. The force is transmitted almost parallel to the axis and the thread is about the same strength as the standard V-thread. Bench vices are usually fitted with spindles having buttress threads. It combines the low frictional resistance characteristic of the square threads with the strength of V- threads.

4. Knuckle Thread

It is also a modification of square thread, *Fig. 18.9 (d)*. It has round top and bottom and can be cast or rolled easily and cannot economically be made on a machine. It is not easily damaged when subjected to rough usage. It is largely used for railway carriage couplings, hydrants, light bulbs, and also on objects made of brittle materials as glass, plastic, porcelain, etc.

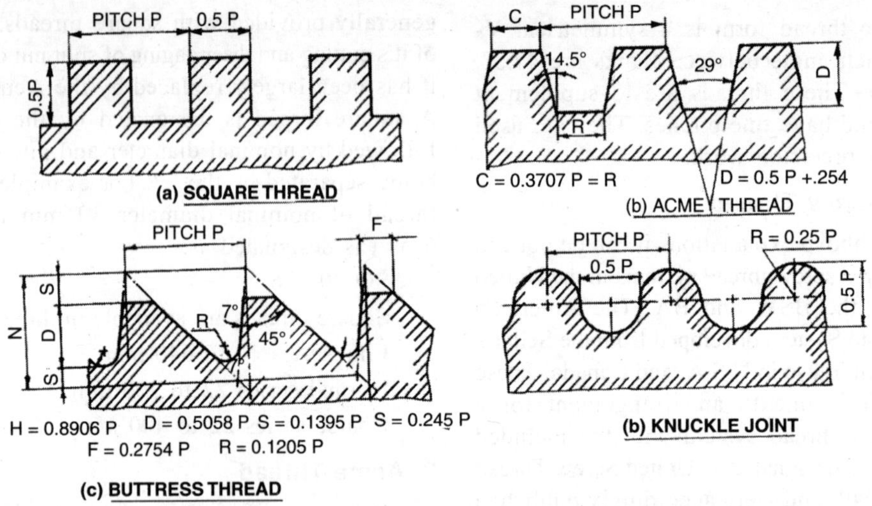

Fig. 18.9. Thread forms of square thread family

18.5. ACTUAL, DETAILED, CONVENTIONAL AND SIMPLIFIED REPRESENTATION OF SCREW THREADS

Figure 18.10 shows the actual representation of a V-thread, *Fig. 18.11* shows actual representation of a single start square thread, whereas, *Fig. 18.12* shows the actual representation of a double start square thread, left hand. It can be observed from these figures that the edges of screw threads are curved lines, known as helices. Such true representation of screw threads, by helical curves, requiring unnecessary time and laborious drafting is rarely used in machine drawing. The detailed representation, closely approximating the actual appearance, is preferred in commercial practice, for it is much easier to represent the helices with slanting lines and the truncated roots and crests, with sharp V. A step-by-step, detailed representation of external V-threads is shown in *Fig. 18.13 (a)* for RH and *18.13 (b)* for LH threads.

Since detailed rendering is also time consuming its use is justified only in those few cases where appearance and permanency are important factors. To save valuable time and expenses in the preparation of drawing the conventional or schematic 'Simplified' representation of threads is adopted. *Fig. 18.18* and *18.19* respectively, show a conventional and 'simplified' representation of external and internal threads, outside or in section.

Figures 18.14 (a) and *(b)* show, the step-by-step, conventional representation of RH and LH V-threads, respectively. Whereas, *Fig. 18.15 (a)* and *(b)* shows conventional and detailed representation of RH and LH square threads.

As the threads occur frequently in drawing, the representation of the threads on a drawing has been so standardized that it saves time and effort, which would otherwise be necessary if actual threads were to be drawn each time. This conventional representation, however, does not give full information about the thread, except its location and a standard form of note is used on the drawing to specify thread specifications. The 'V' thread, as shown in *Figs. 18.14* and *18.18*, is represented by drawing sloping parallel lines, alternately thick and thin, and spaced half pitch apart. The slope of lines is kept equal to half the lead. The thick lines, used to indicate roots, are drawn shorter than thin lines for crests, by a distance equal to the depth of thread on each end.

In a more simplified method two continuous thin lines are drawn parallel to the axis and depth of thread within this, indicating the roots of threads. The length of these thin lines shows the threaded length of the screw, *Figs. 18.17* and *18.19*.

External square thread is represented by sloping parallel lines, spaced half pitch apart, slope being half the lead of thread. The depth of thread is also equal to half the pitch. Only a small portion of the thread may be shown as described above, the

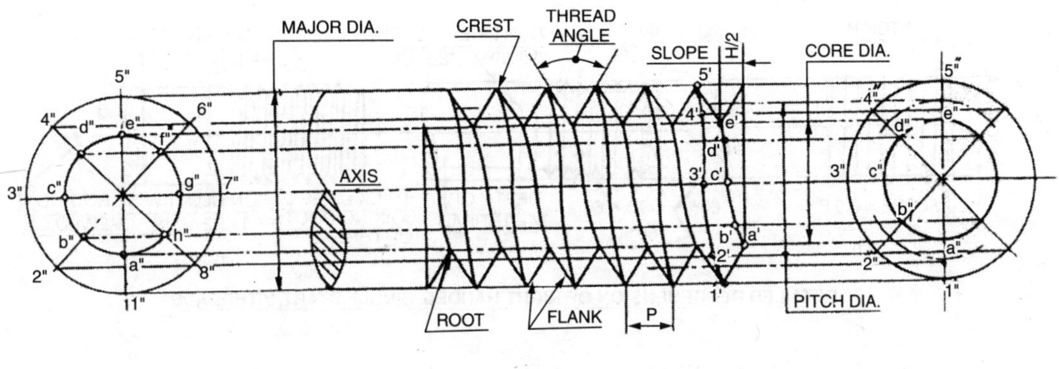

Fig. 18.10. Actual representation of single start V-Threads

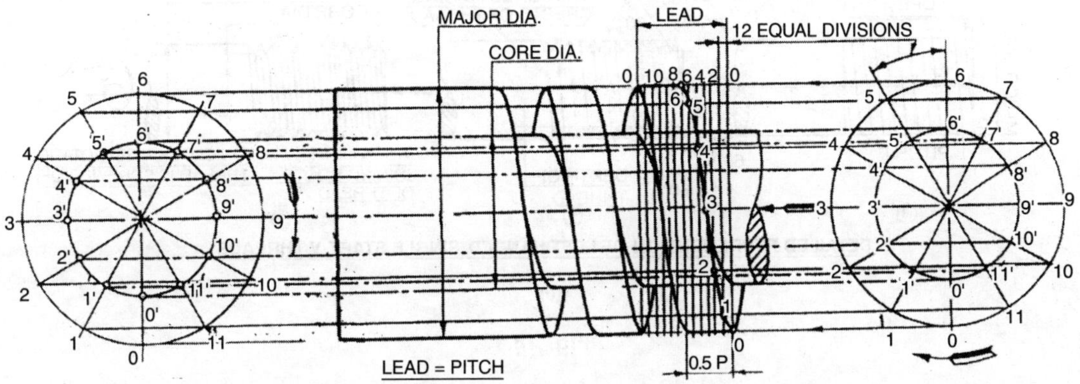

Fig. 18.11. Actual representation of single start square

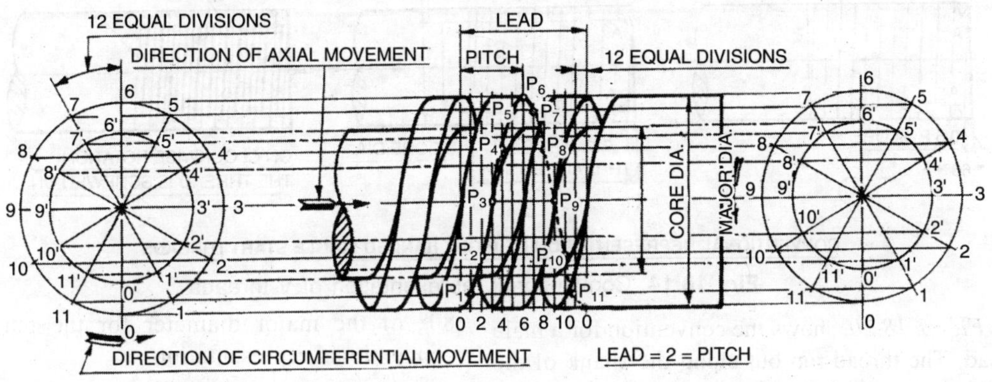

Fig. 18.12. Actual representation of double start square threads LH

remaining portion being shown by two straight lines parallel to the axis. The detailed and conventional representation of a square thread is shown in *Fig. 18.15 (b)* and *(c)*.

To summarize, when the threads are to be shown on a drawing, do not draw their actual shape. When the threads are on two mating parts on assembly to be shown in section, or they are important and permanency is required, show them by detailed representation, otherwise show the threads by conventional, or schematic and simplified methods, wherever possible, to save time and labour and give necessary information by a note.

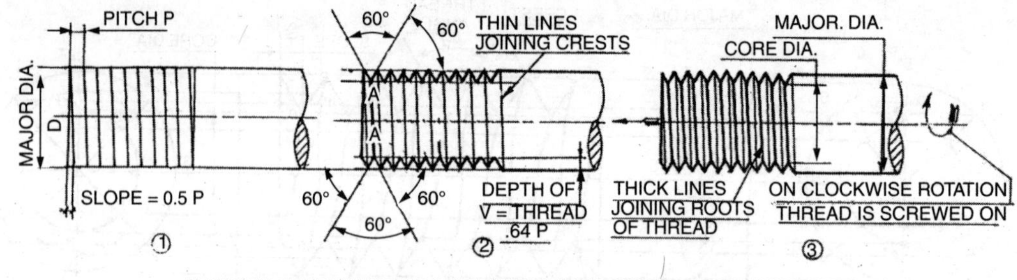

DETAILED REPREHENSION OF RIGHT HANDED, SINGLE START, V-THREADS

(a)

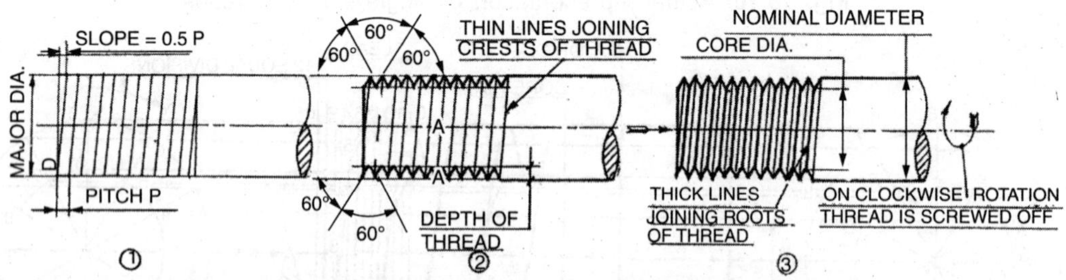

DETAILED REPREHENSION OF LEFT HANDED, SINGLE START, V-THREADS

(b)

Fig. 18.13.

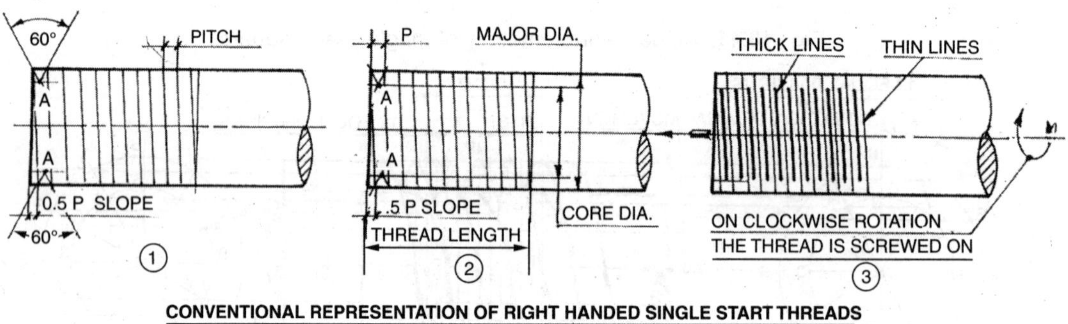

CONVENTIONAL REPRESENTATION OF RIGHT HANDED SINGLE START THREADS

Fig. 18.14. Conventional representation of V-threads

Figure 18.20, shows the convention for a male thread. The thread run out along the shank of the stud is indicated by a line drawn at 30° to the thread axis. The minor diameter of the thread is shown by parallel lines and in the end elevation the projected circle is not continuous. The break in the inside circle distinguishes the end elevation of a male thread from a female thread. Line thicknesses are given for each part of the thread. The actual dimensions of the minor diameter, for any particular thread size, can be approximated at 80° of the major diameter for the purpose of conventional representation.

18.6. THREADS IN SECTION

The detailed representation of internal threads in section, which is used for large diameters, is shown in *Fig. 18.22*. Since the far side of internal threads in section is visible, the crest and root lines incline in the opposite direction to those of an external thread having the same specifications. For smaller sizes the schematic and simplified representation is used as shown in *Figs. 18.17, 18.18, 18.19* and

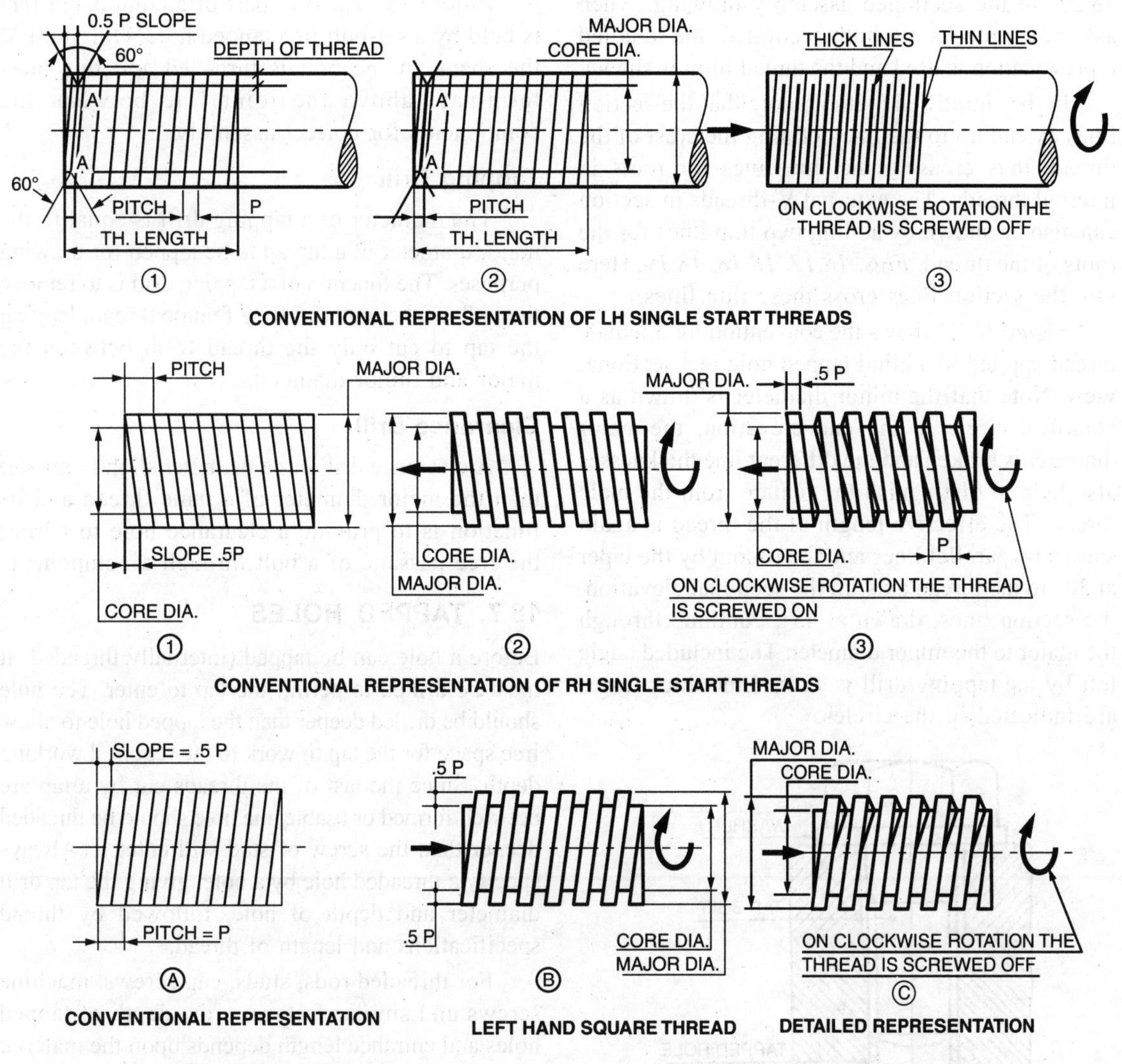

Fig. 18.15. Detailed and conventional representation of square threads

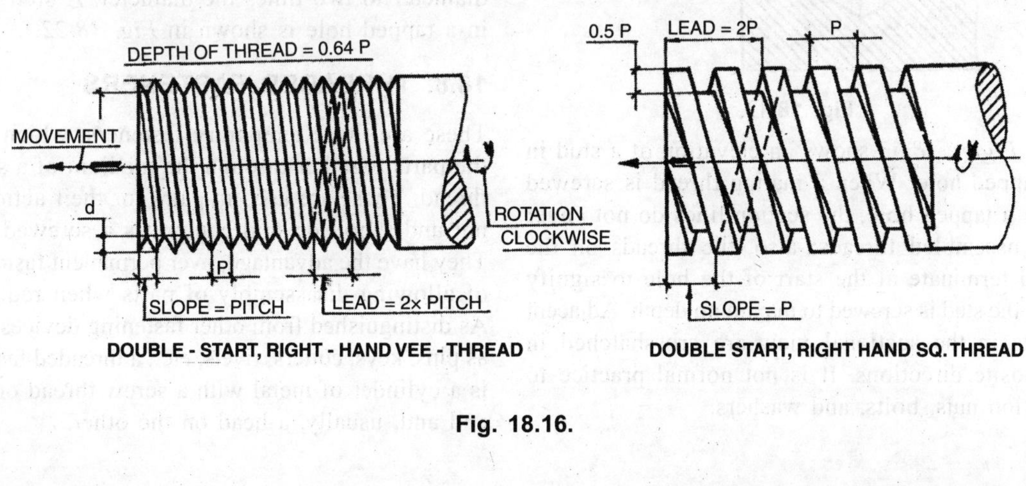

Fig. 18.16.

18.21. In the sectioned assembly drawing, when assembled pieces are both sectioned, the detailed representation is used and the thread form is shown.

In the simplified method, note that the section lines extend up to the line defining the crest of the thread, thus crossing the thin lines for roots in internal threads. The external V-threads in section can also be shown by adding two thin lines for the roots of the thread, *Figs. 18.17, 18.18, 18.19*. Here too, the section lines cross these thin lines.

Figure 18.21 shows the convention for a female thread applied to a blind tapped hole in a sectional view. Note that the minor diameter is drawn as a complete circle in the end elevation, the-major diameter is broken and the different line thicknesses also help to distinguish the female from the male thread. The effective length of the thread is again shown by parallel lines and the runout by the taper at 30° to the thread axis. In the sectional elevation, the section lines, drawn at 45°, continue through the major to the minor diameter. The included angle left by the tapping drill is 120°. Line thicknesses are indicated in the circles.

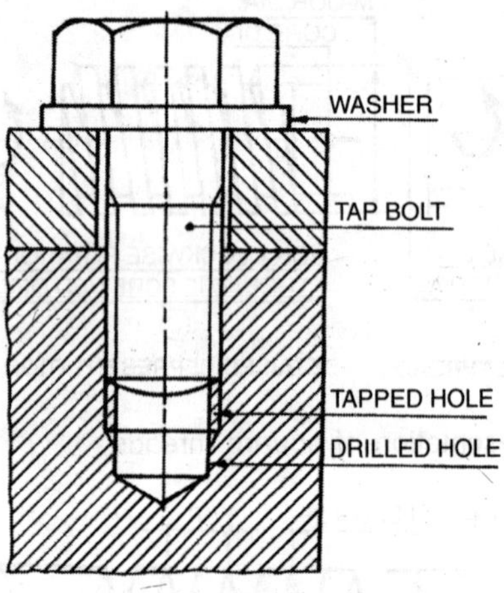

Fig. 18.17.

Figure 18.53 shows an elevation of a stud in a tapped hole. When a mating thread is screwed into a tapped hole, the section lines do not cover the assembled threads; also, the threads on the stud terminate at the start of the hole to signify that the stud is screwed to maximum depth. Adjacent parts in the sectional view are crosshatched in opposite directions. It is not normal practice to section nuts, bolts, and washers.

Figure 18.17 shows part of a component that is held by a set-bolt in a tapped hole. Only part of the shank of the bolt is threaded and this must terminate above the joint line between the components for correct assembly.

Tapping Drill

The diameter of a tapping drill is equal to the major diameter of a thread to be tapped for drawing purposes. The function of a tapping drill is to remove the bulk of the material for a female thread, leaving the tap to cut only the thread form between the major and minor diameters.

Clearance Drill

A clearance drill has a diameter slightly greater than the major diameter of a male thread and its function is to provide a clearance hole to permit the free passage of a bolt through a component.

18.7. TAPPED HOLES

Before a hole can be tapped (internally threaded) it must be drilled to permit the tap to enter. The hole should be drilled deeper than the tapped hole to allow free space for the tap to work to the required working depth. Since the last of the threads cut by a tap are not well formed or usable, the hole should be threaded deeper than the screw of stud will enter it. Always specify a threaded hole by a note, giving the tap drill diameter and depth of hole, followed by thread specifications and length of thread.

For threaded rods, studs, cap screws, machine screws and similar fasteners, the depth of tapped holes and entrance length depends upon the material of the fastener and its diameter. It varies from diameter to two times the diameter. A stud fixed in a tapped hole is shown in *Fig. 18.22*.

18.8. THREADED FASTENERS

These are those temporary fasteners, which hold the parts together through the medium of a screw thread. These are used in pairs for their action. A nut and a bolt or stud comprises a screwed pair. They have the advantage, over permanent fasteners, of allowing disassembly of parts when required. As distinguished from other fastening devices such as pins, keys, cotters, rivets, etc., a threaded fastener is a cylinder of metal with a screw thread on one end and, usually, a head on the other.

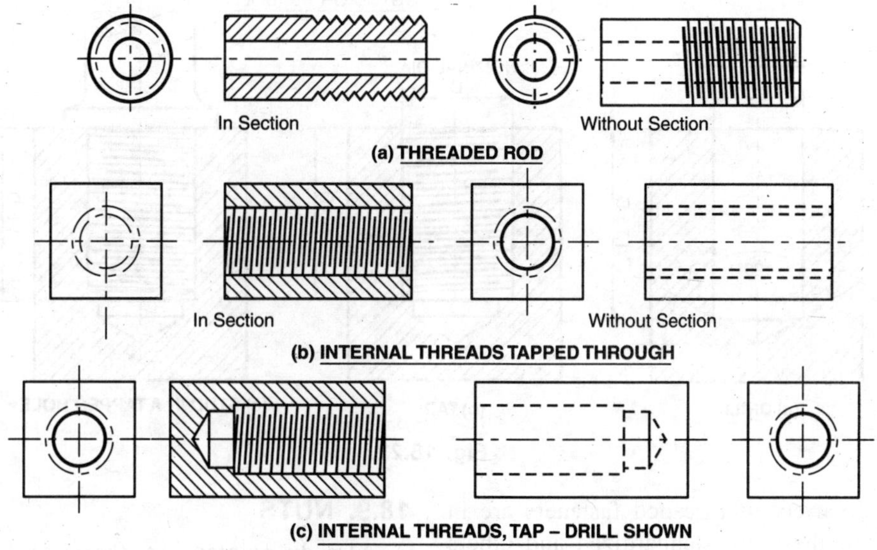

Fig. 18.18. Conventional thread symbols

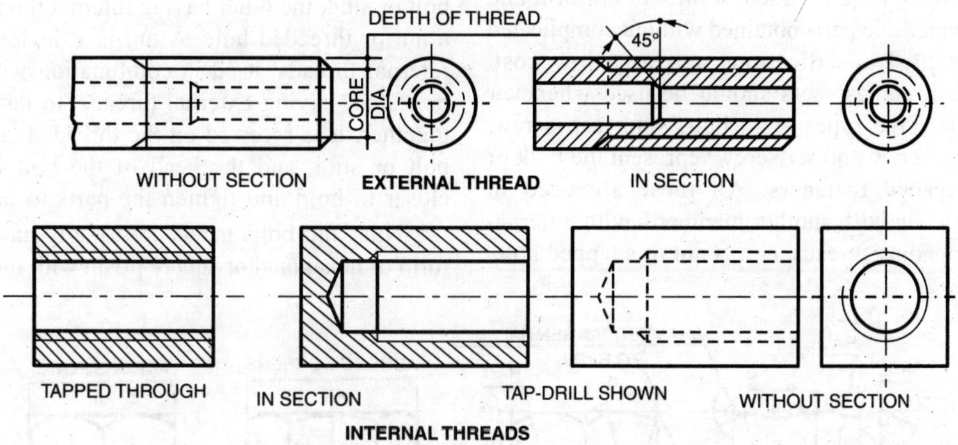

Fig. 18.19. Simplified thread symbols

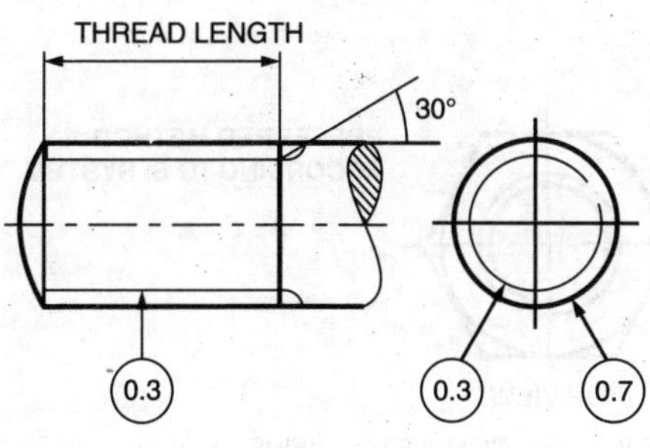

Fig. 18.20.

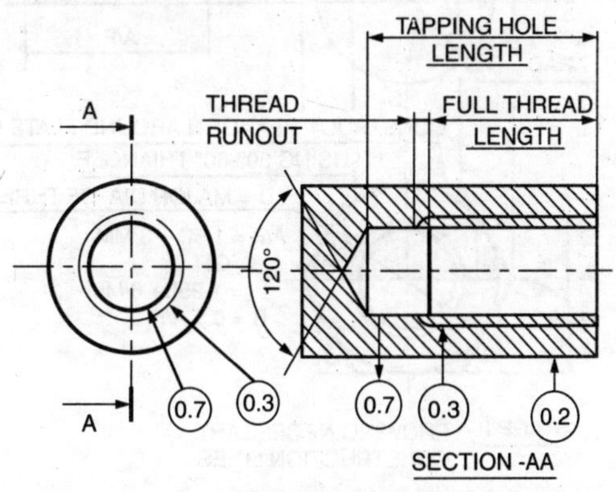

Fig. 18.21.

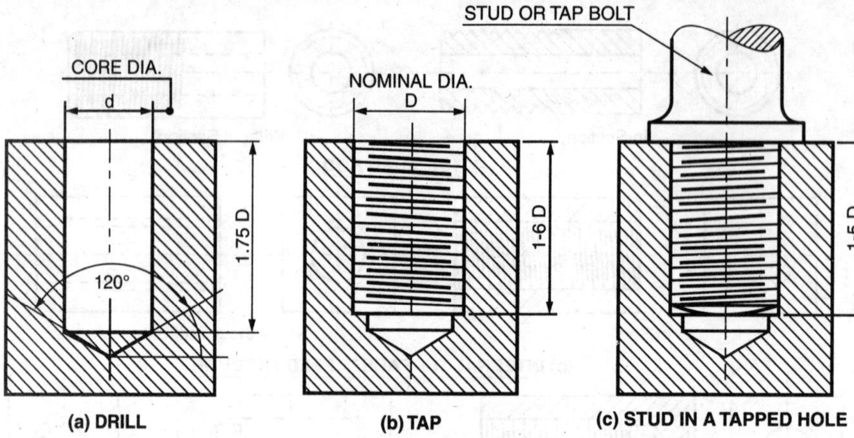

Fig. 18.22.

A wide variety of threaded fasteners are in use. Some of these are standardized and others are made for special use. The standardization of such widely used products results in uniform and interchangeable parts obtained without complicated and detailed specifications and at a low cost. Standardized fasteners should be used whenever possible. Five types, e.g., bolt, stud, cap screw, machine screw and set-screw represent the bulk of the threaded fasteners. All these are used in combination with another member having internal corresponding threads, e.g., a nut or a tapped hole.

18.9. NUTS

As already pointed out, the screwed pieces are used in pairs, one having external threads, e.g., bolt or stud, the other having internal threads, e.g., a nut or threaded hole. A nut is a device, having internal threads, used in combination with a bolt or a stud, having external threads, to fasten parts together. It is screwed on the threaded end of the bolt or stud, and the head of the bolt is drawn closer to hold and tighten the parts to be joined.

Nuts and bolts heads are usually made in the form of hexagonal or square prism with the corners

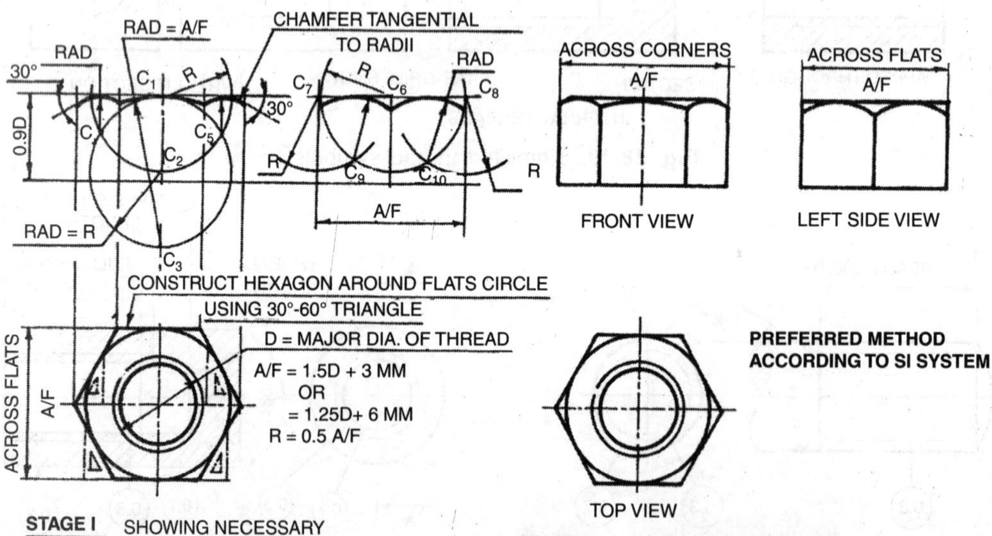

Fig. 18.24. Drawing projections of a Standard hexagonal Nut using First Angle projection method

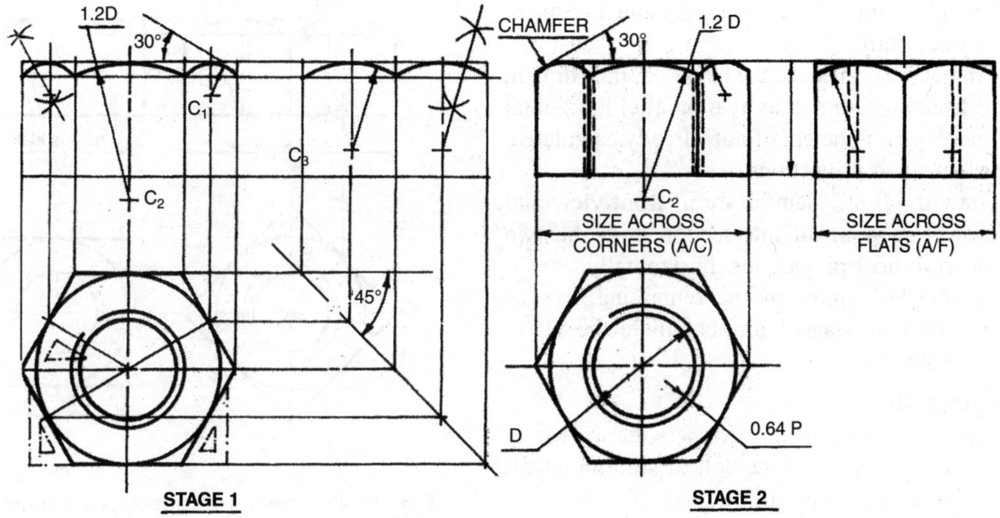

Fig. 18.25. Drawing a Hexagonal Nut (BSW) (First Angle Proj.)

on one end (outer) chamfered to ensure safety of the user. *Figure 18.23 (at the end of this chapter)* and *Fig. 18.24* show the construction of a hexagonal nut, in two stages each, by 130 system method in Third Angle, and First Angle, respectively. The *Fig. 18.25* shows the projections drawn in First Angle using BSW method. Whereas *Figs. 18.26, 18.27* shoe the construction of square nut, by two different methods. Bolts and nuts are drawn to dimensions obtained from standard tables,' but for drawing purposes their simplified (general) proportions are used. Normally, in drawing a nut (hexagonal or square), we know only diameter of the hole in the nut. Other important proportions are usually specified in terms of the nominal (major) diameter (D) of the bolt on which the nut is to be used.

1. Hexagonal Nut

Its drawing is shown completed in two stages by two different methods, in *Figs. 18.23 & 18.24* (ISO system) and *Fig. 18.25* (BSW Method). First method shown in Figs. 18.23/18.24 should be preferred, while second method can still be used for drawing purposes. Nuts are to be used for drawing frequently in drawing, therefore, it is extremely important to have a clear idea of drawing the nuts. The usual proportions, for ordinary hexagonal nuts, are as follows:

Size across flats, A/F = 1.25 D + 6 mm for bolts less than 12 mm in diameter. or = 1.5 D + 3 mm for bolts more than 12 mm in diameter. In general size A/F = 1.5 D.

Size across comers = 1.55 × distance across flats. It can be obtained by construction, or, for drawing, it can be taken approximately = 2 D.

Height of nut = 0.9 D to D, e.g., 0.9 D in general case and D in heavy duty cases.

Radius of front chamfer = 1.5 D in general case, 1.5 D + 3 mm in heavy cases (SI system), (Radius of front chamfer is 1.2 D for BSW hexagonal nuts and is as well fairly accurate for drawing purposes). Chamfer at top = 30° to base of nut.

Hints on Drawing a Hexagonal Nut

1. Work out the required dimensions i.e., nominal diameter of bolt, height of nut, size across flats, radius of front chamfer and radius of crest of the thread, etc.
2. Draw the centrelines, horizontal and vertical, to locate the centre of circles in to view.
3. Commence drawing from top view, as it contains circles. Draw the two circles for roots and crests of threads and also draw the third circle for chamfer on the nut.

4. Outside the chamfering circle, complete the hexagon with 30° – 60° triangle and T- Square or mini-drafter.
5. Project the corners of the hexagon, thus drawn, to front view and draw two parallel horizontal lines, apart by height of nut, already calculated, to cut these projections.
6. Draw the front chamfer arc in front view and project its point of intersection with the two inner vertical projections, horizontally.
7. Locate the centres of the remaining arcs as illustrated in stage 1 and complete the views as in stage 2.

2. Square Nut

The usual proportions for a square nut, in terms of the diameter of the bolt of stud on which it is to be used, are as follows:

Size across flats = 1.5 *D* + 3 mm, for bolts more than 12 mm in diameter. Or 1.5 *D* + 6 mm, for bolts less than 12 mm in diameter.

In general A/F = 1.5 *D*

Size across corners = $\sqrt{2}$ × size across flats. Or A/C = 1.414 × size across flats.

Height or thickness = 0.8 *D* to *D*. In general it is taken as 0.9 *D*.

Radius of front chamfer = 1.5 *D* to 1.75 *D*. In general it is taken as 1.5 *D* and 1.5 *D* + 3 mm in heavy cases (SI system).

Angle of chamfer = 30° to the base of nut, actually it is 25°.

Two methods of drawing a square nut are illustrated in *Fig. 18.26* and *18.27*. The first method should be preferred.

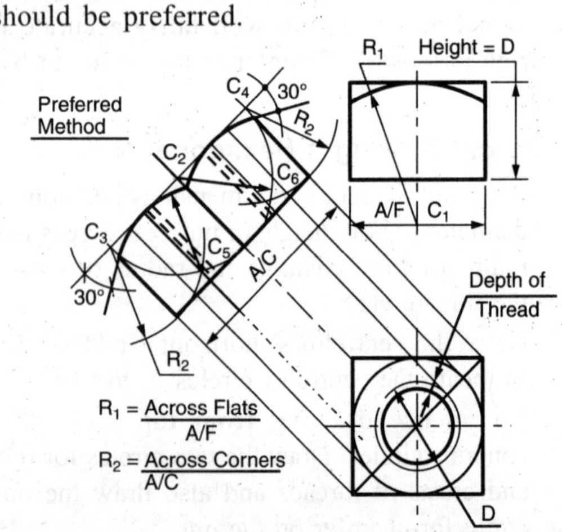

Fig. 18.26. Standard square nut (ISO system) (First mehod)

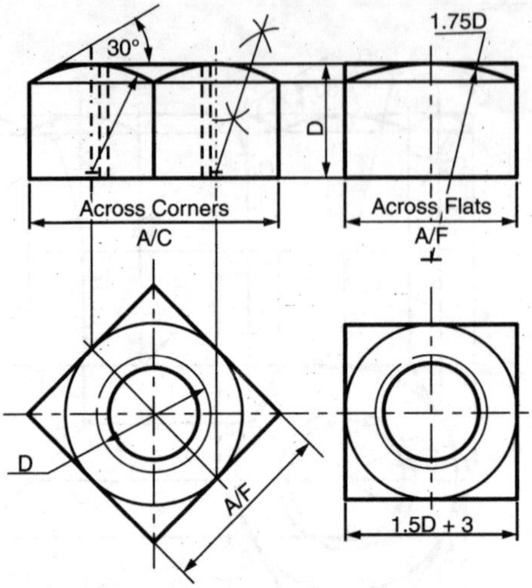

(a) (b)

Fig. 18.27. Method of drawing a square nut (First angle projection) (Second method)

The dimensions given above, for hexagonal and square nuts are approximate, but are satisfactory for drawing work. For accurate work the standard BIS tables should be consulted.

In the front view of a hexagonal nut, three flats are visible and the corners are seen chamfered, whereas in the end view the corners are square. When two or more views of a nut are required, commence with the view containing circles.

When only one view of a nut is required, it is usually drawn across corners as shown in *Fig. 18.27* in front view. BIS recommends that the hexagonal and square nuts should only be drawn across corners to avoid any confusion, *Fig. 18.28*.

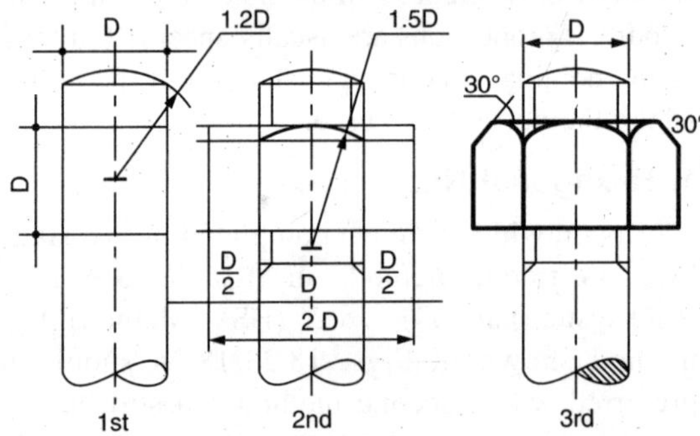

STEPS IN DRAWING ONE VIEW OF HEX. NUT

Fig. 18.28.

Nuts are generally of hexagonal or square forms as already discussed. Other forms of nuts are used for special purposes. Some of these special purpose nuts are briefly described below.

3. Castle Nut

In *Fig. 18.29* are shown two views of a very well known castle nut. It consists of a hexagonal portion with a cylindrical upper part that is slotted in line with the centre of each face. It is used when there is the possibility of a simple hexagonal nut getting loose while in use. The split pin, to lock this nut with bolt, passes through two slots in the nut and the hole drilled in the bolt, after fully tightening the pair, so that a positive lock is obtained unless the pin shears. Its commonly used proportions, in terms of nominal diameter, are given below:

Width across flats	= 1.5 *D*
Total height of the nut, including castle	= 1.2 to 1.25D
Height of the nut only	= 0.8 *D*
Depth of slot in castle	= 0.4 *D*
Width of slot	= 0.25 *D*
Chamfer at the top of castle	= 45°
Number of slots	= 6

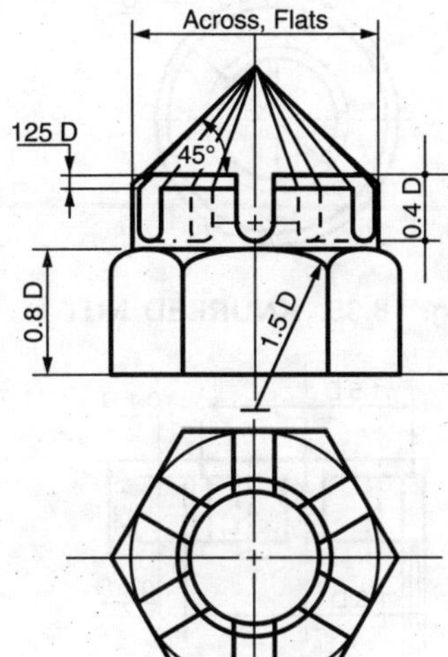

CASTLE NUT

Fig. 18.29.

4. Lock Nut

Figure 18.30 it is used to lock a standard (main) nut, when there is possibility of the nut to get loose, while in use. Height of lock nut varies from 0.5 D to 0.66 *D*; usually it is taken as 0.6 *D*. Other proportions are same as for ordinary hexagonal nut with the difference that it is chamfered from both sides at 30° with the horizontal.

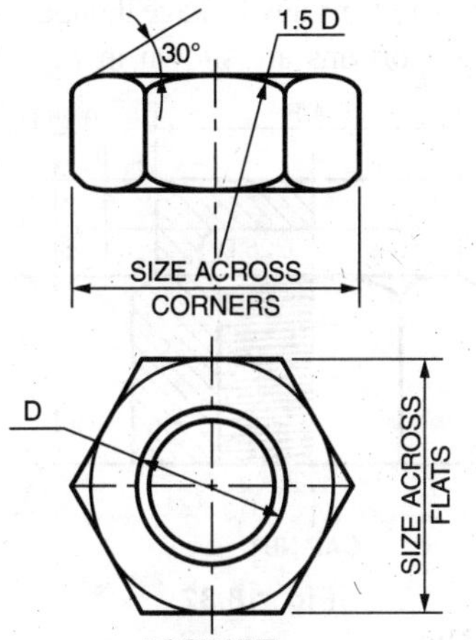

LOCK NUT

Fig. 18.30.

5. Flanged Nut

It has an integral circular flange, at its base, of thickness 0.25 *D* and diameter 2.2 *D*, *Fig. 18.31*. Thus flanged nut, too, is a hexagonal nut with an integral washer. The flange increases the bearing area of the nut, and the bolt can be used in comparatively large sized hole due to the flange.

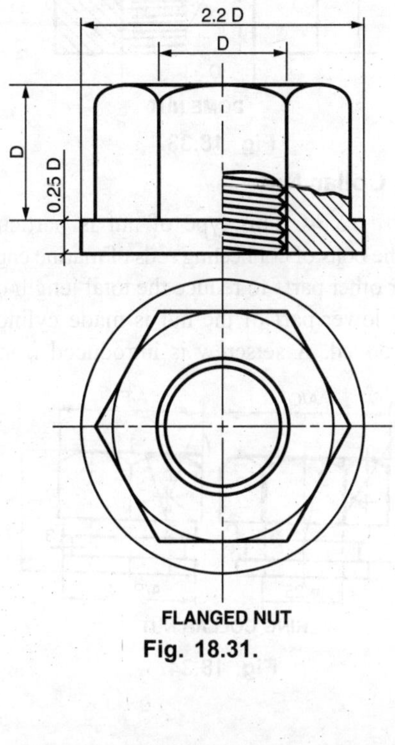

FLANGED NUT

Fig. 18.31.

6. Cap Nut

An integral cylindrical cap is provided at the top of this nut. It protects the end of the bolt against corrosion and prevents any leakage though the threads. The usual proportions are shown in *Fig. 18.32*.

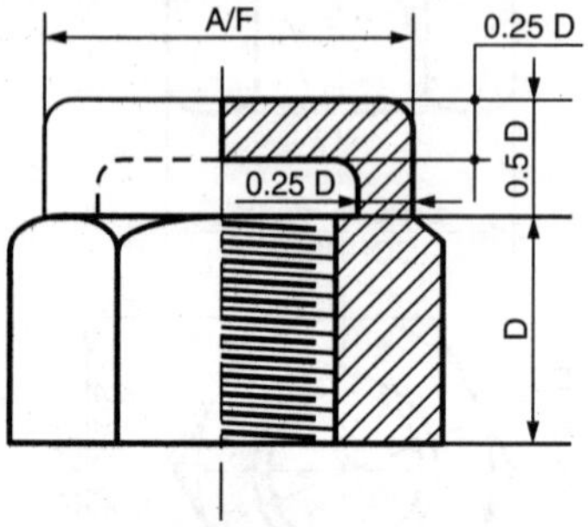

CAP NUT

Fig. 18.32.

7. Dome Nut

A spherical cap is provided on the top surface of the nut, *Fig. 18.31*. It serves the same purpose as the cap nut, with only a change in shape.

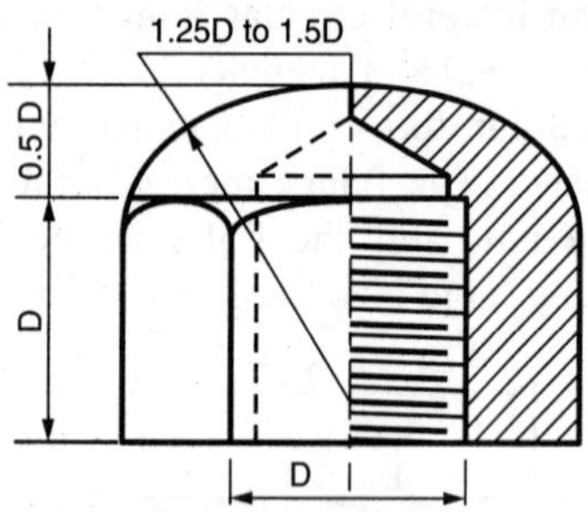

DOME NUT

Fig. 18.33.

8. Ring Collar Nut

Figure 18.34. This type of nut is particularly used for the bolts of connecting rods of marine engines, or similar other parts to reduce the total length of the bolt. The lower part of the nut is made cylindrical and is grooved. A setscrew is introduced from the side, the point of which fits in the groove of the nut, to lock it. The method of locking such a nut is illustrated in locking arrangements.

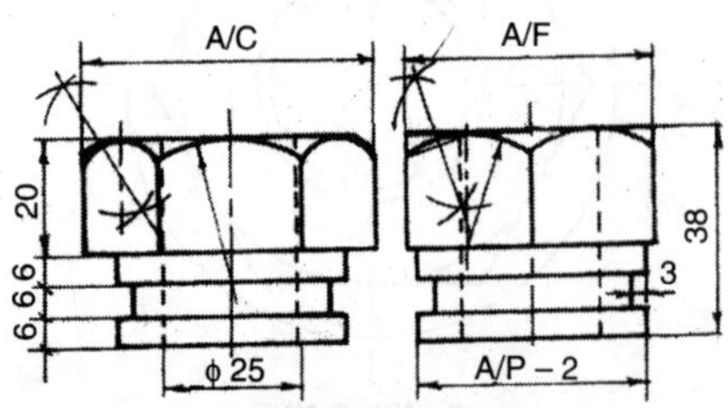

RING COLLAR NUT

Fig. 18.34.

9. Knurled Nut

It is cylindrical nut with knurled curved surface. It is useful where the nut is frequently required to be screwed on and off by hand, *Fig. 18.35*.

10. Fluted Nut

It is cylindrical nut with slots parallel to axis cut on its curved surface, *Fig. 18.36*. These nuts are used in pairs for locking purposes. For this reason it is sometimes called fluted locknut. It is also known as a ring nut.

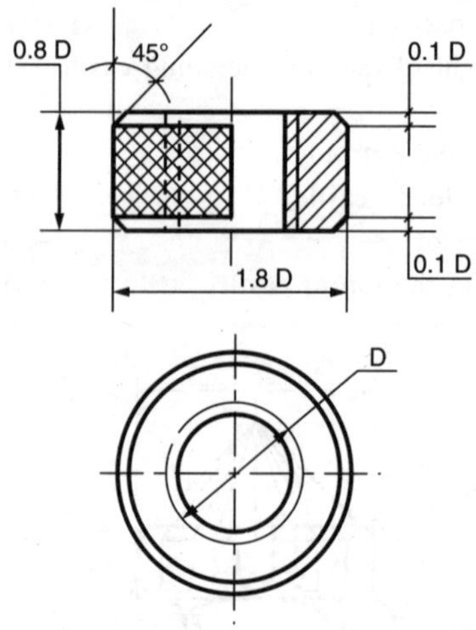

Fig. 18.35. KNURLED NUT

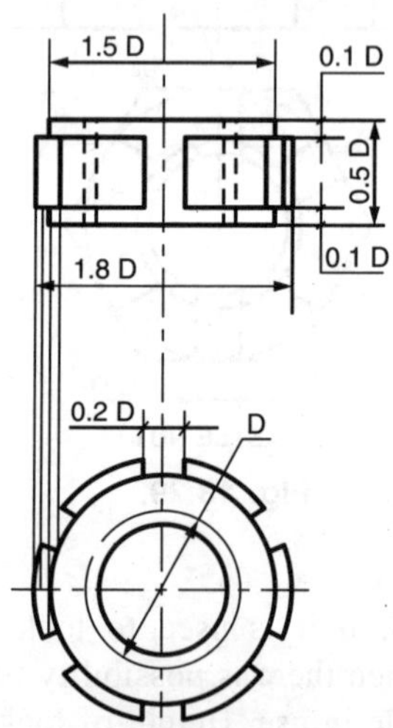

RING NUT

Fig. 18.36.

11. Wing Nut

This type of nut is used for light duty only. It is used wherever the nut is required to be frequently adjusted during use, *Fig. 18.37*. It can easily be tightened or un-tightened by thumb and a finger; i.e., without the use of a spanner.

12. Capstan or Cylindrical Nut

Figure 18.38. It is a circular nut provided with circular holes in the curved cylindrical surface, which are drilled for turning the nut with a Tommy bar. The holes can also be drilled in the top surface of the nut for which a pin spanner is used to turn the nut.

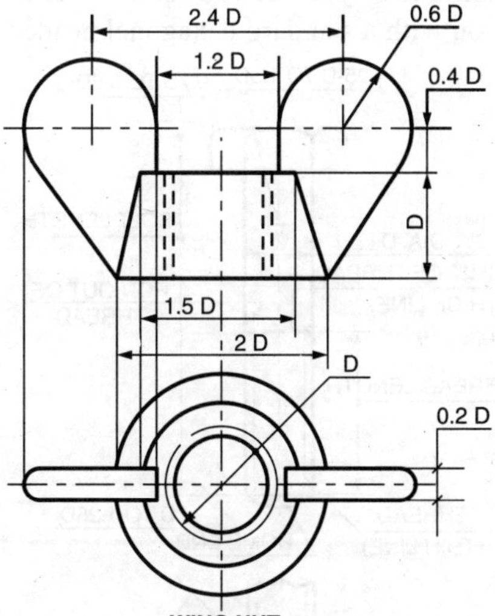

WING NUT

Fig. 18.37.

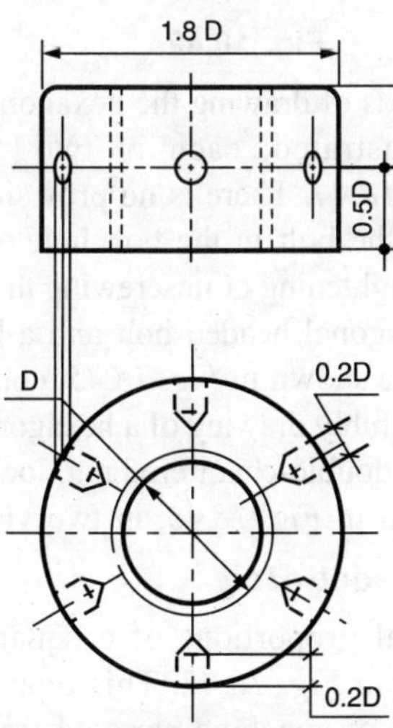

CAPSTAN NUT

Fig. 18.38.

18.9A. WASHER

A washer is placed below the nut, to give a smooth bearing surface, when the seating of the nut is rough or uneven. The pressure of the tightened nuts also spreads over a large area of the washer. Washer also prevents the nut from cutting into the metal and this all allows the nut to be tightened to a greater extent.

1. Plain and Chamfered Washers

The ordinary washers, shown in *Figs. 18.39, 18.40* are simple circular pieces of plate. The dimensions of these types of washers, in terms of *D*, the nominal diameter of the bolts, are as follows:

Diameter of washer = $2D + 3$ mm

Thickness = $0.125\,D$

Angle of Chamfer = $30°$

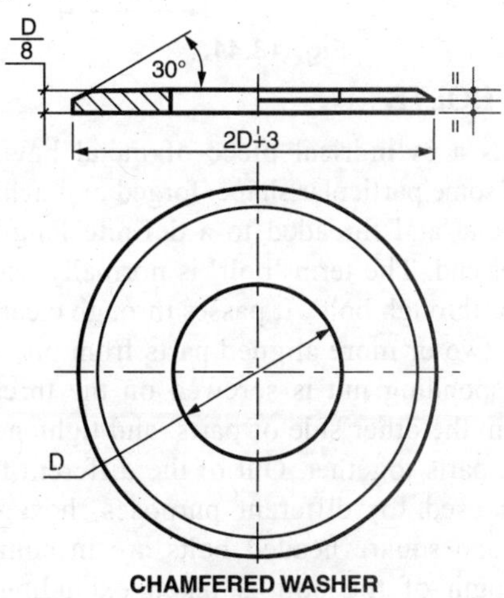

CHAMFERED WASHER

Fig. 18.39.

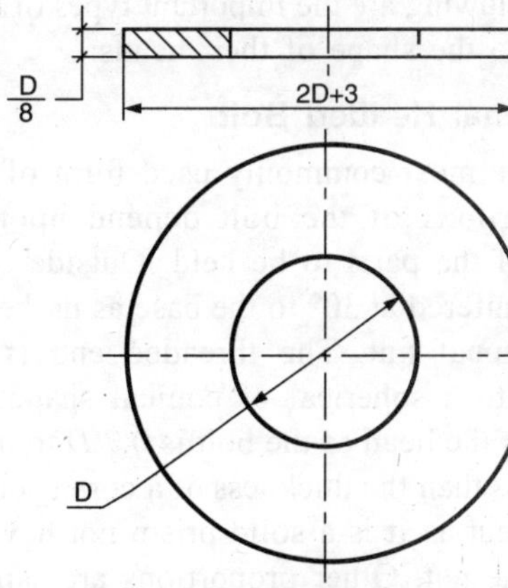

PLAIN WASHER

Fig. 18.40.

2. Spring Washer

It is used as a locking-device for the nuts and is made of spring steel. It is split and sprung open. It is put below the nut and the nut is then tightened. The elasticity of the spring keeps the nut tight on the bolt and does not allow it to get loose. Spring washers are widely used in automobiles or in other moving parts where the nut may work loose due to vibrations and shocks. Spring washers are shown in *Fig. 18.41.*

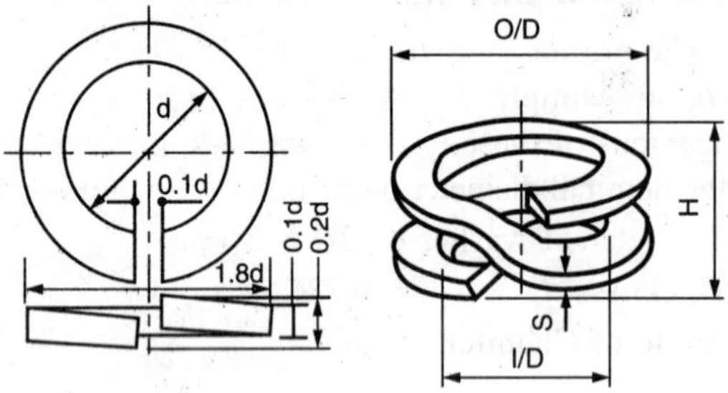

Fig. 18.41.

18.10. BOLTS

A bolt is a cylindrical piece of metal having a head, of some particular shape, forged or machined, on one end and threaded to a definite length on the other end. The term 'bolt' is normally used to denote a 'through bolt'. It passes through clearance holes in two or more aligned parts from one side. A corresponding nut is screwed on the threaded end, from the other side of parts, and tightened to hold the parts together. Out of the different types of bolts used for different purposes, hexagonal headed and square headed bolts are in common use. Length of the bolt is taken excluding the thickness of head.

The following are the important types of bolts, according to the shape of their heads:

1. Hexagonal Headed Bolt

It is the most commonly used form of bolt. The proportions of the bolt depend upon the thickness of the parts to be held. Outside of the head is chamfered at 30° to the base as in the case of a hexagonal nut. The threaded end is also chamfered to a spherical or conical shape. The thickness of the head of the bolt is 0.7 D to 0.8 D, which is less than the thickness of a corresponding hexagonal nut as it is a solid prism not having a hole like the nut. Other proportions are same as in a hexagonal nut.

Three dimensions are usually sufficient for simplified representation of a bolt.

(*i*) The bolt shank diameter, D,

(*ii*) The bolt length, and

(*iii*) The length of threaded portion of the shank.

The standard thread length = $2D + 6$ mm

(for length up to 150 mm)

or = $2D + 12$ mm

(for length over 150 mm)

Unless a special thread length is needed. A thread note, as, for example, M 30 × 3-g, should as well be given.

Figure 18.42 illustrates some important proportions, two types of end, and terms used in connection with a standard hexagonal headed bolt.

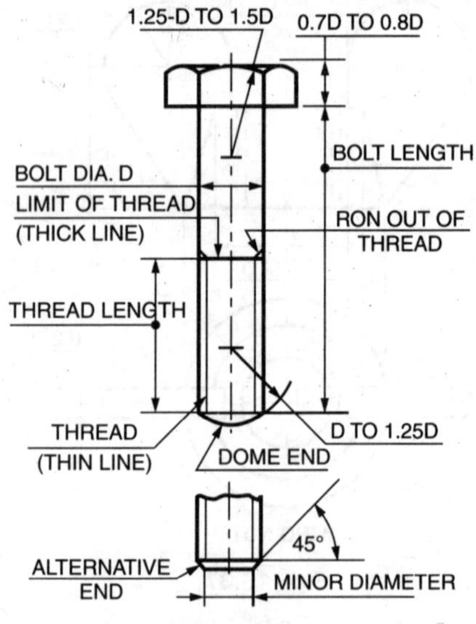

Fig. 18.42.

Two methods of drawing the hexagonal head of a bolt are illustrated, each in two stages, in *Fig. 18.43* and *18.44*. There is no provision to stop the rotation of the bolt in the bolt hole, if it starts rotating, while tightening or unscrewing the nut. Two views of a hexagonal headed bolt and a hexagonal nut assembly are shown in *Fig. 18.45*, completed in two stages. Assembly drawing of a hexagonal headed bolt, hexagonal double-chamfered nut, lock nut, and washer is shown in *Fig. 18.46*, in two views.

2. Square Headed Bolt

The general proportions of a square headed bolt are shown in *Fig. 18.47*. This type of bolt is frequently used on machine parts where the head is to be accommodated in a recess in a member.

The recess, too, is cut square and slightly larger than the head. The provision prevents the rotation of the bolt in bolthole when the nut is screwed on or off.

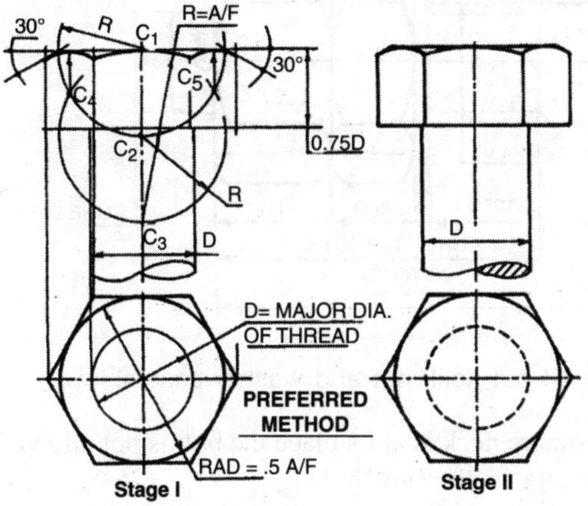

Fig. 18.43. Drawing a Hexagonal headed bolt (First method)

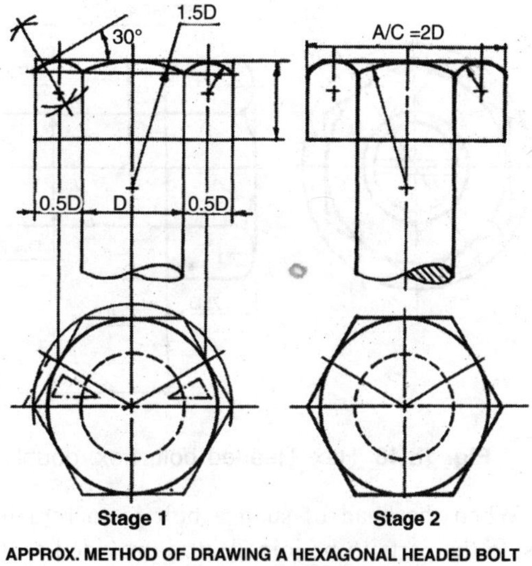

Fig. 18.44. (Second method) (First angle projection)

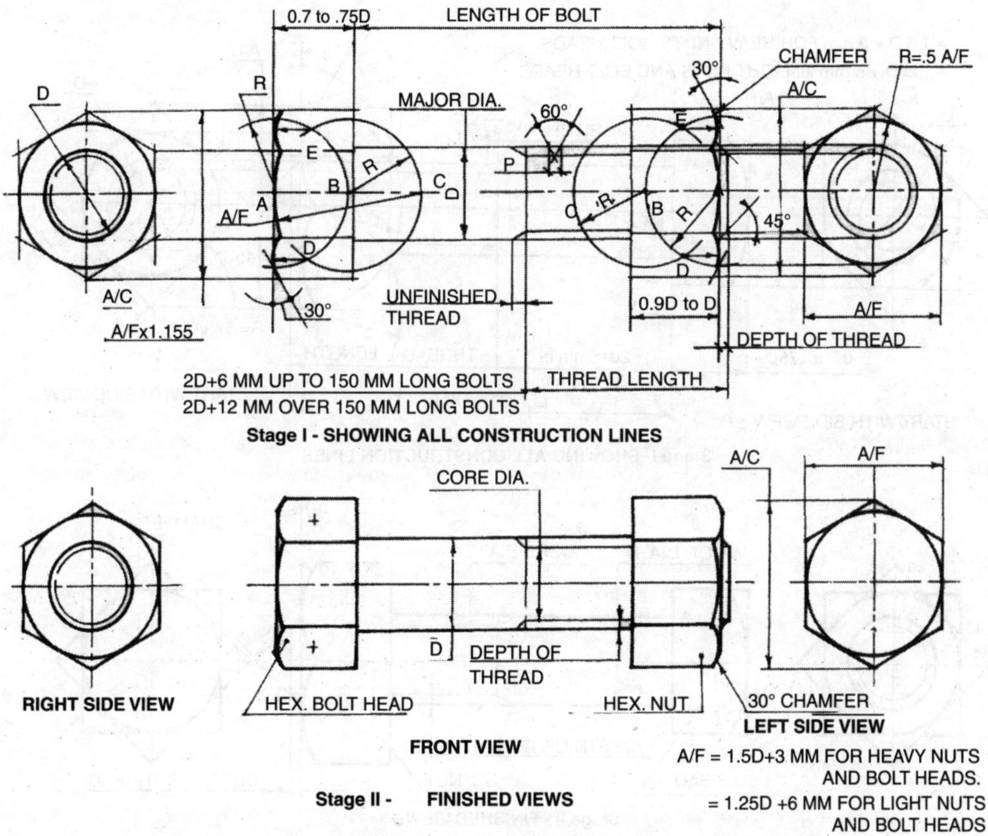

Fig. 18.45. Hexagonal heated bolt and hexagonal nut assembly

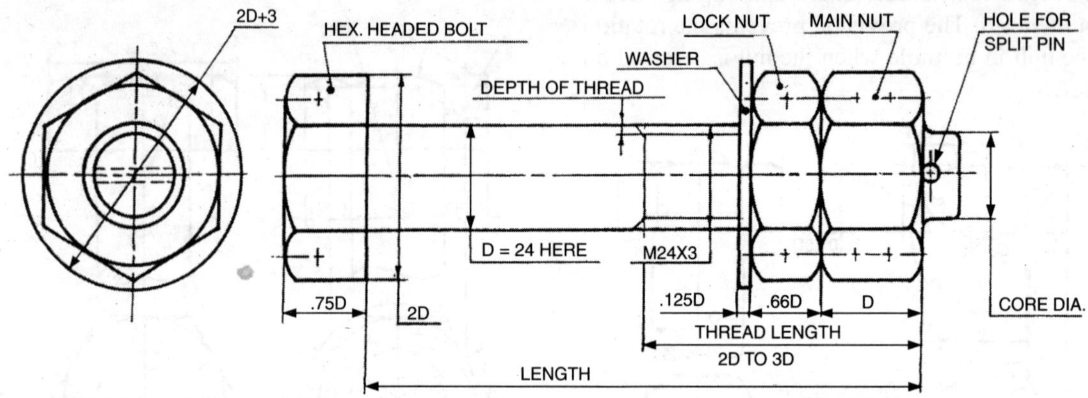

Fig. 18.46. Hex. Headed bolt, hex. double chamfered nut, lock nut and washer assembly

When the head of such a bolt is to remain outside the parts to be joined, it is provided with a neck of square cross-section. The bolt hole is made of the same form to just accommodate this square neck. When the bolt is put in position and the square neck is at its place the bolt is not allowed to rotate in the bolt hole.

An assembly drawing of a square headed bolt with a square nut, drawn in two stages, is shown in *Fig. 18.47*, and a square headed bolt with square

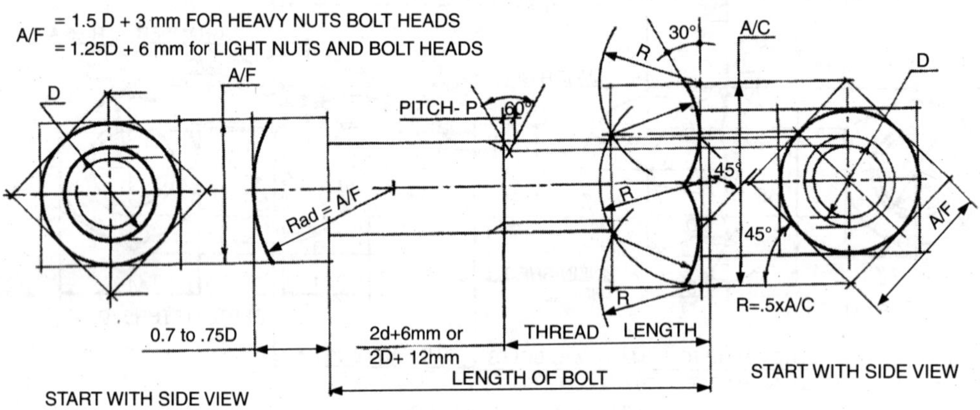

Stage I- SHOWING ALL CONSTRUCTION LINES

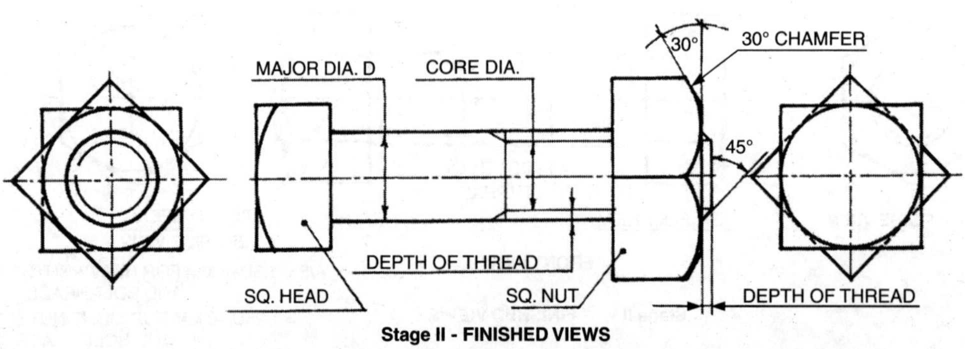

Stage II - FINISHED VIEWS

Fig. 18.47. Square headed bolt and square not assembly

neck, a hexagonal nut and a washer is shown in *Fig. 18.48*, in two views.

3. Cylindrical or Cheese Headed Bolt

The usual proportions of two types of this head are shown in *Fig. 18.49*. The head of this type of bolt is of a cup shape and a snug is forged with the bolt, to fit in a similar recess in the bolt hole, to prevent the rotation of the bolt while screwing the nut on or off. Sometimes the cup headed bolt is provided with a square neck to prevent rotations in its hole.

4. Cup or Round Headed Bolt

Its usual proportions are shown in *Fig. 18.49*. It is the simplest form of a bolt and can be easily manufactured by turning a rod on a lathe or by forging the cylindrical head. It is frequently used in a limited space and where the use of a spanner is to be avoided on the head side. It is used on cross-heads, connecting rods and eccentrics etc. The rotation of bolt in bolt hole is prevented by means of a pin inserted into shank, just below the head. This pin serves as a snug. Sometime a little portion of the head, from below, is chipped off and given the shape of a snug to prevent rotation of the bolt,

5. T-headed Bolt

The usual proportions of a T- headed bolt are shown in *Fig. 18.49*. The head of this bolt is like a T. It is used in machine table. The T- head of the bolt can slide to the required position, through T- slots cut on the machine table. A square neck is provided with the head.

6. Eye Bolt

Head of one type of eyebolt, shown is *Fig. 18.49*, is of the shape of a circular ring of the rectangular cross-section.

Another type of eyebolt, having a circular ring of circular cross-section as head (lifting eye-bolt), is also shown in *Fig. 18.49*. A flat circular portion, integral with the head, is also provided. Below this circular portion, the shank of the bolt is of a cylindrical shape on which threads are cut.

This type of bolt is used for lifting a heavy machine. The bolt is screwed in a threaded hole, provided for this purpose, on the top of the machine, directly above the centre of gravity, so that, while lifting, the machine does not change from its usual working position.

7. Hook Bolt

The shape and usual proportions of a hook bolt are shown in *Fig. 18.49*. In this bolt the head projects only to one side of its shank. The shank of the bolt passes through a hole in one part and the other part, to be jointed, comes under the bolt head and in supported by it. This bolt is also provided with a square neck to prevent its rotation in the bolt hole.

8. Counter-sunk Headed Coach Bolt

Figure 18.49. Where the head of a bolt must not project above the surface of the connected pieces, this form of bolt is used. It may be provided with a snug or a square neck.

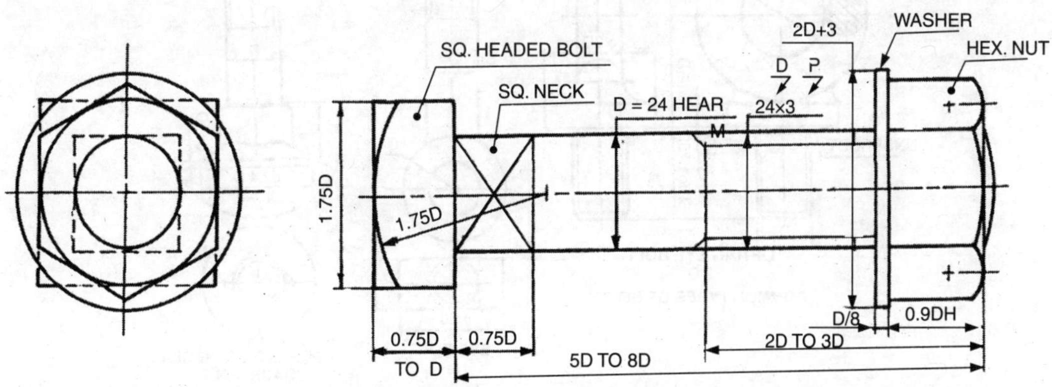

Fig. 18.48. Square head bolt with square neck, washer and hex. Nut assembly (Third Angle projection)

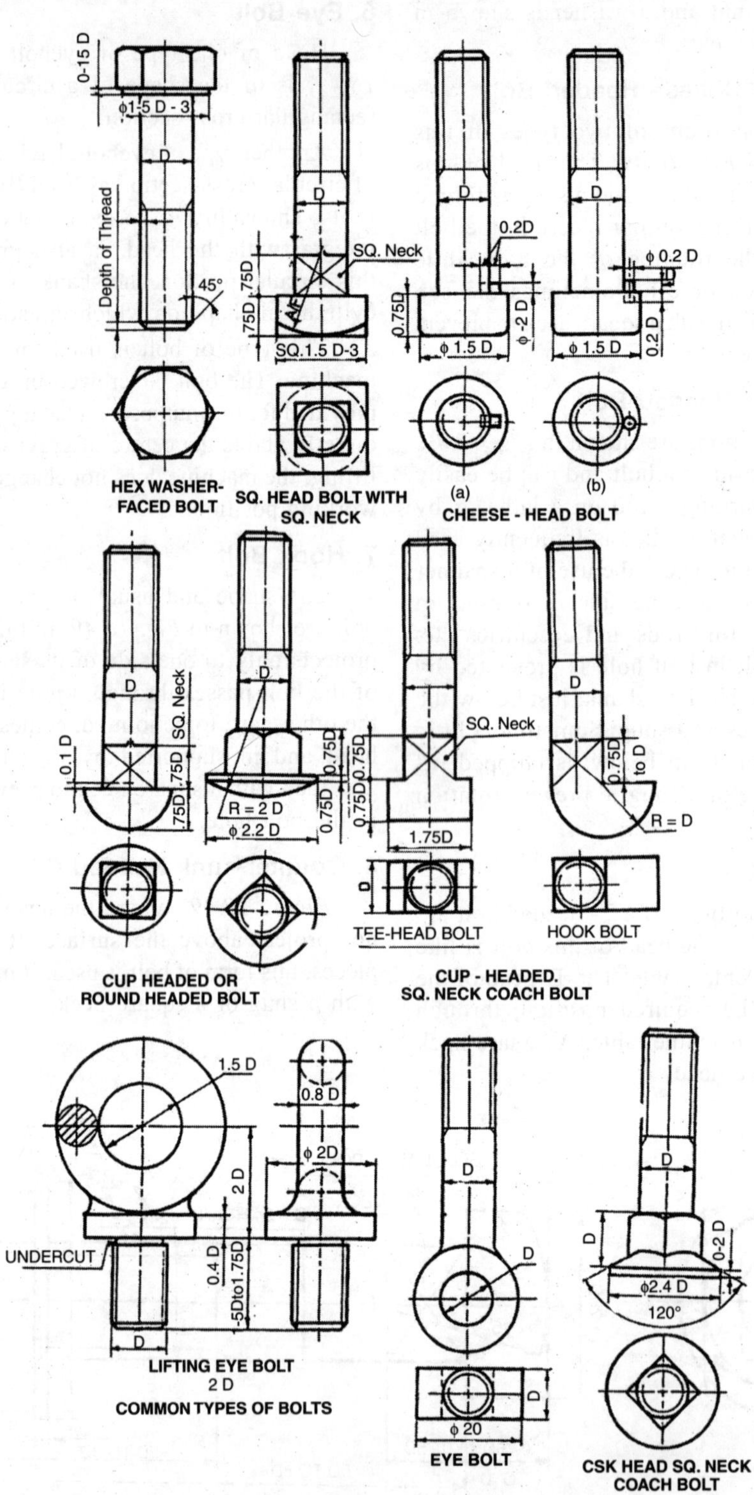
0-15 D
φ1·5 D - 3
D
Depth of Thread
45°
HEX WASHER-
FACED BOLT
D
SQ. Neck
AF
SQ.1.5 D-3
SQ. HEAD BOLT WITH
SQ. NECK
D
0.2D
0.75D
φ -2 D
φ 1.5 D
(a)
D
φ 0.2 D
0.2 D
φ 1.5 D
(b)
CHEESE - HEAD BOLT
D
SQ. Neck
0.1 D
R = 2 D
φ 2.2 D
CUP HEADED OR
ROUND HEADED BOLT
D
SQ. Neck
1.75D
TEE-HEAD BOLT
D
R = D
HOOK BOLT
CUP - HEADED.
SQ. NECK COACH BOLT
1.5 D
0.8 D
φ 2D
2 D
0.4 D
UNDERCUT
D
LIFTING EYE BOLT
2 D
COMMON TYPES OF BOLTS
D
D
φ 20
EYE BOLT
D
D
0-2 D
φ2.4 D
120°
CSK HEAD SQ. NECK
COACH BOLT

Fig. 18.49.

18.11. THROUGH BOLT

When a bolt passes through the holes in the two parts to be joined and is held in place by a nut, it is called a through bolt, *Fig. 18.50.*

18.12. TAP BOLT

When a bolt passes through a hole in one part and is screwed into a tapped hole in the second part, which acts as a nut, thus holding the two parts together, it is called a tap bolt, *Fig. 18.51.*

18.13. STUD or STUD BOLT

Stud is a round bar threaded on both its ends. Some types of studs are shown in *Fig. 18.54*. An undesirable feature of a tap bolt fastening is the tendency to damage the threads in the holes when the bolts are frequently removed and replaced, especially when the screwed holes are in cast iron and aluminum alloys. This disadvantage is overcome by the use of stud bolts, *Fig. 18.52*. Use of a stud is illustrated in *Fig. 18.53*.

It has threads on both ends so that one of the pieces, being held together, must be threaded to replace the head. The stud bolt is screwed tightly into the threaded part by a special locknut device, until it jams. The studs then become an assembly, guide by means of which the other part, which is drilled but not threaded, is easily placed in position, A nut screwed on the other end of the stud bolt holds the two parts together. The end of the stud, which is screwed permanently in the tapped hole, is called **stud end** or **metal end** and the other end on which the nut is screwed is called **nut end.**

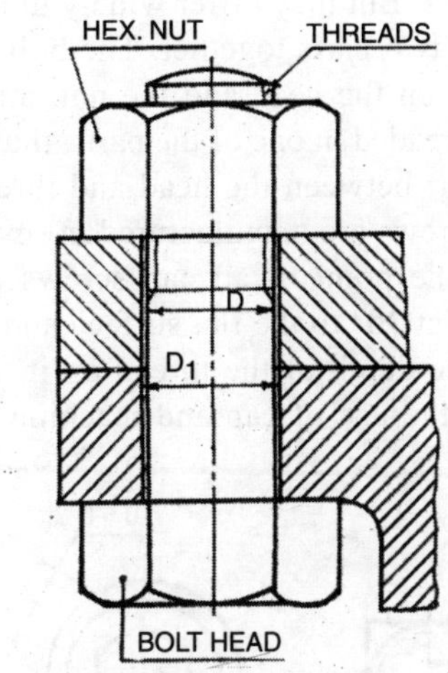

Fig. 18.50. Through bolt

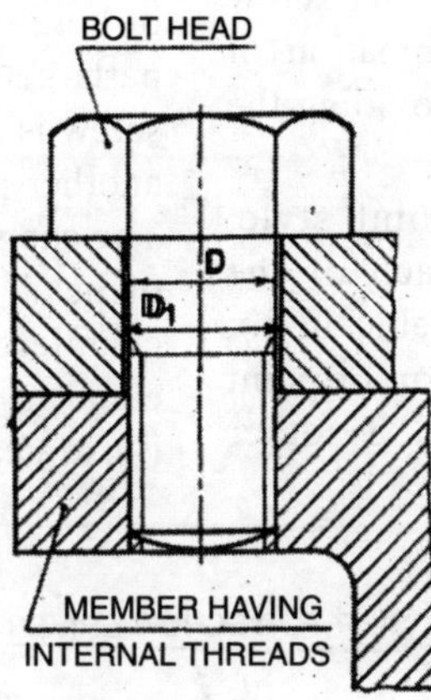

Fig. 18.51. Top bolt

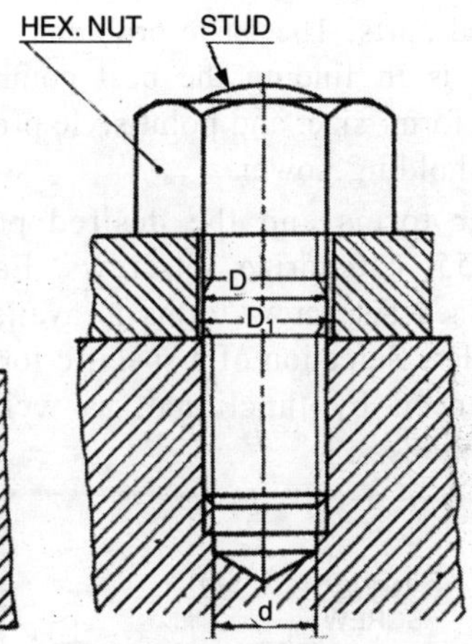

Fig. 18.52. Stud bolt

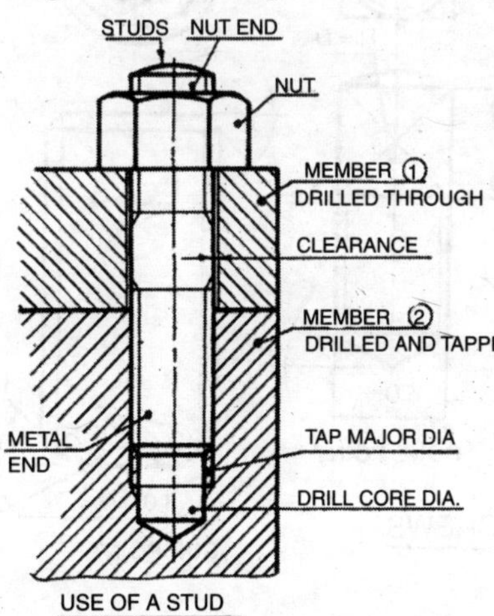

Fig. 18.53.

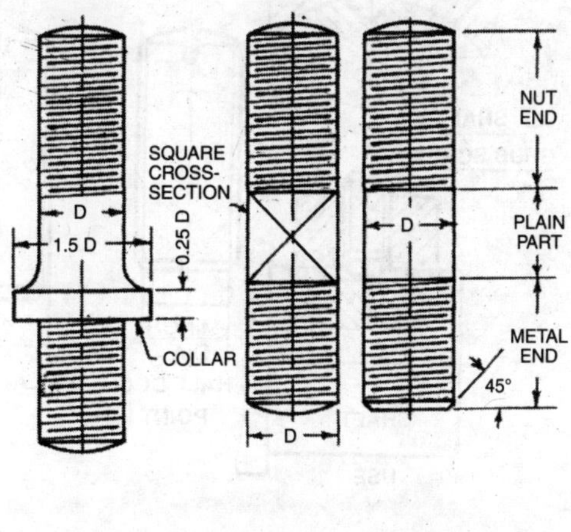

Fig. 18.54. Studies

The stud may have one or two collars. The stud having one collar is called single collar stud and that having two collars is called double collar stud. The collar gives a bearing surface to the stud and supports the adjoining part. The central portion of the stud may be round or square. The square neck, at the centre, facilitates gripping of stud while screwing or unscrewing.

18.14. SET SCREWS

Set screws are used as semi-permanent fasteners to hold a collar, sheave, pulley or gear on a shaft against rotational or translational forces. In contrast to most fastening devices, the set screw in essentially a compression device. Forces developed by the screw point during tightening produce a strong clamping action that resists relative motion between assembled parts. The basic problem in set screw selection is in finding the best combination of setscrew form, size, and point style providing the required holding power.

Their forms and the desired point style, *Fig. 18.55*, categorize setscrews. Each of the standardized set screw forms is available in six point styles. Selection of a specific form or point is influenced by functional, as well as other considerations.

The conventional approach to setscrew selection is usually based on a rule of thumb that the setscrew diameter should be roughly equal to one half the shaft diameter. This rule of thumb often gives satisfactory results but it has a limited range of usefulness. When a setscrew and key are used together, the screw diameter should be equal to the width of the key.

A flat is a slight depression usually cut on a shaft to serve as a surface on which the end of its setscrew can rest when holding an object in place, *Fig. 18.55*.

18.15. CAP SCREWS

Cap screws are similar to bolts in that they have a head on one end and threads on the other, *Fig. 18.56, 18.57*. But they differ widely in the method of holding two parts together. The bolt keeps two parts between the head and the nut, and the cap-screw is threaded in one of the parts, thus clamping another part between the head and threaded part. The cap screws are manufactured in several styles of head. The point of all cap screws is flat and chamfered at 30° to the flat surface and to a depth equal to the depth of the thread on it. *Fig. 18.58* illustrates the use of cap and machine screws.

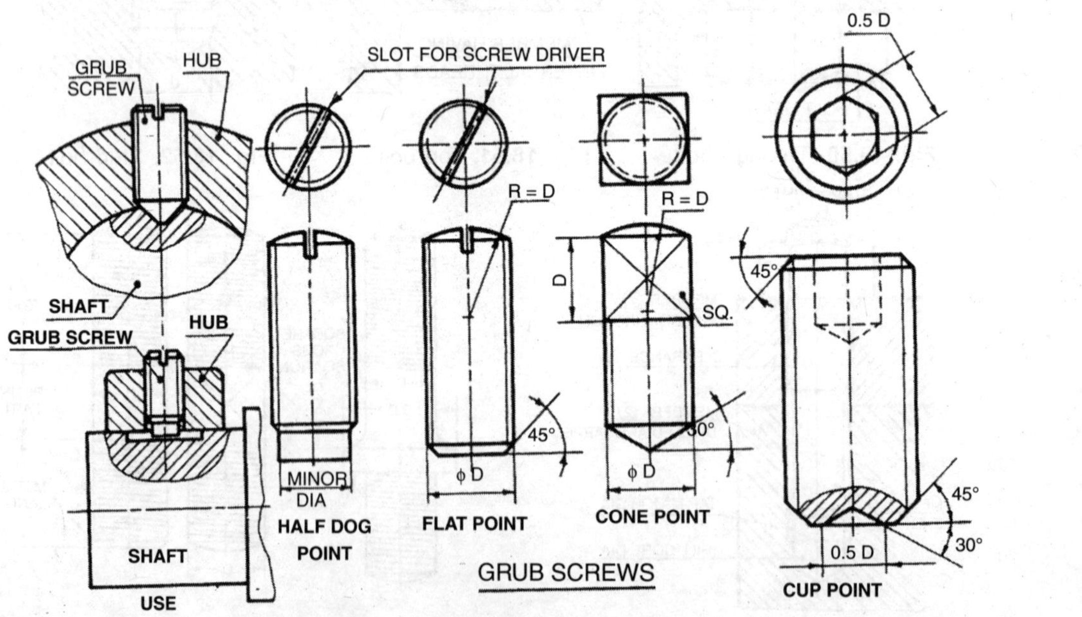

Fig. 18.55. Set screws and their use

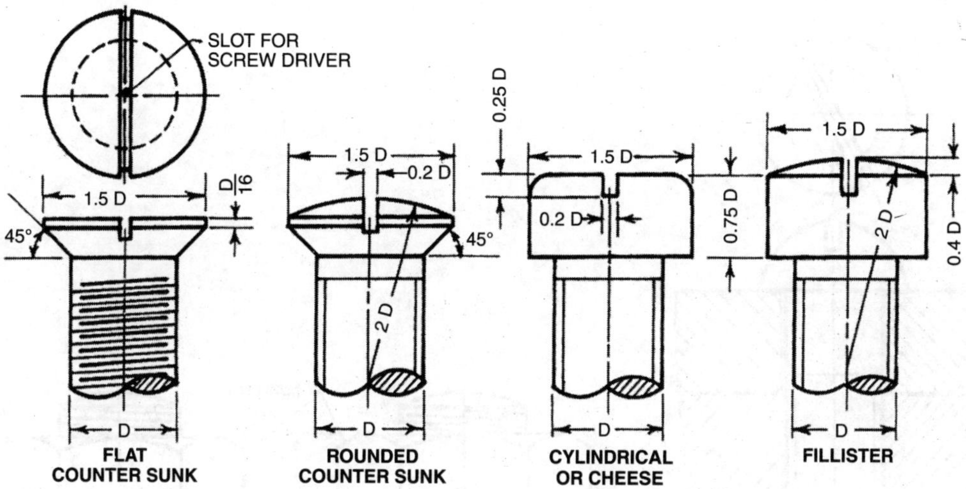

Fig. 18.56.

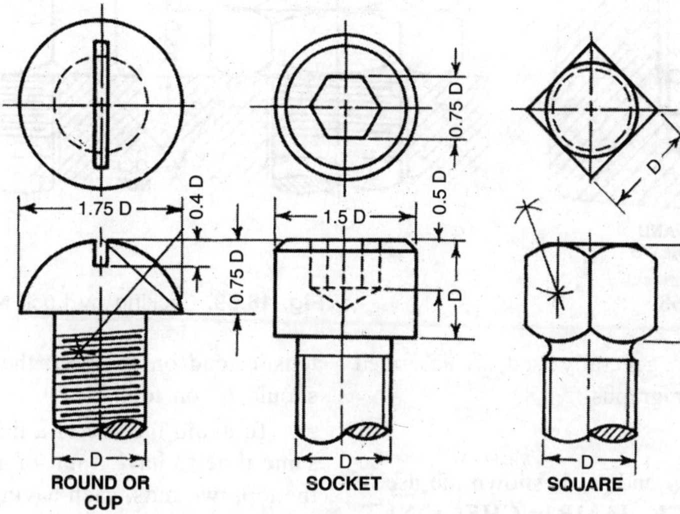

Fig. 18.57. Cap and Machine Screws

18.16. MACHINE SCREWS

These are similar in function and operation to cap screws, but are usually smaller in diameter.

Materials: For general engineering purposes, nuts and screws are made of Mild steel (MS). However, copper and its alloys, aluminium alloys, etc. are also used, for special purposes, in their manufacture.

18.17. LOCKING ARRANGEMENTS

It is frequently necessary to provide some method of preventing the nut from unscrewing and thus leaving a loose connection, which may be dangerous to the machinery. This is particularly true when the screwed fastening is subjected to vibration. To check against this slackening tendency of the nuts and keep them in tight position, some devices called locking devices are employed. Some of these devices are positive locks and some are friction locks. Out of a wide variety of locking devices available,

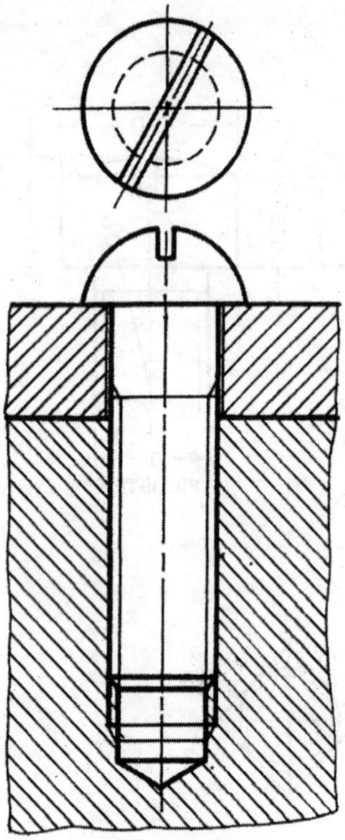

Fig. 18.58.

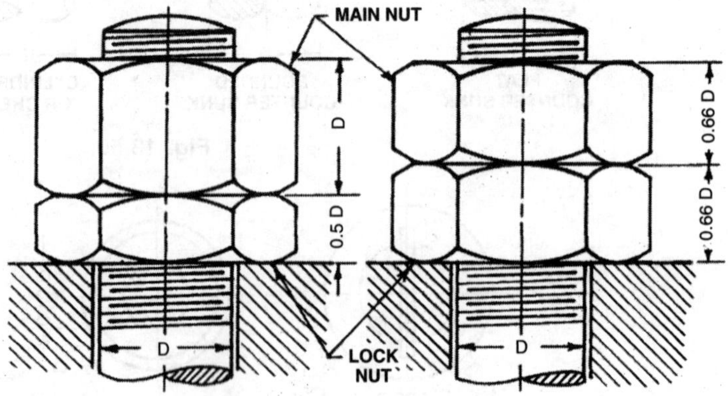

Fig. 18.59. Locking by Lock Nut

some important types, generally used, are described in the following paragraphs.

1. Lock Nut

In *Fig. 18.59 (a)* and *(b)* is shown the use of the well known **LOCK, JAMB** or **CHECK NUT**, chamfered at both ends and about 0.5 to 0.75 of the standard nut thickness. It is the most common frictional locking device.

The lock nut (lower and thinner) is first tightened down on the bolt with ordinary force, and then the upper main nut is tightened down upon it. The upper nut is then held tightly whilst the lower lock nut is slackened back against it. Due to this action the threads in nuts get wedged tightly against those of the bolt. A thin spanner is necessary in slackening back the lock nut. As thin spanners are not always available in workshops, many times the lock nut is placed above the main nut. If the nuts are really tightened down, as they should be, the upper nut carries the whole of the tensile load on the bolt; therefore, the thick nut should be on top.

To avoid the use of a thin spanner and at the same time to have a nut of adequate thickness at the top, two nuts, each having a thickness of 0.66 to 0.75 of thread diameter, *Fig. 18.59 (b)*, are sometimes used.

2. Spring Washer

Figure 18.60 illustrates this very common frictional locking device. The single coil type is illustrated but double coil examples are also seen in practice. Tightening the nut threads against bolt threads, thus increasing the friction between them.

3. Split Pin or Taper Pin

In *Fig. 18.61 (a), (b), (c)* the nut is shown locked by means of **PINS** passing through the bolts. At *(a)* we have a split pin driven through a hole, drilled through, in the centre of the bolt, after fully tightening the nut. It gives positive locking. This is very safe form of locking device. The wire of which the split pin is usually made is of semicircular section, so that when its two ends are carefully bent we obtain a circular section pin with a loop at one end. A disadvantage of this method is that if the nut is removed from the bolt, for some reason, it is very difficult to align the holes in the nut and bolt again to pass the pin. The hole also weakens both details. At (*b*) is shown an arrangement of locking by the use of **SLOTTED NUT**. This is a standard hexagonal nut with a slot cut across the top of each pair of flats. A split pin passes through one slot and the hole in the bolt, the legs of the pin being opened and bent around the faces of the nut. This gives a positive lock. It is sometimes difficult, with this arrangement, to align the slots and the hole in the bolt. Also the hole and slots weaken the bolt and nut. At (*c*) is shown a view of a well-known **CASTLE NUT**. Its shape, construction and general proportions are shown in *Fig. 18.29.* This is a development of the slotted nut. With this nut the length of thread in engagement

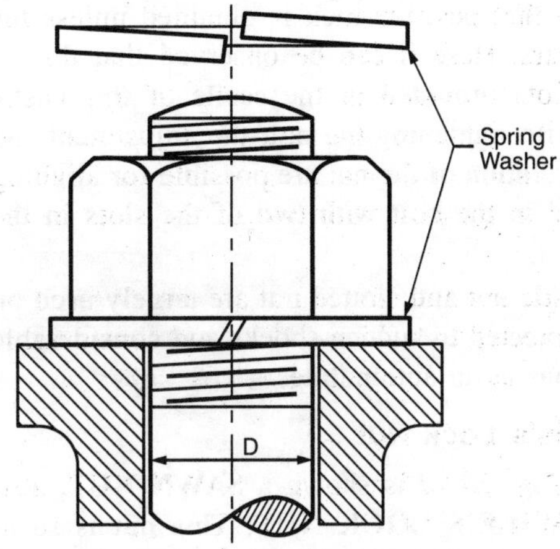

Fig. 18.60. Looking by spring washer

with the bolt thread is not reduced as it is with the slotted nut, and the bent legs of the split pin are safer as they do not stand out on the nut faces. The proportions of both slotted and castle nuts have been standardized. The split pin passes through two opposite slots is the nut and the hole in the

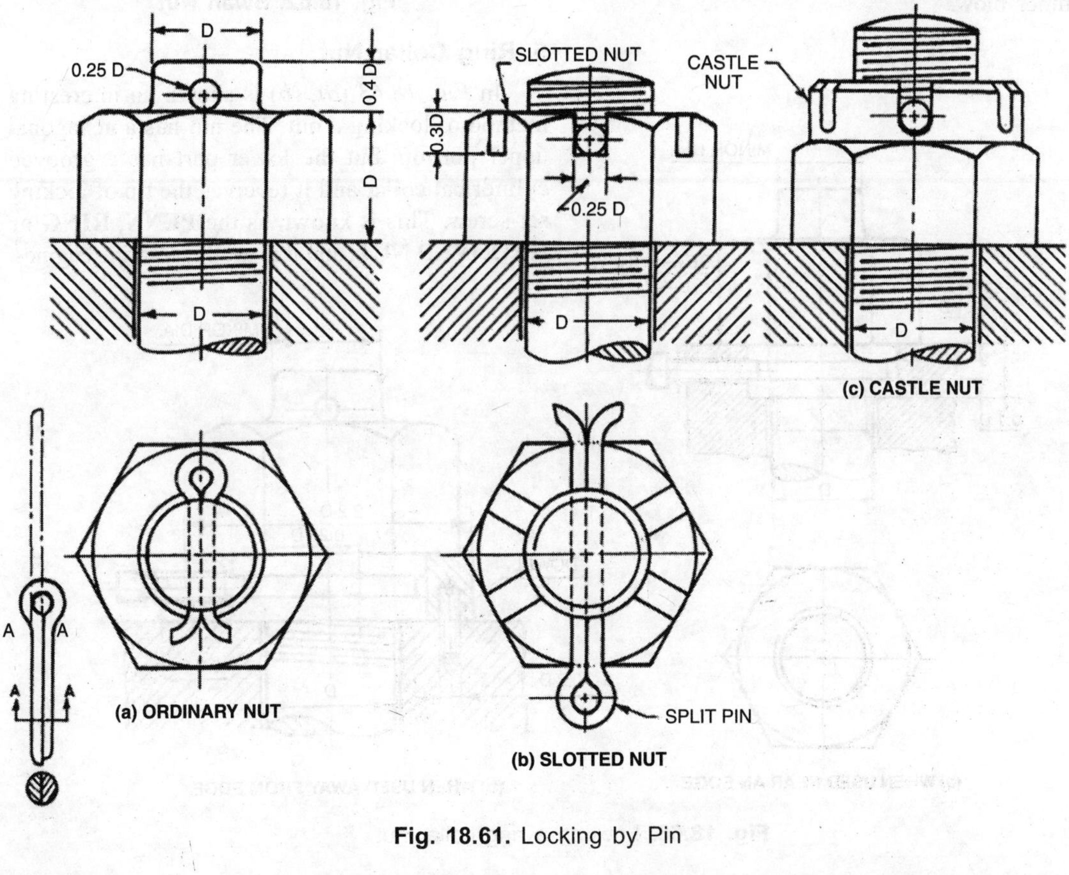

Fig. 18.61. Locking by Pin

bolt, so that positive lock is obtained unless the pin shears. Here it can be observed that due to these slots provided in the castle of this castle nut, while tightening the nut, the adjustments up to 30° rotation of the nut are possible for aligning the hole in the bolt with two of the slots in the nut.

Castle nut and slotted nut are largely used on jobs subjected to sudden shocks and considerable vibrations as in automobiles.

4. Wiles's Lock Nut

In *Fig. 18.62* is shown a **SAWN NUT,** also called **WILE'S LOCK NUT**. This nut is sawn half way though. When the set screw is tightened up (shown loose in figure) the two halves of the nut are pulled together and the nut is locked in position by jamming the threads. The small hole in the lower portion of the nut is tapped, to receive the set screw, and the upper portion is drilled as a 'clearance hole'. Small nuts, with insufficient space to accommodate a set screw, are locked by a hammer blow.

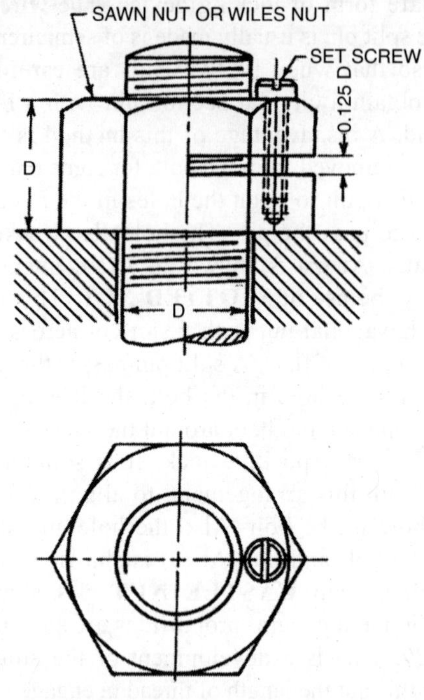

Fig. 18.62. Swan Nut

5. Ring Collar Nut

In *Fig. 18.63 (a)*, (*b*) is shown an interesting method of locking a nut. The nut has a hexagonal upper portion, but the lower part has a grooved cylindrical collar and it receives the tip of locking set-screw. This is known as the **PENN, RING or GROOVED NUT**, and it is used on marine engines.

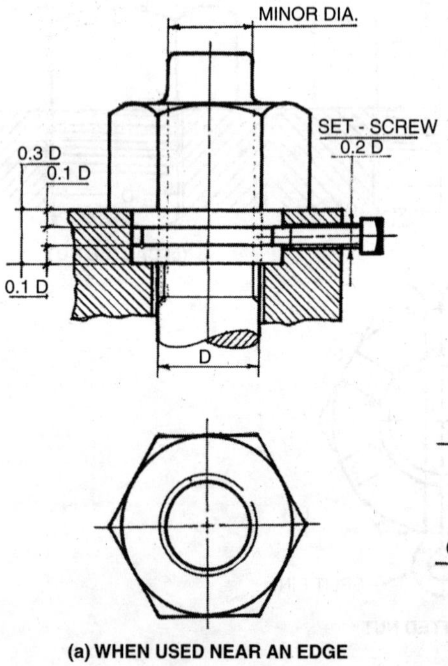

(a) WHEN USED NEAR AN EDGE

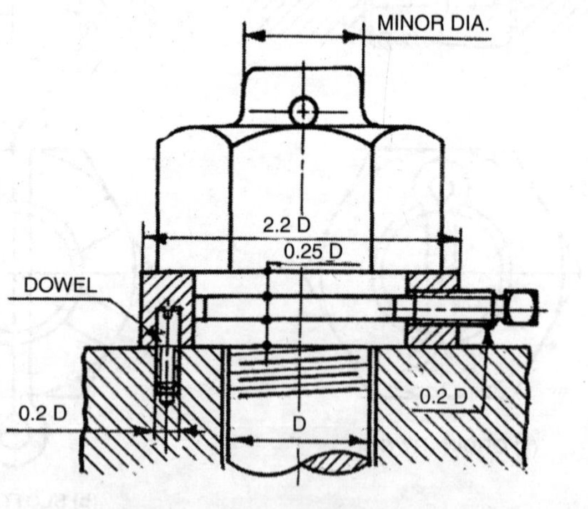

(b) WHEN USED AWAY FROM EDGE

Fig. 18.63. Locking a ring collar nut

Figure 18.63 (*a*) shows its application near an edge. The bolt hole requires counter boring to receive the cylindrical portion of the nut. Away from an edge the design shown in *Fig. 18.63 (b)* is used. The grooved cylindrical portion of the nut passes through the ring, which is retained in position by a dowel. The set screw passes through the ring into groove. A split-pin in often fitted as an additional safeguard.

6. Locking Plate

In *Fig. 18.64* is shown a form of **STOP PLATE** or **LOCKING PLATE.** The nut can be adjusted and subsequently locked through angular intervals at 30° by using such plates.

7. Tab Washer

This also is a form of positive locking device, *Fig. 18.65*. This type, shown here, is suitable for fitting over the edge of flange. The washer is fitted under the nut; one leg is bent over the flange and other up against a flat on the nut. The use of a threaded pin, shown in *Fig. 18.67*, to lock the nut is self-evident.

8. Simmond's Lock Nut

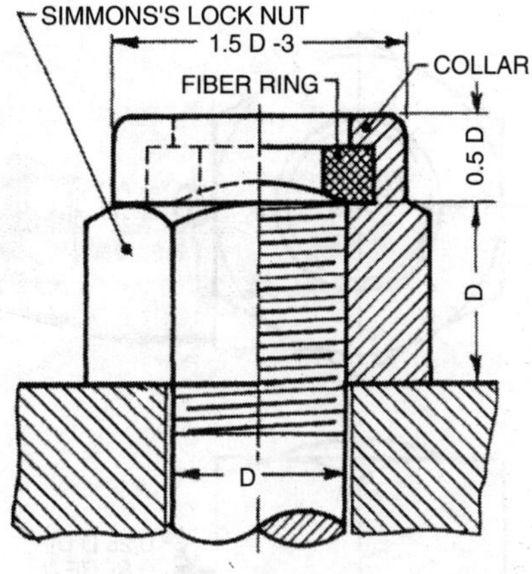

Fig. 18.66. Simmond's Lock Nut

Figure 18.66. In these nuts a collar with an internal diameter smaller than the thread minor diameter is fitted in a cylindrical projection on the top face of the nut. When the nut is screwed on to top bolt, threads force their way though this collar which grips them tightly producing a functional

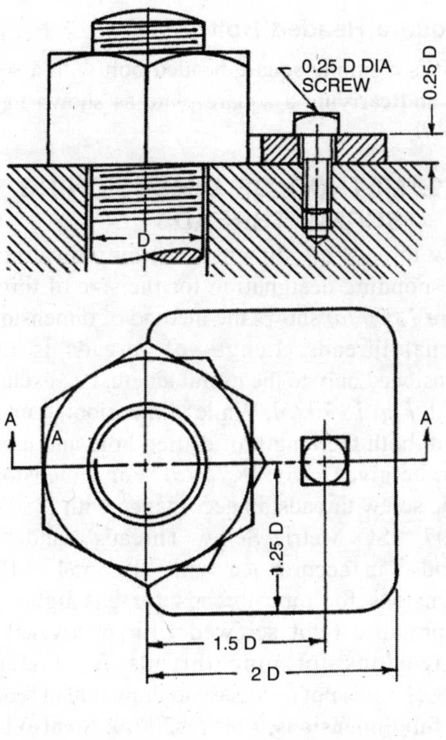

Fig. 18.64. Stop plate or locking plate

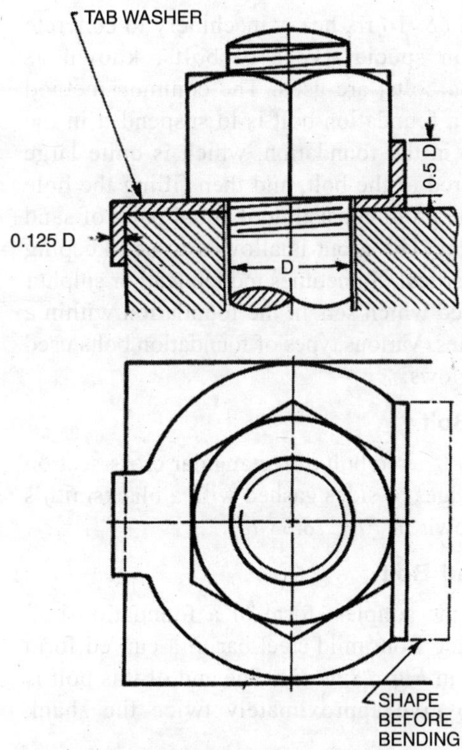

Fig. 18.65. Use of tab washer

lock. The collar may be made of fiber or nylon, the nylon being suitable for use at high temperatures.

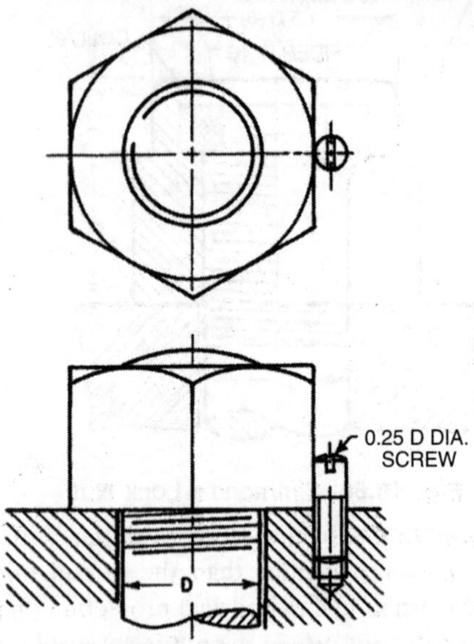

Fig. 18.67. Use of threads pin (Third angle projection)

18.18. FOUNDATION BOLTS

Figure 18.68. To fix heavy machinery to concrete foundation special types of bolts, known as foundation bolts, are used. The common method of fixing a foundation bolt is to suspend it in the hole kept in the foundation, which is quite large as compared to the bolt, and then filling the hole with a fine grout consisting of equal parts of sand and cement. The grout is allowed to set, keeping the bolt vertical. Sometimes molten lead or sulphur is also used which sets in the foundation within a few minutes. Various types of foundation bolts used are as follows:

1. Rag Bolt

It is a tapered bolt of rectangular cross-section with its edges or sides gashed with a blacksmith's set as shown in *Fig. 18.68 (a)*.

2. Curved Bolt

It is the simplest form of a foundation bolt and is made from mild steel bar in a curved form as shown in *Fig. 18.68 (b)*. The end of this bolt is forged to give approximately twice the shank diameter.

3. Lewis Bolt

It is used for a temporary foundation and is shown pictorially in *Fig. 18.69 (b)*. One side of this bolt is straight and other tapered. It is fixed along with a mating key. When the key is driven in position, after putting along the bolt, the tapered side of the bolt is driven below the foundation material. So long as the key is in position the bolt cannot be taken out of foundation. To remove the bolt, take out the key and slide the bolt to space vacated by the key. The tapered side of the bolt then comes clear off the foundation and the bolt can easily be taken out, *Fig. 18.68 (c)*.

4. Cotter Bolt

It is used for fixing heavy machines to the brick or stone foundation. This bolt is secured below the floor or foundation level by means of a cotter as shown at *Fig. 18.69 (c)* and *Fig. 18.68 (d)*.

5. Hoop Bolt

Figure 18.68 (e) shows two views of an eye of hoop bolt forged form a bar. It has a piece of mild steel bar passing through the eye and at right angles to it.

6. Square Headed Bolt

It is a simple square headed bolt with a square neck and carrying a square plate as shown in *Fig. 18.68(f)*.

18.19. METHOD OF DIMENSIONING SCREW THREADS

Screw threads are dimensioned by indicating the corresponding designation for the size of threads. *Figure 18.70 (a)* shows the method of dimensioning external threads. Length of threads is to be dimensioned only to the useful length, i.e., excluding runout, *Fig. 18.70 (b)*. While .dimensioning internal thread, both the length of drilled hole and tapping are to be given, *Fig. 18.70 (d)*. For dimensioning metric screw threads, in accordance with IS:.42318 - 1967 "ISO Metric Screw Threads", and "Pipe Threads" in accordance with "IS: 554 - 1964" Dimensions for pipe threads for gas tight tubes and pressure tight screwed fittings revised and "Dimensions for pipe threads for fastening purposes". It is not necessary to draw thread sections with full dimensions, *Fig. 18.70 (c)*. So also is the case where Indian Standard has been published

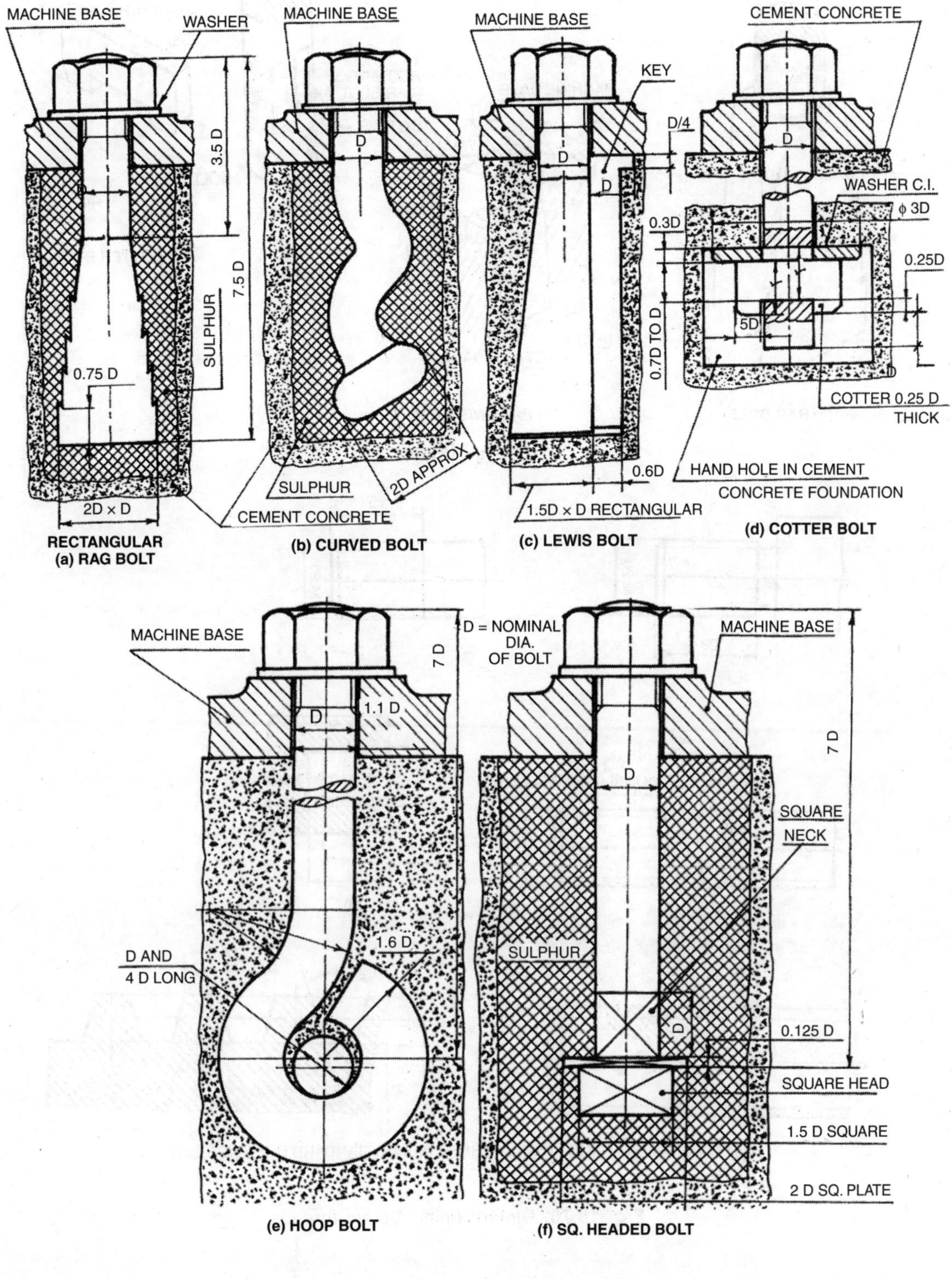

Fig. 18.68. Foundation bolts

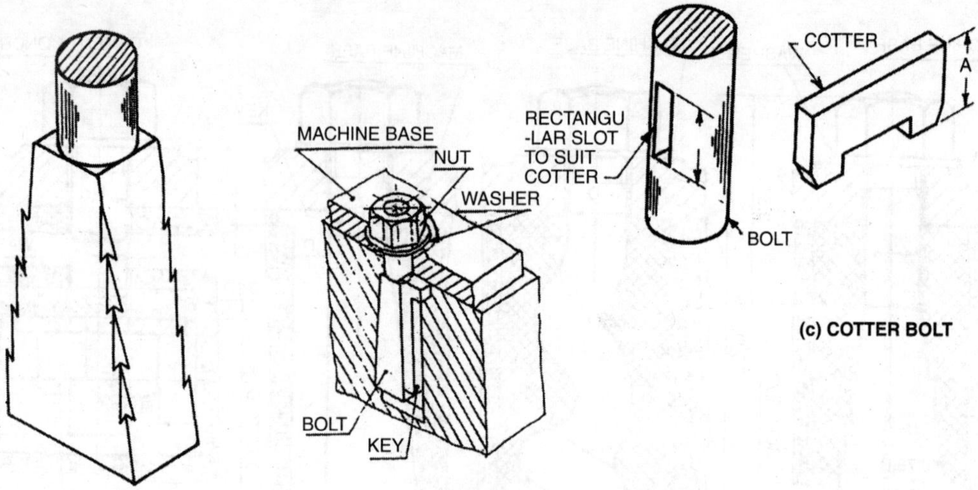

(a) RAG BOLT

(b) LEWIS BOLT

(c) COTTER BOLT

Fig. 18.69.

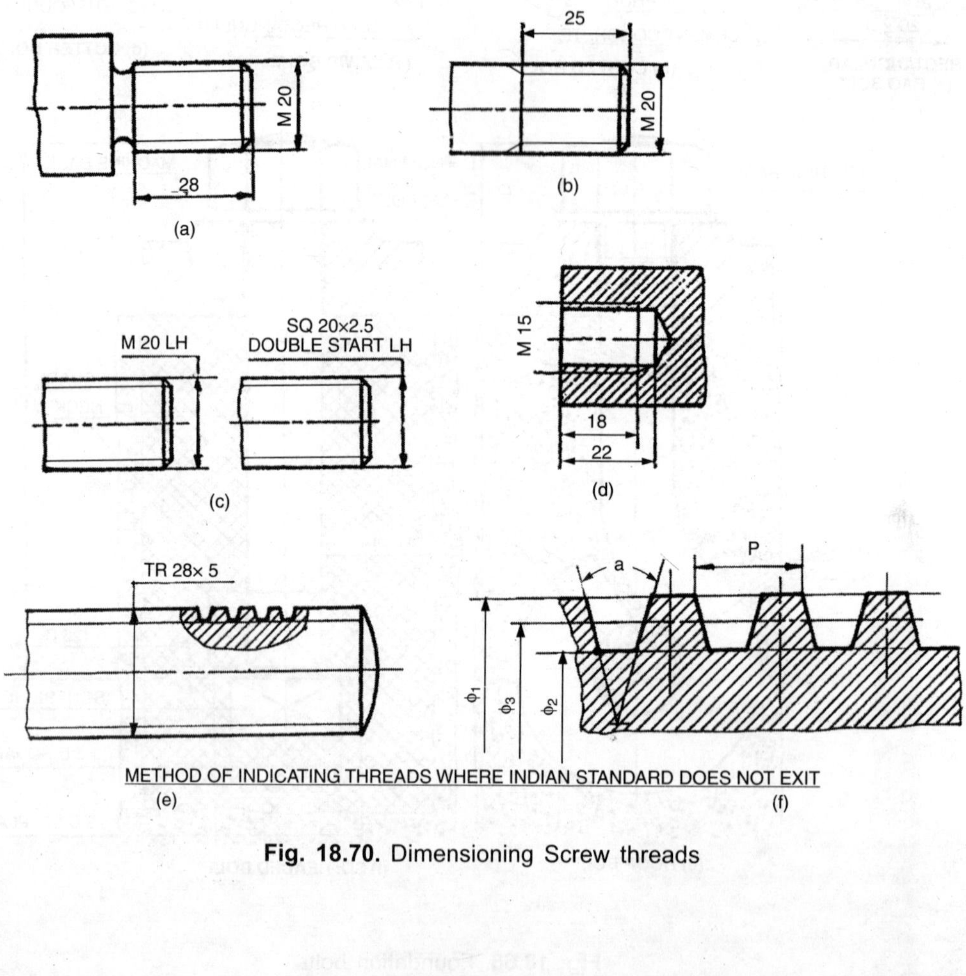

Fig. 18.70. Dimensioning Screw threads

for their types of threads such a square threads etc. However, in cases where IS does not exist, the thread section and all-important dimensions, including the designation, are to be clearly indicated, *Fig. 18.70 (e)* and *(f)*.

All dimensions such as runout, bolt or screw ends etc., which are standardized need not be dimensioned, but indicated with respect to corresponding Indian Standard. Left hand threads and multistart threads shall have the appropriate designation as given in relevant Indian Standard. For example, square threads of 40 mm nominal diameter, 3 mm pitch, double start with left-hand threads (IS: 4694 - 1968) shall be designated as: SQ 40 × 3 DOUBLE START LH.

Conventional symbols for bolts and nuts are shown in *Fig. 18.71* and application of these symbols is shown in *Fig. 18.72*.

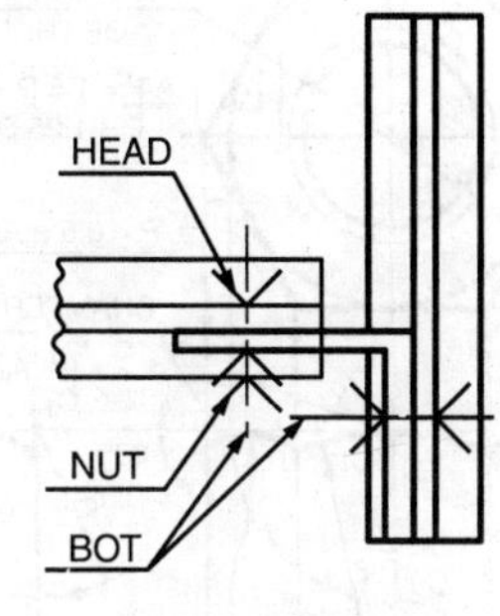

Fig. 18.71.

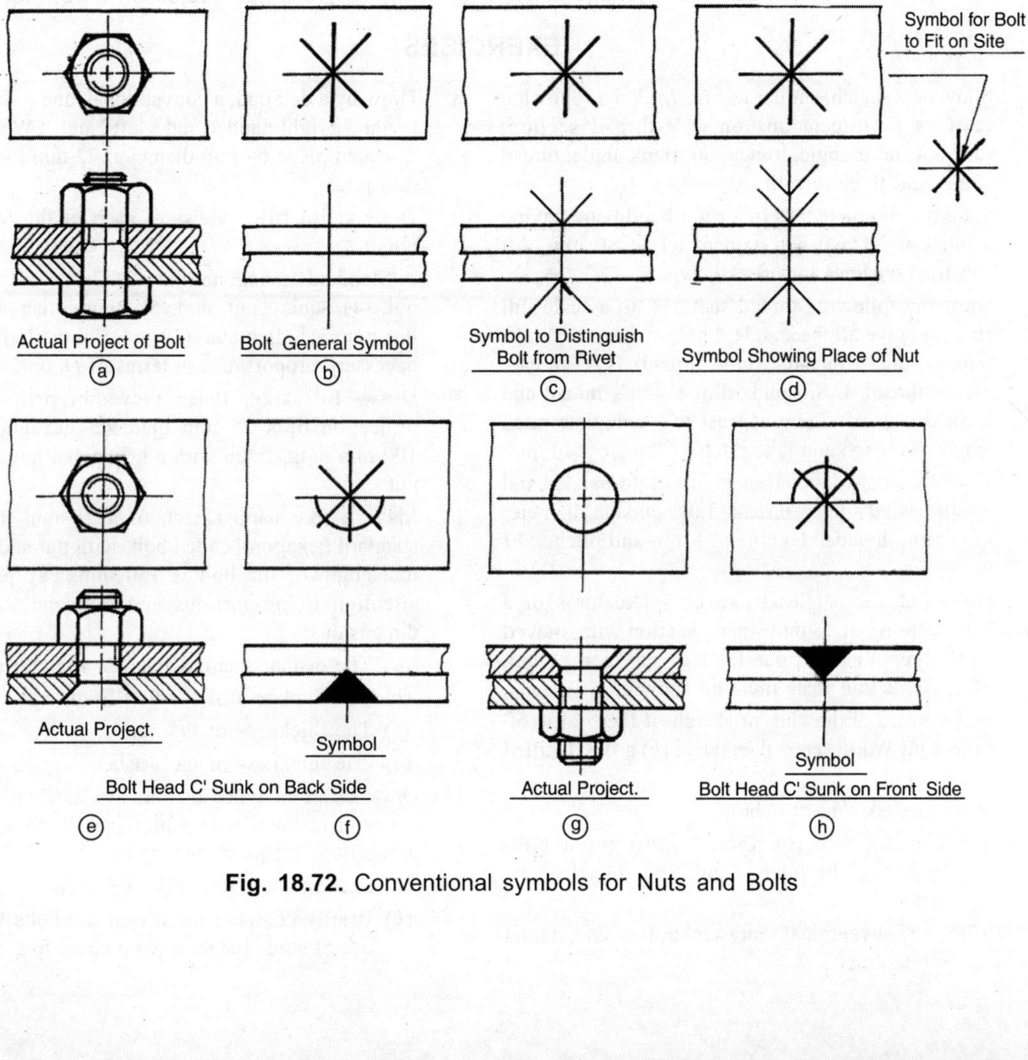

Fig. 18.72. Conventional symbols for Nuts and Bolts

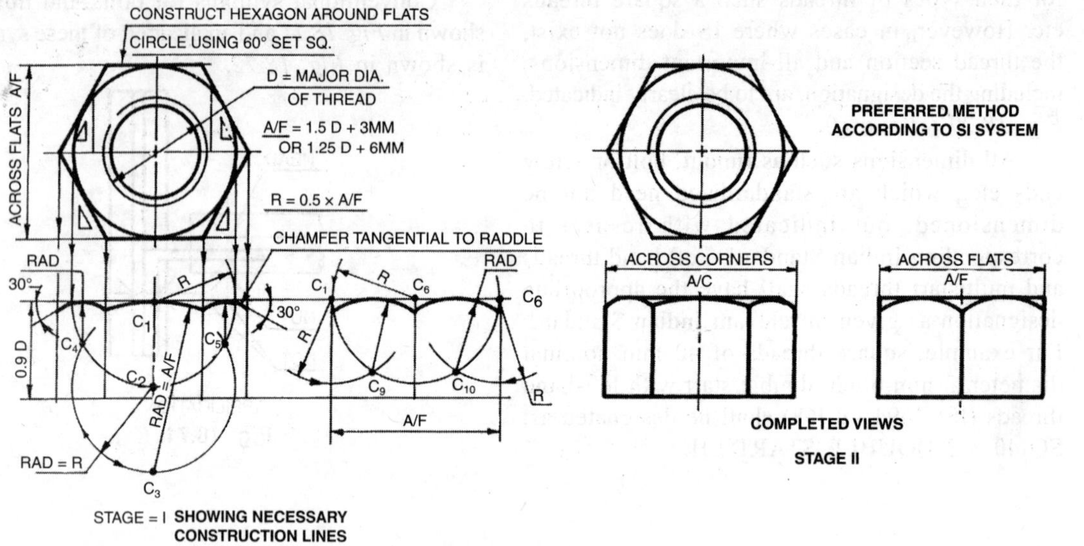

Fig. 18.73. Drawing a standard Hexagonal (ISO system) – Third angle Projection (First Method)

EXERCISES

1. Show on a sketch, similar to *Fig. 18.2*, the following features or dimensions on a V-thread section; fundamental triangle, truncation, flank angle, thread angle, and thread depth.
2. Set out a complete turn of a right hand helix having a pitch of 60 mm and a diameter of 80 mm. All construction lines must be shown.
3. Draw the following thread sections, to a scale full size, and give all the standard proportions: BSW, IS thread, square thread, Acme thread, B.A. thread, Metric thread, U.S. standard of Seller's thread and Knuckle thread. Show at least two complete pitch lengths for each and take pitch of thread = 40 mm.
4. Draw by a conventional method a right-handed and a left-handed square thread. Take outside diameter = 64 mm, threaded length = 72 mm and pitch = 12 mm.
5. Draw full size, at least two complete turns of a helical spring of 18 mm square section wire, moved on a 72 mm diameter bar. Pitch of spring = 72 mm. All construction lines must be shown.
6. Make a careful drawing of sketch of the profile of, (*a*) a Whit-Worth screw thread and (*b*) a B.S. Unified thread form.
7. Make neat sketches to show:
   (*a*) The basic metric thread form for a bolt, indicating the pitch, depth and thread angle, and
   (*b*) One conventional representation of an external screw thread.
8. Draw by a detailed, a conventional, and a simplified method a right-handed and a left-handed Whit-Worth threaded piece 64 mm diameter, 72 mm long and 6 mm pitch.
9. Draw at least two views of each of the following nuts:
   1. Standard Hexagonal nut, (2) Castle nut, (3) Cap nut, (4) Square nut, and (5) Ring-collar nut. Take the nominal diameter (*D*) = 25 mm. Indicate all necessary proportions, in terms of *D*, on the views.
10. Draw, full size, three views in orthographic projection of a 25 mm diameter hexagonal bolt, 100 mm long, fitted with a hexagonal nut and lock nut.
11. Make a free hand sketch of a 25 mm diameter, standard hexagon-headed bolt, with nut and washer the length of the bolt is 150 mm. Pay particular attention to proportions and give the following dimensions:
    (*a*) The distance across flats for the bolt and nut;
    (*b*) The thickness of the bolt head;
    (*c*) The thickness of the nut;
    (*d*) The thickness of the washer.
12. Draw different types or bolt heads. Take nominal diameter of bolts = 25 mm. Give all the necessary proportions taken.
13. (*a*) Sketch a tap bolt and a setscrew.
    (*b*) Briefly compare the use of tap bolts with the use of studs for securing a cover to a cylinder.

14. Sketch freehand, approximately to scale and in good proportions, a sectional view showing a 12 mm dia, stud 72 mm long, screwed 24 mm into an iron casting. What suitable material is used for the stud?
15. Draw two views of a stud fixed in a blind hole in a: (*i*) part of mild steel, part of cast iron. Take diameter of stud = 25 mm and pitch = 3 mm.
16. Sketch at least three different methods of stopping rotation of bolts in their corresponding holes, take diameter of bolt = 25 mm.
17. Sketch three forms of spring washer.
18. Sketch the following: an eyebolt, a hook bolt, a setscrew and a cup headed bolt with square neck.
19. A small shaping machine is to be installed on cement concrete foundation. Make neat sketches showing three alternate types of foundation bolts, which could be used. One should be of a type, which will allow the machine to be moved to a new site, without the need for new bolts.
20. Draw neat freehand sketches in good proportions to indicate clearly at least three different ways in which hexagonal nuts may be securely locked.

QUESTIONS FOR VIVA VOCE

1. What is a fastening? What are its types? Under what circumstances each type is used?
2. What are the two basic types of threads? Where each type is used?
3. Define the terms: fundamental triangle, truncation, depth of thread, pitch, lead, slope, flank, thread angle, crest, root, and axis of a thread.
4. What are a left-handed and a right-handed screw thread?
5. What is a Jielix and pitch of a helix?
6. Which type of thread is used for power transmission or load lifting and why?
7. What is a single start thread?
8. What are pitch, slope and lead in the case of a double-start thread?
9. What are Multi-start threads? Where these are used and why?
10. How threads are cut?
11. What are a through bolt, a tap bolt, and a stud bolt? Where each is preferred?
12. What is a T- bolt and where it is used?
13. What is an eyebolt and for what purpose it is used?
14. What is a setscrew? What is its function?
15. What do you mead by actual, detailed, conventional and simplified thread representations?
16. Why a nut of a bolt head is chamfered? Why sometimes a nut is chamfered at both ends?
17. What are locking devices? Where and why they are used?
18. What are the various methods of stopping the bolts from rotating in bolt holes, while tightening the nuts?
19. What are foundation bolts and where they are used?
20. What are the machine screws? What is the difference between a set-screw, a through-bolt and a machine screw?

❐❐❐

Keys, Cotters And Joints 19

19.1. KEYS

These are temporary fasteners and are chiefly used to hold pulleys, wheels, gears, crank arms, and flanges of couplings etc., on rotating shaft, such that there takes place no relative rotary motion between the pair. A key, as commonly used, is also an agent in the transmission of motion and power. In addition a key is used as a safety feature. Because of its size, when overloading occurs, the key slows or breaks before the actual part or shaft breaks. Keys are always made of steel since they are subject to considerable crushing and shearing loads. They are usually of rectangular section. To accommodate the key, a groove called the "key-seat" is cut in the shaft and a groove called "keyway" is cut in the hub of the part to be mounted on the shaft. Thus after assembly the key is partly in shaft and partly in the hub of the mating part, locking the two members together so that one can not rotate without the other.

Keys are easy to replace and cost little. In assemblies whose parts are subject to sudden loads or accidental stresses, these are designed to fail first and thereby act as a safety device for other members. The main function of a key is to transmit torque between a rotating shaft and a machine element assembled to it.

Figure 19.1, shows pictorially a Gib-head taper key used to secure a round boss to a shaft. This is a rectangular sunk key to be fitted in a key-seat in the shaft as well as a keyway machined in the hub. This key has parallel sides but has a standard taper (1 in 100) on its upper face, like a parallel-sided wedge. The bottom surface of the key-seat machined in the shaft is parallel to the shaft axis. In the mating part (hub or boss) there is a tapered keyway to match the taper on the upper surface of the key.

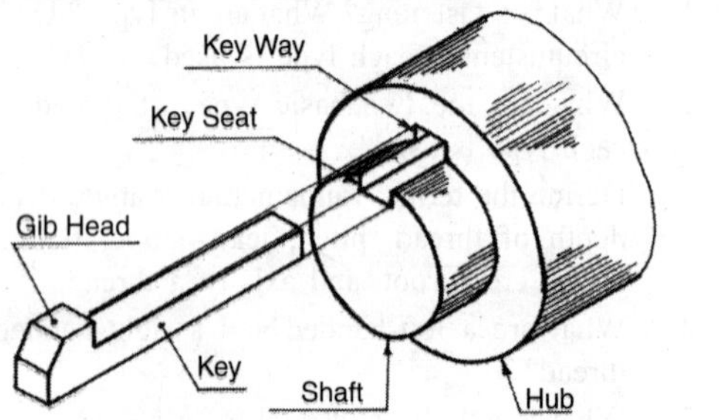

Fig. 19.1. Use of a key

A large number of standard keys are used in engineering applications. The choice of a key depends chiefly upon the load to be transmitted.

19.2. CLASSIFICATION OF KEYS

Keys are broadly classified into the following categories, according to the shape and the duty for which they are used, *Fig. 19.2*:

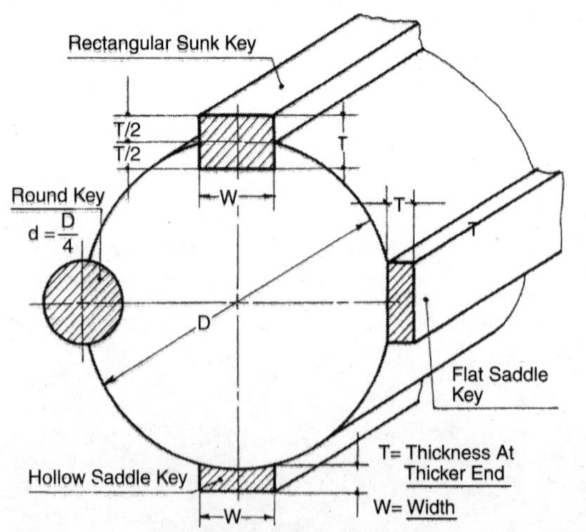

Fig. 19.2. Types of keys

1. **Sunk keys** -Heavy duty keys.
2. **Saddle keys** and **Pin** or **Round keys** - Light duty keys.

19.2.1. Heavy Duty Keys

1. SUNK KEYS

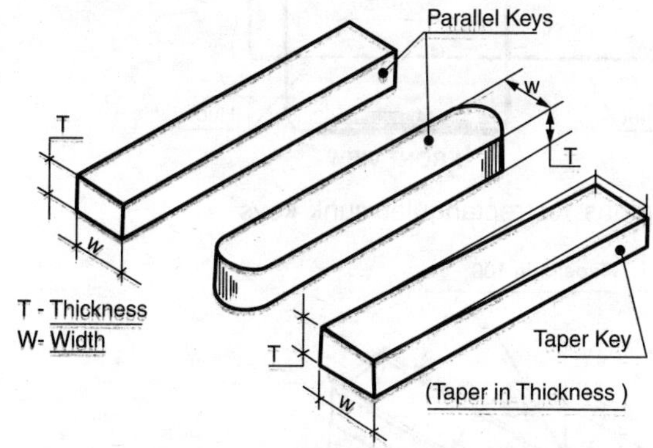

Fig. 19.3. Parallel and taper sunk keys

A key which goes partly in the key-seat, machined in the shaft, and partly in the keyway, machined in the hub, is called a sunk key. Most of the keys can be included in this category. Out of these Rectangular and Square sunk keys are widely used. Both may be parallel (taperless) or taper. The sides of both parallel and taper keys are parallel and taper referring to the relationship between their upper and lower faces, as shown in *Fig. 19.3*. When the section of a key does not change throughout its length, it is called a parallel key and when the thickness of a key reduces along its length, it is called 'taper-key". The sunk (taper) key is strong and reliable for transmitting a turning moment.

Both square and rectangular sunk keys are placed half in the shaft and half in the mating hub when measured at the side and not when measured on the centreline, *Fig. 19.4*.

2. EASY DRAWING PROPORTIONS

For drawing purposes, it is convenient to take width $W = 0.25\ D$; and thickness, T, at the thicker end = $0.66\ W$ for a rectangular key, as shown in *Fig. 19.5*. For a square key take $W = 0.25\ D$ and $T = W$, where D stands for diameter of the respective shafts on which the keys are to be used. When dimensions have to be given to fine limits the draughtsman should refer to BIS standard Code. Standard BIS Tables give the required information for selecting the right key size.

3. GIB-HEAD KEY

It is provided with a gib-head on one end to facilitate withdrawal of the key without disturbing the setting of the hub, which is fastened to the shaft of the mounting of the shaft. This is a sunk key tapered on the upper surface. Its standard proportions are shown in *Fig. 19.6*.

The key is fitted so that there is a clearance, between the inner face of the gib-head and the outside of the hub for inserting an extractor tool of some kind.

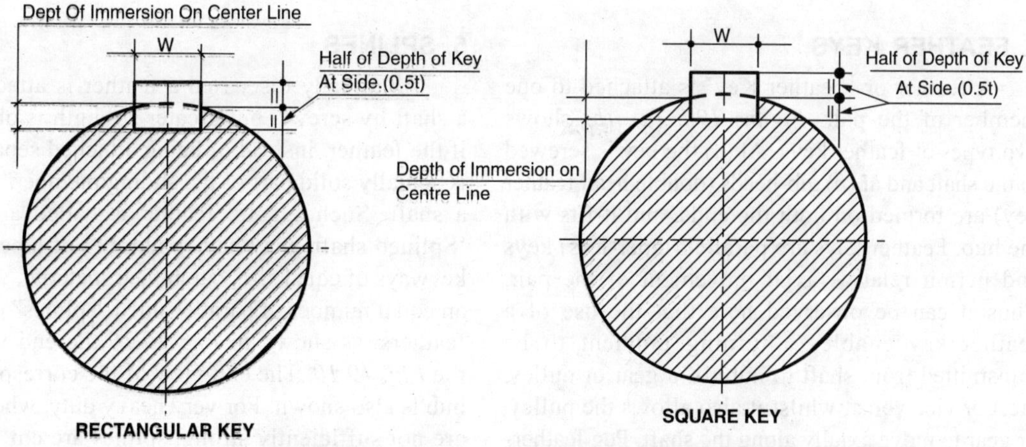

Fig. 19.4. Depth of immersion (sunk keys)

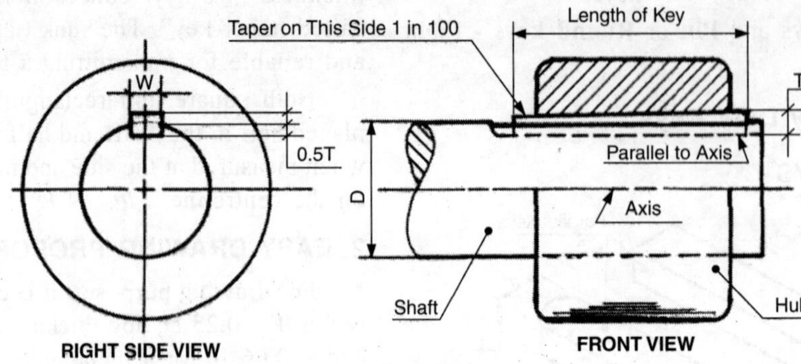

Fig. 19.5. Easy drawing proportions for rectangular sunk keys

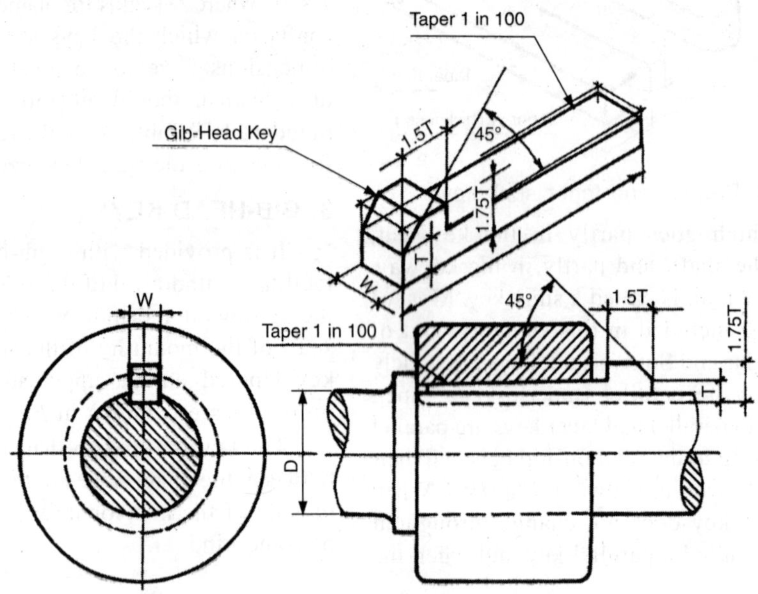

Fig. 19.6. Gib-head key, its use and proportions

4. FEATHER KEYS

A feather or 'Feather Key' is attached to one member of the pair. *Figure 19.8 (a), (b)* shows two types of feather keys. At (a) the key is screwed to the shaft and at (*b*) gib-heads (double head feather key) are formed on both the ends, and it fits with the hub. Feather keys are parallel (taper-less) keys and permit relative axial movement of the pair. Thus it can be observed here that the use of a feather key enables a turning moment to be transmitted from shaft to hub (of a gear or pulley, etc.) or vice versa, whilst it also allows the pulley, or gear to move axially along the shaft. Peg feathers are shown in *Fig. 19.9 (a), (b)*. The pegs fit holes in the hubs.

5. SPLINES

As already discussed a feather is attached to a shaft by screws, but greater strength is obtained if the feather, instead of being attached separately, is actually solid part of the sliding member, usually a shaft. Such a solid feather is called a spline. 'Splined shaft' may have a number of taper sided keyways of equal size, cut in their material, yielding an equal number of parallel sided "splines" or solid feathers, as shown in *Fig. 19.10* the end view of the *Fig. 19.10.* The end view of the corresponding hub is also shown. For very heavy duty, when keys are not sufficiently strong, splines are cut both in the hub and the shaft, arranged so that they fit (one within the other. Splined shafts are used where,

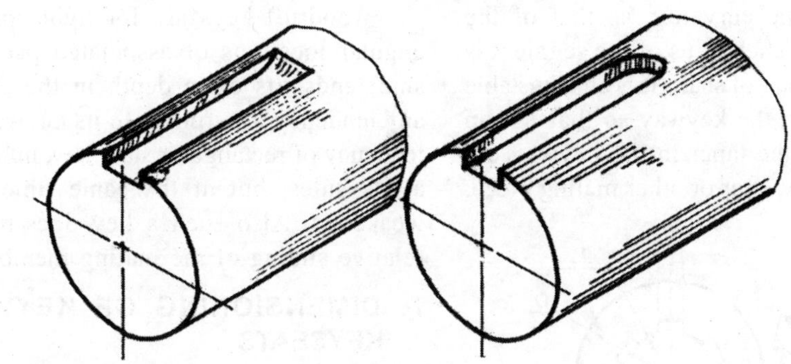

Fig. 19.7. Two forms of keyway

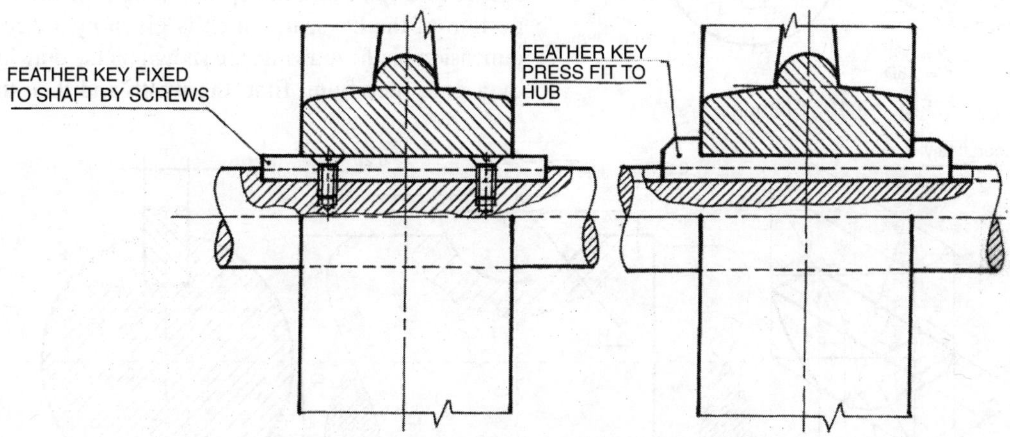

Fig. 19.8. Use of feather keys

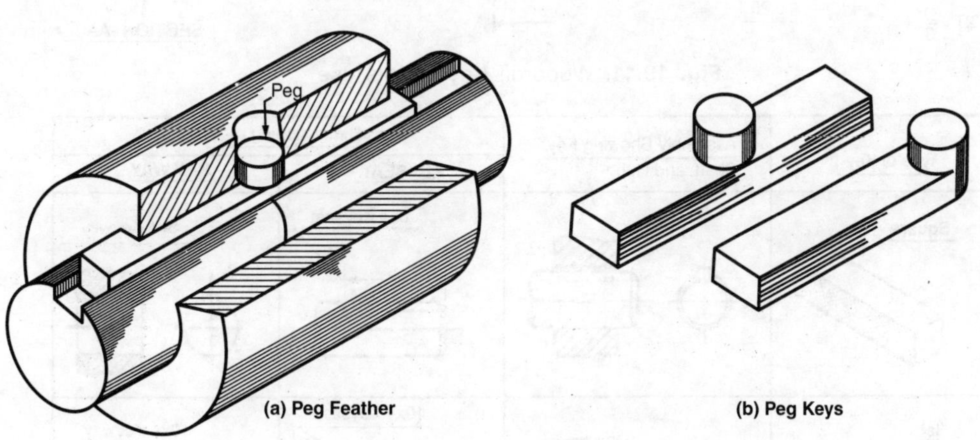

(a) Peg Feather **(b) Peg Keys**

Fig. 19.9. Use of a feather key

in addition to transmitting torque, axial sliding motion is also to be permitted. The splines are usually produced on shafts by means of hobs, but splined holes are generally produced by broaching.

6. WOODRUFF KEY

It is a flat segmental disc with a flat or round bottom. As shown in *Fig. 19.11*, a keyway of segmental shape is cut in the shaft by a milling cutter having the same curvature as that of the mating woodruff key. The radius of the segment is kept equal to the diameter of shaft. It is an adjustable sunk key. It can tilt in the keyway so that its top face can lineup with the taper in the keyway cut in the hub of the pulley, gear or other mating piece.

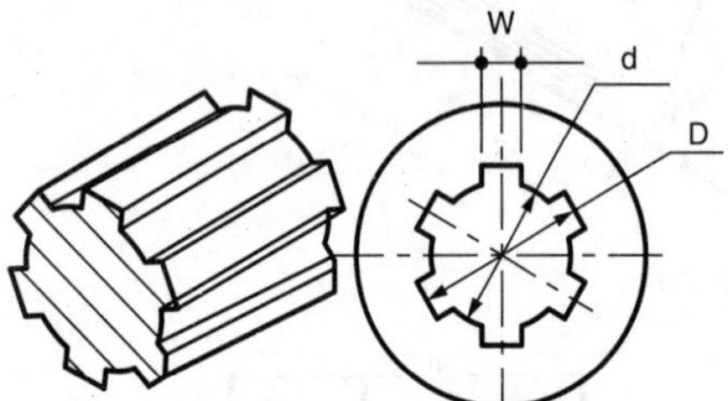

Fig. 19.10. Splines

Woodruff keys are for light applications, or angular locations of associated parts on tapered shaft ends. Its extra depth in the shaft prevents any tendency to turnover in its keyway, a common tendency of rectangular sunk key, unless well fitted at its sides, but at the same time the shaft is weakened. Also such a key does not allow any relative sliding of the mating members.

7. DIMENSIONING OF KEYWAYS AND KEYSEATS

All dimensions of keyways and keyseats for square and flat keys, except the length of the flat portion of the keyseat, which is given by a direct dimension on the drawing, are shown on the drawing by a note specifying first, the width and then the

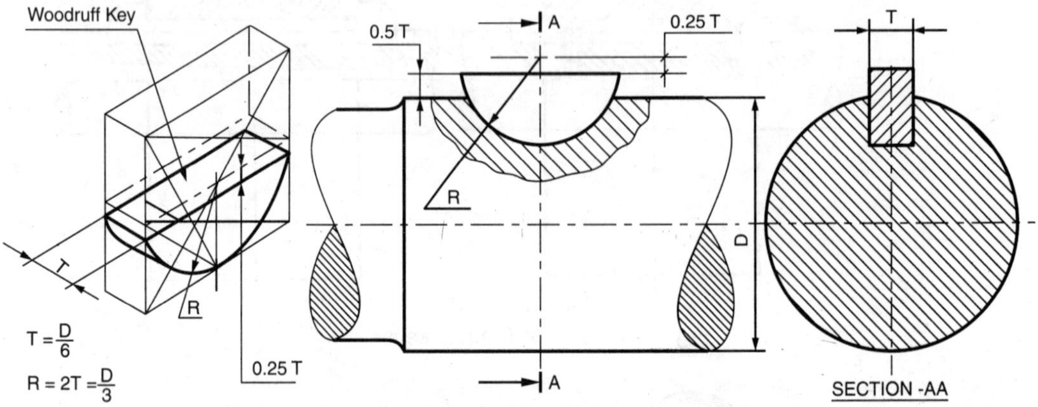

Fig. 19.11. Woodruff Key and its use

Type of Key	Assembly Showing Key. Shaft. and Hub	Dimensioning of KEYSEAT	Dimensioning of KEY WAY
Square		8x4 Keyseat, L	8x4 Keyway
Flat		10x4 Keyseat, L	10x4 Keyway

Fig. 19.12. Dimensioning of Keyways and Keyseats

depth, *Fig. 19.12*. This type of dimensioning is the standard method used for unit production where the machinist is expected to fit the key into keyway and keyseat.

For interchangeable assembly and mass production purposes, keyway and keyseat dimensions are given in limit dimensions to assure proper fits and are located from the opposite side of the hole or shaft, *Fig. 19.13*.

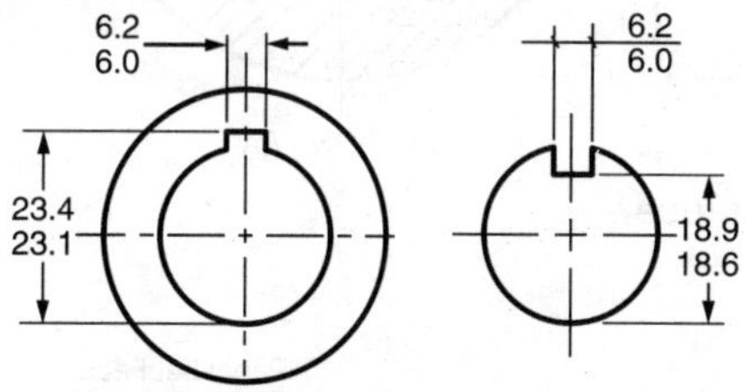

Fig. 19.13. Limit dimensioning of keyseat and Keyway

19.2.2. Light Duty Keys

1. FLAT SADDLE KEY

It is taper key fitting a keyway in the hub and a flat on the shaft. As it is liable to slip on the shaft, it is suitable for light duty only. In *Fig. 19.14 (a)* are shown its different proportions where W = 0.25 *D*; *T* = 0.33 W (minimum).

2. HOLLOW SADDLE KEY

Figure 19.14 (b). It is taper key fitting a keyway in the hub, the bottom of the key being shaped to fit the curved surface of the shaft. As the two parts are held only due to friction, it is liable to slip and is suitable only for the lightest duty, as in the temporary positioning of eccentric on a shaft. Its proportions are: W = 0.25 *D* and *T* = 0.33 *D* (minimum).

3. ROUND KEY or PIN KEY

Figure 19.15. It is circular pin fitted in a hole drilled partly in the end of a shaft and partly in the mating hub, as in fixing a hand wheel to the end of a small shaft. Its proportions are: *W* = 0.166 *D* to 0.25 *D*.

19.3. TANGENTAL KEYING

Here the keys act tangentially in opposite directions and this arrangement is used for heavy drives, which may reverse direction. *Figure 19.16* shows an arrangement of tangential keys. For drawing purpose, we can take W = 0.3 *D*, *T* = 0.1 *D*. The length of the key is, of course, determined by the length of the hole in the part to be keyed to the shaft, and it is recommended that an extra length, for driving the keys, should not be less than the width of a pair of keys.

19.4. COTTERS

Cotters are temporary fasteners and are used to hold two parts together where the parts are subjected to axial forces only. A cotter is a wedge shaped piece of mild steel, often rectangular in section, tapered on one or both sides, *Fig. 19.17*. It is usually driven in perpendicular to the axis of the connected parts. Taper on it is provided to facilitate it's driving into position and withdrawal and for lateral adjustments of connected parts. Usually the taper varies from 1 in 50 to 1 in 25. If the taper provided is in excess (up to 1 in 8) there is possibility of the cotter slipping back. To prevent this slipping back of cotter some locking device, may be a taper pin or a setscrew, should be used.

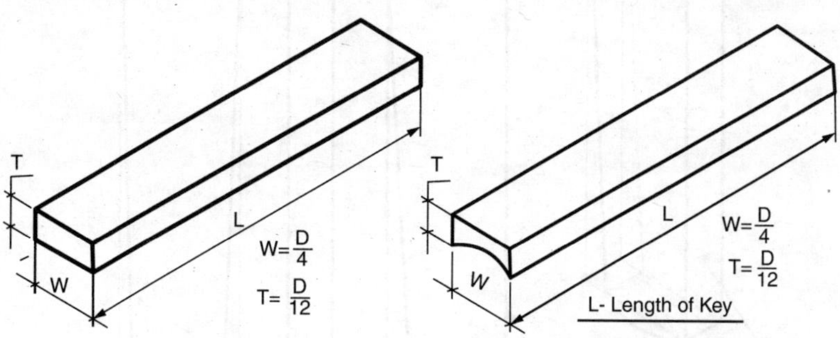

(a) FLAT SADDLE KEY **(b) HOLLOW SADDLE KEY**

Fig. 19.14.

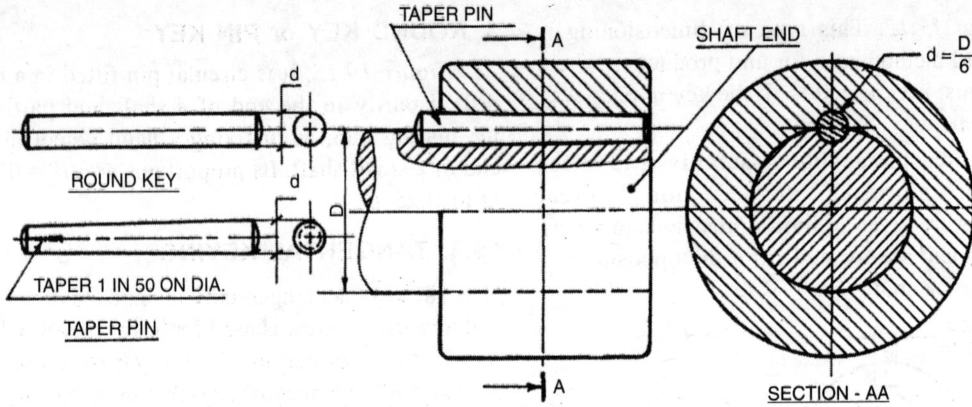

Fig. 19.15. Pin Key and its use

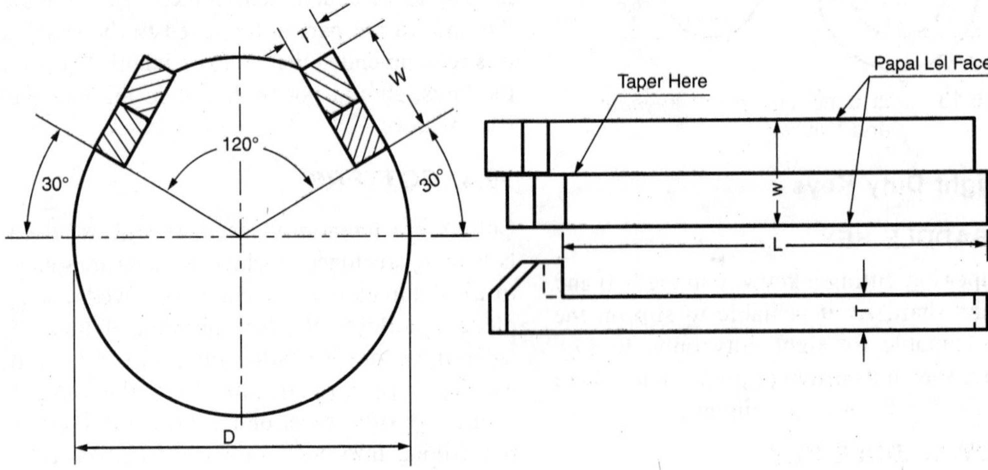

Fig. 19.16. Tangential keying

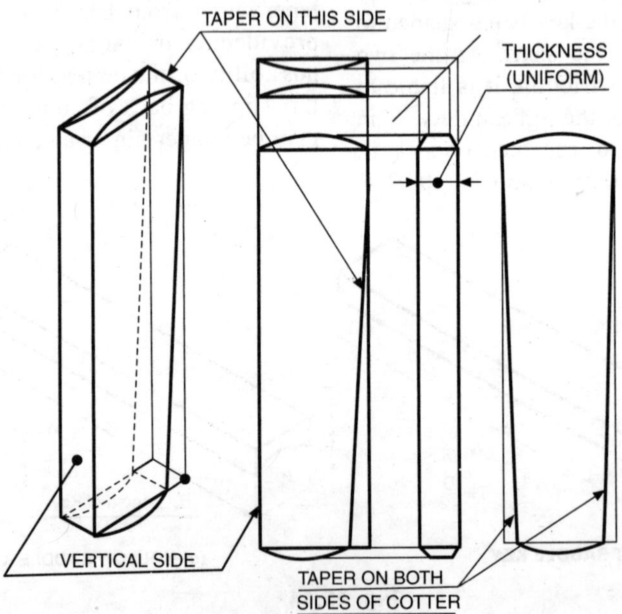

Fig. 19.17. COTTER

19.5. COMPARISON BETWEEN KEYS AND COTTERS

Keys are usually driven in parallel to the axes of shafts, which are subjected to torsional or twisting stress. Cotters are normally driven at right angles to the axes of the connected parts, which are subjected to tensile or compressive stress along their axes. A key resists shear over a longitudinal section, whereas, a cotter resists shear over two transverse sections.

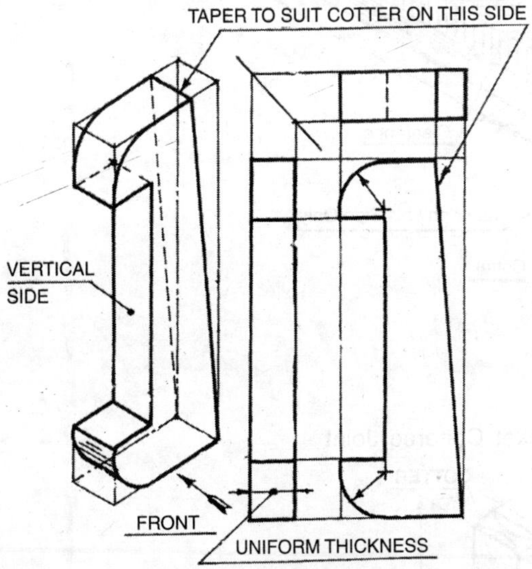

Fig. 19.18. GIB

19.6. COTTERED JOINTS

Some important types of cottered joint, which are of interest from drawing point of view, are discussed below:

1. SPIGOT AND SOCKET COTTERED JOINT

A joint formed by using a cotter is called a cottered joint, *Fig. 19.19*. In the cottered joint, shown pictorially, in *Fig. 19.20*, one rod end is formed into a socket, and the other, called spigot end, is forged to a diameter a little larger than that of the rods. A hole, to suit the dimensions of the cotter is provided in both the socket and the spigot. The usual clearance, when the cotter is driven home, varies from 1.5 to 3.0 mm (3.0 mm here). When the cotter is driven in, it forces the rod more inside the spigot and makes the joint. Note that the spigot end is provided with a collar and a clearance is left between the end of the spigot and the base of cavity in the socket when the joint is made.

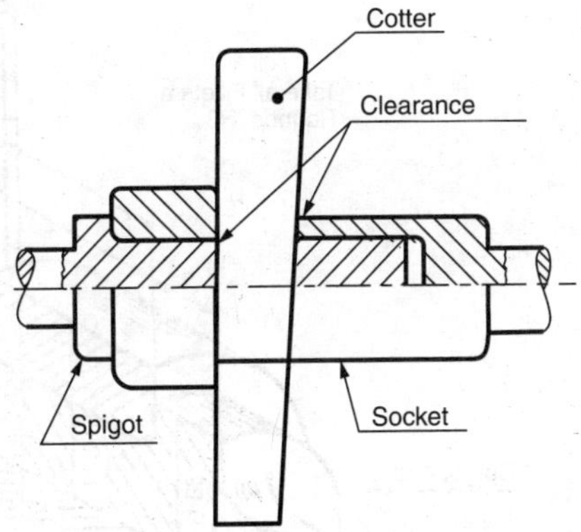

Fig. 19.19. Spigot and Socket Cottered Joint

The dimensions of the rod end and socket end should be such that the area of each across the slot is not less than area of the rod.

2. GIB AND COTTER JOINT

A jib and cotter joint, *Fig. 19.21*, mostly used for square rod ends is drawn pictorially in the form of an exploded assembly. One of the ends is formed into the shape of a fork (or straps). Suitable rectangular section holes, enough to accommodate the gib and cotter, and necessary clearance, are made in both the rod ends. For making assembly, the rod end (left) is first pushed in the fork end (right) till its movement is stopped by contact with the solid part of the fork end. When the two ends are in position, the gib is then inserted in aligned holes (3 mm offset in this case) and after locating it in the proper position, the cotter is driven in to make the joint tight. It should be noted here that the gib is to be placed on the open side of the fork.

3. GIB AND ITS USE

Gib is a piece of mild steel having the same thickness as the cotter, in combination of which it is to be used, *Fig. 19.18*. It has the same taper as on its mating cotter. The two outer lines of the joined gib and cotter are parallel to each other.

A gib is used in a gib and cotter joint to achieve the following purposes :

- When the cotter (without gib) is driven in position, it tends to bend the lower part of the fork-end away from the rod, *Fig. 19.23 (a)*,

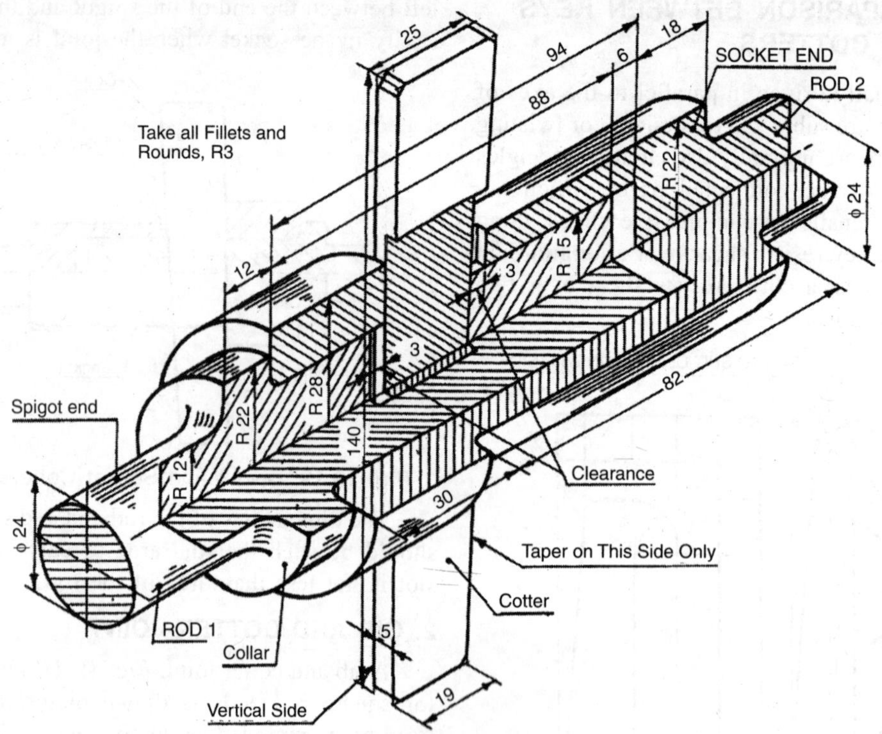

Fig. 19.20. Spigot and Socket Cottered Joint

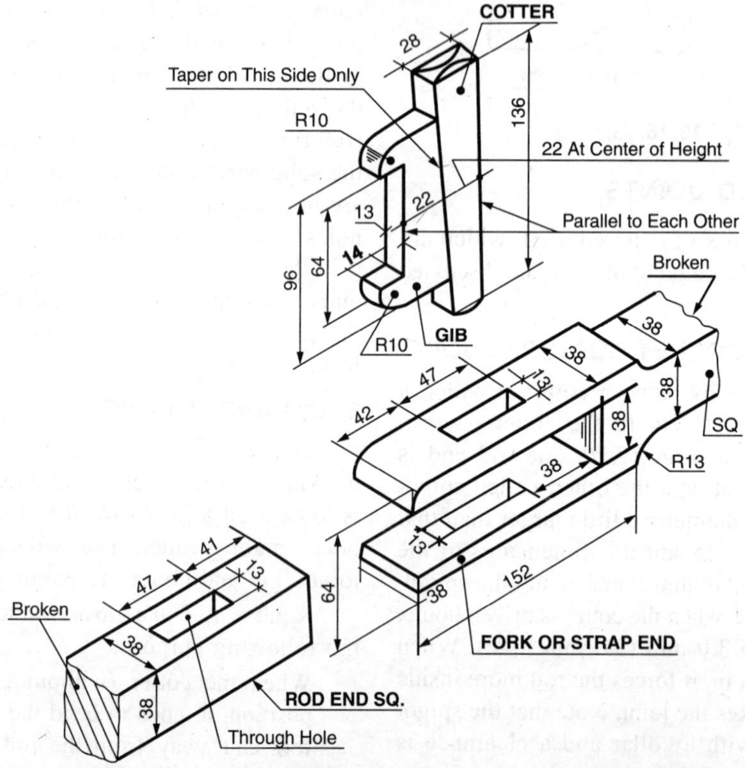

Fig. 19.21. Gib and Cotter for square rods (Exploded details)

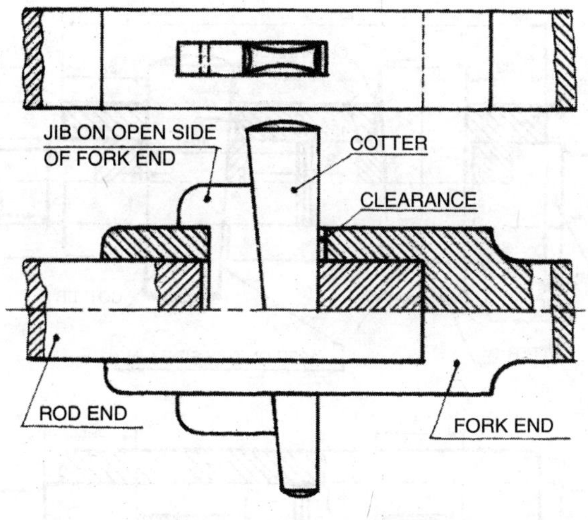

Fig. 19.22. Gib and Cotter Joint (General Arrangement)

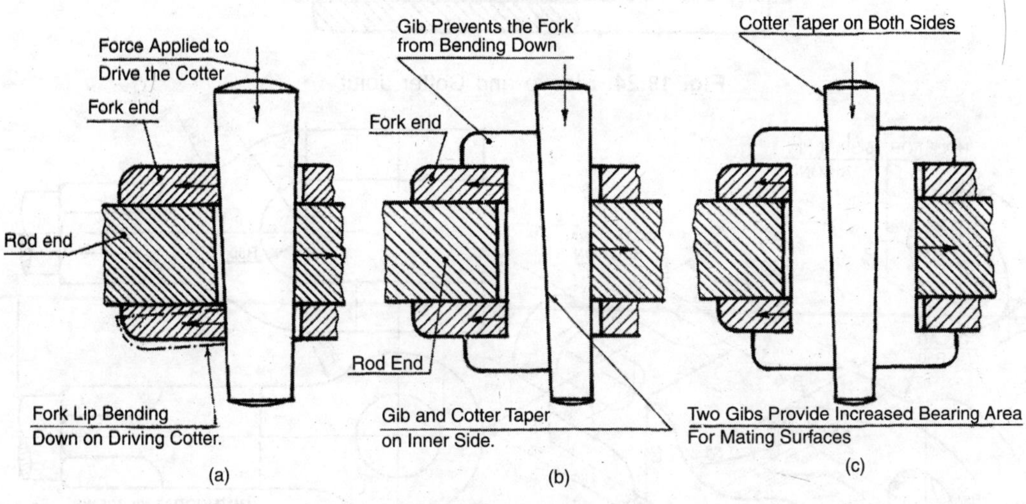

Fig. 19.23. Use of a Gib

and this is prevented by using a gib in combination with the cotter, *Fig. 19.23 (b).*

- To increase the bearing area of contact between the mating surfaces, a gib is used in the joint. When one gib is used, the cotter is cotter used has taper only on one side and the gib is always kept on the outside, and when two gibs are used the cotter is tapered on both sides, *Fig. 19.23 (b), (c).*

4. SLEEVE AND COTTER JOINT

In this joint the rod ends are circular in section, and are connected by using a sleeve or muff and then applying two cotters, one on each rod end, *Fig. 19.24.* The two rod ends are made to meet at the centre of sleeve. When the cotters are driven in, they press the two rod ends against each other and make the joint tight. Note the clearances on the sides of the two cotters. These clearances are important to ensure a tight joint.

19.7. PIN JOINT - KNUCKLE JOINT

Figure 19.25, shows exploded assembly of the most common type of knuckle joint, used to connect two rods, whose axes meet at a point. This joint allows a small angular movement of the rods relative to each other and is capable of transmitting rotary and transverse motions.

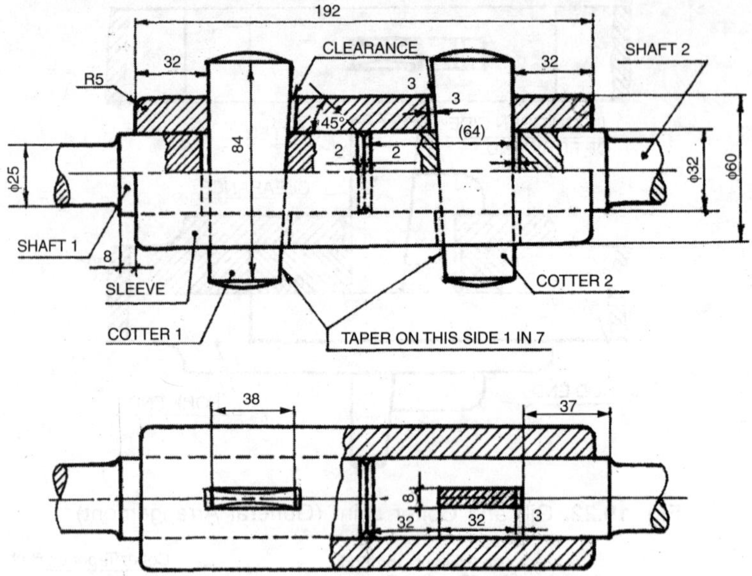

Fig. 19.24. Sleeve and Cotter Joint

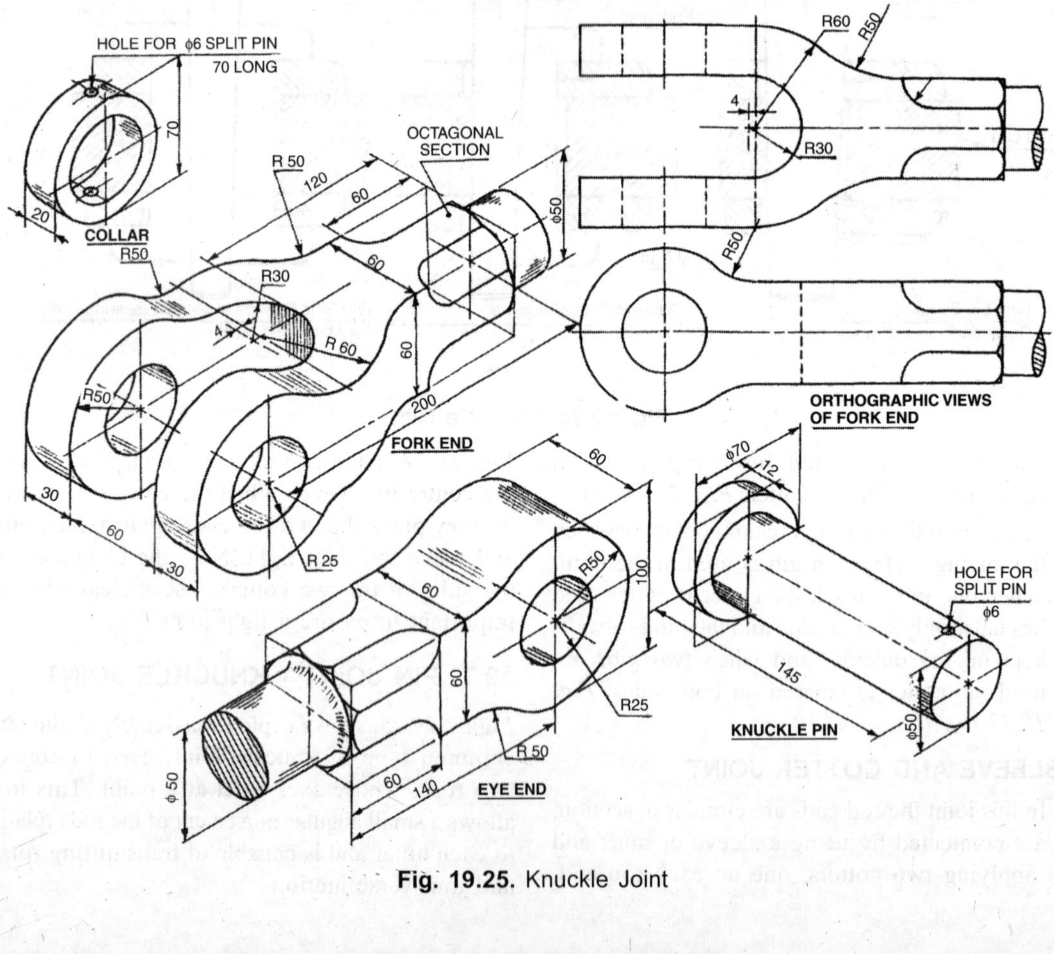

Fig. 19.25. Knuckle Joint

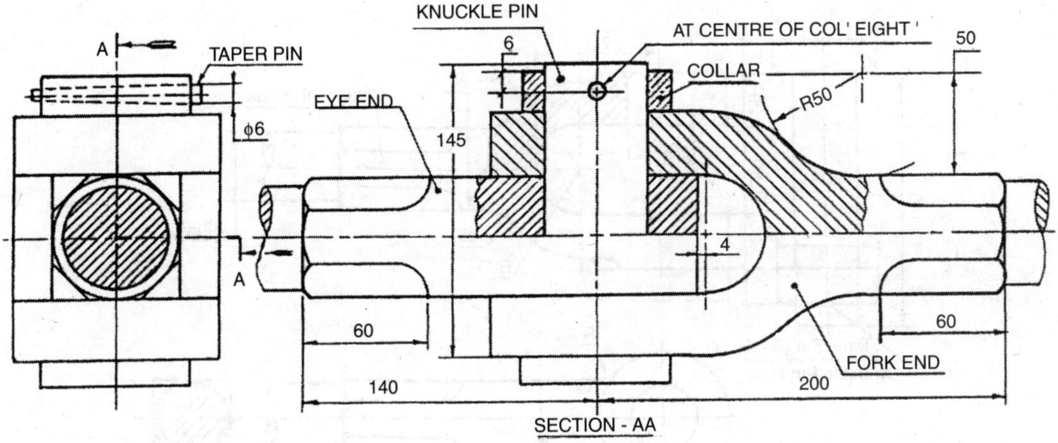

Fig. 19.26. General arrangement of Knuckle Joit (part solution to exercise 14)

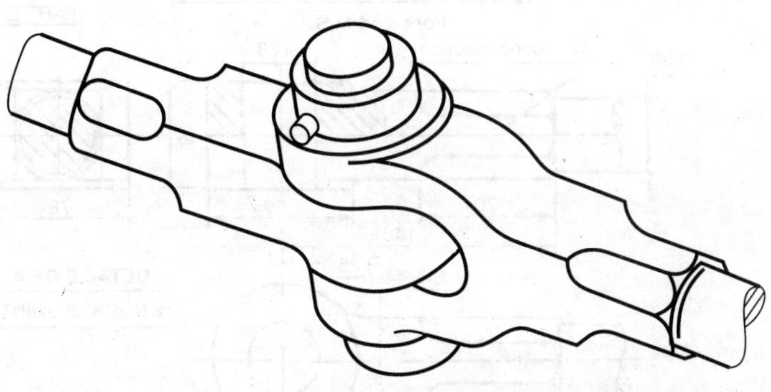

Fig. 19.27. (a) General arrangement of Knuckle joint

One of the rod ends is forged to the shape of a fork (or double eye) and the other rod end is forged to the shape of an eye. The eye end is inserted between the jaws of the fork end and, after aligning the holes in the fork end and eye end; the knuckle pin is passed through them. The knuckle pin has a head at one end and a collar, secured to it by a taper pin, at the other end. The forged fork and eye ends at octagonal in cross-section to a certain length and then square for some length.

Detail drawings of another design of a knuckle joint are given in Fig. 19.27.

EXERCISES

1. Draw, orthographic views of a hub and a shaft with a wood -ruff key in position. Take diameter of the shaft as 40 mm.
2. Sketch proportionately two orthographic views of a gib head key suitable for 50 mm diameter shaft.
3. Draw at least two views of a hub and shaft, f 54 mm, with a rectangular taper sunk key in position.
4. Show clearly, with the help of a proportionate sketch, the difference between a gib-head key and a woodruff key.
5. Make freehand pictorial sketches of the following:
   (*a*) a gib-head key;
   (*b*) a wood-ruff key; and
   (*c*)a splined shaft.
   State at least one application of each type of drive.
6. What is the difference between a key and a feather? Sketch in good proportions:
   1. a rectangular taper sunk key in its shaft,
   2. a feather in its shaft.

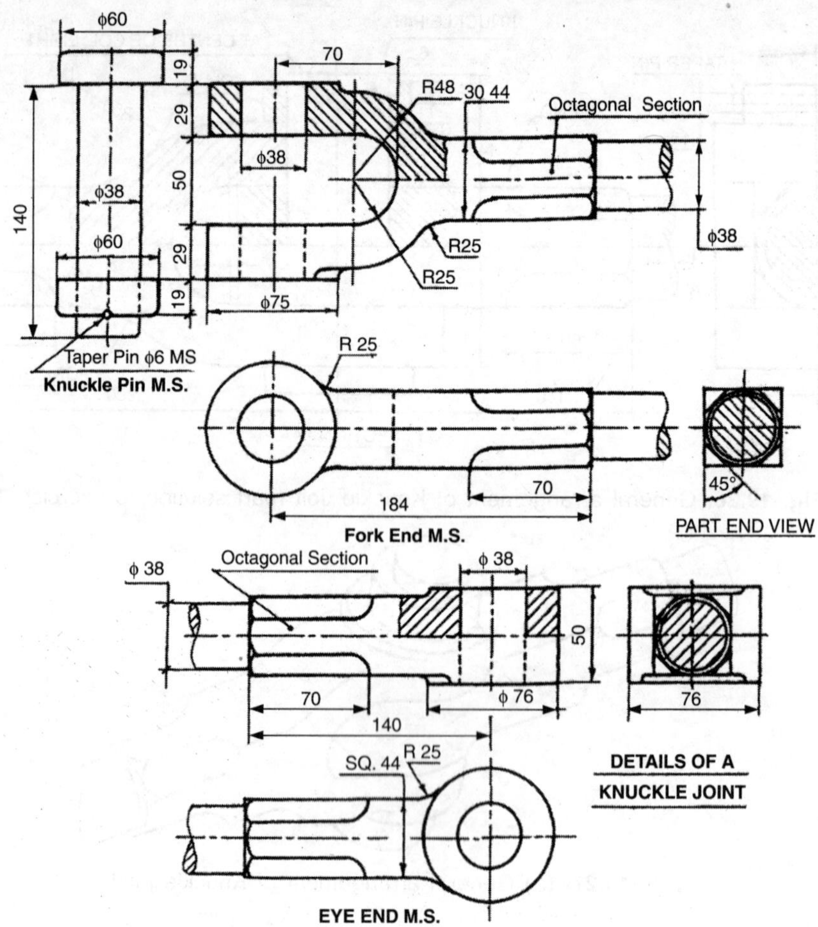

Fig. 19.27. (b) Knuckle Joint (Detail drawings)

7. By means of neat and proportionate sketches show and specify:
   (*a*) any three types of sunk key,
   (*b*) two types of cotter.
8. (*a*) Sketch two forms of saddle key.
   (*b*) How should sunk keys be fitted for transmitting unidirectional torques?
   (*c*) Sketch a woodruff key in a tapered shaft end and briefly state the advantages and disadvantages inherent in the use of these keys.
9. Compare the use of keys with the use of cotters and illustrate your answer by sketching an example of each, when used to fasten parts together.
10. (*a*) Sketch an example illustrating the use of tangential keys.
    (*b*) What is the standard taper for a taper key?
11. From the pictorial view of a spigot and socket cottered joint shown in *Fig. 19.20*, to a scale full size, draw the following orthographic views:
    1. Front view upper half in section;
    2. Right side view.

    Give all the necessary dimensions.
    - Place the cotter centrally.
    - Refer *Fig. 19.19*, for partial solution.
12. *Figure 19.21* shows the exploded assembly of a gib and cotter joint for square rods. Draw, to a scale full size, the following orthographic views of the - assembly:
    - Front view upper half in section.
    - Top view, outside.
    - Right hand side view.

    Give all the necessary dimensions.

DRAWING HINTS

1. To fix the cotter in position there are two alternate ways out:

(*a*) Draw the front view of the cotter separately, and find out the distance along the cotter from its one end where the cotter is 22 mm wide. Then keep that portion of the cotter at horizontal centerline.

(*b*) As given in the solution to this problem, place the cotter centrally about the horizontal centerline. Take the width of the cotter 28 mm at the top and 22 mm at the center. Join the corresponding points as shown and let it be any width, whatever it comes, at the lower end.

2. Refer to front and top views drawn for the joint, *Fig. 19.22*.

13. *Figure 19.24* shows two orthographic views of a sleeve and cotter joint, where the cotters are placed symmetrically about the centerline. Draw the detail drawings of its component parts. Give all the necessary dimensions for each part.

14. *Figure 19.25* shows functionally the exploded assembly of a knuckle joint. To a scale half the full size, draw the following orthographic views of the assembly:
    1. Front view upper half in section.
    2. Top view.
    3. Left side view.

    Give all the required dimensions.

DRAWING HINTS

- As the assembly is symmetrical about the centerlines or axes the drawing should be started by drawing the centerlines.
- The octagonal shapes of the ends can be drawn in front and top views after drawing the corresponding octagon in end view.
- As almost all parts involved project as circles in one of the views, the corresponding circles should be drawn first in their respective views and the projections of the circular parts in the other views should be taken from these circles instead of taking measurements again. It is true for both CAD and manual drafting.
- Before drawing any circle, first find its center by drawing its two centerlines.
- Start all the views simultaneously and complete these by transferring dimensions from one view to the other. Refer *Fig. 19.26* for the part solution of this exercise.

15. *Figure 19.27* shows the detail drawing? of a knuckle joint; draw the following views of its assembly;
    1. Front view - lower half in section.
    2. Right side view.
    3. Top view.

    Dimension the views completely. Pay proper attention to line work.

QUESTIONS FOR VIVA VOCE

1. What is a Gib? Where and why it is used? How it is installed in place.
2. Why it is necessary to include some clearances in cotter joints?
3. Under what circumstances cotter joints are commonly used?
4. What is the difference between a cotter joint and a pin joint? Where each type is preferred?
5. Why a head is made at one end of a Gib-head key?
6. What is the usual taper provided on the face of a sunk key?
7. What is the difference between a saddle key and a sunk key?
8. Under what conditions a woodruff key is used?
9. How a feather key is fitted on its keyway or key-seat? Under what circumstances a feather key is used?
10. What is a cotter and under what conditions of loading it is used?
11. Why a gib is used in combination with a cotter?
12. What type of joint you would recommend for shafts whose axes do not lie in straight line?

❑❑❑

Shaft Couplings 20

20.1. INTRODUCTION

Mild steel shafts are usually available in specified lengths. Sometimes it is inconvenient to use a shaft of large length due to space limitations or transport difficulties. Thus a long line shaft is composed of a number of small shaft pieces, put together end-to-end, joined by couplings, and supported by bearings. Couplings are also used to connect the shafts of two machines so that the power can be supplied directly from one machine shaft to the other machine shaft.

20.2. CLASSIFICATION OF COUPLINGS

Couplings can be classified with respect to the kind of alignment and centre line position of the connecting shafts. For a preliminary examination the shaft couplings may be divided into two main groups, nameely, *rigid couplings* and *non-rigid* or *flexible couplings.*

Rigid shaft couplings connect shafts in rigid alignment. In this type are included flange couplings, muff-couplings, cone or compression couplings and claw couplings.

Nor-rigid or flexible couplings allow slight angular deviations, also permit the axes of the shafts to float or run slightly out of alignment and in some cases to move end wise, whilst still transmiting torque from one shaft to another shaft. Further, the use of a flexible coupling enables us to compensate for the inaccuracy of workmanship in aliging the two connected machines.

A typical example is a direct electric drive from an electric motor to a machine tool, where we have four bearings relatively close together.

20.3. RIGID COUPLINGS

Rigid couplings can be divided, further into the following three types:

1. **FLANGE COUPLINGS:** In this category solid flange coupling, straight or unprotected flange coupling, and protected type flange coupling.
2. **MUFF COUPLINGS:** Solid muff coupling, half-lap solid-muff coupling, and split-muff coupling fall under this category.
3. **FRICTION GRIP or COMPRESSION COUPLINGS.**

20.4. RIGID FLANGE COUPLINGS

1. SOLID or FORGED FLANGES

Refer *Fig. 20.1.* The abutting ends of two steel shafts are upset or flanged from the solid. There is an outstanding cylindrical portion on the right flange, which is called *''spigot or register"* and there is a corresponding recess in the mating flange on the left side with which it *"registers"*. It helps to bring the shafts into line and to maintain alignment. This is called "spigot and socket centring". The machined faces are held together by means of tapered headless bolts, numbering from four to twelve, depending upon diameter of shafts. Such couplings are mostly used in marine shafts or other similar large capacity applications.

These couplings have standard dimensions as given in *Table 20.1.* For the purpose of sketching, for shafts up to 150 mm in diameter, the following usual proportions may be used:

Diameter of shaft = *D* (known)
Outside diameter of flange = 2D + 35 mm
Pitch circle diameter = 1.5 D + 20 mm
Spigot and socket diameter = 0.75 D to 1.4 D

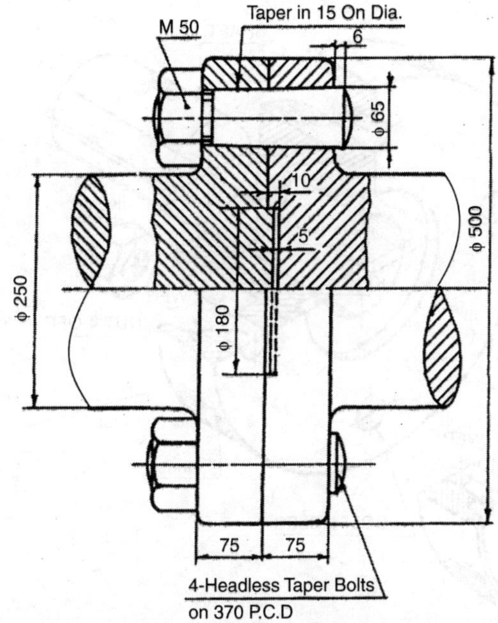

Fig. 20.1. Solid Flanged Coupling

Depth of socket	= 6 to 9 mm
Spigot projection	= Depth of socket – 2 mm
Thickness of flange	= 0.33 D + 3 mm
Number of bolts	= 4 up to ϕ 55 shafts
	= 6 above ϕ up to ϕ 150 mm shafts
Size of bolt	= As per standard *Table 20.1*

2. FLANGE COUPLINGS

The term flange coupling usually applies to a coupling having two separate cast iron flanges. Flange is a cast iron disc with holes to receive the key, bolts and the shaft. Each flange is mounted on a shaft end, and is keyed to it. Flange face is made at right angle to the axis of the shaft. Usually abutting end of one flange has a projected portion and the other flange has a corresponding recess. The keys are inserted from the inside faces of flanges. The two flanges are then pulled together by means of parallel bolts and nuts, Fig. 20.2.

3. UNPROTECTED FLANGE COUPLING

Each shaft is keyed to the boss of flange with a rectangular key and the flanges are held together with the help of bolts, numbering from 4 to 6. The keys are staggered through 90° along the circumference of shafts to divide the weakening effect caused by keyways, *Fig. 20.3*. These flange couplings are drawn to standard dimensions, taken from standard dimensions *Table 20.2*. For the purpose of sketching, the following proportions may be used:

Let D be the diameter of the shafts to be coupled.

Outside diameter of hub	= 1.8 D + 5 mm
Pitch circle diameter	= 2.2 D + 35 mm
Outside diameter of flange	= 2.6 D + 75 mm
Thickness of flange	= 0.33 D + 3 mm
Length of hub	= 1.3 D + 3 mm
Number of bolts	= 4 up to ϕ 55 mm shafts
	= 6 above ϕ 55 mm and up to ϕ150 mm shafts
Size of bolt	= As per standard Table 20.2.

Dimensions for spigot and socket in flanges may be similarly taken as for solid flange couplings.

4. PROTECTED FLANGE COUPLING

Refer *Figs. 20.2, 20.4,* and *20.5*. In this coupling, projections overhang bolt heads and nuts

Table 20.1. Dimensions of Solid Flange Couplings

Shaf Dia.	*Flange Dia.*	*Flange Width*	*Locating Dia.*	*Recess Depth*	*Spigot Depth*	*PCD*	*Bolt size*	*Bolt Hole*	*No.of Bolts*
35	100	17	50	6	4	70	M 10	11	4
45	120	22	60	6	4	85	M 12	13	4
55	140	22	75	7	5	100	M 14	15	4
70	175	27	95	7	5	125	M 16	17	6
80	195	32	95	7	5	140	M 18	19	6
90	225	32	125	7	5	160	M 20	21	6
110	265	36	150	9	7	190	M 24	25	6
130	300	46	150	9	7	215	M 30	32	6
150	335	50	195	9	7	240	M 32	34	6

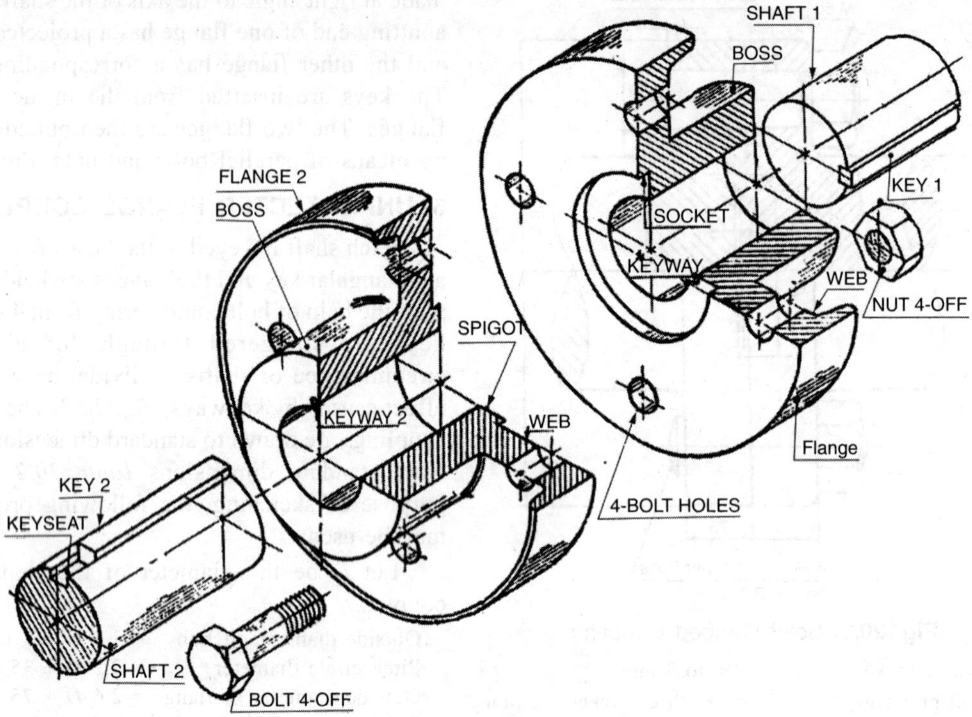

Fig. 20.2. Exploded Assembly of a Protected type Flange Coupling

Table 20.2. Dimensions of Unprotected Type Flange Couplings

Shaft Dia. D	*Hub Dia.*	*PCD*	*Flange Dia.*	*Thickness of Flange*	*Length of Hub*	*Bolt Dia*	*No. of Bolts*
35	68	120	180	15	48	M 10	4
45	84	140	192	18	62	M 12	4
55	106	160	220	22	74	M 14	4
70	132	190	260	26	94	M 16	6
80	150	212	285	30	110	M 18	6
90	170	240	310	33	120	M 20	6
110	200	280	362	40	146	M 24	6
130	240	324	415	46	172	M 30	6
150	275	370	470	54	230	M 32	6

so that they do not catch the fingers or the clothes of workers. That is why this type of coupling is referred to as protected. This is the most commonly used type of coupling. To secure a good alignment of the two shafts, spigot and socket centreing is used, i.e., a circular axial extension on the outside of one flange fits into a corresponding recess in the other flange.

In *Fig. 20.5* are shown two orthographic views of a protected type flange coupling with lower half of front view in section. The flanges are held together with the help of 4 bolts.

Detail drawings of its component parts are given in *Fig. 20.4*. The sunk keys are of rectangular taper type and are driven from the inside faces of the flanges. When the bolts are tightened, the flanges are pulled even more tightly on to the keys. Keys should not project more than 3 mm beyond the bosses of the bosses in the flanges.

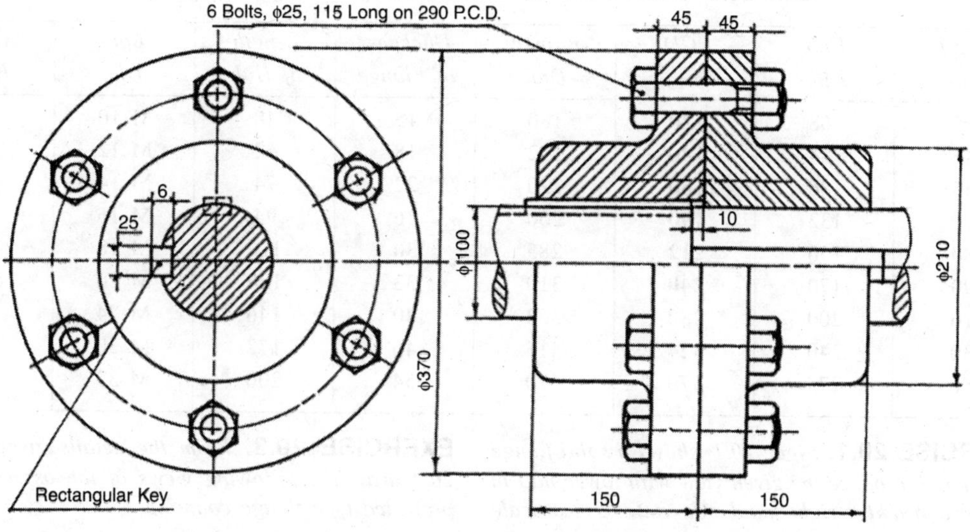

Fig. 20.3. Unprotected type Flange Coupling

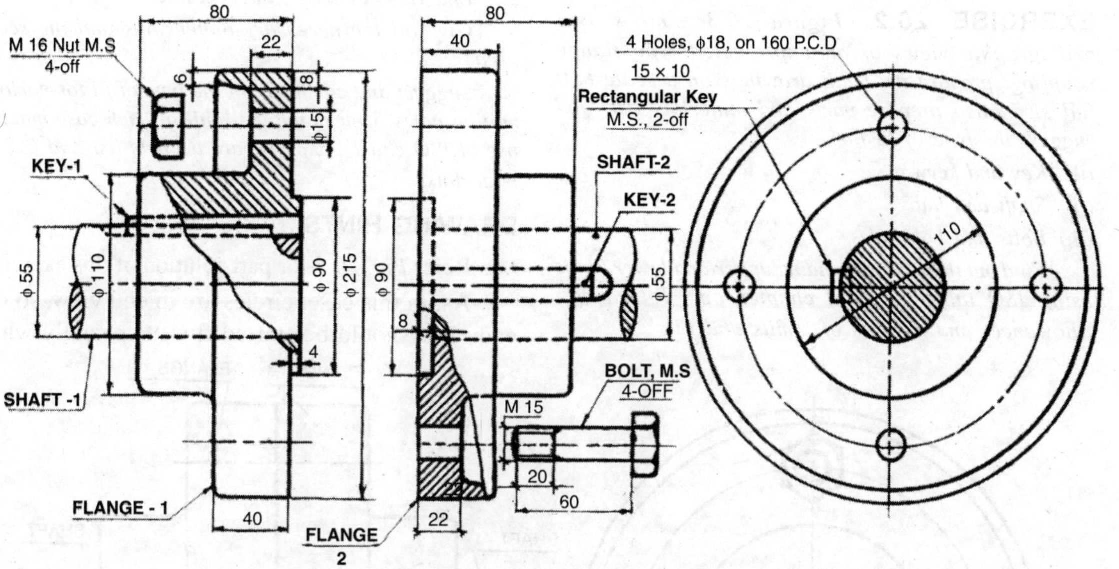

Fig. 20.4. Protected type Flange Coupling details

Keyways should not extend more than 15 mm beyond the ends of the keys.

For drawing these couplings, dimensions should be taken from standard dimensions Table 20.3. For the purpose of sketching, following proportions may be taken:

Let D be the diameter of shafts to be coupled.

Outside diameter of the hub = $1.8\,D + 5$ mm
Pitch circle diameter = $2.2\,D + 35$ mm
Outside diameter of flange = $2.6\,D + 75$ mm
Thickness of flange = $0.333\,D + 3$ mm
Shroud width = Thickness of flange + height of nut + 3 mm
Thickness of shroud = 6 to 10 mm
Length of hub = 1.3 D + 3 mm
Number of bolts = 4 up to ϕ 55 shafts
= 6 above ϕ up to
= ϕ 150 mm shafts

Size of bolt = As per standard *Table 20.3*.

Figure 20.2, shows pictorially, in the exploded form, this design of a flange coupling.

Table 20.3. Dimensions of Protected Type Flange Couplings

Shaft Dia.	*Hub Dia.*	*PCD*	*Flange Dia.*	*Thickness of Flange*	*Length of Hub*	*Bolt Dia*	*No. of Bolts*
35	68	120	180	15	48	M 10	4
45	84	140	192	18	62	M 12	4
55	106	160	220	22	74	M 14	4
70	132	190	260	26	94	M 16	6
80	150	212	285	30	110	M 18	6
90	170	240	310	33	120	M 20	6
110	200	280	362	40	146	M 24	6
130	240	324	415	46	172	M 30	6
150	275	370	470	54	200	M 32	6

EXERCISE 20.1. *Figue 20.1, shows a solid flange coupling. Reproduce the given view with lower half in section and add a right hand side view. Use a scale half full size.*

EXERCISE 20.2. *Figure 20.3, shows two orthographic views of an unprotected type flange coupling; prepare its detail drawings on a scale half full size. Also prepare parts and materials list, and suggest the type of fit for:*

(i) Key and keyway,

(ii) Shaft and hub,

(iii) Bolts and bolt holes.

Find out their corresponding upper and lower limits using ISO table given in chapter on "Tolerances, Allowances and Systems of Limits and Fits".

EXERCISE 20.3. *From the details given in Fig. 20.4 draw the following views of the assembly of a protected type flange coupling:*

1. Front view, lower half in section,

2. End view, looking from left side.

Give all the necessary dimensions and the scale used.

Suggest and comment on the type of fit for various mating parts. Determine the limit in each case making use of ISO chart. Also prepare a Parts-list and Bill of materials.

DRAWING HINTS

- Refer *Fig. 20.5* for part solution of this exercise. As, in this case, circles are in end view, so the end view should be started first. Especially when

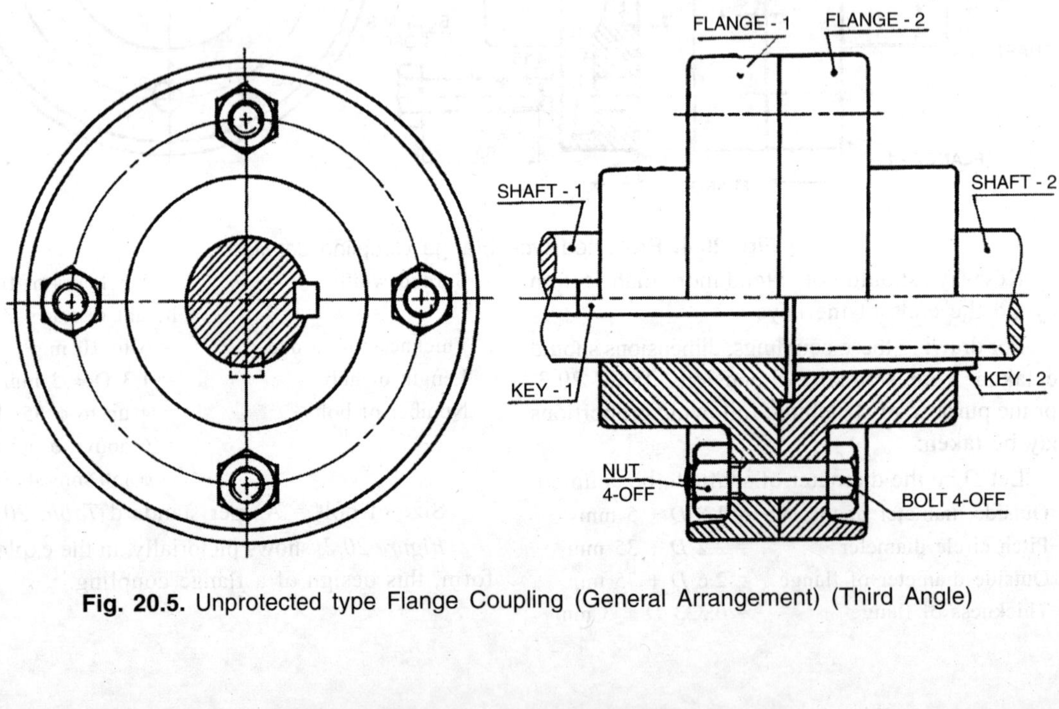

Fig. 20.5. Unprotected type Flange Coupling (General Arrangement) (Third Angle)

drawing the nuts, they should be drawn here in end view.

20.5. MUFF COUPLINGS

These are also called **box** or **sleeve** couplings. A muff coupling consists of a hollow cast iron cylinder (muff) enveloping the abutting ends of the shafts and keyed to both the shafts by a sunk key (usually rectangular taper key). In *Fig. 20.6* are shown two views of a simple muff coupling. Shaft ends are sometimes enlarged by approximately 25% to compensate for the weakening effect of the key way. Sometimes shaft ends are lapped and a common sleeve, over the two ends, is put. A single saddle key (tapered) holds the muff in position. This type of key coupling is called half lap muff coupling, *Fig. 20.7*.

In the above two designs of muff coupling, when the enveloping muff is made in one piece, it is essential to move one of the shafts endwise in order to fix this coupling. Where limitations do not allow this movement, a split muff coupling is used, *Fig. 20.8 (a), (b)* and *Fig. 20.9*. In this case the box, sleeve or muff is made in two halves. The shaft ends are made to abutt each other and a single key is fitted directly in the keyways of both the shafts. One half of the muff is fixed from below and the other half is placed from above. Both halves are held together with the help of mild steel studs (or bolts) and nuts, numbering 3, 4, or 6. The nuts are recessed, in these examples, into the bodies of the muff castings.

EXERCISE 20.4.

1. *Figure 20.8 (a) shows two halves of a split muff coupling. Its two orthographic views are shown in Fig. 20.8 (b). Draw the given views and add an outside top view. There are four studs use full size scale.*
2. *Draw its detail drawings to a scale half full size. Give all the required* dimensions.

EXERCISE 20.5. *Figure 20.9*, shows details of a split-muff coupling. Assemble the parts and draw to a full size scale:

1. *Front view, lower half in section.*

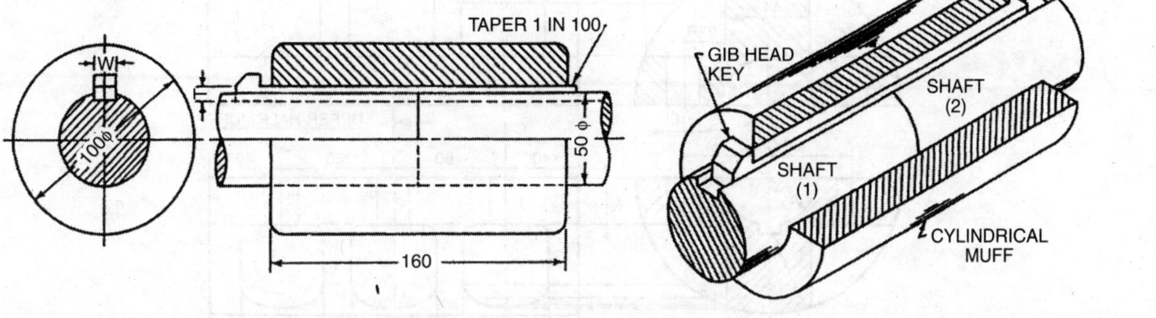

Fig. 20.6. Muff Coupling

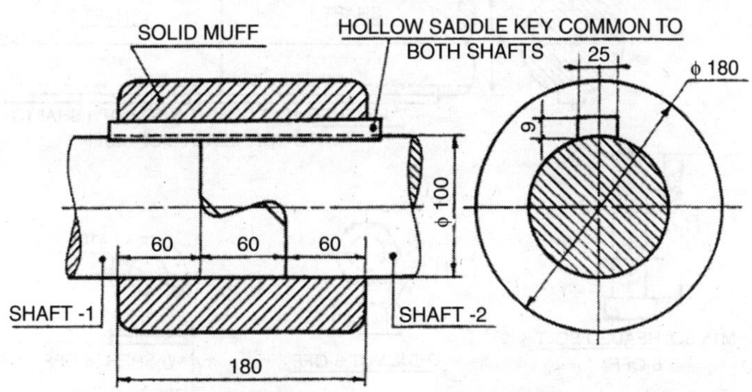

Fig. 20.7. Half-Lap Muff Coupling

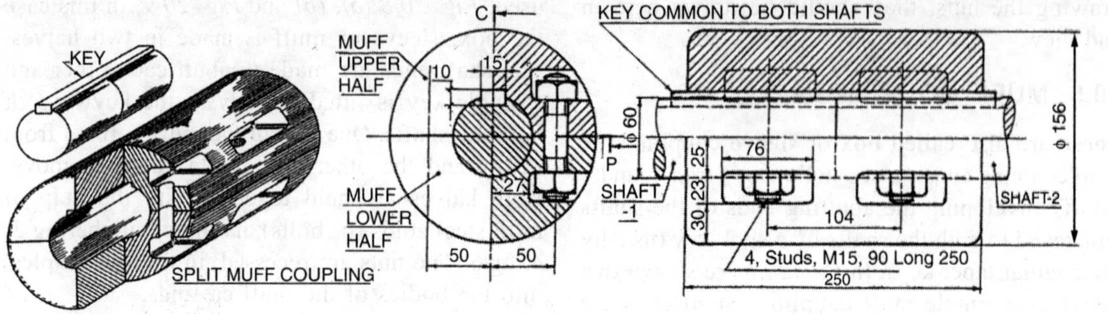

(A) SPLIT MUFF COUPLING (PICTORIAL VIEW)

(B) SPLIT MUFF COUPLING (ORTHOGRAPHIC VIEWS) (Third Angle)

Fig. 20.8.

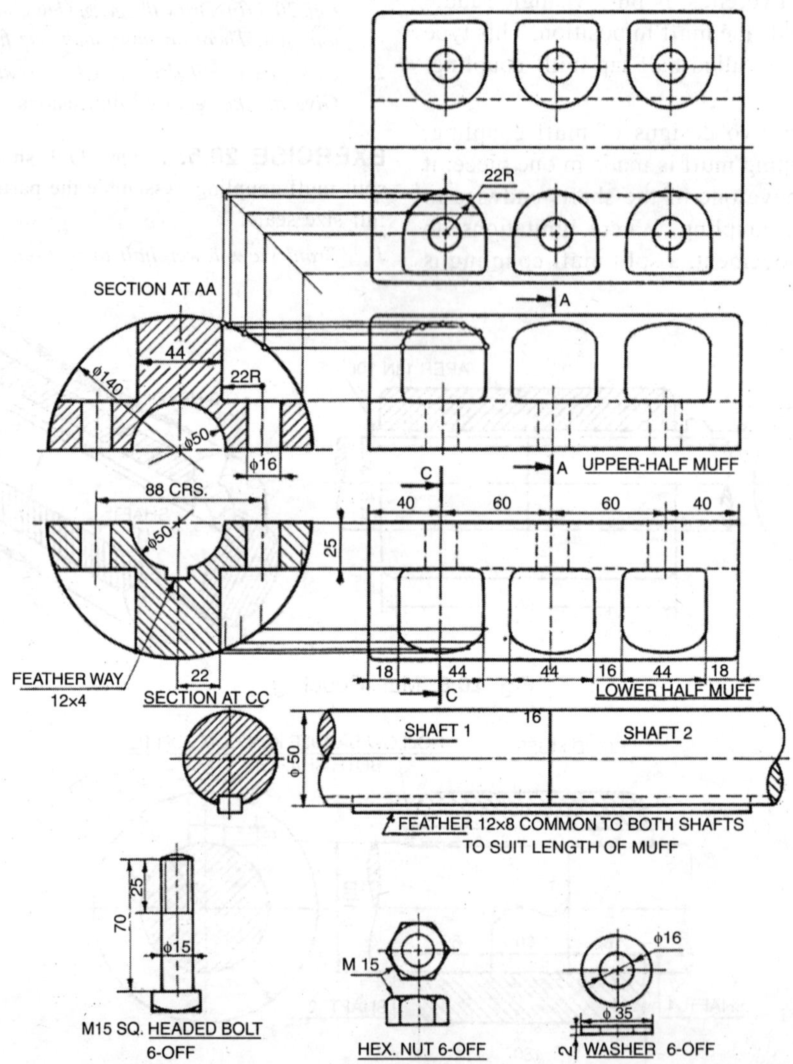

Fig. 20.9. Details of a Split Muff Coupling (Third Angle)

2. *Left side view, left half in section.*
3. *Top view outside.*

DRAWING HINT

Refer *Fig. 20.10*, for the method of projecting holes for recesses in the two halves of the split-muff coupling. It can be observed in top view that the top view of these recesses is partly circular.

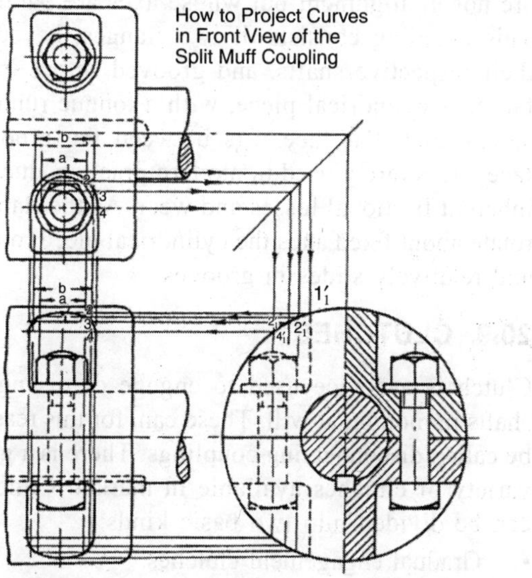

Fig. 20.10. Split Muff Coupling (part solution to exercise 20.5) (Third Angle Projection)

20.6. FLEXIBLE or NON-RIGID COUPLINGS

When shafts are supported on a number of supporting bearings, placed over a comparatively small length, it is very difficult to obtain a perfect axial alignment.

In the case of a drive from a prime mover to an electric generator or from a motor to a pump, four bearings will normally be used in a small space. The flexible couplings are used to join the abutting ends of shafts when they are not in exact alignment and undertake misalignment of the shafts without considerable absorption of power that is being transmitted.

20.6.1. Pin Type Flexible Coupling

Figure 20.11. In pin type flexible couplings, motion from one half of flange coupling is transmitted to the other half by means of pins or bolts, rigidly bolted to one flange and loosely fitting corresponding holes in the other. In various designs this type of flexible coupling is extensively used, especially where the driven units are mounted on a common base plate, and where excessive misalignment is not likely. They can drive in either direction, allow for slight misalignment and end float of shafts, and absorb impulses, shocks and vibrations. The pins are covered with rubber or

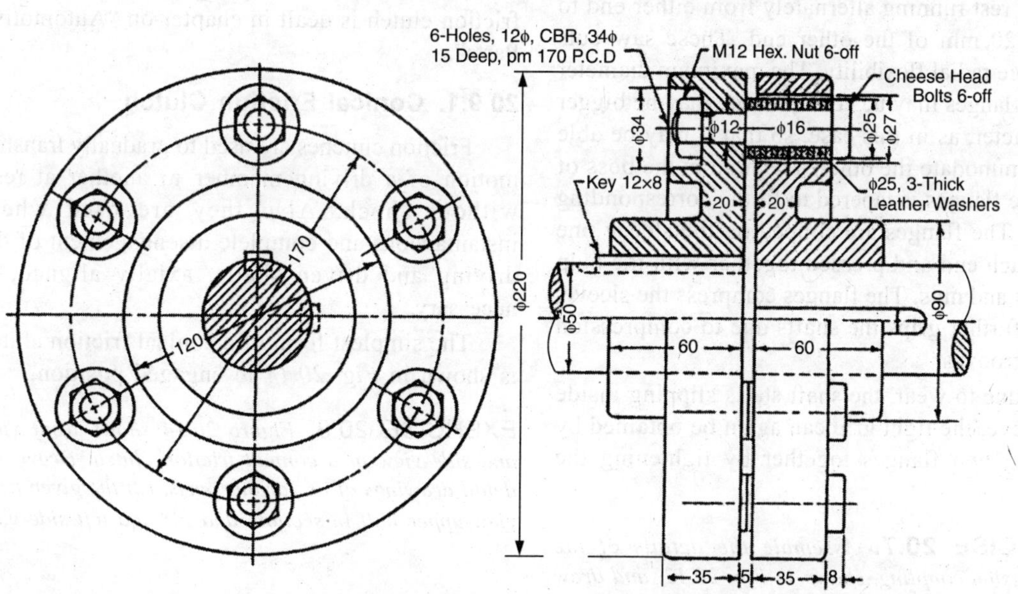

Fig. 20.11. Pin type flexible coupling (Third Angle)

some other suitable elastic, insulating material. Cushion or leather washers are employed on the pins. Since the driving pins are easily removable, either shaft may be moved without shifting it endwise.

EXERCISE 20.6. *Draw the following views of the pin type flexible coupling given in Fig. 20.11. Use a full size scale and give all the necessary dimensions.*

1. *Front view, lower half in section.*
2. *Side view, looking from right side.*

Also prepare a suitable Title-block, Bill of materials, and Parts-list.

20.7. COMPRESSION COUPLINGS

They are rigid couplings, for they connect shafts in rigid axial alignment. The skill and trouble inherent in the correct fitting (mounting) and alignment of keyed couplings render these couplings costlier. Several types of compression couplings are designed which tend to overtake partially the above factors. An example is given in *Fig. 20.12*. It can be observed here that the torque is transmitted not by the bolts and keys, as in other couplings, but by the friction between the conical surfaces in contact.

This coupling consists of a double conical sleeve and two cast iron flanges. The sleeve has six longitudinal saw cuts, one running throughout, and the rest running alternately from either end to within 20 mm of the other end. These saw cuts introduce radial flexibility. The maximum diameter of two flanges may be equal, or one may be bigger in diameter, as in this case, so that it may be able to accommodate the other. The hole in the boss of both the flanges is tapered to fit the corresponding sleeve. The flanges are slipped on the sleeve one from each end and pressed together with the help of bolts and nuts. The flanges compress the sleeve, which further grips the shafts due to compression of its grooves.

If due to wear, the shaft starts slipping inside the sleeve, the tight grip can again be obtained by pressing two flanges together by tightening the bolts.

EXERCISE 20.7. *Assemble the details of the compression coupling, shown in Fig. 20.12, and draw the following views:*

1. *Front view, upper half in section.*
2. *Right end view, outside.*

Use a scale full size, add important dimensions, and show the shafts in position.

20.8. OLDHAM'S COUPLING

Figure 20.13 (a), (b). The special advantage of this coupling is that it can be used for shafts which are not in alignment but whose axes are parallel. This coupling consists of two flanges, keyed to their respective shafts, and grooved across inner faces. A cylindrical piece, with a tongue running across each flat face, fits between the grooved faces. It is rarely used in modern practice, due the inherent frictional losses and wear. As the flanges rotate about fixed axes the cylindrical piece rotates and relatively slides in grooves.

20.9. CLUTCHES

Clutch is a device used to engage or disengage shafts in motion, at will. These can, for this reason, be called disconnecting couplings. There is a great variety of clutches available in market. Clutches can be divided into two basic kinds:

- Gradual engagement clutches.
- Positive engagement clutches.

Here we shall take up only two clutches, one of each kind, namely, conical friction clutch and claw clutch or claw coupling. Another type of a friction clutch is dealt in chapter on "Automotive Parts".

20.9.1. Conical Friction Clutch

Friction clutches are used to gradually transmit motion of a driving member to another at rest, without shock. Also they are used where instantaneous and complete disengagement of the driving and driven shafts, axially aligned, is necessary.

The simplest form of a conical friction clutch is shown in *Fig. 20.14* in engaged position.

EXERCISE 20.8. *Figure 20.14 shows front view and side view of a conical friction clutch. Draw; (1) detail drawings of its various parts, (2) the given front view-upper half in section, and (3) add left side view*

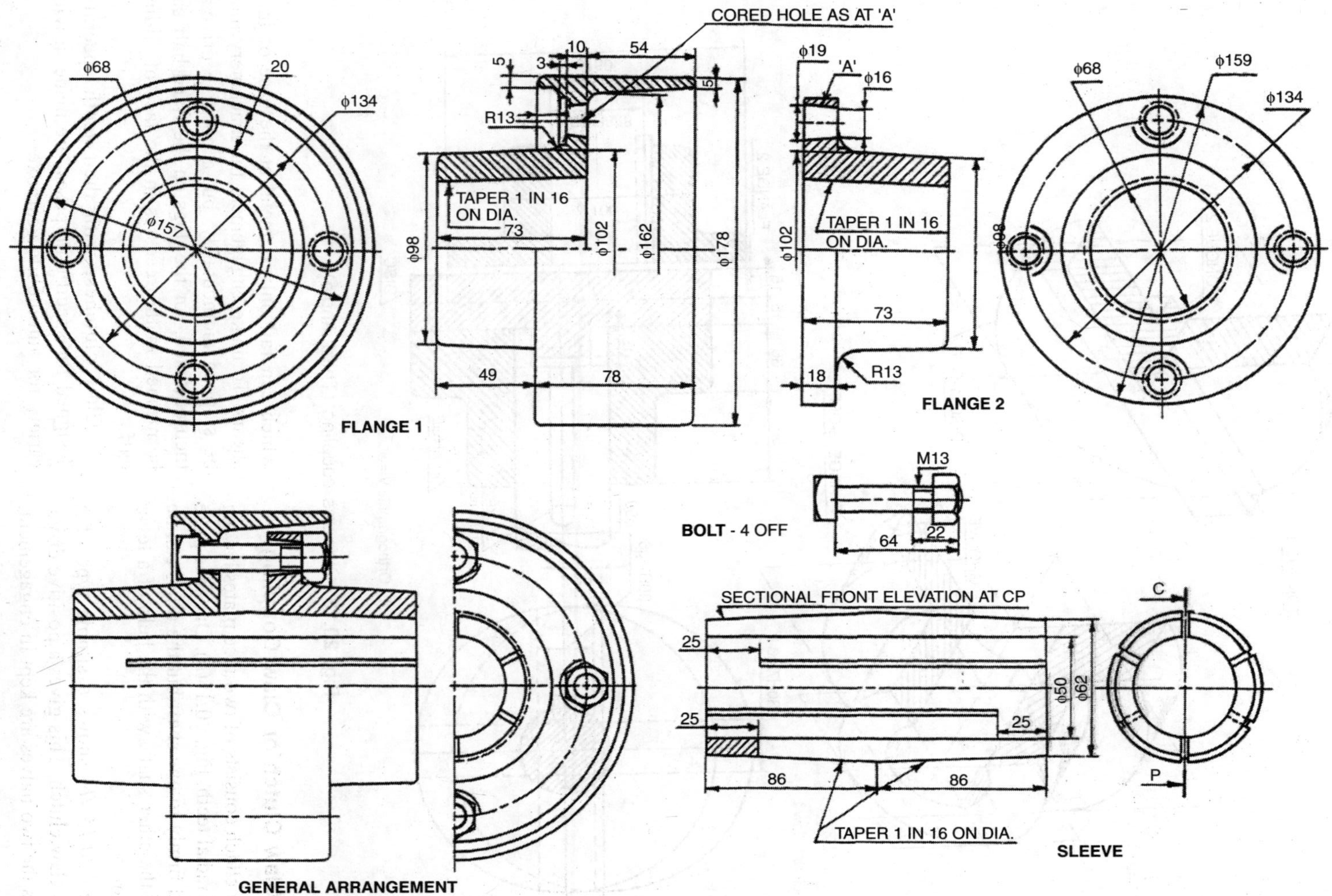

Fig. 2.12. Details of a Compression Couping (Third Angle)

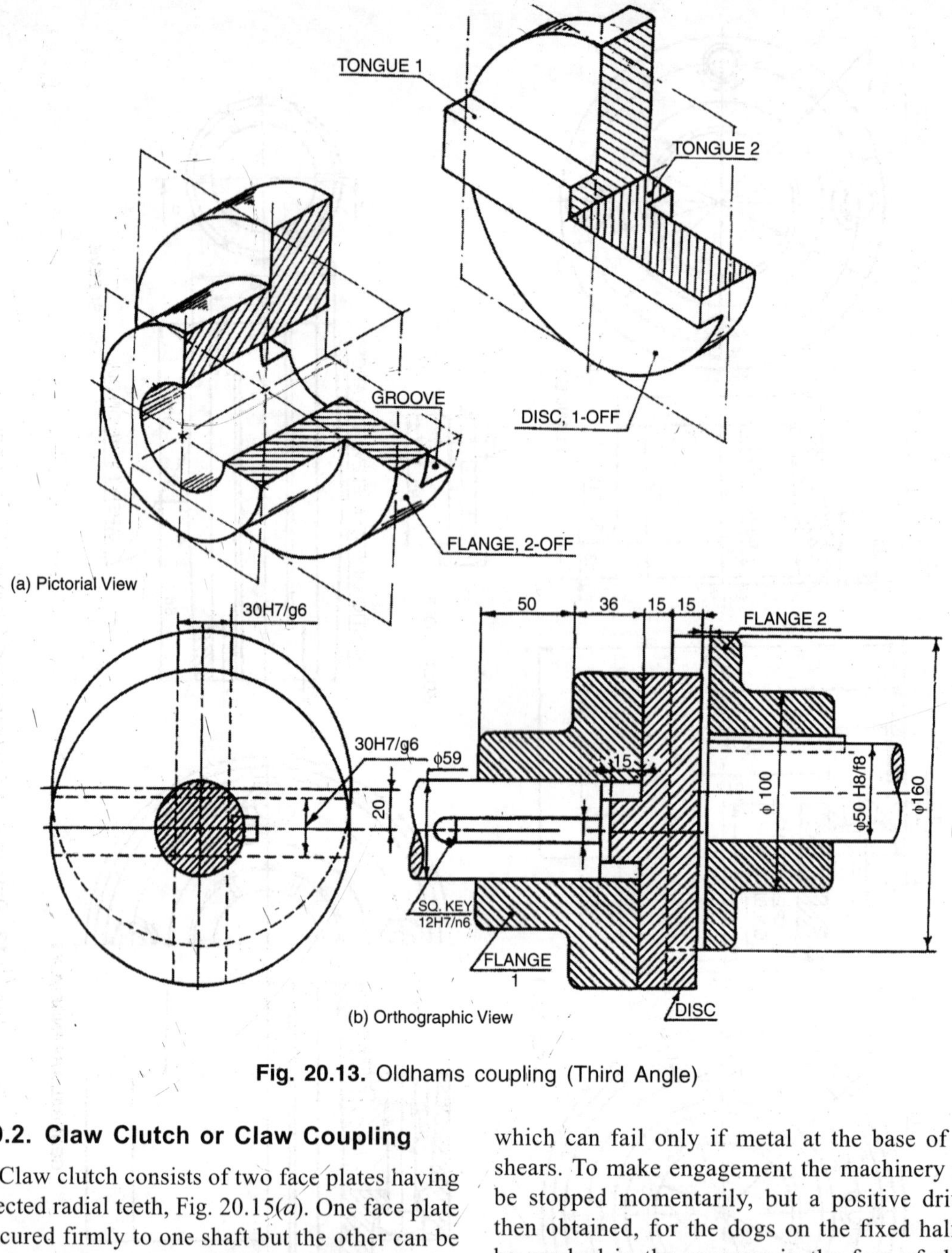

Fig. 20.13. Oldhams coupling (Third Angle)

20.9.2. Claw Clutch or Claw Coupling

Claw clutch consists of two face plates having projected radial teeth, Fig. 20.15(*a*). One face plate is secured firmly to one shaft but the other can be slided on the other shaft, with the help of a lever mechanism.

Figure 20.15 (b) shows a simple type of a three-jaw claw clutch. This gives a positive drive, so long as the two halves are kept in engagement, which can fail only if metal at the base of jaws shears. To make engagement the machinery must be stopped momentarily, but a positive drive is then obtained, for the dogs on the fixed half can be meshed in the recesses in the face of sliding half.

Claw clutches, with their teeth specially designed for specific purposes, are made in wide variety for automobile gearboxes.

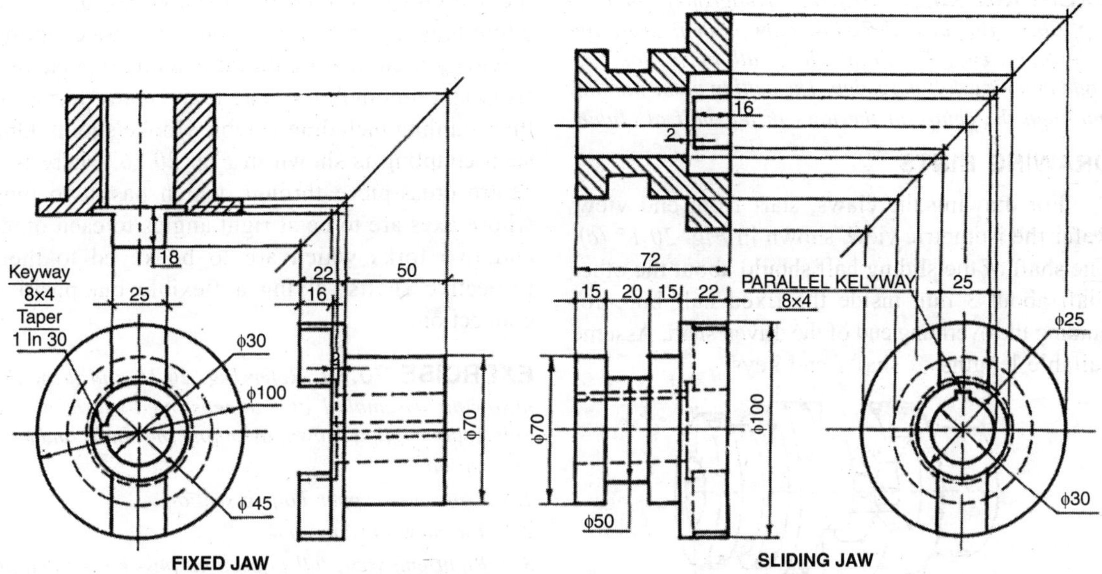

Fig. 20.15. (a) Two jaw claw coupling

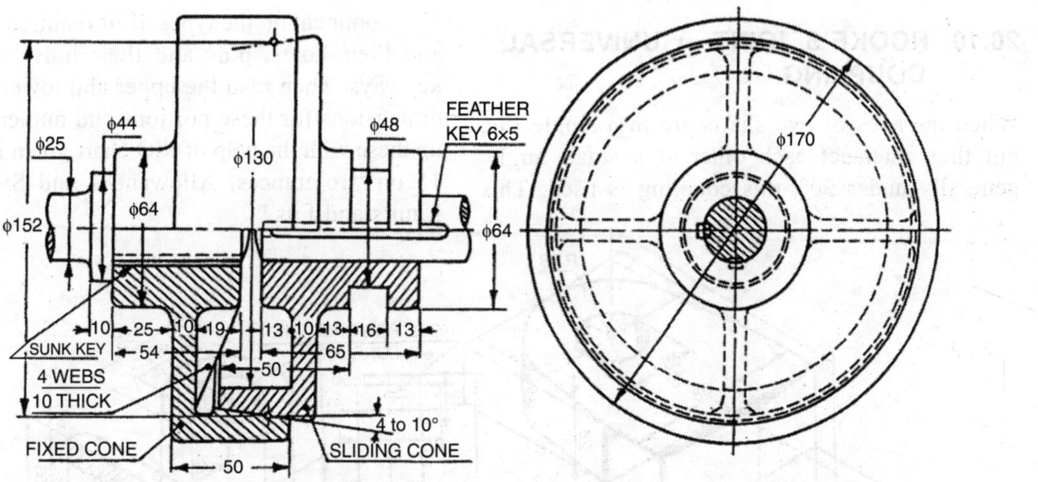

Fig. 20.14. Conical friction clutch

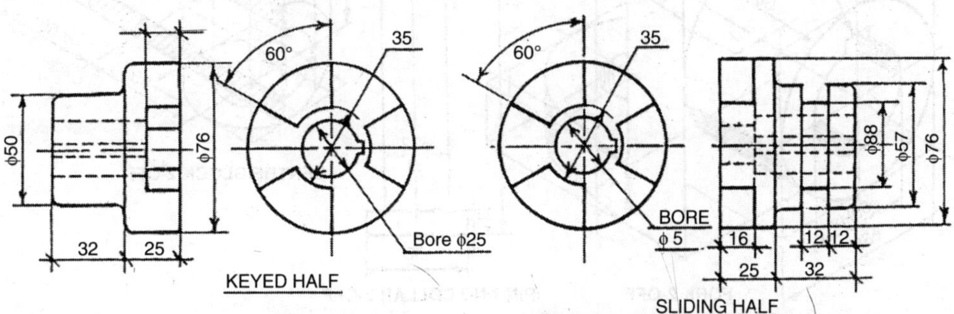

Fig. 20.15. (b) Three jaw claw clutch or coupling

EXERCISE 20.9. *From the orthographic details, Fig. 20.15 (b), of a three-jaw claw clutch, draw the following views, (1) Front view, full in section, (2) Right end view, full in section. The cutting plane passing thorough the centre of the jaws in left half of clutch.*

DRAWING HINTS

For drawing the claws, start from end view. Refer the isometric view, shown in *Fig. 20.15 (c)*. The shaft of the sliding half should about the other shaft about 3 mm inside the fixed half bore for guiding the overhung end of the driven shaft. Assume suitable lengths of shafts and keys.

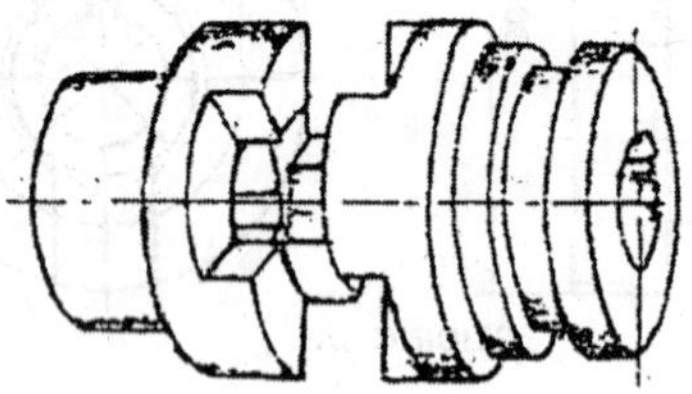

Fig. 20.15. (c)

20.10. HOOKE'S JOINT or UNIVERSAL COUPLING

When the axes of two shafts are in a single line but they intersect each other at a small angle, generally under 30°, this coupling is used. This inclination of the connected shafts can be varied while they are transmitting power. This coupling is widely used in agricultural machinery, machine tools and automobiles. There are many forms of this coupling including Double Hooke's joint. One such coupling is shown in *Fig. 20.16*. There is a centre cross-piece through which pass two pins whose axes are to be at right angles to each other and two forks which are to be keyed to their respective shafts, giving a flexible but positive connection.

EXERCISE 20.10. *Refer Fig. 20.16 and draw the following assembled and dimensioned views of the universal coupling shown, assuming the shafts and keys in position:*

1. *Front view upper half in section.*
2. *Top view outside, and*
3. *Right end view, full in section, passing the cutting plane through the centre of vertical portion of its centre block.*

Comment on the types of fit required for shafts and their holes, pins and their holes, keys and keyways. Then read the upper and lower limits of dimensions for these portions and numerically fix up these with the help of ISO chart given in chapter 15 on "Tolerances, Allowances and Systems of Limits and Fits."

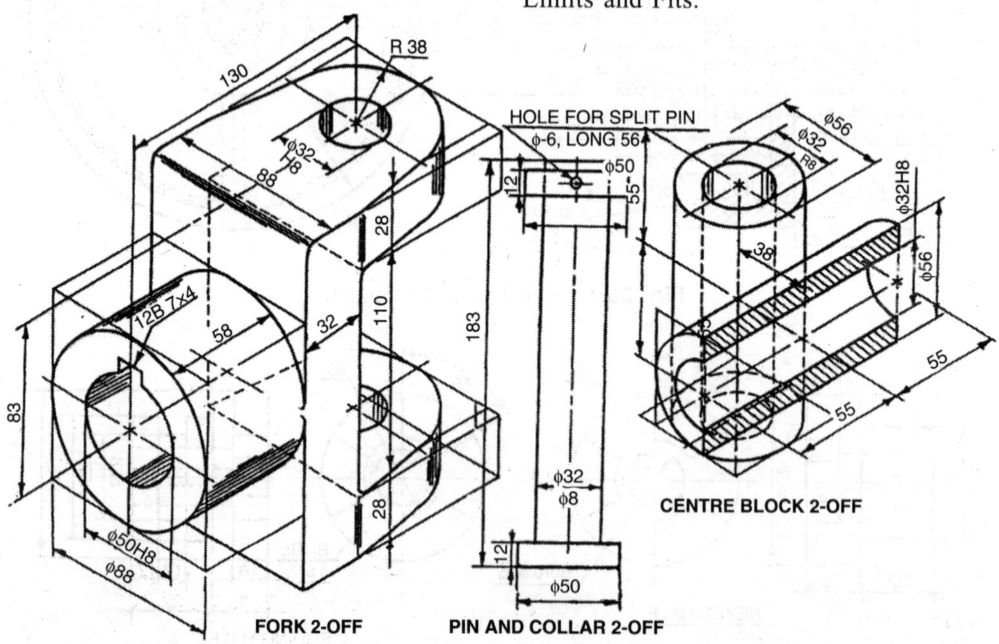

Fig. 20.16. Hooke's Joint or Universal Coupling

Hint: Refer *Fig. 20.17*, for its solution. Draw the views commencing at and building around centre lines.

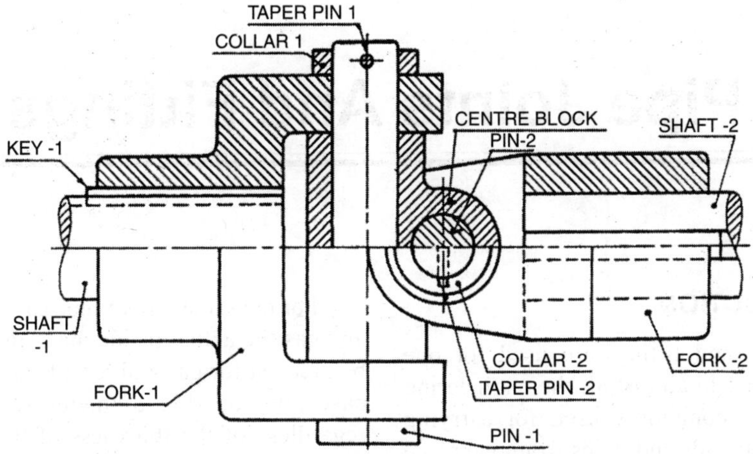

Fig. 20.17. Part solution to exercise 21.10

EXERCISES

1. Make a well-proportional freehand sketch of a simple coupling, which is fastened with standard hexagonal bolts and nuts. Take diameter of shafts, to be coupled, as 30 mm.
2. Briefly state the difference between rigid and non-rigid shaft couplings. Neatly sketch an example of each type.
3. Sketch a half-sectional front view and a side view of a protected type flange coupling.
4. Given the following dimensions of a solid-muff coupling. Make dimensions sketches, approximately half size, consisting of, (1) a front view in half section and (2) a side view with the shaft ends secured by sunk keys:

Outside diameter of muff	90 mm
Diameter of muff bore	40 mm
Length of muff	140mm
Width of keys	10mm

5. Neatly sketch a split muff coupling for 25 mm diameter shafts.
6. (*a*) Name three kinds of rigid shaft couplings? State why they are called "rigid"?
   (*b*) Why are split muff couplings used in preference to solid muff?
7. Make well-proportioned sketches of a typical flexible coupling for 25 mm diameter shafts.
8. Two shafts are to be connected together in line. The connection should fulfill the following requirements:
   (*a*) It must be possible to disconnect the shafts without disturbing either shaft.
   (*b*) It must not have any projecting pieces, which may cause injury.
9. (*a*) Make a proportionate freehand sketch of a simple claw clutch, stating its principal purpose,
   (*b*) Under what circumstances an Oldham's Coupling can be used?
10. Make a freehand-dimensioned sketch of a simple Hooke's joint of Universal coupling, take diameter of shaft as 30 mm.
11. What is spigot and socket centering, as applied to flange couplings? Illustrate by a freehand sketch.
12. Sketch freehand, two/b views of a suitable coupling to joint two 30 mm diameter shafts when their axes are parallel to each other and are 3 mm apart.

❑❑❑

Pipe Joints And Fittings 21

21.1. INTRODUCTION

Pipes, with joints and fitting, are used for one purpose or the other in almost all the engineering projects. They are commonly used for carrying fluids such as waste, oils and gases from one place to another. They are also some times used as structural elements such as columns and beams.

Pipe drafting is a type of assembly drawing which is used to represent pipe layouts. Pipe drafting is used in any industry where it is necessary to transfer fluids. An engineering student should develop knowledge and skills necessary to draw piping systems.

21.2. STANDARDIZATION

Because piping is used almost everywhere, it has been standardized. There are standard sizes of pipes, fittings and valves, which simplify the design, drawing, building and repairing of pipe systems.

21.3. TYPES OF PIPE

The material of a pipe depends upon the purpose for which the pipe is used. A large number of materials, varying from cast iron, steel, brass, copper, lead and aluminium, etc., to wood, concrete, clay, glass, plastic and rubber, etc., are used these days in pipe manufacture. This chapter will deal only with metal and clay pipes, which are commonly used. Steel and wrought iron pipes are in common use as water, steam, oil and gas pipes. These are available in three different weights, known as standard, extra strong and double extra strong.

Metal pipes are specified by their nominal diameter, which differs somewhat from their actual diameter, when it is up to 25 cm. For all pipes over 25 cm diameter, the nominal diameter equals the actual outside diameter; and such pipes are called O.D. (outside diameter) pipes.

For the same size of pipe carrying steam, water and gas the actual inside diameter of the pipe will be greatest for gas and least for steam. In all these cases the outside diameter remains the same regardless of the thickness of the wall.

Steel pipe is used for high temperature and or pressure applications. Its size is specified by its nominal diameter, referred to NPS (Nominal pipe size). This nominal size is roughly the inside diameter (ID) of the pipe. There are ten weights of pipes with different wall thickness and many sizes in each weight.

Cast iron pipe is usually of large diameter and is used only for low pressure applications. A common applicaton for cast iron pipe is large diameter underground sewer pipe.

Copper pipe and tubing is used where corrosion is a problem. Copper is one of the most expensive materials used for piping. It may be used for house plumbing and heating.

Plastic pipe is corrosion resistant, flexible, easily installed and inexpensive. Sprinkler systems, and domestic sewer systems are a common application for plastic pipe.

21.4. PIPE JOINTS

The C.I. pipes are available in lengths up to 5.5 meters only, due to manufacturing, transportation, storing and handling difficulties. In many cases when a pipeline longer then the available lengths are required, two or more such pipes are axially connected, end-to-end.

Pipes are connected by fittings. Fittings have been standardized to ensure that connections will not leak or vibrate loose.

The various methods of connecting pipes depend upon the material of pipes, material to be handled by pipes, pressure, shock and stresses to which the joint may be subjected. Out of the several forms of joints used, some common pipe joints are described here.

21.5. SCREWED JOINT

A short cylindrical sleeve, called a socket or a coupler, threaded from inside, is screwed on to the threaded ends of the two pipes, *Fig. 21.1 (b)*. Pipe threads are tapered at a rate of 1 : 200 to ensure a tight, connection. Tapered pipe threads may be used for all types of pipe. This type of joint is not readily disassembled as flange joints and may be designed to withstand relatively high pressure. To make the joint leak-proof a few strands of jute or hemp, coated with red lead, are wound round the bottom of the thread on each pipe end.

A nipple, which is threaded on the outside, is also sometimes used, *Fig. 21.1 (a)*. It is screwed inside the internally threaded ends of the two pipes. In this case, the passage of the fluid is partially chocked due to the resulting smaller inside diameter of the nipple.

21.6. WELDED JOINTS

Figure 21.2 (a) shows a butt-welded joint, which can be subjected to high pressure, but cannot be disassembled. Copper and brass tubes are usually connected by soldering; these cannot be subjected to pressure and temperature. Fittings are made very precise and the solder flows in to make a good, tight connection, *Fig. 21.2 (b)*. Welded joints are lightweight and less bulky than screwed or flanged fillings.

21.7. CEMENTING

Cementing or gluing is a common method of connecting plastic pipes, *Fig. 21.2 (b)*. It is easy and takes no special tools or skills.

21.8. FLANGE JOINTS

Flanged connections are used to connect large diameter pipes.

In the cast iron pipes the flanges are cast with the pipes and then their faces are machined, while in the case of M.S. and W.I. pipes these are usually screwed on to the pipe ends. The flanges are held together by a number of bolts, with a gasket of soft material in between them to make the joint

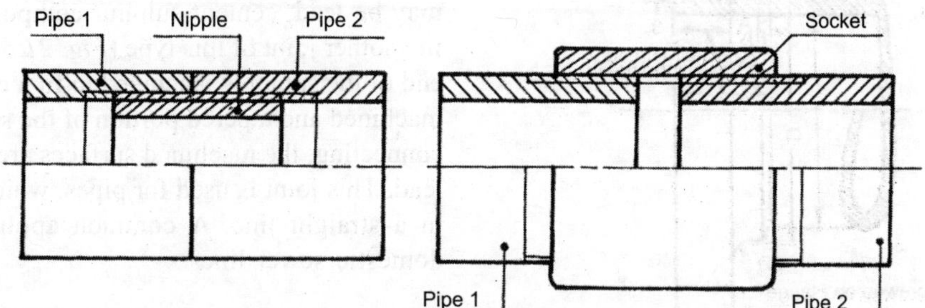

(a) NIPPLE JOINT (b) SOCKET JOINT

Fig. 21.1.

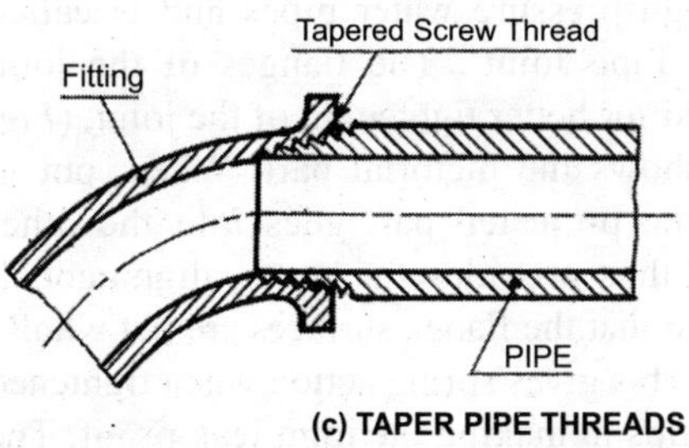

(c) TAPER PIPE THREADS

Fig. 21.1. Screws joint

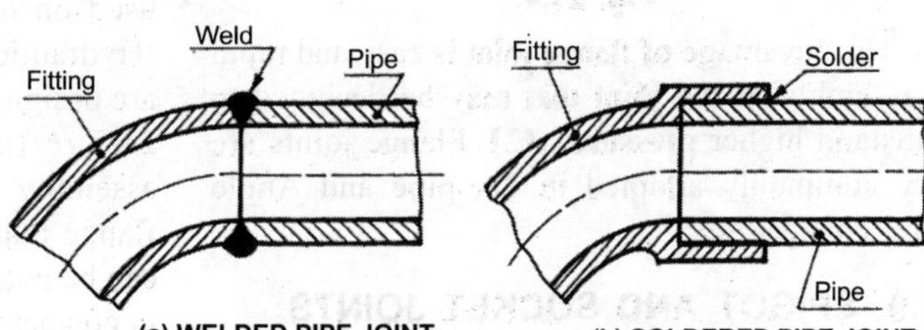

(a) WELDED PIPE JOINT (b) SOLDERED PIPE JOINT

Fig. 21.2. Pipe joints

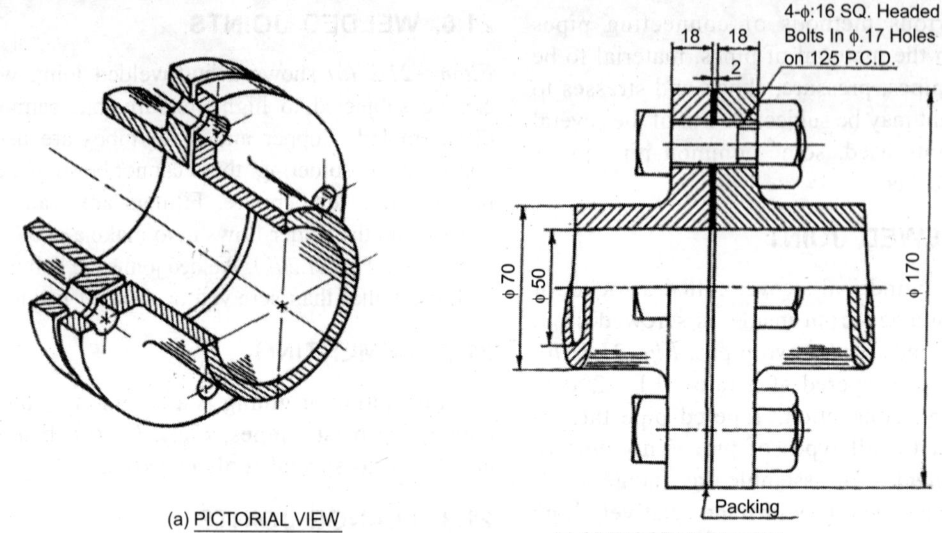

(a) PICTORIAL VIEW

(b) ORTHOGRAPHIC VIEW

Fig. 21.3. Flanged pipe joint

leak proof. A simple type of C.I. flange joint is shown *Fig. 21.3 (a), (b)* and in *Fig. 21.4* is shown a flange screwed to the pipe.

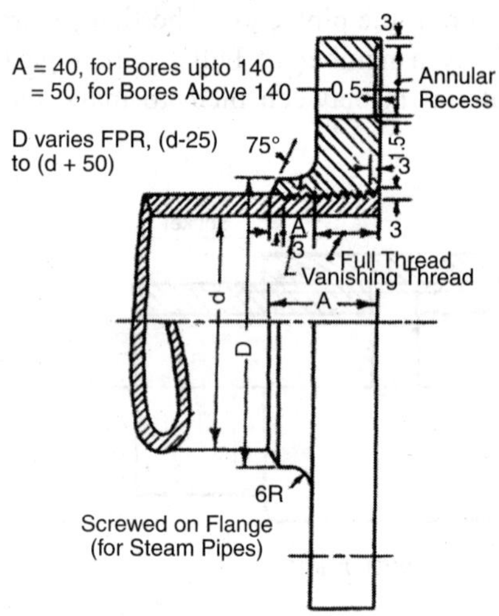

Fig. 21.4.

The advantage of flange joint is easy and rapid disassembly of the joint that may be designed to withstand higher pressures. C.I. Flange joints are very commonly adopted in Tee-pipe and Angle Bend connections.

21.9. SPIGOT AND SOCKET JOINTS

These joints are adopted for connecting cast iron pipes in underground installations, where there is low pressure and little vibrations. As these joints are flexible, they can adopt themselves to the settlement of earth, which takes place due to climatic or other conditions. As shown in *Fig. 21.5*, this joint consists of two main parts, 'Socket' (Bell) and 'Spigot'. After placing the spigot in position in the socket, the remaining space between the two is filled with some packing material, which may be lead, cement sulphur compound and jute. In another joint of this type (*Fig. 21.6*) the enlarged end of the spigot is machined to fit a corresponding machined and tapered portion of the socket. Before connecting, the machined surfaces are painted with lead. This joint is used for pipes, which are strictly in a straight line. A common application is for domestic sewer lines.

21.10. HYDRAULIC PIPE JOINT

Figure 21.7 (a) shows a flanged joint, which is used on high pressure water pipes and is called 'Hydraulic Pipe Joint'. The flanges of the joint are designed for better tightening of the joint. (*Fig. 21.7 (b)*) shows the pictorial part. When, put in assembly the projected part goes into the other flange and thus provides for better alignment. It can be noted that the flange surfaces are not wholly in contact. This gives spring action when tightened and thus helps in making the joint leak proof. The detail of portion 'A' of the joint, illustrating the

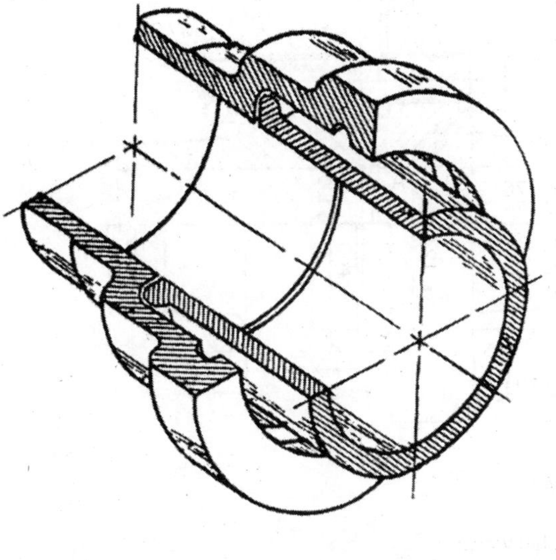

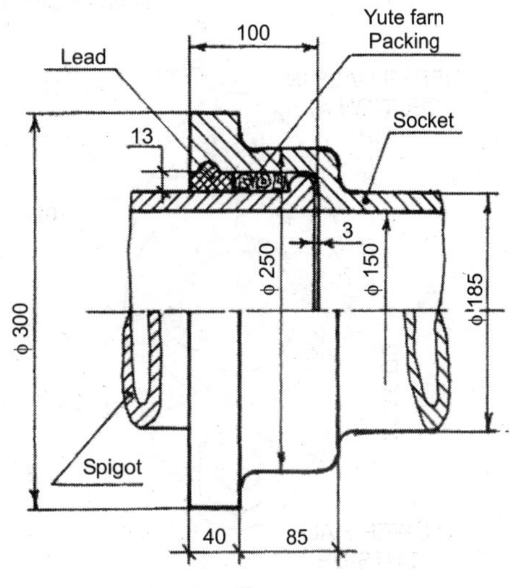

Spigot and Socket Pipe Joint
(with no Direct Pipe Contact)
(b) Orthographic view

Fig. 21.5. Spigot and socket pipe joint (with no direct pipe contact)

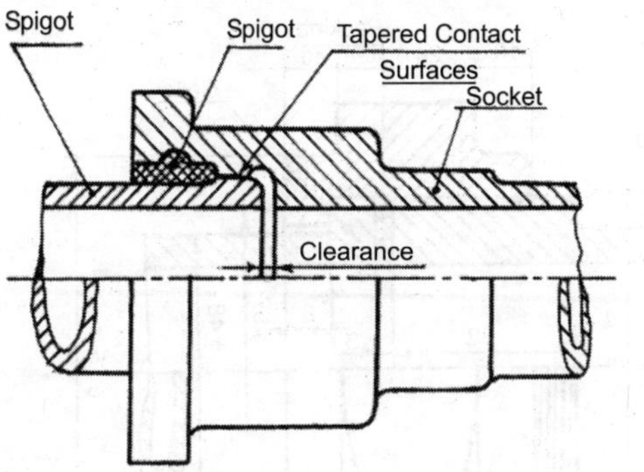

Fig. 21.6. Spigot and socket joint (with direct contact of pipes)

position of the packing washer, is shown on an enlarged scale.

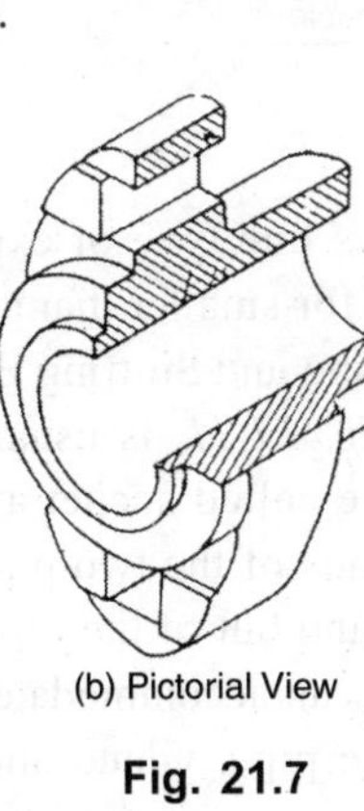

(b) Pictorial View

Fig. 21.7

21.11. SCREWED AND FLANGED UNION (UNION JOINT)

The screwed joints discussed above cannot be disassembled easily. For rapid disassembly, the joint of the type shown in *Fig. 21.8*, called pipe union or union joint, is used. A coupler having screw threads on one side is used to connect the two flanged pipes. The flange on one pipe is threaded and the threaded end of the coupler is screwed on to it. The outer surface of the coupler is usually made hexagonal for using a spanner to tighten the joint. Gunmetal or brass pieces, properly machined, are used as packing materials.

21.12. PIPE FITTINGS

Pipe fittings are used to connect lengths of pipes, to change the direction of pipes, and to change pipe size. Some of the more common pipe fittings are shown in *Fig. 21.9*.

Use of a socket or coupler and a nipple to change the length of pipes is already illustrated in *Fig. 21.1 (a), (b)*.

Pipes are connected or branched off at right angles by means of fittings known as **bends, elbows, tees** and **crosses**, *Fig. 21.9 (a), (b), (c) and (f)*.

Pipes of different diameters are joined by the use of reducing sockets, *Fig. 21.9 (e)*. A plug is used to close the end of a pipe, *Fig. 21.9 (d)*.

(with no direct pipe contact)

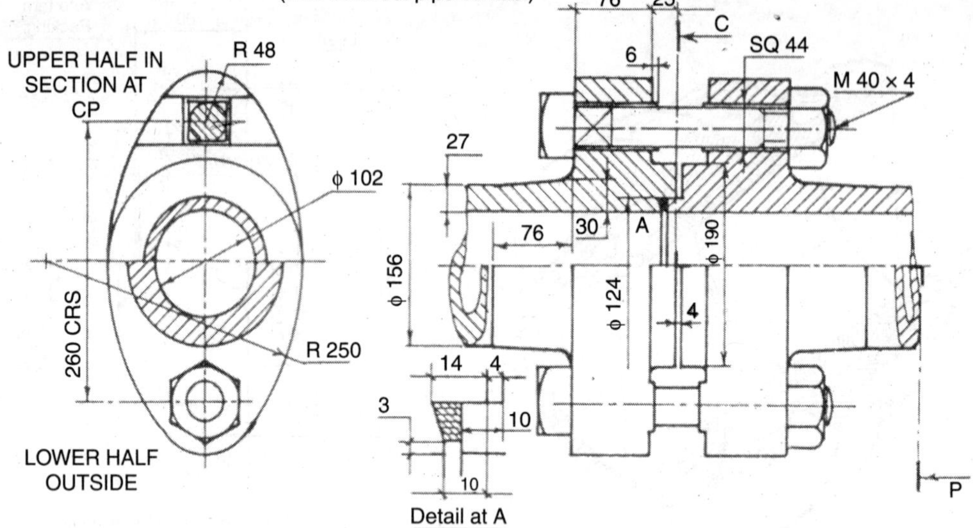

(a) Orthographic Views

Fig. 21.7. Hydraulic pipe joints

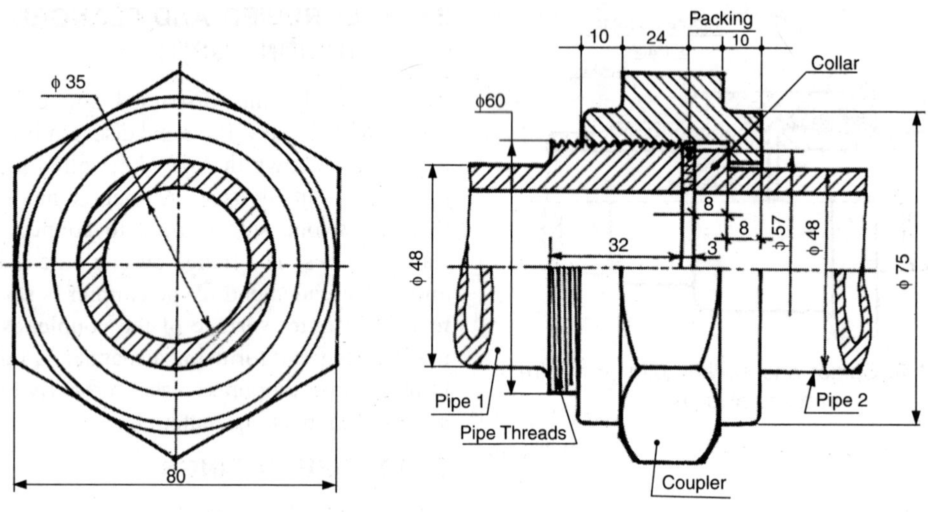

Fig. 21.8. Union joint

When a fitting is called out on a drawing, it is specified according to its nominal size, material and strength factor.

In long pipes carrying live steam, there is too much variation in temperature due to which contraction and expansion of the pipes takes place. This may result in the breakage of some joint or the buldging out of the pipes. To prevent it, various types of expansion take up equipments are used. *Figure 21.10* shows one type of expansion bend, which can be used for smaller diameter pipes. For larger pipes the Gland and Stuffing Box Expansion Joint, shown in *Fig. 21.11*, is usually used. Two parts which may be called socket and sleeve, are connected to the ends of the two pipes. the sleeve is free to move in and out of the turned portion of the socket and thus can accommodate any variation in the length of the pipe. Gland and packing are

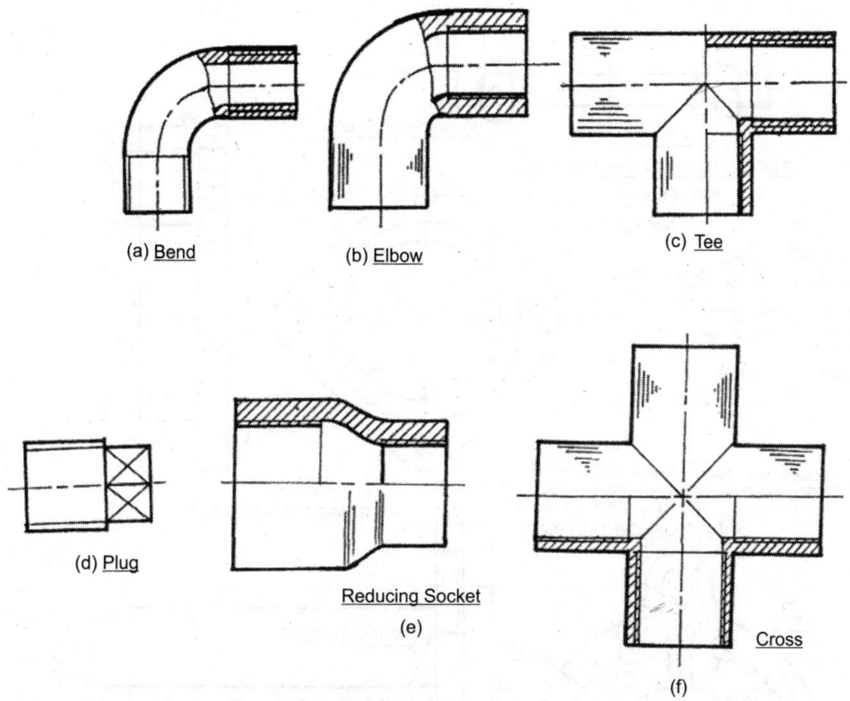

Fig. 21.9. Pipe Fittings

used to prevent the leakage of steam. Guard bolts are provided to prevent the undesired, excessive movement of the sleeve in the socket.

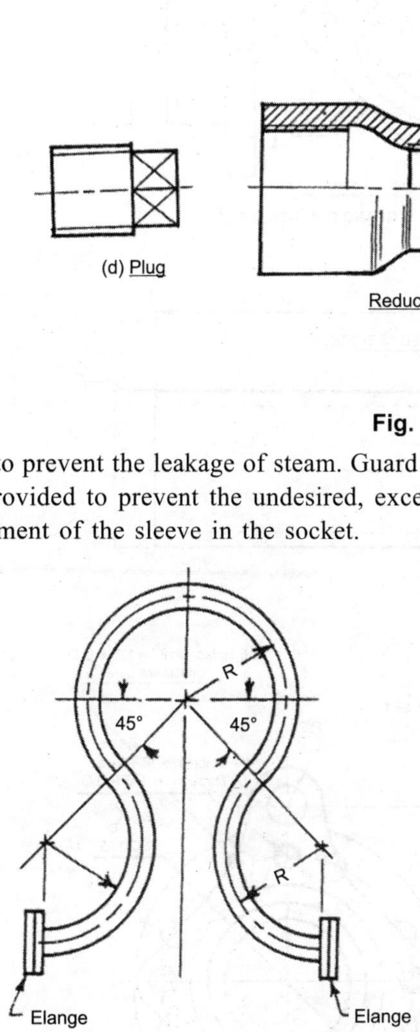

Fig. 21.10. Expansion Bend

Another design of expansion joints is shown in *Fig. 21.12*.

21.13. VALVES

Valves play a major role in pipe-system. They are used to stop or to control the flow of liquids through the pipe lines. There are many types of valves; but the three most common are gate valves, globe valves and check valves. For greater details on valves read chapter 24 on **'Valves'.**

21.14. PIPING DRAWINGS

Piping drawings are system drawings, that is they show how the pipes go together to make up a system. The drawings show the location, size and orientation of the pipe fittings and valves. Because the components of a piping system are standard items, it is unnecessary to detail individual parts. To simplify the drawing scheme symbols and simplified outlines of the components are used. *Fig. 21.13*.

Four types of piping drawings are commonly used. They are:

1. Single line orthographic, *Fig. 21.14 (a)*.
2. Double line orthographic, *Fig. 21.14 (b)*.
3. Single line isometric, *Fig. 21.14 (c)*.
4. Double line isometric, *Fig. 21.14 (d)*.

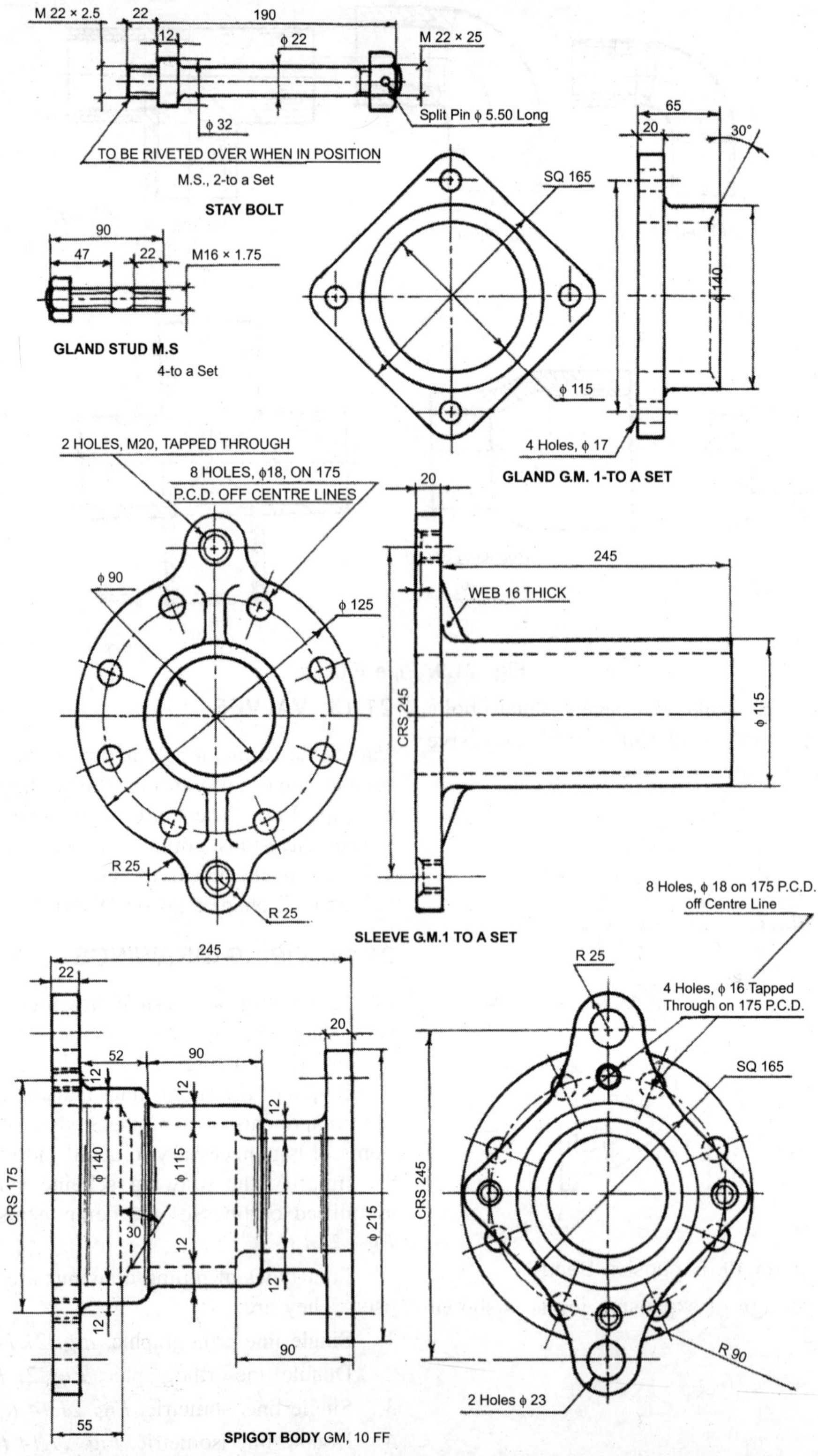

Fig. 21.11. (a) Detain drawing of a 90 mm BORE. Expansion Joint – G.M.

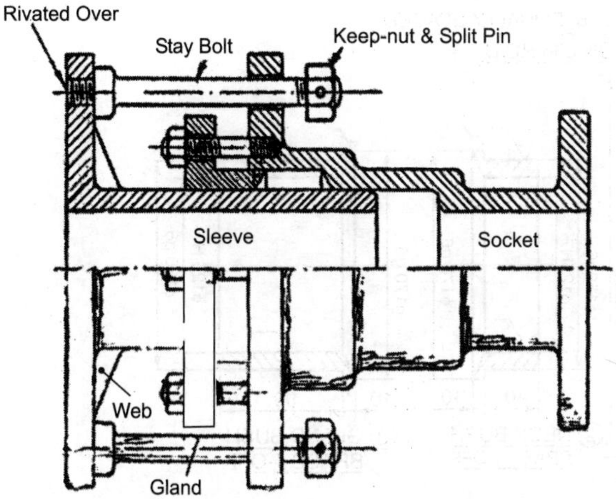

Fig. 21.11. (b) General arrangement of Expansion joint

Isometric piping drawings are laid out and dimensioned in the same manner as any other isometric drawing. *Figure 21.15* illustrates how some of the typical piping symbols are drawn and how dimensions are given on three isometric planes.

21.15. DIMENSIONING

Piping drawings are dimensioned for locational purpose only. Locational dimensions are given to the centre lines of pipes, fittings and valves. The size of the pipe fittings may be shown besides the components by a note or they may be called out in the parts-list. Pipe lengths are usually not shown. They are left up to the pipe fitters.

21.16. SIZES OF STANDARD PIPE

Table 21.1 gives the nominal size, outside diameter and inside diameter of steel pipe as specified in the ISO standard C 65. It should be noted that the nominal size of the pipe is approximately the inside diameter size. Therefore, a pipe referred to as a 6 mm pipe (its nominal size) has an inside diameter of approximately 6 mm (actually 6.5 mm).

Table 21.1.

Nominal Size mm	*Ouside Diameter mm*	*Inside Diameter mm*
6	10.2	6.5
8	13.6	8.0
10	17.1	12.5
15	21.4	16.0
20	26.9	20.0
25	33.8	25.0
32	42.2	31.5
40	48.4	40.0
50	60.2	50.0
65	76.0	63.0
80	88.8	80.0
90	101.1	90.0
100	114.0	100.0
125	139.6	125.0
150	165.2	160.0

EXERCISES

1. Sketch freehand the sectional front view and outside end view of the Socket and Spigot Joint shown in *Fig. 21.5*.
2. *Figure 21.3 (b)* shows an orthographic view of Flanged Pipe Joint, Draw its following views:
   (*a*) Front view full in section.
   (*b*) Outside end view
3. Reproduce the two views of the Hydraulic pipe joint shown in *Fig. 21.7 (a)*.
4. Make proportionate sketches of the following:
   (*a*) Screwed joint
   (*b*) Welded joint
   (*c*) Flanged joint,
   (*d*) Hydraulic pipe joint, and
   (*e*) Expansion joint.
5. Draw two views of a 'Flanged pipe joint' for a pipe of 100 mm diameter and having four bolts securing the flanges together. Give all important proportional dimensions.
6. What is the specific use of an expansion pipe joint?
7. Sketch neatly, a suitable joints for 200 mm diameter C.I. Pipes, to be laid underground.
8. *Figure 21.11* shows the details of an expansion joint. Draw the following assembled views, on a scale half the full size:
   (*a*) Front view upper half in section.
   (*b*) End view looking from left hand side.
9. *Figure 21.12*, shows the detail drawings of an expansion joint. Assemble the parts and draw the following views:
   (*a*) Front view lower half in section.
   (*b*) Right side view
   Give all the necessary dimensions. Also prepare a Title-Block Bill of materials and Part list.

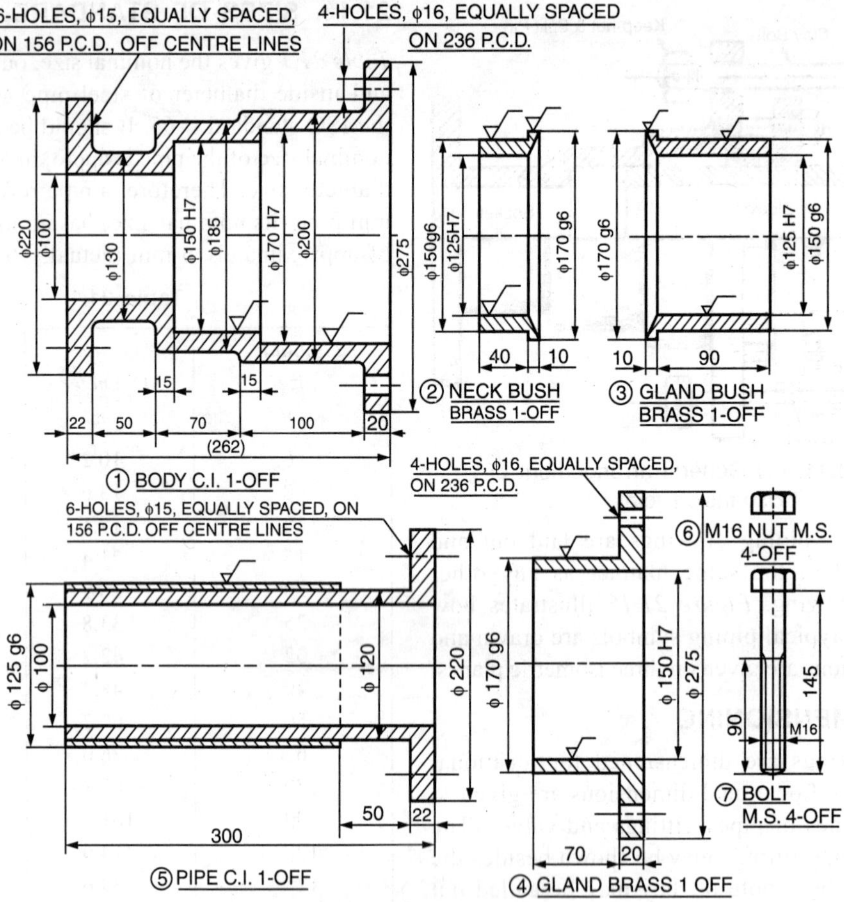

Fig. 21.12. Expansion Joint

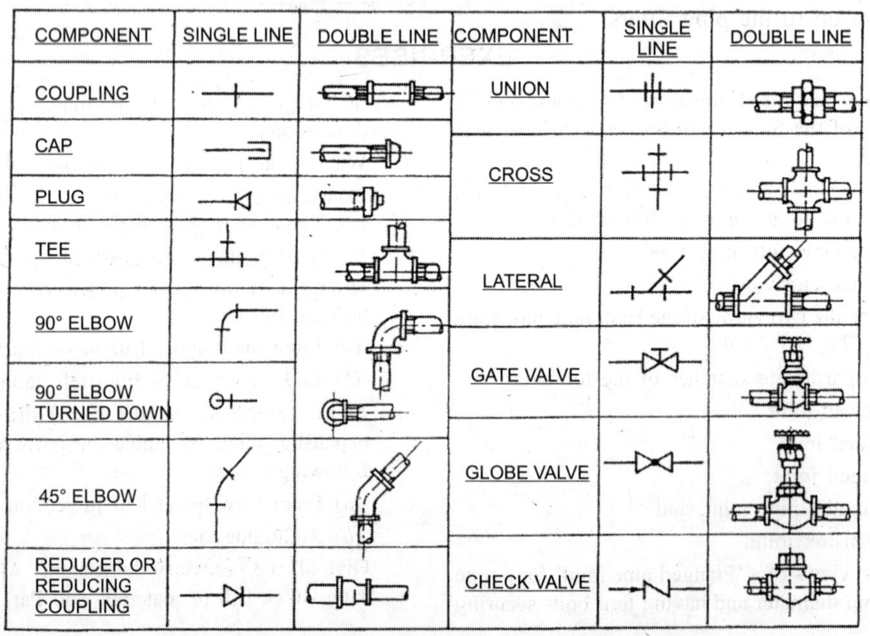

Fig. 21.13. Symbols for screw connection fittings

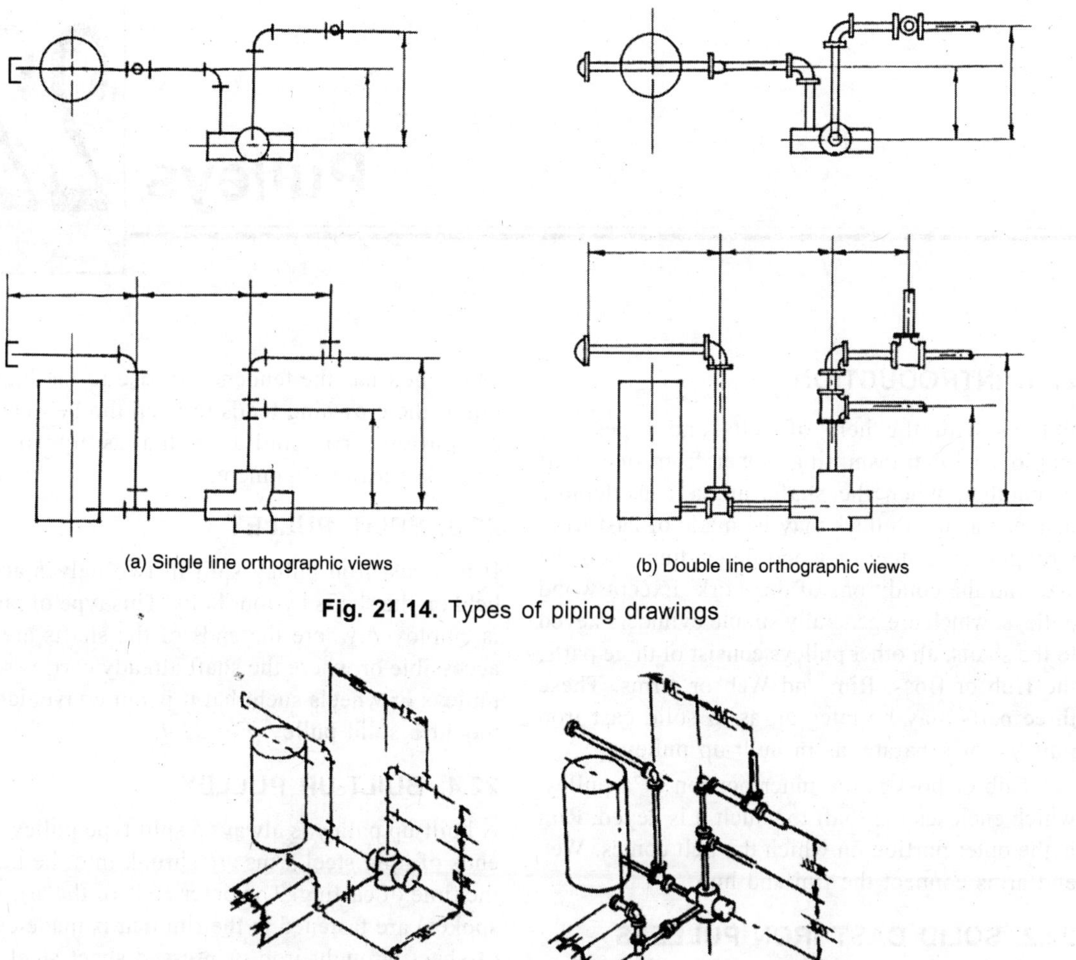

(a) Single line orthographic views

(b) Double line orthographic views

(c) Single line isometric view

(d) Double line isometric view

Fig. 21.14. Types of piping drawings

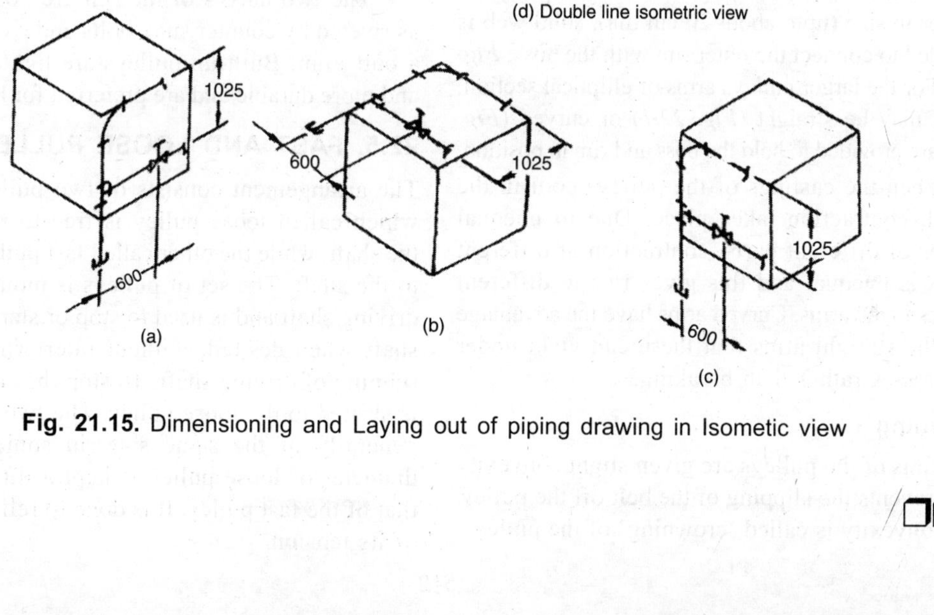

(a) (b) (c)

Fig. 21.15. Dimensioning and Laying out of piping drawing in Isometic view

Pulleys 22

22.1. INTRODUCTION

Pulleys with the help of belts and ropes, are employed for transmitting power from one shaft to another, when the shafts are a considerable distance apart. Pulleys may be made of cast iron, wrought iron, steel or wood depending upon the size and the conditions of the work. Except wood pulleys, which are generally simple cylinders keyed to the shafts, all other pulleys consist of three parts, the Hub or Boss. Rim and Web or Arms. These three parts may be integral, as in solid cast iron pulleys, or separate, as in built-up pulleys.

Hub or boss is the inner portion of a pulley, which encloses the shaft to which it is keyed. Rim is the outer portion on which the belt comes. Web and arms connect the rim and hub.

22.2. SOLID CAST IRON PULLEYS

These are one-piece castings. When the pulley is smaller in size (upto about 20 cm dia), solid web is provided to connect the outer rim with the boss, *Fig. 22.1*. For the larger pulleys arms of elliptical section, which may be straight (*Fig. 22.1*) or curved (*Fig. 22.3*) are provided to hold the boss and rim in position.

When the castings of the pulleys cool in the mould, contraction takes place. Due to unequal masses of different parts, contraction at different places is unequal and this gives rise to different stresses in the arms. Curved arms have the advantage over the straight arms that these can yield under the stresses rather than breaking.

Crowning

Rims of the pulleys are given slight convexity that prevents the slipping of the belt off the pulley. The convexity is called 'crowning' of the pulleys. as the belt has the tendency to rise to the highest point, the crowning tends to keep the belt central on pulley's rim and thus makes up for the irregularities in alignment.

22.3. SPLIT PULLEY

It is a cast iron pulley split in two halves and is held on the shafts by four bolts. This type of pulley is employed where the ends of the shafts are not accessible or where the shaft already carries some pulleys or wheels such that it is not convenient to mount a solid pulley, *Fig 22.4*.

22.4. BUILT-UP PULLEY

A built-up pulley is always a split type pulley. The ends of mild steel arms are shrunk into the hub at the time of casting. The outer ends of the arms (or spokes) are fastened to the rim that is made either of sheet wrought iron or pressed sheet steel.

The two halves of the rim are bolted as well as riveted by counter sunk bolts and rivets forming a butt joint. Built-up pulleys are lighter, stronger and more durable and are preferred for high speeds.

22.5. FAST AND LOOSE PULLEYS

The arrangement consists of two pulleys, one of which called loose pulley is free to rotate about the shaft, while the other called fast pulley, is keyed to the shaft. The set of pulleys is mounted on the driving shaft and is used to stop or start the driven shaft when desired, without interfering with the running of driving shaft. To stop the shaft the belt is shifted to the loose pulley. The two pulleys are generally of the same size, in some cases the diameter of loose pulley is kept a little less than that of the fast pulley. It is done to relieve the belt of its tension.

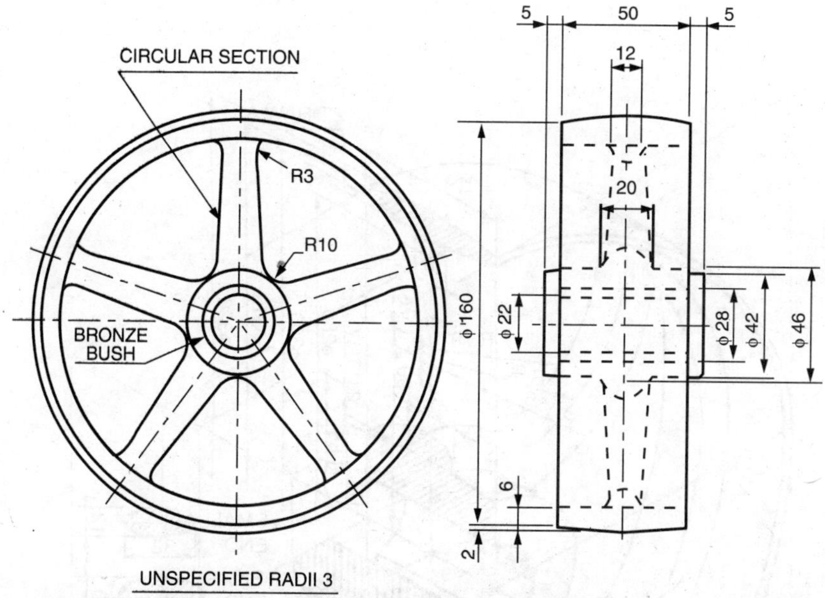

Fig. 22.1. Solid pulley (with straight arms)

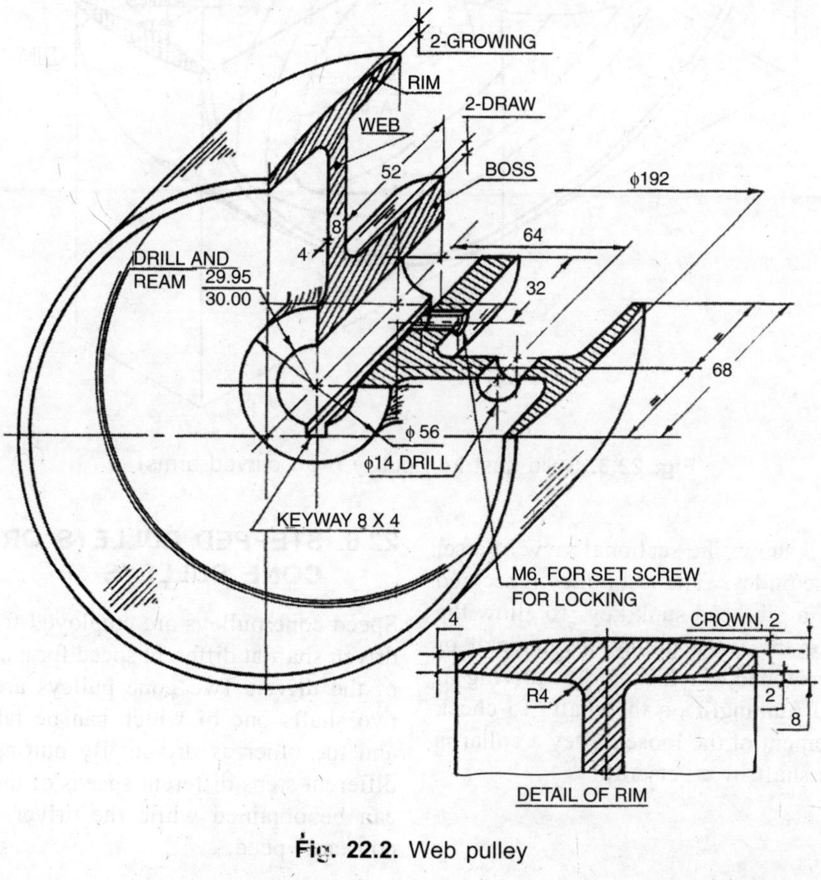

Fig. 22.2. Web pulley

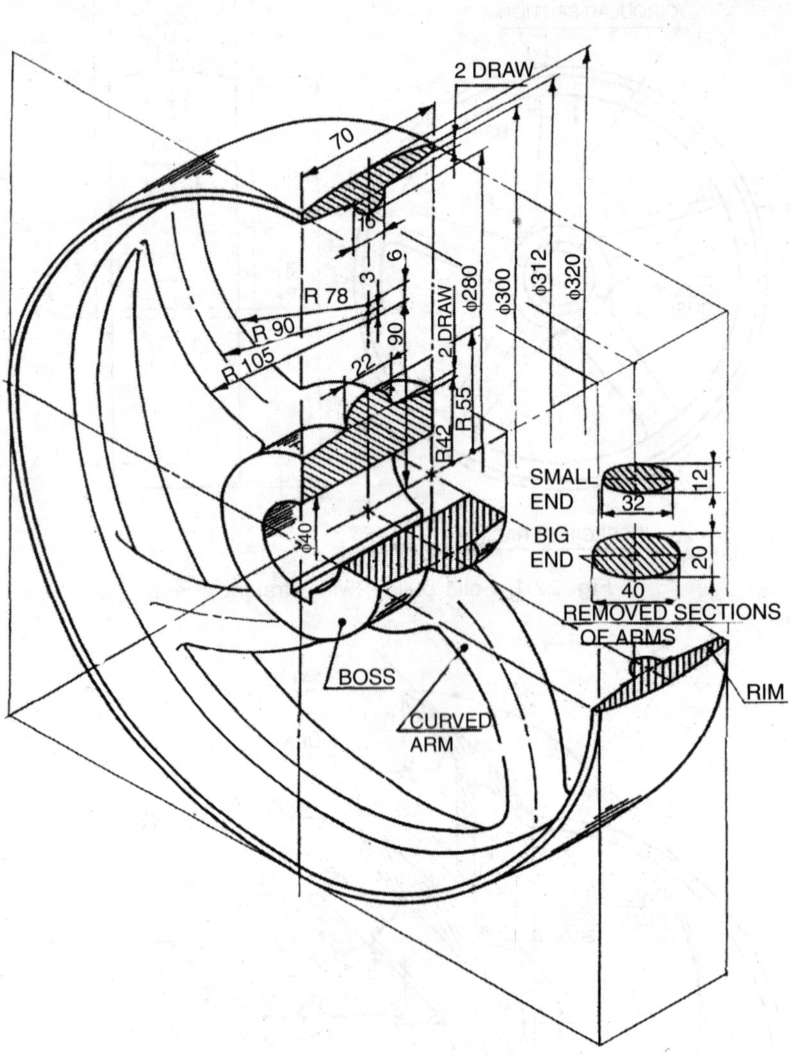

Fig. 22.3. Solid cast iron pulley (with curved arms)

Figure 22.5 shows the sectional view of a set of fast and loose pulleys. The fast pulley is secured to the shaft by a gib-head sunk key. To allow the free running of the loose pulley, a gunmetal or brass bush is provided which should be driving fit in the hub and runningfit on the shaft. To check the axial movement of the loose pulley a collar is secured to the shaft by a setscrew.

22.6. STEPPED PULLEYS OR SPEED CONE PULLEYS

Speed cone pulleys are employed for running the driven shaft at different speed for a constant speed of the driver. Two cone pulleys are mounted to two shafts one of which can be taken as driver and the other as driven. By putting the belt on different steps different speeds of the driven shaft can be obtained while the driver shaft runs at constant speed.

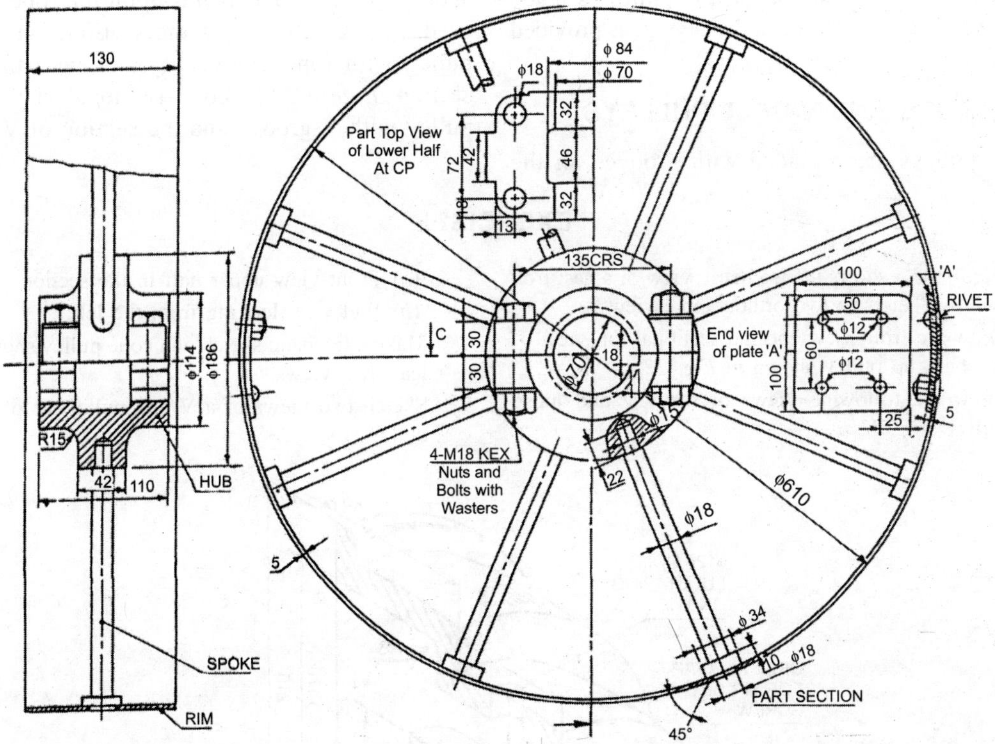

Fig. 22.4. Wrought iron split or built-up-pulley (Third Angle Projection)

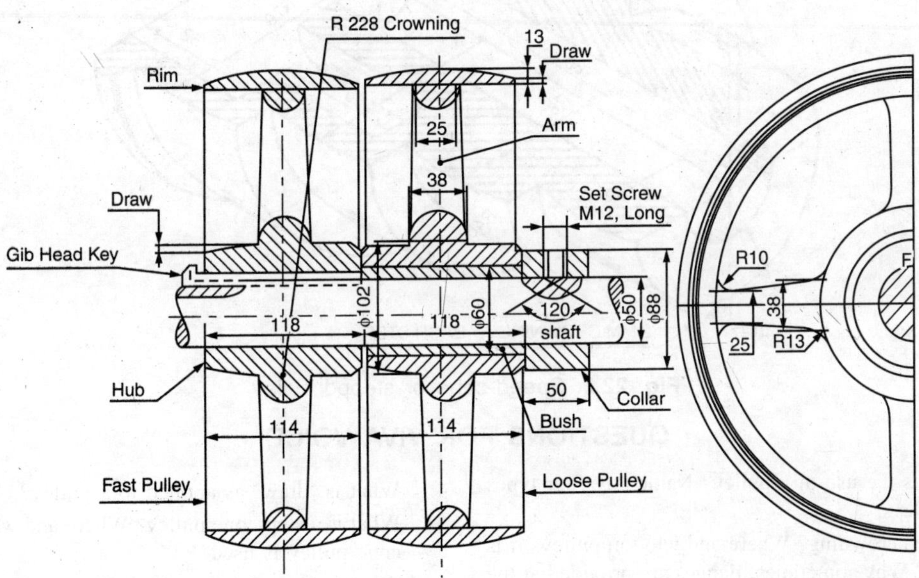

Fig. 22.5. Fast and loose pulleys

Figure 22.6 shows four-speed cast iron, step pulleys. To reduce the weight the pulley is made hollow, as shown in figure, and a keyway is provided in the hub (not shown Fig.)

22.7. ROPE AND V-BELT PULLEYS

These pulleys are provided with grooves on the rim to accommodate the rope or V-belt. The shape and size of groove depends upon the rope or belt used. *Figure 22.7* shows the sectional view of a pulley with rope groove together with the rope section. *Figure 22.8* shows sectional view of the pulley with V-groove and the section of V-belt.

EXERCISES

1. *Figure 22.3* shows the isometric view of a cast iron pulley. Sketch its two orthographic views.
2. Draw the front view and the sectional top view of the built up pulley shown in *Fig. 22.4*.
3. Draw the following views of the fast and loose pulley, *Fig. 22.5*.
   (*a*) Front view upper half in the section.
   (*b*) End view looking from the left.
4. Make a freehand sketch of a cone pulley showing at least two views.
5. Sketch two views of a V-belt pulley, or of a rope pulley.

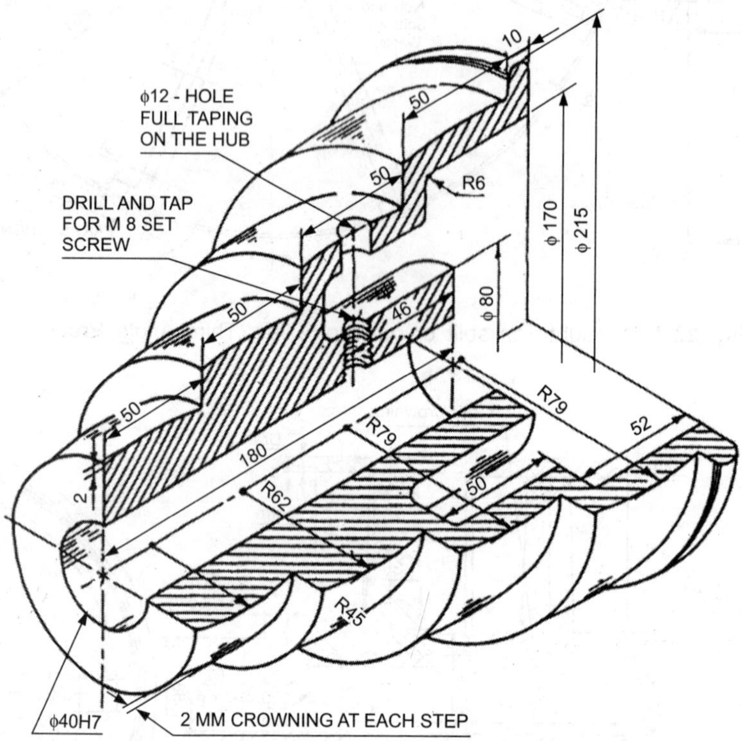

Fig. 22.6. Speed-cone or steppd pulley

QUESTIONS FOR VIVA VOICE

1. What is the use of a pulley? Name any four types of pulleys.
2. What is crowning? Where and why on pulleys it is given? Why sometimes, flanges are provided at the sides of the rim or a pulley in addition of crowning?
3. Why usually in pulleys curved arms are preferred to straight arms?
4. What is 'draw' as applied to a pulley?
5. What is a step-cone pulley? Where and why a step-cone pulley is used?
6. What is fast and loose pulleys? Where these are used.

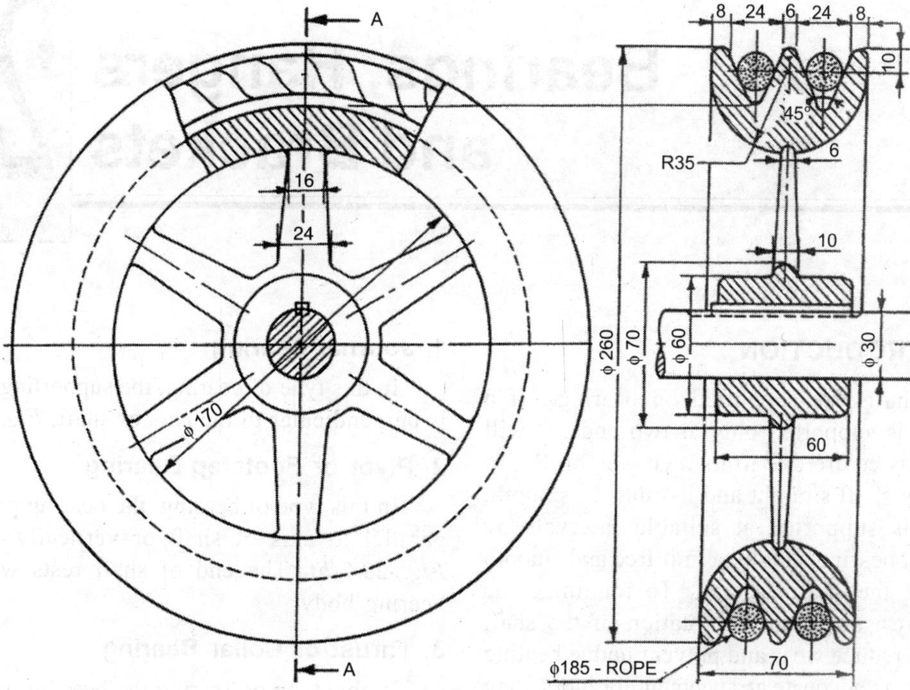

Fig. 22.7. Rope Pulley

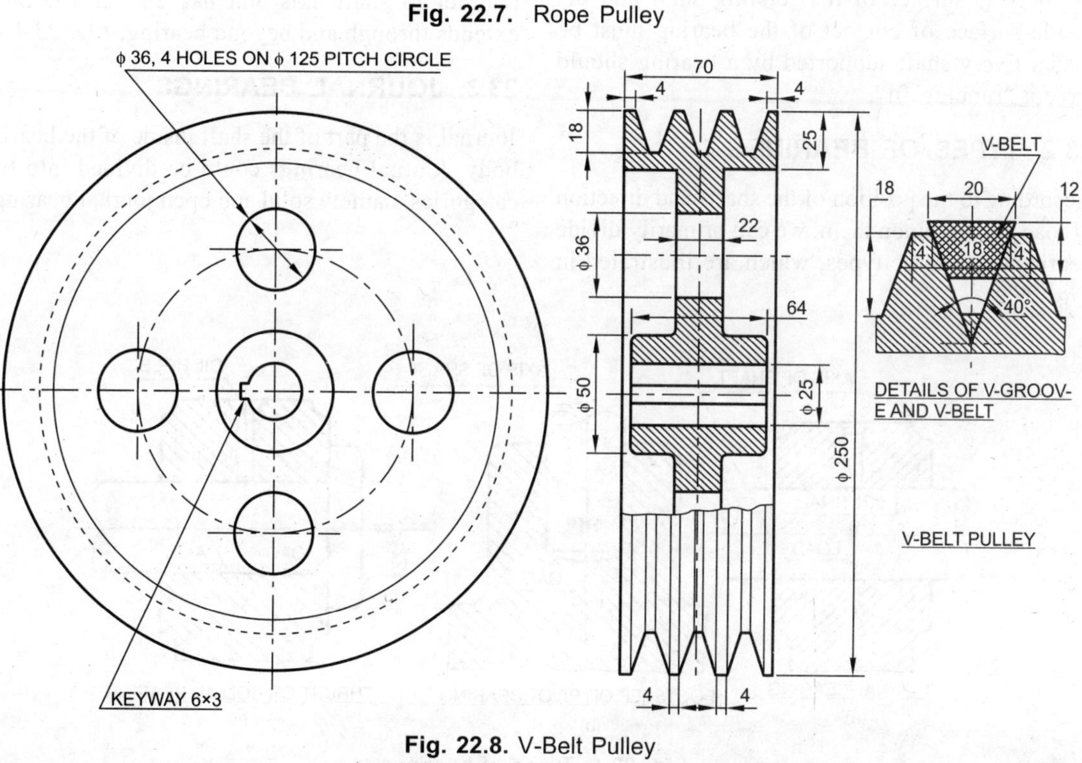

Fig. 22.8. V-Belt Pulley

Bearings, Hangers and Brackets

23

23.1. INTRODUCTION

Rotating shafts are supported on bearings. If a long shaft is supported only at two ends, it will deflect at its centre due to its own weight. So, to have a long shaft straight and its running smooth, the shaft is supported at suitable intervals by bearings. A bearing should permit free and smooth rotation of the shaft as well. To minimize the frictional resistance due to rotation of the shaft and also to reduce wear and prevent undue heating and seizing, an adequate arrangement for lubricating the outside surface of the rotating shaft and the inside surface of contact of the bearing must be made. Every shaft supported by a bearing should have a "running fit".

23.2. TYPES OF BEARINGS

According to the position of the shafts and direction of load coming upon them, we can primarily divide bearings into three types, which are illustrated in *Fig. 23.1*.

1. Journal Bearing

In this type of bearing, the supporting pressure is perpendicular to the axis of shaft, *Fig. 23.1(a)*.

2. Pivot or Footstep Bearing

In this type of bearing, the bearing pressure is parallel to axis of shaft or vertically upwards, *Fig. 23.1(b)*. The end of shaft rests within the bearing body.

3. Thrust or Collar Bearing

In this type of bearing the bearing pressure is parallel to shaft axis and has end thrust. Shaft extends through and beyond bearing, *Fig. 23.1(c)*.

23.3. JOURNAL BEARINGS

Journal is the part of the shaft inside of the bearing body. Journal bearings could be divided into two categories, namely solid and open journal bearings.

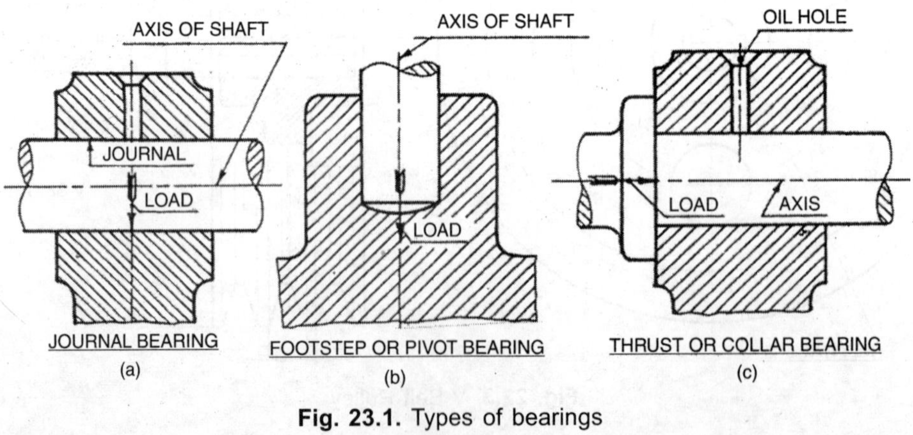

Fig. 23.1. Types of bearings

23.4. SOLID JOURNAL BEARING

This is the simplest type of a journal bearing. A cylindrical block is extended at the base to permit bolting down. This portion is called sole. A hole, equal to diameter of shaft, is drilled through the cylindrical block making the shaft a running fit. The shaft is inserted in the hole from its one end. An oil hole, leading up to inner hole, is provided at the top centre of its body for lubrication, *Fig. 23.2.*

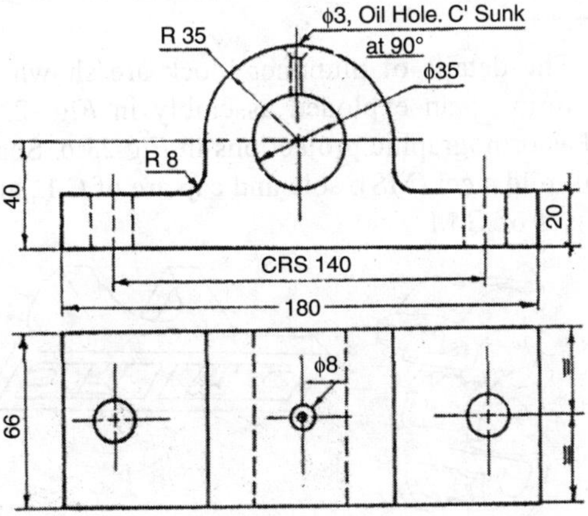

Fig. 23.2. Solid Journal Bearing

This bearing is made of cast iron and is used for shafts, which rotate slowly or at infrequent intervals, carrying light load only. There is no provision for adjustment of wear.

23.5. BUSHED BEARING

The bushed solid bearing, *Fig. 23.3 (a), (b)*, is developed by introducing a bush or brasses made or brass, gunmetal, phosphor bronze or managanese bronze. This bush is replaceable when worn. The outside of the bush is a driving fit or it is prevented from slipping inside the block by using one of the methods described under Article 23.8 in *Fig 23.14.* The shaft is running fit in the brasses, and is passed into the bearing axially. So they are generally used at the end of a shaft.

Exercise 23.1. *Draw the following views of he bushed bearing shown in Fig. 23.3 (a) for 40 mm diameter shaft:*

(a) Front-view, right half in section

(b) Right side view, left half in section

(c) Top view, outside.

Dimension the views completely. What should be the type of fit for bush and body? Determine the limits for the suggested fit, using ISO table given in chapter on 'Tolerances, Allowances and Systems of Limits and Fits.'

23.6. OPEN BEARING

The bearing is open at the top and is used for very long zig-zag shafts. To prevent the shaft from falling out, at high speed, the sides of bearing are kept sufficiently high. The shape of the bush is such that it cannot rotate with the shaft. An open bearing

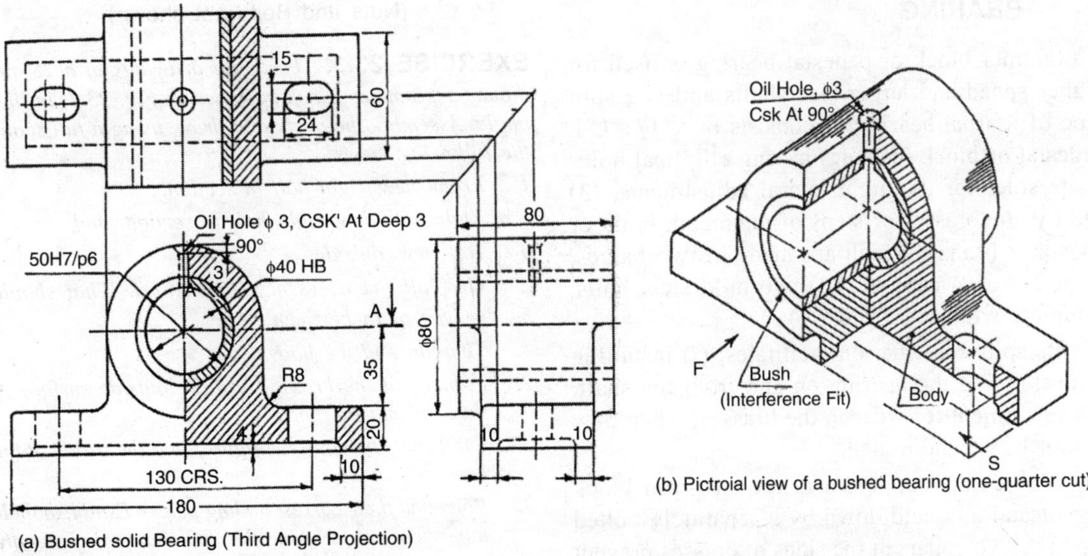

(a) Bushed solid Bearing (Third Angle Projection)

(b) Pictroial view of a bushed bearing (one-quarter cut)

Fig. 23.3.

is shown in *Fig. 23.4*. The shaft rotates in a flanged bush.

EXERCISE 23.2. *Figure 23.4 shows an open bearing in exploded pictorial form. Assemble the parts and draw its; (a) Front view; left half in section, (b) Top view, and (c) Right side view. Suggest the type of fit for bush and body.*

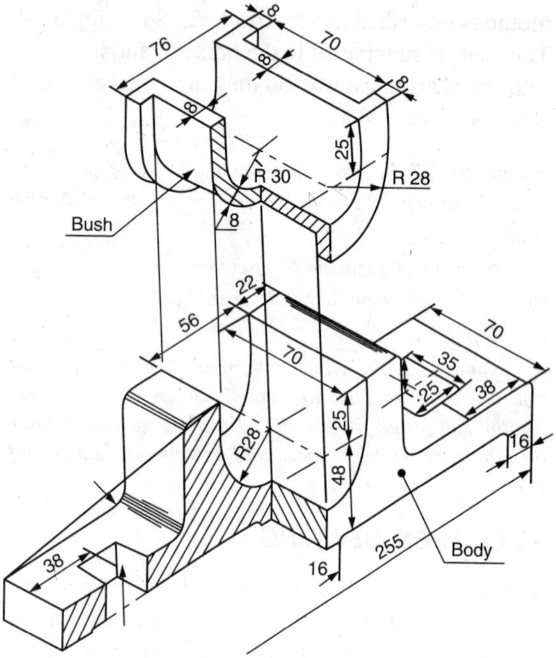

Fig. 23.4. Open bearing

23.7. PLUMMER BLOCK OR PEDESTAL BEARING

A plummer block or pedestal bearing is used for higher speed and larger size shafts and is a split type of journal bearing. It consists of : (1) a C.I. pedestal or block (casting) having elliptical holes in its sole, for its slight lateral adjustments, (3) and (4) the brasses or steps of gunmetal, brass or phosphor bronze, which are made in two halves, (2) a cast iron cap, and (5) two mild steel bolts, complete with nuts (6 and 7).

Its split construction facilitates; (*i*) installing and removing the bearing on and from the shaft, (*ii*) adjustment for wear in the brasses, when they cannot be reconditioned.

Brasses are supported in the cast-iron block or pedestal and held down by a separately bolted cap. The two collars at the sides of brasses prevent the axial movement of the brasses. A snug at the bottom centre of lower brass, fitting in a corresponding hole in the block prevents the rotation of brasses along with the shaft.

The cap while resting on the upper brass fits inside the pedestal but is not supported by it. The square heads of bolts fit in corresponding square recesses in the bolt holes at the bottom of the pedestal, to stop rotation of bolts in their holes while tightening the nuts. Cap is spot-faced around the bolt holes to provide a smooth seat for the nuts.

The details of plummer block are shown in the form of an exploded assembly in *Fig. 23.5* and as orthographic projections in *Fig 23.6*. Shaft is of mild steel (MS); sole and cap are of C.I., and brasses of G.M.

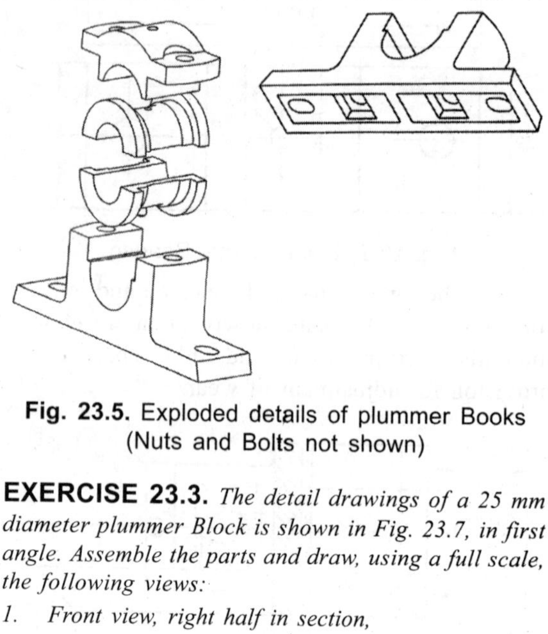

Fig. 23.5. Exploded details of plummer Books (Nuts and Bolts not shown)

EXERCISE 23.3. *The detail drawings of a 25 mm diameter plummer Block is shown in Fig. 23.7, in first angle. Assemble the parts and draw, using a full scale, the following views:*

1. *Front view, right half in section,*
2. *Right end view, right half in section, and*
3. *Top view, outside.*

Give all the necessary dimensions. What should be the type of fit between;

1. *The cap and the body,*
2. *Flanges of the brasses and the outside surface of the cap and the body,*
3. *Shaft (not shown in the figure) and the inside of the brasses,*
4. *Bolts and their corresponding holes? Establish their numerical values using the ISO chart. Also prepare a parts list and billo materials.*

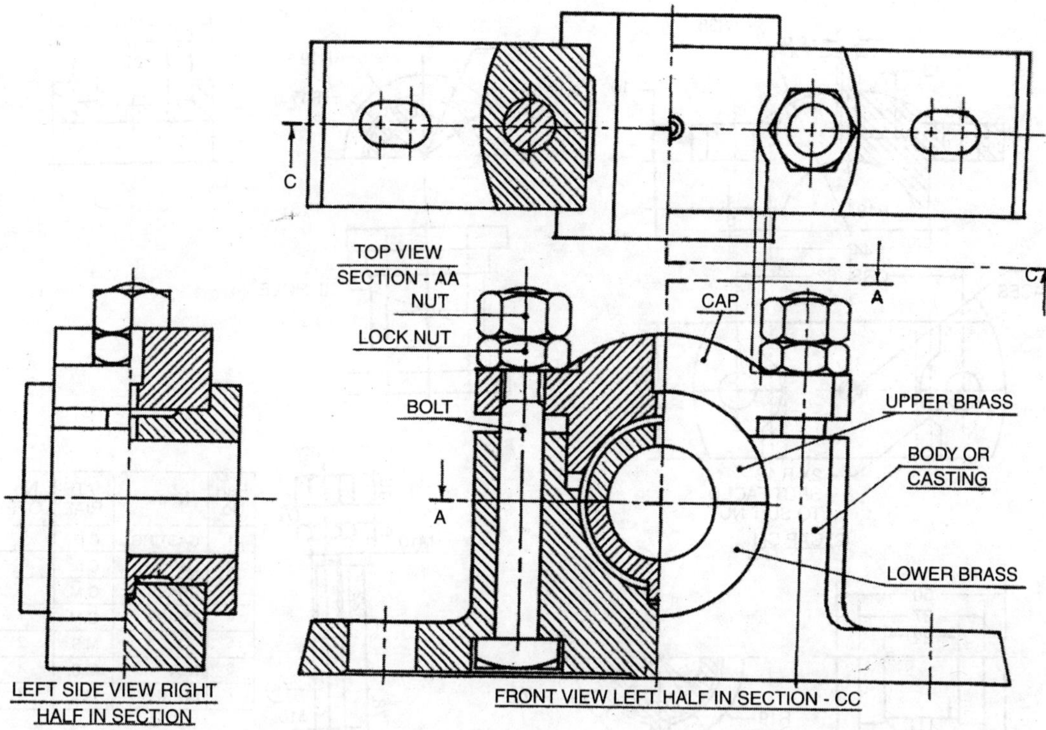

Fig. 23.6. Orthographic Projections of the Plummer Block (Third Angle Projection)

Drawing Hints

Study the details carefully and visualize the position and relation of each detail in the assembly. The assembly is symmetrical about the vertical centre line. After spacing the sheet and drawing centre lines for views, start at the centre of the brasses in the front view and then project these to other views. For holes and nuts start in the top view.

Cap is drawn simultaneously in top and front views. To generalize, we can say that, start for each part from the view, which gives its arcs or circles first. Note that a clearance of 3 mm, in this case will exist between the cap and the main body or block.

EXERCISE 23.4. *Details of a pedestal bearing are shown in Fig. 23.8. Assemble the parts and draw the following views:*

1. *Front view outside*
2. *Left side view–full central section,*
3. *Bottom view–lower half on section, the section plane passing through the centre of the brasses. Suggest the suitable types of fit for mating parts.*

EXERCISE 23.5. *Details of a plummer block are given in Fig. 23.9. Assemble the parts and draw its:*

1. *Front view–left half in section,*
2. *End view (right)–left half in section, and*
3. *Top view–lower half in section.*

EXERCISE 23.6. *Details of a Marine Engine shaft bearing are given in Fig. 23.10 (a). Assemble its parts and draw:*

1. *Front view, right half in section,*
2. *A side view, left half in section, and*
3. *Top view.*

Give all dimensions according to IS: 696-1972 or SP: 46-1988. Also prepare a parts-list and bill of materials.

HINT: For solution refers to its general arrangement given in *Fig. 23.10 (b).*

EXERCISE 23.7. *Details of a Pedestal Bearing are given inFig. 23.11 (b). Assemble the details and draw to a suitable scale A2 drawing sheet:*

1. *Front view–right half in section,*
2. *Right end view–left half in section,*
3. *Top view outside.*

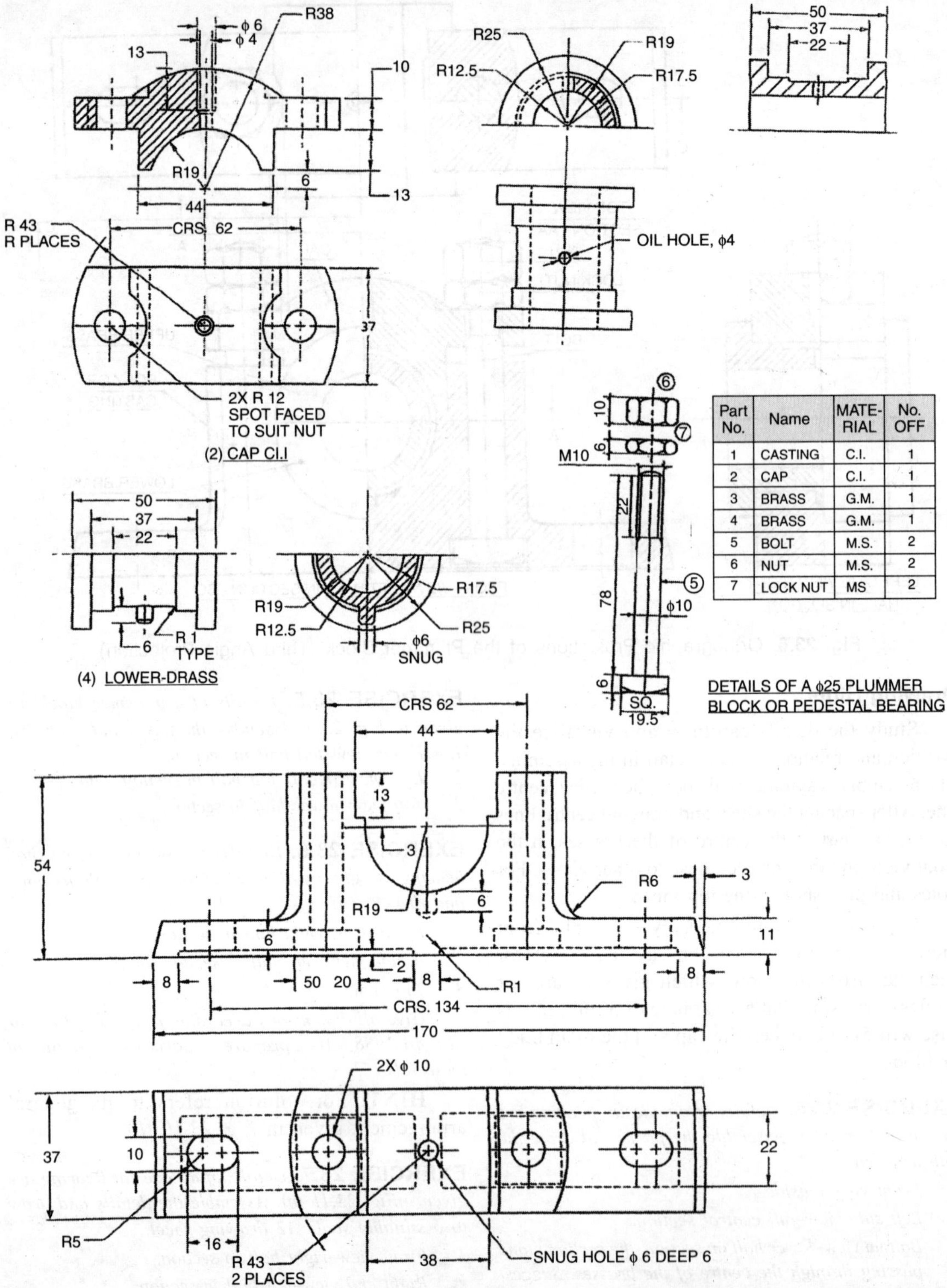

Part No.	Name	MATE-RIAL	No. OFF
1	CASTING	C.I.	1
2	CAP	C.I.	1
3	BRASS	G.M.	1
4	BRASS	G.M.	1
5	BOLT	M.S.	2
6	NUT	M.S.	2
7	LOCK NUT	MS	2

Fig. 23.7. Details of a ϕ 25 mm Plummer Block.

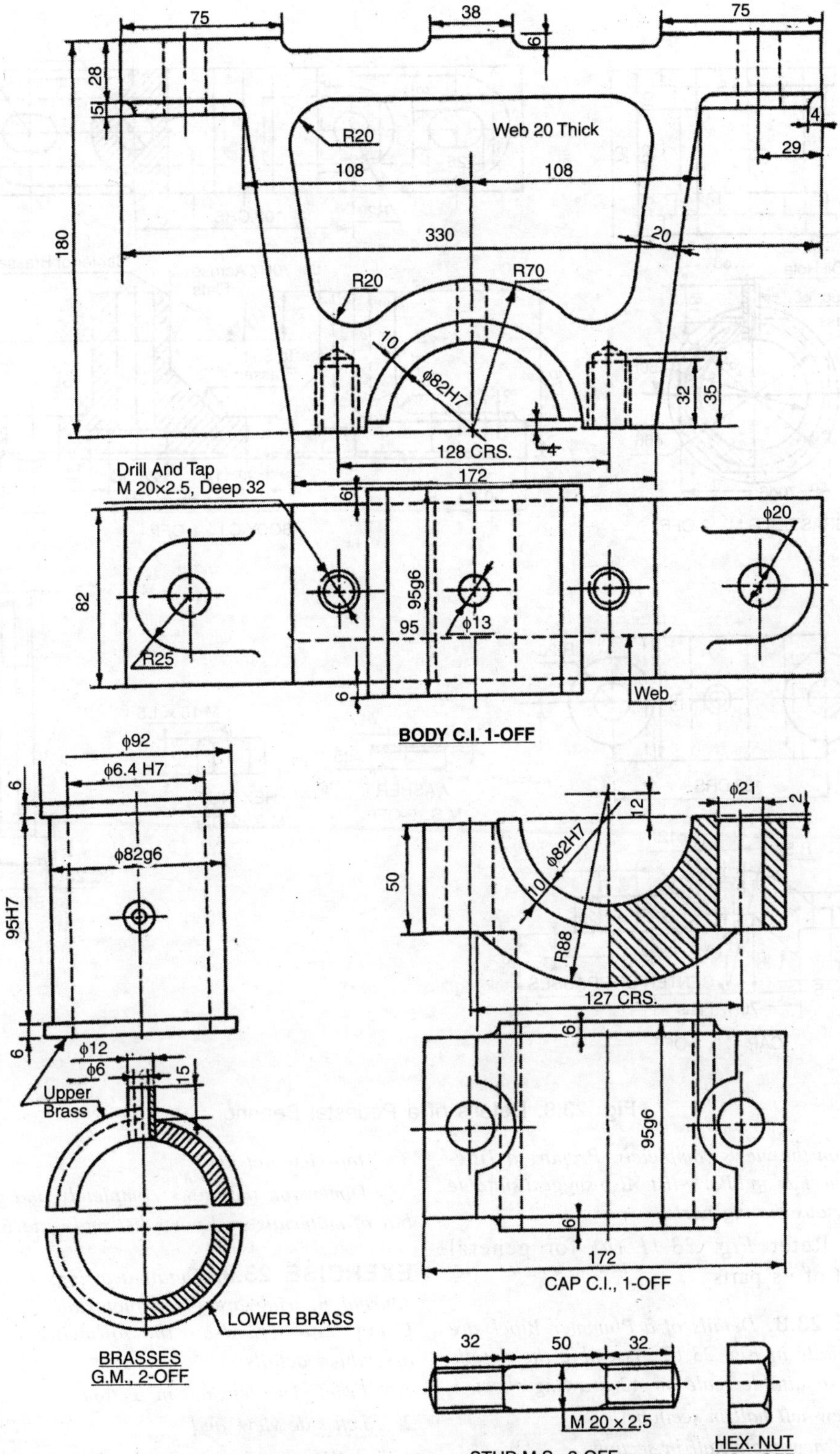

Fig. 23.8. Pedestal Bearing

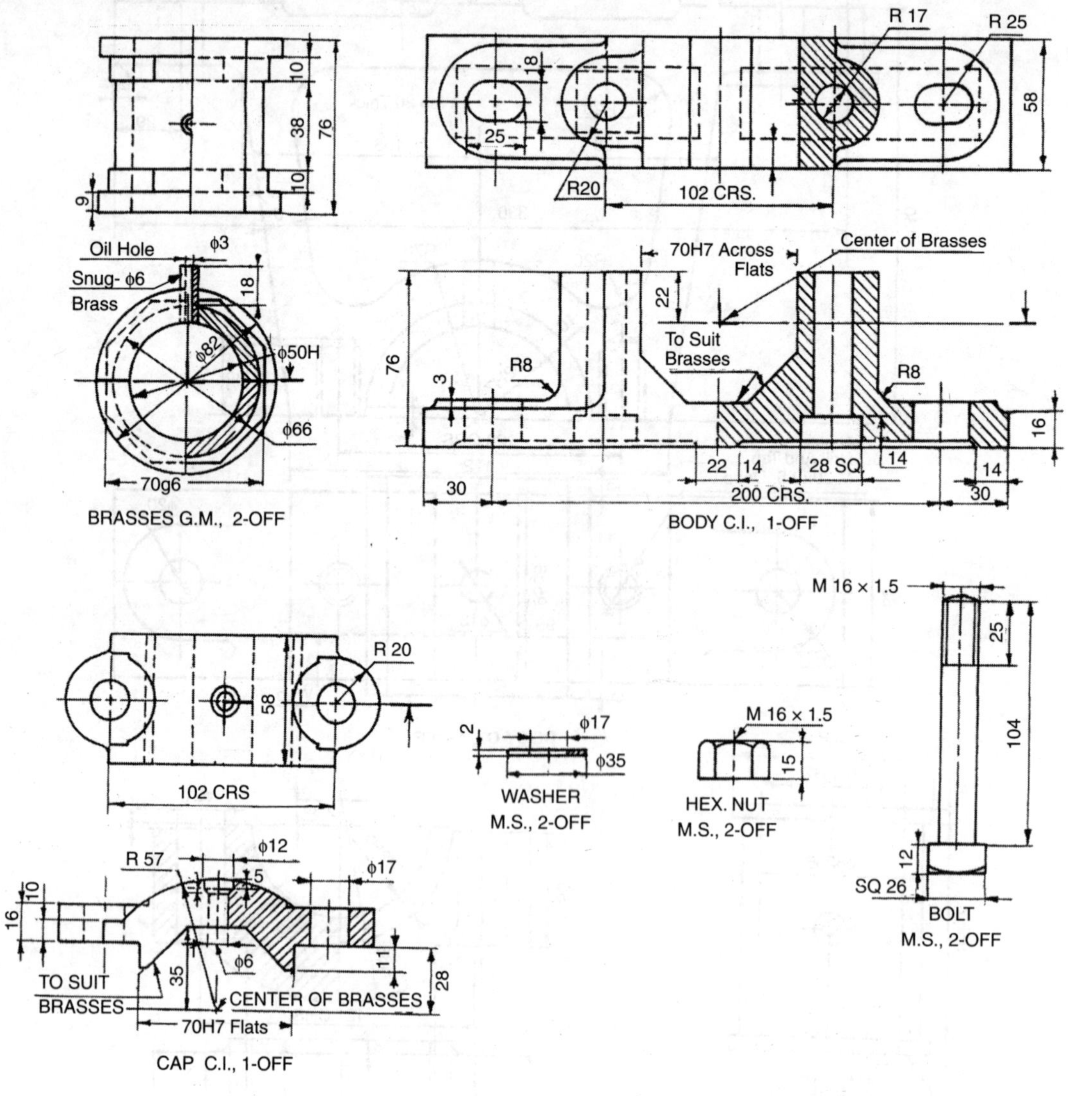

Fig. 23.9. Details of a Pedestal Bearing

Dimension the views completely. Prepare a Title-Block and Materials and Parts-list. Also suggest suitable limits for various mating parts.

HINT: Refer *Fig. 23.11 (a)* for general arrangement of its parts.

EXERCISE 23.8. *Details of a Plummer Block are shown pictorially in Fig. 23.12. Assemble the details and draw to a suitable scale on A2 drawing sheet:*

1. *Front view–left half in section,*
2. *Left side view–right half in section,*
3. *Top view outside.*

Dimension the views completely and prepare the bill of materials and parts-list and Title Block.

EXERCISE 23.9. *Detail drawings of an inverted Journal Bearing are given pictorially in Fig. 23.13. Using scale 1:2 draw the following views of the assembled details:*

1. *Front view, left half in section,*
2. *Left side view, and*
3. *Bottom view.*

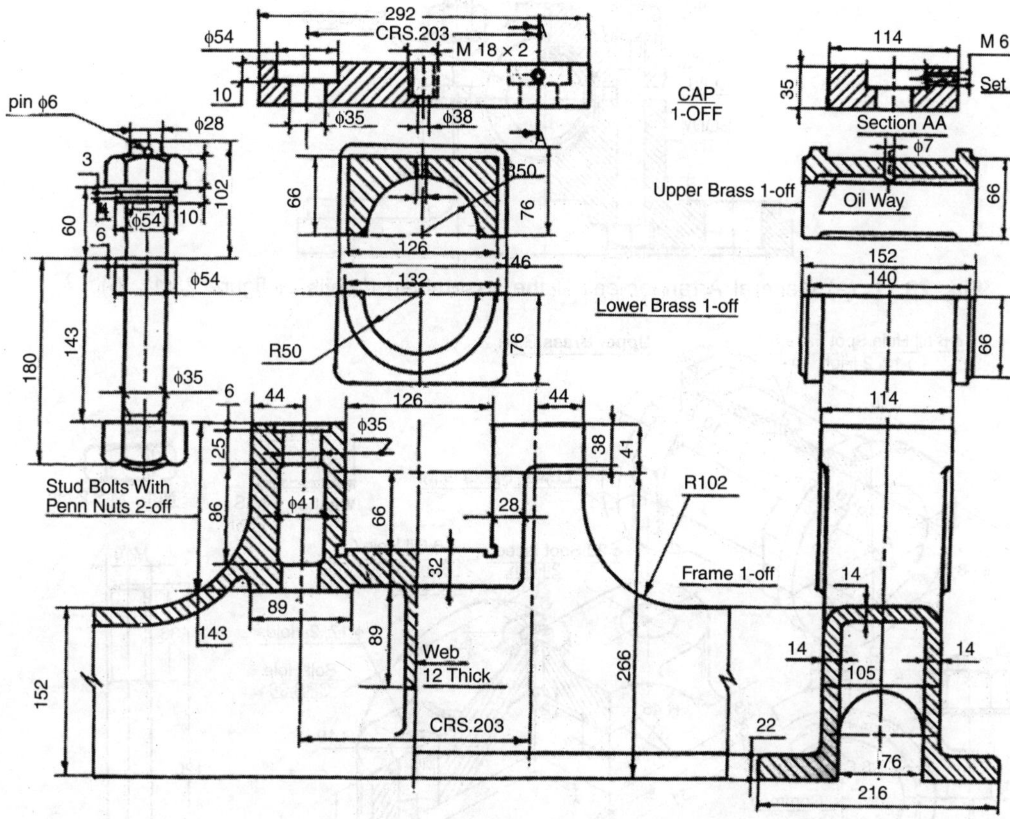

Fig. 23.10. (a) Details of Marine Engine shaft bearing

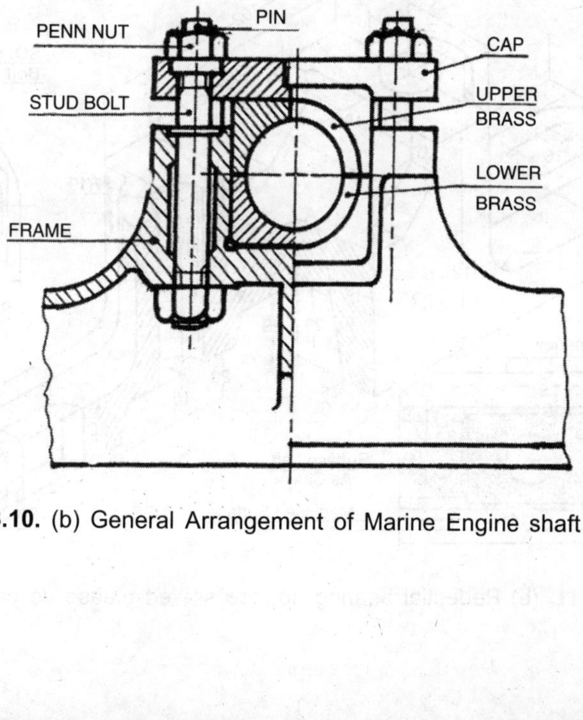

Fig. 23.10. (b) General Arrangement of Marine Engine shaft bearing

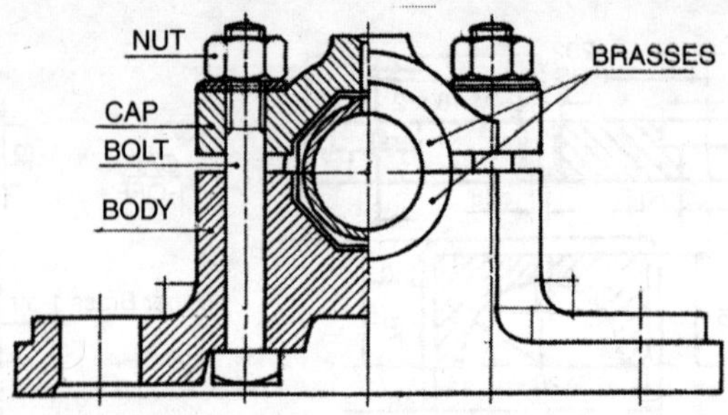

Fig. 23.11. (a) General Arrangement of the assembled details of figure 23.11 (b)

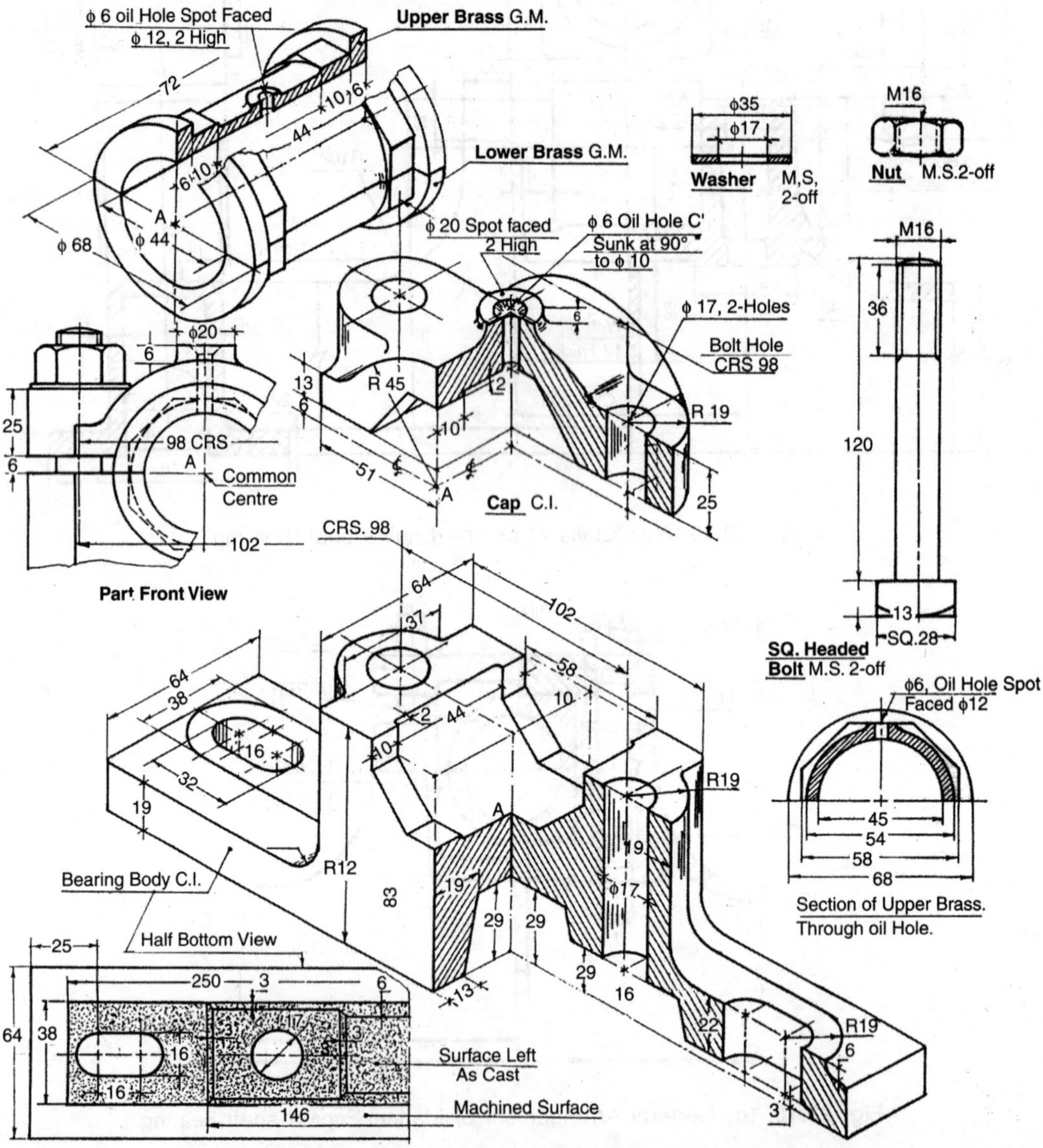

Fig. 23.11. (b) Pedestial bearing no. are shifted please do carefully

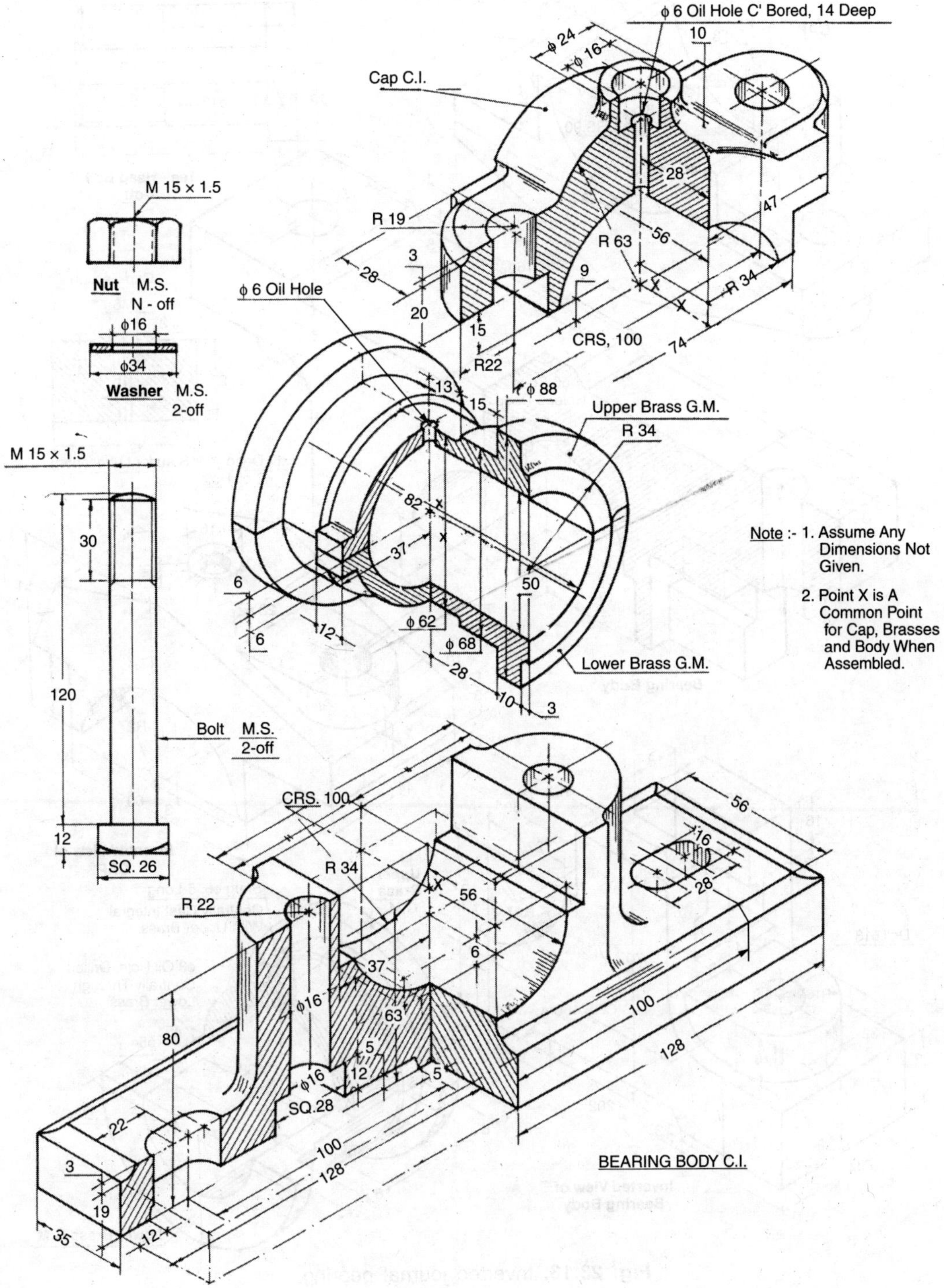

Fig. 23.12. Plummer block

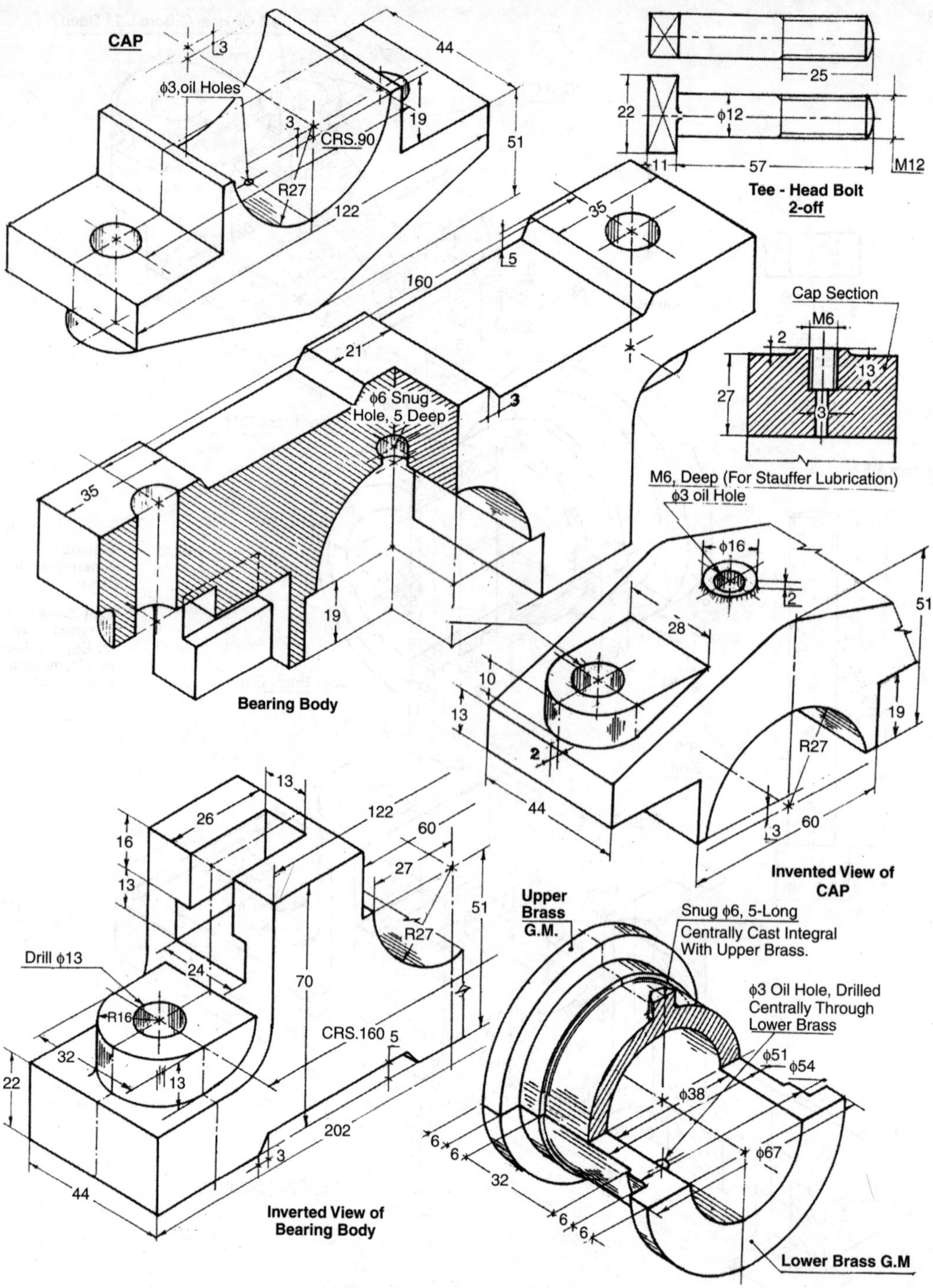

Fig. 23.13. Inverted journal bearing

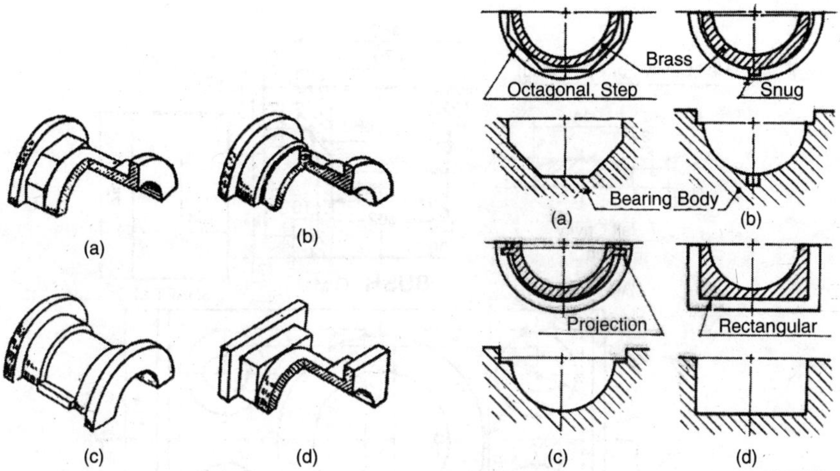

Fig. 23.14. Prevention of rotation of Brasses along with their shafts

Dimension the views completely and prepare the Bill of Materials, Parts-list and Title-Block.

23.8. PREVENTION OF ROTATION OF BRASSES

Many methods are provided to check the rotation of the brasses along with the shaft. Some most common in them, are shown in *Fig. 23.14*.

(*a*) The outsides of the brasses are made octagonal casting, near the collars, which sit in similar shaped steps provided in block.

(*b*) A snug, a pipe, or a register is cast integral with lower brass, which fits in a corresponding hole (cavity) in casting.

(*c*) On the side of brasses, necks are provided and corresponding recesses are provided in the main casting, to fit them.

(*d*) The lower bush is squared except in the middle and fit in the corresponding square shaped casting. The recess, in middle, helps fitting.

23.9. PIVOT OR FOOTSTEP BEARING

This is used for vertical shafts. A gunmetal bush (having collar on top) is placed inside a C.I. Sole. Inside the bush, at the bottom, rests a gunmetal disc or pad. On this disc rests the shaft. The disc is prevented from rotation with the help of a snug fitting in the sole. The block and the bush are recessed to form fitting strips. The sole is also recessed at the bottom to reduce machining. The bush and the pad can be replaced when worn out.

EXERCISE 23.10. *Referring to Fig. 23.15, from the details of a Foot-step Bearing draw the following assembled views:*

1. *Front view–right half in section,*
2. *Top view outside. Give all necessary dimensions and suggest the fits and limits for various mating parts. Prepare a parts list and bill of materials.*

DRAWING HINTS

Refer *Fig. 23.16* for solution, in First Angle. Draw the holes and other circles in top view first. Draw the front view around the vertical centreline.

23.10. SPLIT TYPE FOOT-STEP BEARING

Where, due to some manipulating difficulties, it is not possible or convenient to shift the shaft to replace the worn out pad or brasses, a split type of footstep bearing is used with advantage. It can be installed without disturbing the shaft. *Figure 23.17* shows, in a pictorial form, details of one such bearing.

EXERCISE 23.12. *The details of a split type footstep bearing are shown in Fig. 23.17, in pictorial form. Draw the following views of the assembly:*

1. *Front view, looking in the direction of arrow E.*
2. *Side view looking in the direction of arrow S, and*
3. *Top view, outside.*

Assume suitable size of bolts and nuts for connecting the cap to the body of the bearing. Comment on the type of fit, which you suggest, for various mating parts. Prepare a parts-list and bill of materials.

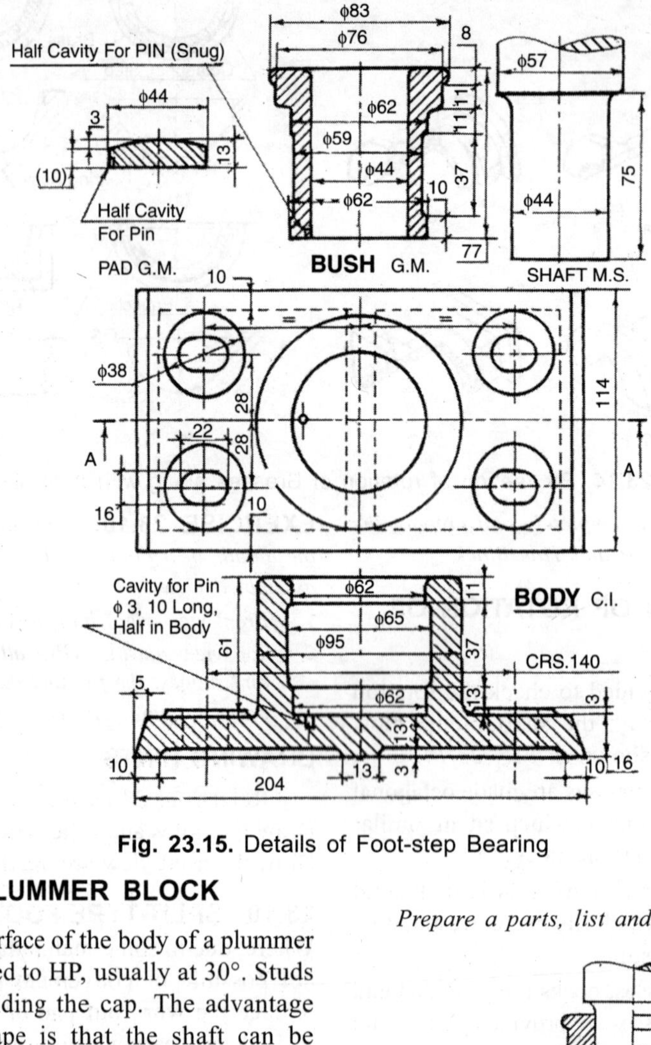

Fig. 23.15. Details of Foot-step Bearing

23.11. ANGLE PLUMMER BLOCK

Sometimes the top surface of the body of a plummer block is made inclined to HP, usually at 30°. Studs replace bolts for holding the cap. The advantage of this, special, shape is that the shaft can be removed, and power transmitted at an angle. One such bearing is shown in *Fig. 23.18 (a), (b)*. The upper brass has an integral snug, which is drilled to serve as oil hole, too. This snug fits in a corresponding cavity, provided in the top of the cap, to stop rotation of brasses with the shafts. It requires sample lubrication and waste oil sumps are cast integral with the body of the bearing.

EXERCISE 23.12. *Details of an angle Plummer Block are shown in Fig. 23.18 (a), in first angle projection. Draw the following dimensioned views of its assembly:*

1. *Front view–right half in section*
2. *Right side view–right half in section; and*
3. *Top view–outside. Use full size scale.*

Prepare a parts, list and bill or materials.

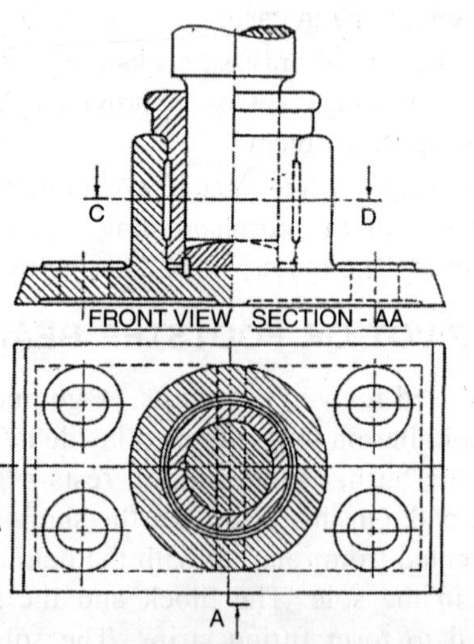

Fig. 23.16. General Arrangement of the Assembled Details

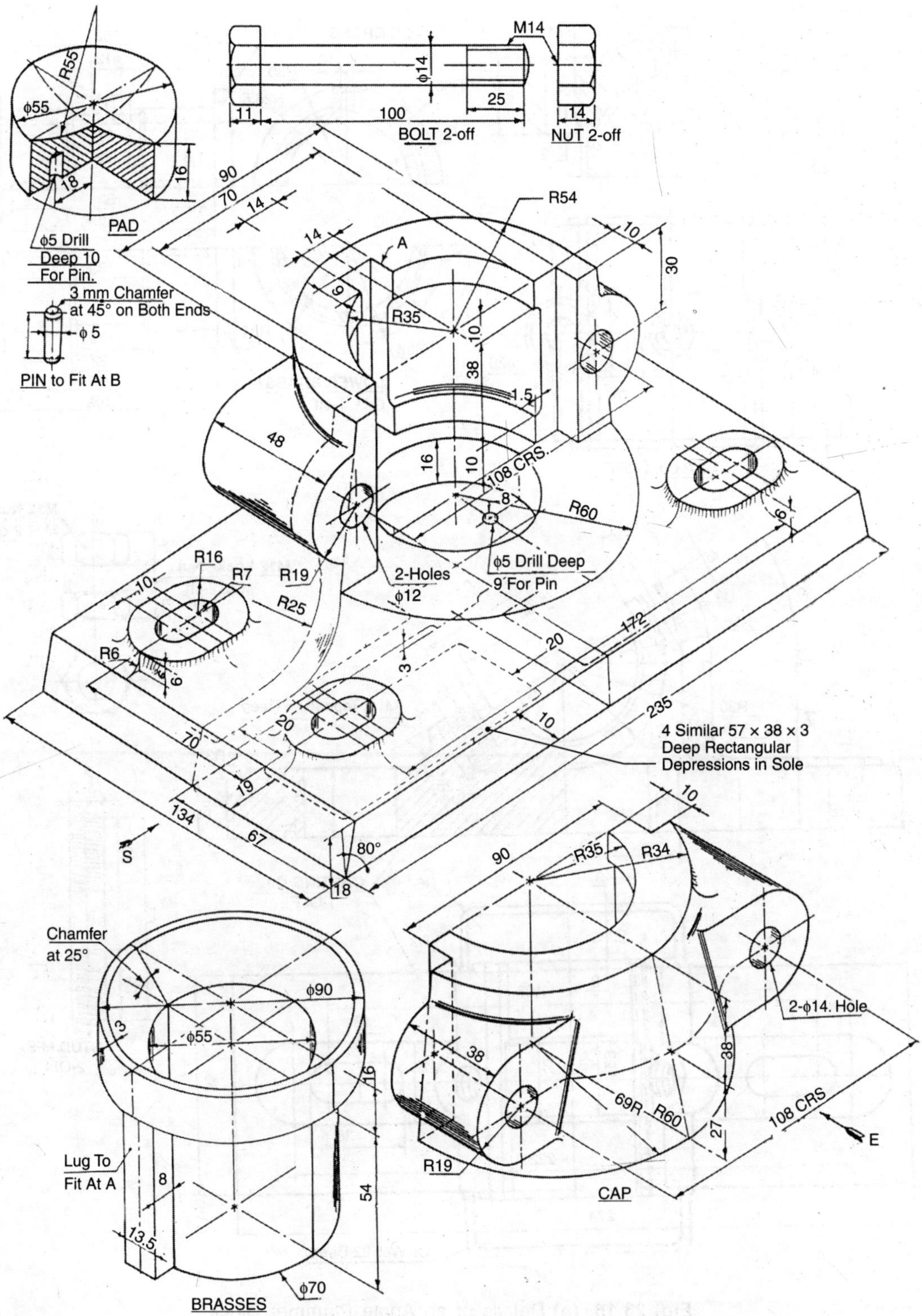

Fig. 23.17. Split Type Foot-step Bearing

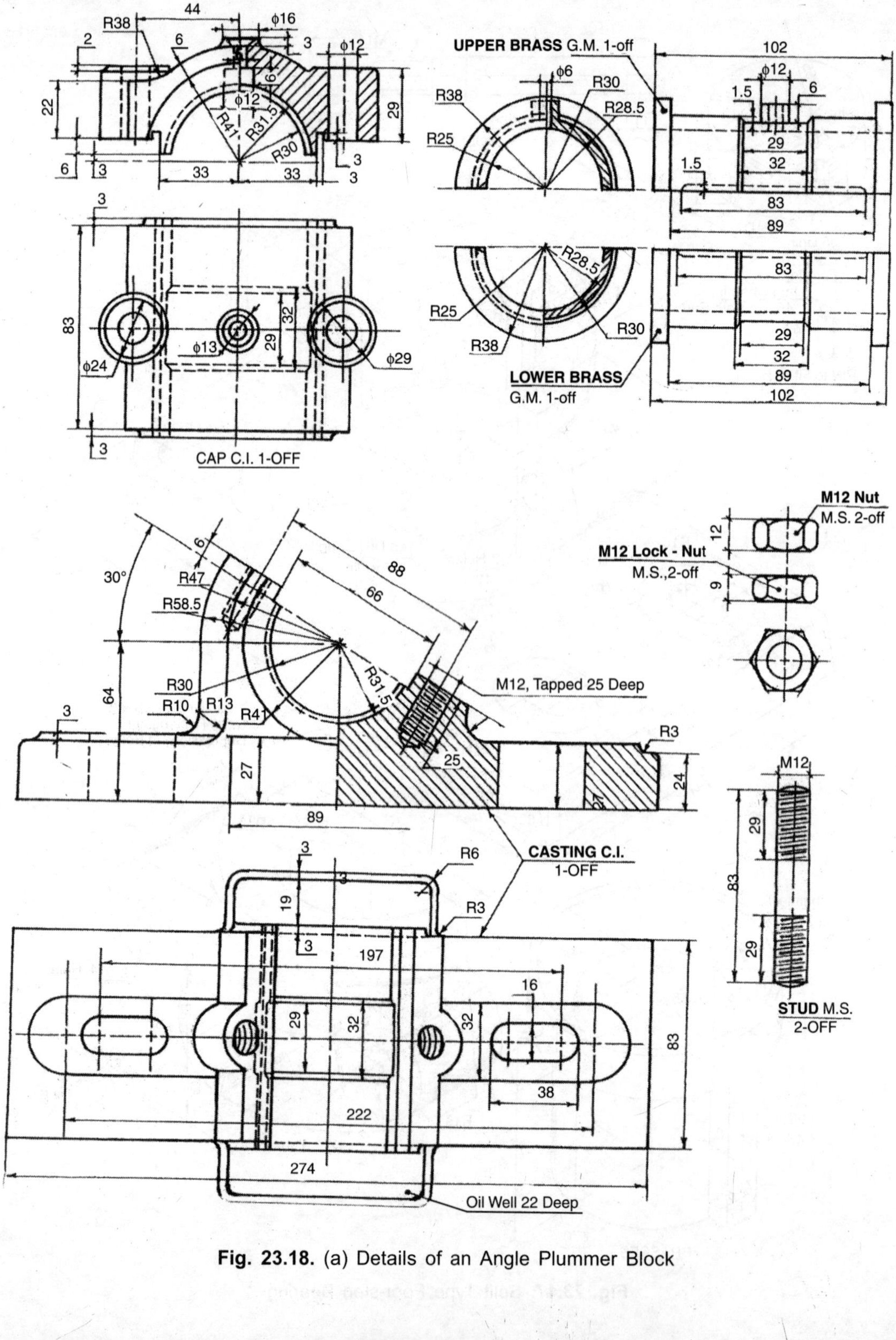

Fig. 23.18. (a) Details of an Angle Plummer Block

23.12. SWIVEL BEARING

Figure 23.19 (a, b). It is a self-adjusting bearing and adjusts itself to the position of the shaft, to be supported, readily. In the upper forked portion two setscrews hold the bearing. The bearing will turn about the conical points of the setscrews to adjust itself according to the position of the shaft. The fork as a whole can be moved up or down by rotating a threaded spindle. The spindle operating in a screwed hole provided in the upper portion of the body. A collar operates on the spindle to the required positon to support the fork. A setscrew, screwed through the threaded hole in the collar, is used to lock the collar in position. Another setscrew, screwed through the threaded hole in the casting, lock the spindle at the desired position. To safeguard the threads on the spindle, pads or discs of some soft material are placed ahead of the screw points.

EXERCISE 23.13. *Details of Swivel Bearing are shown in Fig. 23.19 (a), in First Angle. Assemble the parts and draw, to a scale double the full size, the following views:*

1. *Front view–right half in section*
2. *End view–left half in section, and*
3. *Top view–outside.*

Also prepare a parts-list and bill of materials.

23.13. SELF-ALIGNING BEARING

Such a bearing supports a horizontal shaft, which gets its alignment itself from the bearing. *Figure 23.20 (a), (b)* shows two assembled views of one such bearing. Its detail working drawings are shown in *Fig. 23.20 (a)*. In its body there is a double row ball bearing in which the shaft rotates. The tapering adopter sleeve is tightened upon the shaft by means of slotted nut, which is locked by a tab-locking washer. Two felt pads, one placed on each side of the ball bearing, in main body, allow any fluctuation to the shaft, as well help to retain the lubricant.

EXERCISE 23.14. *Figure 23.20 (a) shows details of a self-aligning Bearing. Assemble the parts and draw to a scale 1 : 2, the following views:*

1. *Front view–outside.*
2. *Left side view–full in section.*
3. *Top view–outside.*

Dimension the views completely, suggest the types fit for various mating parts and fix up the limits using ISO chart.

DRAWING HINT:

Refer *Fig. 23.20 (b)* for solution.

23.14. HANGERS

Sometimes it is required to support a shaft from the ceiling of a room or from a beam. In such cases, shafts are carried by bearings fitted on hangers. Hangers take various shapes but common forms are 'U' shaped, and 'J' shaped, *Fig. 23.21*. In case of a 'U' hanger the bearing is supported between two arms. The soleplate is cast integral with the body. Usually 'T' section is used for J-hanger. The cap is fitted to the sole by two studs and nuts. The brasses are prevented from rotating by snug at the top through which an oil hole is also drilled.

EXERCISE 23.15. *Referring to detail drawings of a J-Hanger given in Fig. 23.21, assemble the parts and draw:*

1. *Front view–left half in section.*
2. *Right end view–showing bearing full in section, and*
3. *Top view.*

HINT: Here it may be noted that sole of the bearing is integral part of the arm.

Lower part of hanger serves as bearing block. The bearing is completed by brasses, a cap and suitable threaded fasteners. There is only one arm supporting the bearing from the ceiling. Hence one side of the bearing is open. The line shaft can be placed into the bearing form this open side and it need not be inserted axially only.

EXERCISE 23.16. *Referring to Fig. 23.22, draw from given drawing of a J-hanger bearing:*

1. *Complete front view–without section,*
2. *End view–section at CC, looking in the direction of arrow, and*
3. *Sectional Top view on AA.*

23.15. WALL BRACKET

A wall bracket holds a bearing, which supports an overhung shaft, which is parallel to and near a wall. The bearing is bolted to the wall bracket, which is bolted to a pillar or a wall. Brackets also

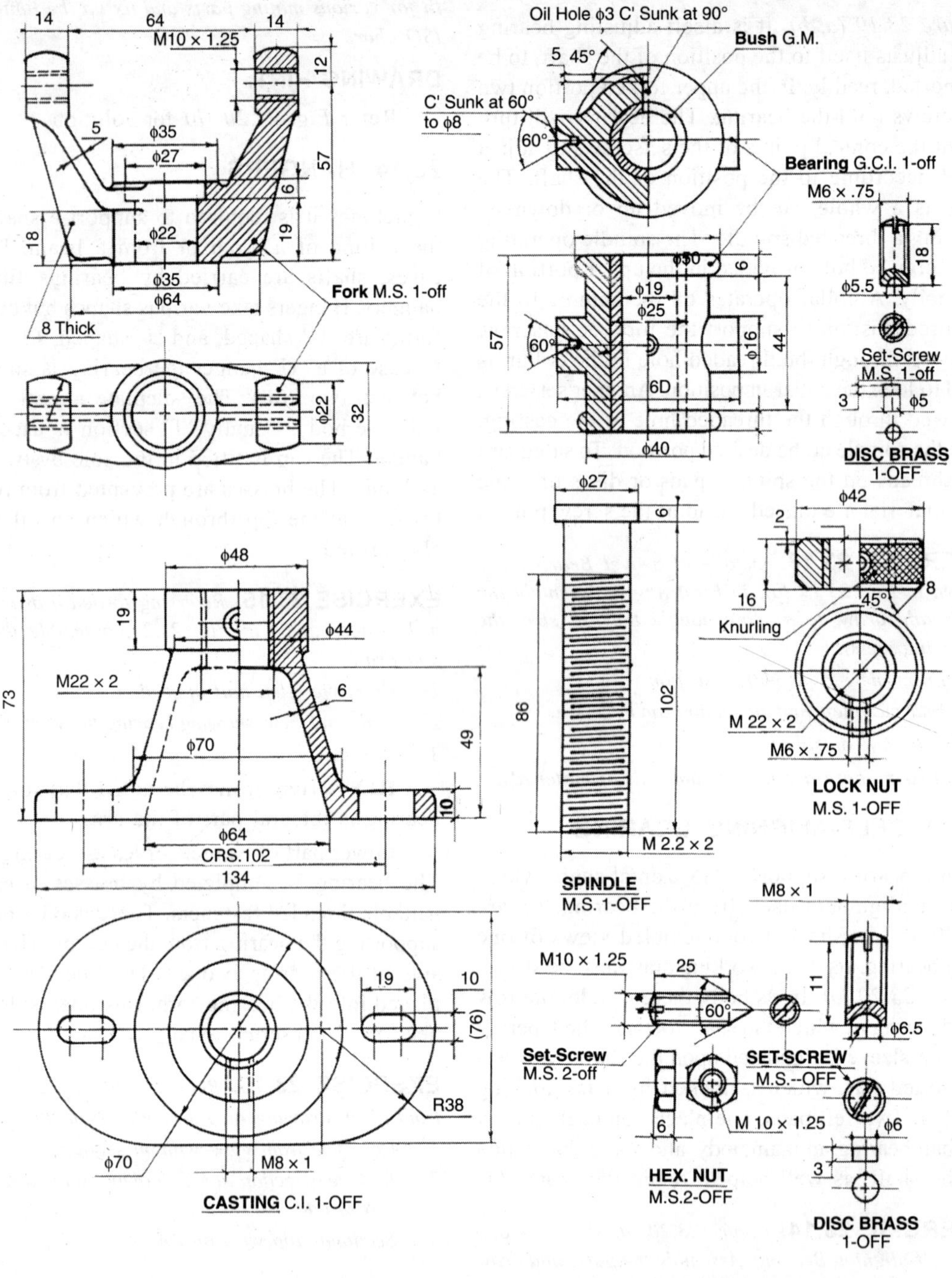

Fig. 23.19. (a) Details of a Swivel Bearing

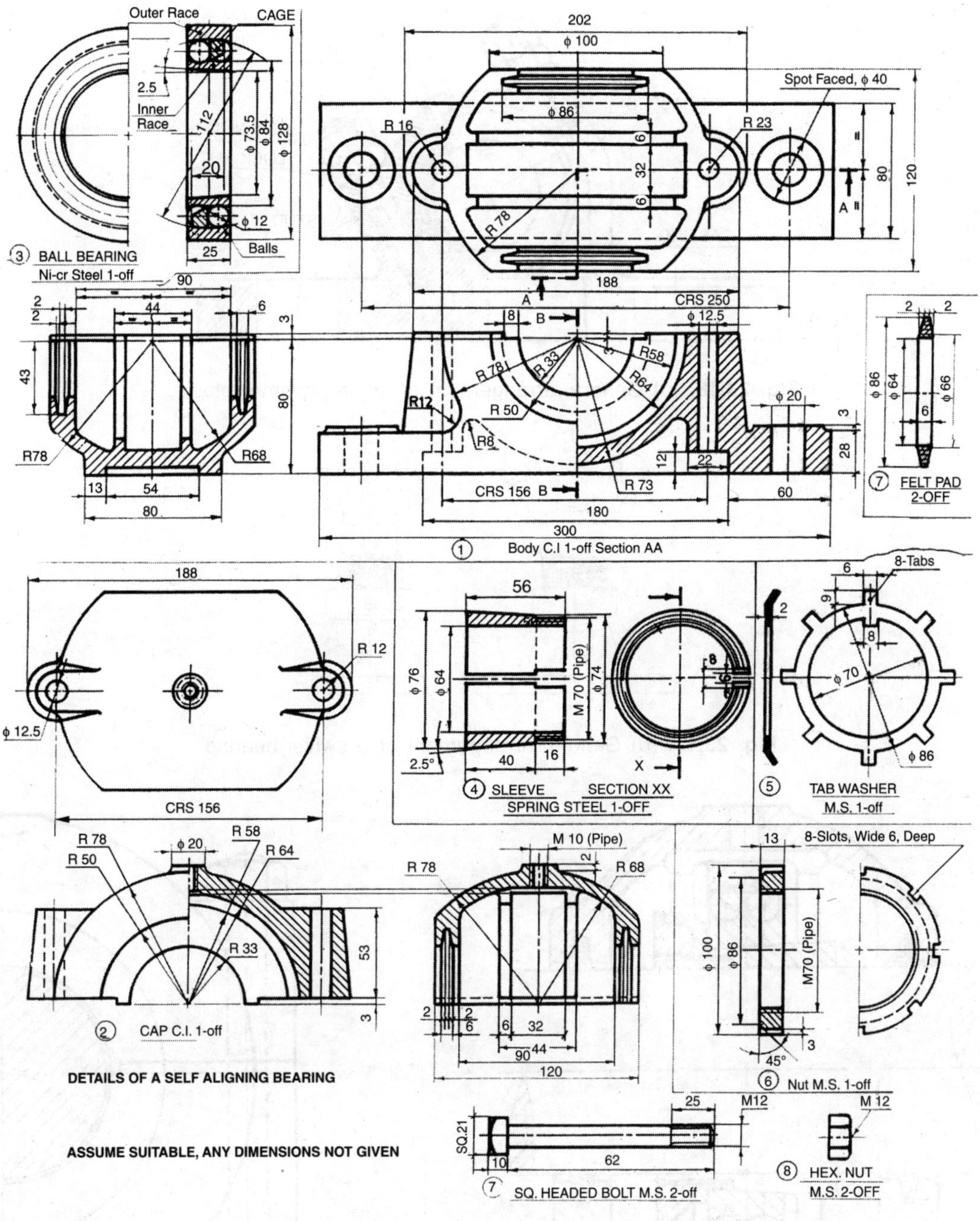

Fig. 23.20. (a)

find their use in holding insulators on electric transmission poles.

There are numerous types of wall brackets but common features of most of the brackets are, a horizontal plate (with holes for holding bearing) cast with a vertical plate and web to support the horizontal plate. Its plate is bolted to the wall or a pillar. A few types of brackets are illustrated in the following figures.

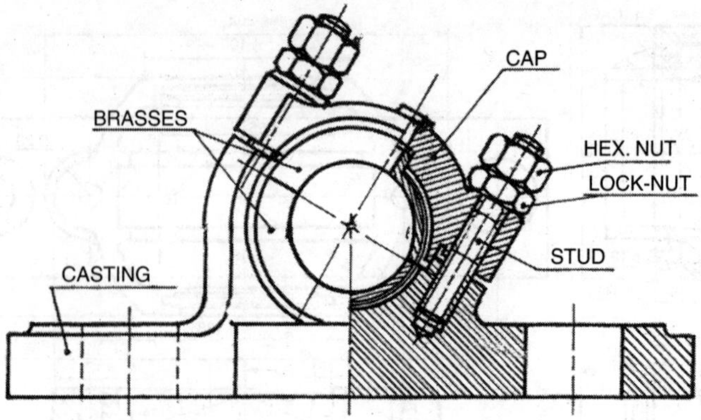

Fig. 23.18. (b) General arrangement of angle plummer block

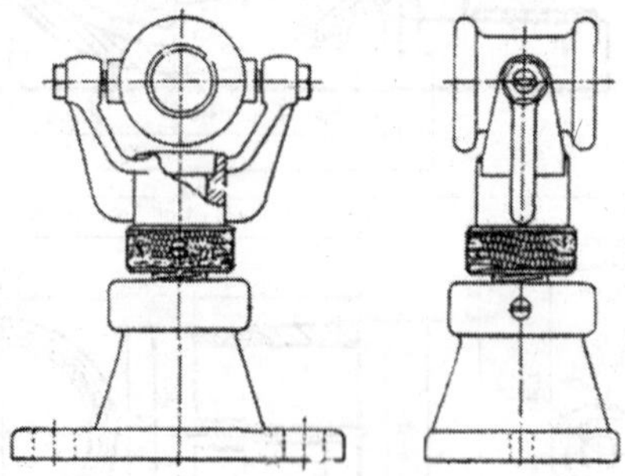

Fig. 23.19. (b) General arrangement of a swivel bearing

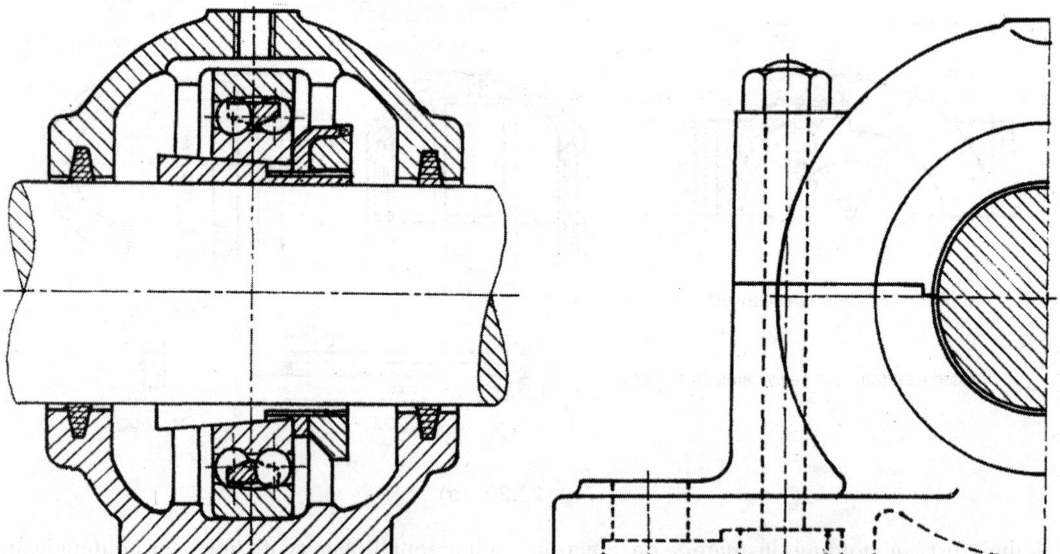

GENERAL ARRANGEMENT OF SELF ALIGNING BEARING

Fig. 23.20. (b)

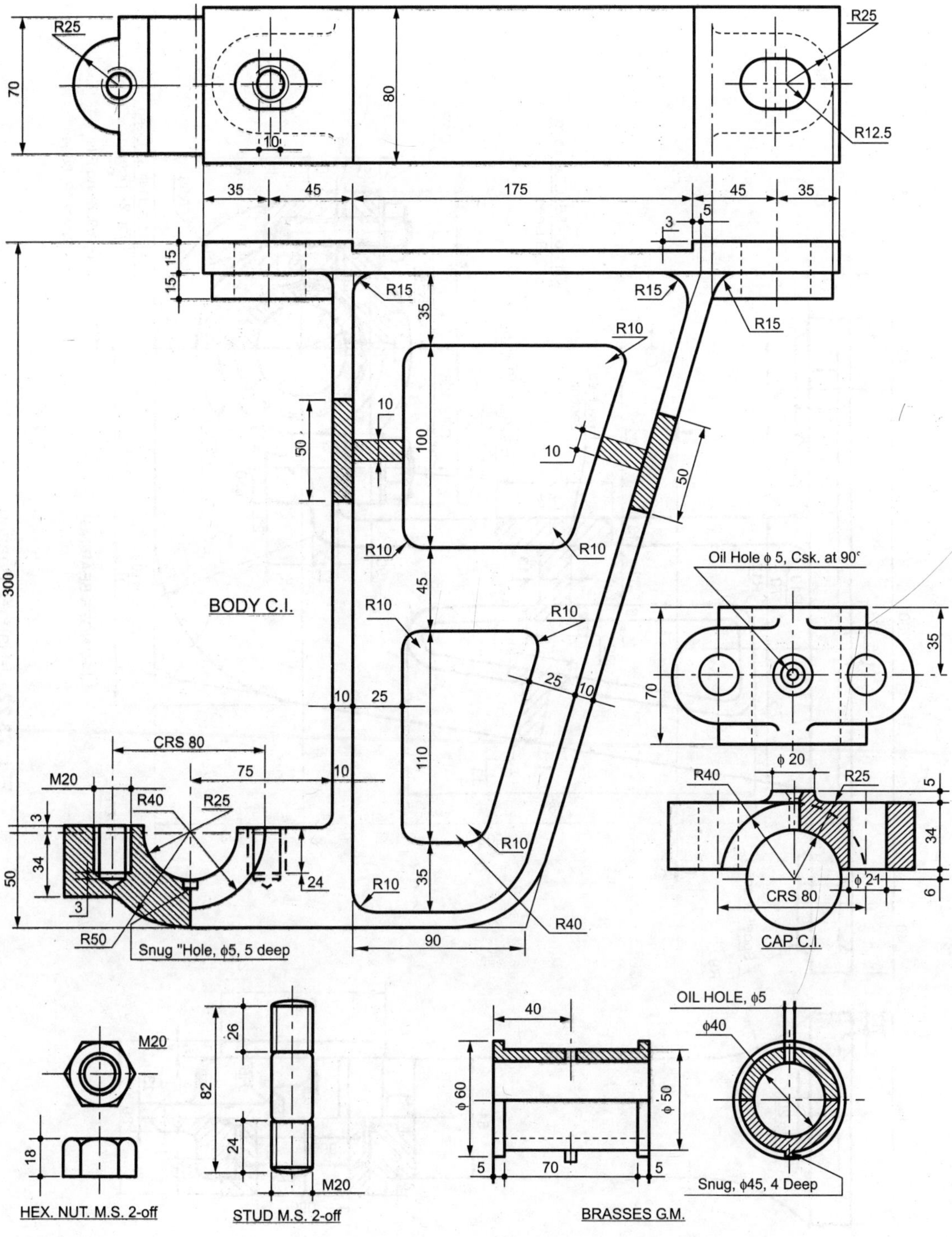

Fig. 23.21. Details of a Hanger Bearing

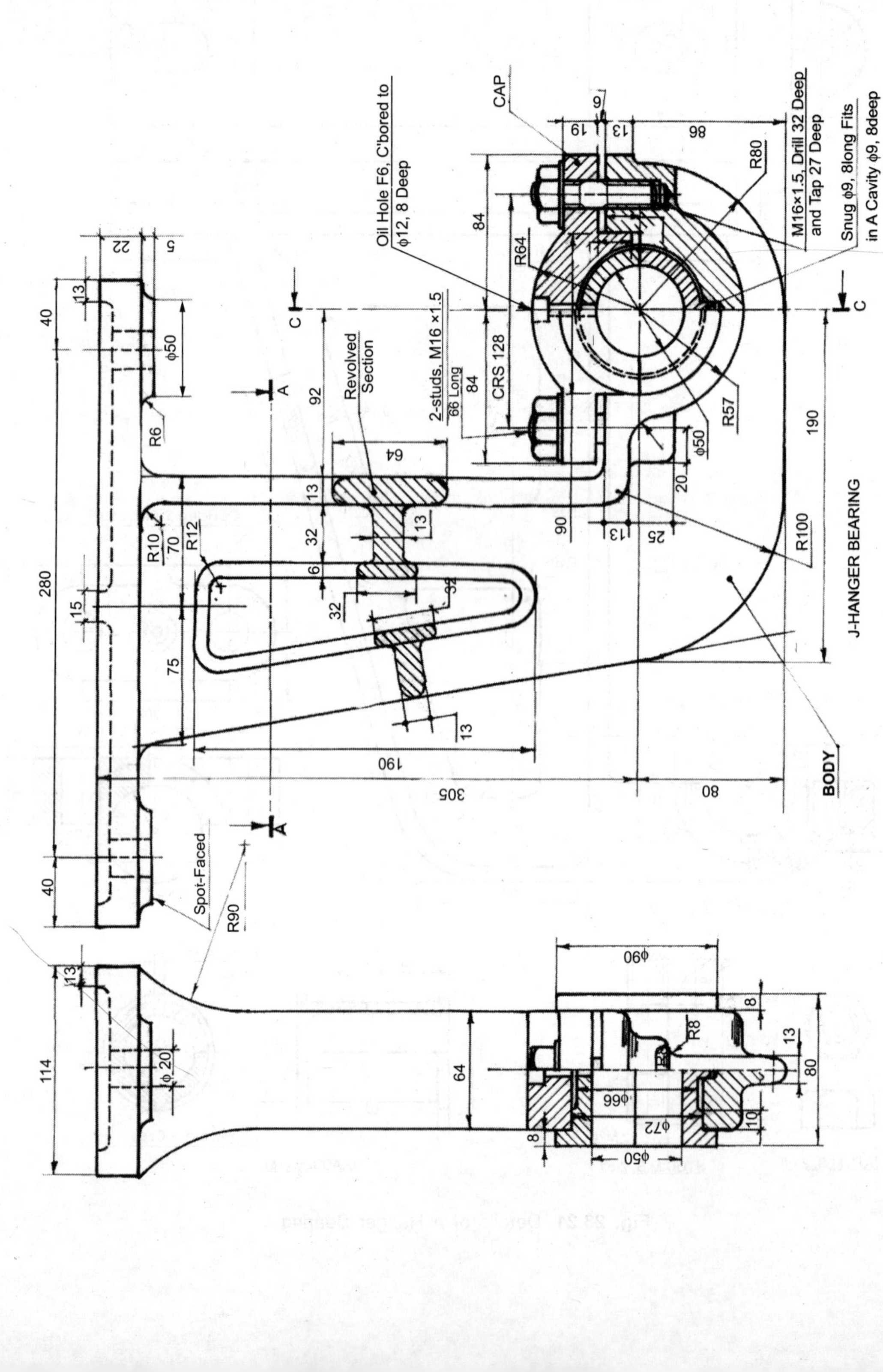

Fig. 23.22. J-Hanger Bearing

EXERCISE 23.17. *In Fig. 23.23 is shown a C.I. Bracket. Draw its:*

1. *Front view – looking in the direction of arrow E,*
2. *End view – looking in the direction of arrow S, and*
3. *Top view.*

EXERCISE 23.18. *Pictorial view of a C.I. Bracket is shown in Fig. 23.24, draw its:*

1. *Front view – looking in the direction of arrow A.*
2. *Left side view, and*
3. *Top view – right half in section.*

EXERCISE 23.19. *Pictorial view of a C.I. Bracket is shown in Fig. 23.25, draw its:*

1. *Front view,*
2. *End view – looking from right, and*
3. *Top view. Also dimension the views completely.*

EXERCISE 23.20. *Referring to Fig. 23.26 draw the three orthographic views of the bracket shown. Also give all the necessary dimensions.*

EXERCISE 23.21. *A wall bracket is shown pictorially in Fig. 23.27. Draw its:*

1. *Front view,*
2. *Left side view, and*
3. *Top view.*

HINT: Note here that neither the supporting plate nor the supporting ribs are placed symmetrically.

23.16. COMBINED WALL BRACKET AND BEARING

It serves the dual purpose of a bracket and a bearing.

EXERCISE 23.23. *Referring to Fig. 23.28 Draw.*

1. *Front view showing in section the portion to the right of centre of bearing brasses,*
2. *Right side view, and*
3. *Top view, right half of bearing in section. Use scale 1 : 2. Give all the necessary dimensions. Also prepare a parts-list and bill of materials.*

23.17. BALL AND ROLLER BEARINGS

Two main types are ball bearings and roller bearings. The ball and roller bearings substitute pure rolling motion for plain sliding motion. Balls make point contact whereas; rollers make line contact with their surface. These bearings have many advantages over plain bearings; namely, coefficient of friction is lower, it can take higher loads and can work at higher speeds, it is compact, lesser wear, and lesser lubricant consumption. The limitations in the use of these bearings are; greater initial cost and that cannot be used in halves.

23.18. BALL BEARINGS

These are of two main types: journal bearings and thrust bearings as shown in *Fig. 24.29 (a, b)*. Its important parts are:

1. CAGE

It is a single piece of sheet iron pressed to suitable shape. It helps to hold the balls. In case of journal bearing the cage is in two halves, and then riveted together, with radial holes to take balls, see *Fig. 23.30 (a)*.

In thrust bearing cage, *Fig. 23.30 (b)*, the holes are drilled, which allow the balls to enter from one side. The projecting central edge is then indented and closed over sufficiently to retain balls.

2. Races or Ring

There are two races, upper and lower, race or outer and inner race, *Figs. 23.33, 23.34*. The lower or inner race is driving fit on the shaft. This is provided with one or two grooved channels on its outer surface to form a track for balls (single row or double row.) The balls and races are both of high carbon chrome steel, hardened and polished.

The outer race is piston fit in its housing and grooved on the inner surface to form a track for balls. Thrust bearing can take thrust loads also. Single row thrust bearing can absorb thrust load in either direction.

The rings and balls are made of high carbon steel, hardened and polished, whereas the cages are of steel or brass.

23.19. ROLLER BEARINGS

Details of one such bearing are shown in *Fig. 23.35*. The construction of this bearing is simiular to that of the one shown in *Fig. 23.20 (a)*. Rollers here replace balls. The rollers are solid cylinders with ends perpendicular to axis. They roll around a

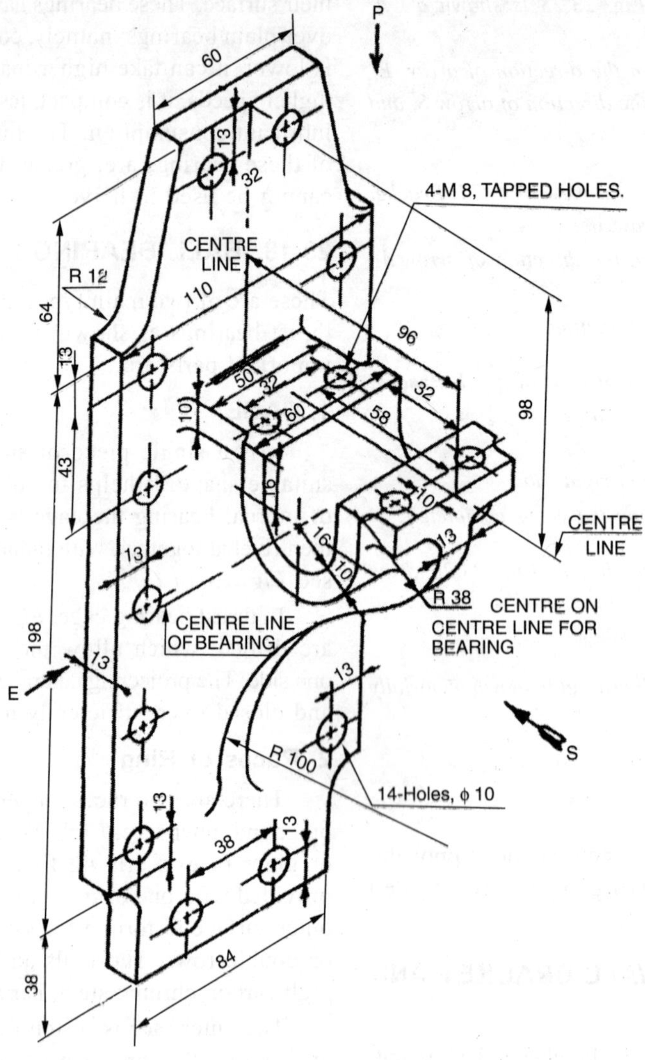

Fig. 23.23. Wall Bracket

grooved channel in the inner race and are spaced and kept parallel to the shaft. They are contained in a cage, (made in two pieces and riveted together, *Fig. 23.31*). The inner race in grooved to form a track for rollers while the outer race is plain so that the rollers can take up their real position. The diameter and length of rollers is usually kept equal. Both the races are secured longitudinally and the revolving race (mostly inner) is an interference fit on the shaft.

The roller bearings can take from 50 to 70% more load than the ball bearings.

The roller bearings are also of many types.

1. Cylinderical Roller Bearing

The bearing described above can take heavy, radial loads but limited thrust load. It is capable of rotating at high speed, used in electric motors, machine tools, motor vehicles. Refer *Fig. 23.32 (a)*.

2. Tapered Roller Bearing

It is suitable for carrying axial and radial loads acting simultaneously and when radial load is more than axial load. It finds it use in wheel hubs, machine tools, gear boxes, traction boxes and axle boxes. Refer *Fig. 23.32 (b)*.

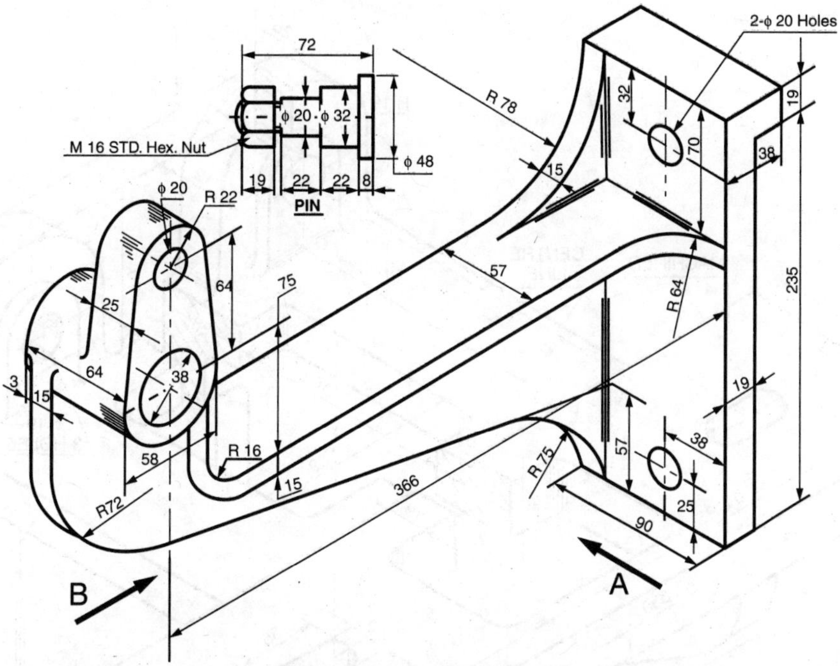

Fig. 23.24. Wall Bracket

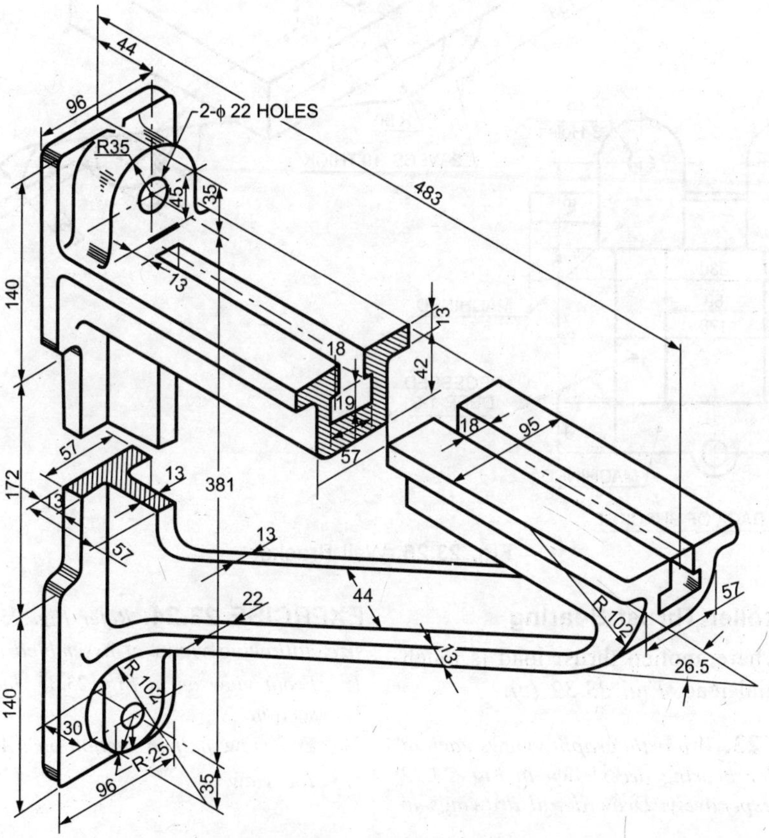

Fig. 23.25. Wall Bracket

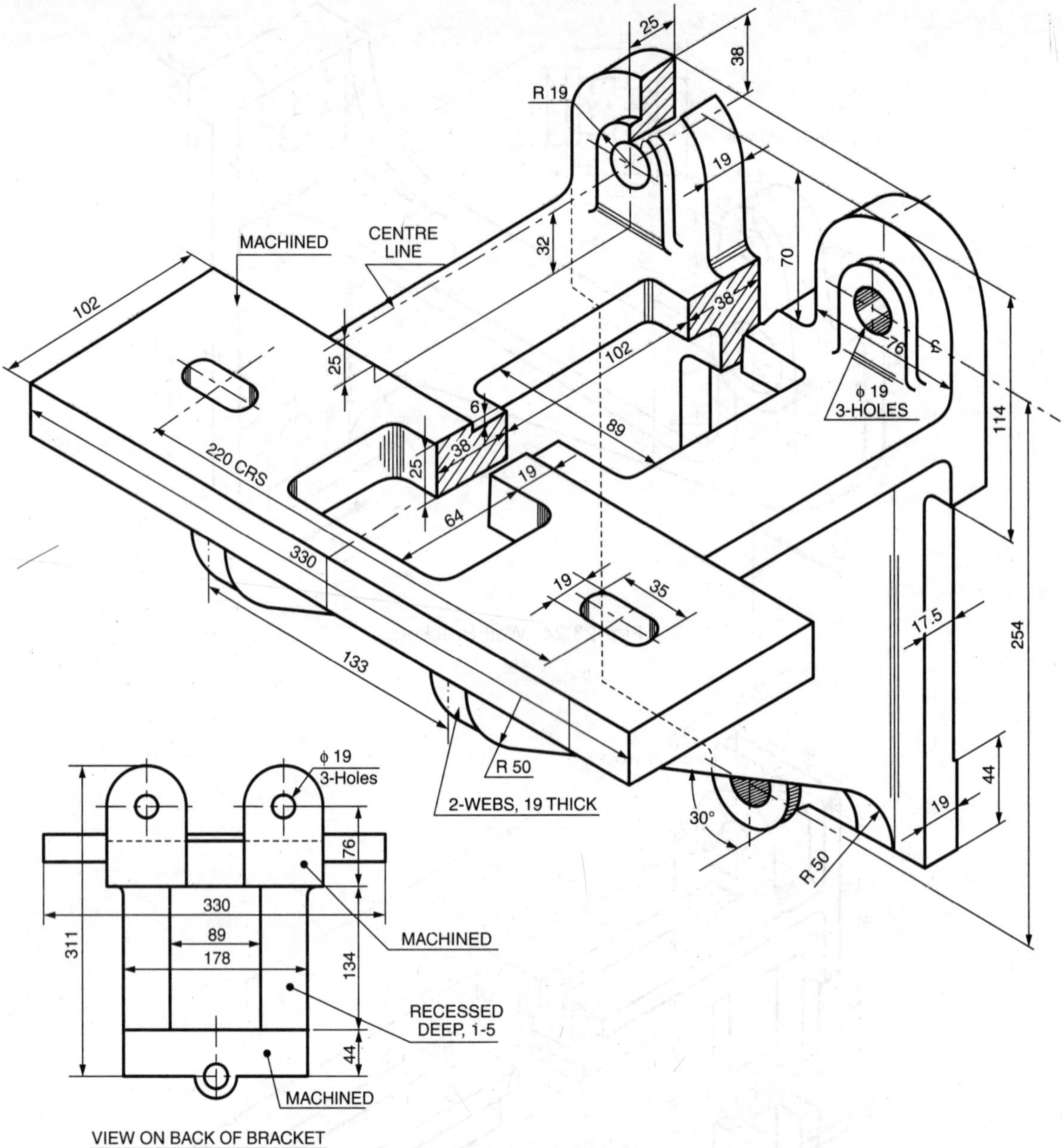

Fig. 23.26. Wall Bracket

3. Spherical Roller Thrust Bearing

It is used where applied thrust load is much greater than radial load, *Fig. 23.32 (c)*.

EXERCISE 23.23. *Two orthographic views each of a ball and a roller bearing are shown in Fig. 23.33 and Fig. 23.34 respectively. Draw detail drawings in each case.*

EXERCISE 23.24. *Refer Fig. 23.35 and draw full size, following views of assembled roller bearing:*

1. *Front view as in Fig. 23.35, but with left half in section.*
2. *End view, half in section on AA.*
3. *Top view.*

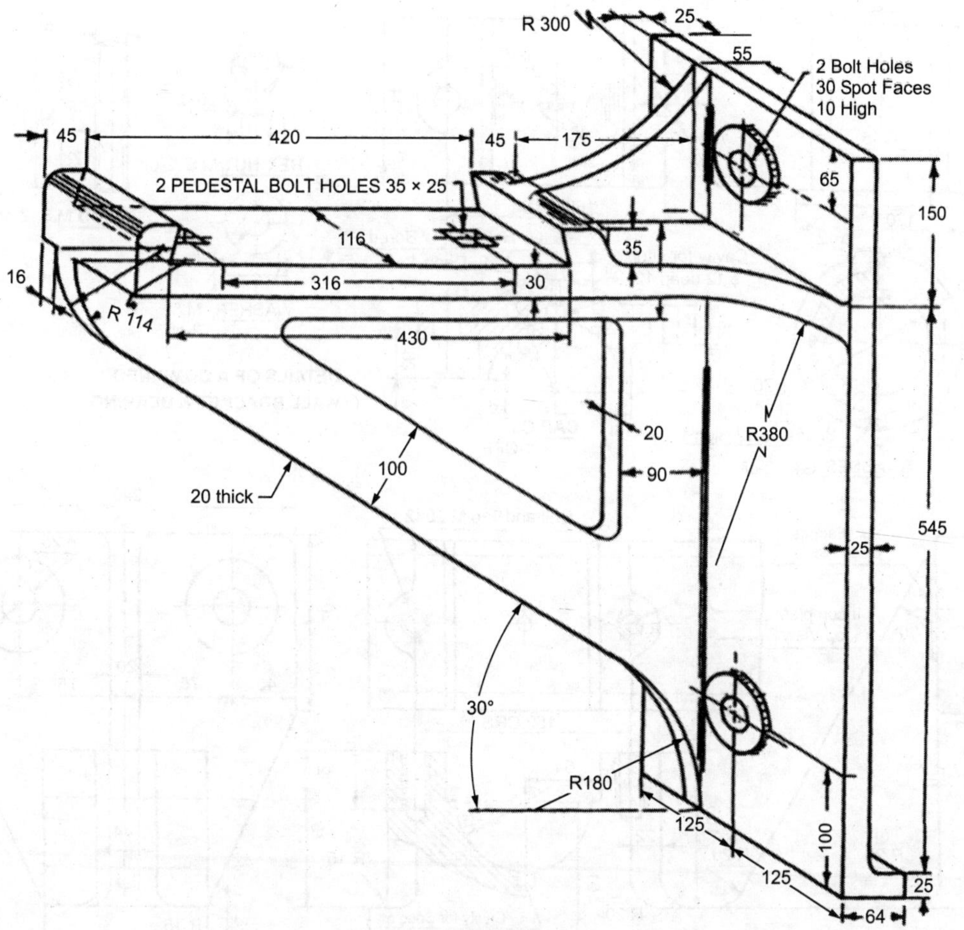

Fig. 23.27.

EXERCISES

1. What is the function of a bearing? What is a journal ? Describe with the aid of proportionate sketches, the three main types of bearings used for supporting shafts.
2. Why ample lubrication of shaft bearing is necessary? Illustrate, by sketches, at6 what point and how the lubricant is fed? How metal-to-metal contact is eliminated? What is meant by seizing of the shafts? What circumstances lead to it?
3. Draw three views of the solid bearing shown in *Fig. 23.2*, each being half in section, passing the cutting planes through the center of bearing hole.
4. (*a*) Why a bush is provided in between the shaft and the body of a bearing? What material is usually selected for a bush?

   (*b*) Make dimensioned sketches of a bushed solid bearing of f 50 mm shaft.

   Illustrate, by drawing neat sketches, at least three different means adopted to stop rotation of brasses in bearing. What is the difference between a bush and brasses? Why the rotation of brasses with shaft is to be prevented?
5. What is a pivot bearing? Sketch freehand, two views of a footstep bearing for supporting 50 mm diameter vertical shaft. Give all important proportionate dimensions. What is the difference between a footstep and a journal bearing? Why a pad or a disc is provided below the shaft end in a pivot bearing?
6. What is the function of a wall bracket? What is the difference between a bracket and a bracket bearing? Illustrate by simple sketches?

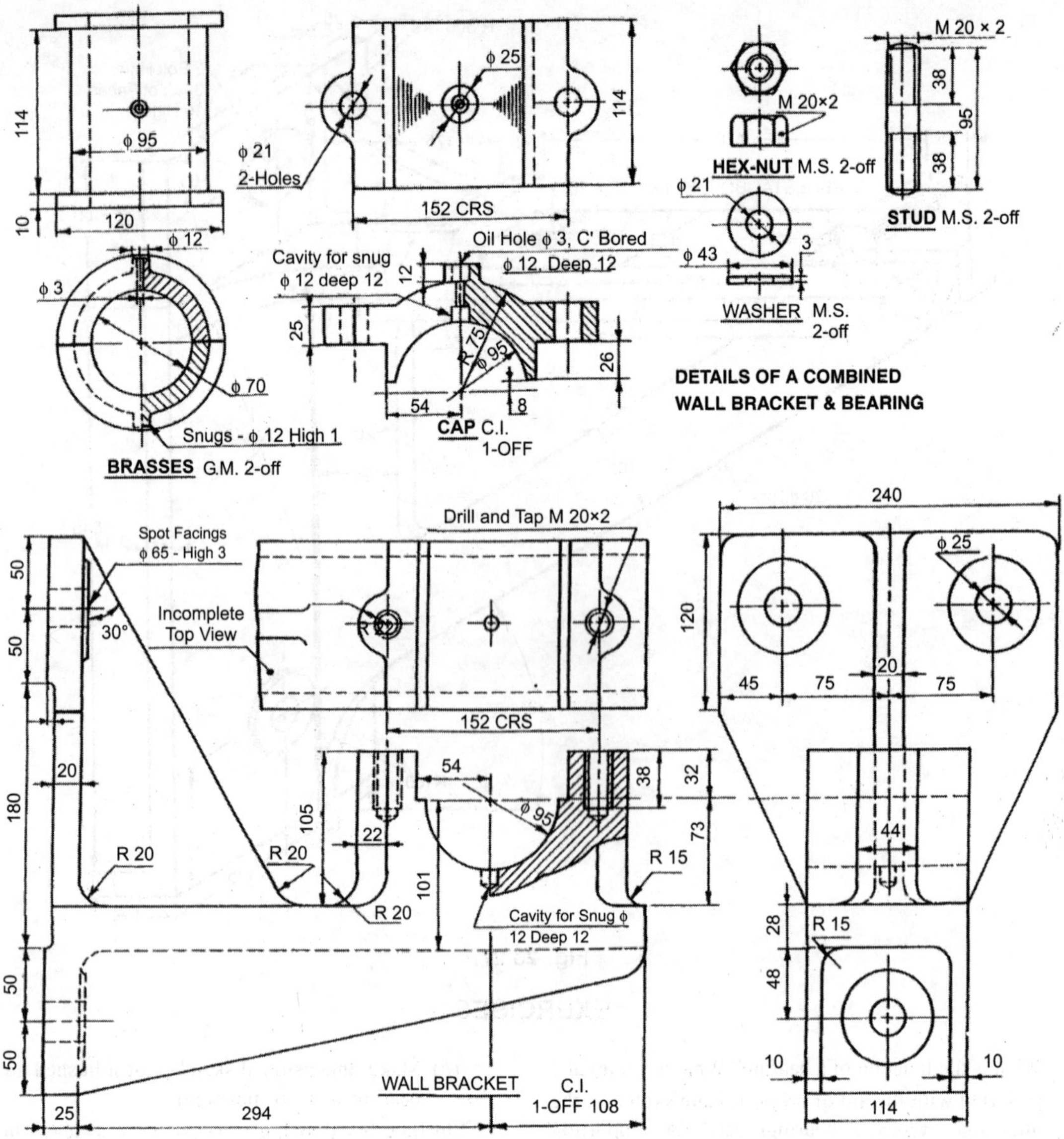

Fig. 23.28. A Combined wall bracket and bearing

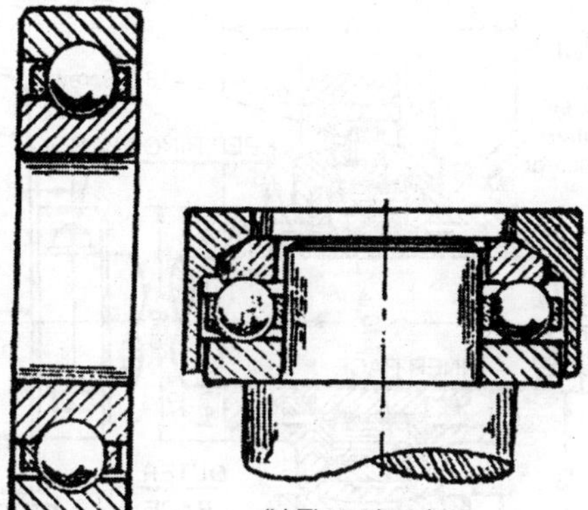

(a) Journal bearing (b) Thrust bearing

Fig. 23.29. Ball bearings

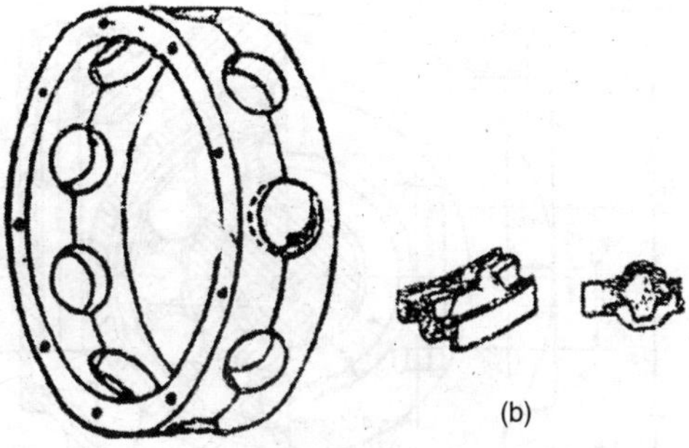

(a) (b)

Fig. 23.30.

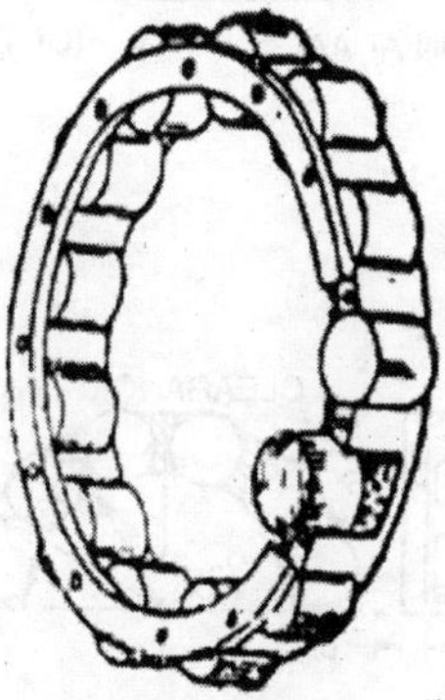

Fig. 23.31.

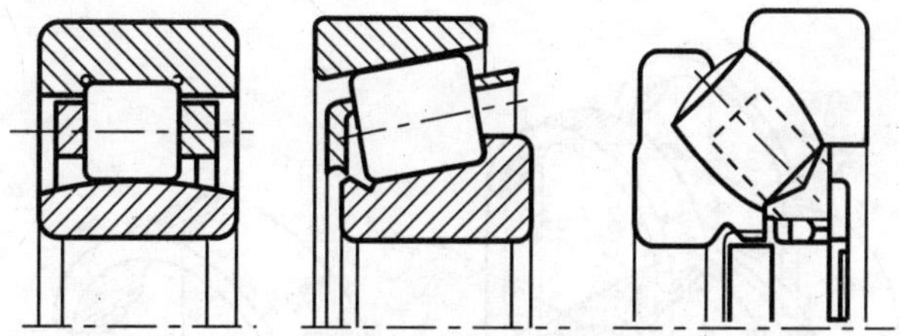

(a) (b) (c)

Fig. 23.32. Types of Roller Bearing

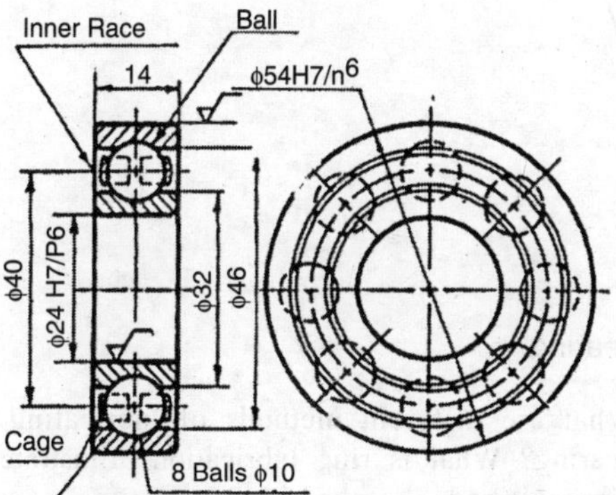

Fig. 23.33. Ball bearing

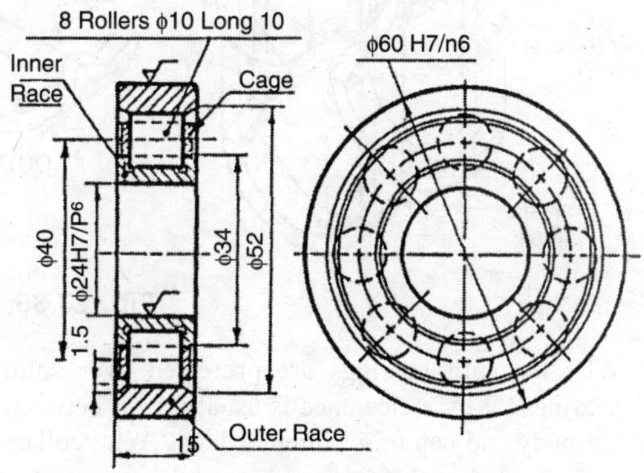

Fig. 23.34. Roller bearing

7. Sketch proportionately, a wall bracket for a line shaft bearing. Take the axis of the line shaft 200 mm from and parallel to the wall.
8. What is an angle plummer block and under what circumstances it is used?
9. Draw, dimensioned proportionate sketches of any one type of hanger, with bearing for a shaft of 25 mm diameter and supporting a 200 mm diameter pulley.
10. What is snug? Illustrate by a sketch its function in a bearing.

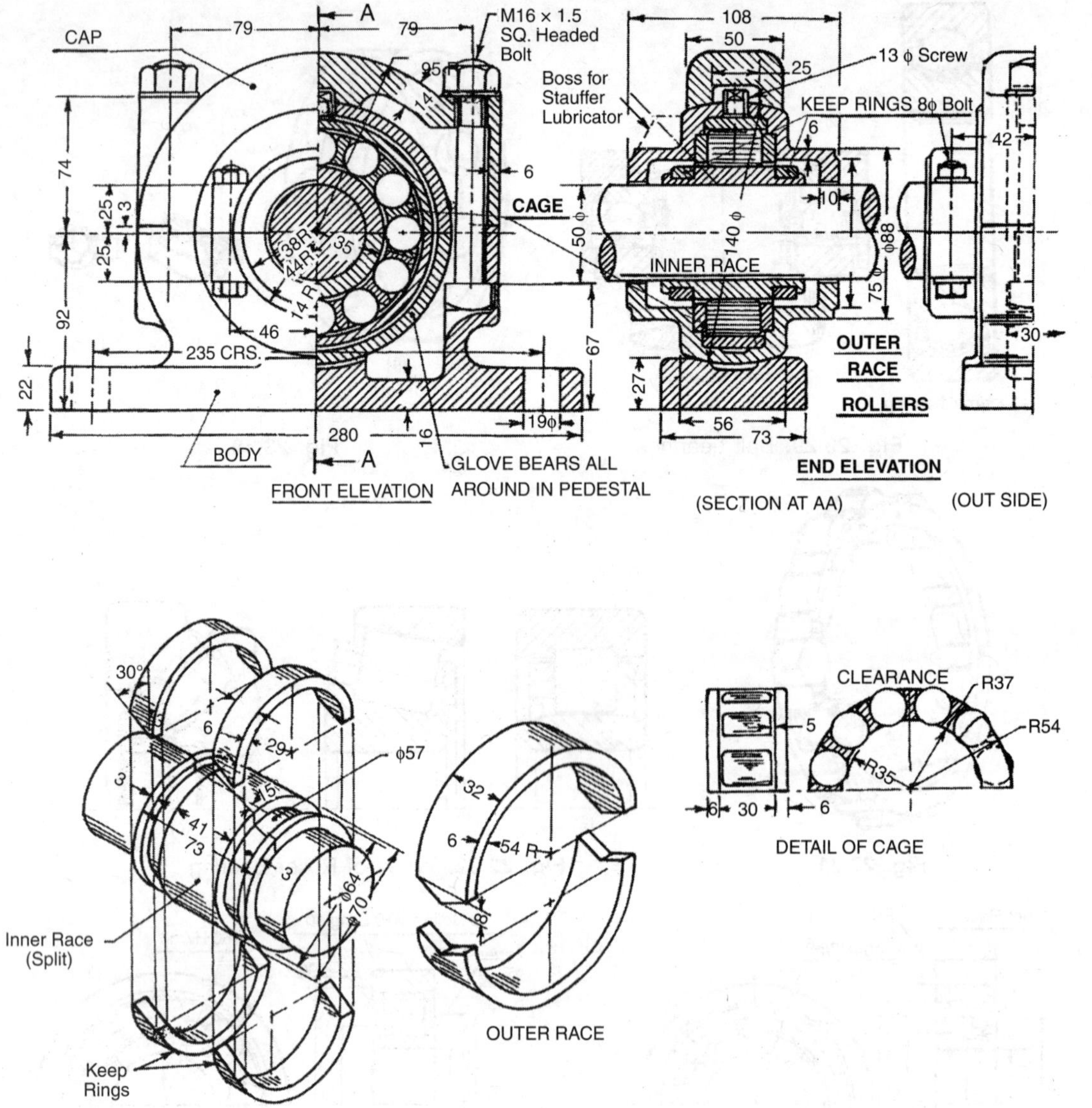

Fig. 23.35. Roller Bearing

11. Why the split bearings are preferred over solid bearings? Why a clearance is usually kept between the body and cap of a Plummer lock? Why collars are provided at the sides of brasses?
13. What is the advantage of using a swivel bearing over a Plummer block? Illustrate, by dimensioned sketches, any one type of swivel bearing.
14. What are different methods of lubricating the bearing? What is ring lubrication? Illustrate by dimensioned sketches any one type of ring lubrication bearing.

Valves And Cocks

24

24.1. INTRODUCTION

Valves are used to control the flow of fluids through the pipes and from the pressure vessels. These are operated by hand, the pressure of the fluid they handle, or some outside mechanism. Valves, which are operated by the pressure of the fluid, are called non-return valves, as they allow the flow of fluid in one direction only and do not allow the fluid flow in the reverse direction, *Figs. 24.1, 24.2.* Relief valves and safety valves, too are operated by the pressure of the fluid. Junction and stop valves, and the cocks are operated by hand.

The material of the valves may not necessarily be the same as that of the pipes to which they are connected. The most common material for a valve in smaller sizes is brass, bronze or gunmetal, and in the large sizes is cast iron. The valves subjected to very high pressure are made of cast steel or cast steel alloys. Numerous types of valves are used in practice. A few of the most common types are discussed in this chapter.

24.2. VALVES AND VALVE SEATS

A valve is usually a gunmetal, brass, bronze or steel circular disc, beveled off at an angle of 45° at its lower edge, and provided with feathers, wings or guides to keep it central in its seat, during operation *(Figs. 24.1, 3, 5, 7, 10, 16)*. A cylinderical, central spindle on the lower surface of the valve also serves to guide the rise and fall of the valve from and onto its seat *(Figs. 24.4, 24.25)*. The recesses, made at the portion of the wings where they join the disc, provide clearance for the cutting tool while preparing the seating surface of the valve. Also, it gives better opening for the fluid. The feathers or wings are made running fit inside the valve seat. The distance by which the valve moves up from its seat is called the lift of the valve.

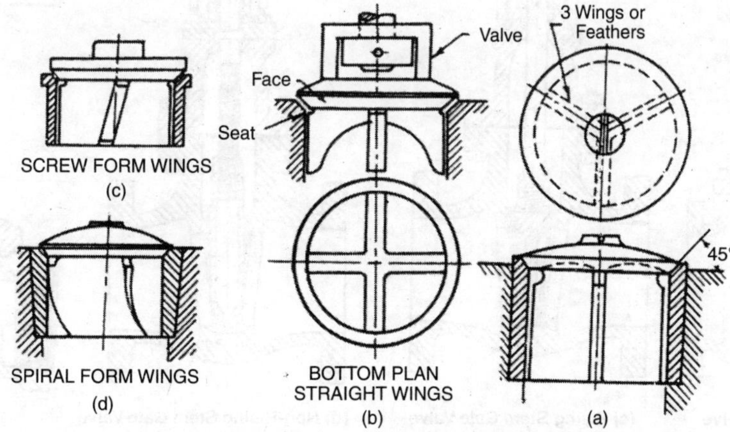

Fig. 24.1. Types of valves

Valve are attached to the spindle by different means *(Figs. 24.3, 24.5, 24.6, 24.10)*.

The valve may have surface (*Figs. 24.1, 3, 4, 5, 6, 7* etc.) or line (*Fig. 24.11*) contact with the valve seat. *Figure. 24.1* shows some types of valves commonly used in practice. At (*a*) is shown a valve with three straight feathers or wings. The valve at (*b*) has four wings and is provided with a recess at the top to receive and hold the lower end of the spindle. At (*c*) is shown a valve with wings of screw form and at (*d*) of spiral form. The upward force of fluid acts upon these curved surfaces and turns the valve a little when it is lifted clear off its seat. These equalizes wear on the valve face.

A valve seat is a G.M. or brass cylinderical bush, screwed, press fit, or held secure in the main body of the valve by cap-screw or set-screw or by some other means so that it does not lift with valve due to pressure of the fluid. It is beveled off at its top edge at an angle of 45°. It is that fixed part on which the valve rests. In small size valves the seat is formed in the passage in the valve body itself. In valves of large diameters, it is made separately and then fitted, by some arrangement, in the passage.

24.3. CLASSIFICATION OF VALVES

Valves may be classified as:

1. Flap Valves or Swing Check Valves– Which bend or turn upon a hinge, *Fig. 24.2 (a)*. The amount of swing given to the valve controls the opening for fluid.

2. Lift Check Valves – which rise and fall in a direction perpendicular to the seat. The lift of the valve is checked by the position of the lower end of the valve spindle, *Figs. 24.2 (b), 24.7* or the projection provided below the cover, *Fig. 24.6*. Such valves are used as feed check valves or hydraulic non-return valves.

3. Gate Valves – rising stem or non-rising stem gate valve *(Fig. 24.2 (c), (d))*. This type of valve is used on pipes conveying water or other liquids. It offers little restriction to straight-line flow. *Fig. 24.2 (c), (d)* illustrate two types of gate valve. At (*c*) as the valve wheel is revolved to open the valve, the stem rises to move the gate out. This type is called rising stem gate value. At (*d*), as the wheel is revolved, the gate moves out along the stem, while the stem does not rise. This type is called non-rising stem valve. In both cases the leakage is prevented by gland and packing.

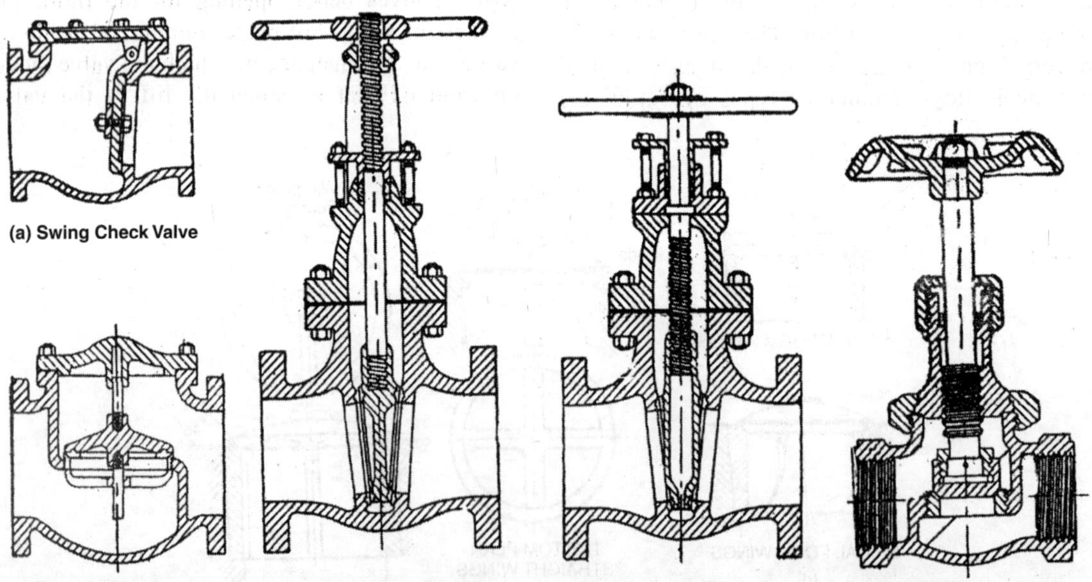

(a) Swing Check Valve (b) Left Check Valve (c) Rising Stem Gate Valve (d) Non-Rising Stem Gate Valve (e) Globe Valve

Fig. 24.2. Types of valves

4. Globe Valves – in which the valve is lifted, perpendicular to the fluid flow, off its seat, by rotation of the head-wheel, that intern operates the spindle. These are used to control the fluid flow for very close regulation, as the throttling, on the steam pipelines. *Figure. 24.3* shows the detail drawings of a globe valve used for regulating steam supply and is usually called "*steam stop valve*". The size of the valve is specified by the size of the passage of steam or the size of the seat in it. As in this case the internal diameter of the valve seat is 125 mm, the valve is specified as 125 mm stop valve.

The cast iron body of the valve is divided into two compartments by a thin web. The horizontal portion of the web is cored out and machined to accommodate the gunmetal valve seat. Brass valve is provided with four feathers, which guide its motion in the seat. The lift of the valve is adjusted by a brass spindle, that works in the threaded part called bridge, which is turn is mounted on the valve body by two studs, called pillars.

24.4. SOME COMMON FEATURES OF VALVES

1. All valves are provided with flanges and are secured to the flanged pipes or seating blocks by bolts.
2. In most valves, the flanges are strengthened by providing ribs.
3. The valves using spindle, to operate the valve and control the amount of flow, are provided with a gland and packing to prevent leakage of the fluid along the spindle.
4. The spindles, generally, have square threads and are rotated by cast iron wheels.
5. The valve seats, which are generally of gun-metal or brass, are either tight fit or are secured by threads into the main body.

24.5. JUNCTION OR STOP VALVE

As already discussed, stop valves are used to control the flow of fluids at high pressure, which may be water or steam. There is no specific difference between the design of a stop valve and a junction valve. When the valve changes the direction of flow through 90°, it may be called a junction valve. As applied to steam supply, the function of a stop valve is to regulate or completely shut off the amount of steam passing through the steam pipe. It is operated by hand. Such a valve when placed directly on a boiler, is called a junction valve, and when placed near a prime mover is called a stop valve.

Figure. 24.5, shows the details of a Stop Valve, which may be called a junction valve, too. Lower flange of the valve is bolted to the mounting plate of the boiler and the other flange is bolted to the pipe flange. It can be noted that:

1. The seat is secured into cast iron body of the valve.
2. A cylindrical sleeve is provided in the centre of the seat, which guides the cylindrical feather of the valve.
3. The gland is held in position by a special type of hexagonal nut, which also holds the spindle. This nut is provided with threads at both ends, on the lower end to suit the threads on the cover; and threads on the upper end to suit the threads on the spindle.

EXERCISE 24.1. *Figure. 24.3 shows the details of a 125 mm stop valve drawn in First Angle. Draw the following assembled views:s*

(a) Front view; full in section,

(b) Right side view outside.

(c) Top view; outside, removing the hand wheel.

Suggest the type of fit for various mating parts and fix the upper and lower limits for them using the ISO chart on limits and fits.

Also prepare a parts list and bill or materials.

EXERCISE 24.2. *Figure. 24.4 shows the exploded assembly of a gunmetal steam stop valve. Draw the following assembled views:*

(a) Front view – full in section,

(b) Top view – outside.

Assume suitable stud bolt and nut sizes. Prepare a parts list and bill of materials.

EXERCISE 24.3. *From the details given in Fig. 24.5, assemble the parts and draw the following views of the 50 mm Stop Valve:*

(a) Front view – full in section,

(b) Right side view, and

(c) Top view.

Also prepare a bill of materials and parts list. Shown suitable nuts and stud bolts in position. Also suggest the types of fit and fix the upper and lower limits for all mating parts.

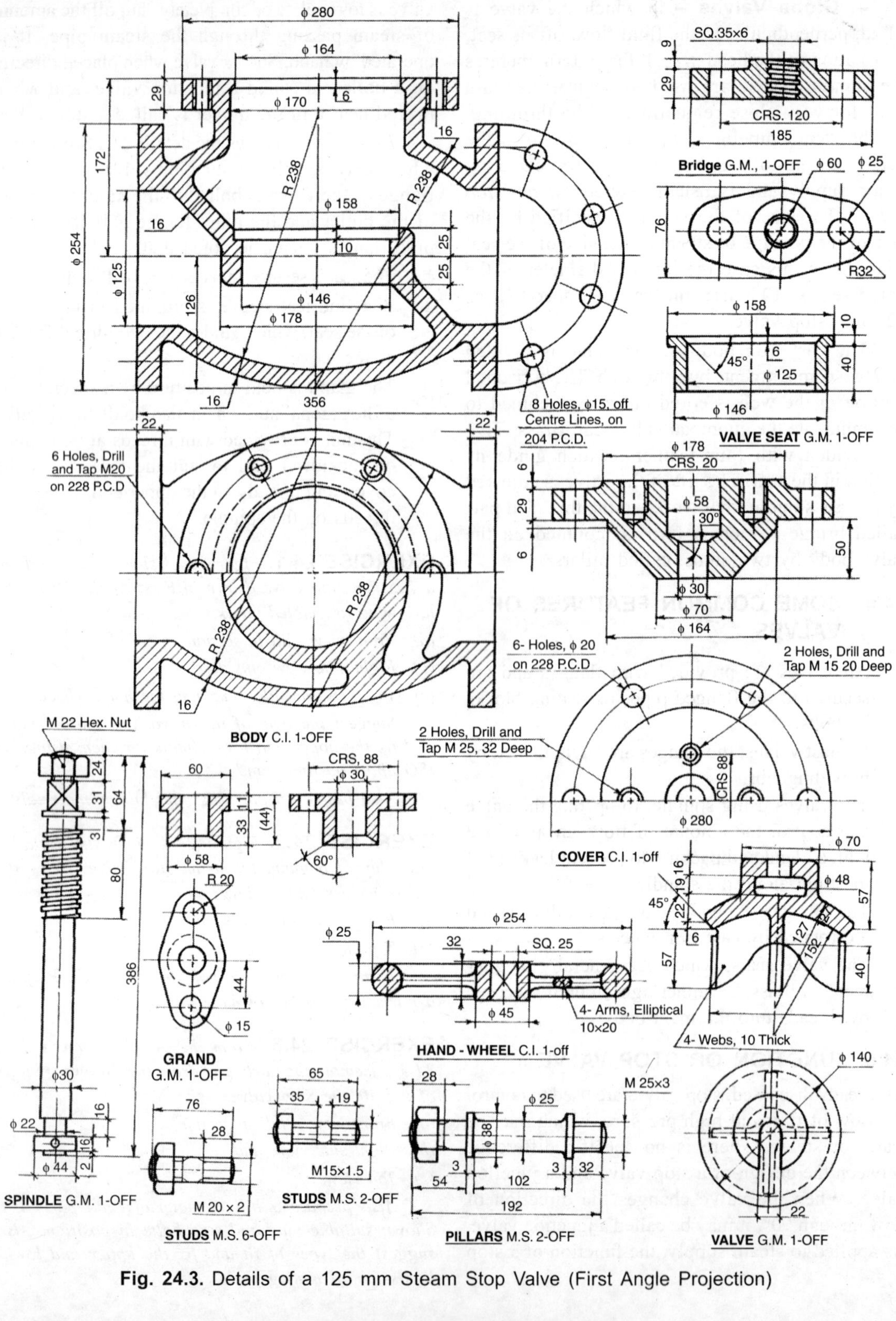

Fig. 24.3. Details of a 125 mm Steam Stop Valve (First Angle Projection)

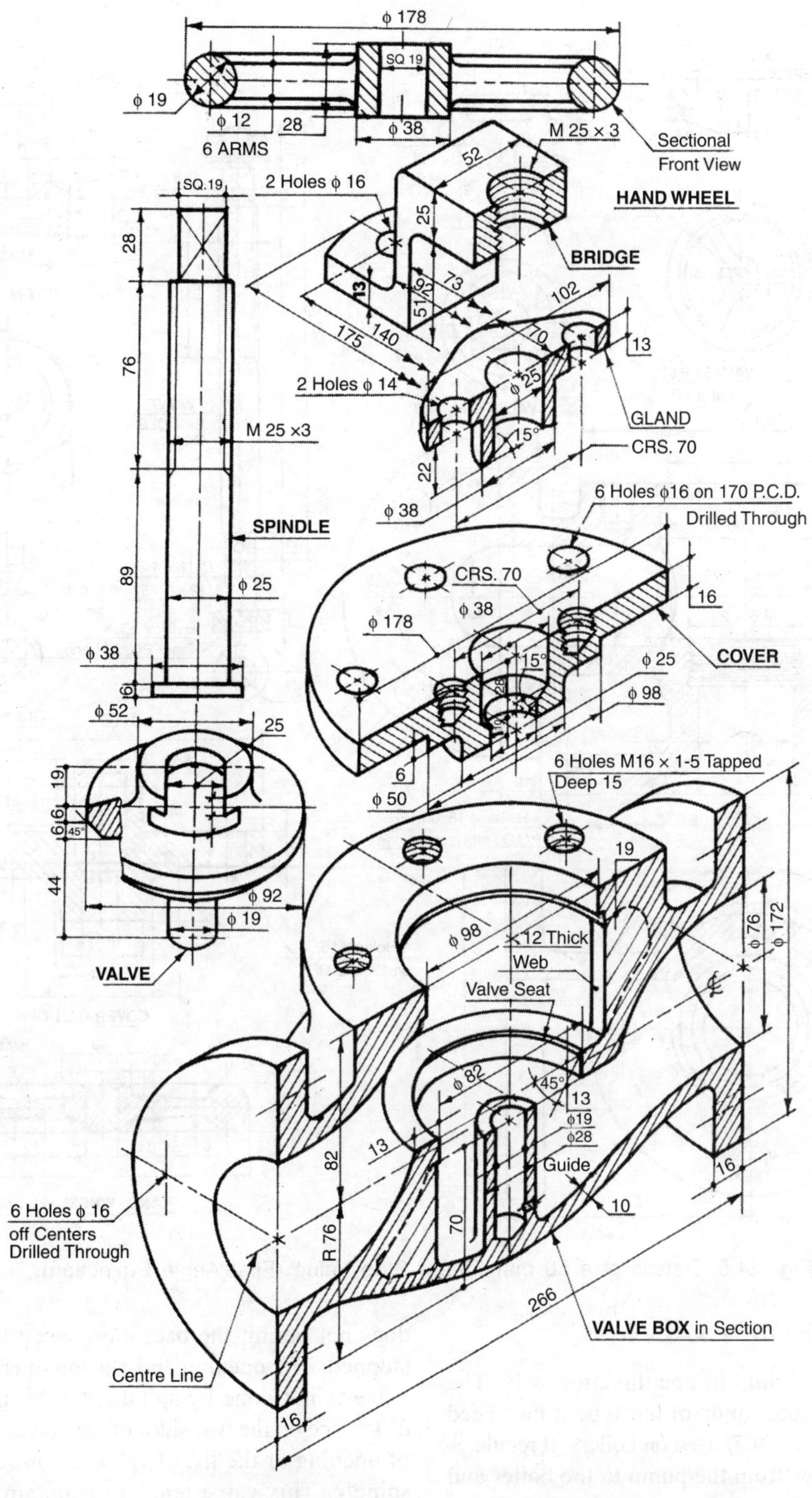

Fig. 24.4. Gun-Metal Steam Stop-Value

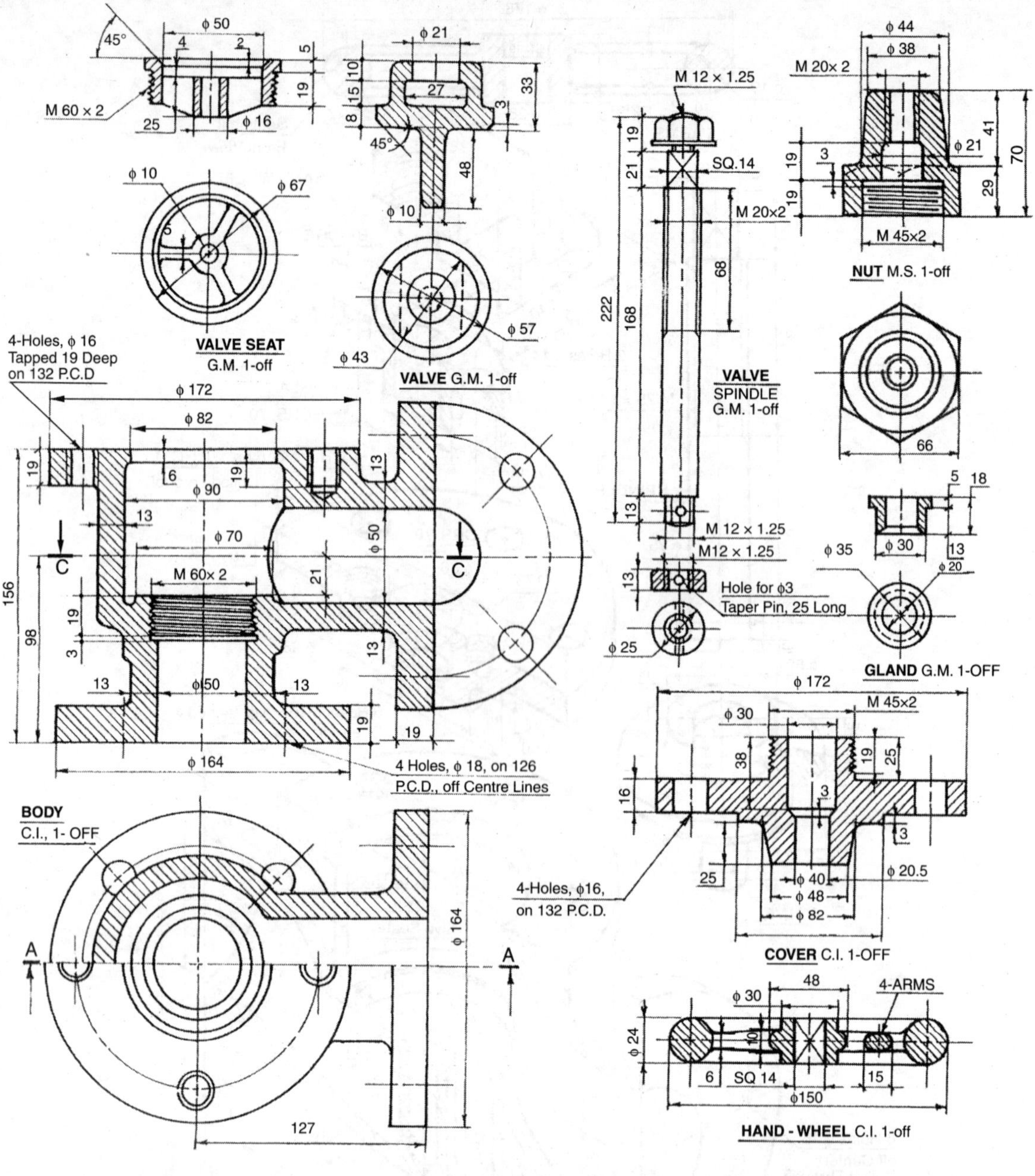

Fig. 24.5. Details of a 50 mm Steam Stop Value (First Angle Projection)

24.6. NON-RETURN VALVE

It allows flow of fluid in one direction only. The most important application of this type is the "Feed Check Valve" *(Fig. 24.7)* used on boilers. It regulates the flow of water from the pump to the boiler and does not permit the back flow when the pump is stopped. The opening and closing operation of the valve is not done by spindle, but by the pressure difference on the two sides of the valve. The amount of opening or the lift of valve is controlled by the spindle. This valve tends to maintain the supply of water, as far as possible, constant and is placed as near to the boiler as possible.

Figure 24.6 shows a type of non-return valve, which is used to control the flow of water and is called the hydraulic non-return valve. The operation is similar to the feed check valve. Only difference is that the spindle is replaced by the projection on valve cover. In this type the opening of the valve cannot be changed.

Figure. 24.7 shows the details of a Feed Check valve. It can be noticed that the spindle cannot be connected to the seat but can act only as a stop to limit the rise of the valve. As no threads are

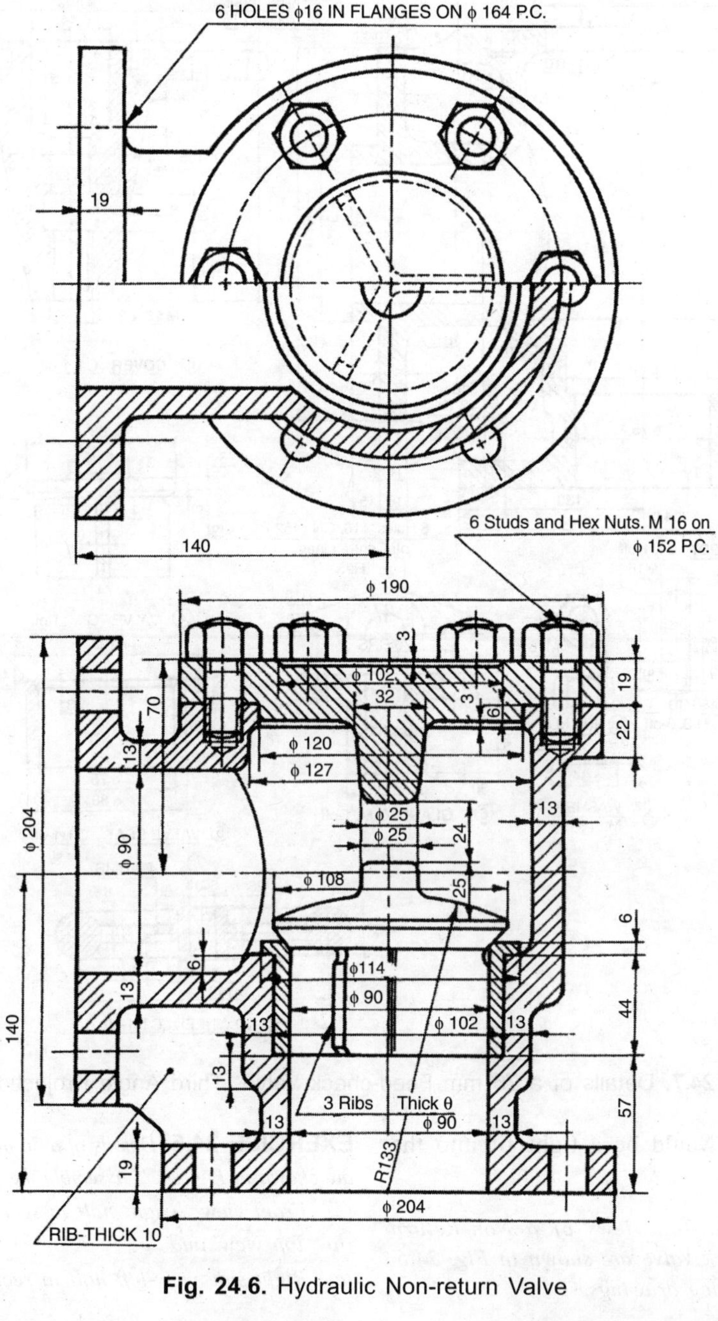

Fig. 24.6. Hydraulic Non-return Valve

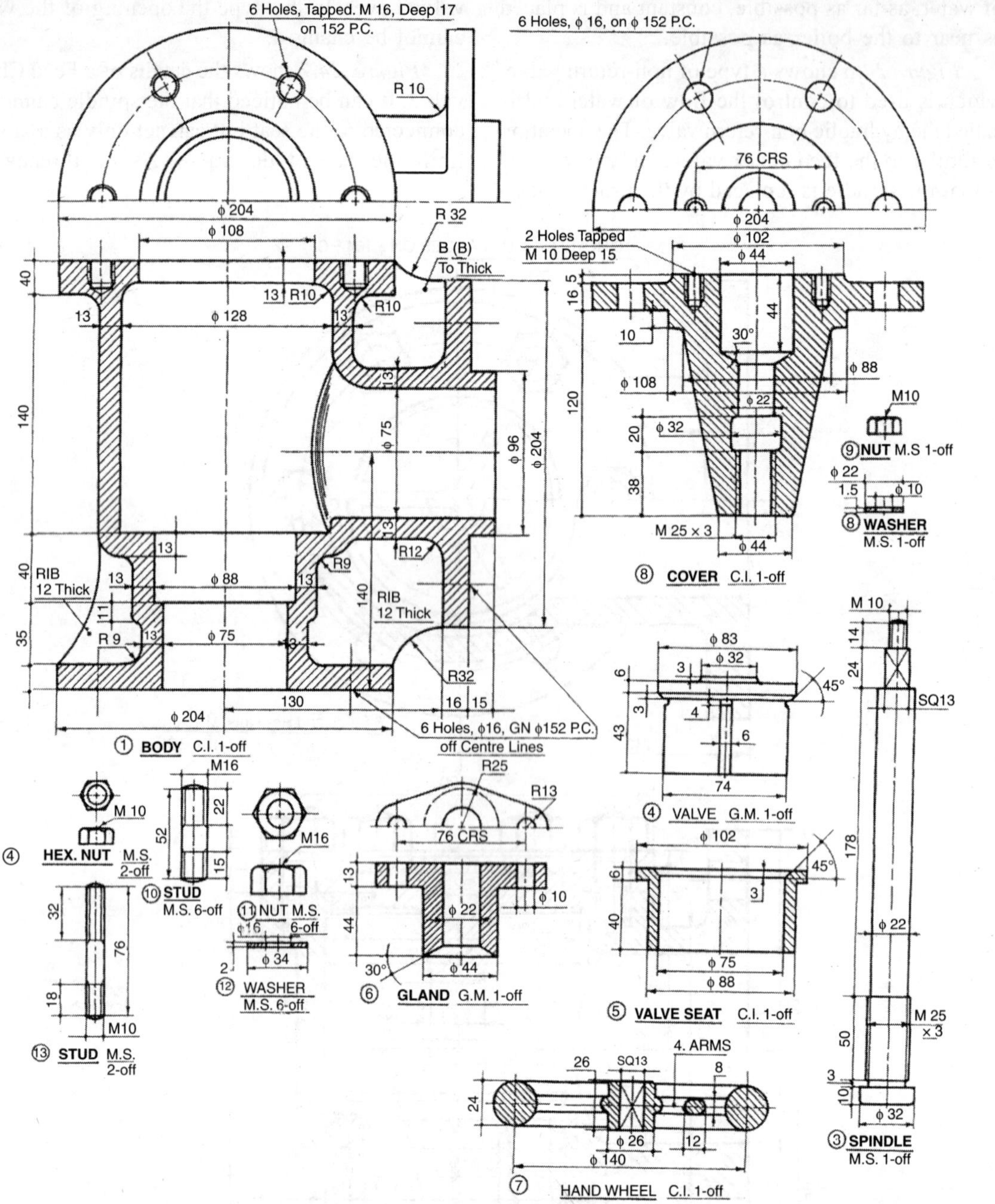

Fig. 24.7. Details of a 76 mm Feed-check Valve (Third Angle Projection)

provided, the seat would be a tight fit into the body.

EXERCISE 24.4. *Top views of a Non-Return Hydraulic Feed Check Valve are shown in Fig. 24.6. Draw its detail working drawings.*

EXERCISE 24.5. *Details of a 76 mm Feed Check Valve are shown in Fig. 24.7. Assemble the parts and draw:*

(a) Front view – right half in section,

(b) Top view, and

(c) Right side view-left half in section.

24.7. BLOW OFF COCK

Blow-off cock is used to drain out or blow-off the water from the boiler shell, when it is to be emptied for checking, repairing or cleaning purposes. In this apparatus the element used to control the flow is called a plug or cock and hence is the name blow off cock, *Fig. 24.8*.

The material of the Blow-off cock is either cast iron or gunmetal. The cock of the blow-off cock is usually a hollow conical part. In some cases it may be hollow cylinderical also. A through hole is cut in the cock which permit water to drain out when in line with the flow. The cock can be rotated by a spanner or hand-wheel. A chisel cut marked on the head of the plug indicates the position of the plug, whether open or close.

As in other valves, gland and packing are used to prevent the leakage of water.

At the bottom of the body, an easing screw in fitted, which is used to remove the cock when it sticks fast with the body.

Figure 24.9 shows another type of blow-off cock, where the two flanges of the C.I. body are at 90° to each other. Except a small change in the design of the body and cock, the remaining is same as in the previous valve. The vertical flange of the body is bolted to the mounting plate of the boiler and the other flange to the pipe, which disposes the water out.

EXERCISE 24.6. *Figure 24.8 shows the details of a Blow-off Cock. Draw the following assembled views:*

(a) Front view – right half in section,

(b) Right side view, and

(c) Top view – outside.

EXERCISE 24.7. *Details of a 75 mm blow-off cock are given in Fig. 24.9. Draw the following assembled views.*

(a) Full sectional elevation, and

(b) Outside plan.

Suggest the types of fit for various mating parts and fix the upper and lower limits of their dimensions using the ISO chart on Limits and Fits.

HINT: *Refer Fig. 24.9 (a) for solution.*

EXERCISE 24.8. *Details of a Plug cock are given in Fig. 24.10 (a). Assemble the details and using a scale 1 : 2 draw the following views:*

(a) Front view – right half in section,

(b) Left side view – outside,

(c) Top view – right half in section.

Taking the section plane along the centreline of the body.

Dimension the views completely and prepare a parts-list and bill of materials. Assume an appropriate size of Nuts, washers (for studs), Joint ring and packing as these parts are not given in details.

HINT: *Refer Fig. 24.10 (b) for part solution.*

24.8. SAFETY VALVES

A safety valve is used to limit the maximum pressure of steam in a boiler. It is generally fitted at the crown of a boiler. When the pressure of steam rises above the specified limits (working pressure), the safety valve opens automatically to allow the steam to escape. Depending upon the method used to balance the steam pressure on the valve, safety valves can be divided into following types:

(*i*) Lever safety valve,

(*ii*) Dead weight safety valve, and

(*iii*) Spring loaded safety valve.

Though there is a large number of safety valve of different types used in practice, only a few, most common, are discussed in this chapter.

24.9. LEVER SAFETY VALVE

Figure 24.11 (a) shows the detail drawings of a lever safety valve, and in *Fig. 24.11 (b)* is shown the sectional pictorial view of the same. The main component parts of the valve are, C.I. valve, lever and dead weight. The lower flange of the body is secured to the seating block of the boiler by bolts, and in the upper flange a brass seat is fitted. The brass valve is kept in position on the seat by a pointed spindle, fitted in the steel lever. The shape of the lever is made such that is curved end fits against the knife edge of the fulcrum, and the other end supports the dead weight.

In the normal position, the moment due to dead weight, about the fulcrum, is more than the moment due to steam pressure. When the pressure exceeds the set limit, the moment due to steam

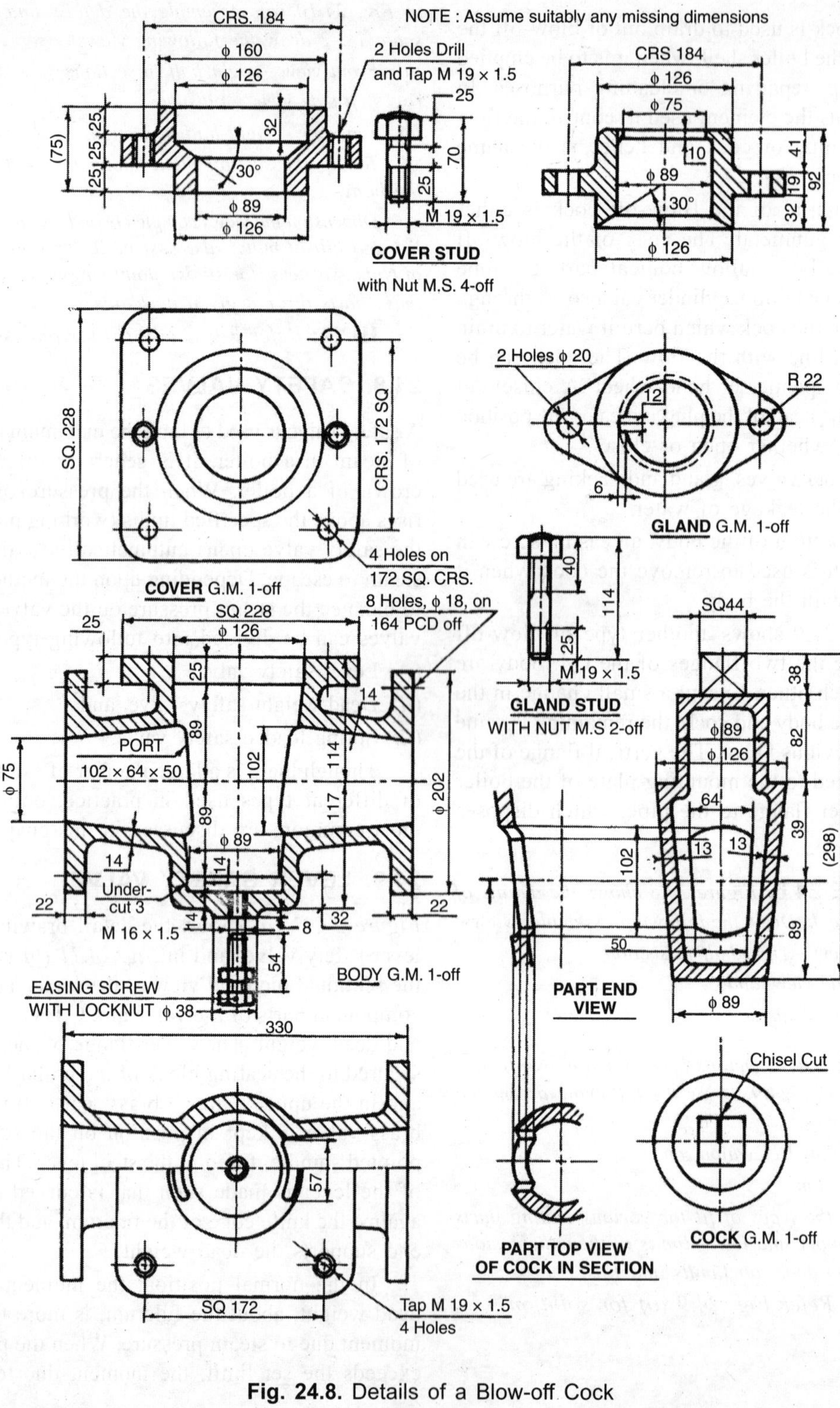

Fig. 24.8. Details of a Blow-off Cock

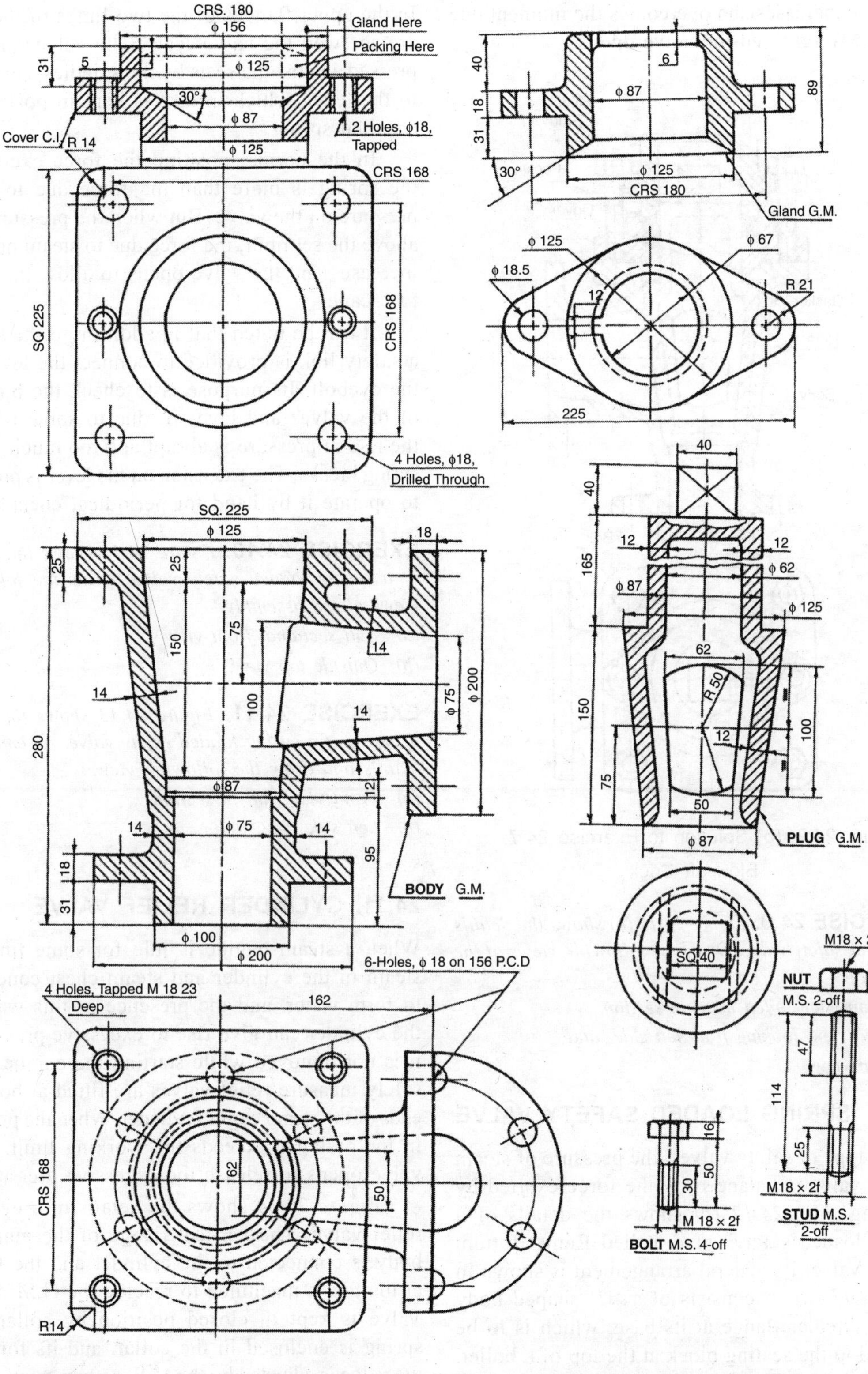

Fig. 24.9. (a) Details of a 75 mm Blow-off Cock (First Angle Projection) Not to Scale

pressure increases and overcomes the moment due to dead weight and the valve opens.

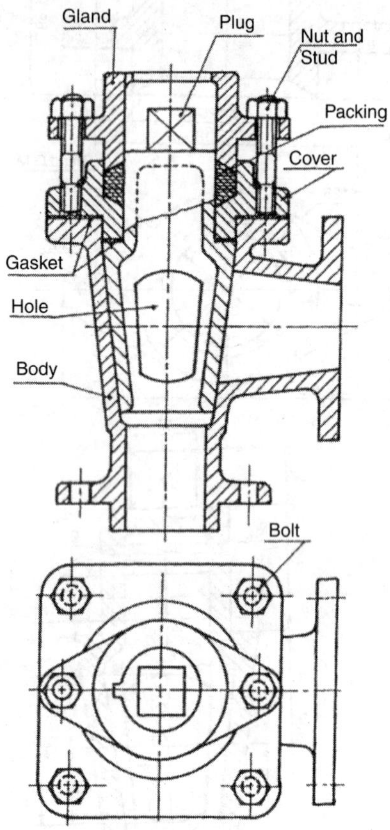

Fig. 24.9. (b) Solution to Exercise 24.7, Blow-off Cock

EXERCISE 24.9. *Figure 24.11 (b) shows the details of a lever safety valve. Draw the following views of the assembly :*

(a) Front view – left hand in section,

(b) End view, looking from left side, and

(c) Top view.

24.10. SPRING LOADED SAFETY VALVE

In this type of safety valves, the pressure of steam on the valve is balanced by the force exerted by a spring. *Fig. 24.12 (b)* shows the details of a spring loaded safety valve, called Rams Bottom Safety Valve. Its general arrangement is shown in *Fig. 24.12 (a)*. It consists of a 'U' shaped body with a circular flange at its base, which is to be secured to the seating block at the top of a boiler. In the upper flanges of the two limbs of the body are screwed the valve seats. The valves are kept pressed against the seats by the spindles, connected to the lever, which in turn is kept in position by a coiled spring.

In the normal position the force exerted by the spring is more than the force due to steam pressure on the valve. But when the pressure rises above the set limit, the force due to steam pressure increases and the valve opens to allow the steam to escape.

It can be noted that in addition to the spring, a safety link is provided to connect the lever and the eyebolt. Its purpose is to check the blow-off of the valves and lever if, due to some reasons, the rise in pressure is abrupt and too much, or the spring breaks, The extension on the lever is provided to operate it by hand for periodical checking.

EXERCISE 24.10. *Figure 24.12 shows the details of a spring loaded safety valve. Draw the following views of the assembly:*

(a) Full sectional front view,

(b) Outside top view.

EXERCISE 24.11. *Figure 24.13 shows the detail drawing of a spring loaded safety valve. Assemble the details and draw the following views:*

(a) Front view, full in section,

(b) Left side view, and

(c) Top view.

24.11. CYLINDER RELIEF VALVE

When a steam engine is idle for some time the steam in the cylinder and steam chest condenses to form water, and the presence of this water in the cylinder can give rise to excessive pressure if it is not removed while starting the engine. As a safety measure relief valves are fitted at both the ends of the steam engine cylinder. When the pressure in the cylinder exceeds the working limit, relief valve opens to release the excessive pressure.

Figure 24.14 shows the details of a cylinder relief valve. The vertical flange of the gunmetal body is connected to the cylinder and the flange at the top is machined to receive the G.M. valve. Valve is kept in closed position by collar. The spring is enclosed in the collar, and its force on the valve is adjusted by the M.S. compression screw.

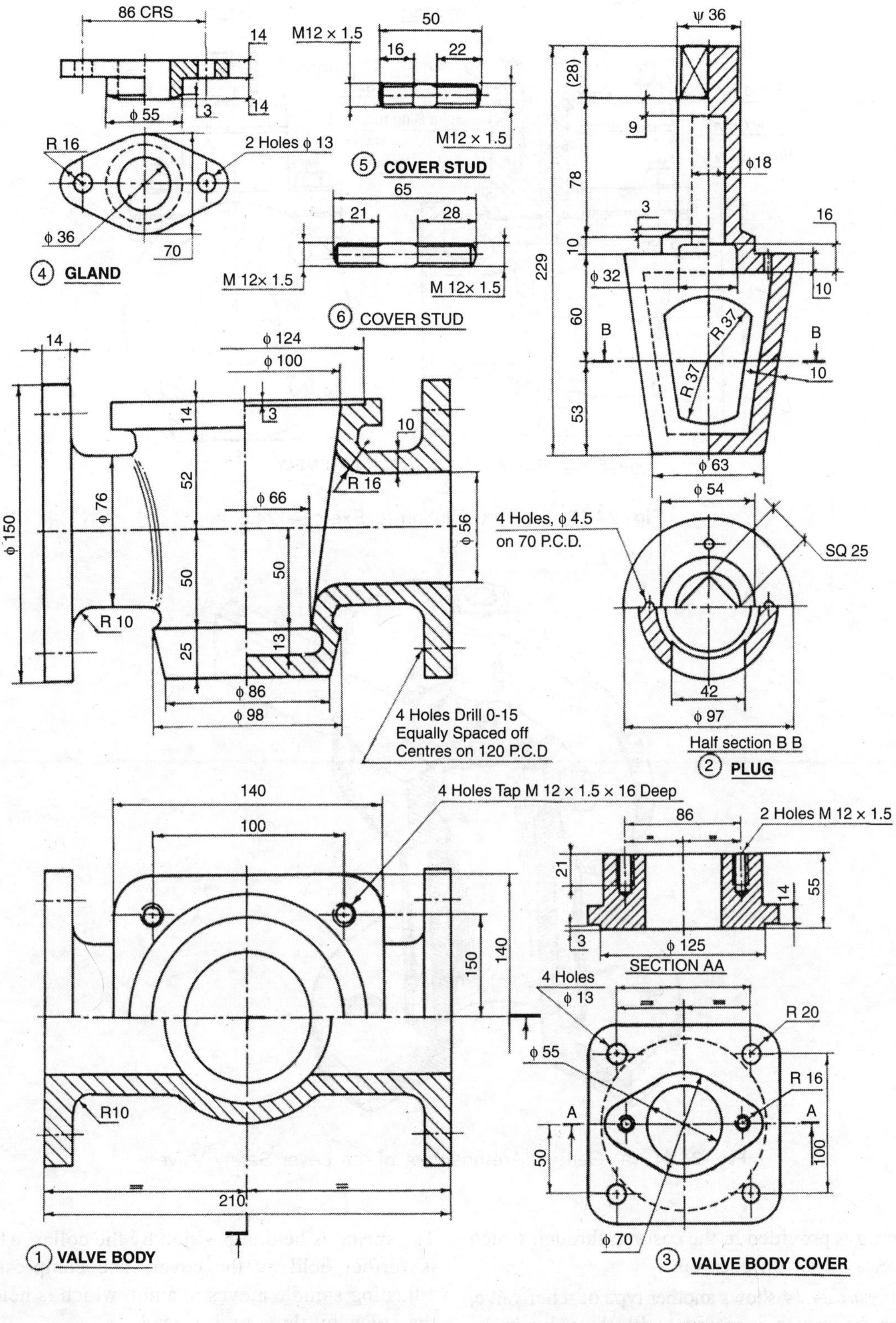

Fig. 24.10. (a) Details of a Plug-cock

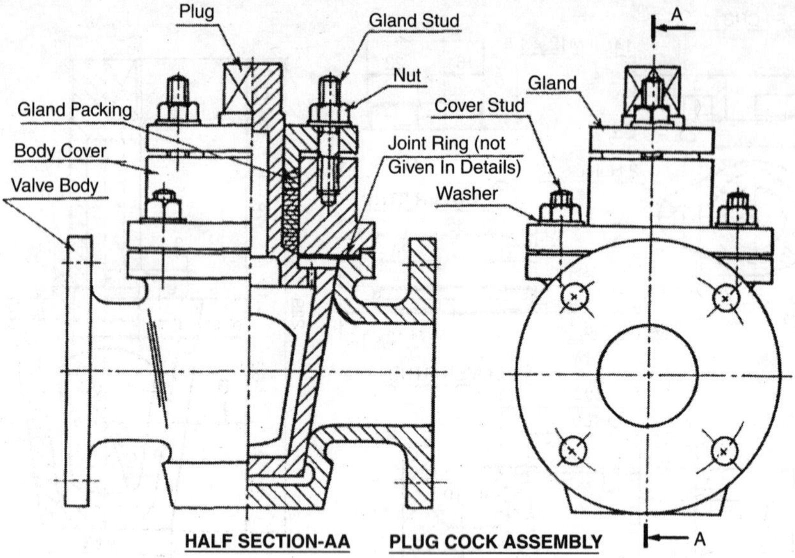

Fig. 24.10. (b) Part solution to Exercise 24.8

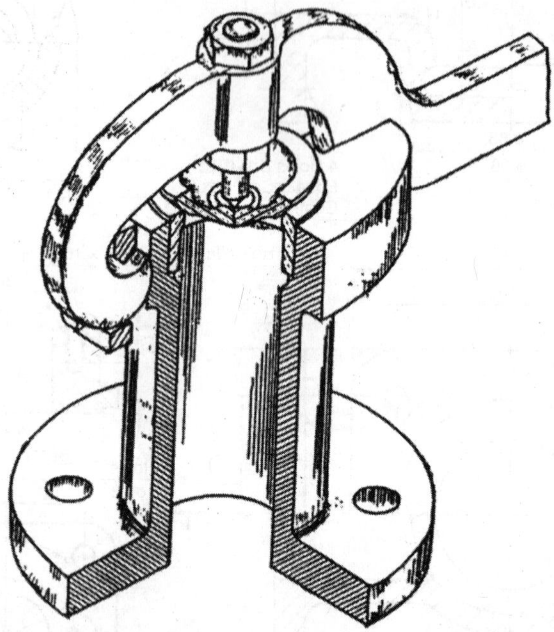

Fig. 24.11. (a) General Arrangement of the Lever Safety Valve

Opening is provided in the casting, through which the released steam goes out.

Figure 24.14 shows another type of relief valve, where the casting is integral with the valve body. The spring is held in position by the collar, which is further held by the cover. The compression adjusting spindle moves in a nut, which is held in the cover by three grub screws.

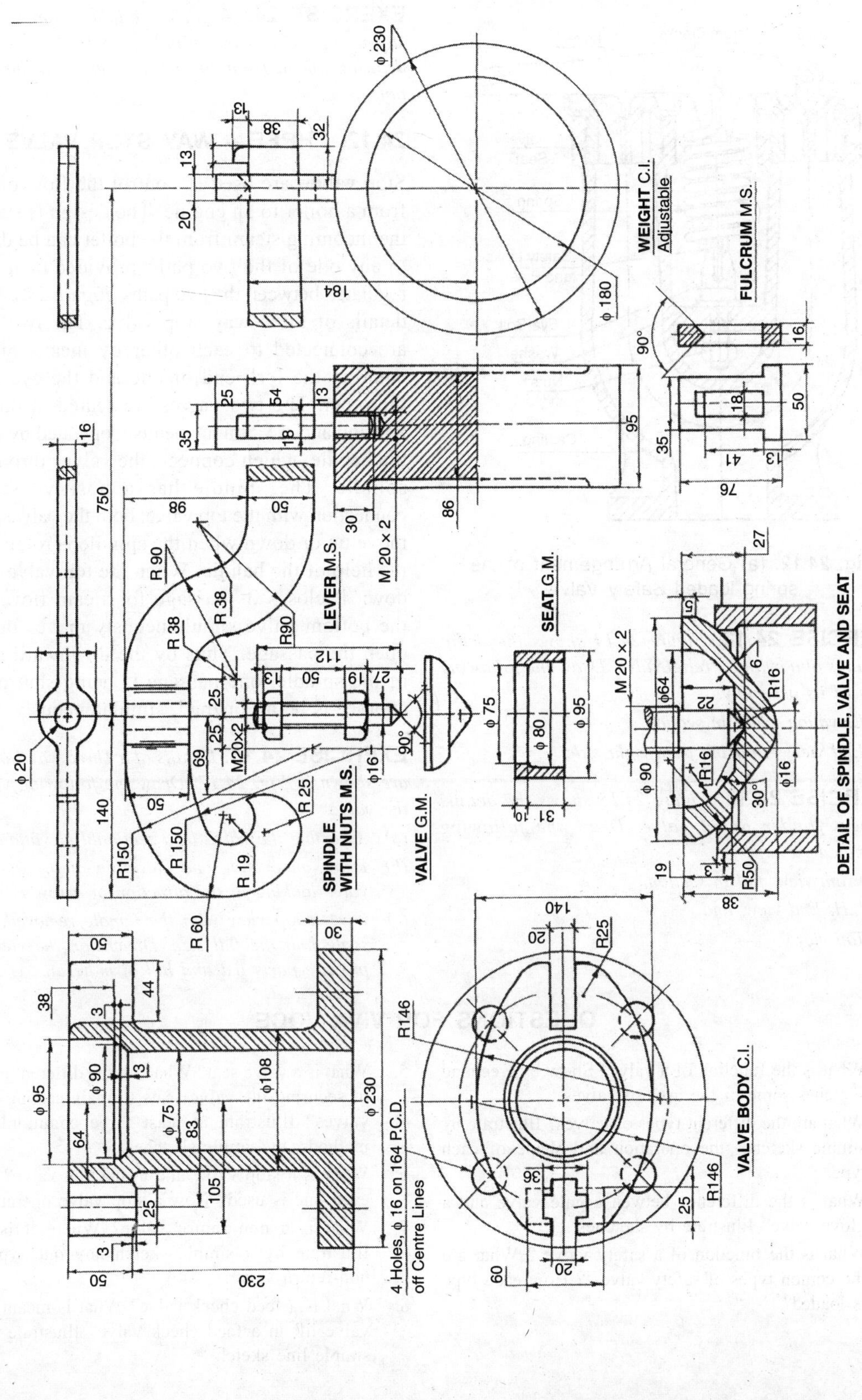

Fig. 24.11. (b) Details of a Lever Safety Valve (First Angle Projection)

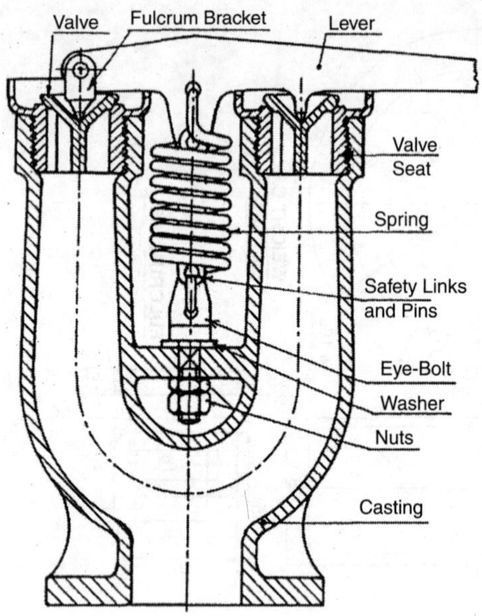

Fig. 24.12. (a) General Arrangement of the spring loaded Safety Valve

EXERCISE 24.12. *Figure 24.14 shows the detail drawings of a cylinder relief valve. Draw the following views of its assembly:*

(a) Elevation – full in section,

(b) End view – looking from right side.

EXERCISE 24.13. *Figure 24.15 shows the details of the Cylinder Relief Valve. Draw the following assembled views:*

(a) Front view, full in section,

(b) L.H. End view, and

(c) Top view.

EXERCISE 24.14. *Full sectional front view of a steam stop valve is shown in Fig. 24.16. Draw its detail drawings. Dimension the views, assume any dimensions not given.*

24.12. THREE – WAY STOP VALVE

Stop valves are used to control the flow of steam from a boiler to an engine. The design is such that the incoming steam from the boiler can be directed to any one of the two paths provided or it can be regulated between the two paths. *Figure 24.17* shows details of three way stop valve. The two valves are connected to each other by means of a pin entering the fork end of one and the eye end of the other. The two valves are seated in the main housing and are simulataneously operated by turning the handle, which connects the valves through the spindle. The spindle has a positive screwed connection with the top valve, both the valves jointly move up or down when the spindle is rotated with the help of the handle. When the top valve moves down it closes its passage for steam flow, while the bottom valve simultaneously moves down to open the passage. Thus by the downward motion of the spindle, one passage is being shut off and another is being opened, simulateneously.

EXERCISE 24.15. *Details of a Three way stop valve are shown in Fig. 24.17. Draw the following views of the valve:*

(a) Assemble the valve in a sectional elevation.

(b) Draw an outside end elevation of the assembled valve looking in the direction of arrow P.

(c) Give a top view with the handle removed. Use a scale half the full size. Dimension the views and prepare parts list and bill of materials.

QUESTIONS FOR VIVA VOCE

1. What is the function of a valve? Show by freehand sketches some of the uses of valves.
2. What are the different types of valves? Illustrate by simple sketches the operation and shape of each type.
3. What is the difference betwen a gate valve and a glove valve? Illustrate by sketches.
4. What is the function of a safety valve ? What are the comon types of safety valves ? How each type is loaded?
5. What is a valve seat? What are the different methods of securing the valve seats to main castings of the valves? Illustrate at least three commonly used methods, by simple sketches.
6. What is a stop valve and a junction valve? Where each one is used? How a stop valve operates?
7. What is a non return valve? Where it is used? Illustrate by a simple sketch any one type of a non-return valve.
8. What is a feed check valve? What is meant by the valve lift in a feed check valve? Illustrate it by a simple line sketch.

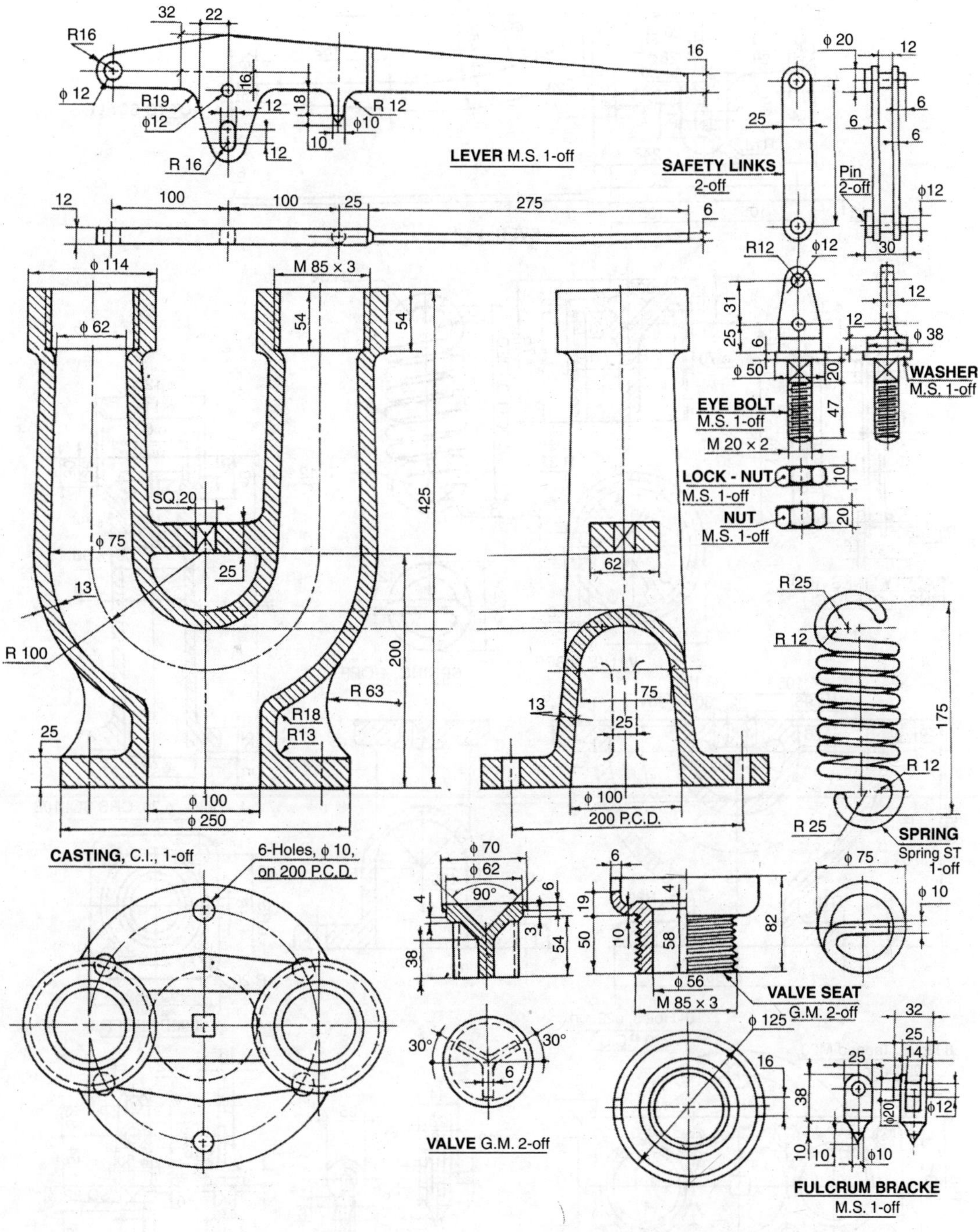

Fig. 24.12. (b) Details of a Spring-loaded Safety Valve

9. What is a blow-off cock and where it is used? Why in certain cases a screw is fitted in the main body, below the cock?

10. What is the function of a relief valve and where it is fitted?

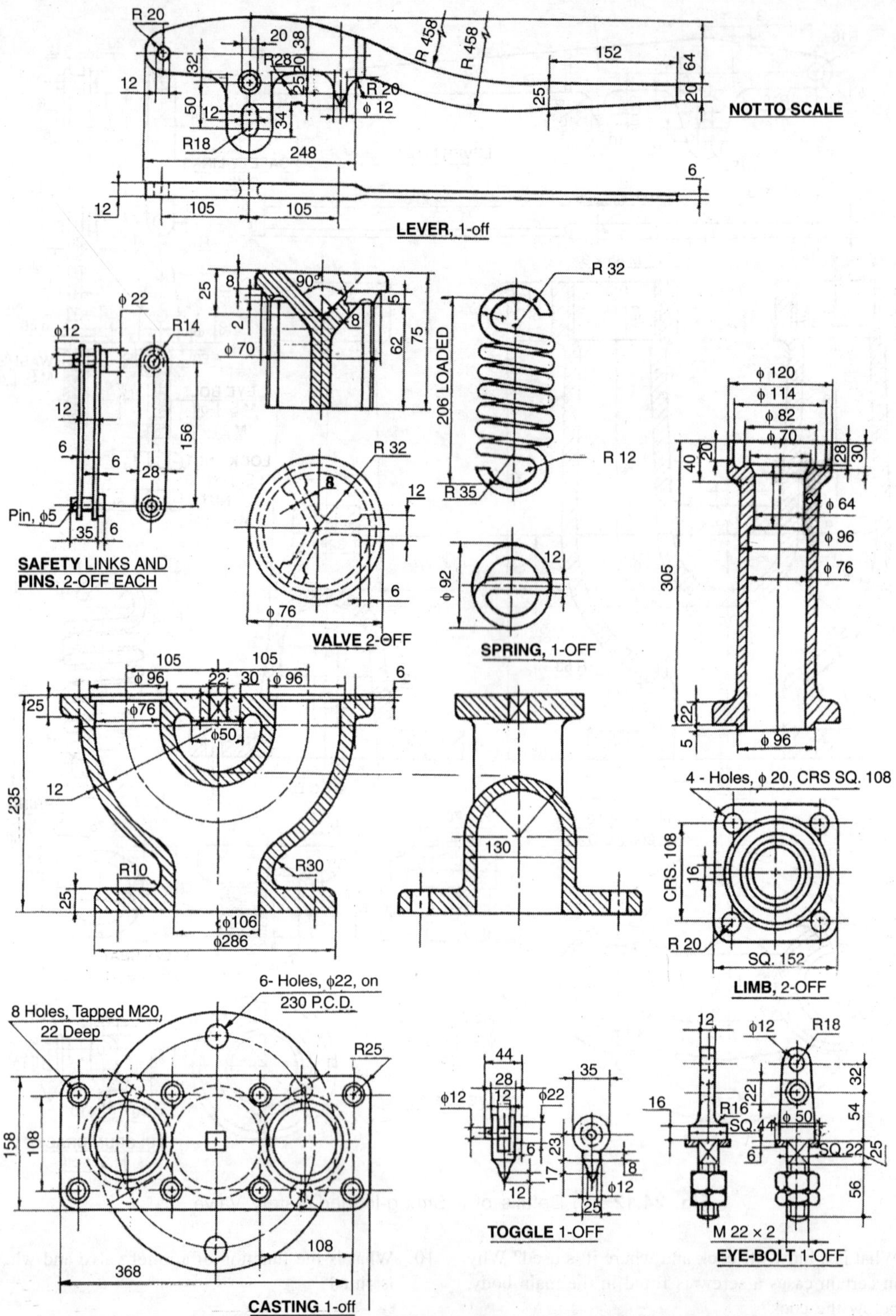

Fig. 24.13. Details of a spring-loaded Safety Valve

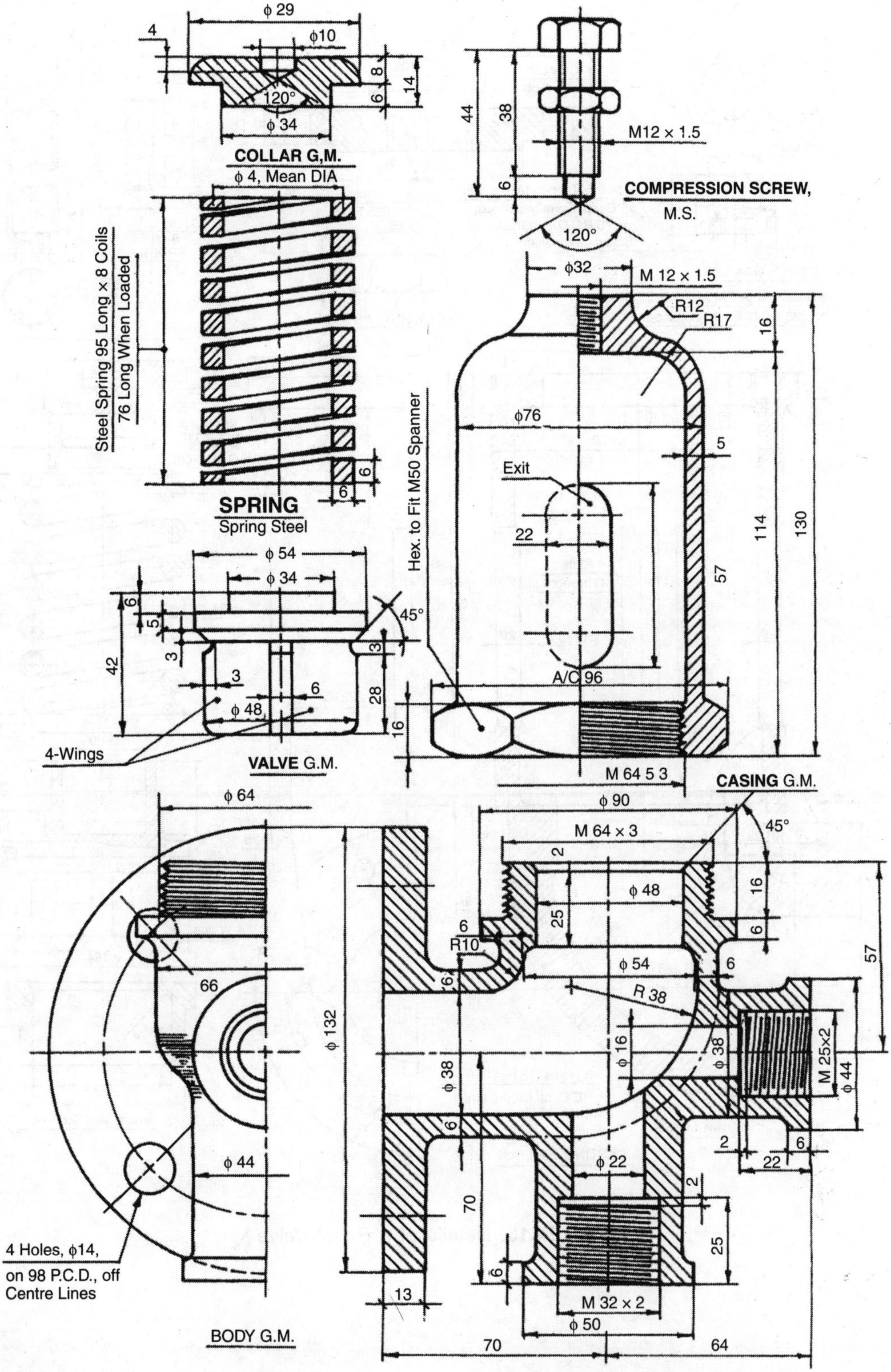

Fig. 24.14. Details of a Cylinder Relief Valve

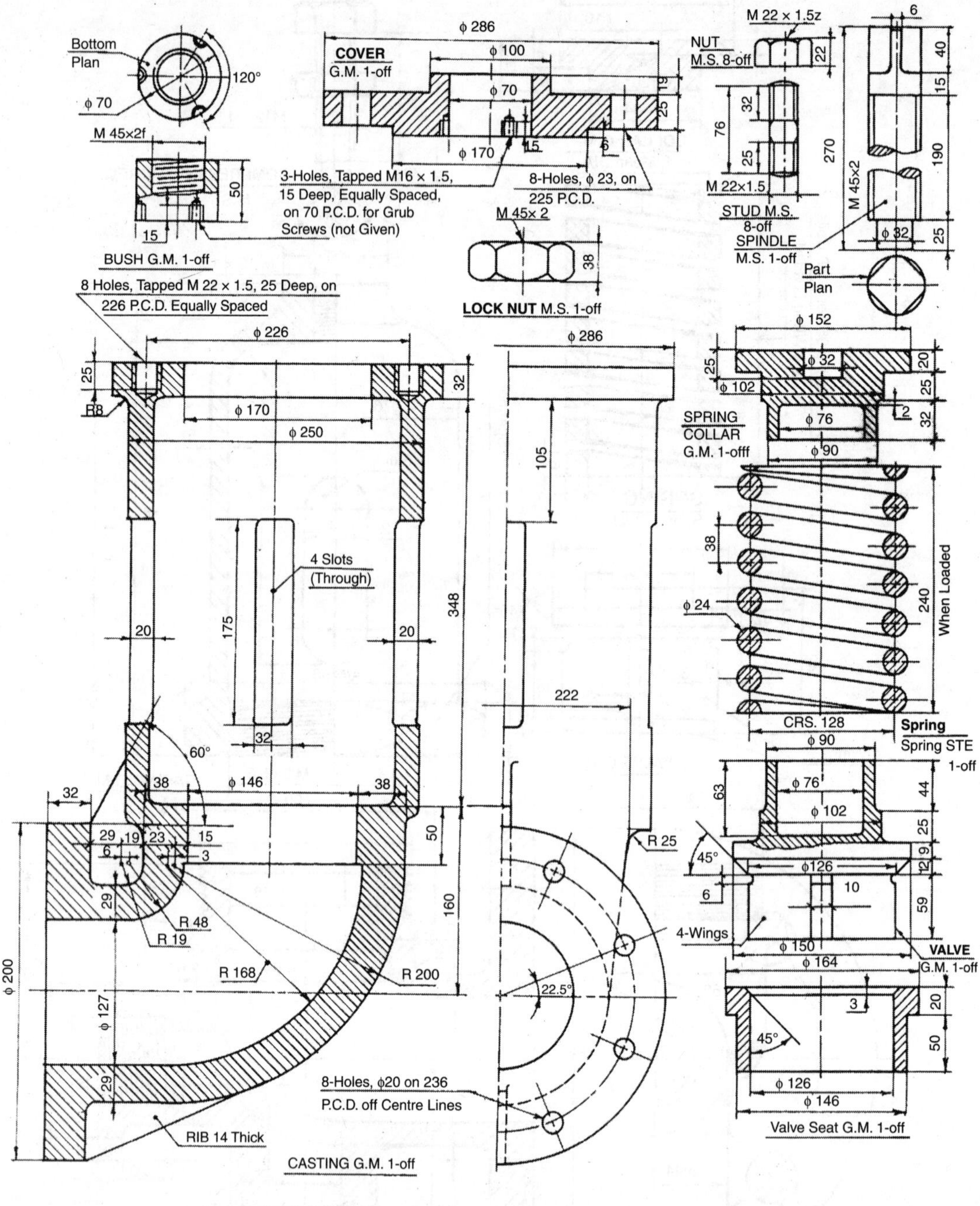

Fig. 24.15. Detaila of a Relief Valve

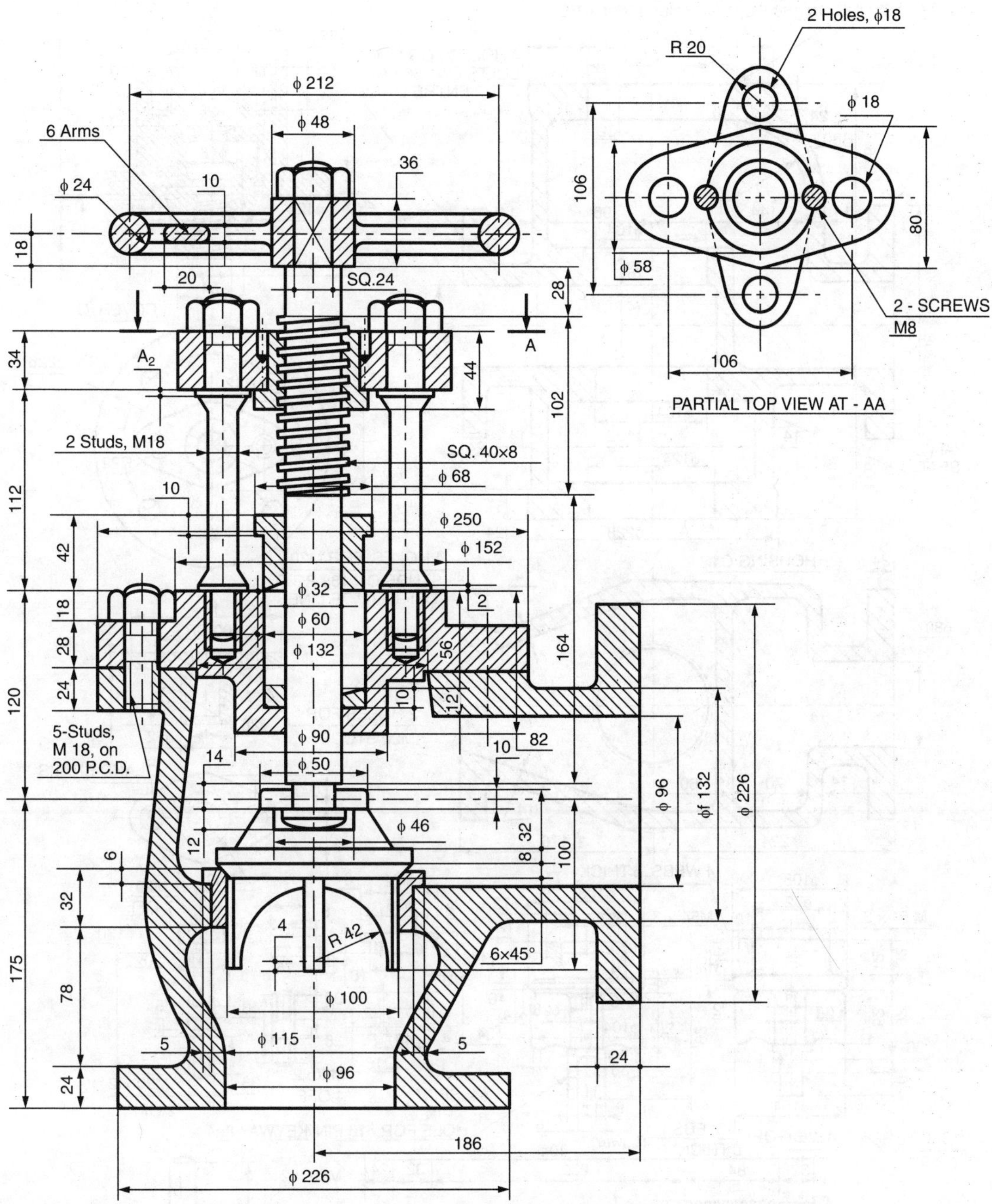

Fig. 24.16 Steam Stop-Valve (Assume, proportionately, any dimensions not given)

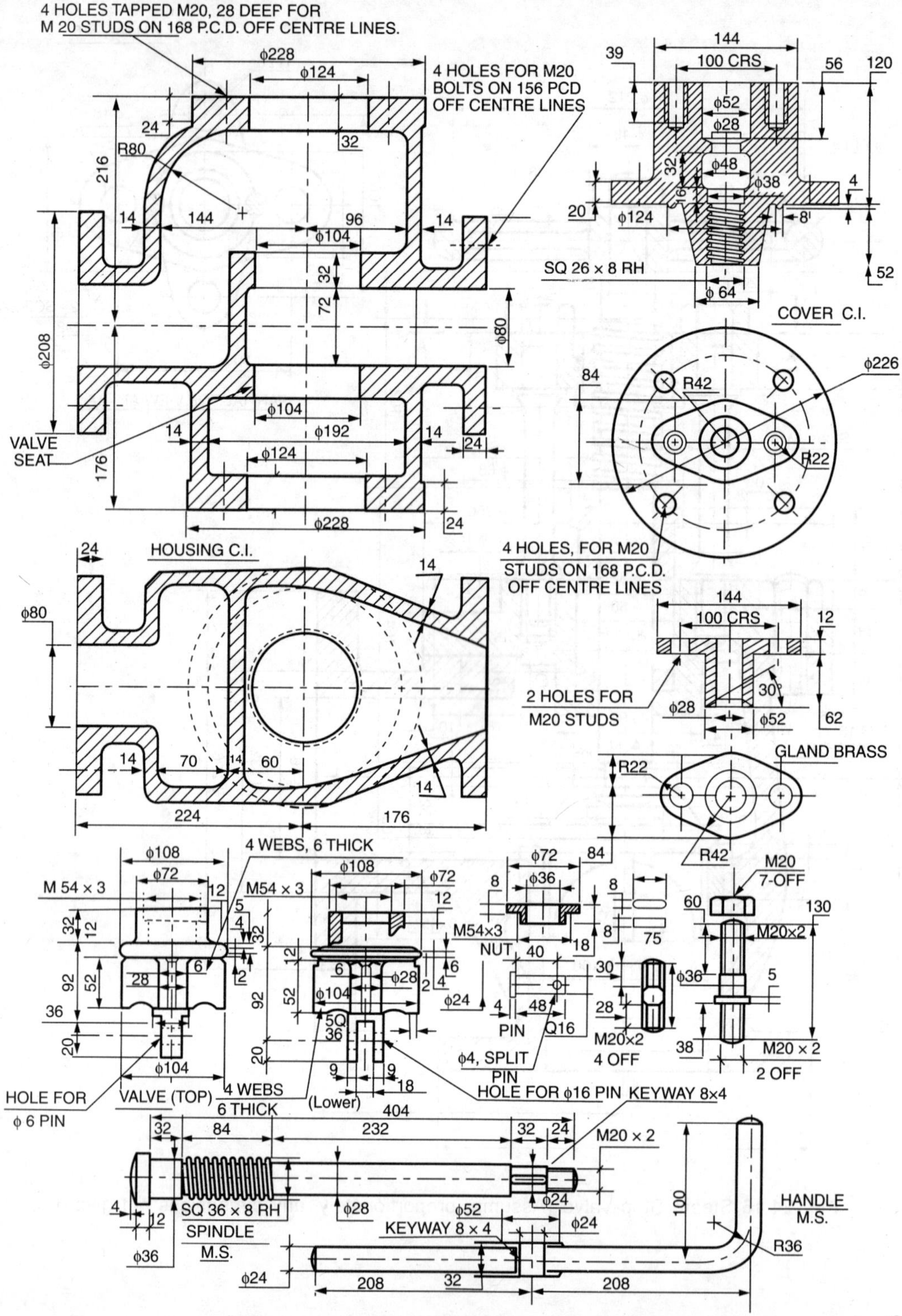

Fig. 24.17. Details of Three-way Stop valve

25 Steam Engine Parts

25.1. INTRODUCTION

The function of an engine, which may be a steam engine or an I.C. engine, is to convert the heat energy in its fuel into mechanical power. In steam engines the fuel supplied to the engine is steam at high pressure, generated in a boiler where the water is heated and converted into steam at the desired pressure, controlled by valves.

To make the students conversant with the function of different parts of a steam engine, line diagram of a simple steam engine is shown in *Fig. 25.1*. The names of important parts are labelled in this diagram. Supply of steam to the cylinder is controlled by the D-slide valve, which slides in the steam chest and gets its motion from the eccentric. Steam expands in the cylinder and makes the piston to slide. This sliding motion of the piston is converted into rotary motion of the crankshaft, through the crosshead and connecting rod. Steam chest, D-slide valve parts, and the valve operating mechanism are not shown in this figure.

25.2. STEAM ENGINE PISTONS

A steam engine piston is a short cylindrical body, which is free to move in a hollow cylinder, in such a way that practically no escape of the fluid, from one side of the piston to the other, takes place. The function of the piston is to convert the pressure of the working fluid into motion. The piston is secured to the piston rod, which passes through the stuffing box in the cover of the cylinder and connects the piston to the corsshead. The crosshead ensures straight-line motion of the piston.

Steam engine pistons are usually made of cast iron or cast steel. *Figure 25.2* shows a "Box type" cast iron steam engine piston. This piston is called box type because it is made hollow to reduce its weight. This piston is provided with grooves in which piston rings are fitted. The outside diameter of the ring is made slightly larger than inside diameter of the cylinder, so that, when put in position, it exerts a little force on the cylinder walls and thus prevents the steam leakage across the piston. In this piston, *Fig. 25.2*, two cast iron piston rings are shown. The rings are held in position in the grooves by dowel pins. To minimize the leakage, through gaps between the ring ends, the

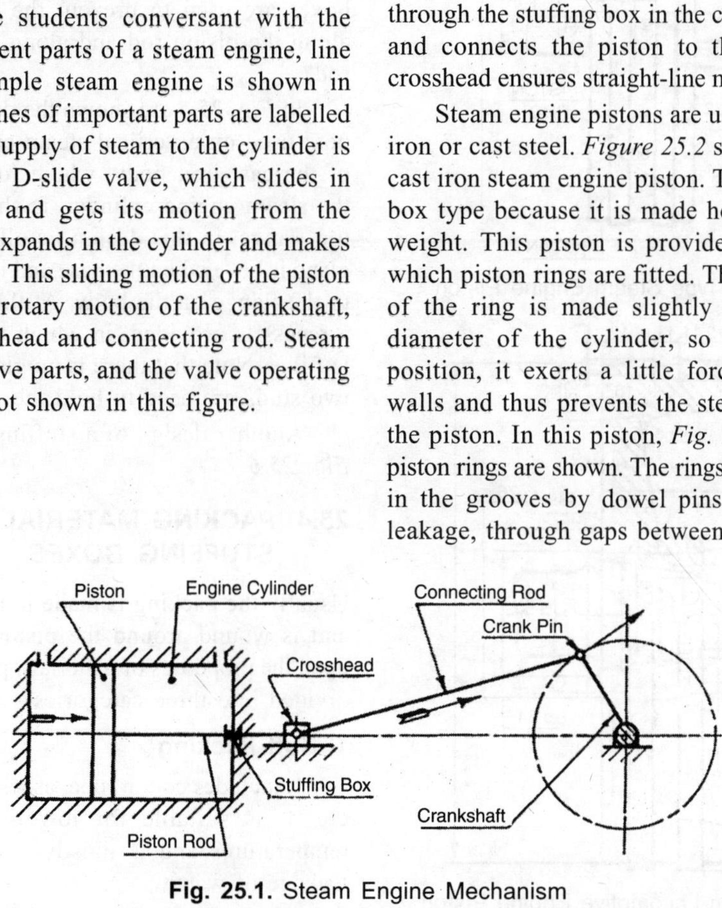

Fig. 25.1. Steam Engine Mechanism

open ends of the rings are not placed in line with each other. Note that the open ends of the rings are not faced parallel, but at an angle to the line of stroke. There are three plugged holes in the piston body. These have no functional purpose, but are kept to remove the core out to hollow the casting at the time of manufacture. The piston is shown secured to the piston rod by a nut and a split pin. The end of the piston rod is tapered to a length somewhat less than the length of the piston. The remaining length is parallel and threaded. This ensures tight hold between the piston and piston rod.

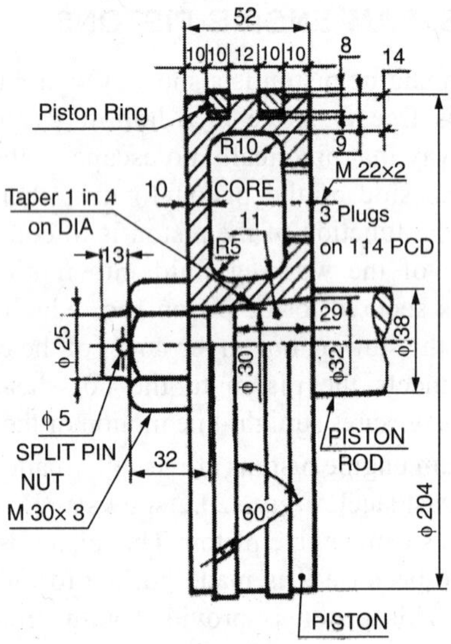

Fig. 25.2. Box type Steam Engine Piston

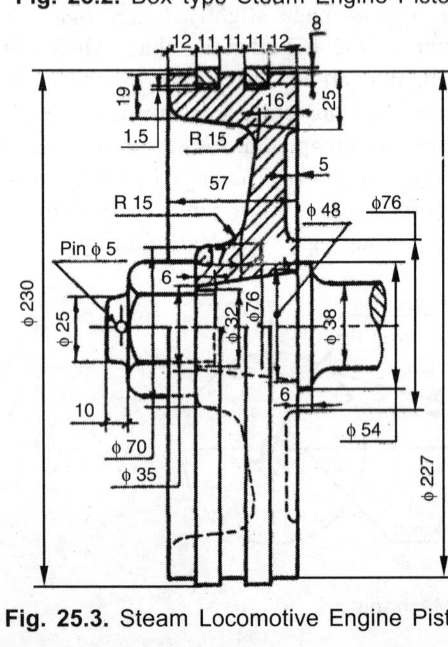

Fig. 25.3. Steam Locomotive Engine Piston

Figure 25.3 shows another type of steam engine piston, which is used on a locomotive. The body of the piston in this case is not hollow, but is just a flanged thin disc. The locomotive pistons are usually made of cast steel.

EXERCISE 25.1. *Referring to Fig. 25.2, draw the half sectional front view and left side view of the box type steam engine piston. Give the necessary dimensions.*

EXERCISE 25.2. *Draw the half sectional front view and an outside end view of the locomotive piston, shown in Fig. 25.3. Assume the rings removed.*

25.3. STUFFING BOX

When a sliding or rotating rod passes through a cylinder carrying fluid under pressure, to prevent the leakage of the fluid along the sliding rod, a stuffing is used. A good stuffing box should allow free movement of the rod, with practically no leakage of the fluid. In steam engines, stuffing boxes are used to prevent the leakage of steam along the piston rod and along the D-slide valve rod.

In *Fig. 25.4* are shown the details of a stuffing box, used on a vertical steam engine. The flange of the cast iron body, which forms the cover of the steam engine cylinder, is shown broken. The top flange, in this design, is elliptical. Gunmetal gland is beveled on the base and its flange is similar to the flange on the body. At the top of the gland a recess is provided, in which the lubricating oil is filled. Note that there is no gland bush and only two studs are used to hold the gland in position.

Another design of a stuffing box is shown in *Fig. 25.6.*

25.4. PACKING MATERIAL FOR USE IN STUFFING BOXES

Usually the packing is made in the form of a rope and is wound around the piston rod. Depending upon the properties of materials, packing are usually divided into three categories:

1. Soft Packing

It includes cotton rope, asbestos, Indian rubber, etc. It is suitable for low pressures and low temperatures and is mostly used to prevent the leakage of water.

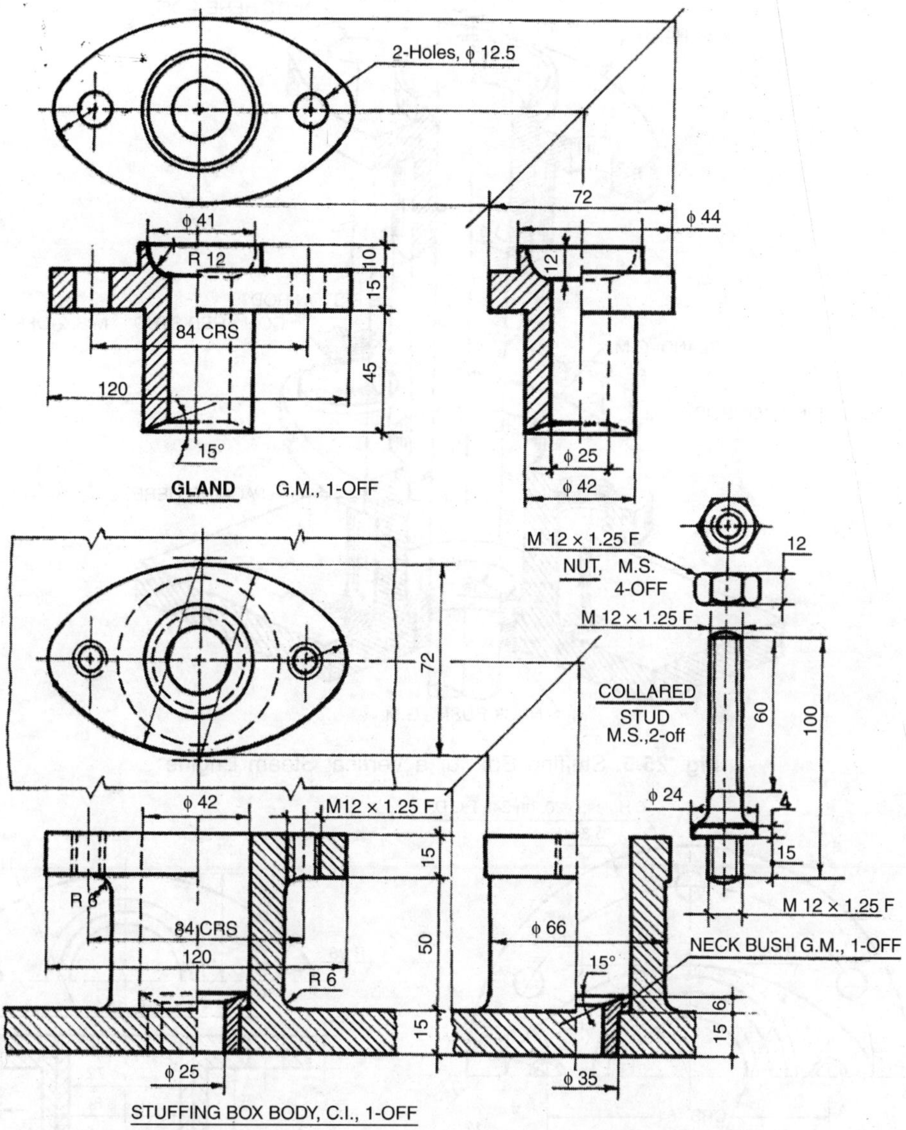

Fig. 25.4. Details of a Stuffing Box for Vertical Steam Engine

2. Semi-metallic Packing

It is a combination of soft packing and metallic packing. It consist of Indian rubber, asbestos, and canvass, etc., lined with metallic cores. It can be subjected to medium pressures and temperatures.

3. Metallic Packing

It consists of soft metal rings, which are pressed against the piston rod by suitable collars and springs. The advantages of metallic packing are:

- It offers minimum of functional frictional resistance to the piston rod and thus absorbs less power.
- Once set properly, no attention other than lubrication is required for a long time.

EXERCISE 25.3. *From the details of a stuffing box given in Fig. 25.4, draw the following views of the assembly:*

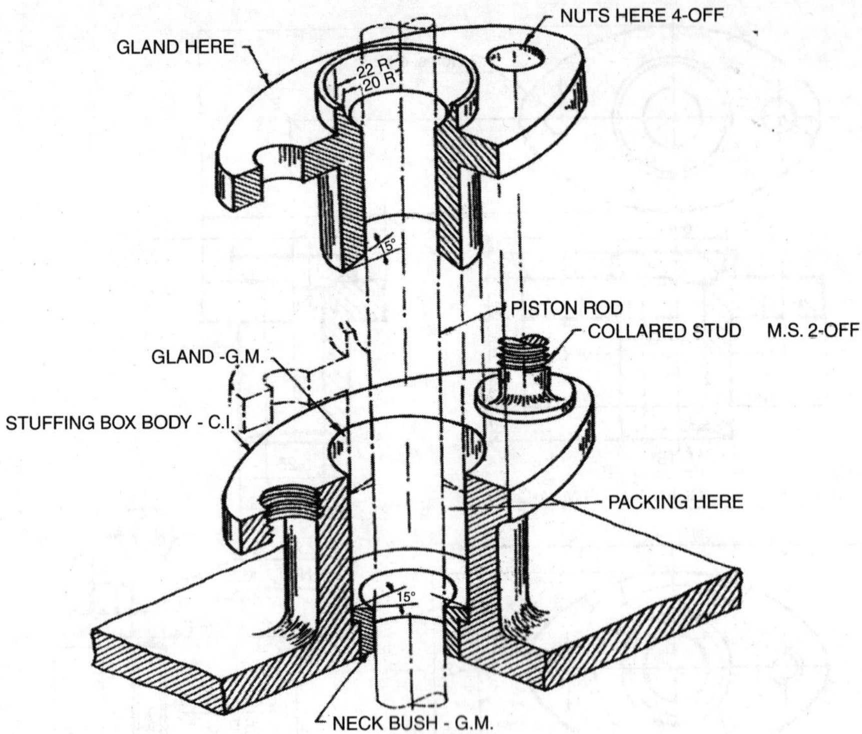

Fig. 25.5. Stuffing Box for a Vertical Steam Engine

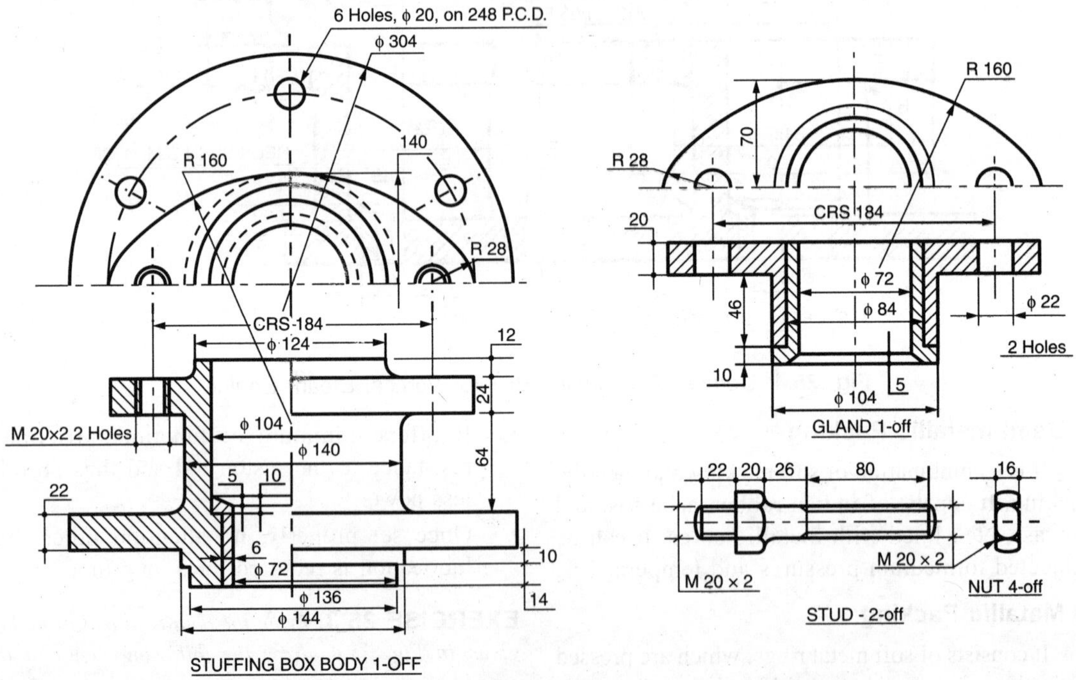

Fig. 25.6. Stuffing Box Details

(a) *Front view – left half in section,*

(b) *Side view, and*

(c) *Top view.*

Suggest suitable fits for the mating parts and then determine the upper and lower limits of size for each fit.

HINT: *Figure 25.5 shows the exploded pictorial view the stuffing box.*

EXERCISE 25.4. *Figure 25.6 shows detail drawings of a stuffing box.*

(a) *Front view – left half in section, and*

(b) *Top view – outside.*

Give all the necessary dimensions. Prepare a suitable drawing sheet format.

25.5. CROSSHEAD

It is that part of a steam engine which connects the piston rod with the connecting rod. Sometimes it is called a Guide Block, too. Parallel bars are provided on the engine frame, which guide the crosshead motion. The function of the crosshead is to transmit the thrust due to obliquity of connecting rod to guide bars and to ensure motion of the piston along a straight line.

The piston rod end may be connected to the crosshead by a screwed union, by a bolted joint, by cottered joint, or by forging the piston rod integral with the crosshead body or with a part of the body. The small end of the connecting rod is made in the from of an eye or double eye, depending upon the design of the crosshead, and is connected to it by a pin.

The material of the crosshead body may be cast iron, wrought iron, mild steel, or forged steel.

Figure 25.7 (b) shows the details of a crosshead used on a vertical steam engine. The main parts are the body, the gunmetal brasses, and rod end and cover plate. The body has a rectangular flange at the base, which can be called a guide plate, shoe, of slipper. See that, in this crosshead, the piston rod is forged with rectangular flat plate, which forms a part of the body. Bearings are made in two halves; one half is circular at the outer periphery and is accommodated in the corresponding space in the body. The other half, which is at the outside, is held in position by the cover plate. The two cover plates and the body are held together by the bolts and nuts. In *Fig. 25.7 (a)* are shown two views of this crosshead. Note that the projection on the central body fits into a corresponding recess in the rod end plate, which ensures proper alignment of the piston rod. Also note that the cover plate does not touch the central body. The clearance left, provides re-matching the bearings to proper size when these become loose on the pin.

Figure 25.8 (a) and *(b)* show the details and general arrangement, respectively, of another crosshead. The design of the crosshead body provides for cottered joint connection with the piston rod and knuckle joint connection with the connecting rod end. A recess is made in the base of crosshead

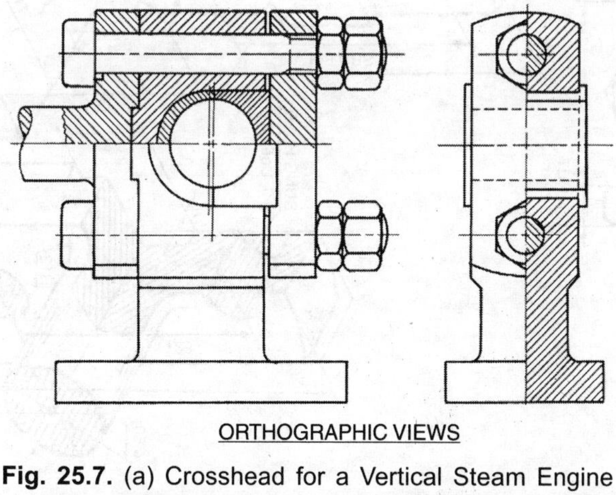

ORTHOGRAPHIC VIEWS

Fig. 25.7. (a) Crosshead for a Vertical Steam Engine

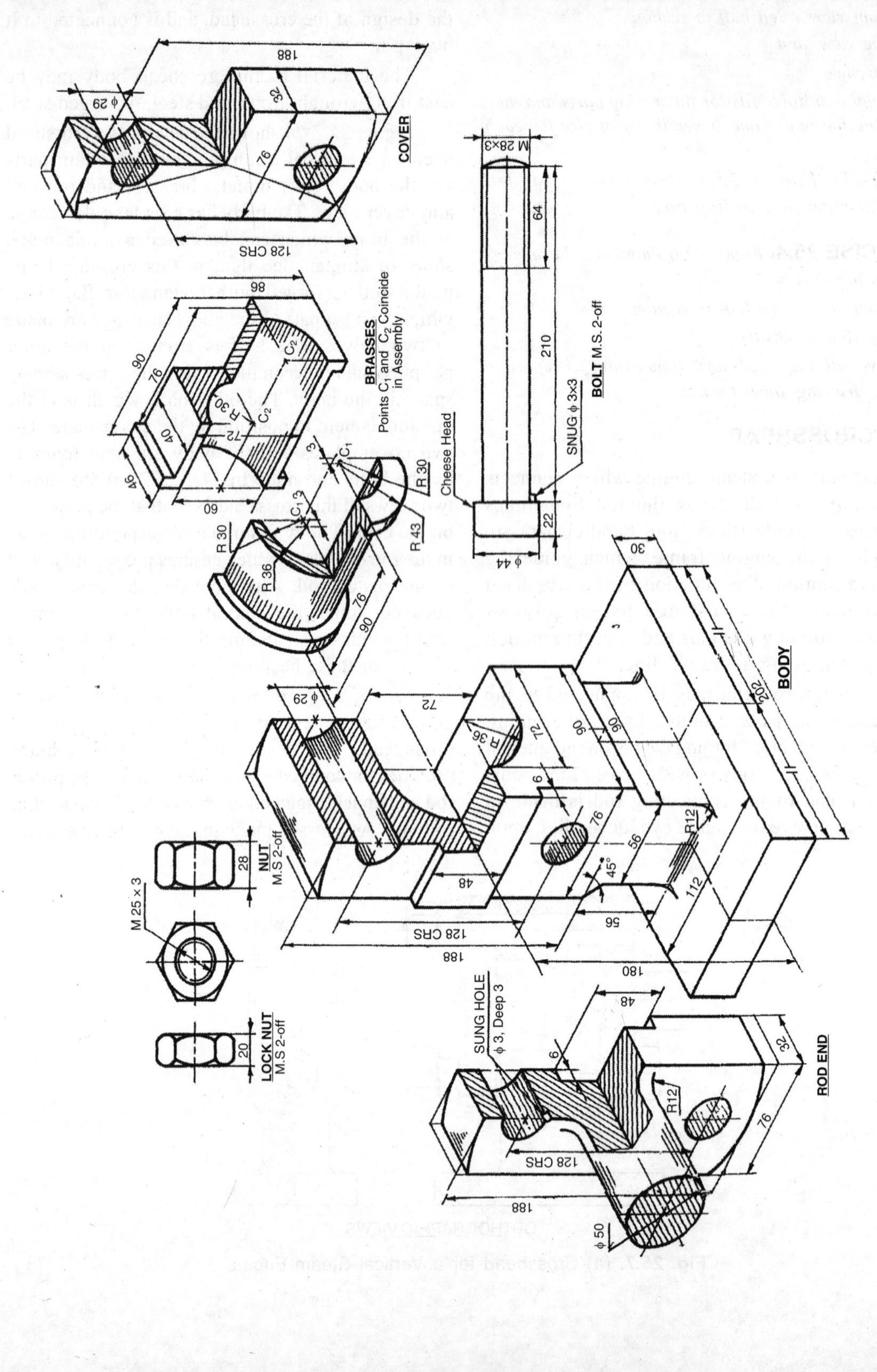

Fig. 25.7. (b) Crosshead for Vertical Steam Engine

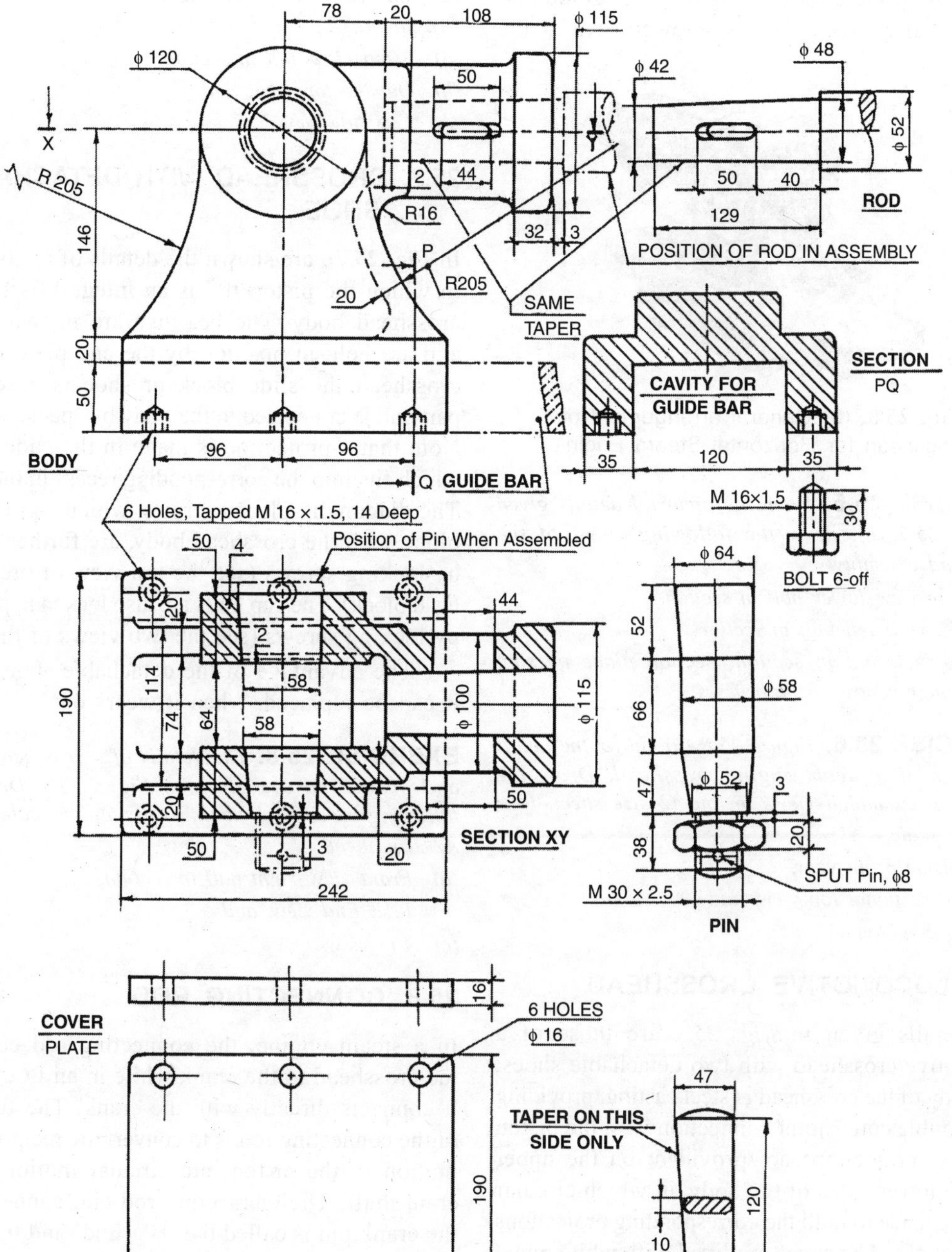

Fig. 25.8. (a) Details of a Crosshead for Horizontal Steam Engine

body, which forms a way for a rectangular guide bar. To hold the crosshead against the guide bar, a rectangular cover plate is screwed to it at the base.

Fig. 25.8. (b) General arrangement of crosshead for Horizontal Steam Engine

EXERCISE 25.5. *From the pictorial details given in Fig. 25.7 (b), draw the following views of the crosshead assembly:*

(a) Front view–lower half in section.

(b) R.H. view–left half in section.

(c) Top view–full in section. Section plane passing through centre of the brasses.

EXERCISE 25.6. *Figure 25.8 (a) shows the detail drawings of a steam engine crosshead. Draw the following assembled views on an A2 size sheet, to a suitable scale:*

(a) Outside Front view,

(b) Full sectional top view, and

(c) Right side view.

25.6. LOCOMOTIVE CROSSHEAD

The details given in *Fig. 25.9* are those of a locomotive crosshead with two detachable shoes. The body of the crosshead is steel casting, providing for a double cotter joint connection with the piston rod. The projections are provided, on the upper and the lower sides of the body, in which circular slots are made to hold the corresponding projections on the shoes. The shoes are lined with white metal pads, and provisions are also made to feed the lubrication oil. to hold the brasses in positions a wedge block is used. By turning the bolt, which passes through the wedge block, the brasses are passed in position.

EXERCISE 25.7. *Assemble the details given in Fig. 25.9 and draw the following views of the Locomotive Crosshead:*

(a) Front view, full in section,

(b) Outside top view, and

(c) Left side view.

25.7. CROSSHEAD WITH DETACHABLE SHOE

In *Fig. 25.10* are shown the details of a crosshead in which the piston rod is an integral part of the crosshead body. The bearings are in two halves and are held in position by the end plate. In this crosshead the slide block or shoe is a separate part and is connected to the body by special screws. Note that a projection is made in the slide block, which fits into the corresponding recess in the body. The diagonally placed screws, which fix the slide block with the crosshead body, are further locked by locking screws (see the top view of the cross-head block). The cap bolts are also locked in position by locking screws (see the two views of the cap).

The advantage of the detachable shoe is that it can be replaced, when it wears out.

EXERCISE 25.8. *The details of a crosshead block and piston rod are given in Fig. 25.10. Draw the following assembled views, to a suitable scale, on an A2 size sheet:*

(a) Front view–right half in section,

(b) R.H. End view, and

(c) Top view.

25.8. CONNECTING ROD

In a steam engine, the connecting rod connects the crosshead to the crank while in an I.C. engine it connects directly with the crank. The function of the connecting rod is to convert the reciprocating motion of the piston into circular motion of the crankshaft. The connecting rod end connected to the crank pin is called the 'Big End' and the other end connected to crosshead wrist pin is called the "Small End". The central body of the connecting rod may be of circular, elliptical, rectangular or I-section, with the dimensions of the cross-section decreasing towards the small end. Connecting rods are generally made of forged steel.

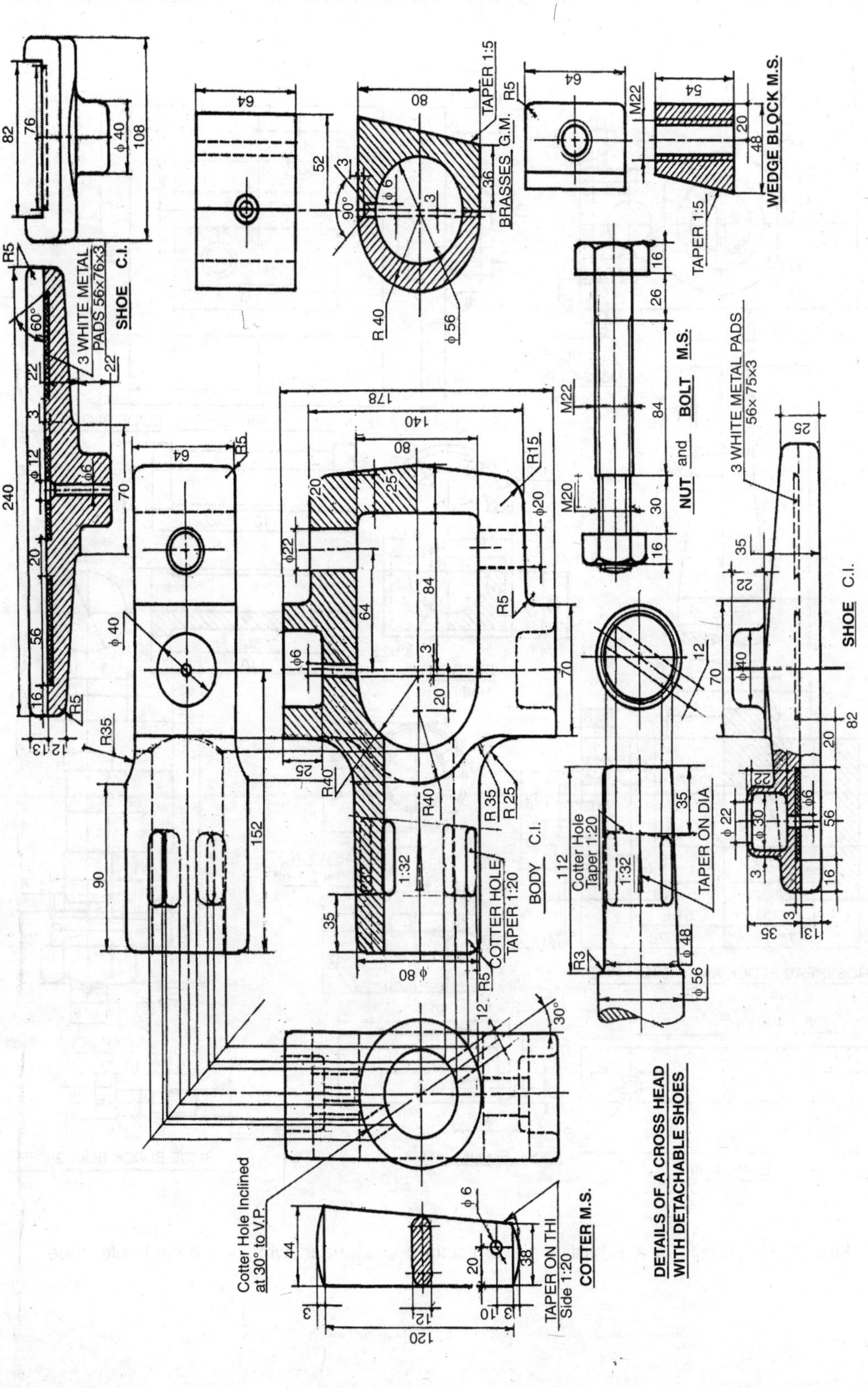

Fig. 25.9. Lo comotive crosshead with two Detachable Shoes (Third Angle Projection)

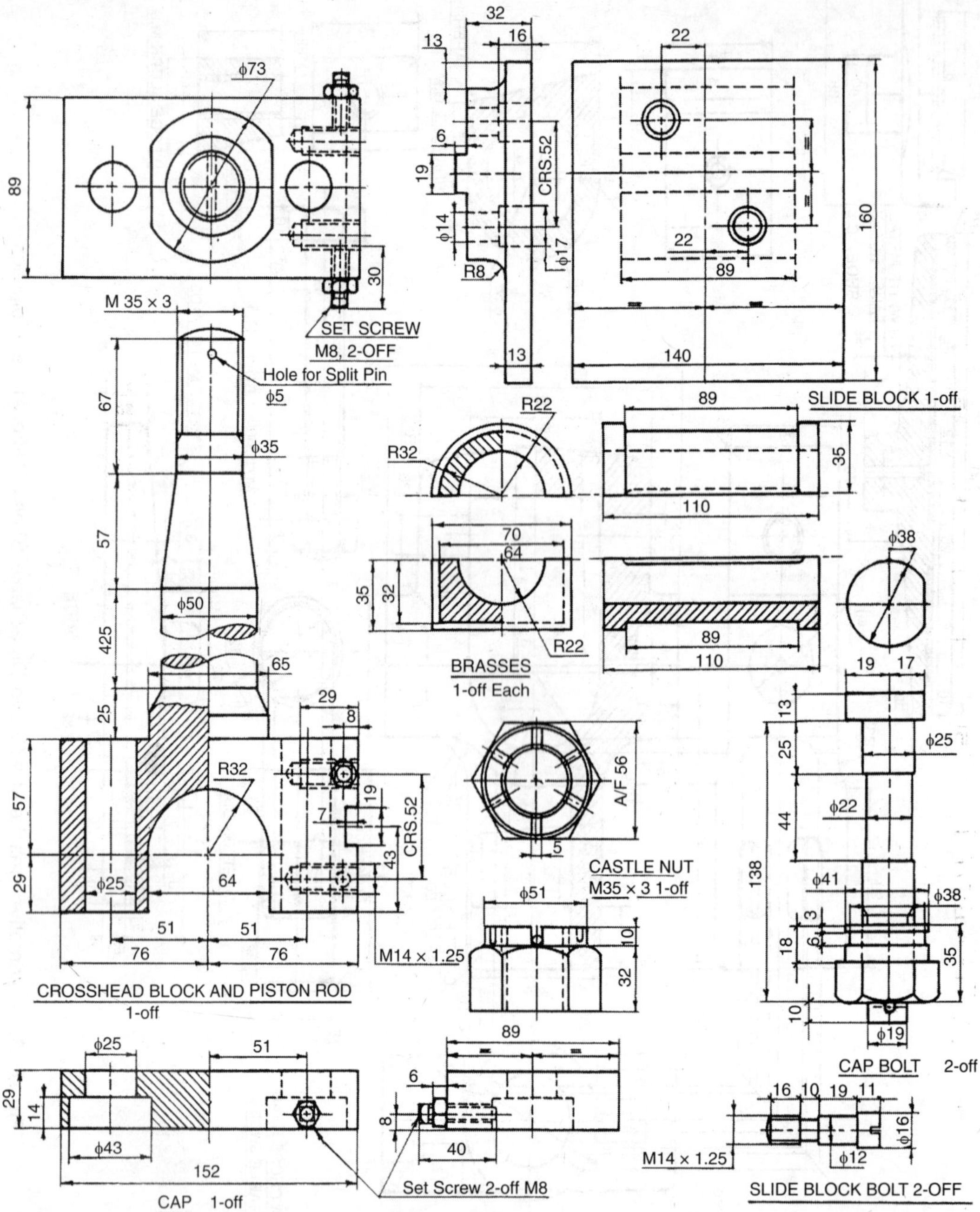

Fig. 25.10. Details of a (Third angle projection) Crosshead with one detachable shoe

25.9. CONNECTING ROD BIG END

From the drawing point of view, this is the only important portion of a connecting rod, as it is an assembly of a number of parts. A few common designs of the connecting rod big ends are included in this chapter.

In *Fig. 25.11 (a)* are shown the pictorial details of Big End (Strap End) of a steam engine connecting rod. Brasses, made in two halves, are provided with flanges to prevent any axial movement. Brasses are held tight against each other by the connecting rod end, which is secured to the strap by two gibs and a cotter. The cotter is locked in position by a setscrew. To lubricate the bearing (Brasses), oil hole is provided through the brasses and the strap. *Figure 25.11 (c)* represents the pictorial view of the assembled joint.

Figure 25.12 shows the sectional pictorial view of another connecting rod Big End, which is almost similar to the one shown in *Fig. 25.11 (a)*. The only difference is that the strap is circular at the end and there is only one Gib used. To match the shape of the strap, one part of the brasses is also made in circular shape.

EXERCISE 25.9. *From the details given in Fig. 25.11 (a) draw the following views of the connecting rod Big End assembly.*

(a) Front view–upper half in section

(b) End view–looking from right side,

(c) Top view–full in section.

Also suggest suitable fits and limits for the mating components. Use ISO chart.

EXERCISE 25.10. *From the pictorial view, shown in Fig. 25.12, draw the following views of the connecting rod Big End:*

(a) Front view–lower half in section,

(b) Top view–front half in section, and

(c) Left end view.

EXERCISE 25.11. *Assemble the details of the connecting rod Big End, given the Fig. 25.13 and draw the following views :*

(a) Full sectional Front view,

(b) Full sectional Top view, and

(c) Outside End view–looking from right.

Also prepare a bill of materials and parts-list for the drawing.

25.10. LOCOMOTIVE CONNECTING ROD END

It is plain strap type connecting rod end provided with safety bolts and is used for locomotive rods, where the forces transmitted are very high. *Fig. 25.14* illustrates the details of one such end. On the strap a projection is made in which there is a provision for fitting the lubricating cup. In addition to the slot for cotter, two holes are made in the strap as well as in the rod end for safety bolts. The brasses which are of the usual shape are lined with Babbitt metal, which adds to lubrication. The brasses are held in position by a rectangular wedge, which is pressed against the brasses by the cotter. Note that in the end of the rod uniform slot is cut, and in the wedge the slot is tapering. These two combine to make the slot for the cotter. Connecting rod end is secured to the strap by two bolts. After driving the cotter in, it is locked in position by two setscrews.

EXERCISE 25.12. *Draw the following assembled views of the Locomotive Connecting Rod End whose details are given in Fig. 25.14.*

(a) Half sectional front view,

(b) Full sectional top view, and

(c) End view–looking from left hand side.

25.11. MARINE ENGINE CONNECTING ROD END

As marine engines are of slow speed and high power type, so all its parts have to be very strong and sturdy. *Figure 25.15* illustrates different parts of the big end of a marine engine connecting rod. Here the connecting rod end consists of a rod, a bearing in two halves, a cover plate, bolts and packings. The bearing halves are manufactured as separate components to facilitate assembly of the heavy parts of the connecting rod of the marine engine.

After the two halves of the bearing have been placed around the crank pin, the cover plate is placed in position and then all these are fastened to the rod end by mean of bolts and nuts. Two snugs, one on each bolt, are provided to prevent rotation of the bolts. Split cotters are used as locking devices for the bolts. Packing strips (made of leather) are usually inserted between the two brass halves. At a later date, if the bearing gets worn out, the bearing halves can be brought still closer by replacing the original packing strips with relatively thinner ones.

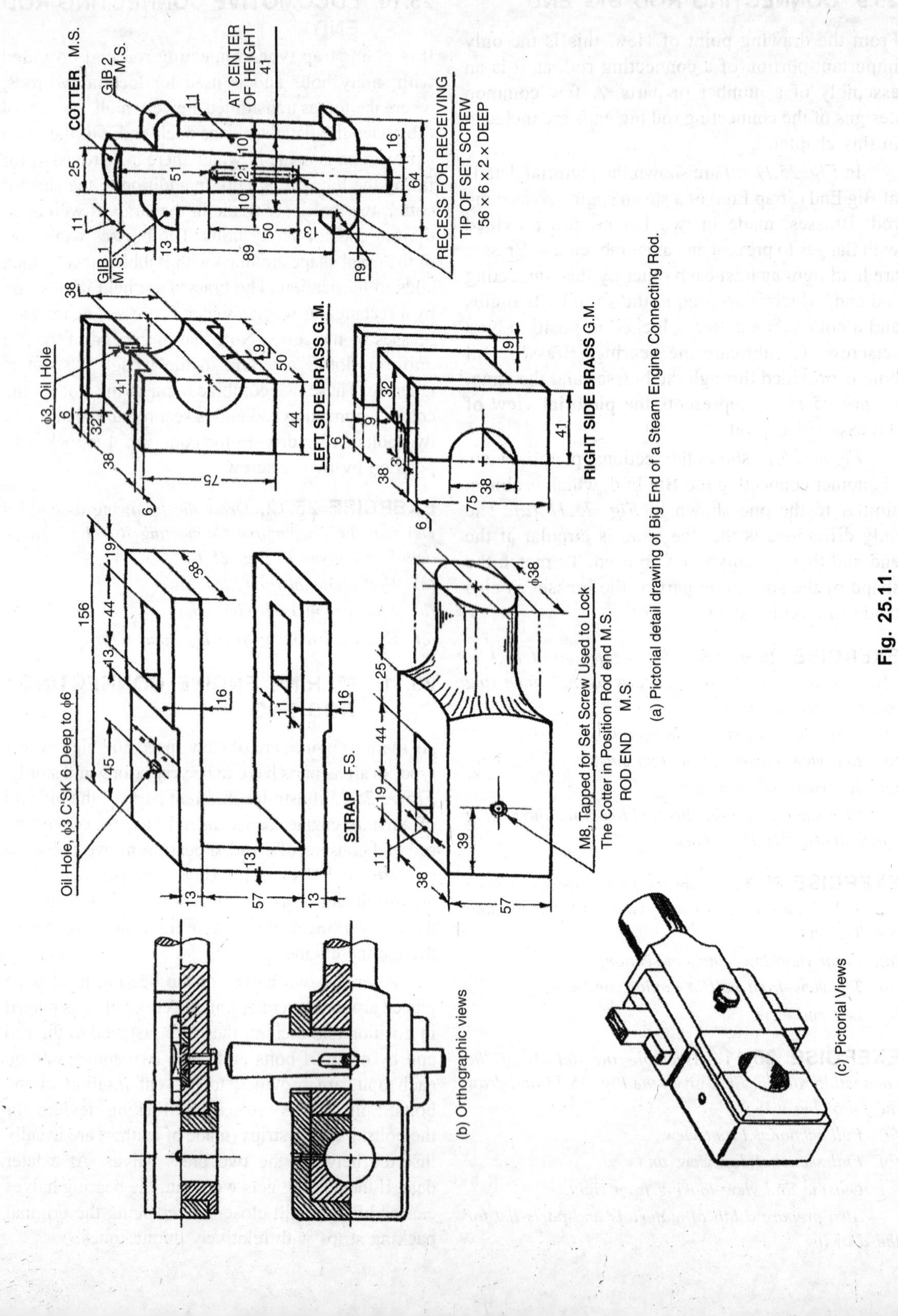

(a) Pictorial detail drawing of Big End of a Steam Engine Connecting Rod.

(b) Orthographic views

(c) Pictorial Views

Fig. 25.11.

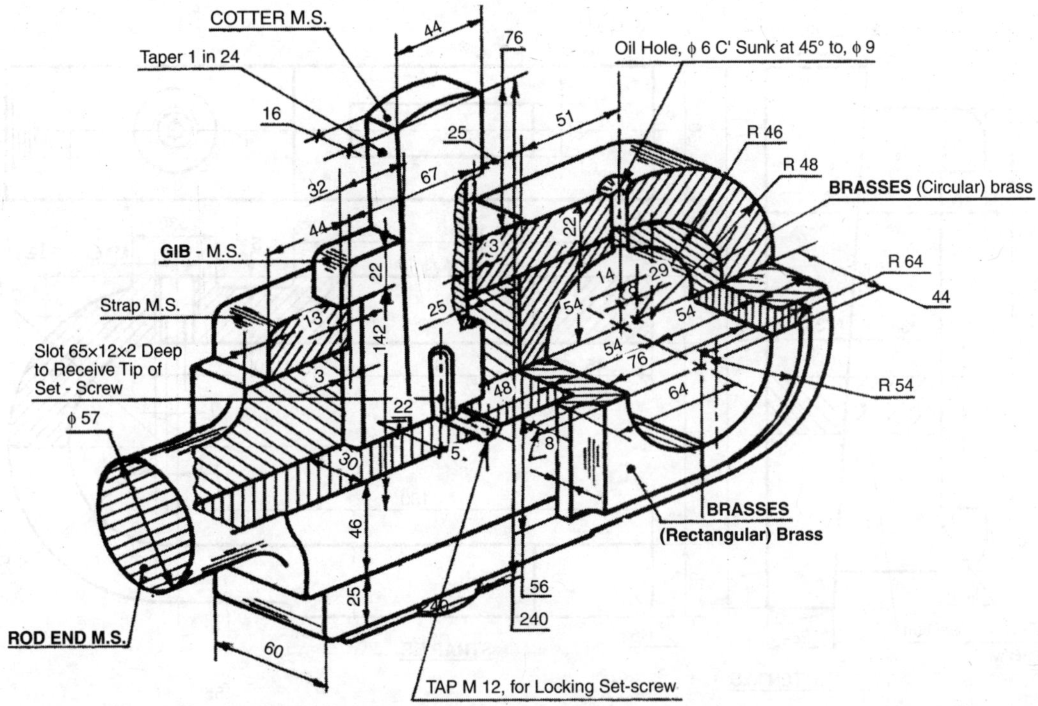

Fig. 25.12. Connecting Rod Big End (Pictorial Assembly)

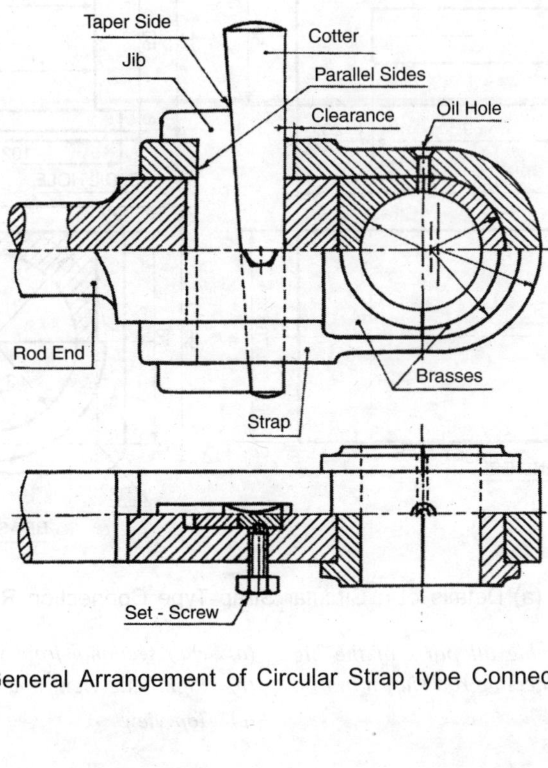

Fig. 25.13. (b) General Arrangement of Circular Strap type Connecting Rod Big End

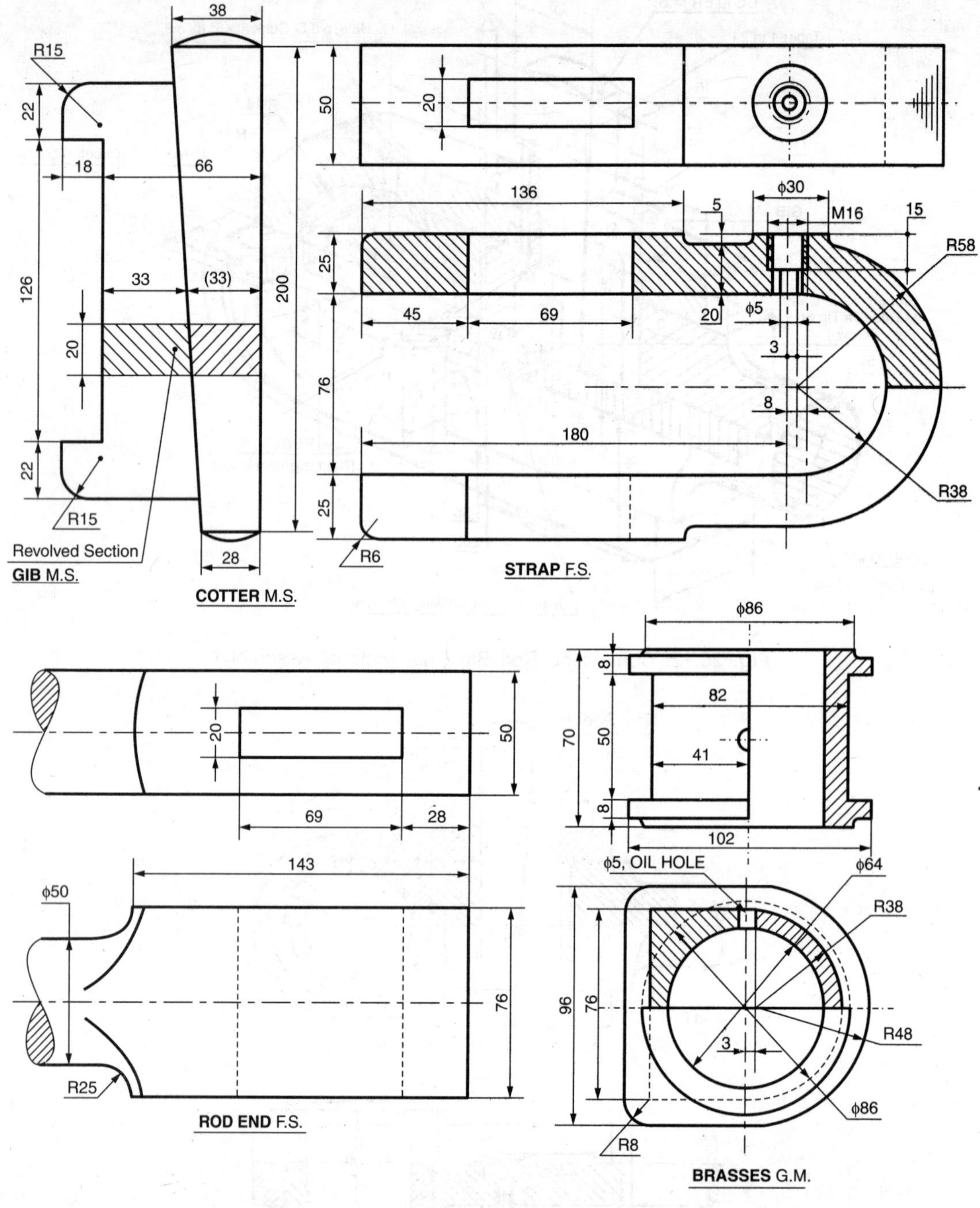

Fig. 25.13. (a) Details of a Circular Strap Type Connection Rod Big End

EXERCISE 25.13. *Assemble all parts of the big end of the marine engine connecting rod shown in Fig. 25.15 and draw its,*

(a) Half sectional front view, upper half in section,

(b) Right side view, and

(c) Top view.

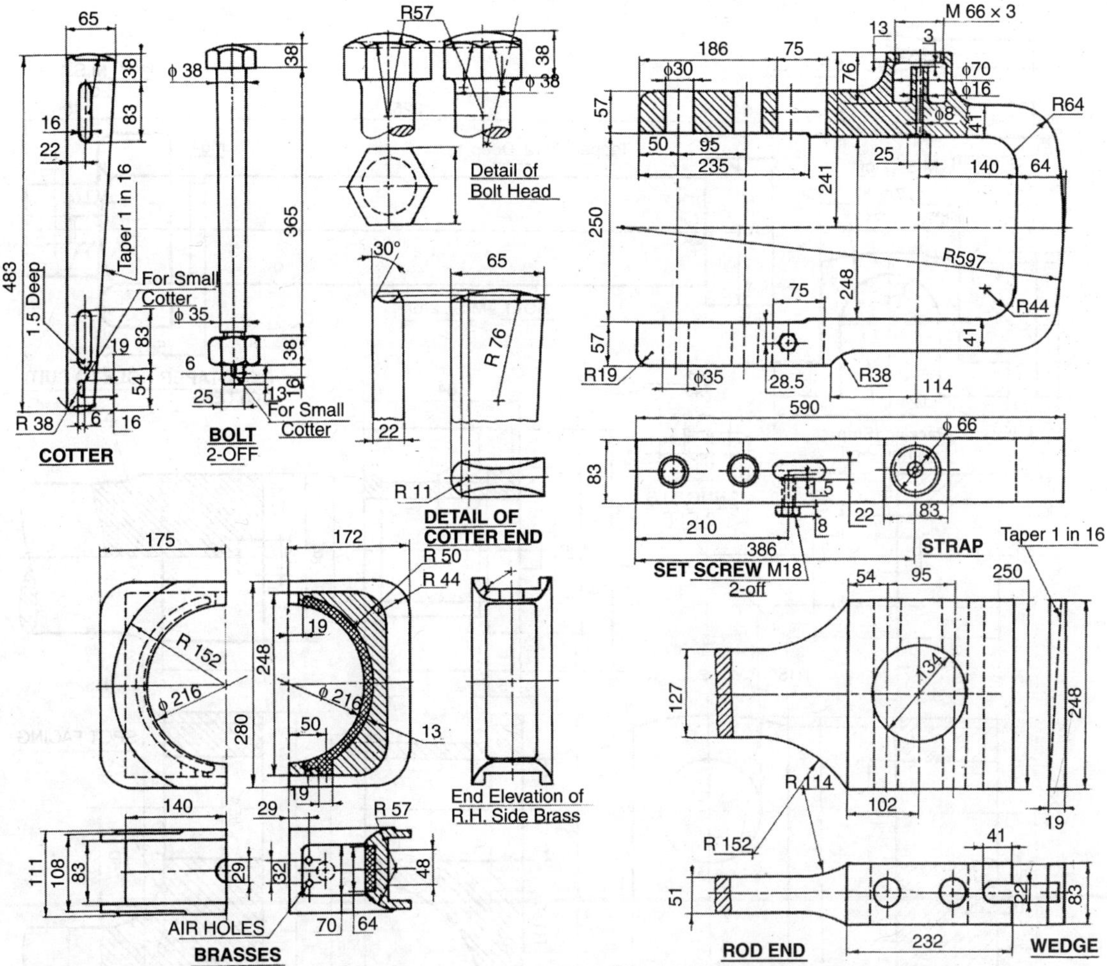

Fig. 25.14. Details of a Locomotive Connecting Rod End

25.12. ECCENTRIC

Eccentric is a machine part, which converts the rotary motion of one link into the sliding motion of the other (Never vice versa). In a steam engine, it converts the rotarty motion of the crankshaft into the sliding motion of the D-slide or piston valve.

Eccentric consists of two main parts, a sheave and straps. Sheave is a circular disc the outer rim of which is projected in the form of the ring. The hole for the shaft is made, not in the centre, but eccentric. Straps are made in two parts, when bolted together, these encircle the eccentric sheave, and a corresponding depression is made in the rim of straps (*Figs. 25.15, 17 and 18*). to one of the straps. Valve rod is studded.

Material of the sheave is usually steel and that of the straps may be gunmetal, brass or some other alloy.

Figure 25.16 shows the details of an eccentric and the eccentric rod end. To connect the rod end, there are two stud holes and one recess in the strap. The projection on the rod end fits into the recess in the straps and helps to achieve better alignment. On one of the straps, arrangement for lubrication is provided. to reduce the weight and for the purpose of balancing, some metal is removed from the sheave. Gunmetal packing is placed in betwen the straps, which can be changed to make the adjustment after wear.

Figure 25.17 shows the details of another eccentric with a little change in the design of eccentric rod and straps.

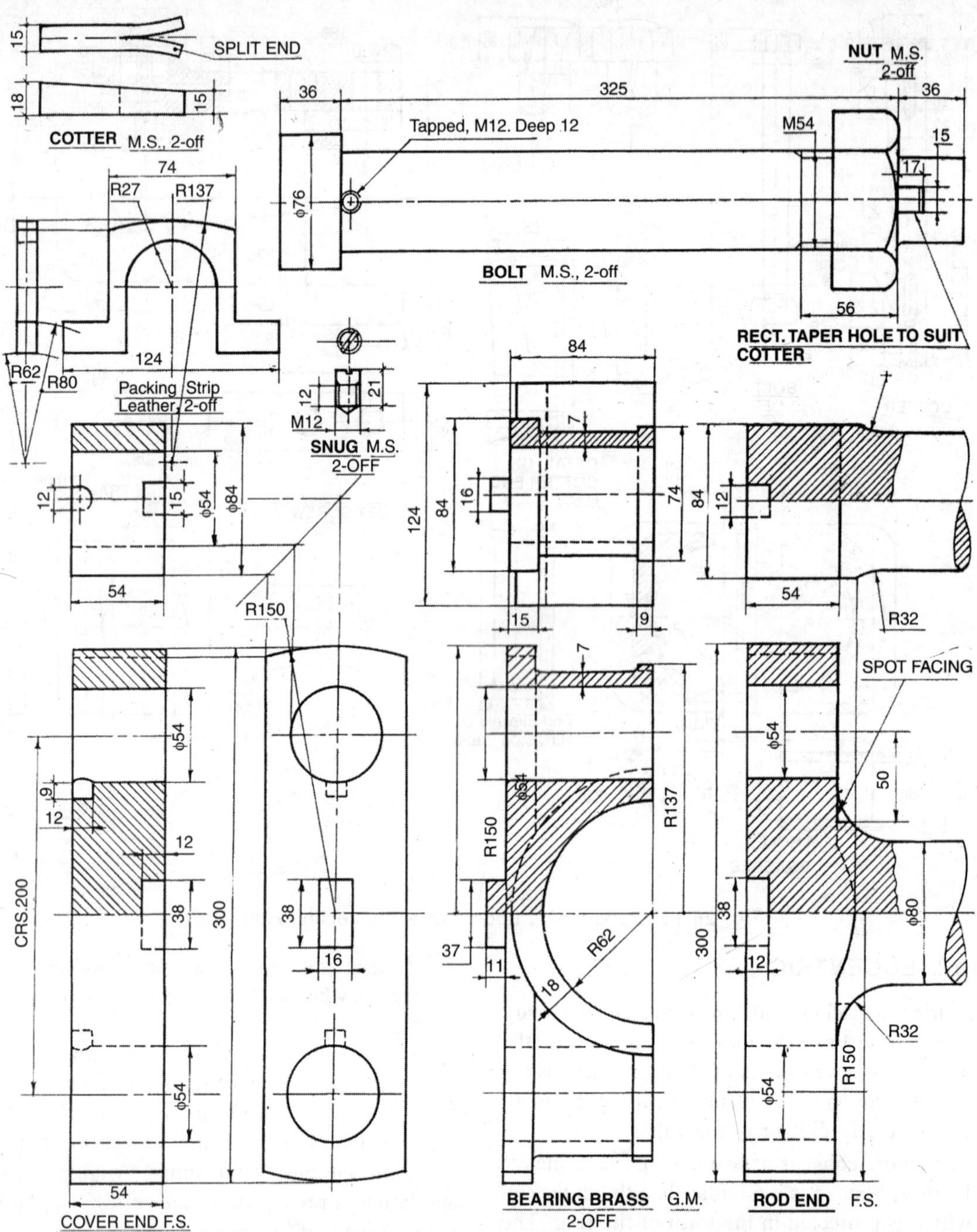

Fig. 25.15. Marine Engine Connecting Rod end details

EXERCISE 25.14. *From the details of the eccentric given in Fig. 25.16, draw the following assembled views:*

(a) *Front view–upper half in section,*

(b) *Top view–full in section, through the centre of the straps, and*

(c) *Left end view.*

EXERCISE 25.15. *From the details given in Fig. 25.17, draw the following views of the eccentric assembly;*

(a) *Front view–upper half in section,*

(b) *Top view–full section and*

(c) *Right End view.*

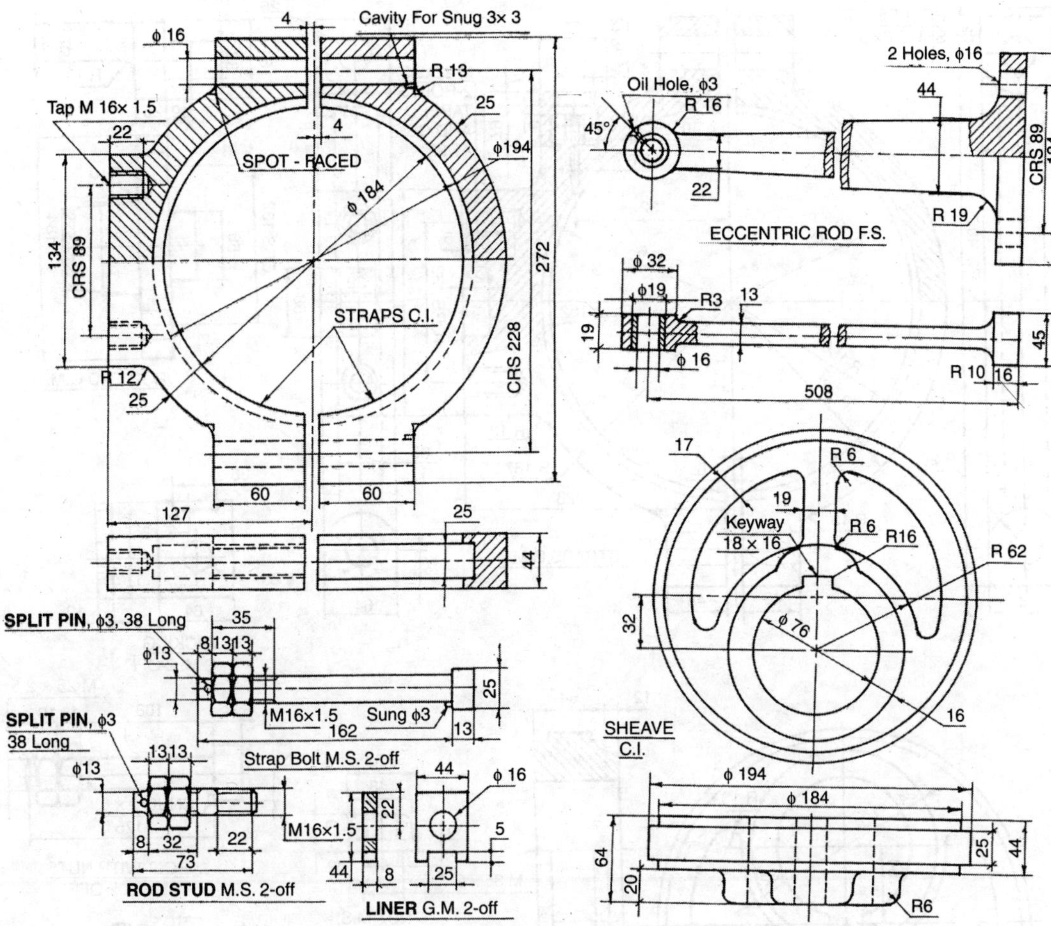

Fig. 25.16. (a) Details of an Eccentric

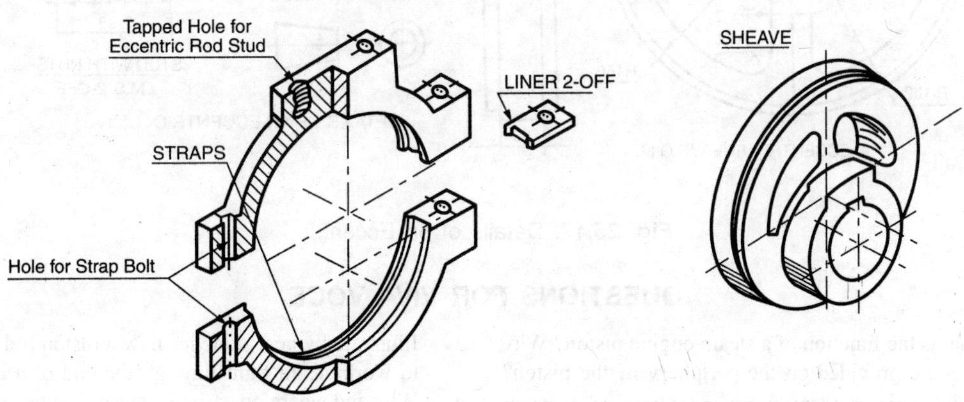

Fig. 25.16. (b) Details of an Eccentric (Pictorial)

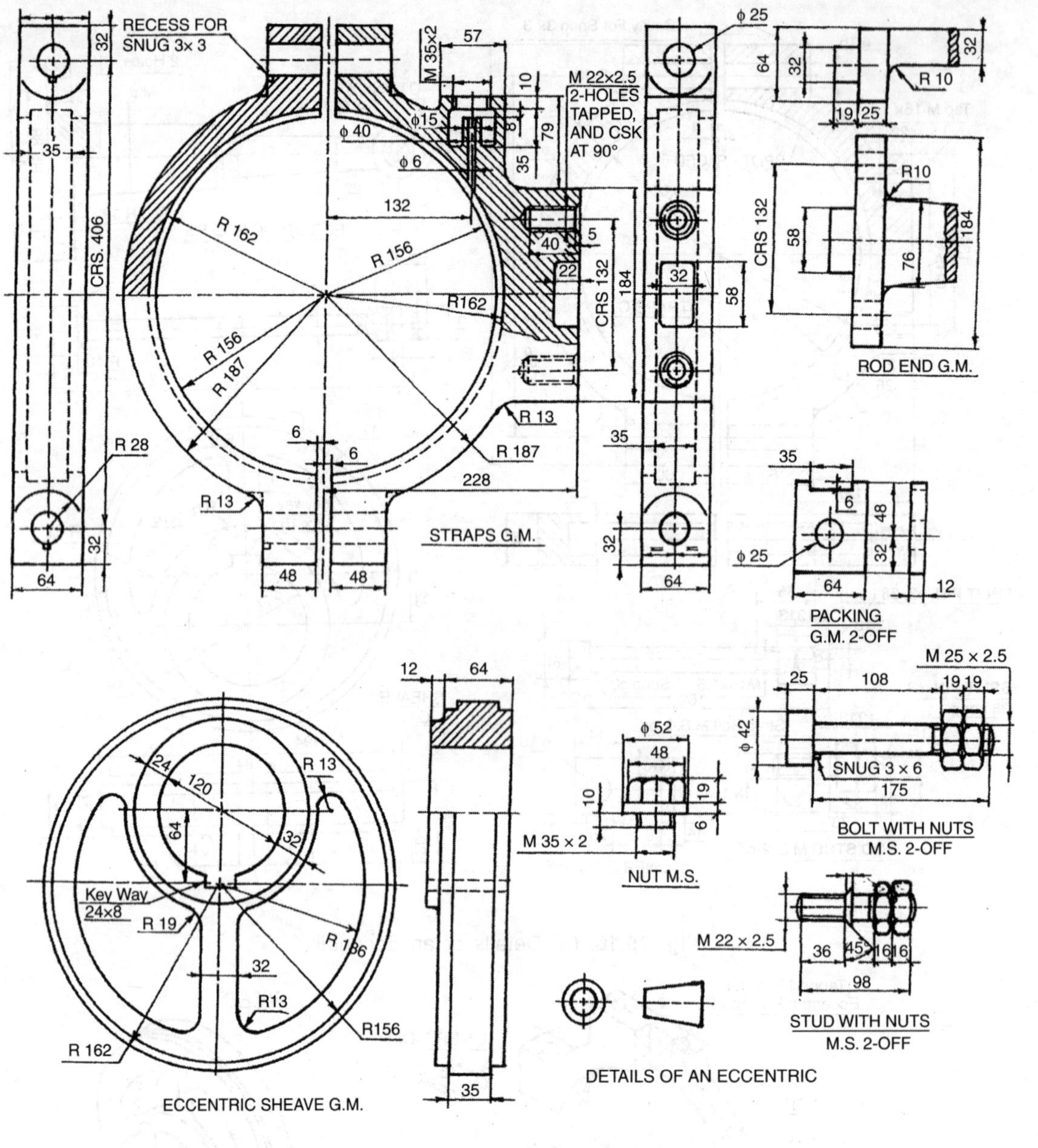

Fig. 25.17. Details of an Eccentric

QUESTIONS FOR VIVA VOCE

1. What is the function of a steam engine piston? Why rings are provided on the periphery of the piston?
2. Sketch proportionately, any one type of a steam engine piston, showing clearly the method of securing it to the piston rod. Also shown the rings in position.
3. Illustrate by neat sketches how a piston rod is made to work steam tight through the end of a cylinder.
4. Why and where, in case of a steam engine, a stuffing box is used?
5. What are the different types of packing materials used in stuffing boxes and where and when each one is used?

6. Why the ends of a gland and a neck bushing of a stuffing box, which are in contact with the packing are beveled? Illustrate the answer with a simple sketch.
7. What is the function of a crosshead in a steam engine? Where it is used and why? Sketch in two orthographic views, a crosshead suitable for a vertical steam engine.
8. Why sometimes a crosshead is made with detachable shoe?
9. What is the function of a connecting rod in a steam engine? What are big end and small end of a connecting rod?
10. Draw a neat and dimensioned sketch, in two orthographic views, of a strap end of a connecting rod of f 50 mm with: (*a*) one jib and a cotter, (*b*) two jibs and a cotter.
11. Why brasses are used in connecting rod ends and why these are made of soft metals?
12. What are a crank and a crankshaft? What are the functions?
13. What is the function of an eccentric in a steam engine? What is the difference between an eccentric and a crank?
14. Illustrate by a suitable sketch, the term eccentricity. as applied to an eccentric, and the valve travel or throw of the eccentric.
15. What is the means adopted to take up any wear in an eccentric?
16. Sketch two views of an eccentric having 50 mm throw.

❑❑❑

Cams 26

26.1. INTRODUCTION

A cam is a rotating device, which transforms rotary motion into reciprocating motion. A rotating cam drives a, spring loaded, follower and gives it some specified type of oscillating motion, according to a predetermined program, in a single plane. Usually this imparted motion is up and down. The type of motion depends upon the shape of the cam and the type of the follower used. Configuration of the cam is analyzed graphically, prior to the preparation of the specifications for its manufacture.

Cams utilize the principle of the inclined wedge, with the surface of the cam, causing a change in the slope of the plane, thereby producing the desired motion. Many types of machinery and engines have valves and mechanisms that are operated by cams, e.g., cam mechanisms in clocks, printing machines, automatic screw cutting machines, internal combustion engines for operating valves, etc., *Fig. 26.1*.

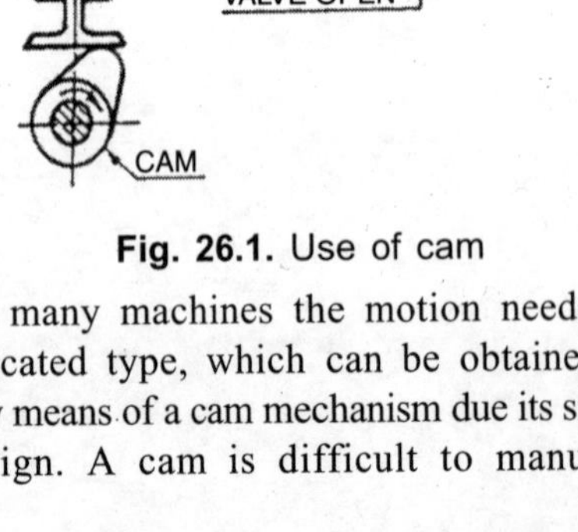

Fig. 26.1. Use of cam

In many machines the motion needed is of complicated type, which can be obtained easily only by means of a cam mechanism due its simplicity of design. A cam is difficult to manufacture, especially when it is to be produced in small quantities. Single cams are often marked out and hand finished, or cut on a milling machine. When produced on mass scale, it is manufactured by a punch press, by die-casting, or by milling from a master cam. A cam profile may be defined by X, Y and Z co-ordinates, so computerized design data can be used to cut the cams on numerically controlled machine tools.

26.2. TYPES OF CAM

Two main types of cam are :

1. Radial, Edge or Plate cams, and
2. Cylindrical cams.

This classification takes into account the relative motion of cam and follower. The follower with a radial cam reciprocates or oscillates in a plane perpendicular to the cam axis, i.e., in the plane of the cam or in a parallel plane, while with a cylindrical cam the follower moves parallel to the cam axis, i.e., generally, in a plane perpendicular to the plane of motion of the cam.

Figure 26.2 (a) Shows the most elementary type of cam. A block with a sloping top face has a knife-edge follower resting on it. When the block is reciprocated on the horizontal plane, the follower is caused to reciprocate, in its guide, in the vertical plane. In *Fig. 26.2 (b)* the cam oscillates about a centre causing the follower to reciprocate up and down, while in *Fig. 26.2 (c)* the cam rotates, giving a reciprocating motion to the follower.

Cylindrical cams are cylinders with grooves cut into their surfaces. As the cam rotates, the groove imparts the designed motion to a follower which is engaged with the groove, *Fig. 26.2 (d)*. The cam may oscillate instead of rotating.

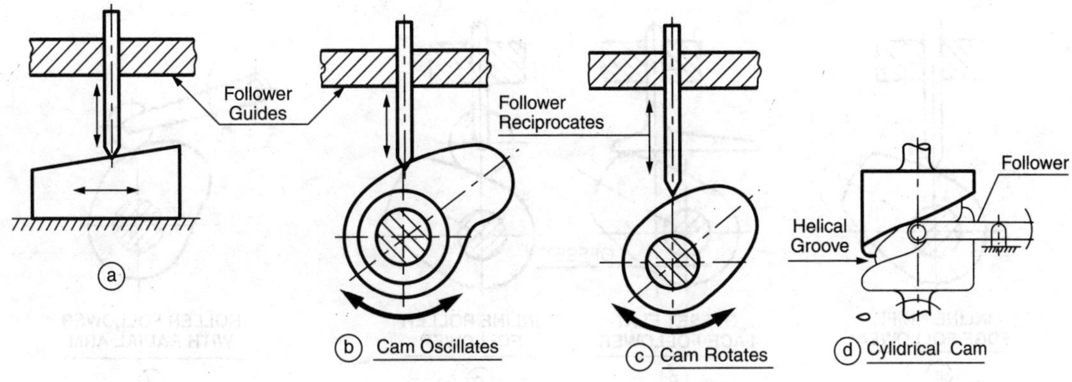

Fig. 26.2. Radial Edge or Plate Cams

26.3. DEFINITIONS

1. Base Circle or Least Radius

Referring to *Fig. 26.3*, it can be observed that the lift of the follower commences at A and continues up to B. From B to C the follower remains at rest, in the lifted position, subsequent to which it descends. This process continues till it comes in contact with the cam at D. The circle drawn with centre O and radius OA or OD is known as base circle and its radius is known as the base circle radius or least radius of the cam.

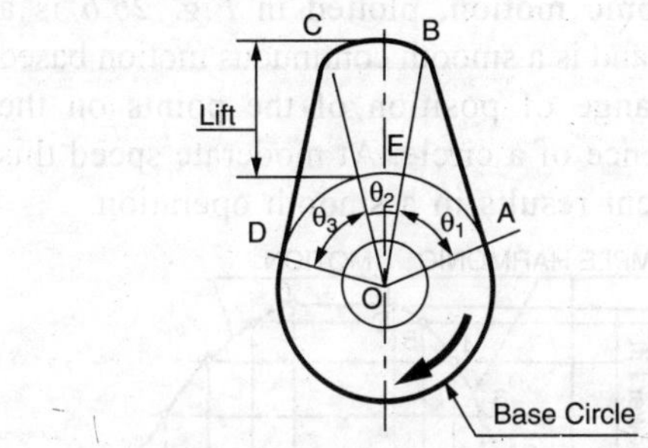

Fig. 26.3.

2. Lift

Difference between maximum distance of the lowest point of the follower from the axis of rotation of the cam and least radius is known as the lift of the follower.

3. Angles of Ascent, Dwell, Descent, and Action

Angle moved by the cam, from the instant the follower begins to rise till it reaches its highest position, is known, as angle **of ascent** and it is equal to the angle AOB, *Fig. 26.3*.

The angle through which the follower remains in the highest position is known as *angle of dwell*. This angle is eq ual to angle BOC in *Fig. 26.3*.

Similarly, angle COD, during which the follower returns to its initial position is known as *angle of descent*. The total angle moved by the cam from the beginning of ascent to the termination of descent is termed, as *angle of action* and it is equal to angle AOD measured in the counter clockwise direction *Fig. 26.3*.

26.4. TYPES OF FOLLOWER

Three basic types of cam follower are: (*a*) the knife, (*b*) the flat face and (*c*) the roller, *Fig. 26.4*. The flat face and knife-edge followers are limited to use with slow moving cams where the minimum of force will be exerted by the friction that is caused during rotation. The roller is the most often used form of followers since it can withstand higher speeds and can transmit greater forces and also used for any application where heat and wear require careful consideration, but they are most expensive.

Followers may oscillate in a straight line or radially and they may be offset from the cam centre of rotation. All these variations give the designer a wide choice for the most suitable arrangement for a given engine or mechanism.

The roller follower causes least wear to the cam but the cam profile must have no concave parts of a smaller radius than the roller radius. Any follower must be kept in contact with its cam and a return spring is generally used in order to maintain continuous contact, *Fig. 26.1*.

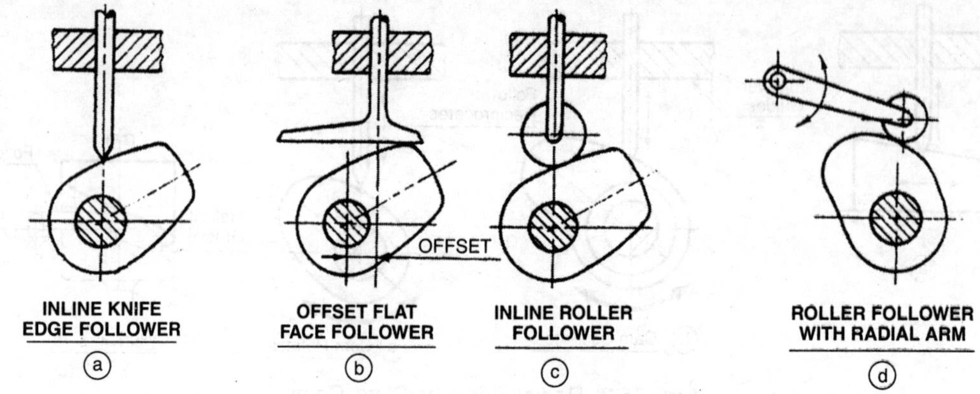

Fig. 26.4. Types of follower

The flat follower motion depnds upon the cam profile, the type of the follower and their relative positions. When studying the motions, which are given to a follower, by the profile of a radial cam, use is made of a displacement diagram of cam graph in which lift of the follower is plotted against the angular displacement of the cam. Construction of a displacement diagram is the first step in cam design. The complete 360° rotation of the cam is laid off in 15° steps (usually) along the horizontal axis to a convenient scale and the lift of the follower is shown on the vertical axis, full size.

Cams are designed primarily to produce the following motions, which may be given to the followers: (*a*) Uniform velocity, (*b*) Simple harmonic motion, (*c*) Uniform acceleration and retardation or gravity motion or (*d*) Combination of these motions. The displacement diagrams for both follower lift and fall, with a dwell period between, are shown in *Figs. 26.5, 26.6* and *26.7*.

(a) Uniform Velocity (Fig. 26.5)

The constant slope line indicates uniform velocity *i.e.,* equal displacements in equal intervals of time for follower lift and fall periods. Sudden changes, as indicated by the corner of the graph, are impractical and would have to be modified by rounding these corners to give more gradual accelerations and retardations. If this is not done large accelerations and retardations are given to the follower, which tends to jump on the cam profile. The radius of the modifying arc can vary upto a radius of one half the total displacement of the follower, depending upon the speed of operation. Usually a radius of about one-third to one-fourth total displacements is best.

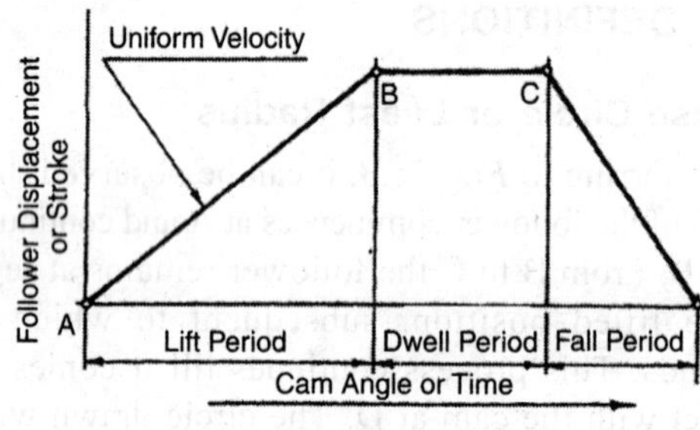

Fig. 26.5. Uniform linear motion

(b) Simple Harmonic Motion

Harmonic motion, plotted in *Fig. 26.6* is a sine curve and is a smooth continuous motion based on the change of position of the points on the circumference of a circle. At moderate speed this displacement results in a smooth operation.

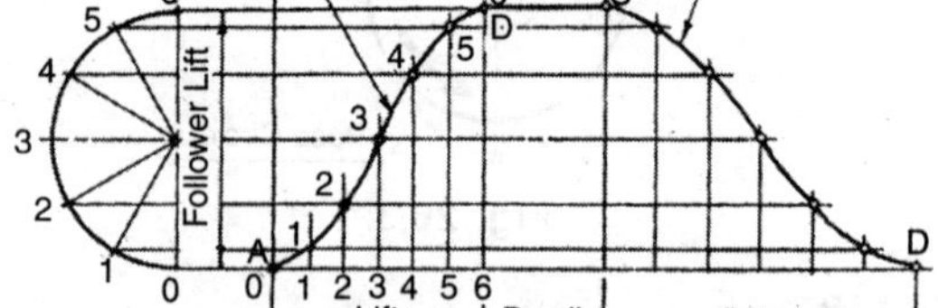

Fig. 26.6. Simple Harmonic Motion

A semi-circle of diameter equal to the follower simple harmonic motion stroke is drawn and divided into a number of equal parts numbered 1, 2, 3 etc,. The corresponding cam angle is divided into same number of equal parts and horizontal projectors through these locate the corresponding points 1, 2, 3, etc. on the displacement diagram curve.

(c) Uniform Acceleration And Retardation or Gravity Motion

Gravity motion, plotted in *Fig. 26.7* is commonly used for high-speed operation. The gravity fall of the follower is designed to conform to the shape of the cam, so that its contact with the surface will provide smooth operation.

The graphs for both parts of *Fig. 26.7* are parabolic in shape. Uniform acceleration is shown between points A and B. Divide the cam displacement angle into a convenient number of parts and vertical height into the same number of parts four here. Radial lines are drawn from the start position A to the four vertical divisions and these intersect with vertical lines from the points, 1, 2, 3 and 4 representing camshaft positions. The curve passes through the intersections. Acceleration of the follower is a maximum at B and it is retarded and stationary at point C. From C to D the follower does not move and it is often called the dwell period. Two semi parabolas are constructed in a rectangle of height equal to the stroke, either directly on the displacement diagram or to one side of it as shown.

26.6. TO CONSTRUCT THE FOLLOWER DISPLACEMENT DIAGRAMS OR CAM GRAPHS FOR GIVEN CAM PROFILES

(a) With an In-line Roller Follower (Fig. 26.8)

The cam dimensions are known and its profile is first drawn to scale. The displacement of the roller centre has to be plotted on a graph of cam angle for one complete revolution. It is usual practice to assume that the cam is stationary and the follower rotates around the cam in the reverse direction to cam rotation. This maintains the same relative motion between cam and follower and simplifies the constructional drawing.

The 360° cam rotation is next divided into a convenient number of equal parts, say twelve or twenty-four. The cam angel base, on the graph, is also divided into the same number of equal parts. The roller is drawn in contact with the profile in each of its angular positions. The follower displacement is given by the distance from the cam centre O to the roller centre.

The construction of one point P_4 is shown. The point corresponds to point 4 of the follower. An arc of radius O4 is swung back to the follower line of action giving point P_4 and by projecting across, to corresponding angle position, P_4 is located. It will be noticed that where the roller centre radius remains constant the displacement curve remains flat, this is called a *period of dwell.* The base of the graph represents angular rotation of the cam but it can also represent the time of one complete cam revolution. The velocity of the follower can be found by measuring the slope of the displacement time curve.

(b) With An Offset Flat Face Followers (Fig. 26.9)

This figure shows the construction of a displacement diagram for a flat face follower. Here

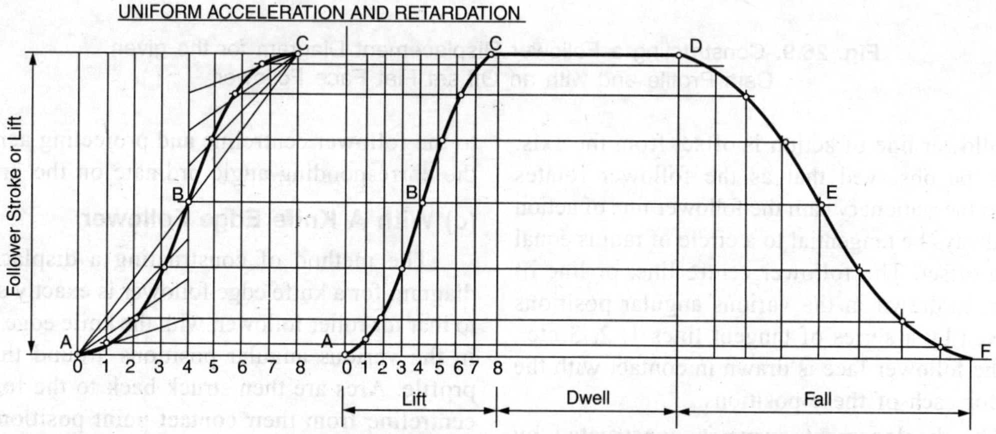

Fig. 26.7. Uniform Acceleration and Retardation or Gravity Motion

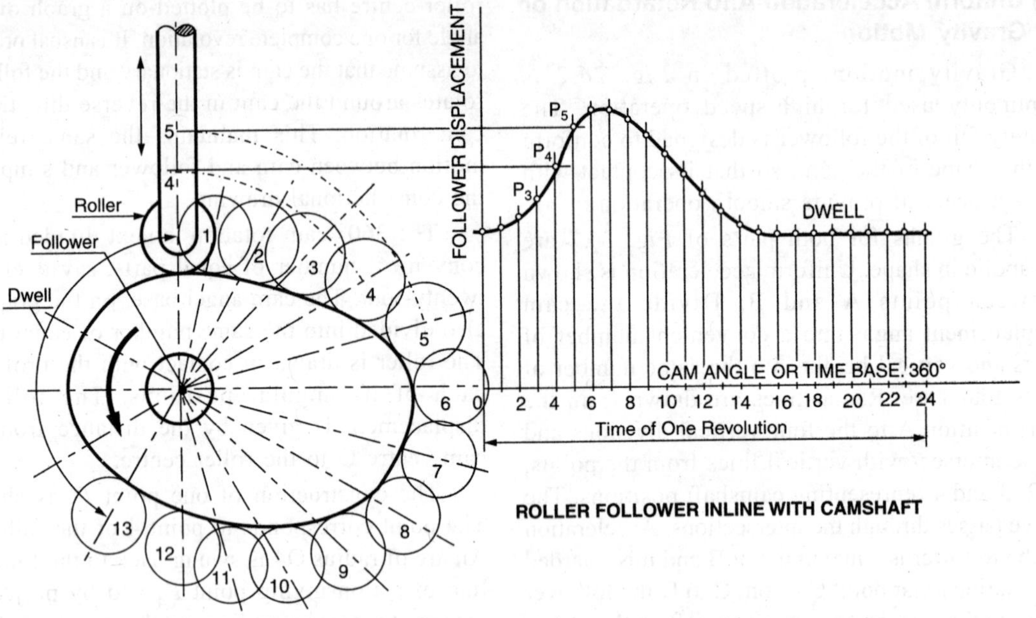

Fig. 26.8. Constructing a Follower Displacement Diagram for the given Cam Profile and with an In-line Roller follower

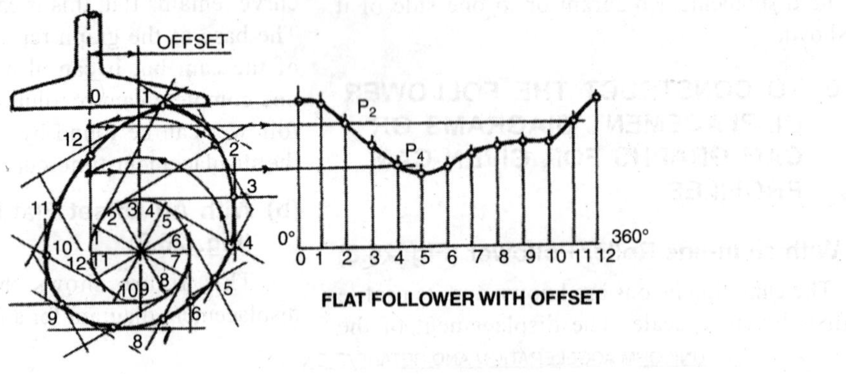

Fig. 26.9. Constructing a Follower Displacement Diagram for the given Cam Profile and with an Off-set Flat Face Follower

the follower line of action is offset from the axis. It can be observed that as the follower rotates around the stationary cam the follower line of action will always be tangential to a circle of radius equal to the offset. The follower centre line, or line of action, is drawn in the various angular positions as given by a series of tangent lines 1, 2, 3 etc., and the follower face is drawn in contact with the cam for each of these positions.

The displacement curve is constructed by striking arcs from the various follower positions to the follower centreline and projecting across to the corresponding angle ordinate on the graph.

(c) With A Knife Edge Follower

The method of constructing a displacement diagram for a knife edge follower is exactly similar to that for roller follower, with the knife edge drawn in the various angular positions around the cam profile. Arcs are then struck back to the follower centreline from their contact point positions.

(d) With An Offset Roller Follower (Fig. 26.10)

(*i*) Start by drawing the cam profile and plot round it a parallel curve representing the path of the roller centre.

(*ii*) Position the line of action of the follower and draw a circle tangent to it using path of the roller centre.

(*iii*) Divide this circle into 15° divisions and draw tangents at each division point, numbering them opposite to the rotation of the cam. Points where these tangents cut the curve, drawn parallel into the cam profile, are positions of the roller centre.

(*iv*) Swing these points around the cam centre to the line of action of the follower and project them into the cam graph to the appropriate displacement line.

(*v*) The cam graph may have a time basis, as shown, instead of an angular displacement basis. To determine the time basis calculate the time taken by the cam to make one revolution and divide it by 24 (*i.e.*, 360/15 = 24). This gives the time for the cam to turn through 15°, enabling the graph to be related to the roller centre positions on the cam.

26.7. TO CONSTRUCT THE CAM PROFILE FOR A GIVEN FOLLOWER MOTION SEQUENCE

This is a design problem where the type of follower motion desired is known. This method is just the reverse of that already described for drawing the displacement diagram for the given cam profile. The follower displacement diagram is drawn from the follower motion data, its base being divided into a convenient number of equal parts and the corresponding cam rotation divided into the same number of equal parts. The cam is again assumed to be stationary and the positions of the follower are drawn around the cam centre, by striking arcs from the follower centreline, according to the projected follower displacement positions. It is to be noted again that to obtain the same relative motion, when the cam is stationary, the follower must rotate round the cam in the oposite direction to the actual cam rotation.

(1) To Construct a Plate Cam Profile With an Inline Knife Edge Follower

The method for drawing the profile of a plate cam to operate a point of knife edged follower is illustrated in *Fig. 26.11*. The same general method is used for all plate cams but it is varied by the type of follower and its position relative to the cam axis.

(*i*) First position the cam centre and the nearest approach of the follower to it. From this point lay off the maximum lift of the follower.

(*ii*) Draw circles through these two points about the cam centre and divide them into 15° divisions.

(*iii*) Construct the cam graph, positioning the angular displacement axis on a level with the lowest position of the follower.

(*iv*) Imagine the cam to be held stationary and the follower to be moved round it. To obtain the same motions as with the cam rotating, the follower has moved in the opposite direction to the cam rotation. When the follower has moved through 15° its height will be given by the point on the cam graph at the 15° station. Project this point to the line of action of the follower, which in this case is the vertical centreline of the cam, and swing it about the cam centre with compass to cut the 15° line on the cam.

(*v*) Succeeding points are obtained in the same way. Where the follower remains at the same height or, 'dwells', the radius of the cam is constant.

(*vi*) When points on all the 15° radial lines have been plotted, a smooth curve drawn through them will be the cam profile.

(*vii*) Complete the cam by drawing an arrow showing the direction of rotation.

(2) To Construct a Radial Cam Profile Using a Flat Follower

The construction of a radial cam to work with a flat follower is shown in *Fig. 26.12*. As before points on the profile of the cam are positioned on the radial displacement lines by projection from the cam graph. these points lie on the face of the flat follower. Lines representing the follower face are drawn through them at right angles to the angular displacement lines. the cam profile must be tangential to these lines.

CAM DATA : Profile shown at (a). Follower offset 25 mm to Left of CAM shaft centre. Rotation of CAM clockwise. Speed of camshaft 15 RPM.

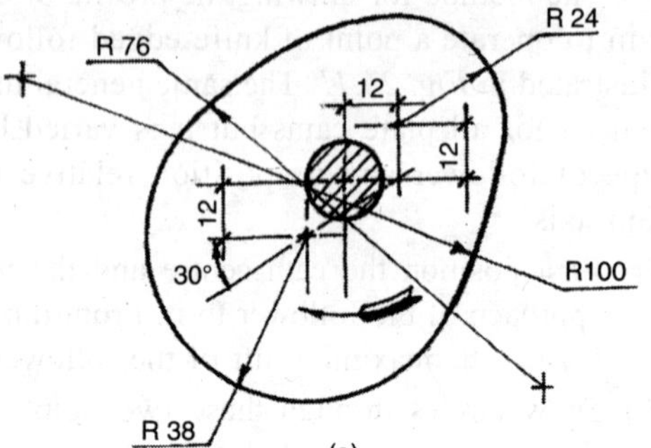

(a)

Stage 1. Redraw cam profile and plot path of roller centre, adjacent to cam profile, at distance X equal to radius of roller.

Stage 2. Draw line of action of roller follower and draw a circle tangential to it. Divide the circle into 24 equal parts, Each part representing 15° of cam rotation.
Draw Tangents to the circle at the division points.

Stage 3. Project centre points of follower to its line of action and then project them to appropriate line on graph.

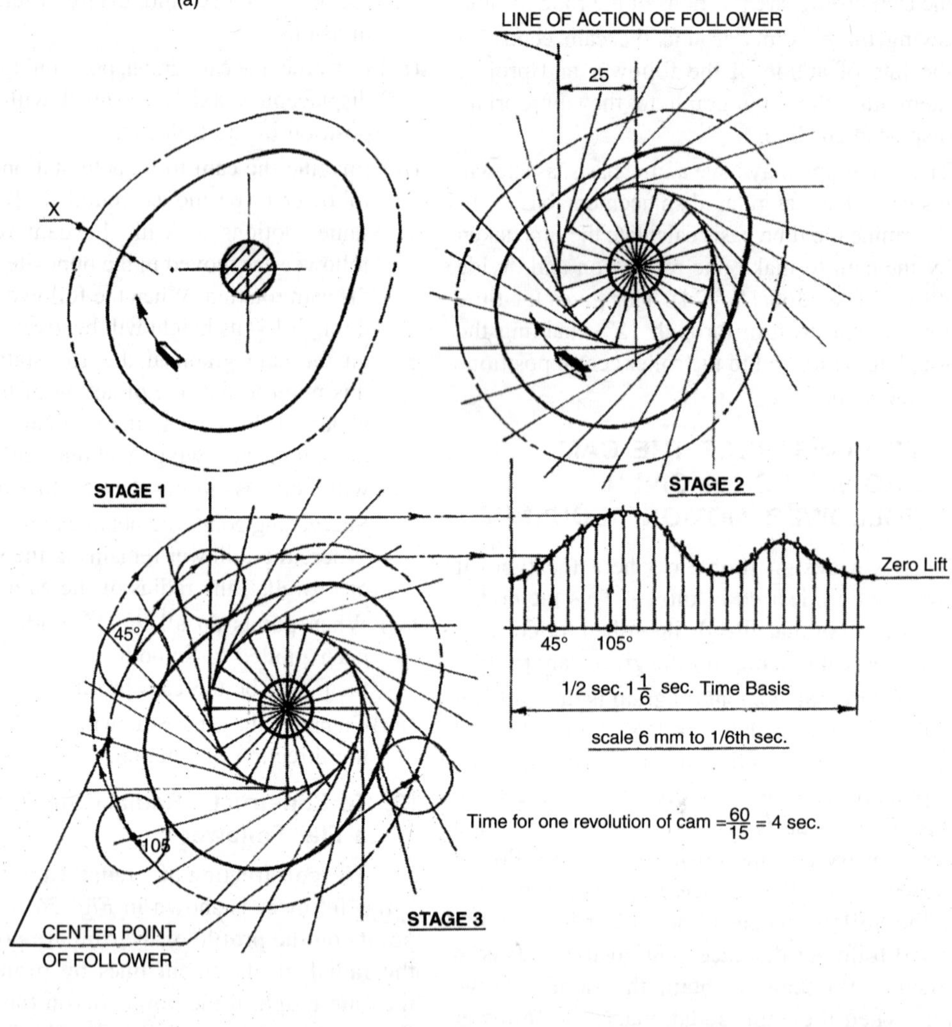

Fig. 26.10. To draw the Cam Graph for a given cam profile with an off-set roller follower

CAM DATA : 0°-180° Lift 25 mm with S.H.M.

240°-360° Fall 25 mm with Acceleration and Retardation at Uniform Rate. Rotation of CAM Anti-clockwise.

Least Thickness of Metal around CAM centre 40 mm. Dia. of Camshaft 24 mm.

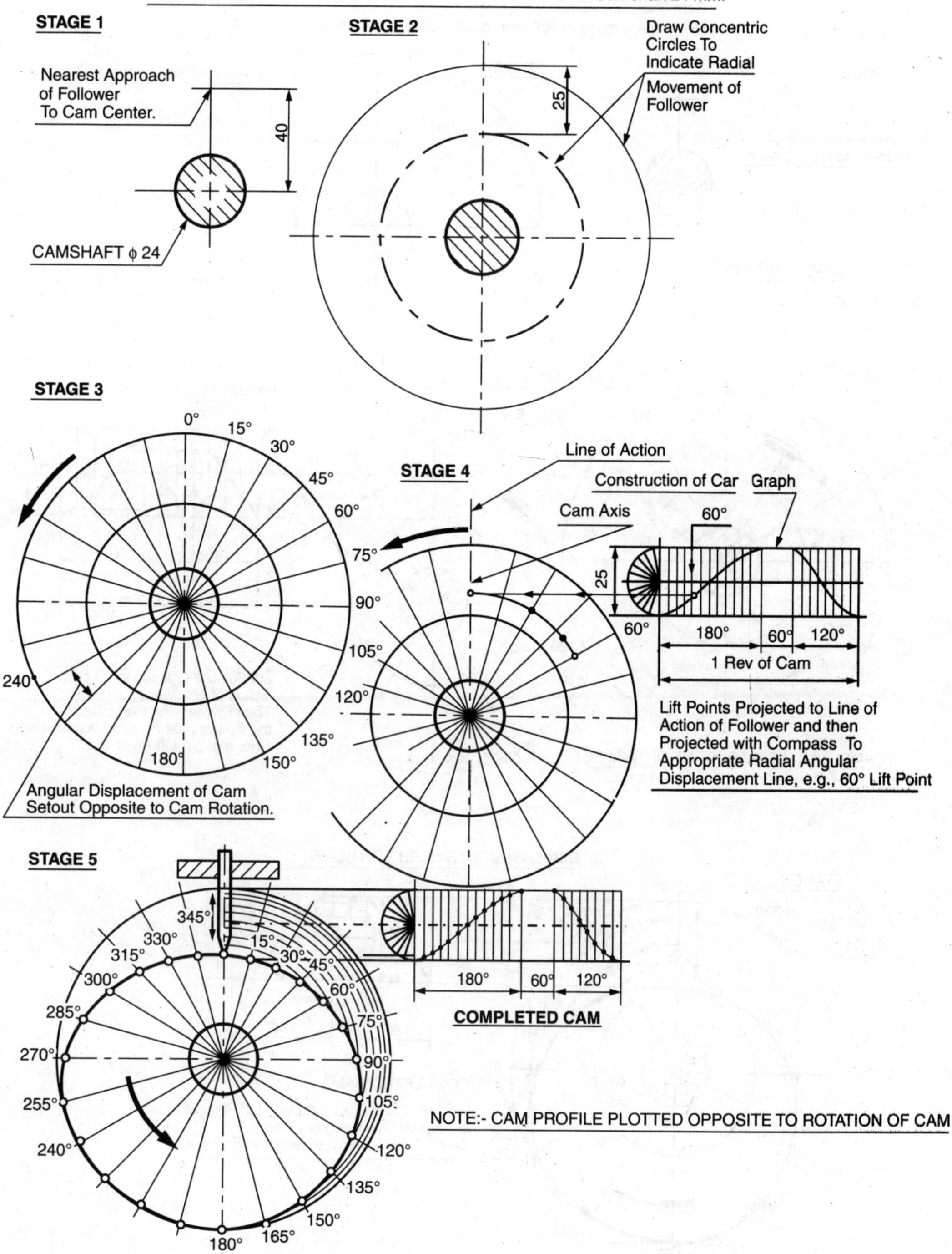

Fig. 26.11. To construct a Plate Cam Profile with an inline kinfe-edge follower.

CAM DATA : 0-120 Lift of 34 mm with Uniform Velocity.
120°-180° DWELL.
180°-360° Fall 34 mm with S.H.M.
Cam Rotation Anti clockwise.
Cam Shaft Dia. 25 mm.
Nearest Approach of Follower to CAM Centre 34 mm.

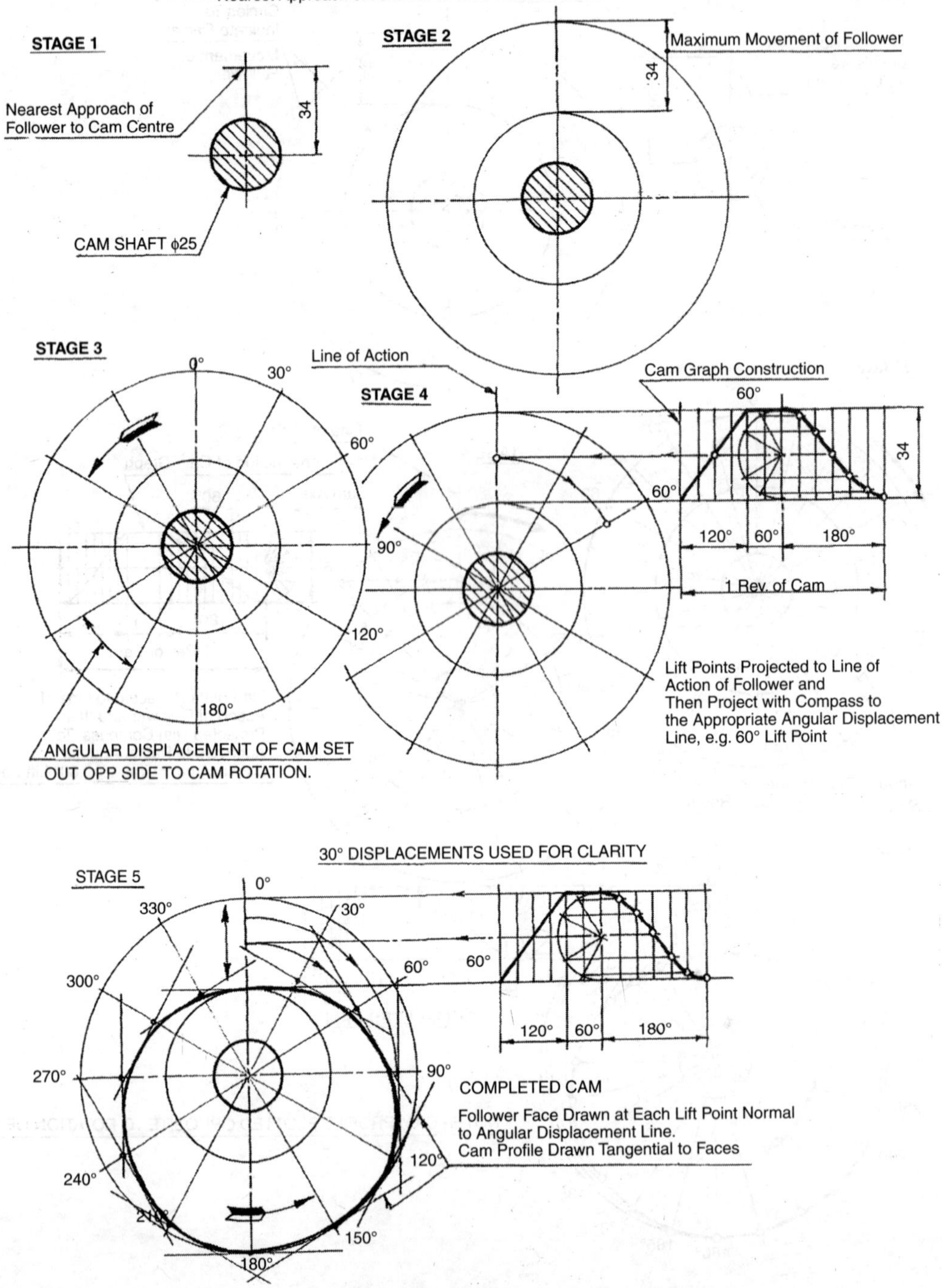

Fig. 26.12. To construct a radial cam profile using a flat follower

It should be noticed that the point of contact between the cam profile and the follower face is, in most cases, a considerable distance from the angular displacement line. This point must be borne in mind when deciding the size of the follower face.

(3) To Construct a Plate Cam Using an Inline Roller Follower

Figure 26.13 shows a plate cam, which gives motion to a roller follower. In this case points projected from the cam graph to the radial displacement lines on the cam are positions of the centre of the follower roller. The roller must be drawn at each of these positions, the cam profile being tangential to the rollers.

Note that in examples involving roller followers the angular displacement axis of the cam graph is positioned on a level with the start position of the roller centre. This ensures that lift points on the cam graph can be projected directly to the cam.

(4) To Construct a Cam Profile With an Offset Roller Follower

The previous examples of cam design have all had inline follower, *i.e.*, the line of action of the follower has been vertically above the cam centre. Sometimes, owing to lack of space for the mechanism, this cannot be arranged and the line of action has to be offset to one side.

Figure 26.14 illustrates the construction of a cam profile with an offset roller follower.

(*i*) Fix the highest and lowest positions of the roller relative to the cam centre.

(*ii*) Draw a circle about the cam centre tangential to the line of action of the follower and divide it into 15° divisions.

(*iii*) At each of these divisions draw tangents numbering them opposite to the direction of rotation of the cam. These tangents represent successive positions of the line of action, as it is moved round the cam.

(*iv*) Construct the cam graph placing the angular displacement axis on the same level as the initial position of the roller centre.

(*v*) Project points on the cam graph to the line of action and then swing them about the cam centre, to the appropriate tangent.

(*vi*) As these points on the tangents are positions of the roller centre, so the roller is drawn at each of these positions and the cam profile is drawn tangent to the rollers.

(5) To Construct a Cam Profile With a Radial Arm Follower

In some applications, for example, automatic machine tools, radial arm follower is more suitable than a reciprocating follower. A radial arm follower is pivoted at one end, the other end working on the cam profile. *Figure 26.15* illustrates the method of constructing the plate cam profile using a radial arm follower.

(*i*) Set out the cam centre, pivot centre and extreme positions of the end of the follower.

(*ii*) Draw circles about the cam centre through the pivot centre and ends of the follower.

(*iii*) Divide the circle through the pivot into 15° divisions, starting from the pivot, and number them opposite to the direction of rotation of the cam.

(*iv*) If the cam is held stationary and the follower moved round it, the pivot centre will move to each of these points in turn. So, with these points as centre draw arcs with a radius equal to the length of the follower arm.

(*v*) The follower is to move outward through 45° with uniform angular velocity during the first half revolution of the cam, that is, the first tweleve 15° divisions. Therefore, divide the 45° movement of the follower into twelve equal parts.

(*vi*) Project these points from the lift line of the follower to the appropriate arcs R_1, R_2, R_3 etc. thus obtaining the first twelve positions of the centre of the roller.

(*vii*) Plot the roller centres for the 30° dwell period and then divide the follower arc into ten equal parts for the return movement. Position the roller centres in the same way as for the outward movement and draw the cam profile tangential to the rollers.

26.8. CYLINDRICAL CAMS

The drawing of a cylindrical cam, illustrated in *Fig. 26.26*, requires a development of the cam

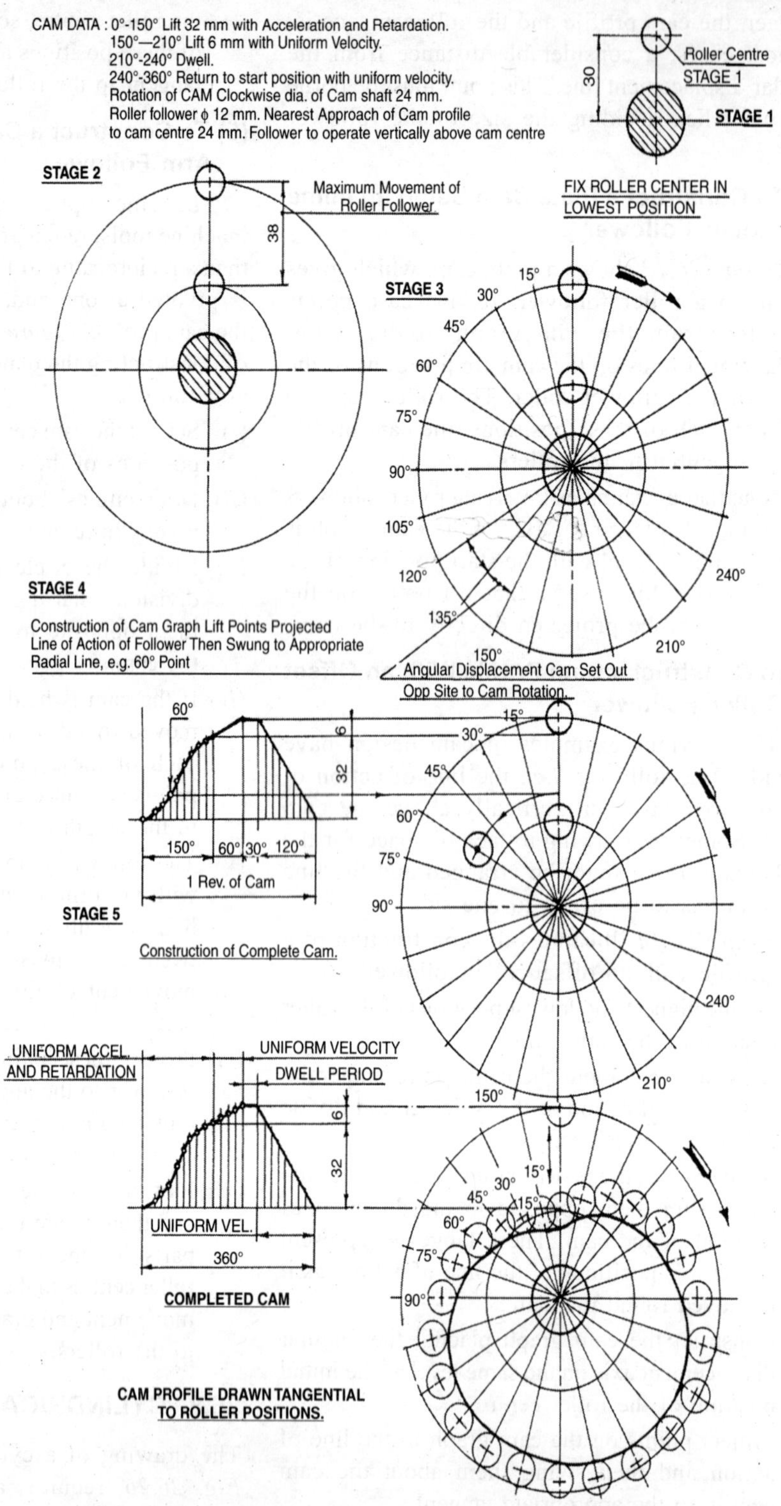

Fig. 26.13. To construct a Cam Profile using an In-Line Roller Follower

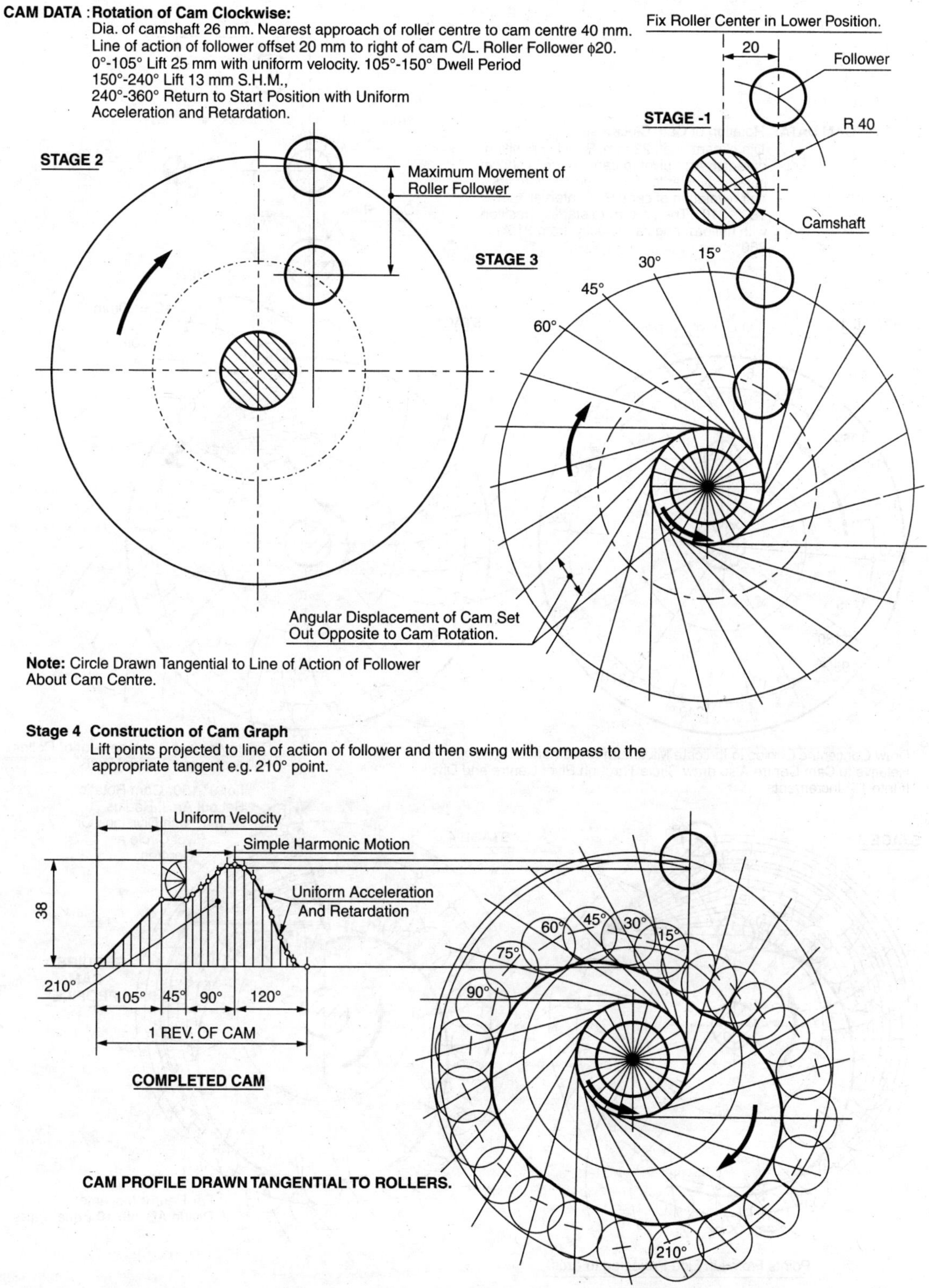

Fig. 26.14. To construct a Plate Cam Profile with an off-set Roller Follower

CAM DATA: Rotation of Cam Clockwise.
Dia of cam shaft 22 mm. Relative position of radial, Arm pivot to cam centre is shown in (a). Follower to lift through 45° with first half revolution of cam. Rest interval 180°–210°. Then return to starting position, with uniform angular velocity, from 210°-360°.

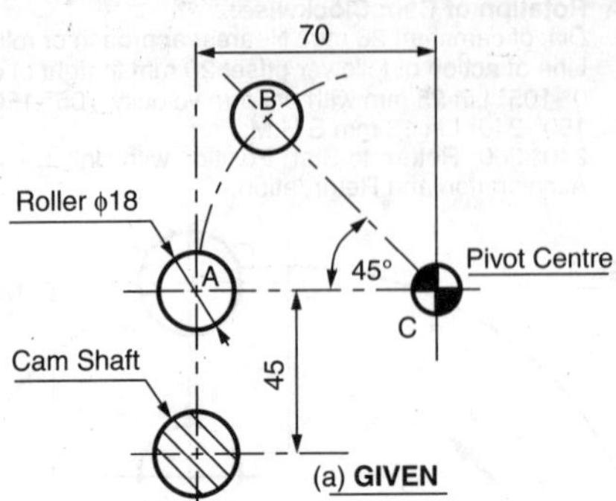

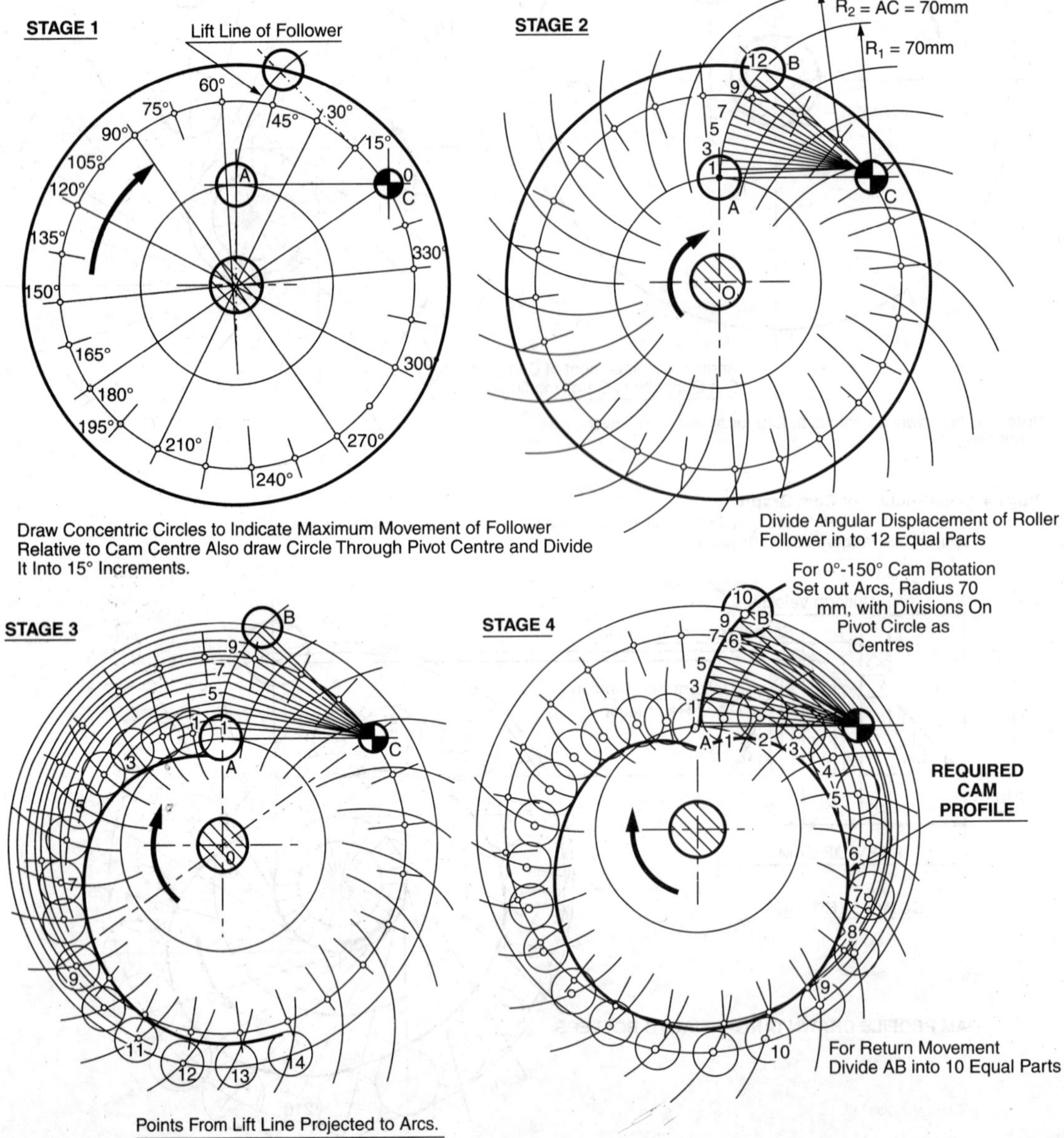

Fig. 26.15. To construct Plate Cam Profile with a Radial Arm Roller ended Follower

surface. From this development a template is made, which is used in the manufacturing of the cam.

1. Start by drawing front and end views of the cam blank, dividing both into a number of equal parts. Twelve divisions have been used in illustration for clarity but the motion of the follower may require 15° divisions.
2. From the front view project the development of the cam surface, dividing it into the same number of equal parts as the front and end views.
3. Layoff on the development the extreme positions of the follower and betwen these points plot the motion of the centre of the follower roller in the same way that motions are plotted on a plate cam graph.
4. Draw the roller at each point where the curve crosses an angular displacement line, and draw the profile of the roller groove tangential to these circles.
5. The elevation can then be completed by projection from the development.

It is usual to use a roller follower to work with a cylindrical cam but the roller must be conical or it will slip on the groove profile. This happens because points on the outside of the cam move faster than points at the bottom of the groove. A cylindrical roller has been used here for simplicity.

ILLUSTRATIVE PROBLEMS

PROBLEM 26.1. *Draw, the profile of an edge cam to give uniform upward motion of 45 mm during the first half of the revolution and again uniform return motion through the remaining half of the revolution. Minimum distance from the cam to the edge of the follower is 50 mm. Diameter of the shaft is 36 mm. The cam makes the wedge ended follower to reciprocate with uniform velocity.*

Construction: (Fig. 26.16) Take the least distance of the cam, OA = 50 mm and its lift AB = 45 mm. Divide the lift AB into six equal parts. With O as centre and OB as radius, draw a circle and divide it into 12 equal parts and join as shown. With O as centre and radius equal to O1 and O11 and similarly draw arcs through 2, 3, 4, 5, and 6 to intersect other corresponding radial lines. Then complete the required profile of the cam by joining smoothly the points of intersection, P_1 to P_{12} so plotted, with French Curves.

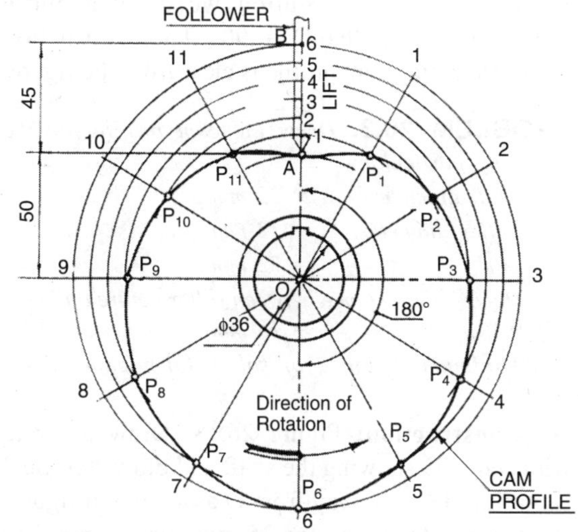

Fig. 26.16.

PROBLEM 26.2. *Design an edge cam to impart a rise of 35 mm at uniform velocity, during 180° of rotation: rest during next 90° and a uniform fall of 35 mm during the remaining 90° and uniform fall of 35 mm during the remaining 90° of revolution. Assume base circle radius to be 35 mm and angular rotation of the cam in anticlockwise direction.*

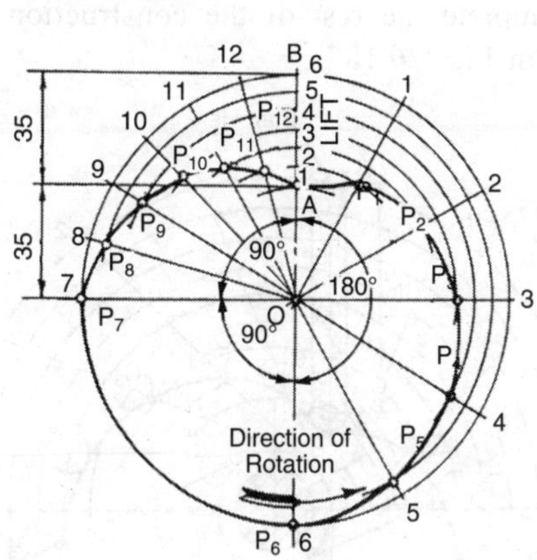

Fig. 26.17.

Construction: Figure 26.17. Through O draw a vertical line OAB and cut OA = AB = 35 mm. Divide AB into six equal parts. Angle BO7 is then divided into six equal divisions and right hand outline of the half cam profile which is a part of an Archimedean spiral, is obtained as in Problem 26.1. After the cam has rotated through 180°, the point 6 on its edge is 70 mm away from centre. As the follower is to remain at that distance during the next 90° of revolution we draw a curve subtending the angle 6 O7 = 90°. The construction of the remaining cam profile is clear from the figure.

PROBLEM 26.3. *Draw the cam profile for the following particular:*

Shaft diameter = *25 mm*

Roller diameter = *12 mm*

Lift = *72 mm*

Horizontal distance between roller centre and cam centre = *18 mm.*

Minimum distance of roller form cam centre = *35 mm.*

Construction: Figure 26.18. Draw a circle with centre O showing the shaft and draw a vertical centreline 18 mm away from O as shown in figure. With centre O and radius 35 mm, strike an arc to cut the centreline to locate the centre, C, of roller. With C as centre of roller, describe a circle of 12 mm diameter. It represents the roller. From C. along the vertical centre line set-off the lift 72 mm and divide it into 8 equal parts. With centre O describe circle through each of these points. Divide the outer circle into 16 equal parts and from each of these points draw tangents to circle of ϕ 25 mm and complete the rest of the construction as is shown in Fig. 26.18.

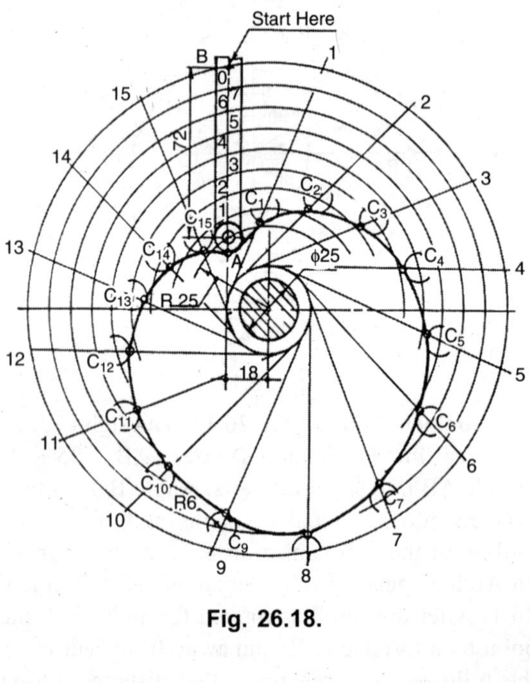

Fig. 26.18.

PROBLEM 26.4. *Determine the profile of a cam, which will allow the follower to oscillat with uniform angular velocity about a fixed centre. Every revolution of the cam completes one oscillation of the follower. Distance between the centre of cam and the roller is 35 mm, Lift = 75 mm, roller diameter = 12 mm, lever FC = 154 mm.*

Construction: (Fig. 26.19) With centre O and radius OF = 140 mm describe a circle. Starting from F divide its circumference into 16 equal parts. From each of these parts in turn, describe arcs passing through O. Locate point C, the centre of roller. Set-off the lift and divide it into eight equal parts. Describe concentric circles through each of the points. The successive points of intersection betwen these circles and the arcs previously drawn are on the locus of the roller centre. Draw the arcs, from the points of intersection of circles and arcs, with radius equal to radius of the roller. A smooth curve touching all these curved arcs gives the profile of the cam.

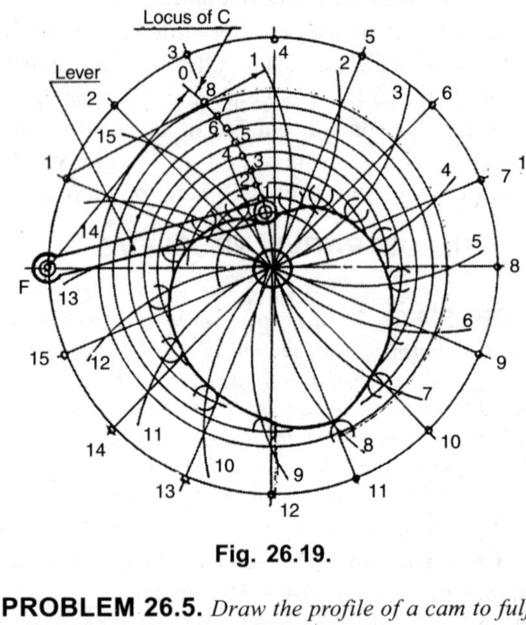

Fig. 26.19.

PROBLEM 26.5. *Draw the profile of a cam to fulfil the following requirements:*

Minimum distance of cam centre to edge = *35 mm*

Lift of follower = *30 mm*

The cam lifts the follower with simple harmonic motion during 120° of the revolution, then remains at rest for the next 60°, then falls with uniform motion during 120° through 15 mm and finally returns to the starting point with another uniform motion.

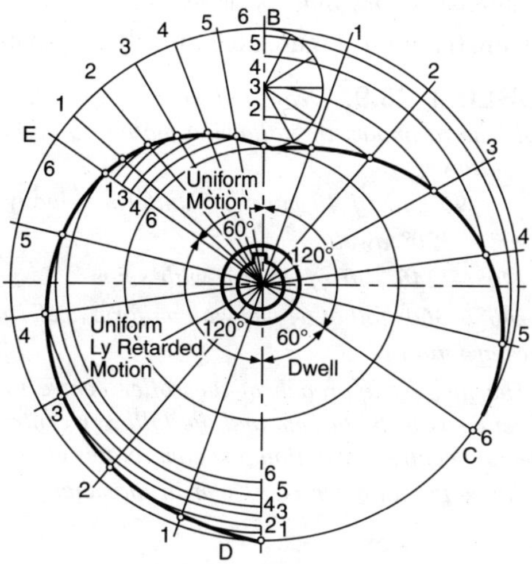

Fig. 26.20.

Construction: Draw the base circle, set-off the lift and mark the outer circle as shown in Fig. 26.20. Now the cam is to lift the follower with S.H.M. during 120° of the revolution. Describe a semi-circle having diameter equal to lift, i.e., 30 mm and divide it is six equal parts. Project these points on the semicircle horizontally to the vertical centreline (lift). Divide the angle of 120° into same number of equal parts, i.e., 6 in this case, and draw this part of the profile as shown in figure. The follower remains at rest for the next 60°. Now the cam falls with uniform motion for the next 120° through 15 mm. This fall of 15 mm is divided into same number of parts (six in this case) and the outer circle is also divided into same number of parts. Points on the fall are transferred to the radials and a smooth curve is drawn as usual. The remaining fall of 15 mm is completed during rest of the revolution, again wih uniform motion. Divide the fall and remaining portion of the outer circle into same number of equal parts (six in this case). Transfer the points on fall to radius and complete the profile of the cam.

PROBLEM 26.6. *Design an edge cam to the following particulars:*

Minimum distance of centre of cam from edge of follower 60 mm, lift 56 mm. The cam is to lift the follower with simple harmonic motion during one half of a revolution, and then allows it suddenly to drop halfway, then to fall with uniform motion during the remaining half of the revolution.

Construction: (Fig. 26.21) Through O draw intersecting centrelines. Mark off OA = 60 mm, AB = 56 mm. On AB describe a semicircle and divide its circumference into six equal/sets. From each of these drop a perpendicular to AB. We thus, divide the circumference into six equal parts. From each of these drop a perpendicular to AB. Thus we divide rise into parts proportional to harmonic motion. This harmonic motion applies to 180° therefore divide the first 180° in the same number of angular parts (six in this case) as was chosen when dividing the lift. Let these radii intersect the arcs from points 1 to 6 (on the lift) in points P_1 and P_6. Draw smooth curve through these points.

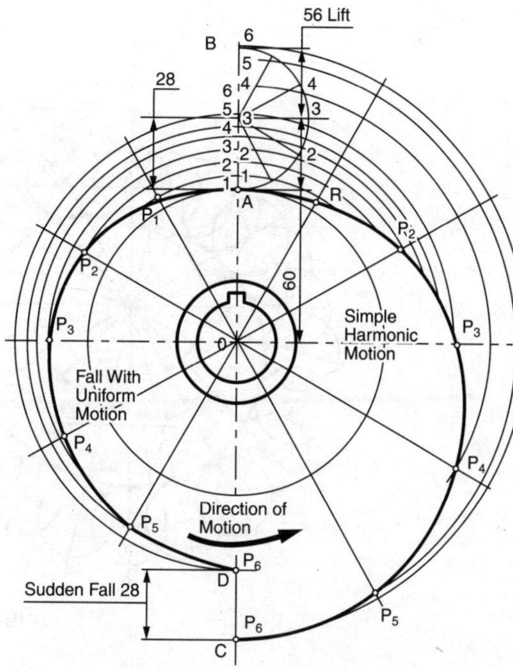

Fig. 26.21.

Mark off CD equal to half AB. The curve form D to A is part of an Archimedean spiral. For uniform motion the follower must rise or fall through equal distances for equal angles of rotation by the cam.

PROBLEM 26.7. *Design a radial cam to impact the following motion to a roller follower whose line of action coincides with the camshaft centre:*

0° – 120° lift of 24 mm with uniform velocity.

120° – 180° dwell.

180° – 360° fall of 24 mm with simple harmonic motion.

The diameter of the roller follower is 19 mm and the cam rotates anticlockwise. The least cam radius is to be 19 mm.

Construction: The Fig. 26.22 is self-explanatory.

PROBLEM 26.8. *Lay out, full size, the profile of a plate cam to give an offset flat follower the specified motion:*

0° – 180° lift of 50 mm with uniform velocity.

180° – 240° dwell.

240° – 360 fall of 50 mm with uniform acceleration and retardation.

The follower line of action is offset by 18 mm to the left of the camshaft centre. The neatest approach of the follower face to the camshaft centre is 36 mm and the cam rotates anticlockwise.

Construction: Figure 26.23 is self-explanatory.

PROBLEM 26.9. *Draw full size the profile of a radial cam to impart the following motion to an offset roller follower:*

0° – 90° lift of 40 mm with uniform velocity.

90° – 120° dwell.

120° – 240° lift of 24 mm with s.h.m.

240° – 360° fall of 64 mm with uniform aceleration and retardation.

The nearest approach of the roller centre to the cam centre is to be 64 mm and the follower centreline of the cam centre of rotation, the cam rotates clockwise and the roller follower is 32 mm in diameter.

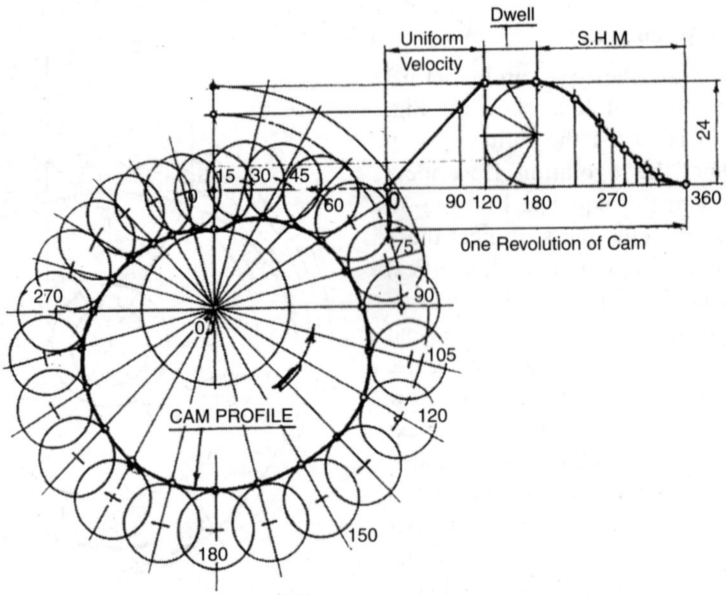

Fig. 26.22.

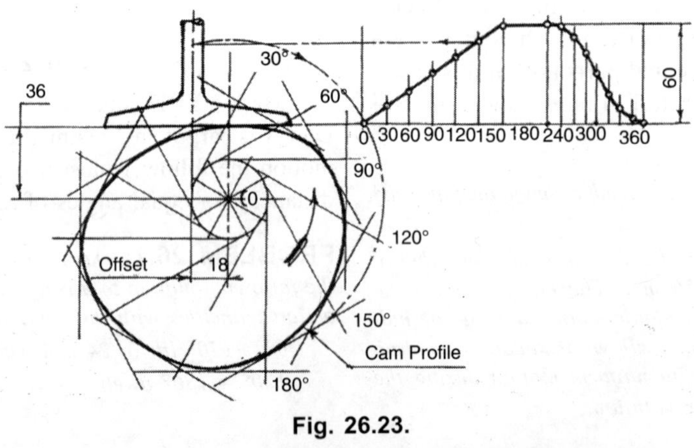

Fig. 26.23.

Construction: *Figure 26.24 self-explanatory.*

PROBLEM 26.10. *Draw the cam profile for a plate cam, using an offset roller follower producing simple harmonic motion through 360°. Take the circle diameter = 60 mm and lift of cam = 30 mm. Assume other necessary dimensions not given.*

Construction: Figure. 26.25. This motion can be plotted directly from the follower rather than on a displacement diagram, since there are no combinations of motion involved. A semi-circle is drawn with its diameter equal to the total motion desired in the follower. In this case the base circle is the centreline of the follower. The centreline of the follower is extended down and a circle is drawn with its centre at the centre of the base circle so that it is tangent to the extension of the follower centreline. The circle is divided into 30° intervals to establish points of tangency for all positions of the follower as it revolves through 360°. These tangent lines can be accurately constructed by drawing them perpendicular to the 30° interval lines extended from the centre of the points on the circumference. The distances are transferred from the harmonic motion diagram to each subsequent tangent line. Distance X is located as an example of this procedure. The circular roller is drawn in all positions and the profile of the cam is constructed to be tangent to the roller at all positions, as shown in Fig. 26.25.

PROBLEM 26.11. *The Fig. 26.26 (a) shows the outline of a cylindrical cam. The roller follower moves through its stroke while the cam rotates once. The follower moves 75 mm to the right with uniform velocity while the cam rotates from 0° to 120°, it then dwell from 120° to 180° and returns to the starting position with S.H.M. from 180° to 360°.*

Develop the cam surface and construct a front view of the cam cylinder showing the follower groove.

Construction: *Figure. 26.26 (b) illustrates the method of construction.*

EXERCISES

1. An inline knife-edge follower for a plate cam is to have the following motion; 0° – 90° dwell, 90° – 120° lift of 30 mm with S.H.M. 120° – 240° dwell, 240° – 360° fall of 30 mm with uniform velocity. The cam is to rotate clockwise and the least cam radius is to be 50 mm. Construct the cam profile full-size.

2. A radial cam, rotating clockwise, operates an offset roller follower and gives it the following motion; 0° – 120° lift of 36 mm with S.H.M. 120° – 180° dwell. 180° – 390° fall of 36 mm uniform acceleration and retardation. Construct the cam profile, full size. If the follower centerline is offset 25 mm to the left of the cam center of rotation. The minimum radial cam thickness is to be 30 mm and the roller radius if 10 mm.

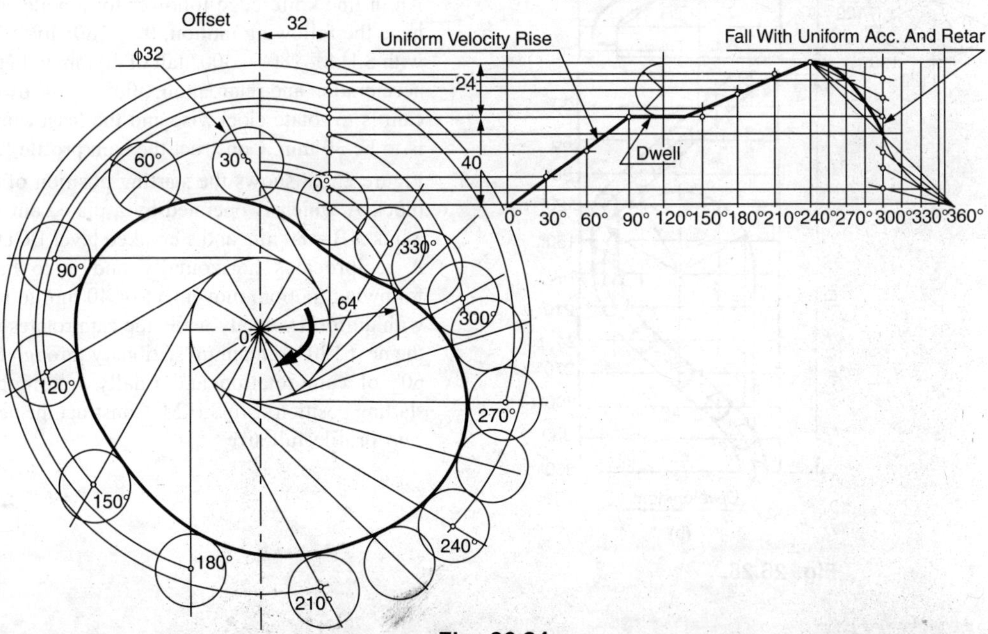

Fig. 26.24.

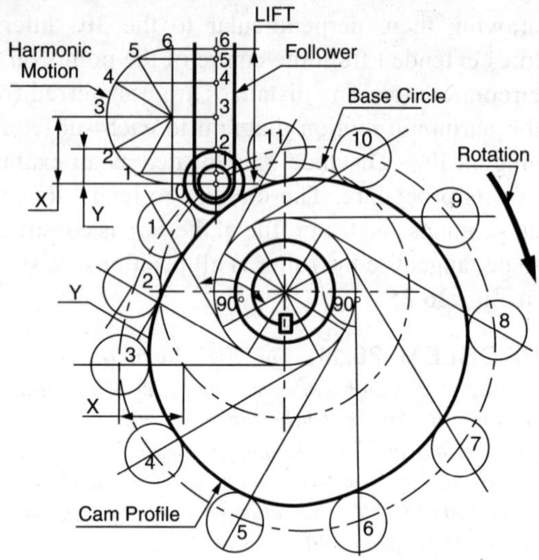

Fig. 26.25.

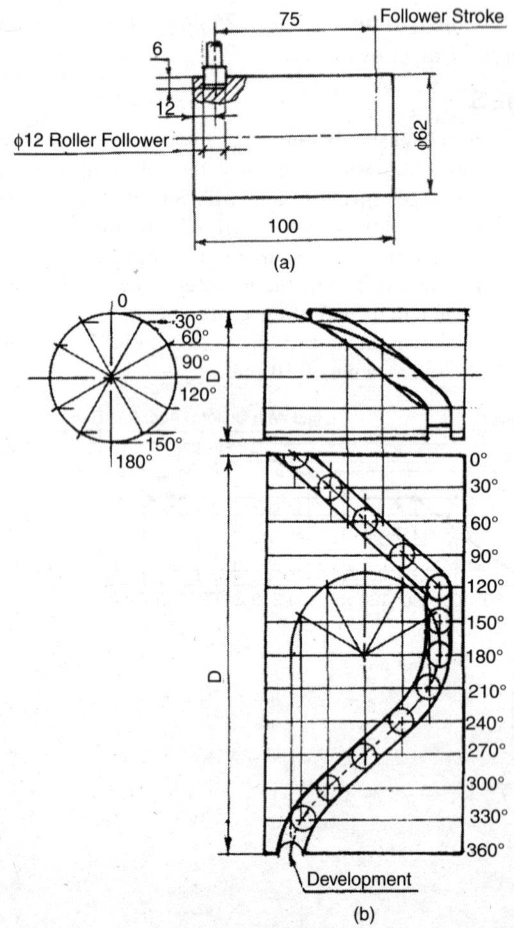

Fig. 26.26.

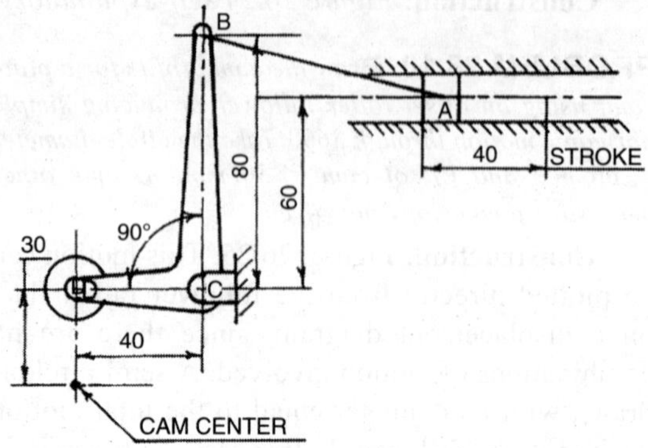

Fig. 26.27.

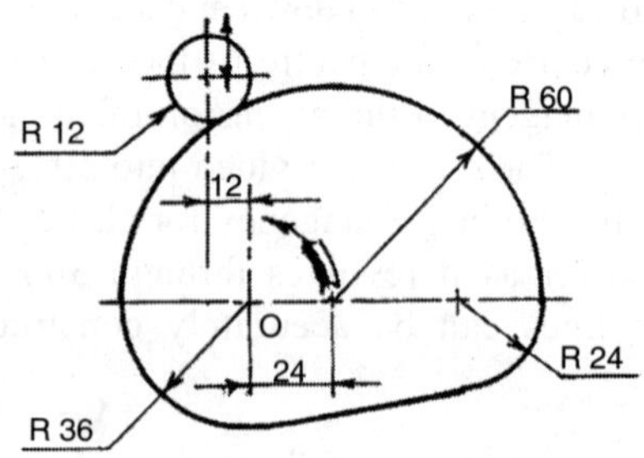

Fig. 26.28.

3. Figure 26.27 shows the profile of a plate cam, which rotates, about center O. The roller follower is offset and the cam rotates anticlockwise at a speed of 2 revolutions/sec. Construct the follower displacement diagram and find the maximum velocity of the cam at which it occurs.

4. An in-line knife-edge follower for a plate cam is to have the following motion; 0° – 180° lift of 40 mm with S.H.M. 180° – 300° fall of 40 mm with uniform acceleration and retardation. 300° – 360° dwell. The cam is to rotate clockwise and the least cam radius is to be 30 mm. Construct the cam profile full size.

5. Figure 26.28 shows the starting position of a guide block A which is operated by a plate cam through a link AB = 80 mm and a cranked lever BCD. Block A reciprocates horizontally and is to have the following motion; movement of 40 mm to the right with uniform velocity while the cam rotates through the next 120°, remaining stationary through the next 60° of cam rotation and finally returning to its starting position with S.H.M. Construct the necessary cam profile full-size.

Toothed Gears

27

27.1. FRICTION DRIVING

When the power and the motion is to be transmitted from one shaft (driver) to another shaft (driven), which is close enough, the purpose can be met with friction gearing, as shown in *Fig. 27.1.* At (*a*) are shown two circular discs, mounted on parallel shafts in the same plane, pressed against each other so that the motion of the driving shaft is transmitted to the driven shaft by means of friction. When the driven shaft offers heavy resistance there is every possibility of 'slip'. This can be overcome to a certain extent by covering the contact surface of the driven wheel with some friction creating material. The driving utility of such friction pairs depends on the coefficient of friction of the driver against the driven. At (*b*) is shown a bevel drive, here the axes of the shafts are again in the same plane, but are mutually at right angles. As both wheels are of the same diameter, these are called miters. It can be observed here that these miters are merely truncated cones, and the apex of each lies on the point of intersection of the shaft axes.

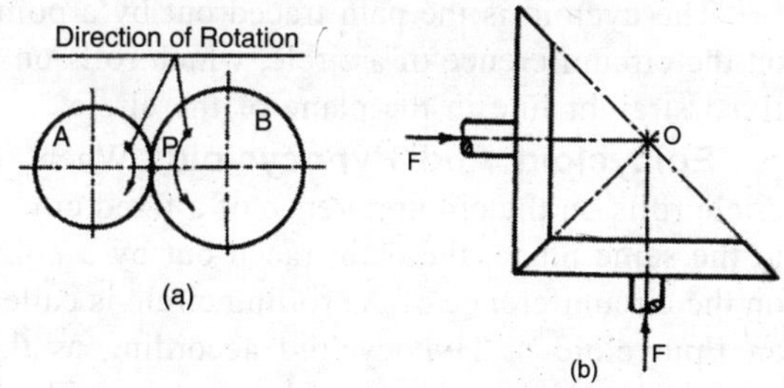

Fig. 27.1. Friction gearing

For light duty and high speeds and where the motion of the shaft is of an intermittent character, friction gearing is satisfactory. The wheel may be thrown out of gear by separating them a small fraction of a centimetre. There is always a certain amount of slip in friction gearing.

27.2. TOOTHED GEARS

Where considerable power has to be transmitted positively (without any slip) and yet at a constant velocity ratio, the smooth rolling discs and cones of *Fig. 27.1* are provided with teeth, which project radically from the circumference. The teeth are formed partly above and partly below the pitch surface of friction gears. The chief function of the teeth is to prevent slip. *Figure 27.2* shows two toothed wheels meshing with each other.

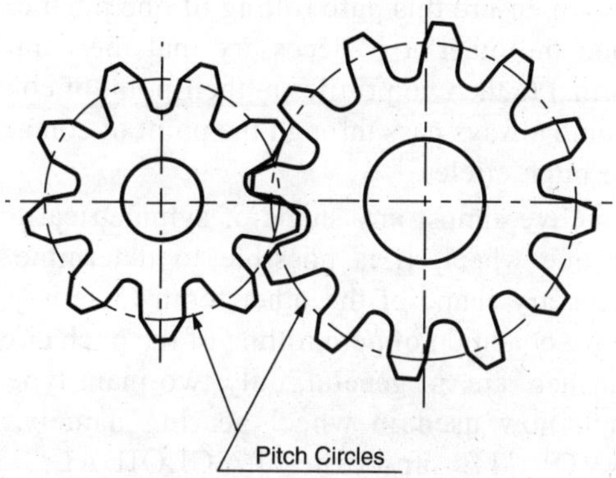

Fig. 27.2. Toothed gears

27.3. CLASSIFICATION OF TOOTHED GEARS

Toothed gearing may roughly be divided into three groups:

(*i*) Gears, which connect parallel shafts, lying in the same plane, comprising spur, helical, double helical or heringbone, treble helical and internal. In case of straight-toothed spur and internal gears, the teeth are parallel with the axis.

(*ii*) Gears connecting shafts, which if produced would intersect at some angle in the same plane, comprising bevel wheels.

(*iii*) Gears connecting shafts at an angle but not lying in the same plane, comprising worms and worm wheels, spiral and skew gears. This group is the least efficient of the three.

27.4. SPUR GEARS

When considering a pair of toothed spur gears we imagine the existence of a pair of rolling cylinders, mating together and having the same velocity ratio as the spur gears. The surface of the smooth cylinders are called pitch surfaces and their outlines or circumferences are known as pitch cricles. Measurements are taken on and from these circles. Thus the diameter of a wheel is the diameter of its pitch circle.

27.5. TOOTH PROFILES

The profile of a pair of teeth in gear must be such that while the driving profile is pressing upon and moving the driven profile the relative motion of the pitch circles of the wheels shall be that of pure rolling.

To ensure this pure rolling of one pitch circle upon the other it is necessary that the common normal to the two profiles at their point of contact should always pass through the point of contact of the pitch circles.

Give almost any shape of symmetrical tooth on one wheel, it is possible to determine the necessary shape of the other tooth to satisfy the above condition of pure rolling of the pitch circles; but there are, in general, only two main types of teeth now used in wheel gearing namely, the INVOLUTE and the CYCLOIDAL. The characteristic appearance of these two types is shown in *Figs. 27.3* and *27.4*. Cycloidal tooth were frequently used in the past as these could easily be cast in a foundry. But the cycloidal tooth is not now used to appreciable extent since the involute tooth is more easily machine-cut, and possesses many valuable advantages over the other:

(*i*) The face and flank form one continuous curve (which may be readily ground);

(*ii*) The pressure angle is constant;

(*iii*) All gears having the same pitch and pressure angle will work together;

(*iv*) If the pitch circles do not exactly touch, the velocity ratio remains constant and the gears work equally well together.

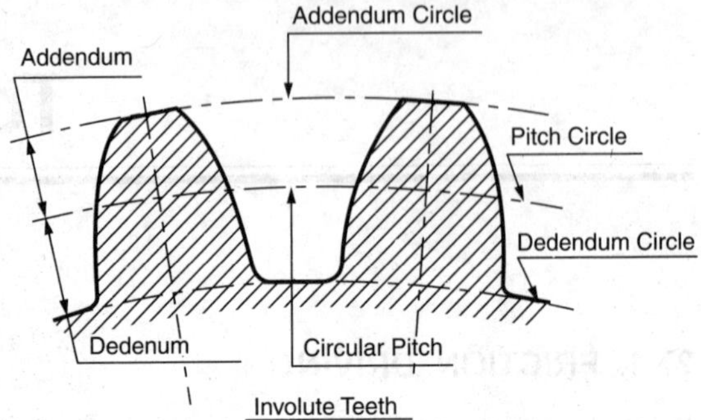

Fig. 27.3.

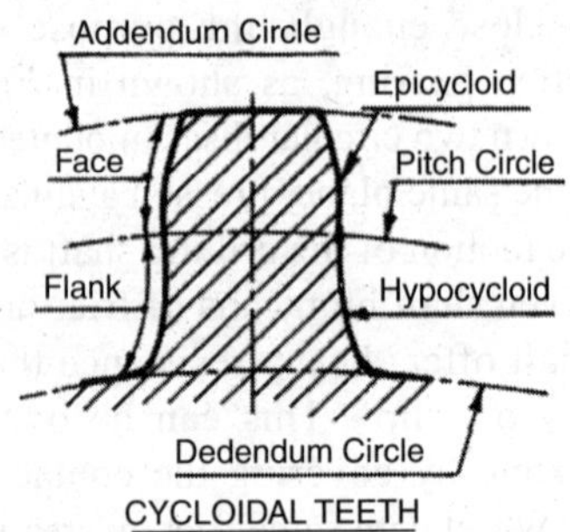

Fig. 27.4.

27.6. CURVES USED FOR THE PROFILES OF TEETH

The cycloidal curves, portions of which go to form the profile of cycloidal teeth, are shown in *Figs. 27.5 and 27.6 (a), (b)* and the involute of a circle, the early part of which forms the working profile of involute tooth is shown in *Fig. 27.7*.

The cycloid is the path traced out by a point on the circumference of a circle, which rolls on a fixed straight line in the plane of the circle.

Epicycloid And Hypocycloid: When a cricle rolls on the circumference of a fixed circle, in the same plane, the path traced out by a point on the circumference of the rolling circle is called an Epicycloid or Hypocycloid according as the circle rolls on the convex or the Concave side of the fixed circle.

The normal through any point on any of the above curves passes through the point of contact of the corresponding portion of the rolling circle with the fixed line *Fig. 27.6 (a)*.

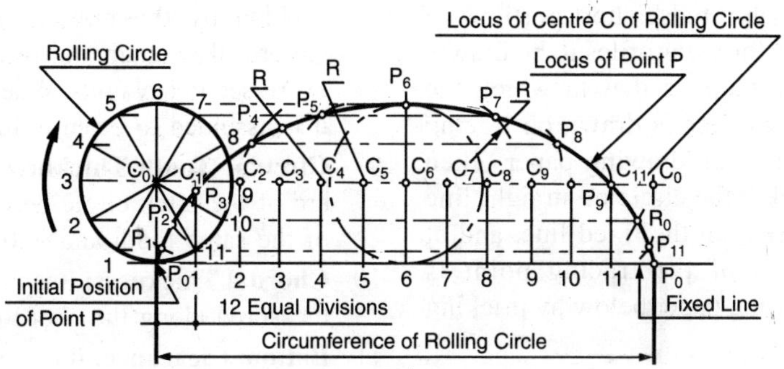

Fig. 27.5. Cycloid

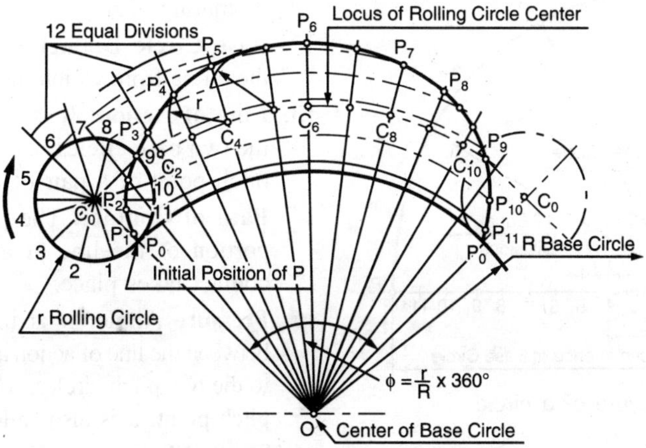

Fig. 27.6. (a) Epicycloid

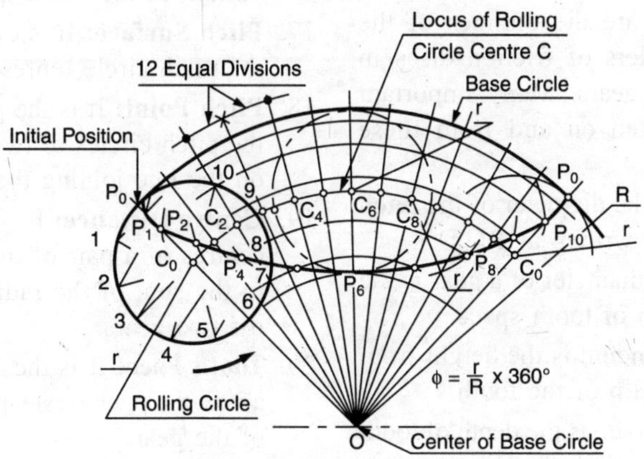

Fig. 27.6. (b) Hypocycloid

27.7. INVOLUTE

The involute of a circle is the path traced out by the end of a straight line rolling on the circumference of the circle, or it is the path traced by an inextensible cord unwound from the circumference of the circle. The straight line or chord is always a tangent to the circle and the normal to the curve at any point is the line passing through the point and tangential to the circle. The figures for these curves are self-explanatory.

When, however, only those short portions of the curves as required for the profiles to be drawn, the better method to adopt is that in which the rolling circle of straight line is drawn on tracing paper and the fixed line on drawing paper. Then with the aid of a pricker the circle or straight line is rolled, in small steps, on the fixed line, and at intervals the position of the tracing point is transferred to the drawing paper below by pricking through.

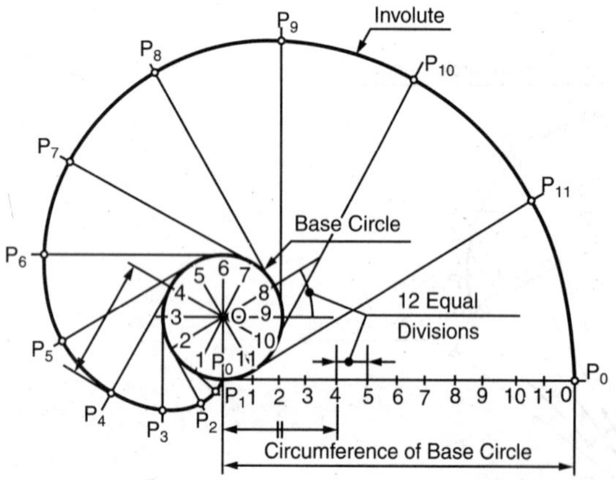

Fig. 27.7. Involute of a circle

27.8. SPUR GEAR NOMENCLATURE

1. **Pitch Circles:** These are the outlines of the imaginary smooth rollers of friction discs in every pair of mating gears. Many important measurements are taken on and from these circles.
2. **Pitch Diameter:** It is the diameter of the pitch circle and is generally written as P.C.D.
3. **Root Diameter:** Root diameter of a gear is the diameter at the bottom of tooth space.
4. **Addendum:** The addendum is the height from the pitch circle to the tip of the tooth.
5. **Dedendum:** The dedendum is the depth of tooth space below the pitch circle and is equal to addendum plus clearance.
6. **Circular Pitch (C.P.):** It is the length of the arc of the pitch circle between similar faces of successive teeth.
7. **Diametral Pitch (D.P.):** Diametral pitch is the number of teeth divided by the pitch diameter. It is ratio not dimension.
8. **Module (m):** The module is the pitch diameter divided by the number of teeth. It is the reciprocal of diametral pitch. Module may be expressed in any units. When no units are stated it is assumed to be in millimeters.
9. **Circular Tooth Thickness:** It is the length of arc of the pitch circle between opposite faces of the same tooth and is 0.5 C.P.
10. **Chordal Thickness:** It is the distance of a tooth measured along the chord of the pitch circle.
11. **Bottom Clearance:** It is the difference between the addendum of a gear and the dedendum of its mating gear.
12. **Base Circle Diameter:** It is the diameter of the circle from which the involute is generated.
13. **Line of Action:** It is the common tangent to the two base circles, which passes throught the pitch point of a pair of mating gears.
14. **Path of Contact:** The path of contact is that portion of the line of action on which tooth contact takes place.
15. **Pressure Angle:** It is the acute angle formed between the line of action and the common tangent to the two pitch circles, which passes through the pitch point. It is also called angle of obliquity.
16. **Tooth Fillet:** It is the rounded corner at the bottom of the tooth space.
17. **Pitch Surface:** It is cylindrical surface, which the pitch circle represents.
18. **Pitch Point:** It is the point of contact between the pitch circles of two gears in mesh and lies on the line joining their centres.
19. **Centre Distance:** It is the distance betwen the centres of a pair of mating gears and is equal to the sum of the radii of the pitch circles of the two gears.
20. **Tooth Face:** It is the side-surface of the tooth above the pitch circle, perpendicular to the plane of the gear.
21. **Tooth Flank:** It is the side-surface of the tooth below the pitch circle perpendicular to the plane of the gear.
22. **Crest of Tooth:** It is the outside surface of the tooth, perpendicular to the plane of the gear.
23. **Root of Tooth:** It is the juncftion of the tooth with the material at the bottom of the tooth surface.

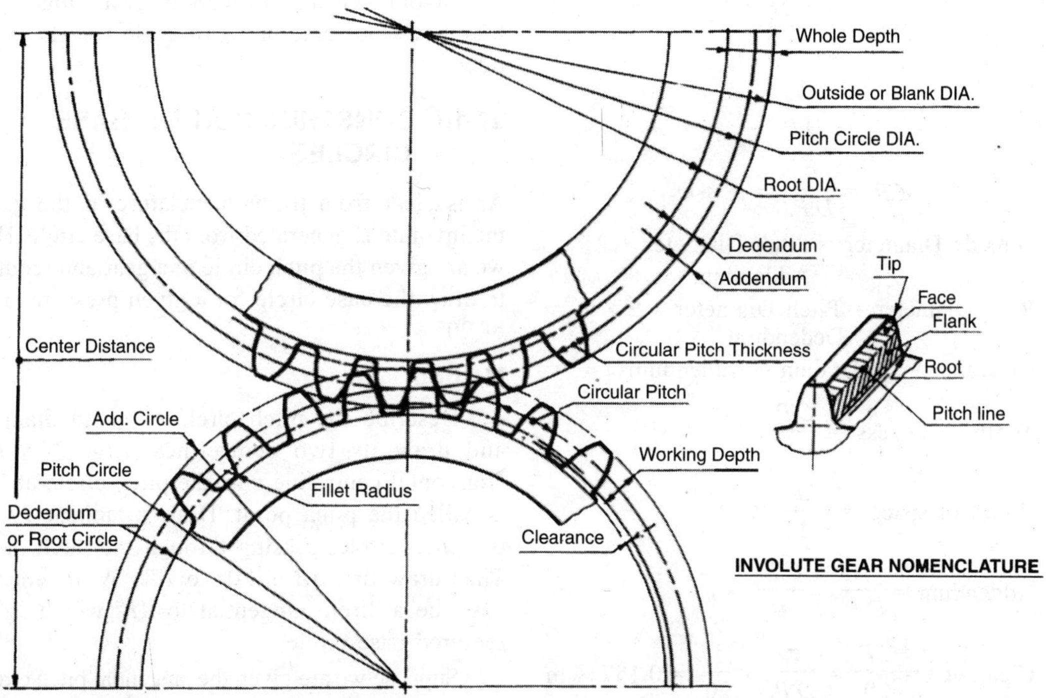

Fig. 27.8. Spur gear nomenclature

24. **Whole Depth:** It is the sum of the addendum and Dedendum of a tooth.
25. **Working Depth:** It is the distance by which a tooth extends into the space of a mating gear and it equal to whole depth minus the clearance, or twice the addendum.
26. **Addendum Circle:** It is the circle, which contains the crests of the teeth. its diameter is called the outside or blank diameter.
27. **Dedendum of Root Circle:** It is the circle, which contains the bottom of the tooth spaces.
28. **Chordal Pitch:** It is the length of the chords of the pitch circle between a point on one tooth and a corresponding point on the adjacent tooth, both on the pitch circle.

27.9. GEAR CALCULATION

$$\text{Circular pitch} = \frac{\text{Circumference of P.C.D.}}{\text{No. of teeth}}$$

$$= \frac{\text{Pitch circle diameter} \times \pi}{\text{No. of teeth}}$$

$$C.P. = \frac{P.C.D. \times \pi}{N}$$

Pitch diameter or number of teeth can also be found out from this relation.

$$\text{Diametral pitch} = \frac{\text{No. of teeth}}{\text{Pitch circle diameter}}$$

$$D.P. = \frac{N}{P.C.D.}$$

Number of teeth or pitch circle diameter can also be found from this relation.

$$\therefore \quad N = P.C.D. \times D.P.$$

$$\text{and} \quad P.C.D. = \frac{N}{D.P.}$$

$$C.P. \times D.P. = \frac{P.C.D. \times \pi}{N} \times \frac{N}{P.C.D.} = \pi$$

$$\therefore \quad C.P. = \frac{\pi}{D.P.}$$

$$\text{and} \quad D.P. = \frac{\pi}{C.P.}$$

Module pitch or metric module is generally used in metric system.

$$\text{Metric Module, } m = \frac{\textit{Pitch Circle Diamenter}}{\textit{Number of teeth}}$$

i.e., $m = \frac{P.C.D.}{N}$

$= \frac{1}{D.P.}$

$C.P. = \frac{\pi}{D.P.} = \pi \times m$

Outside Diameter = Pitch Diameter + 2 × Addendum

Root Diameter = Pitch Diameter – 2 × Dedendum

Clearance = Dedendum – Addendum

Tooth Thickness $= \frac{C.P.}{2}$

Width of space $= \frac{C.P.}{2}$

Addendum $= \frac{1}{D.P.} = \frac{C.P.}{\pi} = m$

Clearance $= \frac{C.P.}{20} = \frac{\pi}{D.P.} \times \frac{1}{20} = 0.157 \times m$

Pressure Angle, θ = 14.5° or 20°. It is usually kept 20°, as this value reduces the possibiolity of interference. In addition, it also gives the tooth a wider root.

Base circle Diameter, B.C.D. = P.C.D. × cos θ as shown in *Fig. 27.9.*

Involute Gears will work correctly together if they have the same pressure angle and diametral pitch.

27.10. CONSTRUCTION OF BASE CIRCLES

As is clear from the nomenclature of the gears, the involute is generated from the base circle. Here we are given the pitch circle of a gear and required to draw the base circle for a given pressure angle of 20°.

Method

Describe the pitch circle of given diameter and draw its two centrelines, *Fig. 27.9 (a)*. Horizontal centreline cuts the pitch circle at P. It is called the pitch point. Draw a tangent TT_1 to the pitch circle, passing through the pitch point. Then draw line *AB* at 20° to TT_1. With centre *O* describe a circle tangential to *AB*, which is the required base circle.

Suppose we are given the tangential pitch circles of two mating gears, as shown in *Fig. 27.9 (b)*, and we are required to draw the base circles. For this purpose draw the two given pitch circles touching at a point P. From P draw a common tangent TT_1. Mark the line of action, *AB*, at an angle of 20° (Pressure angle) to tangent TT_1. Draw the base circles tangential to the line of action *AB*.

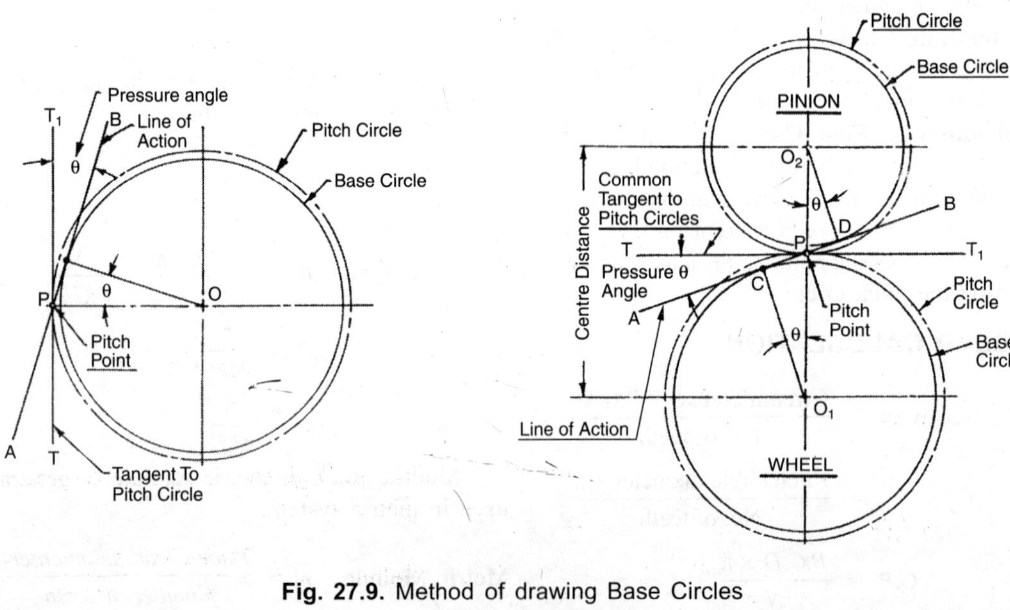

Fig. 27.9. Method of drawing Base Circles

27.11. CONSTRUCTION FOR APPROXIMATE INVOLUTE

Describe the base circle. Cut-off any number of equal spaces along the base circle, *i.e.*, A-1, 1-2, 2-3 etc. Join these points with the centre of the base circle. Draw tangents to the base circle from these points. Now with centre 1 and radius equal to 1A draw an arc to cut the first tangent at 1'. Again with 2 as centre and radius equal to 2 A draw an arc to cut the second tangent at 2' and so on. The curve joining 1', 2', 3' etc. will be the involute curve *Fig. 27.10 (a)*.

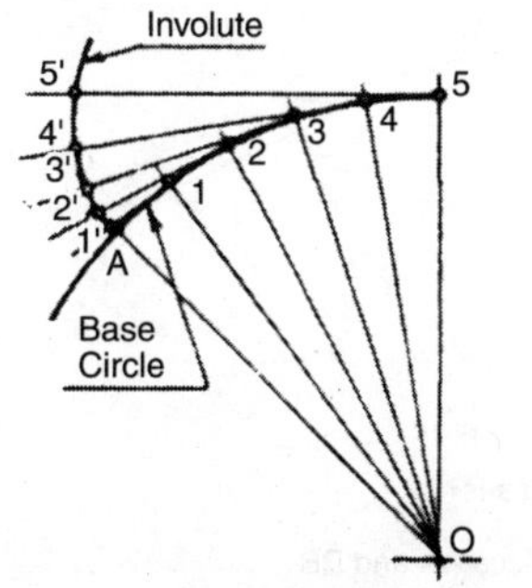

(a) **CONSTRUCTION FOR APPROX. INVOLUTE**

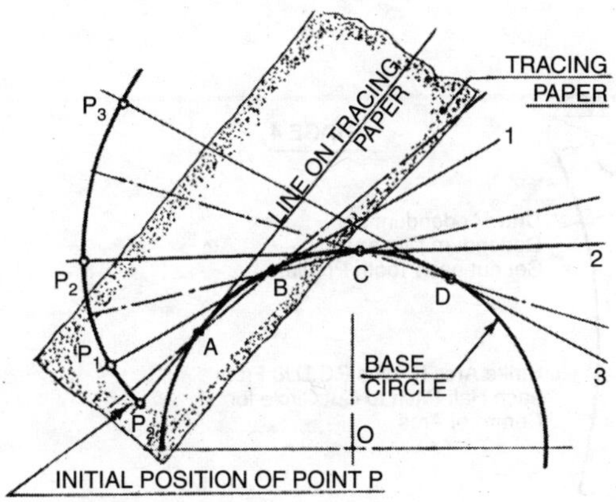

(b) **TRACING PAPER METHOD**

Fig. 27.10. Construction for Approximate Involute.

27.12. APPROXIMATE CONSTRUCTIONS FOR THE INVOLUTE TOOTH PROFILE

Since accurate drawing of actual involute curves on the gear teeth is a lengthy procedure, it is usual to represent those involute curves approximately by circular arcs.

Four such approximate methods are shown here in *Figs. 27.10 (c), 27.11, 27.12* and *27.13*. The first two methods differ only at one point, depending on the number of teeth on the gear. First draw the pitch circle of diameter D and a tangent to it to fix the pitch point. On the pitch circle radius draw a semi-circle of radius D/4 to cut at P an arc of radius D/8 drawn with the pitch point as centre. On this circle lie the centres of the profile arcs of the teeth *Figs. 27.10(c), 27.11*. Then draw the addendum and dedendum circles and from the pitch point mark off tooth widths C.P./2 around the pitch circle. From these points draw arcs of radius D/8 to the circle for the centres of the profile arcs. Then:

1. For a gear of 30 teeth and over (Fig. 27.10(c)) draw the profile arcs radius D/8 with these points as centres from the addendum circle to the Dedendum circle and complete the teeth with fillet radii equal to one-seventh of the widest tooth space.
2. For a gear with less than 30 teeth (Fig. 27.11) draw the profile arcs as in 1 above and draw radial lines tangential to them before adding the fillet radii. This method avoids having teeth, which appear to be excessively undercut.
3. The third method can be used for a gear having any number of teeth and makes use of the base circle. This construction is more suitable for teeth with 20° pressure angle than the methods described in 1 and 2 as it produces teeth which have wider roots, Fig. 27.12.
   - Draw the pitch, addendum and dedendum circles, and the base circle tangential to the line of action.
   - On a convenient radial line, fix a point *A* on the addendum circle and a point *E* on the base circle.
   - Divide *AE* into three equal parts and through *B*, the division nearest *A*, draw a tangent to the loose circle at *D*.
   - Divide *BD* into four equal parts and through F, the division nearest to F draw circle. The centre for *F* is the centre for the profile arc passing through *B*.
   - From *C*, the intersection of this profile arc and the pitch circle, mark off teeth thickness round the circle.

1. CONSTRUCTION OF GEAR TEETH SUITABLE FOR GEARS OF 30 TEETH AND ABOVE PRESSURE ANGLE 14.5°

GEAR DATA–METRIC MODULE (m) 6; NUMBER OF TEETH 30

CALCULATED DATA

PITCH CIRCLE DIAMETER (P.C.D) = N × m = 30 × 6 = 180 mm

ADDENDUM = m = 6 mm

CIRCULAR PITCH (C.P.) = π.m = 3.142 × 6 = 18.852 mm

$$\text{CLEARANCE} = \frac{\text{C.P.}}{20} = \frac{18.852}{20} = .9426$$

DEDENDUM = ADDENDUM + CLEARANCE

= 6 + .9426 = 6.9426 mm

TOOTH THICKNESS = C.P/2 = 9.426 mm

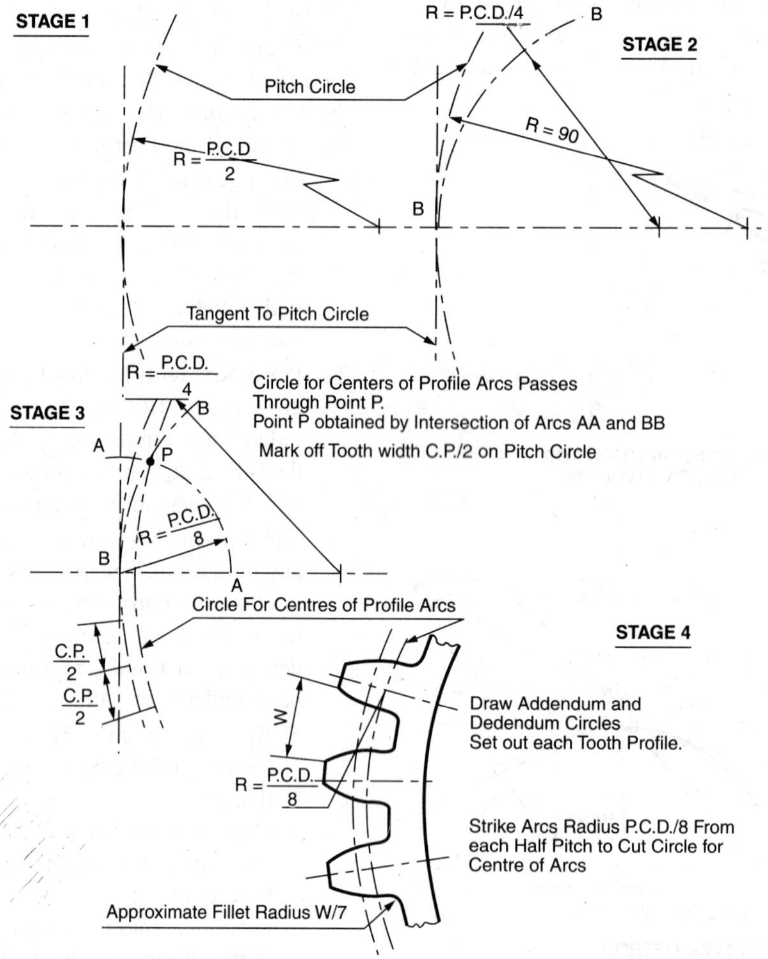

Fig. 27.10. (c) Approximate Construction Method (for 30 teeth and over)

- Draw profile arcs of radius *FB* through these points, centred on the circle through *F*. The fillet radius at the roots of the teeth is again one seventh of the widest tooth space.

4. **Tracing Paper Method:** This method is also very versatile and can be used for gears with any number of teeth.

- With centre *O*, draw the pitch circle. At any point *P* on it draw a tangent TT_1. The point P is pitch point.
- Draw the line of action *AB*, through the pitch point *P*, inclined at pressure angle θ either 14.5° or 20° to TT_1.

2. CONSTRUCTION OF GEAR TEETH SUITABLE FOR GEARS WITH LESS THAN 30 THEETH. PRESSURE ANGLE 14½.

GEAR DATA–MATRIC MODULE 8.5; NUMBER OF TEETH 12
CALCULATED DATA-

PITCH CIRCLE DIAMETER (D here) = metric module (m) × Number of Teeth (N)
= 8.5 × 12 = 102 mm
Addendum = m = 8.5 mm
Circular pitch = πD/N = m × π = 8.5 × 3.142 = 26.707 mm
Clearance = C.P./20 = 26.707/20 = 1.335 mm
Dedendum = Addendum + Clearance = 8.5 + 1.335 = 9.835 mm

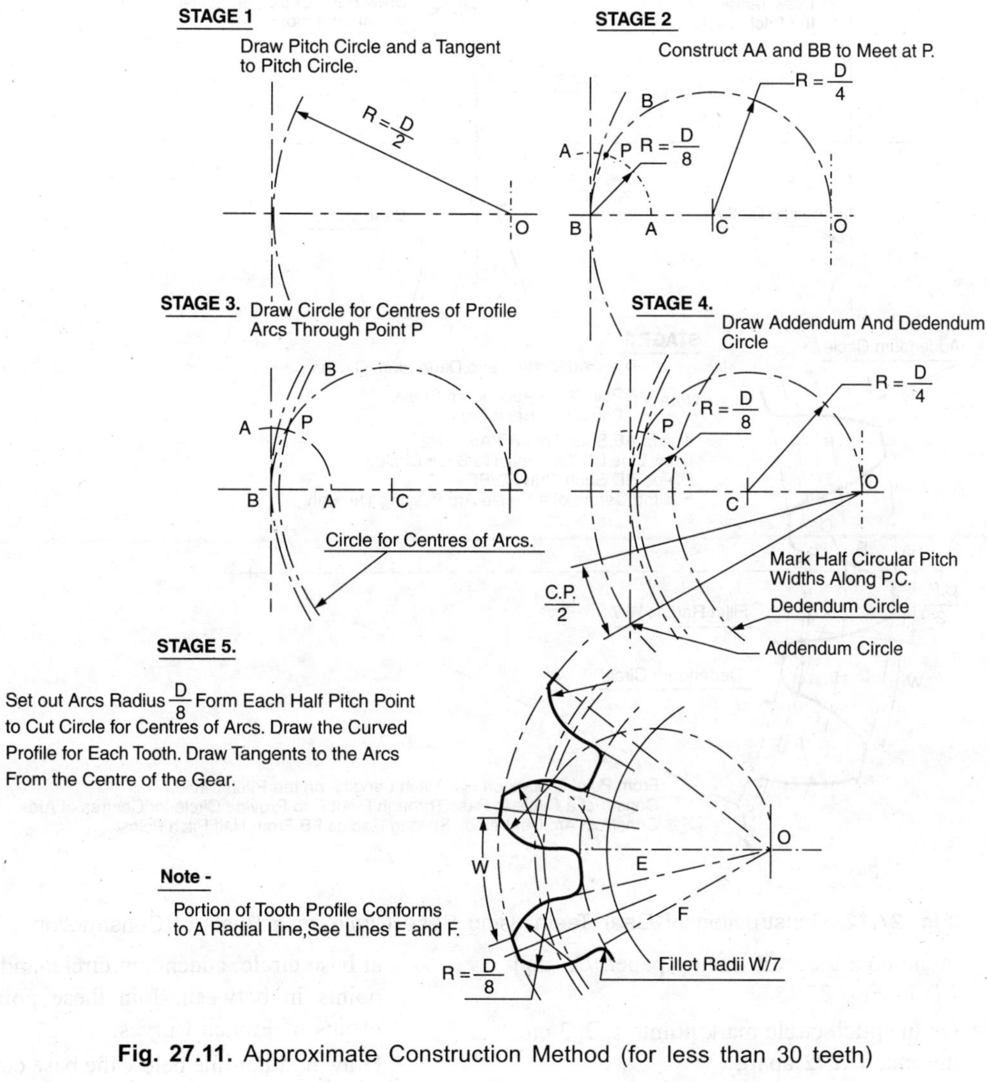

Fig. 27.11. Approximate Construction Method (for less than 30 teeth)

- Through O, draw a line OC inclined at θ to OP, i.e., perpendicular to line of action AB. Then with centre O and radius OC, draw the base circle.
- Draw the addendum and dedendum circles.
- Construct a portion of involute curve from the base circle, shown in *Fig. 27.13* as DF. Trace out this curve and a part of the base

3. Construction of Gear teeth using base circle as a basis of construction:

GEAR DATA-Pressure angle 20°, Metric Module (m)10, number of teeth 25.

CALCULATED DATA-

Pitch Circle Diameter = D = m.N. = 10 × 25 = 250 mm

Addendum = m = 10 mm

Circular pitch = C.P. = $\frac{pD}{N} = \pi \cdot m = 3.142 \times 10 = 31.42$ mm

Clerance = C.P/20 = 31.42/20 = 1.571 mm

Dedendum = Addendum + Clearance = 10 + 1.571 = 11.571 mm

TOOTH THICKNESS = C.P/2 = 31.42/2 = 15.71 mm

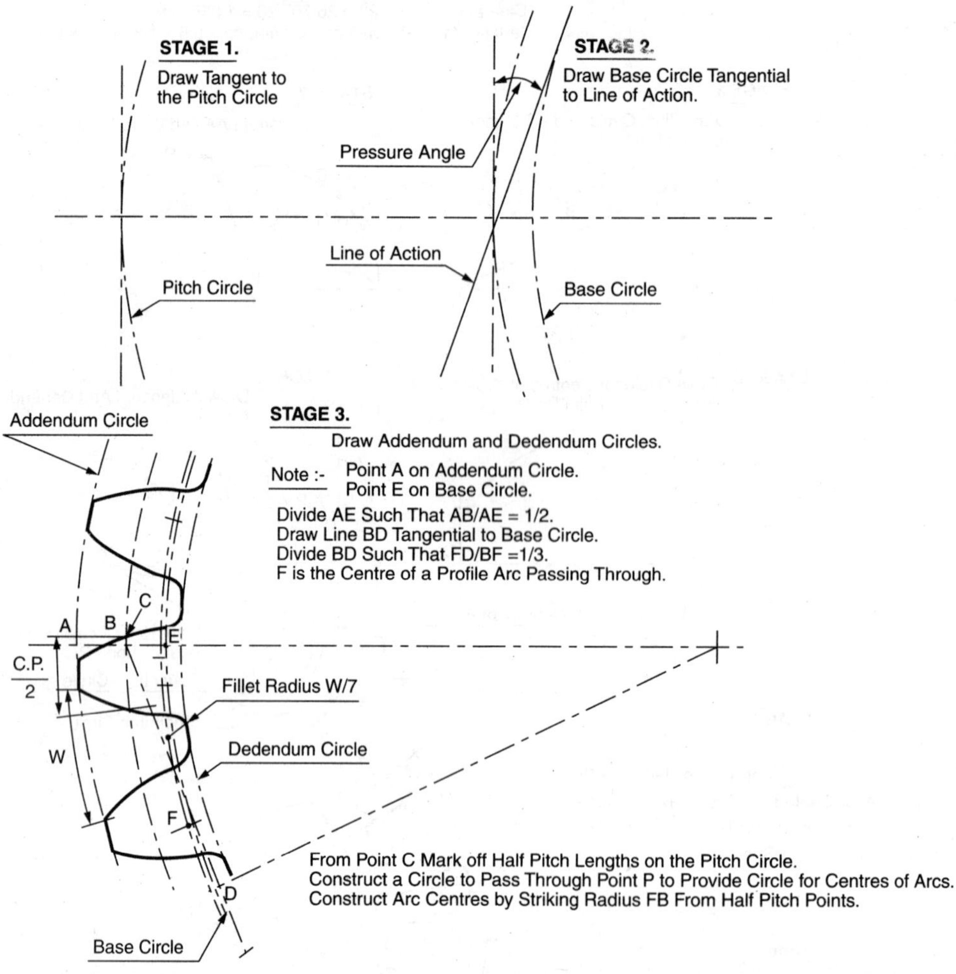

Fig. 27.12. Construction of Gear Teeth using Base Circle as a Basis of Construction

circle on a piece of tracing paper, *DF* and *ED* in *Fig. 27.13*.

- On the pitch circle mark points 1, 2, 3 etc. distance C.P./2 apart.
- Place the tracing paper on the base circle such that the arc ED coincides with it, while the curve DF passes through the point 1 (for example). Prick points along the curve at base circle, addendum circle, and a few points in between. Join these points by means of French Curves.
- Draw a radial line below the base circle up to dedendum circle. Then complete one side of the tooth profile by joining the radial line with the bottom of the tooth space by a fillet of radius r = C.P./8.

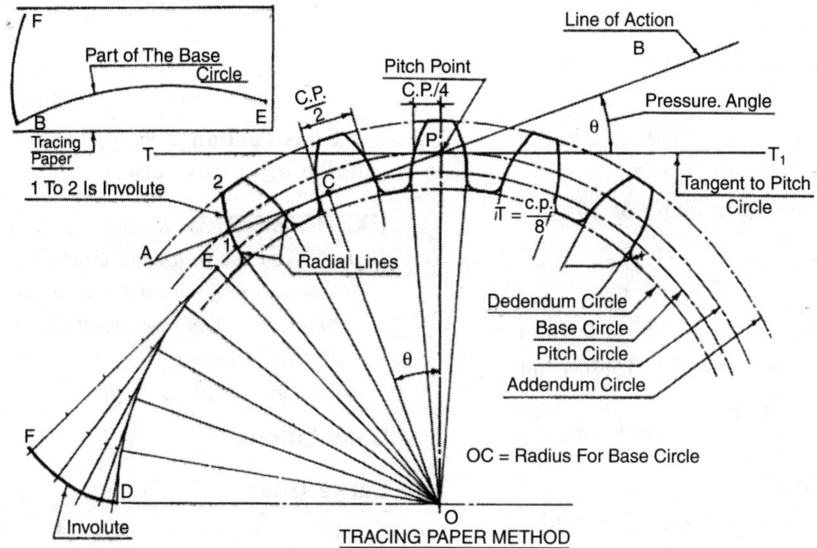

Fig. 27.13.

- Reverse the tracing paper and plot, as explained, the other side the tooth profile, through the adjacent point on the pitch circle (i.e., point 2). In this way the first tooth profile is completed.
- Repeat the procedure for each tooth.

On a working drawing of a spur gear it is usually unnecessary to show more than two or three teeth. The working drawing consists generally of two views, a circular view showning the addendum, dedendum, and pitch circles drawn in chain (dot) lines and a sectional view taken through the centre line of the gear. On the sectional view the section line should not cross the teeth. The data relating to the teeth are generally shown in tabular form on the drawing.

EXERCISE 27.1. *A wheel has 24 teeth of 33 mm circular pitch and pressure angle 20°. Set out the various circles and a few teeth.*

Calculations:

$$\text{P.C.D.} = \frac{\text{No.of teeth} \times \text{C.P.}}{\pi}$$

$$= \frac{24 \times 33}{\pi} = 252 \text{ mm}$$

$$\text{Addendum} = \frac{\text{C.P.}}{\pi} = \frac{33}{\pi} = 10.5 \text{ mm}$$

$$\text{Add circle dia} = P.C.D. + 2 \times \text{ Add.}$$

$$= 252 + 2 \times 10.5 = 273 \text{ mm}$$

$$\text{Clearance} = \frac{\text{C.P.}}{20} \text{ (Approximately)}$$

$$= \frac{33}{20} = 1.65 \text{ mm}$$

$$\text{Dedendum} = \text{Add + Clearance}$$

$$= 10.5 + 1.65 = 12.15 \text{ mm}$$

$$\text{Ded. Circle Dia.} = \text{P.C.D.} - 2 \times 12.15$$

$$= 252 - 2 \times 12.15$$

$$= 277.7 \text{ mm}$$

$$\text{Tooth Thickness} = \frac{\text{C.P.}}{2} = \frac{33}{2} = 16.5 \text{ mm}$$

With these dimensions and after drawing the base circle and approximate involute the rest of the construction can be complete, as shown in *Fig. 27.13*, with the help of the tracing paper trammel.

EXERCISE 27.2. *Draw a gear and pinion in engagement showing few teeth on each, with the following data:*

Pinion:

No. of teeth = 12

Metric module, m = 5

Gear:

No of teeth = 30

Metric module, m = 5

Assume a pressure angle of 20°

Calculations

Pinion:

$P.C.D. = m \times N = 5 \times 12 = 60$ mm

$C.P. = \pi \times m = \pi \times 5 = 15.708$ mm

$$\text{Addendum} = \frac{\text{C.P.}}{\pi} = m = 5 \text{ mm}$$

Add. circle dia

$= P.C.D. + 2 \times \text{Add.}$

$= 60 + 2 \times 5 = 70$ mm

$$\text{Clearance} = \frac{\text{C.P.}}{20} = \frac{15.708}{20}$$

$= 0.785$ mm

Dedendum = Addendum + Clearance

$= 5 + 0.785 = 5.785$ mm

Ded. circle diameter

$P.C.D. - 2 = \times$ Ded.

$= 60 - 2 \times 5.785$

$= 48.43$ mm

Tooth Thickness

$$= \frac{\text{C.P.}}{2} = \frac{15.708}{2}$$

$= 7.854$

Gear : P.C.D. $= m \times N = 5 \times 30$

$= 150$ mm

C.P. $= \pi \times m = \pi \times 5$

$= 15.708$ mm

$$\text{Add.} \frac{\text{C.P.}}{\pi} = m = 5 \text{ mm}$$

Add. Circle Dia.

$= \text{P.C.D.} + 2 \times \text{Add.}$

$= 150 + 2 \times 5$

$= 160$ mm

Clearance $= 0.785$ mm

Dedendum $= 5.785$ mm

Ded. Circle Diameter

$= \text{P.C.D.} - 2 \times \text{Ded.}$

$= 150 - 11.57$

$= 138.43$ mm

Construction : Refer to *Fig. 27.14* for the solution of this exercise.

EXERCISE 27.3. *Layout a pair of Spur Standard Gears, given the following data:*

Number of Teeth on Large Gear = 24

Number of Teeth on small Gear = 16

Metric Modul, m =8

Pressuire Angle ϕ = 20°

Calculations

Large Gear

P.C.D. $= m \times N = 8 \times 24 = 192$ mm

C.P. $= \pi \times m = \pi \times 8 = 25.133$ mm

$$\text{Add} = \frac{\text{C.P.}}{\pi} = 8 \text{ mm}$$

Add. Circle dia. $= P.C.D. + 2 \times 8$

$= 192 + 16 = 208$ mm

$$\text{Clearnace} = \frac{\text{C.P.}}{20} = 1.257 \text{ mm}$$

Dedendum = Add. + Clearance

$= 8 + 1.257 = 9.257$ mm

Ded. Circle Dia. $= P.C.D. - 2 \times$ Ded.

$= 192 - 2 \times 9.257$

$= 173.49$ mm

$$\text{Tooth Thickness} = \frac{\text{C.P.}}{2} = 12.566 \text{ mm}$$

Small Gear

P.C.D. $= 8 \times 16 = 128$ mm

C.P. $= \pi \times 8 = 25.133$ mm

Addendum $= 8$ mm

Add. Circle Dia. = P.C.D. + 2 × Add.

$= 128 + 16 = 144$ mm

Clearance $= 1.257$ mm

Dedendum $= 8 + 1.257 = 9.257$ mm

Ded. Circle Dia. = P.C.D. − 2 × Ded.

$= 109.486$ mm

Construction : Refer *Fig. 27.15* for the solution of this exercise. Here the gears are drawn by using **Base circles**, *i.e.*, **Method 3**.

PINION Drawn by Tracing Paper Method

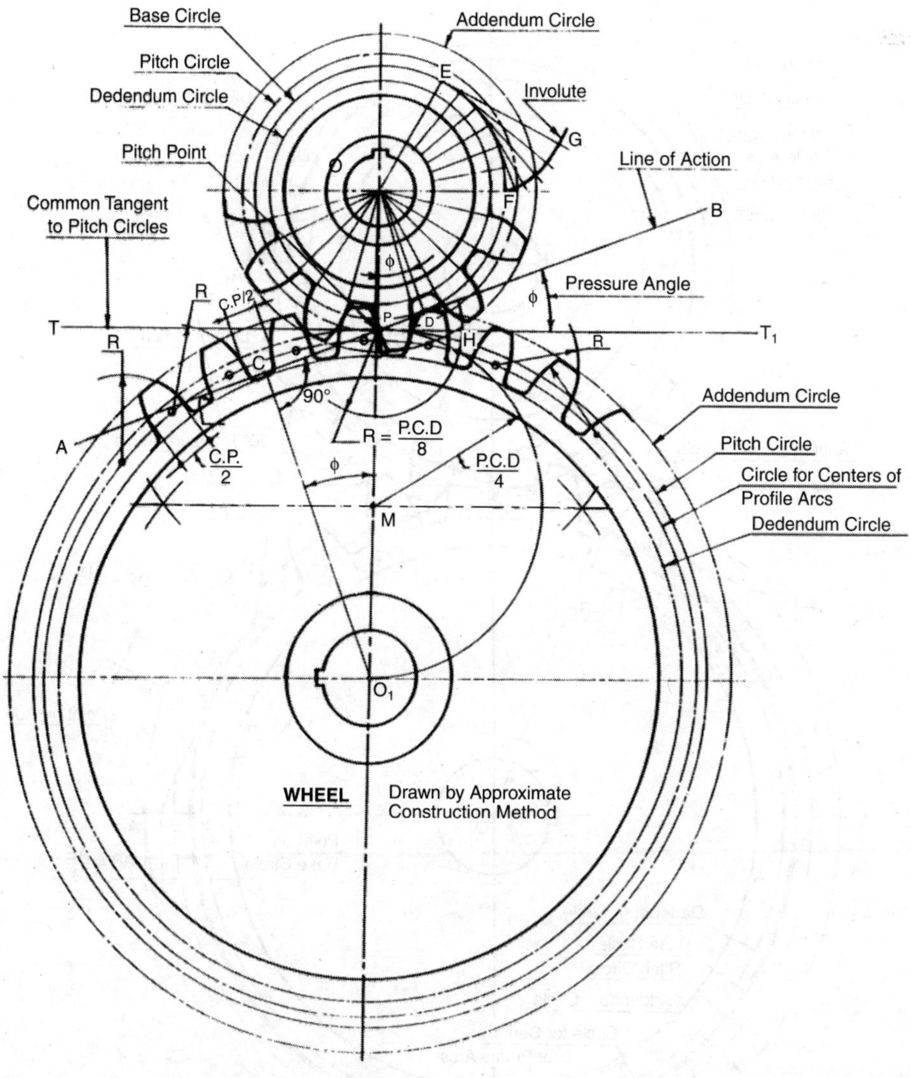

Fig. 27.14. Wheel and Pinion in engagement

EXERCISE 27.4. *Draw the profile of involute teeth, for a gear having 22 teeth and a D.P. = 1.1 taking centimeter as unit length. Assume a pressure angle of 20°.*

Calculations

$$\text{P.C.D.} = \frac{N}{\text{D.P.}} = \frac{22}{1.1} = 20 \text{ cm} = 200 \text{ mm}$$

$$\text{C.P.} = \frac{\text{P.C.D.} \times \pi}{N}$$

$$\text{or} \quad \frac{\pi}{\text{D.P.}} = \frac{\pi}{1.1} = 2.856 \text{ mm}$$

$$\text{Addendum} = \frac{1}{\text{D.P.}} = \frac{1}{1.1} = 0.909 \text{ cm}$$

Addendum Circle Dia.

$$= \text{P.C.D.} + 2 \times 9.09 \text{ cm}$$
$$= 218.18 \text{ mm}$$

$$\text{Clearance} = \frac{\text{C.P.}}{20} = \frac{28.56}{20} = 1.428 \text{ mm}$$

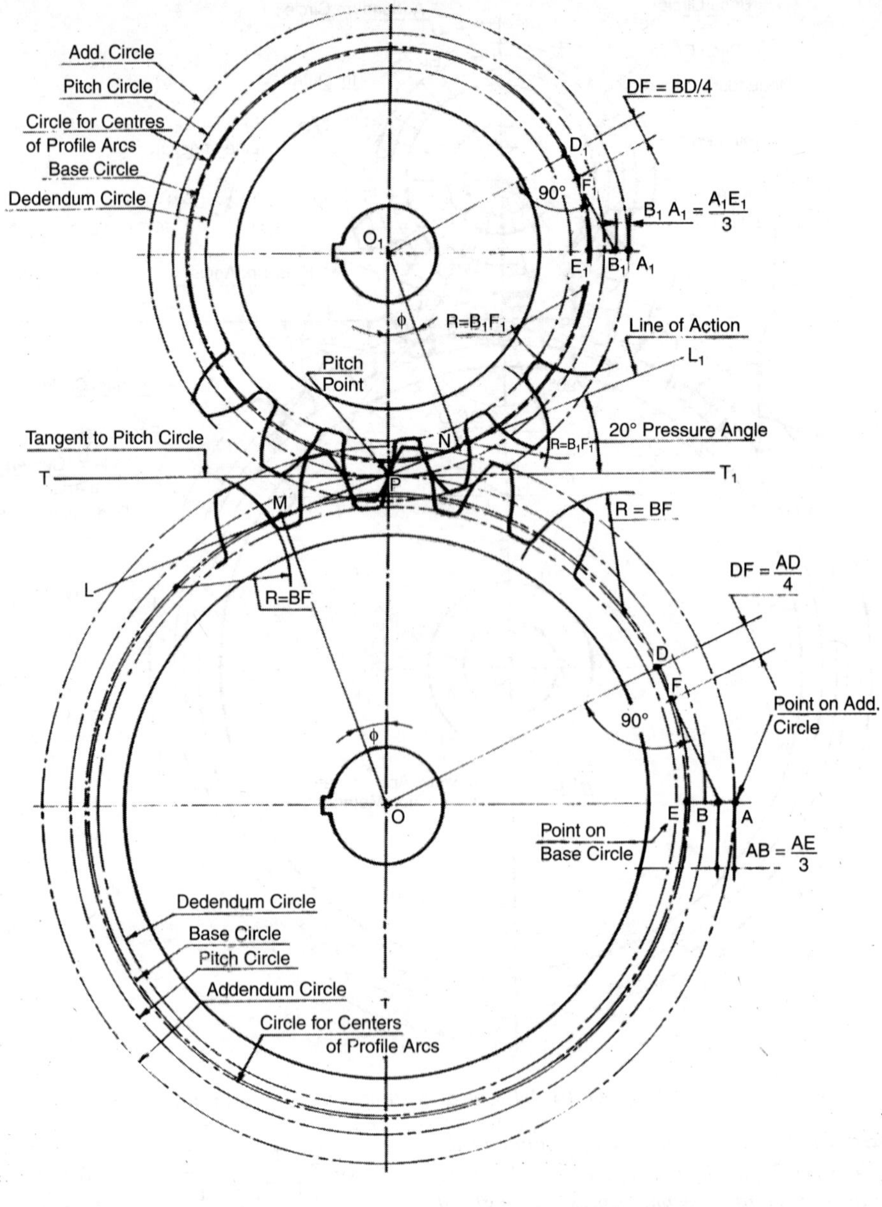

Fig. 27.15. Wheel and Pinion in engagement

Dedendum = Add. + Clearance

$= 9.09 + 1.428 = 10.518$ mm

Ded. Circle Dia.

$=$ P.C.D. $- 2 \times$ Ded.

$= 200 - 2 \times 10.518 = 178.964$ mm

Tooth Thickness

$$= \frac{\text{C.P.}}{2} = \frac{28.56}{2} = 14.28 \text{ mm}$$

Construction : Use the **Tracing Paper Method** of drawing the teeth profile as explained in **Method 4**. Refer *Fig. 27.14* for solutin of this exercise.

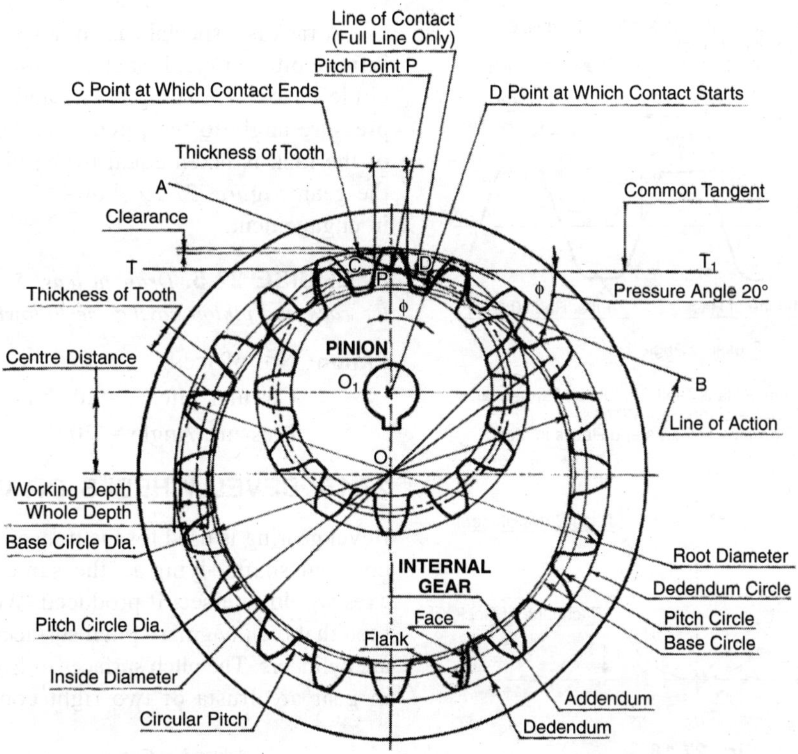

Fig. 27.16. Internal Gear Nomenclature

27.13. INTERNAL GEARS

These are also called *annular gears*. These consist of a *Large Wheel* having teeth cut on the interior surface, in gear with a *Small Pinion*. The gears, which mesh externally, run in opposite directions, *Fig. 27.14*, whereas the internal gears run in the same direction. These gears are used where compact designs are desired. The central distance betwen the parallel shafts, driver and driven, is reduced by the use of these gears. An internal gear and pinion are shown in engagement in *Fig. 27.16*. The particulars of internal gears, shown in *Fig. 27.16* are as follows:

Internal Gear: No. of Teeth = 20

Circular Pitch = 20 mm

Pressure Angle = 20°

Pinion: No. of teeth = 12

Circular Pitch = 20 mm

27.14. INVOLUTE RACKS

Involute racks have straight-sided teeth. The sides of the teeth are normal to the line of action, therefore, they are inclined to the vertical at the pressure angle ϕ.

To draw a standard involute rack, approximately, first set out the pitch line and addendum and dedendum lines, as shown in *Figs. 27.17*, 27.18. Step off on the pitch line the circular thickness, $\frac{\text{C.P.}}{2}$, of the teeth of the mating gear. Through these points, draw the tooth faces at the pressure angle and complete teeth with the tips and roots. The fillet radii are 0.1 of the circular pitch.

Rack Data :- Pressure Angle 20° Metric Module (m) 6, Mesh with 30 Tooth Gear.

Pitch of Rack Teeth = m.π = 6×3.142 = 18.852 mm

Addendum = m = 6 mm

$$\text{Clearance} = \frac{18.852}{20} = 0.9426 \text{ mm}$$

Dedendum = Addendum + Clearance
= 6 + 0.9426 = 6.9426 mm

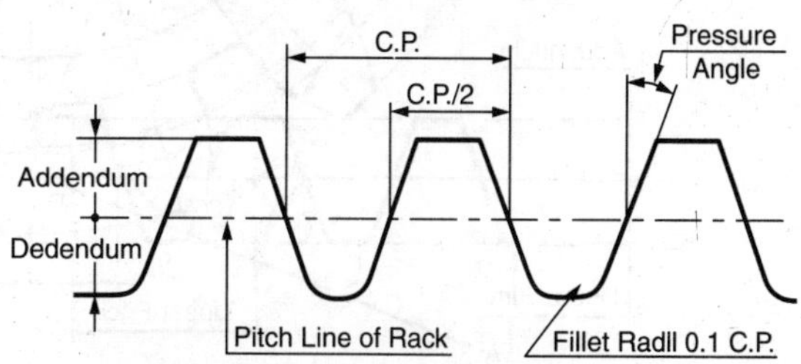

Fig. 27.17. Pressure Angle 20°

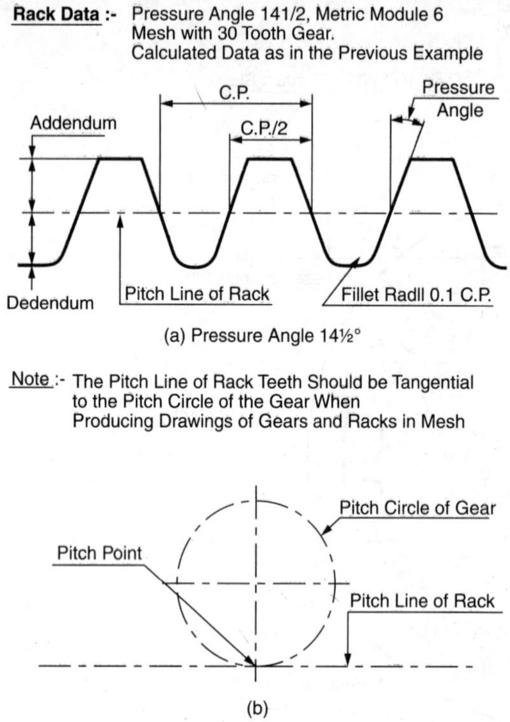

Fig. 27.18.

A rack is a special case of a spur wheel having infinite pitch circle diameter. In that case the pitch circle becomes a straight line and inclined at the pressure angle to the pitch line. The linear pitch of the rack is taken equal to the circular pitch of the gear. *Figure. 27.19* shows "Rack and Pinion" in engagement.

EXERCISE 27.5. *Draw at least 5 teeth of each of the rack and pinion having the following particulars:*

Pinion: No. of teeth = 15
Circular pitch = 20 mm
Pressure Angle = 20°

27.15. BEVEL WHEELS OR GEARS

Bevel gearing is used for the transmission of power between shafts, lying in the same plane, whose axes would intersect if produced. We have already seen that the imaginary pitch surfaces of spur gears are cylinders. The pitch surfaces of two bevel wheels in gear are frusta of two right cones in contact,

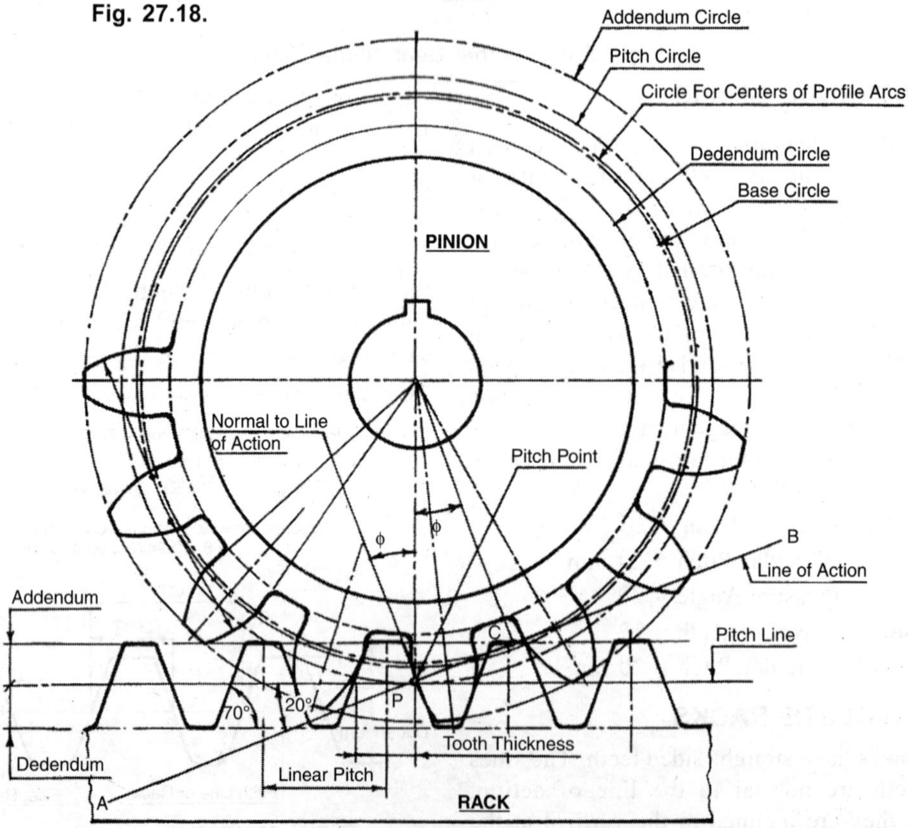

Fig. 27.19. Rack and Pinion in engagement

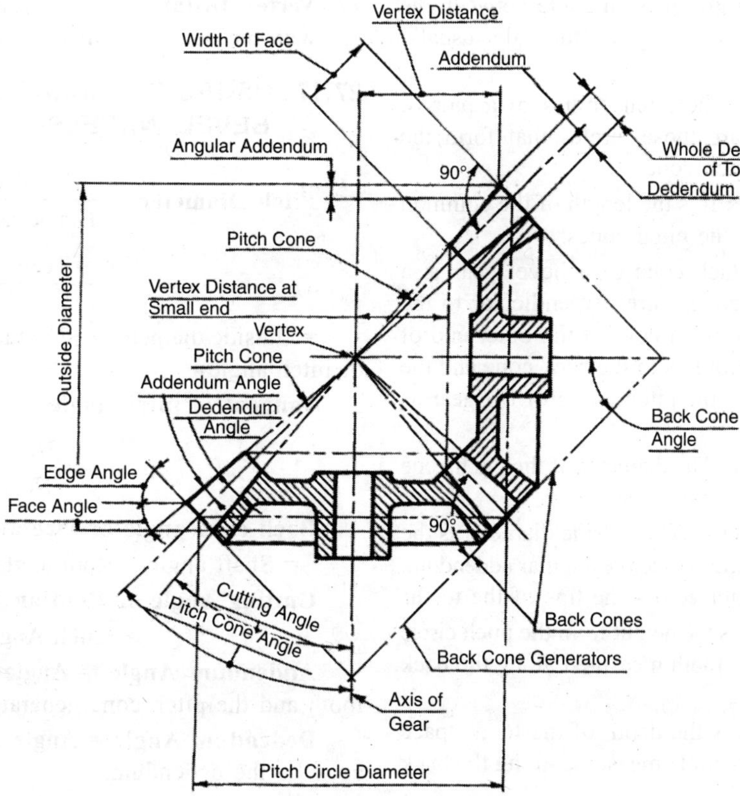

BEVEL GEAR TERMINOLOGY

Fig. 27.20.

having a common vertex O at the point of intersection of the axes of the shafts. The diameters of the bases of the pitch cones are inversely proportional to the number of teeth on the corresponding wheels. As in the case of spur gears, teeth are formed partly below the pitch surfaces. The large and small ends of the teeth lie on the surface of cones with axes coinciding with the axis of the pitch cone and having generating lines at right angles to the generating lines of the pitch cones.

It is the length of the generating lines of back cone, which is the equivalent radius of the pitch circle about which the profiles of the teeth, at the larger end, are generated.

The nominal pitch of teeth of a bevel gear is pitch of the teeth on the base of the pitch cone. The nominal diameter of the bevel gear is the diameter of the base of the pitch cone. All straight lines of the surface of the teeth radiate to the common vertex of the pitch cones, so that the pitch of the teeth diminishes uniformly along the width of the gear.

The teeth on a bevel wheel is on the curved surface of a second cone having the same base as the pitch cone and with its generators at right angles to the generators of the pitch cone. This cone is the back cone of the gear.

If the curved surface of the back cone is viewed normally, the teeth have the same profiles as the teeth on a spur gear. The addendum and dedendum have the same proportions as spur gear teeth but are measures above and below the pitch circle, parallel to the back cone generator.

27.16. TERMS USED IN BEVEL GEARING

1. **Shaft Angle:** Shaft angle is the angle between the intersecting shaft axes. It is equal to the sum of the pitch angles of the two mating wheels.

2. **Pitch Cone Angle:** It is an angle between the axis and the pitch cone generator and is usually denoted by V_P.
3. **Pitch Circles:** The pitch circles of a pair of bevel wheels are those circles that form the bases of the pitch cones.
4. **Cone Distance:** It is the length of the common slant height of the pitch cones.
5. **Back Cone:** Back cone of a bevel gear is a cone whose elements are perpendicular to the elements of the pitch cone at the outer end of the gear. Therefore, both the back cone and the pitch cone have the pitch circle of the gear as base.
6. **Pitch Diameter:** The diameter of the pitch cone is known as the pitch diameter.
7. **Outside Diameter:** The outside diameter is the pitch diameter plus twice the angular addendum. It is the diameter across the tips of the teeth.
8. **Addendum:** It is the height from the pitch circle to the tip of the tooth measured along the back cone generator.
9. **Dedendum:** It is the depth of the tooth space below the pitch circle measured along the back cone generator.
10 **Face Width:** It is the length of the teeth measured along the pitch cone generator. BIS recommends that the face width should not normally exceed one-third of the cone diameter or three times the pitch.
11. **Cutting Angle or Root Cone Angle:** It is the angle between the gear axis and the root cone generator.
12. **Face Angle:** It is the angle between a line at right angles to the axis and the top surfaces of the teeth.
13. **Edge Angle:** It is the angle between a line at right angles to the axis and the generator of the back cone.
14. **Addendum Angle:** It is the angle betwen the pitch cone generator and the top surfaces of the teeth.
15. **Dedendum Angle:** It is the angle between the pitch cone generator and the bottoms of the tooth surfaces.
16. **Angular Addendum:** It is the addendum measured on a line at right angles to the gear axis.
17. **Vertex Distance:** It is the height of a cone with the outside diameter as base.

27.17. USUAL CALCULATIONS FOR BEVEL WHEELS

$$\textbf{Pitch Diameter} = \frac{\text{No. of Teeth}}{\text{Diametral pitch}}$$

$$= \frac{\text{No. of Teeth} \times \text{Circular pitch}}{\pi}$$

= Outside diameter + (2 × addendum × cosine of pitch angle)

Tangent of pitch angle

$$= \frac{\text{No. of Teeth in wheel}}{\text{No. of teeth in pinion}}$$

Back cone angle or face angle

= Shaft angle – Root angle of mating gear

Cutting Angle or Root angle

= Pitch Angle – Ded. Angle

Addendum Angle = Angle between top of tooth and the pitch cone generator

Dedendum Angle = Angle subtended at the apex by the dedendum.

Pressure Angles for bevel gears are usualy taken as 14.5° or 20° as for spur gears.

27.18. APPROXIMATE CONSTRUCTION FOR INVOLUTE BEVEL GEAR TEETH

Teeth is shown in *Fig. 27.21*, drawn in four stages, and is explained below:

1. Draw the gear centreline with the pitch circle diameter crossing it at right angles.
2. On the pitch circle diameter draw the tangent to the pitch cone using the pitch cone angle.
3. Draw the back cone with the generators at right angles to the pitch cone generators.
4. On the back cone set out the addendum and dedendum and join these points to the vertex.
5. Mark off the face width of the teeth on the pitch cone and complete the sectional view of the tooth.
6. Draw lines across the view at right angles to the axis through the tip and root points and points on the pitch cone at each end of the teeth.

7. Then develop the back cone by swinging arcs about O from the addendum, dedendum, and pitch points on the back cone. These arcs fix the addendum, dedendum, and pitch circles of the developed teeth.
8. Using the same construction as used for a spur gear, lay out on these circles the profiles of one or two teeth.
9. In the end view set out the centrelines of the teeth and draw six circles representing the addendum, dedendum, and pitch circles at each end of the teeth.
10. Fix three points on the profiles of teeth at the large end by transferring the widths X, Y, and Z from the developed view of the teeth. Finishing the tooth profiles with French-curves, or with circular arcs may now complete the view.
11. As all straight lines on the teeth are radial, this fact enables three points on the profile of the teeth at the small end to be fixed.
12. The front view can then be completed by projection from the end view. Note that all straight lines on the teeth in this view pass through the vertex if produced.

It can be observed from this illustration that when the back cone is developed a spur gear is formed with a pitch circle diameter equal to twice the length of the back cone generator. This spur gear is called the **virtual spur gear** and is used in the manufacturing calculations for the bevel wheel.

EXERCISE 27.8. *Draw front and top views of a Bevel gear having the following specifications:*

No. of teeth = *25*

C.P. = *25 mm*

Face width = *57 mm*

Pitch Cone Angle = *70°*

Diameter of Pitch Cone

$$= \frac{\text{No. of teeth} \times \text{C.P.}}{\pi}$$

Construction : (Fig. 27.22)

- First draw pitch cone *EDF*, Angle *DEF* = 70° and DF equal to diameter of pitch cone.
- Then describe circle representing end view of the pitch surface.
- On the back cone generator mark addendum and dedendum.
- Angle *FDB* = 90°. Set off *CD* = 57 mm on the pitch cone generator.
- Describe an arc DD_1 taking *B* as centre and radius = *ED*.
- Mark the distance equal to 1/2 × C.P. along the curve DD_1 and draw the radials.
- Now complete the true shape of the teeth on arc DD_1 by approximate construction method, as explained earlier.
- Mark the distances *X* and *Y* on the true shape of one tooth, tips *X* mm wide and root thickness *Y* mm.
- Draw the pitch, addendum, and dedendum circles in top view.
- Mark spaces on the pitch circle and each equal to C.P., i.e., 25 mm.
- Mark *a*, *b*, *c*, *d*, etc. on the addendum, pitch, and dedendum circles.
- Project the points *a*, *b*, *c*, *d*, *e* and *f* from top view to the front view. As the point '*a*' lies dedendum circle so its projection *a will also lie on dedendum circle in front view.*
- Repeat the process for all other points.
- Join *a′*, *b′*, and *c′* with a smooth curve and on the same lines join *d′*, *e′*, and *f′*.
- Other tooth can be drawn in a similar manner.

EXERCISE 27.9. *Figure 27.23 shows two bevel wheels in gear. Some of the dimensions are shown on the drawing and rest of them are as follows:*

No of involute teeth is = *20 and 25*

Circular Pitch = *20*

Angle of Obliquity = *20°*

Addendum = *6.3 mm*

Dedendum = *7.3 mm*

Draw the following views:

1. *The sectional front view, with the left half of the larger wheel in outside view, showing the teeth by projecftion from top view.*
2. *The top views of the pitch cones, with the lower half of the larger wheel showing the teeth in top view.*

Use full size scale.

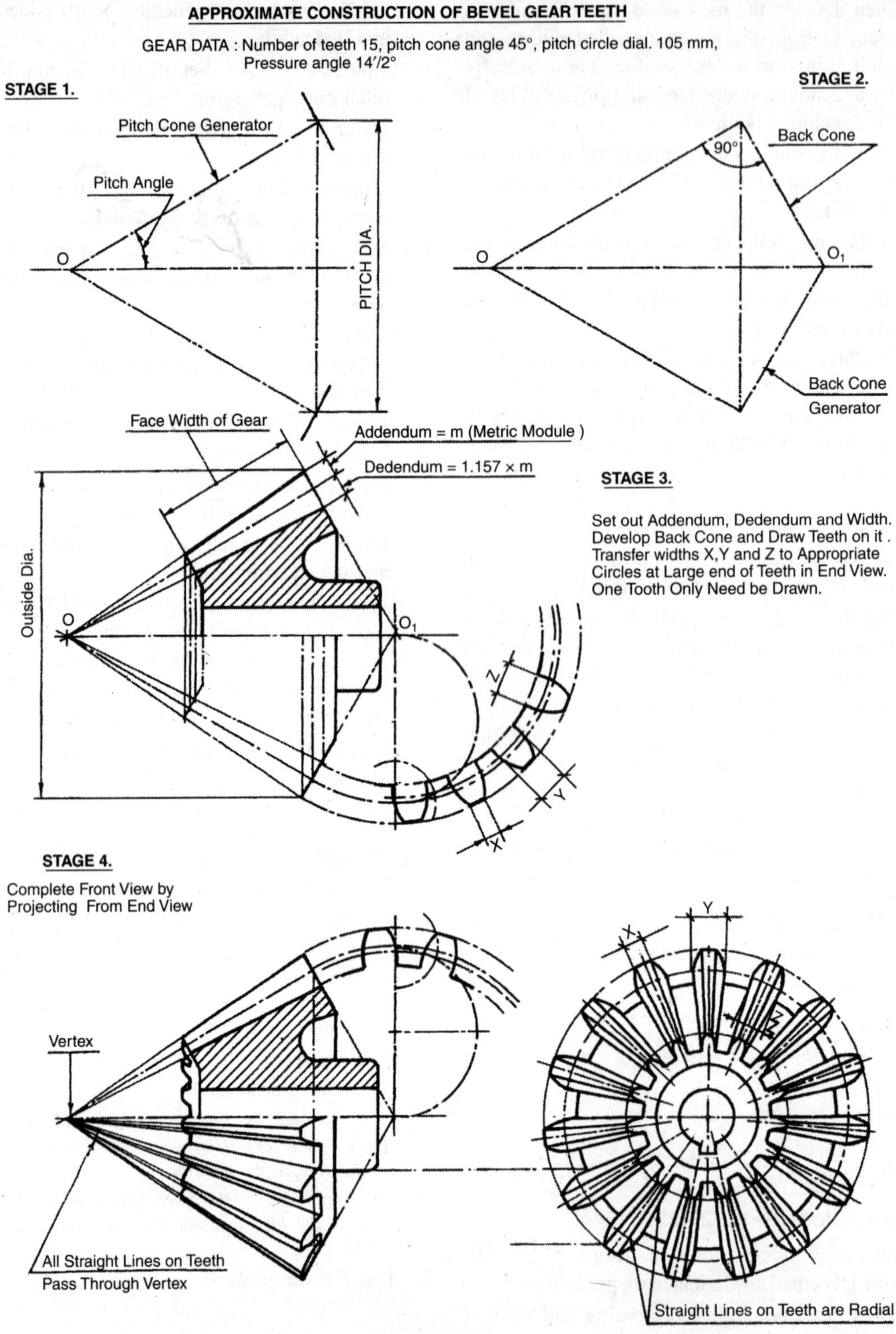

Fig. 27.21. Approximate construction of Bevel Gear Teeth

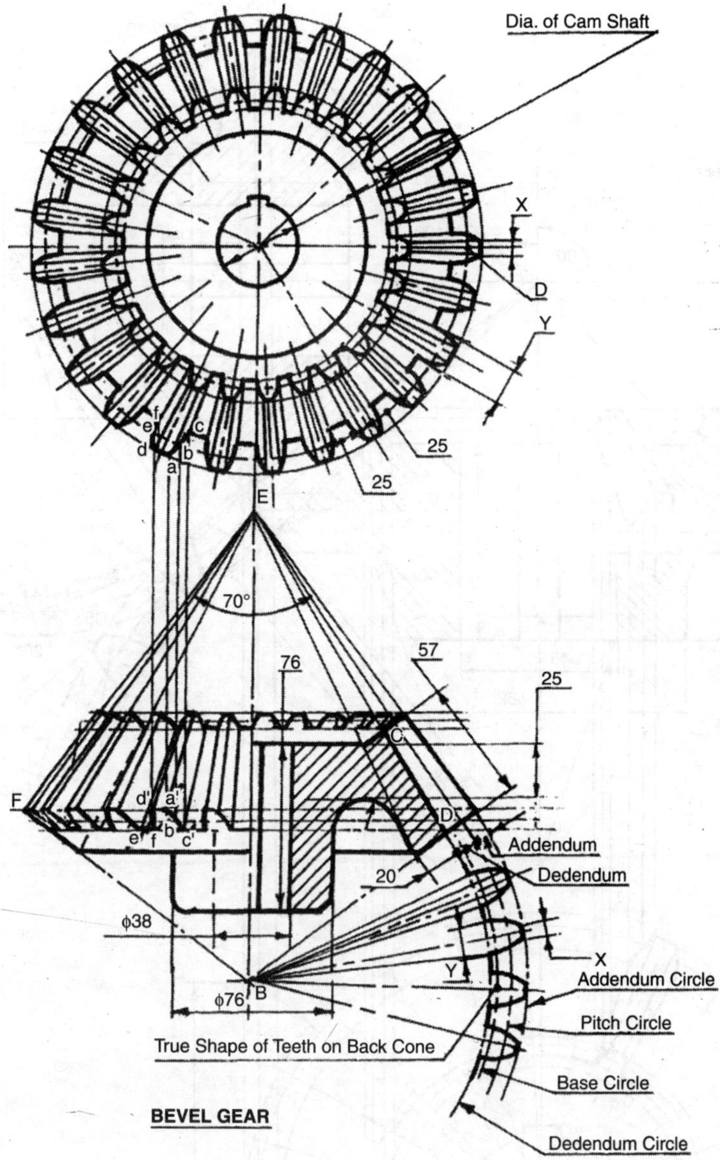

Fig. 27.22.

27.19. CONVENTIONAL REPRESENTATION OF GEARS

Gears in engagement are to be frequently drawn in engineering drawing. As their true projections are very time consuming and serve not much useful purpose, so to simplify the drawing of mating gears BIS has recommended following conventions for drawing such gears. *Fig. 27.24.*

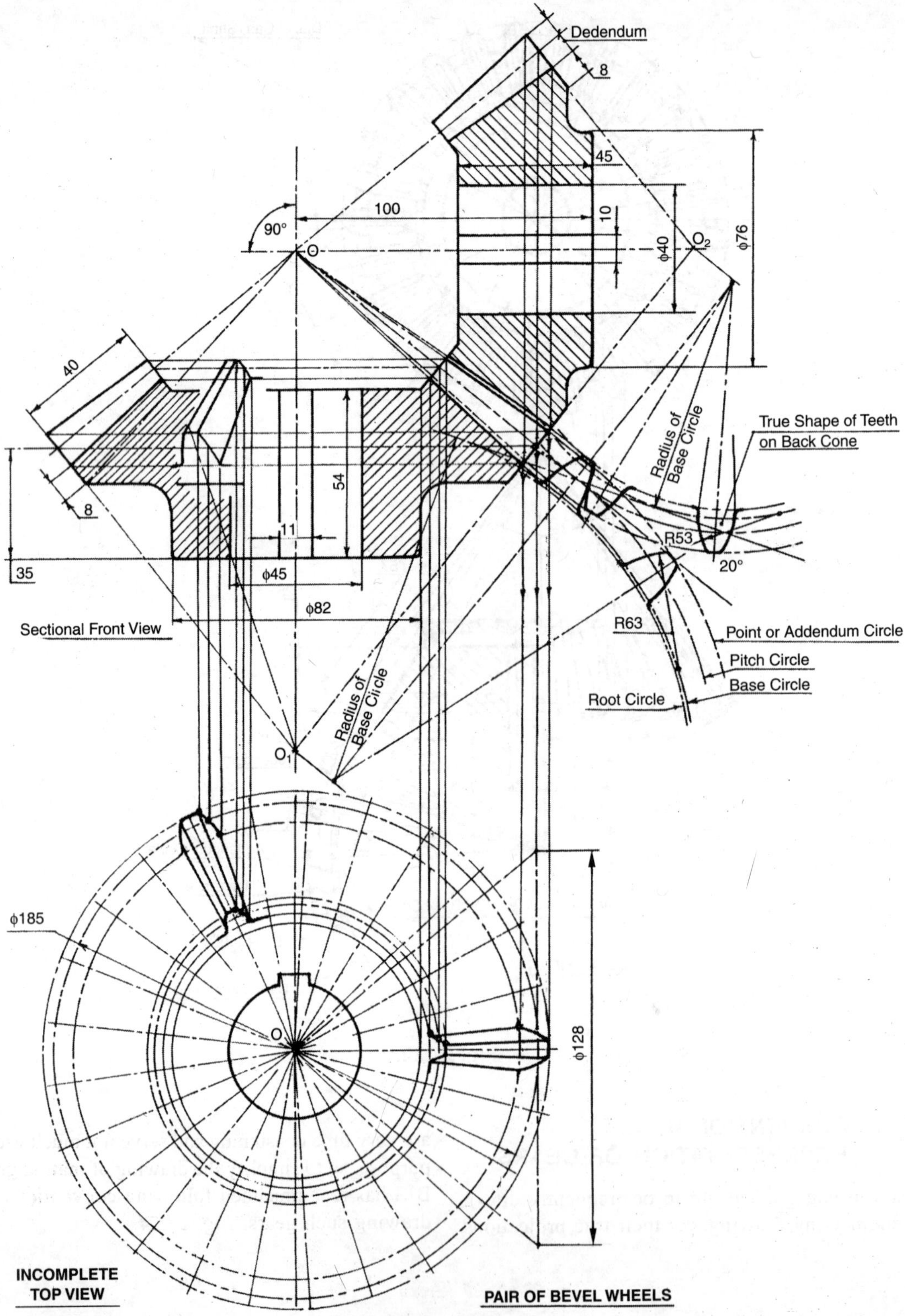
Dedendum
8
45
100
90°
O
10
φ40
O2
φ76
40
54
8
11
35
φ45
φ82
Sectional Front View
Radius of Base Circle
True Shape of Teeth on Back Cone
R53
20°
R63
Point or Addendum Circle
Pitch Circle
Base Circle
Root Circle
Radius of Base Circle
O1
φ185
O
φ128
INCOMPLETE TOP VIEW
PAIR OF BEVEL WHEELS

Fig. 27.23.

Title	Conventional Representation
Spur/1 Helical Gears	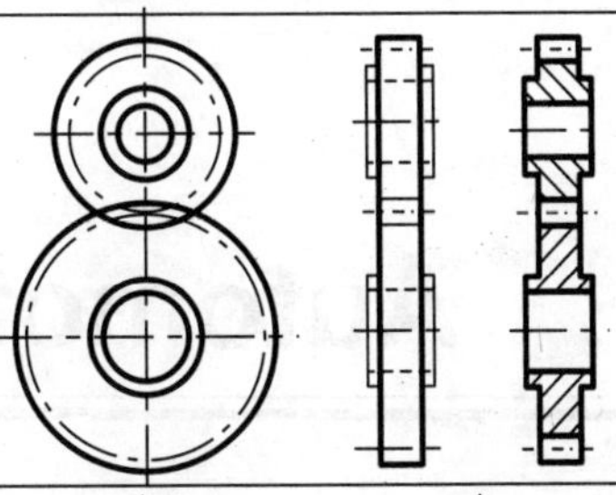
Screw Gears	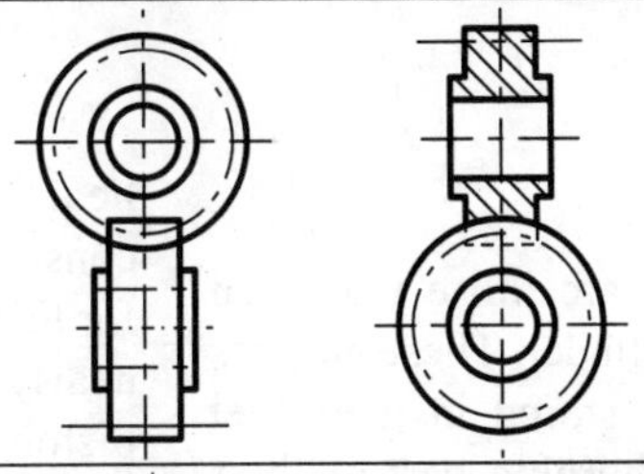
Rack and Pinion	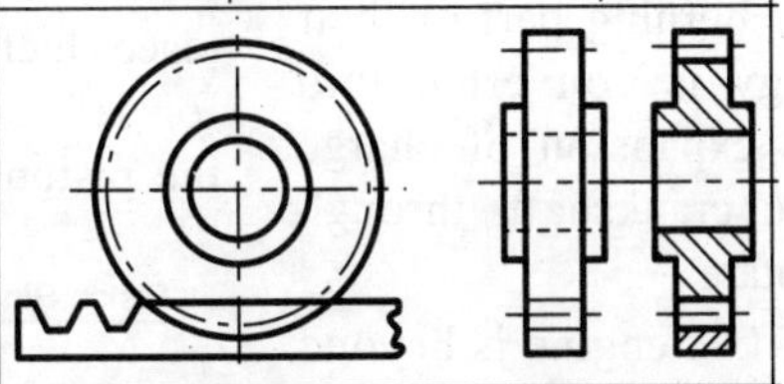
Bevel Gears	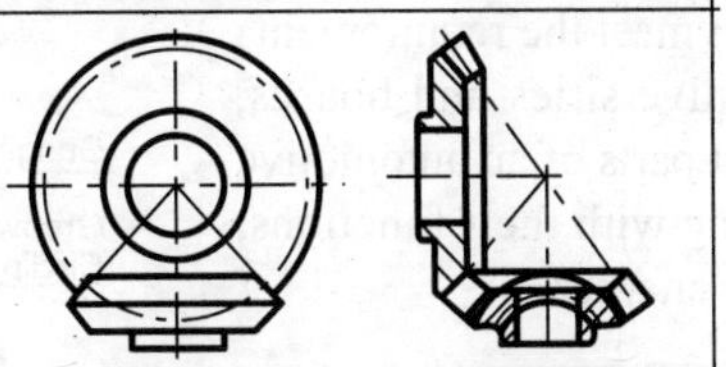
Worm and Worm Wheel	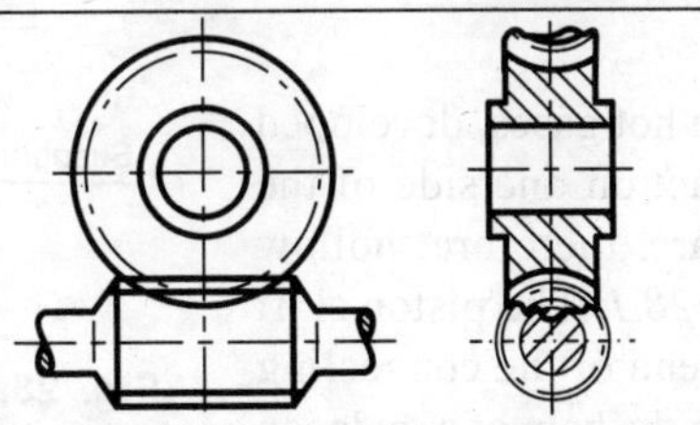

Representation of Gear Assemblies

Fig. 27.24. Conventions for Spur gears, Bevel Gears, and Worm and Wheel in enamement.

Automotive Parts

28

28.1. INTRODUCTION

Internal combustion engines are those that burn their fuel inside the engine cylinder. These engines convert the chemical energy stored in their fuel into heat energy during the burning part of their operation. The heat energy is converted into mechanical energy by the expansion of charge against pistons attached to crankshaft, through connecting rod, that can rotate.

Detailed description of I.C. engine is beyond the scope of this book. But to meet the requirements of syllabi of most of the universities and boards, drawings of some component parts of an automotive engine and power train, along with their functions, have been included in this chapter.

28.2. PISTON

In I.C. engines, high pressure hot gases, developed during combustion of fuel, act on one side of the piston. I.C. engine pistons are, therefore, hollow and have one end open, *Fig. 28.1.* The piston skirt acts as a guide for the small end of the connecting rod, attached to the piston with the help of a gudgeon pin also called a piston pin.

Pistons are slightly smaller in diameter than the bore of the cylinder. The space between the piston and cylinder wall is called the piston clearance. This clearance is necessary because the piston reaches a higher temperature than the cylinder walls, since a cooling fluid surrounding them cools the walls. Clearance is necessary also to provide the space for a film of lubricant between the piston and the cylinder wall.

Piston are made of aluminium alloys, cast steel, cast iron, or chrome nickel. Aluminium alloy piston usually are lighter than other types and are excellent heat conductors. But they expand more and, consequently, require some means of compensating for this characteristic. To achieve this compensation nearly vertical slots are usually cut in aluminium pistons so that expansion, when it occurs, takes place chiefly in the slots, leaving sufficient clearance, for safe opeeation, betwen the cylinder wall and the piston.

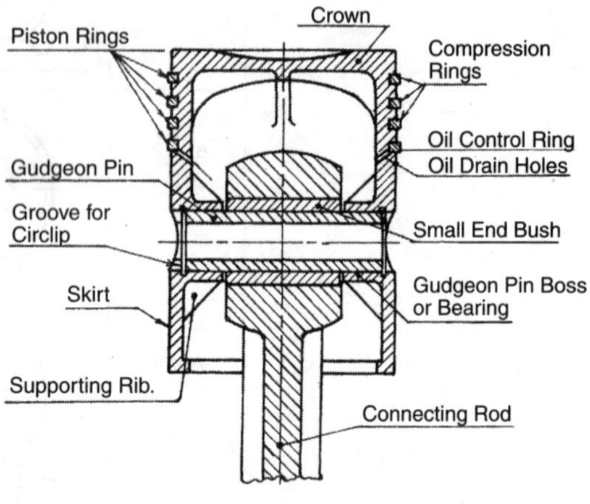

Fig. 28.1. I.C, engine piston anf connecting rod connection

Figure. 28.2 and *Fig. 28.3* shows the construction of two typical aluminium alloy pistons. Grooves are cut in the side of each piston for Three or four piston rings. which are included to seal the compressed and expanding gases above the piston and to control lubricating oil film.

The top is called the head; the part below the ring grooves is called the skirt. The portions of the piston that separate the grooves are called the lands. The piston bosses are the reinforced sections of the piston designed to hold the piston pin or wrist pin.

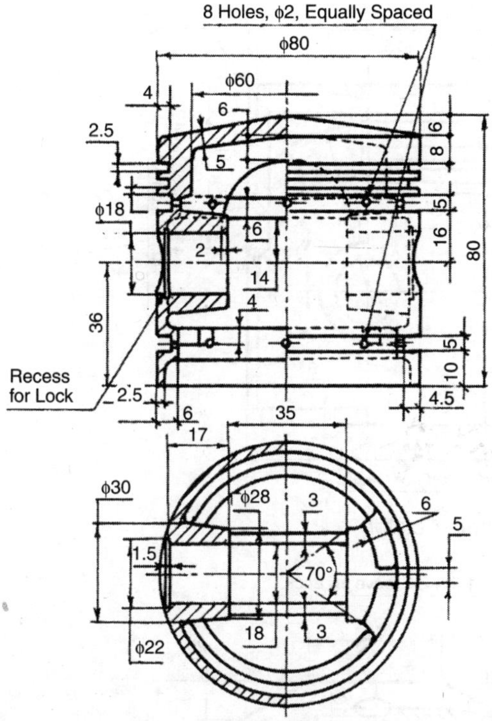

Fig. 28.2. Gasoline engine piston

EXERCISE 28.1. *Two views of a gasoline engine piston are given in Fig. 28.3. Draw its:*

(i) Front view, right half in section.

(ii) Left side view, left half in section.

(iii) Top view, full in section through the piston pin bosses.

EXERCISE 28.2. *Two views of a gasoline engine piston are given in Fig. 28.3. Draw to full size scale:*

(i) Front view–left half in section.

(ii) Left side view, right half in section.

(iii) Top view, full in section through the piston pin bosses.

Assume proportionately, any missing dimensions. Dimension the views completely.

EXERCISE 28.3. *Figure 28.4 shows three views of a two-stroke gasoline engine piston. Reproduce the views, full scale; give all required dimensions with suitable tolerances.*

EXERCISE 28.4. *Figure. 28.5 shows three views of a gasoline engine piston. Reproduce the views, full scale; give all required dimensions with suitable tolerances.*

28.3. CONNECTING ROD

Connecting rod is the connection between the piston and crankshaft. It joins the wrist pin of the piston with the throw or crankpin of the crankshaft. For a given size engine, lighter the connecting rod and the piston the greater the resulting power, and lesser the vibration as a result of the decreased reciprocating weight. The connecting rod usually has an I-beam cross-section. The lower part of the connecting rod is split to permit its being clamped round the crankshaft.

The split big end, usually, incorporates Babbitt bearing. The lining may either be in the form of a separate split shell with a steel backing or it may be spun on the inside of the rod and cap during manufacture of the connecting rod. When a separate split shell liner is used, it is held in place by the connecting rod cap and the bolts that clamp together the split parts of the liner. A dowel pin or lip prevents the liner from turning. In the spun type of liner the Babbitt is reamed to proper size at the factory floor.

Pieces of metal, called shims are sometimes used when spun bearings are employed. These can be peeled or filed thinner to compensate for wear of the bearing.

EXERCISE 28.5. *In Fig. 28.6 details of connecting rod for a gasoline engine, are given. Draw, to a scale double the full size, the following orthographic views:*

(i) Front view, right half in section.

(ii) An outside side view.

(iii) Top view.

Suggest, with reason, the suitable types of fits for the mating parts, where necessary.

28.4. FUEL PUMP AND ATOMIZER

Engines working on constant pressure (Diesel cycle) or Semi-Diesel Cycle, both of which require pure air for compression, must have some extra source of forcing the oil into the cylinder. This purpose is achieved by a suitable arrangement. The two main parts of this arrangement are:

1. **Fuel Pump** and
2. **Atomizer**.

These parts are briefly discussed here. For their detailed description refer some book on diesel engines.

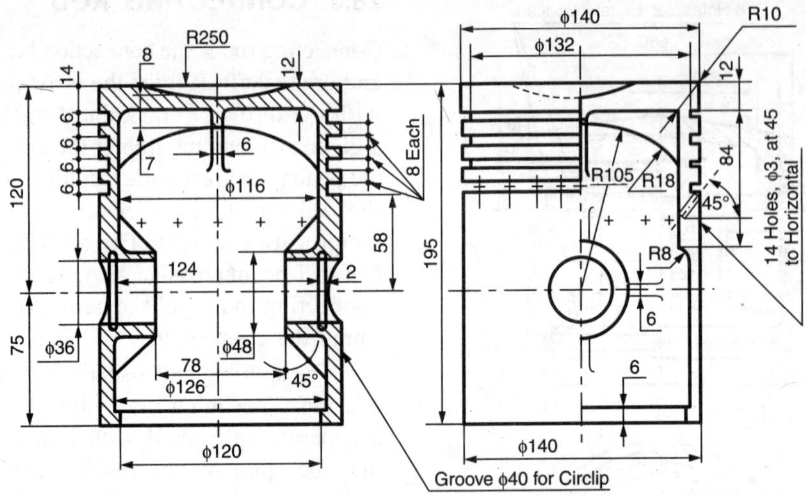

Fig. 28.3. Gasoline engine piston

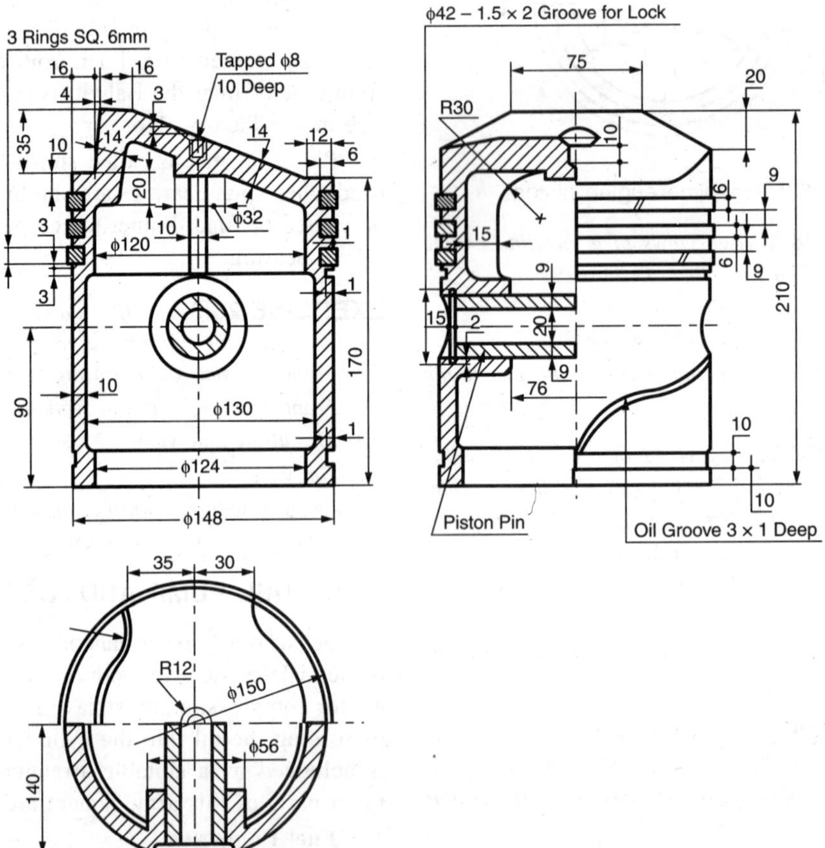

Fig. 28.4. Two stroke I.C. engine piston

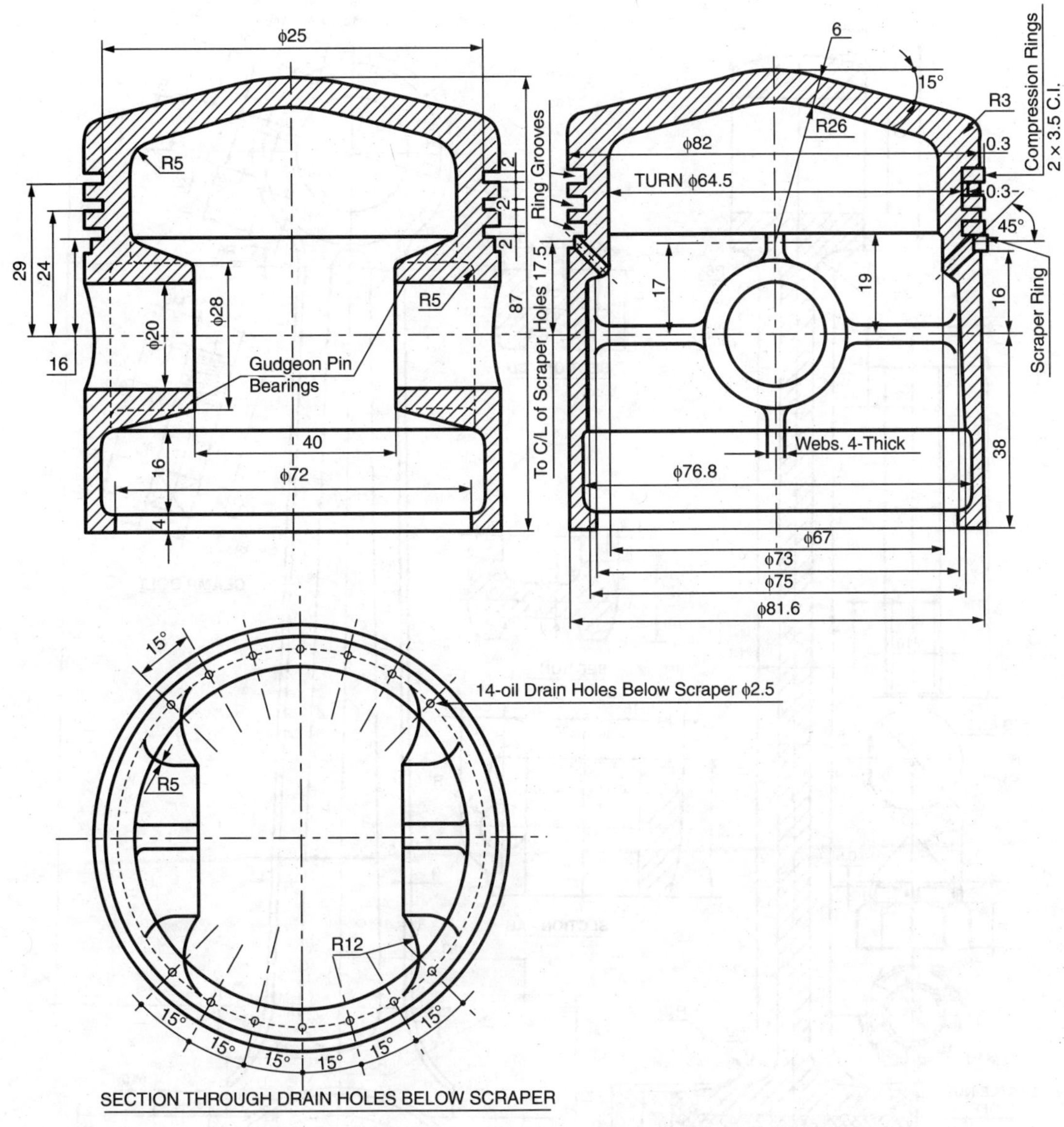

Fig. 28.5. Gasoline engine piston

Fuel Pump

A fuel pump has two main functions to perform:

1. It must start the fuel injection at the proper crank angle, late in the compression stroke of the engine, and
2. It must force through the nozzle, into the cylinder, the exact quantity of the fuel needed to produce the desired power.

EXERCISE 28.6. *Fig. 28.6 (b) shows the details of the component parts of a diesel engine fuel pump. Assemble the parts and draw, to a full size scale, the following views:*

(i) Front view-right half in section:

(ii) Right side view, outside.

Also prepare a bill of materials and parts-list.

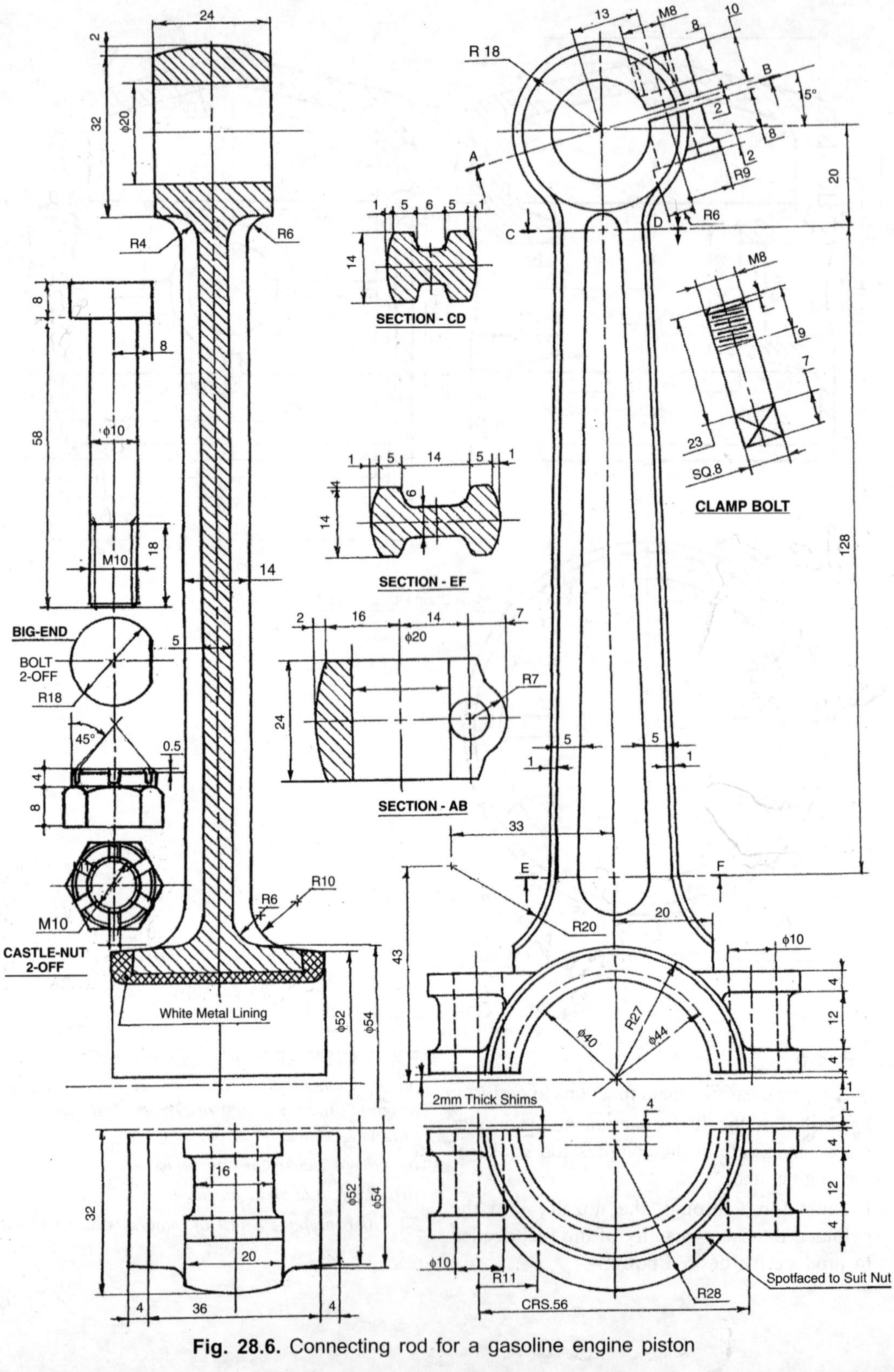

Fig. 28.6. Connecting rod for a gasoline engine piston

DRAWING HINT

Figure 28.7 (a) shows the general arrangement of the assembled details, in the form of front view right half in section.

Fuel Injector or Atomizer

The purpose of the fuel injector, atomizer, or nozzle, as it is called, is to atomize or break the fuel into fine particles and to direct the spray into the combustion chamber so that every fuel particle mixes with the air compressed in the cylinder. In order to achieve this, injection takes place through very fine holes in the nozzle body at a pressure of about 160 kg/cm^2. A fuel pump of the plunger type, already discussed, produces the desired pressure.

EXERCISE 28.7. *Figure 28.8 (b) shows the details of a fuel injection nozzle. Draw to a full size scale, the following views of the assembled details:*

(i) Front view-full in section.

(ii) End view-outside.

Also prepare a bill of materials and parts list. Discuss the types of fit that you will recommend for various mating parts of the assembly.

DRAWING HINT

Refer *Fig. 28.8 (a)* for part solution and general arrangement of the parts in the assembly.

28.5. SPARK PLUG

Spark plugs are used to ignite the compressed charge, in the cylinder of a gasoline engine, by producing a spark by means of electrodes and a high voltage produced through coils and magneto system or a coil and a battery system. The two electrodes of a spark plug are perfectly insulated and separated from each other to stop any short circuiting. A suitable gap is maintained between the two electrodes.

EXERCISE 28.8. *Details of a Spark plug are given in Fig. 28.9. Draw, to a scale 2.5 : 1 the following orthographic views of the assembled parts:*

(i) Front view–right half in section.

(ii) Top view.

Add all the necessary dimensions.

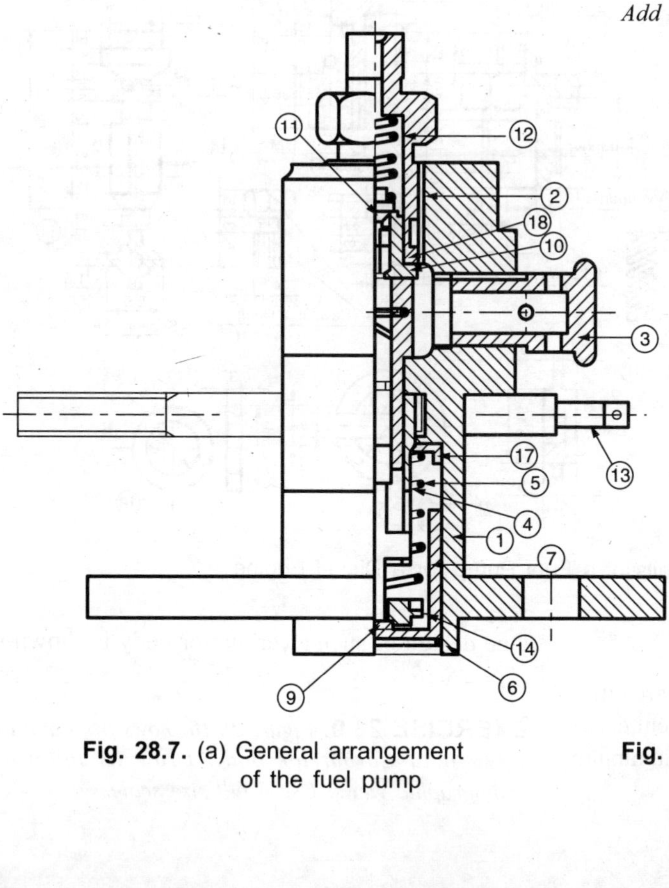

Fig. 28.7. (a) General arrangement of the fuel pump

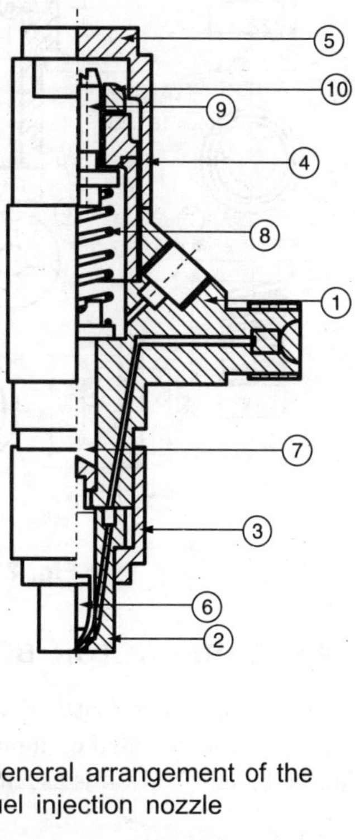

Fig. 28.8. (a) General arrangement of the fuel injection nozzle

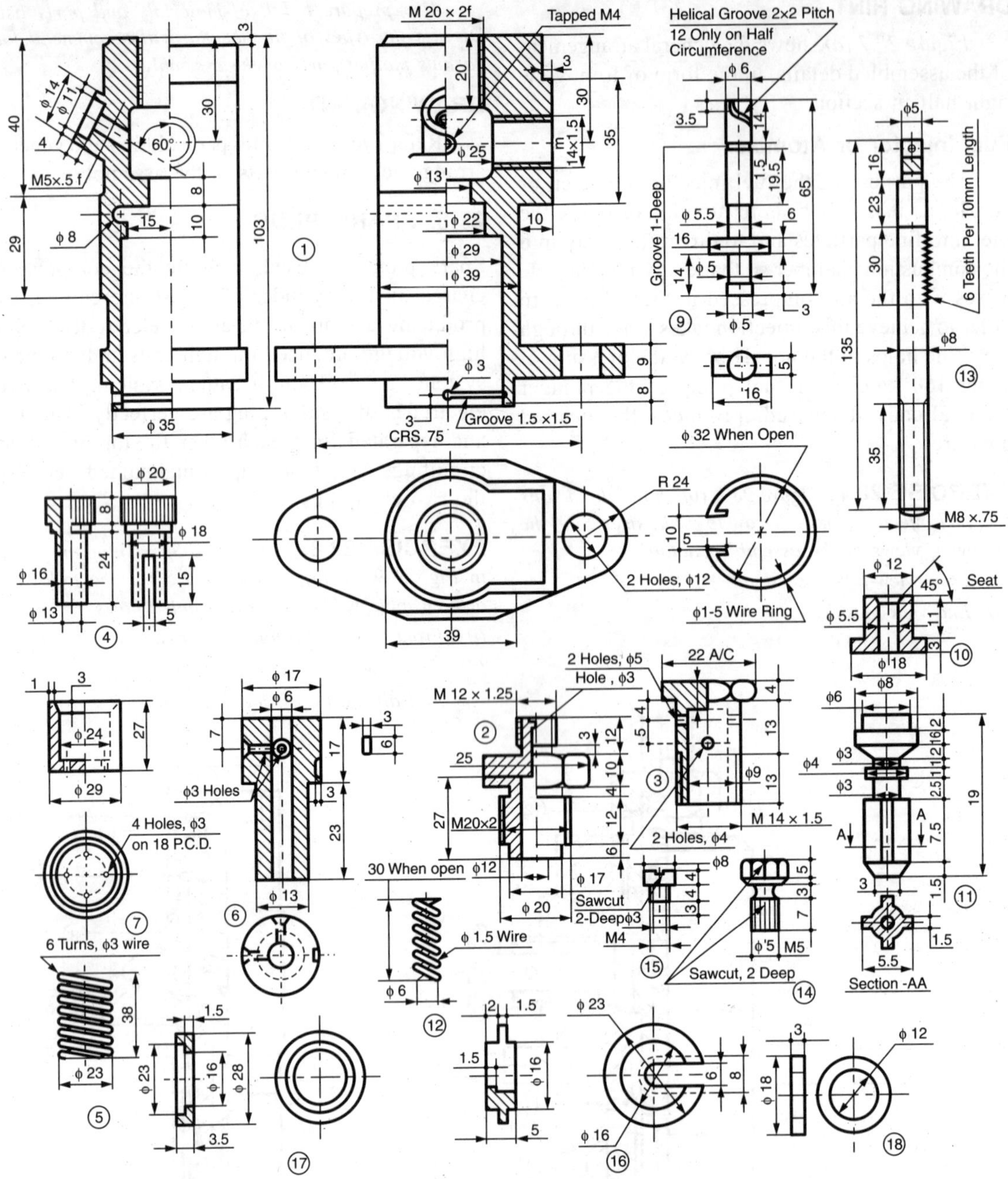

Fig. 28.7. (b) Details of a Fuel Pump for a Diesel Engine

28.6. CARBURETOR BODY

Various component parts of a carburetor are either housed in or mounted on a main component called the carburetor body. It is usually made of aluminium. One design of such a carburetor body is shown in *Fig. 28.10*.

EXERCISE 28.9. *Figure 28.10 shows front and top views of a Carburetor Body. Draw its following orthographic views. Use a full size scale.*

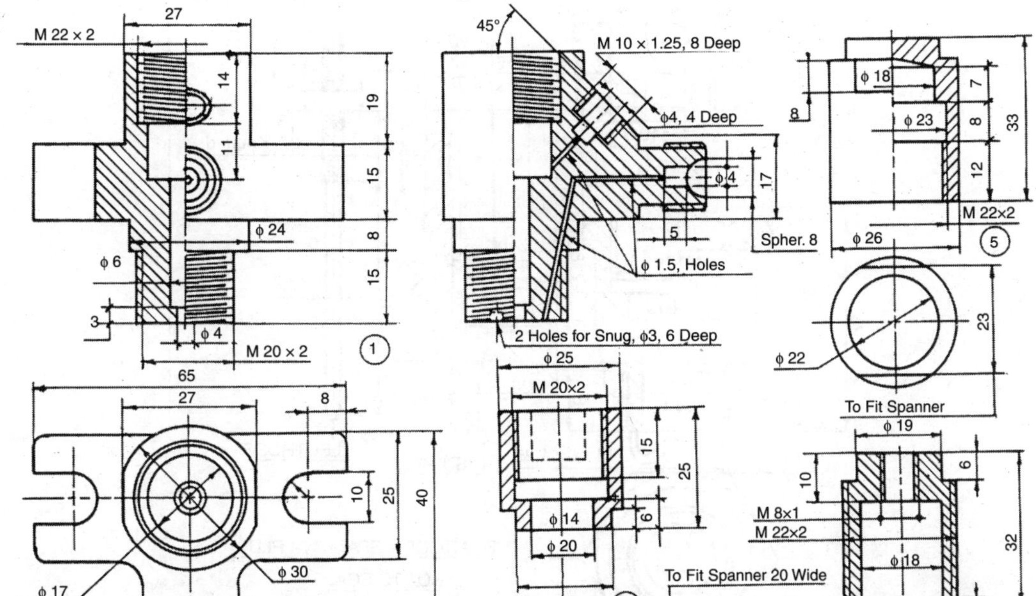

Fig. 28.8. (b) Details of a Fuel Injection Nozzle

(i) *A sectional front view–section AA*

(ii) *Left side view–in section at CC, and*

(iii) *Top view.*

Dimension the views completely, as per latest BIS standard.

EXERCISE 28.10. *Details of a Fuel Injection Nozzle are shown in Fig. 28.11. Assemble the parts and draw, to a scale double the full size, the following views of the assembly:*

(i) *Front view–full in section.*

(ii) *Right end view–outside.*

(iii) *Top view.*

Dimension the views. Also prepare a bill of materials and parts list.

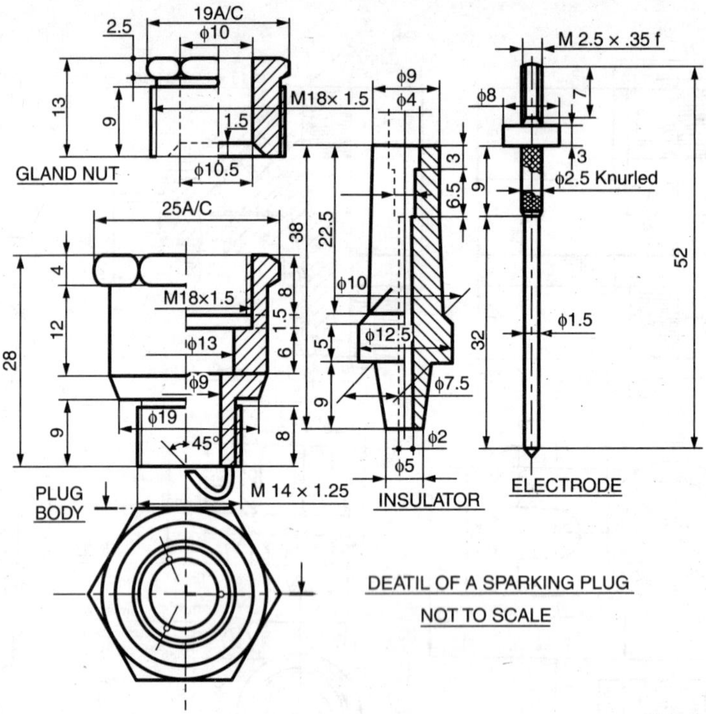

Fig. 28.9. Details of a Spark Plug (DO NOT SCALE)

EXERCISE 28.11. *Figure 28.12 shows the "Crankshaft and Flywheel" of an I.C. Engine. Assemble the parts and draw the following orthographic views:*

(i) Front view–full in section, (the crankshaft need not be in section)

(ii) Side view–outside, looking from the right side.

Dimension the views completely. Also give the two appropriate limits, for dimensions, using ISO chart, where these are required.

28.7. ROTARY GEAR PUMP

A rotary gear pump is a positive displacement type pump. A definite volume, excluding leakages of liquid per revolution of the drive shaft, is pumped from the intake to delivery side, irrespective of the pressure condition in the unit. The gears are usually of equal size. To obtain maximum capacity per revolution of the drive shaft a certain minimum number of involute teeth, that can maintain continuous drive between the gears, are employed.

The function of a rotary gear pump is to supply oil, under pressure, to the various parts of an engine, an automatic transmission, or steering system of an automobile, so as to allow the components to perform their intended functions as designed. For an I.C. Engine, the oil pump is generally located inside the crankcase below the oil level. This is the type almost universally used for pressure lubrication in automotive engines. Its construction is very simple in that it consists of two gears only, which are in mesh with each other. One of the gears is mounted on a fixed spindle and is driven only, whereas the other gear is the driving gear, itself being driven directly by the camshaft through the same gear that drives the distributor shaft. *Figure 28.13* shows the various parts of a rotary gear pump. The oil is transported from the inlet to the outlet side by entraining it in spaces between the gear teeth and the casing and carrying it round the gears from the suction to the delivery side.

Details of another gear pump are shown in *Fig. 28.14*. This gear pump, too, consists of two equal sized involute teeth gears placed in a housing. The base, the body, and the cover, along with the gaskets, are the main parts of the housing. Each gear is integral with its shaft. Each shaft is supported

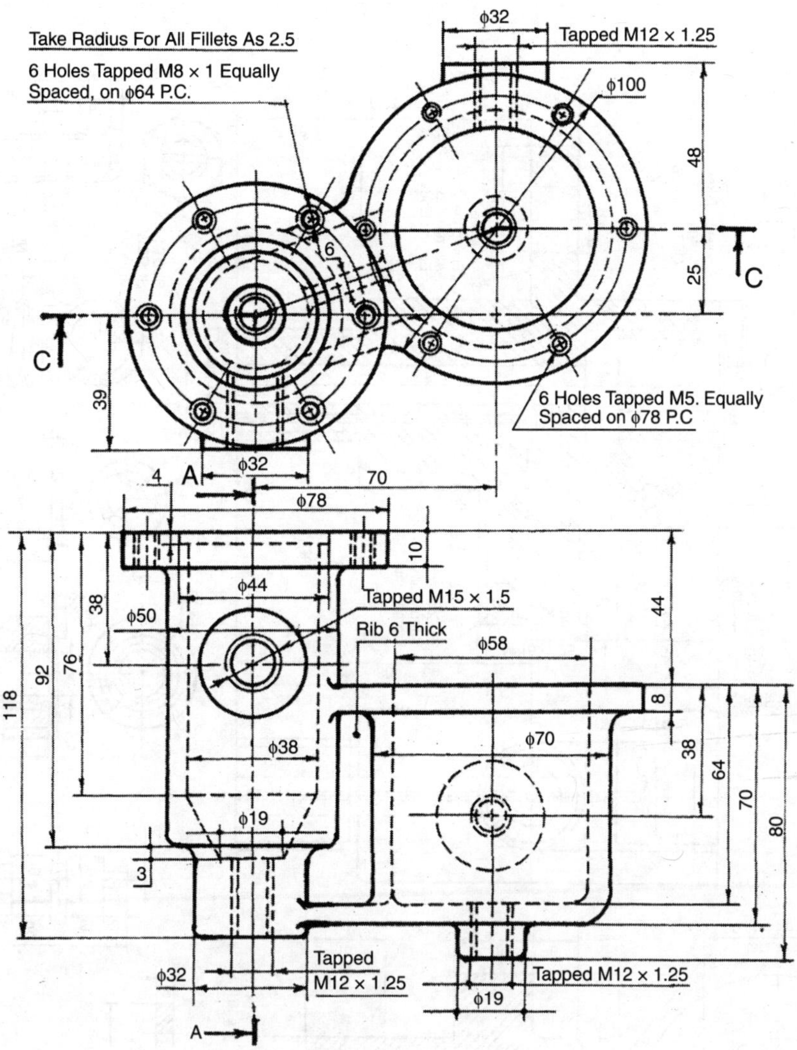

Fig. 28.10. Carburetor Body

in two bush bearings provided in the housing. The pulley is fitted to the drive shaft. In this case the pump gets its drive from an external source through a belt. Where the drive shaft passes through the housing, a gland and a bush are provided the prevent leakage of the fluid (oil).

EXERCISE 28.12. *Figure 28.13 shows the details of a rotary gear pump. Imagine the parts assembled together and draw, to a scale full size, the following views:*

(i) Sectional front view.

(ii) An end view.

Also prepare a parts-list and bill of materials.

EXERCISE 28.13. *Figure 28.14 shows the details of a Rotary Gear Pump Imagine the parts assembled together and draw to a scale full size, the following views:*

(i) Sectional front view.

(ii) An end view.

Also prepare a parts list and bill of materials.

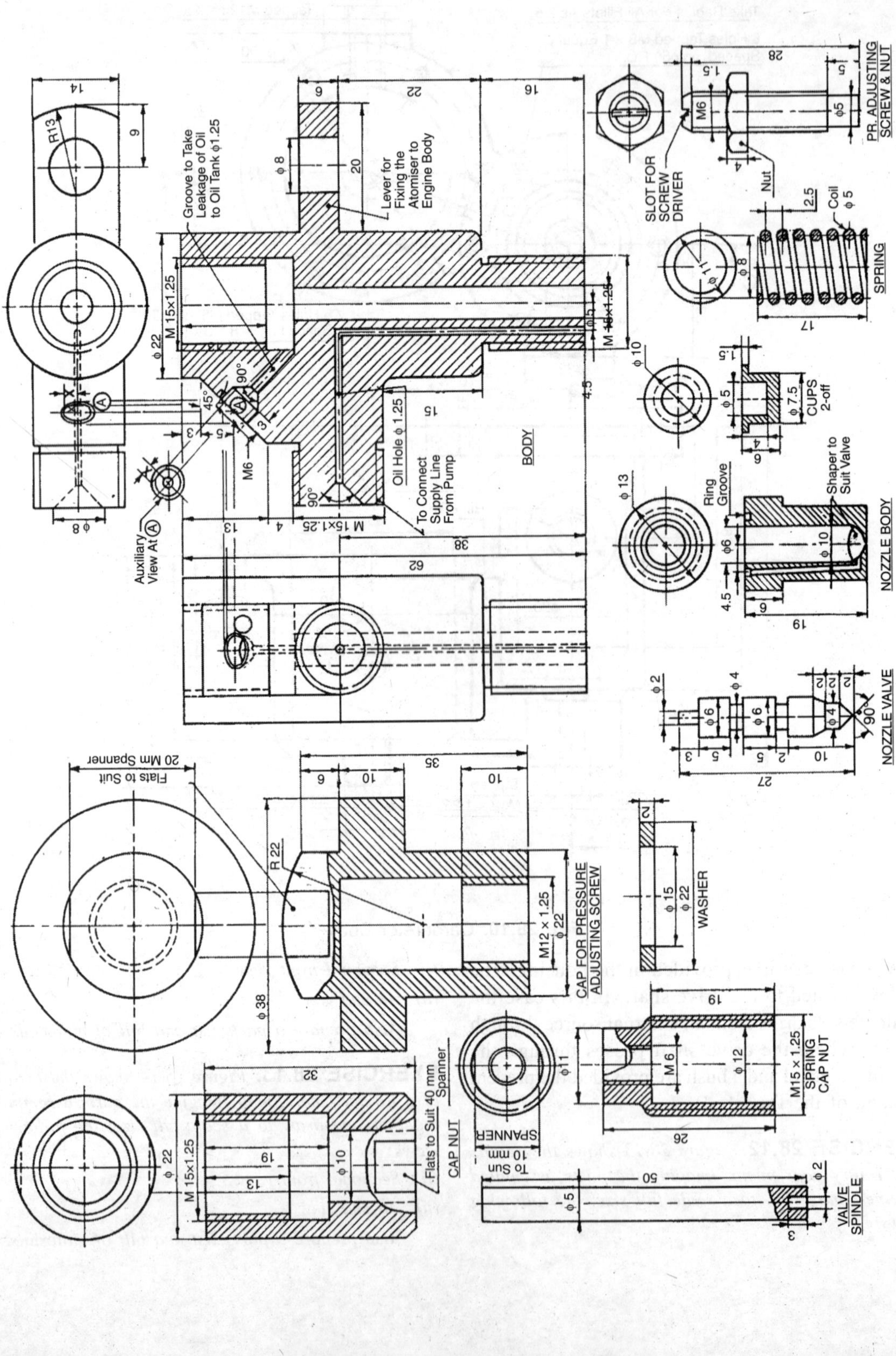

Fig. 28.11. Detail drawings of a Fuel Injection Nozzle

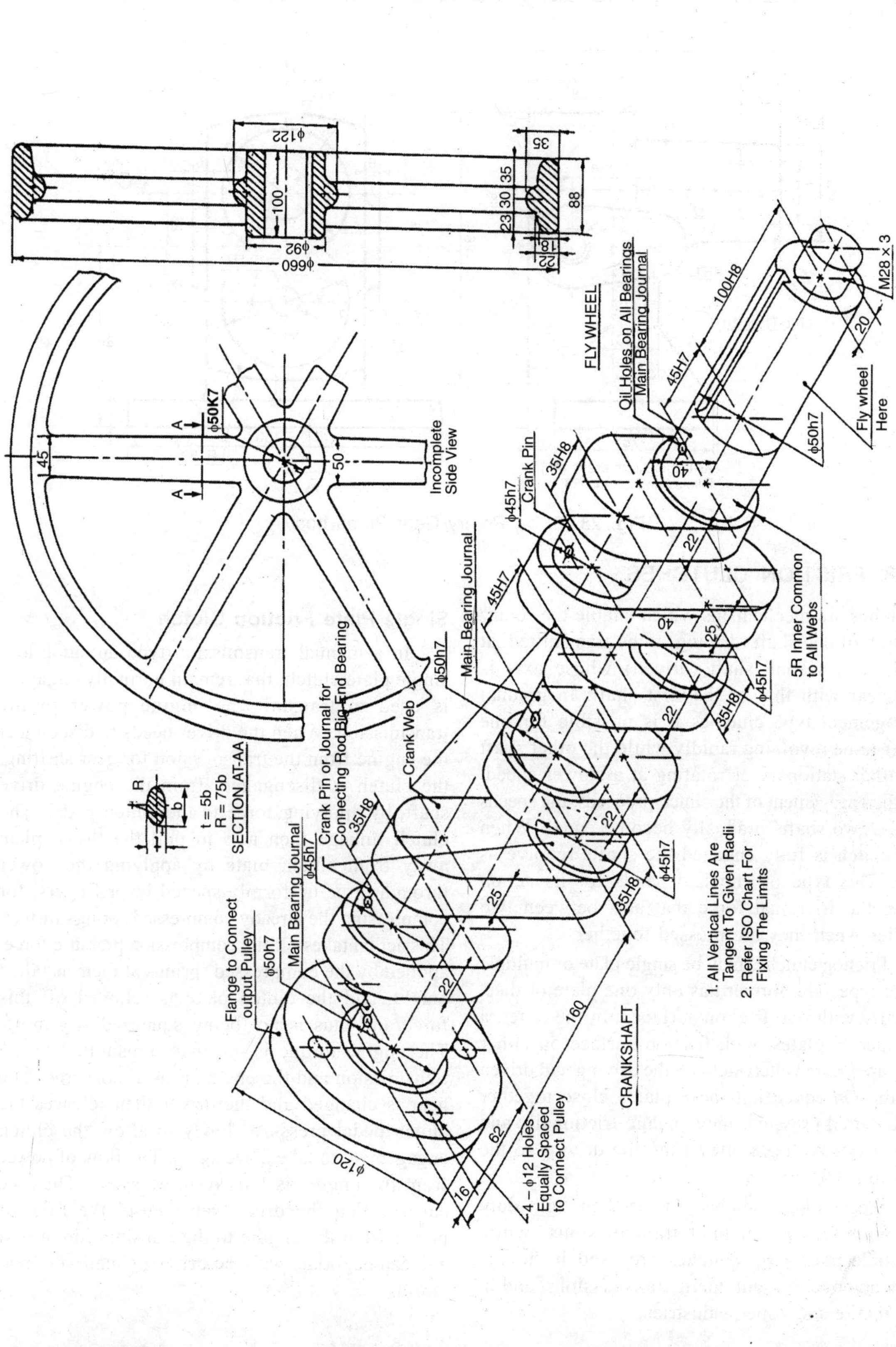

Fig. 28.12 I.C. Engine Crankshaft and Flywheel

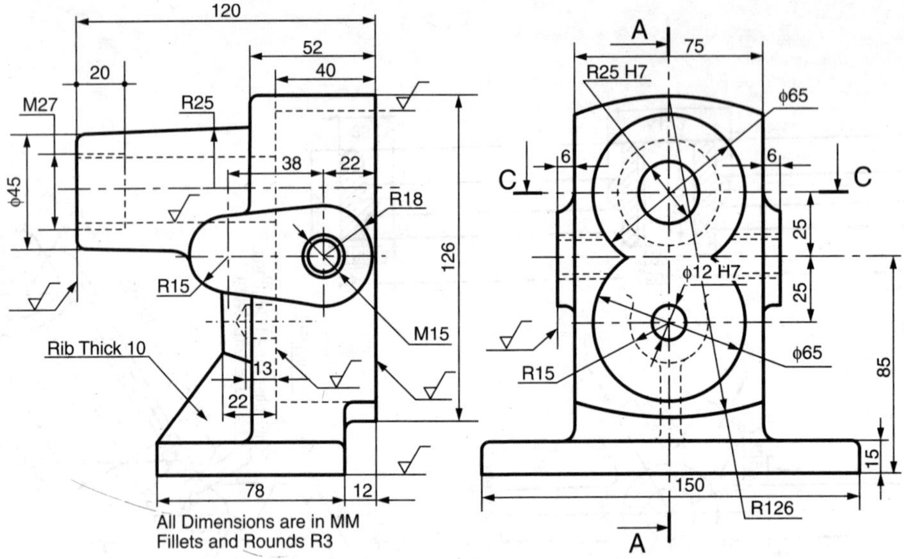

Fig. 28.13. (a) Rotary Gear Pump body

28.8. FRICTION CLUTCHES

Clutches are mechanisms which enable the rotary motion of one shaft (driving) to be transmitted, at will, to a second shaft (driven) whose axis is collinear with that of the first shaft. In gradual engagement type clutches it is possible for one shaft to be revolving rapidly while the other shaft is either stationary or rotating at a slower speed. As the engagement of the clutch proceeds, the speeds of the two shafts gradually become equal. When the clutch is fully engaged the shafts revolve as one. This type of clutch depends for its action upon the frictional force that acts between two bodies when they are pressed together.

Friction clutches may be single plate or multiple plate type. The former has only one plate or disc, usually with two friction surfaces. In the latter, a number of plates, with friction surface on either side are located alternately on the driving and driven shafts. On squeezing these plates close together by external pressure, they engage frictionally and power can be transmitted from the driving to the driven shaft.

Single plate clutches are used in light duty automobiles and manual transmissions, while multiple plate type clutches are used in heavy-duty automobiles, automatic transmissions, and in the textile and paper industries.

Single Plate Friction Clutch

In a manual transmission type automobile a single plate clutch, that remain normally engaged, is used to transmit the engine power to the transmission. When the driver needs to disconnect the engine from the transmission for gear shifting, the clutch is disengaged, from the engine drive shaft, by applying force to the clutch pedal. The clutch linkage then acts to pull the thrust plate away from clutch plate by applying the power through three uniformly spaced lever fingers, for compressing the already compressed springs further. This action takes off the compression pressure force, applied by the compressed springs at their installed height, and the clutch plate is relieved off this force and thus is not being squeezed any more. The engine torque is not then transmitted to the transmission and the engine runs almost free. The gear is changed and the driver then releases the clutch pedal pressure slowly to allow the clutch engagement to take place again. The flow of power from the engine is thus restored again. Thus we can see that the driver can control the flow of power from the engine to the transmission at will and can negotiate with the driving situation in real world.

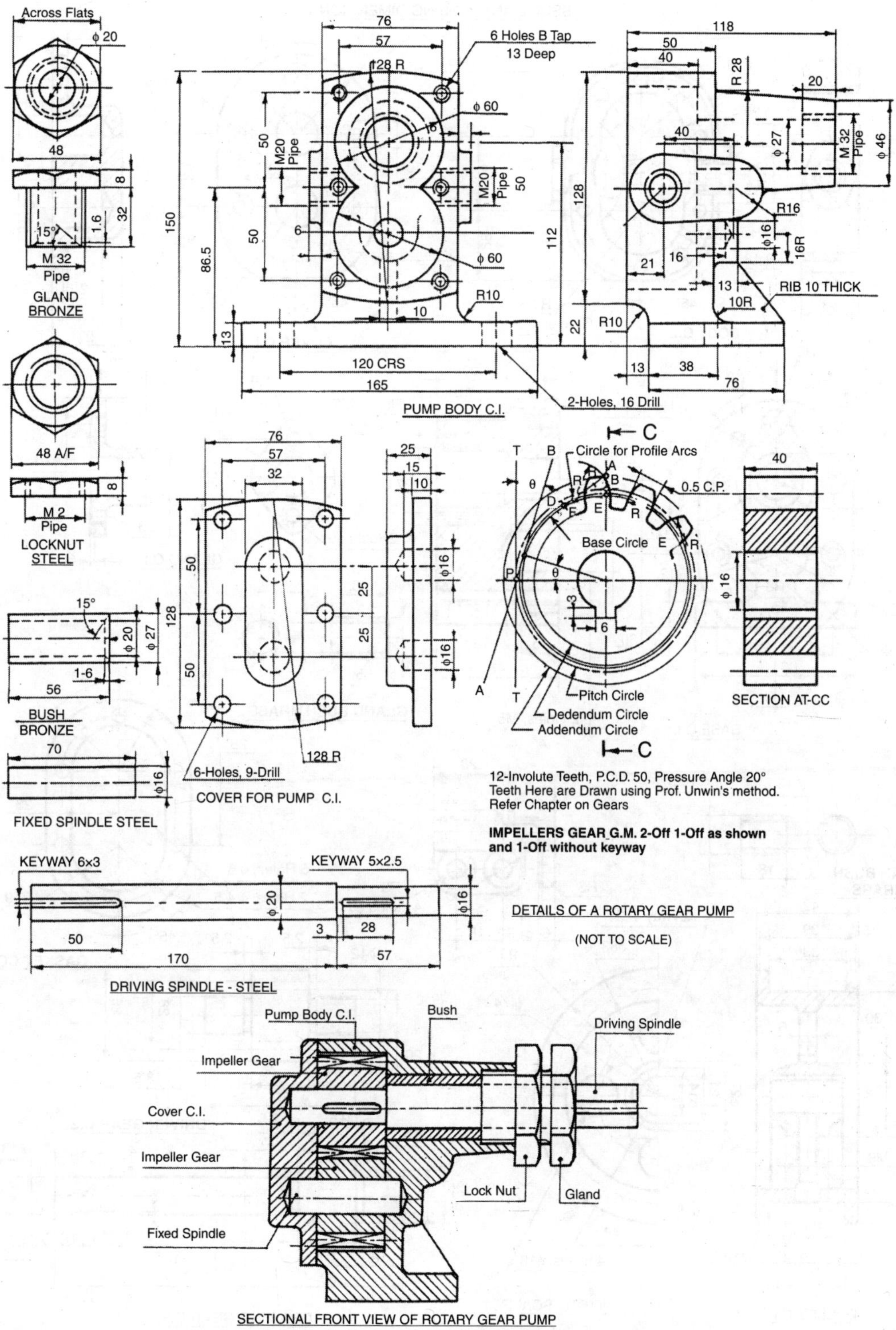

Fig. 28.13. (b) Details of a rotary gear pump

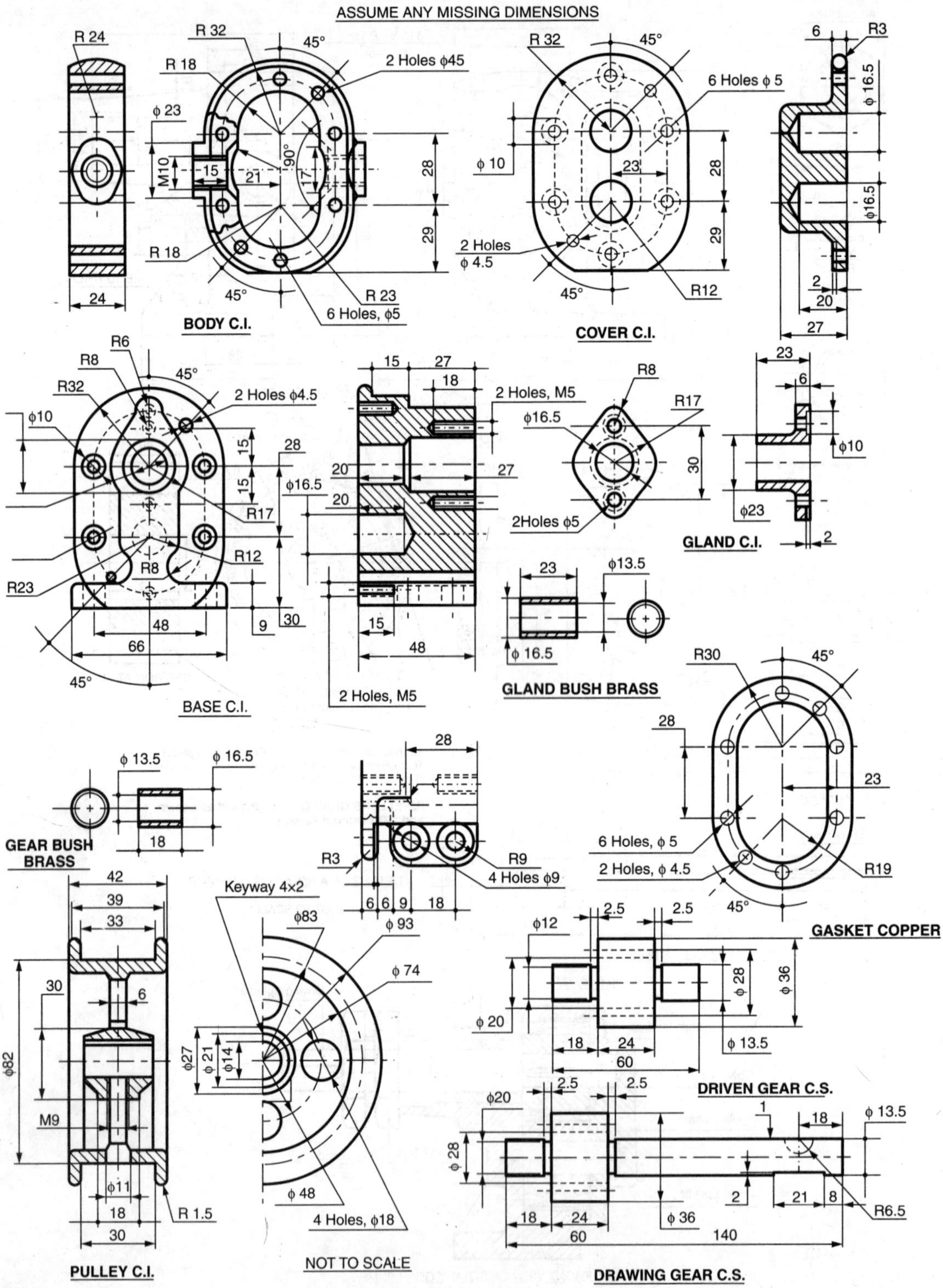

Fig. 28.14. Details of a Rotary Gear Pump

EXERCISE 28.14. *Figure 28.15 shows full sectional front view of the assembled single plate tpe clutch. Most of the parts are round in shape, except the levers and linkage used for declutching. Draw the most appropriate details to match the assembly and meet the functional requirements you want to accomplish. Dimension the details and prepare a parts-list and bill of materials. Assume, suitably, any dimensions and information that is missing and is required to give a complete shape description of the details.*

Multiple Disc Friction Clutch

Figure 28.16 (a) shows the details of a multiple disc friction clutch. The centre block is serrated externally to receive alternate friction discs and is keyed to a shaft. The body of the clutch is serrated on the internal diameter to receive alternate friction discs and is free on the shaft. It is prepared to receive the boss of a driven pulley or wheel. The clutch is operated by means the sliding sleeve, which rotates freely on the shaft with the clutch body. When the sleeve is moved forward by the rotating lever, it exerts pressure on the setting up studs, presses the plates together and completes the drive. The setting up studs are screwed in the first (or striking) friction disc and thus help in obtaining the correct pressure on the driving discs. The front cover is fixed to the shell by means of six small setscrews. The sliding sleeve is connected to the front cover through three jaw levers and three L-levers. The striking or outside plate called

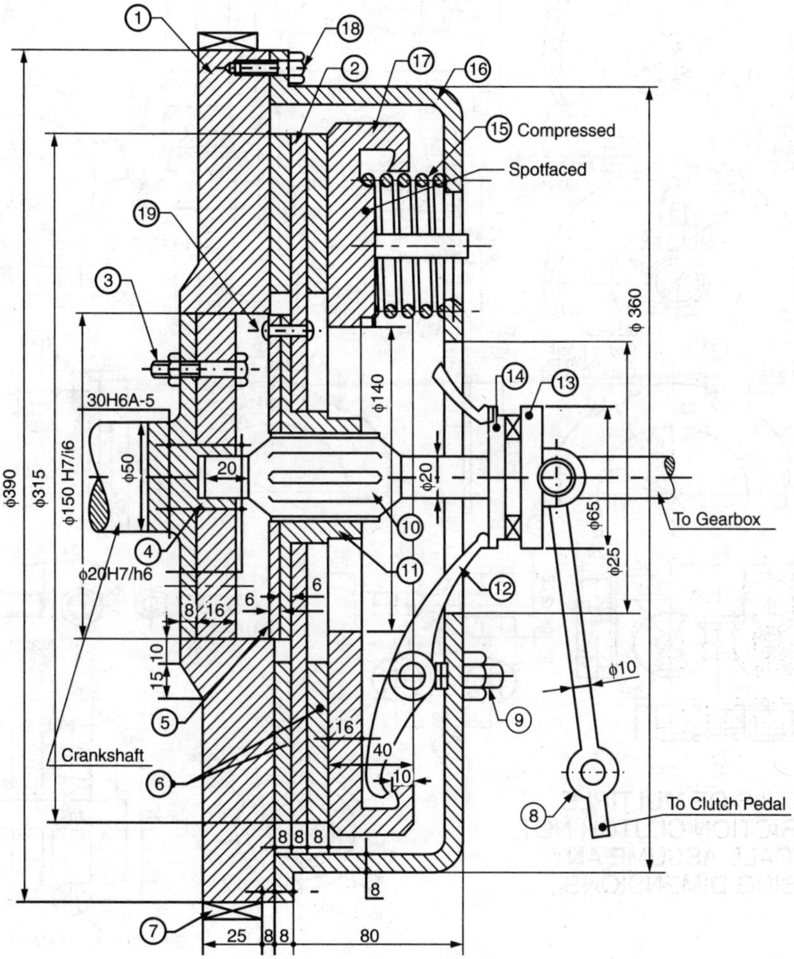

SINGLE PLATE CLUTCH

Fig. 28.15.

the stud carier is provided with the arrangement to carry three studs that are operated by the sliding sleeve through the L-levers.

EXERCISE 28.15. *Detail drawings of a Multiple Disc Clutch are given in Fig. 28.16 (a). Whereas three pins for forks on front cover are to be designed. Assemble the details and draw, using a suitable scale, the following orthographic views:*

Front view upper half in section.

Left side view–outside.

Assume, suitably, any dimensions that are missing. Dimension the views and prepare a parts and materials list.

HINT: Refer *Fig. 28.16 (b)* for part solution of this exercise.

Note: A complete understanding of the function and operating principle of the assembly or sub-

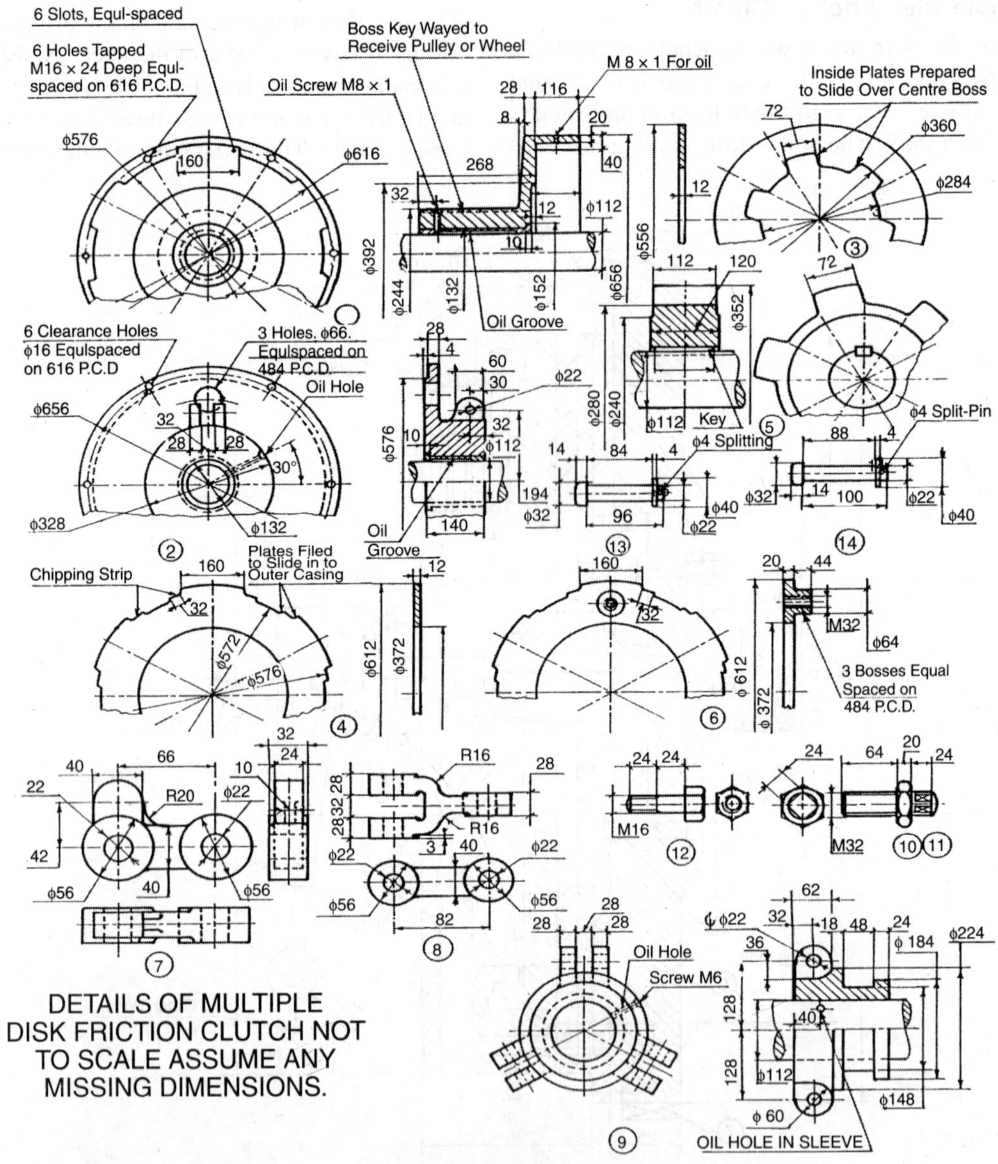

Fig. 28.16. (a)

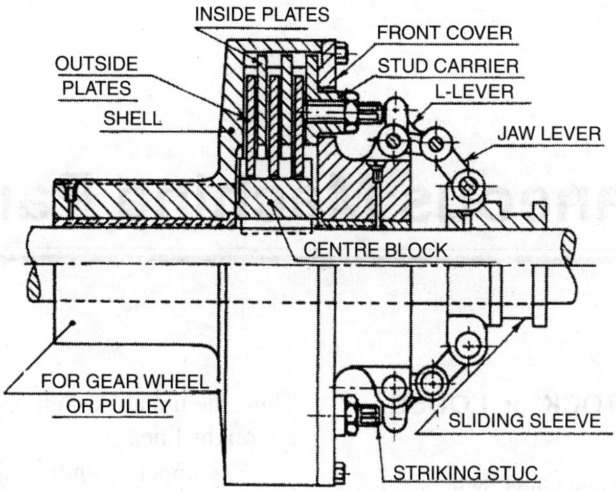

Fig. 28.16. (b) General arrangement of the Multi-Disc Friction Clutch

assembly of the component parts is of a paramount importance to execute the assembly from the detailed components of any system. Therefore, before attempting to complete the assembled views from the details given the drafter must spend some time understanding the above facts about the system of component parts at hand.

Miscellaneous Machine Parts

29

29.1. LATHE TAIL STOCK or LOOSE HEAD STOCK

On a lathe the work, to be machined, is held betwen two centres that are carried by two Head Stocks. Fast Head Stock, which usually carries a dog-chuck, faceplate, or centre, is permanently fixed to the lathe bed on the left hand side. It is sometimes called the driving headstock, too. Loose headstock, commonly called Tailstock, can slide on the lathe bed and can be clamped to it in any position, to accommodate different lengths of jobs.

The detail drawings shown in *Fig. 29.1 (b)* show a tailstock (or loose headstock), used on light duty lathes. Lathe base of the cast iron body is planed and is provided with four feet, which fit into the parallel machined ways of the lathe bed. Thus, the tailstock body is constrained to move in a straight line.

The upper cylinderical part of the body (1) is cast hollow and is machined to receive the barrrel (2). The barrel hole is coned at one end to carry the centre, and on the other end it is provided with threads in which the screw or spindle (3) works. On the right hand end of the body the spindle bearing (5) is fitted, against which rests the collar of the spindle or screw. On the other side is keyed the hand-wheel (7) which is turn is secured in position by a hexagonal nut (10) and washer (not shown in drawing).

The barrel has a keyway or feather way cut in it. Feather (9), when put in its place to fit the keyway cut in the barrel, prevents the rotation of the latter. When the screw of spindle is turned, as

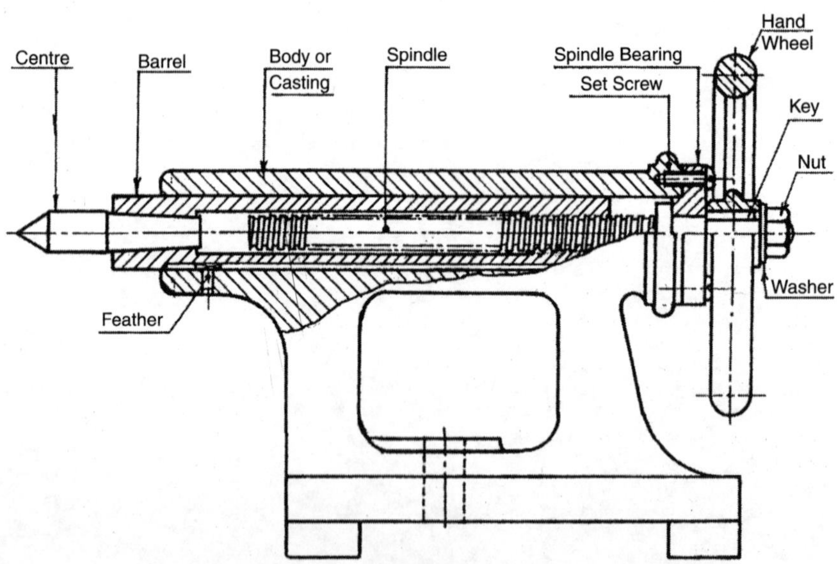

Fig. 29.1. (a) General arrangement of the Lathe Tailstock

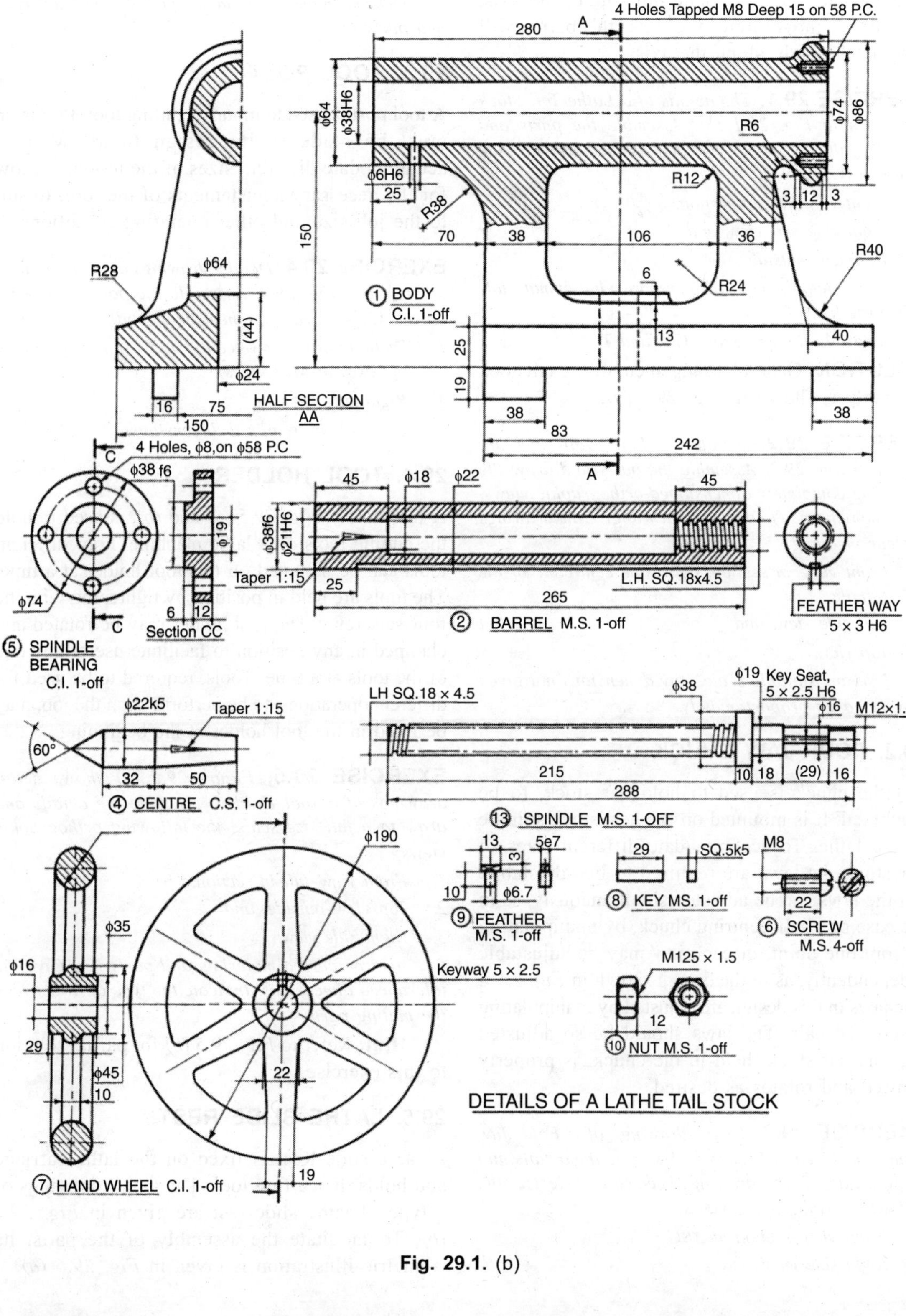
4 Holes Tapped M8 Deep 15 on 58 P.C.
280
A
φ64
φ38H6
φ74
φ86
R6
φ6H6
25
R38
R12
3
12
3
150
70
38
106
36
R40
R28
φ64
6
R24
(44)
① BODY
C.I. 1-off
25
13
40
φ24
19
16
75
HALF SECTION
AA
38
38
150
83
242
4 Holes, φ8,on φ58 P.C
C
φ38 f6
45
φ18
φ22
45
φ19
φ38f6
φ21H6
Taper 1:15
25
L.H. SQ.18x4.5
φ74
265
FEATHER WAY
5 × 3 H6
6
12
C
Section CC
② BARREL M.S. 1-off
⑤ SPINDLE
BEARING
C.I. 1-off
φ19 Key Seat,
5 × 2.5 H6
φ38
LH SQ.18 × 4.5
φ16
M12×1.5
φ22k5
Taper 1:15
60°
215
10
18
(29)
16
32
50
288
④ CENTRE C.S. 1-off
⑬ SPINDLE M.S. 1-OFF
φ190
13
5e7
3
29
SQ.5k5
M8
10
φ6.7
⑧ KEY MS. 1-off
22
⑨ FEATHER
M.S. 1-off
⑥ SCREW
M.S. 4-off
φ35
φ16
Keyway 5 × 2.5
M125 × 1.5
12
29
⑩ NUT M.S. 1-off
φ45
22
10
DETAILS OF A LATHE TAIL STOCK
19
⑦ HAND WHEEL C.I. 1-off

Fig. 29.1. (b)

it is confined, by the collar on one end and hand-wheelon the other, to rotate only about the axis, and not to move along the axis, the barrel will slide in the body along the axis.

EXERCISE 29.1. *The details of a Lathe Tail Stock are given in Fig. 29.1 (b). Assemble the parts and draw the following completely dimensioned orthographic views:*

(i) Front view–full in section,

(ii) Right side end view, and

(iii) Top view, outside

Use a suitable scale. Also prepare a bill of materials and parts-list.

Also add the necessary G.D.and T.

SOLUTION: General arrangement of the tailstock assembly is shown in *Fig. 29.1 (a)*.

EXERCISE 29.2. *Details of a Lathe Tail-Stock are shown in Fig. 29.2. Assemble the parts and draw the following completely dimensioned orthographic views. Give the required G.D. and T that will give the drawings a clear meaning.*

(i) Front view, in section, above the centreline of the spindle,

(ii) Left side view, and

(iii) Top view.

Use any suitable scale. Any dimensions not given may be taken proportionately.

29.2. FOUR-JAW CHUCK

A lathe chuck is used to hold bar stock, to be machined. It is mounted on the headstock spindle of the lathe. To accommodate different sizes of bar stock, its jaws are required to be adjustable. All the jaws can be adjusted simultaneously, as in the case of a self-centring chuck, by manipulating it from one point, or each jaw may be adjustable independently, as in the design shown in *Fig. 29.3*. The jaws in this design are adjusted by manipulating a screwed pair. The jaws should be so adjusted that the bar stock, held in the chuck, is properly centred and rotates as desired.

EXERCISE 29.3. *Detail drawings of a Four Jaw Chuck are given in Fig. 29.3. Assemble the details and draw, on an A2 size drawing sheet, to a scale 1:2, the following orthographic views:*

(i) Front view–section at ABC,

(ii) Right side view, and

(iii) Top view.

Dimension the views, and prepare a bill of materials and parts list.

29.3. TOOL POST

A tool post is used to mount a cutting tool. Provision must be made in its design to allow it to accommodate different sizes of the tool and allow for the necessary adjustements of the tool, to suit to the job size and other operating conditions.

EXERCISE 29.4. *Details drawings of a "Tool post" areshown in Fig. 29.4. Assemble the parts and draw, on an A2 size drawing sheet, using full size scale:*

1. *Sectional front view-section at XX (passing through the axes of M12 holes).*
2. *Right side view.*

Give all the required dimensions.

29.4. TOOL HOLDER

A tool holder, *Fig. 29.5 (a)* and *(b)*, is used to hold the cutting tools on a lathe machine. Four different tools can be mounted on this tool holder at a time. The tools are held in position by tightening with the four setscrews. The tool holder may be rotated and clamped in any position to facilitate use of any one of the tools at a time. Tools, required to be used for different operations, to be performed on the job, may be fixed in the tool holder at the beginning.

EXERCISE 29.5. *Figure 29.5 (a) shows detail drawings of a Tool Holder. Assemble the details and draw to a full size scale, the following orthographic views:*

1. *Front view–full in section,*
2. *Top view–outside, and*
3. *Right side view.*

Also prepare a materials and parts list. Give the toleranced dimensions. Indicate the type of fit between the mating parts.

Hint: Refer to *Fig. 29.5 (b)* for partial solution to this exercise.

29.5. LATHE SLIDE REST

A lathe slide Rest is fixed on the lathe carriage and holds the cutting tool. The detail drawings of a type of lathe slide rest are given in *Fig. 29.6 (b)*. To facilitate the assembly of the parts, its isometric illustration is given in *Fig. 29.6 (a)*.

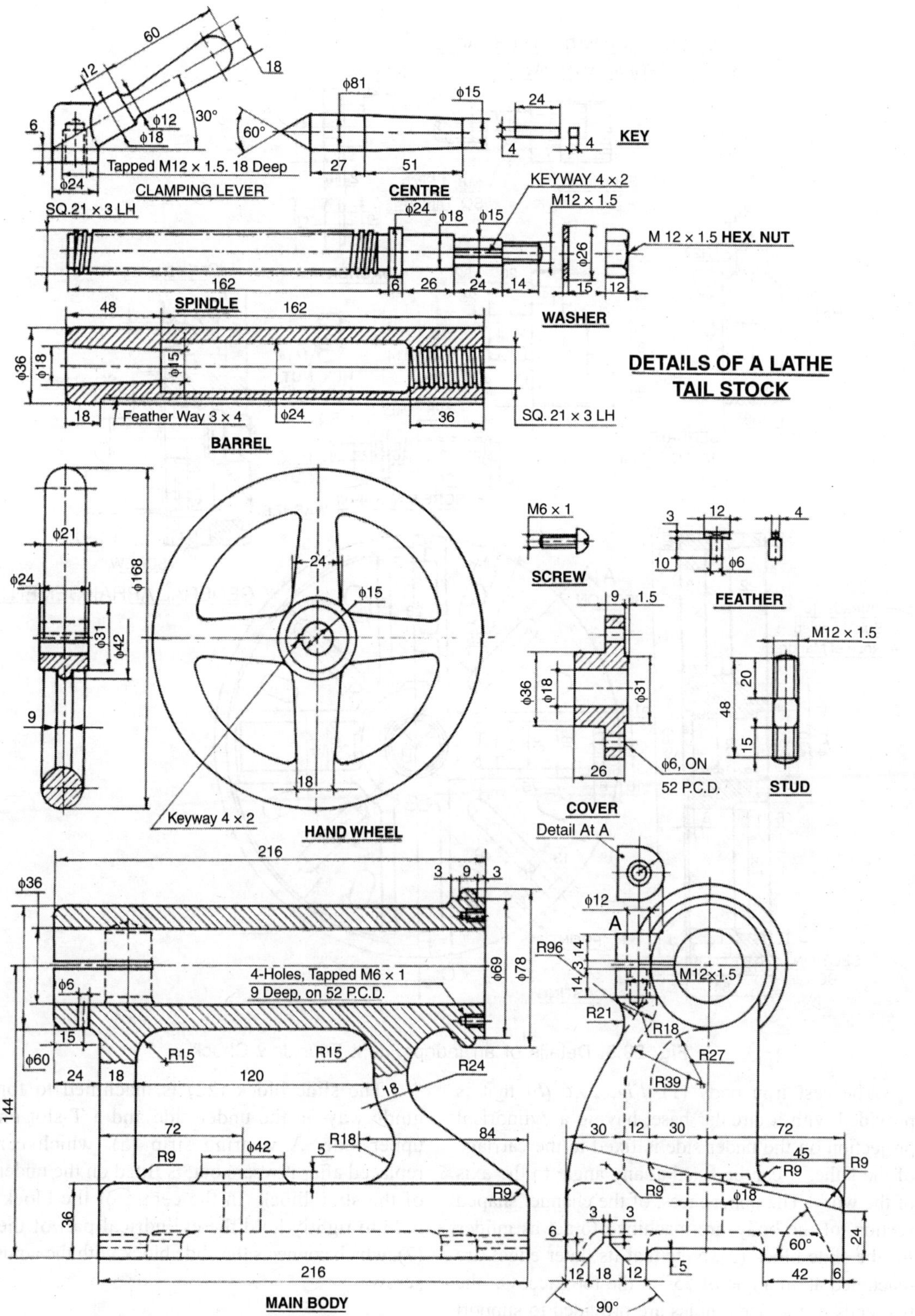

Fig. 29.2. Details of a Lathe Tailstock

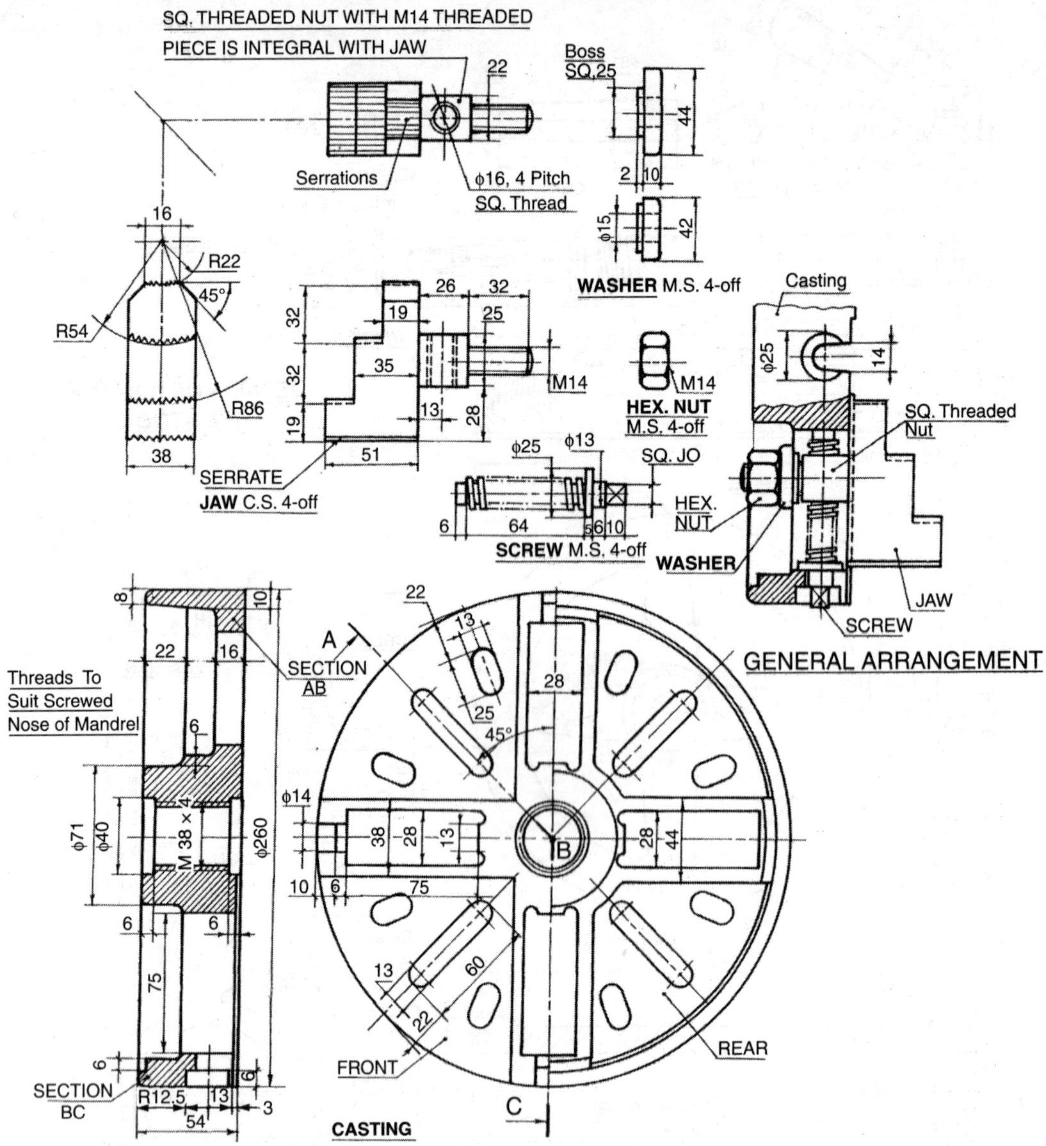

Fig. 29.3. Details of an Independent Four Jaw Chuck

The cast iron body (1), *Fig. 29.6 (b)* that is provided with a circular base, having a cylindrical projection on the under side is fixed to the carriage of the lathe. It can be fixed at any angle to the axis of the work. The upper faces, of the channel shaped section, of the body, are machined to act as guides for the slide block (2). Note that its outer edges are machined at an angle of 55° to the top face. On the two ends of the body, holes are provided to support the adjusting screw or slide screw (9).

The slide block (22) is machined to form a guide way on the under side and a T-slot on the upper side. A wearing strip (5), which can be replaced after it wears out, is fitted on the underside of the slide block. In the centre of the block is a hole to tightly hold the cylindrical part of the nut (7), which connects the slide block with the adjusting screw.

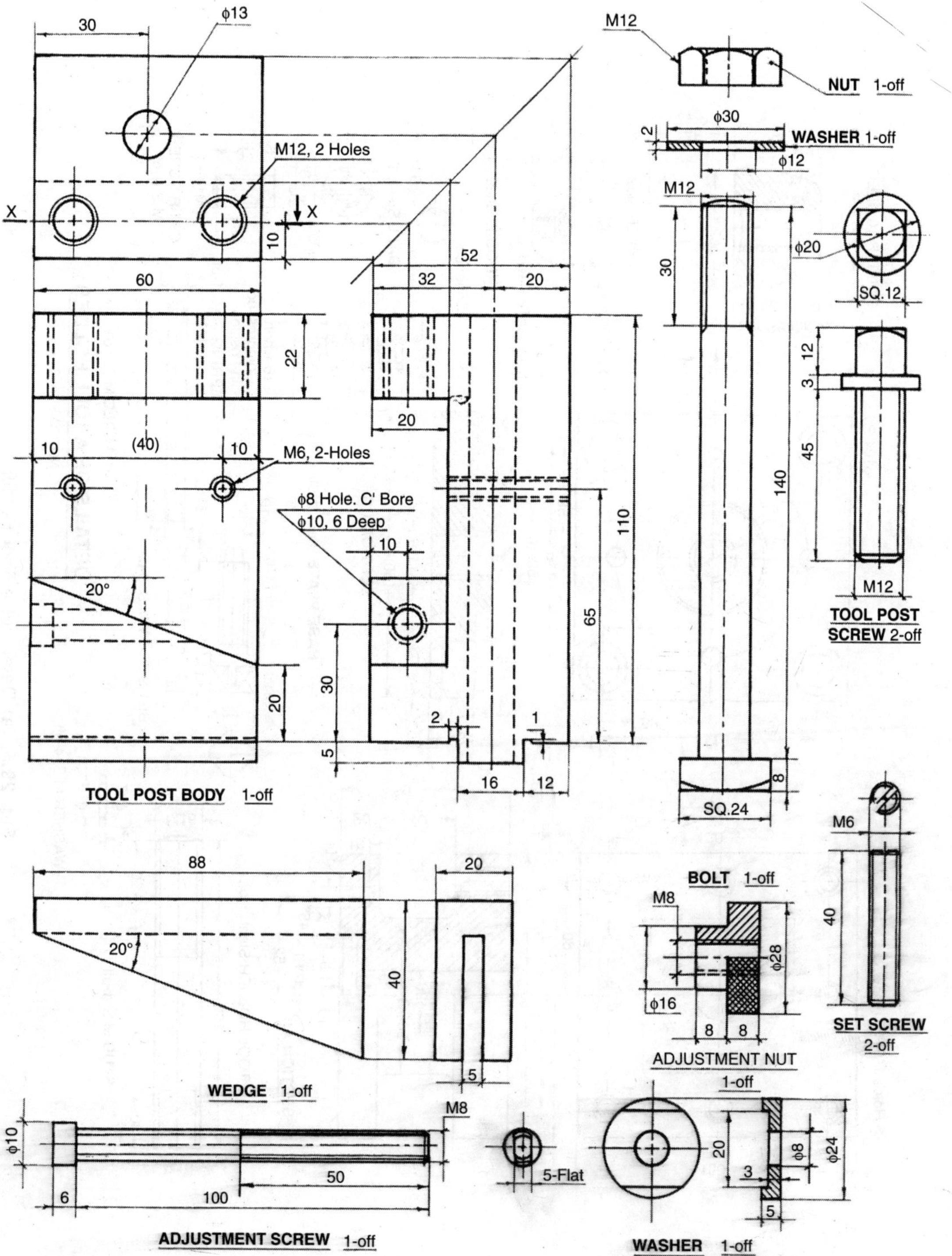

Fig. 29.4. Details of a Tool Post.

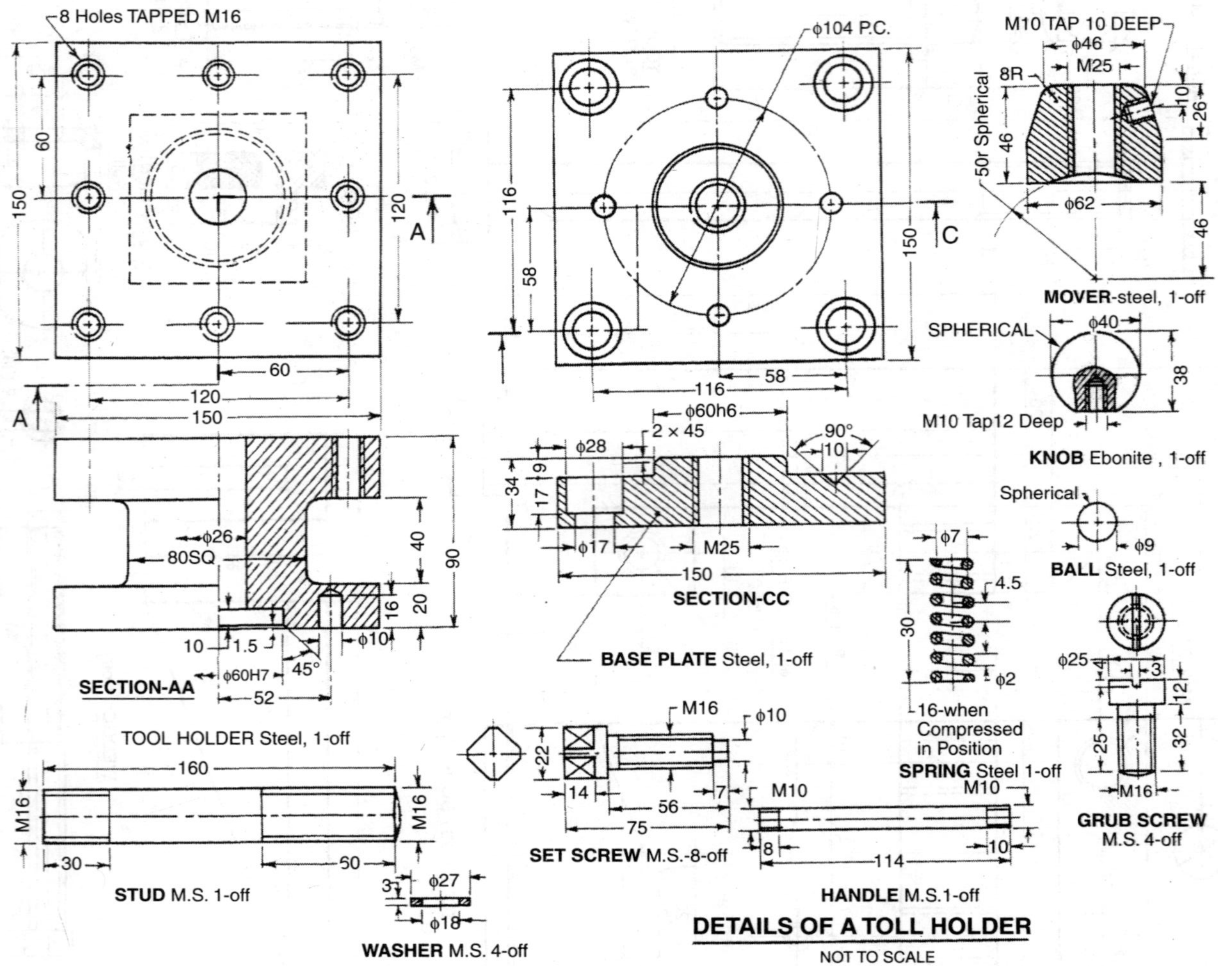

Fig. 29.5. (a) Details of a Tool Holder

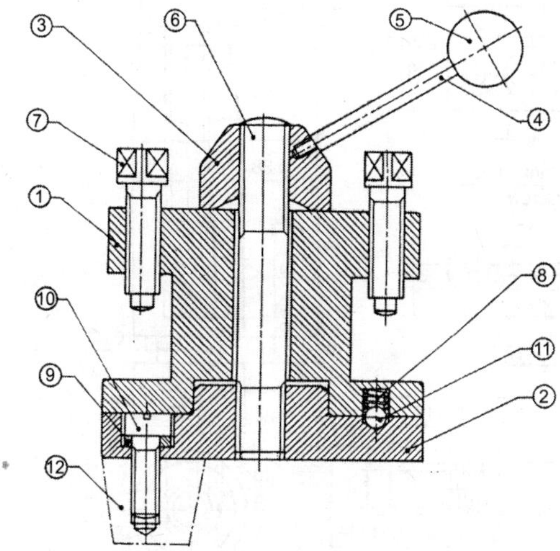

Fig. 29.5. (b) General arrangement of the Tool Holder

In the T-slot block, is slipped the cylindrical collar of the tool holder, and on the washer (4) is held the cutting tool by the screw (8). The washer helps in holding the tool holder tightly in the slide block and provides a good bearing suerface for the tool. On the front face of the body is secured a plate (6), having a stepped hole, to accommodate the collar on the adjusting screw. The plate serves the purpose of a bearing.

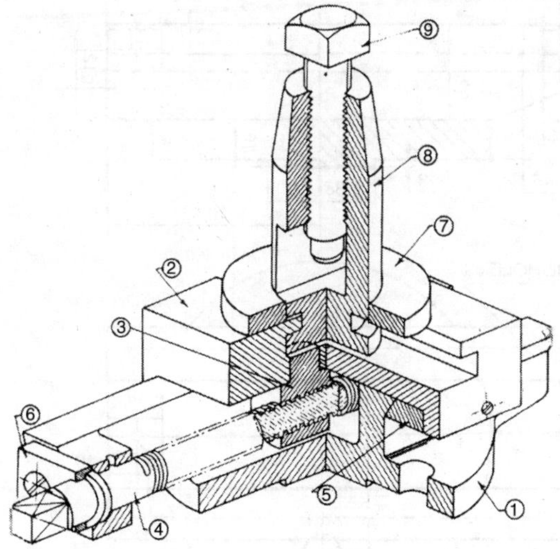

Fig. 29.6. (a) General arrangement of the Slide Rest

EXERCISE 29.6. *Details of a Lathe Slide Rest are shown in Fig. 29.6 (b). Draw the following views of the assembled parts:*

1. *Front view–left half in section*
2. *Right side view–right half in section and*
3. *Top view, outside.*

The outer end of the screw is supported in a hole in the back of the body.

29.6. SCREW JACK

Screw jack is a simple machine, which can raise a very heavy load through some distance by applying comparatively a small force at the end of a long lever. The simple design shown in *Fig. 29.7 (a)* can raise the load through a small distance in a vertical direction, only. These are also designed to raise the load vertically and then to carry it horizontally for a small distance. Small jacks are found in every automobile toolbox, which are used to lift the automobile for changing the wheels.

The screw jack illustrated in *Fig. 29.7 (a)* consists of a C.I. body (1) in the top flange of which is fitted a G.M. Nut (2). In the nut works a M.S. square threaded spindle (3). In the head of the screw is made a through hole, in which the 'Tommy bar' is inserted, for turning the screw. The end of the Tommy bar is knurled to provide a good grip for holding. The C.S. cap (4) is put on the head of the spindle and is free to rotate about the cylinrical projection on the spindle head. The cup is loosely held in position by a screw (6) and washers (5). To achieve a good grip of the load on the cup, the upper face fo the cup is serrated.

The lift of the load is equal to one pitch length, of the threads on the spindle, for every complete turn of the assembly.

EXERCISE 29.7. *From the details of a screw jack, shown in Fig. 29.7 (a), draw the following views of the assembly.*

1. *Front view–right half in section.*
2. *Top view*

Also make a parts-list and bill of materials. Give complete dimensions showing suggested tolerances, wherever necessary.

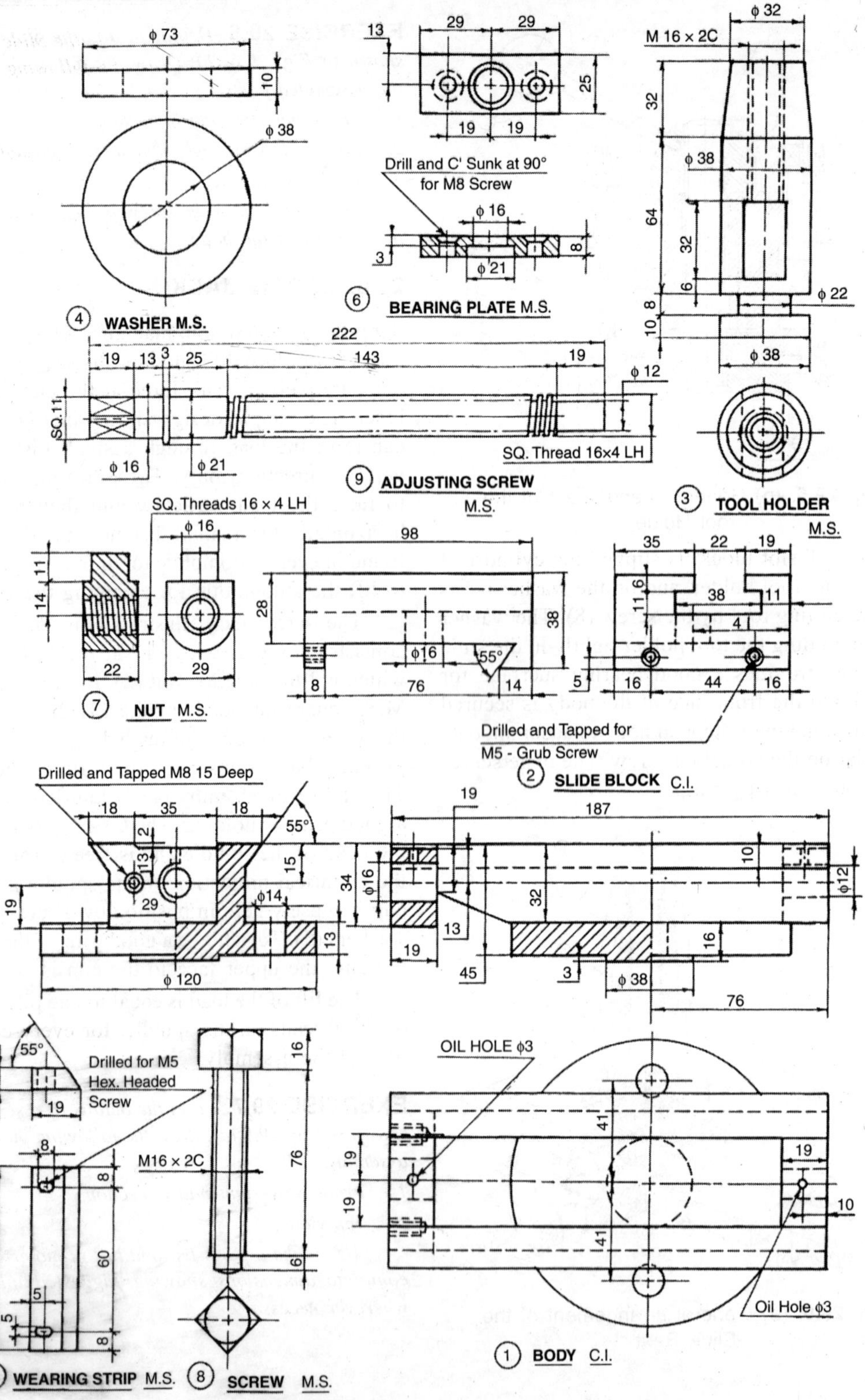

Fig. 29.6. (b) Details of a Lathe Slide Rest

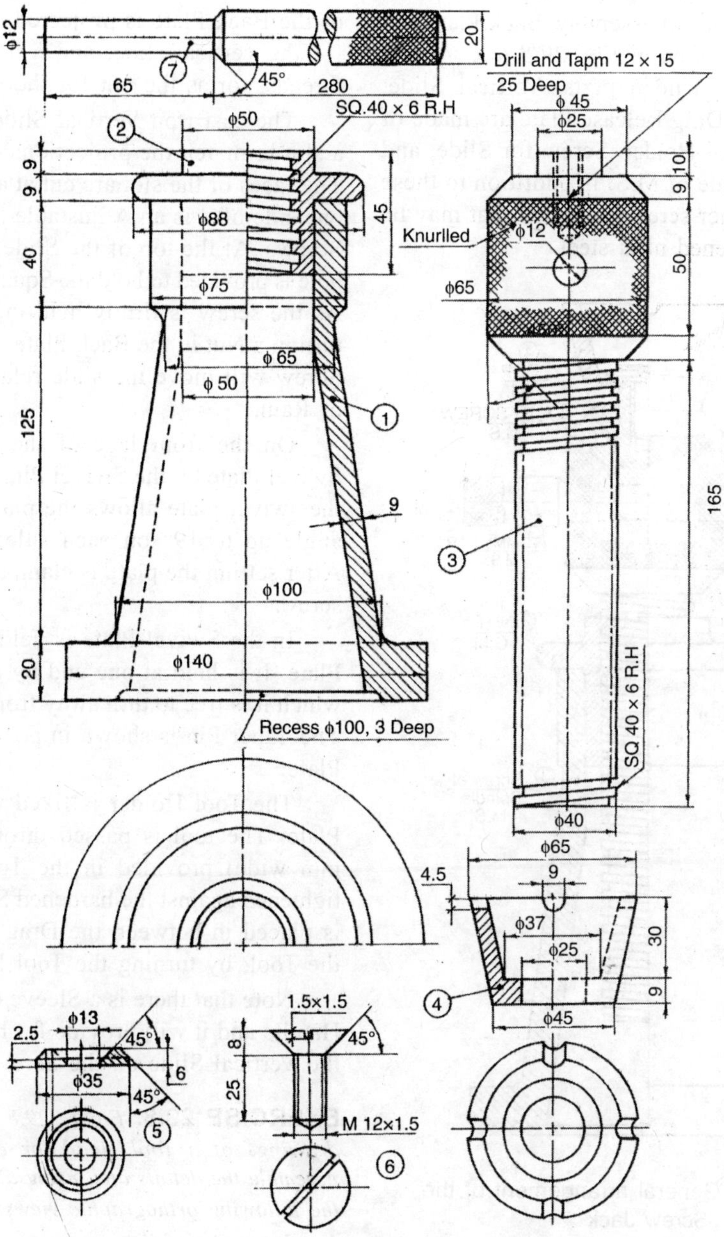

Fig. 29.7. (a) Detail drawings of a Screw Jack

29.7. TOOL HEAD OF A SHAPING MACHINE

Tool Head is fixed on the ram of a shaper and is designed to hold the cutting tool in such a manner that the tool cuts the metal during the forward stroke and drags on the work piece on the return stroke. In *Fig. 29.8 (e)* are shown the details of one such tool head. To facilitate understanding of the assembly and drawing of component parts, isometric illustrations of assembly, Back Plate and Vertical Slide are given in *Fig. 29.8 (a)*, *(b)* and *(c)*, respectively. Its main parts, Vertical Slide, Swivel Plate, and Drag Release Plate are made of C.I., while the Tool Holder, Screw for Slide, and Swivel Pin are made of M.S. In addtition to these there are some other screws and pins that may be made of casehardened mild steel.

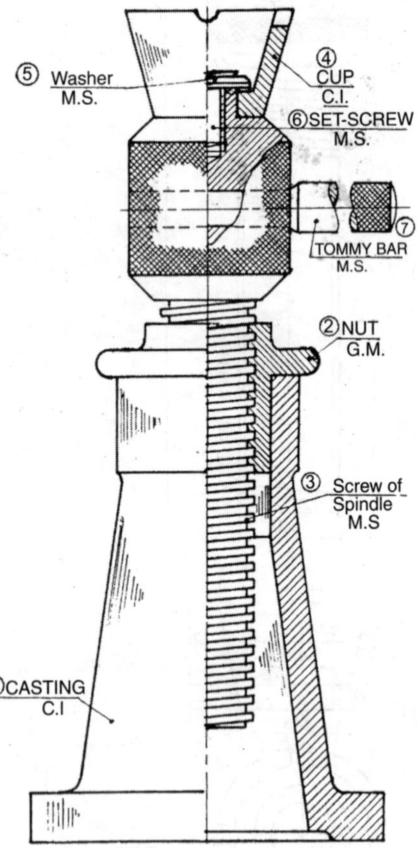

Fig. 29.7. (b) General arrangement of the Screw Jack

The cast iron Back Plate is attached to the circular flange of the horizontal ram of the shaping machine by means of a pin (not shown in details) through the centre of the plate. The pin is provided at one end, with a Hexagonal Nut for adjusting and tightening. The flange on the Ram is graduated so that, by this simple device, the Tool Head may be set quickely, to any desired angle, in a limited range on either side of the vertical. The front face of the Back Plate is projected to form the guides for the vertical slide, and a hole, having square threads, forms the nut for the slide screw.

The cast iron Vertical Slide is provided with a slot to match the projections on the Back Plate. The faces of the slot are cut at an angle of 60°. At one side of it is an Adjustable Strip held by Grub Screws. At the top of the Slide, in a projection, a hole is provided to hold the Square Threaded Screw. As the screw is firmly held in the Slide and can turn in a Nut in the Back Plate, any rotation of the Screw will move the slide relative to Back Plate or Ram.

On the front face of the slide is fixed the Swivel Plate by the Swivel Pin. The curved slot in the swivel plate allows the plate to be set at any angle up to 19° on each side of the centreline. After setting the plate is clamped by the clamping screw.

In the Swivel Plate is held the Drag Release Plate. It is held at one end by a Taper Pin, about which it is free to turn away from the Swivel Plate. The Taper Pin is shown in position in the Swivel Plate.

The Tool Holder is fixed with Drag Release Plate. The tool is passed through the slot (17.5 mm wide) provided in the Tool Holder, and is tightened against the hardened Steel Washer which is placed in between the Drag Release Plate and the Tool, by turning the Tool Holder Screw.

Note that there is a Sleeve on the Slide Screw Handle and it will provide for better tightening of the Vertical Slide on the Screw.

EXERCISE 29.8. *Figure 29.8 (e) shows the detail drawings of a Tool Head for a Shaping Machine Assemble the details and, using a suitable scale, draw the following orthographic views:*

1. *Front view–full in section,*
2. *End view, and*
3. *Top view.*

Give all the necessary dimensions and prepare a bill of materials and parts list.

Hint: Refer *Fig. 29.8 (d)* for part solution of this exercise.

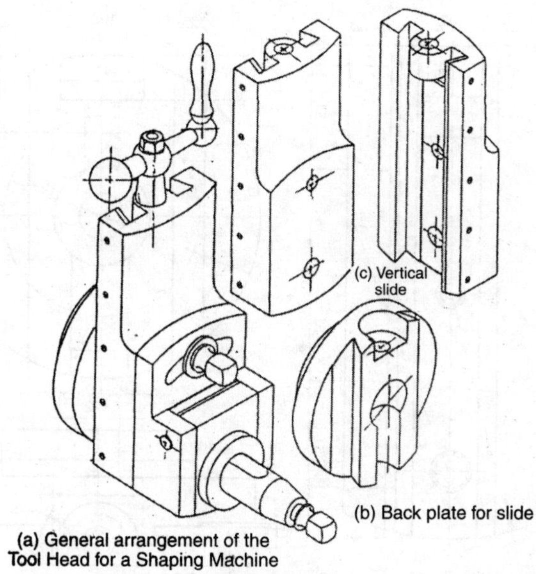

Fig. 29.8. Tool Head for a Shaping Machine

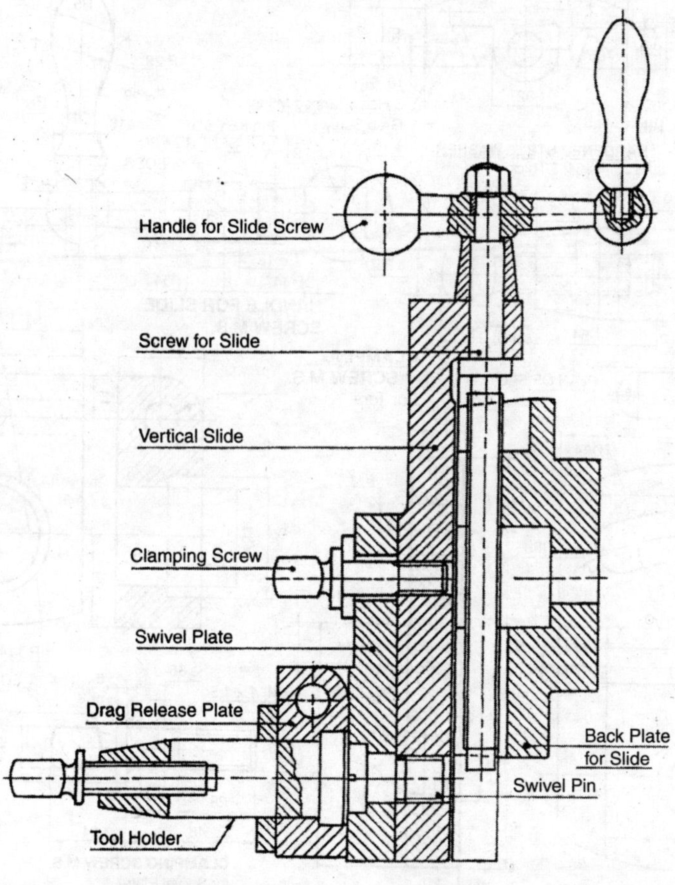

Fig. 29.8. (d) Full sectional front view of the Tool Head

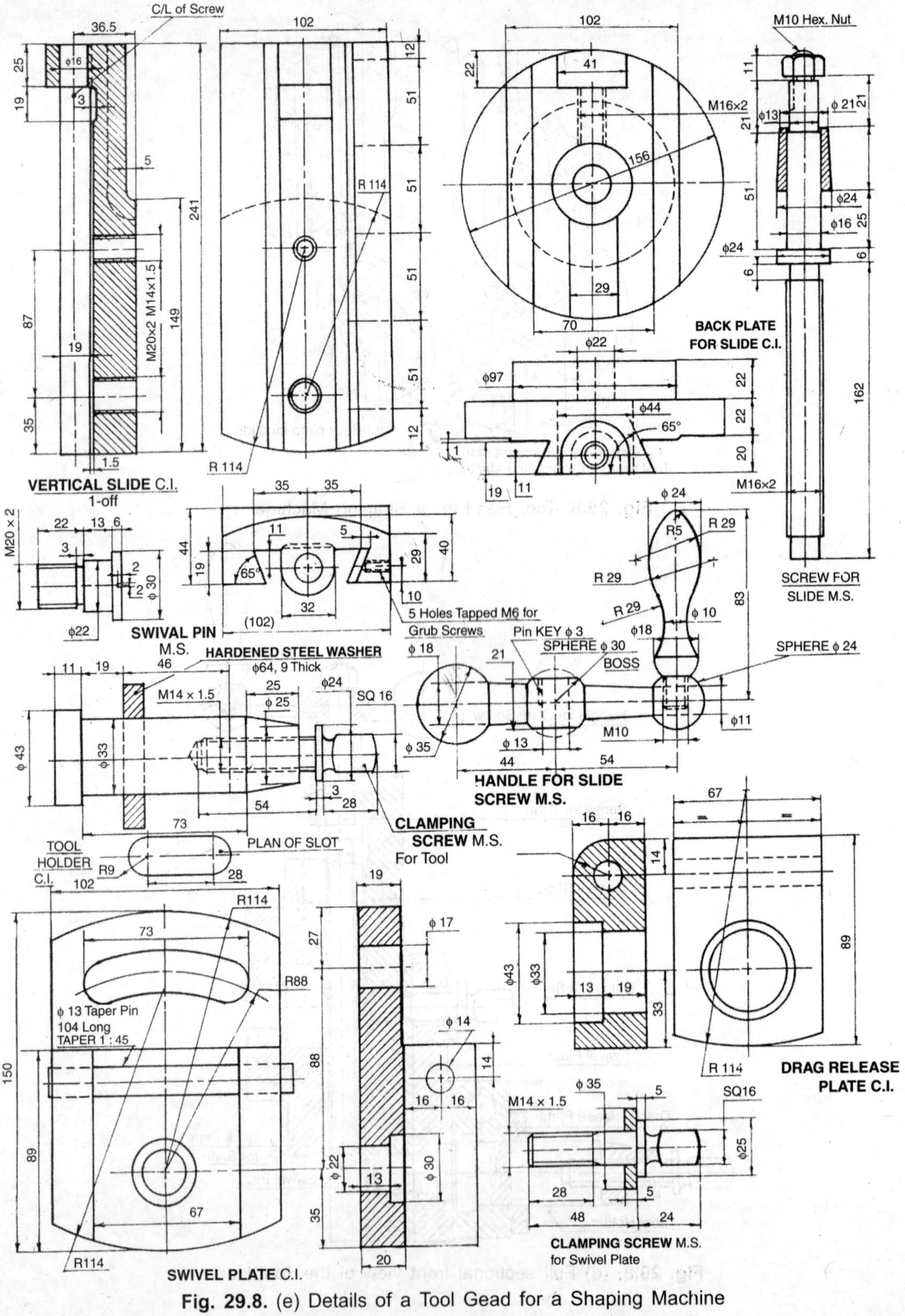

Fig. 29.8. (e) Details of a Tool Gead for a Shaping Machine

29.8. VICES

A **Vice** is the most common device used to hold work, for carrying out various operations, due to its quick loading and unloading capability. It may be fitted on a worktable or on a machine table. These are named accordingly as **Bench Vice** and **Machine Vice**, respectively. For light duty work a small vice may also be held by hand. Such a vice is called **Hand Vice**. Provision is made in the design of each vice to adjust the distance between its jaws for accommodating different sized jobs. Usually a screwed pair is used for this purpose. Each vice has *four* main parts:

1. **Fixed Jaw:** It is usually cast integral with the vice body or base.
2. **Moveable Jaw:** It slides on the ways of the casting and is operated with a screw or spindle.
3. **Screw:** It gives the movable jaw the forward or backward movement and to do so it is rotated by a handle called a screw bar.
4. **Casting:** It constitutes the base of the vice and has ways for the moveable bar.

29.9. PIPE VICE

Pipe Vices are designed specialy afor holding pipes to facilitate operations such as threading or cutting off to length. *Figure 29.9* illustrates various parts of a pipe vice.

To assemble the unit, the screw is screwed into the housing from above. When the circular groove at the end of the screw is in line with the 7.5 mm transverse hole in the housing, the jaw is inserted from below. After correct alignment two small M5 setscrews are inserted into the two holes in the housing, one from either side, and each is screwed into its respective tapped hole in the jaw, (details to be assumed) securing the screw to the jaw. This arrangement allows the jaw to move vertically, without rotation, when the handle is operated and the screw is turning.

The V-shaped base of the housing can accommodate pipes of different diameters. When the screw is turned clockwise, the jaw moves down and grips the pipe. Square threads are used on the screw, as they are stronger. The serrations, provided in the V-shaped end of the jaw, exert effective grip on the pipe surface.

EXERCISE 29.9. *Details of a Pipe Vice are shown in Fig. 29.9. Assemble the details and draw, on an A2 size drawing sheet, the following orthographic views:*

1. *Front view, section through the centre of the threaded hole having square threads,*
2. *Right side view, section through the centre of the square threaded hole, and*
3. *Top view, section through ϕ 7.5 hole.*

Use any convenient scale.

EXERCISE 28.10. *Details of a pipe vice are shown in Fig. 29.10. Assemble the details and draw, on an A2 size drawing sheet, the following orthographic views:*

1. *Front view, section through the M12 hole,*
2. *Left side view, section through the M12 hole,*
3. *Top view section–AA.*

Use full size scale, and dimension the views completely. Assume proportionately any missing dimensions.

EXERCISE 29.11. *Details of a Machine Vice are given in Fig 29.11. Assemble the details and draw, on an A2 size drawing sheet, to a full size scale, the following orthographic views:*

1. *Front view–full central section,*
2. *Right side view–section AA, and*
3. *Top view.*

Give all the necessary dimensions. Also prepare a bill of materials and parts-list. Assume proportionately any missing dimensions.

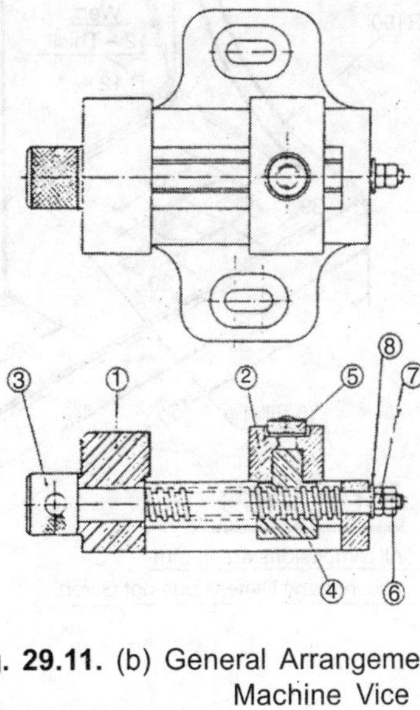

Fig. 29.11. (b) General Arrangement of Machine Vice

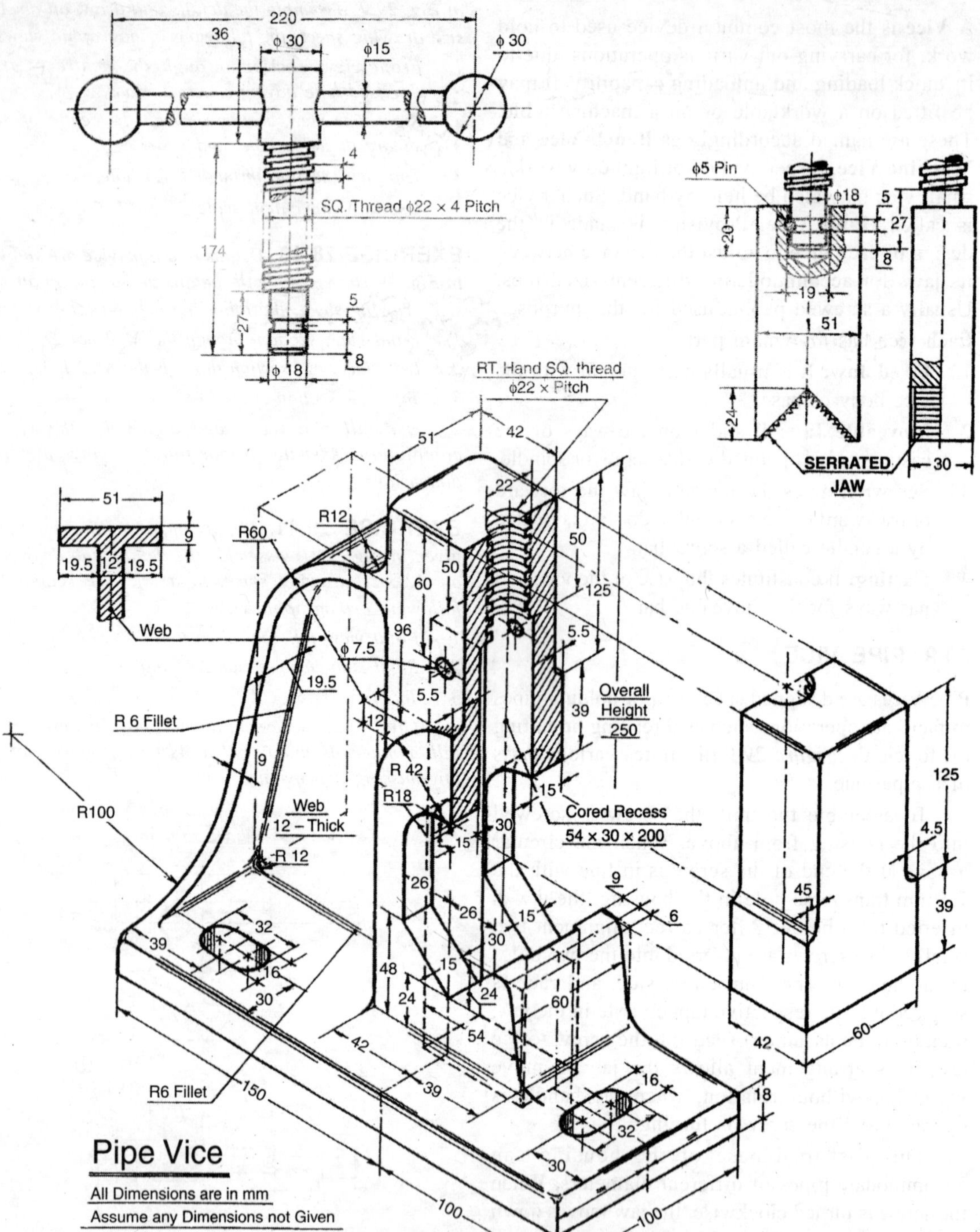

Fig. 29.9. Details of a Pipe Vice

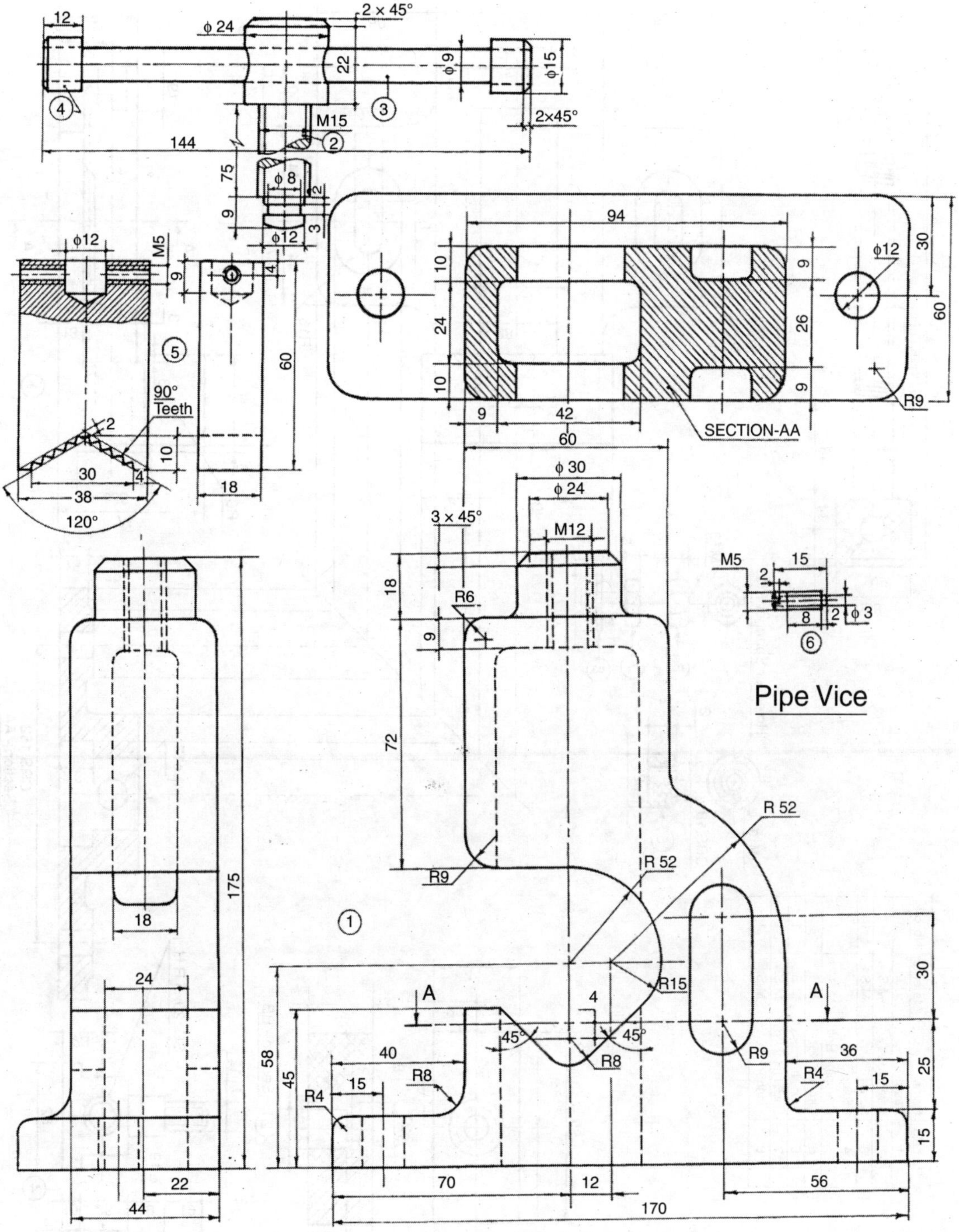

Fig. 29.10. Details of a Pipe Vice

Part No.	Name	Material	No. off
1	Base	C.I.	1
2	Sliding Jaw	M.S.	1
3	Screw	M.S.	1
4	Sliding Jaw Clamping Bolt	M.S.?	1
5	Circular Nut	M.S.	1
6	Hex. Nut	M.S.	1
7	Lock Nut	M.S.	1
8	Washer	M.S.	1

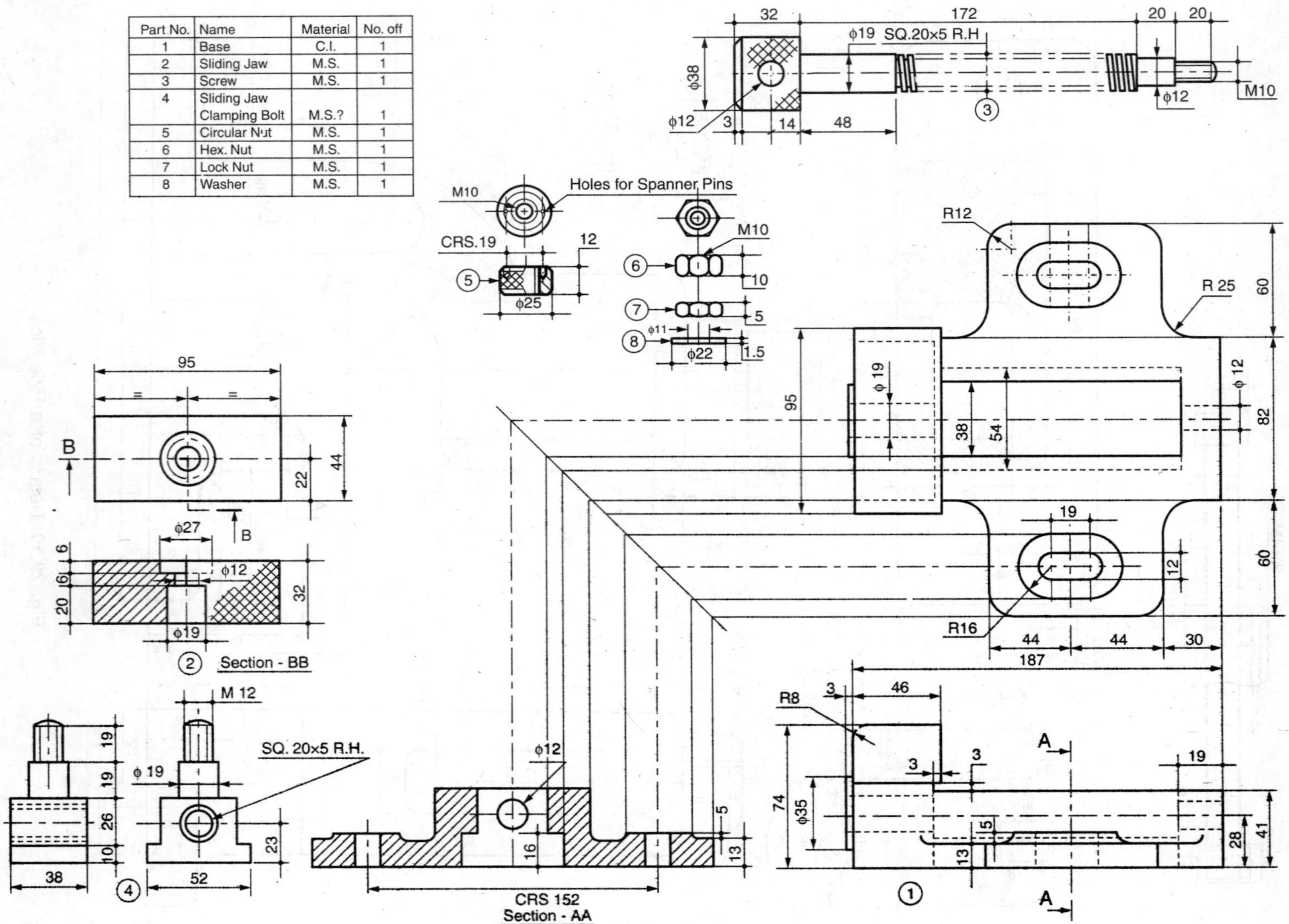

Fig. 29.11. Details of a Machine Vice

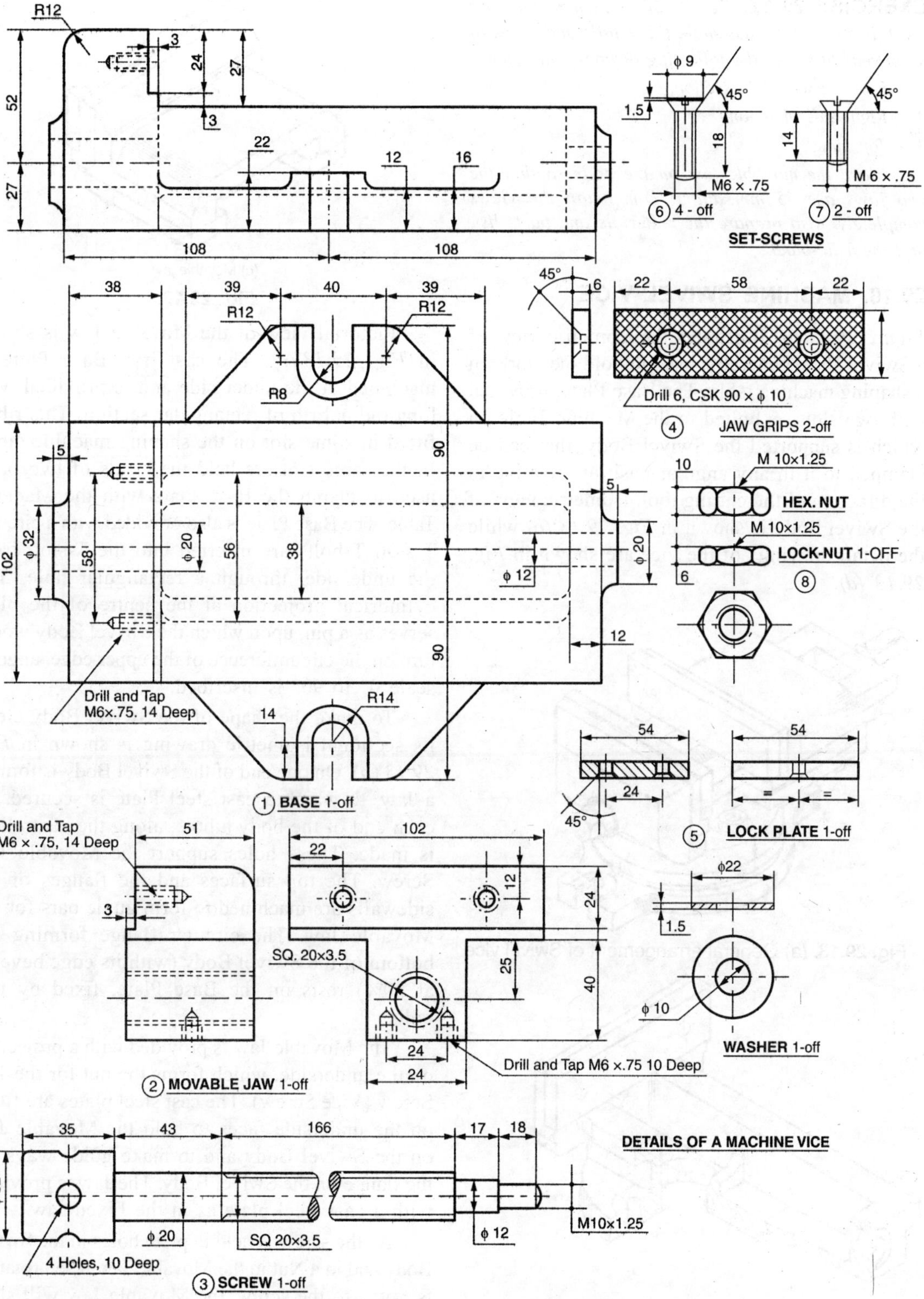

Fig 29.12. Details of a Machine Vice

EXERCISE 29.12. *Details of a Machine Vice are given in Fig. 29.12. Assemble the details and draw, to a convenient scale, the following otrhographic views:*

1. *Front view–full in section,*
2. *Right side view, and*
3. *Top view,*

Place the movable jaw on the body so that the two jaws are 65 mm apart. Dimension the views completely. Also prepare the materials and parts list, and the Title-block.

29.10. MACHINE SWIVEL VICE

Figure 29.13 (a) illustrates the isometric view of a Swivel Vice, generally, used to hold the work on a shaping machine table. The Base Plate, provided with two slots, is bolted to the Machine Table on which is supported the Swivel Body, that can be clamped to it in any angular position, relative to the direction of the cutting tool. Isometric view of the Swivel Body is shown in *Fig. 29.13 (b)* while the detail drawings of the vice are shown in *Fig. 29.13 (d)*.

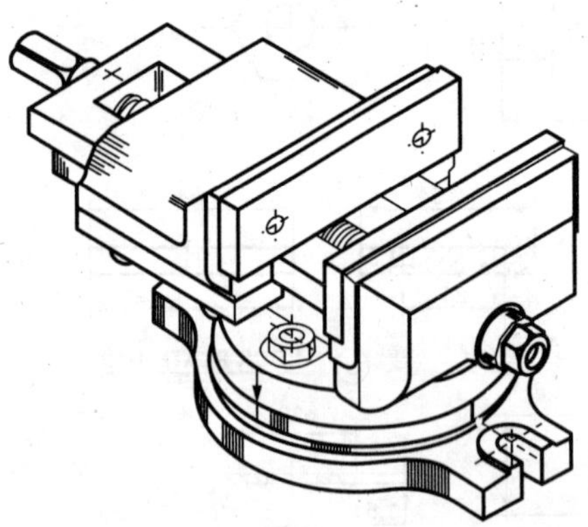

Fig. 29.13. (a) General arrangement of Swivel vice

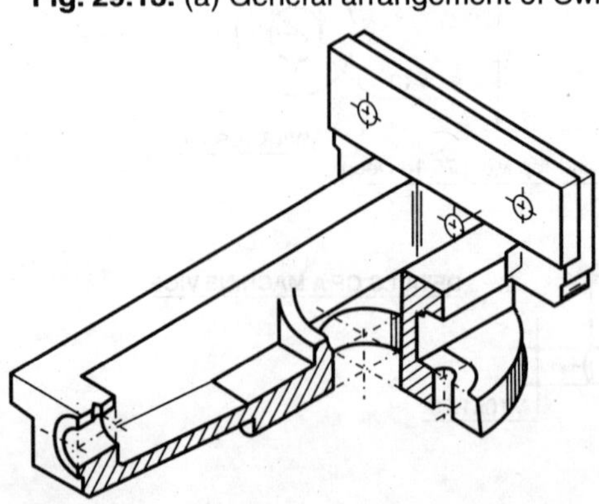

Fig. 29.13. (b) Swivel body and fixed jaw (contd.)

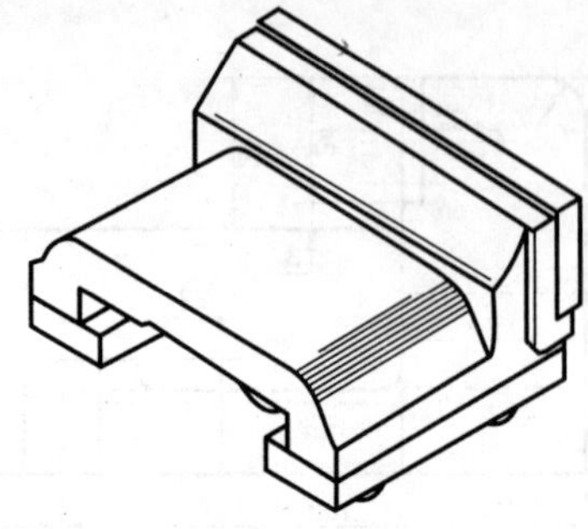

(c) Movable jaw

Fig. 29.13.

Pictorial view of the Movable Jaw is shown in *Fig. 29.13 (c)*. The cast iron Base Plate is machined at the under side and is provided with longitudinal rib of rectangular section. This rib is fitted in some slot on the shaping machine table. In the same slot are held the heads of two bolts used to clamp the Base Plate with the Machine Table. The Base Plate is also provided with a circular T-slot. T-bolts are inserted into the T-slot, from the underside, through a rectangular hole. The cylindrical projection at the centre of the plate serves as a pin, upon which the Swivel Body would turn on the circumference of the upper edge, angular scale 0° to 90° is inscribed.

To make the shape of the Swivel Body clear, its sectional isometric drawing is shown in *Fig. 29.13 (b)*. On one end of the Swivel Body is formed a Jaw, to which a cast steel Plate is secured. At each end of the body a hole, along the centreline, is made. These holes support the Movable Jaw Screw. The top surfaces and the flanges of the sidewalls are machined to form guide bars for the Movable Jaw. The circular flange forming the bottom of the Swivel Body (with its edge beveled at 80%) rests on the Base Plate, fixed by two bolts.

The Movable Jaw is provided with a projection on the underside, which forms the nut for the Jaw Screw (Vice Screw). The cast steel plates are fitted on the underside faces to hold the Movable Jaw on the Swivel Body and to make guide ways for the flange on the Swivel Body. The Jaw is provided with a cast steel plate as in the Fixed Jaw.

As the screw is held in plain holes in the Swivel Body and in a Nut in the Movable Jaw when rotation is given to the screw, the Movable Jaw will slide on the Swivel Body. To prevent the axial movement

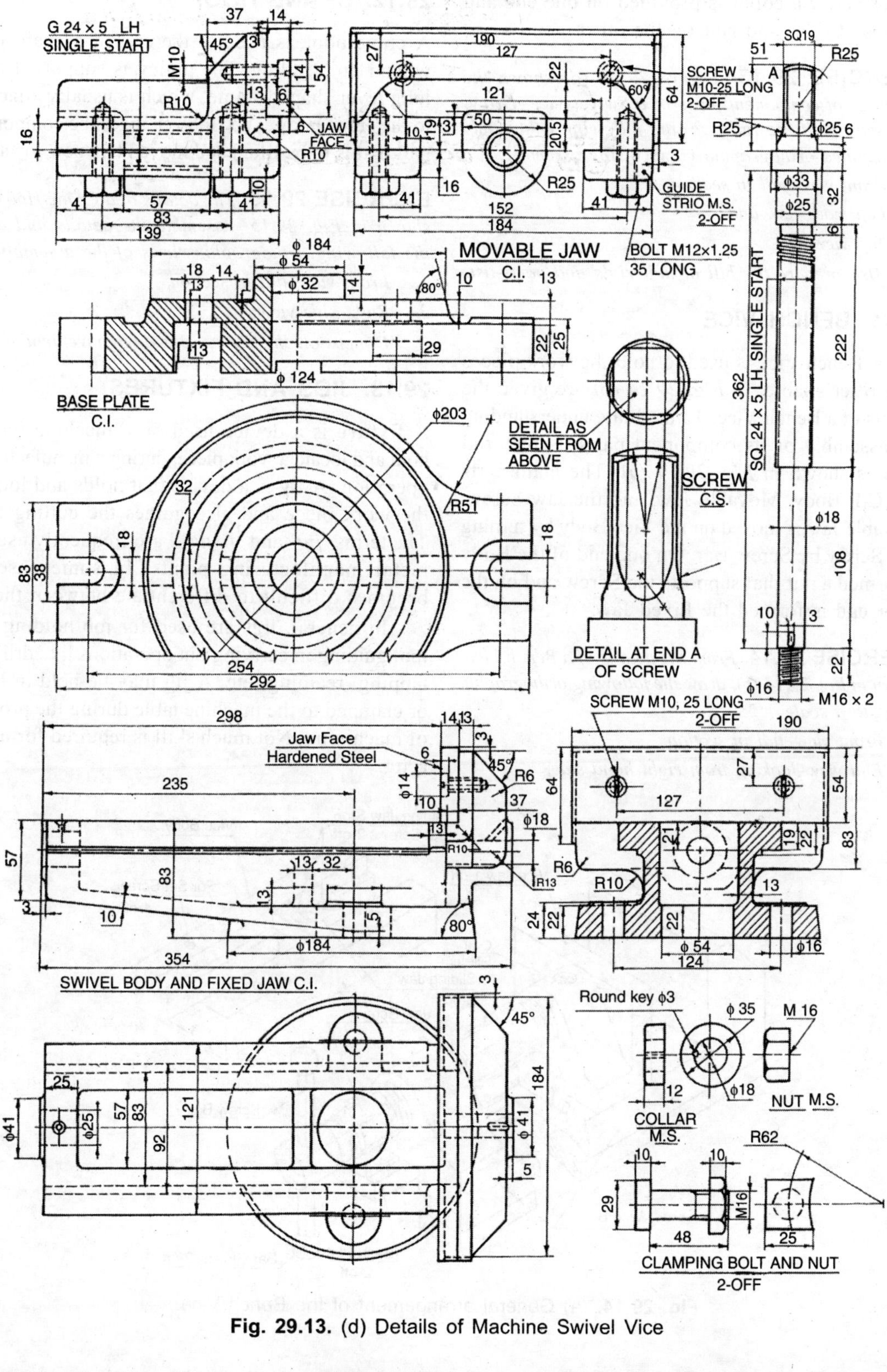

Fig. 29.13. (d) Details of Machine Swivel Vice

of the Screw a collar is provided on one end and a Loose Collar and Nut on the other.

EXERCISE 29.13. *Figure 29.13 (d) shows the drawings of component parts of a Swivel Vice. Draw; to a scale half the full size, the following completely dimensioned orthographic views of the assembly:*

1. *Front view–full in section,*
2. *Left side view, and*
3. *Top view.*

   Also prepare the bill of materials and parts-list.

29.11. BENCH VICE

A Bench Vice is used to hold the work pieces on a fitter's table. In *Fig. 29.14 (b)* are given the details of a Bench Vice. To facilitate understanding the assembly of its component parts its pictorial view is shown in *Fig. 29.14 (a)*. The main parts are: C.I. Body, Movable Jaw, and the Jaw Screw. Movable Jaw is moved on the Vice Body by turning Jaw Screw by Screw Bar. On one end of the Body is formed a nut that supports the Screw and on the other end is formed the Fixed Jaw.

EXERCISE 29.14. *From the details of a Bench Vice, shown in Fig. 29.14 (b), draw the following orthographic views to a scale 1:2:*

1. *Front view–full in section,*
2. *End view–looking from right hand side.*

29.12. CRANE HOOK

A crane hook is used at the end of a chain rope wound on a pulley. The pulley is supported on a long lever shaped frame, which is usually attached to the crane body. The Hook engages the component part that is to be lifted and moved to a new location.

EXERCISE 29.15. *The details of a Crane Hook are shown in Fig. 29.15. Assemble the details and draw the following orthographic views of the assembly:*

1. *Front view–full in section,*
2. *Right side view.*

   Dimension the views. Use any convenient scale.

29.13. JIGS AND FIXTURES

A **fixture** is a device used on a machine tool to hold and locate a work piece, during a manufacturing operation. A **jig** is a device that holds and locates the work piece and also guides the cutting tool. The terms '*jig*' and '*fixture*' are frequently used in metal manufacturing plants. In some cases it becomes difficult to differentiate between them.

In general, jigs are used for job holding and tool guiding in carrying out operations like drilling, tapping, reaming, etc. A jig may be held in hand or clamped to the machine table during the process of machining. Not much skill is required for using a jig.

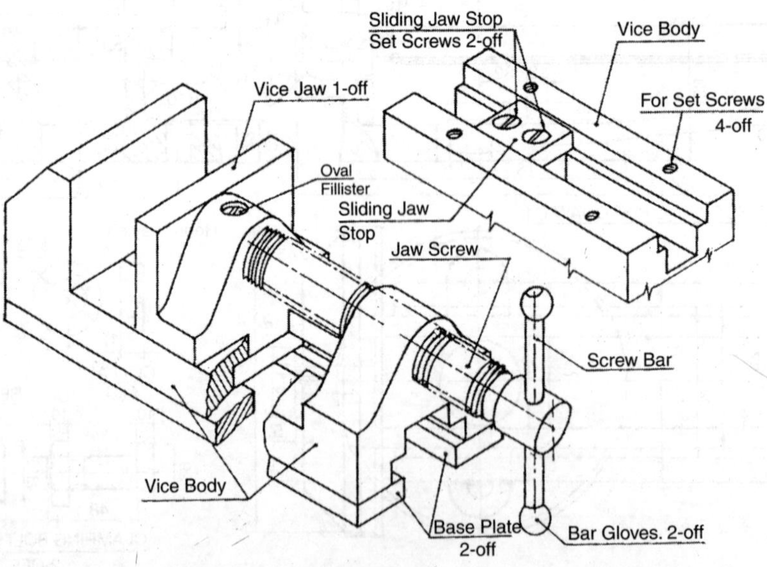

Fig. 29.14. (a) General arrangement of the Bench Vice

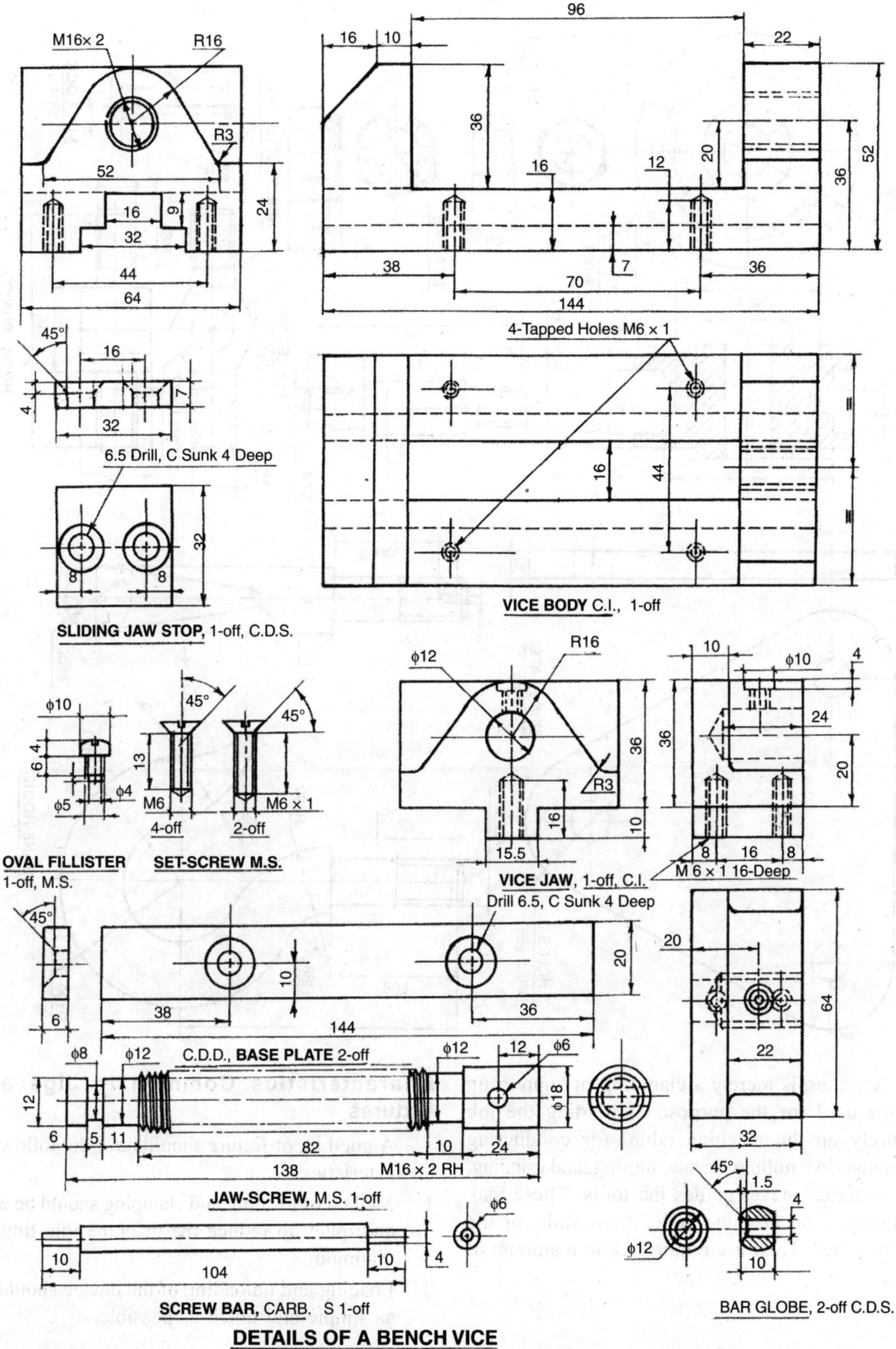

DETAILS OF A BENCH VICE

Fig. 29.14. (b)

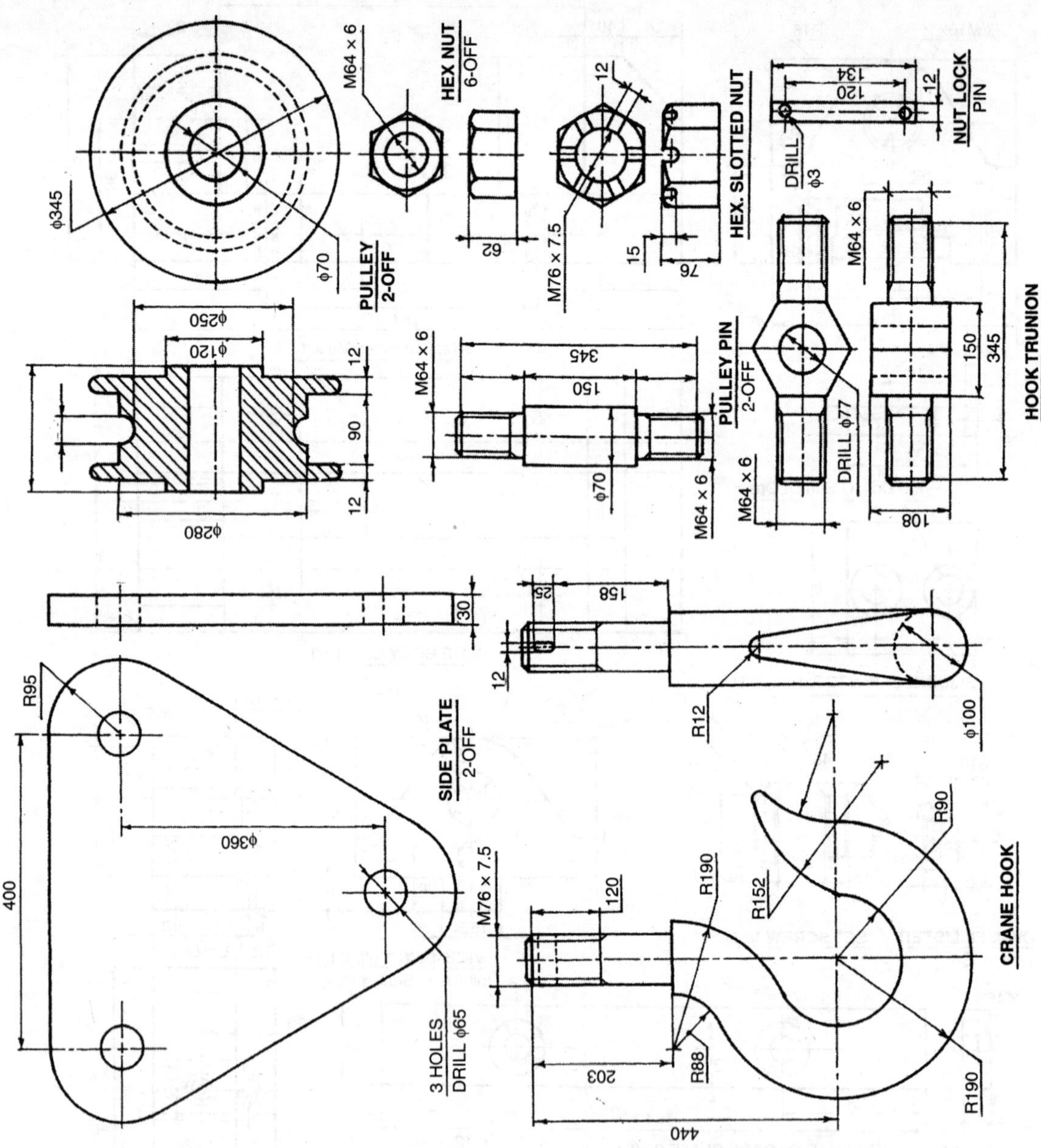

Fig. 29.15. Details of a Crane Hook

A fixture is merely a clamping or supporting device used for the purpose of holding the job securely on the machine table, for conducting operations like milling, boring, turning, and grinding, etc. A fixture never guides the tools. These may be clamped on machine table, cross-slide, or the machine bed. Their use requires certain amount of skill.

Characteristics Common to Jigs and Fixtures

A good jig or fixture should have the following characteristics:

1. Method of locating and clamping should be easy and quick to reduce the machine idle time to minimum.
2. Loading and unloading of the device should be as simple and quick as possible.

3. As cleanliness is of great importance, the device must have adequate provision for machined material chip removal.
4. Locating and supporting surfaces should be of hardened steel and should, wherever possible, be renewable.
5. A jig or fixture should be fool proof. The tool must contain a nest, which should receive and locate the part only in the correct position.
6. Clamps should be positioned to hold the part firmly in the nest and should be strong enough to resist bending.

General Arrangement of the Tumble Jig

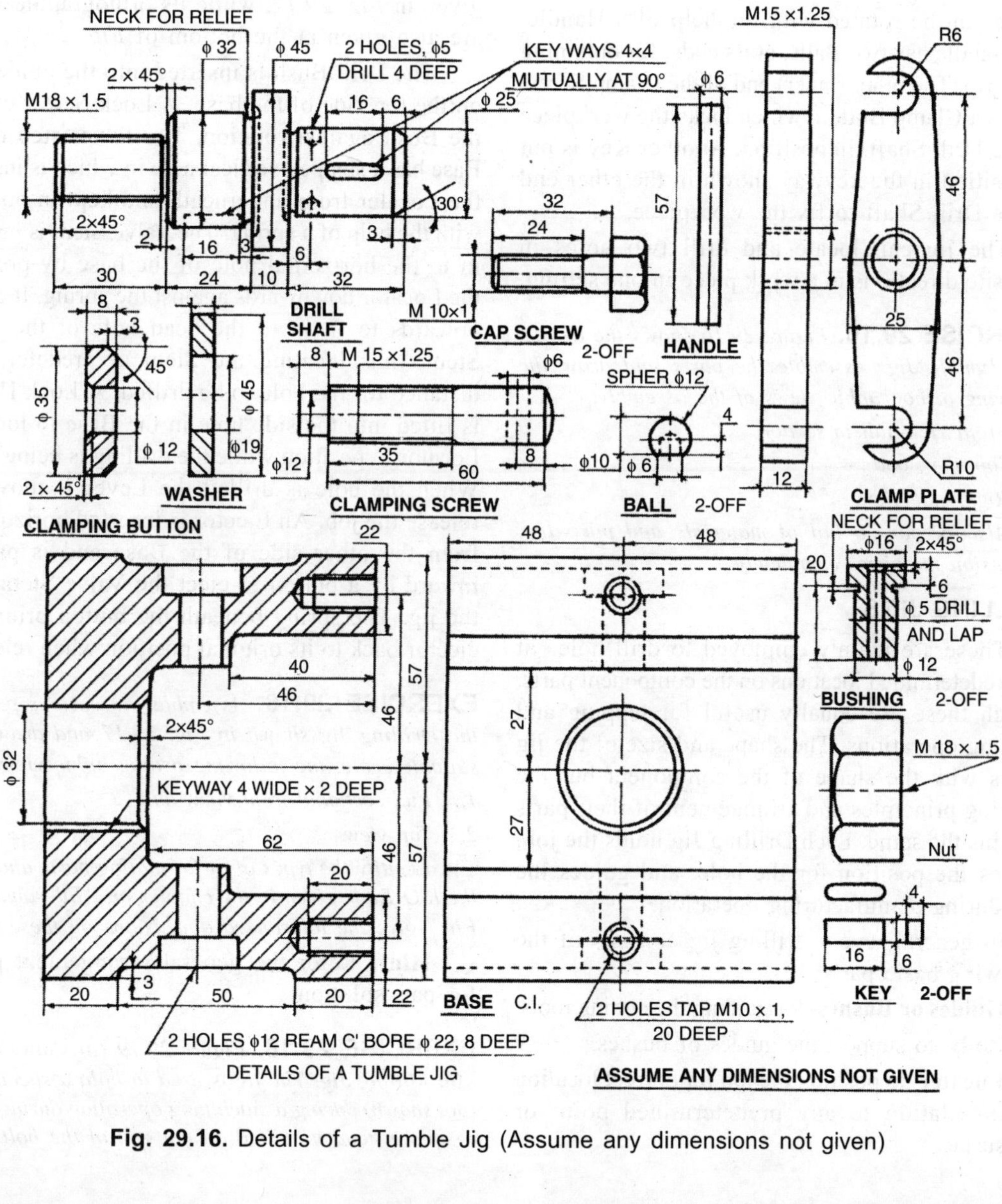

Fig. 29.16. Details of a Tumble Jig (Assume any dimensions not given)

TUMBLE JIG

The details of a Tumble Jig are given in *Fig. 29.16*. The two Bushings are press fitted into the holes (one provided at the top and other at the bottom of the jig base), to guide the drill. The Drill Shaft is inserted into the central horizontal hole of the base and checked against rotation by a key inserted in the aligned keyways cut on the drill shaft and Jig base hole. The Shaft is secured in position with the help of a Washer and Nut. The Clamp Plate is fitted with the mouth of the jig base by two Cap Screws. Through the central hole of the Clamp Plate is screwed a Clamp Screw which can be rotated with the help of a Handle. The Handle has two Balls, at its ends, for obtaining easy grip. The other (inner) end of the Clamp Screw carries a Clamp Button, which locks the workpiece on the Drill Shaft, in position. Another Key is put in position in the keyway shown in the other end of the Drill Shaft to fix the workpiece.

The jig can locate and drill two holes in opposite directions in a work piece in one setting.

EXERCISE 29.16. *Figure 29.16 shows the details of a Tumble Jig. Assemble the parts and draw the following orthographic views of the assembly:*

1. *Front view–full in section,*
2. *Top view, and*
3. *Right side view.*

Also prepare a bill of materials and parts-list. Dimension the views completely.

DRILLING JIGS

These are mainly employed to drill holes at the predetermined locations on the component parts, though these are equally useful for tapping and reaming operations. The shape and size of the jig varies with the shape of the component but the working principles and arrangement of their parts remains the same. Each Drilling Jig holds the job, locates the position for the hole, and guides the tool during manufacturing operation.

In general, every drilling jig consists of the following basic parts:

1. **Guides or Bushes**–for guiding the cutting tools.
2. **Body**–to support the guides or bushes.
3. **Locater**–to position the job for correct location in relation to any predetermined point or surface.
4. **Clamping arrangement**–to lock the job in the desired position.
5. **Base**–to support all the parts of the jig and secure the assembly to the operating part of the machine tool.

DRILLING JIG FOR VALVE STEM HOLE

The valve stem shown at bottom right corner of *Fig. 29.17*, is to have a small hole of $\phi 2$ mm accurately located at a specified distance from one end and running exactly parallel to another, already drilled hole, in head. The detailed drawings of the jig used to drill $\phi 2$ mm hole in the valve stem are given in *Fig. 29.17*, while its orthographic views are also given at the bottom of *Fig. 29.17*.

The Drill Bush is inserted into the guide hole on the top side of the base. A Lock Screw checks the Bush against rotation. The lower step of the Base has a stepped vertical hole in which is inserted the Locater from underneath and kept in position with the help of a spring. The Valve Stem is inserted in to the horizontal hole of the base by pressing the Locator downwards against the spring. It comes upwards to fit in to the head hole of the Valve Stem, thus locating the Stem at predetermined distance for the hole to be drilled. A Lock Handle is fitted into the side hole in the Base to lock the Locator in position while the drilling is being done. When the hole is drilled the Lever is moved to release the job. An Ejector is inserted horizontally from the other side of the Base and is pressed inward by a button to eject the Valve Stem from the jig. The spring beneath the button brings the ejector back to its original position when released.

EXERCISE 29.18. *Assemble the various parts of the Drilling Jig shown in Fig. 29.17 and draw, to a suitable scale, the following orthographic views.*

1. *Full sectional front view, and*
2. *Top view.*

Identify the types of fit for mating parts and using the ISO Table given in the chapter on "Tolerances and Fits" give the toleranced dimensions on these views.

Hint: Refer the general arrangement given, for part solution.

EXERCISE 29.18. *Figure 29.19 (a) shows details of a Milling Jig. The jig is used to hold a special bolt (not shown) during a machining operation during which a square head is milled on the end of the bolt. The*

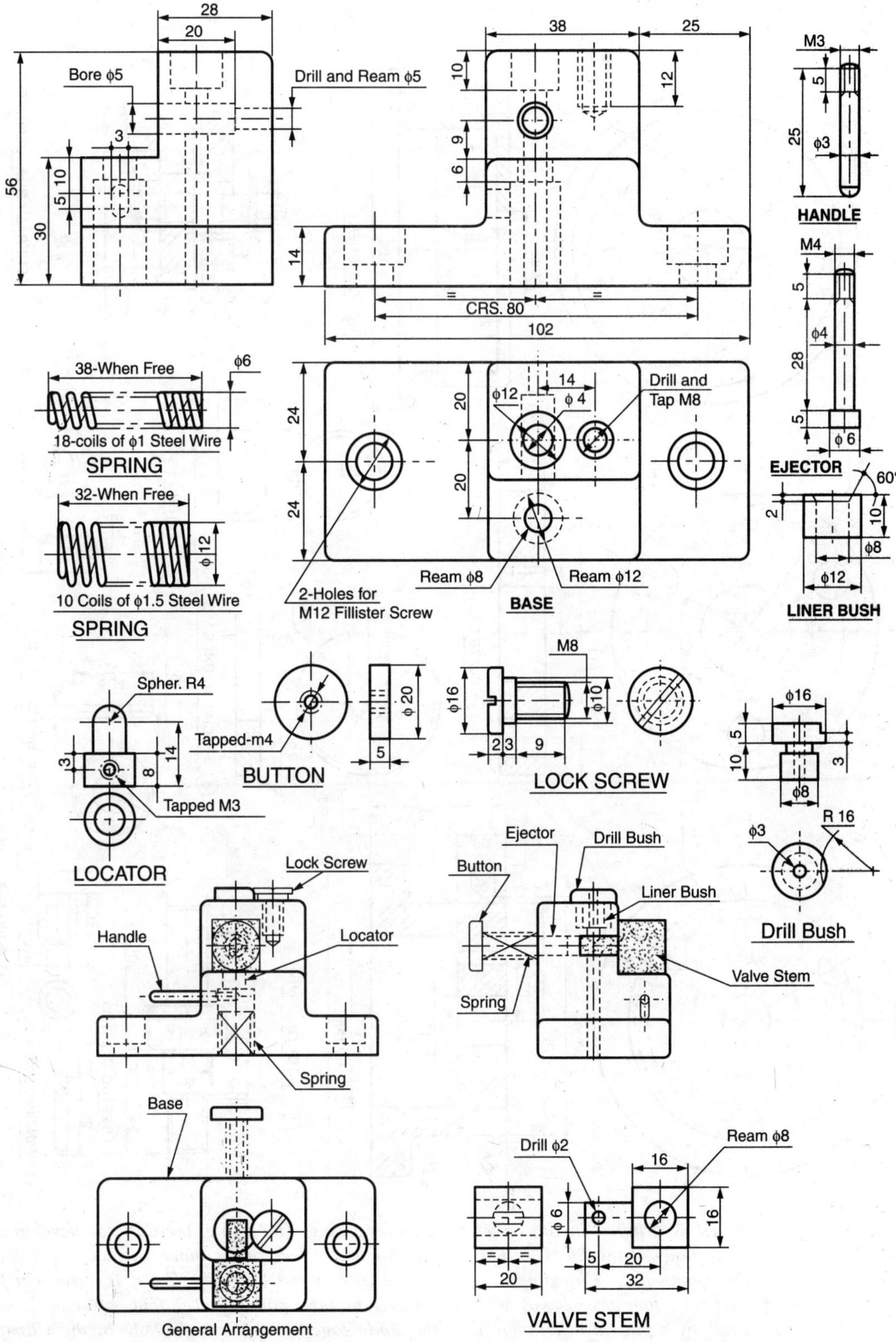

Fig. 29.17. Orthographic views and detail drawings of a Drilling Jig

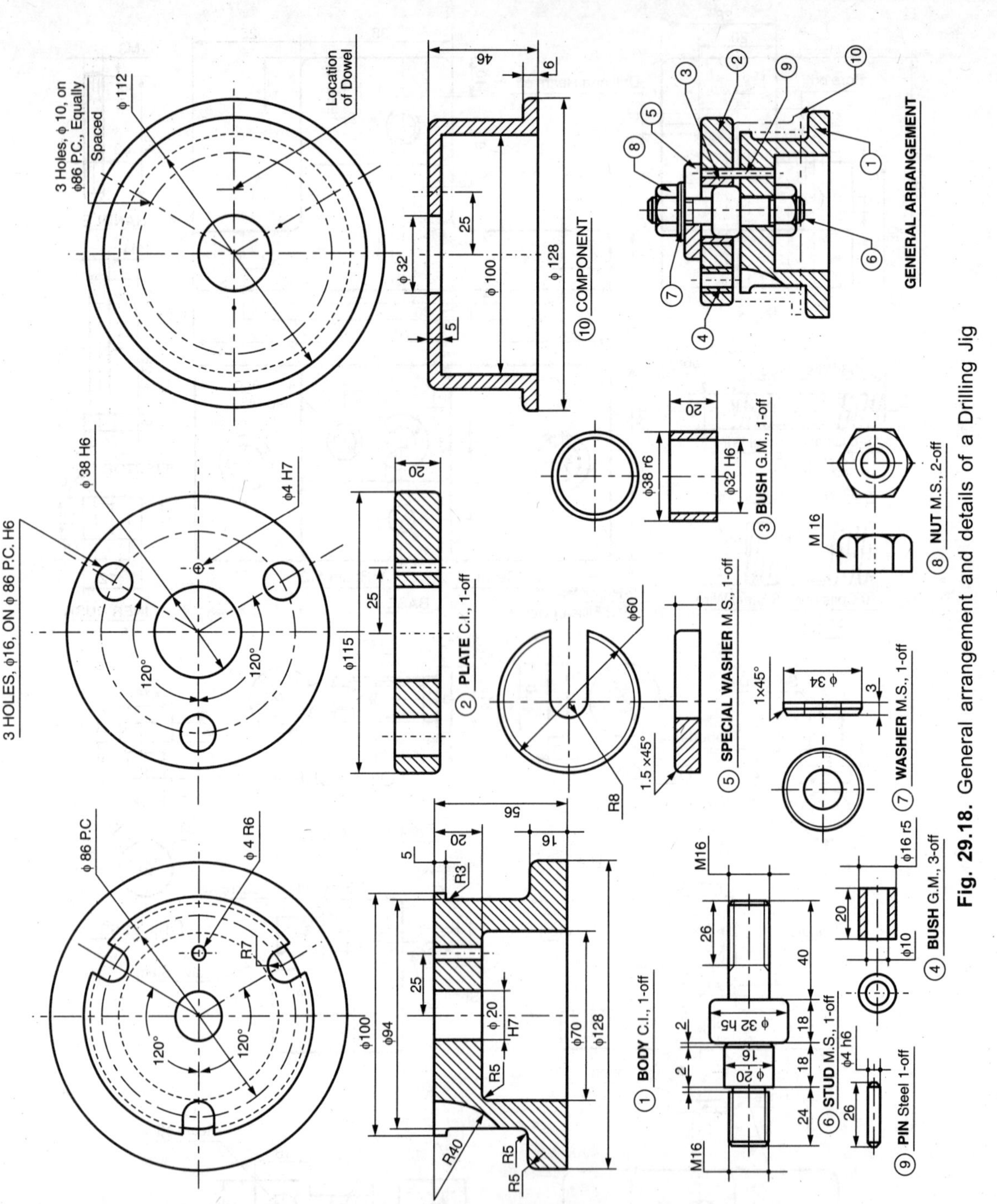

Fig. 29.18. General arrangement and details of a Drilling Jig

component to be machined is gripped in the collet, which is firmly held in the Body of the jig. The Body is secured to the Base by means of the Clamping Ring but is free to rotate into one of the four possible positions, in which, it can be locked, by a pin (also not shown) that engages in mating holes drilled in the Body and Clamping Ring. The grip of the Collet is developed by the 'wedge' effect of its outer surface when forced downwards by the Cap. The Collet is prevented from rotating by the small Key. The four, ϕ10 mm, holes in the Body together with single hole in the Clamping Ring are for locating the special bolt in each of the

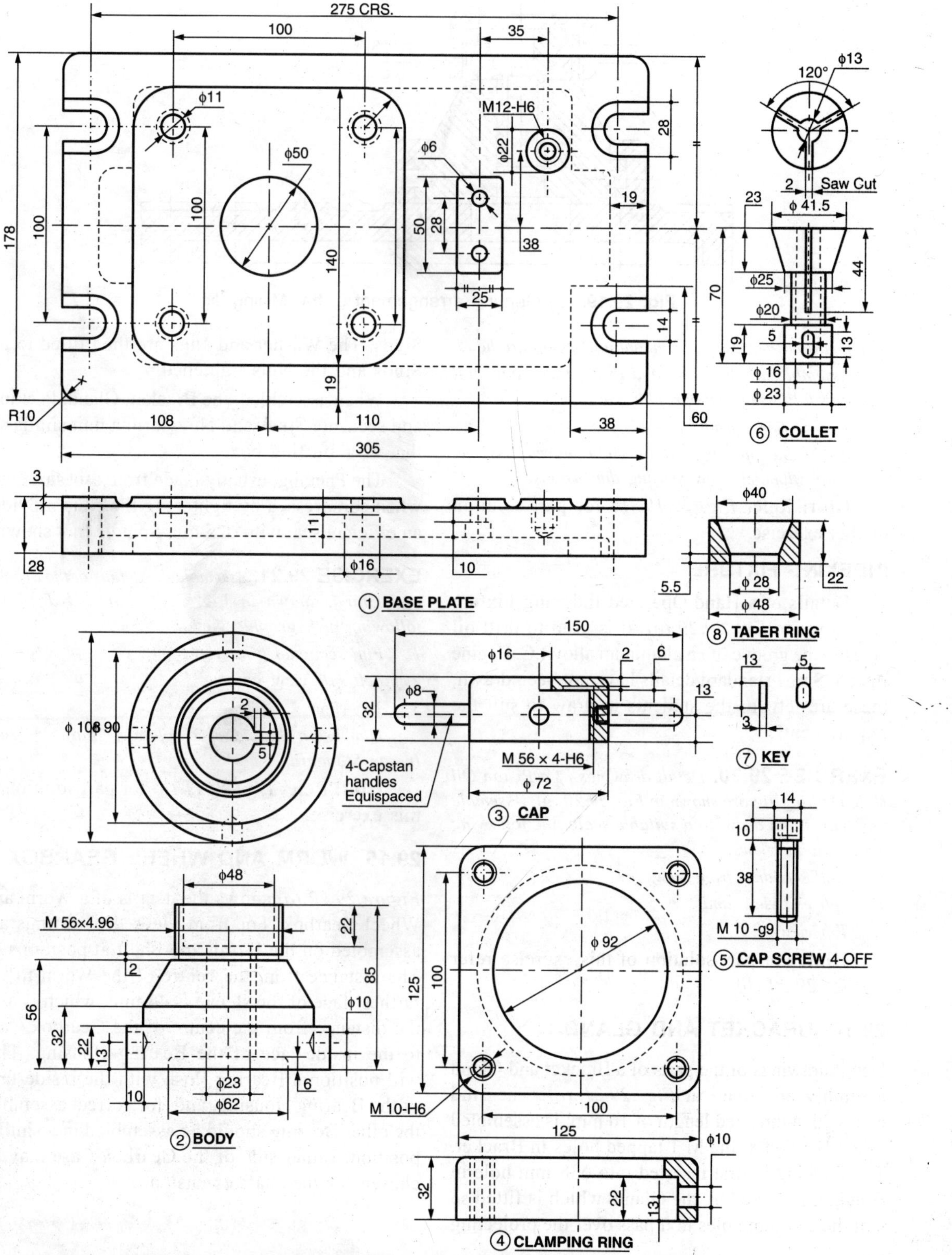

Fig. 29.19. Details of a Milling Jig

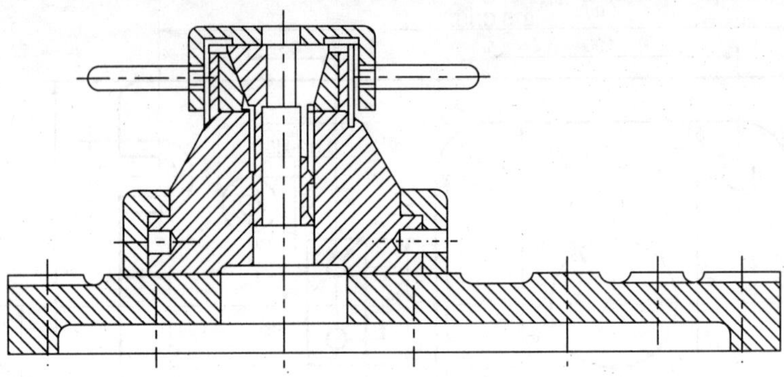

Fig. 29.19. (b) General arrangement of the Milling Jig

four positions necessary for machining its square head.

Assemble all the parts and draw, to a scale full size, the following orthographic views:

1. *Full sectional front view,*
2. *An outside top view, with the cap removed. Assume, proportionately, any missing dimensions.*

Hint: Refer *Fig. 29.19 (b)* for part solution of this exercise.

INDEXING FIXTURE

Details of a Hand Operated Indexing Fixture are shown in *Fig. 29.20 (a)*. It is used to drill oil holes in the groove of an aluminum alloy I.C. Engine Piston. Some standard details have not been drawn; these are left for the students to draw to suit the requirements.

EXERCISE 29.20. *Detail drawings of a Piston Oil Hole Drilling Jig are shown in Fig. 29.20 (a). Assemble the details and draw, to a suitable scale, the following orthographic views:*

1. *Half sectional front view,*
2. *Left end view, and*
3. *Top view.*

Hint: For part solution of this exercise refer to *Fig. 29.20 (b)*.

29.14. BRACKET AND GLAND

Detail drawings of the parts of a Bracket and Gland assembly are given in *Fig. 29.21 (a)*. The stud end, with a threaded length of 16 mm, is assembled in the Bracket at the M 1 tapped holes in Bracket. The Packing is first inserted into ϕ38 mm hole in Bracket, followed by the Gland, which is fitted so that the ϕ11 mm holes in it pass over the projecting Studs. The Washer and Nuts are then fitted to the Studs and the Nuts tightened.

When assembled, the Bracket, Gland, Packing, and securing Studs and Nuts form a unit similar to that of a stuffing box.

The Packing, usually made from asbestos cord, when compressed by tightening the Nuts provides an efficient seal with ϕ25 mm Spindle (not shown).

EXERCISE 29.21. *Assemble the details of the Bracket and Gland, shown in Fig. 29.21 (a) and draw the following orthographic views:*

1. *Full sectional front view,*
2. *Left side view, and*
3. *Top view.*

Dimension the views completely, following latest Indian Standard.

Hint: Refer *Fig. 29.21 (b)* for part solution to this exercise.

29.15. WORM AND WHEEL GEARBOX

Figure 29.22 (a) shows the details of a Worm and Wheel Gearbox. For front view, the bearings are assembled on the Worm and Shaft at positions B. The distance from the centre of the Worm to the farthest face of the sleeve is 44 mm, which is also the distance from the centre of the Gearbox Case to the inside face of the Bearing Housing. This will position a Bearing flush with the inside face of the Bearing Housing, and, for correct assembly, the other Bearing should be assembled in a similar position. Either side of the Gearbox Case may be chosen for the shaft extension.

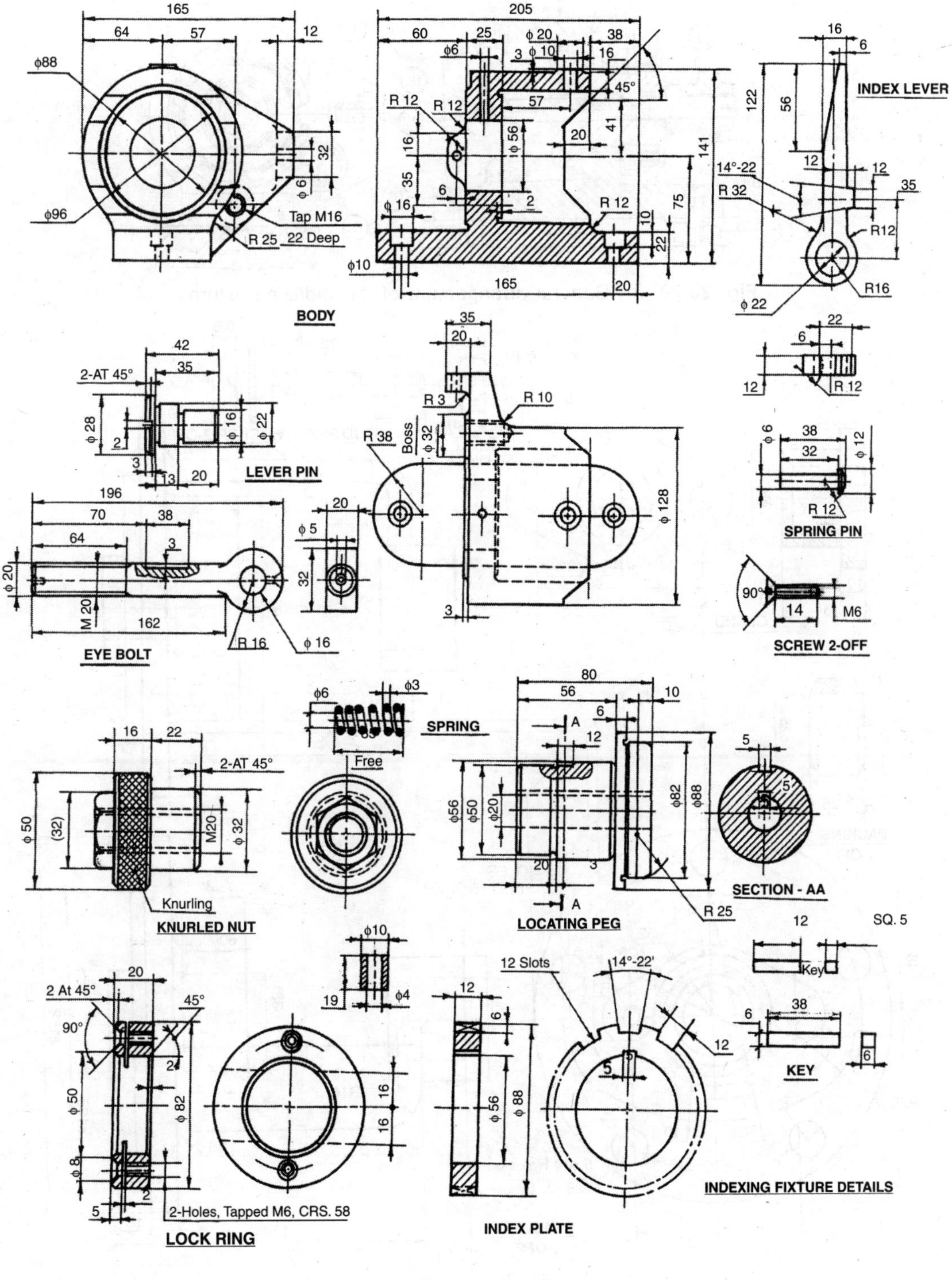

Fig. 29.20. (a) Detail drawings of an Indexing Fixture

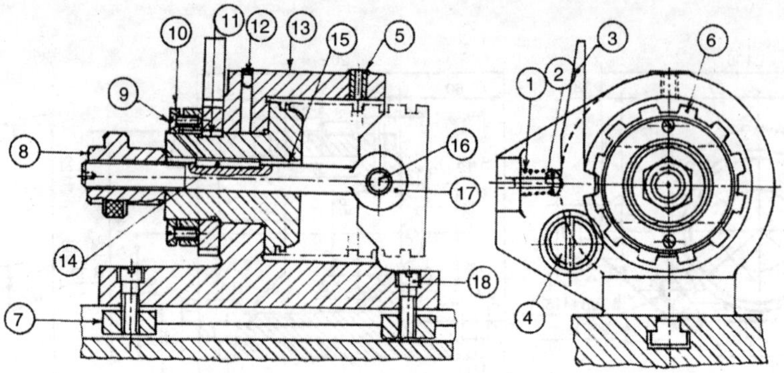

Fig. 20.20. (b) General arrangement of the Indexing fixture

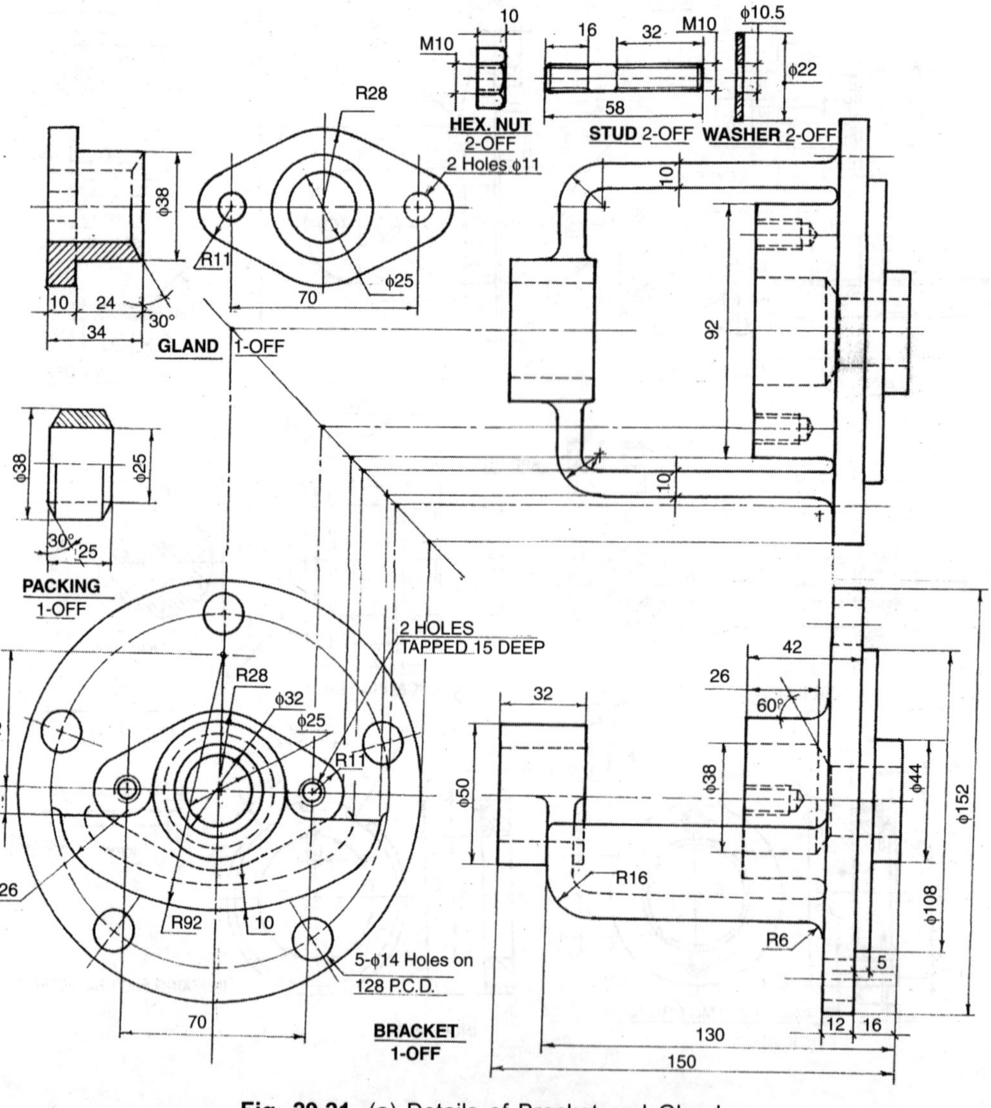

Fig. 29.21. (a) Details of Bracket and Gland

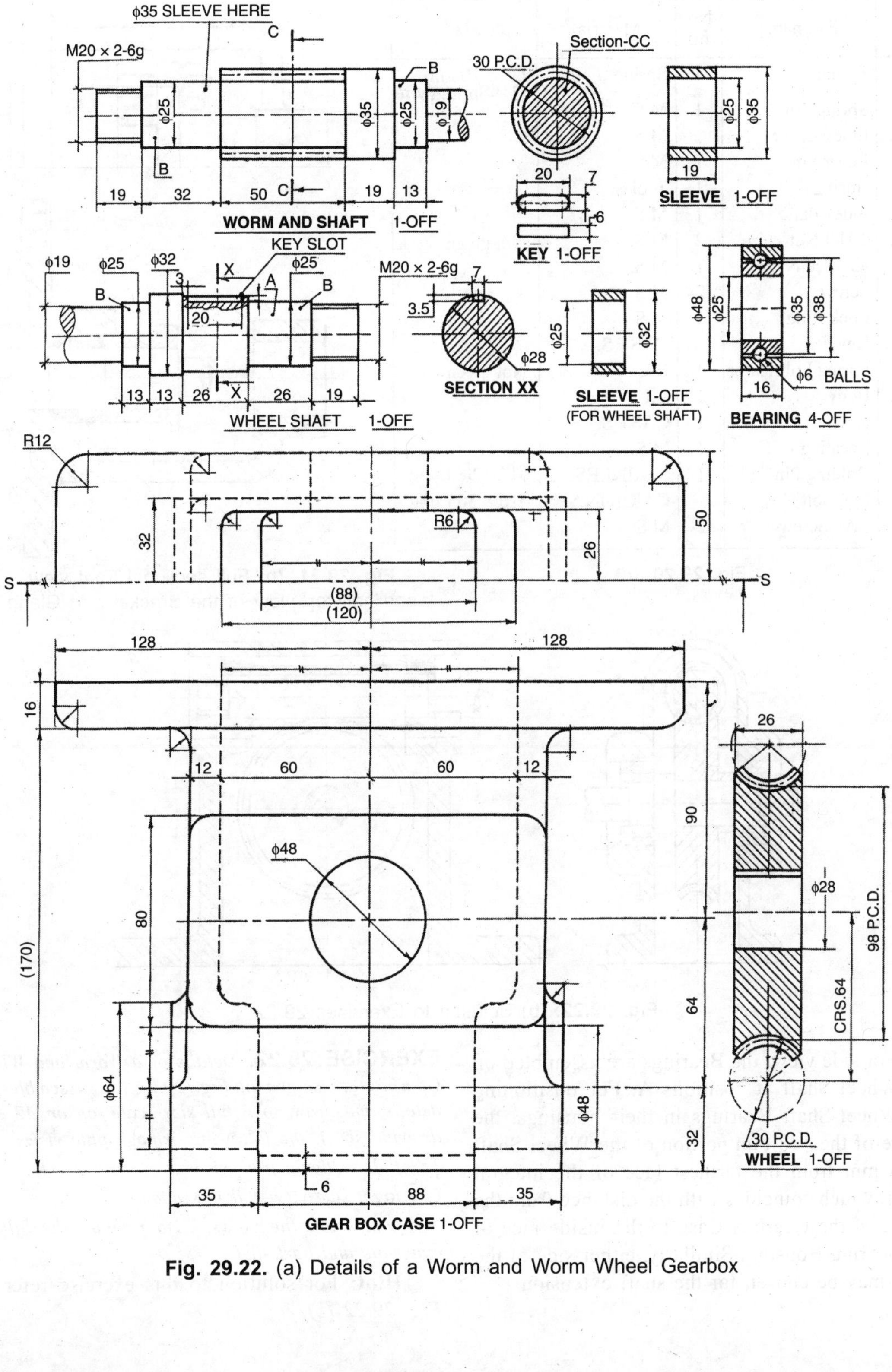

Fig. 29.22. (a) Details of a Worm and Worm Wheel Gearbox

Part no.	Part name	No. off	Material	Size
1.	Spring	1	Spring S.	0.D = 10 mm, φ2 = Spring wire
2.	Spring Pin	1	M.S.	
3.	Index Lever	1	M.S.	
4.	Lever Pin	1	M.S.	
5.	Bush	1	Tool S.	
6.	Index Plate	1	M.S.	
7.	T-slot Nut	2	M.S.	Not given
8.	Knurled Nut	1	M.S.	
9.	Screw	2	M.S.	
10.	Lock Ring	1	M.S.	
11.	Key	1	CAST S.	
12.	Spring Lubricator	1	C.I.	Not given
13.	Body	1	C.I.	
14.	Key	1	CAST S.	
15.	Locating PEG	1	M.S.	
16.	Holding Pin	1	STAINLESS S.	φ14 × 88 Long
17.	Eye Bolt	1	CARBON S.	φ10 × 30 Long
18.	CAP Screw	2	M.S.	

Fig. 29.20. (c)

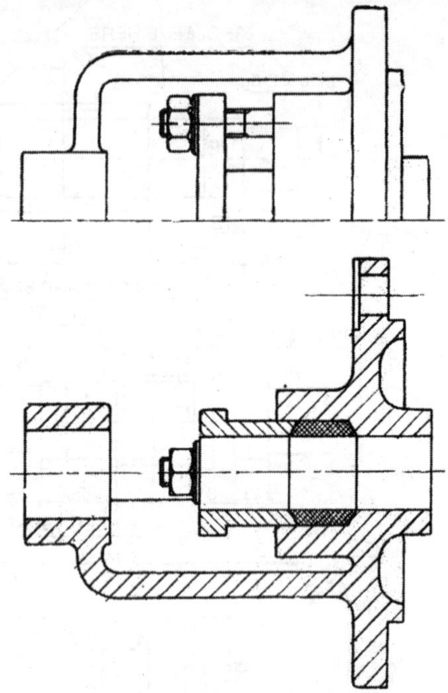

Fig. 29.21. (b) Full sectional front view and half top view of the Bracket and Gland

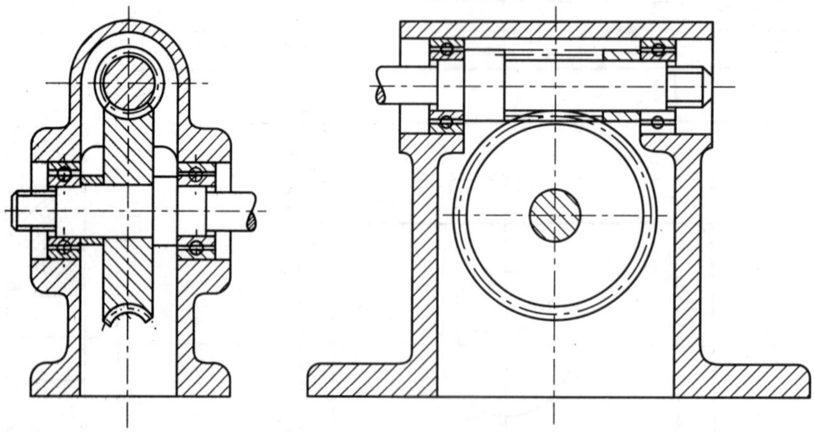

Fig. 29.22. (b) Solution to Exercises 29.22

For side view, the Bearings are assembled on the Wheel Shaft at positions A. For positioning the Wheel Shaft Bearings in their housings, the centre of the φ28 mm portion of the Wheel Shaft is 26 mm from the farthest face of the integral collar, which coincides with the distance from the centre of the Gearbox Case to the inside face of the Bearing Housing. Similarly, either side of the Case may be chosen for the shaft extension.

EXERCISE 29.22. *Details of a Worm and Wheel Gearbox are shown in Fig. 29.22 (a). Assemble the details and draw, to a full size scale, on an A2 size drawing sheet, the following orthographic views:*

1. Full sectional front view,

2. Right side view–full in section.

Dimension the views. Also prepare the bill of materials and parts-list.

Hint: For solution to this exercise refer to *Fig. 29.22 (b).*

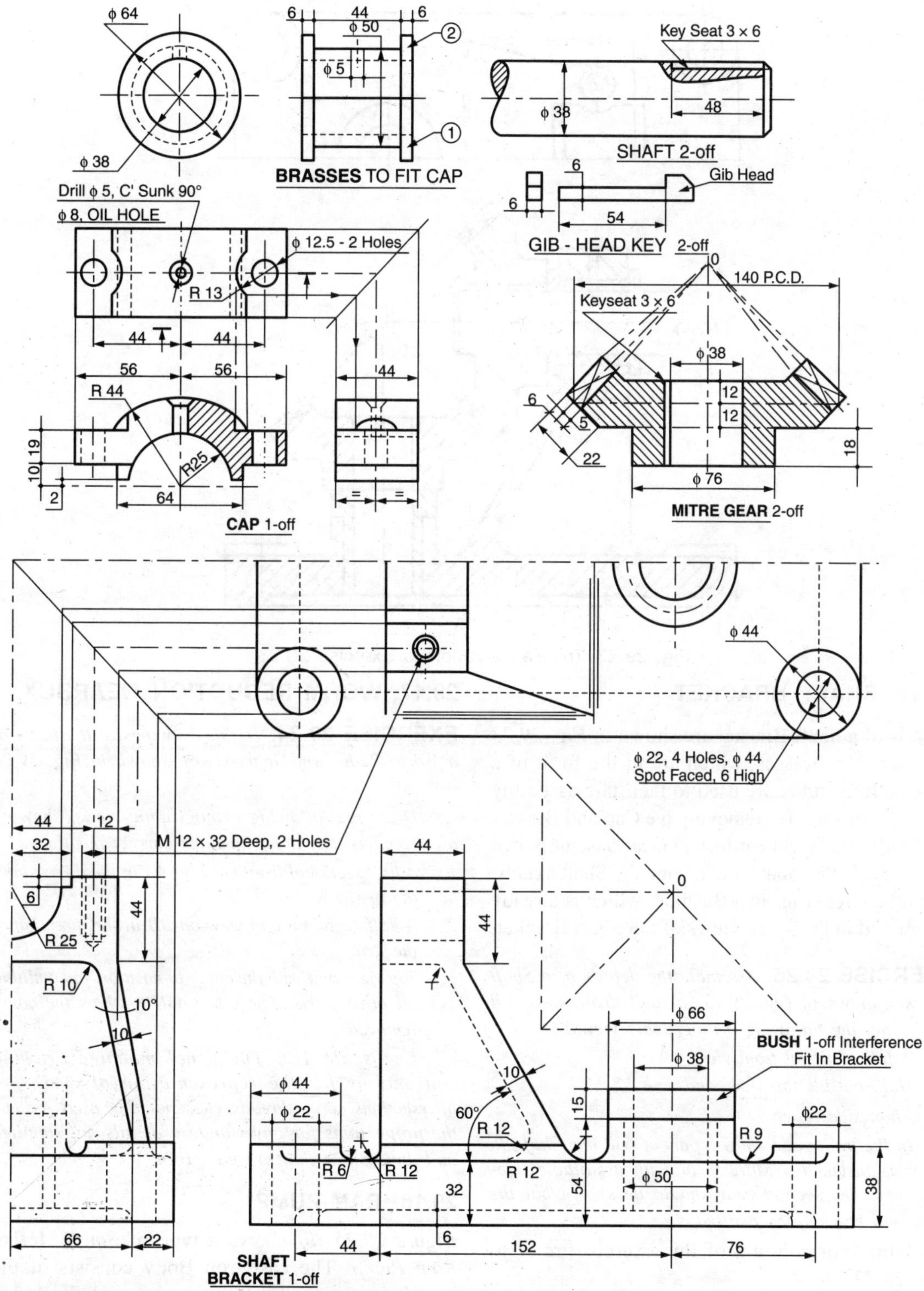

Fig. 29.23. (a) Details of a Shaft Bracket

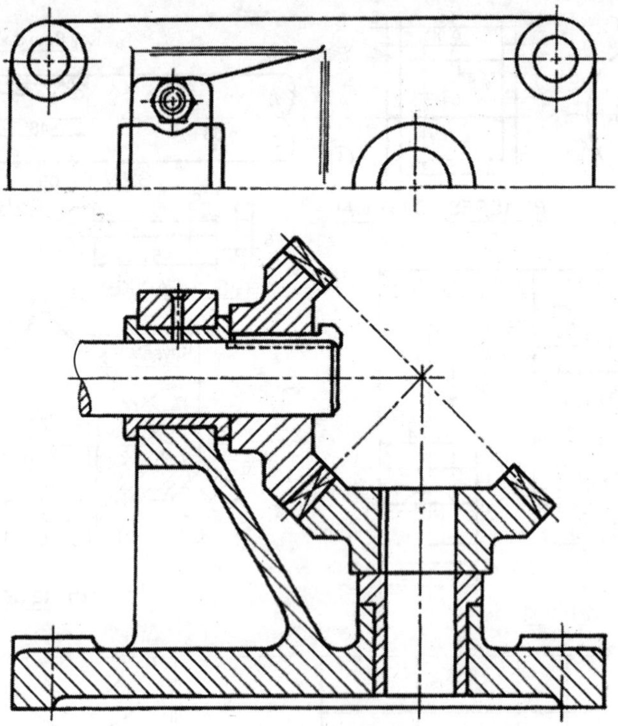

Fig. 29.23. (b) Part solution to Exercise 29.23

29.16. SHAFT BRACKET

Details of a Shaft Bracket are shown in *Fig. 29.23 (a)*. Here the Brasses, which are in the form of a split-collared tube, are used to facilitate assembly and maintenance. By removing the Cap and Brasses the Shaft and Bevel (or Mitre) Gear assembly can be dismantled without withdrawing the Shaft axially. The other Gear runs in a Bushing, which is already assembled in the given views of the Shaft Bracket.

EXERCISE 29.23. *Assemble the details of a Shaft Bracket shown in Fig. 29.23 (a) and draw, to a full size scale, the following orthographic views:*

1. *A full sectional front view,*
2. *Half, outside top view, and*
3. *Right side view.*

In the front view, show a part of φ38 mm Shaft, to which is fastened a Mitre Gear with a suitable Gib-head key. The second gear should be shown but the shaft and key may be omitted.

Hint: For solution of this exercise, refer to *Fig. 29.23 (b)*.

29.17. WORM REDUCTION GEARBOX

EXERCISE 29.24. *Detail drawings of the parts of a Worm Reduction Gearbox are shown in Fig. 29.24 (a), (b).*

Draw the following orthographic views, showing all the parts assembled in their correct positions:

1. *A half sectional front view on AA in the direction of the arrow X,*
2. *A half sectional end view on BB in the direction of the arrow Y,*
3. *Top view, half with the cover in position, but omitting all other parts. The other half to show the cover removed.*

Use scale 1:2. The usual machine drawing conventions for the representation of curves of intersections, screw threads, etc., should be used. Assume the proportions for any standard details not supplied and dimensions that are not given.

29.18. RAM PUMP

Figure 29.25 (b) shows a type of *Single Acting Ram Pump*. The cast iron Body consists of the barrel, in which the Ram works, and the Valve

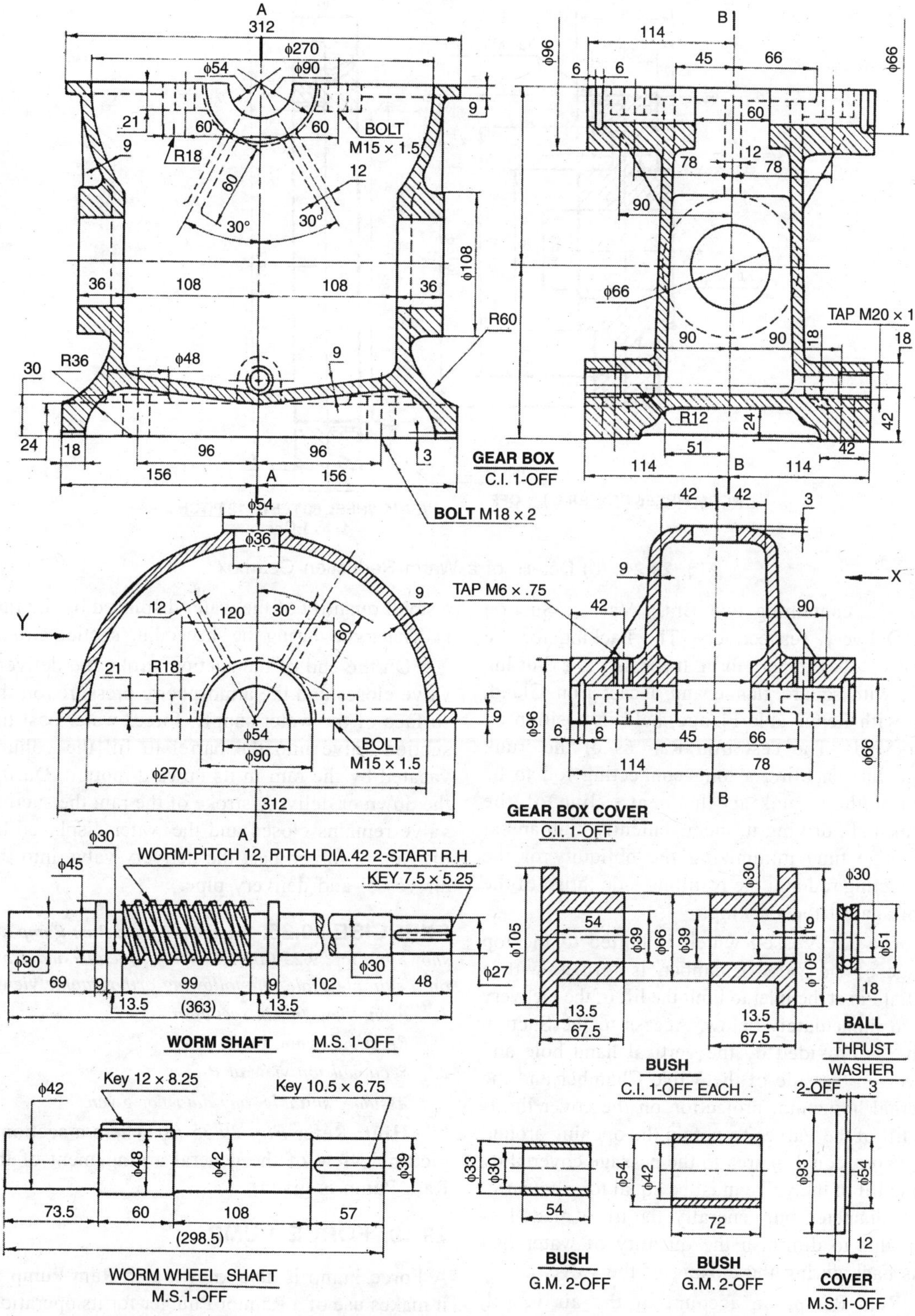

Fig. 29.24. (a) Details of a Worm Reduction Gearbox

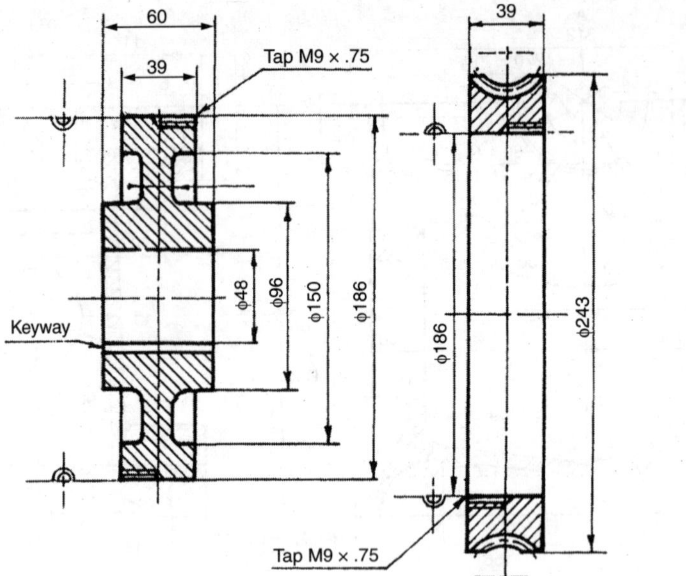

Fig. 29.24. (b) Details of a Warm Reduction Gearbox

Chamber containing two Brass Valves, Suction and Delivery, respectively. The Packing for the Ram is of the ordinary form. It consists of a Stuffing Box, with square flange and a cast iron Gland, also with Removable Flange, held in position by four Studs. The cast iron Ram is of the trunk type, thus enabling a long connecting rod to be used without making the centre line of the Crankshaft, driving it, inconveniently high and at the same time minimizing the obliquity of the connecting rod and the resulting side thrust of the Ram against the Packing.

The Air Vessel, which is bolted to the top flange of the Valve Chamber, is formed with a central Pin at the inlet to limit the lift of the Delivery Valve immediately below. Access to the Suction Valve is provided by the vertical hand hole and Cover on the side of the Valve Chamber and the internal horizontal projection on the cover limits the lift of the Valve. To obtain the opening around the Valve, equal in area to the passage covered by it, the lift of the valve must be equal to one quarter of its diameter, but generally, the lift is made less than this to diminish the quantity of water that runs back during the closing of the valve.

Ram Pump, on account of the air vessel, produces an approximately uniform delivery, since a small quantity of the water displaced by the ram is discharged during the succeeding suction stroke.

On the end of the suction stroke the delivery valve closes and the atmospheric pressure, on the surface of the water supply, forces water past the suction valve into the barrel to fill the volume vacated by the ram in its upward motion. During the down or delivery stroke of the ram the suction valve remains closed and the water displaced by the ram is forced past the delivery valve into the air vessel and delivery pipe.

EXERCISE 29.25. *The details for a Ram Pump are shown in Fig. 29.25 (a). Assemble the details and draw, on a suitable scale, the following orthographic views:*

1. *Front view–full in section,*
2. *Right side view, and*
3. *Sectional top view at P-S.*

Assume, suitably, any data not given.

Hint: Refer *Fig. 29.25 (b)* where a sectional pictorial view of the general arrangement of the Ram Pump is given.

29.19. FORCE PUMP

A Force Pump is also known as a Ram Pump as it makes use of a Ram or Plunger for its operation. Force Pumps are used to force or pump water

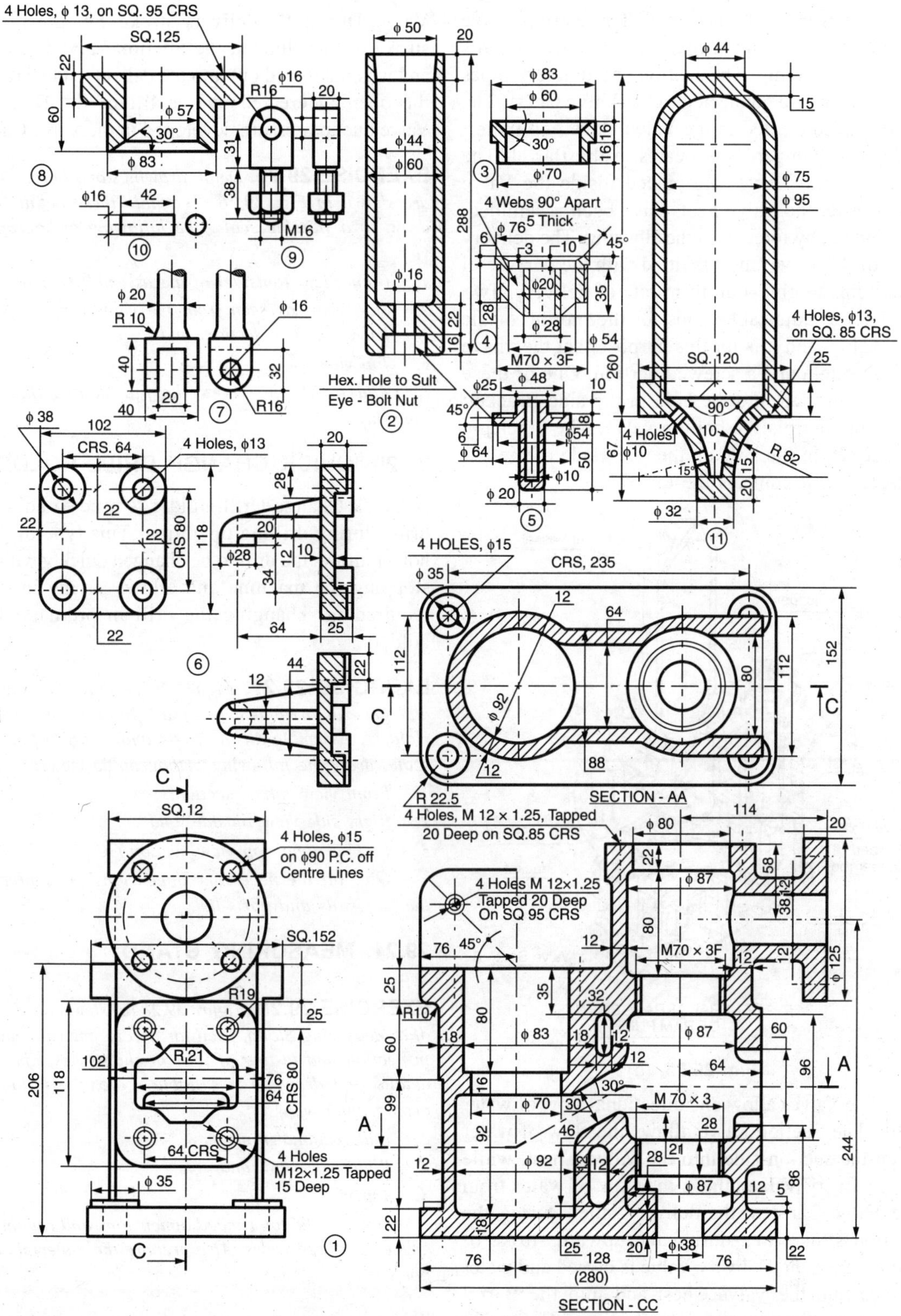

Fig. 29.25. (a) Details of a Ram Pump

from a lower to a higher level. These are positive displacement pumps. *Figure 29.26* shows details of one such pump. The suction and delivery pipes (not shown) are connected to the Valve Chest. The Barrel is connected to the Valve Chest, on its left side. The Plunger reciprocates in the Barrel. The two identical Valves are placed inside the Valve Chest after the Valve Seats have been assembled, in position, by force fit in the Housing. The delivery side of the Valve Chest is fitted with the Air Vessel that helps to give a uniform rate of delivery. Two Pins act as adjustable stops for the motion of the two valves and it is for this purpose that they have a flat bottom and a screwed portion. One of these pins is inserted into the hole in the side Bracket; this Pin controls the bottom valve. The other pin is inserted in the hole on the underside of the Air Vessel and it controls the top valve.

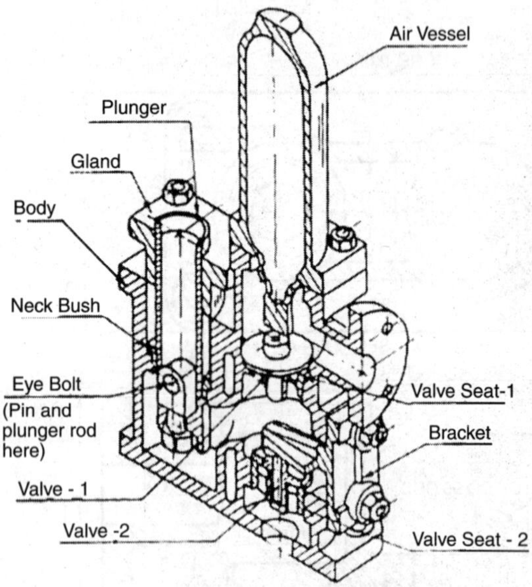

Fig. 29.25. (b)

The Valves allow one directional flow of water only. The Suction Valve allows the flow of water from the suction side into the Valve Chest, while the Delivery Valve allows the flow of water from the Valve Chest to the delivery side. During the suction stroke, i.e., during the upward stroke of the Plunger, the Delivery Valve is closed and water rushes into the Valve Chest and into the Barrel through the opening created by the lift of the Suction Valve. During the delivery stroke, *i.e.,* downward stroke of the Plunger, the Suction Valve is closed and water is forced out into the delivery side through the opening created by the lift of the Delivery Valve due to pressure developed in the Valve Chest.

EXERCISE 29.26. *Detail drawings of a Force Pump are shown in Fig. 29.26. Assemble these details and draw, to a suitable scale, the following orthographic vies:*

1. *Front view, (outside) of the assembled details,*
2. *Side view, as seen from the right side, with the Valve Chest in section, and*
3. *Top view.*

Add to all the views the few details that are "missing".

29.20. QUICK CHANGE DRILL HOLDER

A Quick Change Drill Holder is used to hold the drills during drilling operation. This type of drill holder allows the drill to be changed quickly without stopping the machine and allows saving in time required for changing the drill in ordinary drill holders.

EXERCISE 29.27. *Figure 29.27 shows the details and general arrangement of a 'Quick Change Drill Holder'. Assemble all the parts and, using a full size scale, draw the following orthographic views:*

1. *Front view–full in section,*
2. *Right side view–outside, and*
3. *Full sectional top view–section CC.*

Give all the necessary dimensions. Also prepare the materials and parts-list.

29.21. MEASURING STAND

EXERCISE 29.28. *Figure 29.28 (a) shows the details of a Measuring Stand, used for precise measurements, inspection and testing of jobs in shops. Assemble the details and draw, to a suitable scale, the following orthographic views:*

1. *Full sectional front view,*
2. *Right side view, and*
3. *Top view.*

Give all the toleranced dimensions and roughness grades on these views. Also prepare the materials and parts-list.

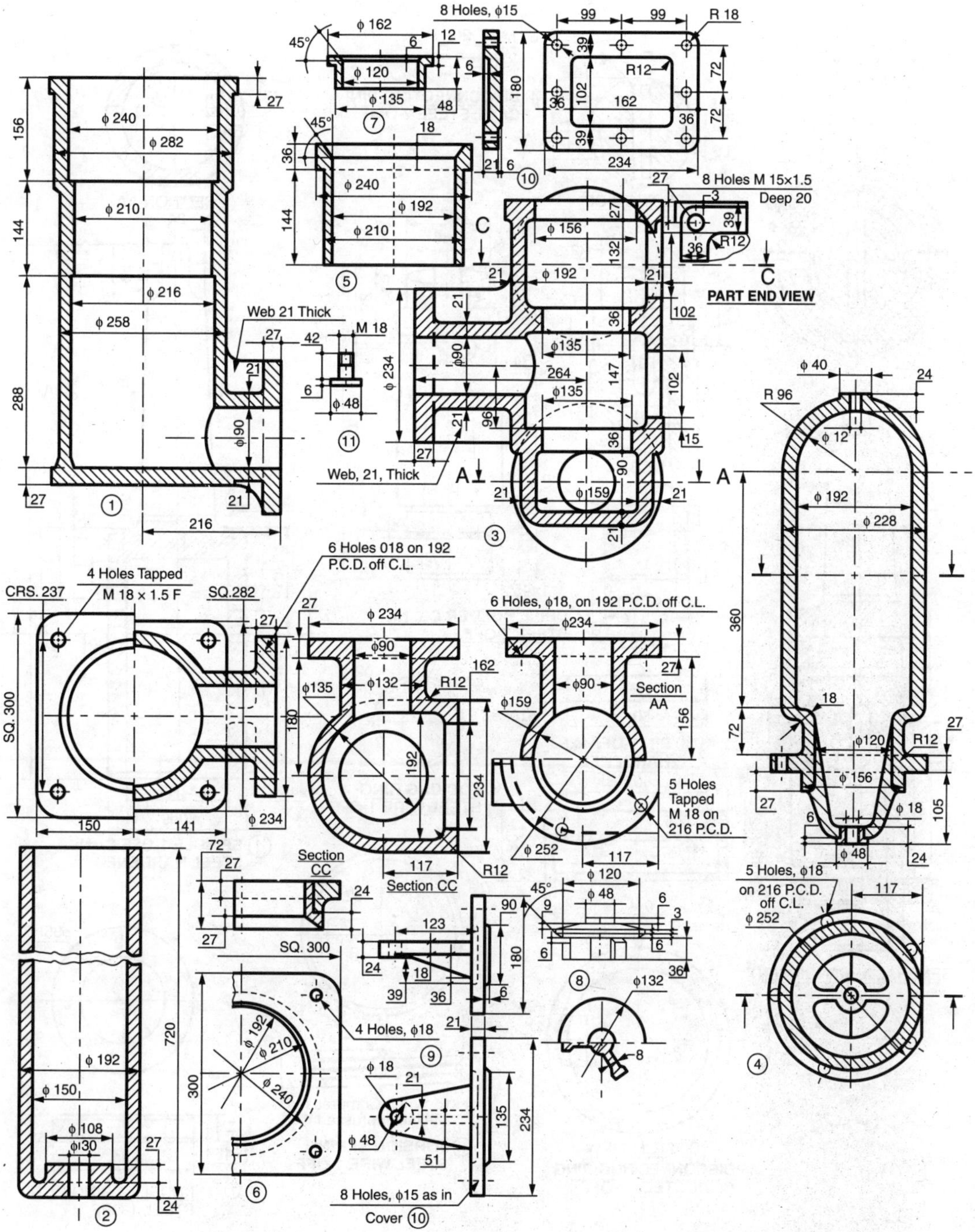

Fig. 29.26. Details of a Force Pump (Not to Scale)

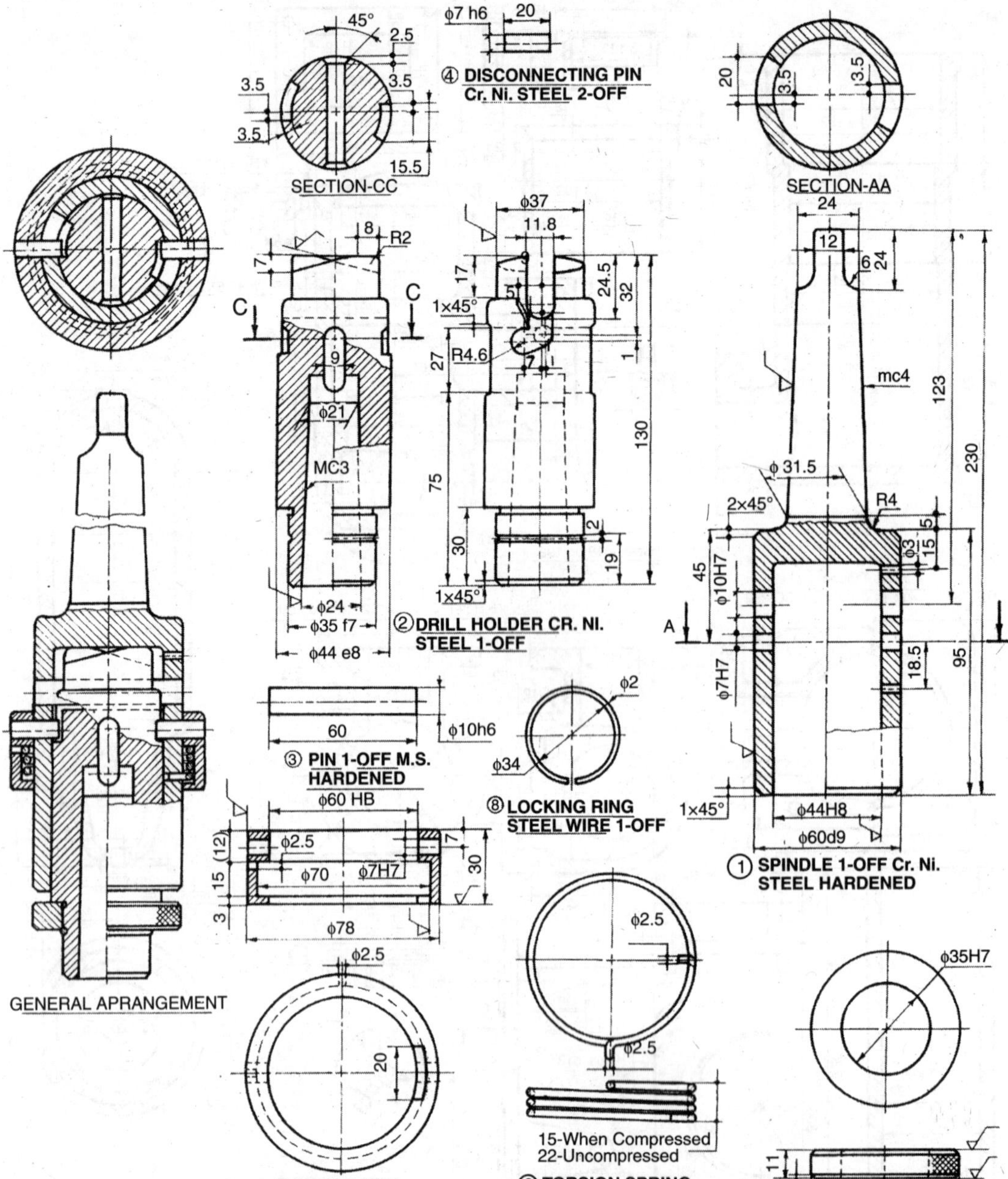

Fig. 29.27. Quick Change Drill Holder

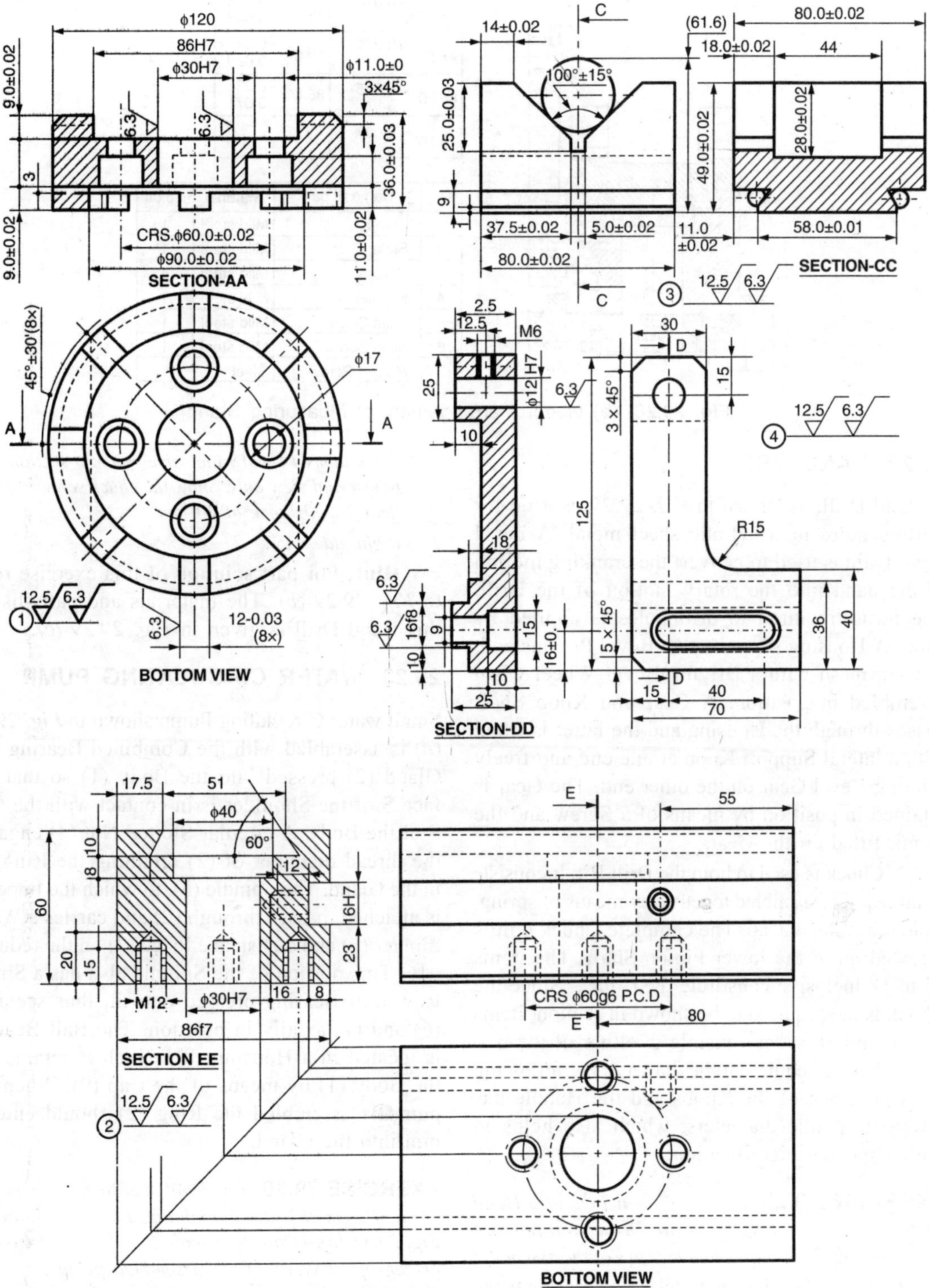

Fig. 29.28. (a) Details of a Measuring Stand

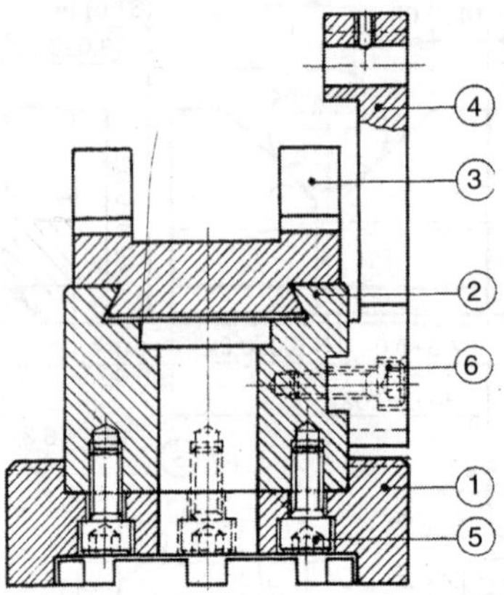

12 H7	+0.018 / 0	16 f8	−0.016 / −0.043
30 H7	+0.021 / 0	60 g6	−0.010 / −0.029
86 H7	+0.035 / 0	86 f7	−0.036 / −0.071
16 H7	+0.027 / 0		

S.No.	Name of part	Material	No.off
1	Stand	Mild steel	1
2	Slide	Mild steel	1
3	V-block	Mild steel	4
4	Bracket	Mild steel	1
5	Allen Screw	Mild steel	1
6	Allen Screw	Mild steel	1
7	M6 Set Screw	Steel	1

Fig. 29.28. (b) General arrangement of Measuring Stand

29.22. HAND DRILL

A Hand Drill, as shown in *Fig. 29.29*, is used for drilling holes in wood and sheet metal. A bevel Gear Train is used to convert the cranking motion of the hand into the rotary motion of the Drill. The thrust required to drill holes is applied by hand. A Housing supports two Bevel Pinions and, in turn, mesh with a Driving Bevel Wheel when assembled in position. A Gear and Knob Shaft passes through the Housing and the latter is fitted with a lateral Support Knob at one end and freely rotating Bevel Gear on the other end. The Gear is retained in position by means of a Screw and the Crank fitted on the Gear.

A Chuck is used to hold the Drill Bit. It consists of three jaws assembled together by means of spring, inside the Chuck Cap. The complete Chuck unit is screwed on to the lower Pinion Shaft. The items 15 to 18 inclusive constitute the Chuck. Since the Chuck is not required to be shown in section, items 15, 16 and 18 are not visible in either of the two views that are required to be drawn in the following exercise. Each of the Knobs and the Handle has different spindle diameters, which also helps in the component identification.

EXERCISE 29.29. *The detail drawings of a Hand Drill are shown in Fig. 29.29 (a). Draw, to a full size scale, the following orthographic views of the assembly:*

1. *Front view full* ***insection,*** *taking the section along the longitudinal centreline, (showing pinions 4 and 5, and Bevel Wheel 6 in section as shown in details). Show items 14 to 18 inclusive, without section, i.e., in external view only. Show the chuck screwed along item 2 a distance of 20 mm.*
2. *Right side view.*

Hint: For part solution of this exercise refer to *Fig. 29.29 (b)*. The materials and parts-list of this Hand Drill is given in *Fig. 29.29 (c)*.

29.23. WATER CIRCULATING PUMP

Small water Circulating Pump shown in *Fig. 29.30 (a)* is assembled with the Combined Bearing and Gland (2) pressed into the Body (1) so that the face S of the Shoulder is in contact with the face S of the Body. A circular Slotted Nut (4) engages the thread diameter of (2) retaining the Ring (3) in the Gland. The Spindle (5), to which the Impeller is attached, passes through (2) and carries a Water Slinger (6) and a Distance Piece (7) on the reduced ϕ16. This portion of the Spindle also has a Single Row Ball Bearing pressed onto it, thus securing (6) and (7) axially in position. The Ball Bearing is located in a Housing (8), which is retained in the Body (1) by means of the Cap (9). When the pump is assembled the Ring (2) should enter 3 mm into the Gland.

EXERCISE 29.30. *The detail drawings of component parts of a small Water Circulating Pump are shown in Fig. 29.31 (a). Draw, to a scale full size, the following orthographic views of the assembled pump :*

1. *A sectional front view on AA as seen in the direction of arrow,*

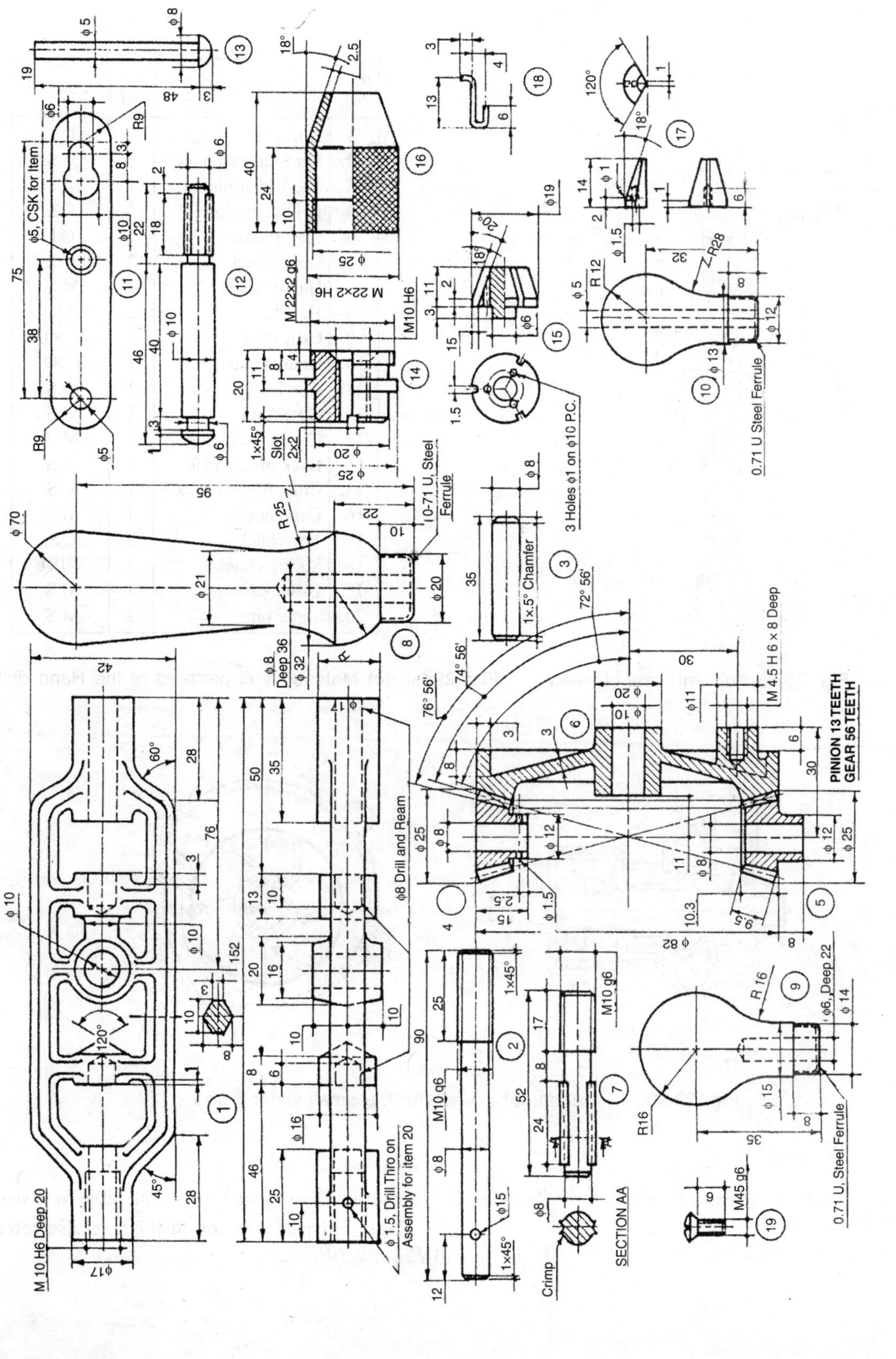

Fig. 29.29. (a) Details of a Hand Drill (Capacity 0–6)

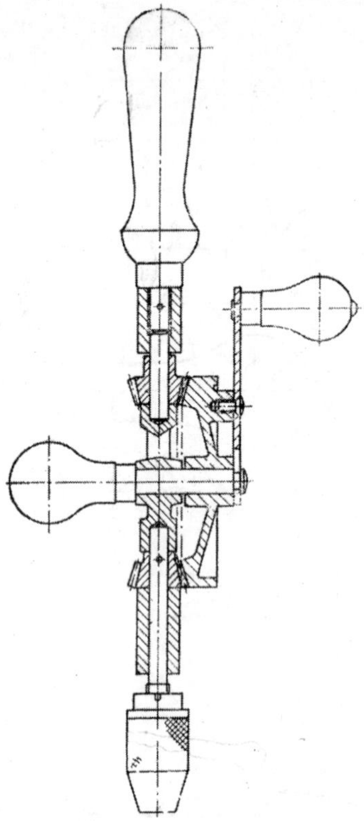

Fig. 29.29. (b) Orthographic view

Item No.	Description	No. off	Material
1.	Housing	1	C.I.
2.	Shaft-driving pinion	1	M.S.
3.	Shaft-idler pinion	1	M.S.
4.	Driving pinion	1	C.I.
5.	Idler pinion	1	C.I.
6.	Gear wheel	2	C.I.
7.	Handle shaft	1	M.S.
8.	Handle	2	WOOD
9.	Gear knob	1	WOOD
10.	Crank knob	1	WOOD
11.	Crank	1	M.S.
12.	Gear shaft	1	M.S.
13.	Crank shaft knob	1	M.S.
14.	Back plate chuck	1	M.S.
15.	Compressor chuck	1	M.S.
16.	Cap-chuck	1	M.S.
17.	Jaw-chuck	3	C.S.
18.	Spring-chuck	3	WIRE
19.	C-Sunk screw	1	M.S.
20.	Taper pins	2	M.S.

Fig. 29.20. (c) Material `and parts-list of the Hand drill

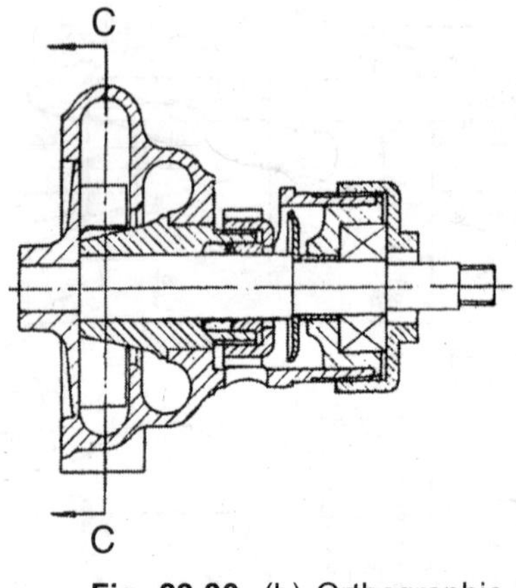

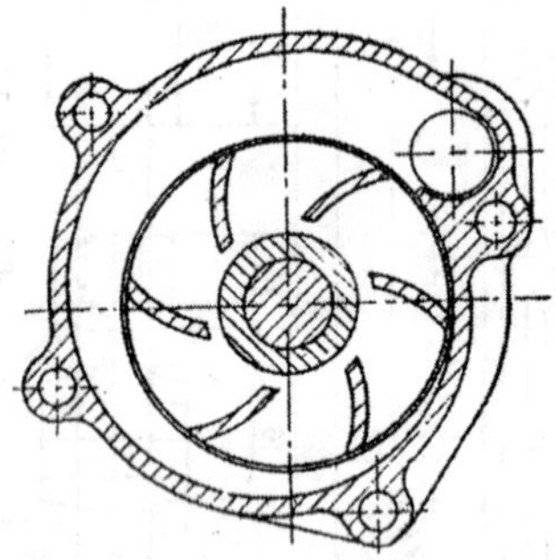

Fig. 29.30. (b) Orthographic views of the small water pump

2. *A sectional side view on CC viewed in the direction of the arrows,*
3. *A complete outside view.*

Small unspecified radii should be taken as 3 mm.

Hint: For part solution to this exercise refer to *Fig. 29.30 (b).*

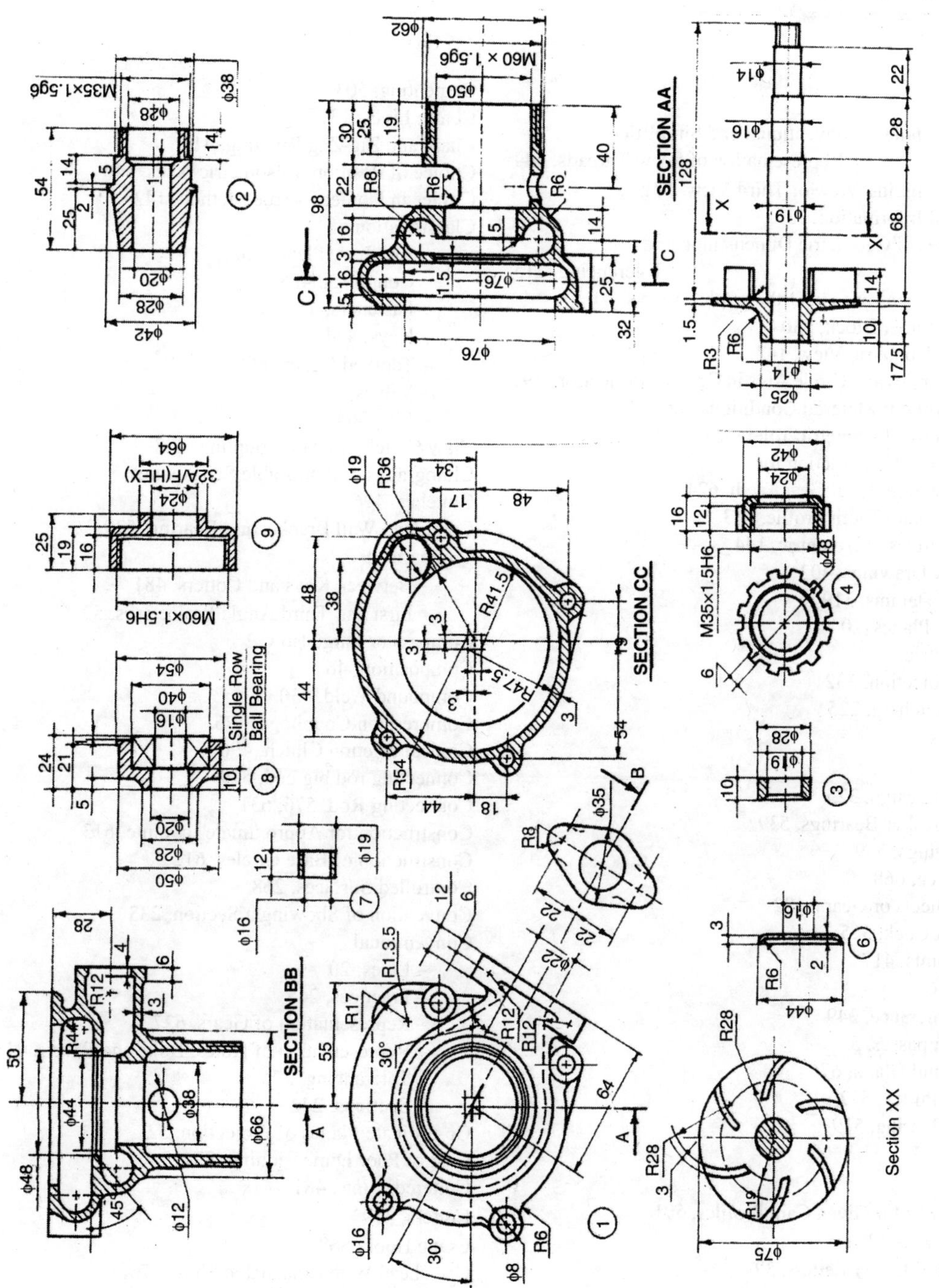

Fig. 29.30 (a) Details of a small Water pump

INDEX

D...

E...

F...

G...

H...

I...

J...

K...

L...

M...

N...

O...

P...

Q...

R...

S...

T...

U...

V...

W...

Win Prizes

ATTENTION READERS

Win Prizes

We request for your frank assessment regarding various aspects of the book as given below :

A TEXTBOOK OF

MACHINE DRAWING

By: P.S. GILL

Please Fill up the Sheet in Neat and Clean Writing

(i) What topics of your syllabus, important for your examination point of view, are not covered in this book ?

(ii) What are the chapters wherein the treatment of the subject matter is not systematic or organised or updated ?

(iii) Have you come across misprints/mistakes/factual inaccuracies in the book? Please specify the chapters and the page numbers.

(iv) What other books on the same subject have you found/heard better than the present book? Please specify in terms of price and quality.

(v) Further suggestions and comments for the improvement of this book.

(PS: If need be please attach a separate sheet to write your views)

OTHER DETAILS

(i) Who recommended you this book ? (Please tick (✓) in the box near the option relevant to you.

Teacher ☐ Friends ☐ Bookseller ☐

....

....

....

....

(ii) Name of the recommending teacher, his designation and address :

....

....

....

....

(iii) Name and address of the bookseller you purchased the book from :

....

....

....

....

(iv) Name and address of your institution (Please mention the University or Board as the case may be)

....

....

....

....

(v) Your name and complete postal address :

....

....

....

....

(vi) Course you are enrolled in :

....

....

....

....

The best assessment will be awarded each month: The award will be in the form of our publications as decided by the Editorial Board.

Please mail the filled up coupon at your earliest to :

S.K. Kataria & Sons®
4885/109, Prakash Mahal, Dr. Subhash Bhargav Lane,
Opposite Delhi Medical Association, Daryaganj, New Delhi-110002
Phones: +91-11-23243489, 43551243; Mobile: +91-9871775858
e-mail: katariabook@gmail.com; katariabooks@yahoo.com
Website: www.skkatariaandsons.com